GERD

This book is dedicated to:

Cherine,
Who was born before esophageal adenocarcinoma was reported;
My children,
Who saw a seven-fold increase in its incidence;
And my grandchildren,
Who I hope will see the day it is eradicated.
With
Gratitude, Respect, and Love.

GERD

A New Understanding of Pathology, Pathophysiology, and Treatment

Parakrama T. Chandrasoma

Academic Press is an imprint of Elsevier
125 London Wall, London EC2Y 5AS, United Kingdom
525 B Street, Suite 1800, San Diego, CA 92101-4495, United States
50 Hampshire Street, 5th Floor, Cambridge, MA 02139, United States
The Boulevard, Langford Lane, Kidlington, Oxford OX5 1GB, United Kingdom

Notices
Knowledge and best practice in this field are constantly changing. As new research and experience broaden our understanding, changes in research methods, professional practices, or medical treatment may become necessary.

Practitioners and researchers must always rely on their own experience and knowledge in evaluating and using any information, methods, compounds, or experiments described herein. In using such information or methods they should be mindful of their own safety and the safety of others, including parties for whom they have a professional responsibility.

To the fullest extent of the law, neither the Publisher nor the authors, contributors, or editors, assume any liability for any injury and/or damage to persons or property as a matter of products liability, negligence or otherwise, or from any use or operation of any methods, products, instructions, or ideas contained in the material herein.

Library of Congress Cataloging-in-Publication Data
A catalog record for this book is available from the Library of Congress

British Library Cataloguing-in-Publication Data
A catalogue record for this book is available from the British Library

ISBN: 978-0-12-809855-4

For information on all Academic Press publications visit our website at
https://www.elsevier.com/books-and-journals

Publisher: Mica Haley
Acquisition Editor: Stacy Masucci
Editorial Project Manager: Sam Young
Production Project Manager: Mohanambal Natarajan
Designer: Christian Bilbow

Typeset by TNQ Books and Journals

Contents

3. Fetal and Postnatal Development of the Esophagus and Proximal Stomach

4. Histologic Definition and Diagnosis of Epithelia in the Esophagus and Proximal Stomach

5. Definition of the Normal State—A Yet Unfinished Saga

6. Definition of the Gastroesophageal Junction

Acknowledgments

This book represents a new pathologic perspective of GERD that is based on the examination of the unique material produced over a period of 25 years by the Department of Foregut Surgery at the Keck School of Medicine, University of Southern California, Los Angeles.

The vision, dedication, and commitment to the academic pursuit of esophageal diseases in that unit are almost entirely those of Dr. Tom DeMeester. His pioneering work in pH testing, manometry, Nissen fundoplication, en-bloc esophagectomy, and LINX established much of the objectivity in the practice of esophageal disease. Many ideas expressed in this book are inextricably linked to his exceptional mind.

I express my deep gratitude to Tom for innumerable hours of discussion and unfettered debate about the most arcane elements of the lower esophageal sphincter and for his and Carol's wonderful friendship.

I also owe my thanks to the exceptional surgeons who were part of his team. The most important of them were the permanent faculty: the incomparable Dr. Cedric Bremner, Dr. Jeffrey Peters, Dr. Steven DeMeester, and Dr. Jeffrey Hagen. Additionally, they include an international cast of fellows that are too numerous to mention individually. These surgeons are now the leaders of the practice of esophageal surgery in all corners of the world. They follow the vision of Dr. DeMeester and are responsible for many of the new advances in the field. Their work represents a significant amount of the evidence that is cited in this book.

Introduction: Discovery: A Path to a New Solution for Gastroesophageal Reflux Disease

This book is aimed at changing the way gastroesophageal reflux disease (GERD) is managed. Its basis is a new understanding of the cellular changes defined by histologic examination. Its primary goal is to develop a method of managing GERD with a view to preventing esophageal adenocarcinoma. Secondary goals are to preventing people from progressing to treatment failure and the misery of GERD where symptoms cannot be controlled. In the process of developing this new understanding, it has become apparent that the goal can be expanded to actually eradicating GERD as a disease.

These are lofty goals emanating from a pathologist. However, the goals and methods are developed from the unique and novel perspective of trying to explain the entire pathogenesis of GERD from the normal state to adenocarcinoma at a cellular level. This must inevitably involve the way the lower esophageal sphincter (LES), which is the mechanism designed to prevent reflux, undergoes damage. It must involve the development of a new ability to measure LES damage by pathology. This it does. This book is about one person's attempt at taking a new diagnostic test for GERD, accurate histopathologic measurement of LES damage, and taking its potential application into the future.

1. STATEMENT OF CONFLICT OF INTERESTS

I have an unusual background. After medical school (a bachelor's degree, MBBS, from Sri Lanka) and internship, I entered a neurosurgery residency. I changed course after the first year to become a lecturer in pathology at my medical school. In the 6 years I practiced medicine in Sri Lanka before immigrating to the United States, I was immersed in an unusual combination of pathology and internal medicine. My study during this time culminated in the highest postgraduate qualification in internal medicine in Sri Lanka (MD in general medicine) and the membership of the Royal College of Physicians in the United Kingdom. This strong clinical background in both surgery and medicine is unusual for any pathologist in the United States.

This introduction is designed to outline the scope of the book and provide information regarding the expertise that allows me to write this book and any conflicts of interest that may induce bias.

It is also designed to show the limitations of my ability to understand GERD from aspects with which I have no direct experience. I do not talk to, evaluate, treat, and follow patients; I do neither endoscopy nor surgery; I never look at radiologic images or interpret pH, impedance, and manometric tests. I have obtained all my information and knowledge on these areas secondhand, largely from a relatively small number of people who I talk to and trust on matters of GERD. I believe they have taught me well and they have my gratitude.

Conflict of interest plagues medicine. Government funding sources are the lifeblood that drives research of many experts in the field who largely hold positions in academic institutions. This research funding is valuable and largely unfettered but requires these experts to produce data to justify and maintain government funding and their position and promotion in the university. Many experts also receive support from private commercial sources, which makes bias difficult to avoid. In general, disclosure of these funding sources is required for physicians presenting at meetings and submitted papers for publication.

The presence of conflict does not (appropriately) disqualify the conflicted expert from a presentation or publication. It is left to the listener and reader to evaluate the conflict that is disclosed and eliminate the possibility of bias. Whether physicians listening to a lecture at a meeting or reading a paper effectively evaluates bias is unknown.

I have never applied for, let alone received, any grant funding. I have never developed any association with a commercial venture of any kind. I own no stock in a pharmaceutical or medical device company except through the mutual funds of my retirement plans and my futile efforts at playing the stock market! I have no, and have never had any, conflicts of interest.

2. PERIOD UP TO 1990: NO INTEREST IN DISCOVERY

I am, and always have been, a general surgical pathologist. Ever since I was hired as a surgical pathologist at the University of Sri Lanka Medical School in 1973, my primary life goals have been to provide the highest level of pathologic diagnosis to patients and to provide education to pathology trainees. I immigrated to the United States in 1978. At the end of my residency, I was hired into the Department of Pathology, University of Southern California (USC), as the Chief of Surgical Pathology at the Los Angeles County + University of Southern California Medical Center. I have worked in this capacity since 1982 and continue to provide high-level general surgical pathology expertise to the largely indigent patients in this hospital. I am not subspecialized in gastrointestinal or esophageal pathology in my daily function as a surgical pathologist.

I run a successful private consultation service that provides expert opinion on difficult problems in all areas of surgical pathology to community pathologists. I also direct the surgical pathology fellowship program at my institution and participate on a daily basis in teaching residents and fellows. I believe I have achieved and continue to further my life's primary goals.

Another major goal in my early career was teaching medical students. In my second year of residency, I was commissioned to write, with Dr. Clive Taylor as a coauthor, what ultimately became the Lange series textbook of Pathology, "Concise Pathology."[1]

This experience changed me. I interacted with Jack Lange's chief medical editor, Jim Ransom, who over a period of 3 years of the most compulsive and critical review tried to teach me the art of medical writing. Whatever clarity exists in my writing I owe to Jim Ransom.

I never had any ambition toward new discovery. I had no interest and made no effort toward basic research. I have always funded my own clinical research. I became a tenured professor at the University of Southern California in 1996. My job is largely in clinical service and teaching. I am under little of no pressure to publish. If not for my discovery, I would have had a relatively short publication list of papers resulting from collaboration with other researchers. I write only what I believe to be new information that is true and important for others to know.

In 1982, I became involved in a new technology to perform needle biopsies of brain lesions. Dr. Michael Apuzzo, a neurosurgeon at USC, had procured one of the prototypic stereotactic biopsy instruments. Over the next decade, I helped him develop a method of pathologic interpretation of the small biopsies he obtained from targeted points within the lesion.[2,3] Without any specialized neuropathology expertise, I found myself being considered a neuropathologist. This reputation persists; I am still a popular neuropathology consultant for pathologists in Los Angeles.

3. THE DISCOVERY: 1990 TO THE PRESENT

My academic interest turned sharply from brain to esophageal pathology in 1990 when Dr. Tom DeMeester took the position of Chairman of Surgery at the University of Southern California. He was recruited to develop a clinical program in foregut surgery at the newly built USC University Hospital. He came with his team of surgeons and staff, including nurses and technicians, who ran his physiology laboratory. His pathologist at Creighton University did not accompany his team to USC, and I was assigned to do his work.

This was an exciting new task for me. My interest in gastrointestinal pathology at that time was only as part of my routine surgical pathology work at the County Hospital. In preparation for doing the pathology for Dr. DeMeester's patients, I read every available textbook and paper on the subject. I thought I was totally prepared by the time I received my first biopsy.

3.1 The Material on Which Expertise Was Developed

Dr. DeMeester brought with him over two decades of experience and research in esophageal diseases. He had pioneered pH testing and manometry and made fundamental changes in the technique of performing Nissen fundoplication. The results of the pH test are expressed as the "DeMeester score." He was a giant in his field.

Incredibly, he had developed a unique biopsy protocol (Fig. 1) that he used for every patient who underwent endoscopy, irrespective of the presence of symptoms or endoscopic findings.

In the endoscopically normal patient (Fig. 1A), this included multiple biopsies from three specimens locations: (1) biopsies across the squamocolumnar junction, (2) retroflex biopsies of the area within 1 cm distal to the squamocolumnar junction, and (3) random biopsies of gastric body and antrum. In patients with a visible columnar-lined esophagus (CLE) (Fig. 1B), additional biopsies were taken at 1–2 cm intervals within the entire columnar-lined segment.

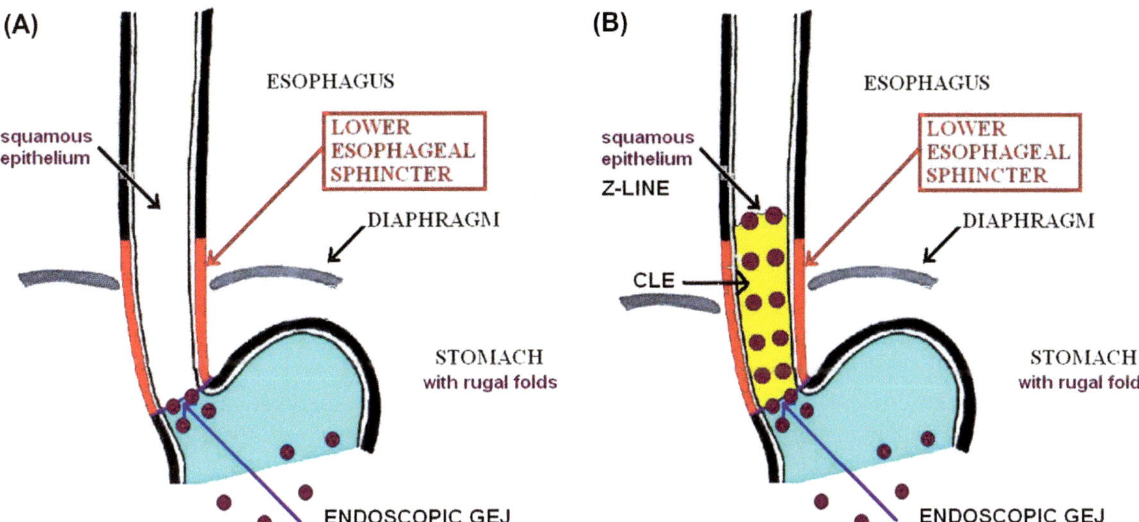

FIGURE 1 (A) Routine DeMeester biopsy protocol for endoscopically normal patients. Biopsy sets (2–4 biopsies at each level) are taken straddling the squamocolumnar junction, within 1 cm distal to the first biopsy on retroflex view, and from the distal body and antrum. (B) Biopsy protocol for patients with a 5-cm segment of endoscopically visible columnar-lined esophagus (yellow). In addition to the biopsy sets in (A), samples are taken at 1-cm intervals distal to the biopsy straddling the squamocolumnar junction to the endoscopic GEJ. All biopsies taken in antegrade view are designated as the distance from the incisor teeth. *CLE*, columnar-lined esophagus; *GEJ*, gastroesophageal junction.

By the time I was assigned to read his biopsies, he and his team were generating biopsies that were done according to this protocol from 2 to 8 patients every day. The pathologic findings were correlated at a weekly meeting with clinical, endoscopic, radiologic (most patients had a video-esophagogram), and physiologic testing (many patients had 24 h pH testing and manometry).

Over the next two decades, Dr. DeMeester's group generated these biopsies continuously at the rate of 800 patients per year for a total of over 15,000 patients. I just happened to be the recipient of this unique pathologic material from the esophagus and stomach. The only reason for my discovery was that it was so obvious in the material that I could not help it; the discovery simply fell into my lap.

Evaluating the biopsies at the endoscopic gastroesophageal junction (GEJ) and a retroflex biopsy from what was called the "cardia" in all people including those with normal endoscopy resulted in the almost immediate realization that the entire literature relating to normal histology in this region was flawed.

The reason why the flaw was not previously recognized is that no one had taken biopsies of normal people in this way. The reason why it is difficult for me to convince people of the discovery is that no one takes biopsies of normal people in this way even today. The American Gastroenterological Association expressly recommends that biopsies should *not be taken* in patients with GERD who do not have an endoscopic abnormality.[4]

I am waging a battle against an unproved dogma that cardiac mucosa is a normal epithelium in this region. Most people arguing on the other side are so steeped in the belief of this false dogma that they refuse to even test the alternative evidence-based truth that cardiac mucosa is an abnormal metaplastic epithelium in the esophagus, not a normal epithelium in the stomach. Whenever anyone has tested this concept, the data have largely supported this concept.

This book is ultimately about that new discovery, its evolution over two decades, the research it has stimulated throughout the world, the present state of its acceptance by the medical community, and its yet unrealized potential value in the future management of GERD.

3.2 Cardiac Epithelium Does Not Exist Normally

The discovery was simple. What was called "cardiac mucosa/epithelium" and believed to normally line the most proximal part of the stomach, which was called the "gastric cardia," was, in fact, neither normal nor gastric. It was *always* an abnormal epithelium that resulted when the squamous epithelium of the esophagus was exposed to gastric acid.

In this book I will use the term "cardiac epithelium" rather than the more commonly used "cardiac mucosa." The reason for this was a question by Tom DeMeester: "Para, why do you use the term cardiac mucosa? Does the mucosa not include the lamina propria and muscularis mucosae? Is it not more accurate to use the term cardiac epithelium?" Yes, Tom, it is.

Unfortunately, like many changes that make sense, this will cause dismay. The short forms for the three epithelial types in CLE will change from CM with intestinal metaplasia, CM, and OCM (oxyntocardiac mucosa) to CE with intestinal metaplasia, CE, and OCE. I will not use these abbreviations in the book. One pleasant side effect of this change is that "nonintestinalized cardiac epithelium" which did not have an abbreviation can now be called "NICE." I will not apologize for this change. Accuracy in terminology is worth the minor confusion it may cause.

I will limit the use of abbreviations in the text of this book to the following: GERD for gastroesophageal reflux disease; CLE for columnar-lined esophagus; LES for lower esophageal sphincter; GEJ for gastroesophageal junction; and SCJ for squamocolumnar junction.

At the time I began to read the biopsies of Dr. DeMeester's patients in 1990, it was believed that cardiac epithelium normally lined the distal 2 cm of the esophagus and the proximal 3 cm of the stomach. As I began seeing specimens from the squamocolumnar junction and within 1 cm distal to it, it became immediately clear that many people had less than 1 cm of cardiac epithelium. Some had no cardiac epithelium at all in the biopsies.

In the correlative weekly conferences, it became clear that the presence of cardiac epithelium in the biopsies was greatest in extent and amount of inflammation in patients with GERD and least in those without GERD. It was therefore obvious to me from the outset that cardiac epithelium was an abnormal epithelium associated with GERD.

The hypothesis that cardiac mucosa at the GEJ was an abnormal epithelium caused by exposure of esophageal squamous epithelium to gastric contents rather than a normal epithelium lining the proximal stomach was first published from our group as an abstract at Digestive Diseases Week by Clark et al. in 1994.[5]

A review of a large series of our patients in 1997 confirmed my observation that the presence of cardiac epithelium in a biopsy of an endoscopically normal patient correlated with the presence of abnormal reflux by a 24-h pH test and associated with abnormalities in the LES.[6]

Publication of the new hypothesis created a considerable amount of interest in the medical community, largely among surgeons. I was invited to a panel discussion on Barrett esophagus at the Digestive Diseases Week and to give the keynote address at the SSAT in 1998. Both these invitations resulted because the faculty of the USC Department of Surgery (Dr. Tom DeMeester and Dr. Jeffrey Peters) had influence regarding the selection of the speaker.

The reaction of the audience to the new information was interest mixed with skepticism and confusion. Histology is not a subject that is quickly assimilated by nonpathology audiences. The keynote SSAT presentation required me to write a review that outlined the new hypothesis.[7]

3.3 Autopsy Studies of the Region

A review of the literature showed that there had never been any previous autopsy study of the GEJ in people without clinical GERD. All the information regarding the normal histology of the region was derived from studies in patients with GERD.

In 1996, I undertook an autopsy study that essentially showed that cardiac mucosa was either absent (Fig. 2) or very limited in length (Fig. 3) in the majority of non-GERD people.[8] This confirmed the hypothesis, at least in my mind. Surely, I thought, a normal epithelium cannot be absent in some people. I knew I had an important new piece of information that needed to be advertised to the world.

Incredibly, our paper reporting the detailed histology of the GEJ region in the first autopsy study ever done was rejected by every medical, gastroenterology, and pathology journal, to which it was submitted from 1996 to 2000[8]

Publication of the paper followed a presentation at the International Academy of Pathology in San Francisco where I was invited to present my autopsy findings. At the end of the conference, the then editor of the *American Journal of Surgical Pathology* came up to me and expressed his interest in the paper and asked me to send it to him. I did not have the heart to tell him that the paper had been summarily rejected previously by the journal he edited.

Although scientific papers are supposed to be evaluated entirely on their content, a paper is much more likely to be accepted when the author has a reputation. This is appropriate because establishing credibility of the data presented in a paper is important and difficult to assess in the review process when the author is unknown. Networking is a critical part of getting known and I had not done that at all. It was largely my fault that the critical information was rejected; it was too radical to be believed when the source was unknown.

3.4 The Normal Histologic State of the Esophagus and Stomach: The Zero Squamooxyntic Gap

If cardiac mucosa was always an abnormal metaplastic epithelium in the esophagus, it *must* follow that the entire normal esophagus was lined by squamous epithelium and the entire normal proximal stomach (including the area adjacent to the junction with esophageal squamous epithelium, fundus, and body) was lined by gastric oxyntic epithelium.

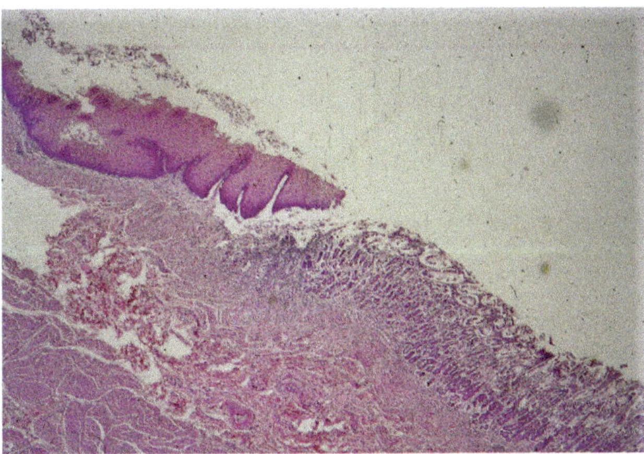

FIGURE 2 Section taken at the squamocolumnar junction at autopsy in a young child who died of a road traffic accident. The esophageal squamous epithelium transitions to gastric oxyntic epithelium characterized by straight tubular glands containing parietal cells. No cardiac epithelium is present. This is a zero squamooxyntic gap.

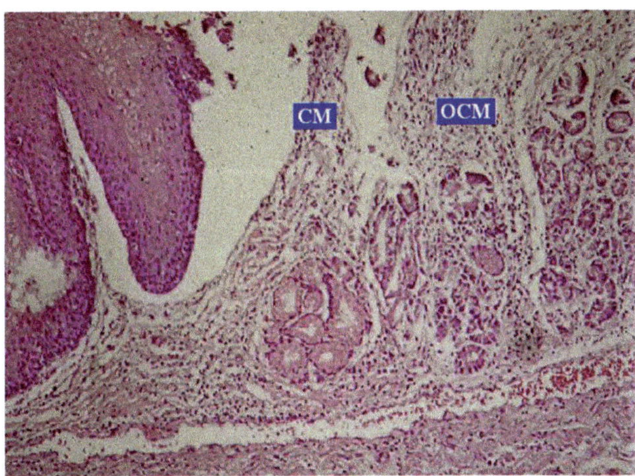

FIGURE 3 Section taken at the squamocolumnar junction at autopsy in a 15-year-old male who died of a drug overdose. The esophageal squamous epithelium transitions to cardiac and oxyntocardiac epithelium. Cardiac epithelium has only mucous cells. Oxyntocardiac epithelium has small round parietal cells with central nuclei in addition to the mucous cells. No gastric oxyntic epithelium is seen in this section. This represents the proximal part of the squamooxyntic gap. *CM*, cardiac mucosa; *OCM*, oxyntocardiac mucosa.

If accurate, this would greatly simplify the understanding of GERD. Of the two epithelia normally present in this region, esophageal squamous epithelium was the only one that was damaged by exposure to gastric juice. Gastric oxyntic epithelium was obviously resistant.

This observation resulted in the development of the concept of the squamooxyntic gap.[9] If squamous epithelium lined the entire esophagus and gastric oxyntic epithelium lined the entire proximal stomach, it meant that the normal GEJ was the junction between the squamous epithelium and gastric oxyntic epithelium. There was no gap between squamous and gastric oxyntic epithelium (Fig. 4A). The normal state was a zero squamooxyntic gap. If cardiac epithelium resulted from the metaplasia of squamous epithelium, it must cause a cephalad migration of the squamous epithelium, separating it from gastric oxyntic epithelium by the intervening metaplastic cardiac epithelium, thereby creating a squamooxyntic gap (Figs. 4B and 5).[9]

3.5 A Cellular Definition of Gastroesophageal Reflux Disease

The squamooxyntic gap composed of cardiac mucosa identifies with absolute accuracy people whose esophageal squamous epithelium has been damaged by exposure to gastric juice. I suggested this as a cellular definition for GERD in 2005 in a review that I was invited to write for *Histopathology*, the British journal for the specialty of surgical pathology.[10]

The gastroenterology community has largely ignored this definition. In many large consensus meetings that were convened to evaluate a definition for GERD, a cellular definition has never even been considered.[11,12] The American College

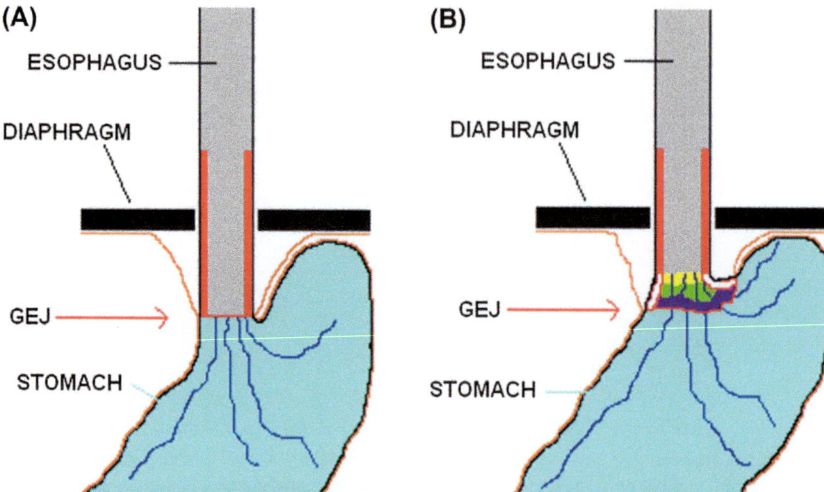

FIGURE 4 (A) Diagrammatic representation of the normal state. The entire esophagus is lined by squamous epithelium (gray), which transitions at the gastroesophageal junction (GEJ) to gastric oxyntic epithelium that lines the entire stomach (blue). The entire LES (red in wall) is intact. The rugal folds reach the GEJ. There is no cardiac mucosa between squamous and gastric oxyntic epithelium, i.e., the squamooxyntic gap is zero. (B) Diagrammatic representation of the majority of people in the world with no visible columnar-lined esophagus at endoscopy. The distal part of the lower esophageal sphincter (LES) has been damaged (white replacing red in the wall) and the squamous epithelium has been replaced by metaplastic columnar epithelium (yellow, green, and purple). This part of the esophagus has dilated to assume the contour of the saccular stomach. The tubular esophagus has shortened and the angle of His has become more obtuse. Rugal folds extend above the true GEJ (marked by the proximal limit of gastric oxyntic epithelium). Note the concordance between the shortening of the LES, dilatation of the distal esophagus, and length of metaplastic columnar epithelium.

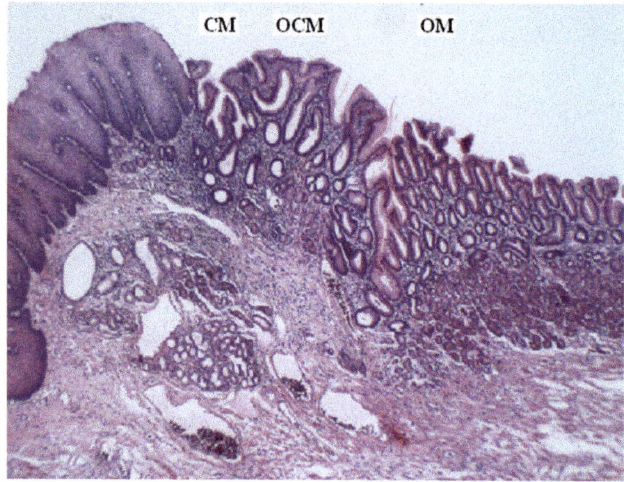

FIGURE 5 Histologic section of a 2-mm squamooxyntic gap in a young child without gastroesophageal reflux disease. The distal limit of squamous epithelium (SCJ) is separated from the proximal limit of gastric oxyntic epithelium (GEJ) by a 2-mm segment of metaplastic cardiac and oxyntocardiac epithelium. Note the disorganized, thin metaplastic cardiac epithelium with chronic inflammation (=reflux carditis) compared with the uninflamed, thicker gastric oxyntic epithelium.

of Gastroenterology's position statement on GERD recommends that the diagnosis of GERD should be made by its symptoms and the empiric proton pump inhibitor (PPI) test, without the use of barium radiographs, upper endoscopy, biopsies, manometry, and pH monitoring.[13]

However, interest in the concept remains. In a recent review that I was invited to write by Dr. Stuart Spechler, the highly influential gastroenterologist, I was encouraged to reiterate the concept and the definition.[14] The mere duration of survival of interest in the hypothesis and the fact that it still has forward momentum suggest it has merit.

3.6 Location of Cardiac Mucosa: Definition of the True Gastroesophageal Junction and the Definition of the Dilated Distal Esophagus

I struggled for many years with an obvious contradiction. Cardiac epithelium that resulted from reflux was found distal to the endoscopic GEJ, i.e., in the proximal stomach (Fig. 4B). This was impossible; GERD could not produce damage in the stomach. Something was awry.

In 2007, I undertook a study on esophagectomy specimens that provided the answer (Fig. 6).[15] The reason for the contradiction was that the universally used definition of the GEJ (the proximal limit of rugal folds as defined by endoscopy and gross examination of specimens) was incorrect. The study showed that the true GEJ was distal to the endoscopic GEJ by 0.36–2.05 cm in the 10 cases in the study.[15] This led to a new definition of the GEJ as the proximal limit of gastric oxyntic epithelium.[15] Endoscopy was not capable of defining the true GEJ because it cannot differentiate cardiac and gastric oxyntic epithelia. This can be done only by histology.

This made logical sense. As the distal squamous epithelium underwent columnar metaplasia by exposure to gastric juice, the squamocolumnar junction would move cephalad. The proximal limit of gastric oxyntic epithelium would remain at the same location, marking the original true GEJ. There would be a gap between the endoscopic GEJ and the true GEJ; in the person without visible CLE, this is the squamooxyntic gap (Fig. 5). The gap is easily identified by the presence of cardiac epithelium (with and without parietal and/or goblet cells) (Fig. 7) between the residual esophageal squamous epithelium and gastric oxyntic epithelium (Fig. 4B).

A finding in this study of esophagectomy specimens that took me by surprise was the anatomic appearance of the damaged distal esophagus that was lined by metaplastic cardiac epithelium. It was no longer tubular. It was dilated and had taken the contour of the stomach. Its mucosal surface had rugal folds (Fig. 6A and B).[15]

The damaged abdominal segment of the esophagus had an appearance that confused physicians throughout history and continues to do so. It exactly resembled everyone's concept of the proximal stomach. It was part of the dilated sac distal to the presently accepted endoscopic GEJ (proximal limit of rugal folds) and the distal end of the tubular esophagus.

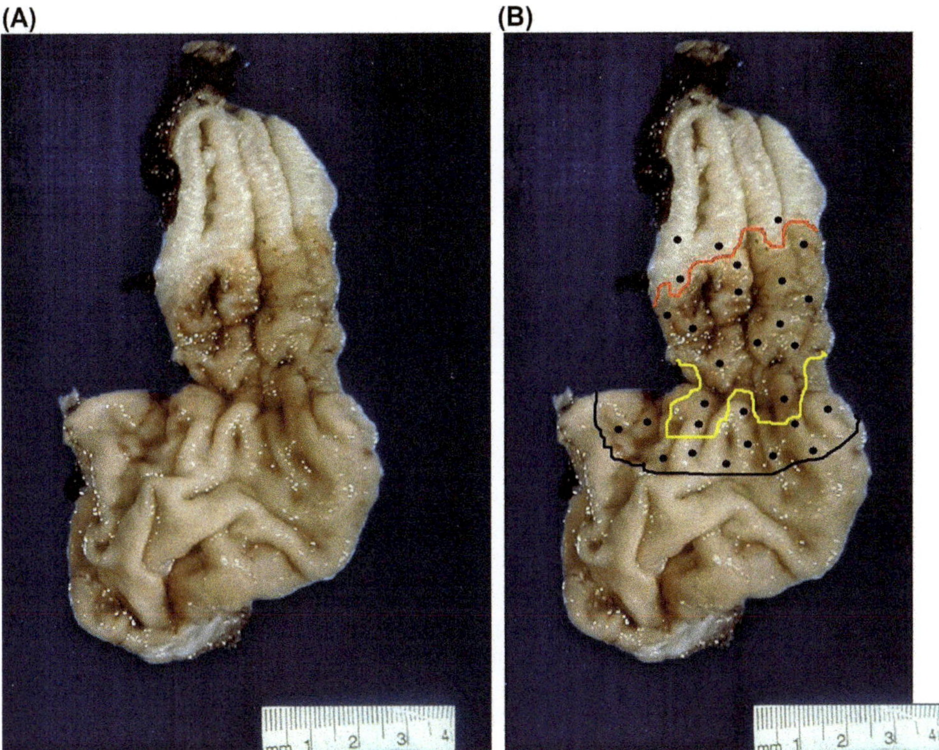

FIGURE 6 (A) Esophagogastrectomy specimen in a patient with an adenocarcinoma arising in a long segment of Barrett esophagus. The specimen shows rugal folds extending to the point where the tubular esophagus flares into what most people think is stomach. (B) Histologic study shows cardiac epithelium with intestinal metaplasia extending from the squamocolumnar junction (*red line*) to the *yellow line*. Cardiac epithelium without intestinal metaplasia extends to the *black line*, which is 2.05 cm distal to the proximal limit of rugal folds and the end of the tubular esophagus. The fact that this 2.05 cm between the presently defined gastroesophageal junction and the proximal limit of gastric oxyntic epithelium (*black line*) is esophagus is proved by the presence of esophageal submucosal glands (*black dots*) up to the *black line*.

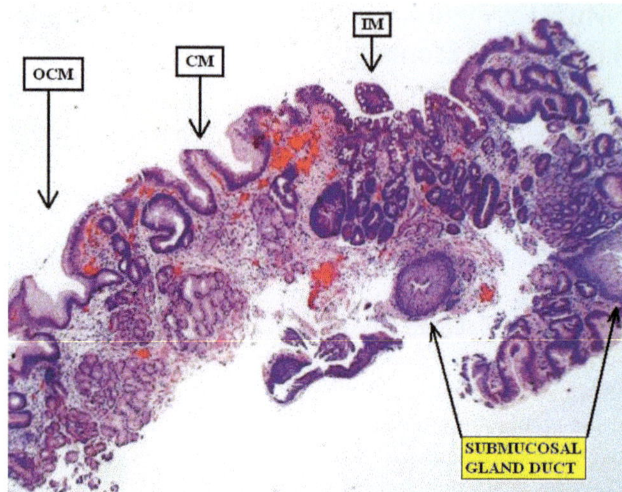

FIGURE 7 A retroflex biopsy taken immediately distal to the squamocolumnar junction in a person with no visible CLE at endoscopy. The biopsy shows oxyntocardiac epithelium (OCM; with parietal cells), cardiac epithelium (CM; mucous cells only), and cardiac epithelium with intestinal metaplasia (IM; with goblet cells). Note the presence of squamous-lined ducts of esophageal submucosal glands. This proves that this biopsy from distal to the endoscopic GEJ is esophageal. *CLE,* columnar-lined esophagus; *CM,* cardiac mucosa; *GEJ,* gastroesophageal junction; *IM,* intestinal metaplasia; *OCM,* oxyntocardiac mucosa.

After trying out many terms for this new entity, I named it the dilated distal esophagus. Credit for this term belongs to Dr. George Triadofilopoulos. Sitting on a balcony in a hotel in Dubrovnik, overlooking the Adriatic Sea, discussing the subject, George complained about the term I had previously used—dilated end-stage esophagus. He said it confused him because his mind immediately went to the esophagus of achalasia of the cardia. After a few more drinks, the term dilated distal esophagus was suggested by George and endorsed by everyone.

The paper that provided an evidence-based definition of the GEJ and clarified the decades-old error wherein the GERD-damaged distal esophagus was deemed to be the proximal stomach was also rejected by multiple gastroenterology journals before being accepted in a pathology journal.[15]

This highlights an important issue about the medical literature. There is no evidence whatsoever to support the present belief among gastroenterologists that the proximal limit of rugal folds is the true GEJ. The proximal limit of rugal folds is simply an endoscopic landmark.[16] The basis for the definition is the opinion of experts.[17] In a world where all experts pronounce that they practice evidence-based medicine, the same experts have no trouble rejecting evidence that contradicts their opinion that is baseless in evidence. It is notoriously difficult to overturn strongly held dogma by evidence.

3.7 Correlation Between the Dilated Distal Esophagus and Damage to the Lower Esophageal Sphincter

Approximately 3 years ago, at a presentation of the hypothesis to a small meeting of community gastroenterologists in San Diego, I was asked a critically important question: "Dr. Chandrasoma, let us assume that you are correct and cardiac epithelium is esophageal and not gastric. What practical value does that information have for me when I treat my patients?"

I tried to tell that gastroenterologist that there surely was scientific value in being able to differentiate between esophagus and stomach when treating a patient with GERD. This was a shallow reply; this new concept would have little value if it did not impact the management of patients with GERD. What did it matter to that gastroenterologist if he was mistaking the GERD-damaged distal esophagus for the proximal stomach if it did not change the way he treated his GERD patient? He would still use PPIs the same way he would have done without that information. The purity of knowledge without a practical value held no attraction to him.

That gastroenterologist was correct. He had given me the reason why there was so much resistance to these new concepts. It was not necessarily that anyone believed I was wrong. It was because they did not find it necessary to change because it made no change in the way GERD was treated.

That question made me delve deeper into what I was seeing. It forced me to find that practical value. In this book, I will present that value in great detail. Ultimately, the power of what I was seeing will lead to a new diagnostic test for GERD based on measuring LES damage histologically. This, in turn, will permit prediction of future LES damage in the patient. The two combined will provide a method of preventing esophageal adenocarcinoma and, if followed aggressively with new technology, allow the complete eradication of GERD as a human disease.

The dilated distal esophagus, which is the area of the GERD-damaged esophagus between the endoscopic GEJ and the true GEJ (the proximal limit of histologically defined gastric oxyntic epithelium), varies in length from 0 to a maximum published length to date of 28 mm and a theoretical length of 35 mm. I will show that the dilated distal esophagus results from damage to the distal esophageal squamous epithelium when it is exposed to gastric contents as a result of gastric overdistension. Such damage ultimately causes cardiac metaplasia, which is associated with loss of tone in the LES.

The length of the metaplastic columnar epithelium in the dilated distal esophagus is equal to the shortening of the abdominal segment of the LES. Measurement of the squamooxyntic gap in the dilated distal esophagus therefore provides a pathological method of assessing the severity of damage of the abdominal segment of the LES.

This has proof. Examination of full thickness sections of the dilated distal esophagus shows the presence of submucosal glands under cardiac epithelium.[15,18] Submucosal glands are limited to the esophagus. They are never found under gastric oxyntic epithelium.

I will show in this book how this assessment of LES damage is achieved practically by biopsy if an appropriate sample can be obtained. This will likely require the development of a new biopsy device that can remove a 20–25 mm vertical piece of mucosa from the region of the dilated distal esophagus.

This has the potential to change the way GERD is defined, assessed, prognosticated, and managed from its present symptom-based method to one that is based on severity of LES damage, the cause of GERD (Table 1). That one question in a small nonacademic meeting had a tremendous impact. I now have an answer to that gastroenterologist whose name I do not know.

The new test will provide an accurate and precise measurement of abdominal LES damage at any point in a person's life, irrespective of whether that person has symptoms or not. Even more important, using an innovative algorithm, a prediction can be made as to future progression of LES damage to the end of the person's expected natural life.

By this, the test will identify which patients will progress in the future to higher stages of abdominal LES damage that are associated with failure of PPIs to control symptoms, Barrett esophagus, and adenocarcinoma. All of these are associated with the severity of reflux into the esophagus, which in turn is dependent on the severity of LES damage. In effect, this book is a root cause analysis of GERD.

This ability to predict future LES damage will allow identification of the minority of patients who will progress to the category of severe GERD with sufficiently abnormal reflux to place the patient at high risk of treatment failure and visible CLE. This will permit early intervention in this minority to prevent this progression of LES damage. Theoretically, this attack of the primary cause of GERD, LES damage, will simultaneously reduce the incidence of adenocarcinoma and treatment failure.

4. ACCEPTANCE OF THE NEW CONCEPTS

The discovery that cardiac mucosa is a GERD-induced metaplastic columnar epithelium in the esophagus rather than a normal proximal gastric epithelium is not yet accepted, although more and more people are beginning to believe that it has substance.

The trend toward acceptance is excellent. When I read the literature in 1990 in preparation for Dr. DeMeester, it was believed that cardiac mucosa normally lined the distal 2 cm of the esophagus and proximal 3 cm of the stomach. Largely because of the data generated in our unit, there is universal agreement that most "normal" (non-GERD) people have less than 0.4 cm of cardiac epithelium in this region.[19] Many persons without GERD have no cardiac epithelium; some have no epithelia other than esophageal squamous and gastric oxyntic in at least a part of the circumference of the squamocolumnar junction.[8]

4.1 The Value of Clinical Research

The concept of medical research has changed. In an effort to prove that medicine is a science rather than an art, physicians have established a hierarchy that defines the value of research. Basic research is now the king; clinical research has been relegated into the background and, in many circles of elite medical scientists, regarded as inferior.

Physician scientists doing basic research in a laboratory have the luxury of developing a hypothesis, planning a scientific experiment, establishing appropriate tests and controls, and then performing a highly specific single test with sophisticated technology in a perfectly reproducible manner. The results are unquestionably true as long as the design and methodology are precise and the researchers are honest and interpret the results accurately and without bias. No defects exist in a well-planned experiment.

Such basic research is beautiful. Truths resulting from experiments come into the mainstream quickly after independent duplication. Funding for basic research is relatively easy to obtain. It is the lifeblood of scientific advancement.

TABLE 1 Stages of GERD Progression From the Point of Damage (Shortening) of the Abdominal Segment of the Lower Esophageal Sphincter (LES)

	AbdLES Damage	Residual AbdLES	Reflux 24-h pH Test	Symptoms	Cellular Changes
Normal	0	35 mm	0	0	0
Phase 1: compensated LES damage	>0–15 mm	<35–20 mm	0 to pH < 4 for <1.1%	0	CLE in DDE 15 mm or less
Phase 2: subclinical GERD	>15–20 mm	<20–15 mm	>0 to pH< 4 for <4.5%	Yes/no; not "troublesome"	CLE in DDE > 15–20 mm NERD Mild EE
Phase 3: clinical GERD, mild	>20–25 mm	<15–10 mm	pH > 0 to pH < 4 for >4.5%	Yes/no; "troublesome"	CLE in DDE >15–25 mm NERD Mild/severe EE
Phase 4: severe GERD	**>25–35 mm**	**<10-0 mm**	**pH < 4 for >4.5%**	**Yes/no; "troublesome"**	**CLE in DDE > 25 mm** **Risk of visible CLE**

AbdLES, abdominal segment of the LES; *CLE*, columnar-lined esophagus; *DDE*, dilated distal esophagus; *GERD*, gastroesophageal reflux disease; *NERD*, nonerosive reflux disease.

The problem with such beautifully controlled basic research is that they usually address very small questions. Experiments are developed to test something that has been conceptualized. Their truths are often divorced from the patient and generated in animals or cell lines. The progress they produce is in small and logical steps, and many such small steps are required to develop a larger truth. However, such a cumulative change is inevitable, orderly, and predictable, albeit slow, and is the way science progresses.

I am not a basic scientist; my research is clinical. I am a busy, practicing academic physician who spends most of my time in patient care and teaching. I do not have the luxury of working entirely in a sterile laboratory that is the domain of the basic scientist. My salary is not supported by a grant. I encounter patients with the disease in which I am interested mixed with thousands of patients with many other diseases. My work can never have that knifelike focus and control of the laboratory experiment.

Clinical researchers have a completely different value to science than basic researchers. We see a problem as a whole in the context of a human population. We encounter a reality that is much broader and deeper. The reality is often precipitated by an unexpected observation, not directed by a prior body of research. We are emotionally affected by seeing patients suffer as a result of their disease and our impotence or failure. We desire to make things better quickly and solve entire problems, not in little steps but in one fell swoop. This is incredibly difficult and therefore rare. But, when it happens, the results are miraculous.

The rising incidence of esophageal adenocarcinoma is the most dramatic change in a disease that I have encountered in my lifetime. When I started looking at pathology slides in 1978 in Los Angeles, almost all esophageal cancers were squamous. I must have seen one adenocarcinoma for every 10 squamous carcinomas. Over the next 35 years, the numbers reversed.

This was partly because of my association with Dr. DeMeester, but a quick search of the literature showed that the increased incidence of esophageal adenocarcinoma was very real and much greater than any other cancer type (Fig. 8).[20]

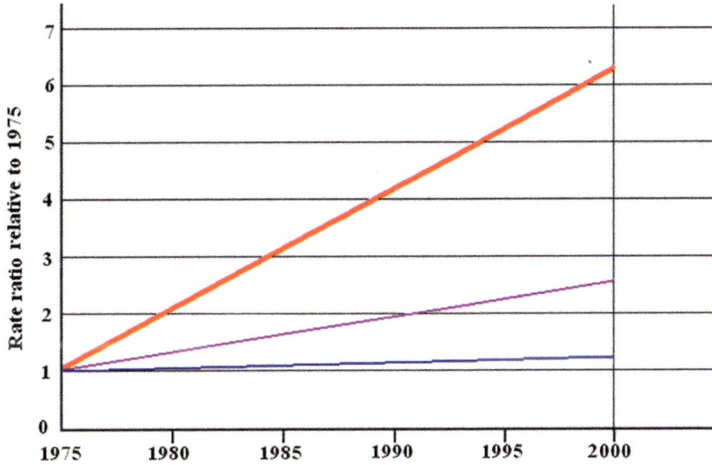

FIGURE 8 The sixfold increase in incidence of esophageal adenocarcinoma from 1973 to 2000 (*red line*) compared with the incidence of breast cancer (*purple line*) and colorectal cancer (*black line*) during the same period.

As a pathologist, any research goal that I would direct my efforts toward would have something to do with trying to impact the incidence of esophageal adenocarcinoma. I have no interest in taking baby steps toward that goal.

Everyone recognizes the difficulties associated with clinical research. The only clinical research that is considered truly scientific is the double-blinded randomized prospective clinical trial. The resources to conduct randomized trials are not available to most physicians like me. They require significant funding, large numbers of patients, which generally means a multicenter trial, a relatively narrow hypothesis, and accurate collection and analysis of critical data. Like basic research, randomized clinical trials usually answer small specific questions. Again, the little steps advance knowledge slowly but in an invaluable way because the results are irrefutable.

The studies done by clinical researchers like me tend to be messy and compare poorly with basic research and randomized clinical trials. The test base often has arbitrary selection criteria and the controls are imperfect based more on availability than design. The study methods frequently lack uniformity. When the data are presented, defects are common. Interpretation is subjective and almost always open to criticism. For these reasons, clinical researchers find it impossible to get published in the best scientific journals. Clinical research dwells in the pages of relatively low-impact clinical journals. All my papers are in pathology journals that are probably never read by gastroenterologists. The data and ideas are hidden in the massive volume of literature that emerges on a monthly basis.

Because of all these limitations, it is easy for my findings and truths to be relegated into obscurity. The scientific community does not need to even refute my findings. They can be simply ignored on the basis that they represent "poor science" or that they are in "low-impact journals." Many physician scientists regard clinical research to be of little value in the new evidence-based world.

This is scientific elitism at its worst. Unlike scientists in laboratories, clinicians are on the front line. They are the only people who can make direct observations about what happens to real people. When clinical research stumbles upon a truth, that truth is often unattainable by any other type of research; not hundreds of basic science experiments and clinical trials. The reason for this is that the truth is such a deviation from accepted dogma that it is rarely a hypothesis for a planned experiment; it must dawn out of nowhere by a clinical observation. Such a truth has the power to suddenly cause new significant direction and change.

Clinicians who keep their eyes and minds open see things that never reach the brains of scientists in isolated ivory towers. Their observations have unique value: Jenner with cow pox, Marshall and Warren with *Helicobacter pylori*, and Semmelweiss with simple handwashing that he observed to decrease the incidence of puerperal sepsis were all clinical researchers. Their observations produce results and change that span centuries. Their research, based on one simple clinical observation, resulted not in small, incremental change but in revolution. These examples have not taught elitist scientists to be careful about belittling clinical research.

Despite the fact that I have found it extremely difficult to get other physicians to agree with me, I am absolutely convinced of my new truths about the pathophysiology of GERD. I also believe that these truths will potentially change the way GERD is fundamentally understood and that it will revolutionize research, treatment, and outcomes.

My published papers have been read, possibly with skepticism but largely without dissent. My lectures have been heard, again largely without dissent. But acceptance of these new truths has been painfully slow.

I used to feel frustrated at the pace of change but now recognize that this is the result of the nature of my research. I expect pathologists to give up their firm dogmas and accept a new truth that they do not encounter because the biopsies they are given do not have the thoroughness of our biopsy protocol. I expect physicians who never look at microscopic slides to understand histologic findings. This is unreasonable.

This is not a new phenomenon. Dr. Barry Marshall, a clinical microbiologist, and Robin Warren, a pathologist, in Australia, who in 1979 discovered the truth of *Helicobacter pylori*, encourage me. The fact that Marshall was led by his frustration at the disbelief of the world of his new truth to swallow a culture of the bacteria to prove his point is discouraging. But his end point of total acceptance is positive. The entry in Wikipedia of what they achieved: "Marshall and Robin Warren showed that the bacterium *Helicobacter pylori* (*H. pylori*) is the cause of most peptic ulcers, reversing decades of medical doctrine holding that ulcers were caused by stress, spicy foods, and too much acid. This discovery has allowed for a breakthrough in understanding a causative link between *Helicobacter pylori* infection and stomach cancer." Two relatively obscure clinical researchers succeeded in overcoming massive establishment resistance to change the world.

Like Dr. Marshall, who swallowed a culture of *Helicobacter pylori* to prove his case, I have had an endoscopic assessment of my LES according to my new protocol. I will describe the process and results in a later chapter. At this time, I will disclose that I know for certain that I am at zero risk for GERD and esophageal adenocarcinoma in my lifetime as a result of my test. Can anyone reading this make this statement with certainty?

4.2 Impossible Contradictions

The acceptance that there is only a very small amount of cardiac mucosa distal to the endoscopic GEJ has created numerous contradictory viewpoints. These contradictions are simply ignored. This is not acceptable. In science, contradictions in viewpoint must either be explained or rejected.

There is strong evidence that the presence of a very small amount of cardiac epithelium at the GEJ is associated with GERD. Oberg et al.[6] showed that people who had cardiac mucosa at the GEJ were more likely to have an abnormal 24-h pH test and LES abnormalities than those without. It has been shown by Glickman et al.[21] that children with >0.1 cm of cardiac epithelium at the GEJ have more reflux esophagitis than those with <0.1 cm. This contradicts the view that 0.4 cm of cardiac mucosa represents normal gastric epithelium. The truth that any cardiac mucosa distal to the squamocolumnar junction is a manifestation of GERD is the only rational explanation for this contradiction.

It is known that cardiac epithelium is frequently present in amounts that are much greater than the 0.4 cm that is considered normal. After all, it was accepted that the proximal 3 cm of the stomach was normally lined by cardiac epithelium. It is true that up to 3 cm of the region distal to the endoscopic GEJ can be lined by cardiac mucosa. What is also true is that this 3 cm is not normal and it is not stomach as is believed. The longest metaplastic epithelium reported is 2.8 cm.[18]

If normal cardiac epithelium in the proximal stomach is <0.4 cm, what is the disease process in the stomach that is defined by cardiac epithelial length from 0.4 to 2.8 cm? There is none. The presence of cardiac epithelium greater than 0.4 cm in length is still regarded as "normal" or simply ignored.

The only reasonable explanation for this is that the length of the dilated distal esophagus is equal to GERD-induced damage of the LES. In fact, dilatation of the abdominal esophagus *only* occurs when the protection of the high pressure zone of the LES is lost. It is not a coincidence that the abdominal segment of the LES has a normal length of approximately 3.5 cm.

Adenocarcinoma arising in the region distal to the endoscopic GEJ is called "adenocarcinoma of the gastric cardia." For decades, this has been classified as a proximal gastric cancer. It is known that adenocarcinoma of the gastric cardia is associated with GERD.[22] How can a gastric cancer be caused by GERD? This has no answer. The reality is that "adenocarcinoma of the gastric cardia" is adenocarcinoma of the dilated distal esophagus. Interestingly, in 2010, the American Joint Committee on Cancer changed the official classification of cancers arising within 5 cm distal to the endoscopic GEJ from gastric to esophageal in their seventh edition.[23]

We now have the contradiction that cancer that is found distal to the endoscopic GEJ will be diagnosed at endoscopy as a gastric cancer by the gastroenterologist, but will be reported as an adenocarcinoma of the esophagus by the pathologist who follows the AJCC classification guidelines. This serious contradiction is ignored.

All these contradictions resolve immediately when the words "gastric cardia" are replaced by "dilated distal esophagus," a condition resulting from columnar metaplasia of the squamous epithelium of the distal abdominal esophagus caused by exposure to gastric contents.

4.3 My Responsibility for the Failure of Acceptance

Ultimately, the responsibility to get these new concepts accepted rests upon my shoulders. My goals and ambitions have not changed to induce a change in my work habits or behavior. I still work hard at providing excellent surgical pathology to my patients at the Los Angeles County Hospital and teaching residents and fellows. I spend most of my time in these endeavors. This is the source of my job satisfaction.

I am not an academic animal. I have never worked in a laboratory. I have never asked anyone to provide financial support for my research. I have never registered for or attended a medical meeting unless I have been invited to present a lecture or participate in a panel. I am an academic recluse by choice. I do not aggressively push these new concepts. I am easily ignored.

I unfailingly attend one conference on the Medical and Surgical Aspects of Diseases of the Foregut held in Hawaii every year. It is a wonderful conference developed by Dr. Tom DeMeester, a renowned esophageal surgeon, and Dr. Don Castell, a renowned gastroenterologist. It has lasted over 30 years.

One of the almost invariable attendees at this conference since I first participated in 1991 is a surgeon from California, Dr. Dennis Wilcox. He faithfully sits in the front row and videotapes the entire conference. A few years ago, he gave me copies of the videos of all the past meetings. I could follow how I presented my discovery through the years. First with hesitancy and then increasing confidence as the ideas became clear in my mind and appeared to have enthusiastic acceptance by some and dissent by none. The concept that I presented has never wavered. It has become stronger and expanded to new areas, and I present it with more conviction with every passing year. The audience skepticism has progressively decreased. Almost everyone who believes the new discovery has heard it at that conference. If the entire world were the audience at that conference over the years, these concepts would have been in the mainstream a long time ago.

I do not belong to any pathology or gastroenterology societies or groups. I do not socialize in this academic world. I regard my reclusive nature as being crucial to the way my thought process developed over the past two decades. I was always independent of mainstream collegial opinion. Medicine, while it claims to be evidence based, practically works through opinions shared at these collegial meetings. This results in a collectivism of thought that impedes new radical ideas. Concepts developed by "consensus" have become valued. I believe that my lack of involvement in these consensus generating groups and meetings played a powerful role in maintaining my ability to develop radical views. Truth is not reached via a vote of the majority. It is a singular fact.

I have only had one academic GI pathologist in the United States who I have ever considered to be a friend. He was Dr. Rodger Haggitt, the renowned Chair of Pathology at the University of Washington. I had met him in a conference in Missilac, France, many years ago. We were the only pathologists at a meeting of clinicians from Europe and the United States. Marooned in a remote town without a car, he spent time with my family during the week driving around the Brittany countryside. It is a rule that anyone who interacts with me when I am with my family members has a higher regard for me than when they deal with me alone.

One of the great losses to the world of GI pathology was the tragic and premature demise of Dr. Haggitt. The last conversation I had with him before his death was shortly after we returned from France, and he had presented at the grand rounds of the Los Angeles County + University of Southern California Medical Center (an invitation that was accepted when I lured him with his favorite wine at a restaurant that served catfish with Ponzu sauce, and the loan of my BMW-M3 with music by Ella Fitzgerald to test drive on Pacific Coast Highway after his lecture!).

I was driving him from his hotel to the meeting. He brought up the subject of my new ideas regarding cardiac epithelium that I had presented at the meeting in France. To the best of my recollection, the conversation went like this:

> *"Para," he said, "Your ideas regarding cardiac mucosa are interesting." "Rodger," I replied, "Does that mean that you do not believe a single word of it?"*
>
> *"Yes," he said, "but what infuriates me is that I have thought about it constantly since France and I still cannot find the flaw in it."*

I am convinced that one result of Dr. Haggitt's death is that over a decade later, the GI pathology world still believes that cardiac mucosa is a normal epithelium in the stomach. I am sure that if he had lived, I would have convinced him of the true nature of cardiac epithelium in very short order and his influence would have facilitated the concept's acceptance. Open minds like his are rare.

4.4 The Reason for This Book

Being trapped in a reclusive world located in the pure bliss of Pasadena, California, perfectly happy doing surgical pathology service work in a teaching hospital, avoiding meetings and people, not having a forum to disseminate my ideas, and having my papers regularly rejected by high impact journals, this book is my preferred and possibly only avenue to publicize this work.

Writing books comes easily to me. It is the lasting effect that Jim Ransom, my superlative editor at Lange Medical Publications, had on me in the early part of my career.

It is only in a book such as this that one has the freedom to present a topic in its entirety. It is only in a book that one can analyze and explain data and conclusions in papers that are seemingly contradictory. It is only in a book that one can develop rational hypotheses for which no evidence exists. It is only in a book like this that one is allowed to dream.

Writing a book is a massive endeavor. It requires critical and deep analysis of all relevant papers. Doing this makes me realize one important fact about the literature: one can almost always trust the data that are presented because researchers are largely honest. However, the interpretation of the data is frequently affected by bias. Reexamination of the data with a different viewpoint commonly leads to conclusions that are different than those presented by the authors. It is only in a book that one has the freedom to present these alternate interpretations of the data.

I have selected papers that I love for detailed review and presentation in this book. The authors of these papers largely had no idea of the new concept that is being proposed when they performed the studies. However, the data that they produced permit me to expand on the evidence base that I have no ability to produce on my own.

As an author, I have the burden of knowing that whatever I write will be in print for eternity. When an individual's written word is examined over the years, one can easily determine which ideas were correct at the time they were written and which were erroneous. It is clear that Allison, in 1948 and 1953,[24,25] knew exactly what he was talking about right from the outset. It is also clear that Norman Barrett was completely wrong in 1950[26] but had the grace to admit his mistake and get on board with Allison in 1957.[27] I will review these wonderful papers in great detail in Chapters 1 and 11.

This book will be an attempt to mimic the classic scientific monographs of the past. One of my greatest life pleasures is to scan the pages of books by Louis Pasteur who described the process of pasteurization in his endeavor to save the wine industry in France,[28] Claude Bernard on the pancreas with the most beautiful hand-drawn medical illustration ever done,[29] and Rudolf Virchow on cellular pathology,[30] the basis of modern medicine. Despite my lack of ability to read French and German, I can follow the logical thought processes of these great minds by simply scanning the pages. The more recent treatise by Harvey Cushing on the pituitary gland is in English and a wonderful experience to peruse.[31] These books show the mastery of an entire subject by one person and a freedom of expression that is unfettered.

It is rare today to find single authored medical books. Even expert physicians do not have the time or inclination. There is little financial gain in writing books. The academic world looks askance at books because they are not peer reviewed. Multiple authors who are invited to write chapters by the editors write the majority of today's medical books. Such books have little cohesion and are largely for reference. They do not read like a story that develops from a beginning and progresses in a logical manner to an end. I hope this book will tell a powerful story that engages the reader.

Journal articles of the past permitted this freedom of author expression. Norman Barrett begins his classical treatise in 1957 (p. 881)[27] that defined and named the CLE as follows: "The ideas discussed here are not based on statistics nor upon a large collection of specimens; they are the results of thinking about a few unusual cases of esophageal disease… Some may be worried because I have changed my opinion relating to certain matters, but progress is not static and there is no subject which does not yield more knowledge as the depths are sounded."

This is different to the present constraints of journal articles: limitation of length, dictatorial rules regarding the way in which data must be presented, inability to express ideas for which there is no evidence, slavish dependence and trust on statistical analysis, and a review process that is not transparent and cannot be challenged. The anonymous power of the reviewer and editor is immense. The author who is not well known or not from a prestigious institution, or expresses a viewpoint that is contrary to the belief of the editor has little chance of acceptance.

If Norman Barrett's paper came into the hands of an editorial review board of any major journal today, that first paragraph would have caused it to be rejected out of hand. Our understanding of CLE that bears his name may have been delayed by decades. I dread to contemplate the wonderful ideas that have never seen the light of day because of the elitism of today's physician scientists that control the literature.

The *New England Journal of Medicine*, review[32] of the first book on GERD that I coauthored with Dr DeMeester[33] states: "The book … demonstrates the advantages of having a limited number of authors and a narrow focus – namely, consistency of style and a cohesive philosophy. There is structural coherence, with a logical progression of the 17 chapters and minimal overlap between them." At the end of a very complimentary review, they write: "The authors are very opinionated, and many readers will not agree with their views, several of which are controversial. Although many of their ideas are indeed provocative and deviate from current consensus, we feel that the authors' vast experience gives their opinions importance – their perspective must be carefully considered."

This book will attempt to similarly be opinionated and provocative and deviate from current consensus, expressing new ideas that will not be believed by many readers. However, the ideas expressed in this book, when compared with my writings dating back to 1997, including two prior books on GERD,[33,34] will show complete commitment and increasing certainty and scope of my hypothesis. The ideas in the first book in 2006[33] that many readers did not believe at the time are now believed by a larger number of people. The new ideas are intended to push the envelop even further with innovative methods of assessing and managing GERD aimed at improving the lot of patients with GERD.

What I hope is that the presentation of the evidence base for the new idea and its logical evolution will be made clear to the few readers who will read this in its entirety with an open mind, evaluate it objectively, understand its truth, and have the influence in the medical world to help effect the necessary changes in our understanding of GERD.

5. THE FUTURE THAT I HOPE FOR

The management of patients with GERD today is based on a symptom-based definition and the control of symptoms as its objective. This would be perfect if the disease had no other manifestations. This is not the case.

Symptoms of GERD result largely from squamous epithelial injury and are controlled successfully, albeit incompletely, by miraculous new drugs that suppresses gastric acid secretion. However, even as it achieves this objective, the management algorithm of GERD has resulted in an ugly new manifestation of the disease that was rare in the 1950s. Adenocarcinoma of the esophagus has become the new face of GERD (Fig. 9). Its meteoric rise in incidence from 1952, when it was first reported,[35] to the present is the most dramatic in the history of medicine. The increase in incidence in the United States from 1973 (the earliest year for which statistics are available) to 2006 is sevenfold (from 3.6 to 25.6/million population).[20]

The management of GERD has not yet adjusted adequately; there is nothing that has been suggested that has any hope of solving the problem. GERD is managed with the same objective as before: the control of symptoms despite the certainty that this management has been accompanied by an increase in the incidence and mortality from adenocarcinoma.

It is irrational to manage a disease with two manifestations with a singular objective aimed at only one. Alkalinization of gastric juice, while it controls symptoms, has no positive impact on columnar metaplasia and its progression to adenocarcinoma.

My hope for the future is that the new ideas in this book will result in a management method that will identify the root cause of both squamous epithelial injury and adenocarcinoma. I will show that this root cause is damage to the LES that permits reflux of gastric juice from the stomach to the esophagus.

The answer to all manifestations of GERD is to attack the root cause by understanding, assessing, and addressing sphincter damage with the objective of preventing reflux. Preventing reflux becomes a new objective that will address both squamous and columnar epithelial complications of GERD that lead to symptoms and adenocarcinoma.

The new method that is described herein follows the fundamental rules of Rudolph Virchow, which is the basis of modern medicine.[30] It takes the disease from its present primitive clinical level of understanding to a new cellular level from the normal state to all end points of the disease. Anytime this happens in a disease, great improvements ensue.

I wish Dr. Rodger Haggitt was alive.

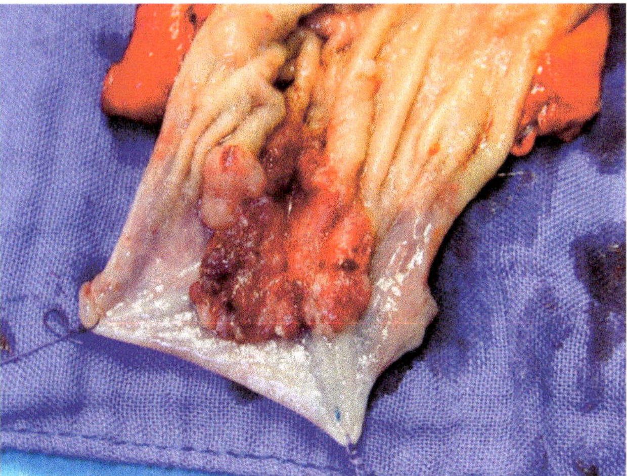

FIGURE 9 Adenocarcinoma of the distal esophagus arising in short-segment Barrett esophagus. The squamocolumnar junction is separated from the proximal limit of rugal folds (endoscopic gastroesophageal junction) by 1–2 cm of flat, salmon pink columnar epithelium. The true gastroesophageal junction (GEJ) (proximal limit of gastric oxyntic epithelium) is a significant distance distal to the endoscopic GEJ by histology. *Photograph courtesy: Dr. Martin Riegler, Reflux Medical, Vienna, Austria.*

REFERENCES

1. Chandrasoma P, Taylor CR. *Concise pathology. A Lange medical book.* Norwalk, Connecticut: Appleton & Lange; 1991.
2. Chandrasoma PT, Smith MM, Apuzzo ML. Stereotactic biopsy in the diagnosis of brain masses: comparison of results of biopsy and resected surgical specimen. *Neurosurgery* 1989;**24**:160–5.
3. Chandrasoma P, Apuzzo MLJ. *Stereotactic brain biopsy.* New York: Igaku-Shoin; 1989.
4. Kahrilas PJ, Shaheen NJ, Vaezi MF. American Gastroenterological Association Institute technical review on the management of gastroesophageal reflux disease. *Gastroenterology* 2008;**135**:1392–413.
5. Clark GWB, Ireland AP, Chandrasoma P, et al. Inflammation and metaplasia in the transitional mucosa of the epithelium of the gastroesophageal junction: a new marker for gastroesophageal reflux disease. *Gastroenterology* 1994;**106**:A63.
6. Oberg S, Peters JH, DeMeester TR, et al. Inflammation and specialized intestinal metaplasia of cardiac mucosa is a manifestation of gastroesophageal reflux disease. *Ann Surg* 1997;**226**:522–32.
7. Chandrasoma P. Norman Barrett: so close, yet 50 years away from the truth. *J Gastrointest Surg* 1999;**3**:7–14.
8. Chandrasoma PT, Der R, Ma Y, et al. Histology of the gastroesophageal junction: an autopsy study. *Am J Surg Pathol* 2000;**24**:402–9.
9. Chandrasoma PT, Wijetunge S, DeMeester SR, et al. The histologic squamo-oxyntic gap: an accurate and reproducible diagnostic marker of gastroesophageal reflux disease. *Am J Surg Pathol* 2010;**34**:1574–81.
10. Chandrasoma P. Controversies of the cardiac mucosa and Barrett's esophagus. *Histopathology* 2005;**46**:361–73.
11. An evidence-based appraisal of reflux disease management – the Genval Workshop Report. *Gut* 1999;**44**(Suppl. 2):S1–16.
12. Vakil N, van Zanten SV, Kahrilas P, Dent J, Jones R, The Global Consensus Group. The Montreal definition and classification of gastroesophageal reflux disease: a global evidence-based consensus. *Am J Gastroenterol* 2006;**101**:1900–20.
13. Katz PO, Gerson LB, Vela MF. Diagnosis and management of gastroesophageal reflux disease. *Am J Gastroenterol* 2013;**108**:308–28.
14. Chandrasoma PT. Histologic definition of gastro-esophageal reflux disease. *Curr Opin Gastroenterol* 2013;**29**:460–7.
15. Chandrasoma P, Makarewicz K, Wickramasinghe K, et al. A proposal for a new validated histologic definition of the gastroesophageal junction. *Hum Pathol* 2006;**37**:40–7.
16. McClave SA, Boyce Jr HW, Gottfried MR. Early diagnosis of columnar lined esophagus: a new endoscopic diagnostic criterion. *Gastrointest Endosc* 1987;**33**:413–6.
17. Sharma P, McQuaid K, Dent J, Fennerty B, et al. A critical review of the diagnosis and management of Barrett's esophagus: the AGA Chicago Workshop. *Gastroenterology* 2004;**127**:310–30.
18. Sarbia M, Donner A, Gabbert HE. Histopathology of the gastroesophageal junction. A study on 36 operation specimens. *Am J Surg Pathol* 2002;**26**:1207–12.
19. Odze RD. Unraveling the mystery of the gastroesophageal junction: a pathologist's perspective. *Am J Gastroenterol* 2005;**100**:1853–67.
20. Pohl H, Sirovich B, Welch HG. Esophageal adenocarcinoma incidence: are we reaching the peak? *Cancer Epidemiol Biomark Prev* 2010;**19**:1468–70.
21. Glickman JN, Fox V, Antonioli DA, Wang HH, Odze RD. Morphology of the cardia and significance of carditis in pediatric patients. *Am J Surg Pathol* 2002;**26**:1032–9.
22. Lagergren J, Bergstrom R, Lindgren A, Nyren O. Symptomatic gastroesophageal reflux as a risk factor for esophageal adenocarcinoma. *N Engl J Med* 1999;**340**:825–31.
23. *American joint commission of cancer staging manual: esophagus and esophagogastric junction.* 7th ed. 2010.
24. Allison PR. Peptic ulcer of the oesophagus. *Thorax* 1948;**3**:20–42.
25. Allison PR, Johnstone AS. The oesophagus lined with gastric mucous membrane. *Thorax* 1953;**8**:87–101.
26. Barrett NR. Chronic peptic ulcer of the oesophagus and 'oesophagitis'. *Br J Surg* 1950;**38**:175–82.
27. Barrett NR. The lower esophagus lined by columnar epithelium. *Surgery* 1957;**41**:881–94.
28. Pasteur L. *Etudes sur la vinaigre, sa fabrication, ses maladies, moyens de les prevenir; nouvelles observations sur la conservation des vins par la chaleur.* Paris: Gauthier-Villars & Victor Masson et fils; 1868.
29. Bernard C. *Memoire sur le pancreas et sur le role du sac pancreatique dans leas phenomenes digestifs, particuleirement dans las digestion des matieres grasses neutres.* From: Academie des Sciences, Supplement aux Comptes Rendus, tome 1. Paris: Bachelier; 1856. p. 379–583.
30. Virchow R. *Die Cellularpathologie: in ihrer Begrundung auf physiologische und pathologische Gewebelehre.* Berlin: Verlag von August Hirschwald; 1858.
31. Cushing H. The pituitary body and its disorders. In: *Clinical States produced by disorders of the hypophysis cerebri.* Philadelphia: J.B. Lippincott; 1912.
32. Book review: Chandrasoma PT, DeMeester TR. *GERD: reflux to esophageal adenocarcinoma.* San Diego: Elsevier Academic Press; 2006. *New Engl J Med* 2007;356:1897–98.
33. Chandrasoma PT, DeMeester TR. *GERD: reflux to esophageal adenocarcinoma.* San Diego, CA: Academic Press, Elsevier; 2006.
34. Chandrasoma P. *Diagnostic atlas of gastroesophageal reflux disease.* San Diego, CA: Academic Press, Elsevier; 2007.
35. Morson BC, Belcher JR. Adenocarcinoma of the oesophagus and ectopic gastric mucosa. *Br J Cancer* 1952;**6**:127–30.

Chapter 1

Definition of Gastroesophageal Reflux Disease: Past, Present, and Future

Gastroesophageal reflux disease (GERD) is a common human disease.[1] Its prevalence is based on how it is defined. Approximately 10% of people in the United States suffer from daily heartburn. 20%–30% have heartburn once a week. Nearly 40% have heartburn at least once a month. The prevalence is increasing all over the world.[2]

This prevalence translates to ~75 million GERD sufferers in the United States when the disease is defined by the presence of symptoms. The very large absolute number of people with heartburn and/or diagnosed with GERD is overwhelming, making GERD one of the most difficult common human diseases to manage.

The enormity of the problem largely dictates the present definition and management algorithms. To avoid overwhelming available resources, the medical establishment defines the existence of GERD at a relatively advanced stage where the numbers are manageable. This causes the many millions of people with mild symptoms to be relegated to self-medication at the pharmacy or management in the nonspecialized environment of primary care without access to endoscopy or specialized diagnostic testing.

This group of patients with mild symptoms has an exceedingly low percentage risk of cancer. However, the many millions of such people translate that minute risk into a large absolute number of adenocarcinomas. These patients present at an advanced stage of cancer, with little or no hope of survival (Fig. 1.1).

Only patients with symptoms that progress beyond the comfort level of a primary care physician reach the care of a gastroenterologist specialized in GERD. This is the point where, from a practical sense, GERD changes from being a mere nuisance that is easily controlled to a problem. At this point, the battle is largely lost. With present management algorithms, adenocarcinoma is inevitable with medical therapy alone in the small minority of people destined to progress to advanced cancer without ever developing an indication for endoscopy.

The only patients who are positively impacted by this management algorithm are those found at endoscopy to have Barrett esophagus. They enter an endoscopic surveillance protocol for early detection and treatment of cancer, which greatly improves survival. Unfortunately, only 15% of all patients developing adenocarcinoma have had an endoscopy with a diagnosis of Barrett esophagus (Fig. 1.1).

The incidence of GERD-induced adenocarcinoma has increased sevenfold from 1975 to 2010.[3] Esophageal adenocarcinoma is a specific complication of GERD. There is no other etiology. In 2016, approximately 20,000 people in the United States will develop GERD-induced esophageal adenocarcinoma, a cancer with an overall mortality of 85%. This number includes adenocarcinoma of the "gastric cardia," which is now classified as an esophageal cancer. The incidence and mortality continues to increase.[3]

Death from complications of GERD was uncommon before the onset of the epidemic of esophageal adenocarcinoma. An increase in the mortality in a disease from near zero to 17,000 patients per year in the United States alone is testament, whatever the reasons, to a dismal failure of management.

Restriction of the definition of GERD to those with troublesome symptoms is appropriate if the only serious problems caused by GERD are symptoms. In the past, when this was largely true, proton pump inhibitors (PPIs) would have been the perfect treatment.

Today, however, the explosion in the incidence of adenocarcinoma makes near total dependence on PPI therapy a very poor management algorithm. By doing nothing to prevent cancer, the medical establishment commits 20,000 people in the United States to develop a lethal cancer annually. We are powerless and without ideas, let alone have a solution to the problem of a cancer that we know has reached epidemic proportions and continues to increase.

GERD. http://dx.doi.org/10.1016/B978-0-12-809855-4.00001-4

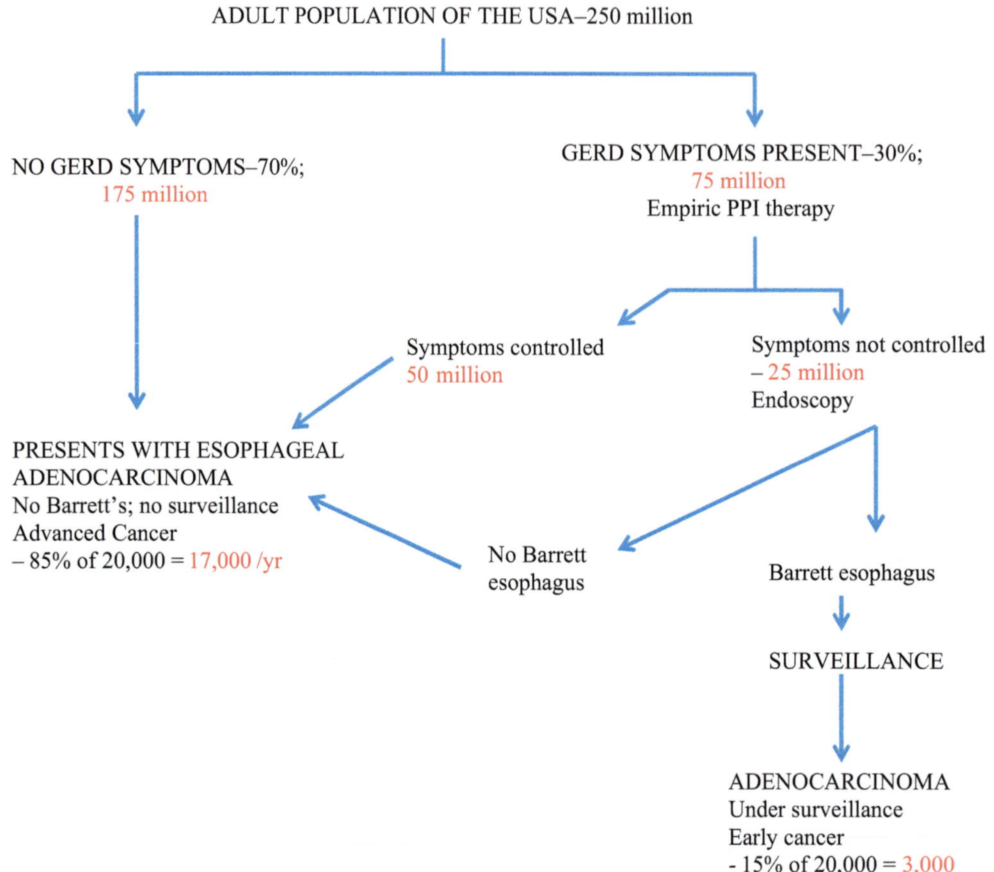

FIGURE 1.1 Algorithm for the present management of gastroesophageal reflux disease (GERD). 70% of the adult population (175 million in the United States) do not have symptoms of GERD. A small minority of these people have asymptomatic cellular progression to Barrett esophagus and adenocarcinoma. 30% (75 million) have symptoms of GERD. These patients are treated (self-medication, primary care physician, or gastroenterologist) empirically with acid-reducing agents including proton pump inhibitors (PPIs). 70% are well controlled for their lifetime; 30 patients fail therapy and undergo endoscopy. Patients found at endoscopy to have Barrett esophagus enter an endoscopy surveillance program aimed at detecting cancer at an early stage. These patients have a much better survival than those developing advanced cancer outside surveillance. Only 15% of patients who develop cancer have been in a surveillance program.

I will show in this book that cellular changes resulting from the basic cause of GERD are almost universally present in the adult population.[4,5] Most patients with minimal histologically demonstrable cellular changes have no symptoms of GERD.[4,5]

The situation is analogous to the relationship between coronary atherosclerosis and ischemic heart disease. Almost all people in the United States have mild coronary atherosclerosis, but very few develop symptoms of ischemic heart disease and a small minority will die of myocardial infarction. Similarly, a minority of patients with cellular changes of GERD will develop symptoms and very few will progress to treatment failure and adenocarcinoma.

To gastroenterologists, GERD is a disease diagnosed by symptoms without any practical method of confirmation at a cellular level. There are no reliable pathologic criteria for diagnosis. Endoscopy and biopsy have little practical value in diagnosis and are not indicated in early GERD.

Patients with symptoms of GERD are treated empirically with acid-suppressive drugs until the disease has progressed to a point of no return.[6] The only reason why this is possible is that GERD, like coronary atherosclerosis, progresses to a severe state only in a small minority of the patients.

70% of patients with cellular changes never develop symptoms that define GERD; a very small percentage in this group progress to cancer. 20%–25% develop symptoms of GERD that are mild enough that PPI therapy controls symptoms beautifully throughout life; a small percentage in this group progress to cancer. 5%–10% with symptoms of GERD progress to treatment failure and disruption of quality of life; a higher percentage in this group progress to cancer. Despite the small percentages in the first two groups, most cancers develop in those two groups because of their very large numbers (Fig. 1.1).

This book proposes a desperately needed effort to alter this management method. I will develop innovative pathologic criteria of early GERD based on measuring lower esophageal sphincter (LES) damage by histology using an appropriate biopsy sample. Management of GERD will have an algorithm based on the ability of the new test of LES damage to

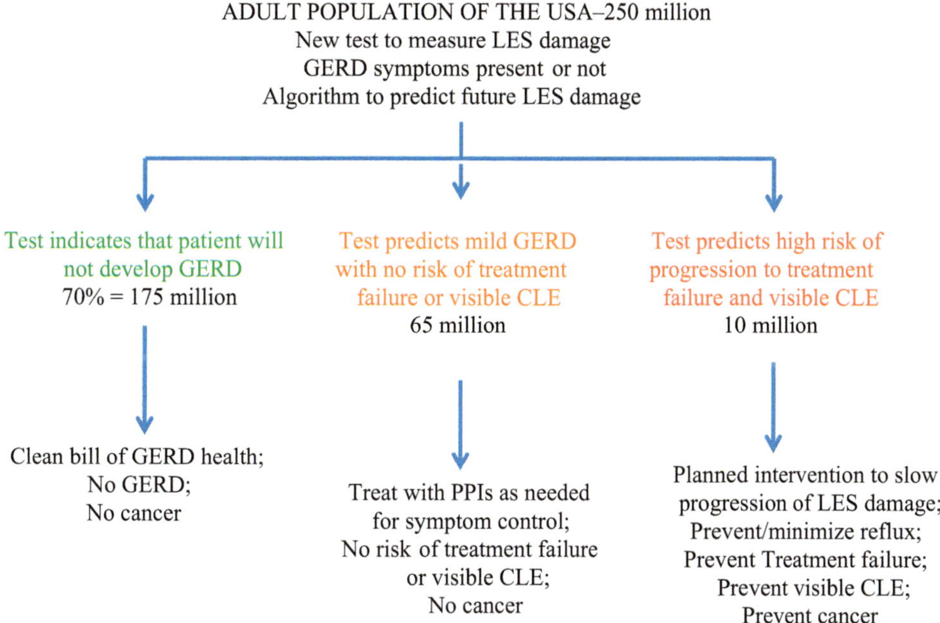

FIGURE 1.2 The proposed new management algorithm for gastroesophageal reflux disease (GERD) based on the new method of measuring lower esophageal sphincter (LES) damage and using the algorithm that will predict future LES damage throughout the person's natural life span. The test can be done at any time during adult life irrespective of presence of symptoms. LES damage severity will be defined as (A) left column: <15 mm during the life of the person. This person will not develop LES damage sufficient to cause GERD; (B) center column: >15–25 mm LES damage during the life of this person. This person is predicted to develop GERD but never progress to severe LES damage sufficient to cause failure of proton pump inhibitor therapy or visible columnar-lined esophagus (CLE). They will be well controlled with PPIs. There is no risk of cancer; (C) right column: This is the minority of persons in whom the test will predict progression to severe LES damage >25 mm in the future. They are at risk for treatment failure and visible CLE at some point in their lives. The numbers are approximate and represent prevalence figures. Once the backlog of present GERD patients is taken care of, the future numbers will be the incidence of GERD, which is 5 per 1000 person years in the United States.[2]

predict progression of LES damage into the future. This will permit stratification of patients into three groups based on the predicted future progressive disease (Fig. 1.2).

The new management algorithm will attack the problem at its root cause, which is LES damage rather than troublesome symptoms. The primary management goal will change from symptom control to prevention of adenocarcinoma. This has the potential to lead to the ultimate control and eradication of GERD-induced esophageal adenocarcinoma.

The availability of a new test for LES damage can produce a fundamental change in the definition, treatment objectives, and management of GERD. It can change the focus of GERD from symptoms to its root cause. The management objective will shift from control of symptoms to control of LES damage. This provides a pathway to eradicate cancer.

1. POTENTIAL CRITERIA TO DEFINE GASTROESOPHAGEAL REFLUX DISEASE

The definition of a disease is a statement of the level of knowledge of that disease. Definition drives diagnosis, treatment, and research. The depth of understanding of the disease and the accuracy of its definition are ultimately gauged by outcomes resulting from the way patients with the disease are treated.

Definition of GERD can potentially use the following different criteria.

1.1 Symptoms of Gastroesophageal Reflux Disease

Symptoms caused by gastroesophageal reflux are presently the chosen criterion for the definition of GERD. Defining GERD by its symptoms drives the management objective of GERD to symptom control and improvement of quality of life. This is pursued with the single-minded goal of reducing gastric acid, mainly by suppression of secretion of acid by the stomach.

GERD is also called "acid reflux," suggesting that acid causes the disease and removal of acid is its cure. This is not correct. No amount of gastric acid will cause GERD in the presence of a competent LES. GERD results from the failure of the sphincter caused by progressive LES damage. Acid-suppressive therapy does not reverse or prevent the progression of

LES damage. It simply changes the pH of the refluxate from strong acid to weak acid (see Chapter 2, Fig. 2.3). It does not therefore prevent the progression of GERD. It converts GERD from a disease in which cellular changes were caused by strong acid reflux in the 1950s to the one that is caused by increasingly weak acid reflux in 2016.

GERD is the result of squamous epithelial damage resulting from exposure to gastric juice. While symptoms and some acid-dependent cellular changes reverse with alkalinization of gastric juice, other cellular events are promoted by alkalinization of the refluxate progress. In particular, PPI therapy does not prevent columnar metaplasia or the progression to adenocarcinoma.

The symptom-based definition will be described in detail in Section 3 of this chapter.

1.2 The Empiric Proton Pump Inhibitor Test

Symptom-based definition is often combined with a therapeutic test to establish the diagnosis.[6] If a given symptom in a patient suspected to be caused by GERD responds to a trial of empiric PPI therapy, the diagnosis of GERD can be confirmed.

Although the empiric PPI therapeutic test is widely used in clinical practice, it is not reliable.[6] A false-positive test may result from a placebo effect or may be due to the existence of a non-GERD cause for the symptom that responded to PPI therapy. A false negative may result because PPI therapy is not absolutely effective in controlling symptoms within the scope of the therapeutic trial. The magnitude of false-positive and false-negative therapeutic tests is unknown.[6]

As a result, an unknown number of patients who are presently being treated with long-term PPI therapy for GERD after a diagnosis based on the empiric PPI test may not have the disease. They are being unnecessarily exposed to the complications of long-term PPI therapy.

1.3 Tests Based on Quantitating Gastroesophageal Reflux

Quantitating gastroesophageal reflux requires the placement of a measuring device in the lower esophagus either at endoscopy or via a nasal catheter to detect reflux over a period of time. This includes tests to monitor pH and impedance in the distal esophagus. While considered by some authorities to be the gold standard for the diagnosis of GERD, these tests have never been used for defining GERD. They are largely used to determine suitability for surgical treatment; to evaluate whether GERD is the cause of possible complications of GERD such as asthma, chronic cough, and pulmonary fibrosis; and to evaluate empirical treatment failures.[6]

They are not indicated as a diagnostic method in the patient with early GERD. The specificity of an abnormal test is high, but the sensitivity is inadequate for use in defining GERD. A negative pH or impedance test does not exclude GERD as the cause of symptoms. Some authorities use the terms "functional heartburn" and "hypersensitive esophagus" for the situation where the patient has typical heartburn and a normal pH or impedance test, but this is not a common practice. It is more common to designate a patient with troublesome heartburn as GERD and not perform the pH or impedance test before treatment.

These diagnostic tests are not used to define GERD in the practical sense. They are not recommended in the initial diagnostic testing of GERD.[6]

1.4 Gross, Radiologic, and Endoscopic Mucosal Abnormalities Caused by Reflux

The presence of grossly and endoscopically visible erosions and ulcers in the esophagus was used historically to define GERD (Fig. 1.3). While these have high specificity, they are relatively insensitive and therefore not suitable for definition presently. Many patients with GERD have no gross or endoscopic abnormalities. Endoscopy is not indicated in the GERD patient who has responded to empiric PPI treatment.[6] Radiologic imaging is of little value in the diagnosis of GERD and is rarely used.

1.5 Cellular (Histopathologic) Changes Caused by Reflux

Presently, histologic criteria are not used in the diagnosis of GERD. The only presently recognized criteria for GERD are changes in the squamous epithelium ("reflux esophagitis"; Fig. 1.4).[7] These changes are not specific for GERD; patients with diseases other than GERD, such as idiopathic eosinophilic esophagitis (Fig. 1.5), can show changes in the squamous epithelium that are similar to reflux esophagitis.[8] The criteria for a diagnosis also have a low sensitivity; many patients with GERD that has been objectively proven by a pH test have no evidence of reflux esophagitis.

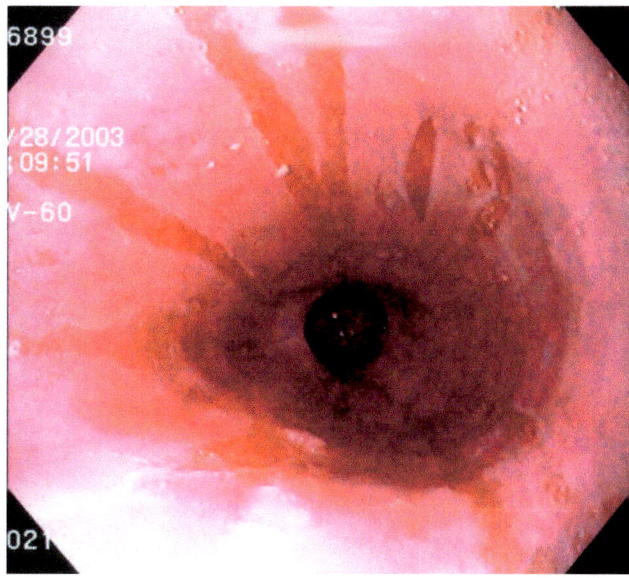

FIGURE 1.3 Erosive esophagitis at endoscopy showing multiple linear erosions.

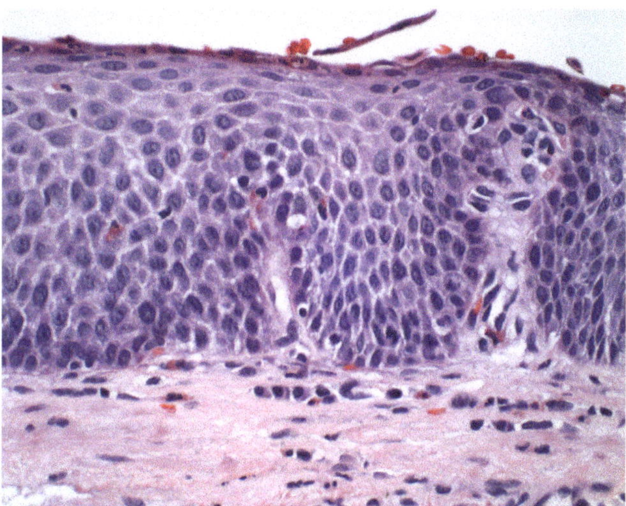

FIGURE 1.4 Histologic changes of reflux esophagitis showing squamous epithelium of normal thickness with basal cell hyperplasia and increased papillary height. The cells show separation due to intercellular edema ("dilated intercellular spaces"), which is associated with increased permeability of the epithelium. There are scattered intraepithelial eosinophils.

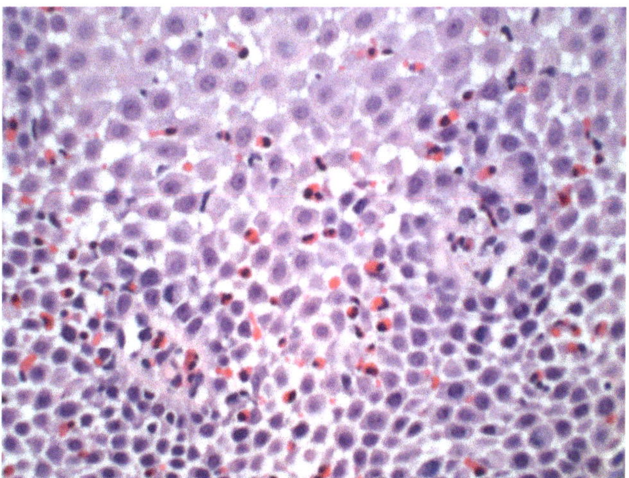

FIGURE 1.5 Esophageal biopsy from the midesophagus of a child with idiopathic eosinophilic esophagitis. This shows basal cell hyperplasia, intercellular edema ("dilated intercellular spaces"), and intraepithelial eosinophils similar to that seen in reflux esophagitis. The number of intraepithelial eosinophils is higher, satisfying criteria for the diagnosis of idiopathic eosinophilic esophagitis.

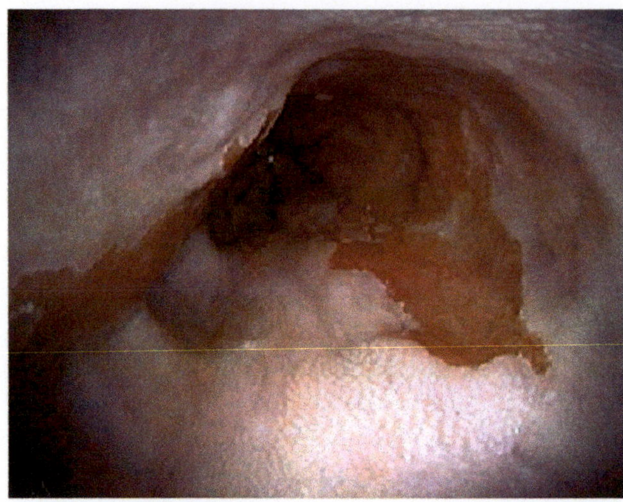

FIGURE 1.6 Antegrade view showing a 6cm segment of visible columnar-lined esophagus (CLE) at endoscopy. This shows a circumferential flat columnar epithelium with a markedly irregular squamocolumnar junction. Note the presence of squamous islands in the CLE. Biopsies showed non-dysplastic intestinal metaplasia in the proximal 2cm with cardiac and oxyntocardiac epithelium in the distal 4cm. *Photograph and data courtesy of Dr. Martin Riegler, Reflux Medical, Vienna, Austria.*

Columnar metaplasia of the esophagus is recognized as being a specific complication of GERD when it is recognized at endoscopy [visible columnar-lined esophagus (CLE)] (Fig. 1.6). Columnar metaplasia results only from exposure of esophageal squamous epithelium to gastric contents. Presently, columnar metaplasia of the esophagus visible at endoscopy is not used as a criterion for the diagnosis of GERD.

When visible CLE is present, biopsies are done; these divide patients into those with and without intestinal metaplasia. Visible CLE with intestinal metaplasia is Barrett esophagus; it is recognized as a premalignant epithelium and its presence is an indication for endoscopic surveillance. Patients with visible CLE who do not have intestinal metaplasia are not placed on surveillance in the United States. Visible CLE is 100% specific for GERD but has a very low sensitivity; only a small percentage of patients with GERD have visible CLE.

1.6 Criteria Based on Molecular (Genetic) Changes

Ultimately, it is probable that GERD will be defined by the molecular changes that are caused by exposure of the esophageal epithelial cells to the various molecules in the gastric juice. The molecular basis of GERD is presently in its infancy and not adequately understood for practical definition. I will discuss the available evidence in Chapter 15.

1.7 Criteria Based on Detecting Lower Esophageal Sphincter Damage, the Cause of Gastroesophageal Reflux Disease (Fig. 1.7)

Presently, there is no method to identify LES damage, the cause of GERD. This does not mean that there is no method; it just means that the method has not been figured out yet. The basic rule of Rudolf Virchow, the father of modern medicine, is that every abnormality must have a cellular basis.[9] There is a cellular basis for LES damage. It just has not yet been recognized.

I will briefly describe this later in the chapter, but this entire book is aimed at recognizing LES damage at a cellular level. When GERD is defined by LES damage, new methods of addressing the disease far more powerful in managing the disease than those existing at present will emerge.

2. PAST DEFINITIONS OF GASTROESOPHAGEAL REFLUX DISEASE BASED ON ULCERS AND EROSIONS

Definition of a disease necessarily changes with time as understanding increases. This is largely a measure of the availability and application of new technology. Advances in radiology, endoscopy, and histopathology; evolution of surgical techniques; and the availability of new drugs have all impacted the definition of GERD at different times. Molecular technology is beginning to do the same, although its practical impact at present in GERD is limited. This is likely to change in the future.

At the **turn of the 20th century**, the inability to examine the esophagus by endoscopy and biopsy prevented the recognition of GERD during life. The entity characterized clinically by heartburn and regurgitation correlated only with the

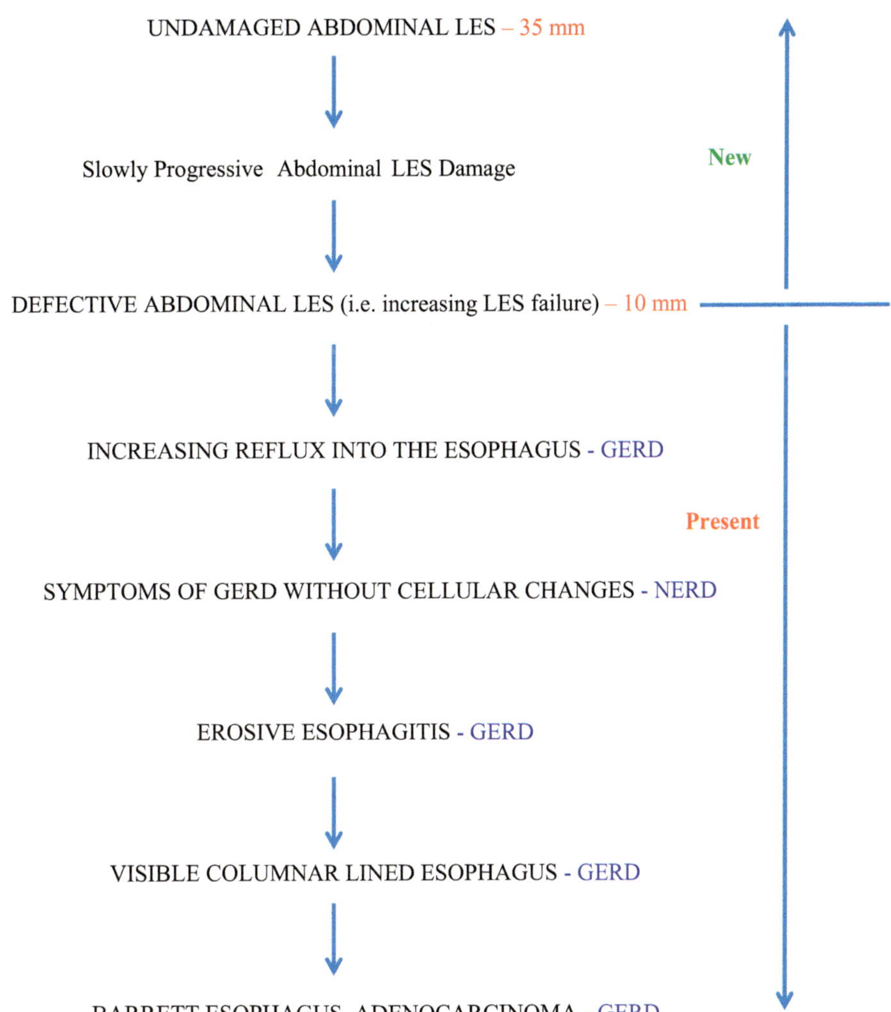

FIGURE 1.7 Criteria used for the definition of gastroesophageal reflux disease (GERD). Present criteria (*red area*) for definition depend on the presence of a defective lower esophageal sphincter (LES) by manometry, sufficient reflux to cause abnormal acid exposure in the pH test, troublesome symptoms, erosive esophagitis, visible columnar-lined esophagus, Barrett esophagus, and adenocarcinoma. These all represent the late stages of LES damage. The new criteria (*green area*) for definition will focus on the early stage of the disease between the normal undamaged LES and the point at which LES failure begins. This is possible because of the new availability of a test to measure LES damage. It is by recognizing this early stage of the disease that the later stages can be prevented.

demonstration of ulcers in the lower esophagus at autopsy. GERD was called "peptic ulcer of the esophagus" and considered to be a rare entity.

In 1906, Tileston[10] separated chronic peptic ulcer of the esophagus from other causes of esophagitis. He identified only 44 cases that had been published from 1839 to that time. The ulcers were large and some perforated into a large vessel, the pericardium, the mediastinum, or pleural cavity, causing death in rare cases. The mucosa adjacent to the ulcers was often "gastric" in type and was considered to be "ectopic." This ectopic gastric mucosa was believed to produce acid that caused the ulcers in the esophagus. In his paper, Tileston suggested that the disease may be more common, quoting that 6 of 4496 autopsies at the Massachusetts General Hospital, Boston City, and Long Island hospitals had noted the presence of peptic ulcer in the esophagus, for a prevalence of 0.13%.[10]

In the 1920s, with increasing availability and improvement in radiological diagnosis and esophagoscopy, surgeons started reporting erosions and ulcers in the esophagus. In 1929, Chevalier Jackson,[11] considered by some authorities to be the father of modern endoscopy, reported esophagitis in 88 of 4000 consecutive endoscopies. There was confusion that the mucosa that was found around the ulcers and erosions was sometimes "gastric" and sometimes squamous. The much higher prevalence of peptic ulcer of the esophagus seen at endoscopy compared with autopsy also caused controversy. Looking back, it is likely explained by the fact that endoscopy is much more sensitive for detecting ulcers than autopsy because rapid postmortem autolysis of esophageal mucosa limits recognition of only larger ulcers at autopsy.

In that era, endoscopy was performed by very few physicians, and the medical establishment back then likely looked on the procedure with suspicion and skepticism. The disease was known as "peptic ulceration of the esophagus" until 1950.

The endoscopic finding of esophagitis with ulceration was confusing until **1948**, when Allison[12] published a detailed study of 63 patients (M:F = 34:29; age range: 22–80, mean age 60 with 56 patients over the age of 50) with peptic ulcer of the esophagus. He writes: "Most of the patients have been observed for some years with repeated radiological and endoscopic examination. Thirteen were diagnosed as suffering from chronic oesophagitis with recurrent acute ulceration, and fifty had chronic ulcers." Allison, in a prior paper in 1946,[13] had described four stages in the natural history of the disease: esophagitis, esophagitis with acute ulceration, esophagitis with chronic ulcer, and healed fibrous stenosis.

In all my reading, Allison's papers are the most detailed, accurate, and illuminating on the subjects of GERD and CLE. They have never been surpassed and are worth recording here in detail. I would recommend to everyone who is interested in this subject to read these wonderful primary papers in their entirety many times over.

His introductory paragraphs to this paper are a thing of beauty: "The patient with peptic ulceration of the lower oesophagus has for a long time suffered his ulcer in a sort of 'no-man's-land' to which the abdominal surgeon could not reach and to which the physician had no access. The radiologist passed it by as quickly as the force of swallowing would take him. The endoscopist saw it as through a glass darkly on the distant boundary of his territory, but was glad to withdraw to more familiar fields. The development of a safe surgical approach to the cardia has helped to revive interest in the condition. It has been looked for with greater care… A lack of detailed investigation in the past led to the easy assumption that the oesophagus suffered acid digestion as a result of congenital maldevelopment. Its failure to reach its full length left the fundus of the stomach in the posterior mediastinum so that the normal barriers to acid reflux were ineffective. Peptic ulcer of the esophagus and congenitally short oesophagus became almost like the Beaver and the Butcher: 'You could never meet either alone!' Although such a congenital abnormality cannot be denied, the burden of this report is that short oesophagus is usually an acquired condition due to defects in the diaphragm which allow a sliding hiatal hernia of the stomach."

Allison carefully documents the clinical features in this patient group: "Dyspepsia in one form or another was present in 74% of patients. The duration varied from 1 to 60 years. It was variously described as heartburn, pain behind the sternum or high in the epigastrium associated with an empty stomach and relieved by alkalis, pain going through between the shoulder blades, or 'wind in the stomach.' Sometimes pain would occur in the night and patients would take alkalis or a glass of milk to the bedside with them. In some a soreness behind the sternum was felt on swallowing. Pain might be worse immediately after food, or it might be relieved by food…. Dysphagia was the symptom that brought most of these patients to hospital. It occurred in 92%. It had been present for between 8 weeks and 50 years…. There were examples of complete obstruction, perhaps lasting a few days and being relieved spontaneously…. Dysphagia was always more marked for solids than for liquids…. 11 of the patients gave a history of bleeding: in one it was slight, but in the other 10 it was severe enough to cause anxiety. No death from bleeding occurred in this series… Loss of weight was in proportion to the amount of starvation…. (Four) people, one man and three women, all of middle age, complained that food came up as easily as it went down…. After a meal the patient only had to lie down, or even to bend down, for the food to flow freely back into the mouth…. In these patients a hernia of the stomach was associated with peptic ulceration of the lower oesophagus with an abnormally patulous cardia. No stenosis was present and the easy regurgitation could be confirmed by screening."

Radiologically: "in all patients, except 2, a 'short oesophagus' has been found and a part of the stomach adjoining the cardia has been demonstrated in the mediastinum. The change from the oesophageal mucosal pattern to that of the coarse gastric folds marks the position of the cardia. If the position of the cardia remains in doubt a Cushing clip may be applied to the junction through an oesophagoscope, and the level of this clip, in relation to the mucosal pattern of the cardia and the diaphragm, can be determined on the x-ray screen."

The endoscopic appearances: "In the lower oesophagus changes in the mucous membrane may be slight and only determined with certainty by histological examination. There may be thickening and congestion of the mucosa, pale patches of leucoplakia, and acute superficial ulceration. This last may be single or multiple small discrete ulcers or extensive denudation."

Allison describes the examination by rigid endoscopy in the normal patient. "During the examination of an anatomically normal oesophagus the pinchcock constriction of the lower end by the diaphragm is noted, after which the patient's head must be depressed and moved to the right to allow the oesophagoscope to pass forward and to the left into the stomach. At a precise moment the lumen of the viscus gapes, gastric mucosa prolapses into the field, and fluid wells up." He contrasts this with the appearance in this abnormal group: "In the patients with short oesophagus the picture is quite different, for in the absence of stenosis the instrument is passed down from the oesophagus to stomach without impediment or deflection. The viscus is lined by esophageal mucosa at one level and by gastric at the next: the level is higher than would be anticipated in the normal, and no pinchcock action of the diaphragm is observed. The level of this change varies with the degree of herniation, but it has been seen at 24 cm from the alveolar margin in a patient in whom the normal level would have been about 39 to 40cm. In these extreme examples there is no difficulty in making a diagnosis, but in the minimal degrees of herniation, where the change in mucosa is found at 38 or 39cm, the actual level of the change is less important than the way in which it occurs. If there is no deflection and the viscus appears to be a direct continuation of the oesophagus lined by gastric mucosa, and if the level of the pinchcock cannot be seen accurately, it suggests that herniation is present. In cases of doubt it may be helpful to apply a silver Cushing brain clip to the junction of gastric and esophageal mucosa and examine the patient radiologically…. In the presence of stenosis the appearances are different. There is usually a moderate collection of mucus (and occasionally food residues) in the oesophagus, but this is never gross as in cardiospasm. The mucosa of the oesophagus shows congestion all the way down, and this becomes more intense at the lower end. About 2 or 3cm above the stenosis the mucous membrane shows acute congestion, often with superficial discrete ulcers or extensive denudations. These last usually have finger-like processes passing up between the pale areas of thickened, sodden esophageal mucous membrane, and below pass into the stricture to be continuous with the chronic ulcer."

The strictures and ulcers in Allison's patients are of varying size and a mixture of active ulceration with inflamed granulation tissue and areas of healing with fibrosis. The mucosa above the ulcer is squamous with changes that include hyperplasia and inflammation. The mucosa below the ulcer is "gastric" with intense chronic inflammation. Back then, anything below the squamocolumnar junction was deemed to be gastric.

It is clear that this patient population is the extreme end of severity of esophagitis defined by the presence of peptic ulceration and strictures.

The squamocolumnar junction in all but 2 of the 63 patients in this study was located in the mediastinum above the diaphragm with the mucosa adjacent to the junction showing "coarse gastric folds." Allison interprets this as a short esophagus with a sliding hiatal hernia. By present understanding, it is likely that most, if not all, these patients had CLE. Allison reports that the position of the squamocolumnar junction varied from 24 to 39cm from the alveolar margin (incisor teeth today), a variation that is similar to that described for CLE today. In 1948, CLE was not recognized. Allison would describe it 5years later, in 1953.[14]

This is an important paper because it describes the natural history of GERD at a time when there were no acid-suppressive drugs. The only available treatment was to neutralize acid with "alkali" and milk. These patients with GERD, seeking treatment in a premier surgical center in England with severe uncontrolled symptoms of GERD, strictures that caused "starvation," and bleeding that caused "anxiety," rarely died of their disease. The larger population of GERD patients with less severe disease without ulcers, strictures, and bleeding do not appear in the literature of this time. They probably suffered on their own, using home remedies such as "alkali" and milk because the physicians of that time had nothing to ease their suffering. The misery index of the disease was probably very high.

With PPIs available today, uncontrolled symptoms, severe strictures causing dysphagia that was present in 92% of patients, and bleeding such as that existed in Allison's era have become uncommon. Patients with GERD now rarely suffer from the misery of those in the era before acid-suppressive drugs were available.

To me, the most important item of data in this paper is that nearly all these patients likely had CLE by Allison's description. Back then, therefore, in the absence of an effective method of suppressing acid, CLE existed, sometimes reaching the upper esophagus 24cm from the incisor teeth. What happened rarely or never was the progression of the CLE to adenocarcinoma. At the time of Allison's paper in 1948, not a single case of esophageal adenocarcinoma had been reported.

In **1950**, 2years after Allison's paper was published, Barrett[15] introduced the term "reflux esophagitis" as the best name "for the lesion so ably described by Allison (1948)."[12] This is the first time that the concept of reflux of gastric juice is put forward to explain acid-induced damage in the esophagus. Previous to this, it was assumed that heterotopic gastric epithelium in the distal esophagus was responsible for acid secretion and ulceration.

Barrett writes[15]: "In the normal person, the oesophagus is completely separated from the stomach by a mechanism at, or immediately above, the oesophagogastric junction, which allows food and liquids to pass down but strictly prevents reflux. This mechanism can become inefficient under many circumstances, the most important of which are some varieties of hiatal diaphragmatic hernia and congenital short oesophagus. It may also be inefficient under conditions which appear, at first sight, to be anatomically normal. Thus Allison has shown that by placing metal clips upon the clear-cut line of demarcation between the esophageal and gastric mucosae at oesophagoscopy, and then screening the patients, the lower parts of the anatomical oesophagus may in fact be lined by gastric epithelium some of which lies above the 'cardiac sphincter'…. In some patients, no harm appears to result from reflux of gastric secretions into the lower gullet, but in others an inflammatory lesion can develop in a few hours…. The mucosa becomes fiery, edematous, and congested; it bleeds easily and may be shed widely or in patches. There is leucoplakia and, in due season, erosion." Barrett goes on to describe the erosions and how they progress to form deeper ulcers and fibrous strictures.

Barrett[15] and Allison[12] establish that reflux esophagitis is an acquired disease resulting from the failure of the valvelike mechanism at or above the squamocolumnar junction that prevents reflux in the normal person.

In 1950, the LES is still a theoretical concept. It was formally defined in the late 1950s. Also in 1950, there is still no recognition of columnar metaplasia of the esophagus. The gastroesophageal junction is equated to the squamocolumnar junction.

When this is located above the diaphragm, it defines the presence of stomach in the thorax with a short esophagus or the presence of congenitally heterotopic gastric epithelium in the lower esophagus. I will review the history of CLE in Chapter 11.

The introduction of the **flexible endoscope in 1957** revolutionized the study of GERD.[16] There was an exponential increase in the number of patients with GERD studied by endoscopy. The method of defining GERD moved rapidly from an examination of excised surgical specimens, which was the basis of Allison's study[12] to endoscopic diagnostic criteria.

Over the next decades, reflux became recognized as being the cause of the vast majority of cases of esophagitis. The symptoms and endoscopic appearances of uncontrolled GERD are, however, probably not described better than in these original papers by Allison[12] and Barrett.[15]

The presence of erosions in the distal esophagus at endoscopy (erosive esophagitis) quickly became the standard definition of GERD (Fig. 1.3). Because of the recognition that GERD was responsible for virtually all cases of erosive esophagitis when there were no obvious other causes such as lye ingestion, impacted foreign bodies, infections, and pill esophagitis, the presence of erosive esophagitis was essentially diagnostic of GERD. The only improvement on this definition until now is the development of endoscopic grading systems of erosive esophagitis by the Savary-Miller and Los Angeles grading classifications.[17]

3. PRESENT (SYMPTOM-BASED) DEFINITIONS OF GASTROESOPHAGEAL REFLUX DISEASE

The next revolution in GERD was the development of drugs that could effectively suppress gastric acid secretion. Beginning with histamine-2 receptor antagonists in the 1960s and progressing to PPIs in the 1980s, physicians were given a new powerful tool for treating patients with GERD.

The most powerful of these drugs, the PPI group, have the ability to maintain gastric pH above 4 for 12–18 h/day at maximum dosage.[18] The ability to achieve such powerful acid suppression was heralded as miracle drugs that would eradicate GERD.

This desire to help people suffering with GERD made erosive esophagitis no longer a suitable method to define GERD. It required endoscopy, which was not feasible because of the large numbers of GERD sufferers. It was also not sufficiently sensitive. Almost half of the patients with typical symptoms of GERD had neither erosive esophagitis at endoscopy nor any specific histologic abnormality if biopsies were done. The designation of "nonerosive reflux disease or NERD" was used for patients with GERD symptoms who were endoscopically normal.

A need arose to develop a symptom-based definition of GERD to replace erosive esophagitis as the defining criterion of GERD. This had the great advantage that it would only need clinical history. No testing was necessary. In their desire to help their patients, physicians could be forgiven for thinking that the only mistake was to not treat someone who may have GERD with these new miracle drugs even if it meant that some people without GERD would be treated unnecessarily. High sensitivity was desirous even if it meant decreased specificity. Acid-suppressive drug therapy, notably with PPI, was considered safe, even when used on a chronic long-term basis. It is relatively recently that significant side effects have been shown to occur with long-term PPI use.

The development of a suitable symptom-based definition of GERD proved to be a challenge because of the large spectrum of symptoms severity and symptom types, typical and atypical, that may occur in GERD.

3.1 Common Unstandardized Definition of Gastroesophageal Reflux Disease in Practice

Most physicians define GERD by the presence of typical symptoms—heartburn and/or regurgitation. Physicians will ascribe numerous atypical symptoms—chest pain, dyspepsia, chronic cough, asthma, hoarseness, etc.—to GERD with a significant interobserver variation.

All patients with symptoms of GERD of sufficient severity are treated empirically with acid-suppressive drugs, usually PPIs. The ease with which any individual physician will attribute a relatively nonspecific symptom such as chronic cough to GERD is very variable. Many patients who have intermittent heartburn whose severity is not sufficient to adversely affect their quality of life are considered to have "episodic heartburn," which is regarded by some authorities as not being GERD. The only definition possible for "episodic heartburn" is the presence of heartburn that the patient and physician agree does not require long term treatment with PPIs. It is highly likely that "episodic heartburn" represents mild GERD that is made to be "not GERD" by defining it away by the requirement of "troublesome" symptoms.

Heartburn is the symptom that is most commonly used for definition of GERD, because it is the commonest symptom of GERD and has a high, but not 100%, specificity. Regurgitation, although more specific for reflux, is much less common and is frequently added to heartburn in the definition ("heartburn and/or regurgitation").

Before the Montreal definition (see below), there was no standardization as to what duration and/or frequency of heartburn was necessary for the diagnosis of GERD. As a result, studies could define the presence or absence of GERD at will based on any frequency of heartburn and/or regurgitation (Table 1.1).[19–22] Patients falling within the definition of GERD in

TABLE 1.1 Definition of Gastroesophageal Reflux Disease in Selected Papers

Reference and Source	Definition
Spechler et al.[19] Harvard	Presence of heartburn equal to or greater than 1 day/week without mention of regurgitation
Goldblum et al.[20] Cleveland Clinic	Presence of heartburn and/or acid regurgitation at least twice per week for at least 6 months
Lagergren et al.[21] Sweden	Presence of heartburn and/or regurgitation once per week
Kahrilas et al.[22] Northwestern University	Presence of heartburn equal to or greater than three times per week controlled by acid suppression

one study could be controls in another study based on the definition used. The literature on GERD is chaotic as a result of a lack of standard definition.

This lack of uniformity was widely known and recognized by the medical community even as it continued to be used in the diagnosis of patients with GERD. The lack of concern regarding this chaos may have been partly because the introduction of PPIs in the late 1980s led to an optimism that GERD would become a disease of the past and strict definition was unnecessary. Unfortunately, this did not happen.

When it was realized that GERD was not disappearing any time soon despite the availability of PPIs, attempts were made to reach a consensus international definition of GERD. Two such major attempts have been made.

3.2 The Genval Working Group Definition, 1999

In 1999, a group of 35 medical doctors (primary care physicians, gastroenterologists, and surgeons) and health economists from 16 countries developed a definition of GERD in the Genval Workshop Report.[23] The core group recognized: "The lack of an explicit definition of GERD has been a problem to date."

They suggested the following definition: "The term 'gastroesophageal reflux disease' should be used to include all individuals who are exposed to the risk of physical complications from gastroesophageal reflux, or who experience clinically significant impairment of health-related well being (quality of life) due to reflux related symptoms, after adequate reassurance of the benign nature of their symptoms."

The last phrase is informative in that this group of international experts thought that emphasizing GERD as a benign disease was so important that it should be stated in the definition. This is a common and unfortunate misconception among physicians. It results in a public perception whereby GERD is treated as nuisance that can be easily treated with pills rather than a premalignant disease. The progression of GERD to Barrett esophagus and adenocarcinoma is a continuum of the same disease. The approximately 20,000 people in the United States who developed adenocarcinoma of the esophagus in 2016 prove that GERD is a premalignant disease.

The complexity and lack of specificity of the Genval workshop definition was possibly the reason that it was rarely used in practice. The main purpose of the workshop was an attempt to standardize the definition of GERD for the purpose of clinical studies and it had some success in this regard. It has now been supplanted with the Montreal definition.

3.3 The Montreal Definition and Classification of Gastroesophageal Reflux Disease, 2006[24]

In 2006, another group of 44 experts from 18 countries, consisting of 38 gastroenterologists, 3 surgeons, 1 family medicine physician, 1 oncologist, and 1 medical researcher, were selected on the basis of expertise, wide geographic distribution, and diversity of views related to GERD. The study was funded by Astra-Zeneca Research and Development; included in the funding was an undisclosed honorarium that was paid to all members of the consensus group.

A statement on sponsor influence on the process was present in the methods: "Ninety-two percent of the participants agreed that the sponsor had not, in any way, influenced their voting." This is an important statement because Astra-Zeneca is a major marketer of PPIs that would benefit significantly from the results of this study. When expressed in an equivalent but reverse manner, this statement shows that 8% (i.e., 4/44) of the participants indicated that the sponsor had influenced their voting in some way. To my ever-suspicious mind, the fact that 4/44 participants who had been paid to participate had this negative impression suggests a powerful sponsor influence.

After a complex system of voting and modification of statements over a period of 2 years, this group developed a new global consensus definition of GERD[24] called the Montreal Definition because it was first presented at a conference in Montreal: "GERD is a condition that develops when the reflux of stomach contents causes troublesome symptoms and/or complications." The degree of agreement with this definition among the experts was high: 81% agreed strongly, 14% with minor reservation, and 5% with major reservation. No expert disagreed.

The group expanded this rather nonspecific statement by listing and classifying all the symptoms and/or complications that fall under this definition (Table 1.2):

From the paper is the intent of the definition: "The language of the definition is designed to allow asymptomatic patients with complications such as Barrett's esophagus to be included in the case-definition of GERD, and be independent of technology used to achieve a diagnosis. For example, patients may be diagnosed based on typical symptoms alone or on the basis of investigations that demonstrate reflux of stomach contents (e.g., pH testing, impedance monitoring) or the injurious effects of the reflux (endoscopy, histology, electron microscopy), in the presence of typical or atypical symptoms."

The result of this statement is that anyone who has not had an upper endoscopy can have GERD. Barrett esophagus and electron microscopic evidence of injury to the esophageal epithelium, e.g., dilated intercellular spaces, can certainly exist in anyone in the asymptomatic population and will not be detected without endoscopy and biopsy. The inclusion of these changes of GERD without a mandate to find them by doing endoscopy has no practical value. It simply provides cover for the inadequacy of the definition. We have shown in previous books that minimal cellular changes of esophageal epithelial injury are present in almost 100% of the population.[25,26]

The Montreal statement covers all injurious effects of reflux irrespective of the presence or absence of symptoms. The problem with this is that the only injurious effects that are now recognized as being specific for GERD are visible CLE and adenocarcinoma. No other endoscopic, histologic, molecular, or ultrastructural feature is considered to be specific for GERD. For example, erosive esophagitis cannot fall within this definition if it is shown that the erosions are the result of herpesviruses or pemphigus vulgaris.

This statement is essentially designed to allow visible CLE and adenocarcinoma to be included in the definition of GERD in the patient without troublesome symptoms.

However, when the presence of cardiac epithelium distal to the squamocolumnar junction is proven to be a specific result of injury to esophageal squamous epithelium caused by exposure to gastric juice, the Montreal definition permits its use as a criterion to define GERD because it is a cellular change in the absence of symptoms.

In addition to the definition, the Montreal group produced numerous statements on which the experts reached agreement (Table 1.3).

The Montreal definition had two significant points of impact on GERD. First, it introduced the concept of "troublesome" symptoms. This removes the physician in clinical practice from requiring any criteria for severity, duration, or frequency of any symptom to diagnose GERD. The consensus groups concluded: "... in clinical practice the determination of whether symptoms were troublesome should be patient-centered without the use of arbitrary cutoffs for frequency and duration."

TABLE 1.2 Syndromes That Fall Under the Montreal Definition of Gastroesophageal Reflux Disease

1. Esophageal syndromes
   a. Symptomatic syndromes:
      i. Typical reflux syndrome: heartburn and/or regurgitation
      ii. Reflux chest pain syndrome
   b. Syndromes with esophageal injury:
      i. Reflux esophagitis
      ii. Reflux stricture
      iii. Barrett's esophagus
      iv. Esophageal adenocarcinoma

1. Extraesophageal syndromes
   a. Established associations:
      i. Reflux cough syndrome
      ii. Reflux laryngitis syndrome
      iii. Reflux asthma syndrome
      iv. Reflux dental erosion syndrome
   b. Proposed associations:
      i. Pharyngitis
      ii. Sinusitis
      iii. Idiopathic pulmonary fibrosis
      iv. Recurrent otitis media

TABLE 1.3 Statements for Which There Was Agreement That Reached the Predetermined Threshold of 67% Among the Experts

Statement	% Agreement	Evidence Grade
1. GERD is a condition that develops when the reflux of stomach contents causes troublesome symptoms.	100	N/A
2. GERD is common and its prevalence varies in different part of the world.	100	High
3. Symptoms related to gastroesophageal reflux become troublesome when they adversely affect an individual's well-being.	100	N/A
4. Reflux symptoms that are not troublesome should not be diagnosed as GERD.	93	N/A
5. In population-based studies, mild symptoms occurring 2 or more days a week, or moderate/severe symptoms occurring more than 1 day a week, are often considered troublesome by patients.	95	Moderate
6. In clinical practice, the patient should determine if their reflux symptoms are troublesome.	100	N/A
7. Heartburn is defined as a burning sensation in the retrosternal area (behind the breastbone).	100	N/A
8. Regurgitation is defined as the perception of flow of refluxed gastric content into the mouth or hypopharynx.	100	N/A
9. Heartburn and regurgitation are the characteristic symptoms of the typical reflux syndrome.	100	N/A
10. Gastroesophageal reflux is the most common cause of heartburn.	100	High
11. Heartburn can have a number of nonreflux-related causes. The prevalence of these is unknown.	98	Moderate
12. The typical reflux syndrome can be diagnosed on the basis of the characteristic symptoms, without diagnostic testing.	100	Moderate
13. Nonerosive reflux disease is defined by the presence of troublesome reflux-associated symptoms and the absence of mucosal breaks at endoscopy.	100	N/A
14. Epigastric pain can be the major symptom of GERD.	91	Moderate
18. Chest pain indistinguishable from ischemic cardiac pain can be caused by GERD.	100	High
22. Esophageal complications of GERD are reflux esophagitis, hemorrhage, stricture, Barrett's esophagus, and adenocarcinoma.	84	
23. Reflux esophagitis is defined endoscopically by visible breaks of the distal esophageal mucosa.	100	N/A
25. Over a 20-year period, the severity of reflux esophagitis does not increase in most patients.	93	Low
26. Although heartburn frequency and intensity correlate with the severity of mucosal injury, neither will accurately predict the severity of mucosal injury in the individual patient.	95	Moderate
31. Dysphagia is troublesome in a small proportion of patients with GERD.	100	Low
32. Persistent, progressive, or troublesome dysphagia is a warning symptom for stricture or cancer of the esophagus and warrants investigation.	98	High
34. Neither the frequency nor the severity of heartburn is useful for prediction of the presence, type, or extent of esophageal columnar metaplasia.	98	Moderate
39. Adenocarcinoma of the esophagus is a complication of GERD.	100	Moderate
40. The risk of adenocarcinoma of the esophagus rises with increasing frequency and duration of heartburn.	96	Moderate

Continued

TABLE 1.3 Statements for Which There Was Agreement That Reached the Predetermined Threshold of 67% Among the Experts—cont'd

Statement	% Agreement	Evidence Grade
41. Long-segment Barrett's esophagus with intestinal-type metaplasia is the most important identified risk factor for esophageal adenocarcinoma.	100	High
42. Chronic cough, chronic laryngitis, and asthma are significantly associated with GERD.	93	High
44. Gastroesophageal reflux is rarely the sole cause of chronic cough, chronic laryngitis, or asthma.	95	Variable

GERD, gastroesophageal reflux disease.
Those statements that are most relevant with the percentage of experts agreeing and evidence grade have been selected.

This meant that, for purposes of clinical management, it was the patient who decided whether he/she had GERD. A patient with heartburn could be diagnosed with and treated for GERD if he/she considered it "troublesome" because it affected quality of life. Another patient with heartburn of similar severity and frequency who did not consider it "troublesome" did not have GERD; he/she had "episodic heartburn," which did not need treatment. The goal of management is therefore clearly the control of troublesome symptoms to return the patient to normal quality of life.

Secondly, the group recognized a broad array of symptoms that include typical and atypical esophageal syndromes and established and proposed extraesophageal syndromes within the definition of GERD (Table 1.2). Because of this, the scope of the diagnosis of GERD expanded greatly. Any patient who was "troubled" by any symptom within these categories who did not have some other obvious cause for these symptoms could be deemed to fall within the definition of GERD. It gave physicians a license to treat chronic cough and asthma empirically with a trial of PPIs if no other cause for these was apparent. Reflux laryngitis became one of the most commonly diagnosed entities among otolaryngologists. If PPI treatment produced a positive response, the patient could be declared as having GERD.

The effect of the definition was a dream come true for pharmaceutical companies who sold PPIs. The use of PPIs increased dramatically; these drugs now represent one of the most widely used and profitable groups of drugs. That the Montreal definition came out of a workshop that was funded by Astra-Zeneca where the participants were paid an undisclosed honorarium and where the sponsors had a stated effect on the voting of a minority of the experts should raise serious conflict of interest concerns.

4. PROBLEMS WITH PRESENT SYMPTOM-BASED DEFINITION

The Montreal definition of GERD, which is widely used, drives treatment, which is the empiric and liberal use of acid-suppressive drugs, mainly PPIs, to control symptoms. It drives research, which for the foreseeable future appears to be almost entirely concentrated on the development of evermore effective acid-suppressive drugs.

The overall effect of treating GERD patients with H_2 receptor antagonists and PPIs since they became available in the 1960s has been extremely positive. The disease that Allison described in 1948[12] has essentially been eradicated. The severity of symptoms and near complete lack of ability to control pain, the severe and deep ulcers, and complex strictures that caused dysphagia in 92% of his 63 patients are a thing of the past. PPIs are miraculous drugs in this regard, improving the quality of life for millions of GERD patients who would be miserable if not for these drugs (Fig. 1.1).

However, PPIs are far from being a perfect treatment for GERD. The initial excitement that they would eradicate GERD has not been realized. The present negative outcomes of GERD management occur despite treatment with PPIs.

The single-minded research focus on producing evermore powerful acid-suppressant drugs has inhibited other avenues of research in GERD. The goal of research even today seems to be based on the idea that all the negative outcomes of GERD therapy will be corrected if only we had better acid-suppressive drugs. This is an incorrect premise. No amount of acid suppression will prevent treatment failure, which is caused by progressive LES damage.

More importantly, the positive effect on symptom control by acid-suppressive drugs has been associated with the powerful negative effect of a large increase in the incidence of GERD-induced adenocarcinoma during the same time frame.[3] No amount of acid suppression will prevent adenocarcinoma in the GERD patient, which is caused by carcinogens in gastric juice that are not impacted positively by removing acid from the equation.

Present negatives of GERD management are largely driven by the focus of the definition of GERD on the presence of "troublesome" symptoms. While symptom severity and duration are known to have a relationship with the development of adenocarcinoma, the majority of patients who develop cancer have not had symptoms of a magnitude that caused them to have endoscopy.

4.1 An Illusion of "Cure" of Gastroesophageal Reflux Disease

An illusion is created by a symptom-based definition of GERD that if the symptoms are controlled, the disease is cured. When the word "troublesome" is added, control becomes easier. It is not necessary to even completely control symptoms; it is only necessary to make these not troublesome to the extent that the patient's quality of life is not impacted.

This goal of management of GERD is not a real cure by any stretch of the imagination. Treatment with acid-suppressive drugs does not have any positive effect on the LES and does not stop reflux. The number and duration of reflux episodes does not decrease. Strong acid reflux is converted to weak acid reflux.[27] This is effective in decreasing heartburn, but does not address the underlying cause.

The illusion of cure that is created by control of symptoms is commercially extremely profitable. The annual revenue to pharmaceutical companies from the sale of PPIs worldwide is estimated to be in the $15–20 billion range. Advertising for these drugs as miracle treatments for heartburn and reflux is everywhere; they encourage the belief that all that is needed for GERD is to pop a pill, purple, or otherwise. Acid-suppressive drugs of all types, including the most powerful PPIs, are available for over-the-counter purchase without prescription and occupy a pride of place in pharmacy shelf space.

The impression is created that these drugs are harmless by FDA approval for over-the-counter use without prescription, an idea that is reinforced by advertising. As a result, many people continue to use these drugs over the long term without seeking physician consultation.

It is theoretically possible that suppression of acid to control heartburn allows overeating and actually removes a natural mechanism of the body to limit the progression of the disease. Like pain limits the movement of an injured joint and controls aggravation of the injury, heartburn may effectively limit overeating that puts further pressure on the LES and causes disease progression. Rather than curing GERD, powerful acid suppression may be facilitating the progression of its cause, LES damage.

4.2 Unnecessary Use of Proton Pump Inhibitors

Statements 10, 11, and 12 in the Montreal workshop document (Table 1.3)[24] demonstrate the impact that the definition has on management. The first two of these state that heartburn is the most common symptom of GERD but that it is not specific for GERD; it can be caused by non-GERD entities in an unknown number of patients. Statement 12 states that GERD can be diagnosed by the presence of typical symptoms without any diagnostic tests. This must mean that an unknown number of patients who have heartburn and will be treated with empirical PPI therapy will have a non-GERD cause for their heartburn. Treatment with PPIs is unnecessary in these patients.

The specificity of the various syndromes for GERD decreases substantially from typical heartburn and regurgitation to extraesophageal syndromes. As more and more of these syndromes fall within the definition of GERD, the frequency with which patients are diagnosed with GERD increases and the likelihood of false-positive diagnosis of GERD increases as does the unnecessary use of PPIs.

The situation is the complete reverse of the first half of the last century, when the definition focused on specificity rather than sensitivity. With only relatively ineffective surgical treatment available, the need was to limit the diagnosis and treatment to the most severely affected GERD patient. Now, the goal is to be overreaching to include all possible people with GERD.

When hospital charts are examined, it is not uncommon to see PPIs listed in the medications used by the patient, often on a chronic basis. In many of these patients, there is no clear indication for use of PPIs. Some have been prescribed the drug for vague dyspepsia; after all, statement 14 of the Montreal document states that epigastric pain can be the main symptom of GERD. Other patients have been prescribed PPIs in an intensive care unit and never come off the drugs. Once started, the tendency seems to be to keep prescribing these drugs.

In a 2003 study of pharmacy billing data for two insurers within a large eastern Massachusetts provider network,[28] 4684/168,727 (2%) of patients were prescribed chronic (more than 90 days) acid-suppressive drugs; 47% were taking H_2 receptor antagonists, 57% PPIs; and 4% both. Diagnostic testing was uncommon in these patients; only 19% of those on chronic acid-suppressive drugs had undergone upper endoscopy within the previous 2 years.

It is likely that present practices of prescribing acid-suppressive drugs have led to significant unnecessary use of PPIs in many patients. With reporting of complications of long-term PPI use, this has recently become a problem that has received considerable attention and causes concern.

Even with a very low percentage risk of complications, the very large population using long-term PPIs translates into a significant increase in hip fractures, *Clostridium difficile* colitis, community-acquired pneumonia, interactions with other drugs such as Plavix, and other reported complications that are difficult to comprehend such as myocardial infarction and dementia. These are largely based on data mining studies that have low credibility, but they are sensationalized on media and Internet sites, and become a reality in the minds of patients taking PPIs. Even if they are groundless, the perception and studies claiming association will result in patient resistance to long-term PPI use. This opens the door to insurance companies that desire to restrict reimbursement for the use of these drugs to control cost.

4.3 Lack of Effort to Prevent Progression to Failure of Symptom Control in Gastroesophageal Reflux Disease

A significant number of GERD patients fail to be satisfactorily controlled with PPI treatment, resulting in a decline in their quality of life and productivity. The rate of satisfactory control of heartburn, the commonest symptom of GERD, is not as good as healing of erosive esophagitis.[6] Heartburn resolution with PPIs in patients with erosive esophagitis is 56% versus 8% for placebo at 4 weeks; resolution of heartburn in patients with negative endoscopy or without endoscopy is 36.7% versus 9.5% for placebo.[6] Control of regurgitation and atypical symptoms is less effective. High-dose PPI treatment is no better than placebo for reflux laryngitis syndrome without frequent heartburn.[29]

Alkalinization of the refluxate with PPIs is effective in controlling heartburn only up to a point of severity of gastroesophageal reflux. Beyond that point, the magnitude of reflux overwhelms the ability of the drugs to control symptoms. Failure of PPIs to control symptoms is inevitable in those patients whose LES damage is destined to progress to a high grade of severity during their lifetime. The only reason why PPI treatment is successful in the majority of GERD patients is that LES damage progresses to this severe grade in only a minority. The majority of patients whose LES damage remains at lower grades of severity throughout their life are well controlled with drugs.

There is no method, at the start of treatment of GERD, to identify those patients who will remain controlled with drug therapy and those who will fail. We treat all patients with GERD in the same way, with acid suppression adequate to control symptoms at one point in time. As the disease progresses, there is often the need for dose escalation in many patients. Failure of therapy to control symptoms adequately is simply an unexpected and unwelcome event that is inevitable in some patients. There is no ability or effort to prevent these negative outcomes. The ability to treat these complications when they occur is ineffective at best.

The patients who progress to treatment failure are already at a significantly advanced stage of LES damage. They are most likely to progress to Barrett esophagus and adenocarcinoma. In fact, the first endoscopy done in this population already has a significant percentage of patients who already have visible CLE. Visible CLE is the first point in the cellular sequence of GERD where medical therapy cannot reverse the change. These patients have entered the irreversible state where prevention of progression to adenocarcinoma is impossible without surgery.

The percentage risk of adenocarcinoma in this population is extremely small, permitting gastroenterologists to minimize and ignore the risk. However, the massive number of patients entering this stage translates the very small percentage risk of adenocarcinoma to a very large and increasing number of people.

4.4 Failure to Prevent Esophageal Adenocarcinoma (Fig. 1.8)

Importantly, when treatment for GERD is started, there is no method of identifying which patients are destined to develop adenocarcinoma. The development of cancer is a totally unexpected adverse outcome in those unfortunate patients. There is little relationship between symptom type or symptom control with drugs and development of cancer.[30] Most patients who develop cancer have disease that is not adequately problematic to have precipitated the need for endoscopy. The increasing incidence of esophageal adenocarcinoma, a lethal complication of GERD, has resulted in a dramatically increasing mortality rate in patients with GERD.[3]

Esophageal adenocarcinoma is a relatively new disease. Morson et al. reported the first case of esophageal adenocarcinoma in 1952.[31] Between 1973, when the increasing incidence of esophageal adenocarcinoma was recognized, and 2006, the incidence of esophageal adenocarcinoma increased sevenfold. It continues to increase.[3]

It is estimated that ~16,000 people in the United States will develop esophageal carcinoma in 2016. Of this, 80%–90% (i.e., ~13,000) will have GERD-induced adenocarcinoma. To this must be added ~8000 people who will develop adenocarcinoma of the "gastric cardia," which is now recognized as an esophageal adenocarcinoma. With an overall mortality rate for esophageal carcinoma of ~85%, this is not an outcome that we should accept. It is probably the greatest failure of modern treatment of any common human disease.

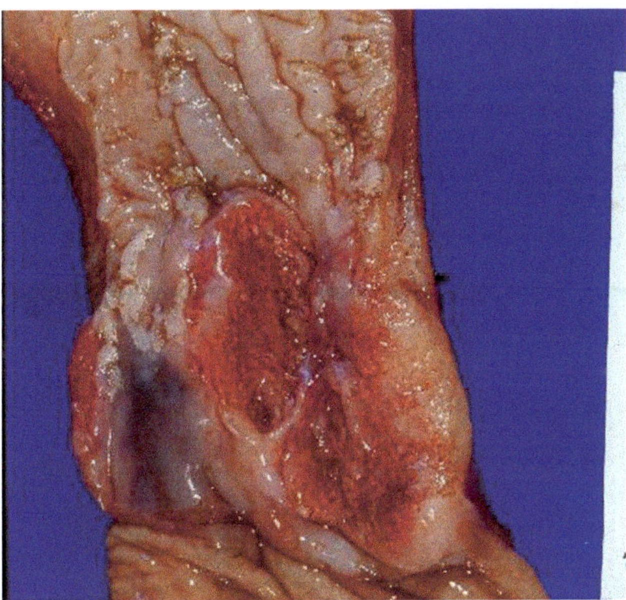

FIGURE 1.8 Adenocarcinoma of the esophagus. This shows an ulcerated mass in the distal esophagus. The tumor is at the junction of squamous and columnar epithelium.

The dramatic increase in the incidence of esophageal adenocarcinoma coincided with the increasing ability of physicians to suppress acid. This should raise questions regarding a relationship between cancer and the use of acid-reducing drugs. The argument put forward by gastroenterologists against this suggestion is that the cancer epidemic preceded PPI use. This is correct; PPI use cannot be the direct cause of this increase in cancer incidence. However, the relationship could be between cancer and the ability to alkalinize gastric contents. Allison[12] reports that his patients used mild alkali in 1948. The progressive refinement of "alkali" with more effective acid-neutralizing agents followed by H_2 receptor antagonists in the 1960s and PPIs in the 1980s has resulted in the material refluxing into the esophagus becoming progressively more alkaline.

There is a logical basis to suggest that acid-reducing drug therapy may be at least partially responsible for the increased incidence of esophageal adenocarcinoma in patients with GERD.[25,26] Acid-suppressive drugs are not carcinogens. However, by changing the milieu of the esophagus during reflux episodes from strong acid to weak acid, they probably induce intestinal metaplasia in CLE. By promoting Barrett esophagus, alkalinization of gastric juice probably has an indirect effect in increasing cancer risk. I will examine the relationship between PPIs and esophageal adenocarcinoma in greater detail in Chapter 12.

The present management of GERD can be described as a method to control the disease symptomatically in a majority of patients while committing a significant minority to the misery of the decreased life quality of failed therapy and increased mortality from cancer. The former positive outcome is wonderful; the latter negative outcome is bad medicine.

4.5 Absence of Alternatives to Treating Gastroesophageal Reflux Disease

The single-mindedness of the pharmaceutical industry to produce evermore effective and evermore profitable acid-suppressive agents has prevented development of drugs aimed at other targets in the pathogenesis of GERD. There are no drugs to effectively enhance the function of the LES. There are no drugs to alter the cellular progression from squamous epithelium to columnar epithelium to intestinal metaplasia to adenocarcinoma.

We are faced with a marked increase in esophageal adenocarcinoma that is not prevented and may be promoted by acid-suppressive drug therapy. We have no good treatment alternatives. Our only defense is to avoid the issue that our treatment may actually be responsible for the cancer. Withdrawal of acid-suppressive drugs from the treatment of GERD is not an option that can even be considered. This will take patients back to Allison's era[12] where patients had to endure the untold misery of uncontrolled symptoms with severe ulceration and strictures.

There are surgical and endoscopic methods of repairing and augmenting the damaged LES, but presently all these have problems with efficacy and complications. They are not done with sufficient frequency to impact the disease significantly. They, however, can have greater relevance in the future if the objectives of management change and treatment is undertaken at an earlier stage of LES damage.

5. A NEW CELLULAR DEFINITION OF GASTROESOPHAGEAL REFLUX DISEASE

In this book, I will develop a new definition of GERD that is logical, scientific, and simple. Simplicity is the basis of precision and accuracy of diagnosis. The definition is based on histologic criteria applied to biopsy samples. Whenever the diagnosis of a disease has shifted from a symptom basis to a pathologic basis, outcomes have improved. This cellular definition can be accurately identified histologically. When it is quantitated, it becomes a new test for LES damage, which will ultimately drive the diagnosis and definition of GERD.

GERD can be conceptually defined by its basic etiology as "an esophageal disease resulting from the exposure of the esophageal squamous epithelium to gastric contents."

The disease results from a failure of the normal mechanism, the LES, which prevents the squamous epithelium of the esophagus from coming into contact with gastric juice. Failure of this process begins the sequence of changes in GERD.

5.1 Pathogenesis of Gastroesophageal Reflux Disease

The anatomic structure of the esophagus and stomach is designed to protect the squamous epithelium that lines the esophagus from gastric juice. The mechanism that protects the squamous epithelium from gastric juice is the LES. The LES is an ~5 cm zone of high pressure (>15 mmHg but usually in the 30–40 mmHg range) in the distal esophagus that effectively prevents reflux of gastric contents from the stomach (normally at a pressure of +5 mmHg in the fasting state) to the esophagus (which has a resting pressure of −5 mmHg in the midthoracic region). A competent sphincter prevents reflux of gastric contents along the natural pressure gradient that exists from the stomach to the esophagus.

In the fasting state, the stomach is collapsed and has a low residual volume. The mucosa is thrown into rugal folds. During a meal, the stomach fills causing it to distend to an extent that is a measure of the rate of filling, the rate of emptying, and the compliance of the wall of the stomach. The stomach is a reservoir. A reasonably sized meal is accommodated easily and without any negative consequence. The stomach distends to within its capacity without an increase in pressure. The rugal folds flatten. There is no pressure exerted on the sphincter from below. The squamous epithelium of the lower esophagus remains sequestered from gastric contents, above the point at which the transition of pH from acid gastric pH (approximately 1–2) changes to the more neutral esophageal pH.

This mechanism gets into trouble when a meal is ingested that causes the stomach to overdistend[32] (see Chapter 8). Gastric overdistension and increased intragastric pressure have an effect on the LES.[5,33] The sphincter becomes partially effaced (shortened), and the squamous epithelium of the distal esophagus moves down to a point below the pH transition point where it is exposed to gastric contents in the distended stomach.[5,33] When this happens, the squamous epithelium of the distal esophagus may undergo damage induced by the acid in the gastric contents.

5.2 Reflux Esophagitis and Reflux Carditis

The pathologic changes resulting from the exposure of squamous epithelium to gastric juice are well known. The initial change is caused by acid in the gastric juice, which damages the squamous epithelium to produce changes of reflux esophagitis (Fig. 1.5). This includes separation of the squamous cells in the epithelium ("dilated intercellular spaces") that causes the epithelium to become more permeable to other molecules in gastric juice (Fig. 1.9).[34,35] Yet undefined molecules in the gastric juice enter the epithelium and interact with the stem cells and proliferative cells in the basal region of the epithelium to induce genetic changes that result in columnar metaplasia (Fig. 1.10).

Inflammation (intraepithelial eosinophils and neutrophils) and reactive changes (dilated intercellular spaces, hyperplasia) in the squamous epithelium are designated as "reflux esophagitis."[7] This is a reversible change that is transient and likely to be patchy. Healing occurs both spontaneously and with acid-suppressive therapy when acid exposure is decreased. A biopsy must be done at the time the injury is present in the area involved. This makes the finding of reflux esophagitis have a very low sensitivity in the diagnosis of GERD. The changes, caused by acid, are also nonspecific. They are manifestations of injury and can be seen in any epithelial disease of the esophagus that causes injury, e.g., eosinophilic esophagitis (Fig. 1.5), chemical (pill, lye) and infection-induced injury, and diseases such as pemphigus vulgaris. Reflux esophagitis is, for these reasons, a very poor diagnostic test for reflux and cannot be used in its definition.

Injury to the epithelium is followed by columnar metaplasia wherein the squamous epithelium changes to cardiac type columnar epithelium.[4,7] This is a highly specific change for GERD because it requires a squamous epithelium whose permeability has been increased by acid exposure *plus* a molecular interaction between cells in the more basal region of the epithelium with a non-H^+ (acid), larger molecule in gastric juice.

Exposure to gastric juice is essential for columnar metaplasia. As such it is not seen in all other conditions causing squamous epithelium damage (allergy, chemicals, lye, infections, etc.), which do not cause the squamous epithelium to be

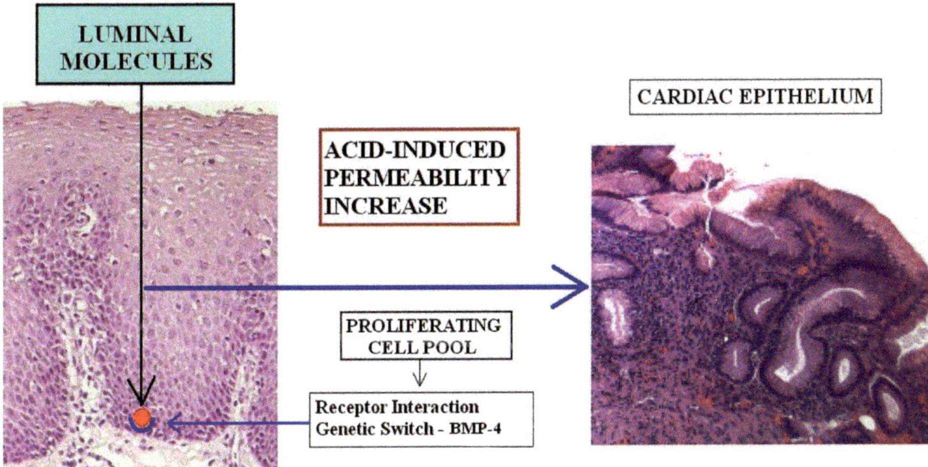

FIGURE 1.9 Diagram showing mechanism of metaplasia of squamous epithelium to cardiac epithelium. The acid-damaged stratified squamous epithelium with dilated intercellular spaces has an increased permeability. Luminal molecules penetrate into the deeper regions of the stratified squamous epithelium. This sets the stage of a genetic change in the cell that induces metaplasia from squamous to columnar (cardiac) epithelium.

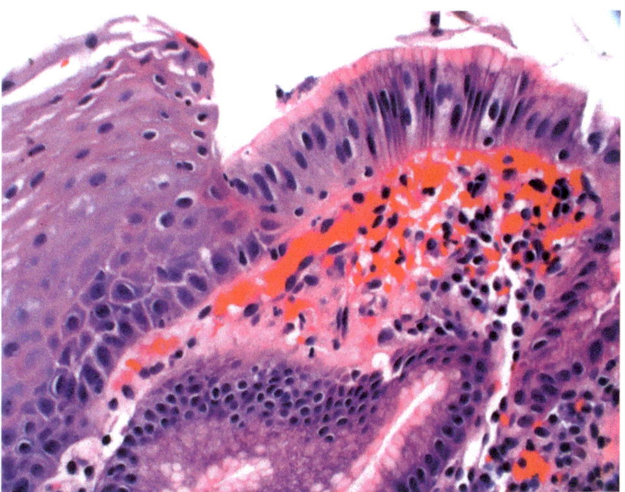

FIGURE 1.10 The zone of transition from stratified squamous to the single layered columnar epithelium. Note the fine line of cleavage in the suprabasal region of the remaining stratified squamous epithelium. This is the expected change resulting from a change in the differentiating signal of the stem/proliferating cell from squamous to columnar. The lack of normal squamous cell attachments will cause separation of the new columnar cell from the superficial squamous cells resulting in the columnar epithelium seen on the right.

exposed to gastric juice. Acid is the key that opens the door in esophageal squamous epithelium and permits access to other molecules in gastric juice.

Reflux esophagitis, with and without erosions, is a reversible change.[36] Columnar metaplasia does not usually reverse once it has occurred. It should therefore be possible to regard reflux esophagitis as an acute change in the squamous epithelium that is completely reversible with PPI therapy and columnar metaplasia as "chronic GERD" that is not reversible. I will show that columnar metaplasia is accompanied by loss of LES tone, which is a second irreversible change in GERD. We have called this "reflux carditis" for over two decades.[37] Only people who accept that cardiac epithelium is never a normal epithelium in the stomach will use this term.

Acute reversible inflammatory changes in the squamous epithelium without chronic GERD are limited to the earliest stage of GERD at the onset of the disease. This is never recognized clinically, endoscopically, or histologically. Columnar metaplasia of the squamous epithelium occurs early. GERD defined by columnar metaplasia is a chronic irreversible disease.

Throughout its course, however, chronic GERD characterized by columnar metaplasia coexists with acute esophagitis in the squamous epithelium. Reflux esophagitis is likely to be the dominant cause of symptoms of GERD. Acute injury of the squamous epithelium is likely to be intermittent ("relapses") and much less common in patients who are being treated with acid-suppressive drugs in adequate dosage.

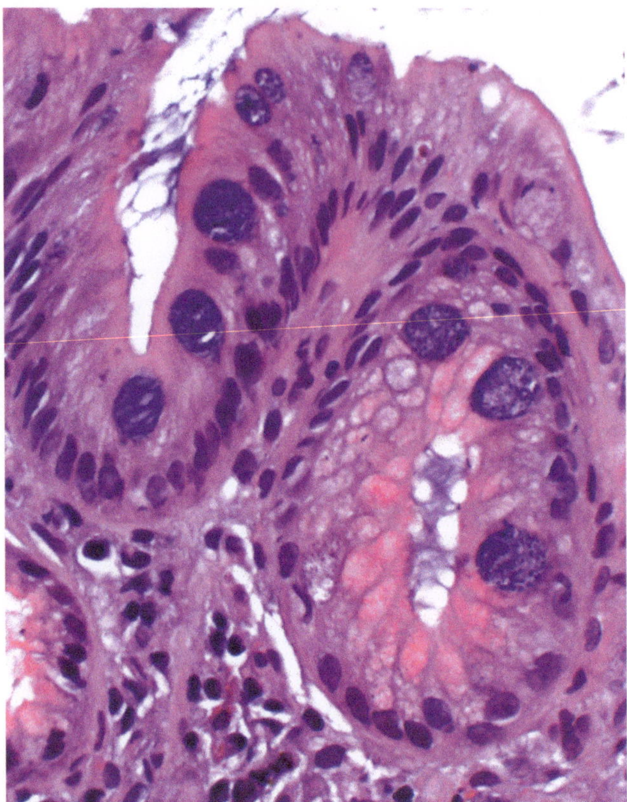

FIGURE 1.11 Cardiac epithelium with intestinal metaplasia, characterized by the presence of goblet cells. This is Barrett esophagus in most of the world when the biopsy is from a segment of visible columnar-lined esophagus. Hematoxylin and eosin stain.

In contrast, columnar metaplastic epithelium progressively increases in amount, replacing squamous epithelium.[38] The columnar epithelium is less sensitive to acid-induced acute inflammation than squamous epithelium. As columnar metaplasia increasingly replaces the squamous epithelium, the quality and severity of pain changes in unpredictable ways in different patients with GERD. This is the reason why the presence of heartburn is a relatively insensitive diagnostic criterion for GERD. Some patients with the most severe GERD with long segments of CLE can be relatively asymptomatic.

Columnar metaplasia of the esophagus continues to evolve by a variable increase in length over time and development of specialized cell types such as goblet cells (Fig. 1.11), representing intestinal metaplasia that defines Barrett esophagus in the United States and much of the world.[38] Barrett esophagus is a premalignant entity that progresses to esophageal adenocarcinoma despite PPI therapy.

I will show in this book that this simple concept of cellular changes that result from exposure of the squamous epithelium to gastric juice can be applied to define and follow the progression of GERD with exquisite accuracy.

5.3 Reasons for Failure to Use Reflux Carditis to Define Gastroesophageal Reflux Disease

Unfortunately, use of a cellular definition based on the presence of metaplastic columnar epithelium is presently not possible because of two fundamental errors in the present understanding: (1) the incorrect definition of the gastroesophageal junction[38] and (2) the dogma that cardiac epithelium is a normal lining of the proximal stomach.[39]

Eradication of these errors results in complete understanding of the pathogenesis of GERD. When this happens, the presence of cardiac epithelium distal to the endoscopic GEJ will be recognized as a 100% specific and 100% sensitive change that defines the presence of chronic, irreversible GERD. It is only when this is recognized that the management of GERD will move forward with significant change.

Chronic GERD characterized by columnar metaplasia of the squamous epithelium of the most distal esophagus defines the presence of GERD. A "normal" (i.e., non-GERD) person differs from a GERD patient in having no metaplastic columnar epithelium between the proximal limit of gastric oxyntic epithelium (the true GEJ) and the squamocolumnar junction.

Over time, and with repeated damage of the squamous epithelium, columnar metaplasia progressively increases in length. This was beautifully described by Hayward, in 1961[40]: "When the normal sphincteric and valvular mechanism in the lower oesophagus and oesophago-gastric junction, i.e. what I call the cardia, fails… reflux from the stomach occurs and acid and pepsin reach the squamous epithelium and begin to digest it…. In quiet periods some healing occurs, and in these periods the destroyed squamous epithelium may re-form, often with leukoplakia, or junctional (=cardiac) epithelium, usually not very healthy looking, may replace it. Where this occurs the area is given considerable protection from future reflux because normal junctional (=cardiac) epithelium resists acid-peptic digestion. Further reflux therefore attacks principally the squamous epithelium higher up. In the next remission it may be replaced by more junctional (=cardiac) epithelium as a further protective reaction. With repetition over a long period the metaplastic junctional (=cardiac) epithelium may creep higher and higher until it reaches the arch of the aorta. It seldom extends higher than this."

Columnar metaplasia of the esophagus can be recognized easily and accurately in biopsies taken from the squamocolumnar junction in patients who are endoscopically normal. Its length distal to the squamocolumnar junction can be measured accurately. This is the crux of the new histologic diagnosis and assessment of the progression of GERD. As Hayward described, columnar metaplasia is progressive, with reflux attacking the squamous epithelium above the CLE causing an increase in the length of CLE. Increase in the length of columnar epithelium is the best method of defining the severity of GERD. The reason for this is that this measurement of columnar metaplasia at the distal end of the esophagus is a measure of LES damage. This is the missing link in the present understanding of GERD.

6. A NEW DEFINITION BASED ON LOWER ESOPHAGEAL SPHINCTER DAMAGE

It is well known that GERD results because of a failure of the LES that is designed to prevent reflux in the normal (non-GERD) patient. A competent sphincter prevents reflux. As LES damage increases, sphincter failure increasingly occurs resulting in reflux into the body of the esophagus. A definition based on the severity and rate of progression of sphincter damage will theoretically be an excellent method of defining GERD (Fig. 1.7).

The LES is defined as a high-pressure zone at the distal end of the esophagus separating the positive pressure in the gastric lumen from the negative pressure in the thoracic esophagus. It can be recognized only by manometry. The LES cannot be recognized at endoscopy, radiology, gross pathologic examination of resected specimens, or autopsy. There is no anatomic counterpart of the high-pressure zone. The high pressure is created by tonic contraction of the smooth muscle of the esophageal smooth muscle wall. The muscle wall in the sphincter zone is anatomically and histologically identical to the muscle wall in the rest of the esophagus.

When LES damage occurs, sphincter pressure is lost. I will show that LES damage begins at the distal end, i.e., at the gastroesophageal junction, and slowly progresses cephalad. The damaged LES is not visible at manometry because it has lost its pressure. In the abdominal esophagus, where LES damage begins, the damaged LES is manometrically identical to the stomach. When manometry is performed, the LES is shorter than it was. However, because the original length of the LES is not known, it is impossible to identify LES shortening. Therefore LES damage cannot be defined manometrically in its early stages.

The other problem with using manometry to define the LES is that presently used high-resolution manometry uses a catheter that has pressure sensors across the LES that are separated by 11 mm. The entire abdominal LES measures 35 mm. Using a system that measures pressure points by sensors separated by relatively high intervals results in a lack of accuracy of measurement. The older, slow, motorized, pull-through method of assessing the LES was more accurate, but has largely become obsolete.

Present manometric assessment is able to identify manometric criteria that are associated with LES failure that is sufficient to produce enough reflux to produce an abnormal pH test. At this degree of LES damage, the patient is likely to have symptoms and a significant prevalence of erosive esophagitis and visible CLE. These manometric criteria are a mean LES pressure <6 mmHg, a total LES length of <20 mm, and an abdominal LES length of <10 mm.

These numbers provide the clue to the massive missing link in the present understanding of GERD. It is generally believed that the LES has a normal abdominal length of 35 mm. The LES becomes incompetent when the abdominal LES becomes less than 10 mm. *What happens between the normal 35 and 10 mm of abdominal LES length?*

At the annual Hawaii conference that I attend, this question was asked about 10 years ago. Dr. Peter Crookes, a brilliant member of our faculty and a good friend, was presenting a talk on the LES and described its shortening in his usually delightfully entertaining manner. A person in the audience raised his hand and asked: "Dr. Crookes, where does the shortened sphincter go?" Peter was perplexed for a moment, but with his extremely quick wit, replied in song: "Oh my darling, Oh my darling, Oh my darling Clementine. Thou art lost and gone forever, dreadful sorrow Clementine." This was a brilliant nonanswer that deflected the question.

That question is still not answered. Presently, the damaged LES simply disappears from the mind. It is invisible at manometry, endoscopy, gross pathology, and present understanding of histology.

Surely, it cannot disappear. It does not. It stays as a damaged distal esophagus between the end of the shortened sphincter and the stomach. The recognition of this one fundamental and obvious fact is the basis of the new understanding of

GERD. This area is dilated, develops rugal folds, and is lined by cardiac epithelium. The length of cardiac epithelium that defines the extent of the dilated distal esophagus is a measure of LES damage.

The objective of this book is to show that correct understanding of the histology of this region permits accurate assessment of sphincter damage. This is based on the interpretation of biopsies taken with a new protocol designed specifically to assess LES damage. This will provide a new method of understanding the pathogenesis of GERD, accurate assessment of GERD severity, and prediction of future progression of GERD.

Such an assessment will transform GERD from its present state of being a disease defined by symptoms and treated to control symptoms, to a disease defined by the severity of LES damage and treated to prevent progression of sphincter damage in the small minority of patients who need such intervention (Fig. 1.2).

Sphincter damage is to GERD what coronary artery narrowing is to ischemic heart disease. We recognize that coronary artery narrowing up to 50%–70% of luminal area is within the reserve capacity of the vessel. Ischemia occurs above this degree of narrowing. I will show that the LES has a similar reserve capacity wherein sphincter competence is maintained with mild damage. Damage has to progress beyond this reserve capacity to cause GERD.

I hope this new assessment of LES damage will do to GERD what coronary angiography did to ischemic heart disease. In the 1970s, ischemic heart disease was diagnosed symptomatically by the presence of angina. It was treated with drugs to control anginal pain. The medical establishment simply waited for patients to progress to myocardial infarction and death. This is similar to the present approach toward the management of GERD, which is diagnosed by the presence of symptoms and treated with the objective of controlling symptoms. The medical establishment simply waits for a minority of GERD patients to progress to treatment failure, adenocarcinoma, and death. These outcomes are inevitable in GERD patients with present management algorithms; there is neither ability nor effort to prevent them.

The increasing mortality trends of ischemic heart disease reversed with the change in the approach to diagnosis and management. The disease was defined by its cause, coronary artery narrowing. Aggressive stress testing followed by coronary angiography permitted identification of patients at high risk for progression. This allowed intervention in these selected patients to prevent progression and reverse coronary artery disease. This new etiology-based management that replaced the prior symptom-based management reduced the incidence of myocardial infarction and death from ischemic heart disease. The outcome, measured by a declining incidence of myocardial infarction by 24% between 1999 and 2008 was impressive.[41]

The new method of assessing of LES damage consists of a new histologic measure of LES damage. I will show that this can be developed into a screening test that can be done on asymptomatic patients or those with early symptoms. Patients can be divided into increasing grades of severity of LES damage that has a high correlation with cellular changes of GERD in the esophagus (Fig. 1.2).

I will show that this new assessment will permit identification of the small minority of patients at risk for progressing to severe GERD in the future, including treatment failure, Barrett esophagus, and adenocarcinoma (Fig. 1.2). Early intervention to prevent or slow progression of LES damage in this selected minority of GERD patients at high risk can theoretically do for GERD what coronary angiography and stress testing did for ischemic heart disease.

It is my hope that in the not too distant future, we will have eradicated GERD as a disease or only have patients with mild GERD that is easily controlled for life with acid-suppressive drugs with no danger of progressing to severe GERD and adenocarcinoma.

The new method is aimed at the minority of patients who are the failures of today's management algorithm. The ability to predict progression of GERD in a defined manner based on progression of LES damage will lead to an opportunity for aggressive intervention in those people at risk and prevent failure of PPI therapy, CLE, and adenocarcinoma (Fig. 1.2).

The new method has a chance of success because it is not designed to change the management of all patients with GERD. Rather, it is targeted to correct the future management failures. The limitation of the number of patients needed to be treated is smaller, albeit still in the millions. The increased cost of treatment in this group will be balanced by the cost saving resulting from the eradication of Barrett esophagus and adenocarcinoma.

In a few decades, if I have been successful, GERD will be defined as follows: "GERD is a disease resulting from progressive damage to the lower esophageal sphincter sufficient to cause LES failure and reflux. The absence, presence, severity, and future progression of GERD is governed by the degree of LES damage defined by histologic measurement."

REFERENCES

1. Dent J, El-Serag HB, Wallander MA, et al. Epidemiology of gastro-oesophageal reflux disease: a systematic review. *Gut* 2005;**54**:710–7.
2. El-Serag HB, Sweet S, Winchester CC, et al. Update on the epidemiology of gastro-oesophageal reflux disease: a systematic review. *Gut* 2014;**63**:871–80.
3. Pohl H, Sirovich B, Welch HG. Esophageal adenocarcinoma incidence: are we reaching the peak? *Cancer Epidemiol Biomarkers Prev* 2010;**19**:1468–70.

4. Chandrasoma PT. Histologic definition of gastro-esophageal reflux disease. *Curr Opin Gastroenterol* 2013;**29**:460–7.
5. Robertson EV, Derakhshan MH, Wirz AA, Lee YY, Seenan JP, Ballantyne SA, Hanvey SL, Kelman AW, Going JJ, McColl KE. Central obesity in asymptomatic volunteers is associated with increased intrasphincteric acid reflux and lengthening of the cardiac mucosa. *Gastroenterology* 2013;**145**:730–9.
6. Kahrilas PJ, Shaheen NJ, Vaezi MF. American Gastroenterological Association Institute technical review on the management of gastroesophageal reflux disease. *Gastroenterology* 2008;**135**:1392–413.
7. Riddell RH. The biopsy diagnosis of gastroesophageal reflux disease, "carditis," and Barrett's esophagus, and sequelae of therapy. *Am J Surg Pathol* 1996;**20**(Suppl. 1):S31–51.
8. Rodrigo S, Abboud G, Oh D, et al. High intraepithelial counts in esophageal squamous epithelium are not specific for eosinophilic esophagitis in adults. *Am J Gastroenterol* 2008;**103**:435–42.
9. Virchow R. *Die Cellularpathologie: in ihrer Begrundung auf physiologische und pathologische Gewebelehre*. Berlin: Verlag von August Hirschwald; 1858.
10. Tileston W. Peptic ulcer of the esophagus. *Am J Med Sci* 1906;**132**:240–65.
11. Jackson C. Peptic ulcer of the esophagus. *JAMA* 1929;**92**:369–72.
12. Allison PR. Peptic ulcer of the oesophagus. *Thorax* 1948;**3**:20–42.
13. Allison PR. Peptic ulcer of the oesophagus. *J Thorac Surg* 1946;**15**:308–17.
14. Allison PR, Johnstone AS. The oesophagus lined with gastric mucous membrane. *Thorax* 1953;**8**:87–101.
15. Barrett NR. Chronic peptic ulcer of the oesophagus and 'oesophagitis'. *Br J Surg* 1950;**38**:175–82.
16. Edmonson JM. History of the instruments for gastrointestinal endoscopy. *Gastrointest Endosc* 1991;**37**:S27–56.
17. Lundell LR, Dent J, Bennett JR, et al. Endoscopic assessment of oesophagitis: clinical and functional correlates and further validation of the Los Angeles classification. *Gut* 1999;**45**:172–80.
18. Katz PO, Castell DO, Chen Y, Andersson T, Sostek MB. Intragastric acid suppression and pharmacokinetics of twice-daily esomeprazole: a randomized, three-way crossover study. *Aliment Pharmacol Ther* 2004;**20**:399–406.
19. Spechler SJ, Wang HH, Chen YY, Zeroogian JM, Antonioli DA, Goyal RK. GERD vs *H. pylori* infections as potential causes of inflammation in the gastric cardia. *Gastroenterology* 1997;**112**:A297.
20. Goldblum JR, Vicari JJ, Falk GW, Rice TW, Peek RM, Easley K, Richter JE. Inflammation and intestinal metaplasia of the gastric cardia: the role of gastroesophageal reflux and *H. pylori* infection. *Gasteroenterology* 1998;**114**:633–9.
21. Lagergren J, Bergstrom R, Lindgren A, Nyren O. Symptomatic gastroesophageal reflux as a risk factor for esophageal adenocarcinoma. *N Engl J Med* 1999;**340**:825–31.
22. Kahrilas PJ, Shi G, Manka M, Joehl RJ. Increased frequency of transient lower esophageal sphincter relaxation induced by gastric distension in reflux patients with hiatal hernia. *Gastroenterology* 2000;**118**:688–95.
23. An evidence-based appraisal of reflux disease management – the Genval Workshop Report. *Gut* 1999;**44**(Suppl. 2):S1–16.
24. Vakil N, van Zanten SV, Kahrilas P, Dent J, Jones B, The Global Consensus Group. The Montreal definition and classification of gastroesophageal reflux disease: a global evidence-based consensus. *Am J Gastroenterol* 2006;**101**:1900–20.
25. Chandrasoma PT, DeMeester TR. *GERD: reflux to esophageal adenocarcinoma*. San Diego: Academic Press; 2006.
26. Chandrasoma PT. *Diagnostic atlas of gastroesophageal reflux disease*. San Diego: Academic Press; 2007.
27. Blonski W, Vela MF, Castell DO. Comparison of reflux frequency during prolonged multichannel intraluminal impedance and pH monitoring on and off acid suppression therapy. *J Clin Gastroenterol* 2009;**43**:816–20.
28. Jacobson BC, Ferris TG, Shea TL, et al. Who is using chronic acid suppression therapy and why? *Am J Gastroenterol* 2003;**98**:51–8.
29. Vaezi MF, Richter JE, Stasney CR, et al. Treatment of chronic posterior laryngitis with esomeprazole. *Laryngoscope* 2006;**116**:254–60.
30. Nason KS, Wichienkuer PP, Awais O, Schuchert MJ, Luketich JD, O'Rourke RW, Hunter JG, Morris CD, Jobe BA. Gastroesophageal refux disease symptom severity, proton pump inhibitor use, and esophageal carcinogenesis. *Arch Surg* 2011;**146**:851–8.
31. Morson BC, Belcher BR. Adenocarcinoma of the oesophagus and ectopic gastric mucosa. *Br J Cancer* 1952;**6**:127–30.
32. DeMeester TR, Ireland AP. Gastric pathology as an initiator and potentiator of gastroesophageal reflux disease. *Dis Esophagus* 1997;**10**:1–8.
33. Ayazi S, Tamhankar A, DeMeester SR, Zehetner J, Wu C, Lipham JC, Hagen JA, DeMeester TR. The impact of gastric distension on the lower esophageal sphincter and its exposure to acid gastric juice. *Ann Surg* 2010;**252**:57–62.
34. Tobey NA, Carson JL, Alkiek RA, et al. Dilated intercellular spaces: a morphological feature of acid reflux-damaged human esophageal epithelium. *Gastroenterology* 1996;**111**:1200–5.
35. Tobey NA, Hosseini SS, Argore CM, Dobrucali AM, Awayda MS, Orlando RC. Dilated intercellular spaces and shunt permeability in non-erosive acid-damaged esophageal epithelium. *Am J Gastroenterol* 2004;**99**:13–22.
36. Malfertheiner P, Nocon M, Vieth M, Stolte M, Jasperson D, Keolz HR, Labenz J, Leodolter A, Lind T, Richter K, Willich SN. Evolution of gastro-oesophageal reflux disease over 5 years under routine medical care – the ProGERD study. *Aliment Pharmacol Ther* 2012;**35**:154–64.
37. Der R, Tsao-Wei DD, DeMeester T, et al. Carditis: a manifestation of gastroesophageal reflux disease. *Am J Surg Pathol* 2001;**25**:245–52.
38. Chandrasoma P, Makarewicz K, Wickramasinghe K, Ma YL, DeMeester TR. A proposal for a new validated histologic definition of the gastroesophageal junction. *Hum Pathol* 2006;**37**:40–7.
39. Chandrasoma P. Controversies of the cardiac mucosa and Barrett's esophagus. *Histopathology* 2005;**46**:361–73.
40. Hayward J. The lower end of the oesophagus. *Thorax* 1961;**16**:36–41.
41. Yeh RW, Sidney S, Chandra M, Sorel M, Selby JV, Go AS. Population trends in the incidence and outcomes of acute myocardial infarction. *N Engl J Med* 2010;**362**:2155–65.

Chapter 2

Present Diagnosis and Management of Gastroesophageal Reflux Disease: The Good, Bad, and Ugly

The present management of gastroesophageal reflux disease (GERD) is dominated by the use of acid neutralizers and acid-suppressive drugs. The popularity of these drugs is shown by the prominent and substantial space they occupy on the shelves of the local pharmacy (Fig. 2.1). This is a testament to the popularity and high profit margin of these drugs. The most powerful of these are available without prescription in package sizes that have large numbers of pills (Fig. 2.2). They are the ones at eye level on the shelves.

Combined with highly visible advertising for these drugs as agents that cure heartburn and "acid reflux disease," the medical establishment gives tacit approval for people to self-medicate themselves without any physician consultation. This includes the government that has approved these powerful drugs for over-the-counter use without prescription.

In a 1998 report, Greenberger[1] estimated that the annual cost of over-the-counter antacids and histamine-2 receptor antagonists in the United States was almost $2 billion with another $6 billion on prescriptions for histamine-2 receptor antagonists and proton pump inhibitors (PPIs). These costs have increased significantly in the past 15 years (Table 2.1).

1. PREVALENCE AND IMPACT OF GASTROESOPHAGEAL REFLUX DISEASE ON PEOPLE AND THE ECONOMY

In a systematic review of the epidemiology of GERD in 2005 using qualified published studies from all over the world, Dent et al.[2] reported an approximate prevalence of 10%–20% for GERD in the Western world (defined as heartburn and/or regurgitation at least 1 day/week). The prevalence was lower (less than 5%) in Asia. The incidence in the Western world was approximately 5 per 1000 person-years. The low rate of incidence relative to prevalence reflects that GERD is a chronic lifelong irreversible disease. The authors state: "The small number of studies eligible for inclusion in this review highlights the need for a global consensus on a symptom-based definition of GORD."

In an update in 2014,[3] the same group reported a range of GERD prevalence of 18.1%–27.8% in North America, 8.8%–25.9% in Europe, 2.5%–7.8% in East Asia, 8.7%–33.1% in the Middle East, 11.6% in Australia, and 23.0% in South America. The incidence was approximately 5 per 1000 person-years in the United States and the United Kingdom. The incidence in pediatric patients aged 1–17 years in the United Kingdom was 0.84 per 1000 person-years.

Of note is that this later review follows the acceptance of the Montreal definition of GERD, causing a change in the GERD definition in this study to include "diagnosis according to the Montreal definition, or diagnosed by a clinician" compared to the prior definition of "heartburn and/or regurgitation on at least 1 day a week" that was used in 2005.

Comparison of the two studies suggested an increase in GERD prevalence since 1995[3] ($P=<.0001$), particularly in North America and East Asia. The authors conclude that the disease burden may be increasing significantly worldwide. The only part of the world where the prevalence is consistently <10% is in East Asia.

The study may underestimate the prevalence of GERD significantly by the definitions used. Limitation to heartburn and/or regurgitation to at least 1 day/week suggests GERD of significant severity. If the definition was a frequency of at least once per month, the prevalence would be much higher, but the accuracy would be lower. The study is correct in limiting the definition to ensure specificity of diagnosis. However, it likely underestimates prevalence.

In Chapter 1, I suggested that when defined at a histologic level based on evidence of lower esophageal epithelial and lower esophageal sphincter (LES) damage, GERD is a universal disease in humans. It is largely subclinical and manifests in selected patients who progress to develop sufficient damage to the LES to become symptomatic and fit into the criteria of definition of GERD that are presently used, as in these studies.

GERD. http://dx.doi.org/10.1016/B978-0-12-809855-4.00002-6

FIGURE 2.1 "Antacid" shelf in local pharmacy enjoys a premium space. The wide array of acid neutralizers and suppressants is available for over-the-counter purchase without prescription. This photograph was taken in 2006. It has pride of place for Prilosec (omeprazole), which is at eye level and occupies a large area.

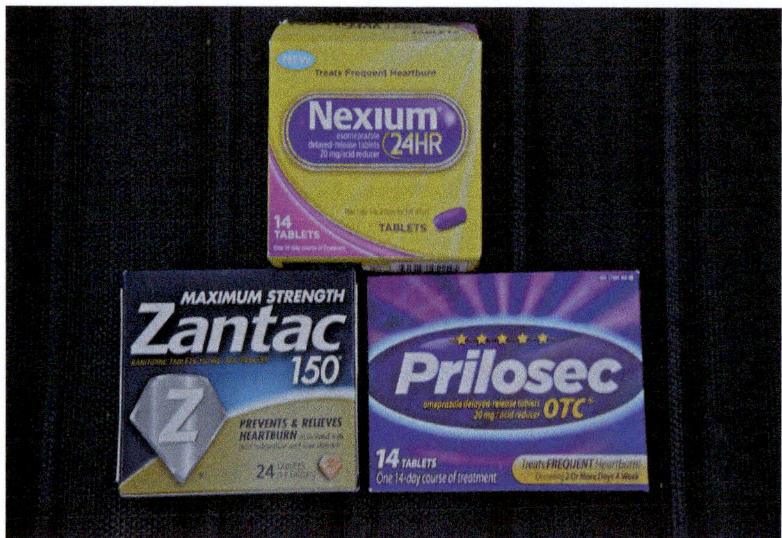

FIGURE 2.2 Today, the pride of place belongs to Nexium (esomeprazole), which is the most powerful acid suppressant. These drugs are now sold in packages that have larger numbers of pills per package, up to a maximum of 45 tablets.

TABLE 2.1 Top Proton Pump Inhibitors Based on US Revenue in 2013

Drug	Revenue (Millions US$)
Nexium	5570
Dexilant	830
Aciphex	692
Omeprazole (Rx)	609
Lansoprazole	289
Prevacid Solu Tab	219
Omeprazole/$NaHCO_3$	114
Pantoprazole	95
Protonix (intravenous)	93
Protonix	66
Total	**8.577 billion**

This does not include H_2 receptor antagonists and acid neutralizers. These data are from an unconfirmed source from the Internet and listed here only to provide a general idea of the widespread use of acid-suppressive drugs.

Toghanian et al.[4] in a study using the National Health and Wellness Survey in Europe and the United States, found that 23% of 116,536 respondents reported symptoms of GERD. 39% (i.e., 9% of this sample) of these were a subgroup that they designated as "disrupting GERD" defined as the presence of GERD symptoms on at least 2 days/week in addition to either nighttime symptoms or the use of prescribed/over-the-counter medication at least twice a week during the past month. This group of patients had higher utilization of health-care resources, poorer health-related quality of life, greater impairments in health-related work productivity and absenteeism (all $P<.05$ vs. nondisrupting GERD), and higher associated total medical costs.

Brook et al.[5] gathered data for 267,269 employees from 2001 to 2004 that included analysis of medical and prescription drug claims, indirect costs for sick leave, short- and long-term disability, and workers' compensation. 11,653 (4.3%) had GERD. GERD was associated with a mean incremental cost of US$3355 per employee of which direct medical costs accounted for 65%, prescription drug costs 17%, and indirect costs 19%. When extrapolated for the entire population of the United States, this amounts to an annual cost of ~$20 billion.

GERD is a very costly disease in terms of reduction in quality of life, cost of treatment with acid-reducing drug use, and indirect costs related to decreased productivity.

2. PRESENT DIAGNOSIS OF GASTROESOPHAGEAL REFLUX DISEASE

GERD is a disease that is diagnosed and managed in ways that are primitive from a scientific point of view. While medicine has evolved rapidly to incorporate new scientific methods, GERD is still diagnosed as it was in the 1960s by the presence of symptoms, syndromes, and complications of varying sensitivity and specificity.[6]

The normal methods we use for the scientific diagnosis of a disease are not applied to GERD. There are no pathologic criteria for diagnosis or exclusion of GERD. Early disease is a mystery at a cellular level because GERD is treated empirically with drugs without scientific investigation. Endoscopic and pathologic study is limited to the more severe disease that fails to be controlled by drugs.[7] Objective pathologic diagnosis is concentrated on the diagnosis of Barrett esophagus, dysplasia, and cancer, which are late complications of GERD.

Treatment guidelines recommend treatment of GERD after an empiric PPI test has been positive.[7] The reasons why no diagnostic testing or endoscopy is recommended at the onset of treatment are as follows:

1. An abundance of confidence that acid-suppressive drugs, particularly the powerful PPIs, are highly effective in treating GERD. This is true in some, but not all patients. Complete control of heartburn is reported only in 56%; control of regurgitation is worse.[7]
2. There is no diagnostic test that can improve the sensitivity of the empiric PPI test that results in the GERD patient being treated. While some authorities believe that the best diagnostic test available for GERD is endoscopy with or without

pH/impedance testing, there is no evidence that this improves the diagnosis of GERD, largely because of their low sensitivity. For this reason, diagnostic testing is not recommended at the onset of treatment.[7]

As the medical establishment has no alternative to the empiric PPI test for early diagnosis of GERD, particularly in the primary care setting where most of the treatment of GERD occurs, it is important to assess the predictive value of a positive and negative empiric PPI test.

The result of a false-positive PPI test is that patients who have no GERD are placed on acid-suppressive medication, often for the long term. Such patients with a false-positive diagnosis of GERD may be subjected to the unnecessary high cost of drugs and risk of adverse drug effects. The number of such patients is unknown although many believe it is significant and costs the health-care system billions of dollars per year.

A recent study analyzed the accuracy of the diagnostic tests (clinical diagnosis and the empiric PPI test) that are commonly used for the initial diagnosis of GERD in the primary care setting.[8] This is a follow-up of the DIAMOND study, a large international study that had shown that the use of the Reflux Disease Questionnaire (RDQ) in patients with upper gastrointestinal (GI) symptoms had moderate and similar accuracy in diagnosis in the hands of both primary care physicians and gastroenterologists when compared with investigation-based criteria of GERD.

308 fully evaluable patients from multiple primary care clinics were chosen for the study. The patients had not used PPIs in the 2 months prior to entry into the study and had never undergone any upper-GI testing. All patients completed the RDQ at entry, and the primary care physician and the study gastroenterologist selected a symptom-based diagnosis from a prespecified list. After this, the patient was asked to identify the most and second-most bothersome upper-GI symptom from a predetermined list of 19 symptoms.

The patients were then placed on a placebo of identical appearance to esomeprazole for 8 days [interquartile range (IQR) 6–9 days] before they were switched to esomeprazole (patients were blinded as to the switch). During the placebo period, all patients had endoscopy and wireless 48-h esophageal pH and symptom association monitoring.

After these studies, the patients received esomeprazole 40 mg daily for 2 weeks (the empiric PPI test). The effect of treatment (placebo and active) on their most and second-most bothersome symptoms was recorded in daily diaries. A positive placebo response was defined as the absence of the most or second-most bothersome symptom during the last 2 days of placebo treatment. A positive PPI test was defined as absence of the most bothersome symptom typical for GERD (heartburn or central chest pain) in the last 3 days of PPI therapy. Other symptom categories were analyzed but were not significantly different.

Based on the outcome of the endoscopy and 48-h pH test with symptom association, all patients were classified into GERD and non-GERD according to the presence of at least one of the four following criteria: (1) reflux esophagitis [any Los Angeles (LA) grade], (2) esophageal pH < 4 for > 5.5% of the time, (3) positive symptom association (95% or greater) for association of symptom with acid reflux, and (4) borderline high esophageal acid exposure (pH < 4 for 3.5%–5.5% of the time) + positive response to the empiric PPI test. Only nine patients fell into the last category.

203 patients were diagnosed with GERD based on these objective criteria. The other 105 patients were classified as "non-GERD." The following similarities and differences were present between patients classified as GERD and non-GERD:

1. Positive response to placebo of the most bothersome symptom (no limitation of the type of symptom) in the GERD and non-GERD groups was 13% and 14%, respectively.
2. Positive response in the empiric PPI test based on the most bothersome symptom when this was heartburn/central chest pain in the GERD and non-GERD groups was 60/81 (74%) and 13/21 (62%), respectively.
3. Positive response of the empiric PPI test based on the presence or absence of endoscopic reflux esophagitis was analyzed for 296 patients. Significantly, more GERD patients with reflux esophagitis had a positive empiric PPI test than non-GERD patients (57% vs. 35%; $P = .002$).
4. The ability of physicians to predict a positive PPI test: During the screening, the primary care physician recorded a most likely symptom-based diagnosis for each patient from a protocol specified list of diagnoses. A positive empiric PPI test was seen in 83/162 (51%) of patients in whom a clinical diagnosis of GERD had been made, compared with 59/134 (44%) of patients who had a non-GERD clinical diagnosis. The correlation between the clinical diagnosis and a positive empiric PPI test was very poor.
5. The study asked the final question: "Does the empiric PPI test provide useful support for the GERD diagnosis?" The PPI test was examined in terms of its sensitivity and specificity and positive and negative predictive value. A likelihood ratio (LR), which is used to evaluate the usefulness of a test in making a diagnosis, was calculated. The calculated LR varied between +2 and −2 for all subgroups, suggesting a very limited diagnostic value of the PPI test.

In their discussion, the authors point to the high quality of their study in establishing the low value of the PPI test in the diagnosis of GERD. They also show that the clinical diagnosis of GERD by primary care physicians also compares poorly with GERD diagnosed by the endoscopic and pH test criteria of the study; of the 308 patients diagnosed with GERD clinically, 105 (33%) did not have GERD according to the criteria in the objective tests.

This suggests that when a clinical diagnosis of GERD is made, it is significantly flawed with a 33% incidence of false-positive diagnosis when compared with objective tests for GERD. The empiric PPI test fails to improve diagnostic accuracy. The net result of this study would be that one-third of all patients being treated for GERD based on the presence of symptoms and an empiric PPI test are being treated unnecessarily because they do not have objective evidence of GERD.

In general, there is a tendency for the more informed gastroenterologists to favor objective criteria for the diagnosis of GERD than the presence of symptoms or the empiric PPI test. Even when typical symptoms are controlled by PPIs, the absence of endoscopic abnormalities and abnormal acid exposure in a pH test usually negates the diagnosis of GERD in the eyes of these experts.

In the discussion of the above paper, the authors address the issue with diagnosis. They write: "We acknowledge that the weakness of the association between proven GERD and a positive empiric PPI test also may reflect the weaknesses of current standards for diagnosing GERD by investigation."

There is no reason to doubt that an abnormal pH test with good symptom association has a 100% specificity for the diagnosis of GERD. It is therefore highly likely that the 60/81 GERD patients who had a positive empiric PPI test had GERD. But the conclusion in the other groups is less clear:

1. Why did 21/81 GERD patients whose most bothersome symptom was heartburn/central chest pain not have a positive empiric PPI test? Was this GERD of such a severity that the symptoms were refractory to PPIs? Or, was it that PPIs have an inherent failure rate. There is some evidence for the latter by the fact that patients with nonerosive GERD have a lower response rate to PPI therapy than those with erosive esophagitis.[7]
2. Why did 13/21 patients who did not have GERD by objective tests have a positive empiric PPI test? Could this be because these patients had GERD but the objective pH tests were not sensitive enough? Or, was this a false-positive empiric PPI test in a person whose symptoms were not caused by GERD?

The importance of these questions is that when patients with symptoms of GERD with a positive empiric PPI test have no esophagitis at endoscopy and pH testing is negative for abnormal acid exposure, some expert gastroenterologists will believe the objective test results and declare that the symptoms are not the result of GERD. These patients run the risk of being labeled as "hypersensitive esophagus" and "functional heartburn" and having their PPIs withdrawn and replaced with drugs that decrease esophageal sensitivity and emotional factors that are possible causes for their symptoms. I have heard gastroenterologists suggest that these patients may do well with antidepressive medication.

It is important that physicians do not use less-than-perfect tests to ascribe emotional reasons to the complaints of their patients. There is no method at present of proving that a person with symptoms and a positive empiric PPI test who does not have objective criteria of GERD does not suffer from GERD. The tests do not have 100% sensitivity.

In summary, it is reasonable to suggest that the modern symptoms-based diagnosis of GERD supplemented by the commonly used empiric PPI test is seriously flawed. The addition of endoscopy and pH testing helps to establish the diagnosis of ~70% of patients with symptoms and a positive empiric PPI test. The other 30% who have a positive clinical diagnosis that is not confirmed by objective testing are a problem. There is no way at the present time to know whether or not they truly have GERD as the cause of their symptoms.

This diagnostic and management algorithm results in the categorization of GERD in the population into well-demarcated groups (Table 2.2). There is movement of patients from one category to the next with the passage of time. However, many patients remain in one of these clinical categories for their entire lifetime.

This highlights all the problems that exist with the present diagnosis and management of GERD.

1. Any adult person in the population who does not have symptoms of GERD may have Barrett esophagus and develop adenocarcinoma. The absence of symptoms therefore does not exclude GERD because the presence of these complications, albeit unknown if endoscopy is not done, satisfies criteria for GERD.
2. Any person with symptoms that are being treated for GERD on the basis of a positive empiric PPI test with long-term PPIs may not have objective evidence of GERD.
3. There is no ability to predict progression from one category to the next until it actually happens. Diagnosis and management are entirely reactive. There is no effort or ability to prevent progression.
4. No person in the population, with or without symptoms of GERD, has a guarantee he/she will not develop esophageal adenocarcinoma in the future.

TABLE 2.2 Different Categories of Gastroesophageal Reflux Disease (GERD) Patients Based on Present Diagnosis of GERD Based on Clinical Symptoms With or Without an Empiric Proton Pump Inhibitor (PPI) Test, Followed by Empiric PPI Treatment to Control Symptoms

Category 1: No Symptoms of GERD
Objective evidence of reflux: pH test (if done) negative>positive (5%)
Visible CLE: Unknown without endoscopy; low but not zero prevalence
Risk of future adenocarcinoma: Very low but not zero
Treatment: None
Category 2: Clinical GERD, Well Controlled With PPI Therapy
Symptoms: Present; controlled with PPI therapy
Empiric PPI test: Positive or negative
Objective evidence of reflux: pH test (if done) positive>negative (30%)
Visible CLE: Unknown without endoscopy; ~5% prevalence
Risk of future adenocarcinoma: Low but higher than in the asymptomatic person
Treatment: PPIs as needed to control symptoms
Category 3: Clinical GERD, Poorly Controlled With PPI Therapy
Symptoms: Present and inadequately controlled with PPI therapy
Empiric PPI test: Positive or negative
Objective evidence of reflux: pH test positive>negative (30%)
Endoscopy (indicated): Normal; erosive esophagitis; visible CLE (10%+prevalence)
Risk of future adenocarcinoma: Low but higher than in both above groups
Treatment: PPIs at maximum dosage and frequency
Category 4: Barrett Esophagus (Visible CLE With Intestinal Metaplasia)
Symptoms of GERD: Can be absent; if present, can be poorly or well controlled with PPIs
Objective evidence of reflux: pH test (if done) positive>negative (< 10%)
Risk of future adenocarcinoma: 0.2%–0.5% per year
Treatment: Endoscopic surveillance and PPI therapy to control symptoms
Category 5: Adenocarcinoma
Symptoms of GERD: Can be absent; if present, can be poorly or well controlled with PPIs; dysphagia
Objective evidence of reflux: pH test (if done) positive > negative (< 10%)
Treatment: Endotherapy when early cancer is discovered during Barrett surveillance (15%); radical with advanced cancer (85%)

Diagnostic testing (endoscopy with or without pH testing and manometry) is indicated only in patients who fail PPI therapy or develop dysphagia.
CLE, columnar-lined esophagus.

3. PRESENT TREATMENT MODALITIES OF GASTROESOPHAGEAL REFLUX DISEASE

The goal of improving the quality of life of GERD sufferers has been achieved to a reasonably high level by present drug treatment. The majority (~70%) of GERD patients are highly satisfied with their treatment. For these patients, PPI therapy has been a godsend. However, GERD is a disease that is beset with many potential long-term future issues even for those whose symptoms are well controlled at a given time.

In many patients, continued control of symptoms with PPIs requires dose escalation after initial satisfactory control. When maximum dose escalation with present drugs has produced no satisfactory relief, the physician has only two options

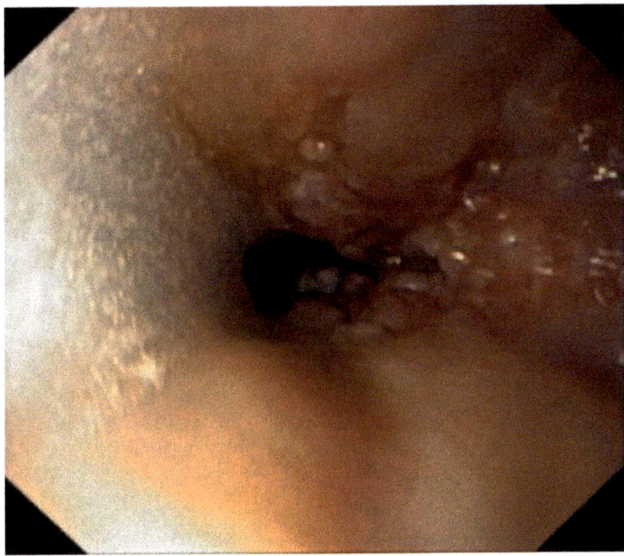

FIGURE 2.3 Antegrade endoscopic view of an early adenocarcinoma of the esophagus detected during surveillance for known Barrett esophagus. *Photograph courtesy of Dr. Martin Riegler, Reflux Medical, Vienna, Austria.*

for the patient: (1) Continue with high-dosage PPI therapy and let the patient tolerate a quality of life that is suboptimal or (2) undergo an antireflux surgical procedure.

When the physician decides that treatment has failed, the patient is referred to a gastroenterologist for endoscopy, often but not always with a pH test. When endoscopy shows erosive esophagitis and the pH test is positive, GERD is confirmed. As I have discussed above, a negative result (normal endoscopy and a normal pH test) is difficult to evaluate because of the less than 100% sensitivity of objective testing.

The only real value of endoscopy is the identification of patients with Barrett esophagus. These patients enter a surveillance program for early detection of dysplasia and adenocarcinoma (Fig. 2.3).

For the others who have no Barrett esophagus, there is no change in the available options. In general, the patient is continued with PPI therapy until the misery of uncontrolled symptoms results in consideration of an antireflux procedure.

Antireflux surgery by Nissen fundoplication has not been proven to be a reliable treatment method. It has a significant failure rate and complications such as dysphagia, gas bloating, and the inability to belch and vomit. These problems prevent referral of patients to surgeons by their treating physicians. Antireflux surgery, for a significant number of patients, is akin to substituting one form of misery by a different form of misery.

The lack of confidence in the outcomes of antireflux surgery is at least partly justified. Failure, which is uncommon in the hands of experts, is too common when inexperienced surgeons perform the procedure. In many cases, the selection of patients for surgery is poorly done, increasing the probability of failure. As a result of these problems, antireflux surgery is performed in only a small percentage of GERD patients with symptoms that disrupt their lives. The often intensely negative impression that gastroenterologists have regarding antireflux surgery turns patients off from this procedure even when it has the potential of improving the quality of their lives. The overall negativity of antireflux surgery in the treatment of GERD on the Internet further reduces the attractiveness of the surgical option.

Currently, ~20,000 antireflux surgeries are performed annually in the United States.[9] This number has declined after PPI therapy became widely available. The annual number of antireflux procedures grew rapidly during the 1990s, peaking at 31,695 (15.7 per 100,000 adults) in 1999. After 1999, the rate has declined steadily, falling ~30% by 2003 to 23,998 (11 per 100,000; $P<.0001$). Use of antireflux procedures fell more precipitously among younger patients (39% for 30- to 49-year-olds vs. 12.5% for those older than 60 years; $P<.0001$) and at teaching hospitals (36% vs. 23% at nonteaching hospitals; $P<.0001$).[9]

The lack of popularity of fundoplication has spawned numerous less-invasive methods of augmenting the LES. These include laparoscopic surgical procedures to implant magnetic rings around the esophagus (LINX), electrical stimulators into the wall (Endo-Stim), and endoscopic procedures using staplers (Transoral incisionless fundoplication) and radiofrequency to cause sclerosis (Stretta). These are variably successful but still performed rarely.

4. EFFECTS OF ACID-SUPPRESSIVE DRUG THERAPY IN GASTROESOPHAGEAL REFLUX DISEASE

4.1 Physiologic Effects

Drug treatment is directed toward neutralizing the normal acid levels in the stomach (pH = 1–2). The holy grail of research in GERD has been to improve our ability to control gastric acid secretion. Like most endeavors on which there is focus, the pharmaceutical research community has been incredibly successful. Modern acid-suppressive drugs, beginning with histamine-2 receptor antagonists in the 1960s to the present PPIs, are increasingly successful in suppressing gastric acid secretion. When used in adequate dosage, the acid-secreting apparatus of the parietal cells can be shut down and gastric pH is maintained at a pH of greater than 4 for nearly 20 h/day. This is remarkable.[10]

Katz et al.[10] studied the effectiveness of gastric pH control by measurement of intragastric pH and pharmacokinetics of twice-daily versus once-daily esomeprazole.

In a randomized, double-blind, three-way crossover study, healthy subjects received esomeprazole 40 mg once daily, 20 mg twice daily, or 40 mg twice daily for 5 consecutive days. Twenty-four-hour continuous ambulatory intragastric pH was recorded on day 5.

Esomeprazole 40 mg twice daily provided a mean of 19.2 h with intragastric pH > 4.0 [80.1% of a 24-h period; 95% confidence interval (CI) 74.5%–85.7%] versus 14.2 h with 40 mg once daily (59.2%; 95% CI 53.7%–64.7%) and 17.5 h with 20 mg twice daily (73.0%; 95% confidence interval 67.4%–78.5%) in 25 subjects. Intragastric pH was maintained >4.0 for a similar percentage of time during active and sleeping periods for all regimens.

The conclusions were that esomeprazole 40 mg twice daily provides significantly greater acid suppression (number of hours in a 24-h period with pH > 4.0) than once-daily dosing and may be a reasonable consideration for patients requiring greater acid suppression for acid-related disease.

The authors' conclusion provides insight into the minds of the most expert physicians who treat patients with GERD. If the symptoms are not controlled with acid-suppressive drugs, the answer is that the dosage of drugs needs to be increased. Or, if this is not effective, push the pharmaceutical industry to develop more powerful acid-suppressive agents. While the authors correctly state that this will be effective only for "acid-related disease," it is difficult to see what symptoms and complications of GERD short of cancer are not improved by acid suppression in the minds of many gastroenterologists. There is practically no medical treatment for GERD apart from acid-suppressive drugs.

Acid-suppressive drug therapy only addresses symptoms and cellular changes in the esophagus caused by acid. Acid suppression does not have any positive effect on the LES, the defective function of which is the cause of reflux. There is little evidence that acid-suppressive therapy has any positive effect on preventing columnar-lined esophagus (CLE) or its progression to adenocarcinoma.

Reflux of gastric juice at a higher pH continues unabated in the patient who is being treated with acid-suppressive drugs. It should be obvious that acid suppression fails to prevent any symptom or complication of reflux caused by any molecule in the refluxate other than the hydrogen ion (H^+).

This was shown in an elegant study by Blonski et al.[11] Combined multichannel intraluminal impedance and pH testing was performed on 70 patients while they were taking a PPI twice daily for 1 week or more and 40 patients who were off all antisecretory therapy for >1 week. Impedance detects reflux episodes that are characterized as either acid (if pH fell to <4 for at least 5 s within 10 s after an impedance detected reflux event) and nonacid (if the pH remained >4 after an impedance detected reflux event).

The authors report the results of testing in the on-PPI and off-PPI groups: "When the total reflux episodes were compared between patients studied on and off PPI therapy, there was a trend (P=0.07) toward higher number of total reflux episodes per hour in patients off PPI (2.3 +/− 0.3 vs 1.7 +/− 0.2) during total monitoring. In contrast, there was no difference (P = 0.9) in the mean number of total reflux episodes per hour during the postprandial interval between patients on (4.5 +/− 0.5) and off (5.1 +/− 0.8) PPI therapy. As expected, patients receiving PPI twice daily had a significantly higher mean number of nonacid reflux episodes per hour during the total monitoring (1.4 +/− 0.1 vs 0.9 +/− 0.1, P = 0.01) and 2-hour postprandial interval (3.7 +/− 0.4 vs 1.8 +/− 0.4, P = 0.0002). Conversely, patients studied off PPI therapy had a significantly higher mean number of acid reflux episodes per hour throughout the total monitoring period (1.4 +/− 0.2 vs 0.3 +/− 0.1, P = 0.0001) and 2-hour postprandial interval (3.2 +/− 0.7 vs 0.9 +/− 0.2, P = 0.0001)."

The authors conclude: "In summary, (1) the number of postprandial reflux episodes is not significantly different on or off PPI therapy; (2) the number of reflux episodes per hour is significantly higher postprandially than during the total monitoring period whether patients are studied on or off PPI therapy; (3) composition of the refluxate (i.e., acid or nonacid) is dependent on PPI therapy. From these data we conclude the PPI therapy has little effect on the frequency of reflux. It simply changes the acidity of the refluxate."

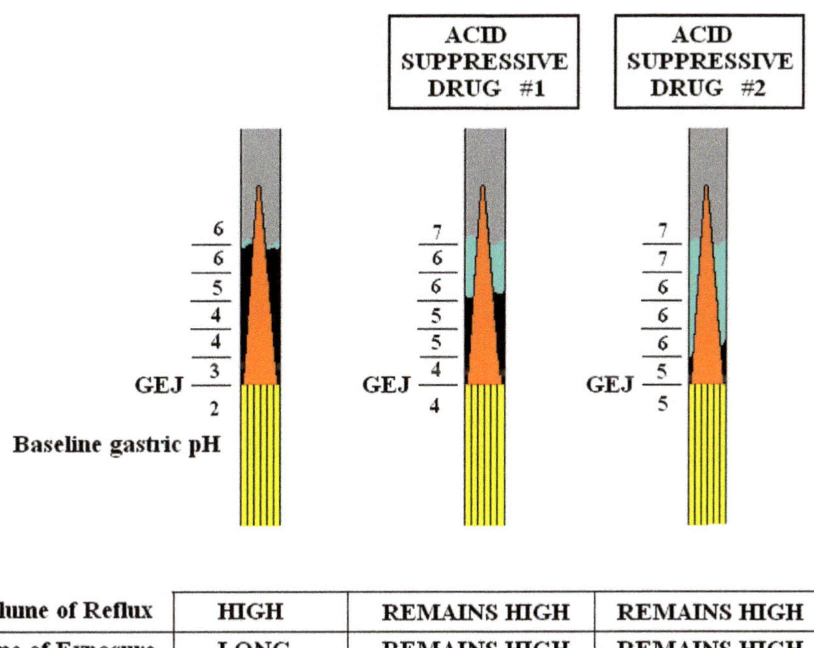

Volume of Reflux	HIGH	REMAINS HIGH	REMAINS HIGH
Time of Exposure	LONG	REMAINS HIGH	REMAINS HIGH

FIGURE 2.4 Diagrammatic representation of a reflux episode in three patients with the identical gastroesophageal reflux disease status (i.e., same lower esophageal sphincter damage with same frequency and volume of reflux). On the left is represented a patient who is not on any acid-reducing medication. The baseline gastric pH is 2. When reflux occurs, a pH gradient is created in the esophagus where the pH increases progressively from 2 at the gastroesophageal junction (GEJ) to the height of the column of refluxed material where the pH is the neutral pH of the esophagus. In the patients who are acid suppressed, gastric baseline pH is increased (shown here as 4 and 5). The entire esophagus is increasingly alkaline during a reflux episode.

It is therefore certain that the only action of PPIs is to increase the pH of gastric contents. The general belief is that when reflux occurs in the GERD patient, the refluxate will have a pH>4 if it occurs during the 12–19h where intragastric pH is >4 in patients on different doses of PPIs and <4 if it occurs during periods where the gastric pH is more acidic. All effects of acid-suppressive therapy will be dependent on the fact that when reflux occurs, the pH of the refluxate is higher than without drugs (Fig. 2.4).

Clarke et al.[12] from the Glasgow group headed by Professor McColl suggested that it is the pH of the refluxed gastric contents rather than the intragastric pH that determines response to PPI therapy. There is no certainty that the pH of the material that refluxes into the esophagus is the same as the intragastric pH recorded by an electrode in the stomach. They showed that a pocket of high acidity (low pH) accumulates at the top of the food column during a meal, contrasting with the higher pH of intragastric contents. This low pH pocket occupies the area immediately below the gastroesophageal junction (GEJ) and likely represents the main component of gastric contents that reflux into the esophagus, particularly in the postprandial phase (Fig. 2.5).

This study using impedance and pH monitoring on and off PPIs changed the viewpoint prevailing at the time that PPIs reduced the amount of reflux. This view resulted from pH monitoring studies that showed a reduction of acid reflux episode numbers in patients on PPIs. Blonski et al.[11] showed that this was the result of a conversion of acid to nonacid reflux episodes rather than a reduction of total reflux episodes.

Increasing the pH of gastric juice has a dramatic effect on the pH of the esophageal body *during* a reflux episode. In the normal person with a gastric pH of 1–2, there is a gradient of pH in the esophagus from strong acid (pH 1–2) in the distal esophagus to the near neutral baseline esophageal pH (6–7) at the top of the column of refluxed gastric juice. In contrast, when acid suppression increases the baseline gastric pH to >4, the entire esophagus is exposed to gastric juice at a higher pH during a reflux episode (Fig. 2.4). This change will tend to suppress cellular changes caused by strong acid (squamous epithelial damage, heartburn) but promote those favored by a pH milieu in the 4–6 range. I will, in Chapter 14, provide powerful evidence that intestinal metaplasia of CLE is promoted by this higher pH 4–6 milieu in the esophagus in a patient with GERD who has visible CLE (Fig. 2.4).

4.2 Clinical Effects

The long experience of treating GERD with ever-more effective acid-suppressive drug therapy makes it important to address the following questions:

1. If a symptom or cellular change has decreased in prevalence and incidence in the period of improving acid-suppressive drug treatment from the 1950s, can the improvement be attributed to alkalinization of gastric juice from pH 1–2 to pH>4?

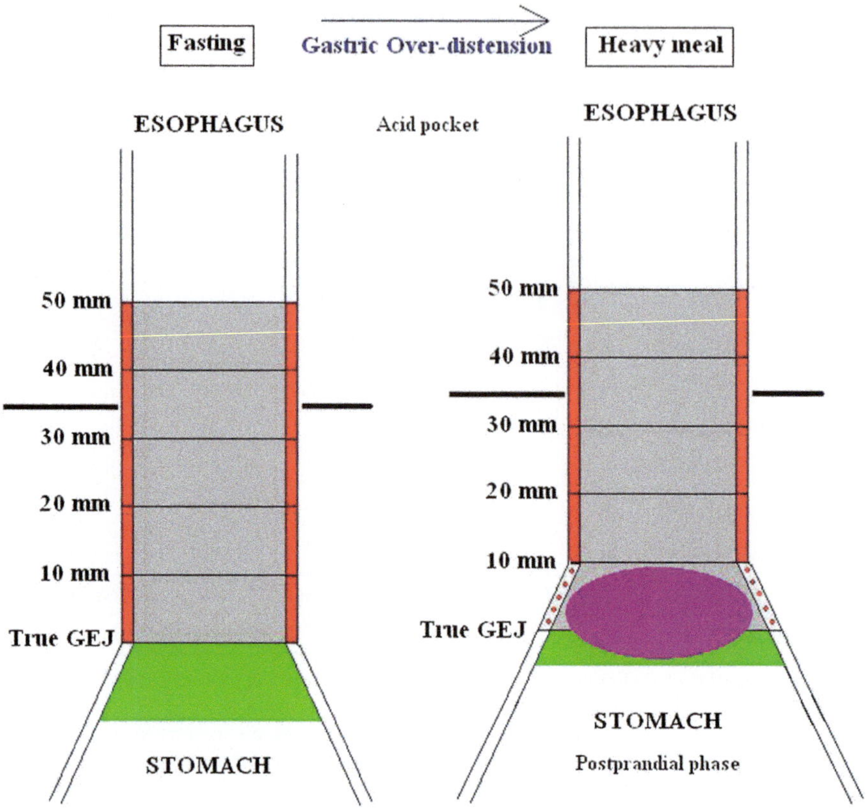

FIGURE 2.5 Characterization of acidity in the distal esophagus and proximal stomach under fasting conditions (left) and in response to a large meal that has caused gastric overdistension (right). In the full state, there is an acid pocket (*purple oval*) at the top of the food column immediately below the gastroesophageal junction (GEJ). If the lower esophageal sphincter fails and reflux occurs, this acid pocket is the first of the gastric contents to reflux into the esophagus.

2. If a symptom or cellular change has increased in prevalence and incidence in the period of improving acid-suppressive drug treatment from the 1950s, can the worsening be attributed to alkalinization of gastric juice from pH 1–2 to pH>4?

4.2.1 Healing of Erosive Esophagitis

The most powerful positive clinicopathological change resulting from acid-suppressive drug therapy in GERD is the positive effect it has had on reversing and preventing squamous epithelial damage. Erosive esophagitis is healed in ~85% of patients with a standard 8-week regimen of once-a-day PPI treatment.

Even in patients who are refractory to this regimen, further treatment with a twice-daily PPI regimen is effective in healing erosions. In a study of 337 patients with refractory reflux esophagitis after the standard 8-week regimen, the rate of endoscopically confirmed healing was significantly higher in those taking rabeprazole at 20mg b.i.d. (77.0%) and 10mg b.i.d. (78.4%) as compared with 20mg q.d. (58.8%).[13]

Statement 8 in the American Gastroenterological Association Medical Position Statement on GERD management[7] addressed the issue of recurrence of erosions. There is good evidence for "long-term use of PPIs for the treatment of patients with esophagitis once they have been proven clinically effective. Long-term treatment should be titrated down to the lowest effective dose based on symptom control." There is fair evidence that the use of less than daily dosing of PPI therapy as maintenance therapy in patients who previously had erosive esophagitis is ineffective in preventing recurrence. There is also good evidence that PPI therapy is more effective than H_2 receptor antagonists in preventing recurrence.

The Pro-GERD study is a large prospective multicenter open cohort study conducted in Europe (mainly Germany). 6215 patients over 18 years of age with the primary symptom of heartburn were recruited from primary care clinics. They could not have been treated continuously with any acid suppressant drug for more than 7 days during the preceding 4 weeks. Upper endoscopy with biopsy was performed and they were classified into nonerosive and erosive disease groups, the latter graded according to the LA classification system.

All patients completed an initial treatment with omeprazole (20 mg o.d. for 4 weeks for patients with nonerosive disease and 40 mg o.d. for up to 8 weeks for patients with erosive disease). Endoscopy was repeated at 4 weeks, and if necessary at 8 weeks, to assess healing of erosive disease and symptom control.

These patients were followed up on a long-term basis under routine care, i.e., at the discretion of their primary referring physician. Routine care may have included treatment with PPIs, H_2-receptor antogonists, or antacids on a regular (at least one dose of PPI every third day) or on-demand basis, and endoscopic intervention when considered necessary.

The patients were followed up regardless of whether or not they responded to initial treatment. Endoscopy with biopsy was repeated at 2 and 5 years, according to the protocol of the study. The results of the 5 year follow-up study of 2721 patients who completed the examination are reported here.

At the initial endoscopy, the distribution of endoscopic changes of these 2721 patients was as follows: nonerosive disease, 1224; erosive disease LA grade A/B, 1044; erosive disease LA C/D, 213. 240 patients who had visible CLE ("Barrett esophagus") are not included in this study.

After the initial treatment, healing was assessed in the nonerosive disease population by symptom control ("symptomatic healing") and by a combination of symptomatic and endoscopic healing in the erosive group. Healing was as follows: 88% in the nonerosive group, 92% in the erosive LA A/B group, and 84% in the erosive LA C/D group.

At 5 years, excluding 241 patients who did not have Barrett esophagus in the initial study but had developed it at 5 years, the status of the esophagus in terms of nonerosive and erosive disease in this population was as follows (Note: the data are presented at 2 and 5 years in the paper, but I have excluded 2 year results and calculated the data at 5 years):

1. Of the 1041 patients with nonerosive disease at baseline, 784 remained nonerosive, 248 had progressed to LA A/B, and 9 to LA C/D erosive disease.
2. Of the 918 patients with LA A/B erosive disease at baseline, 578 now had nonerosive disease, 331 remained LA A/B, and 9 had progressed to LA C/D erosive disease.
3. Of the 188 patients with LA C/D erosive disease at baseline, 94 now had nonerosive disease, 78 had LA A/B, and 16 stayed at LA C/D erosive disease.

The authors summarize these data: "Most patients remained stable or showed improvement in their grade of oesophagitis. However, some patients with milder oesophagitis did progress to more severe grade C/D and also to Barrett esophagus. A multivariate analysis of baseline factors that may be associated with GERD progression to LA grade A/B or C/D after 5 years indicated that a family history of GERD was associated with progression, and that remaining unhealed after baseline treatment predisposed patients to progression. Regular intake of PPI reduced the likelihood of progression compared with on demand PPI or other therapy, although the severity of symptoms at baseline did not seem to be a predictor of progression."

These results are impressive. They show that routine treatment of GERD by primary care physicians over a 5-year period resulted in a reduction in the number of patients with severe (LA C/D) erosive disease from 188 at baseline to 29 at the end of 5 years. Of these 11 stayed at LA C/D throughout the 5 years, 9 each progressed from nonerosive disease and LA A/B disease. That progression was significantly more common in patients on less than regular PPI use suggests that some of the failure to heal and prevent severe erosive esophagitis was the result of inadequate acid suppression.

This study shows that the likelihood and probably speed of healing of erosive esophagitis increases with increasing dosage and frequency of PPI therapy. Coupled with the data that shows that increasing PPI dosage and frequency produces progressively increasing alkalization of gastric juice, this is strong evidence that the cause of erosive esophagitis is the acid in the refluxate. Other molecules in gastric juice such as pepsin and bile are potentially toxic to esophageal epithelium. However, the fact that simple alkalinization of gastric juice is so effective in healing erosive esophagitis indicates that these molecules play a relatively insignificant role in the genesis of erosions.

This is an important logic that needs to be developed and researched, particularly because acid suppression is the only available effective drugs for treating GERD. There is a mindset among physicians that acid in gastric juice is responsible for every pathologic change and symptom in GERD. This is very likely to not be true.

A logical way of assessing the impact of acid in the genesis of any change is to evaluate the effect of different dosage and frequency of acid-suppressive drug therapy on that change. The most pronounced positive effect of acid-suppressive drug therapy is the healing of erosive esophagitis and this is clearly dose dependent.

As efficacy of acid suppression decreases with other symptoms and pathologic changes of GERD, the question must be asked as to why this is the case. One possibility is that a symptom is dependent on the volume and frequency of reflux, which do not change with PPIs, rather than its pH. Regurgitation is much less responsive to PPI therapy, probably for this reason. Another possible answer is that molecules other than H^+ ions in gastric juice are responsible for the changes.

4.2.2 Decrease of Complicated Ulcers and Strictures

In the past, when effective acid suppression was not available, erosive disease led to large and deep peptic ulcers in the esophagus that became complicated by significant hemorrhage, fibrous strictures of the esophagus, and intractable pain. In rare cases, the penetration of the ulcers through the esophageal wall and into the adjacent aorta or bronchus resulted in fatal hemorrhage and infection. It is interesting to note the severity of these complications by evaluating a historical paper by Allison.[14]

This is a detailed report of 63 adults (M:F=34:29; age range: 22–80 years with a mean of 60 years). 13 had chronic esophagitis with acute ulceration and 50 had chronic ulcers.

Pain was present in 74% of patients and was of 1–60 years duration. Pain was of three types: (1) high epigastric discomfort, (2) burning pain behind the sternum, and (3) boring pain radiating to the back, thought to be due to deep penetration of the ulcers. Dysphagia was present in 92% of patients and was the event that caused the patient to be admitted to the hospital. It was 8 weeks to 50 years in duration, more for solids than liquids, and of varying severity. Significant bleeding was present in 11 patients, 10 sufficient to cause anxiety; none were fatal. Radiology and endoscopy showed strictures of varying length associated with ulceration and narrowing in most cases. A postmortem specimen shows a long stricture in the distal esophagus.

There is no doubt that all these complications have become extremely rare. The decrease in the rate of these complications began in the 1960s when histamine-2 receptor antagonists were introduced and have continued to decrease after the introduction of PPI in the late 1980s. The positive change in reducing these complications of GERD is without question the result of effective alkalinization of gastric juice, and the progressive reduction in their incidence is the result of increasing efficacy of acid suppression.

The mortality from these complications of GERD was never high. They were reasons for pain, hemorrhage, and dysphagia that required surgical intervention rather than common causes of death. Their reduction with acid-suppression drug treatment did not have a significant impact in lowering mortality because relatively few people died of GERD complications in the 1900–60 era.

4.2.3 Control of Heartburn

Typical heartburn is the symptom that is associated most specifically with GERD. It is defined "as a burning sensation in the retrosternal area (behind the breastbone)."[15] Many patients have pain that does not precisely satisfy this definition; pain is not "burning" or not exactly "behind the retrosternal area." These are frequently included in "heartburn" or called "atypical heartburn" and subject to the same treatment protocols as typical heartburn.

The greatest problem of treating heartburn with PPIs is that *its efficacy is not sufficient.* While there is voluminous data that prove the efficacy of PPIs in healing erosive esophagitis, the data suggest that the efficacy of PPIs in resolving heartburn is much less.

In the American Gastroenterological Association Institute Technical Review on GERD management,[15] the best analysis of the available evidence is summarized: (1) Patients with esophagitis have a 83% rate of healing at 8 weeks; (2) patients with heartburn who have esophagitis at endoscopy have a symptom resolution rate of 56% at 4 weeks; (3) patients with heartburn who are endoscopy negative or uninvestigated have a symptom resolution rate of 36.7% at 4 weeks, still significantly superior to placebo. Heartburn in patients with nonerosive disease is more resistant to treatment.

These are dismal rates of heartburn resolution with the best available therapy. Many of these patients are improved to some extent and may cease to consider their heartburn not troublesome. However, these data suggest that the general claim that 30% of patients being treated for GERD are dissatisfied with their therapy and that 10% have their quality of life seriously impacted is more likely an understatement than an exaggeration.

If one analyzes these data, it is suggested that heartburn is caused by a mechanism that is different than erosive esophagitis in that it is not controlled to the same extent by withdrawing acid from the refluxate. Heartburn also occurs frequently in patients without erosive esophagitis (nonerosive reflux disease, NERD). The now obsolete Bernstein test that reproduces heartburn when acid is instilled into the distal esophagus in patients who do not have erosive disease suggests that the mechanism of heartburn is independent of erosive disease.

That pain is reduced significantly in the majority of patients treated with PPIs suggests that acid is the most important pain-inducing agent in the refluxate. The significant number of patients whose heartburn is not resolved suggests that other factors are involved, likely related to the fact that reflux of alkalinized gastric contents continues in patients who are on PPI therapy.[11]

The effect of reflux of alkalinized gastric contents on heartburn has been studied by impedance technology in patients who continue to have symptoms while on adequate doses of PPI.[16] It has been shown that the occurrence of symptoms in such patients correlates in many cases with a reflux episode that is above a pH 4 (i.e., weak acid reflux). In such patients,

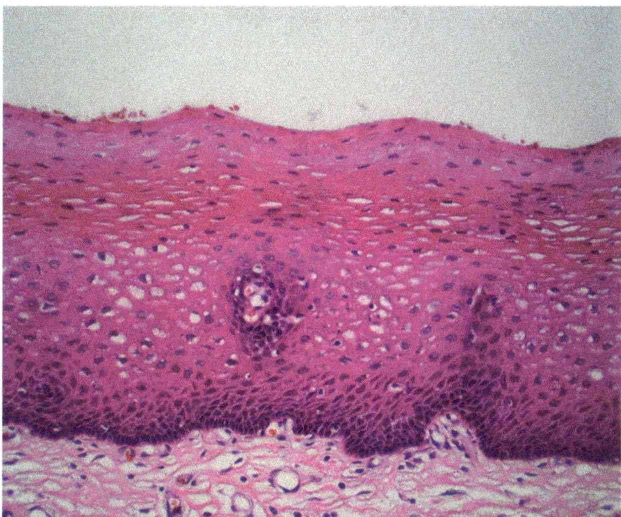

FIGURE 2.6 Section showing normal esophageal squamous epithelium. The cells are tightly apposed to one another without separation. The lack of space between cells makes the normal squamous epithelium impervious to the entry of any molecules from the refluxate into the epithelium. This prevents stimulation of nociceptive nerve endings in the epithelium as well as molecular changes in the proliferative and stem cells that reside in the basal region.

antireflux surgery has been shown to be effective in stopping symptoms.[17] This indicates that, in some patients, PPI may heal erosions but does not completely resolve heartburn. This suggests that when reflux occurs, nonacid molecules in the refluxate can induce pain in patients without erosive disease.

Understanding a reason for the genesis of heartburn by both strong acid and weak acid reflux in patients without erosive disease requires careful study of the squamous epithelium of the esophagus.

The normal squamous epithelium is a stratified epithelium that is impermeable. This means that luminal molecules cannot penetrate into the epithelium. This is largely because of the tight junctions between the squamous epithelial cells (Fig. 2.6).

In a study from Tulane University, Tobey et al.[18] showed by electron microscopic measurement that an increase in the intercellular spaces resulted from acid exposure of esophageal squamous epithelium. Endoscopic biopsies from 11 patients with recurrent heartburn (6 had erosive esophagitis, 5 had a normal appearing squamous epithelium at endoscopy) and 13 control patients (no symptoms or endoscopic evidence of reflux esophagitis) were examined using transmission electron microscopy. Computer-assisted measurement of the intercellular space diameter in the electron photomicrographs was performed in each specimen. The intercellular space diameter was significantly greater in specimens from patients with heartburn (irrespective of whether or not they had endoscopic erosions) than in the control specimens. Space diameters of 2.4 μm or greater were present in 8 of 11 patients with heartburn and in none of the controls. The authors concluded that dilated intercellular spaces are a feature of reflux damage to squamous epithelium.

In a similar study using transmission electron microscopy, Caviglia et al.[19] reported that patients with NERD (typical heartburn with or without an abnormal 24-h pH test) had significantly dilated intercellular spaces than controls (no symptoms, pH study negative). Villanacci et al.[20] developed a reproducible grading system (grades 0–3) for dilated intercellular spaces based on a study of routine biopsies from 21 patients with reflux symptoms. They showed that increasing grade of dilated intercellular spaces correlated with the severity of reflux symptoms and the histological grade of esophagitis. We commonly see dilated intercellular spaces in the squamous epithelium of patients with GERD (Fig. 2.7) but do not use it as a criterion for the diagnosis of GERD because this change is not specific for GERD.

In all these studies, the control population was "normal." These studies therefore provide strong evidence that dilated intercellular spaces result from reflux and are a sensitive indicator of reflux-induced squamous epithelial damage. The test is more sensitive for GERD than a 24-h pH test by data from Caviglia et al.[19] However, no studies have been done to evaluate the specificity of this finding for reflux. This would require comparison with other diseases of the esophagus.

In our experience, patients with eosinophilic esophagitis have markedly dilated intercellular spaces even without symptoms of reflux (see Fig. 1.5; Chapter 1).[21] This suggests that "dilated intercellular spaces" is a marker for nonspecific acute injury of the squamous epithelium rather than a specific marker for GERD. This is my reason for not using it as a criterion for the diagnosis of GERD.

Tobey et al.[22] in a follow-up study showed that exposure of an in vitro model of rabbit squamous epithelium to acid and acid-pepsin damage resulted in an increase in permeability of the epithelium. They showed than luminal hydrochloric acid

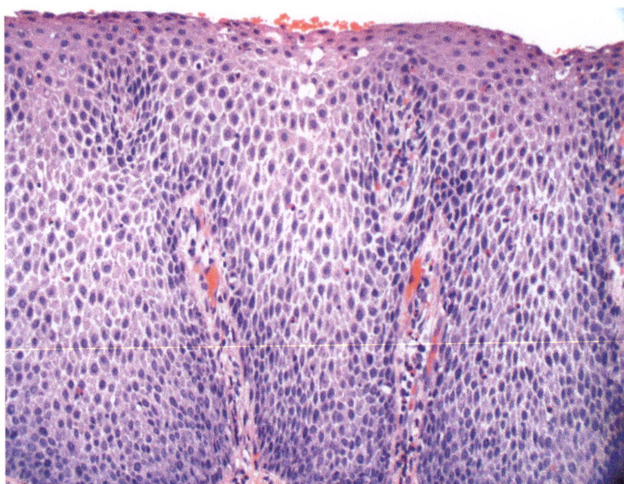

FIGURE 2.7 Esophageal squamous epithelium in a person with gastroesophageal reflux disease. The cells are separated by edema ("dilated intercellular spaces"), resulting in increased permeability. Molecules in the refluxate of varying size can enter the epithelium, stimulating nociceptive nerve endings and interacting with proliferative and stem cells in the basal and suprabasal region to induce genetic switches that result in columnar metaplasia.

(pH 1.1) or a mixture of hydrochloric acid (pH 2.0) plus pepsin (1 mg/mL) for 30 min caused a linear increase in permeability to 4–20kD dextran molecules as well as 6kD epidermal growth factor without gross erosions or histologic evidence of cell necrosis. Transmission electron microscopy documented the presence of dilated intercellular spaces.

In their discussion, Tobey et al.[22] concluded that in noneroded acid-damaged esophageal squamous epithelium, dilated intercellular spaces develop and result in increased permeability. This change in permeability on acid or acid-pepsin exposure is substantial, permitting dextran molecules as large as 20kD (20 angstrom units) and luminal epidermal growth factor at 6kD to diffuse across acid-damaged epithelium and enabling them to access receptors on epithelial basal cells.

It is clear by this experiment that the effect of acid or acid-pepsin on esophageal squamous epithelium is to permit any molecule in the refluxate within the size range of this experiment to diffuse into the epithelium to a varying depth. With severe damage, molecules can reach all the way down to the basal region. Acid is, in effect, a key that causes damage to the squamous epithelium that opens the door to the entry into the epithelium of a multitude of molecules in gastric refluxate.

Tobey et al.[22] postulate in their discussion that epidermal growth factor entering the epithelium may stimulate the basal cells to cause the basal cell hyperplasia that is seen in reflux esophagitis. While this is difficult to believe unless it is shown that epidermal growth factor is commonly present as a free molecule in gastric juice, the concept is valuable. Once the squamous epithelium has been rendered permeable by acid, it is open to attack by every molecule in gastric juice. Like Tobey et al.[22] postulate, large molecules are much more likely than acid to cause pathologic changes that result from interactions with receptors on cell surfaces. They are more likely also to produce genetic changes. The belief that acid is directly responsible for all pathologic changes in GERD is likely to be incorrect.

The question whether this increased squamous epithelial permeability can explain the symptom of heartburn and the variation in the resolution rate of heartburn in different patients depends on understanding the innervation of the squamous epithelium.

The squamous epithelium contains nonmyelinated nerve endings that are invisible in routine microscopy but can be seen with special techniques.[22] Some of these nerve endings in the epithelium likely play a role in the generation of sensory afferent impulses as a result of nociceptive stimuli. Nerve endings also exist in the muscle layer and adventitia of the esophagus. These are probably responsible for perception of discomfort and pain caused by mechanical distention of the esophagus. All sensory afferents pass up the vagus nerve to the nucleus tractus solitarius in the medulla oblongata. A few nociceptive afferents from the muscle wall pass up the sympathetic nerves to the spinal cord and to the thalamus in the lateral spinothalamic tract.

The submucosa and myenteric plexuses also have numerous ganglion cells. With the afferent nerve fibers from the mucosa and muscle wall, these likely form local reflex arcs that probably control peristalsis and maintenance of tonic muscle contraction in the sphincters. There is very little data on these neural mechanisms except to demonstrate that they exist. The esophagus that is separated from all external neural connections still generates peristaltic contraction and maintains sphincter tone.[23]

This excellent study from the Cajal Institute, Madrid, Spain, used the esophagi from ten cats and three rhesus monkeys. The esophagi were impregnated throughout their length by osmium tetroxide–zinc iodide, a technique that permits recognizing the course of the nerve fibers, discerns their connections, and locates their endings.

The nerve fibers reaching the basement membrane region come from the distal part of the submucosal nerve plexus. In the proximity of the submucosal plexus the nonmyelinated nerve fibers are grouped in bundles and surrounded by a common Schwann protoplasm. In the neighborhood of the basal layer, they form a subepithelial plexus consisting of groups of two to three fibers and later completely isolated single fibers, commonly associated with capillaries. They then pass through the basement membrane as single fibers and take a sinuous course upward between the spaces of the epithelial cells.

The density of nerve fibers in the esophageal epithelium is greatest in the upper and lower parts immediately distal to the pharynx and in the abdominal esophagus, respectively. The epithelium of the middle portion of the esophagus is sparsely innervated with only occasional nerve fibers.

The intraepithelial nerve fibers bifurcate within the epithelium and have a beaded appearance because of dilatations. They end freely at variable levels in the spaces between the epithelial cells. While some end in the basal zone and others end in the midregion of the epithelium, most of the fibers pass upward to end near the surface where they are separated from the lumen by the cytoplasm of one or two epithelial cells. The free endings have various shapes, described as resembling buds or buttons, pear-shaped or cup shaped.

In the normal patient without reflux-induced increased permeability of the squamous epithelium, luminal molecules do not enter the epithelium. Intraepithelial nociceptive nerve endings are not stimulated; pain does not occur. When reflux-induced increased permeability is present, molecules of increasing size can enter the epithelium to an increasing depth as the size of the dilated intercellular spaces increases, which in turn is dependent on the amount of damage.

Patients who have typical heartburn without erosive esophagitis can be explained on the basis of increased permeability caused by acid and the presence of nociceptive sensory receptors in the epithelium. Two extremes can be visualized:

1. Mild NERD results from relatively mild acid-induced damage that has resulted in a permeability increase that limits entry to small molecules (such as H^+) into the superficial region of the epithelium, stimulating nociceptive receptors and producing heartburn. When these patients are treated with PPI, rapid resolution of heartburn occurs as acid is removed from the refluxate. The epithelium is still impermeable to other larger molecules. The minor injury heals and the epithelial impermeability returns. This is a patient whose heartburn responds quickly to low PPI doses.
2. In contrast, a patient with a more severe nonerosive injury has a marked increase in permeability that permits deeper entry to H^+ as well as other larger molecules. Nerve endings at all depths of the epithelium are stimulated. The dominant nociceptive receptor stimulus is H^+, but other molecules are also able to induce pain. PPI therapy will produce rapid but partial improvement based on removal of acid. However, complete resolution depends on reversal of damage of the epithelium. If the reversal of damage is incomplete and the epithelial permeability remains abnormal, molecules other than H^+ in the refluxate can continue to stimulate nociceptive receptors. Patients whose heartburn is refractory to PPI therapy belong to this group, which has been called "nonacid or weak acid reflux disease."

The difference between healing of erosive esophagitis and resolution of symptoms with PPI therapy now becomes more understandable. Erosions result from extreme damage and necrosis of squamous epithelium caused by acid. This reverses rapidly when acid is neutralized with adequate doses of PPI. It is probable that a significant time span is necessary for regeneration of intraepithelial nerve endings in the newly regenerated squamous epithelium. Pain resulting from the increased permeability of noneroded squamous epithelium with its innervation intact is a change that is more difficult to reverse.

When assessed by carefully designed questionnaires, ~30% of patients with GERD who are treated with PPIs in adequate dosage are dissatisfied to a significant extent with their quality of life; 10% are seriously dissatisfied. This minority continues to have significant heartburn that disrupts their lives. Their lifestyle, particularly their diet and sleeping habits, is often altered; they fear eating and sleep in awkward positions, sometimes having their sleep interrupted by pain or a sensation of choking. They may need dose escalation; treatment becomes chronic and then lifelong. New symptoms such as regurgitation, chronic cough, and hoarseness may appear and may not be significantly improved by drug therapy.

When one considers actual numbers rather than percentages, the failure rate is staggering. Because approximately 20%–30% of the adult population in the United States and Western Europe suffer from heartburn,[2,3] the number of GERD sufferers is huge.

The 15%–30% of GERD patients who are dissatisfied with their quality of life number in the tens of millions. These people continue to be treated ineffectively with drugs and have reached an uncomfortable brick wall with few good options presented to them other than "keep taking the drugs." Only a few of these will undergo antireflux surgery.[9] Many patients who undergo surgery are helped, but some have complications that create a different set of problems. The poor reputation of antireflux surgery ensures that less than 1% of severe GERD sufferers will proceed to surgery.

The gap between the 30% dissatisfied with their medical treatment and the 1% who undergo successful surgery is the misery index of this disease. These millions of silent GERD sufferers swallow pills every day while enduring significant symptoms. Their lives are dominated by the fear and discomfort of their disease. This should not be acceptable.

4.2.4 The Price of Good Symptom Control

Even as we applaud the successful control of symptoms in millions of GERD patients, an important question must be asked: *Is good control of heartburn positive for the patient in all facets of the disease?*

For those destined not to progress to Barrett esophagus, the answer is absolutely positively "yes." The control of symptoms dramatically improves quality of life even when the control of heartburn is less than complete. Taking drugs on demand or even on a regular basis is a small price to pay for symptom control.

However, to the small minority that is destined to progress to treatment failure, Barrett esophagus, and cancer, good control increases their likelihood of death from the disease. A well-controlled GERD patient who no longer has "troublesome" symptoms does not reach a gastroenterologist and endoscopy is not performed. Progression to Barrett esophagus remains undetected, surveillance does not occur, and the opportunity for early diagnosis and treatment of dysplasia and cancer is lost. It is also possible that good control increases the risk of progression.

In a 2011 study, Nason et al.[24] provided evidence that medically treated patients with mild or absent GERD symptoms have significantly higher odds of adenocarcinogenesis (defined as nondysplastic and dysplastic Barrett esophagus and adenocarcinoma) compared with medically treated patients with severe GERD symptoms. They studied patients referred to gastroenterology clinics from 2004 to 2007 for (1) a clinically indicated upper endoscopy regardless of indication; (2) typical GERD symptoms enrolled in a trial studying small-caliber endoscopic screening without sedation; and (3) patients with atypical GERD (hoarseness, cough, etc.) in otolaryngology clinics and enrolled in a Barrett esophagus prevalence study.

All these patients underwent primary screening endoscopy defined as the first ever esophagogastroduodenoscopy in their life. They completed the validated GERD health-related quality of life questionnaire and reflux symptom index. Symptom severity was designated as "no symptoms," "mild symptoms," and "severe symptoms" based on the questionnaire responses. Barrett esophagus was diagnosed by the histologic presence of unequivocal goblet cells in an endoscopically visible CLE. Intestinal metaplasia of the gastric cardia was not considered Barrett esophagus.

Of the 769 patients studied, 65.4% were males and 94.3% were whites. Typical GERD symptoms were present in 67.1% and 57.2% were using PPI at the time of primary endoscopy. 179 (23.3%) had esophagitis and 365 (47.5%) had a hiatal hernia with a mean/SD of 2.72/1.72 cm.

122 (15.9%) of the patients had esophageal adenocarcinogenesis, 99 (12.9%) with nondysplastic Barrett esophagus, 17 (2.2%) with dysplastic Barrett esophagus, and 6 (0.8%) with esophageal adenocarcinoma.

In the results section of this study:

1. "… the total number of severe GERD symptoms was inversely and significantly associated with the odds of esophageal adenocarcinogenesis among patients using PPIs (OR 0.90; 95% CI 0.84–0.96)." Within this model, no significant association was observed between symptom severity and adenocarcinogenesis in patients not using PPIs. The association was present for both typical and atypical GERD symptoms among PPI users.
2. "The association between the number of severe typical and/or atypical GERD symptoms and esophageal adenocarcinogenesis was then determined while controlling for duration of symptoms and PPI use. Patients with a 10-year or longer history of typical or atypical GERD symptoms were 3-fold more likely to have esophageal adenocarcinogenesis (OR, 3.02; 95% CI 1.70–5.40) compared with a duration of symptoms of less than 10 years. Despite this, the inverse association with the total number of severe GERD symptoms persisted (OR 0.82; 95% CI 0.74–0.92). The relationship was even more pronounced when the analysis was stratified for duration of symptoms of more than 10 years in patients who were using PPIs (OR 0.70; 95% CI 0.58–0.84)."

The authors point out that the findings in this study "highlight potential causes for the ineffectiveness of the current cancer screening guidelines that recommend screening for Barrett esophagus only in patients with poorly controlled GERD."

This study provides evidence that risk of progression to Barrett esophagus in patients with GERD is associated with duration of symptoms, PPI use, and good control of symptoms by PPIs. We reach the uncomfortable conclusion that successful medical treatment of GERD is negative for those patients destined to develop adenocarcinoma. If adenocarcinoma is dependent on an unknown carcinogen in gastric juice, which is likely, there is no way at present to predict who is destined to develop cancer.

Adenocarcinogenesis as defined by the authors to include nondysplastic Barrett esophagus is an accurate concept that is not used in most patients with GERD. The discussion of cancer risk *must* begin at GERD, not at Barrett esophagus as it does with most of the gastroenterology literature. We will show that all the evidence suggests that a significant reason for the increase in adenocarcinoma in the past 60 years is the increase in the conversion of patients with GERD to Barrett esophagus, not to increase in cancer rates in Barrett esophagus. As such, the term adenocarcinogenesis as defined in this paper is an extremely valuable concept.

4.2.5 Control of Other Symptoms of Gastroesophageal Reflux Disease

Regurgitation is the second symptom of the typical GERD syndrome. The Montreal group, after extensive review of the literature, found very little data specifically relating to regurgitation. In most papers, heartburn and regurgitation are lumped together and only heartburn scores used as clinical endpoints. A recent study showed that regurgitation was even less effectively controlled by PPIs than heartburn. The likely reason for this is that regurgitation is a marker for a severely abnormal LES. There are no effective drugs that reverse sphincter dysfunction, and the gastroenterology literature's silence on the topic is in itself a statement of their relative inability to impact this symptom.

There are numerous other GERD syndromes that are defined in the Montreal classification,[6] including reflux chest pain syndrome and extraesophageal syndromes (chronic cough, laryngitis, asthma, dental erosions, etc.). The relationship of GERD to these syndromes is difficult to establish. I have little experience with these syndromes, particularly relating to histopathologic changes, and will not consider them further.

4.2.6 Increase in the Incidence of Barrett Esophagus

There has certainly been an increase in the incidence and mortality from esophageal adenocarcinoma.[25,26] This was first noted in 1975 and has increased sevenfold in the past 40 years.

Esophageal adenocarcinoma is a complication of GERD. It does not result from any other disease. Assuming that adenocarcinoma is preceded by Barrett esophagus, an increase in adenocarcinoma can be explained in one of two ways: (1) The carcinogenicity of gastric juice has increased resulting in an increased carcinogenesis in Barrett esophagus or (2) the carcinogen level has remained constant and the increase in the incidence of adenocarcinoma is the result of an increased incidence of Barrett esophagus, its premalignant state.

In the absence of routine endoscopy of GERD patients, it is difficult to define the exact prevalence of Barrett esophagus in the population and how this has changed in the past century. Despite this, there is reason to believe that the prevalence of Barrett esophagus is increasing.

There is an enormous silence about the reason for the increase in the incidence in Barrett esophagus in the gastroenterology literature. While many experts are concerned, the average primary care physician and the lay public are unaware of the magnitude of the Barrett esophagus problem or the exact reason for the increase in its incidence. The general assumption is that GERD has increased since the 1960s and Barrett esophagus is the result of this increase. Some believe that the obesity epidemic is responsible for the increase in GERD and in turn for Barrett esophagus.

Obesity is unlikely to be the entire reason for the increase in esophageal adenocarcinoma. The epidemic of Barrett esophagus and adenocarcinoma, which were first recognized in the late 1970s, predates the obesity epidemic.

In the 1930–50 era, patients who had long segments of columnar metaplasia in the tubal esophagus were reported. The condition that we call CLE after Barrett[27] coined the term in 1957 existed dating back to the early 20th century. The first clear description is in Lyall's paper in 1937[28] where he reported 8 cases of peptic ulcer of the esophagus. He described his last case as follows: "Close examination of the mucous membrane in the region of the ulcer showed the presence of a remarkable state of affairs. The intact mucosa separating the lateral edges of ulcer was found to be heterotopic gastric mucosa which extended as a tongue-shaped process of well preserved tissue upwards from that of the fundus of the stomach…. The mucosa bore a resemblance to that normally found towards the pyloric end of the stomach."

In the original papers of Barrett[27,29] and Allison and Johnstone,[30] these columnar-lined segments were very long, often extending up to the arch of the aorta. There is strong evidence that the length of CLE correlates directly with severity of reflux disease. As such, it is clear that there were patients with very severe GERD in the first half of the 20th century. Without clear and consistent definition of GERD, the prevalence at different points in history is impossible to ascertain.

What was very different in the 1950s was that patients who had long segments of columnar epithelium rarely had significant amounts of intestinal metaplasia,[29,30] which defines Barrett esophagus (Fig. 2.8). This contrasts with histologic mapping studies in 1976 that show a higher prevalence and extent of intestinal metaplasia within CLE (Fig. 2.9).[31] More detailed mapping studies of the early 21st century show an explosion in the prevalence and extent of intestinal metaplasia within CLE (Fig. 2.10).[32]

The change that has occurred without doubt is not so much an increase in the length of CLE but an increase in the presence and extent of intestinal metaplasia within a given length of CLE. In 1994, Spechler reported that 19.4% of patients

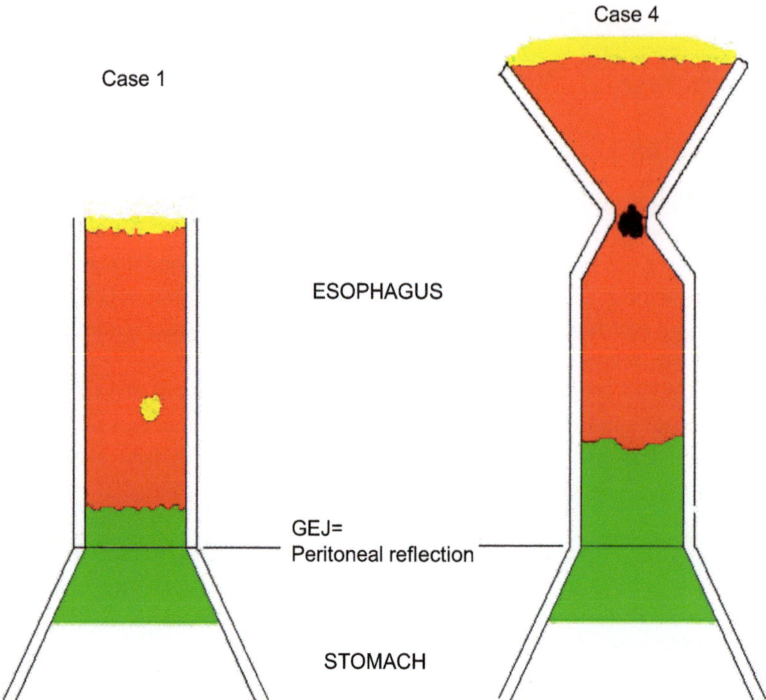

FIGURE 2.8 Diagrammatic representation of two cases in Ref. 30. In case 1, there is a 5-cm segment of columnar-lined esophagus (CLE) above the true gastroesophageal junction (GEJ) (defined by the peritoneal reflection) with a small amount of squamous epithelium (yellow) at the proximal margin. The CLE segment is lined by cardiac epithelium (red) and columnar epithelium with parietal cells (green). There is a squamous epithelial island described within the area of cardiac epithelium. In case 4, there is a 7.5-cm segment of visible CLE with an ulcerated stricture. The CLE segment is lined by cardiac epithelium (red) and columnar epithelium with parietal cells. There is no distinction between oxyntocardiac and gastric oxyntic epithelium; both have parietal cells. No intestinal metaplasia is reported in both these specimens. Intestinal metaplasia was described in two of the other five cases described (not shown).

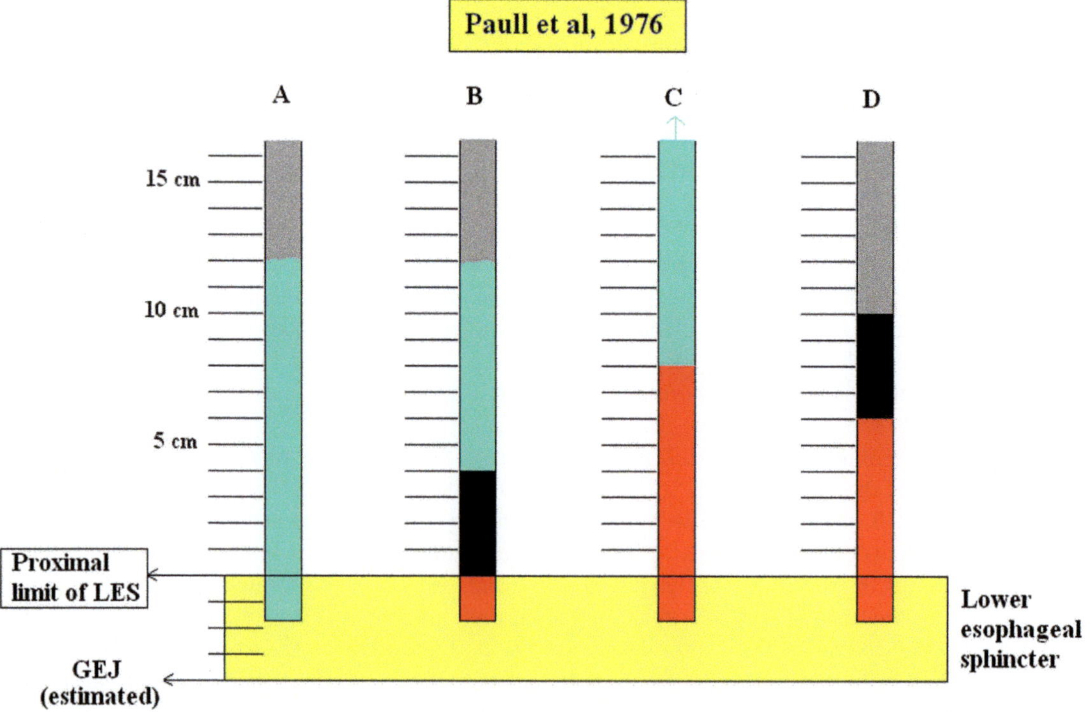

FIGURE 2.9 Diagrammatic summary of the histologic findings in four representative patients demonstrating the heterogeneity of epithelial patterns in the columnar-lined esophagus in the study by Paull et al.[31] The distribution of intestinal metaplasia (blue), cardiac epithelium (black), and oxyntocardiac epithelium (red) is shown. *GEJ*, gastroesophageal junction; *LES*, lower esophageal sphincter.

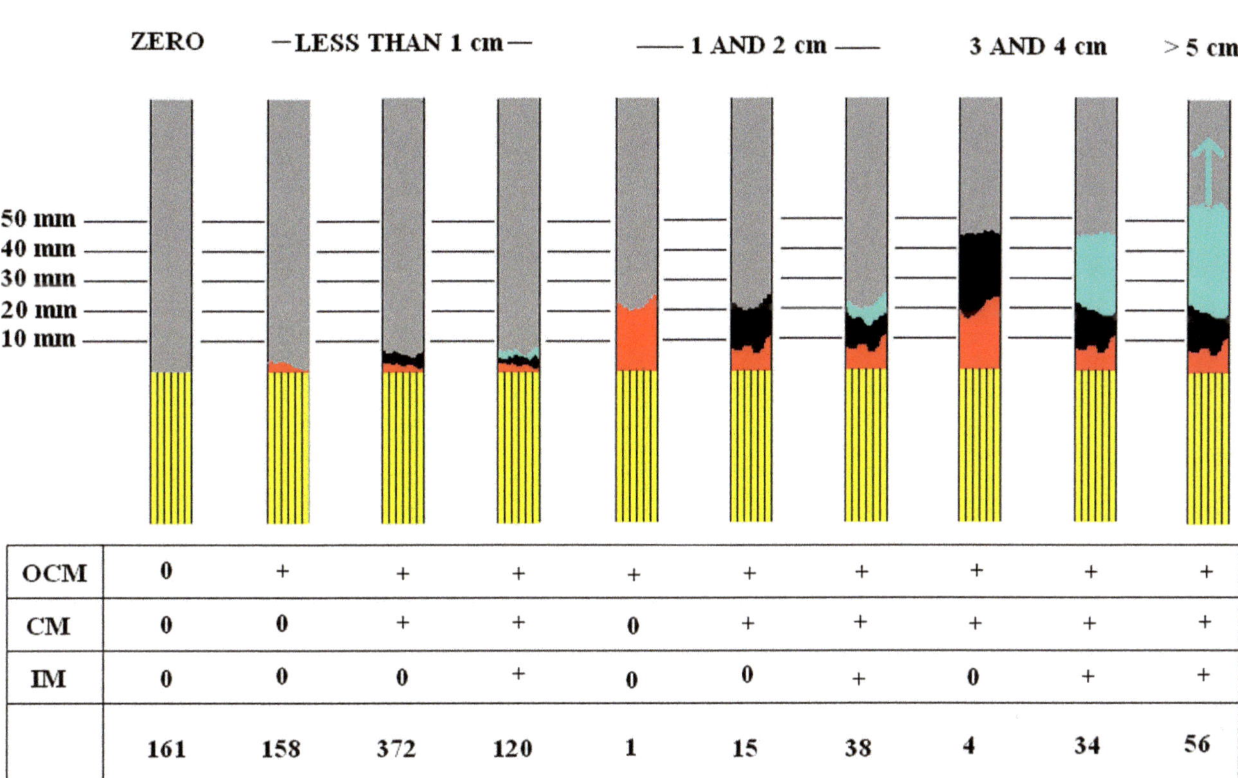

OCM	0	+	+	+	+	+	+	+	+	+
CM	0	0	+	+	0	+	+	+	+	+
IM	0	0	0	+	0	0	+	0	+	+
	161	158	372	120	1	15	38	4	34	56

FIGURE 2.10 Histologic study of 959 patients undergoing biopsy in our unit using the DeMeester biopsy protocol. This shows the increase in prevalence of intestinal metaplasia (blue) as the length of visible columnar-lined esophagus increases. It also shows the proximal distribution of intestinal metaplasia. Cardiac epithelium (black) and oxyntocardiac epithelium (red) are distal to the intestinal metaplasia.

with an endoscopically defined CLE that was less than 2cm had intestinal metaplasia.[33] In the Pro-GERD study with 5years follow-up that was reported in 2012, 50% of patients with a visible CLE <1 cm had intestinal metaplasia.[34]

If there were data in 1950 relating to the prevalence of Barrett esophagus (i.e., CLE with intestinal metaplasia), examination of the studies of Barrett[27,29] and Allison and Johnstone[30] would suggest they had the following features: 100% incidence of erosive esophagitis with ulceration and strictures, significant prevalence of columnar-lined segments >3cm, very low prevalence of nondysplastic Barrett esophagus (i.e., intestinal metaplasia), and a very low incidence of adenocarcinoma.

Although arguable, the critical difference between then and now is likely the prevalence of nondysplastic Barrett esophagus (intestinal metaplasia). When the actual size of the GERD population is taken into account, even a slight increase in the prevalence of Barrett esophagus (visible CLE with intestinal metaplasia) is sufficient to explain the increased incidence of adenocarcinoma that has occurred.

The most obvious thing that is fundamentally different today than it was in the 1950s is our progressively increasing ability to alkalinize gastric juice. If there is no other explanation for the increased prevalence of GERD-induced Barrett esophagus, we must ask the obvious question: *Is Barrett esophagus an unintended consequence of the alkalization of gastric juice that results from treating GERD with acid-suppressive drugs?*

The ability to alkalinize gastric juice began in the 1950s with acid neutralizers and has increased in effectiveness with H_2 receptor antagonists in the 1960s and PPIs in the 1980s. The curve of increasing prevalence of Barrett esophagus likely matches the ability to alkalinize gastric juice closely enough to raise suspicion that there may be a relationship.

Recent evidence in the Pro-GERD study suggests that treating patients with PPIs does not inhibit and may actually increase the occurrence of Barrett esophagus in patients with GERD.

The Pro-GERD study (reviewed partly earlier in this chapter) is a large multicenter study in Europe. Patients with GERD were enrolled into the study and had endoscopic examination at the study centers. They were then sent back to their primary physicians for treatment of their disease. The patients were brought back to the study centers for repeat endoscopy at 2years and 5years.

This report details the progression of GERD without CLE at the index endoscopy to endoscopic CLE with and without intestinal metaplasia at 5years. Of the total of 2721 patients, 240 were excluded in this analysis because they had CLE at the initial endoscopy, leaving 2481 patients.

A total of 241/2481 (9.7%) patients had developed endoscopically visible CLE at 5 years. Data on length of CLE were available for 186 patients; 73% of these had a length of CLE equal to or less than 2 cm (79% for NERD, 73% for ERD LA A/B and 66% for ERD LA C/D). The remainder had a length exceeding 2 cm (Table 2.3).

Intestinal metaplasia was present in 50% of patients with a CLE length of 1 cm or less, 71% with a CLE length of >1–<3 cm, and 100% with a CLE length of 3 cm or more.

The progression to CLE was reported in the subclassifications of nonerosive and erosive disease by endoscopy in the baseline endoscopy (Note: All patients with endoscopic CLE at baseline were excluded, so these were patients who developed these changes during the 5 year period):

1. Patients with NERD at baseline: Of the 1224 patients who were in the NERD category at baseline, follow-up endoscopy at 5 years showed that 72 (5.9%) had an endoscopically visible CLE with 51 (4.2%) confirmed as having intestinal metaplasia on histology (Fig. 2.11).

TABLE 2.3 Progression of Different Gastroesophageal Reflux Disease Categories to Visible Columnar-Lined Esophagus (CLE) at the 5-Year Endoscopy With and Without Intestinal Metaplasia (IM) on Biopsy

	Baseline No.	All Visible CLE	Visible CLE, IM+	Visible CLE, IM−
NERD	1224	72 (5.9%)	51 (4.2%)	21 (1.7%)
ERD LA A/B	1044	127 (12.1%)	85 (8.1%)	42 (4.0%)
ERD LA C/D	213	42 (19.7%)	22 (10.3%)	20 (9.4%)
Total	2481	241 (9.7%)	158 (6.4%)	83 (3.3%)

ERD, erosive reflux disease; *LA*, Los Angeles; *NERD*, nonerosive reflux disease.

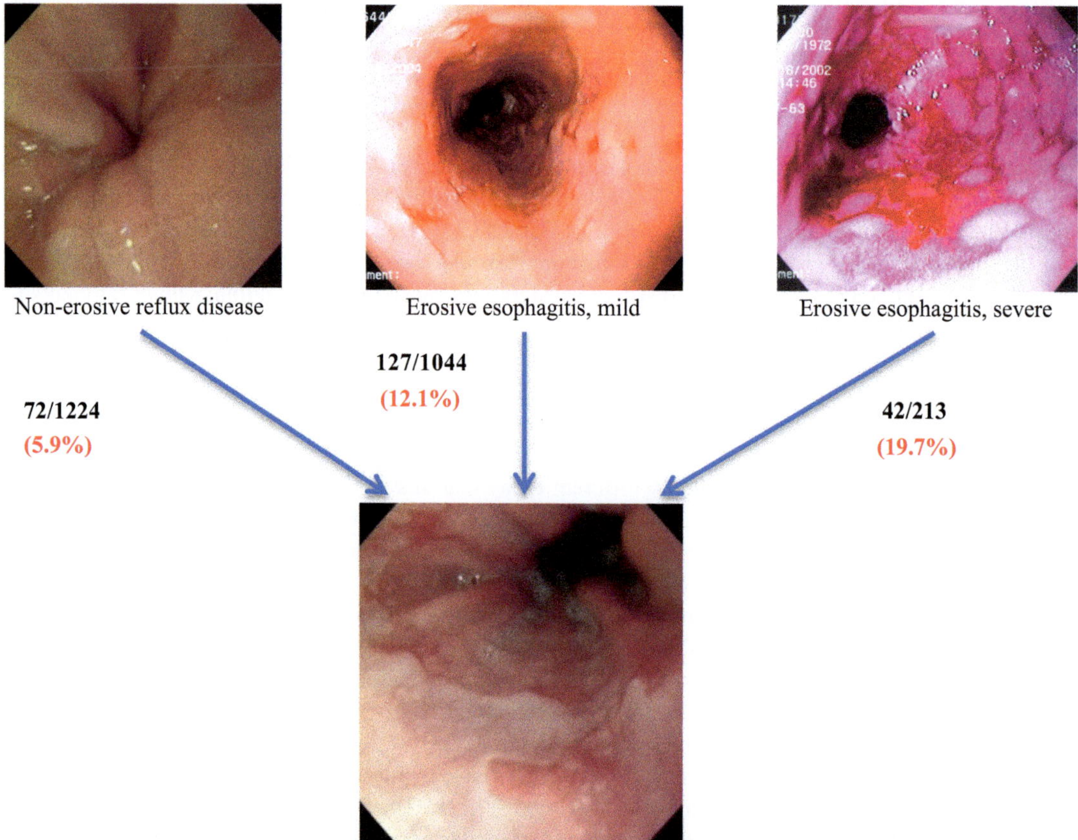

FIGURE 2.11 Progression of patients under treatment for gastroesophageal reflux disease from not having visible columnar-lined esophagus (CLE) at the index endoscopy to being positive for visible CLE within 5 years. The incidence of visible CLE increased progressively in patients who had nonerosive disease (5.9%) to mild erosive disease (12.1%) to severe erosive disease (19.7%) at the index endoscopy.

2. Patients with erosive disease LA A/B at baseline: Of the 1044 patients who were in the erosive disease LA A/B category at baseline, 127 (12.1%) had an endoscopically visible CLE with 85 (8.1%) confirmed as having intestinal metaplasia on histology (Fig. 2.11).
3. Patients with erosive disease LA C/D at baseline: Of the 213 patients who were in the erosive disease LA C/D category at baseline, 42 (19.7%) had an endoscopically visible CLE with 22 (10.3%) confirmed as having intestinal metaplasia on histology (Fig. 2.11).

The study evaluated risk factors predicting progression to Barrett esophagus. This evaluation was based on follow-up questionnaires that were completed by the participants.

The following factors at baseline predicted progression to visible CLE (=endoscopic Barrett esophagus) without designating whether these were positive or negative on biopsy for intestinal metaplasia: (1) presence of erosive esophagitis at baseline; (2) alcohol intake; and **(3) regular PPI therapy**. The odds ratio for progression to Barrett esophagus (endoscopic/confirmed) for regular PPI use was 1.71 (95% CI 1.09–2.67; P=.019), for PPI on demand was 1.53 (95% CI 0.93–2.51; P=.10), and for other non-PPI acid-suppressive drugs was 0.94 (95% CI 0.56–1.57; P=.81).

It must be noted that in the same study regular PPI therapy was protective in the progression within GERD categories with respect to healing of erosive esophagitis compared with on-demand PPI therapy and other acid-suppressive therapy.[34]

The following factors did not significantly influence progression: age, gender, BMI, smoking, GERD in family, duration of GERD, GERD symptoms score at baseline, and *Helicobacter pylori* infection.

This is a prospective study of patients with early GERD that shows with little doubt that the progression to Barrett esophagus is promoted by acid suppression in a dose-related manner. The greater the alkalinization of gastric juice, the greater is the progression to Barrett esophagus.

In the discussion of PPI effect in healing erosive esophagitis, I emphasized that a dose-related effect of PPIs in promoting healing was strong evidence that acid was primarily responsible for erosive esophagitis. The demonstration that the effect of PPIs in promoting Barrett esophagus is dose-related is strong evidence that acid has a protective effect against the development of Barrett esophagus and removal of acid with PPIs promotes Barrett esophagus.

The explanation given by many gastroenterologists to explain this association is that it is the result of indication bias, i.e., patients with more severe reflux received more PPIs, and the reason for the higher incidence of CLE was the more severe reflux rather than the PPIs. This is really a desperate excuse rather than an explanation. The study found no association between the occurrence of CLE and the GERD symptom score at baseline, making indication bias unlikely as the reason for the increased incidence of Barrett esophagus with regular PPI use.

In Chapter 15, I will show how the distribution of intestinal metaplasia within a segment of CLE provides further powerful evidence that the occurrence of intestinal metaplasia is favored by an alkaline milieu in the esophagus.

4.2.7 Increase in the Incidence of Esophageal Adenocarcinoma

Morson et al.[35] reported the first case of esophageal adenocarcinoma in 1952. Allison and Johnstone illustrated the second case in their 1953 paper.[30] Since then, the number of patients developing and dying of esophageal adenocarcinoma has increased dramatically in the United States and Western Europe. An estimated 80% of the 16,000 patients developing carcinoma of the esophagus in the United States in 2010 were patients with adenocarcinoma and therefore GERD-induced. In addition, it is now believed that adenocarcinomas occurring in the GEJ region and "gastric cardia" are actually GERD-induced esophageal adenocarcinomas. This would add approximately another 8–10,000 patients to the annual incidence of GERD-induced esophageal adenocarcinoma. This is the largest single rate of increase in the incidence of any cancer type associated with one etiology that we have ever encountered in the history of medicine.

In a recent update, Pohl et al.[26] suggested that the rate of increase of esophageal adenocarcinoma may be decreasing. They showed that the incidence of esophageal adenocarcinoma increased from 3.6 per million in 1973 to 25.6 per million in 2006, a nearly sevenfold increase. When separated into periods, the annual increase in incidence from 1973 to 1996 was 8.2% and that from 1996 to 2006 was 1.3% per year.

This change was due mainly to a change in the incidence of early stage disease. Regional and distant stage disease continued to increase without significant change. Early stage disease which had increased 10% annually prior to 1999, declined in incidence by 1.6% since then.

The authors, discussing the reason for this, commented that the change may be due to chance and simply reflect annual fluctuation. They did not believe that stage migration due to better diagnostic techniques picking up previously undetected regional and distant disease was likely because there was no acceleration of late-stage disease incidence. They postulated that it was possible that the decreased incidence of early stage disease may be real, but there had been insufficient time for this to translate to a decline in late-stage disease.

Their explanation for a possible real decrease in incidence of early cancers included the following: (1) Risk factors such as GERD and obesity may have exerted their maximum effect; (2) transition of Barrett esophagus to cancer may have declined; and (3) increased endoscopy with detection of Barrett esophagus and surveillance leading to increased detection and successful treatment of dysplasia may have prevented progression to cancer.

It is true that there has been a trend toward a decreasing percentage annual risk of cancer arising in Barrett esophagus in the past decade. In a review article in 2002, Spechler assessed the risk as 0.5% per year.[36] In a review article in 2013, the same author assessed the risk at 0.25% per year.[37]

This 50% decline in risk associated with Barrett esophagus has occurred despite a continuing absolute increase in the incidence of esophageal adenocarcinoma during this period. This percentage decline in the setting of increasing total number can be most easily explained by an increased prevalence of patients with Barrett esophagus. As suggested in the Pro-GERD study, when patients with GERD are acid suppressed with increasing effectiveness, the disease is significantly more likely to progress to Barrett esophagus during a 5-year period.[34]

The temporal increase in the incidence of esophageal adenocarcinoma has probably paralleled the increase in prevalence and extent of intestinal metaplasia within CLE. It was not till 1976 that Paull et al.[31] developed the histologic classification of CLE and recognized intestinal metaplasia as one epithelial type. Because detailed histologic studies of columnar metaplasia are not available before this, the time relationship of the increase in intestinal metaplasia and adenocarcinoma is unknown. Given the lag phase of 10–20 years for the development of cancer in a new target epithelium, one would expect that intestinal metaplasia in CLE was uncommon until the 1950s.

In the only historical paper we can find from before 1976 that has detailed histology and mentions intestinal metaplasia, Allison and Johnstone, in 1953,[30] reported that two of seven patients with long segments of columnar epithelium had goblet cells and one had adenocarcinoma (3/7 or 43%). In contrast, the Pro-GERD study reported that 100% of patients who developed columnar-lined segments greater than 3 cm long in the 5-year follow-up period had intestinal metaplasia.[34] The comparisons are dramatic particularly when indications for endoscopy in GERD were limited to the most severe patients in the 1950s compared to the relatively high frequency with which endoscopy is performed today.

There are few reliable studies that assess the association of PPI use and adenocarcinoma in patients with GERD. This is in contrast to numerous studies that evaluate risk of adenocarcinoma in patients with a diagnosis of Barrett esophagus. If, as is suggested by the Pro-GERD study, the effect of PPI treatment is to increase the incidence of Barrett esophagus in patients with GERD, the cancer-promoting effect of these drugs is indirect by expanding the pool of patients with Barrett esophagus. Studying the risk of cancer in the Barrett esophagus patient population is much less relevant. PPI treatment has already had its negative effect at that point.

I found the following statement, without any supporting data, in a paper by Lagergren et al. in the *New England Journal of Medicine* in 1999, which established symptomatic GERD as a significant risk factor for adenocarcinoma[38]: "The odds ratio for adenocarcinoma was 3.0 (95% CI 2.0–4.6) for symptomatic patients using acid-suppressive medications compared to symptomatic patients not using such medications." When adjusted for severity of symptoms, the odds ratio remained high, at 2.9 (95% CI 1.9–4.6), suggesting that acid-suppressive medication was the primary association rather than the severity of symptoms.

This group follows up this statement with a detailed study in the United Kingdom.[39]

This is a population-based nested case control study using patients registered in the general practitioners research database in the United Kingdom during the period 1994–2001. The study cohort required a patient to be enrolled with a general practitioner for at least 2 years and have at least 1 year of prescription history recorded in the database. The data in this research database have been shown to be of high quality.

Patients were followed up until they developed esophageal or gastric cancer, any other type of cancer (excluding skin cancer), age 85 years, death, or end of study period. Control subjects were selected following a standard incidence density sampling technique.

Drug exposure was classified as (1) never used and (2) ever used with this group being divided into current use (use within the previous year) and past use. Among current users, treatment duration was grouped into <1 year, 1–3 years, or >3 years, the last designated "long-term use." Protopathic bias was adjusted by a 1-year lag time analysis.

In 4,340,207 person-years of follow-up, 2128 patients with a newly diagnosed esophageal or gastric cancer were identified along with 10,000 control patients. After excluding patients due to predetermined criteria, the final case group consisted of 287 patients with esophageal adenocarcinoma, 195 with gastric cardia adenocarcinoma, and 327 with gastric noncardia adenocarcinoma.

A history of GERD and hiatal hernia was associated with an approximate twofold increase in the risk of esophageal adenocarcinoma, did not reach statistical significance for cardia adenocarcinoma, and showed no association with noncardia gastric adenocarcinoma. Gastric ulcer and a history of dyspepsia were associated with both cardia and noncardia gastric adenocarcinoma but not esophageal adenocarcinoma. A history of duodenal ulcer was associated with a twofold increase of noncardia gastric adenocarcinoma.

From the results: "Overall use of H_2 receptor antagonists and PPIs were associated with increased risks of both esophageal and gastric adenocarcinoma…. Among patients with more than three years use of acid suppressing drug use there was a threefold increased risk of oesophageal adenocarcinoma (OR 2.99; 95% CI 1.95–4.59) and a twofold increased risk of gastric non-cardia adenocarcinoma (OR 2.00; 95% CI 1.24–3.20) compared with non-users…. No association was found between long term use of acid suppressing drugs and risk of gastric cardia adenocarcinoma."

Indications for acid-suppressive drug use classified as "esophageal" (reflux symptoms, esophagitis, hiatal hernia, and Barrett esophagus) were more common among cases of esophageal adenocarcinoma (62%) than among cases of gastric cardia (33%) and noncardia adenocarcinoma (25%). "Long term use of acid suppressing drugs due to oesophageal indications carried a greater than fivefold increased risk of oesophageal adenocarcinoma (OR 5.42; 95% CI 3.13–9.39) while no statistically significant increased risk was found in the group of long-term users with other indications." Peptic ulcer as an indication for use of acid-suppressing drugs was associated with a greater than fourfold increased risk of gastric noncardia adenocarcinoma among long-term users. In their analysis, the authors state that the statistical power was not adequate to detect significant associations for gastric cardia cancers.

The authors conclude: "Long term pharmacologic gastric acid suppression is a marker for increased risk of oesophageal and gastric adenocarcinoma. However, these associations are most likely explained by the underlying treatment indication being a risk factor for the cancer rather than an independent harmful effect of these agents per se."

The authors are very careful in their conclusion in not attributing a causative role of acid-suppressive drugs in esophageal adenocarcinoma. They point to the possibility that the effect of increased risk of adenocarcinoma with PPI use may be secondary and the primary reason for using these drugs may be the underlying disease that resulted in their use.

The American Gastroenterological Association Technical Review on the management of Barrett esophagus (from the website, dated 2011)[15] discusses the findings in this paper: "Using the large general practitioners research database in the United Kingdom… Garcia-Rodriguez et al found that patients who were treated with acid suppression for an 'esophageal indication' such as GERD had a significantly increased risk of developing esophageal adenocarcinoma (odds ratio 5.42, 95% CI 3.13–9.39). In contrast, for patients who were treated with acid suppression for a 'gastroduodenal indication' such as peptic ulcer disease, there was no significantly increased risk of adenocarcinoma (odds ratio 1.74; 95% CI 0.90-3.34). The lack of an association with cancer in patients taking PPIs for gastroduodenal disease suggests that the positive association with esophageal disease resulted from confounding by indication. In other words, it was likely the GERD, not the GERD treatment, that increased the incidence of cancer."

While this analysis superficially seems to be reasonable, it has one flaw. The explanation of these findings as a result of confounding by indication is certain only if PPIs act as direct carcinogens to produce esophageal adenocarcinoma. There is absolutely no evidence or suggestion that PPIs are direct carcinogens. If this was the case, the carcinogenity of PPIs would have been detected in the basic safety testing that is required by the FDA before any drug is approved for human use. Also, the epidemic of esophageal adenocarcinoma began long before the late 1980s when PPIs were first introduced into the market.

The action of PPIs in promoting carcinogenesis is much more complex. Let us assume that esophageal adenocarcinoma results from the effect of refluxed carcinogens acting on intestinalized columnar epithelium in the esophagus. Without reflux, the carcinogen in gastric juice will never be delivered to the esophagus. Without reflux, the patient will never develop the necessary target cell (i.e., intestinal metaplasia) in the esophagus. Nothing, including PPI, will influence the risk of adenocarcinoma in that person. PPIs are completely safe until a person has reflux and intestinal metaplasia in the esophagus.

It is only patients who have reflux with intestinal metaplasia in their esophagus that are exposed to carcinogen in gastric juice both before and after treatment with PPI. If they have an increased risk of developing esophageal adenocarcinoma when treated with acid suppression, *it must mean that the acid suppression was the cause of the increased risk from baseline (i.e., the risk in the same patient group without acid suppression.)* The patients who did not have reflux (i.e., patients who were given acid-suppressive drugs for treatment of gastroduodenal indication) would have neither intestinalized columnar epithelium in their esophagus nor exposure to carcinogens in gastric juice. Acid suppression would therefore have no possibility of having any increased risk of esophageal adenocarcinoma. The data are not completely clean only because of the probability that a small percentage of patients in the "gastroduodenal indication" group may have had asymptomatic GERD with esophageal intestinal metaplasia.

The explanation of the data as being the result of "confounding by indication" is not definite or even likely. The findings provide evidence that *acid suppression increases the risk of adenocarcinoma in GERD patients but not in patients without GERD.*

There is a simple test for whether PPIs are direct or indirect promoters of esophageal adenocarcinoma in GERD. The basic action of PPIs is to suppress acid. Its primary effect and mechanism of action is to alkalinize gastric juice. Among drugs that alkalinize gastric juice, PPIs are the most powerful. However, there are other drugs that are capable of suppressing and neutralizing gastric acid that are less powerful. These are acid neutralizers and H_2 receptor antagonists. That they are clinically effective in treating GERD is shown by the fact that they have populated the shelves of the over-the-counter market in pharmacies. These drugs existed in the 1950s before the epidemic of adenocarcinoma began.

If it is the effect of PPIs, i.e., the alkalinization of gastric juice, that is responsible for promotion of carcinogenesis, the time frame of the esophageal adenocarcinoma epidemic can be explained. The critical test for this hypothesis is to see whether non-PPI alkalinizers of gastric juice are associated with an increased risk of cancer in patients with GERD.

I will show in later chapters that the cellular change that is most affected by alkalinization of gastric juice is the development of intestinal metaplasia (i.e., Barrett esophagus) in nonintestinalized cardiac epithelium in the esophagus that has resulted from GERD. The action of PPIs and other alkalinizing agents is to increase the pool of patients with Barrett esophagus. The increased incidence of adenocarcinoma is due to the increase in the risk pool, not necessarily the promotion of carcinogenesis in patients with Barrett esophagus.

I was peripherally involved in an epidemiologic study that was done in LA by Dr. Leslie Bernstein as doctoral thesis project of a graduate student.[40] In this population-based case–control study, patients with incident esophageal adenocarcinoma (n=220), gastric cardiac adenocarcinoma (n=277), or distal gastric adenocarcinoma (n=441) diagnosed between 1992 and 1997 were recruited and matched with 1356 control participants in LA County. Unconditional polychotomous multivariable logistic regression analyses were done to evaluate the association between acid-suppressive drug use and these cancers. The study period was too early for assessing the effect of PPI use, which began in the early 1990s. The study really evaluates antacids and H_2 receptor antagonists.

Among participants who took nonprescription acid-neutralizing agents for >3 years, the odds ratio for esophageal adenocarcinoma was 6.32 compared with never users [95% confidence interval, 3.14–12.69; $P(\text{trend}) < .01$]. Regular use of nonprescription acid-neutralizing agents was not associated with risk of adenocarcinomas of the gastric cardia or distal stomach. Regular use of prescription acid-suppressive drugs was not associated with an increased risk for any of these cancers.

The authors conclude that the increased risk of esophageal adenocarcinoma may represent self-medication for undiagnosed precursor conditions or it may be that nonprescription acid neutralizing drugs, taken without limitation of amount used when symptoms are most intense, may permit alkaline bile reflux into the lower esophagus, thereby increasing esophageal adenocarcinoma risk.

This study is valuable because it suggests a possible association between self-medication with over-the-counter antacids and a risk of esophageal adenocarcinoma. Patients who self-medicate with antacids for symptoms that may or may not be GERD-related number in the millions and represent a group of patients who first seek medical attention with symptoms of advanced adenocarcinoma. They have never had typical GERD symptoms, PPI use, or a diagnosis of Barrett esophagus till they develop cancer.

A recent study by Lada et al.[41] suggested that the fact that patients develop adenocarcinoma without ever having symptoms of GERD may not be correct. Among 345 consecutive patients with esophageal adenocarcinoma undergoing surgery, 64% gave a history of GERD symptoms, 52% had used PPIs, and 34% had a history of Barrett esophagus. This study did not report use of non-PPI acid-suppressive agents by these patients. It is probable that there is likely a large population of GERD patients that may be using over-the-counter acid neutralizers and controlling symptoms without seeking treatment until they develop cancer.

Antacids are generally dismissed as irrelevant because they provide very limited but rapid neutralization of gastric acid within minutes of ingestion. However, their use is frequently on-demand at the time of onset of symptoms or before an expected onset of symptoms before a specific meal type. It is therefore possible that their use coincides with reflux episodes and they produce significant alkalinization at the point of maximum impact. The recent discovery of the acid pocket and its importance in GERD has provided a scientific basis for the efficacy of simple acid neutralizers.[12] That antacids have remained on the market for decades attests to this effectiveness.

None of these studies provide evidence that the use of acid-suppressive medication in patients with GERD is directly causative for esophageal adenocarcinoma. However, what their data show clearly is that the use of long-term acid-suppressive drugs in the treatment of GERD *does not prevent progression to esophageal adenocarcinoma* even if the data do not prove that they promote cancer.

Much more commonly found in the literature are studies assessing cancer risk with use of PPIs in patients with Barrett esophagus. These studies are relatively easy because the entry point is a patient who has an endoscopy and a diagnosis of Barrett esophagus. They are within the purview of academic gastroenterologists and surgeons who are much more likely to perform clinical studies. This is quite different to the GERD patient who either self-medicates with over-the-counter antacids or is treated empirically and often successfully by primary care physicians in the community who are much less inclined to research.

There are few good studies of the association between PPI use and cancer in patients with Barrett esophagus. We will present two studies here that express the opposite viewpoints:

Kastelein et al.[42] published a prospective cohort study in the Netherlands of 540 patients with Barrett esophagus between November 2003 and December 2004. The study group was selected from 756 patients with known or newly diagnosed Barrett esophagus where the endoscopic diagnosis was confirmed histologically by the presence of intestinal metaplasia. Patients with Barrett esophagus shorter than 2 cm, patients with high grade dysplasia and adenocarcinoma, and those who had antireflux surgery were excluded.

216 patients dropped out of the study for a variety of reasons (severe comorbidity, 30; death unrelated to Barrett esophagus, 18; refusal to participate, 89; migration, 10; no follow-up endoscopy, 66; or neoplastic progression to high grade dysplasia or adenocarcinoma within 9 months of inclusion, 3).

At baseline, 460 patients had a prior diagnosis of Barrett esophagus (duration not stated in the study). The other 80 were diagnosed at the initial endoscopy.

Incident cases of high-grade dysplasia or adenocarcinoma were identified during follow-up with surveillance endoscopy and biopsy according to guidelines of the American College of Gastroenterology (nondysplastic Barrett esophagus every 3 years). At each surveillance visit the patients filled a detailed questionnaire which included medication use. Information on medication use was cross-checked with pharmacy records.

At baseline, patients were classified as "current user" of H_2 receptor antagonists or PPIs when they used these drugs at that time for at least 1 month, "former user" when they used these drugs for at least 1 month but not at the time of inclusion, and "nonuser" when they used these drugs for less than 1 month. Of the 540 patients in the study at baseline, 462 (85%) patients were current users, 10 (2%) former users, and 68 (13%) nonusers.

During follow-up, patients were classified as "user" of H_2 receptor antagonists or PPIs according to their exact start and stop dates. Duration was calculated by adding the duration of prescriptions starting from the time of inclusion. Adherence was calculated by dividing the duration of medication use by the duration of follow-up.

The 540 patients were followed for a median duration of 5.2 years (IQR 3.5–5.7). 28 patients developed high-grade dysplasia and 12 developed adenocarcinoma. These were endpoints in the study. The annual incidence of high-grade dysplasia and cancer was 1.6% (95% CI 1.1–2.1) and the annual incidence of adenocarcinoma was 0.5% (95% CI 0.2–0.8). The risk of neoplastic progression increased with age, length of Barrett esophagus, esophagitis, and low-grade dysplasia.

H_2 receptor antagonist use did not affect the risk of neoplastic progression [hazard ratio (HR) 0.83 (95% CI 0.11–6.03)].

PPI use at inclusion was associated with a reduced risk of neoplastic progression (HR 0.43; 95% CI 0.21–0.88). This uses the 68 (13%) of nonusers of PPIs as the reference group for comparison with the 462 (85%) current users. 10 "former users" are excluded from this comparison.

PPI use during follow-up was associated with a reduced risk of neoplastic progression (HR 0.15; 95% CI 0.07–0.66). The risk of neoplastic progression decreased with prolonged PPI use (P=<.001) indicating a duration response relationship. The calculation of hazard ratio uses person-years of follow-up for comparisons. The cohort nonprogressing had 2543 person-years of follow-up and those developing high-grade dysplasia or adenocarcinoma had 104 person-years of follow-up. The follow-up person-years in the group that did not progress were 2488 (98%) among PPI users and 55 (2%) in those who did not use PPIs. In the group that progressed to high-grade dysplasia or adenocarcinoma, the follow-up person-years were 92 (88%) among PPI users and 12 (12%) among nonusers. The comparisons are made between the groups using person-years of follow-up rather than actual number of patients.

As the authors comment in their discussion, "only 8 patients (2%) never used PPI, and 18 patients (3%) used a PPI during part of their follow up period… limits the options for investigating the effect of PPIs."

It should be emphasized that this study has a nonuser of PPI control group of 8/540 patients. While the statistical calculations cannot be questioned, the methodology is unusual. If we conclude that the reason why person-years rather than patient numbers were used in the calculations was that the control numbers were too small to have adequate statistical power, then the study conclusions should be viewed with considerable caution.

Contrast of numbers with the study by Garcia-Rodriguez[39] above is interesting. That study had 4,340,207 person-years of follow-up, 287 patients with esophageal adenocarcinoma, 195 patients with cardia adenocarcinoma, 327 patients with noncardia gastric adenocarcinoma, and 10,000 controls. Their statistical analyses were based on actual patient numbers, not person-years. Despite the large numbers, the power in their study was not sufficient for conclusions regarding the group of cardia cancers. While the study was an epidemiologic retrospective study unlike the prospective study above, it provides some sense of statistical power needed for firm conclusions.

There are other concerns. The differences between current, former, and non-users are defined in the methods. By their definition (see above), a person who has used PPI for at least 1 month is a user and less than 1 month is a nonuser. This is a sloppy definition at best. A person using PPIs for 28 days in February is a user and for 30 days in March is a nonuser. Surely, a study published in a prestigious journal should demand a precise definition in days rather than a single month.

The baseline PPI use on risk of neoplastic progression is divided into no (68), former (10), and current (462) users. The reference group is the 68 patients who were nonusers. When compared against former users (10) the hazard ratio is 1.32 (95% CI 0.29–6.04). Two (5%) former users developed high-grade dysplasia or adenocarcinoma and would have increased the number in the user group from 28 to 30. This is the only place in the data analysis where "former user" is used. The authors open themselves to the criticism that this was a statistical manipulation aimed at reaching statistical significance for the comparison of the nonuser group with the current user group. The reviewer should have asked the obvious question: If the PPI users were divided into nonusers and users with the former and current users combined, would the hazard ratio still have been significant? This question will never be answered and represents a significant weakness in the paper.

The authors state: "In this large prospective cohort study, PPI use was associated with a 75% reduction in the risk of neoplastic progression in patients with Barrett esophagus, independent of age, gender, BE length, esophagitis, histology, and use of other medications." The authors point to the uniqueness of their findings and point to three previous cohort studies and two case control studies that could not provide definite conclusions regarding the protective effect of PPI therapy in patients with Barrett esophagus. They attribute this to defects in those studies. They go on to state: "Although prolonged PPI use is accompanied with considerable costs, it is an effective strategy to prevent neoplastic progression in Barrett esophagus."

This is a remarkably powerful conclusion from a study with a control population of 8 patients without PPI use being compared to a PPI user group of 532 patients as its method of reaching their conclusion.

This study, published in April 2013, was heralded in the gastroenterology world as justification that PPIs have a chemopreventive effect against cancer development in Barrett esophagus. It therefore justified treating all patients with Barrett esophagus with PPIs irrespective of the presence of symptoms with the goal of preventing adenocarcinoma. The following incredible response to this paper in various websites provides an insight as to how news gets to the modern busy physicians of today:

An article by D. Napoli appeared in *Practice Update Frontline Medical News* (March 7, 2013), *Family Practice News* (April 1, 2013), and *Internal Medicine News* (April 1, 2013):

Headline: "PPIs lower progression risk in Barrett's esophagus."
Vitals: "Major Finding: In Barrett's esophagus, use of proton pump inhibitors was associated with a hazard ratio of 0.21 for neoplastic progression."
Data Source: A multicenter, prospective cohort study of 540 patients with known or newly diagnosed Barrett's esophagus.
Disclosures: "None of the authors disclosed any conflicts of interest related to this article. No funding was reported. Best data yet in support of PPIs."

In the first three paragraphs following the text: "Proton pump inhibitor use was associated with a greater than 75% reduction in the risk of neoplastic progression in Barrett's esophagus. Indeed, despite the "considerable costs" of these drugs, 'prolonged PPI use is … justified and feasible in Barrett's patients and should be strongly recommended, in particular in guidelines,' concluded Dr. Florine Kastelein… In what they called 'the first methodologically sound prospective study which shows that PPIs strongly reduce the progression of neoplastic progression in BE…."

PPIs had suddenly developed an evidence base for a new use: they were cancer chemopreventive agents. The limitations of the study, even as stated by the authors: "despite the large sample size, only 8 patients (2%) never used a PPI and 18 patients (3%) used a PPI during part of their follow-up period. Although this reflects clinical practice in Western countries and is representative for a disease in which patients seek to avoid reflux symptoms, it limits the options for investigating the effect of PPIs." The authors go on to state: "A randomized controlled clinical trial would be the ideal way to investigate the effect of PPIs without the risk of confounding. Although not impossible, it will be difficult to perform such a trial, because many Barrett esophagus patients suffer from reflux symptoms without the use of a PPI."

This is not true. It is well known that many patients with Barrett esophagus do not have significant symptoms of GERD if detected by screening rather than endoscopy performed for failure of therapy to control symptoms.[43] The best study to test the impact of PPI treatment in the incidence of adenocarcinoma in patients with Barrett esophagus would be to limit the study to asymptomatic patients newly diagnosed by screening with Barrett esophagus and randomize them into PPI and placebo groups. Such a study is not difficult but has never been done. It is only when data from such a study are available that a decision can be made to use PPIs purely as a cancer-preventive measure in asymptomatic Barrett esophagus patients. The above study does not provide an answer to that question because all patients were symptomatic and PPI use was justified on that basis.

I believe there is a significant bias in the publication of papers in the medical literature. It is easier to get papers published that have appeal to proponents of medical treatment of GERD than papers that contradict the establishment opinion. Reviewers ignore flaws in papers that espouse their opinion and rush to a positive judgment even when the methodology is suspect. When the paper is published, readers assume that the review process has been rigorous and frequently read the abstract and conclusions without critical review. The headline conclusion hits the GERD media outlets and is quoted in doctor's dining rooms. The conclusions become the evidence base that drives treatment of patients. This paper and its response in the news outlets used by many practicing physicians is a good example of this dangerous process.

Individual physicians who prescribe drugs that have potential side effects like PPIs to their patients have a duty to critically review papers without being guided by headlines and conclusions in abstracts. This rarely happens although its importance is taught in journal clubs of every residency training program. The excuse of the practicing physician who does no critical review is that they have no time. This is not true; I took a little over 1 h for my critical review.

The most important statement in the paper of Kastelein et al.[42] is the absence of any evidence before their study that PPIs decrease the incidence of adenocarcinoma in patients with Barrett esophagus. Given the limitations of their own study, the evidence base for the use of PPIs as a cancer prophylaxis in Barrett esophagus is limited at best. Despite this, it is not uncommon for gastroenterologists to prescribe PPIs in high dose to patients with Barrett esophagus with the goal of cancer prevention even when the lack of symptoms does not justify PPIs. This is the result of opinion-based medicine that not infrequently masquerades as evidence-based medicine.

The second study on this topic was by Hvid-Jensen et al.[44] The stated background of this study is that "PPIs may potentially modify and decrease the risk for development of esophageal adenocarcinoma in Barrett esophagus." The stated aim of the study is "to investigate if the intensity and adherence of PPI use among all patients with Barrett esophagus in Denmark affected the risk of oesophageal adenocarcinoma."

In the introduction, the authors put forward a rationale for the cancer-preventive action of PPIs: "In Barrett esophagus, the metaplastic cells have a higher proliferative rate than the normal squamous epithelium. It has been shown, that this activity increases during both persistent and pulsatile acid exposure via mitogen activated protein kinase pathways, transmitting growth regulatory signals in order to enhance proliferation and decrease apoptosis. Inhibiting these pathways, by minimizing acid-induced stimulation, might therefore be beneficial in preventing progression from Barrett's esophagus to high grade dysplasia or oesophageal adenocarcinoma."

They continue: "Apart from relieving symptoms, inhibition of acid production should decrease the reflux of acid into the oesophagus, thereby decreasing the ongoing inflammation, proliferation and risk of dysplasia in the epithelia. Especially in Barrett's oesophagus patients, with most to gain from acid inhibition, this effect is desired. The use of PPIs has risen rapidly—but so have the incidence of oesophageal adenocarcinoma. Studies investigating the potential protective effects of PPIs on Barrett's oesophagus have found some or none protective effect from PPIs. However, the majorities of published studies had methodological limitations, were limited in size and follow-up and relied on selected patient cohorts."

The authors enter into the study with the belief that their findings will provide the evidence that is sorely needed to prove that PPIs decrease cancer risk in Barrett esophagus and justify the use of PPIs in these patients. Their stated bias is to find such evidence.

This is a nested case–control study in a cohort of all patients diagnosed with Barrett esophagus in Denmark from 1995 to 2009. The Pathology Registry was used to identify 9883 patients with Barrett esophagus, defined as presence of specialized intestinal metaplasia in esophageal biopsies. Patients within this cohort who had low- and high-grade dysplasia were identified. Linking with the Cancer Registry, patients in this cohort who developed esophageal adenocarcinoma by the end of 2009 were identified. Patients with a diagnosis of high-grade dysplasia or adenocarcinoma made up to 1 year after the diagnosis of Barrett esophagus were excluded from the cohort.

For each patient with cancer, 10 control subjects were selected from the cohort. These controls had Barrett esophagus, were alive, and had no diagnosis of high-grade dysplasia or adenocarcinoma matched for age and date of diagnosis of Barrett esophagus.

The Danish Prescription Database was used to identify the drug history for all study patients and controls. PPI use was defined as (1) "ever users" who were individuals with >2 prescriptions and (2) "never/rare users" as those with <2 prescriptions during the study period. Ever users were further divided into recent users (>2 prescriptions in the 2 years before the case date) and former users (>2 prescriptions overall, but <2 during the latest 2-year period). Duration of use was classified as short term (<7 years) and long term (>7 years), based on the number of days between the first and last prescription dates. The adherence of use/intensity of PPI therapy was defined as the total number of defined daily dosages divided by the total duration of use. Using this definition, PPI use was classified as low adherence (<75%) or high adherence (>75%).

The findings were: 9883 patients were diagnosed with Barrett esophagus (66.5% males) with a total follow-up time of 66,037 person-years. Median age at diagnosis of Barrett esophagus was 62.6 years (IQR 52.4–72.9) and median follow-up time was 5.7 years (IQR 3.4–9.3). Within this cohort, 140 cases of adenocarcinoma or high-grade dysplasia were identified (median age 67.7 years and median follow-up time 10.2 years) and matched with 1297 controls from the cohort of Barrett esophagus patients.

Among the cases of esophageal adenocarcinoma, 45 (75%) were recent users with 23 (38.3%) being high-intensity users. The corresponding numbers among comparable controls were 341 (61.9%) and 176 (31.9%). Among all cases of esophageal adenocarcinoma, high PPI use adherence was short term in 31.7% and long term in 16.7%. For controls, the corresponding numbers were 32.5% and 11.4%.

When esophageal adenocarcinoma and high-grade dysplasia were combined, 116 (82.9%) of the cases were recent users of PPI and 70 (50%) were high-adherence PPI users. Among the controls, the corresponding numbers were 848 (65.4%) and 589 (45.4%).

The relative risk of esophageal adenocarcinoma or high-grade dysplasia among Barrett esophagus patients using PPI compared to never/rare users was 1.1 (95% CI: 0.4–3.3) in former users, 1.9 (95% CI: 0.7–4.9) in ever users, and 2.1 (95% CI: 0.8–5.6) in recent users. Long-term PPI use yielded a relative risk of esophageal adenocarcinoma or high-grade dysplasia of 2.2 (95% CI: 0.7–6.7) in the low adherence group and *3.4 (95% CI: 1.1–10.5) in the high-adherence group.*

In their discussion, the authors state: "We were not able to prove a preventive effect from proton pump inhibitors, instead we found an increased risk of oesophageal adenocarcinoma and high grade dysplasia related to long-term PPI therapy."

In their discussion they comment on Kastelein et al.[42]: "Irrespective of the strong study design, however, the conclusions may be premature due to the relatively small cohort diluted into several stratified groups and a small control group of non-PPI users."

The authors comment, however, on the common clinical practice of treating Barrett esophagus patients with PPIs: "Previously reflux symptoms have been described as an independent risk factor for oesophageal adenocarcinoma, and oesophageal acid exposure as the prominent factor in malignant transformation from Barrett's esophagus to oesophageal adenocarcinoma. This assumption has justified widespread routine prescription of PPI to Barrett's oesophagus patients, despite present international guidelines recommend PPIs as symptomatic treatment only."

The authors conclude that the study showed no cancer-protective effects from PPIs use. They sound a warning that is justified by their findings: "In fact, high-adherence and long-term use of PPI were associated with a significantly increased risk of adenocarcinoma or high-grade dysplasia. This could partly be due to confounding by indication or true negative effect from PPIs. Until the results from future studies hopefully can elucidate the association further, continuous PPI therapy should be directed at symptom control and additional modalities considered as an aid or replacement."

This conclusion is without hyperbole. They point to the near certainty that PPI use does not prevent progression to adenocarcinoma when used in patients with Barrett esophagus. They provide two possibilities for their finding of an increased risk of adenocarcinoma with increasing PPI dose: a true negative effect of PPIs or a result that is confounded by indication for PPI use. They are much more cautious about their conclusion than Kastelein et al.[42]

Two things are worthy of note regarding the differences between the studies by Kastelein et al.[42] and Hvid-Jensen et al.[44] The first is the reaction. The paper by Kastelein et al.[42] evoked an immediate and highly visible reaction in the Internet websites, providing information to internists and family practitioners, essentially telling these physicians that there was now excellent evidence that treating patients with Barrett esophagus with the objective of reducing cancer risk was appropriate. However, the positive impact of the story for PPI sales makes us suspicious about the origin of these articles in news outlets. It also makes us aware that commercial interests now have numerous avenues to influence patients and physicians with the increasing width of the Internet.

The other question was the journal where Hvid-Jensen's paper was published. It reminds me of a conversation I had with Tom DeMeester about this paper at dinner.

> *"Para," asked Tom, "have you seen that paper from Denmark that showed an increased risk of cancer in Barrett esophagus in patients who used PPIs over the long term?"*
>
> *"Yes," said I, "isn't that the paper that was in that journal called Aliment Pharmacol something?"*
>
> *"Yes," said Tom with a smile. "That is the journal where the really good papers end when they have been rejected by all the big gastroenterology journals when they contain conclusions they do not like!"*

When I was reviewing the paper by Hvid-Jensen et al.,[44] I realized that their prior study using this same excellent and accurate cohort from the Danish Registries was published in 2011 in the *New England Journal of Medicine*.[45] That study demonstrated that the annual risk of esophageal adenocarcinoma in Barrett esophagus was 0.12%, much lower than the generally accepted incidence at the time of 0.5%.[45] This information is appealing to gastroenterologists because it justifies minimizing the risk of cancer to patients who express concern.

It would be interesting to know if the present paper was also submitted to the *New England Journal* and rejected. It would raise questions of possible subject bias and censorship by the highly regarded journal wherein excellent papers that do not fit their view of the subject are more likely to be rejected than less meritorious papers that are concordant with their viewpoint. I think secrecy in this regard can be removed and transparency increased if there is a requirement for authors to state in a footnote, just like their conflict of interest, what journals the paper has been previously reviewed by and rejected.

I will revisit the question of the progression of Barrett esophagus to dysplasia and adenocarcinoma in Chapter 15. Like many other things in GERD, the careful histologic study of how columnar epithelium undergoes intestinal metaplasia, the distribution of intestinal metaplasia, and where dysplasia and adenocarcinoma arise sheds light to this question.

To summarize the literature relating to the association between PPI therapy and neoplastic progression in GERD:

1. There is strong evidence that PPI use in patients with GERD does not decrease the progression to Barrett esophagus and cancer. This is strongly supported by the epidemiology of esophageal adenocarcinoma.
2. In the previous section, we showed that there is strong evidence that regular PPI use increases the risk of progression of GERD without visible CLE to Barrett esophagus.
3. The jury is out on the question of whether PPI use in patients with Barrett esophagus has a positive or negative effect on the risk of cancer. Most of the studies show no significant positive or negative effect. A few show a positive effect and a few show a negative effect.

We are using these acid-suppressive drugs without proof that they are not responsible for the epidemic of GERD-induced esophageal adenocarcinoma. Esophageal adenocarcinoma could be an unintended iatrogenic disease that is more serious than any other in the history of medicine.

This behavior is dangerous. The entire establishment is at risk of litigation if it is proven some day that the use of acid-suppressive drugs was responsible for the epidemic of esophageal adenocarcinoma. Ignorance is not a defense. We are guilty of a failure of requiring *proof of the absence of an association* between use of these drugs and cancer. This is difficult to justify when the increase in our ability to alkalinize gastric juice with these drugs occurred in a similar time frame as the increase in cancer. That we explained this as mere chance, coincidence, or indication bias rather than a causative effect will be considered ridiculous if and when evidence becomes available in the future that the drugs played a causative role.

5. WHAT DOES THE FUTURE WITHOUT ANY CHANGE HOLD?

The management of GERD is in a deep rut.

The rut began in the 1960s when physicians were presented with what, at that time, seemed to be a miraculous gift by the pharmaceutical industry. This was cimetidine, a H_2 receptor antagonist; a drug that could suppress gastric acid secretion. This drug was heralded as causing an end to the misery caused by acid-induced diseases of the foregut: GERD in the esophagus and peptic ulcer disease in the stomach and duodenum. A miracle cure had entered the armamentarium of the gastroenterologist. Nothing else needed to be done to manage these diseases; eradication of these diseases, commonly regarded as being caused by acid, was a done deal.

It took a long time for physicians to recognize that acid suppression did not cure these diseases. Barry Marshall and Robin Warren explained why they did not work to cure peptic ulcer disease. Acid was not the cause of peptic ulcers: *Helicobacter pylori* was. Acid suppression relieved symptoms of peptic ulcer but did not cure it. With the new understanding that *H. pylori* was the primary cause of peptic ulcer, the management changed to focus on eradicating *H. pylori* in the treatment. Combined with acid suppression, this new management resulted in the conversion of peptic ulcer disease from being a life-threatening serious disease to one that rarely causes serious problems today.

When H_2 receptor antagonists failed to eradicate GERD, the medical community believed that this was because the degree of acid suppression produced by these drugs was not adequate. The pharmaceutical industry responded to this perceived need by producing PPIs, a new group of acid suppressants that were far more effective. Omeprazole was introduced in the late 1980s. Again, the euphoria of the medical community was palpable. GERD would now surely become a disease of the past.

For the next 35 years, the medical community has been acting on the belief that PPIs will cure GERD. Their vision of this miracle has been so strong that they have only seen the good that has resulted from these drugs. The good is that millions of GERD sufferers have a rapidly effective mechanism of controlling their symptoms. People with heartburn do not even need a physician's prescription; a trip to the nearby pharmacy is all that is needed. Advertising persuades the population that all they need to do is swallow a pill, purple, or other hued, to both control and prevent heartburn. Take a pill and you can eat anything is the message of some of the commercials.

Slowly, over time, gastroenterologists who are more seriously involved in treating the more complicated GERD patients have come to recognize that there is a bad and ugly element of PPI therapy for GERD.

The bad is that 30% of people with heartburn fail to be controlled in the long term by PPI therapy despite dose escalation and become chronically dissatisfied and disabled by their disease. Regurgitation is not well controlled and a significant number of other symptoms such as chronic cough, hoarseness, and even chronic pulmonary fibrosis can occur in these patients. Treatment failure results in a decline in the life quality of the patients and often has a severe negative impact on job performance and productivity.[4,5]

Gastroenterologists are also the physicians capable of performing endoscopy. They recognize that ~10% of patients with GERD go on to develop Barrett esophagus relatively quickly after initial diagnosis,[34] raising the specter of increased cancer risk. This is another element of the bad element of PPI therapy.

The ugliest element of PPI therapy is that an increasing number of people go on to develop esophageal adenocarcinoma.[25,26] When these occur in the setting of surveillance for known Barrett esophagus, they are detected early and amenable to endotherapy, including mucosal resection and ablation. Enormous effort and expense is required to identify these early cancers with surveillance.

The alternate is worse. 85% of patients who develop GERD-induced adenocarcinoma have never had an endoscopy or a diagnosis of Barrett esophagus. These patients are commonly at an advanced stage of cancer, beyond the scope of endotherapy. They require esophagectomy, chemotherapy, and radiation with an overall mortality of 85%.

Despite the knowledge that there is a bad and ugly component of PPI treatment, the presently recommended management is not designed to do anything other than control symptoms of GERD. Physicians simply accept the bad (treatment failure) and the ugly (adenocarcinoma).

The future as it stands at present has little hope for improvement in terms of reversing the increasing incidence of GERD, Barrett esophagus, and adenocarcinoma. There is no effort at or ability to control or prevent these outcomes.

There is no attempt at or ability to diagnose GERD at an early stage. The requirement of troublesome symptoms delays diagnosis of GERD. The recommendation that endoscopy is indicated only if treatment fails to control symptoms further delays diagnosis to a time when their disease is at or close to irreversibility.

The absolute numbers involved render the medical establishment helpless. 30% of the adult population with heartburn translates to a pool of 75 million people with GERD in the United States. This is too many to make screening for Barrett esophagus cost-effective. Even identifying factors that increase the risk of Barrett esophagus such as severity of symptoms, age over fifty, male sex, Caucasian race, and body mass index do not decrease numbers sufficiently to permit screening without overwhelming the resources available in the health-care system.

There is an atmosphere of helplessness and nihilism that we can do nothing to stop the inevitability of treatment failure and cancer. Present management is based on an irrational hope that PPIs used for symptom control will have a miraculous unintended consequence of preventing cancer. This is not likely.

6. THE FUTURE WITH THE NEW METHOD OF ASSESSMENT OF LOWER ESOPHAGEAL SPHINCTER DAMAGE

We need to stop, reevaluate, and develop a new pathway to manage GERD with the objective of preventing treatment failure and adenocarcinoma.

Physicians have not learned the lesson of peptic ulcer disease. GERD is exactly like peptic ulcer disease. Peptic ulcer disease was thought to be caused by gastric acid; GERD is similarly thought to be caused by gastric acid (GERD is often called "acid reflux disease"). Acid-suppressive drugs were heralded as a cure for peptic ulcer disease; similarly PPIs were heralded as a cure for GERD and believed by many to be so. Acid-suppressive drugs did not eradicate peptic ulcer disease; they certainly did not eradicate GERD.

It was when the cause of peptic ulcer disease was recognized as being *Helicobacter pylori* and steps were taken to address this that success was achieved. It is well known that GERD is primarily caused by LES damage. GERD will be eradicated only when LES damage, the primary cause, is addressed.

At present, damage to the LES, which is the known cause of GERD, is hardly taken into any account in the management of GERD. Manometry, which is presently the only available method of assessment of the LES, is rarely performed and hardly ever used in the diagnosis of GERD.

There are no new drugs in the pipeline that are designed to improve LES function. Surgical and endoscopic methods to augment and repair a damaged LES are not adequately successful and not performed frequently enough to impact the overall outcomes of GERD. They are usually done as a last-ditch resort in patients who have reached the end of the line in medical therapy and so dissatisfied with their quality of life that they are referred for sphincter repair procedures.

We need new avenues of research that will manipulate the cellular events in GERD. This requires new definitions and clear understanding of every step of cellular progression from the squamous epithelial cell through cardiac epithelium, intestinal metaplasia to dysplasia, and adenocarcinoma. We need to understand that the cellular progression is dependent

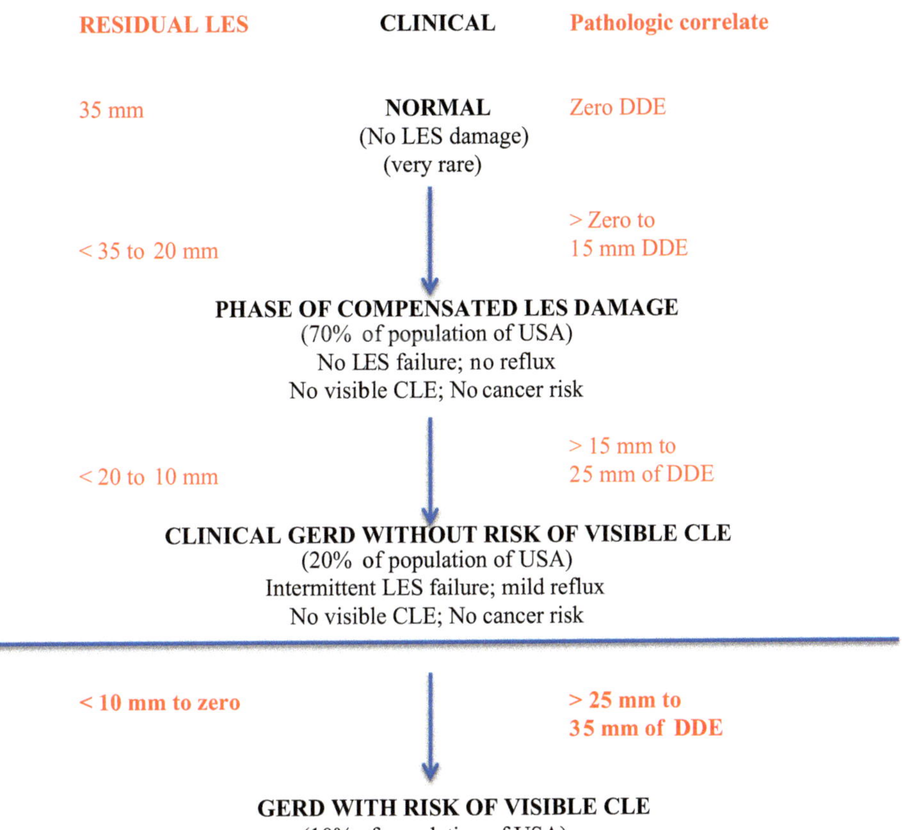

FIGURE 2.12 Progression of gastroesophageal reflux disease (GERD) according to the degree of abdominal lower esophageal sphincter (LES) damage (shortening). The progression is orderly because sphincter failure and reflux is primarily determined by LES damage. The critical point of the disease from a cancer standpoint is shown by the horizontal line. Above the line, patients have no cancer risk and all manifestations of GERD are reversible. Cancer risk begins with the development of visible columnar-lined esophagus (CLE), which is irreversible. *DDE*, dilated distal esophagus.

on progressive LES damage (Fig. 2.12). This new understanding will allow us to answer critical questions about the disease that have not yet been asked:

1. What is the normal state? In Chapter 5, I will define the normal person based on histologic criteria that are defined in Chapter 4. Histology is the only reliable method of defining the normal state.
2. How does the normal patient become the earliest GERD patient and progress to the most severe GERD patient? In subsequent chapters I will take you step by step, cell by cell, millimeter by millimeter through the progression of GERD from beginning (the normal state) to the end in pathologic terms. The end state of GERD in any given person varies with the rate of progression of LES damage and the life span of the person.
3. I will develop a new pathologic method of assessment of LES damage based on a simple biopsy protocol at endoscopy. This will provide a means to define and quantify LES damage, the cause of GERD.
4. I will show that the progression of LES damage into the future life of the patient is predictable by an algorithm that is based on the histologic assessment of LES damage at any given point in the life of a patient and the linearity of progression of LES damage.

The ability to measure LES damage and predict its progression into the future will permit the recognition of the following patient groups.

1. People whose predicted LES damage in their lifetime is so mild that they will never develop LES failure. They will never have reflux. Without reflux, they will never develop visible CLE and therefore will never be at risk for esophageal adenocarcinoma. This is different from the present where a person without symptoms of GERD can harbor Barrett esophagus, dysplasia, and adenocarcinoma. This is 70% of the population in the United States.

If a person has symptoms suggestive of GERD and the test of LES damage shows that the patient's LES is not damaged adequately to cause reflux, reflux can be excluded as the cause of the symptoms. This is far more accurate than the empiric PPI test or clinical diagnosis of GERD. It is much more sensitive than endoscopy and pH testing for the diagnosis of GERD.

2. People whose predicted LES damage in the future will produce mild reflux. Reflux occurs but is so mild that symptoms are not troublesome and easily controlled with PPI therapy. LES damage will never progress to the severe state where there is a risk of visible CLE. This is about 70% of all people who develop symptoms of GERD. They are different than present GERD patients in that the test of LES damage has predicted they will not progress to severe LES damage that is associated with a high risk of visible CLE during their lifetime. They will not be at risk for adenocarcinoma. Progression that is maximally short of severe GERD also means that they will likely be well controlled with PPI therapy all their life.
3. Patients predicted by the new test of LES damage to progress to severe LES damage that places him/her at high risk of future treatment failure, visible CLE, and adenocarcinoma. The algorithm will predict when in the future LES damage will reach that critical point. Intervention to slow or stop LES progression by dietary modification, prokinetic drugs, or endoscopic and surgical methods will theoretically prevent the occurrence of treatment failure, CLE, and adenocarcinoma.

One of the reasons why there is a potential for success of this new proposed management is that it changes the cost structure of treating GERD. The prevention of treatment failure, Barrett esophagus, and adenocarcinoma will result in a huge savings in cost at the bad and ugly end of GERD. More importantly, the human cost of the misery of GERD that disrupts quality of life because drugs cannot control the symptoms adequately, the need for Barrett surveillance, and the suffering associated with the management of esophageal adenocarcinoma will disappear.

The significant added cost will be the upfront need for assessment of the LES for all patients with GERD at the onset of symptoms or before. This is essential to identify people who will progress and intervene in those people to prevent progression to LES damage to a severity that will result in treatment failure, Barrett esophagus, and adenocarcinoma.

The population of 70% of GERD patients who will remain well controlled with PPI therapy all their life can be maintained on drug therapy unless the evolution of methods to prevent progression of LES damage becomes more effective and decreases cost when compared with lifetime PPI therapy.

In addition to the above, the new understanding of the pathogenesis of the cellular events of GERD will provide numerous avenues of research. These are exciting and will stimulate new technologies. GERD research will move from its present rut of trying to create newer and more powerful methods of acid suppression to a new pipeline of technological advancement. Ultimately, these will have the effect of increasing cost-effectiveness of treating GERD. This contrasts with the ever-increasing cost structure at present that is inevitable because of increasing GERD, increasing Barrett esophagus, and increasing adenocarcinoma.

The future is replete with what can be in store for medical device companies in the future. I will list some of the future technologies that will have potential value in the future.

1. A biopsy device that can take a sample of mucosa from the endoscopic GEJ distally to a vertical length of 20–25 mm. This will be essential for accurate measurement of LES damage.
2. New devices that will remove the need of endoscopy to obtain the mucosal sample. A device that can be passed without sedation that will have the ability to visualize the squamocolumnar junction and take the required biopsy will decrease cost and expand the new LES assessment away from the endoscopy suite and into the doctor's office where it can be done safely and at very low cost.
3. New methods of addressing the LES damage. Companies have been trying to develop that magical endoscopic or surgical method to augment or repair the damaged LES for years. These efforts have not resulted in great success. Surgical (laparoscopic) fundoplication remains very expensive and has a need for surgical expertise that is not readily available and not completely successful. LINX and Endo-Stim require less expertise but still has the expense of laparoscopy. Endoscopic fundoplications of many kinds have come and gone. At the present time, Stretta and TIF (transoral incisionless fundoplication) exist in the arena but are not adequately effective.

   It is highly likely that the reason for lack of success is that all these procedures attempt to repair a severely damaged LES (one whose abdominal length is <10 mm).

   What will be required in the future is easier. The goal will be to slow or stop progression of LES damage to <10 mm, not repair an LES that is severely damaged. If the same technologies are used to prevent predicted progression in an LES that measures 20 mm, the task will be easier and very likely to be more effective. The present methods may suffice, but the enormous market created by the new assessment will likely stimulate innovation.

4. Like everyone else in medicine today, the dream of controlling the molecular progression of the disease is attractive. In the first GERD book that I authored with Tom DeMeester in 2006, on page 378, we wrote: "Our dream is that identification of a molecular mechanism that converts cardiac mucosa to oxyntocardiac mucosa will result in a drug that can induce this change. Once a drug is developed, all that will need to be done to prevent intestinal metaplasia will be to screen people for cardiac mucosa and remove them from the reflux-adenocarcinoma sequence by inducing oxyntocardiac mucosa. Or perhaps a drug can be developed to inhibit or reverse the molecular step involved in the conversion of cardiac mucosa to intestinal metaplasia. This will have a similar effect in preventing esophageal adenocarcinoma and appears much more feasible than gene therapy directed at esophageal dysplasia or cancer. There is precedent for such success: identification of the molecular mechanism associated with CD117 and tyrosine kinase receptors permitted the development of Gleevac, a drug directed at this molecular interaction that effectively controls the growth of $CD117^+$ gastrointestinal stromal tumors. Our hope is for a similar drug that can be used at an early metaplastic stage of reflux-induced esophageal carcinogenesis."

Since that time, billions of dollars of research have gone into increasing our knowledge into molecular mechanisms of all diseases including the reflux to adenocarcinoma sequence of GERD (see Chapter 15). There are possible avenues to produce molecules to control progression to adenocarcinoma. However, these address the disease in its evolution.

I now believe that the best way to attack GERD is to address its cause, which is LES damage. Attacking the root of a tree is always a more effective method of killing a tree than lopping off its branches. If reflux can be prevented before it takes hold by ensuring that LES damage is controlled, GERD and all its molecular changes will not occur. The best method of cure is prevention of all cellular and molecular changes by preventing reflux.

REFERENCES

1. Greenberger NJ. Update in gastroenterology. *Ann Intern Med* 1998;**129**:309–16.
2. Dent J, El-Serag HB, Wallander MA, et al. Epidemiology of gastro-oesophageal reflux disease: a systematic review. *Gut* 2005;**54**:710–7.
3. El-Serag HB, Sweet S, Winchester CC, et al. Update on the epidemiology of gastro-oesophageal reflux disease: a systematic review. *Gut* 2014;**63**:871–80.
4. Toghanian S, Wahlqvist P, Johnson DA, Bolge SC, Liljas B. The burden of disrupting gastro-esophageal disease; a database study in US and European cohorts. *Clin Drug Investig* 2010;**30**:167–78.
5. Brook RA, Wahlquist P, Kleinman NL, Wallander MA, Campbell SM, Smeeding JE. Cost of gastro-esophageal reflux disease to the employer: a perspective from the United States. *Aliment Pharmacol Ther* 2007;**26**:889–98.
6. Vakil N, van Zanten SV, Kahrilas P, et al. The Montreal definition and classification of gastroesophageal reflux disease: a global evidence-based consensus. *Am J Gastroenterol* 2006;**101**:1900–20.
7. American Gastroenterological Association Medical Position Statement on the management of gastroesophageal reflux disease. *Gastroenterology* 2008;**135**:1383–91.
8. Bytzer P, Jones R, Vakil N, et al. Limited ability of the proton-pump inhibitor test to identify patients with gastroesophageal reflux disease. *Clin Gastroenterol Hepatol* 2012;**10**:1360–6.
9. Finks JF, Wei Y, Birkmeyer JD. The rise and fall of antireflux surgery in the United States. *Surg Endosc* 2006;**20**:1698–701.
10. Katz PO, Castell DO, Chen Y, Andersson T, Sostek MB. Intragastric acid suppression and pharmacokinetics of twice-daily esomeprazole: a randomized, three-way crossover study. *Aliment Pharmacol Ther* 2004;**20**:399–406.
11. Blonski W, Vela MF, Castell DO. Comparison of reflux frequency during prolonged multichannel intraluminal impedance and pH monitoring on and off acid suppression therapy. *J Clin Gastroenterol* 2009;**43**:816–20.
12. Clarke AT, Wirz AA, Manning JJ, et al. Severe reflux disease is associated with an enlarged unbuffered proximal gastric acid pocket. *Gut* 2008;**57**:292–7.
13. Kinoshita Y, Hongo M, Japan TWICE Study Group. Efficacy of twice-daily rabeprazole for reflux esophagitis patients refractory to standard once-daily administration of PPI: the Japan-based TWICE study. *Am J Gastroenterol* 2012;**107**:522–30.
14. Allison PR. Peptic ulcer of the esophagus. *Thorax* 1948;**3**:20–42.
15. American Gastroenterological Association Institute technical review on the management of gastroesophageal reflux disease. *Gastroenterology* 2008;**135**:1392–413.
16. Sharma N, Agrawal A, Freeman J, Vela MF, Castell D. An analysis of persistent symptoms in acid-suppressed patients undergoing impedance-pH monitoring. *Clin Gastroenterol Hepatol* 2008;**6**:521–4.
17. Frazzoni M, Conigliaro R, Melotti G. Reflux parameters as modified by laparoscopic fundoplication in 40 patients with heartburn/regurgitation persisting despite PPI therapy: a study using impedance-pH monitoring. *Dig Dis Sci* 2011;**56**:1099–106.
18. Tobey NA, Carson JL, Alkeik RA, Orlando RC. Dilated intercellular spaces: a morphological feature of acid reflux-damaged human esophageal epithelium. *Gastroenterology* 1996;**111**:1200–5.
19. Caviglia R, Ribolsi M, Maggiano N, Gabbrielli AM, Emerenziani S, Guarino MP, Carotti S, Habib FI, Rabitti C, Cicala M. Dilated intercellular spaces of esophageal epithelium in nonerosive reflux disease patients with physiological esophageal acid exposure. *Am J Gastroenterol* 2005;**100**:543–8.

20. Villanacci V, Grigolato PG, Cestari R, et al. Dilated intercellular spaces as markers of reflux disease: histology, semiquantitative score and morphometry upon light microscopy. *Digestion* 2001;**64**:1–8.
21. Rodrigo S, Abboud G, Oh D, DeMeester SR, Hagen JA, Lipham J, DeMeester TR, Chandrasoma P. High intraepithelial counts in esophageal squamous epithelium are not specific for eosinophilic esophagitis in adults. *Am J Gastroenterol* 2008;**103**:435–42.
22. Tobey NA, Hosseini SS, Argore CM, Dobrucali AM, Awayda MS, Orlando RC. Dilated intercellular spaces and shunt permeability in non-erosive acid-damaged esophageal epithelium. *Am J Gastroenterol* 2004;**99**:13–22.
23. Rodrigo J, Hernandez CJ, Vidal MA, Pedrosa JA. Vegetative innervation of the esophagus. III. Intraepithelial endings. *Acta Anat* 1975;**92**:242–58.
24. Nason KS, Wichienkuer PP, Awais O, Schuchert MJ, Luketich JD, O'Rourke RW, Hunter JG, Morris CD, Jobe BA. Gastroesophageal reflux disease symptom severity, proton pump inhibitor use, and esophageal carcinogenesis. *Arch Surg* 2011;**146**:851–8.
25. Pohl H, Welch G. The role of overdiagnosis and reclassification in the marked increase of esophageal adenocarcinoma incidence. *J Natl Cancer Inst* 2005;**97**:142–6.
26. Pohl H, Sirovich B, Welch HG. Esophageal adenocarcinoma incidence: are we reaching the peak? *Cancer Epidemiol Biomarkers Prev* 2010;**19**:1468–70.
27. Barrett NR. The lower esophagus lined by columnar epithelium. *Surgery* 1957;**41**:881–94.
28. Lyall A. Chronic peptic ulcer of the oesophagus. A report of eight cases. *Br J Surg* 1937;**24**:534–47.
29. Barrett NR. Chronic peptic ulcer of the oesophagus and 'oesophagitis'. *Br J Surg* 1950;**38**:175–82.
30. Allison PR, Johnstone AS. The oesophagus lined with gastric mucous membrane. *Thorax* 1953;**8**:87–101.
31. Paull A, Trier JS, Dalton MD, Camp RC, Loeb P, Goyal RK. The histologic spectrum of Barrett's esophagus. *N Engl J Med* 1976;**295**:476–80.
32. Chandrasoma PT, Der R, Ma Y, Peters J, DeMeester T. Histologic classification of patients based on mapping biopsies of the gastroesophageal junction. *Am J Surg Pathol* 2003;**27**:929–36.
33. Spechler SJ, Zeroogian JM, Antonioli DA, Wang HH, Goyal RK. Prevalence of metaplasia at the gastroesophageal junction. *Lancet* 1994;**344**:1533–6.
34. Malfertheiner P, Nocon M, Vieth M, Stolte M, Jasperson D, Keolz HR, Labenz J, Leodolter A, Lind T, Richter K, Willich SN. Evolution of gastro-oesophageal reflux disease over 5 years under routine medical care – the ProGERD study. *Aliment Pharmacol Ther* 2012;**35**:154–64.
35. Morson BC, Belcher JR. Adenocarcinoma of the oesophagus and ectopic gastric mucosa. *Br J Cancer* 1952;**6**:127–30.
36. Spechler SJ. Barrett's esophagus. *N Engl J Med* 2002;**346**:836–42.
37. Spechler SJ. Barrett esophagus and risk of esophageal cancer: a clinical review. *JAMA* 2013;**310**:627–36.
38. Lagergren J, Bergstrom R, Lindgren A, Nyren O. Symptomatic gastroesophageal reflux as a risk factor for esophageal adenocarcinoma. *N Engl J Med* 1999;**340**:825–31.
39. Garcia-Rodriguez LA, Lagergren J, Lindblad M. Gastric acid suppression and risk of oesophageal and gastric adenocarcinoma: a nested case control study in the UK. *Gut* 2006;**55**:1538–44.
40. Duan L, Wu AH, Sullivan-Halley J, Bernstein L. Antacid drug use and risk of esophageal and gastric adenocarcinoma in Los Angeles County. *Cancer Epidemiol Biomarkers Prev* 2009;**18**:526–33.
41. Lada MJ, Nieman DR, Han M, Timratana P, Alsalahi O, Peyre CG, Jones CE, Watson TJ, Peters JH. Gastroesophageal reflux disease, proton-pump inhibitor use and Barrett's esophagus in esophageal adenocarcinoma: trends revisited. *Surgery* 2013;**154**:856–64.
42. Kastelein F, Spaander MCW, Steyerberg EW, Bierman K, Valkhoff VE, Kuipers EJ, Bruno MJ. Proton pump inhibitors reduce the risk of neoplastic progression in patients with Barrett's esophagus. *Clin Gastroenterol Hepatol* 2013;**11**:382–8.
43. Rex DK, Cummings OW, Shaw M, Cumings MD, Wong RKH, Vasudeva RS, Dunne D, Rahmani EY, Helper DJ. Screening for Barrett's esophagus in colonoscopy patients with and without heartburn. *Gastroenterology* 2003;**125**:1670–7.
44. Hvid-Jensen F, Pedersen L, Funch-Jensen P, Drewes AM. Proton pump inhibitor use may not prevent high grade dysplasia and oesophageal adenocarcinoma in Barrett's oesophagus. *Aliment Pharmacol Ther* 2014;**39**:984–91.
45. Hvid-Jensen F, Pederson L, Drewes AM, Sorensen HT, Funch-Jensen P. Incidence of adenocarcinoma among patients with Barrett's esophagus. *N Engl J Med* 2011;**365**:1375–83.

Chapter 3

Fetal and Postnatal Development of the Esophagus and Proximal Stomach

The foregut develops from the cephalic end of the primitive endodermal tube that is lined at the earliest stage of gestation by a primitive endodermal stratified columnar epithelium. In the fetal esophagus, this primitive epithelium undergoes transformation to several different types of columnar epithelia before it is replaced by stratified squamous epithelium (Table 3.1).

It is presently accepted (correctly) that the normal fully developed esophagus is entirely lined by stratified squamous epithelium. The presence of columnar-lined esophagus in the adult esophagus is universally accepted (correctly) as an acquired pathologic epithelium induced by exposure of esophageal squamous epithelium to gastric juice.

The combination of the fact that the columnar epithelia that are seen in the fetal esophagus are absent in the fully developed esophagus *must* mean that all the fetal columnar epithelia in the esophagus disappear and are replaced by stratified squamous epithelia (Fig. 3.1).

Therefore, when columnar epithelium is found in the esophagus in late gestation and early postnatal life, it represents residual fetal epithelium that is destined to be replaced by squamous epithelium at full maturation. It is well known that the point of birth does not necessarily coincide with the point where all fetal tissue has been replaced by adult tissue. In general, the likelihood of finding fetal epithelia at birth increases with premature birth occurring at earlier gestational ages.

There are many examples of fetal tissue being present at birth in other organs. Fetal nephroblastic cells dominate the fetal kidney. As development proceeds, this fetal nephroblast progressively decreases in amount even as the kidney increases in size, producing the nephrons composed of glomeruli and tubules that form the adult kidney. At birth and in the first 2 years of life, a rim of fetal nephroblastic cells may still be present in the subcapsular region of the kidney. These fetal nephroblastic cells decrease and finally disappear when renal development is complete. Similarly, fetal columnar epithelium in the esophagus disappears at full development, being completely replaced by squamous epithelium.

Extreme caution is necessary in fetal and perinatal studies to correctly define and name the epithelia that are seen. Fetal esophageal epithelia never contain parietal cells and, with rare exceptions, do not contain goblet cells. They consist entirely of mucous cells. If the definitions that we use for metaplastic esophageal cardiac epithelium are used for fetal columnar epithelia, the entire esophagus will be composed of "cardiac epithelium" from early fetal life. Such terminology must be avoided because it leads to confusion.

Fetal esophageal columnar epithelium must be designated as such (Table 3.1). Cardiac epithelium is a term that must be limited to describe an adult epithelium that results from GERD-induced metaplasia of squamous epithelium. Using adult epithelial definitions for fetal epithelia leads to erroneous conclusions.

Fetal epithelial development in the esophagus is completed at a variable time during the perinatal period, usually just before birth but not infrequently after birth in the first year of life.

In contrast, fetal epithelial development in the stomach is completed to its adult state at around 17 weeks of gestation. Unlike fetal development of esophageal epithelium, which resembles a confused traveler going in various directions until the correct path is decided on, gastric fetal epithelium is decisive in the direction of differentiation from the early stages of fetal life.

In general, it is safe to assume that a child over the age of 2 years has a fully developed esophageal and gastric epithelium and does not have any residual fetal epithelium. Before the age of 2, it is possible to have a confusing mixture of fetal and adult epithelial types (Fig. 3.1).

The fetus grows rapidly throughout its intrauterine life. With increasing age, tissue proliferation keeps pace with the rapid growth. In all developing organs, growth by cell proliferation is accompanied by rapid differentiation of all fetal tissues into adult tissues. The time during gestation when fetal tissues begin and end their differentiation into adult tissues

GERD. http://dx.doi.org/10.1016/B978-0-12-809855-4.00003-8

TABLE 3.1 Epithelial Types in the Fetal Esophagus and Stomach at Different Ages of Gestation

Type	Description	Location	Gestational Age
Fetal esophageal columnar epithelium, type 1	Stratified primitive columnar cells; three cell layers	Entire esophagus	First trimester
Fetal esophageal columnar epithelium, type 2	Stratified ciliated epithelium	Entire esophagus	Second and early third trimester
Fetal esophageal columnar epithelium, type 3	Nonstratified short or tall columnar epithelium	Upper and lower ends	Third trimester and early infancy
Fetal esophageal columnar epithelium, type 4	Intestinal type with goblet cells	Exceptionally seen	Only one case report
Squamous (fetal and adult)	Stratified squamous epithelium	Entire esophagus	22nd week to adult
Superficial (mucosal) esophageal glands	Forms in type 3 columnar epithelium	Entire esophagus	Third trimester; persist under squamous epithelium in adult
Deep (submucosal) esophageal glands	Forms in squamous lined esophagus	Entire esophagus	22nd week to adult
Fetal gastric columnar epithelium, type 1	Stratified primitive columnar cells; three cell layers	Entire proximal stomach	First trimester
Fetal gastric columnar epithelium, type 2 (fetal and adult)	Gastric oxyntic epithelium with parietal cells	Entire proximal stomach	13th week to adult life

The demarcation of the stomach from esophagus in fetal life is accurately defined by the angle of His.

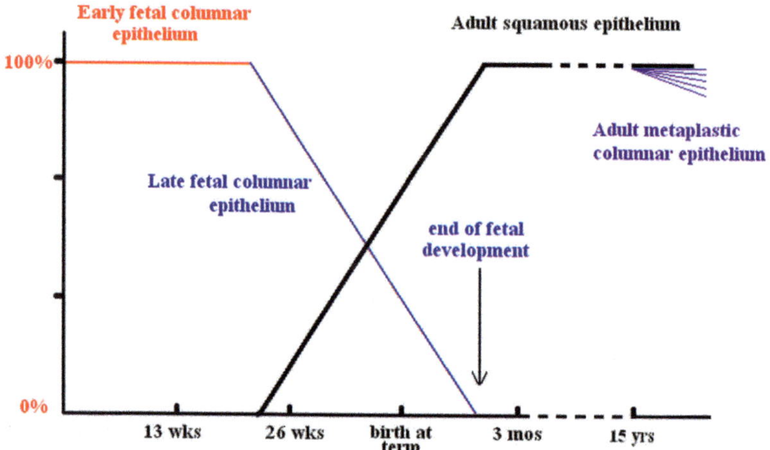

FIGURE 3.1 Diagrammatic representation of the epithelial composition of the esophagus from early fetal life to adulthood. Fetal columnar epithelia of different types line the entire esophagus till the 22nd week of gestation. At this time, stratified squamous epithelium appears in the midesophagus and rapidly spreads in both directions to progressively replace fetal columnar epithelium. At the end of development, which is in the perinatal period, the entire esophagus is lined by stratified squamous epithelium and all fetal columnar epithelia have disappeared. Metaplastic columnar epithelium (purple) commonly begins to develop in the young adult where it is associated with abdominal lower esophageal sphincter damage.

varies. In general though, it is a rule that is applicable to all organs that fetal tissues decrease and adult-type tissues increase in amount relative to gestational age and fetal growth.

One critical test when evaluating tissues in fetuses is: "Does the type of tissue in question decrease or increase in amount with increasing fetal age?" All fetal tissues decrease in extent; all adult tissues increase in extent relative to the growth of the fetus. This test is valuable in assessing data from the studies of fetal columnar epithelia in the esophagus, particularly in the gastroesophageal junctional (GEJ) region.

1. ANATOMIC DEVELOPMENT OF THE FETAL ESOPHAGUS AND STOMACH

The endodermal tube grows rapidly in length during the first trimester. During this time, the endoderm becomes surrounded by mesodermal elements that become organized to form an epithelium-lined tube surrounded by muscle. There is an intimate interrelationship between the maturation of mesenchymal and epithelial elements during fetal life.

In the first trimester, the developing endodermal tube undergoes rotations in its longitudinal axis. During this early gestational period, there is no demarcation between esophagus and stomach.

To assess the development of the fetal stomach, Hawass et al.[1] performed esophageal injections of contrast media into 162 spontaneously aborted normal fetuses from 7.5 to 26 weeks of gestation. The length of esophagus, the fetal trunk, greater and lesser curvature, and vertical and oblique axis of stomach were measured, and the means and standard deviations at each gestational age were calculated.

The demarcation of esophagus from the stomach results in the late first trimester from a much faster growth rate of the greater curvature of the stomach compared with the lesser curvature. This remarkable change occurs in the part of the endodermal tube immediately distal to the point of the future GEJ. It is likely the result of a genetically programmed selective increase in cell proliferation on the greater curvature of the stomach. This results in the outpouching of the proximal stomach on the side of the greater curvature leading to the formation of the gastric fundus and the creation of the acute angle of His at the GEJ.

The relative growth of the endodermal tube and the fetal trunk causes a change in the position of the future GEJ. This moves cephalad in the first trimester and by the 25th week settles into its final position. By this time, the proximal stomach has moved sharply upward from the angle of His at the GEJ to form the gastric fundus, which comes into close apposition with the undersurface of the developing diaphragm. The developing diaphragm separates the esophagus into a long upper thoracic segment and a shorter abdominal segment.

There is no information in the literature regarding the fetal development of the lower esophageal sphincter. This is not unexpected; the LES can only be defined by manometry, and there is yet no method to perform fetal manometry.

2. FETAL DEVELOPMENT OF THE ESOPHAGEAL EPITHELIUM

The epithelium of the entire gastrointestinal tract develops from the endodermal layer of the embryonic disk. As the fetus grows, the endoderm develops a lumen and becomes a tube lined by undifferentiated endodermal stem cells.

The esophagus and stomach develop in the cephalad region of the endodermal tube (the foregut), immediately caudal to the point of origin of the lung bud at the distal end of the developing pharynx. As the cephalic curve of the developing fetus develops, the fetal foregut increases in length rapidly. By the 25th week of gestation, the esophagus can be demarcated from the stomach by the formation of the gastric fundus and the anatomic angle of His at the GEJ.[1] The angle of His is an important landmark. Because of the absence of reflux-induced damage to the esophagus in fetal life, the angle of His is an accurate marker for the fetal and neonatal GEJ.

The rapid growth of the esophagus and stomach continues throughout fetal life into childhood during which both organs increase in size to keep pace with the growth of the baby. The epithelia also grow to keep pace with the increasing surface area of the esophagus and stomach. Along with the rapid growth, the epithelium simultaneously differentiates toward its final adult state at a varying rate in the esophagus and stomach. Once this final adult state is reached, further development consists only of growth.

Material for the study of fetal epithelium is difficult to access, particularly in the highly restricted environment of today. I was given the opportunity to evaluate the material that formed the fetal and perinatal autopsy study from Children's Hospital of Los Angeles.[2] This has been my only direct experience relating to the fetal development of the esophagus and stomach. The information presented here is largely derived from a critical study of the literature.

Fortunately, an extremely detailed study of fetal development of the esophagus by Johns in 1952[3] provides an authoritative and credible information base. Johns studied human embryos ranging in crown–rump length from 3 to 230 mm. In 14 embryos under 42 mm length, the whole embryo was sectioned transversely at 10 mm. In those of greater length, the esophagus was dissected with its attachment to the pharynx above and stomach below and then serially sectioned at 10 mm. Johns also examined serial transverse sections of the esophagus in 4 patients aged 2, 11, 47, and 68 years and every 10th section of a transverse series in 20 other subjects aged 20–60 years.

Johns describes the epithelium in terms or type, length, and thickness for each fetal stage.[3] One should keep in mind that the fetus and esophagus are growing rapidly during this time. Reported measurements must be taken in this context. If the length of a particular epithelial type has remained constant between two stages, it means that the percentage length of this epithelium has decreased in relation to the size of the fetus.

I use a terminology for the fetal epithelia that is purely descriptive and uniquely identifies fetal columnar epithelia, distinguishing them from terms used to describe adult epithelia of the esophagus and stomach (Table 3.1). In general, all fetal columnar epithelia in the esophagus (types 1–4) and the early type 1 epithelium in the stomach disappear when differentiation of the epithelium is complete. These fetal epithelia decrease progressively relative to fetal size throughout gestation. The adult epithelia, stratified squamous in the esophagus, and gastric oxyntic in the stomach appear in the 22nd and 13th week of gestation, respectively. These adult epithelia progressively increase in extent relative to fetal size as they replace fetal epithelia (Fig. 3.1). When fetal development is completed, the entire esophagus is lined by stratified squamous epithelium and the entire stomach is lined by gastric oxyntic epithelium. Once development has completed, the epithelium simply grows to keep pace with the growing child.

Johns' descriptions of the esophageal epithelium at various stages, expressed as size of fetus with gestational age, summarized here, are as follows[3]:

1. *Between 3 and 13 mm (25 days to 6 weeks old)*: The esophageal epithelial lining is stratified columnar in type with a uniform thickness of three cells deep. Oval nuclei are seen throughout the thickness of the epithelium. This is a nonciliated epithelium that is presumably composed of undifferentiated fetal endodermal stem cells.
2. *13–16 mm embryo (6–7 weeks old)*: Vacuolation begins in the cells. This results in spaces in the epithelium that push aside the lumen. This vacuolation increases till the 25 mm embryo stage and decrease in size but can be seen till the 72 mm stage.
3. *23–130 mm embryo (late first to end of second trimester)*: Starting at the 40 mm stage, the superficial layer of columnar cells becomes ciliated. This begins in the midesophagus and extends in both directions. By the 62 mm stage, stratified ciliated columnar epithelium has replaced the primitive columnar epithelium except for a small area at both ends, in which there is a nonciliated epithelium lined by a single layer of columnar cells. Around the 78 mm stage, the nonciliated epithelia at the lower end of the esophagus increase in height and their nuclei assume a basal position. The 130 mm embryo shows the first appearance of an esophageal gland as an outpouching of this tall columnar epithelium. These outpouchings reach the muscularis mucosae. In the late stages of the second trimester, stratified squamous epithelium appears in the midesophagus in patches replacing the ciliated columnar epithelium.
4. *130–230 mm (third trimester)*: Stratified squamous epithelium, which appears first in the midesophagus, progressively replaces the ciliated columnar epithelium. The tall columnar epithelium at the lower end has decreased in length. More mucosal outpouchings have formed from the columnar epithelium and developed into mucous glands in the mucosa. These persist as mucosal mucous glands into adult life. These do not penetrate the muscularis mucosae. The ciliated epithelium progressively disappears except for small patches that may remain until the mid third trimester.
5. *Full term and age 4–60 years*: The lining is entirely stratified squamous. Small tubular outpouchings develop in the squamous-lined esophagus. These extend through the muscularis mucosae to form secretory submucosal glands that contain mucous and serous cells.

Johns' study shows that the esophageal epithelium is somewhat confused as it develops in the fetus, with multiple different types of columnar epithelium appearing at different stages, to be replaced in the third trimester by stratified squamous epithelium. It almost seems that the esophagus develops along the line of the respiratory tract, producing a stratified ciliated epithelium resembling tracheal epithelium, before deciding to form the adult squamous epithelium beginning in the 22nd week.

It also shows that superficial mucous glands (mucosal, above the muscularis mucosae) arise in the early third trimester from the tall columnar epithelium and the deep submucosal glands arise in the perinatal period from stratified squamous epithelium. It is important to note that parietal cells are not seen in any esophageal glands during fetal life.

Not recorded in Johns' paper is the very rare occurrence of goblet cells during fetal development of the esophageal epithelium. This has been recorded in one fetus at 23 weeks.[4] I have also seen photographic material from fetal esophagi collected by Dr. Dorothea Liebermann-Moffett, a friend and surgeon researcher in Germany. These pictures showed the presence of goblet cells in fetal esophageal epithelium.

This does not mean that Barrett esophagus occurs in the fetus. This finding only indicates that there is a genetic signal that is rarely expressed in the fetus that induces goblet cell expression in the esophageal stem cell. This is not expressed in most fetuses and suppressed effectively later in fetal life. Intestinal metaplasia is extremely rare in the pediatric age group. Fetal intestinal metaplasia is not Barrett esophagus. Barrett esophagus is a pathologic state where goblet cells arise in metaplastic GERD-induced cardiac epithelium in postnatal life.

Unfortunately, Johns' study does not address epithelial development in the neonatal period. His study of esophageal epithelium ends at full term where nearly the entire esophagus is lined by stratified squamous epithelium except for a small area of residual tall columnar epithelium at either end.

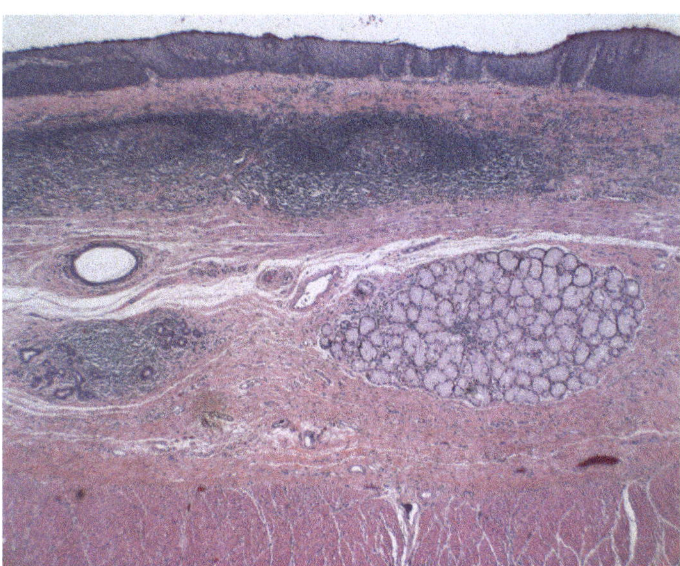

FIGURE 3.2 Adult esophagus lined by stratified squamous epithelium with 2 submucosal glands and a dilated submucosal gland duct cut in cross section. Submucosal glands are specific to the esophagus and are not seen in the stomach. They arise in late fetal life as outpouchings from the squamous epithelium in the esophagus. They do not arise from columnar epithelia.

In the adult state, the entire esophagus is lined by squamous epithelium and contains mucous glands in both the mucosa and the submucosa. The submucosal glands drain to the surface of the squamous epithelium by gland ducts that traverse the superficial submucosa, muscularis mucosae, lamina propria, and the squamous epithelium to open at their surface (Fig. 3.2). All fetal columnar epithelium disappears. The esophageal squamous epithelium extends to the end of the esophagus where it transitions into gastric oxyntic epithelium.

Johns' study does not describe the GEJ region in detail. Studies of the fetal development of the GEJ region will be reviewed below.

3. FETAL DEVELOPMENT OF THE STOMACH AND ITS EPITHELIUM

In contrast to the esophagus, which simply elongates as a tubal structure, the stomach undergoes a dramatic change in the first trimester. There is both clockwise rotation and rapid selective growth in the greater curvature compared to the lesser curvature.[1] This causes the fundus of the stomach to develop as an outpouching from the greater curvature. This begins at the future GEJ from which the gastric fundus moves sharply upward and to the left, forming the angle of His on the side of the greater curvature. The angle of His accurately defines the fetal GEJ. The lesser curvature of the stomach simply passes straight down and to the right toward the antrum.

Unlike the esophagus, fetal development of gastric epithelium is in one direction from the outset. The many studies that provide detail, such as the classical and most detailed by Salenius in 1962,[5] are consistent in the information they provide. I have chosen to review the paper by Menard and Arsenault in 1990[6] because it combines histology combined with cell kinetic studies.

Menard and Arsenault[6] studied 22 fetuses from legal abortions. The body of the stomach was dissected with the pyloric and cardiac ends removed. This study, like most others, does not provide information regarding the GEJ region. Longitudinal strips of gastric wall were cultured in vitro. Radioautography was done to detect radioactive labeling of the cells after the explant was exposed for 2 h to tritiated thymidine. The following observations were made:

1. *8–10 weeks of gestation*: The gastric epithelium is a flat stratified columnar epithelium identical to that seen in the esophagus at this stage. Its thickness averaged 0.029 mm, and cells labeling with tritiated thymidine were abundant (9% of total cell number) and scattered throughout the thickness of the epithelium.
2. *11–17 weeks of gestation*: The stratified columnar epithelium showed the first evidence of invagination into a foveolar pit during this period. Differentiated parietal (oxyntic) cells appeared almost immediately in the foveolar pit. From their first appearance at 13 weeks, the parietal cells progressively increased in number in the deep region of the pit.

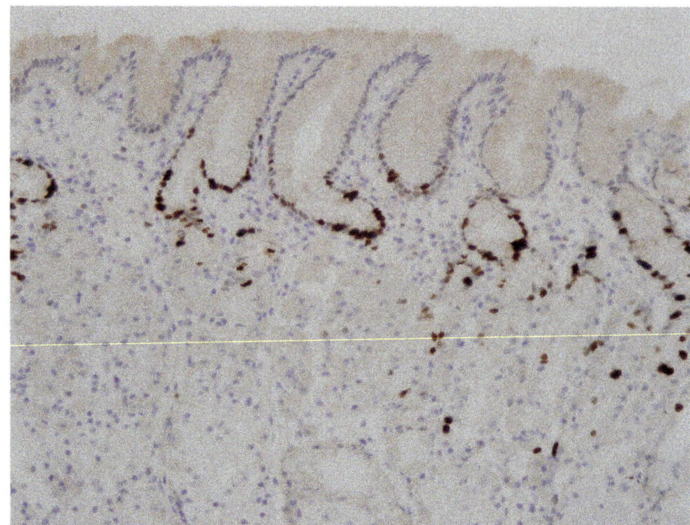

FIGURE 3.3 Immunohistochemical stain for Ki67 showing proliferative zone in adult gastric oxyntic epithelium in the deep foveolar region and isthmus of the gland. The pattern is identical to that established in the fetal stomach from the angle of His distally in the 17th week of gestation.

The stratified columnar epithelium was progressively replaced at the surface by a simple columnar mucous epithelium with basal nuclei. The tritiated thymidine labeling moved away from the surface to the cells lining the foveolar pit. The length of the pit-gland complex invaginating from the surface nearly doubled from 0.050 mm at 11 weeks to 0.097 mm at 17 weeks, typical of a developing adult-type epithelium. At 17 weeks, the thymidine labeling was concentrated in the deep foveolar region and the gland neck region with the surface and parietal cell–containing glandular area deep to the foveolar pit being negative.

Menard and Arsenault[6] commented that the developing stomach epithelium acquired the typical adult proliferative pattern at the beginning of the second trimester. The further development of the stomach during fetal life is represented by growth without change in either the differentiation or proliferation characteristics seen at 17 weeks of gestation. The gastric epithelium is thin at full term and increases rapidly in thickness after birth. This increased thickness is due almost entirely to an increase in the length of parietal cell–containing glands below the relatively short foveolar pit. In the adult gastric oxyntic epithelium, the proliferative zone is exactly the same as it was in the 17-week fetus, in the deep foveolar and isthmus regions. The surface, superficial foveolar region and the long tubular gland below the isthmus are nonproliferative (Fig. 3.3).

All studies of the fetal development of the stomach describe only the transformation of the primitive endodermal tube columnar epithelium into gastric oxyntic epithelium with parietal cells appearing as the first sign of differentiation. No study has shown the transformation of the primitive columnar epithelium to anything resembling cardiac epithelium. This could be either because the primitive columnar epithelium in the stomach does not differentiate into cardiac epithelium or because studies have not been done on the correct area. Menard and Arsenault's study specifically excluded the most proximal region of the stomach.[6]

However, De Hertogh et al.[7] showed that gastric epithelium at the angle of His contained parietal cells (see below). This appears to prove that cardiac epithelium does not develop in fetal life in the proximal stomach.

4. FETAL DEVELOPMENT OF THE GASTROESOPHAGEAL JUNCTIONAL REGION

In contrast to the esophagus and stomach, the literature was silent on the subject of fetal development of the epithelium in the region of the GEJ until the beginning of this millennium. The controversy that our group raised in the 1990s about whether or not cardiac epithelium in the GEJ region was normal or abnormal[8–10] stimulated several studies of fetal development of the GEJ region around 2003.[2,7,11]

The first two autopsy studies of GEJ histology were published in 2000.[12,13] Both of these had significant numbers of young children. These produced significant new data regarding the epithelial composition of the GEJ region. At that time, 2–3 cm of cardiac epithelium was believed to be normally present in the proximal stomach (the "gastric cardia"). Both autopsy studies showed that cardiac epithelial length rarely exceeded 0.5 cm.[12,13] In Chandrasoma et al.,[12] the minimum length of cardiac epithelium was zero. In Kilgore et al.,[13] the minimum length of cardiac epithelium was 0.1 cm. All studies done on people without GERD have shown that the cardiac epithelial length in this population is largely <1.0 cm.

These autopsy studies created a controversy. Chandrasoma et al.[12] concluded from the data that cardiac mucosa was normally absent from the junction. Almost at the same time, Kilgore et al. from the Cleveland Clinic[13] concluded from the data in their study that cardiac epithelium was always present as a normal epithelium in the proximal stomach to an extent that was 0.1–0.4 cm.

At face value, these data strongly suggest that cardiac epithelium does not normally exist at the GEJ. It should be difficult to accept that 0.1 cm of cardiac epithelium represents a normal structure. This can only happen because of the powerful dogma of many decades that dictates that cardiac epithelium is present normally in the proximal stomach.

Many authorities undertook studies based on the hypothesis that the presence or absence of cardiac epithelium at birth would settle this controversy. It was postulated that if cardiac epithelium was present in late fetal life and at birth, it must mean that cardiac epithelium is a normal epithelium. If it was an abnormal epithelium resulting from columnar metaplasia of esophageal squamous epithelium induced by GERD, it should be absent in late fetal life and birth. It is reasonable to assume that damage caused by reflux does not occur in fetal life.

However, this argument is not valid because columnar epithelium is a normal fetal epithelium in the esophagus.[3] If the differentiation of fetal esophageal columnar epithelium to squamous epithelium is not completed until after birth, a residual amount of fetal columnar epithelium will be present in late pregnancy and at birth. Error can arise if this fetal columnar epithelium is confused with cardiac epithelium, particularly because fetal columnar epithelium in the esophagus consists entirely of mucous cells. While the two are different in histologic appearance, they both consist only of mucous cells.

Three studies on the histology of the perinatal GEJ were published in 2003, one from Children's Hospital of Los Angeles,[2] one from Belgium,[7] and one from South Korea.[11]

The findings in these three studies[2,7,11] are similar, but the terminology and conclusions are very different. When the studies of fetal development in the region of the GEJ region are reviewed, two questions need to be addressed:

1. How does the proximal limit of gastric oxyntic epithelium relate to the GEJ (accurately defined by the angle of His in the fetus and neonate)? If gastric oxyntic epithelium with parietal cells is found at the GEJ, it must mean that cardiac epithelium is not present at birth in the proximal stomach.
2. If columnar epithelium without parietal cells is present between gastric oxyntic epithelium and squamous epithelium, how does the distal end of this epithelium relate to the GEJ? If this epithelium is proximal to the angle of His (the GEJ), it must be esophageal, not gastric.

Only the finding of a columnar epithelium without parietal cells distal to the GEJ during fetal life will mean that it is even possible that cardiac epithelium is normally present in the proximal stomach. No study has ever shown this to be the case. When present, nonoxyntic columnar epithelia have always been in the esophagus proximal to the GEJ. Any columnar epithelium found in the esophagus must be a fetal epithelium because the fully developed esophagus does not normally have columnar epithelium.

A second important consideration in evaluating this question is whether the length of columnar epithelium between gastric oxyntic epithelium with parietal cells and squamous epithelium decreases or increases with increasing fetal age. If the columnar epithelium decreases in length, it is likely to be a fetal columnar epithelium being replaced progressively by squamous epithelium (Fig. 3.1).

Of the three papers, only that of De Hertogh[7] relates the epithelial types seen in this region to the GEJ defined by the angle of His. This study therefore provides extremely useful data that permit definition of the location of the different epithelia as distal esophageal or proximal gastric by virtue of their relationship to the angle of His.

This study consists of 24 fetuses ranging from 13 to 23 weeks of gestation, 6 cases from 24 to 41 weeks, and a 7-month-old infant. In the 24 cases of smaller fetuses, the entire esophagus and stomach were sectioned longitudinally. The larger specimens were embedded top-down and sectioned transversely. The number of sections studied varied between 10 (small specimens) and 40 (large specimens).

The authors used the following anatomic definitions:

1. The GEJ is defined as the angle of His (Fig. 3.4). Anything above it is the esophagus and anything below it is the stomach;
2. The length of the abdominal esophagus is the distance from the lower rim of the diaphragm to the angle of His;
3. The length of what De Hertogh calls "columnar-lined esophagus" is the distance from the squamocolumnar junction (SCJ) to the angle of His.

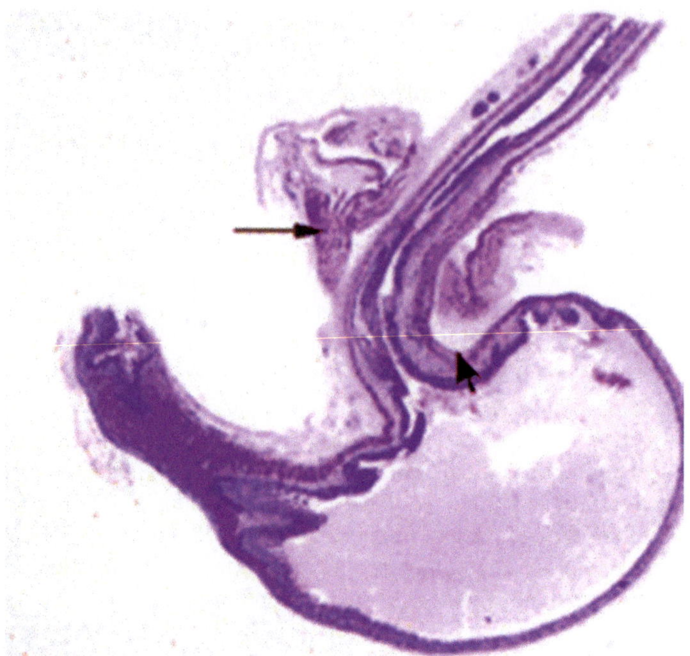

FIGURE 3.4 Autopsy specimen in a fetus showing the esophagus and stomach with the diaphragm in place (*arrow*). The angle of His (*arrowhead*) marks the gastroesophageal junction (GEJ). The area between the two is the abdominal esophagus. The data in the study showed that the epithelium at and distal to the GEJ contained parietal cells from early gestation. Fetal columnar epithelium type 3 occupied the esophagus above the GEJ. This progressively decreased in length with increasing fetal age, indicative of a fetal epithelium. *Reproduced with permission from De Hertogh G, Van Eyken P, Ectors N, Tack J, Geboes K. On the existence and location of cardiac mucosa; an autopsy study in embryos, fetuses, and infants.* Gut *2003;***52***:791–6.*

4. The "cardia" is explained by De Hertogh et al. as follows: "Anatomists have applied the term 'cardia' to that part of the stomach which lies around the orifice of the tubular esophagus. There is no anatomical landmark for the distal margin of the so-defined cardia. Its proximal margin is the gastroesophageal junction which, according to anatomists, is localized at the level of the angle of His." This is accurate.

The use of the angle of His for defining the GEJ is accurate in this fetal population. In adults, the location of the angle of His is altered by GERD-induced anatomic changes and becomes unreliable as a means of defining the GE junction.

The histologic definitions used in this paper are critical to understanding the authors' findings and conclusions. It is unfortunate that they do not cite Johns' paper[3] or use his terminology and definitions of the various epithelia. De Hertogh et al.[7] classify columnar epithelia in the GEJ region into the following types:

1. Primitive esophageal mucosa, consisting of a stratified columnar epithelium with ciliated cells. This corresponds to what I call fetal esophageal columnar epithelium, type 2.
2. Primitive stomach mucosa (PSM), defined as a layer of columnar epithelial cells with no glandular structures. This epithelium was limited to two fetuses of the lowest gestational age (13 and 14 weeks). The fact that this epithelium is not seen in the older fetuses makes it likely that it is the primitive endodermal tube epithelium before differentiation begins (what I call fetal esophageal columnar epithelium, type 1).
3. Cardiac mucosa, which is defined as a lining composed of foveolar and surface epithelium overlying glandular structures containing no parietal cells. The authors define the distal end of cardiac mucosa by the first parietal cell that is encountered. (In my classification of fetal epithelial types, this would correspond to fetal esophageal columnar epithelium, type 3.)

In their results, they state: "Cardiac mucosa (CM) was present at a gestational age of 13 weeks. The mucosal surface and pits were lined by a single layer of tall columnar cells with elongated basally situated nuclei. The apical cytoplasm was filled with mucus, which was lightly eosinophilic in H&E stained sections.... CM was interposed between PSM proximally and mucosa containing parietal cells distally... PSM was present at 13 and 14 weeks only."

De Hertogh et al. illustrate "cardiac mucosa" in a 13-week fetus that shows a nonciliated columnar epithelium with a rudimentary outpouching or gland.

In a 24-week fetus, the "cardiac mucosa" is between stratified squamous epithelium proximally and parietal cell–containing gastric mucosa distally. It is composed of a single undulating layer of tall columnar epithelial cells with basal nuclei and abundant apical mucin.

In the 1-week-old neonate, what De Hertogh calls "cardiac mucosa" is represented by a small mucous gland under the squamous epithelium and a single ductlike structure immediately distal to the squamous epithelium. The stratified squamous epithelium and parietal cell–containing gastric epithelium are within one high power field. In comparison to the 24-week fetus, the "cardiac mucosa" separating squamous from gastric mucosa in the 1-week infant is much smaller in extent.

There is great similarity of the epithelial types described by De Hertogh[10] when compared to the illustrations of Johns.[3] The appearance of "cardiac mucosa" described by De Hertogh et al. in the 13-week fetus is similar to the stratified columnar epithelium described by Johns[3] in the first trimester. The appearance of "cardiac mucosa" described by De Hertogh et al.[7] in the 24-week fetus is similar to the tall columnar epithelium described by Johns[3] in the second and third trimester. The definition of the proximal limit of gastric mucosa by the first parietal cell is accurate; Johns[3] did not find parietal cells in any fetal esophageal epithelium.

The study by De Hertogh et al.[7] showed the relationship of the epithelial types to the GEJ, which was defined as the angle of His. What they call "cardiac mucosa" was located *proximal* to the angle of His, i.e., in the distal esophagus. It was not present in what they defined as the gastric cardia, which was distal to the GEJ. The epithelium at the angle of His in De Hertogh's[7] study was parietal cell–containing gastric mucosa in all their cases. Contrary to the conclusion in their study, the data prove that what they call "cardiac mucosa" is found in the distal esophagus in fetal life. It is not found at or distal to the GEJ (angle of His).

De Hertogh et al.[7] concluded: "If cardiac mucosa develops during pregnancy, it is a normal structure at birth." There are many problems with De Hertogh's conclusion that cardiac mucosa is a normal adult epithelium because it is found in fetal and neonatal life:

1. "Cardiac mucosa" described by De Hertogh et al.[7] is a fetal columnar epithelium (type 3) that is located in the esophagus, not in the stomach. Cardiac epithelium is not normally present in the esophagus in the adult; it results from metaplasia or esophageal squamous epithelium induced by GERD. The presence of cardiac epithelium in the fetal esophagus does not prove that the proximal stomach in the adult is normally lined by cardiac epithelium.

TABLE 3.2 Length (mm; Mean Values for All Ages) of the Abdominal Esophagus (AO), Length of the Columnar-Lined Esophagus (CLO), Length of Cardiac Mucosa (CM), Distance From the CM to the Angle of His (CMH), and Length of Primitive Stomach Mucosa (PSM), as a Function of Age

Gestational Age	No of Cases	AO	CLO	CM	CMH	PSM
13 weeks	2	1.2	0.8	0.2	0.3	0.3
14 weeks	1	0.7	0.9	0.5	0.1	0.3
15 weeks	2	1.6	1.7	1.0	0.7	–
16 weeks	4	1.6	1.9	1.2	0.7	–
17 weeks	1	2.5	2.7	0.9	1.8	–
18 weeks	1	1.7	1.0	1.0	0.0	–
19 weeks	1	2.2	1.8	0.8	1.0	–
21 weeks	3	2.4	2.5	0.8	1.7	–
22 weeks	1	3.3	2.8	0.4	2.4	–
23 weeks	1	0.3	0.3	0.4	0.0	–
24 weeks	2	1.8	1.4	0.4	1.0	–
41 weeks	1	0.6	0.3	0.3	1.3	–
7 months	1	4.9	0.9	0.6	0.3	–

Reproduced from De Hertogh G, Van Eyken P, Ectors N, Tack J, Geboes K. On the existence and location of cardiac mucosa; an autopsy study in embryos, fetuses, and infants. *Gut* 2003;**52**:791–6, Table 2.

2. The distinction between a fetal and adult structure is that a fetal structure is limited to fetal life and decreases to full term and disappears, whereas an adult structure increases from its origin in fetal life to adulthood. The ciliated columnar epithelium is a fetal esophageal epithelium because it is seen in the middle of fetal life and disappears before birth, being replaced by squamous epithelium. Squamous epithelium is an adult epithelium because it arises in fetal life and increases as the fetus matures to the adult state. "Cardiac mucosa" described by De Hertogh et al.[7] is a fetal epithelium. It reaches a maximum length of 1.2 mm at 16 weeks (Table 3.2), is 0.3 mm at 41 weeks, and has virtually disappeared in the 1-week-old neonate.
3. De Hertogh's data prove that what they call "cardiac mucosa" is not present at or distal to the GEJ. The mucosa at the fetal GEJ contains parietal cells, which are found in the fetal stomach from 17 weeks and never found in the fetal esophagus.
4. In Table 3.2, the rapid growth in length of the abdominal esophagus, from 0.6 mm at 41 weeks gestation to 4.9 mm at 7 months, is shown. The epithelial lining of this abdominal esophagus is largely squamous epithelium; the fetal columnar epithelium found between the SCJ and the angle of His (where the first parietal cell is found) is 0.9 mm.

Data by De Hertogh et al. are precisely developed by excellent and careful methods and are accurate. Unfortunately, the data contradict the authors' conclusion that cardiac mucosa is a normal structure at birth because it develops during pregnancy.

The correct conclusions from their data are (1) that there is no cardiac epithelium in the proximal stomach at any time during fetal life, and therefore cardiac epithelium does not exist as a normal structure in the adult proximal stomach and (2) that the columnar epithelium they call "cardiac mucosa" is actually a fetal columnar epithelium (type 3) that progressively decreases with increasing fetal age as it is replaced by adult squamous epithelium.

Accurate terminology and definition is essential. The incorrect conclusion reached by De Hertogh et al.[7] is caused by their use of the incorrect term "cardiac mucosa" to a fetal columnar epithelium that is located in the distal esophagus. If the use of the term "cardiac mucosa" in their paper is replaced by "fetal columnar epithelium type 3," the data lead to the conclusion that there is no cardiac mucosa in the proximal stomach.

The most persuasive finding that shows why "cardiac mucosa" described by De Hertogh et al. is esophageal fetal epithelium, type 3 is its histologic appearance. This is an undulating single-layered columnar epithelium with basal nuclei and apical mucinous cytoplasm that sits on the muscularis mucosae. There are no glands and no inflammatory cells in the lamina propria. The appearance is constant in all the published papers on the subject.[2,3,7,11] It bears no resemblance to metaplastic cardiac epithelium seen in adults with GERD.

The study by Park et al.[11] from Seoul National University coincided in time with De Hertogh's[7] study. It was a mixed retrospective and prospective study of the GEJ region in 15 prenatal autopsies ranging from 18 to 34 weeks and eight postnatal cases from 9 days to 15 years. The authors made no attempt to define the GEJ. They simply measured the distance from the SCJ to the first parietal cell. This is a precise measurement with no possibility of error.

The method by Park et al. measures the gap from a point proximal that is certainly esophageal (squamous epithelium is not seen in the human stomach at any stage) and a point distal that must be stomach (parietal cells do not occur in the fetal esophagus). The measured area must therefore include the GEJ, although the exact location of the GEJ cannot be defined.

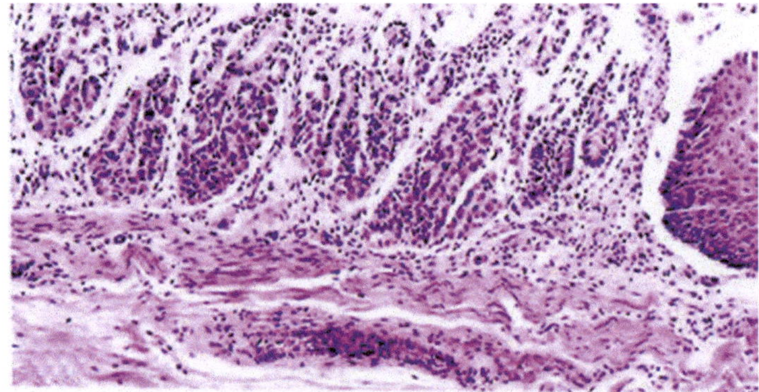

FIGURE 3.5 Sections across the fetal squamocolumnar junction showing a direct transition from squamous epithelium to parietal cell–containing fetal gastric oxyntic epithelium. There is no cardiac epithelium. *Reproduced with permission from Park YS, Park HJ, Kang GH, Kim CJ, Chi JG. Histology of gastroesophageal junction in fetal and pediatric autopsy.* Arch Pathol Lab Med *2003;**127**:451–5.*

Five patients (three prenatal cases 19, 23, and 27 weeks old and two postnatal cases 1 month and 5 years old) had a direct transition from squamous epithelium to parietal cell–containing gastric mucosa (Fig. 3.5). Their illustration of the junction in the 5-year-old child convincingly demonstrates absence of any cardiac epithelium between gastric oxyntic mucosa and squamous epithelium.

In the other 18 patients, Park et al. reported that there was a "transitional zone" between the squamous epithelium and parietal cell–containing gastric mucosa. This transitional epithelium was composed of an undulating tall columnar epithelium with basal cells and abundant apical mucin lining the surface and foveolar region. This is the same epithelium that Johns[3] describes as "tall columnar epithelium" and De Hertogh[7] as "cardiac mucosa." It is what I have called fetal esophageal columnar epithelium, type 3.

The only epithelia found in the distal esophagus and proximal stomach in the late fetal and early postnatal period are (1) stratified squamous epithelium (up to the SCJ), (2) fetal esophageal columnar epithelium of tall columnar type (between the SCJ and the first parietal cell at the GEJ), and (3) parietal cell–containing gastric oxyntic mucosa distal to the GEJ.

Demonstration by Park et al.[11] that the "transitional" columnar epithelium between the SCJ and the first parietal cell is not found in some of the older fetuses and children proves that this is fetal esophageal columnar epithelium. This is the only explanation for its complete disappearance in the perinatal period. Fetal columnar epithelium in the esophagus has been replaced by stratified squamous epithelium.

The third study by Derdoy et al.[2] was also published in 2003, coincidentally with the other two. I was given all the microscopic material for my examination and opinion prior to publication. I was acknowledged in the paper but was not an author because the authors of the paper did not agree with my conclusions, which were conveyed at a meeting prior to submission of the paper for publication. That is certainly their prerogative.

The study by Derdoy et al.[2] was a retrospective analysis of one longitudinal section taken routinely from the SCJ at autopsies performed at Children's Hospital of Los Angeles in fetuses from gestational age 26 weeks to birth and children up to 18 years. These data are from a different population than the other fetal studies. The patients studied were live-born premature babies and children, not aborted fetuses. They frequently had significant diseases that may have required nasogastric intubation. External influences affecting normal development are more likely.

As published, the data were confusing and will not be included here. Their conclusions were similar to those of De Hertogh et al.,[7] suggesting that cardiac mucosa was normally present at birth.

The study, however, gave me a unique opportunity to examine fetal esophagi. One patient (the youngest in the study, 26 weeks old) had ciliated columnar epithelium (fetal esophageal columnar epithelium, type 2). One child had a direct transition from squamous epithelium to parietal cell–containing epithelium without intervening columnar epithelium that did not have parietal cells.

Many of the neonates had an undulating tall columnar epithelium consisting of uniform tall columnar cells with basal nuclei and abundant apical mucinous cytoplasm (Figs. 3.6–3.8). This was located between the distal limit of squamous epithelium and parietal cell–containing gastric oxyntic epithelium. The transitional epithelium (fetal esophageal columnar epithelium, type 3) did not have any glands or parietal cells. There were no inflammatory cells in the lamina propria. This

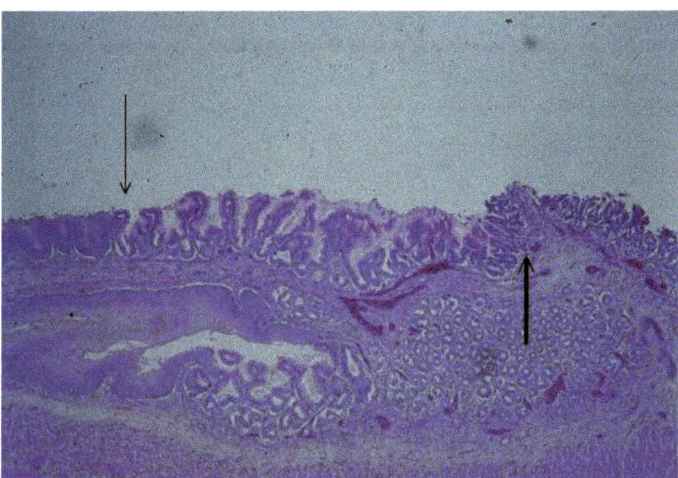

FIGURE 3.6 Tall columnar fetal epithelium in the distal esophagus of a third-trimester fetus between stratified squamous epithelium of the esophagus to the left (*thin arrow*) and parietal cell–containing gastric oxyntic epithelium to the right (*thick arrow*).

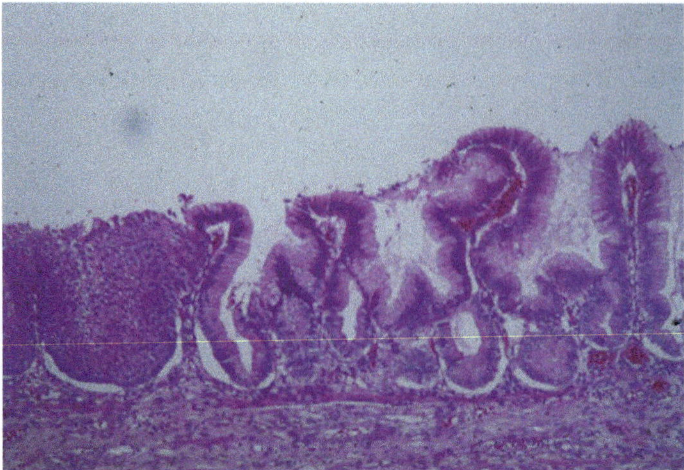

FIGURE 3.7 The transition between squamous epithelium and type 3 fetal columnar epithelium in the distal esophagus. Note the undulating single layer of columnar cells with basal nuclei and apical mucinous cytoplasm sitting on the thin muscularis mucosae.

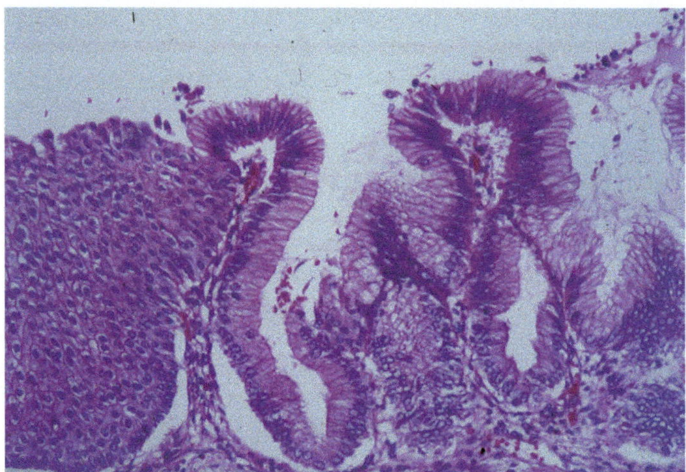

FIGURE 3.8 High power of the fetal squamocolumnar junction with fetal esophageal columnar epithelium, type 3 to the right. Note the complete absence of any inflammatory cells in the lamina propria.

epithelium was identical in appearance to "tall columnar epithelium" described by Johns,[3] "cardiac epithelium" by De Hertogh et al.,[7] and "transitional epithelium" by Park et al.[11]

A columnar epithelium with these histologic features does not occur in adult life. It was never seen in our autopsy study of the GEJ, which had many young patients. Its only similarity to adult cardiac epithelium was that it had only mucous cells.

This fetal columnar epithelium was different to metaplastic cardiac epithelium in the adult in the following ways (Figs. 3.6–3.8): (1) It was very thin and uniform, consisting of a surface epithelium and a very short foveolar pit. (2) It lay directly on the thin muscularis mucosae and had very scanty lamina propria devoid of any lymphocytes or plasma cells. Adult cardiac epithelium is thicker, more disorganized, and always has chronic inflammation in the lamina propria.

In patients of Derdoy et al.,[11] the tall columnar fetal epithelium decreased progressively as the age of the baby increased, a hallmark of fetal epithelium. In contrast, all autopsy studies in adults show that cardiac epithelial length increases with increasing age, a hallmark of an acquired epithelium.

The conclusions that I reached based on my examination of the material in the study by Derdoy et al.[11] were that the esophageal squamous epithelium was separated from parietal cell–containing gastric oxyntic epithelium by a fetal esophageal columnar epithelium. This progressively decreased in its measured extent as the gestational age increased to very small lengths at birth and disappeared completely in one baby.

The authors of the study decided that they would ignore my conclusions and published the paper with their conclusion, which was that cardiac mucosa was normally found in the GEJ region in the perinatal period.

5. SUMMARY OF FETAL EPITHELIAL DEVELOPMENT IN THE ESOPHAGUS AND STOMACH

In the final analysis, the fetal development of the esophagus is simple to understand. The entire endodermal tube in the first trimester is lined by primitive undifferentiated fetal columnar epithelium. This is fetal columnar epithelium, type 1, and is similar in the entire foregut (Fig. 3.9A). Anatomic differentiation is first seen around the 13th week immediately distal to the future location of the GEJ. Rapid cell growth on the greater curvature aspect causes the gastric fundus to protrude upward and to the left and create the angle of His at the GEJ.

Beginning in the 13th week, the primitive undifferentiated columnar epithelium at and distal to the GEJ differentiates into gastric oxyntic epithelium with parietal cells (Fig. 3.9B). The gastric mucosa quickly develops adult characteristics with a mucous cell surface epithelium and foveolar pit and a parietal cell–containing gland deep to the foveolar region. From this point to adult life, this gastric oxyntic epithelium grows in thickness to its adult state.

Proximal to the GEJ, the fetal esophagus changes in the second trimester from primitive undifferentiated columnar epithelium to a ciliated columnar epithelium that quickly lines the entire esophagus. This is fetal esophageal columnar epithelium, type 2. Beginning in the 22nd week, stratified squamous epithelium replaces the ciliated columnar epithelium, which disappears in the third trimester (Figs. 3.9 and 3.10). The squamous epithelium first appears in the midesophagus and progresses to both ends of the esophagus. In the third trimester, a third fetal columnar epithelium is seen in the distal esophagus between the SCJ and the parietal cell–containing gastric oxyntic epithelium (Figs. 3.6–3.8). De Hertogh et al.[7] measured this epithelium; it was 1.2 mm at 16 weeks of gestation, 0.3 mm at 41 weeks, and had virtually disappeared 1 week after birth.

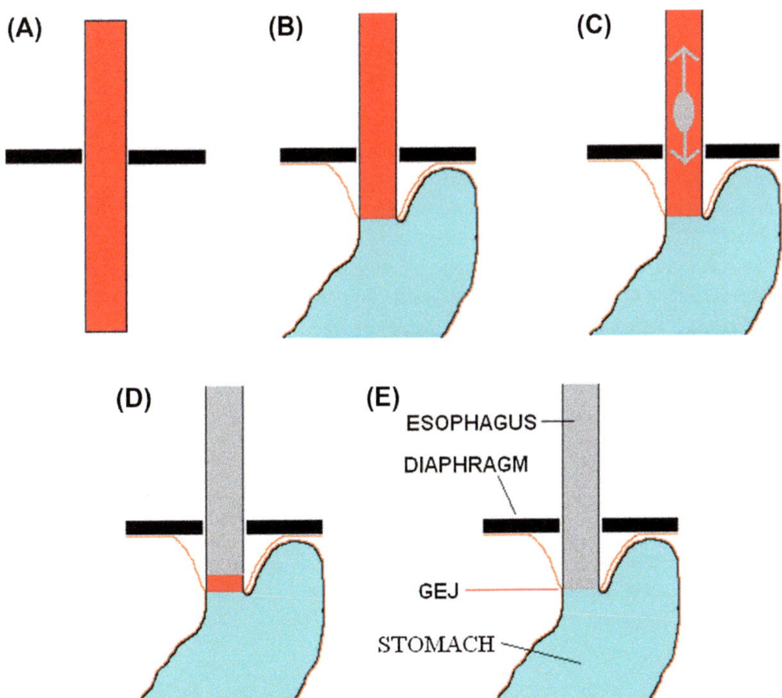

FIGURE 3.9 Progression of fetal (red) and adult-type (gray) esophageal epithelia during development. (A) In the first trimester, the entire foregut (esophagus and stomach) is a tube with no anatomic demarcation and lined by fetal columnar epithelium, type 1. (B) The foregut at 17 weeks. The anatomic demarcation between stomach and esophagus is defined by the angle of His. The stomach has developed into its adult state with parietal cells (blue), already present at the angle of His. The esophagus is lined entirely by fetal ciliated columnar epithelium, type 2. (C) At 22 weeks, squamous epithelium (gray) appears in the midesophagus and progressively replaces ciliated columnar epithelium. (D) In the perinatal period the squamous epithelium lines the entire esophagus except for a small area of fetal columnar epithelium, type 3 (red) at the distal end. (E) At full development in the perinatal period, the entire esophagus is lined by squamous epithelium and the entire stomach by gastric oxyntic epithelium. All fetal columnar epithelia have disappeared. *GEJ*, gastroesophageal junction.

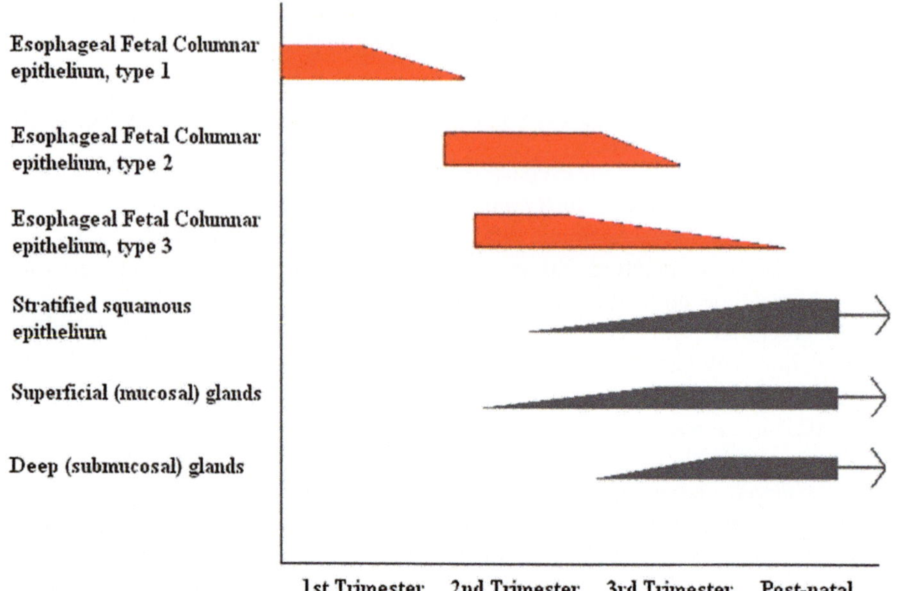

FIGURE 3.10 Diagrammatic representation of the appearance and disappearance of the three different types of fetal columnar epithelia in the esophagus, adult squamous epithelium, and esophageal glands in the esophagus related to gestational and perinatal period.

The end result of fetal development in the esophagus is a complete replacement of all columnar epithelial types by stratified squamous epithelium (Figs. 3.9 and 3.10). This reaches distally to the GEJ where it transitions directly to gastric oxyntic epithelium. The exact time at which this fully developed state is reached varies from before birth to sometime in the first year of life.

REFERENCES

1. Hawass NE, al-Badawi MG, Fatani JA, Meshari AA, Edrees YB. Morphology and growth of the fetal stomach. *Invest Radiol* 1991;**26**:998–1004.
2. Derdoy JJ, Bergwerk A, Cohen H, Kline M, Monforte HL, Thomas DW. The gastric cardia: to be or not to be? *Am J Surg Pathol* 2003;**27**:499–504.
3. Johns BAE. Developmental changes in the oesophageal epithelium in man. *J Anat* 1952;**86**:431–42.
4. Ellison E, Hassall E, Dimmick JE. Mucin histochemistry of the developing gastroesophageal junction. *Pediat Path Lab Med* 1996;**16**:195–206.
5. Salenius P. On the ontogenesis of the human gastric epithelial cells: a histologic and histochemical study. *Acta Anat* 1962;**50**(Suppl. 46):S1–70.
6. Menard D, Arsenault P. Cell proliferation in developing human stomach. *Anat Embryol* 1990;**182**:509–16.
7. De Hertogh G, Van Eyken P, Ectors N, Tack J, Geboes K. On the existence and location of cardiac mucosa; an autopsy study in embryos, fetuses, and infants. *Gut* 2003;**52**:791–6.
8. Clark GWB, Ireland AP, Chandrasoma P, DeMeester TR, Peters JH, Bremner CG. Inflammation and metaplasia in the transitional mucosa of the epithelium of the gastroesophageal junction: a new marker for gastroesophageal reflux disease. *Gastroenterology* 1994;**106**:A63.
9. Chandrasoma P. Pathophysiology of Barrett's esophagus. *Semin Thorac Cardiovasc Surg* 1997;**9**:270–8.
10. Chandrasoma P. Norman Barrett: so close, yet 50 years away from the truth. *J Gastrointest Surg* 1999;**3**:7–14.
11. Park YS, Park HJ, Kang GH, Kim CJ, Chi JG. Histology of gastroesophageal junction in fetal and pediatric autopsy. *Arch Pathol Lab Med* 2003;**127**:451–5.
12. Chandrasoma PT, Der R, Ma Y, et al. Histology of the gastroesophageal junction: an autopsy study. *Am J Surg Pathol* 2000;**24**:402–9.
13. Kilgore SP, Ormsby AH, Gramlich TL, et al. The gastric cardia: fact or fiction? *Am J Gastroenterol* 2000;**95**:921–4.

Chapter 4

Histologic Definition and Diagnosis of Epithelia in the Esophagus and Proximal Stomach

The basis of the new method that I have proposed for the diagnosis and management of gastroesophageal reflux disease (GERD) is based on the histologic assessment of biopsies. It requires definition of the various epithelial types that will permit accurate diagnosis. This is a process that has high precision for the experienced pathologist. With minimal training, interobserver variation is close to zero.

Squamous and columnar epithelia are, with few exceptions, easily distinguished from each other at endoscopy. The transition of squamous to columnar occurs at the squamocolumnar junction (SCJ).

When the SCJ coincides with the endoscopic gastroesophageal junction (GEJ) (the proximal limit of rugal folds), endoscopy is deemed normal.[1,2] Although the proximal limit of rugal folds is universally accepted as the true GEJ, there is no evidence that this is correct.[1,2] This definition is based on one paper with inadequate controls and data (see Chapter 6).[3] It has been proved to be wrong,[4] but the compelling evidence of this opposing viewpoint has been ignored.

When columnar metaplasia is seen above the endoscopic GEJ, biopsies are taken with the sole objective of diagnosing Barrett esophagus, dysplasia, and adenocarcinoma. In these patients, the squamous epithelium transitions at the SCJ to either cardiac epithelium or cardiac epithelium with intestinal metaplasia (IM). Very rarely, it may transition to oxyntocardiac epithelium but never to gastric oxyntic epithelium (GOE) (Fig. 4.1).

In most of the world, the presence of IM in visible columnar-lined esophagus (CLE) is required for a diagnosis of Barrett esophagus. In England and Japan, IM is not a required criterion for the diagnosis of Barrett esophagus.

In persons who are endoscopically normal, the management guidelines recommend that no biopsies should be taken. All columnar epithelia distal to the endoscopic GEJ are, by that definition, proximal gastric epithelia.

In endoscopically normal people, the squamous epithelium may transition into four different columnar epithelial types (cardiac epithelium with IM, cardiac epithelium, oxyntocardiac epithelium, and GOE; Fig. 4.2). When cardiac epithelium with IM is found at the SCJ, the diagnosis of "IM of the gastric cardia" is made. All other columnar epithelial types are regarded by most pathologists as normal epithelia of the proximal stomach.

This present interpretation of endoscopy and histology is completely wrong. The error results in all the early pathologic changes of reflux disease being incorrectly ascribed to normal and abnormal proximal stomach. The management of GERD is so terribly flawed today because the damaged esophagus is seen by the world as proximal stomach.[5,6]

The impetus for the pathologist to make an accurate diagnosis is a clinical demand and a belief that the diagnosis impacts patient care. Pathologists will get on board only when our clinical colleagues understand and accept these histologic concepts and embrace the new histology-based method of managing GERD. Pathologists are dependent on receiving appropriate biopsies.

Without a change in the management guidelines regarding biopsy of endoscopically normal people, pathologists never receive the required specimens. As a result, pathologists never see the material on which the new histology-based method is based. The new method will be adopted only when the incredible resistance to the adoption of this histology-based method is overcome and appropriate biopsies taken.

There is a significant failure of understanding pathology by nonpathologists. It is the natural result of subspecialization in medicine. Before surgical pathologists existed, histology was the province of surgeons. The papers by Allison,[7] Barrett,[8] and Hayward[9] in the 1950–60 era demonstrate a deep knowledge of histology and close interaction with their pathologists. Even relatively recent generations of surgeons actively rotated through the pathology service and were very conversant with histology. The first 2 years of medical school required all students to develop basic skills in microscopic diagnosis.

GERD. http://dx.doi.org/10.1016/B978-0-12-809855-4.00004-X

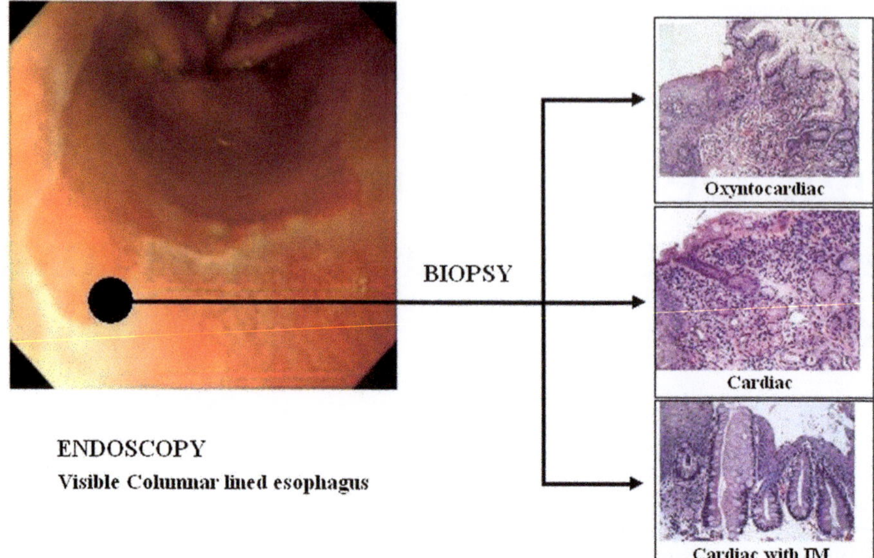

FIGURE 4.1 Biopsy at the squamocolumnar junction in a patient with endoscopically visible columnar epithelium in the thoracic esophagus. The columnar epithelium immediately distal to squamous epithelium can be cardiac epithelium with intestinal metaplasia (IM), cardiac epithelium, or (very rarely) oxyntocardiac epithelium. Gastric oxyntic epithelium is never seen.

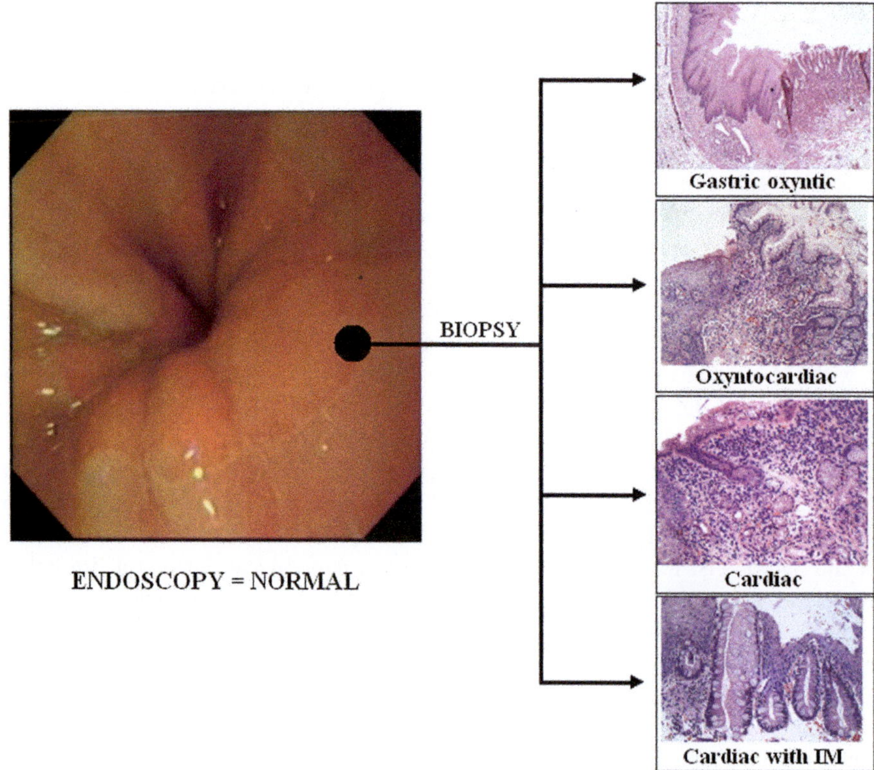

FIGURE 4.2 Biopsy at the squamocolumnar junction in a person who has no visible columnar-lined esophagus. The columnar epithelium immediately distal to squamous epithelium can be cardiac epithelium with intestinal metaplasia, cardiac epithelium, oxyntocardiac epithelium, or gastric oxyntic epithelium. *IM*, intestinal metaplasia.

This has changed in the new generation. A physician can obtain a medical degree today without ever having looked through a microscope. It is a rare surgeon or internist who visits a pathology department today to look at the slides of their patients. Gastroenterology and surgery training programs require pathology education and have pathology questions in the board examination. This does not equate to deep knowledge or even significant ability to understand histology.

1. PRESENT VALUE OF HISTOLOGY IN DIAGNOSIS OF GASTROESOPHAGEAL REFLUX DISEASE

At present, biopsies and histology are used only for the diagnosis of Barrett esophagus, dysplasia, and adenocarcinoma. They have no practical value in the diagnosis of GERD in the patient without visible CLE at endoscopy.

The following reasons exist for the lack of value to histology in the diagnosis of GERD:

1. Histologic changes in the squamous epithelium are neither sufficiently sensitive nor specific for the diagnosis of GERD. The presence of erosions at endoscopy is a sufficiently specific feature of GERD (in the absence of other causes of erosions) to not require histologic confirmation. The detection of highly sensitive features of GERD such as dilated intercellular spaces requires electron microscopy.[10] Its presence, which is simply a general manifestation of epithelial injury, is also likely to be relatively nonspecific for GERD.
2. Biopsy is not recommended in the patient with GERD who does not have a visible CLE.[1] The reason for this is the belief that biopsies taken distal to the endoscopic GEJ represent the proximal stomach and this cannot have relevance in the diagnosis of GERD. The problem, however, is that the true GEJ is not the endoscopically defined GEJ. There is a variable amount of damaged esophagus distal to the endoscopic GEJ. By using the endoscopic GEJ for definition, this damaged esophagus is ignored.

As a result, pathologists all over the world consistently receive endoscopic biopsies at the present time only in the patient who has a visible CLE at endoscopy. This has resulted in the region distal to the endoscopic GEJ being inaccessible to pathologic study. As a result, the evidence base for qualitative and quantitative histologic changes in this region is incredibly small.[11–15] All the evidence supports the fact that there is damaged esophagus distal to the endoscopic GEJ and that its extent correlates with severity of GERD.

The only recognized role the present pathologist has in the diagnosis of GERD is the diagnosis of IM in CLE (which defines Barrett esophagus in the United States), dysplasia, and adenocarcinoma. To the pathologist who is asked to perform only this function, there is no necessity to learn about cardiac and oxyntocardiac epithelia. Careful and accurate diagnosis of these epithelial types is a waste of time for today's pathologist because, except for the presence of IM, it has no clinical significance.

2. A HISTOLOGIC DEFINITION OF THE GASTROESOPHAGEAL JUNCTION

At present, there is no histologic definition of the GEJ where the esophagus transitions to the stomach. The gastroenterology world is managing GERD without clear definition of where the esophagus ends and the stomach begins, or worse, with an incorrect endoscopic definition of the GEJ.

The only histologic definition of the GEJ that has ever existed was proposed by Norman Barrett in 1950,[16] who declared that the esophagus was "that part of the foregut, distal to the cricopharyngeal sphincter, which is lined by squamous epithelium." This was totally incorrect and was made before it was recognized that columnar metaplasia caused a cephalad migration of the SCJ.

In 1957, faced with the evidence of a CLE distal to the SCJ, Barrett withdrew this definition graciously in his landmark paper[8] and accepted the CLE that Allison and Johnstone had described in 1953.[7] The amazing grace and lack of ego to freely admit error by a man who was the leader in his field at that time is what is sorely needed today.

Since then, the GEJ has been defined by endoscopic (the proximal limit of rugal folds)[1–3] or gross anatomic (end of the tubular esophagus)[17] criteria (Fig. 4.3A). Both these criteria used to define the GEJ are reasonably reproducible endoscopic landmarks that can be correlated with gross examination of resected specimens (Fig. 4.3A). Unfortunately, both definitions are incorrect (Fig. 4.3B).

Both the proximal limit of rugal folds and the end of the tube are altered in the patient whose distal esophagus has been damaged by GERD. The damaged distal esophagus dilates and develops rugal folds, separating the endoscopic GEJ from the true GEJ (Fig. 4.3B).[6,18,19] The dilated distal esophagus (DDE), which is distal to the endoscopic GEJ, is lined by cardiac epithelium and is mistaken for proximal stomach by present definitions. The true GEJ can only be defined histologically by the proximal limit of GOE (see Chapter 6).[4]

The incorrect endoscopic definition of the GEJ causes the early pathology of GERD (the DDE lined by metaplastic esophageal columnar epithelium) to be erroneously called proximal stomach and ignored as either normal or irrelevant in the study of the patient with GERD.

The greater the severity of GERD, the greater is the separation of the endoscopic GEJ from the true GEJ and the greater is the error of the present endoscopic definition of the GEJ.[4] This error is corrected by the histologic definition of the GEJ, which is the junction between metaplastic cardiac epithelium (with and without parietal cells) and GOE. This junction is invisible endoscopically and at gross pathologic examination.

The true GEJ is accurately defined by the proximal limit of GOE. When this is accepted, a method to define GERD at a cellular level and accurately measure the severity of GERD emerges.

3. DIAGNOSTIC CRITERIA FOR EPITHELIA IN THE ESOPHAGUS AND STOMACH

There are three basic epithelial types that exist in the esophagus and proximal stomach: normal squamous epithelium of the esophagus, normal GOE that lines the entire proximal stomach, and the abnormal metaplastic esophageal columnar epithelia that replaces the distal esophageal squamous epithelium in GERD. The metaplastic columnar epithelium is cardiac epithelium and its variants (cardiac epithelium with IM and oxyntocardiac epithelium).

Cardiac epithelium and its two variants are *always* located between the distal limit of squamous epithelium (i.e., the SCJ) and the proximal limit of GOE (the true GEJ). They *always* represent columnar metaplasia of esophageal squamous epithelium. As such their presence equates to the distal esophagus.

In the majority of people in the world, there is no visible CLE and cardiac epithelium is limited to the region between the incorrectly defined endoscopic GEJ and the true GEJ (proximal limit of GOE) in the DDE.

I will show later that the length of this cardiac epithelium in the dilated distal esophagus is equal to LES damage involving the abdominal segment of the LES. In a minority of patients with severe reflux, cardiac epithelium and its variants extend into the thoracic esophagus as CLE visible at endoscopy.

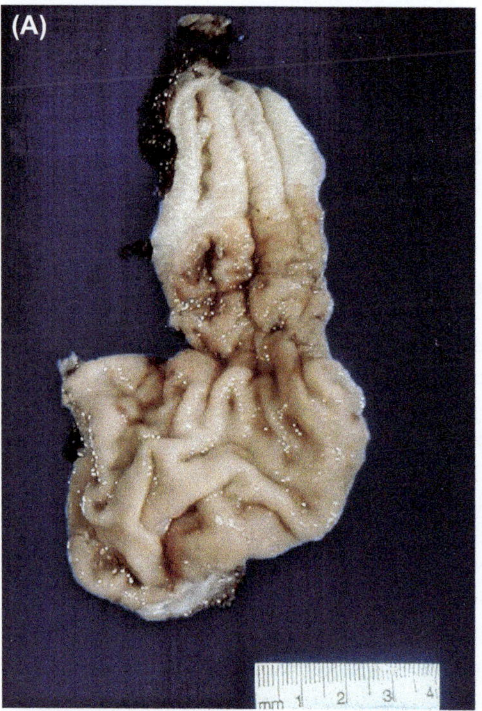

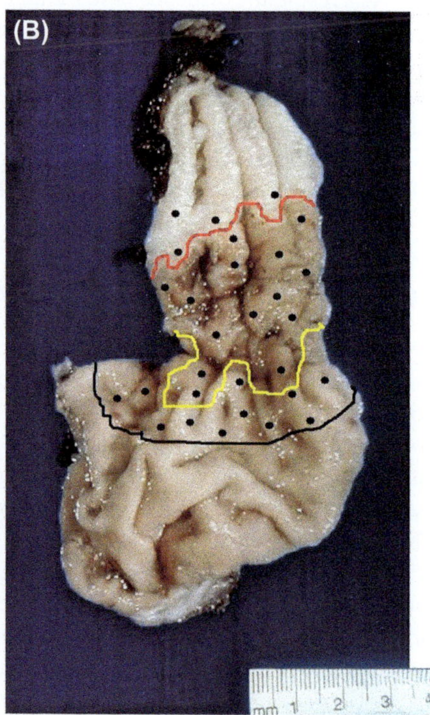

FIGURE 4.3 (A) Resected esophagogastrectomy patient with an ulcerated adenocarcinoma immediately distal to the squamocolumnar junction. The point at which the tubular esophagus flares and the proximal limit of rugal folds (the two presently used definitions of the gastroesophageal junction) end is concordant. (B) Histologic assessment of this esophagus shows a long segment of columnar metaplasia that extends from the squamocolumnar junction (*red line*) to 2.06 cm distal to the end of the tubular esophagus into the area that has rugal folds. This segment of columnar epithelium consists of intestinal metaplasia that extends into the area distal to the tubular esophagus (*yellow line*) and cardiac and oxyntocardiac epithelium that extends to the true gastroesophageal junction (GEJ), which is defined as the proximal limit of gastric oxyntic epithelium (*black line*). The 2.06-cm area between the end of the tubular esophagus and the true GEJ is the dilated distal esophagus. The fact that this is esophagus and not the proximal stomach is proved by the presence of submucosal glands (*black dots*) that are seen only in the esophagus.

All five epithelial types (including the three variants of metaplastic cardiac epithelium) can be defined precisely by simple histologic criteria. The precision of identification of these epithelial types is so high that it can be a powerful basis for histologic diagnosis of the entire spectrum of GERD. This contrasts with the incredible lack of precision in the diagnosis of GERD based on symptoms.[20]

3.1 Stratified Squamous Epithelium

Stratified squamous epithelium is limited to the esophagus in the human foregut. It does not occur in the stomach. This is not universal to all species; ruminating animals have squamous epithelium in the proximal stomach.

The normal stratified squamous epithelium of the esophagus can be reliably differentiated from columnar epithelium at gross examination of specimens (Figs. 4.3 and 4.4) and endoscopy (Fig. 4.5). Histologically, it has multiple layers of epithelial cells with a flat surface and broad rete pegs in its deep aspect (Fig. 4.6). The rete pegs are separated by short papillae that give the deep aspect of the epithelium an undulating appearance.

Stratified squamous epithelium consists of a single basal layer containing stem cells, 2–3 layers of proliferative basaloid cells in the suprabasal region, and larger keratinized cells toward the surface. The esophageal squamous epithelium is nonkeratinizing, i.e., it does not have a stratum corneum.

The overall thickness of the epithelium varies. No normal measured limits exist for squamous epithelial thickness. In the "normal" state, the proliferative basal cell region is less than 30% of the epithelial thickness and the papillary height is less than 60% of the thickness of the epithelium (Fig. 4.6). It is difficult to define "normal" because of the potential frequency of mild unrecognized reflux-induced changes. Squamous epithelial parameters for "normalcy" are adjusted to the upper limit beyond which the epithelium can be recognized as abnormal with high specificity (Fig. 4.7).

It is likely that acid-damaged squamous epithelium often shows changes that cannot be differentiated from "normal." One of these that have been suggested as a sensitive diagnostic marker of GERD is the presence of "dilated intercellular spaces" at the ultrastructural level.[10] The requirement of electron microscopy to accurately detect this change limits its practical usefulness.

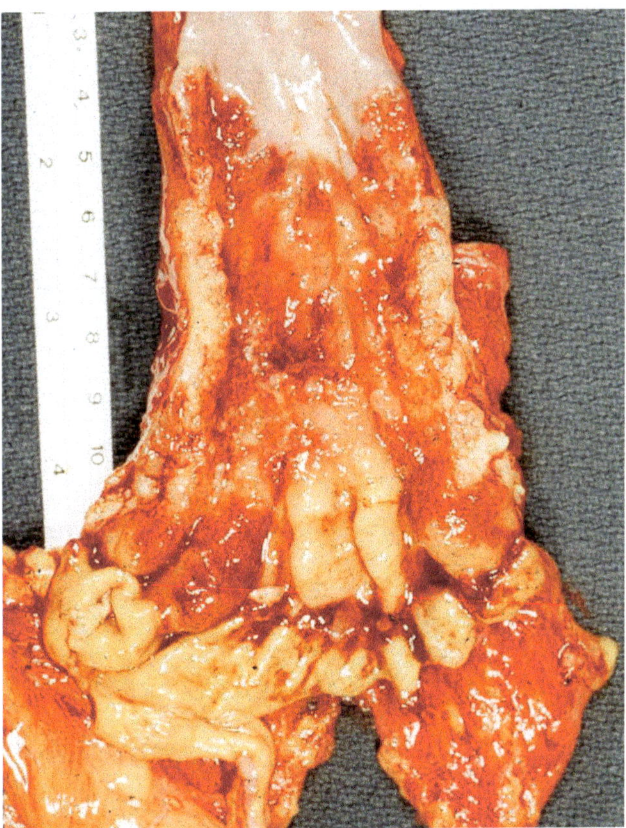

FIGURE 4.4 Esophagectomy specimen showing a long segment of columnar metaplasia involving the thoracic esophagus. The differentiation of squamous from columnar epithelia is easy. The differentiation of different types of columnar epithelium and the distal end of the columnar-lined segment (CLE) is impossible; it requires histology. Note that the distal end of the CLE (i.e., the gastroesophageal junction) is not clearly definable in this specimen.

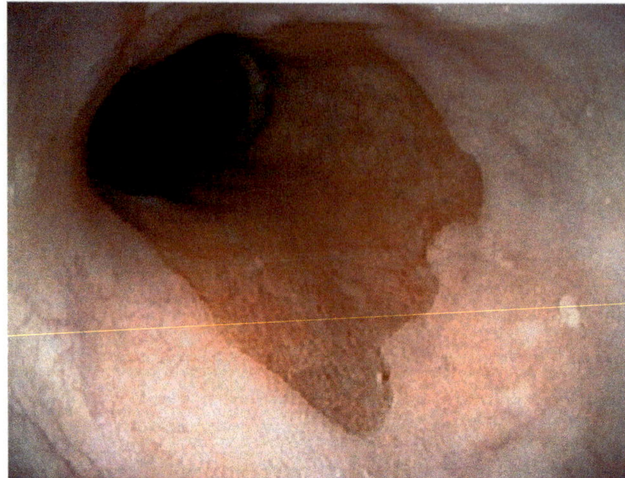

FIGURE 4.5 Endoscopic view of the thoracic esophagus shows visible columnar-lined epithelium, its pink color clearly contrasting with the white appearance of the squamous epithelium.

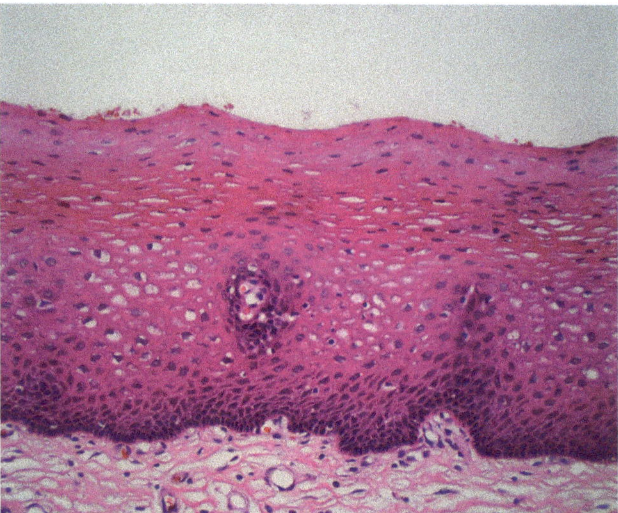

FIGURE 4.6 Histology of normal squamous epithelium. Note the tight cell junctions with no spaces and the thin basal cell zone.

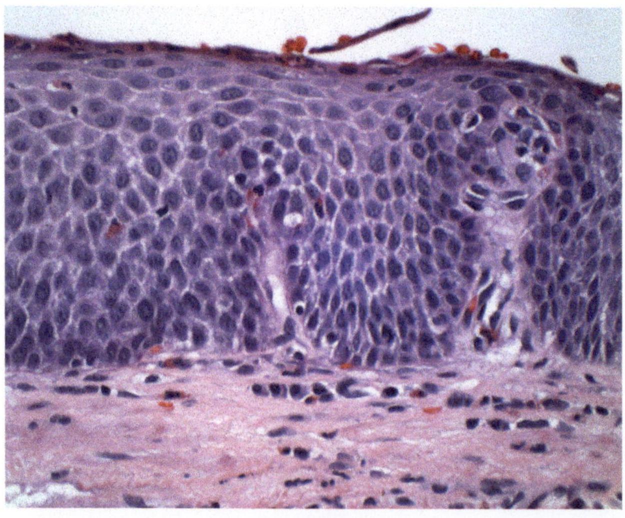

FIGURE 4.7 Reflux esophagitis. This shows basal cell hyperplasia, separation of cell junctions (dilated intercellular spaces), and scattered intraepithelial eosinophils. These findings are not specific for reflux and can be seen in other types of esophageal injury.

The normal state of the squamous epithelium depends on the dynamics of cell loss and replacement. The cells in the basal proliferative zone continually divide and move upward in the epithelium, becoming terminally differentiated keratinocytes that replace the cells that are continually shed at the surface. The keratinocytes have small nuclei and abundant eosinophilic cytoplasm and flatten out at the surface.

In the normal epithelium, the proliferative zone can be identified by immunoperoxidase staining for Ki67 where it is normally seen as 2–3 layers of basaloid cells above the Ki67 negative basal layer (Fig. 4.8). In normal squamous epithelium, the average time taken for a cell to move from the basal zone to the surface is 7–8 days.[21]

Normal stratified squamous epithelium is an excellent barrier that is impervious to luminal molecules.[22] The lack of permeability results from the presence of tight junctions that keep the keratinocytes closely apposed to one another. With reflux-induced damage, intercellular edema results in separation of the squamous cells (Fig. 4.9), increasing permeability of the epithelium to increasingly larger molecules.[22]

Two types of mucous glands are seen normally in the squamous-lined esophagus. There are small mucous glands that resemble minor salivary glands in the mucosal lamina propria between the squamous epithelium and muscularis mucosae (Fig. 4.10). Larger lobulated mucous glands are present in the submucosa deep to the muscularis mucosae (Fig. 4.11). The submucosal glands drain onto the surface by well-formed ducts that traverse the muscularis mucosae, the lamina propria of the mucosa, and the stratified squamous epithelium to open at the surface (Figs. 4.12 and 4.13). The ducts may be lined by columnar epithelium, which can sometimes be ciliated, squamous epithelium, or a mixture of the two. The function of the mucous glands is presumably to secrete mucin that lubricates the surface of the squamous epithelium.

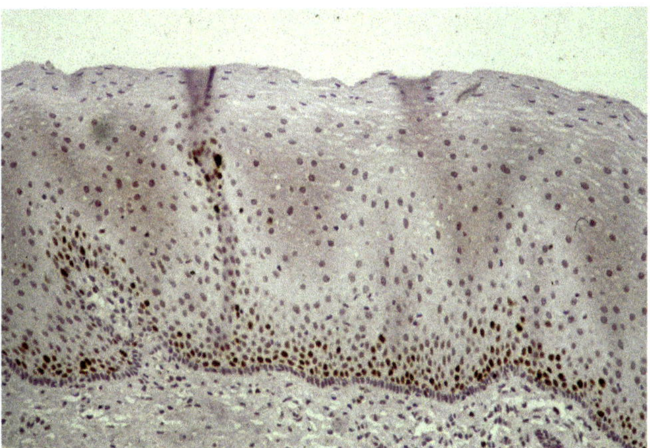

FIGURE 4.8 Immunoperoxidase stain for Ki67 showing positive nuclear staining of a thin proliferative cell zone immediately above the basal layer that contains the stem cells (usually nonproliferative).

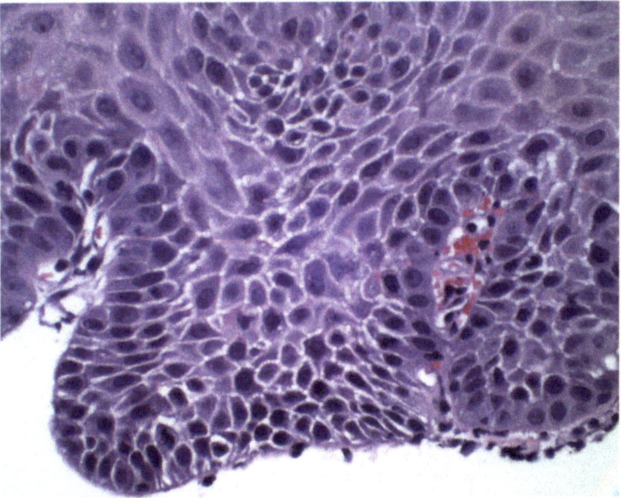

FIGURE 4.9 Squamous epithelium in reflux disease showing intercellular edema causing separation of cells (dilated intercellular spaces). This increases permeability of the epithelium, permitting entry of luminal molecules into the epithelium.

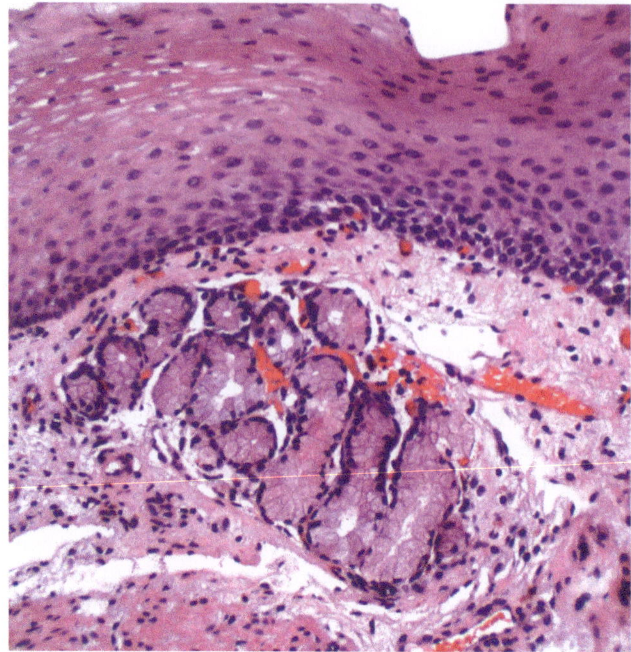

FIGURE 4.10 Normal esophageal squamous epithelium with a mucous gland in the mucosa above the level of the muscularis mucosae.

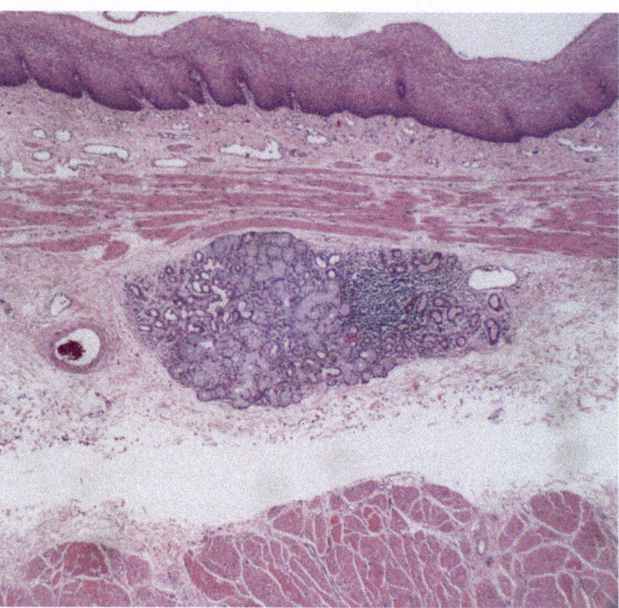

FIGURE 4.11 Resected esophagus, normal area lined by squamous epithelium, showing a submucosal gland between the muscularis mucosae above and the muscularis propria below. Submucosal glands are evidence of esophageal location of this specimen.

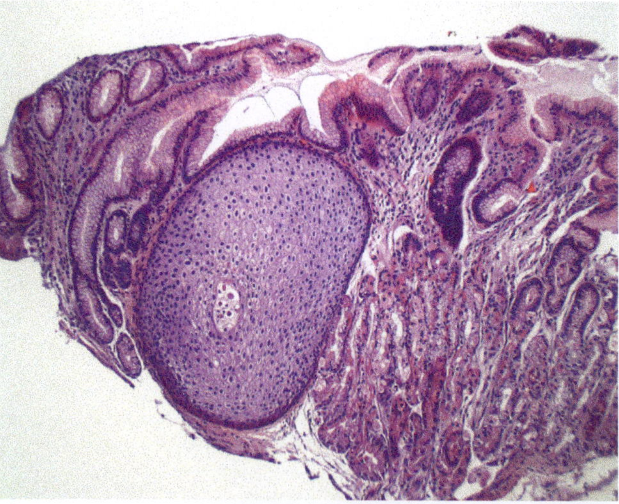

FIGURE 4.12 Squamous-lined gland duct in mucosa lined by metaplastic oxyntocardiac epithelium. The presence of the gland duct is evidence of an esophageal location of this biopsy.

The submucosal glands of the esophagus develop in late fetal life after squamous epithelium has replaced the fetal columnar epithelium of the esophagus.[23] Submucosal glands do not occur in the human stomach where squamous epithelium does not develop. As such, submucosal mucous glands and their gland ducts are specific markers for the anatomic esophagus.[4,7] Submucosal glands are seen only in resection (endoscopic and surgical) specimens. Ducts from the glands are seen in mucosal biopsies.[24] It is important to recognize that only the positive finding has value in localizing location. If a submucosal gland or gland duct is present in this region, the conclusion can be made that the location is the esophagus and not the stomach. Because submucosal glands are distributed variably in number and randomly in distribution in different people, the absence of submucosal glands and ducts is not evidence that the organ is the stomach and not the esophagus.

When columnar metaplasia of the squamous epithelium of the esophagus occurs in patients with GERD, the other anatomic elements of the esophagus apart from the surface epithelium remain relatively unaltered. The mucosal glands and submucosal mucous glands and their ducts remain and serve as evidence of esophageal location of the columnar epithelium (Fig. 4.14).

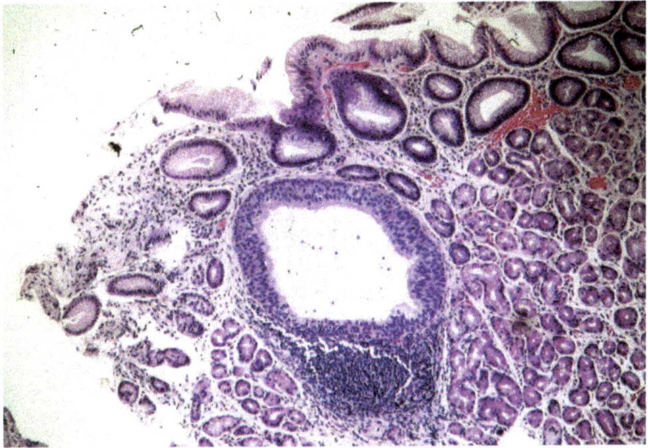

FIGURE 4.13 Columnar-lined gland duct in mucosa lined by oxyntocardiac epithelium. The columnar cells are focally ciliated. The presence of the gland duct is evidence of esophageal location of this biopsy.

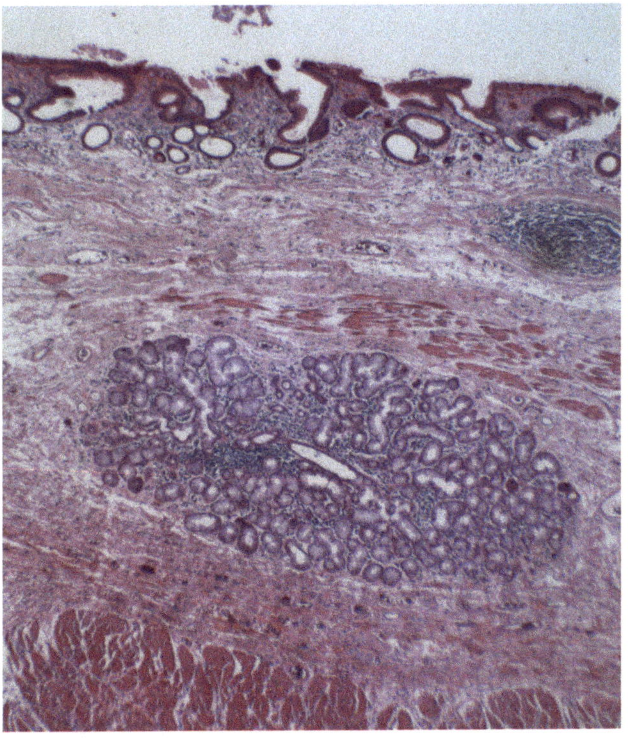

FIGURE 4.14 Esophagogastrectomy specimen showing surface lined by columnar epithelium. A submucosal gland is present, indicating that this is esophageal in location irrespective of whether this is above or below the end of the tubular esophagus.

3.2 Gastric Oxyntic Epithelium

GOE is present in all humans. It is normally limited to the stomach. Rarely, heterotopic GOE is seen in the first part of the duodenum and in some cases of Meckel's diverticulum. It does not occur as a metaplastic epithelium in the esophagus. When parietal cells are encountered in the esophagus, it is in metaplastic cardiac epithelium that has developed parietal cells (i.e., oxyntocardiac epithelium). GOE lines the fundus and body of the stomach and transitions into antral mucosa at a variable point in the distal stomach.

GOE is normally the thickest columnar epithelial type in this region. Normal GOE consists of a surface epithelium and shallow foveolar pit composed of columnar cells with basal nuclei and apical neutral mucin (Fig. 4.15). This surface epithelium and foveolar pit are similar in all types of columnar epithelia in this region. Starting at the base of the foveolar pit and extending down to the muscularis mucosae are long, nonbranching straight tubular glands composed of parietal cells with abundant eosinophilic cytoplasm and basophilic chief cells (Fig. 4.15). The glands extend all the way down to the thin muscularis mucosae. The glands are tightly packed and there is minimal lamina propria apart from the capillary vessels between the glands. The number of lymphocytes in the normal lamina propria is zero to very low. There are normally no plasma cells or eosinophils. The tubular glands of the normal stomach do not have mucous cells below the foveolar pit. Scattered neuroendocrine cells are present in the glands. There are normally no Paneth or goblet cells.

The defining criteria of normal GOE are the presence of straight tubular glands below the foveolar pit that contain parietal (oxyntic) cells but not mucous or goblet cells.

3.3 Metaplastic Esophageal Columnar Epithelia in the Squamooxyntic Gap

The normal state where there is no metaplastic columnar epithelium between the distal limit of squamous epithelium and GOE is very rare (Figs. 4.16A and 4.17). Almost all people damage their distal esophageal squamous epithelium by exposure to gastric juice during their lifetime.

In patients with squamous epithelial damage that has resulted in cardiac metaplasia, the gastric oxyntic mucosa is separated from the squamous epithelium by a squamooxyntic gap composed of cardiac epithelium and/or its variants (Figs. 4.16B and 4.18).

This gap composed of cardiac epithelium (with or without parietal and/or goblet cells) varies from >0 to 25 cm (the entire length of the esophagus). Both extremes of length are extremely rare. The gap progressively increases its length from the normal state of zero depending on the amount of damage caused to the esophageal squamous epithelium by the cumulative lifelong exposure to gastric juice. As cardiac epithelium in the squamooxyntic gap progressively increases in length, the SCJ moves cephalad and becomes increasingly separated from the true GEJ (the proximal limit of GOE).

The distal abdominal esophagus that has been damaged by exposure to gastric juice can be identified histologically by the presence of metaplastic columnar epithelium. When so identified, it is seen that this damaged segment of esophagus is

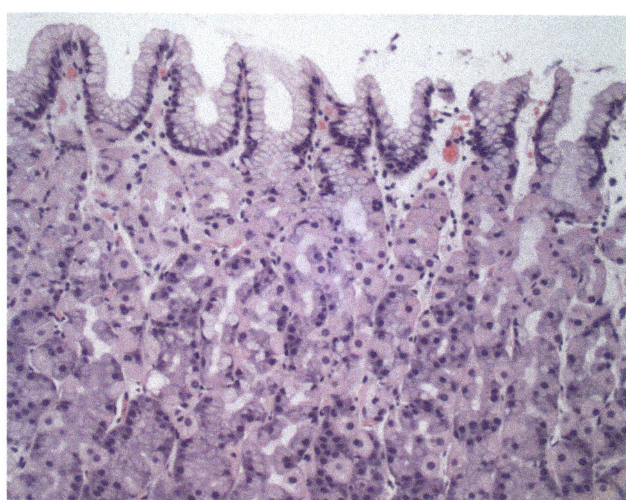

FIGURE 4.15 Gastric oxyntic epithelium, normal, higher power, showing a short foveolar pit lined by mucous cells and the upper region of the gastric gland showing parietal and chief cells without mucous cells. Note the absence of inflammation and scant lamina propria between the tightly crowded glands.

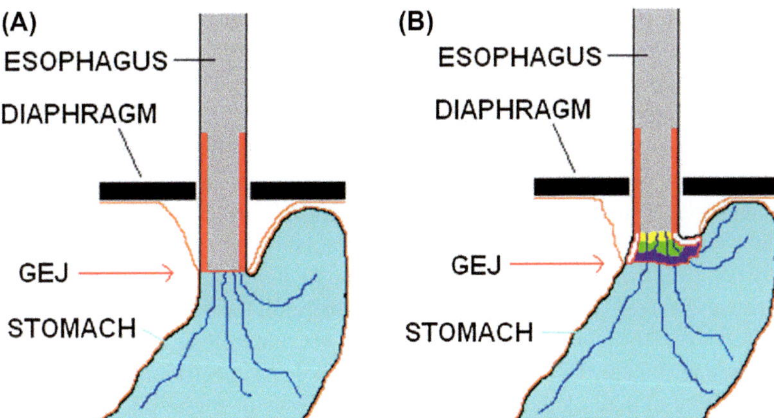

FIGURE 4.16 Two different entities in a person who is endoscopically normal. (A) The squamous epithelium (gray) transitions directly to gastric oxyntic epithelium (blue) with rugal folds. This is the normal state without metaplastic cardiac epithelium. (B) The squamous epithelium (gray) is separated from the gastric oxyntic epithelium (blue) by metaplastic columnar epithelium (green, cardiac epithelium; purple, oxyntocardiac epithelium; yellow, intestinal metaplasia). The end of the tubular esophagus and the squamocolumnar junction has moved cephalad and the metaplastic segment has developed rugal folds. The metaplastic segment is the dilated distal esophagus. It is proximal to the true gatroesophageal junction (GEJ) (proximal limit of gastric oxyntic epithelium) but distal to the endoscopic GEJ. Note that the dilated distal esophagus = length of metaplastic cardiac epithelium = lower esophageal sphincter damage (white replacing the red LES). Note also the shortening of the abdominal esophagus and the more obtuse angle of His.

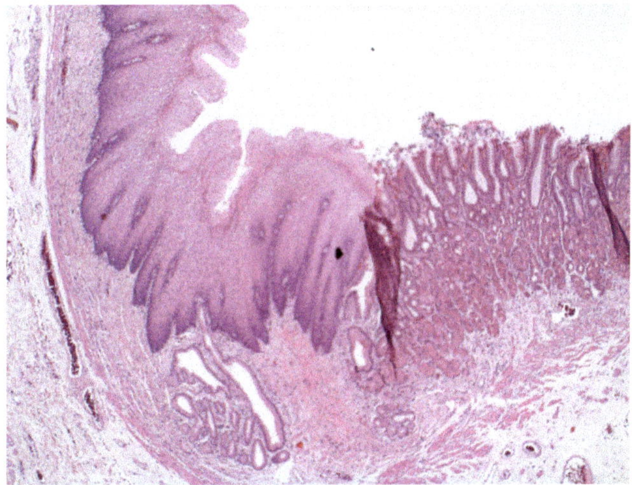

FIGURE 4.17 Section across that squamocolumnar junction in a 67-year-old male who had an esophagogastrectomy for squamous carcinoma. The squamous epithelium transitions directly to gastric oxyntic epithelium. There are two small mucosal glands, one under the squamous epithelium and the other exactly at the distal limit of the squamous epithelium. Both show gland ducts extending up toward the mucosa. This is the normal state shown in Fig. 4.16A.

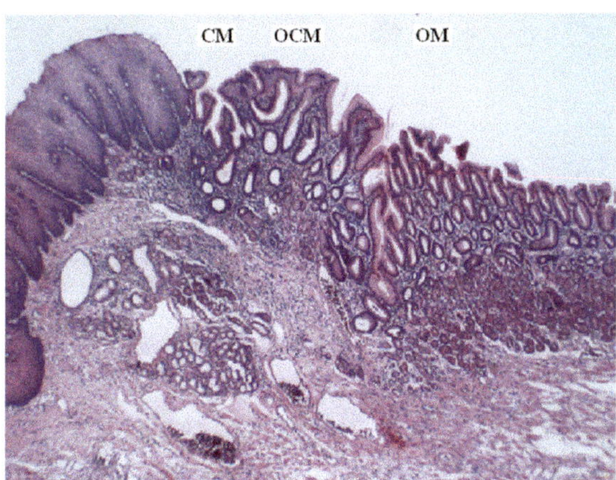

FIGURE 4.18 Section across the squamocolumnar junction (SCJ) in a person whose SCJ is separated from gastric oxyntic epithelium [gastric oxyntic mucosa; true gastroesophageal junction] by a very short (2 mm) segment of metaplastic columnar epithelium consisting of inflamed cardiac epithelium (CM) and oxyntocardiac epithelium (OCM). *OM*, oxyntic mucosa.

dilated and has developed rugal folds (Fig. 4.3A and B). The DDE assumes an appearance that makes it endoscopically and grossly impossible to differentiate from the proximal stomach. It is easy for anyone not looking at the area at a histologic level, which is the majority of people, to fall into the obvious trap.

Metaplastic esophageal columnar epithelium begins as an undifferentiated columnar epithelium with only mucous cells when it arises from metaplasia of esophageal squamous epithelium. This is cardiac epithelium (Fig. 4.19). Initially, when it arises by metaplasia of squamous epithelium, cardiac epithelium consists of either a single layer of mucous cells of a multilayered epithelium composed of a mixture of columnar and basaloid squamous cells. Recapitulating the differentiation of the fetal endodermal tube that is seen in the stomach and intestine, this flat epithelium that is rich in stem cells immediately invaginates into a foveolar pit that can reach a significant length and become tortuous (Fig. 4.20). From the depth of the foveolar pit, glands arise and extend downward. In cardiac epithelium, these glands are also composed of only mucous cells. The cells of the surface layer, foveolar pit, and glands are all mucous cells in cardiac epithelium.

Over time and under different luminal environments, cardiac epithelium may develop one of two mutually exclusive specialized cells (Fig. 4.21). The first of these is the parietal cell, resulting in an epithelium composed of a mixture of mucous and parietal cells (oxyntocardiac epithelium; Fig. 4.22). The second of these is the goblet cell, resulting in an epithelium composed of a mixture of mucous and goblet cells (cardiac epithelium with IM; Fig. 4.23). This evolution of cardiac epithelium has important implications for the progression of columnar epithelia in the esophagus that I will discuss in Chapter 11.

In 1976, Paull et al.[25] developed the histologic classification of metaplastic esophageal columnar epithelium with precise definitional criteria. They studied the columnar epithelium above the upper border of the manometrically defined

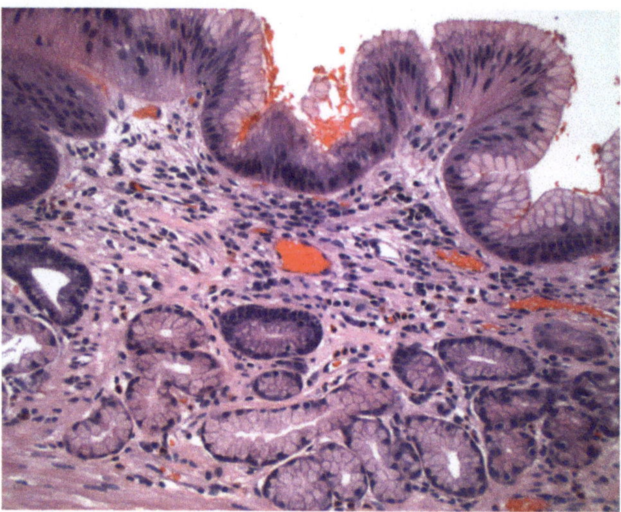

FIGURE 4.19 Cardiac epithelium showing surface epithelium, foveolar pit, and disorganized glands composed of mucous cells only. Note the chronic inflammation and fibrosis in the lamina propria.

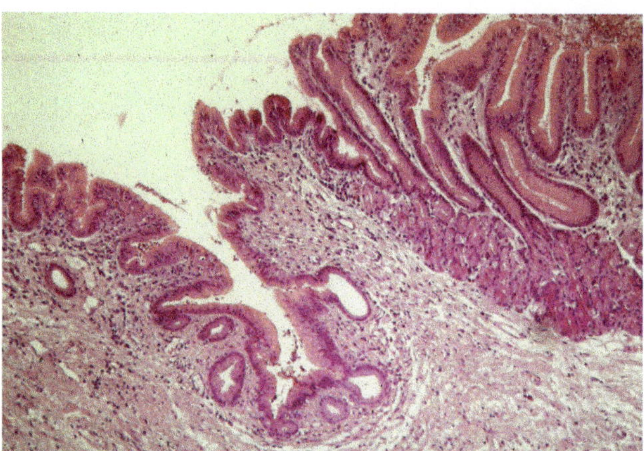

FIGURE 4.20 An evolving zone of metaplastic columnar epithelium. At the left, the epithelium is cardiac with a combination of flat epithelium with rudimentary foveolar pits. Going to the right, the cardiac epithelium becomes thicker and finally develops parietal cells to become oxyntocardiac epithelium.

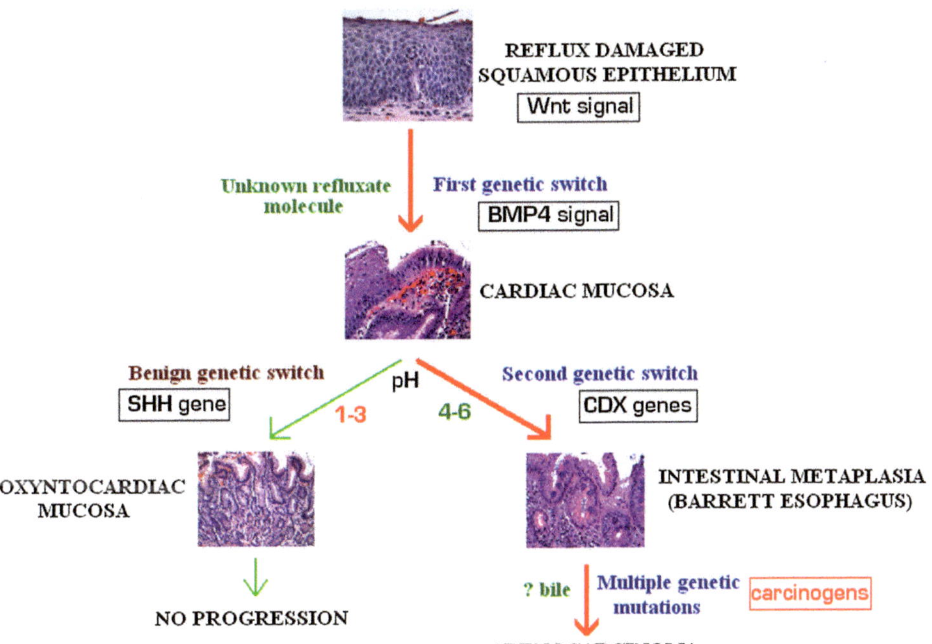

FIGURE 4.21 Evolution of the entire process whereby esophageal squamous epithelium changes due to exposure to gastric contents. The first change is into cardiac epithelium. Cardiac epithelium is a pivot point where it can evolve in two directions, one with parietal cells to form oxyntocardiac epithelium and the other with goblet cells to form intestinal metaplasia (IM). The epithelium at risk for progression to dysplasia and adenocarcinoma is cardiac epithelium with IM that defines Barrett esophagus.

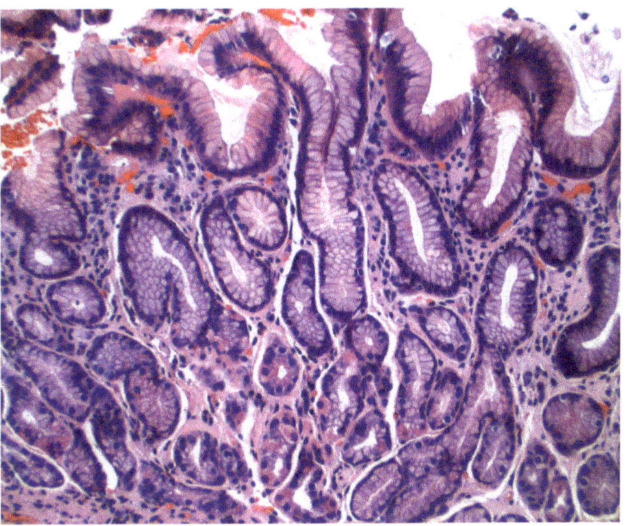

FIGURE 4.22 Oxyntocardiac epithelium showing glands with a mixture of mucous cells with scattered parietal cells, seen as round cells with small central nuclei and abundant eosinophilic cytoplasm.

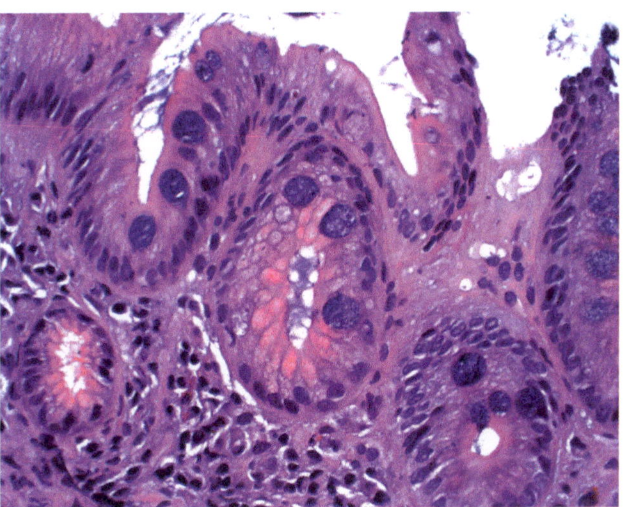

FIGURE 4.23 Cardiac epithelium with scattered goblet cells. These are characterized by a well-defined round cytoplasmic vacuole filled with basophilic mucinous material.

LES in patients with a long segment of CLE. The classification was based on the presence of the three different epithelial types within the foveolar unit. I have proposed slight modifications to the definitional criteria and terminology of Paull et al. The reason why I have modified the definitions is to make them slightly simpler and more reproducible. I changed the terminology because the term cardiac mucosa had largely replaced junctional mucosa of Paull et al. by the turn of the century. However, the definitional criteria developed by Paull et al. were essentially perfect from the outset and have stood the test of time (Table 4.1).

The definition of the three types of metaplastic columnar epithelia in the squamooxyntic gap is applied to each individual foveolar-gland unit. The three epithelial types can coexist in one small area, sometimes in a single biopsy specimen (Fig. 4.24).

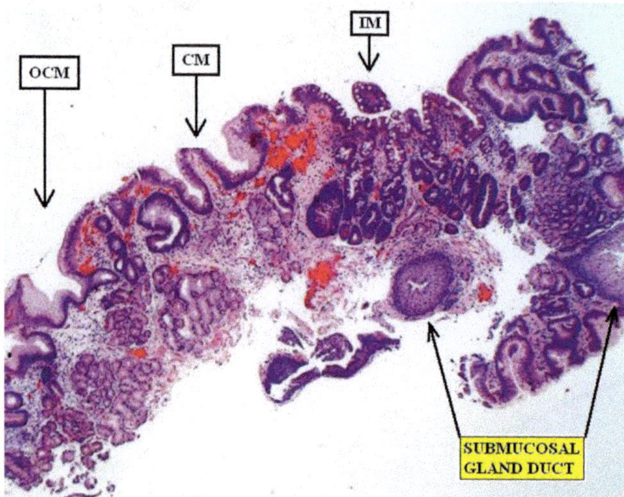

FIGURE 4.24 Biopsy from distal to the endoscopic gastroesophageal junction consisting of oxyntocardiac epithelium to the left, cardiac epithelium in the center, and intestinal metaplasia (IM) with goblet cells to the right. The definitions used to classify the three metaplastic epithelia must be applied to each foveolar/gland unit. Note the presence of squamous-lined gland ducts, indicating esophageal location of this biopsy. *CM*, cardiac mucosa; *OCM*, oxyntocardiac mucosa.

TABLE 4.1 Definitional Criteria and Terminology of Metaplastic Esophageal Columnar Epithelia in the Esophagus

Recommended (Future) Term	Paull et al. Term	Present Term	Paull et al. Criteria	Present Criteria
Metaplastic esophageal columnar epithelium, cardiac type	Junctional epithelium	Cardiac epithelium	Mucous cells+	Mucous cells+
			Parietal cells–	Parietal cells–
			Goblet cells–	Goblet cells–
Metaplastic esophageal columnar epithelium, oxyntocardiac	Gastric fundic-type epithelium	Oxyntocardiac epithelium	Mucous cells+	Mucous cells+
			Parietal cells+	Parietal cells+
			Chief cells+	Goblet cells–
			Goblet cells–	
Metaplastic esophageal columnar epithelium, intestinal	Specialized columnar epithelium	Cardiac epithelium with intestinal metaplasia	Mucous cells+	Mucous cells+
			Parietal cells–	Parietal cells–
			Chief cells–	Goblet cells+
			Goblet cells+	

(1) The only change between the criteria of Paull et al. and present criteria is the removal of the requirement of chief cells from the definition. Parietal cells coexist with chief cells. Because parietal cells are much easier to recognize in routine sections than chief cells, this change makes the definition more precise. (2) The criteria of Paull et al. were developed for epithelia in columnar-lined esophagus in the body of the esophagus above the manometrically defined lower esophageal sphincter. These same definitional criteria are applicable to epithelia throughout the squamooxyntic gap.

3.3.1 Cardiac Epithelium (Metaplastic Esophageal Columnar Epithelium, Cardiac)

Cardiac epithelium (synonyms: junctional epithelium; mucous cell–only epithelium) is defined as an epithelium composed only of mucous cells without any parietal cells or goblet cells (Fig. 4.20). The presence of neuroendocrine cells, chief cells, Paneth cells, and pancreatic acinar cells does not impact the diagnosis of cardiac epithelium.

Cardiac epithelium is present in the squamooxyntic gap, distal to the SCJ and proximal to the true GEJ (proximal limit of GOE). In patients without endoscopic CLE, it is limited to the DDE between the endoscopic GEJ and the true GEJ. In patients with endoscopic CLE, cardiac epithelium is present on both sides of the endoscopic CLE.

Cardiac epithelium exists in a variety of morphologic patterns.[5] Rarely, it consists of a flat epithelium composed of a single layer of mucous cells (surface type) (Fig. 4.25). Surface type cardiac epithelium is likely the first epithelium to form from metaplasia of squamous epithelium. It has all the precious stem cells on the surface. Very quickly, as occurs in the embryologic development of gastric epithelium (see Chapter 3), foveolar pits form, resulting in movement of the stem cells to the deep foveolar region where they are sequestered. Foveolar type of cardiac epithelium is more common. It consists of a surface layer with foveolar pits only, all composed of mucous cells (Fig. 4.26). The foveolar pits can vary in length and become tortuous. In its most developed form, cardiac epithelium has mucous glands beneath the foveolar pit composed of mucous cells only (Fig. 4.27). In all types of cardiac epithelium, only mucous cells are present.

Formation of glands in cardiac epithelium is necessary for specialized glandular cells such as parietal cells to occur (Fig. 4.28). Oxyntocardiac epithelium is therefore always a glandular type of epithelium. The glandular element is beneath the foveolar pit and can be recognized by the fact that it is below the proliferative zone of the epithelium in the deep

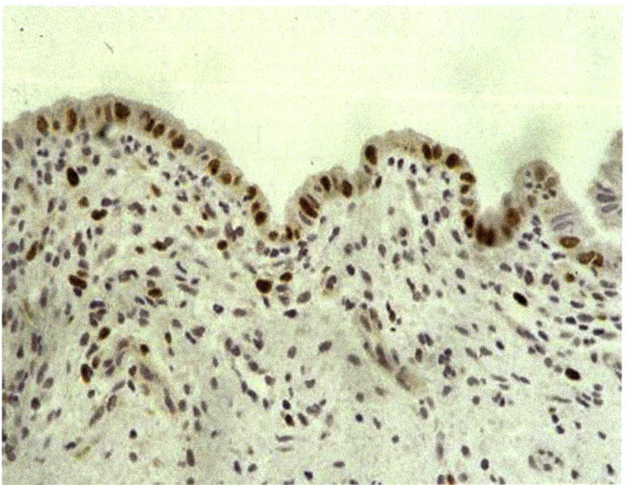

FIGURE 4.25 Cardiac epithelium, surface only type, consisting of a single layer of flat mucous cells. This is a transient epithelium that contains the stem cells on the surface. Ki67 immunoperoxidase–stained section showing nuclear staining of proliferative cells.

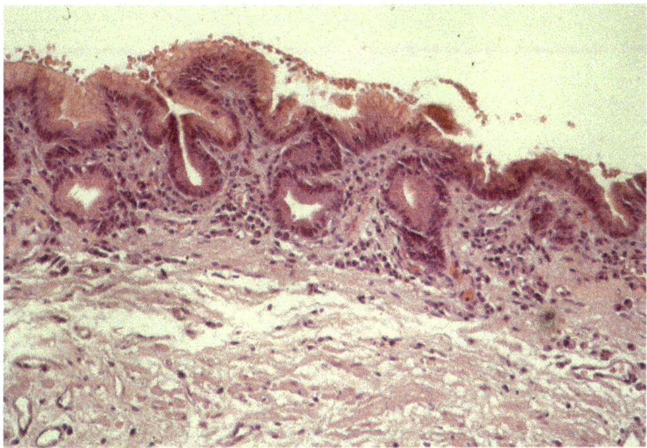

FIGURE 4.26 Cardiac epithelium, foveolar type. The surface epithelium has invaginated to form short foveolar pits. Both surface and foveolar pit cells are mucous cells.

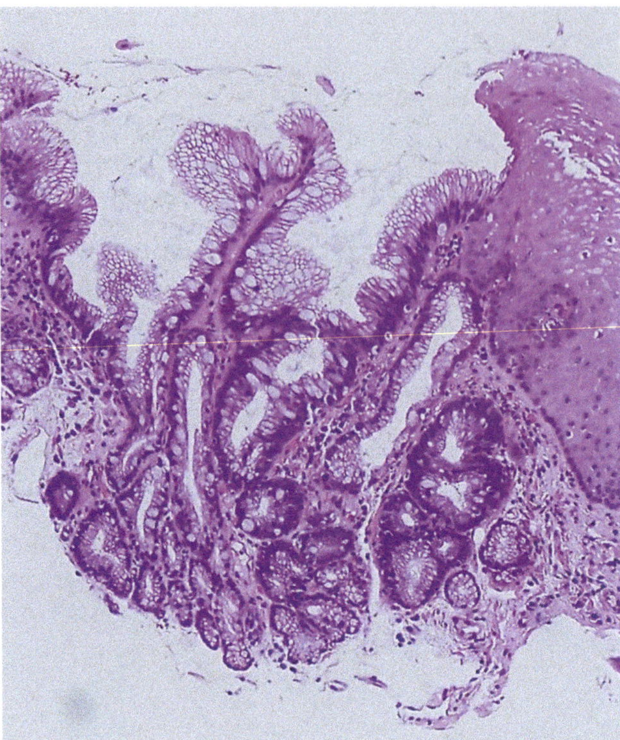

FIGURE 4.27 Cardiac epithelium, glandular type, with mucous cells in the surface, foveolar pit, and in disorganized glands. Note the mucin distension of some of the foveolar cells producing pseudogoblet cells. No definite goblet cells are seen.

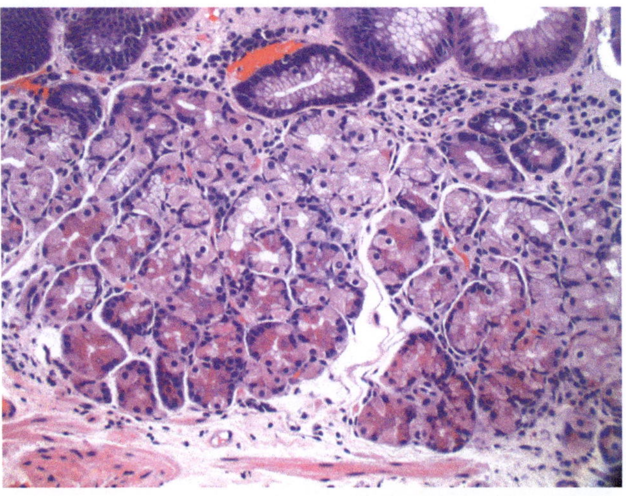

FIGURE 4.28 Metaplastic columnar epithelium glandular type with the presence of parietal and Paneth cells in the glands beneath the foveolar region. The presence of parietal cells makes this oxyntocardiac epithelium. Paneth cells are not used for definition and have no known significance.

foveolar region, well demonstrated in a Ki67 immunostain, which marks proliferative cells (Fig. 4.29). IM generally forms in the surface and foveolar region, but can involve the glands.

In a minority of cases, the cardiac epithelial surface is lined not by the usual single layer of mucous cells but is multilayered.[26] The multilayered epithelium usually consists of 2–5 layers of columnar cells, but may be a mixture of columnar cells and a basal region of basaloid squamous cells.

Cardiac epithelium always shows an expanded lamina propria with inflammatory cells and fibrosis (Fig. 4.30).[27] Hypervascularity and congestion are common. Villiform change of the surface and foveolar hyperplasia characterized by serration and associated with reactive cytologic changes are common (Fig. 4.31). The amount of inflammation, fibrosis, and edema

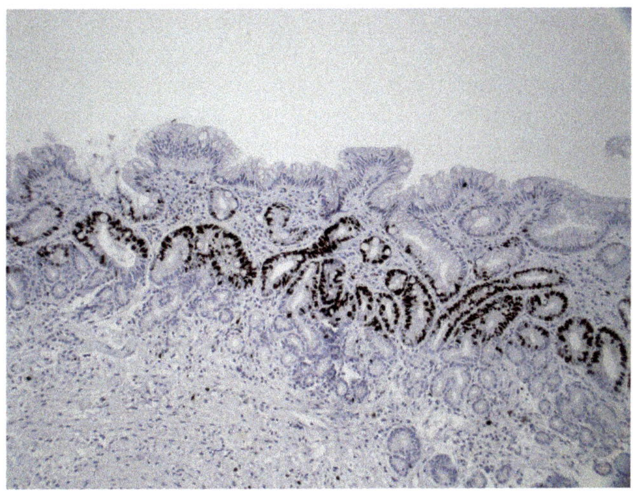

FIGURE 4.29 Metaplastic columnar epithelium, glandular type with intestinal metaplasia. This section is an immunohistochemical stain for Ki67 showing the proliferative zone in the deep foveolar region. The surface and superficial foveolar region and the deep glands are negative.

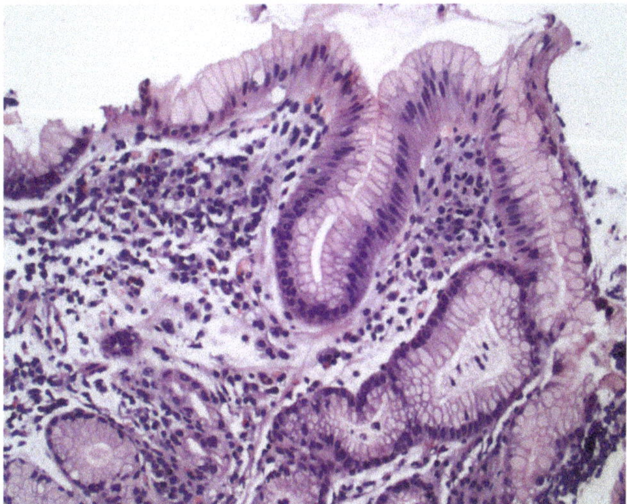

FIGURE 4.30 Cardiac epithelium with chronic inflammation in the lamina propria. Inflammatory cells are lymphocytes with scattered plasma cells and eosinophils. No neutrophils are seen.

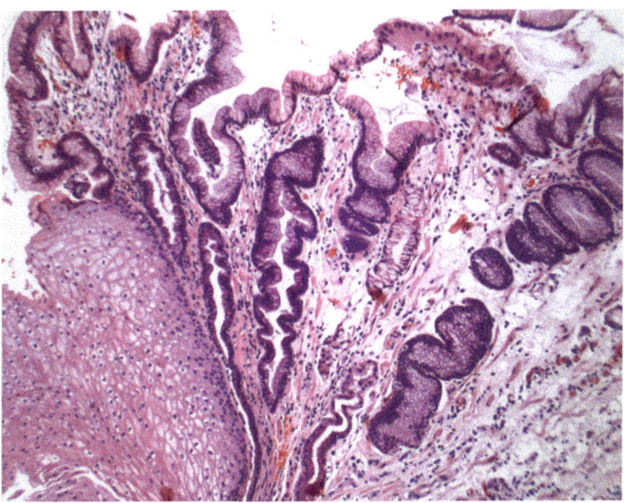

FIGURE 4.31 Cardiac epithelium at the squamocolumnar junction showing foveolar elongation and serration producing a villiform appearance.

of the lamina propria varies. The cells present are small lymphocytes, plasma cells, and eosinophils. Rarely, reactive lymphoid follicles may be present. Neutrophils are generally absent unless there is erosion or secondary infection by *Helicobacter pylori*. Proliferation of smooth muscle fibers of the muscularis mucosae causes thickening and splitting of the muscularis mucosae, irregularity of muscle fibers in its interphase with the submucosa, and upward extension of muscle fibers toward the surface.

Endoscopically, mucosa lined by cardiac epithelium may be flat or have rugal folds. It is indistinguishable from GOE endoscopically. When foveolar hyperplasia, lamina propria inflammation, and edema are severe, small polypoid lesions may be seen. Edematous, inflamed cardiac epithelium with reactive epithelium is the commonest cause of benign polyps in the region immediately distal to the endoscopic GEJ (Fig. 4.32).

Inflammation of cardiac epithelium is different than inflammation in the GOE in gastritis. It has a combination of reactive change and chronic inflammation with absent active inflammation. As such, use of the Sydney grading system for gastritis is not appropriate for assessment of severity of carditis. The Sydney system requires active inflammation as a criterion for the more severe grades. In cardiac epithelium, neutrophils are not seen unless secondary changes such as superinfection with *H. pylori* or surface erosion are present. As such, neutrophils are not predictive of severity of cardiac epithelial inflammation.

One remarkable finding in biopsies in the region of the GEJ is the difference in the amount of inflammation that is present in cardiac and gastric oxyntic epithelia when they coexist in the same or adjacent biopsies.[28] When gastric oxyntic mucosa is normal, as is common in the Caucasian population that has a low incidence of *H. pylori* gastritis, cardiac epithelial inflammation is strikingly greater than that in the immediately adjacent GOE.

3.3.2 Oxyntocardiac Epithelium (Metaplastic Esophageal Columnar Epithelium, Oxyntocardiac)

Oxyntocardiac epithelium (synonyms: gastric fundic-type epithelium; mixed mucous–parietal cell epithelium) is defined as an epithelium that contains glands beneath the foveolar pit composed of a mixture of parietal and mucous cells (Figs. 4.22 and 4.33).

Oxyntocardiac epithelium is seen in the squamooxyntic gap. In patients with very short squamooxyntic gaps, oxyntocardiac epithelium can be the only metaplastic columnar epithelium found in the gap (Fig. 4.33).[29] In patients with longer squamooxyntic gaps, oxyntocardiac epithelium is seen in the distal part of the gap immediately proximal to the true GEJ where it transitions into GOE.

The relative number of parietal and mucous cells varies greatly with predominance of either cell type being possible in the glands. Oxyntocardiac epithelium is differentiated from cardiac epithelium by the presence of parietal cells. It is differentiated from GOE by the presence of mucous cells in the glands under the foveolar region. Care is needed to identify sparse parietal cells to differentiate oxyntocardiac epithelium from cardiac epithelium and sparse mucous cells in the glands to differentiate it from GOE.

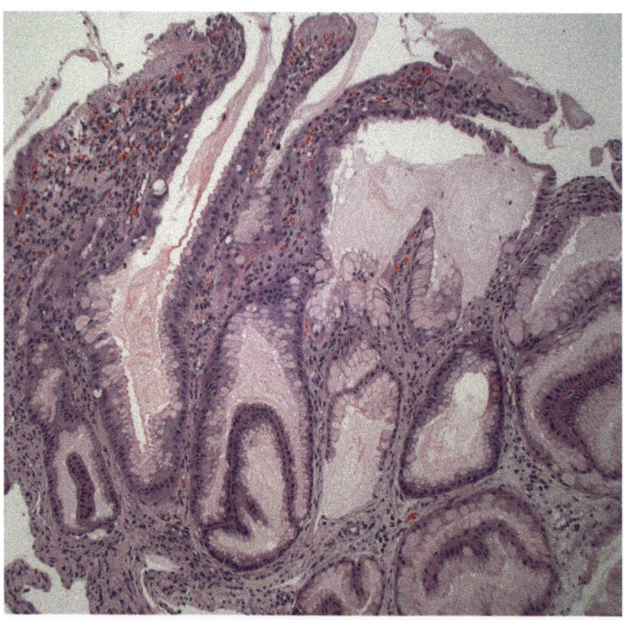

FIGURE 4.32 Cardiac epithelium with marked hyperplasia of mucous cells in the foveolar region with severe inflammation and surface erosion. This was from a specimen labeled "polyp in gastric cardia." Note the mucin-distended pseudogoblet cells.

The glands in oxyntocardiac mucosa vary from lobulated and disorganized resembling cardiac epithelium to less lobulated and straighter, resembling GOE. Inflammation is present in oxyntocardiac epithelium, but is less than that seen with the adjacent cardiac epithelium.[30] Inflammation is similar in nature to that in cardiac epithelium with lymphocytes, plasma cells, and eosinophils and a paucity of neutrophils. Reactive change and proliferative activity are also less than those in adjacent cardiac epithelium.[30]

When cardiac, oxyntocardiac, and normal gastric oxyntic epithelia are present in one biopsy or adjacent biopsies at the same level, the order of epithelia from proximal to distal usually is squamous, cardiac, oxyntocardiac, and gastric oxyntic. The inflammation in oxyntocardiac epithelium is generally less than that in cardiac epithelium and greater than that in GOE (unless the latter is involved by gastritis).

3.3.3 Cardiac Epithelium With Intestinal Metaplasia (Metaplastic Esophageal Columnar Epithelium, Intestinal)

Cardiac epithelium with IM (synonyms: specialized columnar epithelium; specialized intestinal metaplasia; cardiac IM; specialized Barrett epithelium) is defined by the presence of goblet cells in cardiac epithelium. Goblet cells have a well-defined round vacuole of acid mucin that occupies the major part of the cytoplasm. This vacuole pushes the nucleus to the base, flattening it against the basement membrane. Depending on the fixation and hematoxylin type used, the vacuole may have a blue tinge (Fig. 4.23) or appear as a clear round space (Fig. 4.34). Not infrequently, the number of goblet cells is small. In

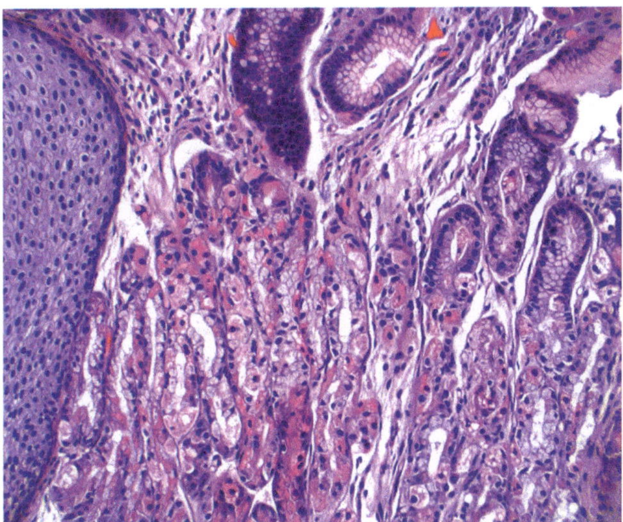

FIGURE 4.33 Biopsy from squamocolumnar junction showing oxyntocardiac epithelium transitioning directly to squamous epithelium without intervening cardiac epithelium. Note that the first foveolar/gland complex adjacent to the squamous epithelium has parietal cells. This transition will mainly be seen in people who are endoscopically normal.

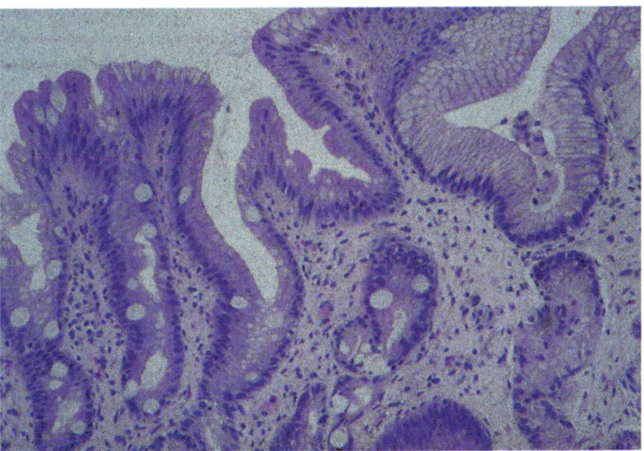

FIGURE 4.34 Intestinal metaplasia in cardiac epithelium. The goblet cells in this case are clear without basophilic staining. The difference in the staining is likely related to fixation and type of hematoxylin used.

these patients, the goblet cells tend to be located immediately adjacent to the SCJ (Fig. 4.35). Biopsy sampling technique is crucial for diagnosis in these patients. If a biopsy straddling the SCJ is not taken, small amounts of IM are easily missed.

Cardiac epithelium with IM is present in a minority of patients with metaplastic esophageal columnar epithelium. Its prevalence in CLE increases with increasing lengths of CLE (Fig. 4.36).[31] When present, it tends to be found preferentially in the most proximal region adjacent to the SCJ (Fig. 4.35).[31] From there, it extends distally to a variable extent within the CLE in a continuous manner, usually without skip areas (Fig. 4.37). It is commonly admixed with nonintestinalized cardiac epithelium at all levels, but in separate foveolar/gland units (Fig. 4.24). It is usually separated from GOE at the true GEJ by nonintestinalized cardiac epithelium and oxyntocardiac epithelium. The order of epithelial types from proximal to distal is

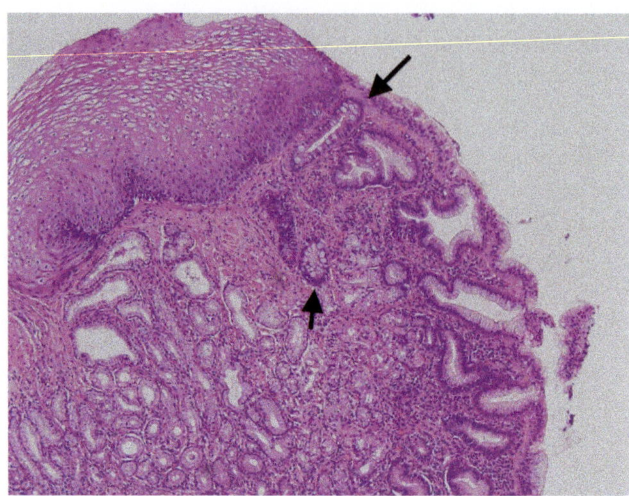

FIGURE 4.35 Cardiac epithelium with a single foveolar unit adjacent to the squamous epithelium showing goblet cells (*arrows*). The diagnosis of intestinal metaplasia is not dependent on the number of goblet cells; one definite goblet cell is sufficient.

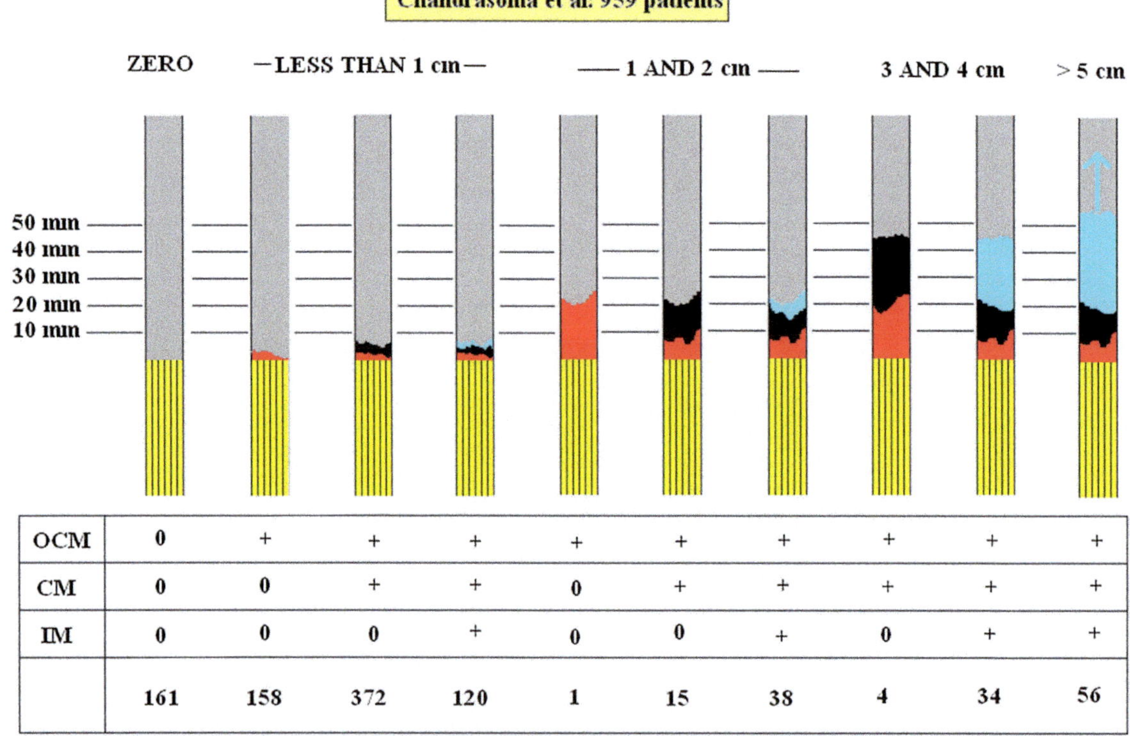

OCM	0	+	+	+	+	+	+	+	+	+
CM	0	0	+	+	0	+	+	+	+	+
IM	0	0	0	+	0	0	+	0	+	+
	161	158	372	120	1	15	38	4	34	56

FIGURE 4.36 Mapping biopsies in 959 patients biopsied in our unit. The prevalence of intestinal metaplasia (IM) increases progressively as the length of columnar-lined esophagus increases. *CM*, cardiac mucosa; *OCM*, oxyntocardiac mucosa.

usually squamous, cardiac with IM, cardiac, oxyntocardiac, and gastric oxyntic (Figs. 4.36 and 4.37). The distribution of the different epithelial types gives insight into their pathogenesis. I will discuss this in Chapter 11.

The diagnosis of IM can be made when one or more definite goblet cells are identified in the routine hematoxylin and eosin stain. All studies in the literature are based on this definition. Special stains and molecular markers are not necessary for confirmation and are not recommended.

The use of special stains such as Alcian blue stain at pH 2.5 and the combined Alcian blue–periodic acid Schiff (PAS) stain is common in practice. These should be discouraged. These stains highlight goblet cells beautifully (Fig. 4.38). However, they also stain nongoblet cells and increase the risk of false positive diagnosis of IM (Fig. 4.39). The definition of a goblet cell is *not* an Alcian blue–positive cell.

Immunoperoxidase stains should also not be used. Differential staining patterns with cytokeratin (CK) 7 and 20, DAS-1, villin, and CDX2 may all be positive in IM[32] but do not replace or enhance the diagnostic accuracy of the routine stain. They can all be positive in cardiac epithelium, usually less intensely than that in IM. In particular, when there are no goblet cells on the routine stain, positivity with any special stain should not be used to make a diagnosis of IM.

There is also no value for special stains that classify the type of acid mucin that is expressed in IM in this setting. Sulfomucins and sialomucins can be differentiated by high iron diamine stain. Mucin patterns are used to differentiate complete from incomplete IM. The vast majority of cases of cardiac mucosa with IM manifest features of incomplete IM, making high iron diamine staining superfluous. High iron diamine is no longer used because of its toxicity to the environment.

All special stains add unnecessary cost without improving the histologic diagnoses made on the basis of a routine hematoxylin and eosin–stained slide.

With very rare exceptions, parietal cells are not present in cardiac epithelium with IM. Cardiac epithelium with and without IM and oxyntocardiac epithelium may coexist within a small area, but it is exceedingly rare for goblet and parietal

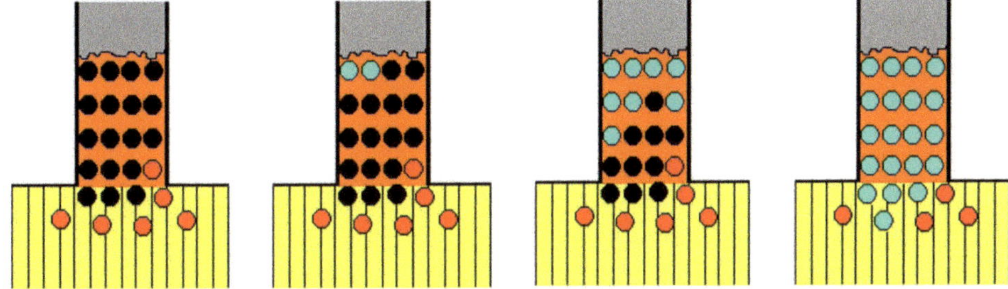

FIGURE 4.37 Diagrammatic representation of 4 different histologic findings in a patient with a visible columnar-lined esophagus measuring 5 cm. On the left, there is only cardiac (black) and oxyntocardiac epithelium (red). Progressing to the right, the extent of intestinal metaplasia (blue) increases, starting at the squamocolumnar junction and progressing caudad. This is the typical distribution of intestinal metaplasia within columnar-lined esophagus.

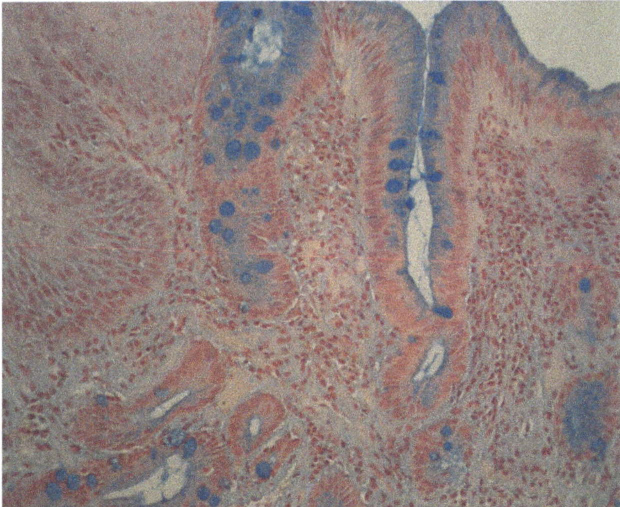

FIGURE 4.38 Alcian blue stain of cardiac epithelium with intestinal metaplasia showing positive staining of goblet cells. Note also that many nongoblet cells show positive Alcian blue staining.

cells to coexist in one foveolar unit. If it is certain that the epithelium being examined is esophageal and not chronic atrophic gastritis (CAG), the presence of goblet cells is given priority and a diagnosis of cardiac epithelium with IM is made.

4. CONTROVERSY REGARDING THE EPITHELIA OF THE ESOPHAGUS AND PROXIMAL STOMACH

4.1 Variation in Terminology

One effect of lack of deep knowledge of histology in this region is that simple concepts appear incredibly complicated. The number of terms that are used in the literature for one epithelial type makes it seem like there are a large number of different epithelia in the esophagus and stomach (Table 4.2). This is particularly true in the literature originating from clinicians without pathologist participation.

This complexity is an illusion. As seen above, the only terms that are required to completely define all epithelia lining the entire normal and GERD-damaged esophagus and proximal stomach to the pyloric antrum are stratified squamous epithelium, metaplastic cardiac epithelium (and its variants), and GOE.

Limiting terminology to these clearly defined histologic terms will improve understanding.

4.2 Points of Universal Agreement Regarding Histology

The following statements regarding histology of this region can be expressed as accepted facts for which there is no controversy at present:

1. The entire normal esophagus is lined by stratified squamous epithelium. The historical claim that the distal 2–3 cm can be normally lined by columnar epithelium has been appropriately rejected. The presence of any columnar epithelium above the endoscopic GEJ (defined as the proximal limit of rugal fold) is now universally considered to be abnormal, irrespective of its histologic type.
2. The esophagus of the patient who has visible CLE at endoscopy is lined by squamous epithelium in its proximal part and metaplastic esophageal columnar epithelium in its distal part. The metaplastic columnar epithelia that can be found are cardiac epithelium with IM, cardiac epithelium, and oxyntocardiac epithelium.
3. The area distal to the endoscopic GEJ can contain cardiac epithelium with IM, cardiac epithelium, and oxyntocardiac epithelium, identical to that in visible CLE. These can be limited to the area distal to the endoscopic GEJ or be associated with visible CLE in the thoracic esophagus.
4. When present, cardiac epithelium (with or without parietal/goblet cells) is *always* located between the squamous epithelium and GOE, i.e., in the "squamooxyntic gap."[19]
5. The body and fundus of the stomach is lined by GOE. This epithelium is present in everyone and extends distally to the pyloric antrum.
6. GOE is never seen proximal to the endoscopic or true GEJ.

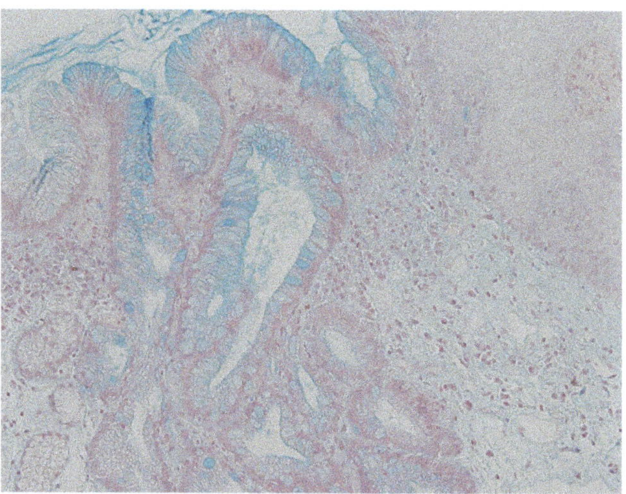

FIGURE 4.39 Alcian blue stain of cardiac epithelium without intestinal metaplasia (IM) showing positive staining of nongoblet cells. Alcian blue positivity in the absence of definite goblet cells does not satisfy criteria for a diagnosis of IM.

4.3 Points of Controversy

The points of controversy and difference of opinion are presented below. The headings reflect my viewpoint.

4.3.1 Cardiac Epithelium = Metaplastic Esophageal Columnar Epithelium, Not Normal Gastric Mucosa

There is controversy regarding the finding of the three nonoxyntic columnar epithelia that define CLE (cardiac, oxyntocardiac, and cardiac with IM) in the area distal to the endoscopic GEJ (Fig. 4.40). Traditional dogma regards this as a normal epithelium of the proximal stomach ("gastric cardiac mucosa").

The situation in Fig. 4.40A is normal. Neither the normal esophagus nor the normal stomach has cardiac epithelium (with and without parietal and/or goblet cells).

In Fig. 4.40B, the following changes, all caused by squamous epithelial damage resulting from exposure of esophageal squamous epithelium to gastric juice, are seen: (1) The distal squamous epithelium of the esophagus has been replaced by cardiac epithelium with or without parietal/goblet cells; (2) The damaged distal esophagus has dilated and developed rugal folds; (3) The endoscopic GEJ is now proximal to the true GEJ; (4) The tubular abdominal segment of the esophagus (i.e., length from diaphragm to endoscopic GEJ) is shortened; (5) The angle of His is more obtuse; (6) The abdominal segment of the LES is shortened to the same extent as the dilated segment of the distal esophagus that is lined by cardiac epithelium with and without parietal/goblet cells.

Every study that has correlated the length of cardiac epithelium with severity of GERD has shown a direct relationship, i.e., increasing length of cardiac epithelium distal to the endoscopic GEJ is associated with increasing severity of GERD.

When found, cardiac epithelium (with and without parietal and/or goblet cells) is *always* abnormal, *always* esophageal, and *always* results from columnar metaplasia of esophageal squamous epithelium as a result of exposure to gastric juice. A finding that has 100% specificity and 100% sensitivity is of extreme value in medicine.

When the fact that cardiac epithelium is never normal and never gastric is accepted, these three epithelial types can collectively be called "metaplastic esophageal columnar epithelia" and classified as cardiac, cardiac with parietal cells

TABLE 4.2 Different Terms Used to Describe Esophageal and Gastric Epithelial Types

Epithelial Type (Recommended Term)[a]	Significance	Different Equivalent Terms Used (Retirement Recommended)
Stratified squamous epithelium	Normal epithelium of the entire esophagus	None
Gastric oxyntic epithelium	Normal epithelium of the proximal stomach	Gastric fundic epithelium Gastric body epithelium
Cardiac epithelium	GERD-induced metaplastic esophageal columnar epithelium[a]	Gastric cardiac epithelium (incorrect) Junctional epithelium Mucous cell-only epithelium
Oxyntocardiac epithelium	GERD-induced metaplastic esophageal columnar epithelium[a]	Gastric fundic-type epithelium Mixed mucous/oxyntic cell epithelium
Cardiac epithelium with intestinal metaplasia	GERD-induced metaplastic esophageal columnar epithelium	Specialized columnar epithelium
		Specialized intestinal metaplasia (SIM)
		Barrett epithelium
		Intestinal metaplasia of the gastric cardia
		SIM of the GEJ

GEJ, gastroesophageal junction; *GERD*, gastroesophageal reflux disease.
The terms "epithelium" and "mucosa" are equivalent terms in this context and used interchangeably. The mucosa is a term that includes the epithelium, lamina propria, and the muscularis mucosae. The epithelium is the only element of the mucosa that varies in the different mucosal types and is used for definition. I will use the term "epithelium" in preference to mucosa.
[a]*These definitions must be applied irrespective of perceived location. They are purely histologic definitions. Anatomically, stratified squamous epithelium is always esophageal, and gastric oxyntic epithelium is always gastric. Confusion arises in terminology that attempts to ascribe cardiac epithelium and its variants to either esophageal or gastric by use of the incorrect definition of the GEJ. These are always metaplastic esophageal columnar epithelia, irrespective of whether they are found proximal or distal to the endoscopic GEJ. They are always proximal to the true GEJ defined as the proximal limit of gastric oxyntic epithelium.*

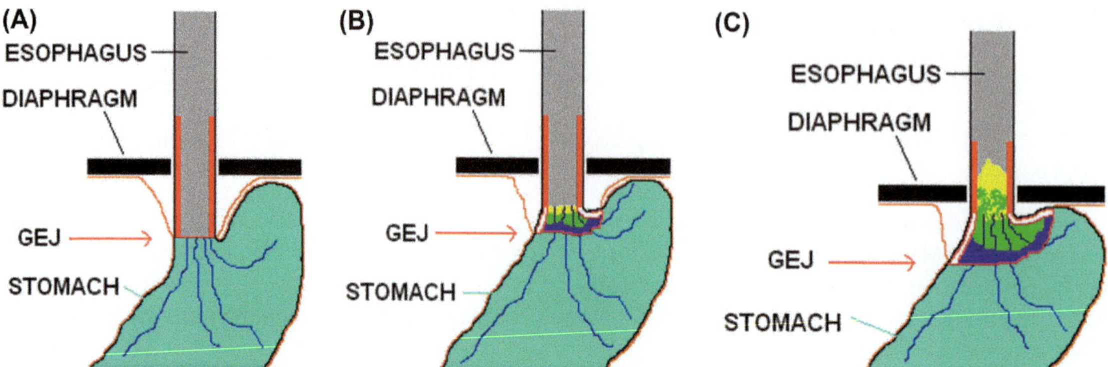

FIGURE 4.40 The relationship of columnar metaplasia in three different situations seen in clinical practice. (A) There is no metaplastic columnar epithelium between the squamocolumnar junction (SCJ) and gastric oxyntic epithelium. This is the rarely encountered normal state. (B) There is metaplastic columnar epithelium between the SCJ (also the end of the tubular esophagus and proximal limit of rugal folds). This is proximal to the true gastroesophageal junction (GEJ) [proximal limit of gastric oxyntic epithelium (blue)] but distal to the endoscopic GEJ. This is the dilated distal esophagus that is mistaken for the proximal stomach. (C) There is metaplastic columnar epithelium above the endoscopic GEJ in the thoracic esophagus. This is Barrett esophagus. Barrett esophagus always continues distal to the endoscopic GEJ to the true GEJ [proximal limit of gastric oxyntic epithelium (blue)]. Note that the dilated distal esophagus and lower esophageal sphincter damage are greater in (C) than (B).

(=oxyntocardiac), and cardiac with IM. In this book, I will continue to use the terms cardiac, oxyntocardiac, and cardiac with IM (or, cardiac epithelium with or without parietal/goblet cells) for these epithelia.

Cardiac epithelium and its variants (oxyntocardiac and intestinal), when found proximal to the endoscopic GEJ represents visible CLE in the thoracic esophagus (Fig. 4.40C). Histologically and in its expression of immunohistochemical and molecular markers, cardiac epithelium above and below the endoscopic GEJ are similar and different than those in GOE.

4.3.2 Cardia = Lower Esophageal Sphincter, Not Proximal Stomach

The term "cardia," though used freely by many authorities, has always been confusing. To experts of the past like Barrett[8,16] and Hayward,[9] the cardia was a term that was synonymous with the lower esophageal sphincter (LES) (as evidenced by the terms "cardiospasm" and "achalasia of the cardia" for a disease of the LES).

Without good reason, the term "cardia" has also been used for an anatomic region of the proximal stomach adjacent to the opening of the esophagus. This is an area supposedly lined by cardiac mucosa that is normally found distal to the endoscopic GEJ. When it shows inflammation, the term "carditis" is used; when it has goblet cells, the term "intestinal metaplasia of the cardia" is used; and when it is the site of a cancer, the term "adenocarcinoma of the cardia" is used. Being located distal to the endoscopic GEJ, "carditis," "intestinal metaplasia of the cardia," and "adenocarcinoma of the cardia" must represent gastric pathology. However, all these entities have been shown to have an association with GERD. The absurdity that GERD can produce gastric pathology is readily accepted in the highest intellectual circles.

The extent of this anatomic "gastric cardia" has not been defined. Based on his selective interpretation of the evidence from two autopsy studies,[29,33] Odze[34] has opined that its maximum extent, defined histologically by the presence of cardiac mucosa, is "less than 0.4 cm (=4 mm)." Sarbia et al.,[15] in a study of esophagectomy specimens, showed that histologic cardiac (including oxyntocardiac) epithelium can extend 2.8 cm distal to the end of the tubular esophagus.

The presence of cardiac epithelium distal to the endoscopic GEJ >4 mm, which is now the accepted maximum length in most pathology circles, is completely ignored. Though abnormal by definition, it is not recognized as a pathologic or clinical entity. All evidence indicates that the presence of increasing amounts of cardiac epithelium distal to the endoscopic GEJ is strongly related to increasing severity of GERD. This fact proves that cardiac epithelium is esophageal, not gastric.

The obvious conflict caused by using the term "cardia" for two completely different areas of the body is somehow easily and inexplicably accepted. "Achalasia of the cardia" is thought of (correctly) as a disease affecting the esophagus and "intestinal metaplasia and adenocarcinoma of the gastric cardia" is thought of (incorrectly) as a disease of the stomach.

The new histology of the region that I have proposed essentially removes the presence of a "gastric cardia" because cardiac epithelium does not line the stomach. When it is accepted that the entire proximal stomach is lined by GOE, the "gastric cardia" disappears; the gastric body lined by oxyntic epithelium extends to the GEJ. The area distal to the endoscopic GEJ that is lined by cardiac mucosa is a GERD-damaged DDE and not part of the proximal stomach.

When this is accepted, the terminology can revert to the accurate use of the term cardia by Barrett[8,16] and Hayward[9] where "cardia" is a synonym for the LES. The LES, which is the distal 4–5 cm of the esophagus, is normally lined by squamous epithelium. When its abdominal segment is damaged by GERD, it dilates and becomes lined by metaplastic cardiac epithelium (Fig. 4.3). This permits recognition of what we know to be true: "intestinal metaplasia and adenocarcinoma of the gastric cardia" are correctly called "intestinal metaplasia and adenocarcinoma of the dilated distal esophagus." There is now a rational explanation as to why these entities are associated with GERD.

The retention of the term "cardia" as being synonymous with the LES, though unnecessary, will facilitate understanding of the histology of this region once the dust clears. The recognition that cardiac epithelium exactly mirrors damage to the LES is critical to understanding that this epithelium is a histologic marker for LES damage (Fig. 4.3). This understanding provides a pathologic method of assessing severity of LES damage that is the basis of the new diagnostic method that I will propose for GERD.

Alternatively, the term cardia can disappear into vocabulary heaven and never be used again. There will then be an esophagus that has a 4–5 cm high-pressure sphincter zone at its distal end and a proximal stomach. Squamous epithelium lines the normal esophagus, including the LES, and GOE lines the proximal stomach. GERD-induced metaplastic columnar epithelia between the two will be first seen as the DDE resulting from LES damage. Clarity will replace confusion. If this happens, I know of no one who will complain. The use of the term "cardia" today is simply a crutch that is used to mask the confusion.

5. PROLIFERATION CHARACTERISTICS OF THESE EPITHELIAL TYPES

The three epithelial types in CLE have, in addition to their cellular composition, differences relating to the amount of inflammation and reactive changes, association with erosion, and proliferative rates of the cells.

The severity of chronic inflammation tends to be the highest in cardiac epithelium with IM, followed by cardiac epithelium with oxyntocardiac epithelium being least inflamed. This suggests a difference in the resistance of the three epithelia to gastric juice because the inflammation is most likely the result of their exposure to gastric acid.

Oxyntocardiac epithelium, which is most distal and therefore subject to more frequent and higher gastric juice exposure than the more proximal epithelia, is the least inflamed. Cardiac epithelium at and immediately distal to the endoscopic GEJ in the DDE tends to have the greatest reactive changes.

We undertook a study[30] to characterize the patterns of Ki67 expression in the different types of columnar metaplastic epithelia that occur in the esophagus as a result of GERD. 26 patients who underwent esophagogastrectomy at the University of Southern California for high-grade dysplasia or adenocarcinoma complicating long-segment Barrett esophagus (LSBE) were selected. 19 patients had nondysplastic IM and all patients showed cardiac and oxyntocardiac epithelia in the tubular esophagus distal to the SCJ. A total of 241 representative noneroded areas, which had a sufficient length of a uniform epithelial type, were selected for study. In squamous epithelium, a minimum of 2 mm was required; in columnar epithelia, three adjacent foveolar-gland units composed of the same epithelial type were required.

We selected 40 areas of stratified squamous epithelium, 43 areas of oxyntocardiac epithelium, 45 areas of cardiac epithelium, 41 areas of cardiac epithelium with IM, 9 areas of low-grade dysplasia, 16 areas of high-grade dysplasia, and 15 areas of invasive adenocarcinoma. 32 areas of normal GOE were taken from near the distal margin of the specimen.

Serial sections were stained with hematoxylin and eosin and for Ki67 by immunoperoxidase. Ki67 (MIB1) is an antigen that is expressed in cells that are in the cell cycle and is therefore a marker of proliferation. Immunoreactivity for Ki67 was defined as strong and complete nuclear staining.

Metaplastic columnar epithelia had the basic structure of gastrointestinal epithelia, being comprised of a surface epithelial cell layer, a foveolar pit region, and glands deep to the foveolar pit. Four areas selected had only a flat surface epithelial layer. All other columnar epithelia had a foveolar region of varying length; three had a rudimentary foveolar pit that precluded division into zones. In all the others, the foveolar pit was divided into superficial, mid, and deep thirds. True glands under the foveolar pit were defined morphologically and by the fact that they were Ki67 negative, similar to the glands in GOE. The Ki67 positivity in each zone was scored as follows: 0 = no Ki67$^+$ cells; 1 = 1%–10% Ki67$^+$ cells; 2 = 11%–25% Ki67$^+$ cells; 3 = 26%–50% Ki67$^+$ cells; 4 = >50% Ki67$^+$ cells.

The 40 areas of squamous epithelium showed a consistent pattern of Ki67 expression in the basal layer and 2–5 suprabasal squamous cell layers. The distribution in normal GOE was used as a control to define normal proliferative activity in columnar-lined epithelia. In all 32 areas of the GOE, Ki67$^+$ cells were restricted to the deep third of the foveolar pit without any Ki67$^+$ cells in the surface layer, the superficial and middle third of the foveolar pit, or the glands below the foveolar pit consisting of parietal and chief cells. Ki67 positivity in the deep third of the foveolar pit was scored as 1 in 1 area, 2 in 10 areas, 3 in 9 areas, and 4 in 12 areas.

We recognized two deviations from this normal pattern of Ki67 expression in the metaplastic CLE (Table 4.3).

5.1 Expanded Proliferation

This was defined as the presence of $Ki67^+$ cells in the middle third of the foveolar pit in addition to the deep third. We quantified this by requiring a combined mid + deep third Ki67 score >5 (normal epithelium had a maximum score of 4).

Oxyntocardiac epithelium in CLE had a pattern of Ki67 expression that resembled normal GOE. Mid-foveolar staining was present in 17/43 (39.5%) of cases. The combined mid + deep third score was >5 in 4/43 (9.3%) cases, not significantly different than that in normal GOE ($P = .21$).

The four cases of cardiac epithelium consisting of a flat surface epithelial layer showed Ki67 positivity in a minority of the surface cells. When a foveolar pit developed in cardiac epithelium, the $Ki67^+$ cells moved to the deep region of the foveolar pit. In the 38 cases of cardiac epithelium with an adequate length of foveolar pit for evaluation, the combined mid + deep third score >5 was seen in 11 (29%) cases, significantly higher than those of both gastric oxyntic and oxyntocardiac epithelium ($P = .02$).

The 41 areas of nondysplastic IM showed a combined mid + deep third score >5 in 26 (63.4%) cases, significantly higher than that of cardiac epithelium ($P = .02$).

5.2 Aberrant Proliferation

This was defined as the presence of $Ki67^+$ cells in the surface epithelial layer and/or the superficial third of the foveolar pit. This was quantified by the combined score for the surface + superficial third of foveolar pit >1.

None of the 43 areas of oxyntocardiac epithelium showed aberrant Ki67 expression.

The four areas of cardiac epithelium composed of only a surface layer showed Ki67 positivity at the surface; 1/3 areas of cardiac epithelium with rudimentary foveolar pits showed surface staining. The other two showed staining limited to the short foveolar pits. The 38 cases of cardiac epithelium with well-formed foveolar pits showed aberrant Ki67 expression (i.e., a surface + superficial foveolar score >1) in five (13.2%) areas. This was significantly higher than that in oxyntocardiac epithelium ($P = .02$). Cellular morphology in the five areas of cardiac epithelium with aberrant Ki67 expression was normal in three and regenerative in two.

7/41 areas of nondysplastic IM showed aberrant expression, similar to that seen in cardiac epithelium ($P = .55$). Cellular morphology in the seven areas with aberrant Ki67 was normal in six and showed regenerative changes in one. IM with low-grade dysplasia showed aberrant Ki67 expression in 3/9 (33%) cases, similar to nondysplastic IM ($P = .51$) and cardiac epithelium ($P = .78$).

Aberrant expression of Ki67 was seen in 14/16 (87.5%) of high-grade dysplasia, and all cases of adenocarcinoma, significantly higher than that in low-grade dysplasia (3/9; 33%) and all nondysplastic epithelia ($P < .001$).

I have found the presence of aberrant expression of Ki67 to be helpful in the differentiation between low- and high-grade dysplasia although I do not believe its reliability is higher than morphologic criteria.

TABLE 4.3 Expanded and Aberrant Proliferative Patterns of Ki67 Expression in the Different Epithelial Types in Visible Columnar-Lined Esophagus

Epithelial Type Number	GOM 32	OCM 43 (%)	CM 45 (%)	IM 41 (%)	LGD 9 (%)	HGD 16 (%)	CA 15 (%)
Expanded proliferation	0	4 (9.3%)[a]	14 (31.1%)[a]	26 (63.4%)[a]	9 (100%)	12 (75%)	15 (100%)
Aberrant proliferation	0	0	10 (22.2%)	7 (15.6%)	3 (33.3%)	14 (87.5%)[b]	15 (100%)[b]

CA, adenocarcinoma; *CM*, cardiac mucosa; *GOM*, gastric oxyntic mucosa; *HGD*, high-grade dysplasia; *IM*, intestinal metaplasia; *LGD*, low-grade dysplasia; *OCM*, oxyntocardiac mucosa.
[a]*Significantly higher versus each previous nondysplastic epithelial type.*
[b]*High-grade dysplasia and adenocarcinoma significantly higher than all previous epithelial types, including low-grade dysplasia (P<.001).*

6. PROBLEMS IN PATHOLOGIC DIAGNOSIS

Diagnostic problems are rarely encountered in the interpretation of epithelial types in the esophagus and stomach if adequate biopsies are taken. The following situations that have potential for error can arise:

6.1 Inadequate Biopsies

Accurate pathologic diagnosis of epithelial type generally requires a full thickness biopsy of the epithelium. Superficial biopsies that are limited to the surface and foveolar region and containing only mucous cells will be identical in cardiac, oxyntocardiac, and gastric oxyntic epithelia and are inadequate for classification. They are sufficient if the epithelium shows goblet cells because that finding alone is diagnostic of IM.

When reporting on a biopsy the pathologist should make a clear statement of inadequacy when this is the case. The main reason for inadequacy is lack of adequate depth. Other causes include poor fixation or improper processing that may render interpretation difficult and therefore liable to error.

6.2 Diagnosis of Intestinal Metaplasia in Cardiac Epithelium

The accurate diagnosis of IM in cardiac epithelium is critical because it defines an increased risk of esophageal adenocarcinoma.[35]

In most cases, the diagnosis of IM is easy because goblet cells are numerous and well defined even if they are present in only a small part of the biopsy. Cases that present difficulty are those where goblet cells are rare and where the features are not typical of goblet cells. Mucin distended cells in cardiac epithelium can mimic goblet cells to the extent that they have been designated pseudogoblet cells (Figs. 4.27 and 4.41). In these cases, there can be significant interobserver variation.

Goblet cells have a single large, round vacuole of acid mucin that distends the cell borders and pushes the nucleus to the base of the cell. The vacuole often has a basophilic tinge if mucinous material has been retained during processing (Fig. 4.34). Otherwise, the vacuole will be empty and appear clear (Fig. 4.35). The cytoplasm around the mucin vacuole is usually dense and deeply eosinophilic (Fig. 4.35). At the luminal aspect, there is usually a deep eosinophilic band of cytoplasm that separates the vacuole from the lumen. Typically goblet cells are found as single cells separated by nongoblet mucous cells.

In contrast, pseudogoblet cells have cytoplasmic mucin that distends the entire cell or form small vacuoles (Fig. 4.41) rather than a definite round mucin vacuole. There is little demarcation of the mucin from the remainder of the cytoplasm. The mucin reaches the surface of the cells. Pseudogoblet cells tend to involve many adjacent cells. They frequently have a lightly basophilic tinge and often stain positively with Alcian blue stain at pH 2.5, though not as strongly as true goblet cells (Fig. 4.39).

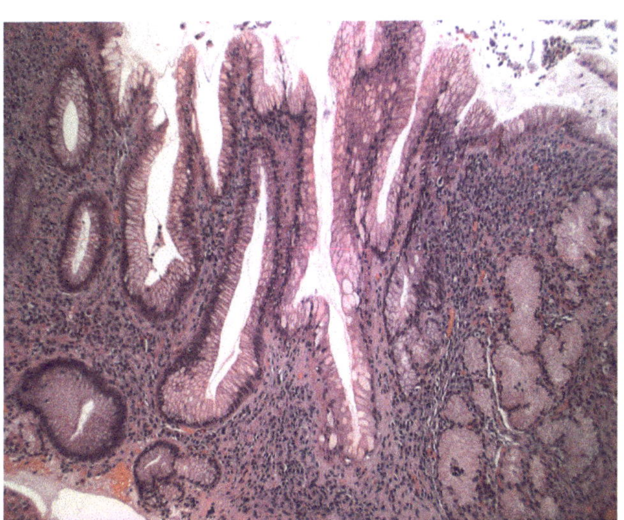

FIGURE 4.41 Pseudogoblet cells in cardiac epithelium. The vacuoles in the cytoplasm are not adequately defined and round to qualify as goblet cells.

Pseudogoblet cells occur frequently in multilayered epithelium and gland ducts that show reactive changes (Fig. 4.42). In these locations, the pseudogoblet cells can have basophilic cytoplasm that stains positively with Alcian blue at pH 2.5, making overdiagnosis of IM a significant risk.

In most cases, the distinction between true goblet cells and pseudogoblet cells is not difficult. Where doubt exists, the pathologist must err on the side of specificity, i.e., any cell that is not definitely a goblet cell is negative. The diagnosis of Barrett esophagus precipitates endoscopic surveillance and creates hardship and cost to the patient. This is justified only with a definite diagnosis.

Special stains make the diagnostic problem worse and result in overdiagnosis of IM. A goblet cell is defined by its morphologic features in routine hematoxylin and eosin stain; it is not an Alcian blue positive cell.

The relationship between pseudogoblet cells and goblet cells is not known. There is no evidence at this time that pseudogoblet cells are a precursor of goblet cells. In my opinion, it is more likely that they represent a reactive change that is not permanent.

6.3 Oxyntocardiac Epithelium Versus Gastric Oxyntic Epithelium

Oxyntocardiac epithelium and GOE both have parietal cells in the glands. The definitional difference is that mucous cells are present in the glands admixed with parietal cells in oxyntocardiac but not GOE.

In most cases of oxyntocardiac epithelium, the presence of lobulation of the glands, numerous mucous cells in the glands, and inflammation and fibrosis of the lamina propria make diagnosis easy (Figs. 4.22, 4.28, and 4.33). As the number of parietal cells in oxyntocardiac epithelium increases, mucous cells decrease, inflammation becomes less, and the lobulation decreases, making distinction more difficult.

In difficult cases, staining for neutral mucin with a PAS stain is helpful. The mucous cells within the gland stain, a bright magenta color that contrast with negative staining parietal cells in oxyntocardiac epithelium (Fig. 4.43), make the identification of small numbers of either of these cell types easier. In contrast, deep magenta staining with PAS is limited to the surface and foveolar cells in GOE where the glands are composed of parietal and chief cells and no mucous cells. Parietal and chief cells do not contain neutral mucin and are therefore negative with PAS (Fig. 4.44). This permits recognition of rare mucous cells in an oxyntocardiac epithelium that resembles GOE.

Another specific difference is the presence of gland ducts of the submucosal glands in the mucosa of oxyntocardiac epithelium (Figs. 4.11 and 4.12).[24] Submucosal glands do not occur in the gastric oxyntic epithelium lining the human stomach and therefore, gland ducts never penetrate the mucosa of GOE.

The distinction between oxyntocardiac and gastric oxyntic epithelia is not critical from a cancer risk standpoint. Oxyntocardiac epithelium is a stable epithelial type that does not develop IM and is not at risk for adenocarcinoma.

However, the distinction has fundamental importance in the definition of the distal limit of the esophagus. Oxyntocardiac epithelium is a metaplastic esophageal epithelium unlike GOE, which is a normal gastric epithelium. Error in differentiating between oxyntocardiac epithelium and GOE leads to an inaccurate measurement of the squamooxyntic gap in the DDE. This is a critical measurement in the new pathologic assessment of LES damage.

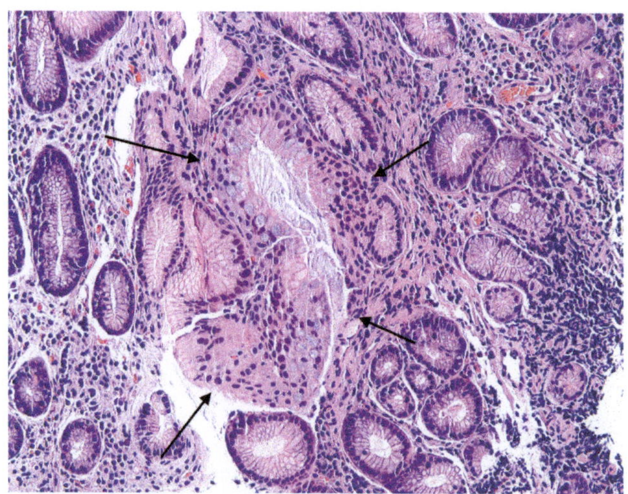

FIGURE 4.42 Pseudogoblet cells in a gland duct within the cardiac epithelium (*arrows* outlining the gland duct). Note the blue tinge in the mucin-distended cells.

6.4 Chronic Atrophic Gastritis Versus Cardiac Epithelium With and Without Intestinal Metaplasia

CAG is a common disorder with a variable geographic distribution. It results from autoimmune gastritis or, much more commonly, infection with *H. pylori*. *H. pylori* infection is relatively uncommon in affluent Caucasian populations of the Western world. Chronic atrophic gastritis is extremely common in populations where *H. pylori* infection is prevalent such as in East Asian countries.

There is an inverse relationship between the prevalence of GERD and *H. pylori* infection in different populations. This translates to a wide difference in the incidence of adenocarcinoma of the esophagus, which predominates in North America and Europe, and adenocarcinoma of the stomach, which predominates in Asia.

CAG caused by autoimmune gastritis primarily involves the parietal cell containing epithelium of the stomach, i.e., proximal to the antrum. In contrast, *H. pylori* preferentially infects the antrum and results in an antral gastritis. As the disease severity increases, it may extend into the proximal stomach resulting in a pangastritis (Fig. 4.45).

When columnar metaplastic epithelium is present in the distal esophagus, *H. pylori* can extend into it causing an increase in the severity of chronic inflammation and the presence of active inflammation.[27] *H. pylori* does not infect esophageal squamous epithelium and is not a cause of columnar metaplasia in the esophagus.

In CAG, there is progressive loss of parietal cells in the glands, associated with chronic inflammation. The glands become shorter and the mucosa thinner. There is frequently hyperplasia of the foveolar region. The atrophic glands

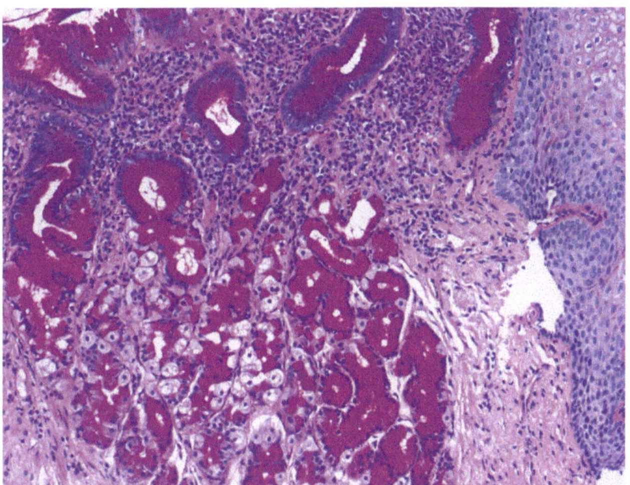

FIGURE 4.43 Oxyntocardiac epithelium stained with digested PAS stain. The mucous cells in the surface, foveolar region, and glands stain a deep magenta color that contrasts with the admixed negatively staining parietal cells in the glands.

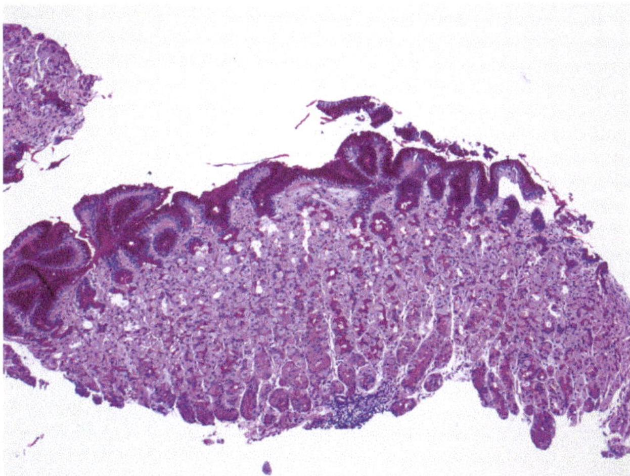

FIGURE 4.44 Gastric oxyntic epithelium stained with digested PAS stain. The positive staining is limited to the surface and the foveolar region that contain cells with neutral mucin. The glands, composed of parietal and chief cells, are negative.

either disappear or undergo metaplasia with replacement of the parietal cells with mucous cells (pseudopyloric metaplasia; Fig. 4.46) or goblet cells (IM; Fig. 4.47). In cases with partial atrophy of the glands, mucous cells may rarely coexist with residual parietal cells in an individual foveolar-gland complex.

This results in conversion of GOE to a columnar epithelium consisting of mucous cells only (which resembles cardiac epithelium), a mixture of mucous and parietal cells (which resembles oxyntocardiac epithelium) or a mixture of mucous and goblet cells (which resembles cardiac epithelium with IM).

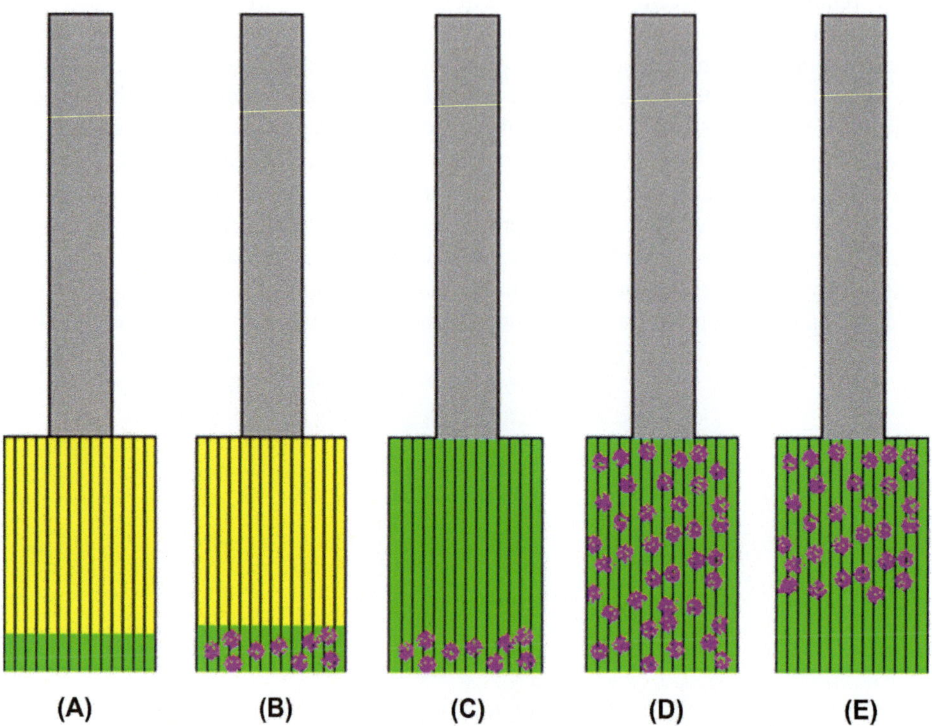

FIGURE 4.45 Diagrammatic representation of different patterns of chronic gastritis. (A) Chronic antral gastritis (green) without atrophy or intestinal metaplasia; the body is normal (yellow). (B) Chronic atrophic antral gastritis with intestinal metaplasia (*purple dots*); the body is normal. (C) Chronic pangastritis with atrophy and intestinal metaplasia involving only antrum. (D) Chronic atrophic pangastritis with intestinal metaplasia involving the entire stomach. (E) Chronic pangastritis with atrophy and intestinal metaplasia that spares the antrum. In all cases shown here, the esophagus is uninvolved (gray squamous epithelium without metaplastic columnar epithelium).

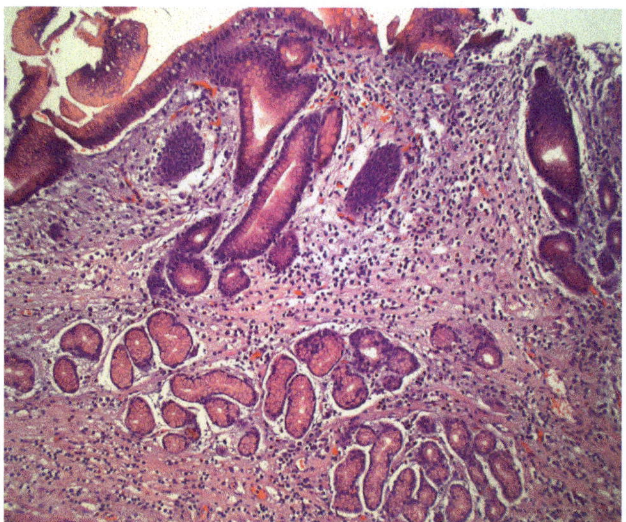

FIGURE 4.46 Biopsy from the gastric body showing chronic gastritis with atrophy (loss of parietal cells) and pseudopyloric metaplasia. There is no intestinal metaplasia. This resembles cardiac epithelium.

It should be understood clearly that while these epithelia may resemble each other, they are completely different. The epithelia resulting from CAG are pathologic changes in GOE unrelated to GERD. Cardiac epithelium with and without IM and oxyntocardiac epithelium are the result of GERD-induced columnar metaplasia of esophageal squamous epithelium.

When these epithelia are found in biopsies taken from an endoscopically visible CLE, there is no difficulty (Table 4.4). The epithelial types observed will be the three types known to comprise CLE (cardiac epithelium with and without IM and oxyntocardiac epithelium).

When there is a coexistence of visible CLE and CAG, the exact classification of biopsies at the GEJ region is usually irrelevant because the two entities are present in more proximal and distal biopsies (Table 4.4).

The only situation that is problematic and important is when a single biopsy is taken at the SCJ in a patient without endoscopic evidence of CLE. In such a case, the differentiation between metaplastic esophageal columnar epithelium with IM limited to the DDE from atrophic pangastritis with IM in GOE extending to the GEJ can be difficult.

This problem is most easily resolved by routinely taking biopsies from the body and antrum in addition to the DDE (Table 4.4). If the distal gastric biopsies are normal, the finding of cardiac epithelium (with or without parietal and/or goblet cells) indicates GERD-induced metaplastic esophageal columnar epithelium. CAG does not occur as an isolated phenomenon limited to the proximal few centimeters of the stomach.

When cardiac epithelium (with and without parietal and/or goblet cells) is found in a biopsy taken from the GEJ in an endoscopically normal patient with CAG in the distal gastric biopsies, the diagnosis is a problem. There is an overlap of features between CAG and metaplastic esophageal columnar epithelia.

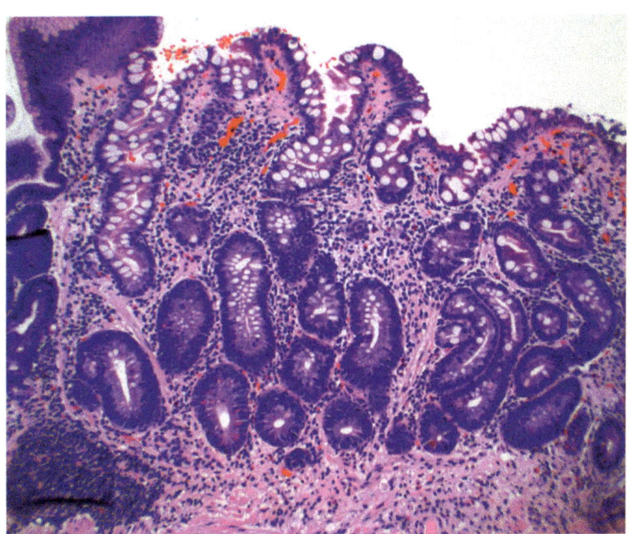

FIGURE 4.47 Biopsy from the gastric body showing chronic gastritis with atrophy (loss of parietal cells) and intestinal metaplasia. This resembles cardiac epithelium with intestinal metaplasia.

TABLE 4.4 Differential Diagnosis of the Presence of Intestinal Metaplasia in a Biopsy Taken From the Region of the Gastroesophageal Junction. The Use of Proximal and Distal Intestinal Metaplasia in the Differential Diagnosis

Diagnosis	IM in Visible CLE	IM Distal to Endo GEJ	IM in the Distal Stomach
LSBE	+; CLE > 3 cm	±	−
SSBE	+; CLE < 3 cm	±	−
CAG	−	±	+
IM in DDE	−	+	−
BE + CAG	+	±	+
IM in DDE + CAG	−	+	+

BE, Barrett esophagus; *CAG*, chronic atrophic gastritis; *CLE*, columnar-lined esophagus; *DDE*, dilated distal esophagus; *endo GEJ*, endoscopic gastroesophageal junction; *IM*, intestinal metaplasia; *LSBE*, long-segment Barrett esophagus; *SSBE*, short-segment Barrett esophagus.

The following differential features in metaplastic esophageal columnar epithelia are helpful in distinguishing it from CAG: (1) The surface epithelium is frequently villiform in metaplastic esophageal columnar epithelia and flat in CAG. (2) The presence of multilayered epithelium and gland ducts indicates metaplastic esophageal columnar epithelia. These are not seen in gastric oxyntic mucosa. (3) Pancreatic metaplasia is seen more often in metaplastic esophageal columnar epithelia than in CAG. (4) Paneth cells are more common in CAG. (5) Neuroendocrine cell hyperplasia favors CAG. Staining for synaptophysin may be helpful. (6) Reactive foveolar hyperplasia with irregular extension of muscle fibers of the muscularis mucosae into the superficial epithelium is more common in metaplastic esophageal columnar epithelia. (7) Neutrophils are rare in CLE without active *H. pylori* infection or erosion. None of these features are definitive.

Immunoperoxidase staining has been attempted to differentiate between cardiac epithelium and CAG with IM. DAS-1 monoclonal antibody is reported to be positive in IM occurring in cardiac epithelium and not in gastric IM in CAG.[36] Staining patterns with CK7 and 20 are reported to be different.[37] Cardiac epithelium with IM shows CK20 staining limited to the surface and CK7 staining in both superficial and deep glands. This contrasts with CAG with IM where there is full thickness staining with CK20 and patchy staining with CK7. These are all unreliable and have generally been abandoned.

The differentiation of IM into complete and incomplete based on staining patterns with high iron diamine stain as a method of differentiating cardiac epithelium with IM from CAG has also been abandoned. Cardiac epithelium with IM in visible CLE is almost always the incomplete type. Gastric intestinal epithelium in CAG can be complete or incomplete.

Except in rare cases where specific criteria of esophageal location are present (see below), I do not attempt to diagnose microscopic CLE in a biopsy from the junctional region in a patient with chronic atrophic pangastritis. It is possible that we miss a rare case of microscopic CLE coexisting with gastritis. We prefer specificity to sensitivity of diagnosis in this situation.

7. EXPRESSION OF OTHER MARKERS IN METAPLASTIC COLUMNAR EPITHELIA

In the discussion of the histologic diagnosis and differentiation of the three columnar epithelial types in CLE (proximal to the endoscopic GEJ in the thoracic esophagus and distal to the endoscopic GEJ in the DDE), I described all the stains used at present in practice. These include hematoxylin and eosin (the routine stain), Alcian blue at pH 2.5, PAS-Alcian blue, and PAS with diastase digestion.

In my practice, I only examine the routine hematoxylin and eosin stain for diagnosis of the different epithelial types. I believe that the use of Alcian blue and PAS-Alcian blue stains greatly increases the risk of overdiagnosis of IM. Lee et al.[37] found acid mucin in nongoblet columnar cells in 71% of esophageal biopsies in patients with CLE stained with Alcian blue at pH 2.5 or PAS-Alcian blue at pH 2.5.

I believe that digested PAS stain has value in differentiating oxyntocardiac from GOE. This is of little practical value at this time. However, it may become essential in the future if the new method of pathologic assessment becomes important. In that setting, it is the best way to demarcate the exact line of the true GEJ, which is the junction between GOE and metaplastic oxyntocardiac epithelium. Digested PAS stain will stain the neutral mucin in the cytoplasm of mucous cells. These are present in the surface epithelium and foveolar pit of both but are found in the glands of oxyntocardiac epithelium but not in the glands of GOE.

The presence of mucin gene products has been suggested as being useful for the diagnosis of IM in cases of doubt. Arul et al.[38] in a study of Barrett esophagus using in situ hybridization and immunohistochemistry showed that the normal esophagus expressed MUC5B in the submucosal glands and MUC1 and MUC4 in the stratified squamous epithelium. In IM, MUC3 and MUC5AC are expressed in the superficial columnar epithelium, MUC6 in the glands, and MUC2 in goblet cells. MUC2 has been reported to be highly sensitive[39] and specific[40] for goblet cells in Barrett esophagus. MUC2 is an intestine-specific mucin gene product whereas MUC5AC and MUC6 are expressed in gastric epithelium.

DAS-1 is a monoclonal antibody developed against a colonic epithelial protein that has been reported to have value in the differentiation of columnar epithelial types in metaplasia of the esophagus and stomach. Glickman et al.[41] reported that DAS-1 antibody stained goblet and nongoblet columnar cells in LSBE (91%), short-segment Barrett esophagus (88%), IM of the GEJ (100%), IM of the gastric antrum (13%), cardiac epithelium at the GEJ (35%), and normal gastric antrum (5%). The antibody therefore had no practical value and has not been used in clinical practice.

Ormsby et al.[42] reported that 29/31 (94%) of Barrett esophagus was characterized by superficial CK20 staining of the surface epithelium and superficial gland and moderate to strong CK7 staining of superficial and deep glands. This pattern was not seen in cardiac epithelium or IM of the stomach. Glickman et al.,[37] however, found this pattern in gastric IM but not in cardiac epithelium. Differential CK7/20 staining was used for a few years but is rarely used today because of its lack of reliability in practice.

In summary, the diagnosis of IM in both the esophagus and stomach is based on the identification of goblet cells in a routine section stained with hematoxylin and eosin. In the absence of definite goblet cells, a diagnosis of IM should not be made. Where there is doubt about the presence of goblet cells, i.e., when pseudogoblet cells are present, the diagnosis of IM should not be made. There is no ancillary method—Alcian blue at pH 2.5, PAS-Alcian blue, CDX2 by immunohistochemistry, DAS-1, differential CK staining—that increases the specificity of routine hematoxylin and eosin identification of goblet cells. In fact, performing Alcian blue and PAS-Alcian blue stains, though done in the majority of pathology laboratories for the routine detection of IM in CLE, has a significant risk of false positive diagnosis of IM. IM is the presence of definite goblet cells, not blue staining cells on Alcian blue. The simplest and cheapest routine hematoxylin and eosin stain is the best.

8. PATHOLOGIC REPORTING OF BIOPSY FINDINGS

There are different ways to report the pathologic findings in the biopsy sample.

The most important element in the pathology report is the epithelial composition of the biopsy. This can be reported comprehensively as all the types of epithelia present with percentages of each type of epithelium. This is not necessary and results in a complicated report with too much information for clinical need.

The method that I use prioritizes epithelial types according to the epithelial type that is the most significant and reports only that epithelium. The priority level is cardiac epithelium with IM, cardiac epithelium, oxyntocardiac epithelium, and GOE. To this is added any pathology that may be present in squamous and gastric oxyntic epithelia if any is present.

The reason for this method of reporting is that, at present, the only clinical relevance of biopsy is the presence or absence of IM. Biopsies are limited to patients with a visible CLE above the endoscopic GEJ. The reason for the biopsy is to see whether IM is present or not. The only answer that is sought in the pathology report is whether or not intestinal metaplasia is present. Similarly if biopsies are performed in endoscopically normal patients, the only finding that has relevance at present is the presence of IM. Most gastroenterologists do not care about the other types of epithelia (cardiac, oxyntocardiac, and gastric oxyntic epithelia) that may be present in the biopsy.

In our unit, the presence of cardiac and oxyntocardiac epithelium without IM has significance because it is recognized as a manifestation of reflux. This is true whether or not the patient has a visible CLE. Because cardiac epithelium is always inflamed, I report this as reflux carditis.

In the future, if the new method of pathologic assessment of LES damage becomes accepted, the biopsy samples must be interpretated to accurately measure the length of cardiac epithelium (with and without parietal and/or goblet cells) between the distal limit of the squamous epithelium and the proximal limit of GOE. This measures damage to abdominal segment of the LES. I will discuss this in Chapter 17.

9. PRECISION AND ACCURACY OF DIAGNOSIS OF COLUMNAR EPITHELIA

The diagnosis of epithelial types using the definitions given above is easy and has a very high interobserver agreement rate. The amount of training required for a high level of proficiency of diagnostic accuracy is very small. Residents in the pathology training program at the Los Angeles County University of Southern California Medical Center become proficient in identifying these epithelia by the end of the first monthly rotation in surgical pathology. Pathologists visiting the Center to learn the method require one 4-h period of looking at slides with me with a relatively painless explanation of the method. Pathologists can also reach proficiency by reading illustrated books on the subject very easily. The reward of a simple diagnostic method is high precision.

A practical anecdotal experience demonstrates this well. Professor Martin Riegler, an esophageal surgeon from Vienna, Austria, became interested in our method of assessing biopsies from GERD patients at the annual Hawaii Foregut Diseases conference conducted by the University of Southern California. On returning to Vienna, he began taking biopsies according to an expanded version of our protocol.[12] He recruited a senior pathologist at the University Hospital, Professor Fritz Wrba, to read these slides. The pathologist educated himself on the criteria based on our papers and read nearly 2000 biopsies from a series of patients.

I was invited to Vienna for a second blind read of these same biopsies. The interobserver variation was miniscule. One patient diagnosed as cardiac epithelium with IM was reclassified as cardiac epithelium with pseudogoblet cells (i.e., no IM). Six cases that were called GOE by Dr. Wrba were reclassified as oxyntocardiac epithelium at the consensus second review of the slides.

There was discussion during this meeting about main source of interobserver variation: a pathologist may frequently miss rare parietal cells in cardiac epithelium. This was not borne out by the study where there was near total concordance in the diagnosis of cardiac and oxyntocardiac epithelia.

Nearly 1500 specimens, which contained cardiac epithelium, were stained by immunoperoxidase with an antibody against parietal cells that was supplied by Dr. John Forte of the University of California at Berkeley. Four cases were seen to have immunoperoxidase positive rare parietal cells. Review of the original biopsy showed that these were present in the deeper sections but not in the original section. The conclusion was that a careful pathologist has an extremely high accuracy in the differentiation of cardiac from oxyntocardiac epithelia. The very small error rate could be almost completely eradicated by the use of one set of deeper sections of all cases that had only cardiac epithelium.

The results of the study based on these biopsies were published and provide independent confirmation of our data.[12] It showed that the histologic classification that we had developed in Los Angeles was duplicated with near perfection by an experienced pathologist in Vienna who had trained himself.

REFERENCES

1. Kahrilas PJ, Shaheen NJ, Vaezi MF. American gastroenterological association institute technical review on the management of gastroesophageal reflux disease. *Gastroenterology* 2008;**135**:1392–413.
2. Sharma P, McQuaid K, Dent J, et al. A critical review of the diagnosis and management of Barrett's esophagus: the AGA Chicago workshop. *Gastroenterology* 2004;**127**:310–30.
3. McClave SA, Boyce Jr HW, Gottfried MR. Early diagnosis of columnar lined esophagus: a new endoscopic diagnostic criterion. *Gastrointest Endosc* 1987;**33**:413–6.
4. Chandrasoma P, Makarewicz K, Wickramasinghe K, Ma YL, DeMeester TR. A proposal for a new validated histologic definition of the gastroesophageal junction. *Hum Pathol* 2006;**37**:40–7.
5. Chandrasoma P. Controversies of the cardiac mucosa and Barrett's esophagus. *Histopathology* 2005;**46**:361–73. [Chandrasoma – histopath].
6. Chandrasoma PT. Histologic definition of gastro-esophageal reflux disease. *Curr Opin Gastroenterol* 2013;**29**:460–7.
7. Allison PR, Johnstone AS. The oesophagus lined with gastric mucous membrane. *Thorax* 1953;**8**:87–101.
8. Barrett NR. The lower esophagus lined by columnar epithelium. *Surgery* 1957;**41**:881–94.
9. Hayward J. The phreno-esophageal ligament in hiatal hernia repair. *Thorax* 1961;**16**:41–5.
10. Tobey NA, Carson JL, Alkiek RA, et al. Dilated intercellular spaces: a morphological feature of acid reflux-damaged human esophageal epithelium. *Gastroenterology* 1996;**111**:1200–5.
11. Jain R, Aquino D, Harford WV, Lee E, Spechler SJ. Cardiac epithelium is found infrequently in the gastric cardia. *Gastroenterology* 1998;**114**:A160. [Abstract].
12. Ringhofer C, Lenglinger J, Izay B, Kolarik K, Zacherl J, Fisler M, Wrba F, Chandrasoma PT, Cosentini EP, Prager G, Riegler M. Histopathology of the endoscopic esophagogastric junction in patients with gastroesophageal reflux disease. *Wien Klin Wochenschr* 2008;**120**:350–9.
13. Glickman JN, Fox V, Antonioli DA, Wang HH, Odze RD. Morphology of the cardia and significance of carditis in pediatric patients. *Am J Surg Pathol* 2002;**26**:1032–9.
14. Robertson EV, Derakhshan MH, Wirz AA, Lee YY, Seenan JP, Ballantyne SA, Hanvey SL, Kelman AW, Going JJ, McColl KE. Central obesity in asymptomatic volunteers is associated with increased intrasphincteric acid reflux and lengthening of the cardiac mucosa. *Gastroenterology* 2013;**145**:730–9.
15. Sarbia M, Donner A, Gabbert HE. Histopathology of the gastroesophageal junction. A study on 36 operation specimens. *Am J Surg Pathol* 2002;**26**:1207–12.
16. Barrett NR. Chronic peptic ulcer of the oesophagus and 'oesophagitis'. *Br J Surg* 1950;**38**:175–82.
17. Association of Directors of Anatomic and Surgical Pathology. Recommendations for reporting of resected esophageal adenocarcinomas. *Am J Surg Pathol* 2000;**31**:1188–90.
18. Chandrasoma PT, Wijetunge S, DeMeester SR, Hagen JA, DeMeester TR. The histologic squamo-oxyntic gap: an accurate and reproducible diagnostic marker of gastroesophageal reflux disease. *Am J Surg Pathol* 2010;**34**:1574–81.
19. Chandrasoma P, Wijetunge S, Ma Y, DeMeester S, Hagen J, DeMeester T. The dilated distal esophagus: a new entity that is the pathologic basis of early gastroesophageal reflux disease. *Am J Surg Pathol* 2011;**35**:1873–81.
20. Vakil N, van Zanten SV, Kahrilas P, Dent J, Jones B, Global Consensus Group. The Montreal definition and classification of gastroesophageal reflux disease: a global evidence-based consensus. *Am J Gastroenterol* 2006;**101**:1900–20.
21. Karam SM. Lineage commitment and maturation of epithelial cells in the gut. *Front Biosci* 1999;**4**:286–98.
22. Tobey NA, Hosseini SS, Argore CM, Dobrucali AM, Awayda MS, Orlando RC. Dilated intercellular spaces and shunt permeability in nonerosive acid-damaged esophageal epithelium. *Am J Gastroenterol* 2004;**99**:13–22.
23. Johns BAE. Developmental changes in the oesophageal epithelium in man. *J Anat* 1952;**86**:431–42.
24. Shi L, Der R, Ma Y, Peters J, DeMeester T, Chandrasoma P. Gland ducts and multilayered epithelium in mucosal biopsies from gastroesophageal-junction region are useful in characterizing esophageal location. *Dis Esophagus* 2005;**18**:87–92.
25. Paull A, Trier JS, Dalton MD, Camp RC, Loeb P, Goyal RK. The histologic spectrum of Barrett's esophagus. *N Engl J Med* 1976;**295**:476–80.
26. Glickman JN, Chen Y-Y, Wang HH, Antonioli DA, Odze RD. Phenotypic characteristics of a distinctive multilayered epithelium suggests that it is a precursor in the development of Barrett's esophagus. *Am J Surg Pathol* 2001;**25**:569–78.
27. Der R, Tsao-Wei DD, DeMeester T, et al. Carditis: a manifestation of gastroesophageal reflux disease. *Am J Surg Pathol* 2001;**25**:245–52.
28. Lembo T, Ippoliti AF, Ramers C, Weinstein WM. Inflammation of the gastroesophageal junction (carditis) in patients with symptomatic gastroesophageal reflux disease; a prospective study. *Gut* 1999;**45**:484–8.
29. Chandrasoma PT, Der R, Ma Y, et al. Histology of the gastroesophageal junction: an autopsy study. *Am J Surg Pathol* 2000;**24**:402–9.

30. Olvera M, Wickramasinghe K, Brynes R, Bu X, Ma Y, Chandrasoma P. Ki67 expression in different epithelial types in columnar lined esophagus indicates varying levels of expanded and aberrant proliferative patterns. *Histopathology* 2005;**47**:132–40.
31. Chandrasoma PT, Der R, Ma Y, Peters J, DeMeester T. Histologic classification of patients based on mapping biopsies of the gastroesophageal junction. *Am J Surg Pathol* 2003;**27**:929–36.
32. Riddell RH, Odze RD. Definition of Barrett's esophagus: time for a rethink – is intestinal metaplasia dead? *Am J Gastroenterol* 2009;**104**:2588–94.
33. Kilgore SP, Ormsby AH, Gramlich TL, et al. The gastric cardia: fact or fiction? *Am J Gastroenterol* 2000;**95**:921–4.
34. Odze RD. Unraveling the mystery of the gastroesophageal junction: a pathologist's perspective. *Am J Gastroenterol* 2005;**100**:1853–67.
35. Chandrasoma P, Wijetunge S, DeMeester S, Ma Y, Hagen J, Zamis L, DeMeester T. Columnar lined esophagus without intestinal metaplasia has no proven risk of adenocarcinoma. *Am J Surg Pathol* 2012;**36**:1–7.
36. DeMeester SR, Wickramasinghe KS, Lord RV, Friedman A, et al. Cytokeratin and DAS-1 immunostaining reveal similarities among cardiac mucosa, CIM, and Barrett's esophagus. *Am J Gastroenterol* 2002;**97**:2514–23.
37. Lee RG. Mucins in Barrett's esophagus: a histochemical study. *Am J Clin Pathol* 1984;**81**:500–3.
38. Arul GS, Moorghen M, Myerscough N, et al. Mucin gene expression in Barrett's esophagus: an in-situ hybridization and immunohistochemical study. *Gut* 2000;**47**:753–61.
39. Chinyama CN, Marshall RE, Owen WJ, et al. Expression of MUC1 and MUC2 mucin gene products in Barrett's metaplasia, dysplasia and adenocarcinoma: an immunohistochemical study with clinical correlation. *Histopathology* 1999;**35**:517–24.
40. Watson C, van de Bovenkamp JH, Korteland-van Male AM, et al. Barrett's esophagus is characterized by expression of gastric-type mucins (MUC5AC, MUC6) and TFF peptides (TFF1 and TFF2), but the risk of carcinoma development may be indicated by the intestinal-type mucin MUC2. *Hum Pathol* 2002;**33**:660–8.
41. Glickman JN, Wang H, Das KM, et al. Phenotype of Barrett's esophagus and intestinal metaplasia of the distal esophagus and gastroesophageal junction: an immunohistochemical study of cytokeratins 7 and 20, Das-1 and 45MI. *Am J Surg Pathol* 2001;**25**:87–94.
42. Ormsby AH, Goldblum JR, Rice TW, et al. Cytokeratin subsets can reliably distinguish Barrett's esophagus from intestinal metaplasia of the stomach. *Hum Pathol* 1999;**30**:288–94.

FURTHER READING

1. Chandrasoma PT, Der R, Dalton P, Kobayashi G, Ma Y, Peters J, DeMeester T. Distribution and significance of epithelial types in columnar lined esophagus. *Am J Surg Pathol* 2001;**25**:1188–93.
2. Theodorou D, Ayazi S, DeMeester SR, Zehetner J, Peyre CG, Grant KS, Augustin F, Oh DS, Lipham JC, Chandrasoma PT, Hagen JA, DeMeester TR. Intraluminal pH and goblet cell density in Barrett's esophagus. *J Gastrointest Surg* 2012;**16**:469–74.
3. Oberg S, Peters JH, DeMeester TR, et al. Inflammation and specialized intestinal metaplasia of cardiac mucosa is a manifestation of gastroesophageal reflux disease. *Ann Surg* 1997;**226**:522–32.
4. Chandrasoma PT, Lokuhetty DM, DeMeester TR, et al. Definition of histopathologic changes in gastroesophageal reflux disease. *Am J Surg Pathol* 2000;**24**:344–51.

Chapter 5

Definition of the Normal State—A Yet Unfinished Saga

Definition of any disease in an organ must begin with the definition of the normal state of that organ in the absence of that disease. Before we can study the pathology and pathogenesis of gastroesophageal reflux disease (GERD), we must answer the following question: "Can we accurately define the normal esophagus and proximal stomach in everyone including fetuses and people with and without esophageal epithelial damage caused by exposure to gastric juice?"

The present definition of GERD essentially precludes accurate definition of the normal state. It equates "normal" with "no-GERD."

According to the Montreal definition,[1] "GERD is a condition that develops when the reflux of stomach contents causes troublesome symptoms and/or complications."

This definition, while it sets criteria to define the presence of GERD, cannot be applied to define the normal state. For example, heartburn that is not troublesome does not meet the definitional criterion for GERD. *Does this mean that all persons not troubled by their heartburn have no GERD? Are these persons normal? Or, have we set our diagnostic criteria to be too exclusive and too specific while sacrificing sensitivity?*

No one will suggest that a person moves from a perfectly normal state to GERD that satisfies the criteria of the Montreal definition in one instant. GERD is a progressive disease that must pass through a pathologic process before it reaches the point where the patient has sufficient reflux to cause troublesome symptoms and/or complications. The question that must be asked and answered is: *Does the patient who does not have GERD by the Montreal definition have any detectable pathologic abnormality related to GERD? Or, does GERD arise directly from the normal state?*

The present answer to this is that there are no detectable pathologic abnormalities in most patients who satisfy the Montreal definition of GERD. They have no abnormality at endoscopy [no erosive esophagitis and no visible columnar-lined esophagus (CLE)]. They have no recognizable abnormality on biopsy (no histologic evidence of reflux esophagitis in the squamous epithelium). The obvious question that now arises is: "Is it reasonable that a disease caused by damage to the squamous epithelium caused by exposure to gastric contents has no cellular abnormality?"

Every available test for GERD has the same problem.

The pH test, which monitors acid exposure and regarded by some to be the gold standard of diagnosis, is regarded as abnormal only when an electrode in the thoracic esophagus detects a pH<4 for 4.5% of the test period. This equates to 64 min/day in a 24-h monitoring period. *Does this mean that if there is reflux sufficient to cause a pH<4 for less than 64 min/day, the person has no GERD? Or, is there a gradation of acid exposure from 0 to 64 min/day that has a significance we are ignoring?*

Manometry, which evaluates the function of the lower esophageal sphincter (LES), defines a defective LES by the following criteria: (1) mean LES pressure<6; (2) total LES length<20 mm; and (3) abdominal LES length<10 mm. These criteria for a defective LES were defined to correlate with the presence of an abnormal pH test. As such, they define an advanced state of LES damage. *These criteria do not define LES damage less than that needed to cause LES failure. Is there a gap between a normal LES and severe LES damage with only the latter being associated with LES failure? Are we missing recognizing this gap between normal and a defective LES?*

Reflection on this situation must lead to the recognition that there is something radically wrong with the way we understand this disease. I will show that this is the case. Because of the use of incorrect definitions of anatomy, histology, GERD, and normalcy, the medical establishment is completely misinterpreting the pathologic manifestations of the early part of the disease.[2]

I will show that *there is a significant gap between the normal state and GERD as it is presently defined by troublesome symptoms, endoscopic changes, histologic manifestations of reflux, pH testing, and manometric criteria. Defining the pathology of this gap is the objective of this book.*

GERD. http://dx.doi.org/10.1016/B978-0-12-809855-4.00005-1

All the present misinterpretations of the pathology of GERD can be traced to one fundamental myth and false dogma: that cardiac epithelium is a normal epithelium in the proximal stomach. Eradication of this dogma leads logically to a recognition of other myths and an understanding of the entire pathophysiology of GERD.

1. NORMAL CARDIAC EPITHELIUM: A MYTH RESULTING IN A PRESENTLY HELD FALSE DOGMA

A myth is a traditional story without a determinable basis of fact. I will trace the story of how the myth that cardiac epithelium is normally present in the esophagus and/or proximal stomach developed.

A dogma is a principle laid down by an authority as incontrovertibly true. The myth that cardiac epithelium is normally present in the distal esophagus and/or proximal stomach created a false dogma that was taught by almost every authority as incontrovertibly true through much of the last century.

When I started looking at Dr. DeMeester's specimens taken according to his biopsy protocol, the accepted dogma regarding cardiac epithelium, following Hayward in 1961,[3] was that when it was found in the distal 20 mm of the esophagus and the proximal 30 mm of the stomach, it was normal (Fig. 5.1).

In the last decade of the 20th century, I proposed that cardiac epithelium was always a columnar epithelium that resulted from metaplasia of esophageal squamous epithelium caused by exposure to gastric juice.[4–7] According to this, cardiac epithelium was normally completely absent in the normal state (Fig. 5.1, middle frame). The suggestion that normal cardiac epithelium was a myth and the universal belief in that myth was a false dogma created a controversy that still remains only partially resolved.[8]

In this chapter I will produce evidence that cardiac epithelium does not exist as a normal epithelium in either the esophagus or the stomach. It is always a metaplastic epithelium resulting from exposure of esophageal squamous epithelium to gastric juice.

Unfortunately, even in this day when we are supposed to practice evidence-based medicine, false dogmas that are fervently supported by experts cannot easily be dispelled. Evidence that does not fit an expert's opinion can be ignored.

Since 1994, there has been some movement toward the truth that cardiac mucosa does not normally exist. However, the myth of normal cardiac mucosa still persists although its accepted normal extent in the region of the gastroesophageal junction (GEJ) has shrunk from 50 mm (Fig. 5.1, left frame) to 0.1–0.4 cm (Fig. 5.1, right frame).[9]

1.1 Early History—Barrett and Allison (1950–60)

From the time when histologic study of the region of the GEJ was performed, a columnar epithelium that was different than the epithelium in the gastric body was noted between the squamocolumnar junction (SCJ) and the proximal limit of gastric

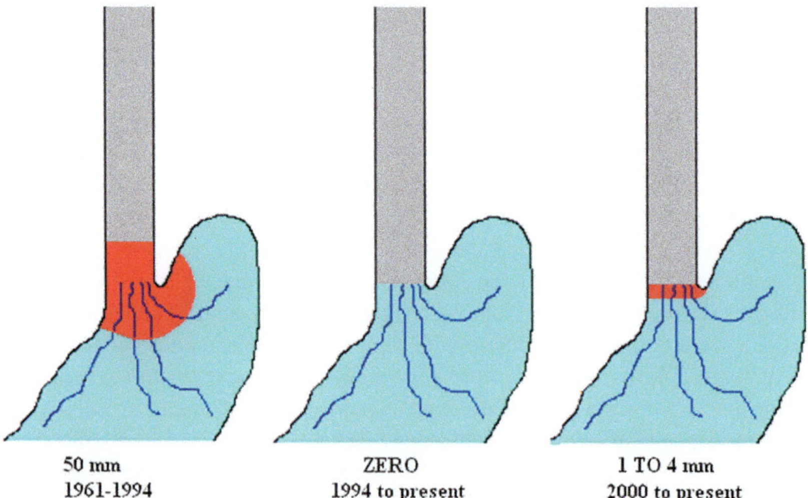

FIGURE 5.1 Historical changes regarding the presence and extent of cardiac epithelium in the gastroesophageal junction (GEJ) region. Left frame: In 1961, Hayward stated that cardiac epithelium (red), when found in the distal 2 cm of the esophagus and proximal 3 cm of the stomach, was a normal finding. Middle frame: In 1994, I proposed that cardiac epithelium was normally absent in this region; the only normal epithelia were esophageal squamous (gray) and gastric oxyntic (blue) epithelia. Right frame: The contrary viewpoint (incorrect) that 1–4 mm of cardiac epithelium was normally present distal to the endoscopic GEJ.

oxyntic epithelium. This epithelium was composed of mucous cells with rare oxyntic cells. It was completely different than gastric oxyntic (fundic) epithelium that lined the body and fundus of the stomach. It was called cardiac (or junctional, depending on the era) epithelium.

This epithelium was found distal to the end of the tubular esophagus in a majority if not all patients in what was believed to be the proximal stomach. No one ever questioned that this was the proximal stomach. It was distal to the end of the tube that defined the esophagus at the time and part of the sac that was the gastric reservoir. It was called "cardiac mucosa/epithelium." That part of the stomach lined by cardiac epithelium became known as "the gastric cardia."

The universally established histology of the normal stomach was that it had three distinct areas: cardiac mucosa proximally, antral mucosa distally, and oxyntic mucosa in between. The initial descriptions of this cardiac epithelium showed it to have glands composed of mucous cells or a mixture of mucous and parietal cells. The myth of normal gastric cardiac epithelium was created with the birth of histology of this region. No one had any doubt that this was true.

This same histologic epithelial type was also reported in the tubular thoracic esophagus in a minority of patients. It was initially described in resected specimens and, when rigid endoscopy was developed in the 1920s, by examination of biopsies. From 1930 to 1953, this columnar epithelium, when seen in the esophagus, was thought, because it was identical to cardiac mucosa in the proximal stomach, to be the result of congenital heterotopia of gastric epithelium. It was considered to be an anatomic curiosity without clinical significance. This was not unreasonable because adenocarcinoma of the esophagus had not been reported then. The idea that cardiac epithelium is harmless is deeply entrenched.

In 1950,[10] Norman Barrett contested the viewpoint that columnar (cardiac) epithelium in what was believed was the tubular esophagus was congenital heterotopic gastric epithelium in the esophagus. By defining (incorrectly) that the esophagus ended at the SCJ, he decreed that what had been called an esophagus containing heterotopic gastric mucosa was actually a tubular intrathoracic stomach associated with a congenitally short esophagus. The esophagus, defined by its squamous epithelial lining, ended at the SCJ, sometimes in the upper region of the tubular structure that everyone had previously called the esophagus.

Actually, Barrett could not have been further from the truth. He misinterpreted the entire CLE that now bears his name as tubular intrathoracic stomach lined by gastric cardiac epithelium.[6]

Barrett's interpretation was quickly disproved in 1953 when Allison and Johnstone[11] showed that what Barrett called the intrathoracic stomach was actually esophagus. They showed that this tubular organ lined by cardiac epithelium was proximal to the peritoneal reflection, an accepted definition of the GEJ (Fig. 5.2). They termed this an "esophagus lined by gastric mucous membrane." By this terminology, cardiac epithelium of the stomach now lined the distal esophagus.

In 1957, Barrett[12] agreed with Allison and Johnstone that what he had called the tubular intrathoracic stomach was actually esophagus lined by columnar epithelium. He made the important recommendation that this entity be called "columnar-lined esophagus" rather than "esophagus lined by gastric mucous membrane." Barrett's reason for the recommended change in terminology is as follows and extremely astute: "This paper concerns a condition whose existence is denied by some, misunderstood by others, and ignored by the majority of surgeons. It has been called a variety of names which have confused the story because they have suggested incorrect etiologic explanations; congenital short esophagus, ectopic gastric mucosa, short esophagus, and the lower esophagus lined by gastric epithelium are but a few. At the present time, the most accurate description is that it is a state in which the lower end of the esophagus is lined by columnar epithelium. This does not commit us to ideas which could be wrong, but it carries certain implications which must be clarified" (p. 881).

In his paper,[12] Barrett discussed three potential causes for the finding of columnar epithelium in the "gullet" (a term he used for the tubular structure that connects the pharynx to the stomach below the diaphragm): (1) a sliding hiatal hernia; (2) the new condition described in the paper where the lower part of the esophagus is lined by columnar epithelium extending upward for a long or short distance above the GEJ; and (3) true ectopic gastric mucosa in the lower esophagus. Barrett so vehemently dismissed the last possibility of ectopic gastric mucosa in the esophagus that this concept was forever dispelled from the subsequent literature.

Barrett's paper eased the confusion somewhat by matching the esophagus to an epithelium that was actually called esophageal rather than gastric. However, the histology remained confusing because the same epithelium, i.e., cardiac epithelium lined both the CLE and the region distal to the tubular esophagus (the "gastric cardia").

However, a major problem remained because of the inability to identify the GEJ. This was where the CLE ended and the stomach began. This is a problem that still exists. It exists because there is still no accurate, evidence-based definition of the GEJ. This issue of the GEJ, however, was unimportant then. CLE was yet believed to be a harmless medical curiosity because adenocarcinoma was still extremely uncommon.

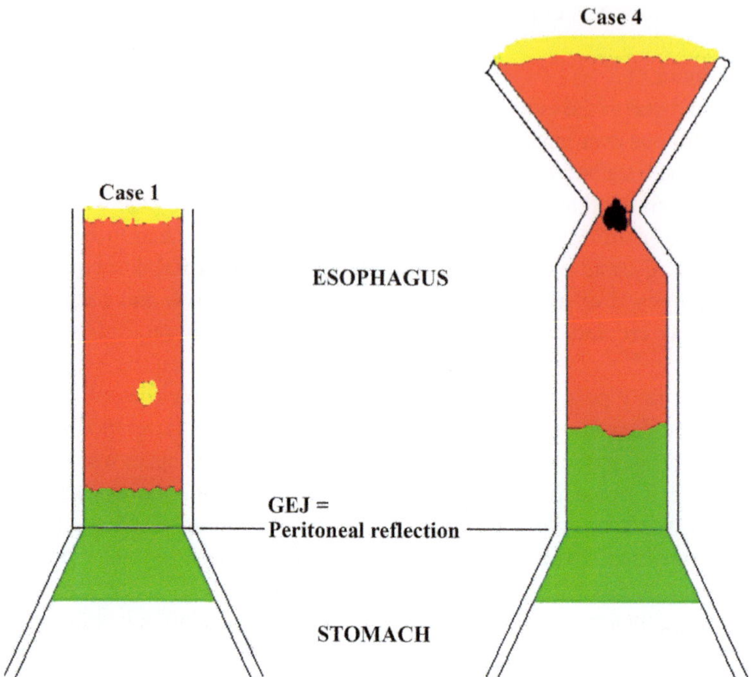

FIGURE 5.2 Diagrammatic representation of two case reports from Ref. 11. Left frame: Case 1, showing a 5 cm segment of columnar-lined esophagus (CLE) in the tubular esophagus composed of cardiac (red) and fundic (green) epithelium. Right frame: Case 4, showing a 7.5 cm segment of CLE with an ulcerated stricture. The CLE consists of cardiac (red) and fundic (green) epithelia. The CLE segment is above the peritoneal reflection [+true gastro-esophageal junction (GEJ)]. Note that any columnar epithelium containing parietal cells was regarded as "fundic" epithelium. They make no differentiation between oxyntocardic and gastric oxyntic epithelia (not defined at the time).

1.2 Short-Segment Barrett Esophagus: Hayward and Spechler (1960–94)

Hayward, in 1961,[3] in a highly influential paper, proposed a new definition for the GEJ: "The oesophagus is defined as the tube conducting food from the throat to the stomach and all of this tube is regarded as oesophagus irrespective of its lining. The lower centimeter or two of this tube is normally lined by columnar epithelium." There was no evidence base for this definition. According to Hayward, the same columnar epithelium that normally lined the distal 2 cm of the esophagus extended across the end of the tube into the proximal stomach for 3 cm (Fig. 5.1, left frame). Hayward suggested that this 5 cm of cardiac epithelium between the squamous epithelium and acid-producing gastric oxyntic epithelium was necessary to act as a buffer protecting the squamous epithelium from acid.[3]

In 1961, flexible endoscopy was still relatively new and likely limited to patients with the most severe reflux disease. The normal extent of cardiac epithelium was largely based on examination of this population. The normal extent of cardiac mucosa in 2 cm in the distal esophagus and 3 cm in the proximal stomach was repeated as absolute truth in every histology textbook in 1991.[13,14]

Spechler et al.[15] published a paper in *Lancet* in 1994, which resulted in the definition of short-segment Barrett esophagus. For patients without a hiatal hernia, the GEJ was defined as the point where the tubular esophagus flared to become a saclike structure. For patients with hiatal hernia, the GEJ was defined as the proximal margin of rugal folds in the hernia pouch. Two biopsy specimens were obtained from the columnar epithelium at the SCJ to seek "specialized columnar epithelium (SCE)" (=cardiac epithelium with intestinal metaplasia). One biopsy specimen was obtained from the squamous epithelium to seek esophagitis.

This was a population where a majority of the subjects had clinical evidence of GERD. 86% of patients had visible CLE > 3 cm and ~60% of those without visible CLE (defined at the time as CLE < 3 cm) had symptoms of GERD.

The patients were divided into two groups by their endoscopic findings:

1. 16 patients with endoscopically apparent Barrett esophagus who had columnar epithelium extending >3 cm from the GEJ. All these patients had SCE (i.e., cardiac epithelium with intestinal metaplasia).
2. 140 patients without endoscopically apparent Barrett esophagus who had <3 cm of CLE. 26/140 of this group had SCE (i.e., cardiac epithelium with intestinal metaplasia); the other 114 did not. By default, these patients must have had cardiac epithelium without intestinal metaplasia.

The data that 26/140 (18.6%) of patients with <3 cm of CLE had intestinal metaplasia were considered important because of the understanding that SCE (cardiac epithelium with intestinal metaplasia) was an abnormal finding. This resulted in a change in the guidelines for biopsy at endoscopy. Biopsy was recommended when any columnar epithelium was seen at endoscopy rather than the prior requirement of >3 cm CLE.

This changed the definition of endoscopic normalcy as an esophagus lined *entirely* by squamous epithelium, at least to precipitate a biopsy. The presence of SCE (i.e., cardiac epithelium with intestinal metaplasia) in a biopsy from any visible CLE resulted in the diagnosis of Barrett esophagus. The term "short-segment Barrett esophagus" was used for the condition where the CLE length was <3 cm. That distinguished this lesion from long-segment Barrett esophagus, which was diagnosed when >3 cm of visible CLE was present.

The study resulted in the present guidelines that recommend that biopsies should be done in patients with any visible CLE even if it is <3 cm.

1.3 The Controversy That Cardiac Epithelium Is Not Normal in the Proximal Stomach (1994 to Present)

The dogma that cardiac epithelium could be found normally up to 3 cm distal to the endoscopic GEJ in the "proximal stomach" was not in dispute in 1994. This area was therefore never subject to routine biopsy, a practice that continues to this day. There are therefore minimal data in the literature regarding the presence, extent, and significance of cardiac epithelium distal to the endoscopic GEJ in normal persons or GERD patients.

Dr. Tom DeMeester moved to the University of Southern California in 1990 as Chairman of the Department of Surgery. He quickly developed a clinical program for the comprehensive study of esophageal disease. I was assigned to read his biopsies and this began a long collaboration that lasts to this day.

Dr. DeMeester's biopsy protocol included taking retroflex biopsies distal to the endoscopic GEJ in all patients, endoscopically normal or not, symptomatic GERD or not (Fig. 5.3). The retroflex biopsies were taken within 1 cm distal to the SCJ. Other biopsies were taken of the SCJ, distal stomach, and visible columnar epithelium above the endoscopic GEJ when such were present.

When I started looking at these biopsies, it was quickly apparent that the accepted 3 cm of normal cardiac epithelium distal to the endoscopic GEJ was a myth. The reason why this was not discovered earlier was that pathologists never received specimens routinely from distal to the endoscopic GEJ. No one had thought biopsies of the proximal stomach had any value in any clinical situation. There was no point taking biopsies of the stomach in the GERD patient, and gastric pathology was more accurately assessed in more distal gastric biopsies.

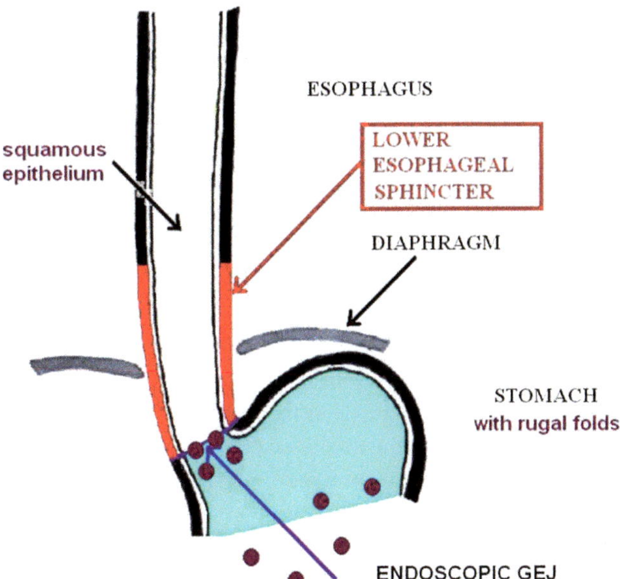

FIGURE 5.3 The routine biopsy protocol for persons who have no visible columnar-lined esophagus at endoscopy. Four-quadrant biopsy sets are taken straddling the squamocolumnar junction, on retroflex view from within 1 cm distal to the squamocolumnar junction, and from the distal gastric body and antrum. *GEJ*, gastroesophageal junction.

At present, the myth still exists but the power of the false dogma that was created by the myth has become much weaker (Fig. 5.1). Most authorities accept that the "normal" extent of cardiac epithelium in the proximal stomach is "less than 0.4 cm."[8] At least, the myth of 3 cm of cardiac epithelium lining the proximal stomach has been reduced to "less than 0.4 cm." However, there are a significant number of experts who seem to be coming around to the correct viewpoint that cardiac epithelium is normally absent.[16]

Also proven but generally ignored is that some people have more than 0.4 cm of cardiac epithelium distal to the endoscopic GEJ.[17,18] This "greater than normal" extent of cardiac epithelium distal to the endoscopic GEJ does not presently indicate any pathologic state.

The remaining mythical 4 mm of cardiac epithelium normally lining the proximal stomach must disappear (Fig. 5.1). When this happens, the presence of cardiac epithelium in any biopsy from this region becomes recognized for what it is: abnormal and esophageal in location. It then becomes necessary to determine the cause of the pathologic abnormality (*columnar metaplasia of esophageal squamous epithelium caused by exposure to gastric juice*) and the reasons for the variation in the length of cardiac epithelium distal to the endoscopic GEJ (*severity of damage to the LES*). This is the entire pathophysiology of early GERD.

2. ERADICATION OF THE MYTH THAT CARDIAC EPITHELIUM NORMALLY LINES THE PROXIMAL STOMACH

The problem with the belief that cardiac epithelium is normally present in the proximal stomach is that all the data that support the myth initially originated from a study of abnormal people, usually with severe GERD. The source of material was specimens resected for intractable esophageal ulcers and strictures caused by GERD or endoscopic biopsies in the days of rigid endoscopy when only the most severe patients were subject to endoscopy.

Hayward's declaration of the normal extent of cardiac epithelium in 1961[3] was based on his observations that his patients frequently had cardiac epithelium in the distal esophagus and proximal stomach. Normalcy was defined by examination of the most abnormal. No autopsy study of people without GERD had been done in the 20th century. This was partly because of lack of interest, and partly because of the difficulties relating to rapid postmortem autolysis of mucosae of the gastrointestinal (GI) tract.

The dogma of normal cardiac epithelium has determined guidelines for biopsy at endoscopy that have been recommended by gastroenterology societies. In the period from 1961 to 1994, the recommendation was to not biopsy patients (even with symptoms of GERD) unless a CLE >3 cm was present. This precluded accumulation of data in all patients with <3 cm of visible CLE.

In 1994, following the study by Spechler et al.[15] (see above), the recommendation changed. Biopsy was recommended in all patients with CLE of any length in the esophagus. This resulted in the accumulation of data regarding the histology of the most distal 3 cm of the tubular esophagus above the endoscopic GEJ.

While these histologic data accumulated in patients with CLE < 3 cm, only those patients with SCE (cardiac epithelium with intestinal metaplasia) were ever analyzed. If there was no intestinal metaplasia, the data were largely ignored. The reason for this was that the only interest that the gastroenterologist had in CLE was its significance as a future risk of adenocarcinoma. Haggitt, in the early 1980s,[19,20] had established that only CLE with intestinal metaplasia was at risk for adenocarcinoma. Columnar epithelia without intestinal metaplasia was generally ignored and never used as a pathologic criterion (except in England, where visible CLE without intestinal metaplasia is diagnosed as Barrett esophagus).

No one ever even considered that cardiac epithelium distal to the endoscopic GEJ was anything other than gastric. There was no disease entity other than adenocarcinoma that was limited to the gastric cardia. Therefore, unless a mass lesion was seen, no biopsy was ever taken from the cardia. When an adenocarcinoma was found, it was without question regarded as an adenocarcinoma of the proximal stomach (=gastric cardia).

Later, when biopsies were taken of the epithelium distal to the normal SCJ, some patients were found to have intestinal metaplasia in cardiac epithelium. This was called "intestinal metaplasia of the gastric cardia." Cardiac epithelium without intestinal metaplasia in such biopsies was normal.

The first time that any systematic histologic examination of this region was done with an open mind was when I was presented with the retroflex biopsies from the patients presenting to Dr. Tom DeMeester's Foregut surgery unit. This was the first time that this region was routinely subject to biopsy in normal or GERD patients. Incredibly, for the first time in history, a unique opportunity to define normalcy arose because of the compulsive biopsy protocol that had been developed by a surgeon. The biopsy material was unique because the protocol ignored the guidelines for endoscopic biopsy recommended by every authority.

Definition of normalcy is often not as simple as it sounds. In the esophagus, where nonneoplastic disease is dominated overwhelmingly by GERD, the definition of normalcy is practically equivalent to defining the absence of GERD. Esophageal diseases unrelated to GERD such as trauma, infections, immune-mediated diseases, neoplasms other than adenocarcinoma, diverticula, and many others are rare and easily excluded. In contrast, GERD is extremely common.

In the study of esophageal disease, the word "normal" is used too easily to describe the situation that is equivalent to "not abnormal." The distinction is important. If a person does not have symptoms of GERD that satisfy the Montreal definition, we should say that the person has no GERD *as defined*, not that he/she has no GERD with the implication that he/she "is normal." It is always possible that the definition is not all inclusive.

In 1994, we proposed that cardiac epithelium, when found in a biopsy distal to the endoscopic GEJ, was never normal and never gastric; it was always an abnormal esophageal epithelium that resulted from columnar metaplasia of esophageal squamous epithelium that had been exposed to gastric contents.[4] The normal state was a complete absence of cardiac epithelium. The entire esophagus was lined by squamous epithelium and the entire proximal stomach by gastric oxyntic epithelium (Fig. 5.1, middle frame).

The hypothesis that our unit proposed in the mid-1990s created a storm of disbelief. There was astonishment bordering on ridicule that anyone could suggest that gastric cardiac epithelium was not normal. Many studies were done to prove the hypothesis incorrect. The power of the dogma instilled a bias toward disproving the hypothesis. Many conclusions drawn from the data in several studies were flawed because of lack of objectivity. Careful examination of the data in these studies, however, provided strong evidence in favor of the absence of cardiac epithelium in the normal state.

2.1 Study of Fetal Development

All organs develop during intrauterine fetal life from an embryonic anlage to its adult state. Development is completed at varying points in either fetal life in utero or postnatal life. Until development is complete, an organ may have a mixture of undifferentiated embryonic (fetal) tissues and differentiated adult tissues.

The normal state of an organ can be defined as the composition of the organ at the end of development where all fetal tissues have differentiated to adult tissues.

The hypothesis that our group put forward in the 1990s resulted in a series of papers that reported the histologic findings in the GEJ region in fetuses and neonates.[21–23] These studies were reviewed in detail in Chapter 3.

The fetal stomach (at and distal to the GEJ as accurately defined in the fetus by the angle of His) is lined by an epithelium that contains parietal cells from the 17th week of gestation and grows to develop into an adult gastric oxyntic epithelium.[19] There is no epithelium in the stomach that resembles cardiac epithelium (i.e., an epithelium consisting of mucous cells without parietal cells) at any time after 17 weeks of gestation when the primitive stratified columnar epithelium of the endodermal tube differentiates into gastric oxyntic epithelium.

The esophagus is lined by primitive endodermal tube stratified columnar epithelium, which differentiates into ciliated columnar epithelium that lines the entire esophagus in the first half of fetal life.[24] At 22 weeks of gestation, stratified squamous epithelium appears in the midesophagus and progressively replaces the ciliated epithelium. A fetal columnar epithelium consisting of mucous cells appears in the third trimester proximal to the angle of His.[21,24] This progressively decreases in length as it is replaced by stratified squamous epithelium.[21,22]

When development is complete, before, at, or in the first year after birth, stratified squamous epithelium has completely replaced the columnar epithelia in the esophagus and extends to the angle of His where it transitions directly into gastric

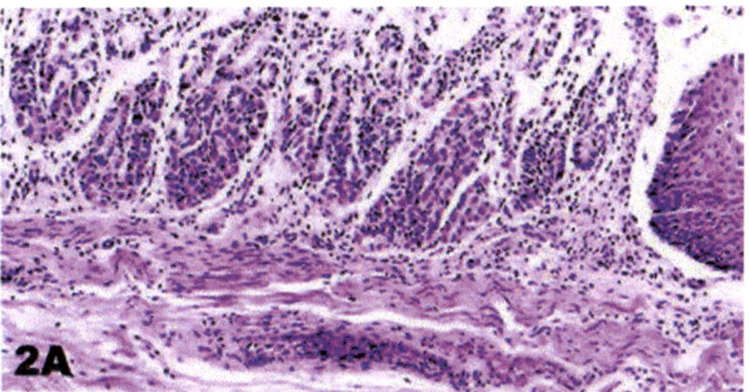

FIGURE 5.4 Section across the squamocolumnar junction in a perinatal autopsy showing a direct transition from squamous epithelium to parietal cell–containing fetal gastric oxyntic epithelium. There is no cardiac epithelium. *Reproduced with permission from Park YS, Park HJ, Kang GH, et al. Histology of gastroesophageal junction in fetal and pediatric autopsy. Arch Pathol Lab Med 2003;**127**:451–55.*

oxyntic epithelium at the GEJ (Fig. 5.4). There are no epithelia in the esophagus and stomach other than stratified squamous epithelium and gastric oxyntic epithelium, respectively.

The fetal studies, interpreted correctly, show that the entire fully developed esophagus is lined by squamous epithelium and that the entire proximal stomach is lined by gastric oxyntic epithelium. One picture in the paper by Park et al.[23] that shows this normal state is worth a thousand words (Fig. 5.4). At the end of development and the beginning of postdevelopmental life, the infant has no cardiac epithelium. This proves that cardiac epithelium is an abnormal epithelium caused by a pathologic event that occurs after birth.

2.2 Autopsy Study of Normal People (i.e., Those Without GERD)

The most time-honored method of establishing normalcy is to study people who have had no evidence of disease during life at autopsy. This is not a foolproof method; findings at autopsy must be interpreted carefully.

When a disease is common in the population, people who have never had clinical manifestations of the disease during life may express that disease at autopsy. For example, careful examination of the prostate in people over 80 years old will more often than not have small foci of well-differentiated prostatic adenocarcinoma. Atherosclerosis of the abdominal aorta is present in small amounts in most people over 50 years old in the United States. These autopsy findings obviously do not make prostatic adenocarcinoma and aortic atherosclerosis normal. The normal state is the absence of these conditions. This is obvious because these conditions are obviously pathological, but more importantly, their presence during life is known to be associated with a higher risk of death than the absence of these entities.

When the normal state at autopsy is the absence of a finding and the normal state is rare i.e., most people have the abnormality, it is important to examine an adequate number of cases and to examine patients who are least likely to have the disease. The likelihood of finding the normal state increases in a population of young people if the abnormal state is an acquired disease. It is also important not to ignore the finding of the negative as an anomaly.

2.2.1 Our Autopsy Studies in Children and Adults

When we developed the hypothesis that cardiac epithelium was not a normal epithelium, my first reaction was to go to the literature in search of studies of the histology of the region of the GEJ and proximal stomach at autopsy, resected specimens, or endoscopic biopsies. I was amazed to find that there were none. The dogma of normal cardiac epithelium in the proximal stomach was so powerful that it did not need to be supported by evidence.

I immediately decided to undertake a study of autopsies of the region. It is an uncommon opportunity in medicine to be the first to do a relatively simple and obvious study that the entire medical establishment had failed to do for a century.

The autopsy protocol at the LAC+USC Medical Center required the taking of one vertical section across the GEJ in all cases. Examination of autopsy reports showed that the protocol was not always followed. However, I was able to find the slides across the GEJ in over 500 autopsies. I immediately realized why no one had ever done this study before. Rapid autolysis, particularly of the columnar epithelium distal to the GEJ, often made it impossible to characterize the type of epithelium. There were only 72 cases where the histology was adequately preserved to accurately define the epithelial type. Three of these seventy-two patients had "esophagitis" recorded in their medical record. The age and histologic findings in these 72 patients are shown in Table 5.1.

The results were astounding and exciting. In 21/72 (29%), only stratified squamous epithelium and gastric oxyntic epithelium were present (Fig. 5.5). There was no intervening cardiac epithelium (with or without parietal cells). Even in the

TABLE 5.1 Age and Epithelial Type at the Gastroesophageal Junction in a Single Section Taken Routinely at Autopsy at LAC+USC Medical Center in 72 Patients

Age (Years)	No. of Patients	Patients Without CE	Patients Without CE or OCE
0–19	11	9	9
20–39	15	10	5
40–59	26	19	4
>60	20	15	3
Total	**72**	**53**	**21**

CE, cardiac epithelium; *OCE*, oxyntocardiac epithelium.

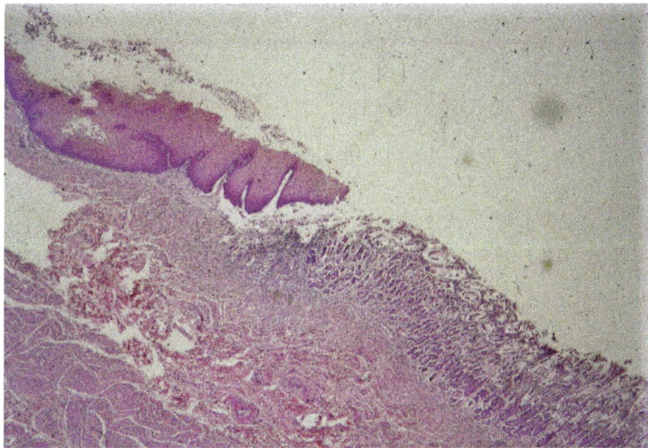

FIGURE 5.5 Section across the squamocolumnar junction at autopsy showing a direct transition from esophageal squamous epithelium to gastric oxyntic epithelium. The gastric epithelium shows the orderly straight tubular glands typical of gastric oxyntic epithelium. There is no cardiac or oxyntocardiac epithelium.

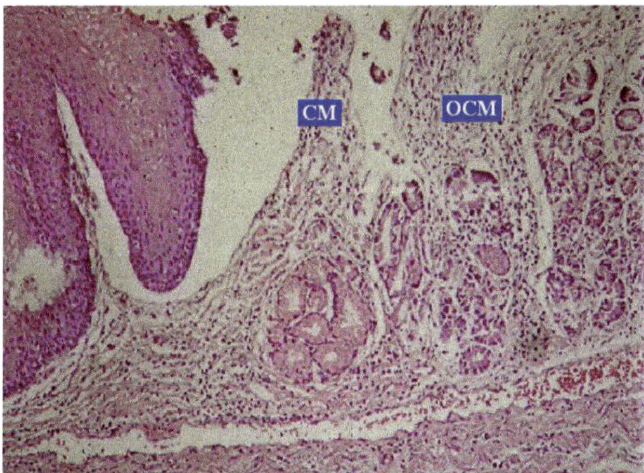

FIGURE 5.6 Section across the squamocolumnar junction at autopsy showing transition from squamous epithelium to cardiac and oxyntocardiac epithelia. *CM*, cardiac mucosa; *OCM*, oxyntocardiac mucosa.

partially autolyzed material, it was easy to recognize the well-organized, uniformly thick gastric oxyntic epithelium with its straight tubular glands, the absence of mucous cells in the glands, and the paucity of inflammatory cells. Its distinction from the more disorganized cardiac epithelium was obvious (Fig. 5.6).

9/11 (82%) patients under the age of 19 years had no cardiac epithelium (with or without parietal cells). The prevalence of cardiac epithelium (with or without parietal cells) increased with age. However, people without cardiac epithelium (with or without parietal cells) were found in all age groups. The oldest person with no cardiac epithelium (with or without parietal cells) was 87 years. All three patients who had a diagnosis of "esophagitis" in their medical record had cardiac epithelium at the GEJ.

53/72 (74%) patients had no pure cardiac epithelium. Of these, 32/72 (44%) patients had oxyntocardiac epithelium between the squamous epithelium and gastric oxyntic epithelium.

I attempted to measure the cardiac and oxyntocardiac epithelium between the squamous epithelium and gastric oxyntic epithelium (Table 5.2). This was not possible because of tangential sectioning in 11 patients. Of the 61 patients in whom measurement was possible, 21 had 0 mm, 25 had >0–5 mm, 12 had 5–10 mm, and 3 had >10 mm (measured at 11, 11, and 15 mm) cardiac epithelium (with or without parietal cells). The three patients with a recorded history of "esophagitis" had cardiac epithelial (with or without parietal cells) lengths measured at 3.7, 5.8, and 11 mm. Although the number of patients (3) and the method of obtaining clinical information (autopsy summary and SNOMED code) were not scientifically sound for any conclusion, the fact that all 3 patients with "esophagitis" were among the 19 patients with pure cardiac epithelium was interesting.

To a clinical researcher with a new radical hypothesis, the data from this study essentially proved the hypothesis that cardiac epithelium (with or without parietal and/or goblet cells) was normally absent in the GEJ region and proximal stomach.

TABLE 5.2 Measured Length of Cardiac Epithelium (With or Without Parietal Cells) Between the Distal Limit of Squamous Epithelium and Gastric Oxyntic Epithelium. The Measurement Was Not Possible in 11 Patients Because of Tangential Sectioning of the Section That Resulted in Improper Orientation

Number of Patients	Length of Cardiac Epithelium (With or Without Parietal Cells)
21	0
25	>0–5 mm
12	5–10 mm
3	>10 mm (11, 11, and 15 mm)
Total of 61 patients	**0–15 mm**

No goblet cells were seen in any of the patients.

The absence was proven in 21/72 (29%) patients and the maximum measured length of cardiac epithelium (with or without parietal cells) in 72 patients was 15 mm.

At the time of this study, Oberg et al.[25] from our unit had shown in clinical studies that the presence of cardiac epithelium at the GEJ in endoscopically normal people correlated with symptomatic GERD and was associated with a greater likelihood of an abnormal 24-h pH test and criteria of a defective LES. We had an explanation for why small amounts of cardiac epithelium (with or without parietal cells) were found in the majority of patients at autopsy; they had subclinical-GERD.

I rushed to publish this with the enthusiasm of Archimedes shouting "Eureka!" I was naïve. *The New England Journal of Medicine*, *Lancet*, every gastroenterology journal, and every pathology journal rejected this paper out of hand. The reviewer comments were incomprehensible. Their purported reasons for rejection were long and complicated. They made no sense. The first ever autopsy study of the GEJ region that conclusively disproved the dogma at that time that up to 30 mm of cardiac epithelium (with or without parietal cells) normally existed in the proximal stomach was not worthy of publication anywhere in an expansive medical literature that is replete with minutiae.

The reason for rejection was simply disbelief that the entire medical establishment could have been so wrong for so long. All the reasons given were invalid. As a young researcher without much standing in the world of gastroenterology and GI pathology, I was lost and dejected.

Among the repeated reasons for rejection by reviewers were that the study used retrospective material that could not be trusted because of the unstandardized way the sections were taken, that the single vertical section examined did not permit assessment of the entire circumference of the GEJ, and that autolysis made the findings unreliable.

These were all invalid objections. The absence of cardiac epithelium cannot be explained by inadequate sampling. Sampling is only required to avoid a false negative of a positive finding, not a negative finding. Standardization is not necessary to show a negative finding. We had already excluded all cases where autolysis had made histologic interpretation unreliable.

With an intense desire to publish these data, I undertook an additional prospective autopsy study to address these issues. I coopted Dr. Mark Taira, a recent resident at LAC+USC Medical Center who was doing a forensic pathology fellowship at the Los Angeles Coroner's office. After getting the study approved, Dr. Taira prospectively harvested the entire circumference of the GEJ region (25 mm of proximal stomach including the SCJ). We selected 18 patients with a bias toward young age, the smallest time lapse between death and autopsy (to minimize autolytic changes), and no evidence of any esophageal disease during life in the medical record. The selection bias was designed to maximize the probability that we would encounter patients with no cardiac epithelium.

The harvested specimen was pinned out for optimal fixation for 2 days and then sectioned completely in vertical orientation providing the entire SCJ and the adjacent 25 mm distal to the junction for histologic examination.

The data from the prospective study were incredibly detailed but simply confirmed the conclusions of the retrospective study (Table 5.3). This was not unexpected. It is impossible, assuming correct methodology and adequate material for any study to contradict these conclusions. The absence of cardiac epithelium in some people cannot be disproved by the demonstration of its presence in others. The 0–15 mm length of cardiac epithelium (with and without parietal cells) can be disproved, but only at its upper end by finding longer segments, not at its minimum of zero.

It is highly unlikely that cardiac epithelium (with or without parietal and/or goblet cells) normally exists if its minimum extent is zero. The body does not create a tissue that is normally absent in some people.

TABLE 5.3 Detailed Data in the Prospective Autopsy Series (n = 18).

Sex	Age	Cause of Death	No Sections Without CE/OCE[a]	Maximum Length of CE (mm)	Mean Length of CE (mm)	Maximum Length of CE + OCE (mm)	Mean Length of CE + OCE (mm)
M	3	Drowning	0/7	2.75	0.704	2.875	1.371
M	11	Stab injury	5/8	0.25	0.036	0.475	0.161
M	12	Gunshot wound	2/8	0	0	1.375	0.639
F	14	Drug overdose	7/10	0	0	4.00	0.70
M	15	Trauma	0/7	1.00	0.442	3.20	2.283
M	17	Gunshot wound	2/9	0.575	0.072	3.00	1.119
M	18	Gunshot wound	5/12	0	0	4.25	0.708
M	20	Electrocution	5/11	0	0	1.50	0.25
M	20	Gunshot wound	2/10	0	0	7.25	1.858
F	21	Blunt trauma	0/9	0	0	4.95	1.536
F	21	Gunshot wound	3/8	0	0	2.375	0.783
M	22	Gunshot wound	0/7	0	0	3.925	1.919
M	25	Gunshot wound	0/13	0.375	0.077	5.375	3.273
M	26	Head trauma	0/12	0	0	2.55	1.106
M	29	Drug overdose	0/7	0.275	0.055	8.05	6.945
M	41	Head trauma	0/8	2.375	1.038	4.80	2.719
M	49	Drug overdose	0/10	2.625	0.735	7.50	4.715
F	61	Blunt trauma	2/9	0	0	1.00	0.40

CE, cardiac epithelium; *OCE*, oxyntocardiac epithelium.
[a]*Number of sections without CE or OCE/total number of sections taken from the full circumference of the specimen across the gastroesophageal junction.*

Cardiac epithelium without parietal cells was present in some part of the circumference in only 8/18 (44%) patients and absent in all sections taken from the entire specimen in 10/18 (56%) patients. All eight patients with cardiac epithelium showed an absence of cardiac epithelium in two to eight vertical sections taken from the GEJ indicating that this epithelium was absent in some part of the GEJ circumference. The maximum length of cardiac epithelium varied from 0 to 2.75 mm with a mean length of 0.704 mm; 14/18 persons had a combined cardiac and oxyntocardiac epithelial length of <5 mm.

In the 10 patients who did not have cardiac epithelium, oxyntocardiac epithelium was present from a minimum of 0.25 to a maximum of 1.919 mm. The eight patients with cardiac epithelium all had oxyntocardiac epithelium located distal to cardiac epithelium (Fig. 5.6), extending to its transition point with gastric oxyntic epithelium. The combined maximum cardiac and oxyntocardiac epithelial length in the eight patients with cardiac epithelium ranged from 0.475 to 8.05 mm. In general, the maximum squamooxyntic gap tended to be greater in patients who had cardiac epithelium than those without.

Therefore, all patients in this study had some separation of squamous epithelium from gastric oxyntic epithelium, but only in a part of the circumference of the GEJ in 9/18 (50%) patients. In these patients, there was a direct transition from squamous epithelium in two–seven vertical sections taken from the circumference (Fig. 5.5).

I submitted the revised paper with the added data of the prospective study that completely addressed the reviewers' major reasons for initial rejection. All the journals again rejected the paper. There were new objections but these were so bizarre that I could not address them. I had no option but to shelve the paper.

I emphasize that these measurements are in millimeters. The data are astonishing because the accepted length of normal cardiac epithelium in the proximal stomach was 3 cm (=30 mm). In reality the measurements of the maximum extent of cardiac epithelium (with and without parietal cells) in these completely studied GEJ regions are <2 mm in 4 patients, <5 mm in 15 patients, and a maximum of 8.05 mm in the entire population of 18 patients. This was revolutionary information. The dogma that 30 mm of cardiac epithelium normally existed in the proximal stomach was so ingrained that it could not be shaken by any amount of data. Rejection of the evidence with no valid reason was preferable to the reviewers than rejection of the dogma.

2.2.2 The Importance of Oxyntocardiac Epithelium

At that time, it was suggested to me by many colleagues that I was making a mistake by including oxyntocardiac epithelium within the definition of abnormal epithelia. Certainly, if I had only considered pure cardiac epithelium as being abnormal, the data were more convincing. 71% of patients in the retrospective study and 56% of patients in the prospective study did not have any cardiac epithelium. Surely, cardiac epithelium could not be a normal epithelium.

Adding oxyntocardiac epithelium made the presence of cardiac and/or oxyntocardiac epithelium universal; it was present in all patients in the prospective study with complete circumferential examination of the region. I had to show that epithelia that were universally present were not normal. This was much more difficult and much more easily rejected.

It was also more convenient in terms of definition if I had limited the discussion to cardiac epithelium. Cardiac epithelium was easily distinguished from oxytocardiac epithelium because it had no parietal cells. The distinction was clear and easily understood. Oxyntocardiac and gastric oxyntic epithelium both had parietal cells (Fig. 5.7A and B). It was difficult to make even pathologists understand the difference. Trying to convince nonpathologists was near impossible.

I steadfastly, and possibly too stubbornly, refused to exclude oxyntocardiac epithelium from the definition of metaplastic esophageal columnar epithelia. The reason for this was initially based on my desire to be scientifically pure. In my examination of thousands of slides, I knew that oxyntocardiac epithelium evolved from metaplastic cardiac epithelium. I could easily distinguish it from gastric oxyntic epithelium. It was esophageal, not gastric. It was one of the three epithelia that Paull et al.[26] had found in visible CLE in the esophageal body above the LES. I could not give into the temptation of not including it simply for the sake of making my argument easier.

Ultimately, this decision was absolutely correct. I will show that the critical histological point that must be recognized is the point where metaplastic esophageal oxyntocardiac epithelium transitions to gastric oxyntic epithelium. This is the true GEJ. The oxyntocardiac epithelium is part of the dilated distal esophagus and therefore has to be included in the measurement that will ultimately quantitate LES damage. It is critical to the new method of understanding and management of GERD based on histology.

It is worth noting that many studies that evaluated the hypothesis did not follow my requirement that oxyntocardiac epithelium should be included with cardiac epithelium. Many studies limit the definition to cardiac epithelium and use the first parietal cell distal to the SCJ as the distal point of measurement. This is not accurate for defining the entire extent of the GERD-damaged esophagus. The entire extent of metaplastic columnar epithelium is defined by cardiac epithelium with and without parietal and/or goblet cells, not simply pure cardiac epithelium.

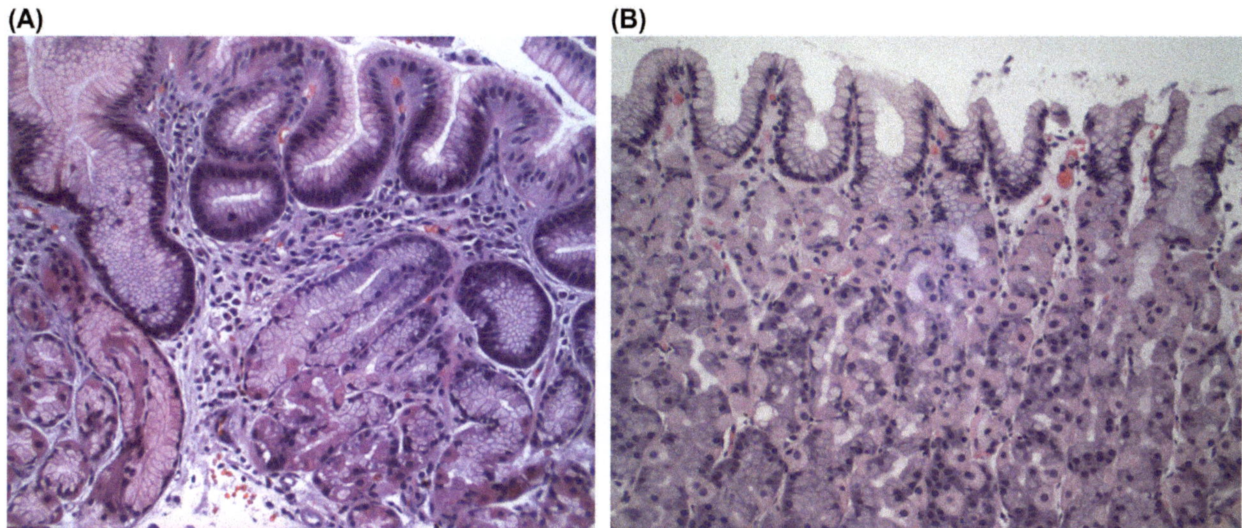

FIGURE 5.7 (A) Oxyntocardiac epithelium showing lobulated glands composed of numerous parietal cells (round with light pink cytoplasm), mucous cells (vacuolated cytoplasm), and scattered Paneth cells (red cytoplasm). Note the presence of mild chronic inflammation in the superficial lamina propria. (B) Gastric oxyntic epithelium showing the straight (not lobulated) tubular glands with parietal cells but no mucous cells. There is no inflammation in the lamina propria.

The true GEJ is not the distal limit of cardiac epithelium. It is the distal limit of oxyntocardiac epithelium where it transitions to the stomach (gastric oxyntic epithelium). Using only cardiac epithelium would commit the medical establishment to repeating the same mistake of the last 50 years, i.e., thinking that the esophagus ends more proximally than the true GEJ.

2.2.3 The Hypothesis Does Not Die

At this point, with the rejection of the paper for the second time, the hypothesis was in danger of descending into oblivion. It did not. I like to think this was because the truth was too powerful. However, I know it was because I had acquired a few believers who were influential and gave me opportunities to, as Dr. Tom DeMeester said, "spread the gospel."

I continued to present this concept at the annual USC Foregut conference in Hawaii. There were a significant number of attendees who heard, understood, and recognized the potential significance of the concept. This resulted in me being invited to present my data at many meetings. I was invited by Dr. Ellie Klinkenberg-Knoll, a leading gastroenterologist in Amsterdam, to a meeting of European gastroenterologists in Vienna to celebrate the 10th anniversary of the release of omeprazole. Dr. John Chang invited me to give the International Academy of Pathology meeting in Hong Kong, allocating an entire day to the topic. Dr. Peters, a senior surgeon in the USC Foregut unit for which I did the pathology, invited me to a panel discussion he organized for the Surgical Society of the Alimentary Tract. Dr. DeMeester invited me to give the state-of-the-art address at the 1998 meeting of the Society for Surgery of the Alimentary Tract.

The various talks resulted in an invitation to write a review article on Barrett esophagus in 1997, which had a summary of the yet unpublished autopsy findings.[5] In light of the repeated rejection of my autopsy studies, these review articles were my first opportunity to put to paper the hypothesis that cardiac epithelium was an abnormal epithelium in the distal esophagus caused by exposure of esophageal squamous epithelium to gastric acid and was never a normal epithelium of the proximal stomach.

The abstract of the review paper in 1997 stated[5]: "A novel pathophysiology of Barrett's esophagus and a new method of assessing biopsy specimens in patients with gastroesophageal reflux disease (GERD) are presented. This is based on the observation in autopsy studies of patients without GERD that the squamous epithelium of the esophagus transitions directly to fundic mucosa in many people and that the cardiac mucosa is of very short length in others. Available evidence suggests that what is termed gastric cardiac mucosa is in reality an abnormal mucosa resulting from metaplasia of the squamous epithelium of the esophagus as a result of GERD. The severity of GERD correlates with the length of metaplastic cardiac mucosa and further changes occurring in it, permitting development of a system that provides good correlation between biopsy histology and severity of GERD. Intestinal metaplasia ("Barrett's esophagus") always occurs in this metaplastic cardiac mucosa. The recognition of this new pathophysiology of Barrett's esophagus permits identification of the entire sequence whereby GERD leads to adenocarcinoma: GERD→cardiac metaplasia of squamous epithelium→reflux carditis→intestinal metaplasia→dysplasia→adenocarcinoma."

TABLE 5.4 Interpretation of Biopsies in Chandrasoma (1997)[5]

Finding	Reflux	GERD	Risk of IM	Risk of Adenocarcinoma
Fundic mucosa only[a]	0	0	0	0
Cardiac-fundic mucosa only[b]	+	±	0	0
Cardiac mucosa <1 cm long; no IM	+	±	+	0
Cardiac mucosa <1 cm long; IM+	+	±	N/A	+
Cardiac mucosa 1–2 cm long; no IM	++	+	+	0
Cardiac mucosa 1–2 cm long; IM+	++	+	N/A	+
Cardiac mucosa >2 cm long; no IM	+++	++	++	0
Cardiac mucosa >2 cm long; IM+	+++	++	N/A	++

GERD, gastroesophageal reflux disease as indicated by symptoms, endoscopic features of reflux esophagitis, and abnormal pH scores; *IM*, intestinal metaplasia.
[a]*Fundic mucosa, gastric oxyntic epithelium.*
[b]*Cardiac-fundic mucosa, oxyntocardiac epithelium.*

In the paper is a table that shows the "interpretation of changes in biopsies taken immediately distal to the squamocolumnar junction, irrespective of any endoscopic or clinical data" (Table 5.4).

When a hypothesis is evaluated over the long term, the likelihood that the hypothesis is correct is predictable by its longevity and the direction of change of the prior opinion held by the medical establishment.

At the time the above review paper was published, the universal belief was that the proximal 3 cm of the stomach was normally lined by cardiac epithelium. I was suggesting that 0–3 cm of cardiac epithelium provided a grading of the severity of GERD. I will show in this book that this concept has not changed at all; the only change has been the realization that the reason for this correlation is that the length of cardiac epithelium up to 3 cm is a measure of abdominal LES damage.

Since 1997, all the papers that I have written and the talks I have given have reiterated this basic hypothesis with new ideas that have expanded the scope of the hypothesis.

In contrast, the reaction of the medical establishment has been one of slow but inexorable capitulation. Between 1994 and 2015, the purported "normal" extent of cardiac epithelium in this region has shrunk from 30 mm to "1–4 mm.[8]" This miniature dogma persists despite convincing evidence that some people have no cardiac epithelium (with and without parietal/goblet cells) (Fig. 5.5).

A significant and increasing number of people accept that all cardiac epithelium is abnormal. Most people who still cling to the dogma that there is 1–4 mm of normal cardiac epithelium in the proximal stomach do so in silence and increasing discomfort. No one will confront me with their opposite viewpoint, but will express their certainty that I am wrong when they are among their cobelievers, causing the dogma to persist. I am ever willing to challenge anyone to argue with me in public. It does not happen.

These are all things that are predictive of a hypothesis that is correct.

2.2.4 Other Autopsy Studies of Normal People Without Gastroesophageal Reflux Disease

The hypothesis generated a rush to perform studies about this subject. Several studies on fetal and neonatal subjects were published in 2003 (see Chapter 3). However, only one other autopsy study from Cleveland Clinic was published. This was in children.

The objective of the study by Kilgore et al.[9]: "It is unclear whether the gastric cardia is present from birth or is metaplastic and develops as a result of gastroesophageal reflux disease. To this end, we evaluated the histology of the entire esophagogastric junction in consecutive pediatric autopsies to determine the presence and extent of cardiac mucosa."

30/33 consecutive pediatric (<18 years old) autopsies were evaluated, including 25 forensic autopsies. Autolysis was the reason for exclusion of three cases. None of these patients had GERD, Barrett esophagus, or *Helicobacter pylori* infection, and none were on acid-suppressive medication. The mean age was 6.3 years (range: 16 days–18 years); 21 were males; 16 were white, and the rest were black.

All cases had a regular Z-line at the esophagogastric junction (EGJ) (defined by the proximal margin of gastric folds and peritoneal reflection, which were concordant). The entire EGJ was sectioned longitudinally and examined histologically. The 3.5 cm sections were composed of 0.5 cm squamous epithelium and 3.0 cm columnar epithelium distal to the Z-line. These sections were stained with hematoxylin and eosin and Alcian blue/periodic acid–Schiff (PAS). In addition, a section of the antrum was taken in each case. The EGJ and antral sections were stained for *H. pylori* with Giemsa stain.

The slides were evaluated by two pathologists looking for: "the presence of cardiac-type mucosa, characterized by unequivocal PAS-positive mucous glands arranged in lobular configuration. The length of the cardiac-type mucosa, the distance between the most distal portion of the squamous mucosa and the identification of the most proximal parietal cell, was measured with a micrometer and recorded. In addition, an attempt was made to measure the distance between the most distal aspect of the cardiac-type mucosa to the point at which the mucosa was composed entirely of fundic-type glands, that is, the so-called "transition" zone composed of a mixture of mucous glands and parietal cells."

They report in their results: "Cardiac mucosa was present in all specimens, with a mean length of 1.8 mm (range 1.0–4.0 mm) and in all cases was found on the gastric side of the EGJ (Figs 2 and 3). There was no significant association between patient age or gender and length of cardiac mucosa."

> *In all cases, an attempt was made to identify and measure the transition zone with an admixture of cardiac and fundic-type glands. In three cases, autolysis precluded adequate histological evaluation of this mucosa. Of the remaining 27 cases, there was an abrupt transition between cardiac-type and fundic type mucosa in 11 cases (37%). In 12 cases (40%) there was an admixture of these two types of glands, which measured <1.0 mm, and in four cases (13%), this mucosa measured >1.0 mm (Fig 4).*

The authors reached two conclusions based on their data. The first, stated in the abstract, was that: "In an unselected pediatric patient population with little or no propensity for gastroesophageal reflux disease, a short segment of cardiac mucosa was consistently present on the gastric side of the esophagogastric junction, independent of gender or age. These results support the concept that the gastric cardia is present from birth as a normal structure."

The second conclusion is in the discussion: "… even if one included the length of oxynto-cardiac mucosa with cardiac mucosa, the longest combined length of these two mucosae in our study was 8 mm. These data emphasize the fact that this zone is extremely small, regardless of whether one includes an admixture of these two types of mucosa."

The study by Kilgore et al.[9] was the only other autopsy study of the histology of the GEJ region in normal children and adults that has been published in any major journal. This is interesting because our study and the Cleveland Clinic study contradicted one another on the question of the presence or absence of cardiac epithelium. One would have expected other studies to provide independent data to support one or the other viewpoint. This did not happen.

2.2.5 Comparison of the Two Autopsy Studies

The above two studies are the only two studies in the literature on the histology of children and/or adults without clinical GERD during life. The two studies agree on the fact that, in this population, the maximum amount of cardiac plus oxynto-cardiac epithelium was found to be very small; 8.05 mm in our study[7] and 8 mm in Kilgore et al.[9]

The two studies are very different in their conclusion regarding the presence of cardiac epithelium. Our prospective study and the study by Kilgore et al. had a complete examination of the entire circumference of the region distal to the normal SCJ. Cardiac epithelium (not including oxyntocardiac epithelium) was absent in 10/18 (56%) patients in our study and present in all the patients in the study by Kilgore et al. The difference of the minimum extent of cardiac epithelium in the two studies was minute; 0 mm in our study[7] and 1 mm in Kilgore et al.[9]

This 1 mm, however, resulted in two completely different conclusions: We concluded that cardiac epithelium did not exist as a normal epithelium at the junction and therefore was an abnormal epithelium.[7] Kilgore et al.[9] concluded the reverse: "These results support the concept that the gastric cardia is present from birth as a normal structure."

When two studies produce a contradictory result and conclusion, there are two ways to decide which is correct.

The first is by duplication of data from other studies. While there were no other autopsy studies of normal patients, there were two studies where the presence of cardiac epithelium at the GEJ was evaluated by endoscopic biopsy.[27,28] These two are detailed in the next section. They both show the absence of pure cardiac epithelium distal to a normal SCJ, contradicting the study by Kilgore et al.[9]

The second method is to demonstrate the negative finding. The finding of a parietal cell–containing epithelium immediately adjacent to the squamous epithelium essentially proves that cardiac epithelium is not universally present. In my experience, it is not rare to find a direct transition of squamous epithelium to oxyntocardiac epithelium even in clinical biopsies if the normal endoscopic GEJ is sampled (Fig. 5.8). Marsman et al.[28] reported a direct transition from squamous to oxyntocardiac epithelium in 38% of patients.

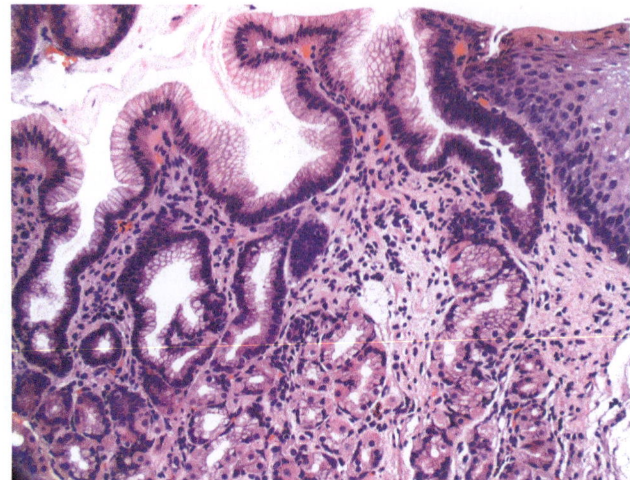

FIGURE 5.8 Section across the squamocolumnar junction in an endoscopic biopsy showing a direct transition of squamous epithelium (right) to oxyntocardiac epithelium consisting of glands with a mixture of parietal (bright pink) and mucous (light pink) cells. There is mild chronic inflammation in the lamina propria.

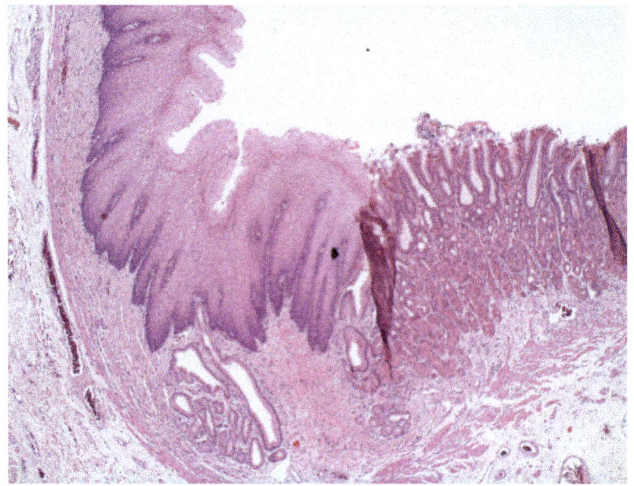

FIGURE 5.9 Direct transition of squamous epithelium to gastric oxyntic epithelium. Section taken from the squamocolumnar junction of a resection specimen in a 67-year-old patient with squamous carcinoma. Note the gland duct of a submucosal gland draining into the squamous epithelium at the junction.

A single illustration showing squamous epithelium transitioning directly to gastric oxyntic epithelium proves that both cardiac and oxyntocardiac epithelium are absent (Fig. 5.5). This is seen in autopsies and very rarely in clinical biopsies or resection specimens (Fig. 5.9). Any number of illustrations showing the presence of cardiac epithelium at the SCJ cannot overturn the proof of the demonstration of the absence of cardiac epithelium (with and without parietal cells) in one patient.

The way our autopsy paper was finally published was very interesting. I had written the 1997 review and given many talks at pathology meetings about the hypothesis. The paper, as stated earlier, had been rejected on two occasions (before and after the prospective part of the study) by many journals, including the *American Journal of Surgical Pathology*.

As a result of the interest created, I was invited to a symposium on cardiac epithelium at a major pathology meeting in San Francisco in 1999. I was to present my hypothesis and autopsy data, followed by Dr. John Goldblum with data from the Cleveland Clinic autopsy study, in which he was the senior author. Neither study had yet been published. This was followed by an audience question session and a summation by Dr. Henry Appelman from Michigan, who is a well-recognized GI pathology expert.

At the end of the meeting, the editor of the *American Journal of Surgical Pathology* came up to me and asked me to resend the paper to that journal, specifically addressed to him. It was rapidly reviewed and accepted, being published in March 2000.[7]

There was one interesting audience question during the meeting. A pathologist came up to the microphone, introduced himself (I do not remember his name), stating that he was from Boston. He said that he had done a similar autopsy study and that his findings supported Dr. Goldblum's. He had found cardiac epithelium in 80% of his patients. Nothing that I have ever heard demonstrates the power of a dogma more than this statement. If his study had shown that 80% of patients had cardiac epithelium, it meant that it was absent in 20%, essentially proving my hypothesis. The fact that 80% was closer to 100% (the prevalence of cardiac epithelium in the study by Kilgore et al.[9]) than the 44% in our study with cardiac epithelium does not mean that his data support Dr. Goldblum's study. It supports ours.

Kilgore et al.[9] in their discussion, offer the following criticism of our paper: "… there is no mention of how the authors localized the EGJ, nor is there much detail regarding the relationship of the squamocolumnar junction relative to the EGJ. Thus, it is unclear whether the cardiac-type or oxynto-cardiac-type mucosae were found within the distal esophagus or proximal stomach. If these mucosae were on the gastric side of the EGJ, then their results were similar to ours. However, without knowing the precise location of these mucosae, it is difficult to compare the results of their study with ours."

This statement is another unbelievable example of the power of dogma. Kilgore et al.[9] believe that the GEJ can be defined at autopsy with a precision that permits them to state with certainty that the 1 mm of cardiac epithelium they find distal to the SCJ is gastric. Their definition of the GEJ is "the peritoneal reflection and the proximal margin of the gastric folds." The transference of the point of the peritoneal reflection, an external landmark, to the mucosal surface has an inherent error based on the angle of sectioning from the serosa to the mucosa. The proximal limit of gastric folds is an endoscopic landmark that is used for defining the GEJ at endoscopy. It is an opinion-based definition that has no evidence whatsoever in support. The gastric folds gradually fade into the flat mucosa without a sharp point and it is well known at endoscopy that there is a 0.5–1 cm error in demarcating its exact location. To claim that 1–4 mm of cardiac epithelium can be placed accurately in the stomach (normal) or distal esophagus (abnormal, metaplastic) based on their imprecise definition of the GEJ is unreasonable and defies logic. It is also wrong. I will show in Chapter 6 that the true GEJ is distal to the proximal limit of rugal folds by the length of cardiac epithelium (with or without parietal and/or goblet cells) distal to the proximal limit of rugal folds.

The entire basis of our hypothesis is that cardiac epithelium (with or without parietal and/or goblet cells) in this region represents metaplastic esophageal epithelium, irrespective of the perceived location of the GEJ at either endoscopy or pathologic examination of resected or autopsy specimens. The true GEJ cannot be defined at either endoscopy or gross pathologic examination. It can be defined only by histology as the transition between cardiac epithelium (with or without parietal and/or goblet cells) and gastric oxyntic epithelium.

If the location of the EGJ is removed from the equation, the two autopsy studies are exactly comparable, leading to the conclusion that the length of cardiac epithelium (with or without parietal cells) is 0–8 mm in our study[7] and 1–8 mm in Kilgore et al.[9] In both studies, this pure cardiac epithelium and oxyntocardiac epithelium (called "transitional epithelium" in Kilgore et al.[9]) lie between the distal limit of squamous epithelium (esophageal) and gastric oxyntic epithelium (called "fundic-type mucosa" in Kilgore et al.). This is therefore the epithelium lining the squamooxyntic gap in both studies. There is no controversy about this. I will show that the squamooxyntic gap is *always* esophageal, *never* gastric.

The main question that arises is whether the presence of cardiac epithelium that has a length of 1–4 mm in this region justifies the conclusion of Kilgore et al.[9] that the gastric cardia is present from birth as a normal structure. It certainly does not. The only conclusion that can be drawn is that 1–4 mm (mean 1.8 mm) of cardiac epithelium was present in this region in 30 children with a mean age of 6.3 years (range 16 days–18 years) who had autopsies performed at Cleveland Clinic. No conclusion can be drawn as to whether this 1–4 mm (mean 1.8 mm) is normal or abnormal. The absence of a history of GERD or acid-suppressive drug use in the medical record is not reliable evidence of the absence of pathology related to GERD.

While there is strong evidence to support that the absence of cardiac epithelium is the normal state, the discrepancy of the minimum length of cardiac epithelium (zero in our study[7] and 1 mm in the study by Kilgore et al.[9]) is difficult to understand. The probability that all 30 children in the study by Kilgore et al. had cardiac epithelium with complete examination of the GEJ region when 56% of patients in our study had no cardiac epithelium is very low if the findings are accurate in both studies.

Careful evaluation of the paper by Kilgore et al.[9] shows the following problems that call their claim that cardiac epithelium was found in all their patients into question: (1) The definition of cardiac epithelium used was different than in our study and the accepted norm that had been established in 1976 by Paull et al.[26] In our study, cardiac epithelium is defined by the presence of mucous cells only, without parietal and goblet cells in a routine section stained with hematoxylin and eosin. In Kilgore et al.[9] the method of evaluation for cardiac epithelium was by: "looking for the presence of cardiac-type mucosa, characterized by unequivocal PAS-positive mucous glands arranged in lobular configuration." (2) The paper is illustrated by three figures of the histology. Figure 3 is a high-power photograph of cardiac epithelium showing glands with only mucous cells in a hematoxylin- and eosin-stained section. Figures 2 and 4, to my eyes, provide a possible solution to the discrepancy in the data. Figure 4 shows transitional mucosa (=oxyntocardiac epithelium) with an admixture of PAS-positive mucus-secreting glands staining a deep magenta color and fundic-type glands characterized by parietal cells that are negative for PAS and have small nuclei surrounded by cytoplasm in this section. This is correctly called oxyntocardiac epithelium. Their Figure 2 also shows a mixture of PAS-positive mucous cells and smaller cells that are PAS-negative that are similar to those illustrated in Figure 4. If this is true, the epithelium illustrated immediately adjacent to the squamous epithelium is oxyntocardiac epithelium, not cardiac.

The data in the study by Kilgore et al. stating that cardiac epithelium is present in all people are an outlier. There are studies in patients who have undergone endoscopy with extensive biopsies at a normal endoscopic GEJ that show cardiac epithelium without parietal cells is frequently absent,[27,28] concordant to the data in our autopsy study.[7]

2.2.6 *Summary of the Data From the Two Autopsy Studies*

The data from the two autopsy studies establish without any controversy that the maximum length of pure cardiac epithelium in these two populations with a bias toward younger age is less than 4 mm. The maximum length of cardiac + oxyntocardiac epithelium in this population in the two studies is 8 mm. This explodes the myth that up to 30 mm of cardiac epithelium (with and without parietal cells) normally lines the proximal stomach, i.e., the "gastric cardia."

The data from the two studies also establish that the minimum extent of pure cardiac epithelium is 1 mm in the study by Kilgore et al.[9] (the mean length in the 30 patients was 1.8 mm) and zero in the study by Chandrasoma et al.[7] The minimum length of cardiac + oxyntocardiac epithelium was not recorded in Kilgore et al.[9] and was zero in Chandrasoma et al.[7]

The present controversy is therefore whether the minimum amount of cardiac epithelium (with and without parietal cells) is 0 or 1 mm. To believe that there is a normal cardiac epithelium that can be as short as 1 mm despite the proof that some people do not have this same cardiac epithelium is illogical at best. It is difficult to believe that this is an argument that has not been resolved by reasonable people for 15 years.

It is the power of dogma.

3. STUDY OF THE REGION DISTAL TO THE ENDOSCOPIC GASTROESOPHAGEAL JUNCTION IN RESECTED SPECIMENS

There are two studies that have carefully looked at the histology of the region distal to the endoscopic GEJ in resected specimens. These have reasonably concordant findings and provide an accurate definition of the histology of the region in patients who are older and have had resections of the esophagus for severe disease.

Chandrasoma et al.[17] is a study of 10 resected esophagogastrectomy specimens selected because they had a well-defined transition from the tubular esophagus to the saccular "stomach" and where this transition point coincided with the proximal limit of "gastric" rugal folds. The 10 patients were aged 47–65 years; there were 7 males; 8 patients underwent the surgery for esophageal adenocarcinoma and the other 2 for squamous carcinoma. Five patients had definite CLE in the tubular esophagus measuring 20 mm or more; three others had possible CLE in the tubular esophagus measuring <5 mm. The two patients with squamous carcinoma did not have CLE.

Full-thickness vertical sections were taken from the entire circumference of the end of the esophagus straddling the proximal limit of the rugal folds. The sections extended from the proximal limit of rugal folds to 25–30 mm distal to this point. The sections were stained with hematoxylin and eosin and examined by the three pathology authors (PC, KM, YM).

All patients had cardiac and oxyntocardiac epithelium distal to the endoscopic GEJ. Intestinal metaplasia was present in this area in all eight patients with Barrett esophagus and one of two patients with squamous carcinoma. The cumulative length of cardiac epithelium (with and without parietal/goblet cells) in the saccular part of the specimen distal to the GEJ (end of tubular esophagus and proximal limit of rugal folds) is shown in Table 5.5.

The data are an incredible contrast to the autopsy data where the minimum length of cardiac + oxyntocardiac epithelium was 0–1 mm and the maximum length was 8 mm.

In this population of older patients with esophageal disease, the length of cardiac epithelium distal to the GEJ ranged from 3.1 to 20.5 mm. Both patients with length < 5 mm had squamous carcinoma, a disease unassociated with GERD. The eight patients with adenocarcinoma, which is a specific complication of GERD, had lengths varying from 10.3 to 20.5 mm (Fig. 5.10).

Another highly significant finding in this study was that the area distal to the GEJ that was lined by cardiac and oxyntocardiac epithelium had submucosal glands under these epithelia (Fig. 5.11). The submucosal glands were never found under gastric oxyntic epithelium. Submucosal glands are recognized as a marker for esophageal location. The concordance of the length of cardiac epithelium (with and without parietal and/or goblet cells) and the presence of submucosal glands is powerful evidence that this region is esophageal (see Chapter 6).

TABLE 5.5 Cumulative Length of Cardiac Epithelium (With and Without Parietal/Goblet Cells) in the Saccular Part of the Specimen Distal to the Gastroesophageal Junction (End of Tubular Esophagus and Proximal Limit of Rugal Folds) in 10 Esophagogastrectomy Specimens

Diagnosis	CLE Length in Tubular Esophagus (in mm)	Length of CM + OCM + IM Distal to Proximal Limit of Rugal Folds (in mm)
Squamous carcinoma	0	3.1
Squamous carcinoma	0	4.3
Adenocarcinoma	30	10.3
Adenocarcinoma	55	10.5
Adenocarcinoma	<5	11.0
Adenocarcinoma	45	11.3
Adenocarcinoma	<5	13.9
Adenocarcinoma	<5	16.0
Adenocarcinoma	20	16.8
Adenocarcinoma	50	20.5

CLE, columnar-lined esophagus; *CM*, cardiac mucosa; *IM*, Intestinal metaplasia; *OCM*, oxyntocardiac mucosa.

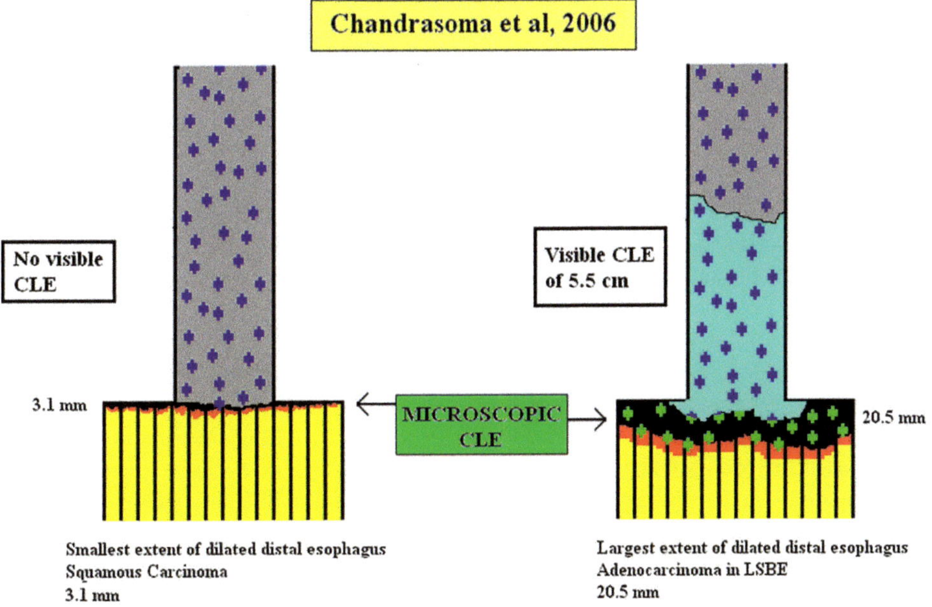

FIGURE 5.10 Diagrammatic representation of the findings at esophagectomy in a patient with resection for squamous carcinoma (left) and a patient with adenocarcinoma (right). The patients represent the minimum and maximum length of cardiac epithelium (with and without parietal and/or goblet cells) between the endoscopic gastroesophageal junction and the proximal limit of gastric oxyntic epithelium. *CLE*, columnar-lined esophagus. Intestinal metaplasia = blue; cardiac epithelium = black; oxyntocardiac epithelium = red.

The second study,[18] which preceded our study of resected specimens presented above, was stimulated by the controversy raised by our hypothesis: "Traditionally, cardiac mucosa (CM) is considered as a zone of approximately 1–2 cm length that separates the most distal portion of esophageal squamous epithelium from the acid-producing fundic mucosa. However, in the meantime it has been suggested that CM is not a normal structure but arises secondary to GERD." The authors cite the papers of Oberg et al.,[25] Chandrasoma et al.,[7] and Kilgore et al.[9] as representing the two sides of the controversy.

They write: "… both biopsy- and autopsy-based investigations have inherent technical disadvantages: sampling error in the first and autolysis in the second type of study. In the current study we therefore investigated the entire GEJ of 36 surgical

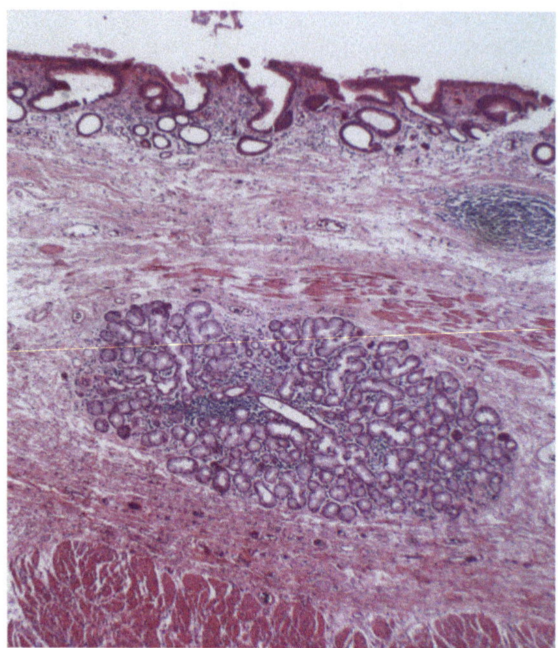

FIGURE 5.11 Cardiac epithelium overlying submucous glands in a resection specimen. This section was taken distal to the endoscopic gastroesophageal junction, proving that this is from the esophagus and not stomach.

esophagogastrectomy specimens which had been resected for squamous cell carcinoma of the upper or middle esophagus. The aim of the study was a detailed examination of the presence and extent of CM in a cohort of adult patients not biased for the presence of GERD."

The age range of the 36 patients was 24–82 years (median 55 years); 30 were males. Cases where the distal margin of the tumor was <5 mm from the GEJ were excluded. None of the patients had CLE in the tubular esophagus. The GEJ was identified as the junction between brown-red gastric mucosa and gray esophageal mucosa. The entire GEJ was sectioned such that the majority of the section was distal to the squamous epithelium. The length of CM was determined as the distance between the distal end of the squamous epithelium and the most proximal parietal cell in all sections. The length of oxyntocardiac mucosa (OCM), characterized by glands that contain a mixture of mucous cells and parietal cells (they cite the definitions used in our autopsy study), was measured.

CM was present in the entire circumference of the GEJ in 20 cases, in parts of the circumference in 15 cases, and entirely absent in 1 case. No patient showed a transition of squamous epithelium directly to gastric oxyntic epithelium in any part of the circumference of the GEJ. Squamous epithelium was separated by either cardiac epithelium or oxyntocardiac epithelium from gastric oxyntic epithelium, i.e., all patients had a squamooxyntic gap whose minimal extent was 1 mm.

An important finding of the study: "Intraesophageal location of CM and/or OCM could be verified histologically in 9 of 36 cases. Thus, in eight cases CM/OCM was situated over submucosal glands (illustrated in their Fig 1a). In the ninth case, CM/OCM was situated over squamous epithelium lined ducts, a finding that was also present in three of the eight aforementioned cases." This means is that in these five cases, tissue taken distal to the GEJ can be proven to be esophageal and not gastric. A similar finding was reported in our study.[17] In addition, we showed that submucosal gland ducts are sometimes present in mucosal biopsies distal to the endoscopic GEJ.[29] We will discuss this further when we reconsider these studies in our discussion of the definition of the GEJ in Chapter 6.

The minimal and maximal lengths of CM, OCM, and CM + OCM in the study by Sarbia et al.[18] are shown in Table 5.6.

This study has a different population than our study of resected specimens. This study is biased toward excluding patients with GERD while patients with adenocarcinoma resulting from GERD dominated our study. It is difficult to compare the two studies meaningfully. However, if one looks at the median length of CM + OCM, ~ 50% of the patients were under 10 mm. However, the maximum length of CM + OCM in this study (28 mm) was greater than the maximum length in our study (20.5 mm). This 28 mm CM + OCM length in the gap between the distal limit of squamous epithelium and proximal limit of gastric oxyntic epithelium is the maximum that I have seen reported in a person without visible CLE.

The authors report that "a statistically not significant trend for increase of minimal length of CM, OCM and the sum of both was found in the presence of gastroesophageal reflux disease." Their final conclusion is very careful and

TABLE 5.6 Minimal and Maximal Length of Cardiac Mucosa (CM), Oxyntocardiac Mucosa (OCM), and CM+OCM (They Call This Junctional Mucosa) in 36 Gastroesophageal Resection Specimens

	Range (mm)	Median (mm)
CM—minimal	0–3[a]	1
CM—maximal	1–15	5
OCM—minimal	1–7	2
OCM—maximal	2–24	7
CM+OCM—minimal	1–12	4
CM+OCM—maximal	5–28	11

[a]*One patient who did not have any CM was excluded in this table by the authors; I have taken the liberty to add this.*

accurate: "In conclusion, the high variability in length, … and the frequent extension into the esophagus suggest that CM/OCM is a dynamic structure that probably mirrors the influence of underlying esophageal disease."

The conclusion suggests strongly that cardiac epithelium is not a normal structure in the proximal stomach although the tissue is distal to the GEJ; it is a manifestation of GERD-induced damage of the esophageal epithelium.

3.1 Summary of the Two Studies

The two studies are essentially concordant. They both show that there is a significantly greater length of cardiac epithelium (including oxyntocardiac epithelium) distal to the endoscopic GEJ than found in the two autopsy studies.

They both show that the available evidence suggests that the area distal to the endoscopic GEJ, to the extent that it is lined by cardiac epithelium (with or without parietal and/or goblet cells), is esophageal and not gastric. The presence of submucosal glands in this esophagus represents proof of this fact.

They both show a trend toward an association between increasing length of cardiac epithelium and GERD. The number of patients in both studies is too small for this trend to reach statistical significance.

If the present guidelines for endoscopic biopsy are followed, the only patients in the four studies of autopsy and resected specimens who will be studied at endoscopy will be the five patients with definite CLE in the esophagus. Biopsy is not recommended in the other patients who do not have CLE at endoscopy. If all the patients in these studies had undergone endoscopy during life, all patients in both autopsy studies,[7,9] all patients in the study by Sarbia et al.,[18] and the 5 patients in the study by Chandrasoma et al.[17] who did not have CLE would not have undergone biopsy. The patients undergoing biopsy would not have had biopsies taken distal to the endoscopic GEJ unless there was a visible neoplasm in that area.

All the data that are represented by these four studies are therefore completely hidden in the endoscopic study of patients with GERD. No biopsies are ever taken and pathologists are blind to these changes. There is no interest among the massive population of gastroenterologists to study the relevance of these changes. There is no opportunity for pathologists to ever see what I am talking about. The only reason why they ever came to light was because Dr. Tom DeMeester followed his own biopsy protocol and allowed me to see them.

4. ENDOSCOPIC BIOPSY IN ASYMPTOMATIC VOLUNTEERS

Asymptomatic volunteers are a population that is closest to patients at autopsy without evidence of GERD during their lifetime. Only one study and a personal communication exist in the literature.

Caution is necessary when data from such studies are evaluated. Asymptomatic for GERD does not necessarily mean absence of GERD. The reason is that symptoms of GERD are not concordant with severity of disease. Manifestations of severe GERD such as Barrett esophagus can exist in people who do not have GERD symptoms.

4.1 The Glasgow Volunteer Study

This is an elegantly designed study that provides incredibly valuable data that help establish the normal baseline as well as provide a basis for the earliest changes in the pathogenesis of GERD.

I will directly quote from this paper because of the perfection of some of the writing. "The first aim of our current study was to examine the association between central obesity and the length of non-acid-secreting columnar epithelium (cardiac mucosa) laying between esophageal squamous mucosa and acid-secreting gastric mucosa."

> *Study participants were healthy volunteers recruited by word of mouth and newspaper advertisement. Those who had ever taken proton pump inhibitors or ever attended primary or secondary care with reflux symptoms were excluded. All subjects were screened for* H. pylori *by urea breath test and those testing positive or who had a past history of the infection were excluded. Subjects who were found to have hiatus hernia during the study protocol were excluded from the current analysis.*

Obesity was assessed by waist circumference and magnetic resonance imaging study of abdominal fat distribution. Patients were divided into two groups, small and large waist circumference. The study had 51 volunteers (M:F = 26:25). The age range was 21–73 years with a median of 45 years in the 24 volunteers with small waist circumference and 46 years in the 27 volunteers with large waist circumference. The patient population was thus significantly older than the populations in the two autopsy studies.

At upper GI endoscopy, biopsy specimens were taken across the SCJ using a pair of jumbo forceps with a jaw span of 8 mm. The biopsies were taken perpendicular to the junction and targeted to include just enough squamous epithelium to confirm position and maximize the amount of columnar epithelium distal to the SCJ. Up to three biopsies were taken to optimize the chances of capturing the full span of cardiac mucosa in a single biopsy, allowing accurate measurement of cardiac mucosa. Biopsies were also taken of the body and antrum. The biopsy handling was optimized to minimize sample contraction and preserve original length as much as possible. The methodology is as good as it can get.

> *The cardiac mucosa was considered fully measurable when consecutive squamous, cardiac, and cardio-oxyntic mucosal types were present in the same biopsy specimen. The cardiac mucosa was defined as epithelium devoid of parietal cells and consisting of mucous-secreting cells. The demarcation between squamous and cardiac epithelium was clear and the distal limit of cardiac mucosa was taken as the appearance of parietal cells.*

The definitions used for histologic classification of the epithelia are those that are described in Chapter 4. The measurement being aimed for is the length of cardiac mucosa. This is not the full squamooxyntic gap because oxyntocardiac epithelium is excluded in the measurement. This is reasonable because the size of the biopsy limits the length that can be measured to 8 mm and biopsies may not reach the end of oxyntocardiac epithelium where the epithelium transitions to gastric oxyntic epithelium. In the autopsy studies reviewed above,[7,9] the cardiac plus oxyntocardiac epithelial length reached a maximum of 8 mm.

Thirty-five patients had a fully measurable cardiac mucosa (at least one biopsy containing the full length of cardiac mucosa from the squamous epithelium to the first parietal cell).

In the 17 volunteers in the large waist circumference group, mean cardiac mucosal length was 2.5 mm [interquartile range (IQR), 0.8 mm], significantly greater than the 1.75 mm (IQR, 1.1 mm) in the 18 volunteers with small waist circumference ($P = .008$). By bivariate analysis, cardiac mucosal length was correlated positively with increasing age ($R = 0.455$, $P = .006$).

This study is perfect in its methodology and permits comparison with other studies where the definitions are applied with care. The only regret when I first read this paper is that it did not mention whether or not there were any patients who did not have any cardiac mucosa in their biopsies. I obtained this information by a personal communication with Professor McColl and Dr. Derakhshan. In answer to my question, they replied that all patients in their series had cardiac epithelium and that the smallest length of cardiac epithelium was 0.5 mm (i.e., 500 µm) in a 36-year-old volunteer with a waist circumference of 88 cm, which was at the high end of the small waist circumference group.

It is remarkable to compare these data relating to those in the prospective part of our autopsy study[7] (Table 5.7). The two studies are exactly comparable in their definitions and measurement methodology insofar as cardiac epithelial length is concerned. Both studies show that the length of cardiac epithelium in the population under study is extremely small (less than 5 mm). In general, the autopsy study shows smaller lengths of cardiac epithelium. The differences can be explained by the fact that the 18 patients in the autopsy study had an age range of 3–61 years with a median age of 20 years and was a much younger population than in the volunteer study (age range 21–73 years with a median around 45 years). In both studies, the length of cardiac epithelium increased significantly with increasing age.

This study establishes the following evidence base with the proviso that it stands alone and has not been duplicated:

1. The length of cardiac epithelium is similar in this population of volunteers without any clinical evidence of GERD to those of an autopsy population without any clinical evidence of GERD. This establishes the feasibility of transferring autopsy data to clinical practice.

TABLE 5.7 Comparison of Length of Cardiac Epithelium in Two Studies Using Identical Definitions and Measurement Methodology

	Autopsy Study[20] (mm)	Small Waist Circumference	Large Waist Circumference
Mean length of cardiac epithelium	0.57	1.75 mm	2.50 mm
Minimum length of cardiac epithelium	0	0.05 mm[a]	Unknown
25th percentile of cardiac epithelial length	0	0.64 mm	1.70 mm
Maximum length of cardiac epithelium	2.75	Unknown	Unknown
75th percentile of cardiac epithelial length	1.00	2.85 mm	3.30 mm

[a]*Data not in the paper; obtained by personal communication with Professor McColl.*

2. The length of cardiac epithelium is significantly correlated with increasing age, suggesting that cardiac epithelium is an acquired epithelium. The slightly higher length of cardiac epithelium in this study compared to the autopsy studies can be reasonably explained on the basis of the higher age of patients in this study.
3. In the volunteers, there was no patient with cardiac epithelium absent in contrast with the autopsy study. This is unexpected. Marsman et al.[28] reported that the epithelium adjacent to squamous epithelium was cardiac in only 62% of their patients.
4. There is a significant difference between cardiac epithelial lengths in volunteers with small compared to large waist circumference. This provides evidence that central obesity is associated with an increase in the length of cardiac epithelium. This again points to cardiac epithelium being acquired.

With such small gaps of nonoxyntic squamous epithelium in either people without symptoms of GERD at autopsy or volunteers at endoscopy, it is reasonable to suggest that the normal length of cardiac epithelium is zero. The lines of evidence for this are as follows: (1) A zero squamooxyntic gap has been described and illustrated by multiple sources[7,23]; (2) The squamooxyntic gap has been shown to increase with both age and waist circumference. Extrapolation back to the age where the esophageal epithelium reaches full development and ideal waist circumference is likely to bring the observed small lengths of the squamooxyntic gap to a baseline of zero. (3) Studies in fetuses and neonates show that all fetal columnar epithelia disappear at full development leaving an esophagus lined entirely by squamous epithelium and the proximal stomach lined by gastric oxyntic epithelium. The stomach never develops a columnar epithelium without parietal cells at any time in fetal life.

4.2 A Personal Study

When it was time for me to have my screening colonoscopy at age 56 years, I sought a gastroenterologist who was willing to add an upper GI endoscopy to the procedure and perform a series of biopsies as instructed by me. I have never had any typical symptoms of GERD. I have had rare episodes of chest pain for several years that last a few minutes and disappear spontaneously. I have ignored these although I have wondered whether they could be an atypical symptom of GERD.

Assuming he would find a normal endoscopic appearance, which turned out to be correct, I instructed my gastroenterologist to take a four-quadrant biopsy that straddled the SCJ and extended 5 mm distal to the SCJ and multilevel four-quadrant biopsies at 5 mm intervals (i.e., one standard biopsy forceps interval) distal to the SCJ to a distance of 20 mm (i.e., 5–10 mm, 10–15 mm, 15–20 mm). The biopsies were processed routinely. They provided a mapping of the histology of this region with the objective of providing a measurement of the dilated distal esophagus.

The biopsies showed 0.2 mm of cardiac epithelium in one quadrant of my SCJ and ~4 mm of oxyntocardiac epithelium distal to the SCJ. The squamous epithelium was normal; the cardiac epithelium showed moderate chronic inflammation and reactive changes; the oxyntocardiac epithelium showed mild chronic inflammation. There was no intestinal metaplasia. A small amount (20%) of the biopsy at the SCJ consisted of gastric oxyntic epithelium. Cardiac and oxyntocardiac epithelia were present only in the first biopsy level between the SCJ and 5 mm. All biopsies distal to 5 mm only had gastric oxyntic epithelium, which was histologically absolutely normal without any inflammation.

I call myself a "4-mm-squamooxyntic-gap" man. My status is equivalent to the people without GERD at autopsy.[7,9] The 0.2 mm length of cardiac epithelium is comparable to the shortest encountered length of cardiac epithelium in the asymptomatic volunteers in Robertson et al.[30]

I will show in a later chapter that these data are predictive of two facts: (1) I will never develop sufficient LES damage to cause clinical GERD in my lifetime. (2) I will never progress to develop CLE in the tubular esophagus or esophageal adenocarcinoma in my lifetime.

Please do not tell me that my short length of cardiac and oxyntocardiac epithelium with significant inflammation is normal. It is evident that 4 mm of my most distal squamous epithelium has been damaged by exposure to gastric juice and undergone columnar metaplasia. I will show that this is equivalent to 4 mm of lower esophageal sphincter damage. These histologic findings identify a stage between the normal state and clinical GERD as it is presently defined. I am neither normal (by virtue of the presence of abnormal metaplastic epithelia in the dilated distal esophagus) nor do I have GERD by present criteria of the Montreal definition.

5. ENDOSCOPIC BIOPSY STUDIES IN CLINICAL PATIENTS

There are a very small number of studies that have produced data regarding the prevalence of cardiac epithelium distal to the endoscopic GEJ.

The reason for this is that there has never been any protocol to study the region immediately distal to the endoscopic GEJ before Dr. DeMeester routinely took retroflex biopsies in all his patients, irrespective of whether the patient had GERD or not and whether endoscopy was normal or not.

Despite all the evidence over more than two decades that this biopsy protocol provides valuable information that has resulted in numerous new ideas regarding GERD, no gastroenterologist that I know has followed suit. They still have implicit faith that cardiac epithelium normally lines the proximal stomach and that the GEJ is accurately defined by the proximal limit of rugal folds.

All the evidence of the autopsy and resection studies reviewed above has ever persuaded any gastroenterology association to change their recommendations for endoscopic biopsy or any gastroenterologist to ignore the guidelines and change their behavior. They have ignored the evidence. As a result, there has been no progress in this area.

From the outset, I knew that gastroenterologists must be my primary target audience. As long as they remained resistant, nothing would happen. Although surgeons were more likely to accept the new hypothesis, they had largely relinquished the performance of endoscopy. Gastroenterologists are responsible for the vast majority of upper endoscopies that are done in the United States. The critical requirement to prove the hypothesis and take it to its next level was to convince gastroenterologists to take appropriate biopsies. Without biopsies, there could be no confirmation. Pathologists had no means of testing the hypothesis without appropriate samples.

Unfortunately, gastroenterologists were the most resistant to the new concept. Even those who were interested enough to listen and understand were not persuaded to take the next step of obtaining biopsies to test the concept by going against the guidelines of the major associations. The resistance still exists like a brick wall.

Gastroenterologists have a history of resistance to important new concepts; the recent history of Dr. Barry Marshall and the intense resistance to his ideas of *H. pylori* that led to the point of driving Dr. Marshall to swallow a culture of the bacterium to prove his case.

The only people who were persuaded to take biopsies according to our protocol were a few surgeons who did their own endoscopy. They mostly ran into the resistance of the pathology community who did not accept the new concept. Only a very small number of surgeon–pathologist groups embraced the new concept. The most prominent of these was Dr. Martin Riegler, a surgeon in Vienna who was an attendee at the conference in Hawaii (see below).

One gastroenterologist who was an exception to the rule was Dr. Stuart Spechler. He was one of the participants at a panel discussion on Barrett esophagus in 1998 where I presented my autopsy data. After the meeting, he came up to me and told me that he was very interested in what I had said. He said that he found it difficult to believe the data and that he intended to test my hypothesis.

He was true to his promise. In the next year's *Digestive Diseases Week*, his test of the hypothesis was presented as an abstract by Jain et al. entitled "Cardiac epithelium is found infrequently in the gastric cardia." The methodology of the study was as follows: "At endoscopy, both the squamocolumnar junction (the Z-line) and the anatomic esophagogastric junction (EGJ) were identified and biopsy specimens were taken from the columnar epithelium at each of 4 quadrants in the following locations: 1) the Z-line, 2) the EGJ, 3) the gastric cardia 1 cm below the EGJ, and 4) the gastric cardia 2 cm below the EGJ. Cardiac epithelium was identified in specimens with mucus-secreting glands devoid of parietal and chief cells, fundic epithelium in specimens with abundant parietal and chief cells, and intestinal metaplasia in specimens with goblet cells."

TABLE 5.8 Histologically Defined Epithelial Types in Multilevel Biopsies Taken at 1 cm Intervals From the Endoscopic Esophagogastric Junction (EGJ) to a Point 2 cm Distal to It.

Biopsy Site	No. of Patients	Cardiac Epithelium	Fundic Epithelium	Intestinal Metaplasia
EGJ	31	11 (35%)	21 (68%)	10 (32%)
1 cm below EGJ	29	4 (14%)	23 (79%)	7 (24%)
2 cm below EGJ	30	1 (3%)	30 (100%)	2 (7%)

The above methods are precise except for one thing. There is a gap between "cardiac epithelium," which has zero parietal cells, and "fundic epithelium" with abundant parietal cells. This leaves a gray area where the epithelium is mainly composed of mucous cells with a few parietal cells. This gray area is oxyntocardiac epithelium and it does not fit into either cardiac or fundic epithelium as defined.

The study had 31 patients. The clinical background is not stated in the abstract. Their endoscopic findings are as follows: "The Z-line and EGJ were located at the same level in 25 patients, whereas in 6 the Z-line was located >1 cm proximal to the EGJ."

They report the results for the epithelial types found in at least one of the four biopsy specimens obtained at the various locations (Table 5.8).

These data, which was published before our autopsy study, can be interpreted only with regard to cardiac epithelium, which was present in at least one of four-quadrant biopsies at the GEJ in only 11/31 (35%) of patients. The prevalence of cardiac epithelium decreased rapidly to 4/29 (14%) at 1 cm and 1/30 (3%) at 2 cm distal to the GEJ. The data clearly support the concept that cardiac epithelium is absent in a majority (65%) of patients and present to a variable extent that rarely (14%) reaches 1 cm distal to the GEJ and very rarely (3%) reaches a point 2 cm distal to the GEJ.

Unfortunately, the histologic methods do not permit the differentiation of oxyntocardiac from gastric oxyntic epithelium at the different levels. All that can be said is that parietal cell–containing epithelium (either oxyntocardiac or gastric oxyntic) is found in the majority of patients at the EGJ (68%) and 1 cm distal to it (79%), reaching 100% at 2 cm distal to the GEJ.

The authors conclude: "Both cardiac epithelium and intestinal metaplasia are found with decreasing frequency as the biopsy sites move distally down the gastric cardia. Even at the GEJ, cardiac epithelium is found in only a minority of patients. These findings challenge the traditional view of a gastric cardia lined predominantly by cardiac epithelium."

Unfortunately, these data never progressed from this abstract to a published paper. The clinical background of these patients remains unknown.

The study by Marsman et al.[28] is from the Academic Medical Center in the Netherlands, one of the premier study centers for GERD. In the introduction, the authors accurately outline the problem as it existed in September 2003, when it was accepted for publication: "Traditionally, it was believed that the cardia is a congenital part of the stomach, based on several studies in mammalians and limited studies in humans. This hypothesis has been challenged by the group of DeMeester. Both in an endoscopic and in an autopsy study, it was noted that cardiac mucosa was not uniformly present at the EGJ. It was suggested that cardiac-type mucosa in the most proximal part of the stomach is a metaplastic lesion that develops due to GERD. In contrast to these findings, a pediatric autopsy study identified cardiac mucosa at the EGJ in all specimens, suggesting that the gastric cardia is a congenital part of the stomach (Kilgore et al.[9] study reviewed above). To contribute to this important debate about the histological definition and possible etiology of cardiac mucosa, we evaluated the presence of cardiac mucosa in a random group of patients who presented at our endoscopy unit."

The methods in this study are impeccable. The study consists of 253 unselected adult patients. The location of the SCJ and EGJ (defined as the most proximal margin of the gastric folds) was documented. Fifty-five patients in whom there was cephalad displacement of the SCJ (i.e., visible CLE) were excluded from this analysis. Only the 198 patients with a normal SCJ seen as a straight line between the white and red mucosa, coinciding with the EGJ, were selected. Study protocol biopsies were taken 2 cm above the SCJ and two biopsies from the gastric cardia just below the SCJ. For purposes of this study, only biopsies that encompassed the epithelial SCJ on histological examination were selected. The pathologic assessment of histologic epithelial types uses the diagnostic criteria and definitions used in one of our papers.

In 63/198 patients without a visible CLE, the epithelial junction between squamous and columnar epithelium was actually present in at least one of the two biopsies taken from the EGJ. In the other 135 patients, the actual histologic junction was not present in any biopsy, the squamous and columnar epithelium being in separate pieces of tissue.

TABLE 5.9 Type of Columnar Epithelium in Biopsies Taken From the Esophagogastric Junction, Which Did (in 63 Patients) or Did Not (in 135 Patients) Include the Squamocolumnar Junction (SCJ) on Histological Examination

Type of Epithelium	Biopsy With Histologic SCJ (n = 63)	Biopsy Without Histologic SCJ (n = 135)
Cardiac epithelium	39 (62%)	2 (1%)
Oxyntocardiac epithelium	24 (38%)	43 (32%)
Oxyntic mucosa (fundic type)	0 (0%)	90 (67%)

TABLE 5.10 Comparison Between the Data in the Study by Marsman et al. and Those of Chandrasoma et al. and Kilgore et al. Relating to the Type of Epithelium Present Immediately Distal to the Squamous Epithelium

	Cardiac Epithelium	Oxyntocardiac Epithelium	Gastric Oxyntic Epithelium
Kilgore et al.	30/30 (100%)	0	0
Chandrasoma et al. (prospective study)	8/18 (44%)	10/18 (56%)	0 (in entire circumference); 50% in part of circumference
Chandrasoma et al. (retrospective study)	19/73 (26%)	32/73 (44%)	21/73 (29%)
Marsman et al.—63 patients with histologic SCJ	39/63 (62%)	24/63 (38%)	0
Marsman et al.—135 patients without histologic SCJ	2/135 (1%)	43/135 (32%)	90/135 (67%)[a]
Jain et al.	11/31 (35%)[b]	Data not available	Data not available

Sampling of the esophagogastric junction decreases from top to bottom. Kilgore et al. and the prospective study by Chandrasoma et al. have vertical sections across the entire circumference of the SCJ at autopsy; the retrospective study by Chandrasoma et al. has a single vertical section across the SCJ at autopsy; in Marsman et al. 63 patients have one to two biopsies across the SCJ.
SCJ, squamocolumnar junction.
[a]*In Marsman et al. 135 patients have biopsies "just below the SCJ" without a histologic SCJ in the sections.*
[b]*In Jain et al. patients have biopsies at the GEJ without mention of the presence of the histologic SCJ.*

In 39/63 (62%) of the patients who had the histologic SCJ in the EGJ biopsies, this showed pure cardiac mucosa adjacent to the squamous epithelium. In the other 24/63 (38%) of patients, these biopsies contained oxyntocardiac mucosa adjacent to the squamous mucosa (Table 5.9). None showed a transition from squamous to gastric oxyntic epithelium.

In the 135 patients in whom the EGJ biopsy (which was taken "just below the SCJ") did not show the histologic SCJ, only 2 (1%) patients had cardiac mucosa, 43 (32%) had oxyntocardiac mucosa, and 90 (67%) had gastric oxyntic mucosa. The extremely low prevalence of cardiac epithelium in EGJ biopsies that were done "just below the SCJ" with the biopsies not straddling the SCJ suggests that these patients either did not have any cardiac epithelium or had a very short length of cardiac epithelium. If the data were given for the entire group of 198 patients, cardiac epithelium was present in 41/198 (21%) of patients. This item of data is not given in the results.

The demographic, clinical, and endoscopic data of these two groups were not significantly different. Severe GERD symptoms were present in 54/198 (27%) patients overall with 15/63 (24%) in the first group and 39/135 (29%) in the second group. This patient group can therefore be regarded as being selected to have a low prevalence and severity of GERD, i.e., closer to the normal state than established GERD.

The intent of this study was to contribute to the debate of whether cardiac epithelium was a normal gastric epithelium or an abnormal metaplastic esophageal epithelium. The data produced show concordance in this regard with our autopsy study[7] and contradicts Kilgore et al.[9] (Table 5.10).

Kilgore et al.[9] is the outlying study with all patients having cardiac epithelium immediately distal to the SCJ. The study by Robertson et al.[30] in asymptomatic volunteers is the only other study where biopsies straddling the histologic SCJ showed cardiac epithelium in all persons. In all the other studies, oxyntocardiac epithelium was present immediately distal to the SCJ in a significant number of patients without intervening pure cardiac epithelium. Patients with a direct transition from squamous epithelium to gastric oxyntic epithelium were seen only in part of the circumference of the SCJ in 50% in the prospective study by Chandrasoma et al. and 29% of the single section in the retrospective study[7] (Fig. 5.5).

The way the data are presented is similar in all the papers except Marsman et al.[28] They divide their initial population of 198 patients into those with (n=63) and without (n=135) the histologic SCJ in the biopsy.

Their conclusion: "Cardiac mucosa was uniformly present adjacent to the squamous epithelium at the EGJ. This argues against the hypothesis that the gastric cardia is an acquired metaplastic lesion." It is accurate to combine cardiac and oxyntocardiac because the latter is a variant of the former and is included in metaplastic columnar epithelia. However, the statement that the uniform presence of cardiac epithelium at the GEJ argues against the gastric cardia being an acquired metaplastic lesion is not correct (see below).

6. MULTILEVEL ENDOSCOPIC BIOPSIES DISTAL TO THE ENDOSCOPIC GASTROESOPHAGEAL JUNCTION

Apart from the abstract by Jain et al.[27] with scanty data, there is only one formal study where the area distal to the endoscopic GEJ was systematically evaluated with a multilevel measured biopsy protocol. This was done in a series of 102 patients in Vienna, Austria.[31] Dr. Martin Riegler, the lead investigator, had attended the USC Foregut conference in Hawaii and completely understood the value of pathologic assessment of this region. He and I had many hours of discussion during the week-long meeting, often on the beaches of the Hawaiian islands in the most wonderful conditions.

On returning to Vienna, he coopted a senior pathologist (Professor Fritz Wrba), overcoming the latter's resistance, and did this study, which involved nearly 2000 biopsies in these 102 patients with clinical evidence of GERD. The biopsy protocol identified the endoscopic GEJ (which was called level zero). He took four-quadrant biopsies proximal to this at 5 mm intervals until the SCJ was reached (called +5, +10, etc.) and similar biopsies distal to this for 15 mm (called −5, −10, −15). All biopsies from one level were processed as separate specimens and examined histologically by Dr. Wrba. Dr. Wrba had educated himself on the interpretation of histology from our papers and used the same histologic classification. The same group had studied the histopathology of this region with a lesser biopsy protocol previously.[32]

After the study was completed, Dr. Riegler invited me to Vienna to review the specimens. He demonstrated his biopsy protocol to me in the endoscopy suite and sat me down with his group for hours on end looking at all his biopsy specimens. Dr. Wrba and the entire clinical team participated in this review, but I was kept blind to the original histologic interpretation. My interpretation was noted without discussion and compared with Dr. Wrba's original diagnosis after the review was over. There was almost total concordance of the pathologic diagnoses. Out of nearly 2000 biopsies, three cases that Dr. Wrba had called "gastric oxyntic epithelium" were called "oxyntocardiac epithelium" and one case he had called "intestinal metaplasia" was called "cardiac epithelium with pseudogoblet cells" by me.

The findings in this multilevel biopsy study provided invaluable data regarding the distribution of cardiac and oxyntocardiac epithelium as well as the prevalence and distribution of intestinal metaplasia in this region. The study and data were both unique and, in my opinion, of great value. This paper was rejected out of hand by all major gastroenterology journals and finally published in a relatively low-impact surgical journal in Europe. The gastroenterology establishment successfully buried this information from the eyes of much of the medical world.

Ringhofer et al.[31] provide data that show the variation in the length of metaplastic cardiac epithelium (with and without goblet cells) proximal and distal to the endoscopic GEJ. The data show that, in this population of patients with clinical evidence of GERD, (1) the metaplastic epithelium is one continuous segment from the distal limit of squamous epithelium to gastric oxyntic epithelium extending distal to the endoscopic GEJ; (2) patients with visible CLE above the endoscopic GEJ *always* have similar columnar epithelia distal to the endoscopic GEJ; (3) all patients without visible CLE have these epithelia distal to the endoscopic GEJ; and (4) when all five epithelial types are present, the order of the epithelia from proximal (i.e., the SCJ) to distal is squamous epithelium, cardiac epithelium with intestinal metaplasia, cardiac epithelium, oxyntocardiac epithelium, and gastric oxyntic epithelium.

These data strongly suggest that columnar metaplasia begins distal to the endoscopic GEJ and progressively increases in length, first entirely distal to the endoscopic GEJ and then, in a minority of patients, extending above the endoscopic GEJ to produce CLE that is visible at endoscopy.

This paper is more completely reviewed in Chapter 17.

7. WHAT IS THE SIGNIFICANCE OF CARDIAC EPITHELIUM DISTAL TO THE ENDOSCOPIC GASTROESOPHAGEAL JUNCTION?

The authors of several studies rush to judgment to designate the high prevalence of cardiac epithelium as normal. Kilgore et al.[9] used their finding of cardiac epithelium at the GEJ in all their patients to conclude: "In an unselected pediatric patient population with little or no propensity for gastroesophageal reflux disease, a short segment of cardiac mucosa was consistently present on the gastric side of the esophagogastric junction, independent of gender or age. These results support the concept that the gastric cardia is present from birth as a normal structure."

Marsman et al.[28] came to a similar conclusion: "Cardiac mucosa was uniformly present adjacent to the squamous epithelium at the EGJ. This argues against the hypothesis that the gastric cardia is an acquired metaplastic lesion."

These are false conclusions even though their claim that cardiac epithelium was universally present in their population is not disputed. The universal presence of a finding does not mean that it is a "normal structure" or that it "is present from birth" or "that it argues against the hypothesis that the gastric cardia is an acquired metaplastic lesion."

Let us consider the following situation. I undertake a study (autopsy or clinical) of the abdominal aorta in 100 young Americans. I find that all subjects have atherosclerosis in their abdominal aorta ranging from 1 to 4 mm plaque. Am I allowed to say that this amount of atherosclerosis "is a normal structure," "is present from birth," or "that it argues against the hypothesis that it is an acquired lesion?" Obviously not.

Universality does not prove normalcy. A finding can be universally present because it is extremely common. If an abnormal finding has a prevalence of 99.9% (i.e., it is absent in 1 out of 1000 people), a sample size of 100 is unlikely to find that person. However, if one person does not have that finding, then that sample size is adequate to prove that the finding is not universally present.

Conversely, the absence of a structure deemed to be normal in a rare person does not mean that the structure is not normal. For example, the human head is a normal structure that is absent in rare anencephalic babies. The test here is that the absence of the structure is associated with a pathologic finding compared with the presence of that structure. Anencephaly is not compatible with life.

The reasons why the presence of 1–4 mm of cardiac epithelium in the proximal stomach is not normal are as follows: (1) Cardiac epithelium is not seen in fetuses distal to the angle of His (GEJ)[21]—see Chapter 3. (2) It is absent in rare autopsies in neonates (Fig. 5.4)[23] and people of all ages at autopsy[7] (Fig. 5.5) and in resected specimens (Fig. 5.9). (3) It increases with age and obesity.[30] (4) Its presence is associated with reflux and LES abnormalities.[25] (5) Increase in its length is associated with increasing GERD.[33,34]

Cardiac epithelium is therefore more like atherosclerosis than the human head. Its absence represents the normal state and its presence represents the pathologic state. Increase in its length represents increasing severity of the pathologic state. The pathologic state that it represents results from exposure of esophageal squamous epithelium to gastric contents that result in cardiac metaplasia. The near universal presence of cardiac epithelium at the SCJ only indicates that the pathologic state is exceedingly common.

The hypothesis that cardiac epithelium was an abnormal state associated with GERD was first presented in the Hawaii Foregut Surgery conference in 1992 and then as an abstract at Digestive Diseases Week in 1994.[4] This was followed by an important study by Oberg et al.[25] from our unit.

This study was possible because of the biopsy protocol that was used in our unit. Every patient undergoing upper endoscopy who has no visible CLE has three sets of four-quadrant biopsies: (1) from the SCJ, which is coincidental with the endoscopic GEJ in the endoscopically normal person, (2) a retroflex biopsy from the region within 1 cm distal to the SCJ, and (3) from the body and antrum of the stomach. At the time, the biopsies from around the SCJ did not attempt to straddle the junction.

A total of 334 patients who had, in addition to the biopsies, a 24-h pH test and manometry were selected. 246/334 (74%) the patients had cardiac and/or oxyntocardiac epithelium at or immediately distal to the SCJ, and the other 88 (26%) had neither; their biopsies contained only squamous and gastric oxyntic epithelia. The 246 patients with cardiac and/or oxyntocardiac epithelium in their biopsies had a significantly higher probability of an abnormal pH test, a hiatal hernia, erosive esophagitis, and criteria for a structurally defective LES than the 88 patients without cardiac and/or oxyntocardiac epithelium (Table 5.11).

These data prove that the presence of cardiac and/or oxyntocardiac epithelium distal to the SCJ in an endoscopically normal person is associated with GERD. The absence of these epithelia, seen in a significant minority of these patients, is associated with a more physiologic state.

Glickman et al.[33] studied 74 children (median age 13 years; range: 1 day to 18 years) with the following aim: "... to evaluate the morphologic features of the cardia in a pediatric population and to determine the significance of inflammation in this region by correlating the pathologic features with clinical and endoscopic data." They cite papers that show a correlation between inflammation of the gastric cardia and both GERD and *H. pylori* infection and aim to test this as well.

The patient selection criteria for the study were as follows: "(1) patient age younger than 19 years at the time of biopsy, (2) absence of endoscopically apparent columnar-lined esophagus (Barrett's esophagus), (3) clinical and endoscopic data available for review, and (4) the presence of squamocolumnar junctional (SCJ) epithelium in the biopsy specimen. The gastroesophageal junction (GEJ) was defined endoscopically as the proximal limit of the gastric rugal folds; in this study group it coincided with the distal limit of the squamous-lined tubular esophagus. Thus, biopsies that straddled the SCJ

TABLE 5.11 Comparison of pH and Manometric Data Between 246 Patients Who Had Cardiac and/or Oxyntocardiac Epithelium in the Gastroesophageal Junction Region Compared With 88 Patients Without These Epithelia. All Patients Did Not Have Visible Columnar-Lined Esophagus at Endoscopy

	Patients With No CE and/or OCE (n = 88)	Patients With CE and/or OCE (n = 246)	*P* value
% time pH<4	1.1 ± 4.6	6.0 ± 7.4	<.01
LES pressure (mmHg)	13.2 ± 12.8	8.0 ± 8.0	<.01
Abdominal LES length (cm)	1.6 ± 1.1	1.0 ± 1.2	<.01
% defective LES	27.2	62.3	<.01

CE, cardiac epithelium; *LES*, lower esophageal sphincter; *OCE*, oxyntocardiac epithelium.

represented the distal esophagus and proximal stomach ("cardia")." The faith in the false dogma that the proximal limit of "gastric" rugal folds accurately defines the true GEJ is absolute. This faith clouds the authors' interpretation of the data. It is important to evaluate the data with the assumption that the true GEJ cannot be defined at endoscopy. If this is done, the interpretation of data changes and becomes objective.

All mucosal biopsies (average 1.2 biopsy fragments per patient; range 1–4) from the SCJ stained with hematoxylin and eosin were examined. Columnar epithelia located within 1.0 and >1.0 mm from the SCJ were evaluated for (among other things) the type of glandular epithelium ("pure mucous glands—mucous cells only," "pure oxyntic-type glands—mixture of parietal and/or chief cells," and "mixed mucous/oxyntic glands—mixture of mucous cells and parietal and/or chief cells"), severity and type of inflammation, presence of active esophagitis in squamous epithelium, and presence of *H. pylori* infection.

The glandular types have been given new terms that have never been used previously. Their "pure mucous glands—mucous cells only," "pure oxyntic-type glands—mixture of parietal and/or chief cells," and "mixed mucous/oxyntic glands—mixture of mucous cells and parietal and/or chief cells" are synonymous with cardiac, gastric oxyntic, and oxyntocardiac epithelia, respectively, in our terminology. Harvard and Boston have every right to be a world unto themselves, but us mere mortals need to decipher their language!

Indication for upper endoscopy was varied. Symptoms related to GERD (heartburn, regurgitation, dysphagia) were present in 40%. The use of H_2 receptor antagonists and PPIs was recorded in 38% and 21%, respectively. Histologically, active esophagitis was present in 28 (38%) and chronic gastritis in 9 (13%) of 69 patients who had biopsies taken from the distal stomach. *H. pylori* infection was present in 5/74 (7%) and this correlated strongly with the presence of antral gastritis. The fact that *H. pylori* was absent in 69 patients makes it irrelevant as a cause of pathologic changes in these patients. It is likely that this is the normal prevalence of *H. pylori* in this population.

The way the histologic features in the cardia are presented is also unique, making it very difficult to decipher the length of the different epithelial type. They do not provide a simple measurement from the squamous epithelium to the first parietal cell (the squamoparietal cell gap) or from the squamous epithelium to gastric oxyntic epithelium (the squamooxyntic gap).

Of the 74 patients, 60 (81%) had pure cardiac epithelium within 1.0 mm from the SCJ; the remaining 14 (19%) had oxyntocardiac epithelium. None of the patients had gastric oxyntic epithelium within 1 mm of the SCJ. Of a subset of seven patients who did not have upper GI symptoms, esophagitis, or gastritis, six had pure cardiac epithelium and one had oxyntocardiac epithelium within 1.0 mm from the SCJ.

Of the 74 patients, the type of columnar epithelium located >1.0 mm from the SCJ is described only in 59 patients. There is no explanation about why these data are not available for the other 15 patients. I can only assume that the total length of the columnar epithelium distal to the SCJ in the biopsy did not exceed 1.0 mm in these 15 patients. Of the 59 patients with >1.0 mm of columnar epithelium, 32 had pure cardiac epithelium, 21 had oxyntocardiac epithelium, and 6 had gastric oxyntic epithelium. Again, I am left to assume that this is the composition of the epithelial types seen >1.0 mm distal to SCJ rather than in the entire length of columnar epithelium. Otherwise, six patients cannot have only gastric oxyntic epithelium because no patient had gastric oxyntic epithelium within 1.0 mm of the SCJ.

Statistical analysis showed: "… patients who had pure mucous glands (i.e. cardiac epithelium) located both within and >1 mm from the SCJ showed a significantly … increased prevalence of active esophagitis (55% vs 21%, p = 0.04) compared with patients who had a combination of pure mucous glands within 1 mm of the SCJ and either mixed mucous/oxyntic glands (i.e. oxyntocardiac epithelium) or pure oxyntic glands (i.e. gastric oxyntic epithelium) >1 mm from the SCJ."

A significantly greater proportion of patients who had eosinophils in cardiac epithelium had active esophagitis in comparison with those without eosinophils (54% vs. 18%, $P=.002$).

The finding that cardiac epithelium extending >1 mm distal to the SCJ had a significantly higher association with active esophagitis than those patients with cardiac epithelium limited to within 1 mm from the SCJ is strong evidence that cardiac epithelium is a metaplastic epithelium associated with GERD rather than a normal gastric epithelium.

The overall discussion seems to agree, but not exactly. The authors begin by considering the data in the study that may suggest cardiac epithelium is normal: "… of the seven patients who presented for endoscopy without upper gastrointestinal tract symptoms and showed no evidence of esophagitis or gastritis histologically, nearly all (6/7) had mucous glands with a mild degree of inflammation in their SCJ biopsies. These results suggest that the anatomic cardia may normally be composed of a small amount of mucous glands in the majority of young individuals and that this mucosa often shows some degree of inflammation. Unfortunately, because of the retrospective nature of our study and the lack of a well-defined 'control' group, we cannot determine with certainty whether significant inflammation in the cardia may be a 'normal' finding or, as we think, more likely a result of either GERD or *H. pylori* infection, even in patients of very young age. Thus, we agree with other authors [they cite 4 papers, all from our group] that in many instances inflammation in the cardia may be a consequence of physiologic, or perhaps, pathologic reflux in pediatric individuals without gastritis, as has been proposed in adults. Our data showing that both active esophagitis and increased inflammation in the cardia correlated positively with a longer length of pure mucous glands (i.e. cardiac epithelium) in the SCJ region… are evidence in support of this theory."

After a discussion of the controversy resulting from the two autopsy studies, these authors state: "Nevertheless, despite the fact that that interpretation of their results is somewhat limited by the absence of a non-GERD (control) group of patients, some of the other data presented by Chandrasoma et al in adults showing the presence of inflammation in most biopsies from the cardia region are similar to our data presented here in pediatric patients. Both our studies suggest that increasing lengths of mucous-type glands (i.e. cardiac epithelium) or mixed mucous/oxyntic-type glands (i.e. oxyntocardiac epithelium), in the proximal stomach (cardia) may be metaplastic in origin and possibly derived as a consequence of GERD."

The final summary statement in both discussion and abstract is not definite and somewhat convoluted in its thought process: "In summary, our study shows that a small amount of pure mucous-type glands (i.e. cardiac epithelium) is present in the cardia in pediatric patients, a finding that supports a congenital origin for this type of epithelium. Also, carditis may be the result of either GERD or *H. pylori* infection. However, our finding of an association between length of mucosa occupied by pure mucous glands and active esophagitis suggests that injury and repair related to GERD may contribute to the expansion of the zone occupied by mucous glands in the GEJ region."

It is important to understand what is being said because most readers will take home the final summative message in the paper. According to Glickman et al.[33] there is a small amount of normal cardiac epithelium in the proximal stomach. There are no data to support this. The only statistically significant finding was that the prevalence of active esophagitis, which existed in both groups, was greater (55%) in the group with >1 mm cardiac epithelium than those with <1 mm (21%). Given the low sensitivity of active esophagitis in the diagnosis of GERD, this datum would suggest that the length of cardiac epithelium is an exquisitely sensitive histologic marker for GERD, not that cardiac epithelium is a normal epithelium at the GEJ.

Then, the authors try to explain how GERD causes an expansion of the zone of cardiac epithelium in the proximal stomach distal to the GEJ. While it seems obvious that this expansion must involve the esophagus, the authors in their discussion and conclusion suggest an alternate explanation: "Although, like Chandrasoma et al, we suspect that longer lengths of pure mucous glands (i.e. cardiac epithelium) in the SCJ region represent proximal extension (metaplasia) into the distal esophagus, based solely on our data we cannot rule out the possibility that this finding represents distal extension of mucous glands (i.e., cardiac epithelium) into the proximal stomach." It is being suggested that GERD may cause an increase in the amount of cardiac epithelium in the proximal stomach. The only reason for this is the authors' total belief in the accuracy of the endoscopic definition of the GEJ. They leave open the possibility that cardiac epithelium may extend *distally* from the GEJ downward. The power of this dogma is incredible.

The truth is much simpler and more logical. The endoscopic definition of the GEJ is incorrect. The true GEJ is the proximal limit of gastric oxyntic epithelium. Damage of the squamous epithelium resulting from exposure to gastric contents causes cardiac metaplasia. In this study, those patients with cardiac epithelial length starting from zero had increasing evidence of active esophagitis and inflammation in cardiac epithelium. As the length of cardiac epithelium increased from <1 to >1 mm, the prevalence of reflux esophagitis increased from 21% to 55%. The risk of cardiac epithelium <1 mm was not zero. There is no more powerful proof that cardiac epithelium at the SCJ is an abnormality associated with GERD.

Occam's razor can be invoked here. There are two alternate explanations for the data in this paper based on two different definitions of the GEJ. When the GEJ is defined as the proximal limit of rugal folds, GERD-induced cardiac metaplasia extends upward into the esophagus as CLE and downward into the proximal stomach as an "expansion" of the gastric cardia. When the GEJ is defined histologically as the proximal limit of gastric oxyntic epithelium, GERD-induced cardiac metaplasia progresses only upward into the esophagus as CLE. The obviously simpler explanation, which is that GERD causes pathology in the esophagus and not the stomach, is the second.

Chandrasoma et al.[34] is a study of 71 patients with CLE from our unit that has as its aim the correlation of the length of CLE with the severity of reflux. It differs from most other similar studies in two respects: (1) It includes a biopsy taken of the region within 10 mm distal to the endoscopic GEJ, allowing the measurement of CLE to include that part of the CLE distal to the endoscopically visible CLE. All other studies of visible CLE assume that the distal limit of CLE is the endoscopic GEJ. (2) It quantifies reflux by a 24-h pH study which was available in 53/71 patients, not symptoms of GERD.

The biopsy protocol is the standard used in our unit (see Introduction). The histologic definitions are as we have defined them (see Chapter 4). The length of CLE was calculated by their presence at the measured biopsy levels. The CLE length was >20 mm when biopsies at a minimum of three levels, 3 cm or more apart, were positive for cardiac epithelium (with and without parietal/goblet cells). The CLE length was <20 mm when biopsies were positive for cardiac epithelium (with and without parietal/goblet cells) in 1 or 2 levels.

Pure cardiac epithelium was present in 68/71 (96%) patients and was associated with intestinal metaplasia in 42 patients. The length of cardiac epithelium (with and without parietal/goblet cells) was >20 mm in 22 patients and <20 mm in 49 patients. The cardiac epithelium (with and without parietal/goblet cells) extended into the retroflex biopsy taken distal to the endoscopic GEJ in all cases. The length of separation of squamous epithelium from gastric oxyntic epithelium was dramatically different than in the autopsy studies.

If these data are interpreted in relation to the endoscopic GEJ, cardiac epithelium (with and without parietal/goblet cells) is present in both the stomach and esophagus. The cardiac epithelium in the esophagus will be recognized as visible CLE at endoscopy. The cardiac epithelium in the retroflex biopsy will not be biopsied, but if it biopsied, it will be regarded as normal proximal gastric epithelium. The single retroflex biopsy does not permit accurate assessment of length of cardiac epithelium, but Sarbia et al.[18] and Chandrasoma et al.[17] have shown that the extent of cardiac epithelium distal to the endoscopic GEJ can exceed 20 mm.

If the data are interpreted in relation to the proximal limit of gastric oxyntic epithelium (the true GEJ), it will be seen that a single column of CLE separates the true GEJ from the cephalad-displaced SCJ. Glickman et al.[33] showed that column of cardiac epithelium (with and without parietal/goblet cells) is associated with GERD at <1 mm and that the association becomes significantly greater with length > 1 mm.

The present study shows that increasing lengths of cardiac epithelium (with and without parietal/goblet cells), beginning at the true GEJ (proximal limit of rugal folds), are associated with increasing severity of reflux as assessed by a 24-h pH study (Table 5.12).

The severity of reflux in the 15 patients with cardiac epithelium (with and without parietal/goblet cells) >20 mm (median 30.6: IQR 16.1, 41.1) was highly significantly (P = .001) greater than in the patients with <20 mm (median 7.1; IQR 5.2, 12.1). The severity of reflux in the patients with intestinal metaplasia was significantly greater than those without IM (P = .001).

TABLE 5.12 Correlation Between Reflux (as Assessed by a 24-h pH Test in 53 Patients and Expressed as % Time the pH Was <4 in the Esophagus) and Length of Cardiac Epithelium (With and Without Parietal/Goblet Cells) 20 mm or Less versus >20 mm

	No. of Patients	Median	Range	IQR (Q_L,Q_U)
All patients	53	10.4	1.1–54.3	5.7, 20.5
CE>20 mm	15	30.6	3.6–54.3	16.1, 41.1
CE<20 mm	38	7.1	1.1–39.7	5.2, 12.1
CE<20 mm, IM+	15	10.5	1.8–39.7	6.6, 20.5
CE<20 mm, IM–	23	6.1	1.8–54.3	8.5, 33.7
IM+	30	16.5	1.8–39.7	6.6, 20.5

CE, cardiac epithelium with and without parietal cells; *IQR*, interquartile range. All patients with CE>20 mm in length had intestinal metaplasia (IM).

8. REVIEWS OF THE MEANING OF CARDIAC EPITHELIUM DISTAL TO THE ENDOSCOPIC GASTROESOPHAGEAL JUNCTION

The controversy regarding cardiac epithelium, its true location (i.e., distal esophagus or proximal stomach), and significance (normal or abnormal) persisted. Two reviews of this subject were written in 2005, 3 years after the paper by Glickman et al.[33] was published. The first was in the *American Journal of Gastroenterology* by Robert Odze,[8] the senior author of the paper by Glickman et al.[33]

Odze[8] is an excellent review that recognizes the problem: "... studies based on mucosal biopsy sampling are fraught with limitations, particularly related to the difficulty in differentiating the distal esophagus from the proximal stomach at the time of endoscopy."

His summation, which is the take-home message, is as follows: "The gastroesophageal junction (GEJ), which is defined as the point where the distal esophagus joins the proximal stomach (cardia), is a short anatomic area that is commonly exposed to the injurious effects of GERD and/or *Helicobacter pylori* infection. These disorders often lead to inflammation and intestinal metaplasia of this anatomic region. The true gastric cardia is an extremely short segment (<0.4 mm) [this is a typographical error, it should read 0.4 cm or 4 mm] of mucosa that is typically composed of pure mucous glands, or mixed mucous/oxyntic glands that are histologically indistinguishable from metaplastic mucinous columnar epithelium of the distal esophagus. In patients with GERD, whether physiologic or pathologic, the length of cardia-type epithelium increases and extends proximally above the level of the anatomic GEJ into the distal esophagus. Columnar metaplasia of the distal esophagus represents a squamous to columnar metaplastic reaction that develops from an esophageal stem cell...."

The practical impact of this review was to perpetuate the dogma that cardiac epithelium normally lined <4 mm of the proximal stomach (cardia). This statement belies the data in Glickman et al.[33] from Dr. Odze's unit where patients with >1 mm of cardiac epithelium had significantly more GERD than those with <1 mm. Cardiac epithelium of any length is associated with GERD by these data; it cannot be normal. This statement emanating from the recognized expert from Harvard is repeated in all conferences by GI pathology experts and accepted without question in most GI pathology and gastroenterology circles as the truth. No argument is tolerated. That statement gives cover to anyone who wants to ignore the reality; unfortunately, this is the majority of gastroenterologists and pathologists.

I wrote the other review by invitation of the editor of *Histopathology*, a British journal.[35] My summary was as follows: "Confusion regarding the diagnosis of Barrett's oesophagus exists because of a false dogma that cardiac mucosa is normally present in the gastro-oesophageal junctional region. Recent data indicate that the only normal epithelia in the oesophagus and proximal stomach are squamous epithelium and gastric oxyntic mucosa respectively. When this fact is recognized, it becomes easy to develop precise histological definitions for the normal state (presence of only squamous and oxyntic mucosa), metaplastic oesophageal columnar epithelium (cardiac mucosa with and without intestinal metaplasia, and oxyntocardiac mucosa), the gastro-oesophageal junction (the proximal limit of gastric oxyntic mucosa), the oesophagus (that part of the foregut lined by squamous and metaplastic columnar epithelium), reflux disease (the presence of metaplastic columnar epithelium), and Barrett's oesophagus (cardiac mucosa with intestinal metaplasia). It is also possible to assess accurately the severity of reflux, which is directly proportional to the amount of metaplastic columnar epithelium, and the risk of adenocarcinoma which is related to the amount of dysplasia in intestinal metaplastic epithelium present within the columnar lined segment of the oesophagus. Histopathological precision cannot be matched by any other modality and can convert the confusion that exists regarding diagnosis of Barrett's oesophagus to complete lucidity in a manner that is simple, accurate, and reproducible."

In that review is a table that uses the recommended definitions of histology and the GEJ (proximal limit of gastric oxyntic epithelium) to classify all people regarding GERD and risk of adenocarcinoma (Table 5.13).

This review was essentially buried and largely ignored, not reaching acceptance by gastroenterologists or pathologists. The ability to use histology to define these groups was not recognized. It did not result in any change in the recommendations for biopsy at endoscopy that is essential to put these definitions into practical use. Without adequate biopsies, including the routine biopsy of GERD patients without visible CLE, this classification cannot be applied. Only patients in grades 1b, 1c, 2b, 2c, and 3 are subject to biopsy. The only value of these is to identify intestinal metaplasia, dysplasia, and adenocarcinoma.

There is a good reason why this method is ignored. The subclassification of patients without endosopic CLE into groups 0, 1a, and 2a is only of academic value at present. All these patients, if symptomatic, will be treated with acid suppression as needed with the objective of controlling symptoms. Endoscopy is indicated only when treatment fails and then only to identify people with Barrett esophagus.

The alternative to PPI therapy is antireflux surgery, which is rarely preformed. There is even the possibility that the concept that cardiac epithelium at the GEJ is not normal is accepted in many minds, but this does not lead to any change in

TABLE 5.13 **Classification of the Population Into Groups Based on Their Risk of Developing Adenocarcinoma of the Esophagus**

Grade 0: No Risk of Adenocarcinoma (55%–65%)
Definition: No cardiac or intestinal epithelium in any biopsy.
Group 0a: Normal; only squamous and gastric oxyntic epithelia.
Group 0b: Compensated reflux: metaplastic columnar epithelium consisting only of oxyntocardiac epithelium.
Grade 1: Reflux Disease (GERD) (30%–45%)
Definition: Cardiac epithelium (reflux carditis) and oxyntocardiac epithelia present; no intestinal metaplasia.
Group 1a: Mild GERD; no CLE at endoscopy.
Group 1b: Moderate GERD; endoscopy shows CLE measuring 20 mm or less.
Group 1c: Severe GERD; endoscopy shows CLE measuring >20 mm.
Grade 2: Barrett Esophagus (5%–15%)
Definition: Intestinal metaplasia (goblet cells) present in cardiac epithelium.
Group 2a: Microscopic Barrett esophagus; no CLE at endoscopy.
Group 2b: Short segment Barrett esophagus; endoscopy shows CLE measuring 20 mm or less.
Group 2c: Long segment Barrett esophagus; endoscopy shows CLE measuring >20 mm.
Grade 3: Neoplastic Barrett Esophagus
Definition: Dysplasia or adenocarcinoma present.
Group 3a: Low-grade dysplasia.
Group 3b: High-grade dysplasia.
Group 3c: Adenocarcinoma.

The percentages are for the United States and Western Europe. Adequate biopsy sampling is essential.
CLE, columnar-lined esophagus; *GERD*, gastroesophageal reflux disease.

treatment and therefore is best hidden. Like the proverbial ostrich with its head in the sand, ignorance is better than knowledge that has no practical impact.

I will show in later chapters that this is the fundamental reason why there is a failure of management of GERD at present. Cancer prevention begins only when a patient is identified as having a visible CLE with intestinal metaplasia at endoscopy. This is a failed patient in terms of cancer prevention. Progression to cancer in this group is only a matter of chance. It is not positively impacted by medical therapy.

I will also show that the changes in early GERD detected by histology in groups 0, 1a, and 1b, who represent ~80% of people who do not have a visible CLE at endoscopy, have potential to change management. This is because these changes provide a pathologic assessment of LES damage that will be the main focus of this book. The ability to quantify LES damage provides a potential method for predicting progression of GERD and recognizing those patients at risk to progress to visible CLE. This opens the door to a new management aimed at preventing visible CLE. If this can be achieved, esophageal adenocarcinoma will become preventable.

9. DOES CARDIAC EPITHELIUM HAVE FEATURES OF COLUMNAR-LINED ESOPHAGUS OR GASTRIC EPITHELIUM?

A study of the histochemical and molecular features of cardiac epithelium can also shed light on the controversy whether this epithelium occurring distal to the endoscopic GEJ is gastric or esophageal.

Cardiac epithelium without goblet cells in the distal esophagus in visible CLE expresses histochemical and molecular markers that are similar to those seen in intestinalized cardiac epithelium in Barrett esophagus. This suggests that molecular changes precede the morphologic appearance of goblet cells in esophageal metaplastic columnar epithelium.

This study by DeMeester et al.[36] examines the expression of cytokeratins and DAS-1 in biopsies that show cardiac epithelium with and without intestinal metaplasia obtained from the GEJ region.

The controversy as to the etiology and significance of intestinal metaplasia occurring at the GEJ is discussed: "This condition, called intestinal metaplasia of the cardia (CIM), is defined by the presence of an endoscopically normal appearing GEJ and a biopsy that shows goblet cells indicative of intestinal metaplasia in cardiac mucosa. A histologically identical appearing biopsy taken from an endoscopically visible segment of columnar mucosa within the esophagus would confirm a diagnosis of Barrett esophagus. Consequently, because the histology of CIM and Barrett's esophagus are identical, it is the endoscopist, not the pathologist, that determines whether the patient has CIM or Barrett's esophagus.... Whether CIM, like Barrett's esophagus, is a manifestation of reflux disease, or, like GIM (gastric intestinal metaplasia in the antrum and body), is a result of some type of gastric insult is unclear and debated."

Tissue blocks from 50 patients evaluated at the University of Southern California for foregut symptoms were obtained on the basis of histology showing one of the following: cardiac mucosa, Barrett esophagus, cardiac intestinal metaplasia (CIM), gastric intestinal metaplasia (GIM), or normal antral, fundic, or squamous mucosa. Barrett esophagus was defined as any length of endoscopically visible columnar mucosa in the esophagus with intestinal metaplasia on biopsy. CIM was defined as intestinal metaplasia in cardiac mucosa on biopsies from an endoscopically normal GEJ. GIM was defined as intestinal metaplasia in antral or fundic/body mucosa.

Histologic slides were stained by immunoperoxidase technique for CK7, CK20, and DAS-1. The CK7/CK20 was assigned as having a Barrett's or non-Barrett's type staining pattern as described by Ormsby et al.[37] DAS-1 staining was graded as trace, 1+, or 2+.

A non-Barrett's-type staining pattern with CK7/CK20 was seen in all normal fundic and antral mucosa as well as all biopsies with GIM. A Barrett's-type pattern was seen in 85% of biopsies with cardiac epithelium and all Barrett's esophagus. 7/9 (78%) of biopsies with CIM showed a Barrett's-type pattern of staining.

Immunostaining with DAS-1 antibody was negative or trace in all biopsies showing normal fundic and antral mucosa and in 94% of biopsies with cardiac mucosa. Staining > 5% for DAS-1 was seen in all 10 biopsies of Barrett esophagus, 9/10 (90%) biopsies of CIM, significantly more than the 4/10 (40%) biopsies with GIM.

Immunohistochemistry is not practically effective and therefore not presently used routinely to define the origin of cardiac epithelium or CIM as being esophageal or gastric. There is, however, no argument that intestinal metaplasia in visible CLE represents Barrett esophagus and cardiac epithelium without intestinal metaplasia in visible CLE is esophageal. Similarly, there is no argument that intestinal metaplasia occurring >5 cm distal to the end of the tubular esophagus in the distal stomach is GIM.

Evidence that cardiac epithelium shows a CK7/CK20 staining pattern that resembles Barrett esophagus and is different than that seen in distal gastric body and antral epithelium suggests that cardiac epithelium is more likely to be esophageal than gastric.

Similarly, evidence that CIM shows a CK7/CK20 staining pattern that resembled Barrett esophagus and is different from that seen in GIM suggests that CIM is more likely to be esophageal than gastric.

The findings with DAS-1 staining are somewhat different. DAS-1 seems to be specific for esophageal intestinal metaplasia, being significantly more expressed in Barrett esophagus and CIM than in GIM. This again suggests that CIM is esophageal rather than gastric. DAS-1 is not expressed in normal gastric body and antral epithelium and similarly not expressed in cardiac epithelium. This likely means only that DAS-1 is expressed only when there is intestinal metaplasia in cardiac epithelium.

Liu et al.[38] designed a study to compare the molecular phenotype of columnar epithelia in visible CLE with and without intestinal metaplasia, as defined by the presence of goblet cells. The study is designed to test the hypothesis put forward by this group that esophageal columnar epithelium without goblet cells may have malignant potential. I will examine this hypothesis in a later chapter. At this time, I will review the data in this paper that suggest that cardiac epithelium in the esophagus is similar to cardiac epithelium with intestinal metaplasia and different to gastric epithelium.

Sixty-eight patients with visible CLE at endoscopy were studied. Nineteen biopsies from the gastric corpus from 19 patients were used as controls. 22/68 patients with visible CLE without goblet cells had short segments (<3 cm) of columnar metaplasia. The other 46 patients who had goblet cells identified in one or more biopsies had varying densities of goblet cells. In 32 of these patients, biopsies with and without goblet cells were analyzed. A comparison of DNA content abnormalities was made between the different study groups, and controls, and also between the nongoblet cell–containing and goblet cell–containing epithelium in patients with columnar metaplasia with goblet cells.

DNA content was measured by the integrated optical density (IOD) value in high fidelity DNA histograms created by the Automated Cellular Imaging System. Comparison of the test cells with the internal control mean IOD value of stromal cells [assigned a DNA index (DI) of 1] was used to identify aneuploidy (defined as a DI > 1.1). The heterogeneity index (HI) (normal < 13) and histograms that revealed cells with a DI content of >5N (5N-EC) were also noted.

The 19 control specimens of gastric mucosa showed a mean DI of 1.02 (diploid), zero aneuploidy, normal HI with no elevation of HI, and zero 5N-EC.

The 46 patients with columnar metaplasia that contained goblet cells showed a mean DI of 1.15±0.12 with 54% of biopsies showing aneuploidy, a mean HI of 18.2±2.1 with elevated HI in 100%, and 5N-EC in 7 biopsies (15%).

The 22 patients who had columnar metaplasia without goblet cells had a mean DI of 1.13±0.16 with 64% showing aneuploidy, a mean HI of 18.4±3.3 with 100% showing an elevated HI, and 5N-EC in 4 biopsies (18%).

In the patients who had biopsies of esophageal columnar epithelium with (n=46) and without (n=54) goblet cells, aneuploidy was present in 54% versus 50%, elevation of HI in 100% versus 98%, and 5N-EC in 15% versus 13%.

The results showed that the DNA content parameters tested were statistically similar in nongoblet cell–containing and goblet cell–containing metaplastic esophageal columnar epithelium, both groups being significantly higher than the control normal gastric epithelium.

The intent of the study was to show that the presence of similar DNA abnormalities in esophageal columnar epithelium with and without goblet cells provides evidence that nongoblet columnar epithelium (which, histologically, is cardiac epithelium) is at the same risk for malignancy as that with intestinal metaplasia. The authors' conclusion: "These findings suggest that metaplastic non-goblet columnar epithelium of the esophagus may have neoplastic potential."

What the study also shows, however, is that cardiac epithelium without goblet cells shows similar DNA abnormalities as intestinalized cardiac epithelium in Barrett esophagus and is completely different than gastric epithelium.

This study uses cardiac epithelium in visible CLE above the endoscopic GEJ. Whether these data can be extrapolated to cardiac epithelium in the dilated distal esophagus distal to the endoscopic GEJ (i.e., "gastric cardia") is not known. However, at a histologic level, cardiac epithelium with or without goblet cells that is found on either side of the GEJ is identical. There is no reason to believe that the histologic similarity is not matched by similar DNA content abnormalities.

Derakhshan et al.[39] frame the issue in the introduction of their 2015 study as follows: "The cardiac mucosa between gastric oxyntic and oesophageal squamous epithelium is nearly always inflamed. This carditis has been attributed to either *Helicobacter pylori* infection or acid reflux, but is also present in most people without *H. pylori* infection or symptoms or signs of reflux. Cardiac mucosa may also show intestinal metaplasia (IM) in the absence of *H. pylori* infection or GORD. IM at the cardia was observed in 16% of Caucasians having screening colonoscopy."[40]

Chandrasoma proposes that cardiac mucosa is a response to reflux causing columnar metaplasia of the squamous mucosa of the most distal oesophagus, and is therefore a form of ultra-short Barrett's oesophagus. In neonates, the cardiac mucosa is <1 mm but lengthens with age. However, mechanisms by which the cardiac mucosa may expand in subjects without reflux is unclear.

We recently studied the cardiac mucosa in 58 healthy volunteers without reflux disease or H. pylori *infection and found it inflamed in all subjects. In addition, the length of the cardiac mucosa was positively associated with age and central obesity. None of the subjects had evidence of traditional reflux 5 cm above the upper border of the lower oesophageal sphincter (LOS), but central obesity was associated with gastric acid extending closer to the upper border of the LOS. Based on these observations, we hypothesized that acid reflux onto oesophageal squamous mucosa normally within the distal sphincter was causing columnar metaplasia promoted by central obesity. As there was no evidence of traditionally defined trans-sphincteric reflux, we refer to this phenomenon as intra-sphincteric reflux.*

To test the hypothesis that much of the cardiac mucosa in subjects without traditional GORD may be established by columnar metaplasia, we have compared immunophenotyping of cardiac mucosa with other upper GI mucosae, that is, oesophageal squamous mucosa, gastric body and antral mucosa in the same healthy volunteers and with classical Barrett's mucosa with and without intestinal metaplasia in other subjects.

Detailed data are presented regarding inflammatory cell infiltration and reactive change in the cardiac mucosa. Cardiac mucosa had a similar intensity of inflammatory infiltrate to non-IM Barrett's and greater than any of the upper GI mucosae.

The immunostaining pattern of cardiac mucosa most closely resembled non-IM Barrett's showing only slightly weaker Cdx2 immunostaining. In distal esophageal squamous mucosa, expression of markers of columnar differentiation [trefoil factor family 3 (TFF-3) and liver-intestine cadherin] was apparent and correlated with central obesity. The expression of TFF-3 in distal esophageal squamous mucosa correlated with proximal extension of gastric acidity within the region of the LES.

The expression of Cdx2, villin, TFF-3, and LI-cadherin was significantly less in oxyntocardiac epithelium compared with cardiac epithelium.

The conclusion: "These findings are consistent with expansion of cardiac mucosa in healthy volunteers occurring by squamo columnar metaplasia of distal esophagus and aggravated by central obesity. This metaplastic origin of expanded cardia may be relevant to the substantial proportion of cardia adenocarcinomas unattributable to *H. pylori* or trans-sphincteric acid reflux."

In their discussion, the authors address the possible implication of expression of TFF-3 and LI-cadherin in distal esophageal squamous epithelium exposed to gastric acid: "We also observed evidence of expression of TFF-3 and LI-cadherin within the distal oesophageal squamous mucosa consistent with early columnar metaplasia or at least a wavering commitment to glandular differentiation. TFF-3 and LI-cadherin especially were expressed by a small number of basal cells and immediately suprabasal cells. This could indicate transient expression by stem cells or early daughter cells, not maintained in later generations of transit-amplifying cells. If such glandular differentiation was maintained, rather than switched off, it would represent a possible pathway to glandular metaplasia potentiated by obesity."

This is a logical suggestion. Partial expression of these factors limited to the stem cells and their daughters would result in maintenance of squamous phenotype. Complete expression, as shown by marked increase in their levels in cardiac and Barrett epithelia, would complete the columnar metaplastic process.

An examination of the immunostaining data also shows that there is a substantial and highly significant increase in the expression of Cdx2, villin, TFF-3, LI-cadherin, MUC2, and MUC5ac in Barrett's mucosa with intestinal metaplasia compared with both cardiac mucosa at the SCJ and non-IM Barrett's mucosa. This finding indicates a significant molecular event that drives intestinal metaplasia in cardiac epithelium.

In a similar study in our unit of Cdx2 expression in squamous and different columnar epithelia in patients with Barrett esophagus, Vallbohmer et al.[41] reported quantitative Cdx2 gene expression in cardiac epithelium, squamous epithelium, and intestinalized Barrett epithelium.

The differentiation of the epithelium of the foregut is driven by expression of specific genetic signals. This process is dynamic and liable to change with activation and suppression of these differentiation signals. In fetal life, the esophagus is initially lined by stratified columnar epithelium and subsequently differentiates into the normal squamous epithelium. In the fetal stomach, the primitive stratified columnar epithelium differentiates into a glandular epithelium containing parietal and chief cells.

During adult life, this dynamic process results in the metaplasia of esophageal squamous epithelium to columnar epithelium. This is a highly specific change when esophageal squamous epithelium is exposed to gastric juice. I will consider the genetic signals involved in differentiation of esophageal epithelial cells in Chapter 15.

The acid-induced metaplastic process in the esophagus results initially in the formation of cardiac epithelium, a simple columnar epithelium composed of only mucous cells. Metaplastic cardiac epithelium in the esophagus can evolve into two other epithelial types by a second step that appears to be mutually exclusive. It can develop parietal cells forming oxyntocardiac epithelium or develop goblet cells forming intestinal metaplasia. The latter event defines Barrett esophagus when it occurs in a biopsy taken from visible CLE above the endoscopic GEJ. When goblet cells develop in cardiac epithelium below the GEJ, it is presently called "CIM."

This study evaluates the possible role of Cdx2 in the genesis of intestinal metaplasia in cardiac epithelium. Cdx2 is a homeobox gene that is an important transcriptional regulator of embryonic differentiation and maintenance of normal colonic and small intestinal epithelium. Two prior studies[42,43] have shown Cdx2 expression in all patients with intestinal metaplasia in CLE and 30%–38% of patients with cardiac mucosa.

In this study,[41] biopsies obtained at endoscopy were immediately snap-frozen in liquid nitrogen. Sections cut from these were classified into the following five groups based on histologic examination: (1) normal squamous epithelium of the esophagus taken 3 cm above the SCJ in 62 patients; (2) cardiac epithelium taken immediately distal to the SCJ in 19 patients; (3) oxyntocardiac epithelium taken immediately distal to the SCJ in 14 patients; (4) intestinal metaplasia within cardiac epithelium from the distal CLE in 15 patients with Barrett's esophagus; and (5) duodenal mucosa from 26 patients as an intestinal columnar control. Multiple samples from these sites were taken from some patients. Patients with Barrett esophagus only had biopsies taken of the Barrett's segment.

Samples were dissected from the slides using a scalpel if the epithelial type was homogeneous or by laser-capture microdissection. The dissected samples were processed by a method that permitted quantification of Cdx2 mRNA. Table 5.14 shows the Cdx2 expression levels in the five histologic groups.

Cdx2 expression was significantly higher in duodenal mucosa than all other groups, in Barrett's intestinal epithelium over cardiac, oxyntocardiac and squamous groups, and cardiac/oxyntocardiac over normal squamous epithelium. There was no significant difference between cardiac and oxyntocardiac epithelial groups.

The study shows that Cdx2 is essentially absent in squamous epithelium, is expressed significantly more in cardiac epithelium, and then shows a marked increase in expression in intestinal metaplasia. The results are "in keeping with the concept that columnar metaplasia of esophageal squamous epithelium is a multistep process. The first step is the transformation of squamous mucosa to nonintestinalized columnar epithelium, known as cardiac mucosa… further expression of Cdx2 is associated with goblet cell differentiation in cardiac mucosa."

TABLE 5.14 Cdx2 mRNA Expression Levels in the Different Histologic Groups

Patient Group	No. of Patients	Cdx2 × 100/β-actin mRNA Expression Median (25th–75th Percentile)
Normal squamous epithelium	62	0.01 (0.01–0.05)
Cardiac epithelium	19	0.4 (0.3–0.71)
Oxyntocardiac epithelium	14	0.76 (0.28–1.14)
Barrett esophagus (intestinal metaplasia)	15	6.72 (3.97–8.08)
Duodenum	26	39.64 (25.98–55.29)

mRNA, messenger RNA.

The expression of Cdx2 at an intermediate level between squamous epithelium and intestinal metaplasia provides evidence that cardiac epithelium is esophageal. It is to be emphasized that cardiac epithelium in this study was taken from biopsies *distal* to the endoscopic GEJ.

The authors point out that simple expression of Cdx2 in these epithelia does not necessarily mean that Cdx2 is the genetic signal involved in the metaplastic process: "Based on our study results we can only speculate about the role of Cdx2 in the development of Barrett's esophagus. Although Cdx2 may be a critical and necessary gene driving the process of intestinalization, it may also represent a biomarker gene merely indicating the presence of intestinal metaplasia."

This is true of any molecular marker associated with these epithelial types. Cardiac epithelium with and without intestinal metaplasia is a metaplastic epithelium derived from exposure of squamous epithelium to gastric juice. This results in injury and inflammation associated with the metaplasia. Molecular changes in the epithelia may be associated with injury and inflammation. Proof of the exact role in these changes requires further study, preferably prospective with long-term follow-up.

It is interesting that this same quantitative expression of a potential signaling gene (TFF-3) was suggested by Derakhshan et al.[39] (see above) as the mechanism for transformation of squamous epithelium to cardiac epithelium. The concept that quantitative rather than qualitative expression may be involved in the metaplastic process should induce caution in the interpretation of simple expression of any gene in this process.

10. RELATIONSHIP OF CARDIAC EPITHELIUM TO GASTROESOPHAGEAL REFLUX DISEASE AND GASTRITIS

There is a perception that inflamed cardiac epithelium distal to the SCJ in endoscopically normal people can result from both GERD and *H. pylori* gastritis. This is not correct. Carditis (i.e., cardiac epithelium) *always* results from metaplasia of esophageal squamous epithelium that has been damaged by exposure to gastric contents.

In the majority of people, the metaplastic cardiac epithelium is limited to the area distal to the endoscopic GEJ. Most studies of cardiac epithelium show the almost invariable presence of inflammation and reactive changes such as foveolar hyperplasia and lamina propria edema. The inflammation in cardiac epithelium usually consists of lymphocytes, plasma cells, and eosinophils, and it varies in severity from mild to severe. The proliferative index of the epithelium, as assessed by Ki67 (MIB-1), mirrors the inflammatory reaction.[44]

Inflamed cardiac epithelium distal to the endoscopic GEJ ("carditis") is commonly an isolated abnormality. Der et al.[45] showed that 111/141 (79%) of patients with carditis had no inflammation in their distal gastric biopsies. Lembo et al.[46] confirmed that carditis is an entity that is limited to a small area immediately distal to the SCJ in endoscopically normal patients.

Cardiac epithelium always results from damage to the squamous epithelium of the esophagus when it is exposed to gastric juice. It begins at the true GEJ where the squamous epithelium meets the normal gastric oxyntic epithelium and extends cephalad as the squamous epithelium progressively undergoes cardiac metaplasia.

Once cardiac epithelium has been produced as a result of metaplasia of esophageal epithelium, it becomes susceptible to secondary infection by *H. pylori*. *H. pylori* most commonly affects the antrum but can extend proximally all the way to the SCJ. When it does so, both gastric oxyntic epithelium and metaplastic cardiac epithelium distal to the SCJ can become infected. Infection of cardiac epithelium results in an increase in the severity of inflammation and the presence of numerous neutrophils in cardiac epithelium. Spechler et al.[47] showed that inflammation in carditis was more severe in the 21% of patients whose carditis was associated with *H. pylori* compared to the 79% whose carditis was not associated with gastritis. Der et al.[45] also showed the relationship between acute inflammation in carditis and *H. pylori* infection.

This explains the relationship between carditis and gastritis. Carditis is caused by metaplasia of esophageal squamous epithelium and is an isolated finding at the SCJ in patients without gastric pathology. When *H. pylori* infection occurs in the stomach and extends proximally, there can be secondary infection of cardiac epithelium resulting in an increase in the inflammation with the addition of neutrophils in the inflammatory infiltrate. *H. pylori* is not a primary cause of carditis; it cannot produce cardiac metaplasia in squamous epithelium.

This issue is discussed in greater detail in Chapter 8.

11. RELATIONSHIP OF CARDIAC EPITHELIUM TO VISIBLE COLUMNAR-LINED ESOPHAGUS

Visible CLE is the result of damage of the squamous epithelium of the thoracic esophagus caused by a sufficiently severe amount of transsphincteric reflux of gastric contents into the esophagus. This is associated with frequent failure of the LES that causes an abnormal pH test. LES failure is strongly associated with permanent LES damage from its original length (~35 mm for the abdominal segment) to <10 mm (for the abdominal segment), which defines a defective LES.

Visible CLE is *always* associated with cardiac epithelium distal to the endoscopic GEJ (end of the tubular esophagus or, in patients with hiatal hernia, the proximal limit of rugal folds). The length of cardiac epithelium distal to the endoscopic GEJ is usually longer in patients with visible CLE than those without, frequently extending >15 mm to a theoretical maximum of 35 mm (see Chapter 11).

With the present perspective where the GEJ is defined endoscopically, this creates a confusing situation. It appears as if GERD with visible CLE has two separate pathologic entities: (1) the visible CLE above the endoscopic GEJ and (2) cardiac epithelium distal to the endoscopic GEJ in what is believed to be the proximal stomach.

Impossible to explain and conveniently avoided in the gastroenterology literature is the fact that the length of cardiac epithelium distal to the endoscopic GEJ is greater in patients with severe GERD (especially those with visible CLE) than mild GERD. To people who believe that the endoscopic GEJ is the proximal limit of rugal folds and/or the end of the tubular esophagus, this must mean that in some way gastric cardiac epithelium expands distally as GERD increases. This makes no sense.

When the perspective is adjusted to the true GEJ (the proximal limit of gastric oxyntic mucosa), the significance of cardiac epithelium becomes crystal clear. Cardiac epithelium increases progressively from the true GEJ (Fig. 5.12). In the early stage of the disease, the length of cardiac epithelium is short. This correlates with a competent LES whose damage is within its reserve capacity. As LES damage increases, cardiac epithelial length increases cephalad, displacing the SCJ (and the endoscopic GEJ) cephalad. In this early stage of the disease, cardiac epithelium is limited to the dilated distal esophagus

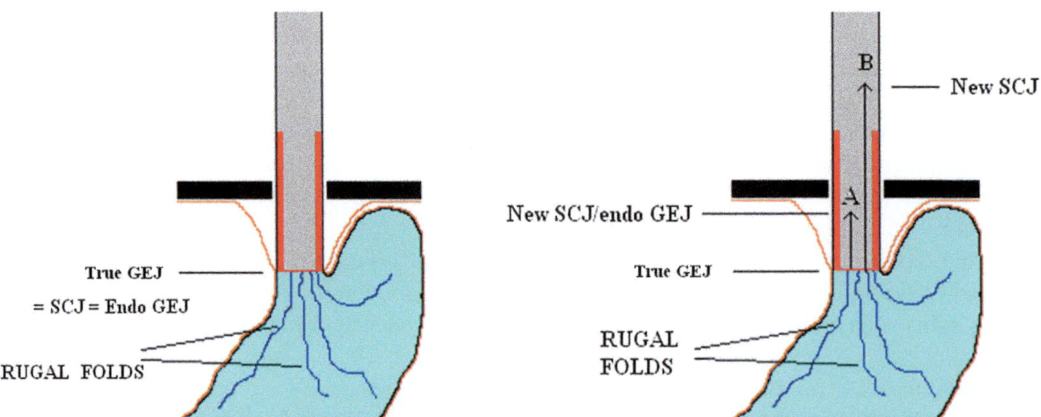

FIGURE 5.12 Diagrammatic representation of how cardiac epithelium develops, always as a result of columnar metaplasia of esophageal squamous epithelium. This always begins at the true gastroesophageal junction (GEJ) and progresses cephalad. On left is the normal state without cardiac epithelium. On right, the *arrows* show the two stages of development of cardiac epithelium. A: Cardiac epithelium limited to the dilated distal esophagus [from the true GEJ to the cephalad-migrated squamocolumnar junction (SCJ) and endoscopic GEJ]. B: Cardiac epithelium extending into the thoracic esophagus as visible columnar-lined esophagus. A results from slow cardiac metaplasia secondary to repeated gastric over-distension in early gastroesophageal reflux disease (GERD). B results from abnormal reflux into the thoracic esophagus in severe GERD.

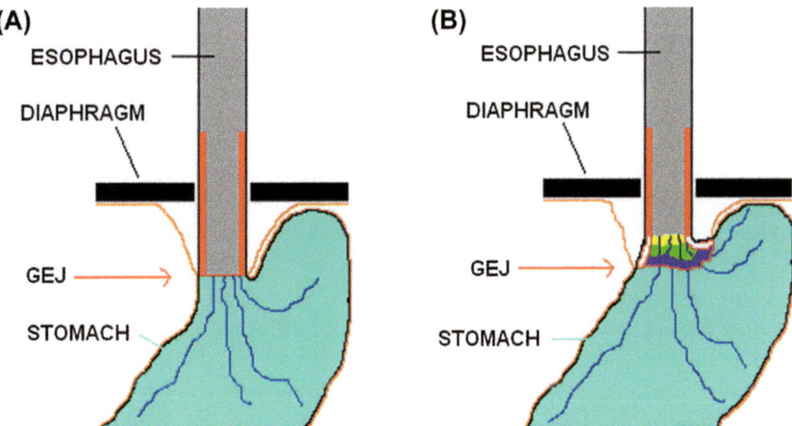

FIGURE 5.13 Diagram of (A) the normal state without any cardiac epithelium and (B) the abnormal state where cardiac epithelium with intestinal metaplasia (yellow), cardiac epithelium (green), and oxyntocardiac epithelium (purple) are interposed between the squamous epithelium and gastric oxyntic epithelium distal to the endoscopic gastroesophageal junction (GEJ). The pathologic anatomy of cardiac metaplasia in the abdominal esophagus is the dilated distal esophagus.

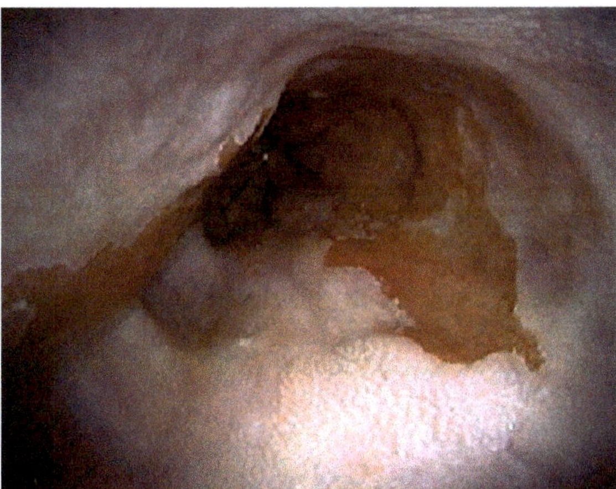

FIGURE 5.14 Antegrade endoscopic view of a long segment of visible columnar-lined esophagus in the thoracic esophagus. *Photograph courtesy of Dr. Martin Riegler, Reflux Medical, Vienna, Austria.*

(A in Fig. 5.12). If the length of cardiac epithelium is <15 mm, the patient has a competent LES and there is no reflux. The patient does not have GERD and endoscopy is normal. However, histology will show the abnormal cardiac epithelium varying in extent from >0 to 15 mm (Fig. 5.13). In this stage, the SCJ and the endoscopic GEJ have both moved cephalad to the same extent creating a gap with the true GEJ that is equal to the length of cardiac epithelium (Fig. 5.13).

With severe LES damage (seen as cardiac epithelial length > 15 mm, and usually >25 mm), the LES starts becoming incompetent. As LES damage progressively increases, transsphincteric reflux progressively increases, the squamous epithelium in the thoracic esophagus undergoes columnar metaplasia, and visible CLE develops (Fig. 5.14). Cardiac epithelium extends into the thoracic esophagus as visible CLE (B in Fig. 5.12), continuous with the cardiac epithelium in the dilated distal esophagus. In this stage, the SCJ moves cephalad from the endoscopic GEJ (which marks the proximal limit of the dilated distal esophagus) creating the gap that is seen as the visible CLE (Fig. 5.14).

The lengthening of cardiac epithelium is always in one direction: beginning at the true GEJ and extending cephalad. The columnar epithelium between the true GEJ (the proximal limit of gastric oxyntic epithelium) and the SCJ is the same cardiac epithelium (with and without parietal and/or goblet cells) irrespective of its length and whether it is endoscopically visible or not.

The most convincing evidence for such a process is the histologic mapping of the epithelial distribution of the segment of cardiac epithelium (with and without parietal and/or goblet cells) between the distal limit of squamous epithelium and the true GEJ (proximal limit of gastric oxyntic epithelium). This is what I call the squamooxyntic gap.[48] When all three epithelial types are present in this segment, cardiac epithelium with intestinal metaplasia is proximal, cardiac epithelium is intermediate, and cardiac epithelium with parietal cells is distal. This distribution is true whatever may be the length of the segment of metaplastic columnar epithelium.

12. CONCLUSION

I have reviewed much of the available literature that was generated as a result of the 1994 hypothesis that cardiac epithelium did not exist as a normal epithelium in either the stomach or the esophagus. We suggested that cardiac epithelium was an abnormal epithelium resulting from columnar metaplasia of the squamous epithelium of the esophagus due to exposure to gastric contents.

There are very few studies that have measured the length of cardiac epithelium (with and without parietal/goblet cells) distal to the SCJ in people without evidence of CLE at endoscopy. Whenever measurements have been made, the length of cardiac epithelium (with and without parietal/goblet cells) has correlated with increasing likelihood of GERD, either by the presence of active esophagitis or by an abnormal 24-h pH test.[25,33,34]

This permits us to define the normal state as an esophagus lined entirely by squamous epithelium. This transitions at the GEJ to gastric oxyntic epithelium (Fig. 5.13A). This point, in the normal state, is also the SCJ, the end of the tubular esophagus, the point at which the tubular esophagus flares into saccular stomach, the proximal limit of gastric rugal folds, the distal end of the abdominal segment of the LES, the peritoneal reflection, the point at which the two-layered muscle of the esophageal wall transitions to the more complex muscle wall of the stomach, and the angle of His. Every definition ever used to define the GEJ is absolutely correct in the normal state.

Cardiac epithelium (with and without parietal/goblet cells) does not exist in the normal state (Figs. 5.4, 5.5, and 5.9). It develops when esophageal squamous epithelium is exposed to gastric juice. The damage caused to the squamous epithelium results in columnar metaplasia, i.e., the development of cardiac epithelium. As damage progresses, the length of cardiac epithelium (with and without parietal/goblet cells) increases as more and more squamous epithelium undergoes columnar metaplasia. This is easy to understand.

What is confusing are the other things that occur when the squamous epithelium undergoes columnar metaplasia. The segment of esophagus where squamous epithelium has undergone columnar metaplasia loses its sphincter pressure and dilates (Fig. 5.13B). When this happens the following occurs: (1) the SCJ moves cephalad; (2) the dilated segment of CLE functions as part of the reservoir (i.e., it becomes gastricized) and develops rugal folds; (3) the endoscopic GEJ moves cephalad, initially retaining its relationship with the displaced SCJ (Fig. 5.12); and (4) the angle of His becomes more obtuse (Fig. 5.13B). Every definition of the GEJ that has been used in the past and to the present has undergone change, moving cephalad. Only the proximal limit of gastric oxyntic epithelium and the peritoneal reflection remain unchanged.

The endoscopist faced with this abnormal state created by GERD sees a SCJ coincident with the proximal limit of rugal (not "gastric" rugal) folds at the end of the tubular esophagus where it flares into the saccular organ. The dilated distal esophagus, which has developed rugal folds, is mistaken for proximal stomach. The trap has been set and for decades every endoscopist has fallen into it. The only unchanged marker of the GEJ is the proximal limit if gastric oxyntic epithelium. This is an epithelium that is designed to resist gastric juice. This point cannot be detected at endoscopy, it needs biopsy.

The trap that everyone has fallen into has created two false dogmas: (1) the GEJ can be defined at endoscopy by the proximal limit of rugal folds and (2) cardiac epithelium is normal in this region.

These false dogmas have prevented recognition of the dilated distal esophagus as the earliest change in GERD. The eradication of these two false dogmas leads to a new understanding of the pathogenesis of GERD, a method of assessing LES damage histologically, and ultimately a new way of diagnosis and management of GERD that has the potential to prevent treatment failure and adenocarcinoma of the esophagus. This is what we will explore in the succeeding chapters.

In an editorial to the paper by Derakhshan et al.[39] (reviewed above), Spechler writes: "Traditional teaching holds that the normal stomach has three types of mucosae. The oxyntic mucosa of the gastric body and fundus has glands with parietal cells that secrete acid, and chief cells that secrete digestive enzymes. The mucosa of the antrum is comprised of mucus-secreting cells and endocrine cells that produce gastrin, which regulates acid production by the oxyntic mucosa. Finally, the most proximal portion of the stomach (the gastric cardia), a region with ill-defined borders, allegedly is lined by 'cardiac mucosa' comprised almost exclusively of mucus secreting cells. Cardiac mucosa has been assumed to function as a buffer zone, preventing the damage that might result if the acid sensitive squamous mucosa of the oesophagus joined directly with the acid-secreting oxyntic mucosa of the gastric body. This dogma went unchallenged until 1997, when Chandrasoma proposed that cardiac mucosa is not a normal structure, but rather is acquired when GORD causes columnar metaplasia of squamous epithelium in the distal oesophagus. Furthermore, he contended that cardiac mucosa evolves into the intestinal metaplasia of Barrett's oesophagus in the setting of persistent GORD. Recent studies by investigators in Glasgow provide support for Chandrasoma's proposals."

TABLE 5.15 Progression From "Normal" Through a Phase of "Pre-GERD" Through Increasing Severity of Clinical GERD Correlated With Abdominal Lower Esophageal Sphincter Damage Measured by Cardiac Epithelial Length in the Dilated Distal Esophagus

Feature	Normal	Pre-GERD	Mild Clinical GERD	Severe GERD
Cardiac epithelial length in DDE	0	>0–15 mm	>15–25 mm	>25–35 mm
Abdominal LES damage	0	>0–15 mm	>15–25 mm	>25–35 mm
Manometric abdominal LES length	35 mm	20–<35 mm	10–<20 mm	<10 mm–0
Acid exposure = pH test (% time pH < 4)	0	0%–1.1%	1.1%–4.5%	>4.5%
Endoscopy	Normal	Normal	Normal or mild erosive esophagitis	Severe erosive esophagitis; risk of visible CLE
Biopsy changes	Zero	Cardiac epithelium < 15 mm limited to DDE	Cardiac epithelium 15–<25 mm limited to DDE; reflux esophagitis	Cardiac epithelium > 25 mm in DDE + visible CLE
Symptoms	Zero	Zero to nonspecific dyspepsia	Zero to mild, controlled symptoms	Zero to severe symptoms

CLE, columnar-lined esophagus; *DDE*, dilated distal esophagus; *GERD*, gastroesophageal reflux disease; *LES*, lower esophageal sphincter.

The hypothesis is now at a point where multiple investigators have tested the hypothesis and produced a critical mass of supportive evidence. There is a likelihood that the hypothesis will be recognized as true, albeit in the time it takes for the medical establishment to move. When this happens, recommendations for biopsy of the dilated distal esophagus will be made. This will result in a spate of new information regarding the significance of cardiac epithelium distal to the endoscopic GEJ. This information will prove the expanded hypothesis that I will detail in the subsequent chapters.

This chapter opens the door to a revolutionary new way of understanding GERD as a disease. At present, we do not define the normal state. We simply define GERD, the disease entity, by criteria that are dependent on the presence of an advanced stage of the disease. Absence of these criteria means only that the patient does not have GERD by these criteria. Unfortunately, this means that there is no diagnostic test for an earlier stage of GERD.

"No-GERD" is not equivalent to "normal." A person does not progress from "normal" to GERD defined by the presence of troublesome symptoms and complications such as erosive esophagitis and Barrett esophagus in one moment of time. The progression from normal to GERD must involve an intervening stage of "pre-GERD" or "subclinical-GERD" that we do not presently recognize.

The definition of the normal state as the absence of cardiac epithelium provides the basis for recognizing the "pre-GERD" or "subclinical-GERD" pathologic change. Evidence suggests that persons diagnosed with GERD have >15 mm of cardiac epithelium distal to the endoscopic GEJ. The progression from 0 to 15 mm cardiac epithelium is the pathologic abnormality that defines "pre-GERD" or "subclinical-GERD."

This progression of measurable change in cardiac epithelial length from 0 to 35 mm provides a basis for evaluating the same concept of "pre-GERD" or "subclinical-GERD" abnormality as it relates to the pH test, manometric measurement of the abdominal LES, cellular changes by endoscopy and biopsy, and symptoms. Let us assume that "normal" represents the highest level of normalcy that has been observed and attempt to correlate cardiac epithelial length with these criteria that are used to assess severity of GERD (Table 5.15).

I will show in the succeeding chapters that this "pre-GERD" or "subclinical-GERD" phase involves decades of slowly progressive lengthening of cardiac epithelium in the distal esophagus. This correlates with damage of the abdominal segment of the LES, which is the primary cause of GERD. In turn, this change correlates with every feature of GERD (Table 5.15). The least reliable correlates are symptoms and symptom control, which are the present criteria of diagnosis of GERD. The ability to measure LES damage by a new pathologic test opens the door to potential methods of preventing complications of GERD, and ultimately GERD itself.

REFERENCES

1. Vakil N, van Zanten SV, Kahrilas P, Dent J, Jones R, The Global Consensus Group. The Montreal definition and classification of gastroesophageal reflux disease: a global evidence-based consensus. *Am J Gastroenterol* 2006;**101**:1900–20.
2. Chandrasoma PT. Histologic definition of gastro-esophageal reflux disease. *Curr Opin Gastroenterol* 2013;**29**:460–7.
3. Hayward J. The lower end of the oesophagus. *Thorax* 1961;**16**:36–41.
4. Clark GWB, Ireland AP, Chandrasoma P, et al. Inflammation and metaplasia in the transitional mucosa of the epithelium of the gastroesophageal junction: a new marker for gastroesophageal reflux disease. *Gastroenterology* 1994;**106**:A63.
5. Chandrasoma P. Pathophysiology of Barrett's esophagus. *Semin Thorac Cardiovasc Surg* 1997;**9**:270–8.
6. Chandrasoma P. Norman Barrett: so close, yet 50 years away from the truth. *J Gastrointest Surg* 1999;**3**:7–14.
7. Chandrasoma PT, Der R, Ma Y, et al. Histology of the gastroesophageal junction: an autopsy study. *Am J Surg Pathol* 2000;**24**:402–9.
8. Odze RD. Unraveling the mystery of the gastroesophageal junction: a pathologist's perspective. *Am J Gastroenterol* 2005;**100**:1853–67.
9. Kilgore SP, Ormsby AH, Gramlich TL, et al. The gastric cardia: fact or fiction? *Am J Gastroenterol* 2000;**95**:921–4.
10. Barrett NR. Chronic peptic ulcer of the oesophagus and 'oesophagitis'. *Br J Surg* 1950;**38**:175–82.
11. Allison PR, Johnstone AS. The oesophagus lined with gastric mucous membrane. *Thorax* 1953;**8**:87–101.
12. Barrett NR. The lower esophagus lined by columnar epithelium. *Surgery* 1957;**41**:881–94.
13. De Nardi FG, Riddell RH. Esophagus. In: Sternberg SS, editor. *Histology for pathologists*. 2nd ed. Philadelphia: Lippincott-Raven Publishers; 1997. p. 461–80.
14. Owen DA. Stomach. In: Sternberg SS, editor. *Histology for pathologists*. 2nd ed. Philadelphia: Lippincott-Raven Publishers; 1997. p. 481–93.
15. Spechler SJ, Zeroogian JM, Antonioli DA, et al. Prevalence of metaplasia at the gastroesophageal junction. *Lancet* 1994;**344**:1533–6.
16. Spechler SJ. Cardiac mucosa: the heart of the problem. *Gut* 2015;**64**(11):1673–4.
17. Chandrasoma P, Makarewicz K, Wickramasinghe K, et al. A proposal for a new validated histologic definition of the gastroesophageal junction. *Hum Pathol* 2006;**37**:40–7.
18. Sarbia M, Donner A, Gabbert HE. Histopathology of the gastroesophageal junction. A study on 36 operation specimens. *Am J Surg Pathol* 2002;**26**:1207–12.
19. Haggitt RC, Tryzelaar J, Ellis FH, et al. Adenocarcinoma complicating columnar epithelium-lined (Barrett's) esophagus. *Am J Clin Pathol* 1978;**70**:1–5.
20. Haggitt RC, Dean PJ. Adenocarcinoma in Barrett's epithelium. In: Spechler SJ, Goyal RK, editors. *Barrett's esophagus: pathophysiology, diagnosis and management*. New York: Elsevier Science Publishing; 1985. p. 153–66.
21. De Hertogh G, Van Eyken P, Ectors N, et al. On the existence and location of cardiac mucosa; an autopsy study in embryos, fetuses, and infants. *Gut* 2003;**52**:791–6.
22. Derdoy JJ, Bergwerk A, Cohen H, et al. The gastric cardia: to be or not to be? *Am J Surg Pathol* 2003;**27**:499–504.
23. Park YS, Park HJ, Kang GH, et al. Histology of gastroesophageal junction in fetal and pediatric autopsy. *Arch Pathol Lab Med* 2003;**127**:451–5.
24. Johns BAE. Developmental changes in the oesophageal epithelium in man. *J Anat* 1952;**86**:431–42.
25. Oberg S, Peters JH, DeMeester TR, et al. Inflammation and specialized intestinal metaplasia of cardiac mucosa is a manifestation of gastroesophageal reflux disease. *Ann Surg* 1997;**226**:522–32.
26. Paull A, Trier JS, Dalton MD, et al. The histologic spectrum of Barrett's esophagus. *N Eng J Med* 1976;**295**:476–80.
27. Jain R, Aquino D, Harford WV, et al. Cardiac epithelium is found infrequently in the gastric cardia. *Gastroenterology* 1998;**114**:A160 (Abstract).
28. Marsman WA, van Sandyck JW, Tytgat GNJ, et al. The presence and mucin histochemistry of cardiac type mucosa at the esophagogastric junction. *Am J Gastroenterol* 2004;**99**:212–7.
29. Shi L, Der R, Ma Y, et al. Gland ducts and multilayered epithelium in mucosal biopsies from gastroesophageal-junction region are useful in characterizing esophageal location. *Dis Esophagus* 2005;**18**:87–92.
30. Robertson EV, Derakhshan MH, Wirz AA, Lee YY, Seenan JP, Ballantyne SA, Hanvey SL, Kelman AW, Going JJ, McColl KE. Central obesity in asymptomatic volunteers is associated with increased intrasphincteric acid reflux and lengthening of the cardiac mucosa. *Gastroenterology* 2013;**145**:730–9.
31. Ringhofer C, Lenglinger J, Izay B, Kolarik K, Zacherl J, Fisler M, Wrba F, Chandrasoma PT, Cosentini EP, Prager G, Riegler M. Histopathology of the endoscopic esophagogastric junction in patients with gastroesophageal reflux disease. *Wien Klin Wochenschr* 2008;**120**:350–9.
32. Lenglinger J, Ringhofer C, Eisler M, Sedivy R, Wrba F, Zacherl J, Cosentini EP, Prager G, Heafner M, Riegler M. Histopathology of columnar lined esophagus in patients with gastroesophageal reflux disease. *Wien Klin Wochenschr* 2007;**119**:405–11.
33. Glickman JN, Fox V, Antonioli DA, Wang HH, Odze RD. Morphology of the cardia and significance of carditis in pediatric patients. *Am J Surg Pathol* 2002;**26**:1032–9.
34. Chandrasoma PT, Lokuhetty DM, DeMeester TR, et al. Definition of histopathologic changes in gastroesophageal reflux disease. *Am J Surg Pathol* 2000;**24**:344–51.
35. Chandrasoma P. Controversies of the cardiac mucosa and Barrett's esophagus. *Histopathology* 2005;**46**:361–73.
36. DeMeester SR. Cytokeratin and DAS-1 immunostaining reveal similarities among cardiac mucosa, CIM, and Barrett's esophagus. *Am J Gastroenterol* 2002;**97**:2514.
37. Ormsby AH, Goldblum JR, Rice TW, et al. The utility of cytokeratin subsets in distinguishing Barrett's related oesophageal adenocarcinoma from gastric adenocarcinoma. *Histopathology* 2001;**38**:307–11.

38. Liu W. Metaplastic esophageal columnar epithelium without goblet cells shows DNA content abnormalities similar to goblet cell containing epithelium. *Am J Gastroenterol* 2009;**104**:816–24.
39. Derakhshan MH. In healthy volunteers, immunohistochemistry supports squamous to columnar metaplasia as mechanism of expansion of cardia, aggravated by obesity. *Gut* 2015;**64**:1705–14.
40. Gerson LB, Shetler K, Triadofilopoulos G. Prevalence of Barrett's esophagus in asymptomatic individuals. *Gastroenterology* 2002;**123**:461–7.
41. Vallbohmer D, DeMeester SR, Peters JH, Oh DS, Kuramochi H, Shimizu D, Hagen JA, Danenberg KD, Danenberg PV, DeMeester TR, Chandrasoma PT. Cdx-2 expression in squamous and metaplastic columnar epithelia in the esophagus. *Dis Esophagus* 2006;**19**:260–6.
42. Phillips RW, Frierson Jr HF, Moskaluk CA. Cdx2 as a marker of epithelial intestinal differentiation in the esophagus. *Am J Surg Pathol* 2003;**27**:1442–7.
43. Groisman GM, Amar M, Meir A. Expression of the intestinal marker Cdx2 in the columnar-lined esophagus with and without intestinal (Barrett's) metaplasia. *Mod Pathol* 2004;**17**:1282–8.
44. Olvera M, Wikramasinghe K, Brynes R, Bu X, Ma Y, Chandrasoma P. Ki67 expression in different epithelial types in columnar lined esophagus indicates varying levels of expanded and aberrant proliferative patterns. *Histopathology* 2005;**47**:132–40.
45. Der R, Tsao-Wei DD, DeMeester T, et al. Carditis: a manifestation of gastroesophageal reflux disease. *Am J Surg Pathol* 2001;**25**:245–52.
46. Lembo T, Ippoliti AF, Ramers C, Weinstein WM. Inflammation of the gastroesophageal junction (carditis) in patients with symptomatic gastroesophageal reflux disease; a prospective study. *Gut* 1999;**45**:484–8.
47. Spechler SJ, Wang HH, Chen YY, Zeroogian JM, Antonioli, Goyal RK. GERD vs *H. pylori* infections as potential causes of inflammation in the gastric cardia. *Gastroenterology* 1997;**112**:A297.
48. Chandrasoma PT, Wijetunge S, DeMeester SR, Hagen JA, DeMeester TR. The histologic squamo-oxyntic gap: an accurate and reproducible diagnostic marker of gastroesophageal reflux disease. *Am J Surg Pathol* 2010;**34**:1574–81.

Chapter 6

Definition of the Gastroesophageal Junction

To begin a rational study of gastroesophageal reflux disease (GERD), we must clearly demarcate the esophagus from the proximal stomach by accurately defining the gastroesophageal junction (GEJ). If we fail to do this, we may potentially ascribe pathology in the esophagus to the proximal stomach.

The present failure in accurately defining the GEJ is self-evident. When an adenocarcinoma arises in an area that is distal to the endoscopic GEJ (Fig. 6.1), it will be designated "adenocarcinoma of the gastric cardia" at endoscopy. Evidence exists that "adenocarcinoma of the gastric cardia" is associated with GERD.[1] This is not rational. Adenocarcinoma of the stomach cannot be caused by reflux of gastric contents into the esophagus.

Furthermore, if that "adenocarcinoma of the gastric cardia," diagnosed at endoscopy using the endoscopic definition of the GEJ, is resected and given to a pathologist, it will be classified by the criteria of the seventh edition of the American Joint Committee on Cancer (AJCC) as an esophageal adenocarcinoma.[2] This was a classification change that occurred in the seventh edition in 2010. In the AJCC sixth edition, the same tumor was classified as gastric carcinoma. The confusion is profound but accepted in the highest intellectual circles.

1. ANATOMY OF THE ESOPHAGUS AND PROXIMAL STOMACH

The normal esophagus is normally a tubular structure with two sphincters at either end. Both sphincters are part of the esophagus; the esophagus extends from the proximal end of the upper esophageal sphincter to the distal limit of the lower esophageal sphincter (LES).

The esophagus can be divided anatomically into cervical, thoracic, and abdominal segments. The cervical segment extends from the inferior border of the pharynx to the thoracic inlet; the thoracic segment from the thoracic inlet to the diaphragmatic hiatus; and the abdominal segment from the diaphragmatic hiatus to the GEJ.

The LES straddles the diaphragmatic hiatus and can be divided into thoracic and abdominal segments. The entire abdominal esophagus is the abdominal segment of the LES, which extends from the diaphragmatic crus to the peritoneal reflection at the GEJ.

The abdominal esophagus is surrounded by a conical mass of loose fibroadipose tissue that extends from the GEJ below to the diaphragm above. It is separated from the peritoneal reflection by this connective tissue, which contains the inferior phrenoesophageal ligament, an ill-defined condensation of collagen.

Unlike the upper esophageal sphincter, which is definable by an anatomically visible ring of muscle, the lower sphincter is difficult to define anatomically in gross specimens, at surgery and at endoscopy. It is generally defined by manometry, which recognizes the sphincter as a high-pressure zone between the stomach distally and the esophagus proximally.

One of the recognized early changes in GERD is a "shortening of the LES" beginning distally. Because it is defined by manometry, "shortening" equates to loss of normal pressure of the sphincter, beginning at its distal end. When the sphincter "shortens," its pressure becomes indistinguishable from gastric baseline pressure. As such, the manometric distal limit of the sphincter does not define the end of the esophagus (i.e., the GEJ) in patients with GERD. The end of the esophagus is distal to the manometric end of the LES by the amount of pathologic shortening of the LES.

There is excellent evidence that the primary reason for GERD is a failure of the LES mechanism. The present inability to define and quantitate damage and shortening of the sphincter is a fundamental problem in the study of GERD.

LES damage begins at the distal end of the esophagus and progressively moves proximally. If this is true, the change in the length of the abdominal segment of the LES will provide a more sensitive assessment of LES damage than a change in its total length. Ayazi et al.[3] and Robertson et al.[4] demonstrated that LES shortening in asymptomatic subjects results

GERD. http://dx.doi.org/10.1016/B978-0-12-809855-4.00006-3

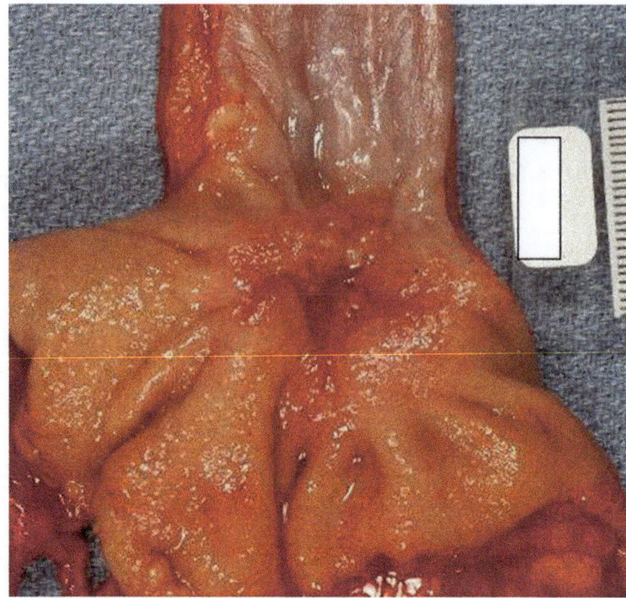

FIGURE 6.1 Adenocarcinoma of the dilated distal esophagus. This is distal to the end of the tubular esophagus in what appears to be the proximal stomach ("gastric cardia"). The rugal folds have been largely obliterated by the tumor. Histologic examination showed the presence of metaplastic cardiac epithelium at the distal edge of the tumor. This indicates that this tumor is entirely within the dilated distal esophagus, i.e., proximal to the true gastroesophageal junction, which is histologically defined as the proximal limit of gastric oxyntic epithelium.

primarily from shortening of the abdominal segment, confirming that LES damage begins at the distal end of the esophagus and extends cephalad.

1.1 The Proximal End of the Esophagus

The proximal end of the esophagus is easy to recognize anatomically; it is where the inferior constrictor muscles of the pharynx decussate and surround the top of the esophagus. The decussating muscle forms a well-defined ring of muscle, that is, the upper esophageal or cricopharyngeal sphincter. The proximal end of the esophagus is the upper limit of the cricopharyngeal sphincter.

There is no histologic difference between the mucosal lining of the pharynx and esophagus; both are lined by nonkeratinizing stratified squamous epithelium. However, the sensitivity of the epithelium is different. The pharynx is sensitive to material within it; the esophagus is less so.

The presence of food in the pharynx stimulates the swallowing reflex that converts the upper aerodigestive tract from its resting state that is designed to ventilate the lungs to a state where ventilation is temporarily suspended to allow pharyngeal contraction to push the food bolus into the esophagus. Initiation of the swallowing reflex causes relaxation of the upper and lower esophageal sphincters, converting the pharynx, esophagus, and stomach into an unobstructed passage. This sphincter relaxation is coordinated with a peristaltic contraction that propagates the food bolus from the pharynx into the stomach. In normal swallowing, a single primary peristaltic wave causes the bolus of food to traverse the full length of the esophagus. Passage of the food bolus across both sphincters causes the sphincters to regain their tone and prevents backflow of food.

The presence of food in the esophagus stimulates peristalsis. If food remains in the body of the esophagus during swallowing, it stimulates a secondary peristaltic wave that pushes the food into the stomach. When reflux occurs and gastric contents regurgitate into the esophagus, clearing of this material probably occurs through a secondary peristaltic wave.

The muscle and innervation of the pharynx and esophagus is also different. The pharyngeal wall is exclusively voluntary skeletal muscle innervated by the glossopharyngeal nerve. The esophagus behaves like involuntary muscle; its wall contains a mixture of skeletal and smooth muscle in a variable part of the upper third of the esophagus and becomes entirely smooth muscle in its lower two-thirds. The esophagus is primarily innervated by the vagus nerve.

There are pathologic processes that involve the upper esophageal sphincter that cause motility abnormalities. Theoretically, in patients with GERD, failure of the upper esophageal sphincter may permit refluxate that enters the esophagus to reach the pharynx. Regurgitated material in the pharynx may produce extraesophageal syndromes of GERD such as laryngitis, cough, asthma, dental caries, sinusitis, otitis media, and pulmonary fibrosis. While these are all of significance

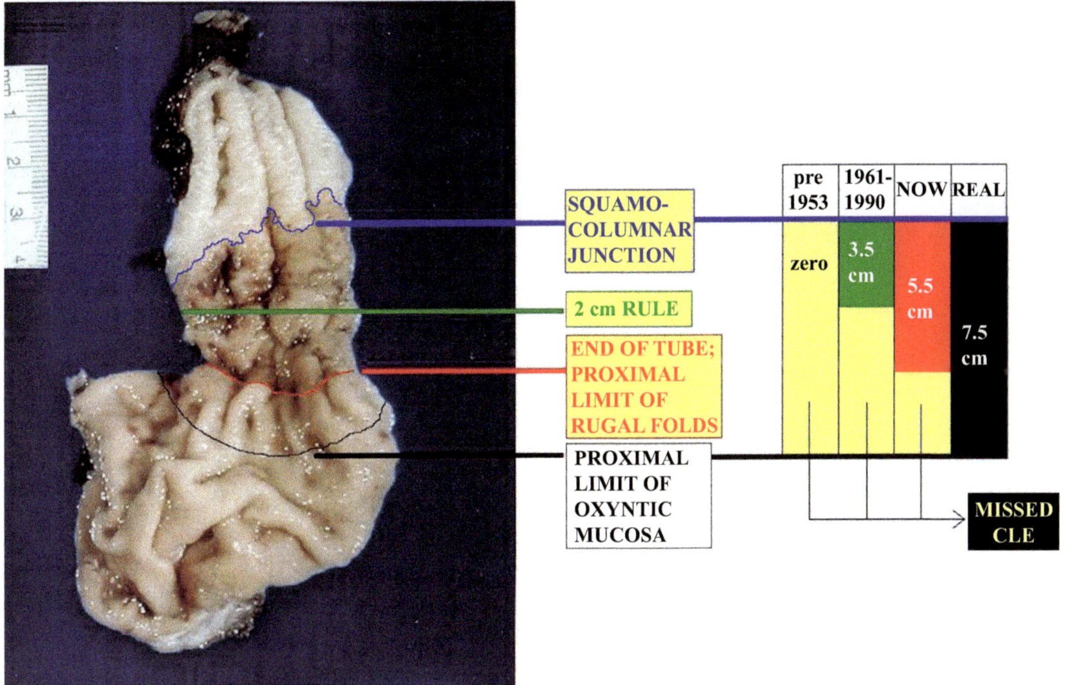

FIGURE 6.2 Historical progression of definition of the gastroesophageal junction (GEJ) in this patient with a 7.5 cm length of cardiac epithelium (with parietal and goblet cells) between the squamocolumnar junction (SCJ) and proximal limit of gastric oxyntic epithelium. In 1950, Barrett defined the GEJ as the SCJ (*blue line*), mistaking the entire segment as the stomach. In 1961, Hayward defined the presence of 2 cm of cardiac epithelium in the distal esophagus as normal (*green line*), ignoring short segment Barrett esophagus. Presently, the definition of the GEJ by the proximal limit of rugal folds (*red line*) misinterprets the dilated distal esophagus as proximal stomach. It is only when the true GEJ is accepted as the proximal limit of gastric oxyntic epithelium (*black line*) that the entire segment of columnar-lined esophagus is recognized.

in the patient with GERD, they are outside the scope of this book. I will not discuss the upper esophageal sphincter or extraesophageal GERD syndromes.

1.2 The Distal End of the Esophagus (Gastroesophageal Junction)

Defining the distal end of the esophagus in anatomic, histologic, and physiologic terms has been a comedy of errors for a century (Fig. 6.2). For many diseases occurring in the area around the GEJ, we remain confused whether the disease affects the esophagus or proximal stomach (Table 6.1).

One interesting method of evading the confusion has been the development of a "GEJ region or zone." In the minds of many people, this is the equivalent of a third organ inserted between the esophagus and stomach. Lagergren et al.,[1] in their historic paper on the relationship between symptomatic GERD and adenocarcinoma of the esophagus and "gastric cardia," as well as Siewert's widely used classification of adenocarcinomas of the GEJ,[5] used the concept that there is a GEJ zone (see below).

The GEJ is neither a zone nor a third organ. The GEJ is an imaginary line drawn between the end of the esophagus and the beginning of the stomach. It has no length. It gives rise to no disease. It only exists as a zone because of our inability to define whether a point within this region is the esophagus or stomach. It will disappear when we define the GEJ accurately (Table 6.1).

Regarding GERD, recognition of a pathologic lesion as esophageal or gastric should not be a problem because a disease produced by the exposure of esophageal squamous epithelium to gastric contents cannot produce disease in the stomach. For this reason, the association of "intestinal metaplasia of the gastric cardia" and "adenocarcinoma of the gastric cardia" with GERD should be problematic. Yet many expert gastroenterologists accept these entities, terminology, and contradictions without question.

A disease believed to be gastric that is limited to the most proximal 3 cm of the stomach in a person whose gastric body and antrum are normal should be regarded with suspicion. Inflammation limited to the cardia ("carditis") and intestinal metaplasia of the "gastric cardia" are very commonly isolated to this area and fall within this dubious category.[6] When the

TABLE 6.1 Definitions of the Gastroesophageal Junction (GEJ): Past, Present, and Future; Validated or Not

Period	GEJ Definition	Validation	Clinical Impact
Past			
<1950	End of the tube	None	CLE misinterpreted as heterotopic gastric epithelium in the esophagus
1950[7]	Distal limit of squamous epithelium, i.e., SCJ	None	CLE misinterpreted as tubular intrathoracic stomach; pathology in CLE mistaken as gastric pathology
1953 Allison and Johnstone[10]	Peritoneal reflection; only in resected specimens	+	Perfect interpretation of CLE; epithelium at GEJ was gastric oxyntic epithelium
1961 Hayward[12]	End of the tube	None	Short segment CLE designated as "normal"; histology in distal esophagus and proximal stomach identical
Present			
1987 McClave[14]	Proximal limit of gastric rugal folds	None	Dilated distal esophagus misinterpreted as proximal stomach
Future			
2006 Chandrasoma[20]	Proximal limit of gastric oxyntic epithelium	+	Entire extent of CLE recognized; dilated distal esophagus equated to LES damage and early pathology of GERD

CLE, columnar-lined esophagus; *GERD*, gastroesophageal reflux disease; *LES*, lower esophageal sphincter; *SCJ*, squamocolumnar junction.

pathologic anatomy of this region is understood, it will become apparent that they are both pathologic changes in the distal esophagus, not the stomach.

One obvious reason for confusion and error that must be considered is that we may be using an incorrect definition of the GEJ. This has precedent in history (Fig. 6.2). In 1950, Barrett[7] defined the GEJ as the squamocolumnar junction (SCJ). Patients who had long segments of columnar-lined esophagus (CLE) were, by this definition, deemed to have a long segment of tubular intrathoracic stomach associated with a congenitally short esophagus.

When there is a historical precedent for a >10 cm error in defining the GEJ in a 25-cm-long esophagus, we should be cautious. Errors of this magnitude indicate a difficult problem and difficult problems tend to result in repeated errors.

At present, two criteria are used to define the GEJ, irrespective of whether the person has a visible CLE or not (Fig. 6.3):

1. Endoscopically, the GEJ is defined as the proximal limit of "gastric" rugal folds. This takes precedence over the point of flaring of the tubular esophagus, particularly when there is a hiatal hernia. In persons without visible CLE, the proximal limit of rugal folds is at the same level as the horizontal SCJ (Fig. 6.4).
2. At gross pathologic examination of esophagectomy specimens, a horizontal line across the point at which the tubular esophagus flares into the saccular stomach is used to define the GEJ (Fig. 6.3).[8] When there is no CLE in the tubular esophagus, this is concordant with the horizontal SCJ and the proximal limit of rugal folds (Fig. 6.5).

In a person without visible CLE, therefore, the end of the tubular esophagus, proximal limit of rugal folds, and the SCJ are in the same position. It is easy to believe that this line is the true GEJ. This is indeed the belief of virtually every endoscopist and pathologist who is faced with this situation.

The obvious, however, is not necessarily true. There is no evidence whatsoever that any of these three criteria, i.e., the proximal limit or rugal folds, end of the tubular esophagus, or the SCJ, are the true GEJ. In fact, in a person with GERD, all these three elements have been displaced cephalad. In the patient with a damaged LES, the true GEJ, which is the proximal limit of gastric oxyntic epithelium, is more distal to all these criteria that presently define the GEJ at endoscopy and gross examination.

The interpretation that this line is the true GEJ is accurate only in a very rare (less than 0.1% of the population) person without any cardiac epithelium distal to the SCJ (Figs. 6.6A and 6.7A). This is a person without any LES damage. The endoscopic and gross definition of the position of the GEJ is therefore accurate in only 0.1% of the population; that is, it is almost always incorrect.

In a person with cardiac epithelium distal to the endoscopic GEJ, the presence of cardiac epithelium is evidence of columnar metaplasia of esophageal squamous epithelium as a result of exposure to gastric contents. The area lined by cardiac

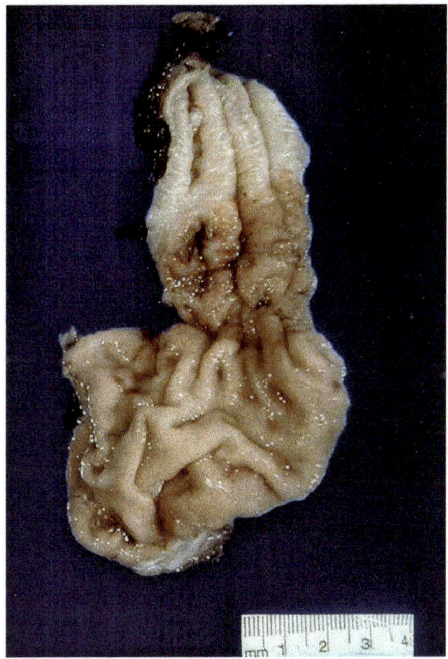

FIGURE 6.3 Esophagogastrectomy specimen in a patient with adenocarcinoma arising in Barrett esophagus. The tumor is seen as an ulcer distal to the squamocolumnar junction. The end of the tubular esophagus and proximal limit of rugal folds (the two presently used definitions of the gastroesophageal junction) are at the same level.

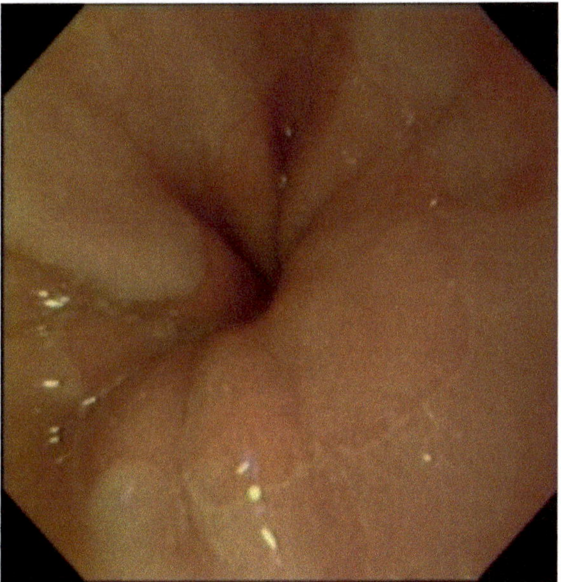

FIGURE 6.4 The normal squamocolumnar junction (SCJ) at endoscopy. The rugal folds extend to the SCJ, which is seen as a horizontal line between the squamous epithelium and columnar epithelium distally.

epithelium is equal to LES damage and represents the dilated distal esophagus. The true GEJ is more distal, separated from the endoscopic GEJ by the variable length of cardiac epithelium between the SCJ and the proximal limit of gastric oxyntic epithelium (Figs. 6.6B and 6.7B). This is the dilated distal esophagus. This can vary from >0 to 35 mm; 35 mm is the length of the undamaged abdominal LES.

This is the error made at endoscopy and gross examination by every gastroenterologist and pathologist. The dilated distal esophagus, being distal to the endoscopic definition of the GEJ, is mistaken as proximal stomach (Fig. 6.7B). It is not subject to biopsy or any kind of study. The only time it is taken seriously is in the person who has a visible lesion in that area. When the visible lesion shows pathology, it is mistakenly designated as pathology in the gastric cardia, e.g., adenocarcinoma of the "gastric cardia" rather than adenocarcinoma of the dilated distal esophagus (Fig. 6.1).

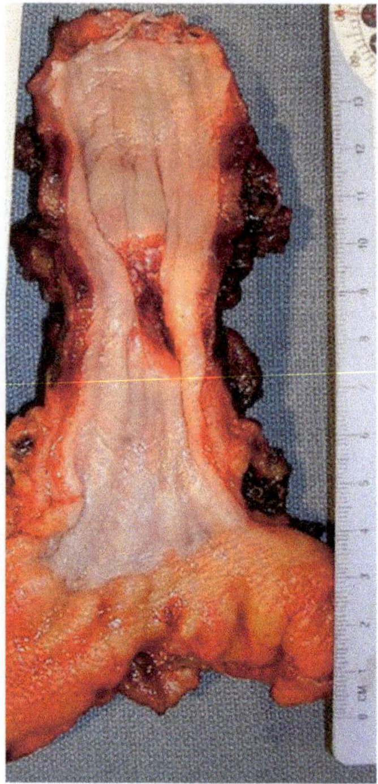

FIGURE 6.5 Esophagectomy specimen showing an ulcerated benign stricture in the midesophagus. The entire tubular esophagus is lined by squamous epithelium. At the point of flaring of the esophagus, the squamocolumnar junction (SCJ) appears as a horizontal line. The columnar epithelium distal to the SCJ shows ill-defined rugal folds.

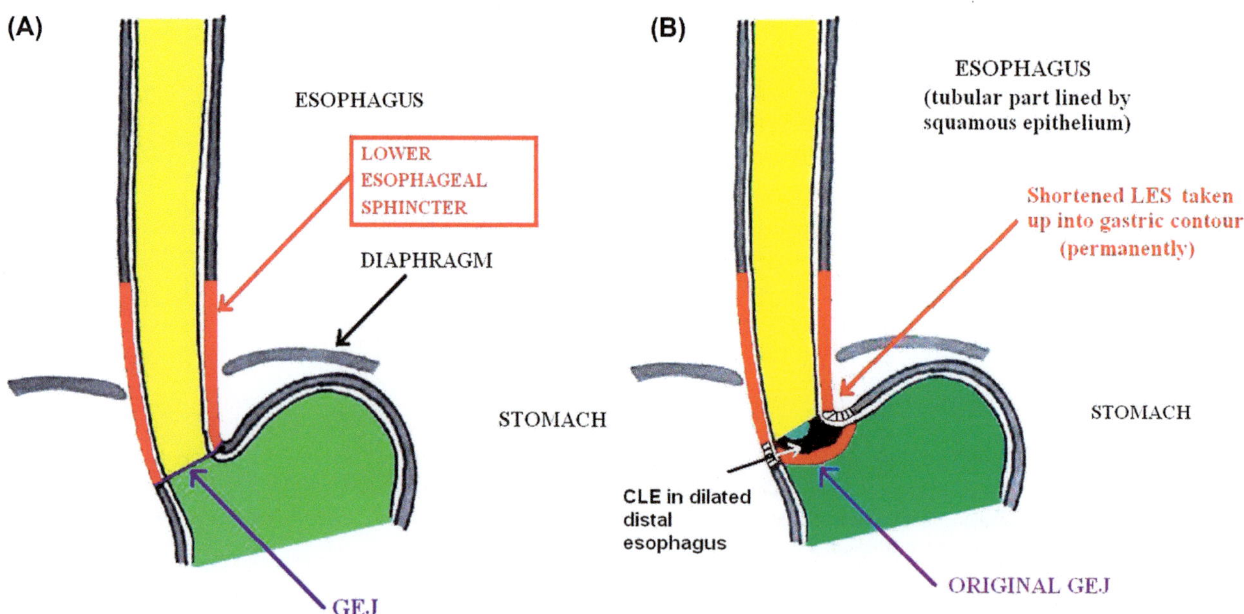

FIGURE 6.6 (A) The normal state. The entire esophagus with an intact lower esophageal sphincter (LES) (*red muscle wall*) is tubular and lined by squamous epithelium (yellow). This transitions at the point of flaring of the tube to gastric oxyntic epithelium (green) with rugal folds. (B) The abnormal state of early abdominal LES damage. The tubular esophagus and LES (*red wall*) have shortened by the extent of the damaged abdominal segment of the LES (*black-striped wall*). Cardiac epithelium (black; with goblet cells—blue; with parietal cells—red) has replaced the distal esophageal squamous epithelium. The part of the distal esophagus that has lost LES tone is dilated, has developed rugal folds, and has the contour of the stomach. The endoscopic gastroesophageal junction (GEJ) has moved cephalad by the extent of abdominal LES damage. The true GEJ is the proximal limit of gastric oxyntic epithelium (green). This remains unchanged.

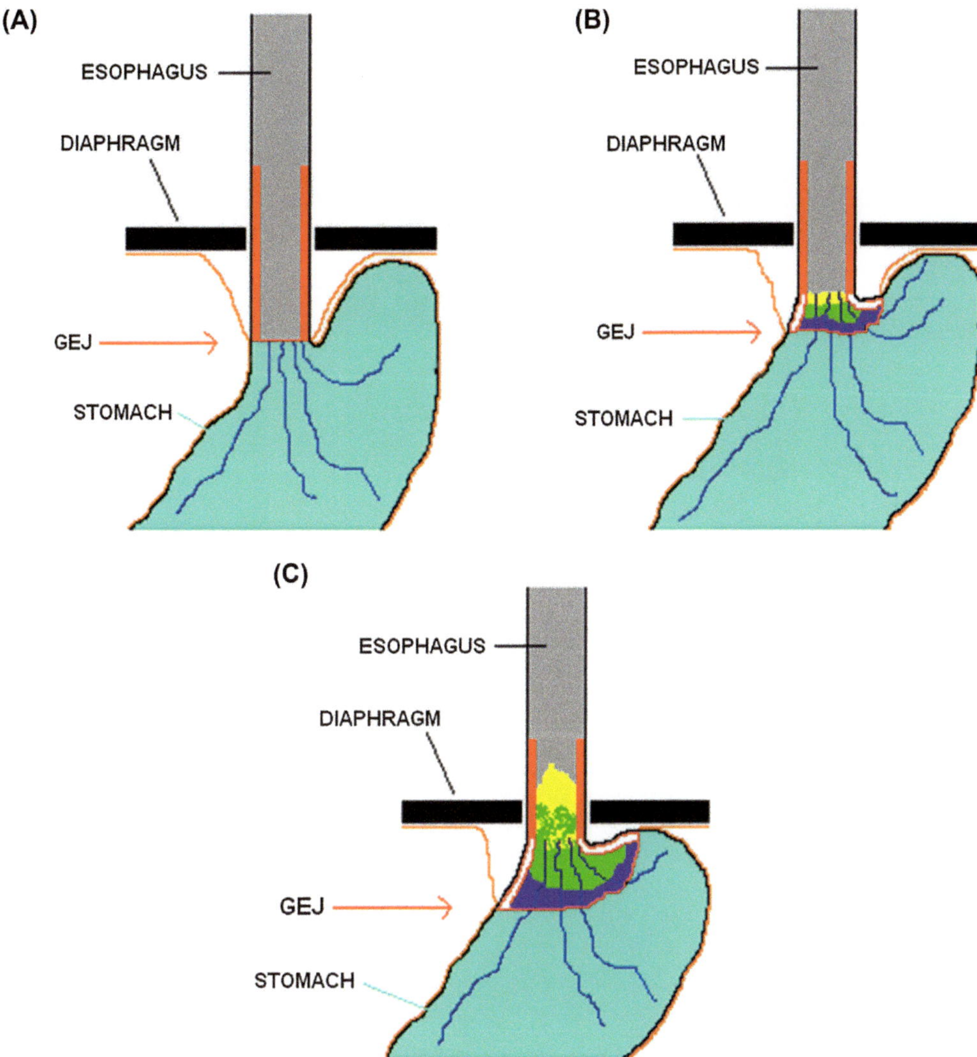

FIGURE 6.7 Diagrammatic representation of: (A) The normal state without cardiac epithelium between the esophageal squamous epithelium (gray) and gastric oxyntic epithelium (blue). The lower esophageal sphincter (LES) (*red line in wall*) is intact. (B) Early gastroesophageal reflux disease with a dilated distal esophagus (yellow-green-purple) interposed between the cephalad-migrated squamocolumnar junction and proximal limit of gastric oxyntic epithelium. Note that the residual abdominal LES and tubular abdominal esophagus have shortened. (C) Visible columnar-lined esophagus involving the thoracic esophagus. Note that the dilated distal esophagus has increased in length and the LES shortened. The length of the dilated distal esophagus is equal to LES damage. *GEJ*, gastroesophageal junction.

The error (i.e., the amount of separation of the endoscopic GEJ from the true GEJ) is small (less than 15 mm) in persons without GERD. It increases progressively in persons with increasing severity of GERD (15–25 mm). It is greatest in the person with visible CLE (25–35 mm) (Fig. 6.7C).

The endoscopist is staring at the pathology of early disease where the abdominal LES has undergone variable damage and believes he/she is looking at the proximal stomach. It is an error made by everyone in the past century.

A visible CLE is seen as a separation of the proximal limit of rugal folds from the frequently serrated SCJ (Z-line) by a variable length of flat, salmon-colored columnar epithelium. This can be a short segment of visible CLE (Figs. 6.8 and 6.9) or a long segment (Figs. 6.10 and 6.11). In such a patient, the endoscopic GEJ is maximally separated from the true GEJ (the proximal limit of gastric oxyntic epithelium). The reason for this is that visible CLE is associated with severe GERD, which is the result of severe (>25 mm) abdominal LES damage, which in turn is associated with the longest dilated distal esophagus (Fig. 6.7C).

In many people who have resections, the anatomy of the region is distorted (Fig. 6.12). The rugal folds are not well defined and often distal to the point of flaring of the tubular esophagus. This and the frequent presence of ulceration, hiatal

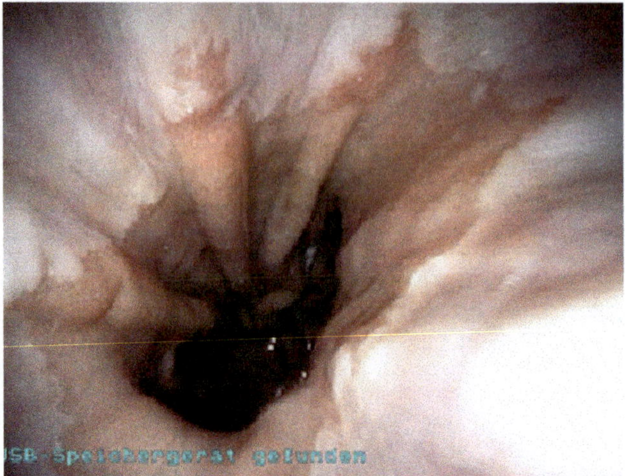

FIGURE 6.8 Antegrade endoscopic view of a 1 cm length of visible columnar-lined esophagus. This appears as a flat, salmon pink epithelium between the proximal limit of rugal folds and the serrated squamocolumnar junction. Biopsy showed cardiac epithelium without intestinal metaplasia. *Photograph courtesy of Dr. Martin Riegler, Reflux Medical, Vienna, Austria.*

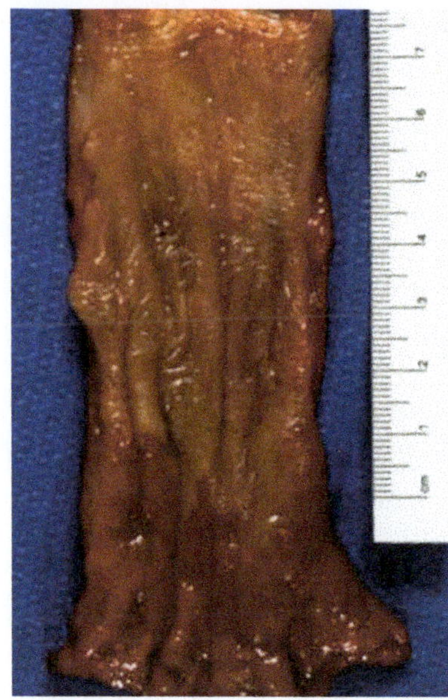

FIGURE 6.9 Resected specimen showing a short segment of flat columnar epithelium in the thoracic esophagus between the irregular squamocolumnar junction and the proximal limit of rugal folds.

hernia, tumors, and strictures make it impossible in such cases to define the GEJ accurately. In such cases, any attempt to define pathology as esophageal or gastric using the above standard definitions of the GEJ is futile. Only histologic definition of the junction between cardiac epithelium (with and without parietal and/or goblet cells) and gastric oxyntic epithelium provides clarity.

The confusion resulting from the use of these definitions of the GEJ should give us pause. When one considers that both these definitions of the GEJ are not evidence-based but simply the opinion of experts, the pause should be prolonged because similar experts in the past have made serious errors.

1.3 Lack of a Histologic Definition of the Gastroesophageal Junction at the Present Time

It is important to recognize that there is presently no accepted histologic definition of the GEJ. The only histologic definition of the GEJ that ever existed was that of Barrett who, in 1950,[7] equated the GEJ with the SCJ (see below). His rationale was that the esophagus should be defined by its epithelium and only squamous epithelium lined the esophagus.

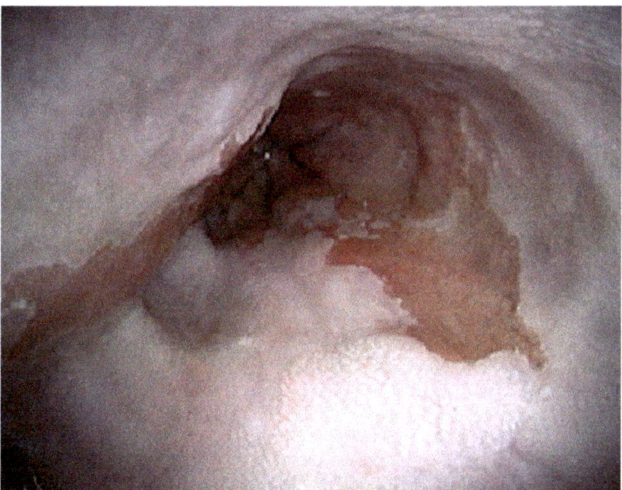

FIGURE 6.10 Antegrade endoscopic view of a 10cm length of visible columnar-lined esophagus. This appears as a flat, salmon pink epithelium between the proximal limit of rugal folds and the serrated squamocolumnar junction. Biopsy showed cardiac epithelium with intestinal metaplasia. *Photograph courtesy of Dr. Martin Riegler, Reflux Medical, Vienna, Austria.*

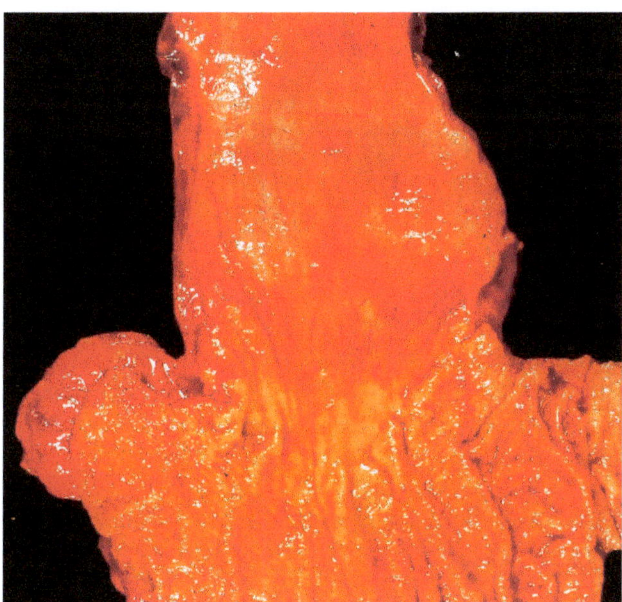

FIGURE 6.11 Resected specimen showing a long segment of flat columnar epithelium in the thoracic esophagus above the rugal folds that are located at the point of flaring of the tubular esophagus. The squamocolumnar junction is not seen in this photograph.

Barrett appropriately changed his mind in 1957[9] as a result of the powerful evidence from Allison and Johnstone in 1953[10] that what he had called a tubular intrathoracic stomach was actually a CLE. This change of mind removed the only histologic definition of the GEJ that has existed.

Barrett was not incorrect in his idea that the esophagus and stomach should be defined by their epithelia. He just chose the wrong epithelium (squamous) to define the esophagus. This is because all columnar epithelium in the esophagus was believed to be heterotopic gastric epithelium at that time. If he had known that GERD resulted in columnar metaplasia, he would likely have correctly chosen gastric oxyntic epithelium to define the stomach.

The correct histologic definition is the complete opposite of what Barrett proposed. The distal limit of squamous epithelium is not the end of the esophagus, i.e., the GEJ. It is clear that everything distal to it is not the stomach, as Barrett initially suggested in 1950. The region distal to it from the SCJ to the proximal limit of gastric oxyntic epithelium that is lined by cardiac epithelium (with and without parietal and/or goblet cells) is metaplastic esophageal columnar epithelium (Figs. 6.2 and 6.7).

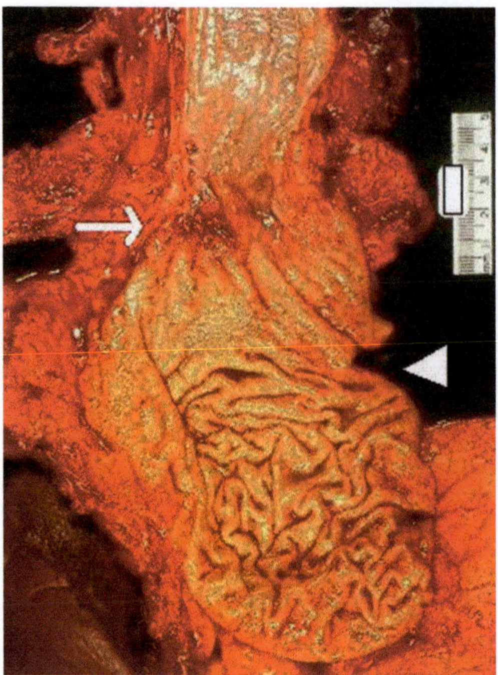

FIGURE 6.12 Markedly distorted anatomy of the distal esophagus. There is an ulcerated carcinoma immediately distal to the squamocolumnar junction (SCJ) at the point of flaring of the tubular esophagus (*arrow*). There is a dilated segment of relatively flat columnar epithelium from the SCJ to the area that shows defined rugal folds (*arrowhead*). The true gastroesophageal junction (GEJ) (proximal limit of gastric oxyntic epithelium) was distal to the proximal limit of rugal folds. The region between the SCJ and the true GEJ is the dilated distal esophagus.

The correct histologic interpretation is that the proximal limit of gastric oxyntic epithelium marks the beginning of the stomach, i.e., the GEJ. This point remains unchanged in GERD because GERD does not affect the stomach. *Everything proximal to the proximal limit of gastric oxyntic epithelium is the esophagus.*

Esophageal epithelium consists of only squamous epithelium in the normal state. In patients in whom the squamous epithelium has undergone columnar metaplasia, the metaplastic cardiac epithelium begins at the true GEJ (the proximal limit of gastric oxyntic epithelium) and extends cephalad. Metaplastic cardiac epithelium (with and without parietal and/or goblet cells) first forms the dilated distal esophagus distal to the endoscopic GEJ and then the visible CLE proximal to the endoscopic GEJ (Fig. 6.7).

The proximal limit of gastric oxyntic epithelium remains in its original position at the true GEJ. It is the most proximal point of the foregut that is not affected at all by exposure to gastric contents.

2. CRITERIA AVAILABLE TO DEFINE THE GASTROESOPHAGEAL JUNCTION

There are many criteria that can potentially define the GEJ. The important thing is to recognize that the definition of the GEJ cannot change in the normal and abnormal state or with different perspectives. If, for example, the proximal limit of rugal folds or the SCJ correctly defines the GEJ, this point must remain constant in normal patients as well as those with severe GERD.

The GEJ is presently defined in different ways by different physicians based on their perspective. Whatever perspective is used, the definition of the GEJ must remain constant. If the definition cannot be applied with certain modalities of observation, the correct reaction is to say that the GEJ cannot be defined with that modality.

For example, if the proximal limit of rugal folds correctly defines the GEJ, the surgeon cannot define the GEJ at surgery unless intraoperative endoscopy is performed or there is an evidence base that has established that the endoscopic GEJ is concordant with some point that is visible externally.

Similarly, if the peritoneal reflection correctly identifies the GEJ, it cannot be defined at endoscopy unless there is a reliable endoscopic landmark that is concordant. If both are used to define the GEJ, they must be at the same location in healthy persons and patients with GERD.

Pathologists are largely blind as to the anatomic location when they interpret biopsies and are presently dependent on the endoscopist for information regarding location of the biopsy, particularly around the GEJ.[11] When they evaluate esophagectomy specimens, pathologists have to deal with an organ freed from its usual tethers and distorted by pathology in the region of the GEJ.

Every definition that has ever been suggested accurately defines the GEJ in the normal state (Fig. 6.7A). These include the following:

1. The SCJ,[7] which can be defined endoscopically and histologically with high precision.
2. The end of the tube where it flares into the saccular stomach,[12] presently used by pathologists and at endoscopy.
3. The angle of His.[13]
4. The proximal limit of gastric rugal folds,[14] which is the presently used definition at endoscopy.[15] This is a useful landmark at endoscopy because it can be identified with reasonable precision. However, it has not been validated as being the true GEJ.
5. The distal limit of the palisading longitudinal blood vessels. This is used mainly by Japanese endoscopists.[16] It is rarely used in the United States.
6. The distal limit of the LES at manometry.
7. The peritoneal reflection, which is sometimes used by surgeons,[10] but lacks value because it cannot be identified at endoscopy.
8. The proximal limit of gastric oxyntic mucosa.[17,18] This is the definition that I have suggested. I will develop this histologic definition in this chapter. It is the correct definition of the GEJ.

When the abdominal LES undergoes damage as a result of cardiac metaplasia of distal esophageal squamous epithelium, many of these landmarks become altered and no longer represent the true GEJ (Fig. 6.7B). Understanding which of these landmarks remain constant markers of the GEJ when there is abdominal LES damage is critical in avoiding error. We will show that all of these criteria change in patients with LES damage, except the location of the proximal limit of gastric oxyntic mucosa (Fig. 6.7B) and, probably, the peritoneal reflection.

It follows from this that the only modality that is capable of accurately defining the true GEJ from the luminal aspect is histology. The reason for this is that the only way metaplastic cardiac epithelium can be distinguished from gastric oxyntic epithelium is histology.

All definitions used for the GEJ at endoscopy, radiology, and gross examination of the mucosa are incorrect in patients with GERD. None of these modalities can distinguish gastric oxyntic mucosa from other types of columnar epithelia that occur in the distal esophagus. Without histology, every physician who deals with GERD is blind to the true GEJ. Incredibly, they make decisions and develop guidelines for management of GERD without accurate knowledge of where the esophagus ends.

3. HISTORICAL DEFINITIONS OF THE GASTROESOPHAGEAL JUNCTION

The instinctive definition of the distal end of the esophagus is that it is the end of the tubal structure where it flares into the saccular stomach. This was probably the unwritten and assumed definition of the GEJ from the time when the question arose. It is still believed by many people to be true if there is no hiatal hernia (Table 6.1).

In the middle of the 20th century, Norman Barrett attempted to develop a formal histologic definition of the esophagus, GEJ, and stomach. The reason for Barrett's desire to define the esophagus was the recognition that some patients had a columnar epithelial lining in a variable part of the distal tube. The finding was considered important at the time because it was associated with peptic ulcers that resembled gastric and duodenal ulcers.

In his first classic treatise in 1950, Norman Barrett[7] defined the esophagus as follows: "The esophagus is that part of the foregut distal to the cricopharyngeal sphincter that is lined by squamous epithelium." This was a precise definition that had essentially no interobserver variation. The distal limit of squamous epithelium where it transitioned to columnar epithelium (SCJ or Z-line) was clearly identifiable at endoscopy (Fig. 6.4), gross pathologic examination of a resected specimen (Figs. 6.3 and 6.5), and histology.

According to this definition of the GEJ, the SCJ was the end of the esophagus irrespective of anything else. The tubular structure lined by columnar epithelium distal to the SCJ was, according to Barrett, a tubular intrathoracic stomach resulting from a short esophagus.

Barrett's reasoning for defining the GEJ as the SCJ was: "the emphasis is on the nature of the secretions produced into the lumen, rather than upon the musculature or the serous covering of the walls. A piece of gullet lined by gastric mucosa, whatever the external or superficial appearances may be, functions as stomach – secreting hydrochloric acid and digestive juice – and is heir to the ailments which afflict the stomach."

Barrett could not have been further from the truth. While this is obvious today because we know that Barrett's intrathoracic tubular stomach is CLE, we make the same mistake again, albeit less obvious. Today, we mistake the metaplastic cardiac epithelium, that is, the dilated distal esophagus as proximal stomach. The error is more subtle, but identical nevertheless.

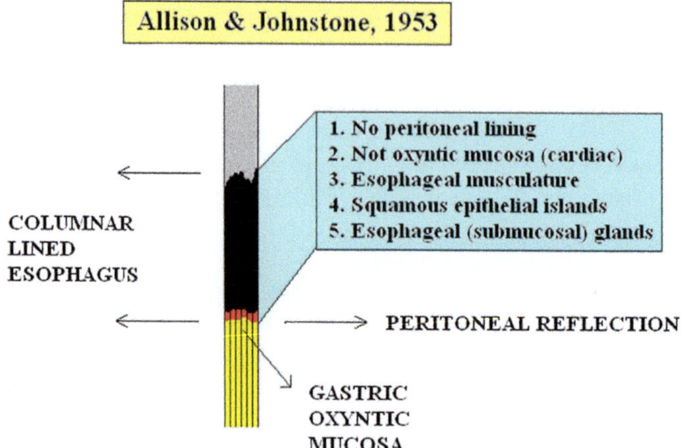

FIGURE 6.13 The evidence that Allison and Johnstone used in 1953 to prove that the tubular structure distal to the squamocolumnar junction that Barrett had called a tubular intrathoracic stomach was the esophagus. Squamous epithelium = gray; cardiac epithelium = black; oxyntocardiac epithelium = red; gastric oxyntic epithelium = yellow.

In 1953, Allison and Johnstone[10] published what was probably the most accurate article ever written on the subject. This provided proof that what Barrett called a tubular lined stomach was actually the esophagus. The evidence presented by Allison and Johnstone that proved that the columnar-lined tubal structure in the thorax was the esophagus rather than a tubular stomach were the following (Fig. 6.13): (1) It was not lined by peritoneum, i.e., it was above the peritoneal reflection, and therefore could not be the stomach. (2) It had islands of squamous epithelium within the columnar-lined segment. (3) It had submucosal glands, which were normally seen in the esophagus but not in the stomach. (4) The muscle wall had two muscle layers characteristic of the esophageal wall rather than the three-layered gastric muscle wall. Remarkably, what is considered by everyone as the "gastric cardia" today has islands of squamous epithelium[19] and submucosal glands.[20,21]

The evidence presented by Allison and Johnstone was powerful. In 1957, in his second classical treatise on the subject, Barrett[9] agreed with Allison and Johnstone that the SCJ was not the GEJ and that what he had designated the tubular stomach in 1950 was actually a CLE.

It is a lesson to all of today's physicians that Barrett, despite his preeminent position in the field, had the grace, openness, and courage to admit his mistake when faced with Allison and Johnstone's evidence.

Between 1960 and 1970, CLE was recognized to be the result of GERD-induced metaplasia of the squamous epithelium of the esophagus. Attempts were made to define the limits of CLE. The proximal limit presented no problem; it was the SCJ. The distal limit, which was the true GEJ, created problems in definition.

In 1961, Hayward[12] formally suggested the first practical method of defining the GEJ at endoscopy. Going back to the instinctive definition, he defined the GEJ as the point at which the tubal esophagus flared into the saccular stomach. "The oesophagus is defined as the tube conducting food from the throat to the stomach and all of this tube is regarded as oesophagus, irrespective of its lining."

Extending his definition of the GEJ to the histology he observed in his patients, Hayward stated that the distal 2 cm of the tubal esophagus was normally lined by cardiac epithelium and that this epithelium extended into the proximal stomach. His illustration suggests that 2–3 cm of stomach could be lined by cardiac epithelium (Fig. 6.14). Because the cardiac epithelium on either side of Hayward's GEJ was identical, there could be no histologic definition of the GEJ.[12]

4. PRESENT DEFINITION OF THE GASTROESOPHAGEAL JUNCTION

Hayward's definition of the GEJ as the end of the tube is still used by gastroenterologists when the anatomy is not distorted by a sliding hiatal hernia. It is also recommended for pathologists in the dissection of resected specimens by the Association of Directors of Anatomic and Surgical Pathology.[8] A line drawn across the point at which the tube flares into the saccular stomach is the GEJ in an esophagogastrectomy specimen.

There were problems, however, with using the end of the tube as the endoscopic marker of the GEJ. The frequent presence of a sliding hiatal hernia in patients with CLE seriously limited the ability to accurately define the end of the tube. This resulted in many patients not having a clear definition of the GEJ, making it impossible to apply the 2-cm rule of Hayward that CLE could only be diagnosed if it was >2 cm above the GEJ.

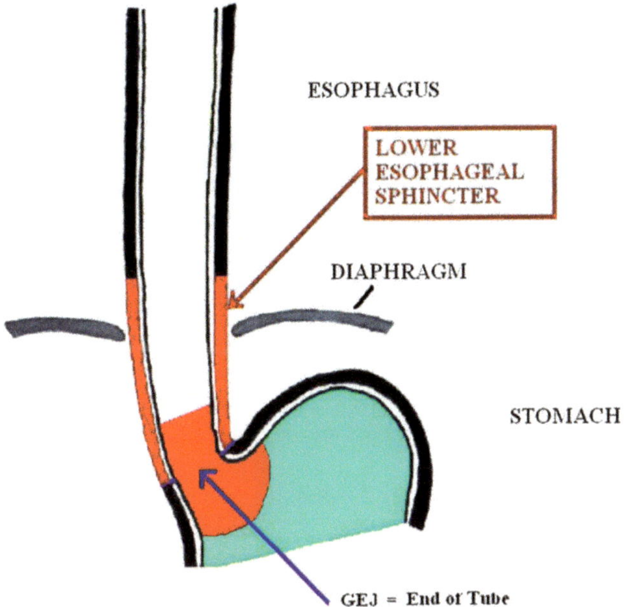

FIGURE 6.14 Diagrammatic representation of the anatomy according to Hayward in 1961. The gastroesophageal junction (GEJ) is defined as the end of the tube. Cardiac epithelium (red) was, according to Hayward, normally present in the distal 2cm of the esophagus and proximal 2–3cm of the stomach.

In 1987, McClave et al.[14] reported that the proximal limit of "gastric" rugal folds was an endoscopic landmark that could be identified with some precision (Figs. 6.3, 6.4, 6.8, and 6.9).

It is clear in the introduction of the paper that CLE "develops as a chronic sequel of gastroesophageal reflux." According to the authors: "Strictly defined, columnar lined esophagus is a columnar lined intrathoracic structure that is otherwise anatomically and functionally the esophagus… (without) set guidelines or specific criteria for making the diagnosis of early columnar lined esophagus by endoscopic criteria alone."

They go on to state: "Documentation that biopsies are obtained precisely from the esophagus at endoscopy is difficult because of alteration or variation in location of the usual endoscopic anatomic landmarks, such as the diaphragmatic hiatus and the squamocolumnar junction."

At he time this paper was published in 1987, it was accepted that the distal 2–3cm of the esophagus was normally lined by gastric cardiac epithelium. A biopsy from the distal esophagus that had columnar epithelium therefore did not mean the patient had CLE. The biopsy had to be >2 (or 3)cm above the GEJ. This was not a problem in the patient without a hiatal hernia, because the GEJ was marked by the end of the tube. However, distortion of anatomy when a sliding hiatal hernia was present left the endoscopist with no reliable marker for the GEJ. As such, the 2-cm rule could not be applied. The authors were examining the possibility of using the landmark of the proximal limit of "gastric" mucosal folds to define the GEJ.

The study is designed to answer two questions: "(1) whether the proximal margins of the gastric folds can serve as a readily identified, fixed anatomic landmark to aid in biopsy site selection for diagnosis of early stages of CLE and (2) whether the relationship between the gastric folds and the squamocolumnar mucosal junction could be designated as a specific diagnostic criterion for CLE."

The study consisted of 18 patients with gastroesophageal reflux and a presumptive diagnosis of CLE made elsewhere during earlier endoscopic evaluation. Four patients without clinical evidence of esophageal disease acted as controls. At endoscopy, measurements were made from the incisor teeth to the hiatus, the gastric folds, and the squamocolumnar mucosal junction. Hiatal hernia was diagnosed in patients if the proximal margin of the gastric folds was more than 2cm above the diaphragmatic hiatus. Biopsies were taken from the proximal margin of the gastric folds present in the hiatal hernia pouch, from any presumed segment of CLE and from the most distal extent of the squamous mucosa.

The four control patients had their SCJ within 2cm (within 1cm in 3/4) above the proximal limit of the gastric folds. This was within the normal state as defined by Hayward. It is interesting that all four control patients had what now would be called short segment CLE. Three controls also had a hiatal hernia and one had mild esophagitis. Biopsies of the proximal margin of gastric folds showed fundic or junctional-type mucosa in all controls. This terminology is following Paull et al.[22]; junctional type is cardiac epithelium and fundic-type is oxyntocardiac epithelium.

All 18 study patients had proximal displacement of the SCJ at endoscopy. This was circumferential in 7/18 and characterized by fingerlike extensions of orange-red mucosa in 11/18, the latter producing a markedly irregular SCJ. The distance between the proximal margin of gastric folds and the SCJ was >2 cm in all patients and >3 cm in 12/18. The proximal margin of gastric folds was seen within a hiatal hernia pouch in all patients. Biopsy showed differences in measurement compared with the endoscopy, probably related to difficulty in taking biopsies from the exact measured location. Columnar epithelium was present >2 cm from the proximal margin of gastric folds in 17/18 patients and >3 cm in 16/18. One patient had squamous and not columnar epithelium in the biopsies at >2 cm from the proximal margin of the folds. Although endoscopy had indicated columnar epithelium >2 cm, biopsy did not confirm this. The possibility of sampling error was considered as a cause for the false-negative biopsy.

The epithelial types in the 17 patients with columnar epithelium >2 cm above the top of the gastric folds were specialized columnar epithelium (=intestinal metaplasia) in 14, junctional (cardiac) in 5, and fundic-type (oxyntocardiac) in 3. The number of biopsies taken was not stated, but the findings suggest a mosaic of the three epithelial types with intestinal metaplasia being common. Intestinal metaplasia was the most proximal type of mucosa in all but one of the 14 patients. Biopsies taken from the proximal margin of gastric folds showed fundic-type (oxyntocardiac) or junctional (cardiac) type epithelium in all patients.

In their discussion, the authors emphasize the great difficulty associated with accurate anatomic identification of the esophagogastric muscular junction at endoscopy in a disease setting that alters the usual landmarks. They describe this problem well: "Endoscopic landmarks that have been employed for identification of the approximate level of the esophagogastric muscular junction are the diaphragmatic hiatus, a Schatzki or B ring, and the squamocolumnar mucosal junction. A Schatzki or B ring is never present in patients with columnar lined esophagus. The squamocolumnar mucosal junction is not reliable in designating the esophagogastric junction in columnar lined esophagus because it can be obliterated by inflammation (esophagitis) and because it migrates orad as part of the pathophysiologic process. The diaphragmatic hiatus is not a reliable reference point for proximal measurement of biopsy sites or squamocolumnar mucosal junction location even in normal patients because of variable movement that occurs with a sliding hiatal hernia." They state that use of any of these criteria results in potential error.

The authors suggest that the proximal margin of gastric folds is a reliable and fixed marker and provides a point of reference for endoscopic diagnosis of CLE. They point to the data in the study where 17/18 patients with endoscopic columnar epithelium >2 cm from the proximal margin of rugal folds had columnar epithelium on biopsy, confirming the diagnosis of CLE (with the 2-cm rule). They suggested a new definition of CLE, which "should be considered at endoscopy when either the squamocolumnar mucosal junction is located or columnar epithelium is obtained by biopsy at a site >2 cm above the proximal margin of the gastric folds located within a hiatal hernia pouch."

This definition of CLE still retained the Hayward rule that columnar epithelium normally lined the distal 2 cm of the tubal esophagus. It should be noted that the four control patients in the study by McClave et al had columnar lining <2 cm above the proximal margin of gastric folds with biopsies showing fundic or junctional epithelium at the gastric folds. This was, to these authors, the normal state. The recommendation then was that patients with <2 cm columnar epithelium in the distal esophagus should not undergo biopsy. With this new definition of the GEJ, the recommendation became better defined as there was a reliable landmark from which to make the endoscopic measurement.

McClave's definition of the GEJ by the proximal limit of gastric rugal folds[14] rapidly gained acceptance. This is easy to understand. To an endoscopist who had no ability to identify the distal limit of a CLE in a patient with a sliding hiatal hernia, the new definition must have been a huge improvement. It was like getting a rudder to a rudderless ship. The requirement of proof and evidence in this setting is not likely to have been stringent.

Careful critical review of the paper by McClave et al.[14] is essential to see if there is an evidence base that justifies the use of the proximal limit of rugal folds as the GEJ. That it is an endoscopic landmark that is fixed and reliable is what the study by McClave et al set out to evaluate. The paper provides excellent evidence that the proximal limit of the gastric folds is a reliable and constant marker that can be used for the purpose of measuring the cephalad displacement of the SCJ. Whether 0 (at present), 2 (Hayward[12]), or 3 (Spechler et al.[23]) cm is used to define the extent of cephalad displacement that we consider abnormal, the proximal margin of rugal folds is an excellent marker to define the distal limit of the measurement.

However, its accuracy in defining the true GEJ does not stand up to critical examination. To use the landmark to get one's bearings is no problem. However, to ascribe a significance of substantial consequence that is not accurate, i.e., the definition of the GEJ, to the landmark creates serious error of diagnosis and interpretation. If, as it turned out, the true GEJ was distal to the proximal limit of rugal folds, that part of the distal esophagus between the endoscopic GEJ as defined by McClave and the true GEJ would be misinterpreted as proximal stomach. This is the present serious error that is made.

In the abstract of McClave et al,[14] the following is stated: "The proximal margin of the gastric folds in a hiatal hernia pouch provide a fixed, reproducible, anatomic landmark at endoscopy, which designates the junction of the muscular wall of the esophagus and stomach and permits one to predict the expected normal location of the squamocolumnar mucosal junction."

In the discussion, the authors correctly point out that other endoscopic landmarks such as the SCJ, Schatzki ring, and diaphragmatic hiatus are not reliable in defining the esophagogastric muscular junction. They then state: "Thus, the proximal margin of the gastric mucosal folds provide the only reliable endoscopic landmark in columnar lined esophagus for designating the junction of the muscular wall of the esophagus and stomach."

This is an incredibly unjustified leap of faith. McClave et al. provide absolutely no data in their study that correlate the presence of the proximal margin of gastric folds to the GEJ which, then, was defined as the muscular junction of the esophagus and stomach.

As evidence that the proximal limit of gastric folds correlates with the muscular esophagogastric junction, the authors cite a previous study in their discussion: "Anatomic studies have shown previously that the proximal margin of the gastric folds closely approximates the esophagogastric mural junction or the muscular junction of the tube (esophagus) and sac (stomach)." This is the only citation they provide to justify their claim.

The citation refers to an excellent paper from the University of Washington in 1966 where the senior author was the highly regarded Dr. Nyhus, Professor of Surgery,[12] that I review in detail in Chapter 10. The most important thing about this paper relating to the present discussion of the GEJ is that it makes absolutely no mention of gastric mucosal folds and their relationship to the muscular junction of the esophagus and stomach. The question of mucosal folds is mentioned once in the introduction: "The mucosa and submucosa in the lower esophagus are thrown up into redundant folds and in some series, marked circular folds have been noted in this area." The only mucosal folds mentioned in this paper are esophageal folds, not gastric, and these were not correlated with anything. The paper by Bombeck et al was published in 1966, 21 years before McClave et al. assigned any significance to the proximal limit of gastric rugal folds.

There is absolutely no evidence that the proximal margin of the mucosal folds has any relationship to the true GEJ. This is just another endoscopic landmark like the SCJ and the point of flaring of the tubal esophagus into the saccular stomach. The rugal folds are as unreliable as a marker for the GEJ as all other endoscopic markers.

The use of the proximal margin of gastric folds is reasonable as an endoscopic landmark to make decisions. However, to ascribe to this landmark a significance that is wrong, i.e., that it accurately defines the GEJ, has created significant error in diagnosis since 1987 when the paper was published to the present time.

The first time the question of the definition of the GEJ was examined critically was 17 years later in the 2004 Chicago workshop of the American Gastroenterology Association.[24] This was a consensus meeting of 18 physicians (15 gastroenterologists, 2 surgeons, and 1 pathologist). The gastroenterologists were among the most highly regarded in the world, the two surgeons were John Hunter from Portland Oregon and Lars Lundell from Sweden, and the pathologist was John Goldblum from Cleveland. All participants had expertise in Barrett esophagus.

The group achieved consensus on 42 statements regarding Barrett esophagus. The nature of evidence for each question was graded I to V (I being best) and the level of support among the group was graded A to E (A being best). In the final analysis, the nature of evidence for the 42 questions (I being best) was graded I for 2 statements, II for 9 statements, III for 14 statements, IV for 7 statements, and V for 9 statements.

Statement 7 was: "The proximal margin of the gastric folds is a reliable endoscopic marker for the gastroesophageal junction." The nature of evidence supporting this was given a grade IV ("Opinions of respected authorities based on clinical experience, descriptive studies, or reports of expert committees"; Grades I–III were for statements based on some evidence).

Support for the 42 statements by members of the group was divided into five grades; grades A–C were "accept (completely, with some reservation, with major reservation)" and grades D and E were "reject (with reservation, completely)." The workshop was divided into subgroups that assessed the evidence and discussed this with the group before assigning a subgroup grade. Each member then voted his level of support on each statement anonymously. The latter is presented as a percentage of members voting for each category.

The subgroup support grade for statement 7 was C. (There is poor evidence to support the statement, but recommendations may be made on other grounds.) When the members voted anonymously for statement 7, 78% accepted with some reservation and 22% accepted with major reservation.

No one rejected this statement. Unanimous absence of rejection was noted for 16/42 statements. Of these, the grade for the subgroup support was A for 12 statements, B and C for 2 statements each. The only statement other than 7 that had no rejections among the members that received a subgroup support grade of C was statement 9: "The normal appearing and normally located squamocolumnar junction should not be biopsied." This also had a grade IV for nature of evidence.

In the discussion of statement 7: "All workshop members were in agreement that a reliable endoscopic determination of the GEJ is needed to make a diagnosis of Barrett esophagus and to record the length of columnar epithelium in the esophagus. A widely used endoscopic marker of the GEJ is the proximal margin of gastric rugal folds. Other markers of the GEJ include the level at which the tubular esophagus meets the wider, sac-like stomach. Given its widespread use by experts in the field, the group uniformly favored using the proximal margin of the gastric folds to identify the GEJ but recognized that there are scant data that validate it. No studies have compared the upper margin of the gastric folds with other markers, such as the level at which the tubular esophagus meets the wider sac-like stomach. However, the group agreed that this still represented the best landmark recognized at endoscopy."

In the current (2011) American Gastroenterology Association guidelines on management of Barrett esophagus,[25] the definition of the GEJ is similar: "The issue of what endoscopic landmark best identifies the GEJ (i.e., the level at which the esophagus ends and the stomach begins) is likely to remain controversial because currently available data do not support a universally accepted definition. A majority of published studies on Barrett's esophagus published over the past 20 years have used the proximal extent of the gastric folds as the landmark for the GEJ. In the absence of compelling data for the use of alternative markers, we advocate the continued use of this landmark."

I would agree that the proximal limit of gastric rugal folds is the *best* landmark compared to others available at endoscopy to define the GEJ. However, the best is simply not good enough if it associated with significant error. The important fact that emerges from this is that the true GEJ cannot be defined at endoscopy. To attempt the impossible is a guarantee of error.

There is no evidence whatsoever that the proximal limit of rugal folds accurately defines the GEJ. This should result in two things: (1) recognition that there is the possibility of error relating to the endoscopic definition of the GEJ, which has no evidence base, and (2) a desperate and active search to find an accurate definition of the GEJ because, as stated, this is essential to make a diagnosis of Barrett esophagus.

Unfortunately, neither the recognition of the possibility of error nor the desire to find an accurate definition of the GEJ exists in the endoscopy community. They continue to use the proximal limit of gastric folds to define the GEJ, use the data derived from this definition as if it was set in stone. The AGA declaration that "a majority of published studies on Barrett esophagus published over the past 20 years have used the proximal extent of the gastric folds as the landmark for the GEJ" should not be a reason for continuing to use this criterion. It should be a reason for calling into question all the data from published studies on Barrett esophagus in the past 20 years. When one looks at the confusion that surrounds this literature and the poor management outcomes for Barrett esophagus and adenocarcinoma in the past 20 years, critical examination is actually essential.

The problem inherent in all statements emanating from the gastroenterology world is their belief that the only definition that can be developed must be endoscopic. The endoscope is an instrument of relatively low optical resolution. It is like asking a pathologist to make a diagnosis of a neoplasm by gross examination without using the microscope. The endoscopist lacks the microscopic point of view.

The only way to accurately define the GEJ at endoscopy is to use histology. I will show below that the true GEJ is accurately defined histologically by the proximal limit of gastric oxyntic epithelium. The microscope, when combined with adequate samples obtained at endoscopy in a manner that is based on correct understanding of the anatomy and histology of this region in GERD, provides accurate definition of the GEJ at a cellular level, not at the optical resolution of the endoscope. Once the GEJ is defined accurately and with precision, we can set about making accurate diagnosis of GERD and CLE. Until then, the present chaos will reign.

To use histology in diagnosis is not a strange suggestion. In virtually all diseases of the gastrointestinal tract other than GERD, diagnosis is based on histologic criteria.

5. THE GASTROESOPHAGEAL JUNCTIONAL ZONE

One consequence of the inability to define the GEJ accurately is the existence of the concept of a "GEJ zone." This is an incorrect concept. The GEJ is an imaginary line drawn across the distal end of the esophagus and the proximal limit of the stomach. It is a constant line that does not change in health and disease.

The extent of the suggested gastresophageal zone varies in different studies. Lagergren et al., in 1999[1] established the relationship between GERD and adenocarcinoma. In this paper, the definition of adenocarcinoma of the esophagus was a tumor centered 2 cm and above the GEJ. Tumors centered 2 cm proximal and 3 cm distal to the GEJ (in essence, a GEJ zone) were classified as adenocarcinoma of the "gastric cardia."

The same GEJ zone concept is central to Siewert's well-accepted classification of junctional adenocarcinoma into three types[5]: Type I: Distal esophageal, centered above the GEJ; Type II: Junctional, centered in the GEJ; and Type III: Subcardiac, below the GEJ. Siewert's GEJ is determined by a combination of "endoscopic and radiologic criteria." These criteria are not further defined.

The GEJ zone has been used for classification of tumors in this region by the AJCC publications that are used for pathologic diagnosis of tumors. The AJCC definition of the GEJ zone is for tumors with an epicenter that is 5 cm on either side of the GEJ. There was a significant change in the classification of tumors in this zone from the sixth to seventh edition in 2011 (Fig. 6.12).[2,26]

In the AJCC sixth edition,[26] adenocarcinomas that had an epicenter that was within 5 cm of either side of the GEJ (defined as the end of the tube) were classified as gastric cardiac (proximal gastric) carcinoma. In the seventh edition,[2] this definition changed. The following characteristics require a tumor within this GEJ zone to be now classified as esophageal: (1) tumors with the epicenter in the distal esophagus that involve the GEJ and (2) tumors with the epicenter within the proximal 5 cm of the stomach if they extend upward to involve the GEJ or distal esophagus. Gastric adenocarcinoma is defined as tumors that (1) have the epicenter beyond 5 cm from the GEJ and (2) have their epicenter within 5 cm of the GEJ but do not extend upward to involve the GEJ.

This change in the seventh edition represents a movement toward the correct interpretation of the location of adenocarcinoma in this region.[27] It is known that many "proximal gastric" adenocarcinomas are associated with GERD, and their designation as esophageal in origin is likely accurate. The effect of this change is that many tumors previously designated gastric will now be esophageal. This can be expected to result in a sudden increase in the incidence of esophageal adenocarcinoma as this classification goes into effect.

The concept of a GEJ "zone" is really an excuse for an inability to define the GEJ accurately by endoscopic and radiologic criteria. It seemingly creates a third organ between the esophagus and the stomach. The confusion this creates is enormous. When we learn to define the GEJ accurately at a histologic level, the GEJ zone will disappear.[27] All pathology will be clearly esophageal or gastric.

6. ACCURATE DEFINITION OF THE GASTROESOPHAGEAL JUNCTION

As we develop an accurate definition of the GEJ, several basic questions need to be answered.

6.1 What Maintains the Esophagus as a Tube?

The esophagus has most of its extent in the thoracic cavity. This is an environment of negative pressure. The mean pressure of the thoracic cavity is approximately −5 mmHg, approximately equal to the intraluminal pressure in the esophagus. The lumen of the esophagus becomes pressurized during swallowing but otherwise remains at negative intraluminal pressure. Apart from obstructive lesions such as achalasia, there is no situation where the esophageal luminal pressure exceeds intrathoracic pressure. This ensures that the thoracic esophagus maintains its tubal shape unless there is esophageal obstruction.

The esophagus passes through the diaphragmatic hiatus and enters the abdomen. The abdominal extent of the esophagus is ~3.5 cm and ends at the GEJ. The right side of the esophagus passes straight down at the junction to continue as the lesser curvature of the stomach. On the side of the greater curvature, the peritoneum lined stomach passes upward to the undersurface of the diaphragm and forms an acute angle (angle of His) with the tubal esophagus at the GEJ. This is the point of flaring of the tubular esophagus.

The LES straddles the diaphragm and has a thoracic and abdominal component. At manometry, the thoracic and abdominal segments of the esophagus are separated by the respiratory inversion point where inspiration and expiration have a different effect on the pressure tracing. The entire abdominal segment of the esophagus is part of the LES. In the normal state, the end of the esophagus is also the distal limit of the LES.

The pressure in the abdomen is normally +5 mmHg. This is also the intraluminal pressure in the stomach. The pressure in the region of the LES is usually >10 mmHg and is maintained in the resting state by tonic contraction of the muscle of the sphincter. This sphincter pressure is greater than the esophageal intraluminal pressure and serves to maintain the esophagus in a collapsed state and in a tubular shape (Fig. 6.15). During swallowing, the sphincter relaxes, the esophageal luminal pressure increases, and the food bolus passes into the stomach. The esophagus dilates as required to permit passage of the food bolus. When the swallow is completed, the sphincter pressure returns and the esophagus returns to its normal tubular shape.

During a meal, the stomach progressively increases in volume, slowly distending to accommodate the meal. Intragastric luminal pressure increases, causing the stomach to dilate, changing from its collapsed state. The mucosal rugal folds flatten. The LES prevents reflux of gastric contents into the esophagus. The pressures in the abdominal segment of the sphincter are maintained at their normal level. The esophagus remains a tubal structure. We will show later that when there is overdistension of the stomach with a heavy meal, the increased gastric pressure can cause temporary dynamic alterations in the distal region of the abdominal sphincter.

Damage to the distal part of the LES is an early change in GERD. This is characterized by "LES shortening." This term is based on the manometric decrease in sphincter length. Pathologically, sphincter shortening simply means that tonic contraction of the muscle in the most distal esophagus has ceased and that there is no pressure exerted by the sphincter in that area.

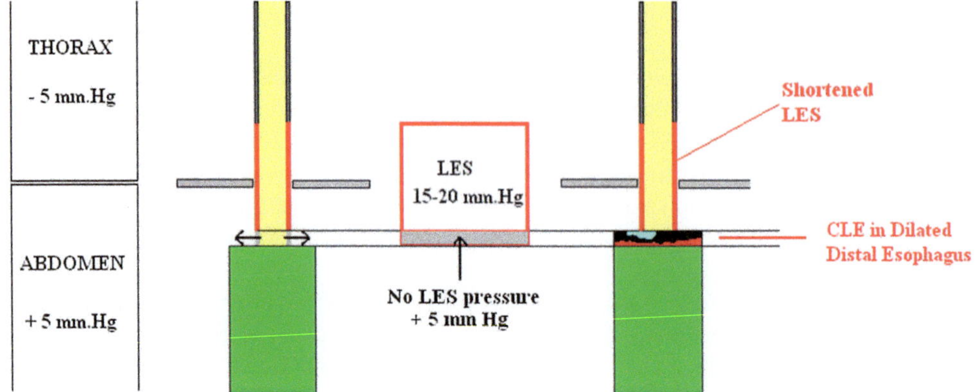

FIGURE 6.15 Pressures in the esophagus, lower esophageal sphincter (LES) zone, and stomach in the normal state and with abdominal LES damage of 5 mm. This shows that the intraluminal pressure in the abdominal esophagus and stomach is positive. The tubular shape of the abdominal esophagus is dependent on the abdominal LES pressure being higher than the intraluminal pressure. When the abdominal LES is damaged, sphincter tone is lost and the intraluminal pressure leads to dilatation of that part of the esophagus. This is associated with cardiac metaplasia (blue, black, and red). *CLE*, columnar-lined esophagus.

This means that the intraluminal pressure in the distal esophagus (+5 mmHg) is now greater than sphincter pressure. The part of the distal abdominal esophagus that has lost LES pressure is now identical to the stomach in its pressure in the fasting state, i.e., has a pressure of +5 mmHg. As the stomach fills and intragastric pressure increases, the part of the distal esophagus that has lost sphincter pressure distends with the stomach (Fig. 6.15).

With time, this part of the esophagus becomes permanently dilated and takes on the contour of the stomach, especially on the greater curvature (Fig. 6.7B). This is the dilated distal esophagus.[7] The abdominal segment of the tubal esophagus shortens. The angle of His becomes more obtuse. The dilated esophagus, now acting as a part of the reservoir, distending with meals, and collapsing in the fasting state, develops mucosal folds, which are a feature of any organ that acts as a reservoir. When the distal abdominal esophagus loses LES pressure, it ceases to be part of the tube that conveys food from the pharynx to the stomach. It becomes part of the reservoir that distends to accommodate the food. It becomes gastricized and resembles the stomach. But, while it functions like the stomach, it is not stomach. It is the dilated distal esophagus.

Therefore, it is the tonic pressure in the abdominal segment of the LES that maintains the tubular shape of the esophagus. When LES pressure is lost, that part of the esophagus ceases to be a tube and ceases to function as the esophagus transmitting food from the mouth to the stomach. The functional esophagus ends at the end of the tube. The GERD-damaged dilated distal esophagus is now functionally part of the gastric reservoir. It has become "gastricized" manometrically, endoscopically, and anatomically (Fig. 6.7B and C) and misinterpreted as proximal stomach. It can only be recognized histologically. The reason for this is because abdominal LES damage is *always* associated with columnar metaplasia of the distal esophageal squamous epithelium. The dilated distal esophagus ends at the proximal limit of gastric oxyntic epithelium, the true GEJ.

6.2 What Are Mucosal Rugal Folds?

Rugal folds are a feature of the mucosal lining of all visceral organs that act as reservoirs, e.g., the gall bladder, urinary bladder, stomach. Rugal folds are a mechanism to permit temporary distension by filling of the organ without a stretching stress on the mucosal lining. The presence of rugal folds in the region of the GEJ only means that the area that contains rugal folds functions as a reservoir, dilating to accommodate a meal and collapsing when empty. Rugal folds are not dependent on the type of epithelium.

The abdominal esophagus, when it dilates because of loss of pressure in the LES, becomes part of the reservoir that accepts a meal. With time, repeated distension with a meal and collapse with gastric emptying can be expected to generate mucosal rugal folds. Increase in the surface area of an epithelial surface occurs by a well-recognized process that is similar to increased surface area of skin that develops when a tissue expander is placed under it in plastic surgical procedures.

6.3 What Methods Can Be Used to Define the Gastroesophageal Junction Accurately?

When trying to solve a problem in medicine, it is rare to find a situation where one cannot, by a diligent search and careful examination of the literature, find a method that has been used effectively in the past. In the case of the solving the problem of accurate definition of the GEJ, it seemed that the best bet was to look carefully at the 1953 paper by Allison and

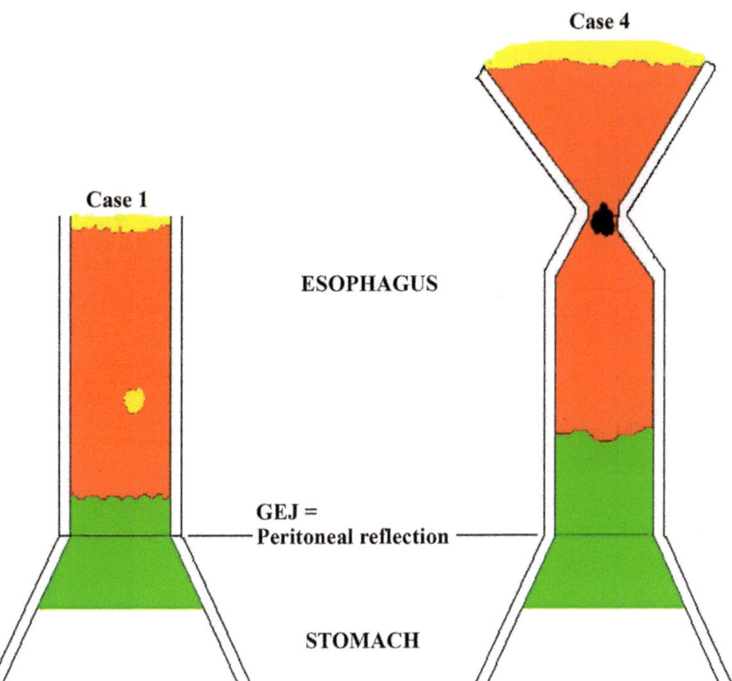

FIGURE 6.16 Diagrammatic representation of Case 1 (left) and Case 4 (right) in Allison and Johnstone's paper in 1953. In both specimens, there is a long segment of columnar epithelium above the peritoneal reflection [true gastroesophageal junction (GEJ)] to 5 and 7.5 cm proximally to the squamo-columnar junction (yellow). The columnar epithelium is described as cardiac epithelium (red) and fundic epithelium (green). The fundic epithelium has parietal cells; there is no differentiation between oxyntocardiac and gastric oxyntic epithelium in 1953. No intestinal metaplasia is seen in either patient.

Johnstone.[10] The evidence that was presented in that paper was so powerful that it made Norman Barrett reverse an opinion that he had expressed about the definition of the GEJ in his 1950 paper within 4 years.[7,9] The evidence was so persuasive that it solved the problem of the structure that looked like the esophagus on the outside but the stomach on the mucosal side. It resulted in the birth of the CLE, an outcome that was perfect.

Allison and Johnstone used four reliable criteria to distinguish the esophagus from the stomach[10] (Fig. 6.13). These were (1) the lack of a peritoneal covering of the esophagus; (2) differences in musculature of the wall; (3) presence of submucosal glands only in the esophagus; and (4) the presence of squamous islands in esophageal columnar epithelium. These criteria, except for the presence of squamous islands in the mucosa could not be evaluated by endoscopy or mucosal biopsy. Squamous islands were unreliable because they were not seen in every case.

The solution that we sought was in the pathologic examination of the resected specimen in Case 1 of this paper.[10] This provides the perfect method of examination of an esophagectomy specimen (Fig. 6.16).

Case 1, described by the pathologist, Dr. D.H. Collins, was an esophagogastrectomy comprising 5 cm of the esophagus and 7 cm of the stomach: "The esophagus was separated with a knife from the stomach *along the line of peritoneal reflection.* A vertical slice was then made through the center of the reconstituted specimen, i.e., up the posterior wall, and three vertically continuous blocks were prepared, and a horizontal block then made across the extreme upper limit of the esophageal mucosa." By this method, everything above the point of separation of the organs is esophageal and everything below is the stomach *if the definition of the* GEJ *used (the peritoneal reflection) was correct.*

In the description of the pathologic findings, the pathologist reports that the epithelium at and below the peritoneal reflection is gastric oxyntic epithelium and the epithelium in CLE above it is cardiac in type. We will consider this further later in this chapter.

7. DEFINITION OF THE TRUE GASTROESOPHAGEAL JUNCTION

When I recognized that cardiac epithelium distal to the endoscopic GEJ was esophageal rather than gastric, I knew that the endoscopic GEJ must be incorrect. However, all I had examined were endoscopic biopsies taken according to Dr. DeMeester's protocol. This provided me with one biopsy set taken in retroflex view from within 1 cm distal to the endoscopic GEJ in all patients. I was dealing with the presence or absence of epithelial types in this biopsy. My mind was tuned to qualitative analysis of the findings. *What was the meaning of the presence of cardiac epithelium compared to its absence?* The establishment resistance to the suggestion that cardiac epithelium was an abnormal esophageal rather than a normal gastric epithelium made me focus all my attention to trying to convince people of this fact.

It took an inexcusably long time for my mind to get to the next obvious step of quantitating cardiac mucosa. This may have been because of my preoccupation with trying to convince people that cardiac mucosa was abnormal.

Unfortunately, Dr. DeMeester did his surgery at the USC University Hospital. I worked at the Los Angeles County hospital. While he sent me all his biopsies for reporting, the esophagectomy specimens that were done by his team were examined at the University Hospital. I did not see those cases on a routine basis.

Endoscopic biopsies in Dr. DeMeester's biopsy protocol did not provide a perspective of length of cardiac epithelium distal to the endoscopic GEJ. In 2005, I had been invited to Vienna by Dr. Martin Riegler to provide a second pathologic review of his patients. He had performed multilevel biopsies at the endoscopic GEJ and at 5, 10, and 15 mm distal to the endoscopic GEJ.[28] These biopsies had shown me that there was a difference in the length of cardiac epithelium with and without intestinal metaplasia and oxyntocardiac epithelium in this region.[28]

It raised the question: "What was the significance of the variation in the length of cardiac epithelium (with and without parietal and/or goblet cells) distal to the endoscopic GEJ?"

It took more than 10 years after I had understood that cardiac epithelium was an abnormal metaplastic esophageal epithelium that was found distal to the endoscopic GEJ for me to change from a qualitative to quantitative perspective. I am disgusted with myself that I wasted 10 years of my life.

Quantitation of cardiac epithelial length was the answer that was critical. This could not be done accurately with Dr. DeMeester's endoscopic biopsy protocol or even in Dr. Riegler's multilevel biopsies; it required the study of esophagectomies. It was an unfortunate quirk of circumstances of the practice locations of Dr. DeMeester's team and I that had delayed my examination of esophagectomy specimens, but this was no excuse. If I had thought of it, the esophagectomies were there for me to examine.

Before I undertook a study of esophagectomy specimens to answer these questions, I had decided to follow the methods of Dr. Collins in his dissection of Case 1 for differentiating the stomach from the esophagus. In our gross examination, we found it very difficult to see the exact point of the peritoneal reflection and found it impossible to identify the point of transition of the muscle of the esophagus and stomach with certainty. We did not see squamous epithelium in the saccular segment distal to the tubular esophagus in any of the cases. We therefore concentrated on mapping the submucosal glands as our main criterion for differentiating the esophagus from the stomach. Submucosal glands are a feature of the esophagus. They are formed after the esophagus has developed stratified squamous epithelium in the perinatal period (see Chapter 3)[29]; they are not seen in the stomach, which never develops squamous epithelium in the human.

Our hypothesis, based on our prior experience with biopsies from this region and our autopsy study, was that the esophagus would extend distally beyond the end of the tube and proximal limit of rugal folds. If this was true, there would be submucosal glands in the saccular segment. We would also map and measure the epithelium distal to the end of the tube in mucosa that had rugal folds.

The aim of the study was to correlate the location of the following structural elements in the region of the GEJ in esophagectomy specimens: surface epithelial type, distal limit of submucosal glands, the point of flaring of the tubal esophagus, and the proximal limit of rugal folds. By this correlative analysis, we believed we would find an accurate definition of the GEJ.

We selected 10 esophagectomy specimens that satisfied the following criteria: (1) absence of distortion of the GEJ region by tumor; (2) presence of a well-defined point of flaring of the tubal esophagus that coincided with the proximal limit of rugal folds (Fig. 6.3); (3) presence of grossly normal mucosa at and distal to the end of the tube; and (4) microscopically normal gastric oxyntic mucosa at the distal margin.

Eight patients had Barrett esophagus (one with high-grade dysplasia and seven with adenocarcinoma in the tubal esophagus) and two patients had squamous carcinoma of the esophagus. All grossly visible tumors were proximal to the end of the tubular esophagus and did not involve the area being studied.

After gross examination and taking appropriate sections from tumor for staging, each specimen was separated by a horizontal cut made at the end of the tubular esophagus where it flared out into the saccular structure (Fig. 6.17A). Our selection criteria made this line coincident with the proximal limit of rugal folds. In cases where there was a columnar-lined segment in the tubular esophagus, this segment was cut in completely with horizontal sections at 0.5 cm intervals (Fig. 6.17A). Random sections were also taken from the more proximal squamous-lined tubular esophagus. The proximal part of the saccular structure distal to the tubular esophagus was cut in completely using vertical sections passing from the end of the tubular esophagus to the maximum distance limited by the size of standard cassettes for histologic processing (~25 mm). Random sections were taken from the more distal pouch and both surgical margins.

The multiple full thickness vertical sections were examined histologically and the lining epithelial types were carefully mapped. The vertical length of the different epithelial types in the 25 mm vertical sections taken distal to the tubular

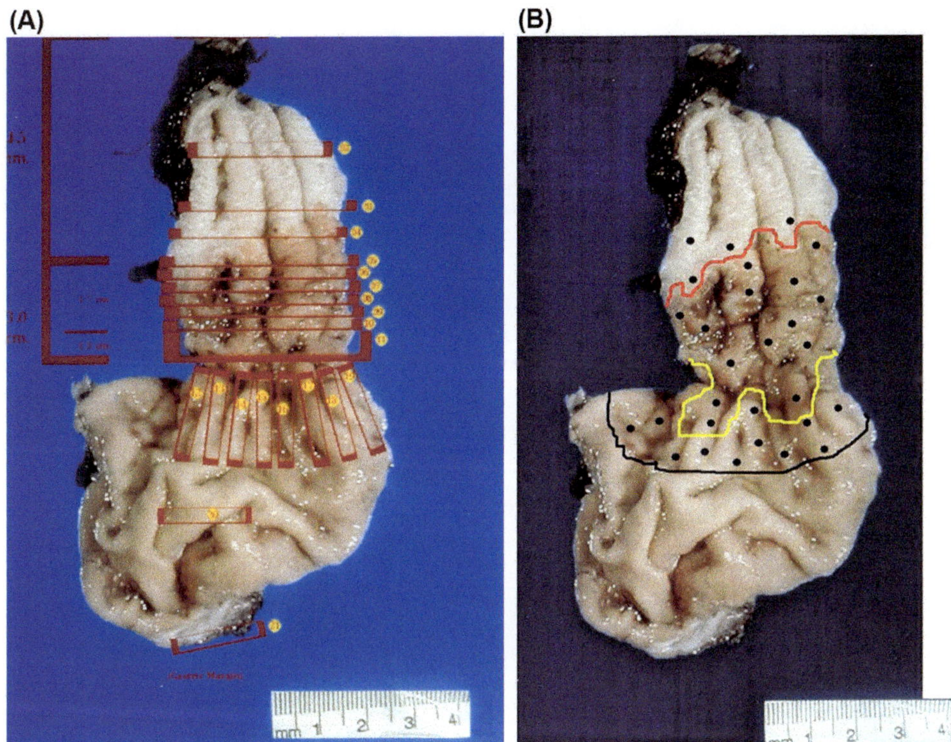

FIGURE 6.17 (A) Sectioning of the esophagogastrectomy specimen shown in Fig. 6.3. A horizontal cut is made at the end of the tubular esophagus and proximal limit of rugal folds (present definitions of the gastroesophageal junction). The entire circumference of the 25-mm area distal to this is sectioned vertically. The tubular esophagus is sectioned by horizontal sections. (B) Results of histologic examination of the specimen. There is a long segment of cardiac epithelium with intestinal metaplasia extending from the squamocolumnar junction (*red line*) to a point beyond the end of the tube and proximal limit of rugal folds (*yellow line*). The proximal limit of gastric oxyntic epithelium is marked by the *black line*; it is 2.05 cm distal to the end of the tubular esophagus. The area between the *yellow and black lines* is composed of cardiac epithelium and oxyntocardiac epithelium without intestinal metaplasia. *Black dots* represent submucosal glands and show that the extent of cardiac epithelium distal to the end of the tubular esophagus is concordant with the extent of submucosal glands. This is proof of the dilated distal esophagus.

esophagus were measured with an ocular micrometer and the location of the junction between oxyntocardiac and gastric oxyntic epithelia was noted. This produced a histologic map of each specimen (Figs. 6.17B and 6.18).

The location of submucosal glands in the tubular esophagus and saccular structure were noted. The location of the glands in the vertical sections distal to the end of the tube was measured from the end of the tube and the type of epithelium overlying the gland was recorded.

Submucosal glands were present in the tubular esophagus in all 10 cases (Fig. 6.17B). The number of glands varied considerably between patients. When the tubular esophagus was lined by columnar epithelium, the number of glands present was similar to that seen in squamous-lined esophagus in the same patient.

Submucosal glands were present in the vertical sections distal to the tube in eight patients under cardiac epithelium (Fig. 6.19). The two patients who did not have glands in the vertical sections distal to the end of the tube were (1) one patient who had only one submucosal gland in the entire esophagogastrectomy specimen and (2) one patient with squamous carcinoma who had a mean cumulative length of nonoxyntic columnar epithelium of 0.31 cm distal to the end of the tube. The number of submucosal glands in the saccular segment varied between 0 and 66 in the 10 patients, the variation being a function of the number of glands present in the esophagus in that patient and the mean cumulative length of cardiac epithelium with and without intestinal metaplasia and oxyntocardiac epithelium distal to the tubular esophagus in the saccular segment.

The epithelium in the tubular segment of the esophagus was only squamous in the two patients with squamous carcinoma. They had no columnar lining in the tubal esophagus. In the eight patients with neoplastic Barrett esophagus, the columnar epithelium in the tubular esophagus was largely cardiac epithelium with intestinal metaplasia and cardiac epithelium in the distal region of the tube.

All patients had cardiac epithelium (with and without intestinal metaplasia) and oxyntocardiac epithelium in the vertical sections from the saccular segment between the end of the tube and gastric oxyntic epithelium (Table 6.2; Fig. 6.18).

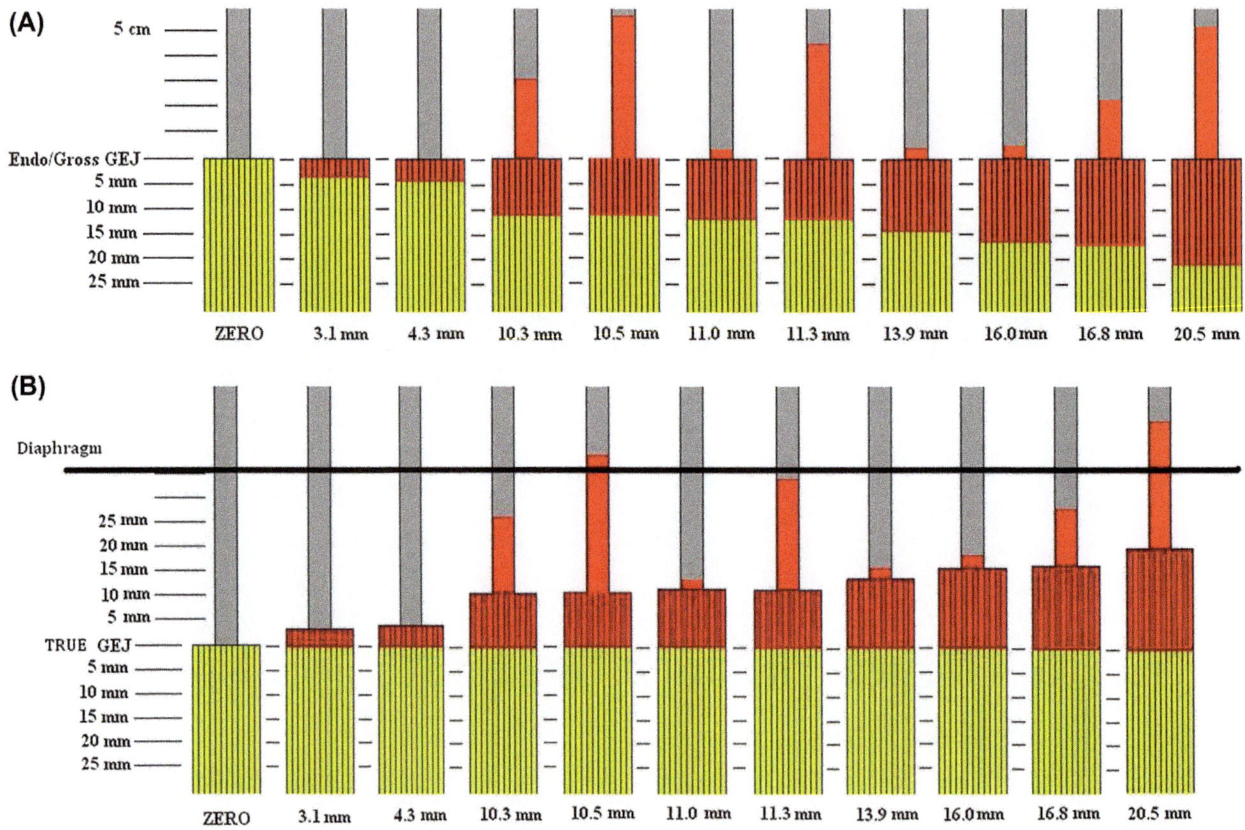

FIGURE 6.18 Diagrammatic representation of the 10 esophagectomy specimens. The normal state (not in the study) is shown on the left. (A) The 10 specimens lined up with the endoscopic gastroesophageal junction (GEJ). They are arranged according to increasing length of dilated distal esophagus composed of cardiac epithelium (red with vertical lines) distal to the endoscopic GEJ. The patients with 3.1 and 4.3 mm of dilated distal esophagus without columnar-lined esophagus (CLE) above the endoscopic GEJ were patients with squamous carcinoma. The other eight patients had visible CLE with intestinal metaplasia (red in tubular esophagus) above the endoscopic GEJ. Note that the dilated distal esophagus appears as proximal stomach distal to the endoscopic GEJ. (B) Same specimens lined up with the true GEG (proximal limit of gastric oxyntic epithelium—yellow). The diaphragm is shown as a black line 35 mm above the true GEJ. The dilated distal esophagus (red with vertical black lines representing rugal folds) is now correctly depicted above the true GEJ. Note that the length of the abdominal esophagus is a constant 35 mm and has a variable length of tubular esophagus (gray for squamous and red for CLE) and dilated distal esophagus.

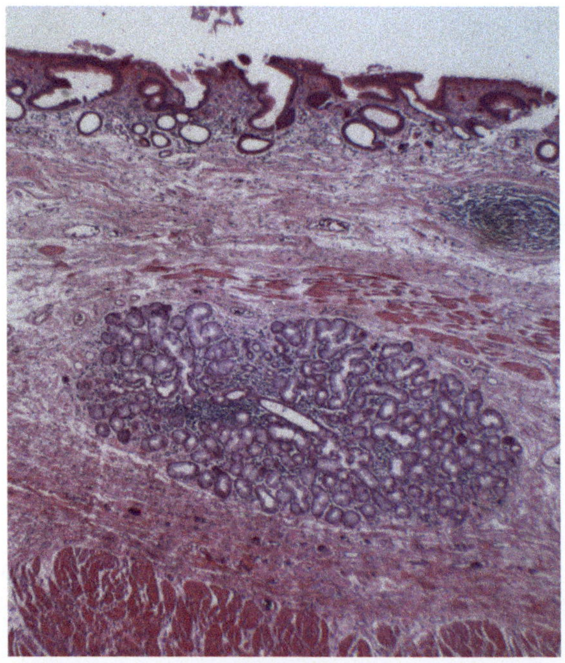

FIGURE 6.19 Section from the area distal to the end of the tubular esophagus and proximal limit of rugal folds showing submucosal glands beneath cardiac epithelium. This is proof that this area is the esophagus when it is lined by cardiac epithelium (with and without parietal and/or goblet cells).

TABLE 6.2 Cumulative Length of Cardiac Epithelium (With and Without Parietal/Goblet Cells) in the Saccular Part of the Specimen Distal to the Gastroesophageal Junction (GEJ) (End of Tubular Esophagus and Proximal Limit of Rugal Folds) in 10 Esophagogastrectomy Specimens

Diagnosis	CLE Length in Tubular Esophagus (in mm)	Length of CM + OCM + IM Distal to Proximal Limit of Rugal Folds (in mm)
Squamous carcinoma	Zero	3.1
Squamous carcinoma	Zero	4.3
Adenocarcinoma	30	10.3
Adenocarcinoma	55	10.5
Adenocarcinoma	<5	11.0
Adenocarcinoma	45	11.3
Adenocarcinoma	<5	13.9
Adenocarcinoma	<5	16.0
Adenocarcinoma	20	16.8
Adenocarcinoma	50	20.5

CM, cardiac epithelium; *IM*, intestinal metaplasia; *OCM*, oxyntocardiac epithelium.

Oxyntocardiac epithelium was the metaplastic epithelium that transitioned to gastric oxyntic epithelium at the GEJ in all cases. The transition point showed inflammation in oxyntocardiac epithelium and uninflamed gastric oxyntic epithelium.

The two patients with squamous carcinoma had squamous epithelium to the end of the tube, i.e., they had no visible CLE in the thoracic esophagus. They had cardiac and oxyntocardiac epithelium distal to the end of the tube with an aggregate length of 0.31 and 0.43 cm (Fig. 6.18). One had a small focus of intestinal metaplasia adjacent to the SCJ.

The eight patients with columnar lining in their tubular esophagus had these three epithelial types distal to the end of the tube in the saccular segment to a aggregate length of 1.03–2.05 cm (Fig. 6.18). All eight patients with Barrett esophagus had intestinal metaplasia extending distal to the end of the tube.

The epithelial types overlying submucosal glands in the saccular segment included cardiac epithelium (eight cases), cardiac epithelium with intestinal metaplasia (five cases), and oxyntocardiac epithelium (three cases). The density of submucosal glands distal to the end of the tube was similar to that in the tubular esophagus in areas of the pouch that was lined by cardiac epithelium with and without intestinal metaplasia. It was slightly less in areas lined by oxyntocardiac epithelium, which was in all cases the most distal part of the saccular segment.

No submucosal glands were seen under gastric oxyntic epithelium, either in the vertical sections of the proximal saccular segment or in the random sections taken from the distal region of the saccular segment of the esophagogastrectomy specimen.

There was a very close correlation between the most distal submucosal glands, which were within 0.5 cm proximal to the proximal margin of gastric oxyntic epithelium in the vertical sections of the proximal saccular segment (Figs. 6.17B and 6.20). Note that the proximal limit of gastric oxyntic epithelium is also the distal limit of oxyntocardiac epithelium. The junction between the esophagus and proximal stomach is the histologic junction between metaplastic oxyntocardiac epithelium and gastric oxyntic epithelium.

There is excellent evidence that submucosal glands are a reliable marker for the esophagus. In the study of fetal esophageal development, Johns showed that submucosal glands develop only after the fetal columnar epithelium in the esophagus has been replaced by squamous epithelium.[29] The submucosal glands begin as outpouchings from the squamous epithelium, not columnar epithelium. In our present study, the submucosal glands were limited to the proximal part of the saccular structure distal to the end of the tube where it was lined by nonoxyntic columnar epithelium. This area had well-formed rugal folds. Submucosal glands were never seen under gastric oxyntic epithelium.

The data in this study prove that esophageal tissue is present distal to the end of the tube in a dilated saccular segment that is easily mistaken for proximal stomach. The extent of this esophageal tissue in the saccular region distal to the end of the tube can be measured by the aggregate length of cardiac epithelium with and without intestinal metaplasia and oxyntocardiac epithelium. Another way of saying this is that the extent of esophageal tissue is the distance between the end of the

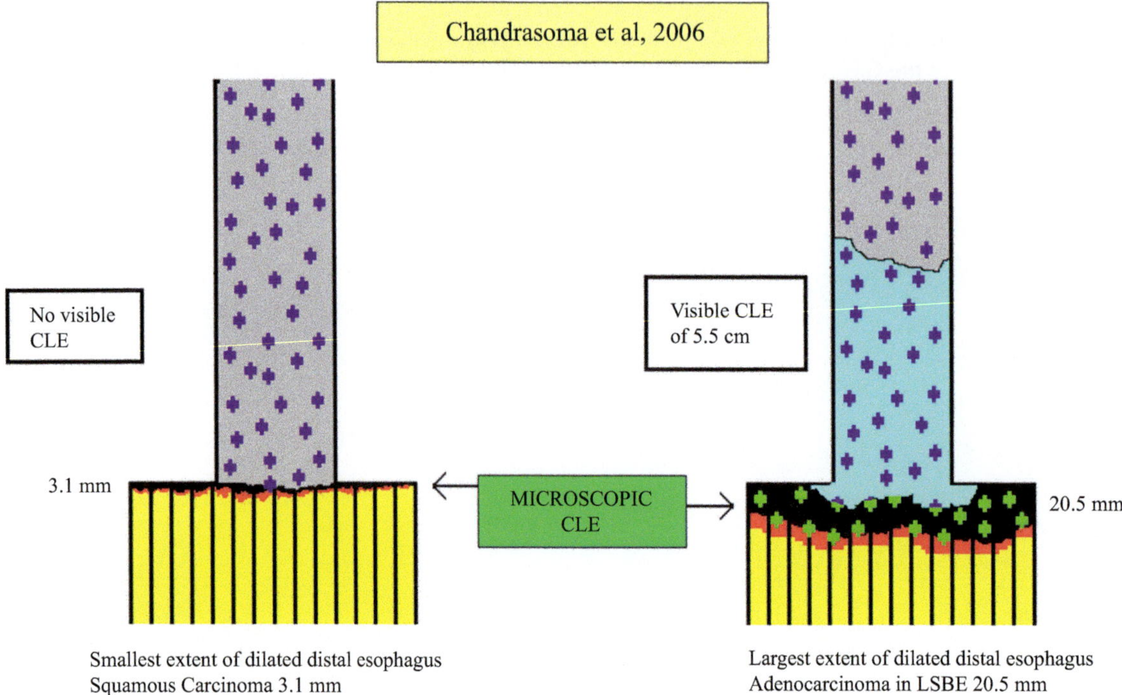

FIGURE 6.20 Details of the two patients at the two extremes of length of dilated distal esophagus. This shows concordance of the cardiac epithelium (blue, black, and red) with the presence of submucosal glands (*colored stars*). The intestinal metaplasia (IM) extends distal to the endoscopic gastroesophageal junction into the dilated distal esophagus in the patient with visible columnar-lined esophagus (CLE). IM is absent in the patient with squamous carcinoma.

tube and the proximal limit of gastric oxyntic epithelium. *This proves that the true GEJ is accurately defined histologically by the proximal limit of gastric oxyntic epithelium.*

The area between the endoscopic GEJ (presently defined by the proximal limit of rugal folds) and the true GEJ (proximal limit of gastric oxyntic epithelium) is the dilated distal esophagus (Fig. 6.7). It is dilated (because it has lost its normal sphincter pressure) and has mucosal rugal folds. It can be recognized by the fact that it is lined by metaplastic cardiac epithelium (with and without parietal and/or goblet cells) and proven to be esophageal by the presence of submucosal glands. *Critically important and of immense value to diagnosis is the fact that the length of the dilated distal esophagus = damage to the abdominal LES = the length of cardiac epithelium (with and without parietal and/or goblet cells) distal to the endoscopic GEJ.*

What is not seen in this study is the normal state without GERD-induced damage (Figs. 6.6A and 6.7A). The two patients with squamous carcinoma (which is a disease that is not associated with GERD) had the entire tubular esophagus lined by squamous epithelium, would have had a normal GEJ appearance at endoscopy, and had esophageal tissue in the saccular segment for 0.31 and 0.43 cm. These patients were 56 and 47 years old. By our definitions they had asymptomatic GERD-induced damage that had produced an aggregate of 0.31 and 0.43 cm shortening of the LES (Fig. 6.18). This corresponds closely to the "normal" state seen in autopsy studies of people without symptoms of GERD during life,[30] in neonatal autopsies,[31] and in asymptomatic volunteers.[4]

8. CORROBORATION OF PROXIMAL MARGIN OF GASTRIC OXYNTIC EPITHELIUM AS THE TRUE GASTROESOPHAGEAL JUNCTION

In a normal patient without any cardiac metaplasia of esophageal squamous epithelium, the GEJ is the junction between the distal limit of squamous epithelium and the proximal limit of gastric oxyntic epithelium.

In a patient whose distal esophageal squamous epithelium has undergone cardiac metaplasia as a result of exposure to gastric contents, the GEJ is at the junction between the distal limit of cardiac epithelium (usually of oxyntocardiac type) and gastric oxyntic epithelium that lines the stomach.

The length of esophagus that is presently mistaken as proximal stomach by the incorrect endoscopic definition of the GEJ is the entire pathology of early GERD. Its length represents the dilated distal esophagus, the length of microscopic cardiac epithelium (with and without parietal and/or goblet cells) distal to the endoscopic GEJ, and the amount of shortening of the abdominal segment of the LES.

The concept that cardiac epithelium is normal is such a powerful dogma that no one has ever thought it needed to be proved. When what is so obviously true is false, there is little or no corroborative evidence for the new truth. Such is the case with cardiac epithelium. The review of virtually the entire literature relating to histology of this region reveals only glimpses of corroborative evidence.

8.1 The Epithelium at the Peritoneal Reflection Is Gastric Oxyntic Epithelium

Allison and Johnstone's paper of 1953[10] provides detailed descriptions of seven cases. In cases 1, 4, and 5, it is clearly stated that the epithelium at and below the peritoneal reflection is gastric fundal epithelium (Fig. 6.16). Dr. D.H. Collin's histologic description of the specimen in Case 1 that was divided at the peritoneal reflection (their GEJ) into an upper esophagus and lower stomach is as follows: "Histology: *The stomach below the anatomical junction with the esophagus is lined by gastric mucosa of fundal type…* Cardiac glands and cardiac gastric mucosa do not appear until 0.6 cm up the anatomical oesophagus, and esophageal glands appear at 2 cm up the oesophagus underlying columnar cell mucous membrane. The first islet of squamous epithelium appears at 3 cm up the anatomical esophagus, but predominantly gastric mucosa continues to the upper limit of the specimen. The transverse sections across the uppermost part of the removed mucosa shows alternations of squamous and gastric mucosa. The rather villous type of cardiac mucosa, its lack of depth, and a diffuse fibrosis of the submucosa in the zone for 2 to 4 cm. above the stomach orifice, suggest healing of previous shallow erosions."

Allison and Johnsone were trying to prove that columnar epithelium above the GEJ was esophageal ("the esophagus lined by gastric mucous membrane") contradicting Barrett's viewpoint that this was an intrathoracic stomach. They, and Dr. Collins, had no doubt that the peritoneal reflection was the GEJ. Their data show clearly the epithelial types in the esophagus and stomach.

Dr. Collins showed that the epithelium at and below the GEJ (peritoneal reflection) was *gastric oxyntic epithelium.* Cardiac epithelium was found above the GEJ starting at 0.6 cm. Their descriptions make it clear that they are describing cardiac and oxyntocardiac epithelium. The description of the esophagectomy specimen in Case 4 is similar (Fig. 6.16).

Although Allison and Johnstone state that submucosal glands were present under cardiac epithelium, their study does not correlate the extent of cardiac epithelium with that of the submucosal glands. In their minds, this is not necessary to demonstrate that Barrett's intrathoracic tubular stomach was actually esophagus.

Allison and Johnstone make no mention of the point of flaring of the tube, mainly because most of their patients had a hiatal hernia. They also make no mention of the mucosal rugal folds, which had no relevance at that time.

Unlike in our study, Dr. Collins' separation of the specimen at an accurate definition of the GEJ (the peritoneal reflection) separated the specimen cleanly into the stomach lined by gastric oxyntic epithelium all the way from the distal margin of the specimen to the GEJ and the esophagus lined by metaplasic cardiac epithelium from the SCJ to the GEJ (Fig. 6.16).

In our study, where we separated the specimen horizontally at the incorrect definitions of the GEJ, i.e., the point of flaring of the tube and the proximal margin of rugal folds, the distal saccular segment consisted of a variable length of the esophagus (0.31–2.05 cm) before reaching the true GEJ. Although we could not accurately define the peritoneal reflection, we can extrapolate Dr. Collins' superior technique and reasonably conclude that the proximal limit of gastric oxyntic epithelium is at the level of the peritoneal reflection.

8.2 Dilated Distal Esophagus Recognized by the Insertion of the Lower Leaf of the Phrenoesophageal Ligament Distal to the Gastroesophageal Junction

In this excellent autopsy study by Bombeck et al. from University of Washington in 1966,[24] the insertion of the lower leaf of the phrenoesophageal ligament to the esophagus was assessed. This ligament is not well defined, but they were able to identify it sufficiently in 20 patients. The lower leaf of the phrenoesophageal ligament must insert at a point either at or above the peritoneal reflection.

They report that the lower leaf of the phrenoesophageal junction inserted into the esophagus at a variable point below the SCJ with the lowest point of insertion being 3.0 cm below the junction *onto the fundus of the stomach.* If the true GEJ is defined by the peritoneal reflection, this indicates that the true GEJ is 3 cm below the SCJ at a point on the "gastric fundus" distal to the point of flaring of the esophagus to the stomach. This suggests that the distal esophagus is dilated distal to the end of the tube and is mistaken for the proximal part of the gastric fundus. This is the dilated distal esophagus.

8.3 The Epithelium at the Angle of His in Fetuses Is Gastric Oxyntic Epithelium

The histologic study of the fetal esophagus and the stomach by De Hertogh et al.[13] included vertical sections through the entire esophagus and stomach. The point of the angle of His was well defined and, in the fetus, accurately represented the GEJ (Fig. 6.21). The epithelium at and below the angle of His was gastric oxyntic epithelium. This proves that there is no cardiac epithelium in the proximal stomach at any time during fetal life. The entire fetal stomach from the angle of His distally is lined by fetal epithelium containing parietal cells from the 17th week of gestation (see Chapter 3). Parietal cells have never been reported in the fetal esophagus.

8.4 Squamous Epithelium Is Present Distal to Endoscopic Gastroesophageal Junction

One of the criteria used by Allison and Johnstone[10] to differentiate the esophagus from the stomach was the presence of squamous epithelial islands in CLE. This was based on the clear knowledge at the time that the human stomach did not have squamous epithelium.

In our study, we did not find islands of squamous epithelium in an area that was distal to the proximal limit of rugal folds in the saccular segment of the specimen. However, we have seen cases where squamous epithelium has been present in retroflex view in areas distal to the endoscopic GEJ (Fig. 6.22). Fass and Sampliner[19] have demonstrated the presence of islands of squamous epithelium distal to the endoscopic GEJ in what they call the proximal stomach. This paper will be reviewed in Chapter 10.

Following Allison and Johnstone's lead,[10] a general rule can be established: anywhere in what is questionably distal esophagus or proximal stomach, the presence of a combination of squamous and columnar epithelium is proof that the location is esophageal.

8.5 Adenocarcinoma in Gastric Cardia After Ablation for Neoplastic Barrett Esophagus

There are consequences to error. One of the negative consequences is in patients with neoplastic Barrett esophagus who are treated with radiofrequency ablation. When these patients are assessed endoscopically prior to ablation, the assumption is made that the esophagus ends at the endoscopic GEJ. Ablation, when it is done, commonly stops distally at the GEJ.

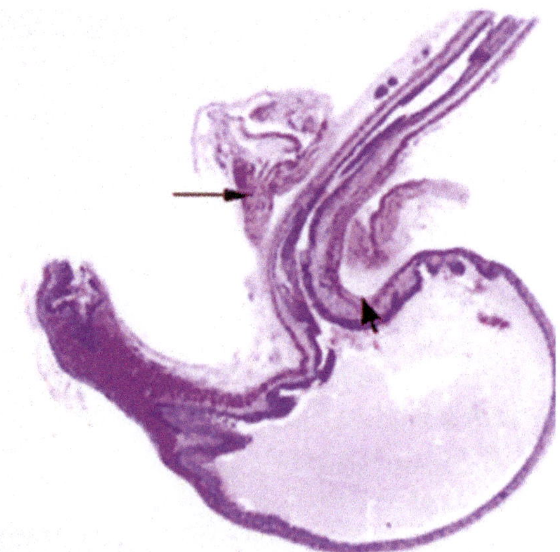

FIGURE 6.21 Autopsy specimen in a fetus showing the esophagus and stomach with the diaphragm in place (*arrow*). The angle of His (*arrowhead*) marks the gastroesophageal junction. The area between the two is the abdominal esophagus. The data in the study showed the epithelium at and distal to the gastroesophageal junction contained parietal cells from early gestation; there was no cardiac epithelium. *Reproduced with permission from De Hertogh G, Van Eyken P, Ectors N, Tack J, Geboes K. On the existence and location of cardiac mucosa; an autopsy study in embryos, fetuses, and infants.* Gut *2003;**52**:791–6.*

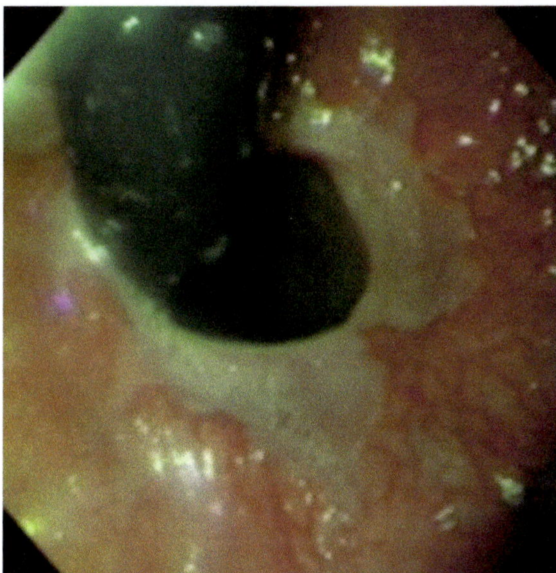

FIGURE 6.22 Retroflex view of the area distal to the endodscopic gastroesophageal junction showing irregular tongues and small islands of squamous epithelium admixed with cardiac epithelium. Squamous epithelium is never seen in the stomach in humans. This proves that this area is the esophagus.

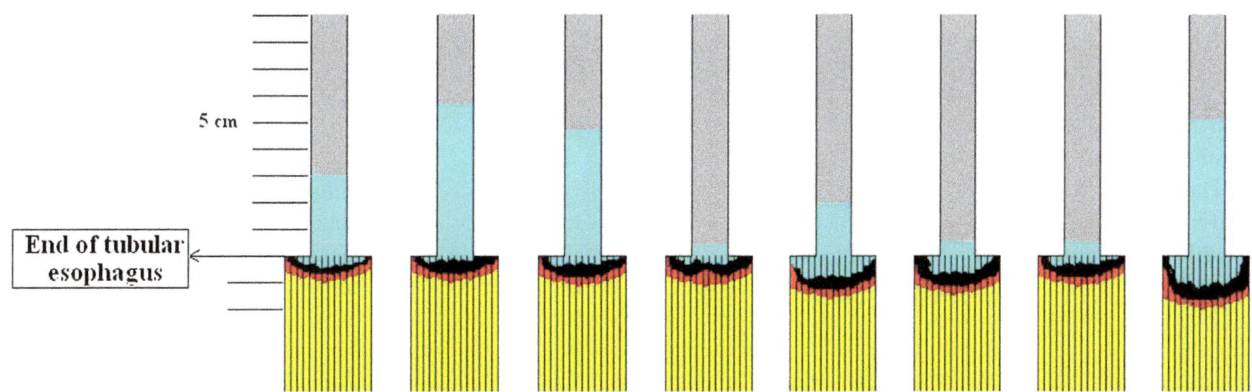

FIGURE 6.23 Histologic features of the columnar epithelium between the squamous epithelium (gray) and gastric oxyntic epithelium (yellow) in the eight patients with adenocarcinoma. This shows that the distribution of the epithelia are squamous epithelium (gray), cardiac epithelium with intestinal metaplasia (IM) (blue), cardiac epithelium (black), oxyntocardiac epipthelium (red), and gastric oxyntic epithelium (yellow). This is evidence that the entire segment of columnar epithelium above the true gastroesophageal junction (GEJ) is one continuous segment of columnar-lined esophagus (CLE). Note that the IM in this CLE segment extends distal to the endoscopic GEJ into the dilated distal esophagus.

We have shown in our esophagectomy study[20] that intestinalized columnar epithelium in CLE above the endoscopic GEJ frequently extends distal to the proximal limit of rugal folds and the end of the tube into the dilated distal esophagus (Fig. 6.23). This will remain undetected by preablation endoscopic biopsies that stop at the endoscopic GEJ.

In one of the patients in our study, there was an adenocarcinoma in the tubular esophagus for which an esophagogastrectomy was performed. Our studies showed that there was a dilated distal esophagus measuring 2.05 cm in length distal to the endoscopic GEJ. Histologically, intestinal metaplasia extended into this segment and a focus of intramucosal adenocarcinoma that had not been detected preoperatively was found therein (Fig. 6.24).

If the present endoscopic definition of the GEJ is wrong, the expectation will be that the intestinal metaplasia in the dilated distal esophagus that has not been ablated will progress in some cases to dysplasia and adenocarcinoma. This will be designated adenocarcinoma of the gastric cardia when it really represents adenocarcinoma in the dilated distal esophagus.

Sampliner et al.[32] have reported that high-grade dysplasia and adenocarcinoma of the gastric cardia occur after ablation of Barrett esophagus. This paper will be reviewed in Chapter 10.

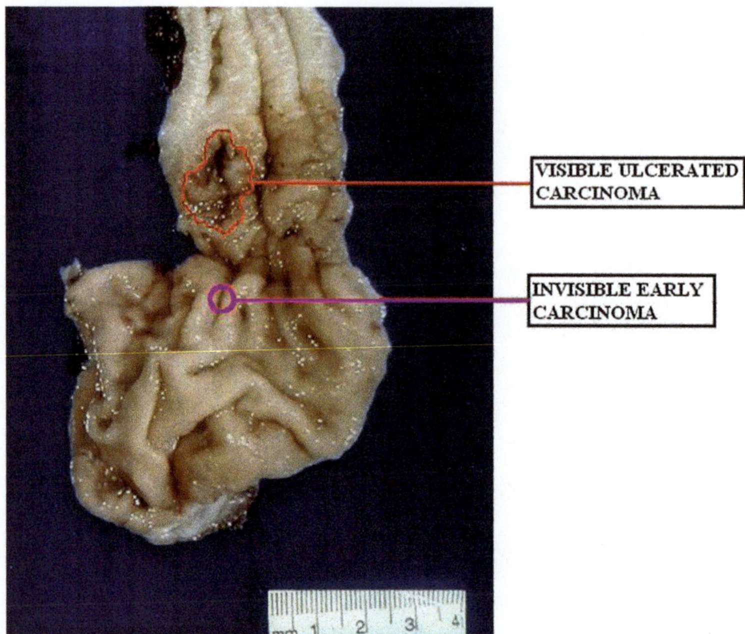

FIGURE 6.24 The patient shown in Fig. 6.3, showing a preoperatively undetected second early intramucosal adenocarcinoma in the dilated distal esophagus in addition to the larger adenocarcinoma in the thoracic esophagus that was the reason for the resection.

9. ENDOSCOPIC LOCALIZATION OF THE TRUE GASTROESOPHAGEAL JUNCTION

The true GEJ can be determined with accuracy in resection specimens where a vertical section extending distally for 3.5 cm from the end of the tube or the proximal limit of rugal folds will invariably reach the proximal limit of gastric oxyntic epithelium, permitting direct localization of the GEJ.

The method of localization of the GEJ at endoscopy needs a biopsy protocol that permits localization of the proximal limit of gastric oxyntic epithelium. This can be adjusted to satisfy the need for accuracy, but it is possible to develop a protocol that permits accuracy within 0.5 cm.

The most reliable endoscopic landmark for reference is the proximal limit of rugal folds. The first set of biopsies is taken at this level. For accuracy within 0.5 cm, sets of biopsies must be taken at 0.5 cm intervals distal to the first set. The most proximal biopsy that shows gastric oxyntic epithelium is the location of the true GEJ. I will describe this in Chapter 17. Unfortunately, even a multilevel biopsy protocol will not be able to accurately measure the dilated distal esophagus. A new biopsy device that can take a vertical sample of mucosa measuring 25–35 mm is necessary. This will provide a vertical section of the entire dilated distal esophagus making its measurement accurate to within a micrometer.

10. CONCLUSION

Accurate definition of the GEJ histologically by the proximal limit of gastric oxyntic epithelium results in a clarity of understanding of the normal state and the entire pathology of GERD. It leads to accurate definition of the normal esophagus and stomach (Fig. 6.7A).

It identifies a stage of pathology of GERD that precedes LES failure and the occurrence of reflux into the esophagus. This stage is characterized by the presence of cardiac epithelium (with and without parietal and/or goblet cells) limited to the dilated distal esophagus (Fig. 6.7B). This area is between the true GEJ (proximal limit of gastric oxyntic epithelium) and the endoscopic GEJ (end of tubular esophagus and proximal limit of rugal folds).

It shows the final change of severe GERD characterized by the presence of visible CLE above the endoscopic GEJ in the thoracic esophagus (Fig. 6.7C) in the minority of patients who develop Barrett esophagus. This is associated with a large extent of dilated distal esophagus distal to the endoscopic GEJ.

With this new understanding of normal anatomy and pathologic changes of GERD as defined by histologic criteria, clear and accurate evidence-based definitions of the normal esophagus and proximal stomach and the entire pathologic spectrum of GERD can be reached (Table 6.3).

TABLE 6.3 New Definitions Based on the Histologic Definition of the Gastroesophageal Junction (GEJ) by the Proximal Limit of Gastric Oxyntic Epithelium

Entity	Definition
The true GEJ	The proximal limit of gastric oxyntic epithelium
The endoscopic GEJ	The proximal limit of rugal folds. This is not the true GEJ. It is a useful and reliably identifiable endoscopic landmark
Normal columnar epithelium	The only normal columnar epithelium in the foregut is gastric oxyntic epithelium. This lines the entire proximal stomach. No columnar epithelium is normally present in the esophagus
Metaplastic columnar epithelia in the esophagus	Cardiac epithelium with and without intestinal metaplasia and oxyntocardiac epithelium; always caused by metaplasia of squamous epithelium as a result of exposure to gastric juice
Metaplastic columnar epithelia in the stomach	Pseudopyloric metaplasia of gastric oxyntic epithelium and intestinal metaplasia of oxyntic and antral epithelium; always caused by chronic gastritis
The normal esophagus	The tubular structure lined by squamous epithelium extending from the pharynx to the proximal limit of gastric oxyntic epithelium (true GEJ)
The GERD-damaged esophagus	The tubular structure lined by squamous epthelium and metaplastic columnar epithelia extending from the pharynx to the proximal limit of gastric oxyntic epithelium (true GEJ)
The visible columnar-lined esophagus	The tubular structure lined by metaplastic columnar epithelium extending from the squamocolumnar junction to the endoscopic GEJ
The dilated distal esophagus	The saccular structure lined by metaplastic columnar epithelium between the endoscopic GEJ and the proximal limit of gastric oxyntic epithelium (true GEJ)
The stomach	That part of the foregut lined by gastric oxyntic and antral epithelium, normal and pathologic, distal to the proximal limit of gastric oxyntic epithelium (true GEJ)

GERD, gastroesophageal reflux disease.

REFERENCES

1. Lagergren J, Bergstrom R, Lindgren A, Nyren O. Symptomatic gastroesophageal reflux as a risk factor for esophageal adenocarcinoma. *N Engl J Med* 1999;**340**:825–31.
2. American Joint Commission of Cancer Staging Manual: esophagus and esophagogastric junction, 7th ed., 2010.
3. Ayazi S, Tamhankar A, DeMeester SR, et al. The impact of gastric distension on the lower esophageal sphincter and its exposure to acid gastric juice. *Ann Surg* 2010;**252**:57–62.
4. Robertson EV, Derakhshan MH, Wirz AA, Lee YY, Seenan JP, Ballantyne SA, Hanvey SL, Kelman AW, Going JJ, McColl KE. Central obesity in asymptomatic volunteers is associated with increased intrasphincteric acid reflux and lengthening of the cardiac mucosa. *Gastroenterology* 2013;**145**:730–9.
5. Siewert JR, Stein HJ. Classification of adenocarcinoma of the oesophagogastric junction. *Br J Surg* 1998;**85**:1457–9.
6. Der R, Tsao-Wei DD, DeMeester T, et al. Carditis: a manifestation of gastroesophageal reflux disease. *Am J Surg Pathol* 2001;**25**:245–52.
7. Barrett NR. Chronic peptic ulcer of the oesophagus and 'oesophagitis'. *Br J Surg* 1950;**38**:175–82.
8. Association of Directors of Anatomic and Surgical Pathology. Recommendations for reporting of resected esophageal adenocarcinomas. *Am J Surg Pathol* 2000;**31**:1188–90.
9. Barrett NR. The lower esophagus lined by columnar epithelium. *Surgery* 1957;**41**:881–94.
10. Allison PR, Johnstone AS. The oesophagus lined with gastric mucous membrane. *Thorax* 1953;**8**:87–101.
11. Odze RD. Unraveling the mystery of the gastroesophageal junction: a pathologist's perspective. *Am J Gastroenterol* 2005;**100**:1853–67.
12. Hayward J. The lower end of the oesophagus. *Thorax* 1961;**16**:36–41.
13. De Hertogh G, Van Eyken P, Ectors N, Tack J, Geboes K. On the existence and location of cardiac mucosa; an autopsy study in embryos, fetuses, and infants. *Gut* 2003;**52**:791–6.
14. McClave SA, Boyce Jr HW, Gottfried MR. Early diagnosis of columnar lined esophagus: a new endoscopic diagnostic criterion. *Gastrointest Endosc* 1987;**33**:413–6.
15. Kahrilas PJ, Shaheen NJ, Vaezi MF. American Gastroenterological Association Institute technical review on the management of gastroesophageal reflux disease. *Gastroenterology* 2008;**135**:1392–413.
16. Palisading vessels.
17. Chandrasoma P. Controversies of the cardiac mucosa and Barrett's esophagus. *Histopathol* 2005;**46**:361–73.
18. Chandrasoma PT. Histologic definition of gastro-esophageal reflux disease. *Curr Opin Gastroenterol* 2013;**29**:460–7.

19. Fass R, Sampliner RE. Extension of squamous epithelium into the proximal stomach: a newly recognized mucosal abnormality. *Endoscopy* 2000;**32**:27–32.
20. Chandrasoma P, Makarewicz K, Wickramasinghe K, Ma YL, DeMeester TR. A proposal for a new validated histologic definition of the gastroesophageal junction. *Hum Pathol* 2006;**37**:40–7.
21. Sarbia M, Donner A, Gabbert HE. Histopathology of the gastroesophageal junction. A study on 36 operation specimens. *Am J Surg Pathol* 2002;**26**:1207–12.
22. Paull A, Trier JS, Dalton MD, Camp RC, Loeb P, Goyal RK. The histologic spectrum of Barrett's esophagus. *N Eng J Med* 1976;**295**:476–80.
23. Spechler SJ, Zeroogian JM, Antonioli DA, Wang HH, Goyal RK. Prevalence of metaplasia at the gastroesophageal junction. *Lancet* 1994;**344**:1533–6.
24. Bombeck CT, Dillard DH, Nyhus LM. Muscular anatomy of the gastroesophageal junction and role of phreno-esophageal ligament. Autopsy study of sphincter mechanism. *Ann Surg* 1966;**164**:643–54.
25. ASGE Standards of Practice Committee, Evans JE, Early DS, et al. The role of endoscopy in Barrett's esophagus and other premalignant conditions of the esophagus. *Gastrointest Endosc* 2012;**76**:1087–94.
26. AJCC. 6th ed.
27. Chandrasoma PT, Wickramasinghe K, Ma Y, DeMeester TR. Adenocarcinomas of the distal esophagus and "gastric cardia" are predominantly esophageal adenocarcinomas. *Am J Surg Pathol* 2007;**31**:569–75.
28. Ringhofer C, Lenglinger J, Izay B, Kolarik K, Zacherl J, Fisler M, Wrba F, Chandrasoma PT, Cosentini EP, Prager G, Riegler M. Histopathology of the endoscopic esophagogastric junction in patients with gastroesophageal reflux disease. *Wien Klin Wochenschr* 2008;**120**:350–9.
29. Johns BAE. Developmental changes in the oesophageal epithelium in man. *J Anat* 1952;**86**:431–42.
30. Chandrasoma PT, Der R, Ma Y, et al. Histology of the gastroesophageal junction: an autopsy study. *Am J Surg Pathol* 2000;**24**:402–9.
31. Park YS, Park HJ, Kang GH, Kim CJ, Chi JG. Histology of gastroesophageal junction in fetal and pediatric autopsy. *Arch Pathol Lab Med* 2003;**127**:451–5.
32. Sampliner RE, Camargo E, Prasad AR. Association of ablation of Barrett's esophagus with high grade dysplasia and adenocarcinoma of gastric cardia. *Dis Esophagus* 2006;**19**:277–9.

FURTHER READING

1. Sharma P, McQuaid K, Dent J, Fennerty B, Sampliner R, Spechler S, Cameron A, Corley D, Falk G, Goldblum J, Hunter J, Jankowski J, Lundell L, Reid B, Shaheen N, Sonnenberg A, Wang K, Weinstein W. A critical review of the diagnosis and management of Barrett's esophagus: the AGA Chicago Workshop. *Gastroenterology* 2004;**127**:310–30.

Chapter 7

The Normal Lower Esophageal Sphincter

The lower esophageal sphincter (LES) is the barrier that prevents reflux of gastric contents into the esophagus. Failure of the LES is accepted universally as the cause of gastroesophageal reflux disease (GERD). Unfortunately, at present, the study of the LES is very limited in the management of the patient with GERD. The only method available for definition of LES characteristics is manometry. There are no manometric criteria for the diagnosis of GERD. The only value of manometry at present is to identify a defective LES that likely requires surgical repair to control reflux.[1]

Different physicians see the LES with varying viewpoints.

Gastroenterologists have an interest in the LES from an academic standpoint. Practically, they have little interest in the sphincter because its assessment is not used for diagnosis of GERD and there are no drugs that can provide a sustained improvement of LES function.

A few academic gastroenterologists with specific focus on GERD have studied the manner in which the LES responds to gastric distension.[2,3] They have concluded that the main mechanism of reflux is the occurrence of transient LES relaxation.[2] While there is little evidence in support, most gastroenterologists believe that these are mediated by a neural mechanism wherein gastric distension activates proximal gastric tension receptors and causes the LES to relax inappropriately independent of swallowing. No such neural mechanism has been demonstrated. There are no drugs that have an action on that neural mechanism with any significant sustained clinical effect on the sphincter.

Surgeons have an interest in identifying criteria for an incompetent LES. These criteria are of value in defining patients who are likely to benefit from antireflux surgery. These criteria are also useful in identifying an LES that is likely to be responsible for GERD symptoms. However, these criteria are designed to be specific for GERD; they detect severe LES damage that is almost always associated with significant GERD. They are not sufficiently sensitive for practical diagnosis of GERD.

Many researchers including gastroenterologists, surgeons, and medical device companies have spent considerable effort in attempting to develop technology to augment or repair the failed LES in patients with severe GERD who have failed medical therapy. Surgical fundoplication, now done laparoscopically, has the highest success rate in experienced hands but has significant complications. Less invasive LES augmentation procedures such as LINX (a magnetic ring placed around the distal esophagus by laparoscopy),[4] and EndoStim (which implants an electrical stimulator in the LES at laparoscopy[5]), and endoscopic procedures such as Stretta (which uses radiofrequency to cause LES fibrosis),[6] transesophageal incisionless fundoplication (which uses a stapling device to "tighten" the defective LES),[7] have had varying degrees of success. The frequency of use of all these techniques at present is not of sufficient magnitude to impact GERD management to any significant extent.

Radiologists do not see the LES, and imaging is not used for assessment of LES damage. The main value of radiology in patients with GERD is in the detection of hiatal hernia and exclusion of achalasia and motility disorders that may produce symptoms that mimic GERD.

Pathologists must see the LES but do not recognize it in resected specimens and during autopsy. They have no method of differentiating the LES from the remainder of the esophagus and stomach on either side. The LES is not associated with thickening of the muscle as is the case in sphincters elsewhere in the body.

At present, when pathologists examine the esophagus, they will regard the point at which the esophagus flares into the saccular organ as the gastroesophageal junction (GEJ).[8] I will show that this is an error created by the dilatation of the abdominal esophagus that results from loss of LES tone when the sphincter is damaged.[9]

At present, there are no pathologic criteria to define a normal LES or recognize an abnormal LES at gross examination or by histology. The only pathologic diagnostic criterion related to the LES is the absence of ganglion cells in the esophagus in patients with achalasia. There are no pathologic criteria at present to define LES damage that is responsible for GERD.

The goal of this book is to develop a histologic method of assessment of LES damage. After I do this, I will pursue this new test of LES damage by pathology into new diagnostic methods for GERD. I believe this has the possibility of transforming the management of GERD in the future with the goals of preventing severe GERD that cannot be controlled adequately with proton pump inhibitor (PPI), Barrett esophagus, and adenocarcinoma.

GERD. http://dx.doi.org/10.1016/B978-0-12-809855-4.00007-5

1. PHYSIOLOGY OF DEGLUTITION (SWALLOWING)

The esophagus and stomach are designed to serve the vital function of ingestion and preparation of food for digestion. Food is ingested intermittently as meals where a large volume of food is eaten during a relatively short time period. Meals are usually separated by several hours where little or no food is ingested allowing the stomach to empty.

The esophagus is a highly efficient conduit for the transmission of food from the pharynx to the stomach. It is ~25 cm (15–40 cm from the incisor teeth) long and extends from the pharynx to the stomach, traversing the thoracic inlet, posterior mediastinum, and the diaphragmatic hiatus to join the stomach at the GEJ (Fig. 7.1).

The cricopharyngeal sphincter, formed by the decussating fibers of the inferior constrictor muscle of the pharynx, is at the upper end of the esophagus. Malfunction of the cricopharyngeal sphincter may be relevant in laryngopharyngeal reflux disease. This will not be addressed in this book.

The stomach is a reservoir that accepts the meal and retains it for several hours, slowly emptying to present it to the small intestine in a manner that facilitates digestion. In the fasting state, the stomach is empty and collapsed with the mucosa thrown into rugal folds. The mucosal rugal folds flatten as the stomach fills during a meal and become folded when the stomach empties. Rugal folds are designed to accommodate expansion without tension. They are a feature of the mucosal lining of all viscera that dilate and collapse, e.g., urinary bladder, gall bladder, and small and large intestines. The intraluminal pressure in the stomach is +5 mmHg, similar to intraperitoneal pressure and the pressure of all viscera in the peritoneal cavity.

The intraluminal pressure in the abdominal LES is also +5 mm, but this is masked by the higher pressure of the LES that normally surrounds the entire abdominal segment of the LES (Fig. 7.1).

This is in contrast to the negative pressure of 5 mmHg in the intrathoracic esophagus. The high pressure of the thoracic segment of the LES also masks this negative pressure above the diaphragm. The high-pressure LES zone therefore acts as a barrier that prevents flow of gastric contents from the stomach to the esophagus along this pressure gradient.

When a meal is ingested, food is processed into a bolus in the mouth and propelled voluntarily by the tongue into the pharynx. This begins the swallowing reflex, driven by the lower cranial nerves and the neuromuscular complex of the esophageal wall, mediated in the deglutition center in the brain stem.

The entry of food, which is normally alkaline, into the stomach quickly neutralizes the resting low pH in the stomach. This alkalinization of gastric contents stimulates the G cells in the antrum to secrete gastrin, causing parietal cells to secrete acid. This acidifies the food column from the periphery inward.

In particular, a strong acid pocket develops at the top of the food column.[10] This creates no problem with a normal meal when the stomach is not filled to capacity because the acid pocket is at a distance from the GEJ. In a stomach that is filled to capacity, however, this acid pocket comes to be located immediately distal to the GEJ (Fig. 7.2).

2. THE NORMAL ESOPHAGUS AND STOMACH DEFINED BY HISTOLOGY

In Chapter 5, I defined the normal state as an esophagus lined entirely by stratified squamous epithelium and a proximal stomach lined entirely by gastric oxyntic epithelium (Fig. 7.3). Cardiac epithelium is not present either in the esophagus or stomach in the normal state. The traditional view that a mucus-secreting epithelium is necessary to act as a buffer between

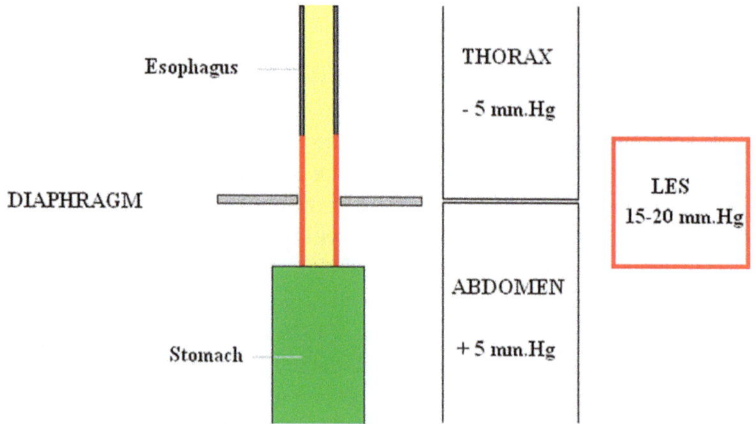

FIGURE 7.1 Diagrammatic representation of the esophagus and stomach with the pressures in this area. The stomach is entirely in the abdomen and is at a positive pressure of 5 mmHg. The esophagus passes from the thorax into the abdomen through the diaphragmatic hiatus. Intrathoracic pressure is negative 5 mmHg. The lower esophageal sphincter (LES) is a high-pressure zone (15 mm or more) that straddles the diaphragm with ~35 mm of its overall 50 mm length being intraabdominal (shown in red). The entire abdominal esophagus is protected by the abdominal LES.

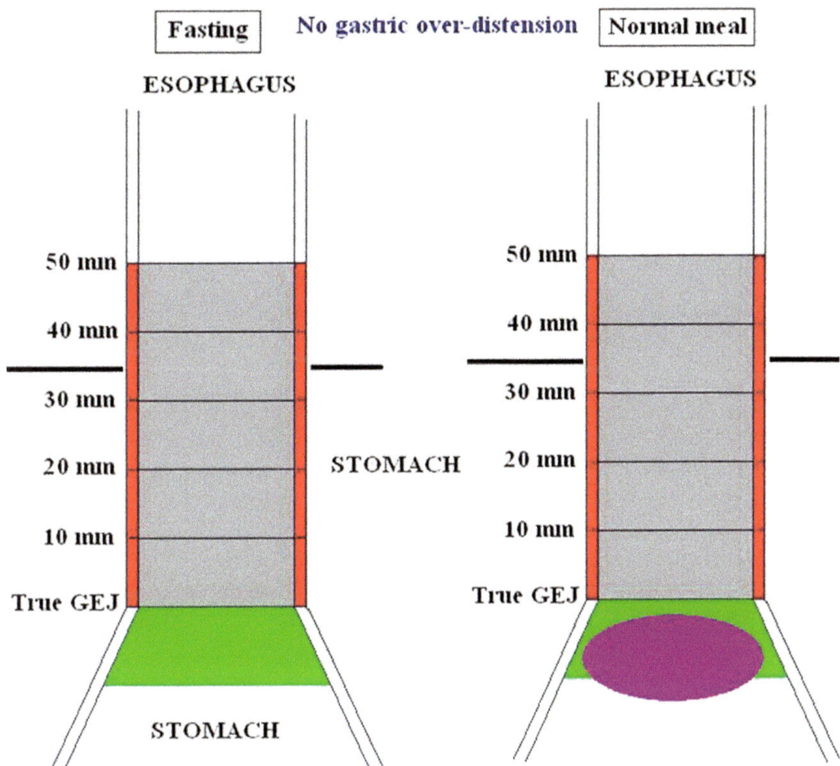

FIGURE 7.2 Diagram showing the status of the full stomach in a normal person at the end of a meal that has been accommodated by the stomach without overdistension. The anatomy is unaltered except for the collapsed stomach becoming full. The lower esophageal sphincter (LES; *red line* in wall) retains its normal length in the absence of pressure from below. The squamous epithelium (gray) is at the distal limit of the LES and therefore not exposed to gastric contents. The food in the stomach has neutralized the resting gastric acid. However, the acid secretion by the mucosa during the gastric phase of acid secretion collects at the top of the food column as an acid pocket (purple). In the full stomach, the acid pocket sits below the gastroesophageal junction (GEJ); this causes no damage when this area is occupied by gastric oxyntic epithelium (green).

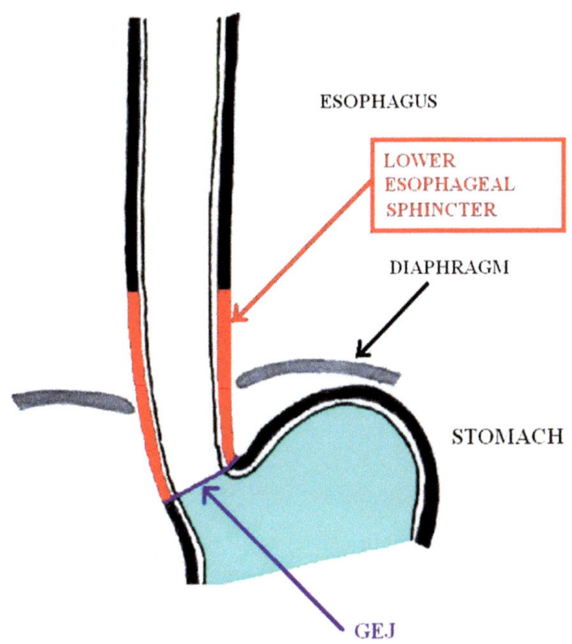

FIGURE 7.3 Diagram of the normal state. The entire esophagus is lined by squamous epithelium (white), the entire stomach by gastric oxyntic epithelium (blue), and the lower esophageal sphincter is undamaged (*red line* in wall). There is no cardiac epithelium. *GEJ*, gastroesophageal junction.

the acid-producing gastric oxyntic epithelium and the squamous epithelium of the stomach is incorrect. A normally functioning LES prevents esophageal squamous epithelium from being exposed to gastric contents.

In the normal person, the GEJ is marked by all the following things: (1) the squamocolumnar junction (SCJ), (2) the distal end of the LES, (3) the proximal limit of rugal folds, (4) the peritoneal reflection, (5) the angle of His, and (6) the proximal limit of gastric oxyntic epithelium. Every definition that has ever been used for the GEJ is accurate in a normal person (Fig. 7.3).

All these criteria except the proximal limit of gastric oxyntic epithelium and the peritoneal reflection move cephalad when the squamous-lined distal esophagus is damaged by exposure to gastric juice. Damage to the LES begins distally and progresses cephalad.

Careful analysis of the literature shows why it has been difficult to convince the medical establishment of the features of the normal state. The normal state is extremely rare. I have encountered this normal state in the retrospective component of our autopsy study (see Fig. 5.5 in Chapter 5)[11] and in the study of fetuses and neonates of Park et al.[12] (see Fig. 3.5 in Chapter 3).

In all my experience looking at thousands of slides from the GEJ, I have encountered the normal state in one esophagogastrectomy specimen in a 67-year-old male with squamous carcinoma of the midesophagus (Fig. 7.4). I have never seen a direct transition from squamous to gastric oxyntic epithelium in any endoscopic biopsy. This is to be expected because endoscopy is not performed until late in the course of GERD and biopsy is not done in endoscopically normal persons.

3. THE MEANING OF CARDIAC EPITHELIUM

The normal state described above is extremely uncommon.

Review of all the papers that have looked at the GEJ region specifically for the presence of cardiac and oxyntocardiac epithelium with adequate sampling (at least one section straddling the SCJ) reveals that almost everyone has these abnormal metaplastic epithelia (Table 7.1).

These data show that virtually all people in the United States and Europe likely have metaplastic cardiac epithelium (with or without parietal cells) at their SCJ.[11] The data also show that this amount can sometimes involve only part of the circumference of the SCJ and is frequently very short (often <5 mm) in people without symptoms of GERD.[11]

If my suggested definition of GERD (a disease characterized by the presence of damage to the esophageal squamous epithelium by exposure to gastric juice) is accepted, nearly everyone in the Western world has GERD. The definition is aimed to be 100% sensitive. It is different to other definitions, such as the Montreal definition, which are designed to be clinically relevant and practical and provides a rational basis for management of GERD. While they succeed in this, the dependence on symptoms for definition results in the failure to recognize the early stages of GERD that are not associated with symptoms.

This is a mistake. One should not ignore the early pathogenesis of GERD for the sake of practicality. One should actively pursue the early pathogenesis of GERD because understanding it can provide valuable keys to earlier diagnosis and provide an opportunity to prevent severe GERD.

It is rare in medicine to have a definitional criterion that is both 100% sensitive and 100% specific. Cardiac metaplasia of the squamous epithelium is produced only by exposure of the squamous epithelium to gastric contents. It is not seen in any other esophageal disease. While the qualitative presence of cardiac epithelium has no value in the diagnosis of GERD,

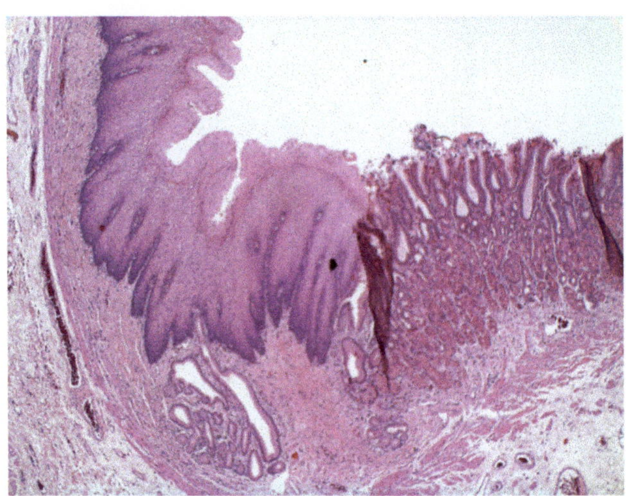

FIGURE 7.4 The normal state by histology where esophageal squamous epithelium transitions directly to gastric oxyntic epithelium. There is no cardiac epithelium.

TABLE 7.1 Prevalence of Cardiac and/or Oxyntocardiac Epithelium Immediately Distal to the Squamous Epithelium

	Cardiac Epithelium	Oxyntocardiac Epithelium	Gastric Oxyntic Epithelium
Kilgore et al.[28]	30/30 (100%)	0	0
Chandrasoma et al.[11] (prospective study)	8/18 (44%)	10/18 (56%)	50% in part of circumference
Chandrasoma et al.[11] (retrospective study)	19/73 (26%)	32/73 (44%)	21/73 (29%)
Marsman et al.[29]	39/63 (62%)	24/63 (38%)	0
Robertson et al.[3]	51/51 (100%)	0	0

Sampling of the esophagogastric junction decreases from top to bottom. Kilgore et al. and the prospective study of Chandrasoma et al. have vertical sections across the entire circumference of the squamocolumnar junction (SCJ) at autopsy; retrospective study of Chandrasoma et al. has a single vertical section across the SCJ at autopsy; Marsman et al. have 1–2 biopsies across the SCJ; Robertson et al. have two biopsies straddling the SCJ.

TABLE 7.2 New Classification of GERD Based on the New Definition

	Cardiac Epithelium Distal to Endoscopic GEJ	AbdLES Damage	Reflux 24 hr pH Test	Symptoms/ Endoscopy
Normal	0	0	0	0
Phase 1: compensated LES damage	>0 to <15 mm limited to DDE	>0–15 mm	0 to pH <4 for 1.1%	0
Phase 2: subclinical GERD	>15–20 mm limited to DDE	>15–20 mm	>0 to pH <4 for <4.5%	Yes/no; not "troublesome" NERD
Phase 3: clinical GERD, mild	CLE in DDE >20–25 mm	>20–25 mm	pH <4 for >4.5%	Yes/no; "troublesome" NERD/mild EE
Phase 4: severe GERD	CLE in DDE >25 mm	>25–35 mm	pH <4 for >4.5%	Yes/no; "troublesome" Severe EE/ visible CLE

AbdLES, abdominal segment of the LES; *CLE*, columnar-lined esophagus; *DDE*, dilated distal esophagus; *GERD*, gastroesophageal reflux disease; *LES*, lower esophageal sphincter. GERD is a disease caused by damage to the abdominal segment of the LES. Damage to the abdominal LES = length of cardiac epithelium (with and without parietal and/or goblet cells) between the endoscopic gastroesophageal junction (GEJ) and true GEJ (proximal limit of gastric oxyntic epithelium). No GERD by present definition = green; increasing severity of clinical GERD = orange and red.

I will show that the quantitative length of cardiac epithelium (with and without parietal and/or goblet cells) has immense unrecognized value in diagnosis.

Cardiac epithelium begins at zero (normal), slowly increases in the dilated distal esophagus (DDE) as abdominal LES damage progresses before LES damage is sufficiently severe to cause reflux into the thoracic esophagus.[9] Reflux into the thoracic esophagus is a prerequisite to the development of visible columnar-lined esophagus (CLE) therein.

It is not the presence of cardiac epithelium in the gap between the SCJ and the proximal limit of gastric oxyntic epithelium that is important. It is its length. The reason for this is that there is a concordance between the length of cardiac epithelium and damage to the abdominal LES. Cardiac metaplasia of squamous epithelium and damage to the abdominal LES is the first event in the pathogenesis of GERD (see Chapter 8). It causes profound, albeit subtle and confusing, anatomic abnormalities (see Chapter 10) that have caused and continue to cause serious errors in interpretation.

Table 7.2 represents a classification of GERD based on the new definition that I have proposed where GERD is defined by the presence of cardiac epithelium distal to the SCJ and the severity of GERD by increasing length of cardiac epithelium (with and without parietal and/or goblet cells) (Fig. 7.5). I am presenting these data in the chapter on the LES because I want to emphasize that cardiac metaplasia of the abdominal esophageal squamous epithelium is exactly concordant with LES damage. As such, when reading Table 7.2, whenever it says cardiac epithelium absent, it means that there is no LES damage and

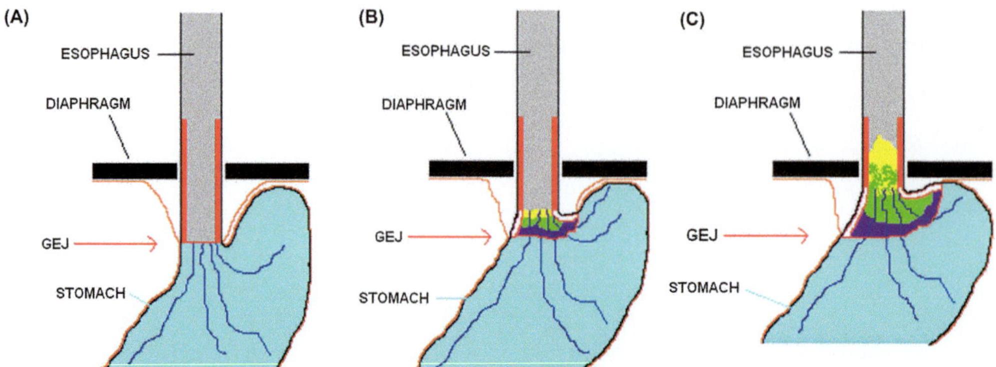

FIGURE 7.5 Diagram of progression of lower esophageal sphincter (LES) damage and its main correlations with cellular changes in the esophagus. (A) The normal state. (B) The first stage of LES damage that has resulted in the formation of a dilated distal esophagus. This is seen as cardiac epithelium (yellow–green–purple) interposed between the cephalad-displaced squamous epithelium (gray) and the true gastroesophageal junction (GEJ) represented by the proximal limit of gastric oxyntic epithelium (blue). The LES damage (white replacing red in wall) is concordant in length with the dilated distal esophagus (=length of cardiac epithelium between the squamocolumnar junction and true GEJ). (C) The final stage characterized by visible columnar-lined esophagus. This is associated with more abdominal LES damage as indicated by a greater length of dilated distal esophagus.

when it says cardiac epithelium present distal to the endoscopic GEJ, it means abdominal LES damage is present. More importantly, I will show that the length of cardiac epithelium distal to the endoscopic GEJ is equal to the amount of abdominal LES damage.

This classification recognizes three important things that represent serious problems in the present management of GERD: (1) Symptoms of GERD or their ease of control with PPIs do not predict progression of GERD at a cellular level. Patients with few or no symptoms can progress to Barrett esophagus and adenocarcinoma; (2) The cellular changes of GERD correlate very well with the severity of abdominal LES damage, which can be measured by the length of cardiac epithelium distal to the endoscopic GEJ in the DDE; and (3) Abdominal LES damage can be recognized by measuring the length of cardiac epithelium in the DDE *before* the onset of clinical GERD.

Visible CLE in the thoracic esophagus represents an irreversible change in the esophageal body. It is not reversed by either medical therapy or antireflux surgery. It can be reversed only by some type of surgical or radiofrequency ablation. Progression of visible CLE to Barrett esophagus (defined as the presence of intestinal metaplasia) is definitely not prevented by medical therapy. The progression of Barrett esophagus to adenocarcinoma is also not prevented by PPI therapy. There is a controversy as to whether the use of PPIs in Barrett esophagus increases or decreases the risk of progression to cancer.[13,14] While there is evidence that surgery decreases the risk of progression of visible CLE to adenocarcinoma, surgery is not yet regarded to be a cancer preventive procedure in Barrett esophagus; it is only recommended to control symptoms.[15]

If the objective of GERD management is the prevention of visible CLE, and if this is successful, adenocarcinoma will be largely prevented. At present, there is absolutely no effort to prevent progression of GERD to CLE. It is simply allowed to happen. When it has occurred, the progression of CLE to adenocarcinoma cannot be prevented by any therapy and no effort is made in this regard. The only thing that can be done is surveillance to detect dysplasia and early cancer that will occur in a minority of these patients.

This classification also directs our attention to two categories of clinical GERD: mild, where the patients remain easily controlled in the long term with medical therapy and severe, characterized by treatment failure, decline in quality of life, Barrett esophagus, and adenocarcinoma. The best criterion to differentiate mild and severe GERD is the presence of visible CLE. The reason for this is that visible CLE is not reversed by medical therapy for GERD. All cellular changes before visible CLE including severe erosive esophagitis are reversible.

This emphasis on visible CLE leads to another simple goal of management: to identify who will progress from mild to severe GERD and take measures to prevent this from happening. At this time, this is impossible because there is no way to predict which patients will progress from mild to severe GERD.

The presence and severity of symptoms is not a reliable indicator of the severity of GERD. The most serious complications of GERD, Barrett esophagus, and adenocarcinoma, can occur in people without symptoms and in patients whose symptoms are well controlled with acid-suppressive drugs.[16]

We have abundant proof that the present management where the focus and target are symptom control and maintenance of quality of life is a failure. This is certainly true if failure is defined as the incidence of adenocarcinoma, which has increased dramatically in the past 60 years and continues to increase.[17]

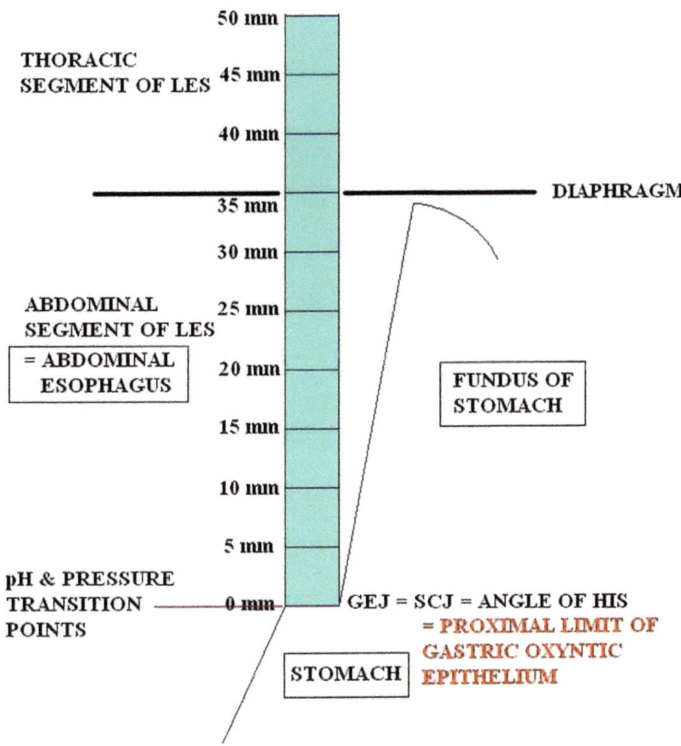

FIGURE 7.6 Diagrammatic representation of the normal lower esophageal sphincter (LES). This measures 50mm with a 35mm abdominal segment distal to the diaphragm. The entire normal LES is lined by squamous epithelium and transitions to gastric oxyntic epithelium at the gastroesophageal junction (GEJ). This is also the point at which the pressure changes from LES to gastric pressure and where the pH of the gastric interior steps up to the neutral pH in the abdominal esophagus. *SCJ*, squamocolumnar junction.

I will show that the method of achieving the goal of preventing progression from mild to severe GERD is to develop a histologic assessment of LES damage that will first accurately differentiate mild from severe GERD. Second, this assessment of LES damage will permit prediction of progression of GERD to severe GERD in the future at an early stage of LES damage at and even before the onset of symptoms of GERD.

The ability of the new assessment to identify at an early stage which patients are at high risk of progression to severe GERD will provide an opportunity to intervene selectively in this minority of patients. Intervention to prevent progression of LES damage in these patients will prevent severe GERD that is characterized by the failure of PPI to control symptoms, Barrett esophagus, and adenocarcinoma. For the first time, this provides a vision of a future where the increasing incidence of adenocarcinoma in GERD patients will reverse.

4. THE NORMAL LOWER ESOPHAGEAL SPHINCTER

The smooth muscle of the distal 4–5cm of the esophagus is a high-pressure zone that forms the LES. The LES straddles the diaphragmatic hiatus and consists of a thoracic segment above and an abdominal segment below the diaphragm. The entire abdominal esophagus, which measures ~3.5cm and ends at the GEJ, is the abdominal segment of the LES (Fig. 7.6).

There is no anatomically recognizable thickening of the muscle wall of the distal esophagus corresponding to the LES. Careful dissection shows a complex arrangement of the muscle fibers at the GEJ that may be representative of the LES.[18] The high pressure of the LES is a function of tonic contraction of the smooth muscle wall of the distal esophagus. LES damage cannot presently be defined by endoscopic, radiologic, gross, or histologic criteria. This is a significant problem in the diagnosis and assessment of GERD because it prevents correlation of the severity of the disease with the degree of LES damage, which is its cause.

In the fasting state between meals, the esophagus is empty and collapsed. It is closed at its proximal end by the cricopharyngeal sphincter, which prevents air from entering the esophagus during ventilation. The intraluminal pressure of the thoracic esophagus is equal to intrathoracic pressure (−5mmHg) and varies with respiration like all other intrathoracic organs.

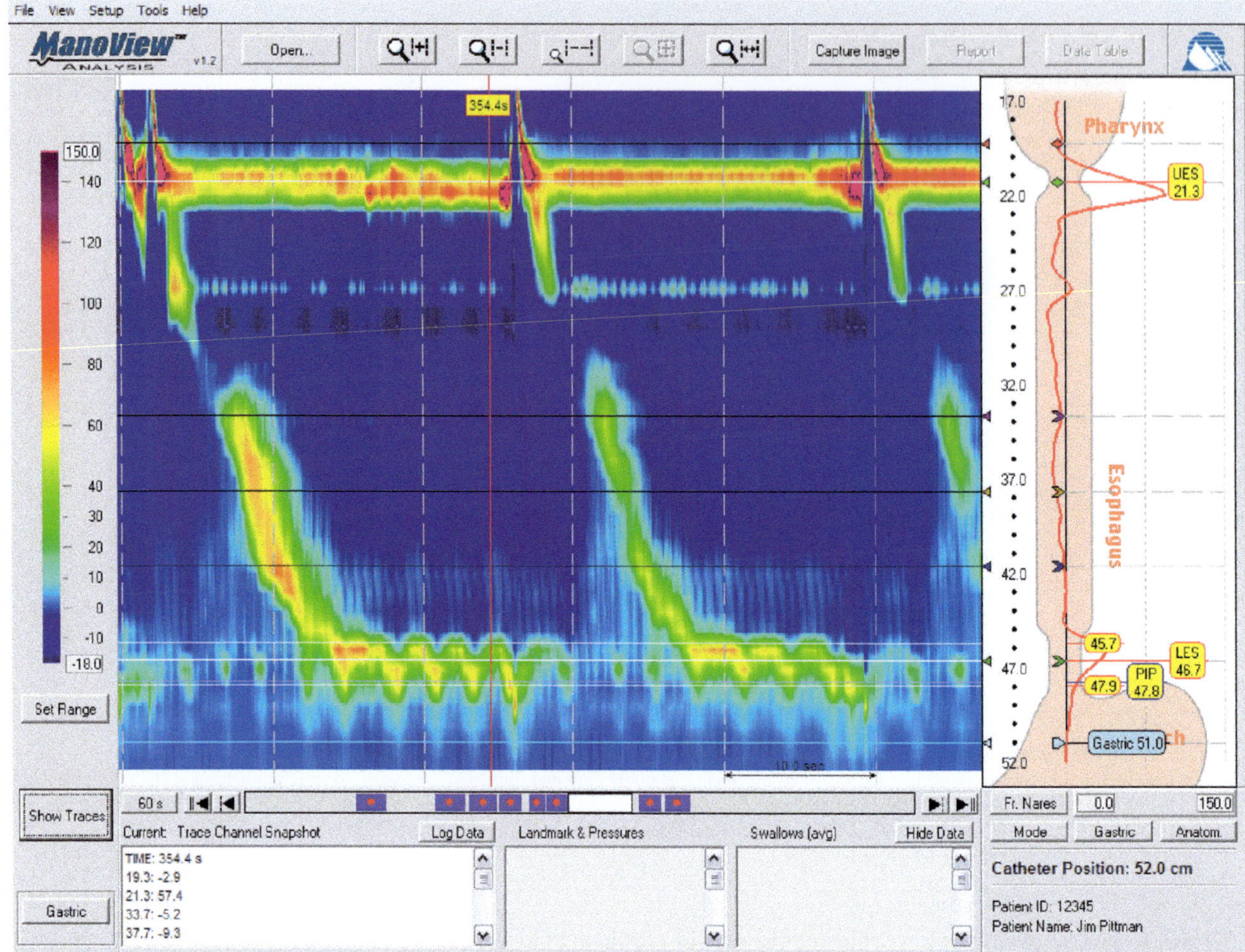

FIGURE 7.7 High-resolution manometry (color code for pressure on the left) showing the high-pressure zone of the lower esophageal sphincter (LES) and relaxation during swallows. The LES is shown in the anatomic depiction of the waveform on the right as extending from 45.7 cm (from the nares) to 51.0 cm. The abdominal length from the pressure inversion point at 47.9–51.0 cm is 31 mm. The LES separates the higher pressure of the stomach from the lower pressure (deeper blue) in the thoracic esophagus.

The intraluminal pressure of the abdominal segment of the esophagus is the pressure exerted by the LES (usually >15 mmHg), which occupies the entire abdominal esophagus to the GEJ (Fig. 7.1). The LES is contracted (i.e., at a high-resting pressure) when food or liquid is not being swallowed and between swallows during a meal, closing the esophagus from the stomach (Fig. 7.7).

Swallowing causes a coordinated relaxation of the upper and lower esophageal sphincters and a peristaltic wave developing in the pharynx that passes rapidly down the entire length of the esophagus. This propels the food bolus into the stomach through the relaxed sphincters (Fig. 7.7).

As the high-pressure propulsive peristaltic wave traverses the upper esophageal sphincter, this closes to shut off the esophagus from the pharynx. As it traverses the LES, this closes to shut off the esophagus from the stomach. The primary peristaltic wave usually propels the food bolus into the stomach. In cases where this fails, a secondary peristaltic wave is generated in the body of the esophagus to complete the propulsion of the food into the stomach.

The LES is normally open only during swallowing, during the time there is a proximal positive pressure created by a peristaltic wave. The LES is normally closed during all times except swallowing (Fig. 7.7). Between swallows, there is a normal pressure gradient from the stomach (+5 mmHg) to the esophagus (−5 mmHg). The LES is designed to permit entry of food into the stomach during a swallow and prevent reflux into the esophagus during the periods between swallows. The only time the LES relaxes physiologically other than during a swallow is in belching, which permits venting of swallowed gastric air into the esophagus, and during vomiting.

The stomach, acting as a reservoir, accommodates the food delivered by swallowing. The stomach can accept a normal-sized meal without any difficulty, filling like a collapsed balloon to capacity without a significant increase in intragastric

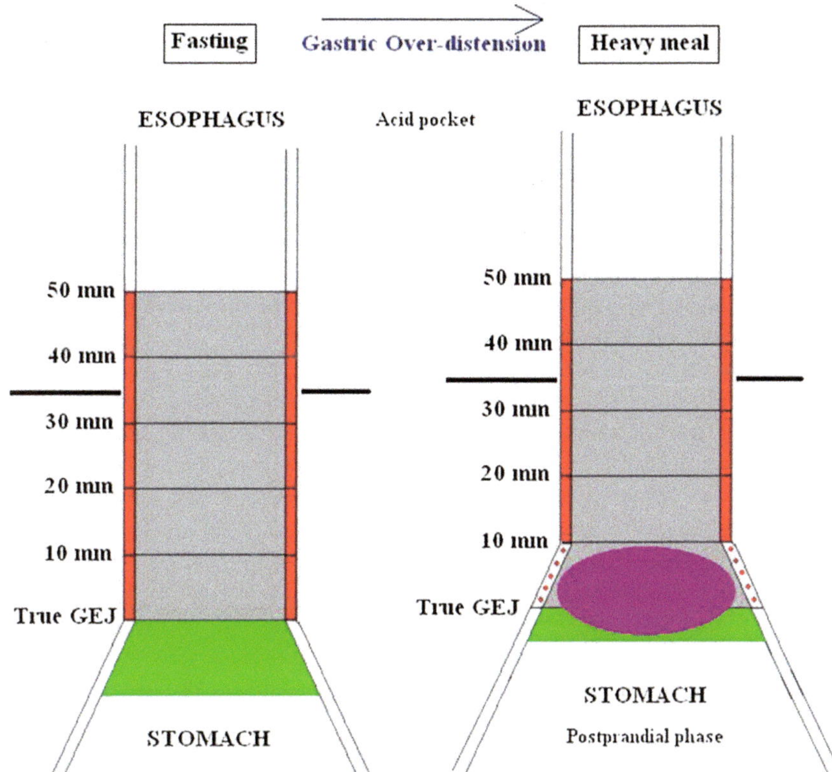

FIGURE 7.8 Diagram showing the status of the full stomach at the end of a meal that has caused gastric overdistension. The lower esophageal sphincter (LES) has been effaced (white with *red dots* in the wall) at its distal end causing dilatation of that part of the abdominal esophagus. The manometric length has shortened. The squamous epithelium (gray) is below the pH transition point and is exposed to the acid pocket (purple) that sits below the gastroesophageal junction (GEJ). This causes squamous epithelial damage during this temporary phase of dynamic shortening of the LES, which eases with gastric emptying.

pressure or wall tension. The muscle wall of the stomach normally contracts to propel food through the pyloric sphincter into the duodenum. As long as there is no gastric overdistension, there is little or no upward pressure on the LES, even during the postprandial period. The entire esophagus retains its tubular shape, the LES retains its normal length and position, and the squamous epithelium is at or above the pH transition point, i.e., the point where the pH changes from the strong intragastric pH to the near neutral esophageal pH (Figs. 7.2 and 7.6).

The LES is a beautifully designed sphincter, perfectly capable of performing its function to prevent reflux of gastric contents into the thoracic esophagus as long as the person ingests meals of reasonable size. Problems arise when excessive eating causes gastric overdistension.

Gastric overdistension during a meal causes dynamic shortening of the LES, moves the esophageal squamous epithelium below the pH transition point exposing it to the acid pocket that is present in this region (Fig. 7.8).[3] This happens in a person with a normal LES as well as a person with LES damage. It is the mechanism of the beginning of the process that leads to GERD as well as the method whereby LES damage progresses (see Chapter 8).[2,3,19]

5. MANOMETRIC ASSESSMENT OF THE LOWER ESOPHAGEAL SPHINCTER

The existence of a barrier at the GEJ was suspected ever since physicians recognized that reflux of gastric contents into the esophagus did not occur despite the fact that there was a pressure gradient from the stomach to the esophagus. The early problem was that no anatomic sphincter could be convincingly demonstrated at the GEJ.

In an analysis of the prevailing opinions up to the mid-1950s, Marchand[20] writes: "The pleuroperitoneal pressure gradient constantly tends to displace stomach contents into the oesophagus, yet continuous regurgitation is in some way prevented. Currently, there is common accord that this is mainly accomplished by the oblique entry of the oesophagus into the stomach which acts as a flap-valve when the stomach distends. Barrett (1952) and Donnelly (1953) are of the opinion that the right crus of the diaphragm compresses the terminal oesophagus on deep inspiration and so reinforces the resistance offered to regurgitation. Johnstone (1951) believes that a sphincter at the gastro-oesophageal junction aids in preventing reflux. Allison (1951), whilst considering the diaphragm to be important, agrees that an intrinsic mechanism contributes to continence of the cardia."

It is interesting to see the names of the same physicians—Barrett and Allison (two surgeons) and Johnstone (radiologist)—who were instrumental in defining reflux esophagitis and the CLE between 1948 and 1957.[21–24] These physicians were giants in their field back then. It is also interesting to see that Allison and Johnstone, who first correctly defined the CLE in 1953,[24] again had the opinion that was closest to what ultimately proved to be the truth.

The development of esophageal manometry in the late 1950s confirmed that there was a high-pressure zone in the distal esophagus that acted as an intrinsic sphincter.

The LES is detected by manometry as a high-pressure zone that separates the negative pressure in the esophagus at its proximal (thoracic end) and the positive pressure in the stomach at its distal end (Fig. 7.7). The upper limit of the LES is marked by an increase in pressure >2 mmHg from the esophageal baseline pressure. The distal limit of the LES is marked by a decline in pressure of the high-pressure zone by >2 mmHg to the gastric baseline pressure.

The LES is divided into a thoracic and abdominal segment by the diaphragmatic hiatus (Figs. 7.1 and 7.6). The junction of the thoracic and abdominal segment of the LES is recognized by a change in the pressure variation during respiration from thoracic in type (decreases in inspiration) to abdominal (increases in inspiration). This point in the manometric tracing is called the respiratory inversion point. The length of the abdominal segment of the LES is the distance from the respiratory inversion point to the distal end of the LES (Fig. 7.7).

The length of the LES during fetal development and the growth of the individual are unknown. The initial total length of the undamaged LES is believed to be ~ 50 mm (Fig. 7.6), with the majority of the normal LES being below the diaphragm (i.e., the abdominal segment).

Studies in "normal" (i.e., asymptomatic) volunteers show a range of abdominal LES length ranging from 10 to 35 mm.[25] There are no data regarding the individual variations in the initial length of the undamaged LES in different people.

Manometry only measures the normally functioning LES. When the LES is damaged, it loses its tonically maintained high pressure. At the distal end, loss of pressure results in the damaged LES becoming indistinguishable from the stomach at manometry, i.e., the damaged distal abdominal LES becomes manometrically "gastricized" (Fig. 7.9).

The original anatomic LES, which is defined by its high pressure, never changes its length. LES damage, which is usually the result of factors exerted from the stomach below, begins at the distal end and progresses cephalad. LES damage is characterized by loss of the high-pressure zone, beginning at its distal end and progressing to a varying length depending on the amount of LES damage. Each person has a different rate of progression of LES damage that determines ultimate severity of LES damage.

From a manometric standpoint, the damaged LES is shortened. The part of the original LES that has been damaged and lost its high pressure is not recognized by manometry. It is distal to the residual functional LES that remains as a high-pressure zone at manometry (Fig. 7.9). Though not recognized, it is part of the original LES and therefore part of the esophagus.

The LES can be completely defined by the following mathematical formula:

Original LES length = Residual (manometric) LES length + length of LES damage

The damaged LES distal to the manometric functional LES is completely misunderstood at present. It is regarded as the proximal stomach (or gastric cardia) by gastroenterologists, surgeons, radiologists, and pathologists. It is the correction of this error that permits a new method of assessment of the damaged LES by pathology.

The damaged LES forms what I have called the DDE (Fig. 7.10). This is easy to understand when one considers the pressures involved. In the normal state, the high pressure of the abdominal LES masks the intraluminal pressure in the abdominal esophagus. Being below the diaphragm, the abdominal esophagus is an abdominal viscus and will have an intraluminal pressure of +5 mmHg. When LES pressure is lost in the distal esophagus, this positive intraluminal pressure will cause this part of the abdominal esophagus to dilate. This is accentuated by the fact that this sphincter-less abdominal esophagus will have no protection from increases in intragastric pressure that occurs repeatedly with eating. As a result, the part of the LES that loses pressure becomes a DDE lined by metaplastic cardiac epithelium (Fig. 7.10).

Therefore, when a manometric tracing is examined, it must be understood that the distal end of the LES in the pressure tracing is not necessarily the end of the anatomical esophagus. In a person with LES damage, there is a DDE that is manometrically indistinguishable to the stomach. This could vary in length from >0 to over 30 mm (Fig. 7.11). Care is necessary, therefore, to make a diagnosis of a hiatal hernia in manometric tracings with a severely damaged LES (Fig. 7.9).

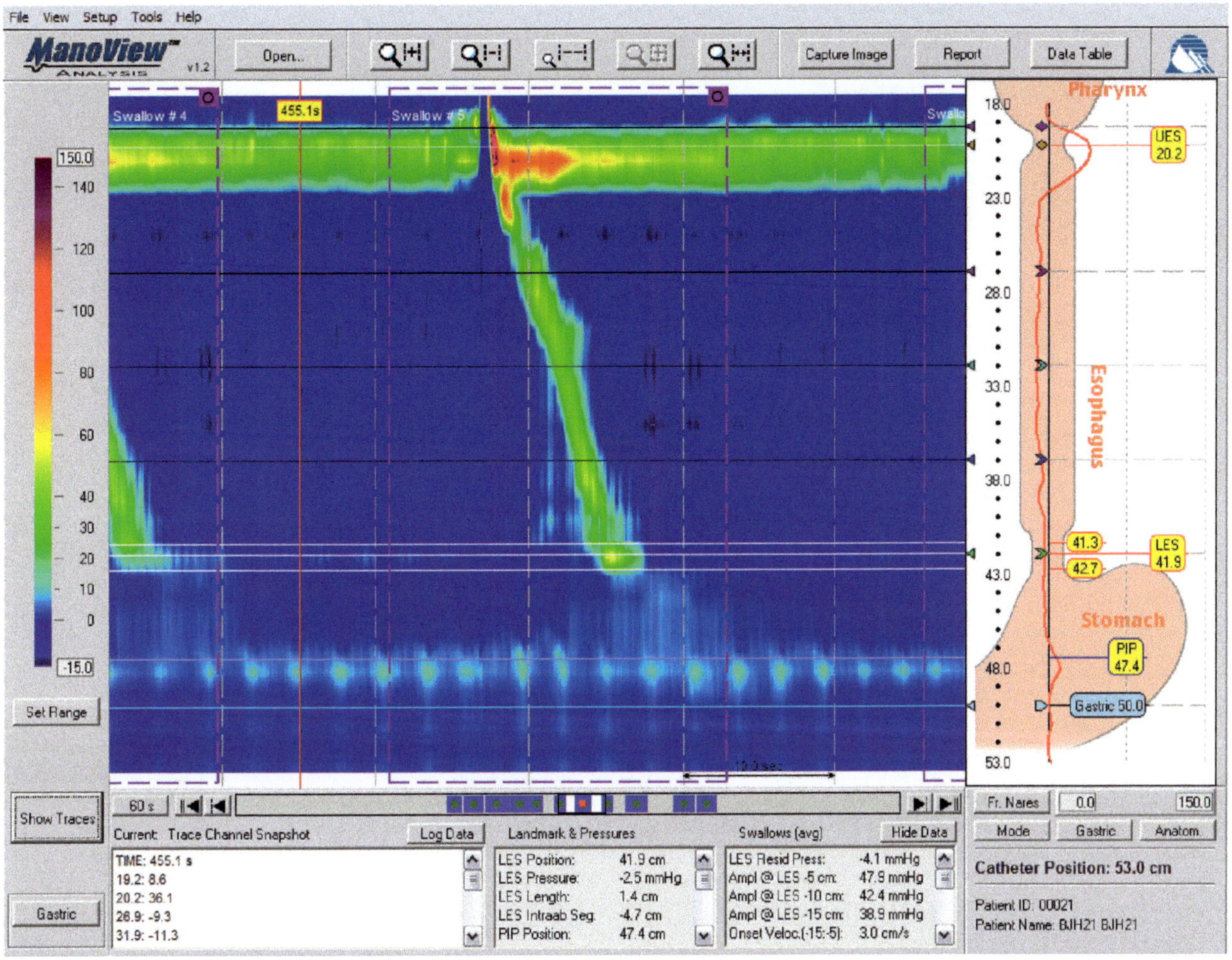

FIGURE 7.9 High-resolution manometry of a defective lower esophageal sphincter (LES) with complete destruction of the abdominal LES. The peristaltic wave ends at 43 cm from the nares, which is the end of the tubular esophagus. The LES extends from 41.3 to 42.7 cm (total length 1.4 cm with zero abdominal length). Some part of the area from 43 cm (end of the tubular esophagus) to 35 mm distal to the diaphragm is represented by the dilated distal esophagus.

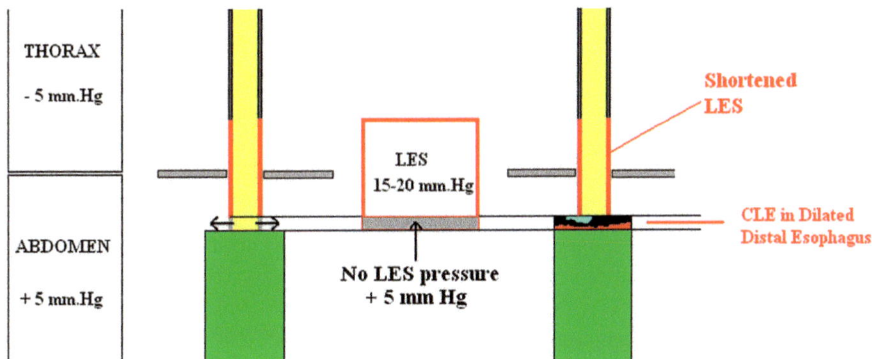

FIGURE 7.10 Diagrammatic representation of the lower esophageal sphincter (LES) showing LES damage of 5 mm at its distal end. This has caused cardiac metaplasia (blue, black, and red) and the dilated distal esophagus that measures 5 mm in length. The residual functional LES is shortened by 5 mm at manometry. The anatomic esophagus ends 5 mm distal to the distal end of the manometric LES. *CLE*, columnar-lined esophagus.

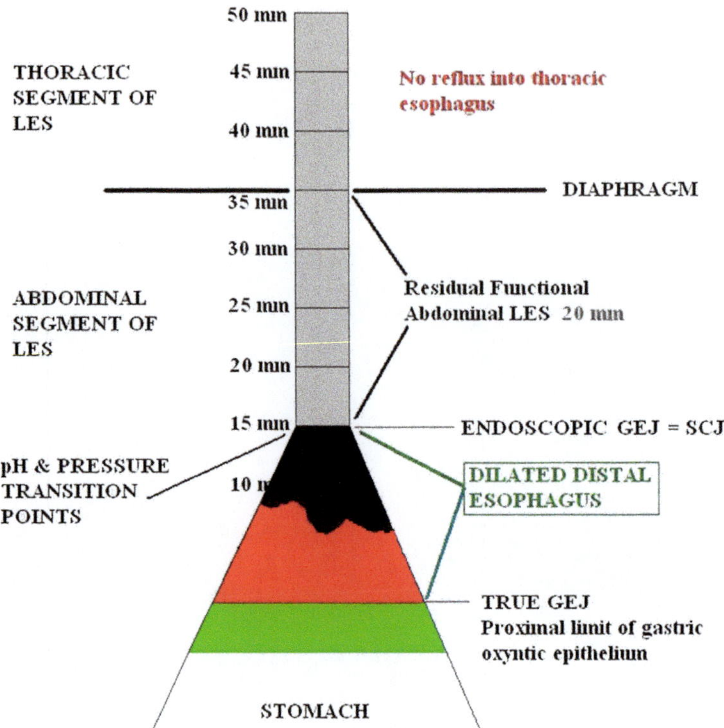

FIGURE 7.11 Diagrammatic representation of the lower esophageal sphincter (LES) showing LES damage of 15 mm at its distal end. This is associated with cardiac metaplasia (black and red) of the dilated distal esophagus that has resulted from the 15 mm of abdominal LES damage. The residual abdominal segment of the LES is now 20 mm in length at manometry. The anatomic esophagus that includes the dilated distal esophagus ends 15 mm distal to the distal end of the manometric LES. *CLE*, columnar-lined esophagus; *GEJ*, gastroesophageal junction; *SCJ*, squamocolumnar junction. Squamous epithelium = gray; cardiac epithelium = black; oxyntocardiac epithelium = red; gastric oxyntic epithelium = green.

6. MECHANISMS OF LOWER ESOPHAGEAL SPHINCTER RELAXATION IN THE PHYSIOLOGIC STATE

LES relaxation is an essential normal function. It is designed to allow swallowed food to enter into the stomach without any obstruction and to allow venting of the stomach when it is overdistended with air swallowed by aerophagy during meals or otherwise. The LES also "relaxes" during vomiting. This is different than the relaxation that occurs during swallowing and belching.

6.1 Lower Esophageal Sphincter Relaxation During Swallowing

The swallowing (deglutition) reflex is initiated by the entry of the food bolus into the pharynx. The propulsion of the bolus from the mouth is a voluntary action produced by contraction of the tongue muscles. Food in the pharynx stimulates afferent endings therein and results in the almost immediate relaxation of both the upper and lower esophageal sphincters while at the same time initiating a propulsive peristaltic wave (Fig. 7.7). The coordinated wave begins in the pharynx and traverses the entire esophagus, pushing the food bolus into the stomach through the relaxed sphincters.

The coordinated peristaltic wave and sphincter relaxation requires the central control of the deglutition center in the brain stem as well as the local neuromuscular complex of the esophageal wall. Diseases involving the brain stem such as bulbar and pseudobulbar palsy interfere with this process. Similarly, diseases involving the neuromuscular complex of the esophagus such as achalasia and scleroderma interfere with peristalsis and relaxation of the LES.

In the normal state, however, the LES relaxes completely at the time the deglutition reflex is initiated. When the peristaltic wave begins in the pharyngeal muscle, the pharynx, the esophagus, and the stomach essentially convert into a single open cavity without barriers created by high-pressure sphincter zones. As soon as the peristaltic wave traverses the LES, pushing the food bolus before it, the LES returns to its resting high-pressure state (Fig. 7.7). If food remains in the esophagus after a swallow, it stimulates the generation of a secondary peristaltic wave in the body of the esophagus. This propels the food into the stomach through an LES that relaxes with the secondary wave.

6.2 Lower Esophageal Sphincter Relaxation During Belching

McNally et al.[26] in 1964 were the first to study the effect of distending the stomach with air. When air was injected into the stomach, there was a slight initial rise in the intragastric pressure, after which the intragastric pressure plateaued in the face of continued injection of air. This suggested that the stomach had a mechanism whereby it relaxed to accommodate the air.

At inconsistent time intervals after gastric pressure had plateaued, either simple gastroesophageal gas reflux or belching (gastroesophagopharyngeal gas reflux) occurs. During simple gas reflux into the esophagus, the LES opened widely to establish a common gastroesophageal cavity. The air entering the esophagus was quickly emptied by a secondary peristaltic wave that returned the air to the stomach and reestablished LES pressure. The person did not experience any sensation during these simple gas reflux episodes into the esophagus.

Belching, on the other hand, appeared to require contraction of the somatic musculature, usually superimposed on an antecedent or simultaneous relaxation of the LES, resulting in expulsion of the air from the esophagus into the pharynx as a belch. The exact mechanism whereby distension of the stomach with air caused relaxation of the LES was not determined.

6.3 Lower Esophageal Sphincter Relaxation During Vomiting

Vomiting is mediated by the central vomiting center as a result of a myriad of stimuli. During vomiting, contraction of the gastric muscle results in a rapid increase in gastric wall pressure. This overcomes the resistance of the LES and causes vomiting. This is not a primary physiologic relaxation of the LES; it is a forced opening of the LES by an increased intragastric pressure created by the vomiting mechanism. This centrally mediated vomiting mechanism causes muscular contraction of the gastric wall irrespective of the presence of gastric contents. When the stomach is empty, there is no reflux; this is called retching. Vomiting is commonly preceded by the sensation of nausea that does not accompany belching or pathologic relaxation of the LES in GERD.

7. MECHANISM OF MAINTAINING LOWER ESOPHAGEAL SPHINCTER HIGH PRESSURE

The high-pressure zone of the LES is normally maintained by tonic contraction of the esophageal smooth muscle in the sphincter zone. There are few or no data on the exact mechanism of how this tonic contraction of the smooth muscle of the sphincter is maintained.

When there is no explanation or study on a subject, the most rational course is to look at other areas of the body for a similar phenomenon. The best understood mechanism of how muscle tone is maintained is in skeletal muscle. Muscle tone is an inherent feature of skeletal muscle. It is maintained by a reflex arc that is controlled by the alpha motor neuron in the anterior horn of the spinal cord. The reflex arc has an afferent limb, which is a sensory nerve that arises in stretch receptors in skeletal muscle known as muscle spindles. These synapse directly with the alpha motor neuron in the spinal cord. The reflex arc is completed by efferent motor fibers that go back to the muscle and produce motor unit contraction that maintains normal tone.

Muscle tone in skeletal muscle is lost when this lower motor neuron reflex is interrupted at any point in the reflex arc. The pyramidal and extrapyramidal tracts that descend from the brain influence muscle tone in a highly complex manner resulting in alterations in muscle tone in diseases of the motor cortex, cerebellum, and the extrapyramidal centers in the brain stem. However, as long as the reflex arc is maintained, muscle tone is not lost.

It is known that tone in the LES is independent of external innervation because it is retained in the completely denervated esophagus. The mechanism must therefore be a local reflex arc in the wall of the esophagus (Fig. 7.12). The central control of the reflex arc must reside in the ganglion cells in the myenteric and submucosal plexuses. The effector mechanism of this reflex must be a relay of impulses from the ganglion cells to the smooth muscle of the LES. The afferent arc must involve sensory impulses of some kind.

There are no data to indicate that this reflex arc exists in the esophageal wall. If there is such a reflex, there are no data regarding the afferent end of the reflex arc or how this is stimulated normally.

The logical, but not the only possible, origin for receptors for the afferent limb of this reflex arc is the squamous epithelium.

The excellent study by Rodrigo et al. from the Cajal Institute, Madrid, Spain, used the esophagi from 10 cats and 3 rhesus monkeys.[27] The esophagi were impregnated throughout their length by osmium tetroxide–zinc iodide, a technique that permits recognizing the course of the nerve fibers, discerning their connections, and locating their endings.

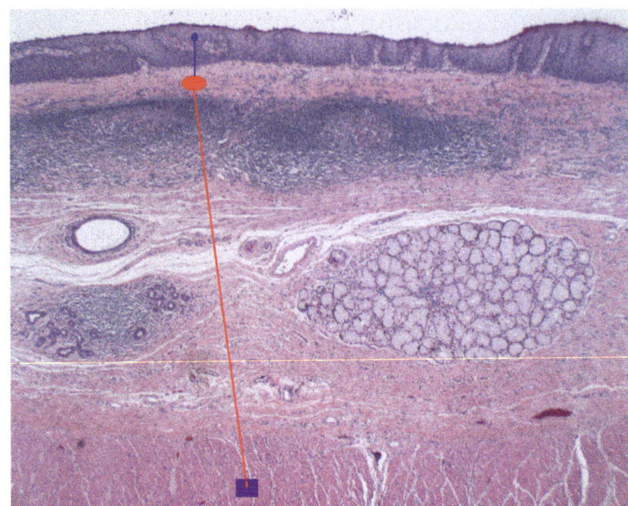

FIGURE 7.12 The hypothesized mechanism whereby the lower esophageal sphincter (LES) tone is maintained normally. Afferent receptors in nerve endings within the squamous epithelium (blue) pass along afferent nerves to the submucosal and/or myenteric plexus ganglion cells. From these neural centers, an efferent stimulus passes to the smooth muscle of the wall producing the contraction that maintains LES pressure.

The nerve fibers reaching the basement membrane region come from the distal part of the submucosal nerve plexus. In the proximity of the submucosal plexus the nonmyelinated nerve fibers are grouped in bundles and surrounded by a common Schwann protoplasm. In the neighborhood of the basal layer, they form a subepithelial plexus consisting of groups of 2–3 fibers and later completely isolated single fibers, commonly associated with capillaries. They then pass through the basement membrane as single fibers and take a sinuous course upward between the spaces of the epithelial cells.

The density of nerve fibers in the esophageal epithelium is greatest in the upper and lower parts immediately distal to the pharynx and cardia, respectively. The epithelium of the middle portion of the esophagus is sparsely innervated with only occasional nerve fibers.

The intraepithelial nerve fibers bifurcate within the epithelium and have a beaded appearance because of dilatations. They end freely at variable levels in the spaces between the epithelial cells. While some end in the basal zone and others end in the mid region of the epithelium, most of the fibers pass upward to end near the surface where they are separated from the lumen by the cytoplasm of one or two epithelial cells. The free endings have various shapes, described as resembling buds or buttons, pear-shaped or cup-shaped.

In Chapter 2, I considered the likelihood that some of these intraepithelial nerve endings were responsible for perception of pain in GERD. However, it is possible that some of these nerve endings also represent the origin of the afferent limb of the reflex arc that is responsible for maintaining tone in the LES.

I emphasize that there is no evidence in support of this suggestion. However, in the absence of any studies whatsoever, I should be allowed the liberty to present a reasonable hypothesis. Hopefully, this will be an impetus for further studies.

My bias toward suggesting this mechanism is based on my observation that columnar metaplasia of the distal esophagus is associated with shortening of the LES to the exact extent of the columnar metaplasia. The length of cardiac epithelium (with and without parietal and/or goblet cells) is concordant with the length of the DDE, which results from loss of LES pressure.[9] This suggests a direct mechanism of LES damage rather than one that is indirect, such as local inflammation associated with squamous epithelial damage.

If it is true that the local reflex arc is dependent on afferent impulses from nerve endings in the squamous epithelium, the effect of columnar metaplasia has the effect of interrupting the afferent arc of the reflex. The removal of all but the basal region of the epithelium during columnar metaplasia will cause removal of all endings in the mid- and superficial part of the squamous epithelium.

This would explain the concordance that we have shown exists between the length of cardiac epithelium (with and without parietal and/or goblet cells) distal to the endoscopic GEJ, shortening of the abdominal segment of the sphincter, and the DDE.

That columnar metaplastic epithelia in the esophagus are less sensitive to acid proves that its sensory innervation is less than that of squamous epithelium. It is not unreasonable to suggest that a similar decline in nerve endings responsible for the afferent limb of the reflex arc is responsible for the loss of muscle tone in the region that has undergone columnar metaplasia.

8. PATHOLOGICAL LOWER ESOPHAGEAL SPHINCTER RELAXATION

The relaxation of the LES at any time other than during swallowing, belching, and vomiting represents pathologic LES relaxation. Removal of the barrier normally created by the high pressure of the LES during such times results in the flow of gastric contents into the thoracic esophagus along the pressure gradient that exists between the stomach (+5 mmHg) and esophagus (−5 mmHg).

Pathologic relaxation of the LES is equivalent to LES failure. This occurs intermittently in patients with GERD leading to reflux episodes that expose the thoracic esophagus to gastric contents. The number and duration of reflux episodes are very variable.

In general, the more frequent and greater duration the reflux episodes, the more severe the cellular changes of GERD. The high correlations between severity of LES damage, LES failure, the occurrence of transsphincteric reflux of gastric contents into the thoracic esophagus, and cellular damage in the epithelium of the thoracic esophagus make a new method of evaluation of LES damage an extremely powerful tool in the diagnosis of GERD.

The complete failure of present medical management to address LES damage is the main reason behind the failure of present GERD management. It can be addressed only by methods that assess LES damage accurately. I will discuss pathologic LES relaxation in detail in Chapter 9.

REFERENCES

1. DeMeester TR, Wernly JA, Bryant GH, Little AG, Skinner DB. Clinical and in vitro analysis of determinants of gastroesophageal competence. A study of the principles of antireflux surgery. *Am J Surg* 1979;**137**:39–46.
2. Kahrilas PJ, Shi G, Manka M, Joehl RJ. Increased frequency of transient lower esophageal sphincter relaxation induced by gastric distension in reflux patients with hiatal hernia. *Gastroenterology* 2000;**118**:688–95.
3. Robertson EV, Derakhshan MH, Wirz AA, Lee YY, Seenan JP, Ballantyne SA, Hanvey SL, Kelman AW, Going JJ, McColl KE. Central obesity in asymptomatic volunteers is associated with increased intrasphincteric acid reflux and lengthening of the cardiac mucosa. *Gastroenterology* 2013;**145**:730–9.
4. Ganz RA, Peters JH, Morgan S, et al. Esophageal sphincter device for gastroesophageal reflux disease. *N Engl J Med* 2013;**368**:719–27.
5. Noar M, Squires P, Noar E, et al. Long-term maintenance effect of radiofrequency energy delivery for refractory GERD: a decade later. *Surg Endosc* 2014;**28**:2323–33.
6. Trad KS, Turgeon DG, Deljkich E. Long-term outcomes after transoral incisionless fundoplication in patients with GERD and LPR symptoms. *Surg Endosc* 2012;**26**:650–60.
7. Rodriguez L, Rodriguez P, Gomez B, et al. Long-term results of electrical stimulation of the lower esophageal sphincter for the treatment of gastroesophageal reflux disease. *Endoscopy* 2013;**45**:595–604.
8. Association of Directors of Anatomic and Surgical Pathology. Recommendations for reporting of resected esophageal adenocarcinomas. *Am J Surg Pathol* 2000;**31**:1188–90.
9. Chandrasoma P, Makarewicz K, Wickramasinghe K, Ma YL, DeMeester TR. A proposal for a new validated histologic definition of the gastroesophageal junction. *Hum Pathol* 2006;**37**:40–7.
10. Clarke AT, Wirz AA, Manning JJ, et al. Severe reflux disease is associated with an enlarged unbuffered proximal gastric acid pocket. *Gut* 2008;**57**:292–7.
11. Chandrasoma PT, Der R, Ma Y, et al. Histology of the gastroesophageal junction: an autopsy study. *Am J Surg Pathol* 2000;**24**:402–9.
12. Park YS, Park HJ, Kang GH, Kim CJ, Chi JG. Histology of gastroesophageal junction in fetal and pediatric autopsy. *Arch Pathol Lab Med* 2003;**127**:451–5.
13. Kastelein F. Proton pump inhibitors reduce the risk of neoplastic progression in patients with Barrett's esophagus. *Clin Gastroenterol Hepatol* 2013;**11**:382.
14. Hvid-Jensen F. Proton pump inhibitor use may not prevent oesophageal adenocarcinoma in Barrett's oesophagus. *Aliment Pharmacol Ther* 2014;**39**:984.
15. Chang EY, Morris CD, Seltman AK, et al. The effect of antireflux surgery on esophageal carcinogenesis in patients with Barrett's esophagus. A systematic review. *Ann Surg* 2007;**246**:11–21.
16. Nason KS. Gastroesophageal reflux disease symptom severity, proton pump inhibitor use, and esophageal carcinogenesis. *Arch Surg* 2011;**146**:851–8.
17. Pohl H, Sirovich B, Welch HG. Esophageal adenocarcinoma incidence: are we reaching the peak? *Cancer Epidemiol Biomark Prev* 2010;**19**:1468–70.
18. Stein HJ, Liebermann-Meffert D, DeMeester TR, Siewert JR. Three-dimensional pressure image and muscular structure of the human lower esophageal sphincter. *Surgery* 1995;**117**:692–8.
19. Ayazi S, Tamhankar A, DeMeester SR, et al. The impact of gastric distension on the lower esophageal sphincter and its exposure to acid gastric juice. *Ann Surg* 2010;**252**:57–62.
20. Marchand P. The gastro-oesophageal 'sphincter' and the mechanism of regurgitation. *Br J Surg*.
21. Allison PR, Johnstone AS, Royce GB. Short esophagus with simple peptic ulceration. *J Thorac Surg* 1943;**12**:432–57.
22. Allison PR. Peptic ulcer of the oesophagus. *Thorax* 1948;**3**:20–42.

23. Barrett NR. Chronic peptic ulcer of the oesophagus and 'oesophagitis'. *Br J Surg* 1950;**38**:175–82.
24. Allison PR, Johnstone AS. The oesophagus lined with gastric mucous membrane. *Thorax* 1953;**8**:87–101.
25. Zaninotto G, DeMeester TR, Schwizer W, Johansson K-E, Cheng S-C. The lower esophageal sphincter in health and disease. *Am J Surg* 1988;**155**:104–11.
26. McNally EF, Kelly JE, Ingelfinger FJ. Mechanism of belching: effects of gastric distension with air. *Gastroenterology* 1964;**40**:254–9.
27. Rodrigo J, Hernandez CJ, Vidal MA, Pedrosa JA. Vegetative innervation of the esophagus. III. Intraepithelial endings. *Acta Anat* 1975;**92**:242–58.
28. Kilgore SP, Ormsby AH, Gramlich TL, et al. The gastric cardia: fact or fiction. *Am J Gastroenterol* 2000;**95**:921–4.
29. Marsman WA, van Sandyck JW, Tytgat GNJ, et al. The presence and mucin histochemistry of cardiac type mucosa at the gastroesophageal junction. *Am J Gastroenterol* 2004;**99**:212–7.

Chapter 8

The Pathogenesis of Early GERD

The importance of defining the normal state is that it permits a study of how the normal state transforms to the abnormal state. I have defined the normal state by histology as an esophagus lined entirely by squamous epithelium and the proximal stomach by gastric oxyntic epithelium[1] (Chapter 5). In this normal state, the entire extent of esophageal squamous epithelium is normally protected by the lower esophageal sphincter (LES), preventing its exposure to gastric contents.

Histology is the only way to define the normal state. Histologic definition has 100% specificity and 100% sensitivity in defining the normal state (absence of metaplastic columnar epithelium) and recognizing gastroesophageal reflux disease (GERD) from its earliest point (the presence of *any* metaplastic columnar epithelium). Histology is also the only accurate way to follow progression of GERD.

In this chapter, I will describe the mechanism by which the earliest stages of the disease that leads to the future clinical entity of GERD is generated and how this differs from the normal state. The changes that occur with each millimeter of damage produce an increasingly pathologic state that can be followed cell by cell, millimeter by millimeter at a cellular level. This results in a conversion of a disease defined by symptoms that occur at an advanced stage to one that is understood from its inception and throughout its course.

The reason why this pathogenesis is easy to understand is that the only normal epithelium of this region that is damaged by exposure to gastric juice is the esophageal squamous epithelium; gastric oxyntic epithelium is resistant. Therefore, any change that occurs in the region of the gastroesophageal junction (GEJ) that is associated with reflux in any way must be a change in the esophagus.

The earliest stage of GERD that is seen in asymptomatic people is one that is defined by the presence of cellular changes that represent a deviation from the normal state without there being transsphincteric reflux into the body of the esophagus. The basic cellular change is conversion of esophageal squamous epithelium to metaplastic cardiac epithelium.

This early cellular change of cardiac metaplasia is associated with LES damage. Initially, LES damage is within its reserve capacity prior to the point at which LES failure occurs. I will call this the phase of compensated LES damage. Without reflux into the body of the esophagus, the person has no symptoms of GERD or any cellular changes in the thoracic esophagus. The observed pathology is in the area distal to the endoscopic GEJ in the dilated distal esophagus (DDE). The changes are primarily histologic and recognizable only by microscopy.

As GERD is presently defined, there is a large gap between the normal state and the onset of GERD as defined by the presence of troublesome symptoms. Troublesome symptoms require significant transsphincteric reflux that results from frequent episodes of LES failure. This is LES damage that is beyond its reserve capacity. This gap between a normal LES and one that is so severely damaged to cause reflux sufficient to cause clinical GERD is the phase of compensated LES damage. It is presently ignored and misunderstood.

The failure to recognize this early pathologic change is responsible for the failure of management of GERD. When GERD is diagnosed by the criteria of the Montreal definition,[2] it is already at an advanced stage with a significant risk of progression to severe GERD (= failure of medical therapy, Los Angeles grades C/D erosive esophagitis, Barrett esophagus, and adenocarcinoma). By the time a patient is diagnosed with GERD, the opportunity for early diagnosis and prevention of progression has disappeared.

By the Montreal definition,[2] a person is normal until he/she develops troublesome symptoms of GERD. From a symptom standpoint, everything that is not "troublesome" is not GERD. If the patient has heartburn that is not troublesome, the patient may be designated as having "episodic heartburn." If the patient has heartburn without objective evidence of reflux (e.g., in a pH test), the designations "functional heartburn" and "hypersensitive esophagus" may be used.

Some asymptomatic patients, if they have endoscopy, may have complications such as Barrett esophagus that are included in the definition of GERD.[3] Therefore, the lack of troublesome symptoms does not exclude even the most severe manifestations of GERD such as Barrett esophagus and adenocarcinoma.

GERD. http://dx.doi.org/10.1016/B978-0-12-809855-4.00008-7

The onset of GERD is not the beginning of troublesome symptoms. It is the onset of LES damage. While persons without symptoms can have the most severe cellular changes of GERD, a person without LES damage can have no manifestations of GERD. This simple difference has profound implications.

In most chronic, progressive diseases, there is a subclinical phase where pathologic evidence of disease exists without symptoms. No such subclinical asymptomatic phase is recognized in GERD. There are no presently recognized reliable histopathologic criteria for the diagnosis of early GERD. The changes that exist are misunderstood and ignored. The only recognized pathologic abnormalities in GERD are reflux esophagitis in squamous epithelium and metaplastic columnar epithelium that is visible at endoscopy.[4] These have low sensitivity and are practically useless as a means of early diagnosis of GERD.

If endoscopy is done, the presence of GERD is denoted by the presence of erosive esophagitis (Fig. 8.1) and visible columnar-lined esophagus (CLE) of varying lengths, divided into short segment (<3 cm) (Fig. 8.2) and long segment (Fig. 8.3). If these changes are not present, endoscopy is normal. A patient with GERD, defined by the presence of troublesome symptoms, is frequently normal at endoscopy. This is called nonerosive reflux disease (Fig. 8.4). It is difficult to believe that an esophagus damaged by exposure to acid for many years shows absolutely no recognizable pathologic abnormality.

CLE is presently defined as the presence of visible columnar epithelium between the endoscopic GEJ and the proximally displaced squamocolumnar junction (SCJ) (Figs. 8.2 and 8.3). Histologically, CLE in the esophagus consists of cardiac epithelium (with and without parietal and/or goblet cells) (Fig. 8.5).

These same epithelia are seen distal to the endoscopic GEJ (Fig. 8.5). They are presently believed to line the proximal stomach ("gastric cardia"). In Chapter 5, I have provided strong evidence that these epithelia are abnormal metaplastic epithelia in the distal esophagus resulting from exposure of squamous epithelium to gastric contents.

Cardiac epithelium (with and without parietal/goblet cells) distal to the endoscopic GEJ is therefore identical to the metaplastic columnar epithelium that constitutes visible CLE in the tubular esophagus. Presently, cardiac epithelia (with and without parietal/goblet cells) distal to the endoscopic GEJ are regarded as normal proximal gastric epithelia (the "gastric cardia").

If a 24-h pH test is done, abnormal reflux is defined as the presence of acid exposure (i.e., pH<4) in the distal esophagus for >4.5% of the 24-h monitoring period. This is 64 min of the day. Anything below this exposure is a "normal" 24-h pH test. A significant number of patients with GERD defined by the presence of troublesome symptoms have a normal 24-h pH test.

A competent LES is designed to completely prevent reflux. If this is true, a normal pH test should show zero acid exposure in the esophagus. The 4.5% amount of acid exposure was established as normal by taking the fifth percentile

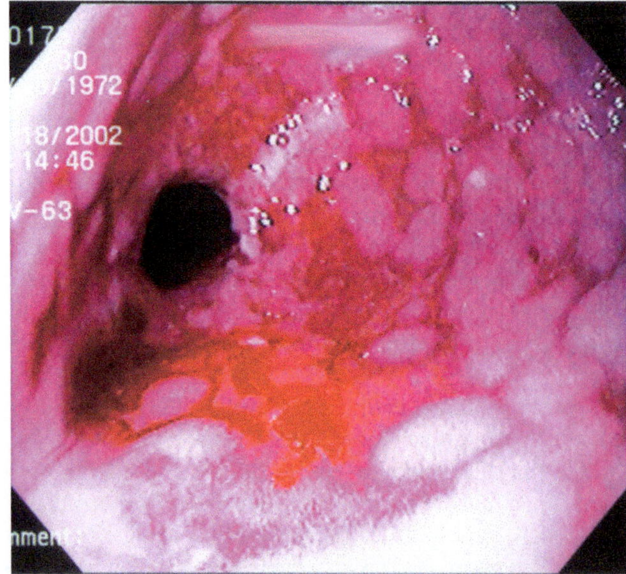

FIGURE 8.1 Severe erosive esophagitis at endoscopy. In these cases, the presence of hidden columnar metaplasia is possible, requiring reexamination after the patient has been treated with proton pump inhibitors to induce healing of erosive esophagitis.

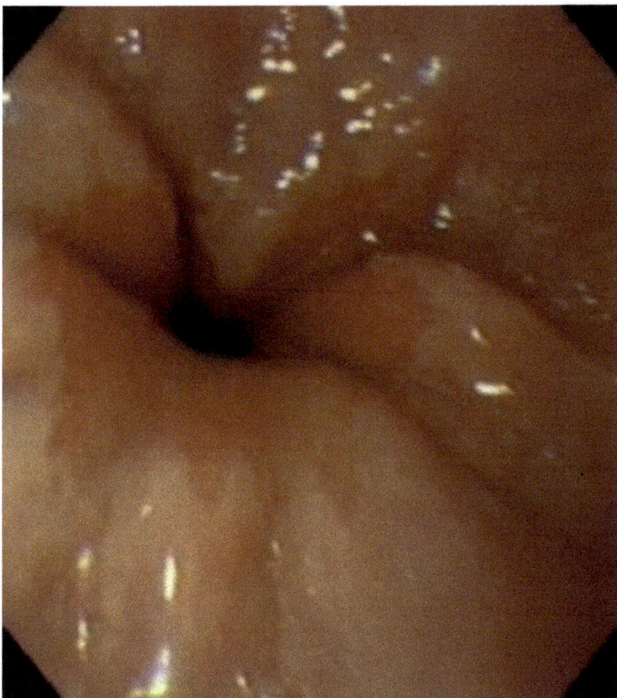

FIGURE 8.2 Short segment of visible columnar-lined esophagus at endoscopy showing short tongues of columnar epithelium extending proximally from the squamocolumnar junction, causing serration of the junction. There is significant interobserver variation in making a diagnosis of minimal (<1 cm) segments of visible columnar-lined esophagus.

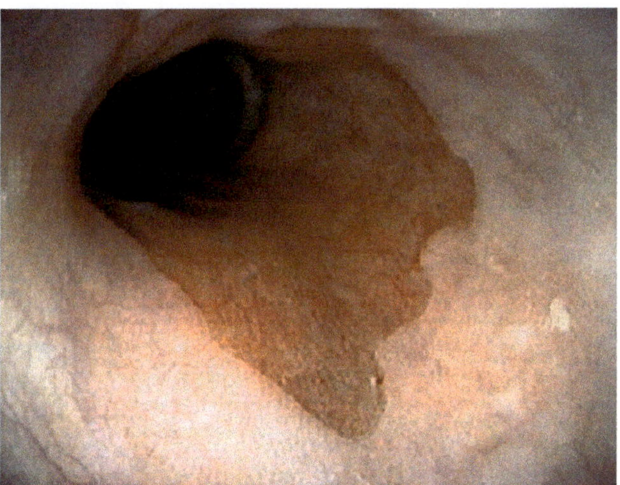

FIGURE 8.3 Antegrade endoscopic view of the proximal part of a 10-cm length of visible columnar-lined esophagus. This appears as a flat, salmon pink epithelium between the proximal limit of rugal folds and the serrated squamocolumnar junction. Biopsy showed cardiac epithelium with intestinal metaplasia. *Photograph courtesy of Dr. Martin Riegler, Reflux Medical, Vienna, Austria.*

value in volunteers where it had the greatest discriminatory value in separating people with and without symptomatic GERD.[5] However, there is no guarantee that all asymptomatic volunteers are normal. I will show that even though asymptomatic volunteers do not have symptoms of GERD, they commonly have pathology defined by histologic changes.

The distinction between symptomatic and asymptomatic GERD is best achieved by >4.5% duration of pH<4 in the esophagus. Complete absence of acid exposure in a 24-h pH test will differentiate normal from abnormal (i.e., patients with histologic evidence of LES damage but not of sufficient severity to cause LES failure) in the new definitions of GERD that I will develop.

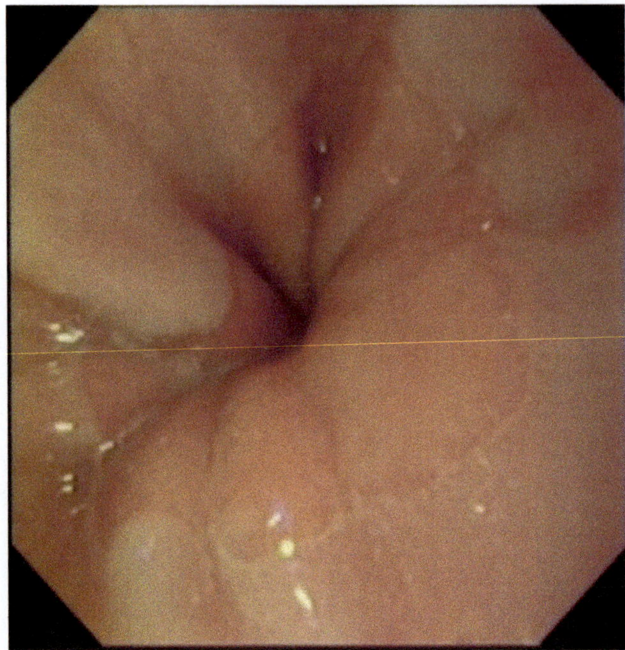

FIGURE 8.4 Normal appearance at endoscopy showing a horizontal squamocolumnar junction (Z-line) at the point of the proximal limit of rugal folds. If this patient has clinical gastroesophageal reflux disease, this represents "nonerosive reflux disease."

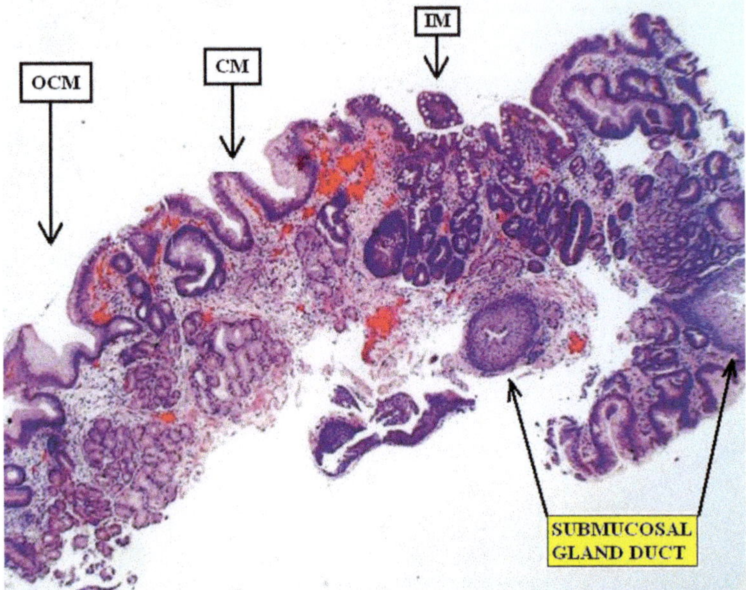

FIGURE 8.5 Retroflex biopsy of the area within 1cm distal to the endoscopic gastroesophageal junction in a patient with gastroesophageal reflux disease. This shows oxyntocardiac mucosa (OCM), cardiac mucosa (CM), and cardiac epithelium with goblet cells indicating intestinal metaplasia (IM). This is the usual distribution of epithelia in columnar-lined esophagus of any length: IM is most proximal and OCM is most distal (to the left in the figure). Note the presence of submucosal gland ducts which established that this biopsy, although taken distal to the endoscopic GEJ, is esophageal in location.

If a patient has manometry, an LES that is considered normal will have a pressure >15 mmHg, an overall length of around 5 cm, and an abdominal LES length of around 3.5 cm (Fig. 8.6A). A mechanically defective LES is defined by the presence of a mean LES pressure <6 mmHg, an overall length <2 cm, and an abdominal LES length <1 cm (Fig. 8.6B).[6] There are no criteria for the diagnosis of an LES that is damaged until the criteria for incompetence are reached. Manometry defines only a defective LES, not a damaged LES that is abnormal but not defective.

All definitions of GERD, endoscopy, acid exposure, and LES parameters by manometry are designed to detect failure of the LES that causes transsphincteric reflux into the body of the esophagus. The goal of these criteria is the high specificity of diagnosis of GERD. They are all designed to identify patients who have troublesome symptoms and complications

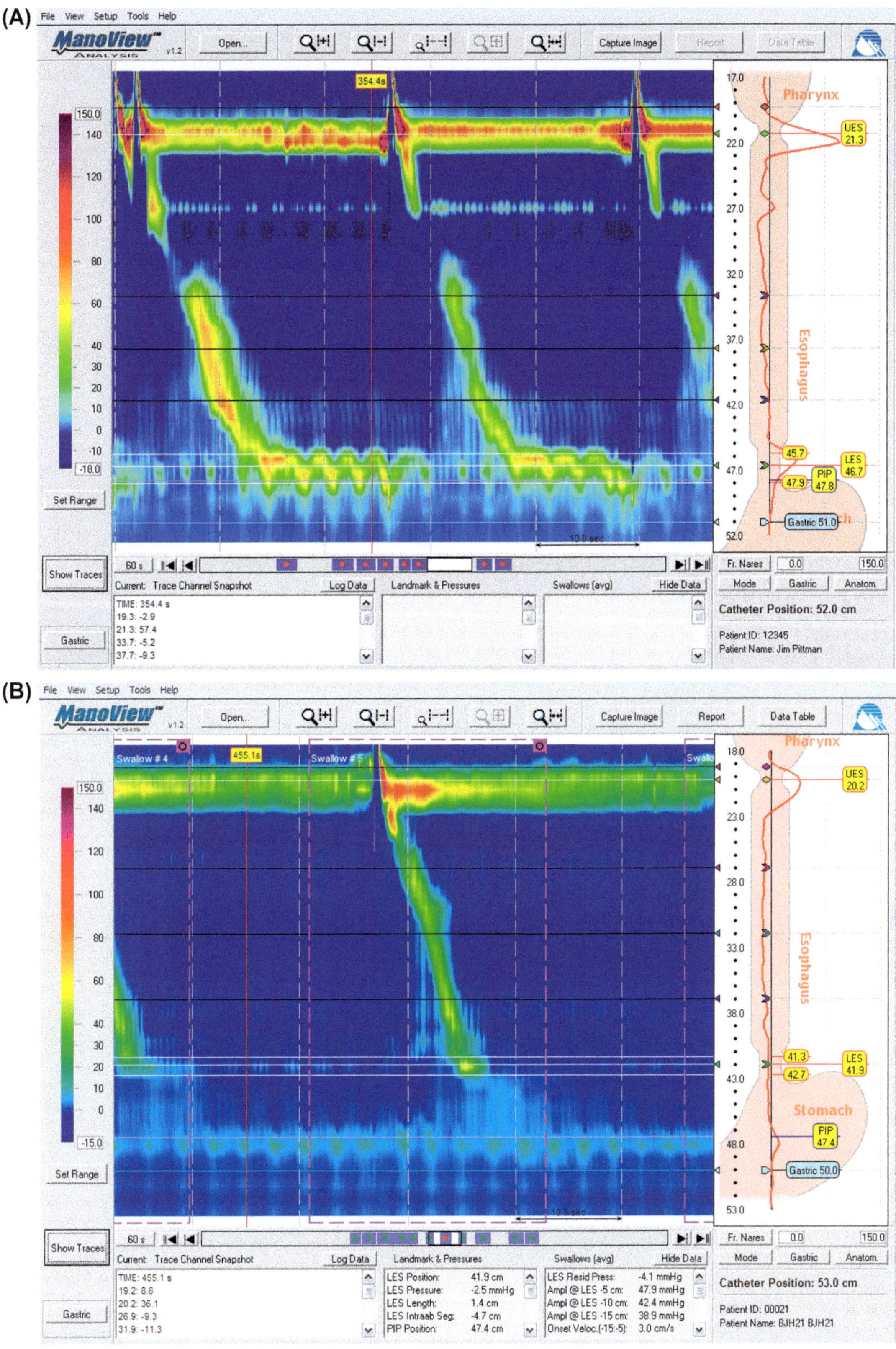

FIGURE 8.6 (A) High-resolution manometry showing a normal lower esophageal sphincter (LES) with the normal high-pressure zone of the LES and relaxation during swallowing. The LES is shown in the anatomic depiction of the waveform on the right as extending from 45.7 (from the nares) to 51.0 cm. The abdominal length from the pressure inversion point at 47.9–51.0 cm is 31 mm. (B) High-resolution manometry of a defective LES with complete destruction of the abdominal LES. The diaphragmatic pinch is at 47.4 cm. The functional LES extends from 41.3 to 42.7 cm (total length 1.4 cm with zero abdominal length). The area distal to the distal limit of the LES likely represents a hiatal hernia associated with a long dilated distal esophagus.

TABLE 8.1 Presently Accepted Criteria That Define "Normal" and "Gastroesophageal Reflux Disease (GERD)"

Presently Accepted Normal = No GERD	GERD
Absence of troublesome symptoms of GERD	Presence of troublesome symptoms of GERD
Endoscopy: absence of erosive esophagitis; absence of visible CLE	Endoscopy: presence of erosive esophagitis and/or visible CLE
Normal histology, i.e., only normal squamous epithelium above endoscopic GEJ; no visible CLE	Reflux esophagitis in squamous epithelium; metaplastic columnar epithelium in visible CLE
24-h pH test with esophageal acid exposure (pH<4) for <4.5% of 24-h testing period	24-h pH test with esophageal acid exposure (pH<4) for 4.5% or more of 24-h testing period
LES pressure 6 mmHg or greater	LES pressure <6 mmHg
Overall LES length 20 mm or more	Overall LES length <20 mm
Abdominal LES length 10 mm or more	Abdominal LES length <10 mm

CLE, columnar-lined esophagus; *GEJ*, gastroesophageal junction; *GERD*, gastroesophageal reflux disease; *LES*, lower esophageal sphincter.

TABLE 8.2 Suggested Criteria for Definition of Normal, the Phase of Compensated Lower Esophageal Sphincter (LES) Damage (pre-GERD), Mild GERD, and Severe GERD

	Normal	Phase of Compensated LES Damage (Pre-GERD)	Mild GERD	Severe GERD
Symptoms of GERD	Zero	Zero	Present; well controlled by PPIs	Present; often poorly controlled by PPIs
Endoscopy	Normal	Normal	Normal or erosive esophagitis (grade A/B)	Erosive esophagitis (grade C/D); visible CLE; neoplasia
Histology	No metaplastic CE	Metaplastic CE limited to dilated distal esophagus	Metaplastic CE limited to dilated distal esophagus	Metaplastic CE in dilated distal esophagus and visible CE
24-h pH test (% time pH<4)	Zero	>0%–1%	1%–4.5%	>4.5%
LES pressure (mmHg)	>15	<15 to >6	<6 in postprandial phase	<6 in fasting state
Overall LES length (mm)	50	>20–50 in postprandial phase	<20 in postprandial phase	<20 in fasting state
Abdominal LES length (mm)	35	>10–35 in postprandial phase	<10 in postprandial phase	<10 in fasting state

The gap between normal and GERD that falls under "no GERD" at present is exposed.
CE, cardiac epithelium; *CLE*, columnar-lined esophagus; *GERD*, gastroesophageal reflux disease; *PPI*, proton pump inhibitor.

of GERD as required in the Montreal definition.[2] This design ensures the failure to recognize early disease in subclinical GERD. All these tests lack adequate sensitivity for early detection of GERD. They commit us to late diagnosis and consequent management failure.

We need to define criteria at endoscopy, the pH test, and manometric testing of the LES that identify early deviation from the normal state. The aim should be to reach 100% sensitivity. In the case of GERD, this is practically possible only by histology.[1] In medicine, increasing sensitivity of a test is generally associated with decline in specificity. The histologic diagnosis of GERD that is 100% sensitive is also 100% specific. We have the perfect test, but we do not use it because it is ignored and misunderstood.

Table 8.1 shows the difference between people who have no GERD and patients with GERD as presently defined.

In contrast, Table 8.2 shows a gap between people who are normal and patients with GERD, characterized by a subclinical ("pre-GERD") stage of the disease between normal and what is presently defined as GERD that is completely ignored.

This gap represented by subclinical GERD is extremely large. When it is considered from the point of view of the abdominal segment of the LES, the normal (initial) length is ~35 mm and the criterion for incompetence is <10 mm. There has to be a 70% decline in the length of the abdominal LES for it to become incompetent.

70% is an interesting number. It seems to be the reserve capacity of many vital structures in the human body. Humans can maintain normal function at rest with 30% of lung tissue, renal tissue, small intestine, and coronary artery luminal size; a 70% reduction can be compensated. The 70% decline in abdominal LES function before the LES fails represents its reserve capacity. We need to identify LES damage during the long period during which it exhausts its reserve capacity, not wait until it fails. I will show that the pre-GERD phase of compensated LES damage commonly lasts many decades.

This gap of presently undiagnosed GERD is a treasure trove. When understood, it will provide the means for new approaches to manage GERD with a view to preventing progression to severe GERD characterized by the failure of proton pump inhibitors (PPIs) to control the disease, severe erosive esophagitis, Barrett esophagus, and adenocarcinoma. Severe GERD correlates best with a defective LES, and it is by concentrating on the progression of LES damage, which is the cause of disease, that we can effectively control GERD.

Table 8.1 suggests that there is a sudden transition from no GERD to GERD when certain definitional points are reached. From a definitional standpoint, this point is reached when the patient decides his/her symptoms of GERD are troublesome. The diagnostic studies for GERD (24-h pH, impedance, LES manometry) have criteria for "normal" that correlate with the clinical onset of GERD. It is no surprise that all these diagnostic tests with their presently defined normal values have little use in the diagnosis of GERD and are not recommended before empiric therapy has failed in the GERD patient.[7]

Table 8.2, on the other hand, shows that the transition from normal to severe GERD is a slow progression of abdominal LES damage that causes shortening from its initial length of ~35 mm to its level of defined incompetence at <10 mm. As LES length decreases slowly over the years, it progressively results in increasing acid exposure in the esophagus from zero to levels that are within the normal range for the test long before symptoms of GERD appear.

I will show that LES damage can be measured by histology because the length of cardiac epithelium (with and without parietal/goblet cells) distal to the endoscopic GEJ in the DDE is concordant with the shortening of the abdominal segment of the LES.

Histologic measurement of LES damage is fundamental to the study of early GERD. Manometric measurement of the residual functional LES and pH monitoring to determine esophageal acid exposure lack the required accuracy and sensitivity. Histologic measurement has an accuracy that is within millimeters with present technology but has the potential of accuracy within micrometers with new biopsy technology that can easily be developed.

1. DEFINITION OF THE SQUAMOOXYNTIC GAP

The squamooxyntic gap is a concept that I developed to more easily explain the progression of GERD.[8] It is defined histologically as the gap that is present between the distal limit of esophageal squamous epithelium (the SCJ) and the proximal limit of gastric oxyntic epithelium. It is measured either directly in a longitudinal section taken at autopsy or in a resected specimen, or by endoscopic biopsies taken at measured intervals.

The normal squamooxyntic gap is zero; that is, there is no cardiac epithelium (with and without parietal/goblet cells) between esophageal squamous epithelium and gastric oxyntic epithelium (Fig. 8.7A).

The squamooxyntic gap begins when the most distal esophageal squamous epithelium undergoes columnar metaplasia. Cardiac epithelium is generated from squamous epithelium, moving the SCJ cephalad and separating it from the proximal limit of gastric oxyntic epithelium (the true GEJ). This is the squamooxyntic gap. It is first composed of a single columnar epithelial cell measuring less than 50 μm.

As damage to the squamous epithelium occurs by exposure to gastric contents during sequential episodes of eating sufficient to produce gastric distension, the length of columnar metaplasia increases slowly, cell by cell, millimeter by millimeter, progressively increasing the length of the squamooxyntic gap.

The squamooxyntic gap can be measured histologically with absolute accuracy. Its proximal limit is the SCJ, which is well defined at endoscopy, gross examination, and microscopic examination. Its distal limit is the junction between oxyntocardiac epithelium and gastric oxyntic epithelium. The distal limit of the gap is only definable by histology.

The maximum extent of the squamooxyntic gap can extend from the true GEJ (proximal limit of rugal folds) to involve almost the entire length of the esophagus. The squamooxyntic gap is a continuous segment but develops in two phases by two completely separate pathogenetic mechanisms:

1. The squamooxyntic gap in the DDE is formed by columnar metaplasia of the abdominal segment of the esophagus (= abdominal segment of the LES). This results from exposure of the squamous epithelium to gastric contents during meals resulting in gastric overdistension, resulting in the DDE. It is a very slowly progressive process. The DDE is distal

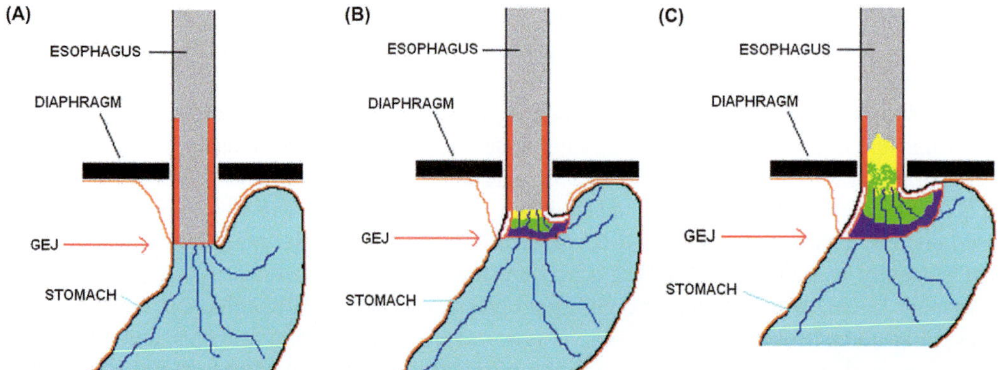

FIGURE 8.7 Diagrammatic representation of the squamooxyntic gap. In (A) there is no gap between squamous epithelium (gray) and gastric oxyntic epithelium (blue). This is the normal state. The lower esophageal sphincter (LES) is normal (red in wall of esophagus). (B) There is a gap between the squamous epithelium that has moved cephalad and gastric oxyntic epithelium, which remains in its original location. The gap is composed of cardiac epithelium with intestinal metaplasia (yellow), cardiac epithelium (green), and oxyntocardiac epithelium (purple). Note that the gap is associated with shortening of the abdominal esophagus and abdominal segment of the LES (white replacing the red LES). (C) The patient with visible columnar-lined esophagus. The gap between squamous epithelium and gastric oxyntic epithelium has increased to involve the thoracic esophagus above the endoscopic gastroesophageal junction (GEJ). The dilated distal esophagus has increased in length with more shortening of the abdominal esophagus and abdominal LES (white replacing more of the red LES). The angle of His is now virtually a right angle.

to the endoscopic GEJ and presently misinterpreted as proximal stomach. Simply put, the phase of compensated LES damage that is the precursor lesion of clinical GERD is mistaken as normal proximal stomach (Fig. 8.7B).

2. The squamooxyntic gap extends from the DDE when columnar metaplasia occurs in the thoracic esophagus. This forms the endoscopically visible CLE (Fig. 8.7C). Visible CLE results from severe transsphincteric reflux caused by frequent failure of the LES. When the LES fails, gastric contents reflux in progressively increasing frequency and volume along the pressure gradient that exists from the stomach (+5 mmHg) to the midesophagus (−5 mmHg). This part of the squamooxyntic gap develops more rapidly.

In the adult population of the United States, only very rare people will have a zero squamooxyntic gap[9]; that is, everyone has some columnar metaplasia of the distal squamous epithelium of the esophagus. Approximately 85% of patients will have a squamooxyntic gap limited to the DDE (Fig. 8.7B).[8] The length of this gap will vary from >0 to approximately 15–20 mm. The other 15% will have a visible CLE in the tubular esophagus at endoscopy, associated with a DDE that is usually >20 mm (Fig. 8.7C).

2. DEFINITION OF THE DILATED DISTAL ESOPHAGUS

The DDE is that area distal to the endoscopic GEJ that is lined by cardiac epithelium (with and without parietal/goblet cells).[10] Its proximal limit is the endoscopic GEJ. In a person with normal endoscopy, this is also the SCJ. Its distal limit is the true GEJ, which is the proximal limit of gastric oxyntic epithelium. It is the part of the squamooxyntic gap that is mistaken for "gastric cardia" by the present endoscopic definition of the GEJ.

The DDE develops as a result of loss of the high-pressure zone in the abdominal segment of the LES. This begins distally and is associated with columnar metaplasia of distal esophageal squamous epithelium.

I have suggested in Chapter 7 that the mechanism of maintenance of the high-pressure zone in the LES is dependent on a local neural reflex arc in the esophagus that maintains tonic contraction of the smooth muscle of the LES (see Fig. 7.12). This is mediated by afferents in the squamous epithelium that relay impulses to ganglion cells in the myenteric and submucosal plexuses, which send effector signals to the smooth muscle to maintain tonic contraction.

Columnar metaplasia of the squamous epithelium changes the intraepithelial innervation and alters the afferent input, interrupting the reflex and causing loss of muscle tone. LES pressure drops to zero signifying sphincter damage. This would explain the observed concordance between the length of cardiac epithelium (with and without parietal/goblet cells) distal to the endoscopic GEJ and the length of the DDE.[11]

When LES pressure is lost, irrespective of its mechanism, the intraluminal pressure of the distal esophagus that has lost LES tone becomes equal to the baseline gastric intraluminal pressure (+5 mmHg). The lack of LES muscle tone results

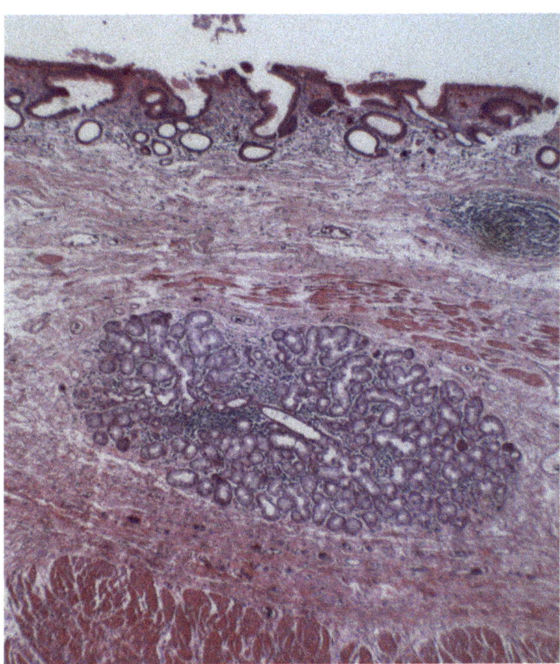

FIGURE 8.8 Cardiac epithelium distal to the end of the tubular esophagus in an esophagectomy specimen. The presence of submucosal glands confirms that this is esophageal in location and in the dilated distal esophagus. Rugal folds were present in the mucosa lined by cardiac epithelium.

in the failure to resist increased intraluminal pressure or increased intragastric pressure associated with gastric distension. Whenever the stomach distends, the part of the distal esophagus that has lost sphincter pressure dilates with the stomach.[1]

When tone is lost in the abdominal segment of the LES, that part of the esophagus ceases to function as a tube that transmits food to the stomach; it now functions as part of the reservoir distal to the conduit. As part of the reservoir, the damaged distal esophagus dilates with gastric filling and collapses with gastric emptying.

Over time, it develops rugal folds in its metaplastic columnar lining. It now exactly resembles the proximal stomach at endoscopy. The DDE is different from the proximal stomach in the following respects (Fig. 8.7B): (1) It is lined by cardiac epithelium (with and without parietal/goblet cells), which is metaplastic and esophageal; (2) its muscle wall is esophageal, not gastric and represents the abdominal LES that has shortened; (3) its submucosa contains esophageal mucous glands that drain to the surface of the metaplastic columnar epithelium through ducts (Fig. 8.8); and (4) it is above the peritoneal reflection.

Anatomically, the length of the abdominal segment of the esophagus has become shorter, ending closer to the diaphragmatic hiatus, and the angle of His has become more obtuse.[12]

These changes in the damaged distal abdominal esophagus are hidden to endoscopic view. They can only be detected by the use of appropriate biopsies to detect the metaplastic cardiac epithelium. These biopsies are not recommended by gastroenterology association guidelines and not done by gastroenterologists.[7] They were part of the routine biopsy protocol used by Dr. Tom DeMeester that I had the privilege and luck to examine.

In a way, this pathologic change can be regarded as a compensatory mechanism whereby the distal esophagus sacrifices itself to increase the capacity of the reservoir to allow a higher intake of food. Like many physiologic compensations, the negative impact is permanent columnar metaplasia and permanent shortening of the LES.

The length of the DDE resulting from LES damage is equal to the length of the squamooxyntic gap distal to the end of the tubular esophagus and the proximal limit of rugal folds (the present endoscopic GEJ). Measuring the length of the squamooxyntic gap distal to the endoscopic GEJ gives an accurate measure of damage to the abdominal segment of the LES.

This is the new tool in the armamentarium to investigate the pathogenesis of GERD from its outset where a single columnar cell replaces the most distal squamous cell in the esophagus. We suddenly have micrometer-level accuracy to understand progression of GERD.

There is direct evidence of dilatation of the distal esophagus that correlates with the severity of GERD. Korn et al.[13] measured the circumference of the esophagus at surgery at the point of the peritoneal reflection. This showed a progressive increase in circumference from normal to patients with increasing GERD. This paper is reviewed completely in Chapter 10.

I have used several terms in the past for the DDE. The first term I used was "gastricization" of the distal esophagus.[14] This was meant to show that when the abdominal segment of the LES loses its pressure, that part of the esophagus takes on the anatomic and functional characteristics of the stomach. It dilates, develops columnar epithelium by metaplasia of the original squamous epithelium, and functions as part of the gastric reservoir. It no longer functions as the esophagus. This was a good term. I thought it would resonate with people because it resembled the familiar "thyroidization" of the end-stage kidney or "bronchiolization" of the abnormal air spaces in honeycomb lung. I was wrong. I should have realized that the problem was not the term but the resistance to the concept that prevented acceptance.

In the GERD book that Dr. DeMeester and I authored in 2006,[15] we used the term "dilated end-stage esophagus." This was supposed to indicate that the loss of sphincter pressure was in effect an end stage for the esophagus where it ceased to have its normal function of transmission of food into the stomach and instead dilated and functioned as part of the reservoir that accepted a meal. This was also a good term. Again, we thought it would resonate with the reader because of similar terminology to describe end-stage kidney and lung.

However, on a balcony overlooking the Adriatic Sea in a hotel in Dubrovnik, sipping a drink in the most salubrious climate, Dr. George Triadafilopoulos suggested to us that while he did not disagree with the concept, he thought the term was confusing. In his mind, dilated end-stage esophagus was the term he would use to describe the sigmoid esophagus that characterizes the end stage of achalasia. A few drinks later, George, almost like Norman Barrett suggesting that Allison's term "esophagus lined by gastric mucous membrane" should be replaced with "columnar-lined esophagus," came up with the term "dilated distal esophagus." We agreed. It is nice to give credit to the person who came up with the term that we now use. Dr. Triadafilopoulos is also a rare gastroenterologist who agrees at least in part with our concept.

3. THE MECHANISM OF DEVELOPMENT OF THE SQUAMOOXYNTIC GAP IN THE DILATED DISTAL ESOPHAGUS

I will now explore the mechanism whereby the normal state is converted to the earliest manifestations of GERD. The evidence base that exists is sparse, largely because most physicians have no interest in studying people without clinical GERD. However, those studies that have been done on asymptomatic volunteers paint a clear picture of the earliest stages of GERD.

The early studies used asymptomatic volunteers to define the correlation of GERD with acid exposure in the distal esophagus in the 24-h pH test and parameters of the LES in manometric tests that were developed in the 1960s.[5,6] These tests failed to define the mechanism of early GERD, largely because the lack of symptoms of GERD resulted in these people being classified as normal. In the absence of a "pre-GERD" concept, the range of manometric findings and acid exposure that these asymptomatic volunteers expressed was regarded as normal individual variation rather than an expression of pathology.

Histology could not be used to study the mechanism of early GERD because it was, and still largely is, completely misunderstood. Most physicians believe that cardiac epithelium (with and without parietal cells) exists normally between the esophageal squamous epithelium and gastric oxyntic epithelium. Although the extent of this "normal" cardiac epithelium has shrunk to "less than 4 mm,"[16] in the "gastric cardia," most physicians readily accept that up to ~3 cm of cardiac epithelium can be normally found distal to the endoscopic GEJ (the proximal limit of rugal folds and/or the end of the tubular esophagus). The presence of >4 mm of cardiac epithelium (with and without parietal/goblet cells) distal to the endoscopic GEJ is not considered to be pathological. This, being distal to the endoscopic GEJ, is regarded as the proximal stomach and therefore unrelated to GERD.

Removal of this false dogma is the key to complete understanding of the pathogenesis of GERD from the point of its first change from the normal state. The suggestion by our group[17] that cardiac epithelium was a metaplastic esophageal rather than a normal gastric epithelium led to a small number of important studies designed to test this hypothesis.

3.1 Exposure of Esophageal Squamous Epithelium to Acidic Gastric Contents

An experimental human study by Ayazi et al.[18] hypothesizes: "Sphincter damage occurs when gastric distension causes progressive and prolonged effacement of the abdominal portion of the LES, causing prolonged exposure of the effaced mucosa and sphincter to acid gastric juice." The aim of the study was to assess the effect of gastric dilatation on LES length and resting pressure in 11 normal subjects and to evaluate the degree to which the mucosa and sphincter are exposed to acid gastric juice.

The study used a slow motorized puller device and a transnasal catheter with eight pressure sensors placed 10 mm apart to obtain a continuous pressure profile of the LES with precise measurement of the total and abdominal LES length. This was done in all subjects after an overnight fast. After the baseline measurement was made, the manometry catheter was placed back into the stomach and positioned such that the catheter straddled the LES with at least one sensor in the stomach.

A standard pH catheter was passed transnasally into the stomach and withdrawn at 5 mm increments every 30 s. This allowed localization of the pH step-up point defined by a rise from baseline gastric pH to a pH > 4. The location of this point was measured from the nares and related to the position of the LES.

The stomach was then distended gradually by insufflating air through the central lumen of the manometry catheter. Air boluses of 50 mL were injected every 30 s until a total of 750 mL was infused. The locations of the upper and lower borders of the LES were measured in reference to the nares after each 50 mL air bolus without moving the pressure catheter. After a total of 750 mL of air was infused, the pH step-up point was again measured and related to the position of the LES.

Continuous measurement of the LES during gastric distension showed a strong inverse correlation between LES overall length and gastric volume; that is, there was a progressive decrease in overall LES length as the volume of insufflated air increased.

Gastric distension moved the location of the lower border of the LES proximally from a median of 44.5 to 42.8 cm from the nares ($P = .001$). The location of the upper border of the LES did not change significantly. This resulted in a reduction of the overall LES length from a median of 34 to 26 mm, almost entirely because of a reduction in the length of the abdominal segment of the LES from a median of 26 to 14 mm ($P = .001$).

Gastric distension resulted in a modest decrease in the median resting pressure of the LES from 27.4 to 23.4 mmHg ($P = .02$). None of the patients showed a complete loss of LES pressure with 750 mL of air insufflation.

The pH step-up point moved proximally with gastric distension from a median of 44 to 42.5 cm from the nares ($P = .001$). The gastric distension shifted the pH step-up point cephalad a median of 10 mm (range 6–60 mm; the LES opened completely in one subject in whom the pH transition point moved into the esophagus).

This study localizes the lower border of the LES as it shortens as well as the pH step-up point with gastric distension. In the fasting state the lower border of the manometrically defined LES is at a median of 44.5 cm from the nares. This point, in a person who is endoscopically normal, would be the SCJ. When the stomach distends, the lower border of the functional LES moves cephalad by a median of 17 mm (from 44.5 to 42.8 cm), almost entirely due to a shortening of the abdominal segment of the LES by a median of 12 mm (from 26 to 14 mm).

It is important to understand what is happening here from the histologic standpoint. In the fasting state, the lower border of the LES (lined entirely by squamous epithelium) is at 44.5 cm (median) from the nares. The pH step-up point in the fasting state is 44 cm (median) from the nares. Given that these numbers are medians, it is reasonable to believe that in the individual person, the pH step-up point is exactly at the point of the lower border of the LES. This is also the SCJ in the normal person. As such, the squamous epithelium is not exposed to gastric contents in the fasting state.

When the stomach is distended with air, the manometric measurement of the LES shortens by 17 mm from its position at 44.5 cm from the nares in the fasting state to 42.8 cm in the distended state. The pH step-up point also moves from a point 44 cm (median) from the nares in the fasting state to 42.5 cm in the state of gastric distension. Again, the pH step-up point is concordant with the lower border of the functional LES in the distended state (42.8 vs. 42.5 cm from the nares).

What is critical to understand is that shortening of the LES by 17 mm results in the distal 17 mm of squamous epithelium lining the abdominal LES being exposed to gastric juice because the pH step-up point has moved cephalad by an amount that is equal to the dynamic shortening of the LES during gastric overdistension (Figs. 8.9 and 8.10).

This should be easily understood by gastroenterologists who perform endoscopy. If the junctional region is viewed in a retroflex view, the SCJ can be seen to descend as the stomach is insufflated with air. This is because gastric distension effaces (i.e., shortens) the lower part of the abdominal LES that is lined by squamous epithelium. The SCJ therefore appears to descend into the stomach.

It is the terminology here that is confusing. The LES never "shortens"; it retains its original length and is lined by squamous epithelium. With gastric distension, the high-pressure zone that defines the LES manometrically shortens from below. The squamous epithelium lining the lower part of the abdominal LES comes to lie below the pH step-up point and is exposed to gastric juice for the duration of the time the stomach remains distended.

In a separate study from the same unit, Tamhankar et al.[19] showed that after gastric distension produced by ingestion of a large fatty meal, the squamous epithelium can remain exposed to gastric juice for up to 150 min.

The study by Ayazi et al.[18] provides valuable data regarding the pathogenesis of early GERD in asymptomatic subjects. In the following study from Glasgow,[20] the placement of a clip to mark the SCJ provided additional insight into the process.

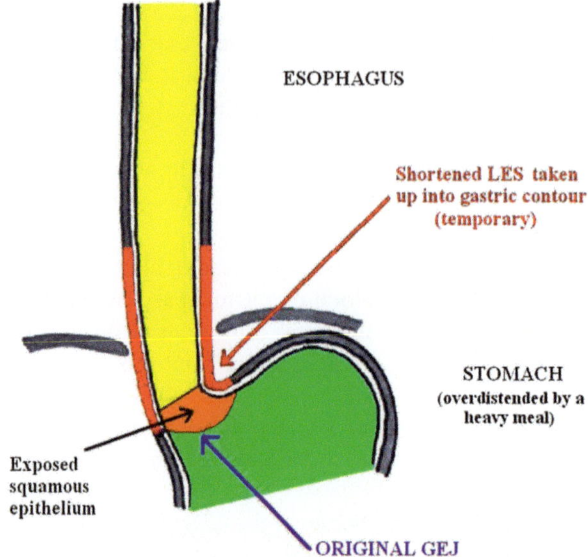

FIGURE 8.9 Dynamic shortening of the lower esophageal sphincter (LES) during gastric overdistension. The effaced LES has moved distally into the contour of the stomach on the greater curvature. This has caused the distal end of the squamous epithelium (shown as orange) to descend below the pH transition point.

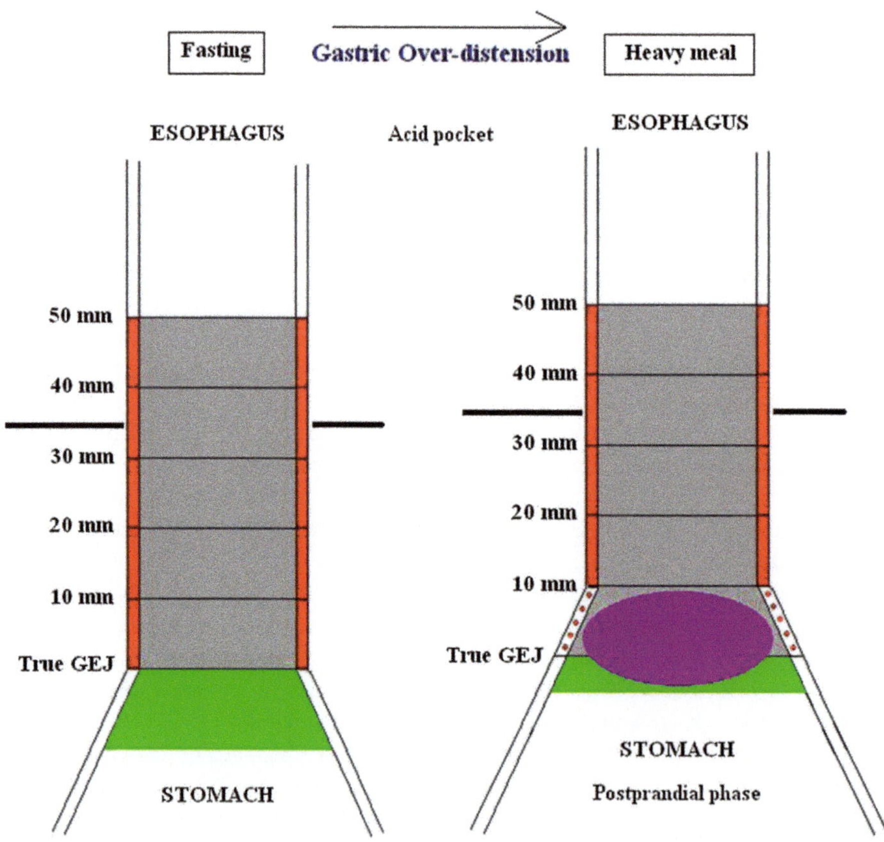

FIGURE 8.10 Dynamic effacement of the lower esophageal sphincter (LES) (shown as white wall with red dots) that has moved into the contour of the stomach with the squamocolumnar junction, exposing the distal part of the squamous epithelium (gray) to the acid pocket (purple). The manometric abdominal LES has shortened by the amount of effaced LES. *GEJ*, gastroesophageal junction.

Robertson et al.[20] is an elegant study using sophisticated technology on asymptomatic volunteers. The study group is divided into 24 with small waist circumference (12 females; median age 45 years, range 21–73 years) and 27 with large waist circumference (13 females; median age 46 years, range 21–71 years) (P = .610). The body mass index was a median of 23.6 kg/m^2 (range 16.7–26.7 kg/m^2) in the small waist circumference group and 30.5 kg/m^2 (range 25.3–43.3 kg/m^2) in the large waist circumference group (P = <.001).

The patients underwent MRI assessment of fat distribution. They then had upper endoscopy with biopsies across the SCJ using a large capacity 8 mm forceps. After biopsy, the SCJ was marked by two endoclips.

The patients were tested with the combined pH and manometry catheter after a 12 h fast with the subjects upright for 15 min and supine for 15 min. Subjects then consumed a standardized meal (battered fish and chips with 150 mL of water) over a 20-min period and were asked to eat until full. After the meal pH and manometric testing was done for 45 min with the patients upright and 30 min in the supine position. In each of the four periods of recording (fasting upright, fasting supine, postprandial upright, and postprandial supine) fluoroscopic screening was done to visualize the endoscopically placed clips.

pH recordings were taken using a high-resolution custom-made probe composed of 12 pH electrodes. The most distal electrodes were placed at the tip of the catheter and the other 11 were 3, 5, 6.1, 7.2, 8.3, 9.4, 10.5, 11.6, 12.7, 13.8, and 17.2 cm proximal to it. The catheter was placed so that the most proximal electrode was 5 cm proximal to the upper border of the LES, and the pH electrodes at 1.1 cm intervals straddled the LES. Acid exposure was the mean pH calculated for each of the 12 pH sensors for each phase of the study.

The pH transition point was defined by the index sensor recording a decrease in mean pH of at least one unit from proximal to distal and corrected for the 1.1 cm distance between the sensors. This is the point of transition from gastric pH to esophageal pH. Acid exposure at the pH transition point was recorded as the mean percent time pH<4. The location of the pH transition point was determined relative to the upper border of the LES.

Manometry was done using a high-resolution system that had 36 circumferential sensors placed at 1.0 cm intervals. The lower border of the LES was defined as the most distal position where the pressure increased to more than 2 mmHg above gastric baseline. The upper border of the LES was defined by a change in pressure to within 2 mmHg of intrathoracic pressure.

The results of the histologic and manometric studies are shown in Table 8.3. These data are used in the study to compare the small and large waist circumference groups in this study. Selected elements of the table that are significant for the discussion are reproduced because the absolute numbers provide invaluable data.

TABLE 8.3 Summary of Differences Between Large and Small Waist Circumference (WC) Groups for Cardiac Mucosal Length and Manometric Measures

		Small WC	Large WC	*P* Value
Cardiac mucosal length, mm		1.75	2.5	.008
(Overall)LES length, cm	FU	6	4	.016
	PPU	4	3	N/S
Length of distal component of LES, cm	FU	2.82 (2.27–3.38)	2.12 (1.63–2.61)	N/S
	PPU	2.36 (1.91–2.81)	1.69 (1.25–2.13)	.026
LES pressure mmHg vs. Intragastric pressure	FU	21.9	18.1	N/S
	PPU	9.1	11.2	N/S
tLESRs, n	FU	1	0	N/S
	FS	0	0	N/S
	PPU	7	6	N/S
	PPS	1	1	N/S

FS, fasting supine; *FU*, fasting upright; *LES*, lower esophageal sphincter; *PPS*, postprandial supine; *PPU*, postprandial upright; *tLESR*, transient lower esophageal sphincter relaxation.

TABLE 8.4 Mean Acid Exposure Time (% Time pH<4) in Sensors Located 5 cm Above the Upper Border of the Lower Esophageal Sphincter, Sensors Detecting the pH Transition Point, and Sensors 1.1 cm Proximal and 1.1, 2.1, and 3.1 cm Distal to the pH Transition Point

Total			Small WC		Large WC	
Sensor Location (cm)	**FU**	**PPU**	**FU**	**PPU**	**FU**	**PPU**
Proximal						
5	0.1	1.4	0.2	1.7	0.1	0.9
1.1	3.9	4.7	3.6	4.8	4.2	4.5
pH Transition Point						
0	37.9	58.5	40.7	61.2	35.4	55.0
Distal						
1.1	71.0	71.2	68.6	70.3	73.2	72.2
2.1	81.1	65.3	83.4	65.2	79.9	65.5
3.1	85.1	49.0	82.3	42.2	87.5	56.8

Only upright position data are given for the sake of simplification; the supine data are not greatly different.
FU, fasting upright; *PPU*, postprandial upright; *WC*, waist circumference.

Biopsies in 35 volunteers (17 in the large waist circumference group and 18 in the small waist circumference group) were adequate for measuring cardiac mucosal length. The length of cardiac epithelium (measured from the squamous epithelium to the first parietal cell, thereby excluding oxyntocardiac epithelium) was 0.25 cm [interquartile range (IQR) 0.8] in persons with large waist circumference, significantly longer than 0.175 cm (IQR 1.1 mm) in the small waist circumference group ($P = .008$). Cardiac mucosal length was correlated positively with age ($R=0.455$, $P = .006$), with intraabdominal fat and total fat. However, on correcting for age, the association with intraabdominal fat was lost. On regression analysis the independent predictors of greater cardiac mucosal length were waist circumference ($P = .035$) and age ($P = .046$).

The study data are analyzed from the point of view of differences between the small and large waist circumference groups. Significant differences between the groups were found for intragastric pressure, total LES length, and postprandial length of the distal (abdominal) LES. No differences were found for distal LES length in the fasting state, LES pressure, distance from nares to pressure inversion point, and transient LES relaxation frequency.

Acid exposure in the proximal electrode located 5 cm above the upper border of the LES (the usual site of electrode placement in the standard 24-h pH test) was in a range of 0%–0.2% time pH<4 in the fasting phase of both groups. In the postprandial phase, the acid exposure ranged from 0% to 1.7% time pH<4. These are all within the normal range for the pH test. This confirms that this population of volunteers had very low levels of free reflux into the body of the esophagus (Table 8.4).

When these esophageal acid exposures are compared with LES parameters, there is an excellent correlation. The length of the distal (abdominal) segment of the LES was a mean of 2.5 cm (25th percentile 2.2 cm) in the small waist circumference group and a mean of 2.1 cm (25th percentile 1.6 cm) in the large waist circumference group in the fasting state. It decreased to 2.1 cm (25th percentile 1.75 cm) in the small waist circumference group and 1.5 cm (25th percentile 1 cm) in the large waist circumference group in the postprandial phase.

The data show that there was a temporary additional abdominal LES shortening by manometry of around 0.5–0.8 cm resulting from ingesting a full meal.

Only the large waist circumference volunteers in the postprandial phase approached the <1 cm value that correlates with LES incompetence. 25% of this group had a postprandial abdominal LES length <1.06 cm and was therefore in danger of postprandial reflux. The total mean acid exposure in this group was 0.1% (very near zero) in the fasting state. Although they did not have symptoms, there was a small amount of acid exposure (1.4%) that was seen in the postprandial phase. These levels were all within the presently defined normal value (4.5%) for the pH test. Caution is necessary in interpretation; these acid exposures were for 30 and 75 min in the fasting and postprandial phases, not for 24 h.

The data also provide a wonderful view of the pH gradient across the GEJ where the pH changes from strong acidic in the proximal stomach to become suddenly less acidic at the pH transition point. Proximal to the pH transition point, the pH

TABLE 8.5 Distance of pH Transition Point From the Squamocolumnar Junction (SCJ) (Demarcated by a Clip Placed at Endoscopy in This Study) in the Small and Large Waist Circumference (WC) Groups

State/Position of Patient	Small WC (cm)	Large WC (cm)
Fasting, upright	0.56	−0.36
Fasting, supine	0.28	0.05
Postprandial, upright	−0.20	−1.03
Postprandial, supine	−0.18	−0.81

Negative values indicate that the SCJ is distal to the pH transition point and therefore exposed to the acid pocket in the proximal stomach.

shows a marked change from 1.1 to 5 cm. The electrode at 1.1 cm proximal to the pH transition point is still within the high-pressure zone of the LES, and the fact that the pH at that point is more acidic than at 5 cm proximal suggests that some reflux episodes are limited to the intrasphincteric part of the distal esophagus, never reaching the level of the electrode at 5 cm.

Also clearly seen in the sensors at 1.1, 2.1, and 3.1 cm below the pH transition point is that in the fasting state, the acidity progressively increases from proximal to distal. This pattern is inverted in the postprandial phase. The presence of an acid pocket at the height of the food column causes the pH in the electrode at 1.1 cm distal to the pH transition point to be much more acidic than the more distal electrodes, which are likely in the interior of the food column where the acid is buffered.

The pH transition point (i.e., the point, irrespective of the true GEJ, where the acidic gastric pH changes to the more neutral esophageal pH of the distal limit of the *functional* LES) was more proximal within the LES in the group with a large waist circumference than in the small waist circumference group. This difference was not significant during fasting where pH transition point was 4.1 (IQR 2.3) cm distal to the upper border of the LES in the small waist circumference group compared with 2.65 (IQR 2.6) cm in the large waist circumference group ($P = .27$).

However, when the stomach was overdistended by a heavy meal, the sphincter shortened further, causing the pH transition point to move more proximally. In the postprandial (supine) phase, this proximal movement of the pH transition point in the people with large waist circumference exposed the squamous epithelium of the esophagus to gastric acid (Figs. 8.9 and 8.10). The proximal limit of the pH transition point was now *proximal to the SCJ* in this group of patients (Table 8.5). In this position, the distal esophageal squamous epithelium in both groups of volunteers is exposed to the acid pocket that is present in the region between the pH transition point and 1.1 cm distal to it (Fig. 8.10). This explains how a heavy meal causes changes in the dynamics of the LES and exposes the distal squamous epithelium to acidic gastric contents.

This study provides valuable insight into the changes that occur in a group of people who are close to normal. These asymptomatic volunteers are not normal from my viewpoint. All these patients had a small squamooxyntic gap composed of cardiac epithelium in the 2.5 mm range plus an unknown amount of oxyntocardiac epithelium. They had an abdominal LES length that trended toward being shorter in the group with large waist circumference and a pH study that was normal but >0, with acid exposure increasing in the postprandial phase. This entire group therefore fits well into our "subclinical GERD" or "pre-GERD" group where LES damage is within the reserve capacity of the LES. Groups with central obesity were further along in their progression of LES damage than those with small waist circumference. Being of identical age as a group, it is not unreasonable to suggest that the large waist circumference group would be more likely to develop symptomatic GERD than those with small waist circumference. Even without the availability of a measured squamooxyntic gap, the ability to predict the future starts to become a reality.

Robertson et al.[20] state in their discussion: "The main finding of this study was that in asymptomatic volunteers without a hiatal hernia, the length of cardiac mucosa lying between esophageal squamous mucosa and gastric oxyntic mucosa increases with waist circumference and age…. (this) suggests that some or all of this mucosa is acquired throughout life. This possibility was suggested originally by Chandrasoma, who proposed that cardiac mucosa developed by gastroesophageal reflux, inducing squamocolumnar metaplasia of the distal esophagus."

Although our subjects with a large waist circumference did not have traditional reflux, they had a number of abnormalities of the GEJ and intrasphincteric acidity, which could explain the lengthening of the cardiac mucosa (CM).

The proximal extension of gastric acid within the LES provides a plausible explanation for the lengthening of the cardiac mucosa. Exposure of the most distal esophageal squamous mucosa to gastric acid will damage it and with time cause it to be replaced with columnar mucosa suited to an acidic environment.

These are inevitable conclusions resulting from active research into the hypothesis that cardiac epithelium is metaplastic and not a normal gastric epithelium. The data from the study also lead to the realization that there is a subclinical phase of the disease with changes within the LES zone of the esophagus before "traditional" transsphincteric reflux causes the disease that is presently recognized as GERD.

3.2 Reversible Squamous Epithelial Damage Caused by Exposure to Acidic Gastric Contents

The studies reviewed above demonstrated the mechanism whereby distal esophageal squamous epithelium is exposed to the proximal acid pocket in the stomach during a heavy meal. This is an artificial model that is created in the experimental setting to maximize the likelihood of this mechanism being revealed.

Most meals in the life of an individual are within the limits wherein the meal is efficiently received by the gastric reservoir without significant gastric overdistension, increase in intragastric pressure, or changes in the LES. The squamous epithelium is not exposed to gastric contents during these meals.

In Robertson et al.,[20] the increase in the intragastric pressure from the fasting upright state to the postprandial stage after a meal of battered fish and chips eaten until the persons were full was 9.0–10.7 mmHg in the small waist circumference subjects and 11.1–11.6 mmHg in the large waist circumference persons. With more normal meals of moderate volume, the likelihood is that the rise in intragastric pressure would be less. More difficult to measure is the amount of distension of the proximal stomach that is responsible for the deformation of the distal LES.

Even with an unusually heavy meal that causes the changes seen in these studies, the duration of the changes is usually short. Gastric emptying will rapidly decrease gastric overdistension and restore the LES to its normal length and position. However, Tamhankar et al.[19] showed that the squamous epithelium may remain exposed to acidic gastric juice for as long as 150 min following a heavy meal with high fat content.

The exposure of the esophageal squamous epithelium to acidic gastric contents during heavy meals is likely to be within the range of the normal squamous epithelium to resist damage. Even when there is damage, it is likely to be mild and reversible when the epithelium is restored to its normal location where it is protected from gastric acid contents by the LES.

The injurious effect of acid on the squamous epithelium of the esophagus is well-documented. Tobey et al.[21] showed by electron microscopic measurement that an increase in the intercellular spaces between cells resulted from acid exposure of esophageal squamous epithelium (Fig. 8.11). Endoscopic biopsies from 11 patients with recurrent heartburn (6 had erosive esophagitis, 5 had a normal appearing squamous epithelium at endoscopy) and 13 control patients (no symptoms or endoscopic evidence of reflux esophagitis) were examined using transmission electron microscopy. Computer-assisted measurement of the intercellular space diameter in the electron photomicrographs was performed in each specimen. The intercellular space diameter was significantly greater in specimens from patients with heartburn (irrespective of whether or not they had endoscopic erosions) than in the control specimens. Space diameters of 2.4 μm or greater were present in 8 of 11 patients with heartburn and in none of the controls. The authors concluded that dilated intercellular spaces are a feature of reflux damage to squamous epithelium.

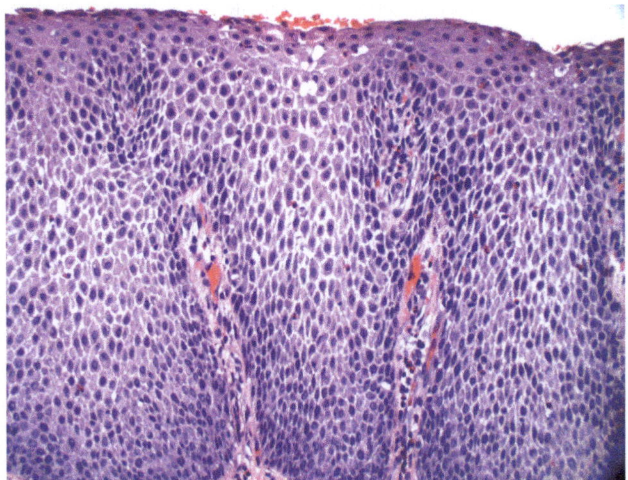

FIGURE 8.11 Squamous epithelium in reflux esophagitis showing marked separation of squamous epithelial cell junctions ("dilated intercellular spaces") resulting from acid-induced damage. This involves almost the full thickness of the epithelium.

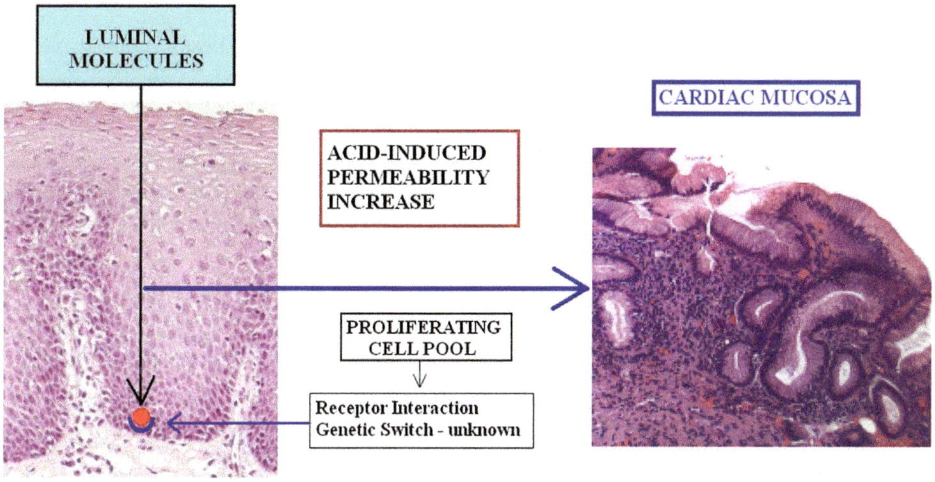

FIGURE 8.12 Mechanism of columnar metaplasia. The increased permeability of squamous epithelium resulting from dilated intercellular spaces allows luminal molecules entering the esophagus during reflux episodes to enter the damaged squamous epithelium. With severe damage, these molecules reach the basal region of the epithelium and interact with the proliferative cells, causing a change in the genetic differentiating signal. This causes the squamous epithelium to change to cardiac epithelium. The nature of the genetic change is yet uncertain.

Tobey et al.[22] in a follow-up study showed that exposure of an in vitro model of rabbit squamous epithelium to acid and acid–pepsin damage resulted in an increase in permeability of the epithelium. They showed that luminal hydrochloric acid (pH 1.1) or a mixture of hydrochloric acid (pH 2.0) plus pepsin (1 mg/mL) for 30 min caused a linear increase in permeability to 4kD to 20kD dextran molecules as well as 6kD epidermal growth factor without gross erosions or histologic evidence of cell necrosis. Transmission electron microscopy documented the presence of dilated intercellular spaces.

In their discussion, Tobey et al.[22] concluded that in noneroded acid-damaged esophageal squamous epithelium, dilated intercellular spaces develop and result in increased permeability. This change in permeability on acid or acid–pepsin exposure is substantial, permitting dextran molecules as large as 20kD (20 Å) and luminal epidermal growth factor at 6kD to diffuse across acid-damaged epithelium and by so doing enable them to access receptors on epithelial basal cells (Fig. 8.12).

Entry of acid into the epithelium to a depth sufficient to stimulate nociceptive receptors in the superficial region of the epithelium (see Chapter 2) will result in pain.

The amount of damage caused to the squamous epithelium by exposure to the acid pocket at times of gastric overdistension will depend on the amount of epithelium exposed, the pH of the acid pocket, the duration of contact, and the resistance of the squamous epithelium to acid-induced injury. In most situations, the likelihood is that overdistension of the stomach is not severe, gastric emptying occurs rapidly, and the squamous epithelial acid exposure is insufficient to overcome its innate resistance. The result is either no or transient minimal damage to the squamous epithelium. With reversal of sphincter shortening as the stomach empties, exposure stops and any mild epithelial damage can heal. The resistance of the squamous epithelium and the ephemeral nature of intermittent exposure to acid are designed to prevent permanent damage to the epithelium.

However, when the distension is severe and emptying is delayed by dietary elements such as high fat content, exposure can be prolonged.[19] More importantly, when gastric overdistension is produced repetitively and at intervals that are too short to permit the epithelium to heal completely, squamous epithelial damage becomes cumulative.

Progressive increase in intercellular gaps first results in pain. This is the pain that arises in the most distal esophageal epithelium in the distal region of the LES. It is not difficult to believe that the pain will not be typical heartburn. Rather, it is likely to be epigastric in location and described as "mild and transient postprandial dyspepsia" rather than "heartburn" and not recognized as being related to the damage of the distal squamous epithelium.

The studies by Tobey et al.,[21,22] which included in vitro studies, showed that acid can directly cause damage to the esophageal squamous epithelium. Souza et al.,[23] in a recent study, summarize the current belief of the pathogenesis of reflux esophagitis in their introduction: "For the past 4 decades, most studies on the pathogenesis of reflux esophagitis have focused primarily on how abnormalities in antireflux mechanisms allow acidic gastric juice to reflux into the esophagus. The concept that the resultant cellular injury is caused by the direct toxic effects of hydrogen ions on the esophageal squamous epithelial cells has been widely accepted and has not been challenged explicitly. Pepsin is known to contribute to the reflux injury, presumably because pepsin degrades the junctional proteins that join cells together and because pepsin might attack vital proteins on the surface of the epithelial cells. Pepsin digestion of the junctional proteins increases mucosal permeability, which facilitates the entry of hydrogen ions into the mucosa and renders cells in the deep layers of the esophageal squamous epithelium vulnerable to acid attack."

> *It has been widely accepted that reflux esophagitis develops from a caustic, chemical injury that starts at the luminal surface of the squamous epithelium, and progresses through the epithelium and lamina propria into the submucosa. The acid-induced death of surface cells is assumed to stimulate a proliferative response in the basal cells that renew the squamous epithelium. This has been presumed to result in hyperplasia of the basal cell layer and the papillae, histologic findings that are considered characteristic of reflux esophagitis. Finally, acid-induced epithelial injury and cell death is assumed to promote an inflammatory response, which is manifested histologically by inflammatory cells infiltrating the damaged squamous epithelium.*

The authors then describe how they were surprised that it took weeks for a rat model designed to produce gastroduodenal reflux into the esophagus to develop erosive esophagitis. The authors ask the following question: "Chemical caustic injuries develop rapidly. If reflux esophagitis is indeed a caustic chemical injury, then why should there be such a long delay in the development of esophagitis after surgical induction of reflux?"

The authors sought to answer this question using a rat model of reflux esophagitis (esophagogastroduodenostomy) and nonneoplastic esophageal squamous (NES) cell lines.

Control (sham operated) animals showed no inflammation in any layer of the esophagus. In the test animals, esophageal inflammation was observed at day 3, most prominent in the submucosa, increasing to reach significant levels in the lamina propria by week 1 and the epithelium by week 3. From weeks 3 to 8, inflammation stayed constant in all layers. Inflammatory cells at day 3 were only $CD3^+$ T lymphocytes; neutrophils appeared at day 7 and eosinophils were not seen during this period. Basal cell hyperplasia appeared at week 1, papillary hyperplasia at week 2, and erosions at week 4.

Interleukin-8 (IL-8) was detected in the submucosa, lamina propria, and within the squamous epithelium by week 2. By week 5, there was strong IL-8 immunoreactivity in the lamina propria and epithelium, particularly in the spaces between cells in the intermediate zone (stratum spinosum; between basal layer and surface). By week 7, IL-8 was detected in the cytoplasm of mature squamous cells as well as in the intercellular space.

In the NES cell lines, there was a significant increase in IL-8 secretion after 2 days of exposure to acidic bile salt media. Significantly increased secretion of interleukin-1β (IL-1β) was detected at day 4. The squamous cells exposed to acidic bile media for 4–5 days caused a significant increase in the migration rates of T lymphocytes and neutrophils compared with controls. The studies suggested that IL-8 increased migration of both T cells and neutrophils, whereas IL-1β had no effect on T cell migration but increased neutrophil migration.

In their discussion, the authors suggest that their findings argue against reflux esophagitis being a direct caustic injury resulting from exposure of the squamous epithelium to acid. They suggest that, at least in this rat model, reflux esophagitis develops primarily as an immune-mediated (T lymphocyte and IL-8) injury. In this new concept of the pathogenesis of reflux esophagitis, "reflux of gastric juice stimulates the esophageal squamous epithelial cells to secrete chemokines that attract inflammatory cells, which ultimately damage the esophageal mucosa."

The possibility that there is an indirect mechanism of epithelial damage that is precipitated by exposure of the squamous epithelium to acid might explain the failure of acid-suppressive drug therapy to control symptoms and heal erosive esophagitis in some patients and why protracted treatment is often necessary.

The study by Dunbar et al. [24] extends that rat model of an indirect mechanism of histologic changes in the squamous epithelium in GERD to humans. 12 patients (11 men; mean age 57.6 years, SD 13.1) at the Dallas Veterans Affairs Medical Center who had reflux esophagitis successfully treated with PPIs were subject to this study. The authors hypothesized that "acute reflux esophagitis could be induced by briefly interrupting PPI therapy in patients with severe erosive esophagitis successfully treated with PPIs. The aim of this study was to evaluate the histologic features of esophageal inflammatory changes in acute GERD."

A total of 215 patients were identified as Los Angeles grade C (LA-C) esophagitis. Of these, 16 were not excluded for a variety of reasons and agreed to participate in the study. These 16 were treated with PPIs for 1 month or more. On study day 1, patients took their morning PPI and completed the GERD Health-Related Quality of Life (HRQL) questionnaire [a validated instrument for GERD symptom severity; score range: 0 (no symptoms) to 50 (worst symptoms)]. Esophageal manometry and 24-h pH and impedance monitoring were performed with a pH electrode placed 5 cm above the LES. Patients took a PPI that evening and the next morning, when esophagoscopy was performed using both high-definition white light endoscopy and confocal laser endomicroscopy. Patients with LA-B, LA-C, or LA-D esophagitis were not eligible for further study; three patients with LA-B esophagitis were excluded and another was withdrawn for an adverse event.

The remaining 12 patients with LA-A or no esophagitis had four esophageal biopsies obtained 1–3 cm proximal to the SCJ for histological evaluation. PPIs were then stopped; patients were given antacids for heartburn. On day 9, esophagoscopy was performed for LA grading and biopsy. On day 15, patients completed another GERD-HRQL questionnaire and had pH and impedance monitoring. The next day, esophagoscopy was performed for LA grading and biopsy and patients resumed PPI therapy. During the second and third esophagoscopy, care was taken to avoid taking biopsies from prior biopsy sites or mucosal breaks.

Confocal laser microscopy with fluorescein enhancement was used to measure intercellular spaces and capillaries in the epithelium. Biopsies were scored on a 0–3 scale (0 = absent; 1 = mild; 2 = moderate; 3 = severe) for type and degree of epithelial inflammation (lymphocytes, eosinophils, neutrophils), basal cell and papillary hyperplasia, and spongiosis (dilated intercellular spaces).

GERD-HRQL symptom scores increased from a median of 2 at baseline taking PPIs to 11.5 at week 1 after stopping PPIs ($P = .008$). Acid exposure (% time pH<4) increased from a median 1.2% at baseline to 17.8% at week 2 after stopping PPIs ($P = .005$). At baseline, 11 patients had no esophagitis and 1 had LA-A esophagitis. At week 2 after stopping PPIs, five patients developed LA-C esophagitis. Esophagitis increased significantly from baseline to week 1 ($P = .006$) and from week 1 to week 2 ($P = .02$).

Histology showed minimal inflammation at baseline with significant increase at weeks 1 and 2 after stopping PPIs. $CD3^+$ T lymphocytes were the predominant inflammatory cell type at all time points; eosinophils and neutrophils were few in number in the absence of erosion. Confocal laser endomicroscopy showed an increasing width in both intercellular spaces and capillaries at weeks 1 and 2 compared with baseline.

In their discussion, the authors suggest that the pattern of injury seen in the present study is not likely to be caused by the direct effect of acid and pepsin starting at the surface and progressing to the submucosa. The findings are consistent with those described in the rat model in the study reviewed above, where the squamous epithelial injury is mediated via activation of cytokines such as IL-8 and IL-1β. The authors suggest that PPIs, which have been shown to inhibit secretion of IL-8, may exert their healing effect in the squamous epithelium by this antiinflammatory effect in addition to a direct acid-reducing effect.

This study did not evaluate interleukin expression in the squamous epithelium in these subjects.

3.3 Irreversible Columnar Metaplasia of Esophageal Squamous Epithelium

If the epithelial damage progresses, the second manifestation of pathologic change, which is columnar metaplasia, will occur (Figs. 8.12 and 8.13).

Until columnar metaplasia occurs, the patient has no GERD in the new definition I have suggested. The mild changes caused in the squamous epithelium by intermittent acid exposure cause no permanent damage because they spontaneously reverse. Columnar metaplasia, an irreversible change, heralds the onset of recognizable chronic GERD.

Recent studies have shown that the squamous epithelium exposed to acid may show molecular changes that are commonly seen in metaplastic cardiac epithelium.[25] This suggests that there is a lag phase between molecular change in

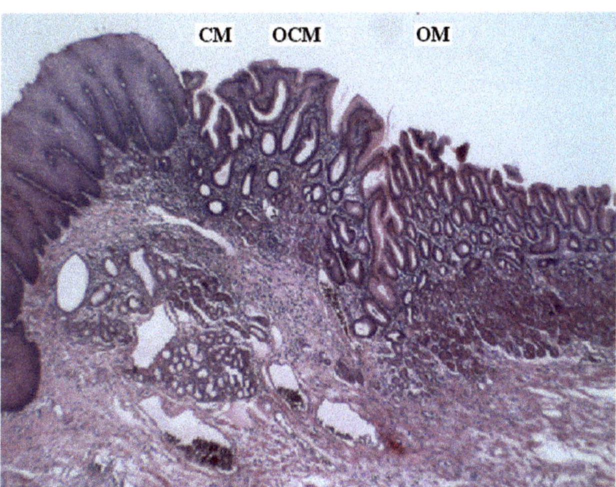

FIGURE 8.13 Sections across the squamocolumnar junction showing a very short length of cardiac mucosa (CM) and oxyntocardiac mucosa (OCM) between the squamous epithelium and gastric oxyntic mucosa (OM). Note the submucosal glands under the squamous epithelium.

squamous epithelium and its metaplasia into columnar epithelium. The new definition will not detect this molecular change that may indicate a commitment to columnar metaplasia in the future. However, whether this actually occurs is undetermined. The failure to recognize this possibility in our new system has no practical negative impact.

Columnar metaplasia requires more substantial damage to the squamous epithelium. This is most likely when repetitive and frequent exposure to acid prevents healing of the damaged epithelium.

Severe damage results in increased permeability that allows larger molecules to reach the proliferative zone of the epithelium near its base. Receptor interactions between cells and molecules cause genetic changes only when they occur in proliferative cells (Fig. 8.12).

The genetic change leading to columnar metaplasia is an activation of the signal that causes the proliferative cells to differentiate toward a columnar cell rather than a squamous cell. There is evidence that this is an activation of bone morphogenesis protein 4.[26] I will discuss the molecular changes in GERD in Chapter 15. It is unlikely that a small ion such as H^+ will be able to activate a genetic pathway. Acid, though, is the key that increases the permeability of the epithelium to provide access to critical larger molecules in the gastric juice to interact with the suprabasal proliferative cells.

The first columnar epithelium to develop is cardiac epithelium composed of nonspecialized mucous cells only. There is evidence that the metaplasia may involve suprabasal proliferative cells in the squamous epithelium in a process called transdifferentiation rather than the basally located stem cells. Metaplasia in suprabasal squamous cells would potentially lead to a multilayered epithelium with a columnar cell at the surface and the residual basal zone cells underneath. Multilayered epithelium with these features is seen in metaplastic columnar epithelium.[27]

The cardiac epithelium when it develops from the basal region of squamous epithelium commonly consists of a single layer of surface cells. Like the primitive endodermal epithelium, this flat epithelium dips down into foveolar pits of increasing length. Glands arise from the deep aspect of the foveolar pits to form the fully developed glandular epithelium. Pure cardiac epithelium is composed only of mucous cells in the surface, foveolar pits, and glands.

When columnar metaplasia occurs, the nociceptive endings of the squamous epithelium disappear, and the quality of pain changes to one that is not as severe and is even less typical for GERD because columnar epithelium is less sensitive. For all these reasons, the initial damage to the most distal squamous epithelium is highly unlikely to be diagnosed as GERD in adults. The pain if present at all is likely to be atypical, probably epigastric and transient.

Endoscopy will be normal because it is not sensitive enough to detect a small gap between the distal squamous epithelium and gastric oxyntic epithelium. The 24-h pH test will be normal because there is no free reflux. Even highly sophisticated high-resolution manometry is not sensitive enough to detect a minimal shortening of the LES in an individual patient. The only way for this diagnosis to be made is by a biopsy at the SCJ, which will show the histologic gap between squamous and gastric oxyntic epithelia.

Glickman et al.[27] showed that children with 0.1 cm or more of cardiac epithelium have more reflux esophagitis than those with <0.1 cm. This is unquestionably GERD if the definition of GERD is damage to squamous epithelium caused by exposure to gastric contents. It can be diagnosed easily and with incredible accuracy by biopsy.

I have a clear knowledge of my eating habits during the first 56 years of my life. While not a glutton, I have eaten my share of heavy meals laden with fat during the first 56 years of my life. Despite this, my squamooxyntic gap was 4 mm consisting largely of oxyntocardiac epithelium with 0.2 mm of cardiac epithelium in one quadrant distal to the SCJ. I would have been a small waist circumference volunteer if I had participated in the study of Robertson et al.[20] I have a smaller length of cardiac epithelium than the smallest length in that study, which, by personal communication with Professor McColl, was 0.5 mm in a patient in the small waist circumference group. The large waist circumference group ate sufficiently in their lives to develop their state of central obesity. This amount of eating correlated with a mean length of cardiac epithelium of only 2.5 mm during their mean 45 years of existence. This is evidence that there is significant resistance to columnar metaplasia by the pathogenetic mechanism that causes cardiac metaplasia of squamous epithelium.

The squamooxyntic gap begins slowly and increases in length very slowly. In my case, the rate of increase of the length of cardiac metaplasia was 4 mm from age 15 years (assumed as the age of the beginning of my established dietary habit) to age 56 years was 1 mm per decade (or 100 μm/year). I will show in Chapter 18 that calculation of the rate of shortening of the abdominal LES is potentially a valuable predictor of future severe GERD.

The first event of GERD has now occurred. The normal patient has been converted to the earliest GERD patient. This is defined by the presence of any cardiac epithelium between the distal limit of squamous epithelium and the proximal limit of gastric oxyntic epithelium. The squamooxyntic gap has been generated. The abdominal segment of the sphincter has been permanently shortened by an amount that is equal to the measurable length of the histologic squamooxyntic gap in the DDE.

4. A HUMAN EXPERIMENT OF THE PATHOGENESIS

Patients who undergo esophagogastrectomy with anastomosis of the distal margin of the resection consisting of gastric body with oxyntic epithelium to the proximal esophageal margin consisting of squamous epithelium represent a human experiment to study the progression of GERD-induced mucosal changes. Because the resection has removed the LES in its entirety, these patients almost invariably have gastroesophageal reflux. Because the entire GEJ including all possible cardiac epithelium has been removed by the surgery, a mucosal junction between esophageal squamous epithelium and gastric oxyntic epithelium has been created with the exact neo-GEJ marked accurately by the anastomotic line. After surgery, the patient has a zero squamooxyntic gap without sphincter protection. Sequential examination of the changes in the epithelia at the new GEJ provides an excellent method of studying epithelial changes resulting from reflux.[28–30] These studies are also important because the exact onset of reflux (i.e., the date of surgery) is known.

The excellent study by Dresner et al.[28] looked at the postoperative course of 40 patients treated for Barrett's adenocarcinoma and high-grade dysplasia (26 patients) and squamous neoplasia (14 patients) with an intrathoracic esophagogastrostomy (i.e., gastric pull-up) after subtotal esophagogastrectomy. The authors state that this is "an ideal model with which to study the early events in the pathogenesis of Barrett's metaplasia from normal human squamous epithelium subjected to duodenogastro-oesophageal reflux" (p. 1121).

The presence of duodenogastroesophageal reflux was verified by combined 24-h ambulatory pH and bilirubin monitoring in 30 patients. Serial endoscopic assessment and systematic biopsy at the esophagogastric anastomosis was done over a 36-month period. The patients had a total of 130 (median 3, range 1–8) endoscopic examinations.

The authors' definition of Barrett esophagus is based on histology and is both clear and precise: "The definition of Barrett's mucosa included both specialized and non-specialized columnar epithelium from the oesophageal remnant. The former epithelium was identified by the presence of goblet cells and is referred to as 'intestinal metaplasia-type' Barrett's mucosa. Non-specialized epithelium, which has no goblet cells and has features similar to those of the mucosa of the gastric cardia or fundus, is referred to as 'cardiac-type' or 'fundic-type' Barrett's mucosa respectively" (p. 1121–1122).

At the end of the study, 7 patients had normal squamous epithelium, 14 had reflux esophagitis of varying grades, and 19 patients had esophageal columnar epithelium above the anastomotic line. Endoscopically, the columnar epithelium was seen as single tongues (6), multiple tongues (2), noncircumferential patches (5), and circumferential segments (6). The measured extent of columnar epithelium ranged from 0.5 to 3.5 cm (5–35 mm) (median 1.9 cm). Coexistent reflux esophagitis in the squamous epithelium was present in 14 of the 19 patients with columnar metaplasia. At the end of the study period of 36 months, 10 of the 19 patients with columnar epithelium had cardiac-type epithelium, and 9 had intestinal metaplasia-type in 9.

At the index endoscopy, 9 patients showed columnar metaplasia. In the other 10 patients, columnar metaplasia developed over the next 6–36 months. In these patients, there was evidence of reflux esophagitis in the squamous epithelium before the development of columnar metaplasia. The initial detection of columnar metaplasia in the esophagus was made at a median of 14 (range 3–118) months after surgery. Of the nine patients with columnar epithelium at the index endoscopy, seven had cardiac-type mucosa and two had intestinal metaplasia. All 10 patients who had only squamous epithelium at the index endoscopy and developed columnar metaplasia during follow-up progressed sequentially from cardiac-type epithelium to intestinal metaplasia. Of these 10 patients, 7 had developed intestinal metaplasia by the end of the study period. The median time for development of cardiac-type mucosa was 14 (range 3–22) months, whereas intestinal metaplasia was first detected significantly later at a median of 27 (range 11–118) months. The gastric epithelium below the anastomosis showed: "gastric body epithelium in all cases, with various degrees of quiescent and active gastritis."

The incidence of columnar metaplasia was similar in patients who had the surgery for squamous cell carcinoma and adenocarcinoma of the esophagus. Patients who developed esophageal columnar epithelium had significantly higher acid ($P = .015$) and bilirubin ($P = .011$) reflux compared with patients who had only squamous epithelium at the end of the study period.

The authors concluded that severe duodenogastroesophageal reflux occurring after subtotal esophagogastrectomy caused a sequence of changes in the esophageal squamous epithelium that showed a temporal progression from reflux esophagitis→columnar metaplasia of cardiac type→columnar metaplasia of intestinal type. This change began immediately above the anastomotic line with the columnar metaplasia involving a range of 0 (21 patients) to 3.5 cm.

The most important finding in this study is the most obvious: that all the changes in reflux involve esophageal squamous epithelium. Gastric oxyntic epithelium is immune to damage caused by gastric contents and remained unchanged after surgery.

This study proves that expansion of cardiac epithelium in these postesophagogastrectomy patients begins at the newly created GEJ (the junction between the SCJ and the proximal limit of gastric oxyntic epithelium) and progressively increased in a cephalad direction in a manner that is directly related to exposure of esophageal squamous epithelium to gastroduodenal contents.

The only differences between the findings in this study and what happens in a person without esophagectomy are that after this surgery the patients have no LES, the new GEJ is in the thorax, and the new GEJ is indelibly marked by the anastomotic line. As cardiac epithelium develops as a result of reflux, it is clear that the expansion occurs in a cephalad direction with the distal end being at the anastomotic line. This line is also the point of the proximal limit of rugal folds, which remains as the permanent marker of the new GEJ. Because the new GEJ is in the thorax, the entire esophagus has an intraluminal pressure that is negative. This keeps the esophagus as a tube without dilatation or development of rugal folds even without a sphincter.

The same histologic events happen in the person without surgery. The histology is identical in that there is a progressive cardiac metaplasia of squamous epithelium in a cephalad direction beginning at the proximal limit of gastric oxyntic epithelium, which remains as a constant marker of the true GEJ. In the person without esophagectomy, cardiac metaplasia of esophageal squamous epithelium is associated with loss of the abdominal segment of the LES and dilatation of the damaged esophagus with development of rugal folds.

The other important finding in the study is the rate of progression of squamous epithelial damage in these patients. The situation that exists at the GEJ is not the same as it is in the normal person. The remnant esophagus is separated from the positive pressure part of the stomach in the abdomen by a segment of the stomach of variable length that has been pulled up into the thorax for the anastomosis. How this impacts reflux is not known; it may depend on the length of the intrathoracic segment of the stomach, its shape, and other factors. The severity of reflux varied significantly in different patients after esophagectomy. The study showed that increasing reflux of acid and bilirubin was associated with columnar metaplasia suggesting that the variation in the severity of reflux may explain the variation in histologic changes and progression.

Although the changes superficially seem surprisingly small, it must be recognized that these patients showed a dramatically high incidence of columnar metaplasia (19/41 or 46%) with 9/41 (22%) developing intestinal metaplasia within 3 years of onset of reflux. This is much greater than the incidence of these complications in the Pro-GERD study where patients without surgery were followed from the onset of GERD for a period of 5 years; only 9.7% developed Barrett esophagus.[31] The difference is clearly associated with the more severe reflux in the patients after esophagectomy. In the Pro-GERD study,[31] patients whose index endoscopy had evidence of Los Angeles grade C/D erosive esophagitis, which is associated with severe reflux, had an incidence of Barrett esophagus of 19.7% within 5 years, similar to this esophagectomy population.

The study by Oberg et al.[29] is from a selected group of 32 patients out of 142 consecutive patients who underwent esophagectomy and gastric tube reconstruction between December 1989 and October 1997. By March 2000, 82 (58%) had died and 60 were still alive. All surviving patients were invited to take part in the study.

32 patients (18 men, median age 64 years; range 46–84 years) agreed to the study. They were evaluated by manometry (standard, stationary), simultaneous pH and bilirubin monitoring, and endoscopy with biopsy. Symptoms of heartburn (defined as a burning sensation in the chest or throat) and regurgitation were graded from 0 (absent) to 3 (severe).

The presence of metaplastic columnar epithelium in the esophagus was suspected when circumferential areas or tongues of pink glandular mucosa extended from the anastomosis into the squamous epithelium. This was measured from the highest point of the SCJ to the anastomosis. CLE was confirmed by histology of multiple biopsy samples taken from the metaplastic columnar mucosa. The histologic classification followed our system of definition of cardiac, oxyntocardiac, and specialized (i.e., cardiac with intestinal metaplasia). Dr. Oberg had done a fellowship at the University of South California (USC) before returning to Sweden and was very conversant with the histologic method.

Indications for esophagectomy were adenocarcinoma in Barrett esophagus (16; 50%), squamous carcinoma (6), achalasia (6), esophageal ulceration without Barrett esophagus (2), failed antireflux surgery (1), and lye injury (1). The median time after esophagectomy was 4.9 years (range 3.0–10.4 years). During the study, 10 (31%) patients were receiving no acid-suppressive therapy and 22 (69%) were on acid-suppressive drugs (8 on a once-daily dose of PPI, 12 on twice-daily PPIs, and 2 on H_2 receptor antagonists).

15 (47%) patients had segments of metaplastic columnar mucosa above the anastomosis. The median length of CLE was 0.5 cm (range 0.5–4 cm). The prevalence of CLE was significantly higher in patients with a preesophagectomy diagnosis of Barrett esophagus compared with those without. CLE was found in 11/16 (69%) patients with preoperative Barrett esophagus and 4/16 (25%) of those without ($P = .032$).

The length of CLE correlated significantly with the degree of esophageal acid exposure, but the presence of abnormal bilirubin exposure was unrelated to the presence of CLE. No difference was seen in percent time pH<4, number of reflux episodes, and length of reflux episodes between patients with and without CLE. The prevalence of metaplasia did not change with increasing time after esophagectomy.

Histologic examination showed cardiac epithelium in 9 (60%) patients and oxyntocardiac epithelium in 3 (20%). Intestinal metaplasia was found in 3 patients 8.5, 9.5, and 10.4 years after esophagectomy. The median postoperative period was significantly longer (P = .004) in patients with intestinal metaplasia compared with those without. All patients with intestinal metaplasia had abnormal esophageal acid and bilirubin exposure, but the combined acid and bilirubin exposure was not significantly higher in these patients compared with those with nonintestinalized CLE. No patient in this study developed dysplasia or adenocarcinoma.

This study shows a similar pattern of development of CLE above the anastomosis with development of cardiac epithelium followed by intestinal metaplasia in a minority of patients. The incidence of intestinal metaplasia in this study is lower and at a time that is much later after surgery than in Dressner et al.[28] The relationship between the occurrence of metaplasia and acid and bilirubin exposure was also much less in this study. It is possible that these differences are related to different surgical technique in the two institutions. In both studies, there was no progression of CLE to dysplasia or adenocarcinoma.

In a similar study from the Foregut Surgery Unit at USC, Lord et al.[30] showed similar findings. This was a retrospective study of 100 patients with esophagectomy, 20 of whom had biopsies taken from the remnant esophagus above the anastomotic line. 10 (50%) patients developed cardiac epithelium with 4 (25%) showing intestinal metaplasia and dysplasia in postoperative biopsies. Adenocarcinoma was present in 1 patient. The cytokeratin 7/20 pattern and DAS-1 staining of the cardiac epithelium in these patients were similar to Barrett mucosa.

This study has one difference than the studies by Dresner et al.[28] and Oberg et al.[29]; it had one patient developing adenocarcinoma in the remnant esophagus. It is uncertain whether this represented a recurrence of the carcinoma or a de novo carcinoma arising in the metaplastic epithelium.

5. A NEW HISTOLOGY-BASED DEFINITION OF GASTROESOPHAGEAL REFLUX DISEASE: REFLUX CARDITIS

A new cellular definition for GERD now becomes possible: "GERD is defined by the presence of cardiac epithelium (with and without parietal and goblet cells) between the SCJ and proximal limit of gastric oxyntic epithelium." This is a practical definition based on histology that replaces the etiology-based definition that I previously suggested: "GERD is a disease characterized by squamous epithelial damage caused by exposure to gastric contents."

By choosing cardiac epithelium (with and without parietal and/or goblet cells), I am excluding the irrelevant, reversible, transient squamous epithelial damage that occurs prior to columnar metaplasia. Being irreversible, cardiac metaplasia is reliably found with adequate biopsy sampling unlike the more patchy histologic changes of reflux esophagitis in squamous epithelium.

This definition of GERD has absolute specificity. Columnar metaplasia of esophageal squamous epithelium requires two steps:

1. Squamous epithelial damage that results in dilated intercellular spaces that increase epithelial permeability to large molecules. This can happen in many different esophageal injuries, chemicals, infectious agents, and allergy (eosinophilic esophagitis) in addition to GERD.
2. Entry of large molecules into the damaged squamous epithelium that have the ability to react with the proliferative cells in the epithelium to induce the change in the differentiating signal in the cell (Fig. 8.12). As far as is known, a large molecule with this ability exists only in gastric juice. No other esophageal disease has this capability. This makes cardiac metaplasia 100% specific for GERD.

This definition also has absolute sensitivity. It can be detected by histology at the earliest stage with a length that can be measured within an accuracy of micrometers (Fig. 8.13). When it is present in small amounts, the patient has no symptoms and no abnormal reflux in a 24-h pH test. The abdominal segment of the LES is shortened by the length of cardiac epithelium (Fig. 8.14). However, manometry cannot detect this because it measures the functional residual sphincter, not damage to the sphincter.

GERD now has a cellular definition based on a simple histologic finding that has 100% sensitivity and 100% specificity. This has been ignored until now because of the belief that cardiac epithelium is a normal epithelium in the proximal stomach. Its acceptance opens the door to the future that I will explore in the succeeding chapters.

6. THE DIAGNOSIS AND SIGNIFICANCE OF REFLUX CARDITIS

The new definition of GERD has value not only because it is theoretically correct but also because it translates into a diagnostic test for GERD that has 100% sensitivity for GERD. The presence of cardiac epithelium (with and without parietal and/or goblet cells) in a biopsy taken anywhere from the esophagus or which is believed to be from the proximal stomach

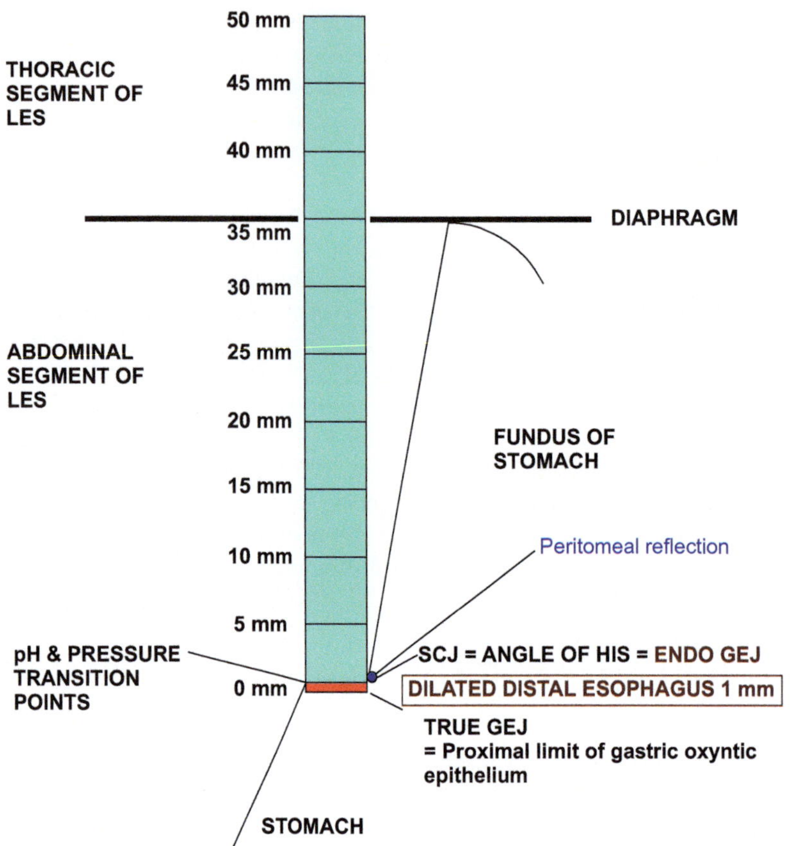

FIGURE 8.14 Diagrammatic representation of the lower esophageal sphincter (LES) showing replacement of the most distal 1 mm of the squamous epithelium by cardiac epithelium (red). The squamocolumnar junction (SCJ) has moved cephalad by 1 mm and the area lined by cardiac epithelium is dilated. *GEJ*, gastroesophageal junction.

represents the DDE. It is diagnostic of abdominal LES damage. This is true irrespective of whether the endoscopist designates whether the biopsy is taken proximal or distal to the endoscopic GEJ.

The term "reflux carditis" was used in the interpretation of biopsies in the Foregut Surgery Unit at USC since ~1994. My naïve expectation was that the power of the data that cardiac epithelium was abnormal would result in total and rapid acceptance of the term. That did not happen. As far as I know, I am still the only person using the term.

A study by Der et al.,[32] which was done in our unit around the time our autopsy study was published in 2000,[9] was an attempt to provide data that would contrast the relative rarity of pure cardiac epithelium at autopsy with its invariable presence in GERD patients.

We introduce the problem: "Histological examination of the gastroesophageal junctional region at autopsy and by extensive biopsy sampling shows a marked variation in the type of glandular mucosa that are found. These can be classified into oxyntic mucosa (OM), cardiac mucosa (CM), oxyntocardiac mucosa (OCM), and intestinal metaplastic mucosa (IM). When anatomic definitions of cardia are used, based on tissue accessed distal to a perceived gastroesophageal junction (GEJ), the variation in histology of this region makes it impossible to know which type of mucosa is involved. There is the possibility that reports of carditis based on an anatomically defined cardia actually use evaluations of inflammation of oxyntic mucosa rather than CM.... When the term "carditis" is restricted to inflammation of histologically defined CM, irrespective of its exact location in the gastroesophageal junctional region, its definition is highly reproducible and has no ambiguity.... The current study was designed to evaluate the relationships between the presence of CM, inflammation of CM (carditis), gastroesophageal reflux as assessed by a 24-hour pH test, gastritis as assessed by biopsy of the distal stomach and *Helicobacter pylori* infection."

The study consisted of 141 patients (79 men; median age 54 years, range 18–89 years) who had histologic cardiac epithelium in biopsy samples from the GEJ region. Each patient underwent endoscopy with systematic biopsy of the distal stomach (antrum in all patients plus body/fundus in 121 patients), a retrograde biopsy of the region of the proximal limit of the gastric rugal folds, and the squamous epithelium immediately above the SCJ. When visible CLE was present, additional

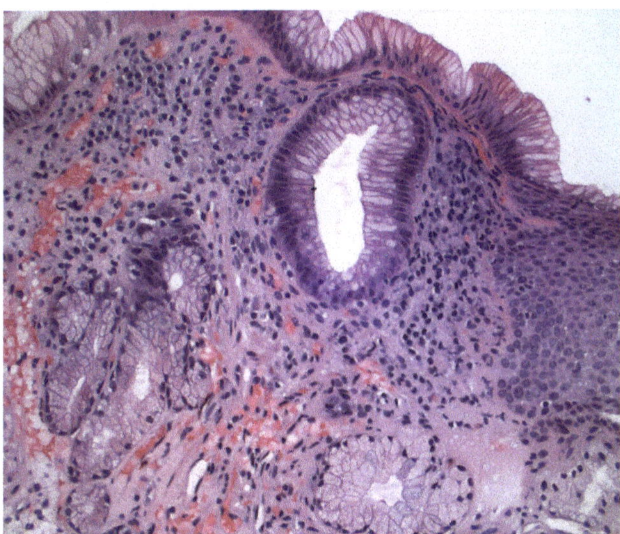

FIGURE 8.15 Sections across the squamocolumnar junction in an endoscopic biopsy showing transition of squamous epithelium (on right) to cardiac epithelium consisting of mucous cells only. The cardiac epithelium shows significant chronic inflammation.

biopsies were taken from this epithelium. 105 of these 141 patients underwent 24-h pH monitoring. Abnormal acid exposure was defined as a pH<4 for >4.5% of the 24-h period.

Columnar epithelia were evaluated for inflammation and *Helicobacter pylori*. Inflammation, when present in gastric mucosa in the distal gastric biopsies, was classified as gastritis and graded as "present" or "absent." More detailed grading did not provide any significant statistical correlations. Inflammation in cardiac epithelium was called "carditis" and classified as "acute" defined by the presence of neutrophils and "chronic" by the presence of plasma cells, lymphocytes, and eosinophils in the lamina propria (Fig. 8.15). Acute inflammation was classified as "present" or "absent." Chronic inflammation was graded on a 0–3 scale based on a subjective assessment of the overall number of inflammatory cells present in all samples with CM. Grading was as follows: 0 = no inflammation; 1+ = chronic inflammatory cells not significantly expanding the interglandular spaces; 2+ = chronic inflammatory cells focally expanding interglandular spaces; and 3+ = chronic inflammatory cells diffusely expanding interglandular spaces. *H. pylori* infection was evaluated in Giemsa-stained sections.

All 141 patients showed evidence of significant inflammation in the cardiac epithelium (Fig. 8.15). Chronic inflammation was present in all biopsy samples of cardiac epithelium, with only 12/141 (9.2%) classified as 3+. Acute inflammation was present in cardiac epithelium in 26/141 (18.4%) patients; all these patients had associated 2+ or 3+ chronic inflammation.

30/141 (21%) of the patients had distal gastritis; the other 111 (79%) had an isolated carditis. A remarkable finding in these patients was the marked difference between the amount of inflammation in cardiac epithelium and gastric oxyntic epithelium when both epithelial types were present. Gastric oxyntic epithelium was commonly devoid of any inflammation.

The type and severity of inflammation in cardiac epithelium had a variable association with distal gastritis (Table 8.6). The presence of acute inflammation in cardiac epithelium was a strong predictor of the coexistence of distal gastritis. 14/26 (54%) of patients with acute inflammation in cardiac epithelium had distal gastritis compared with 16/115 (14%) patients with no acute inflammation in cardiac epithelium ($P = .001$). 10 of the 26 (39%) patients with acute inflammation in cardiac epithelium had evidence of *H. pylori* gastritis.

H. pylori infection was present in only 20/141 (14%) of the patients; 20/30 patients with distal gastritis had *H. pylori* infection. 17 of the 20 patients with *H. pylori* infection had a pangastritis; 15 of these patients had *H. pylori* in cardiac epithelium. The other 3 patients had *H. pylori* limited to the antrum. The 14% prevalence of *H. pylori* in this population is likely to be not significantly different than the prevalence of this infection in the general population in Los Angeles.

The majority of patients with carditis had neither *H. pylori* infection (121/141 or 86%) nor gastritis (111/141 or 79%). In these patients carditis was an isolated finding in the region of the GEJ.

Carditis was associated with an abnormal 24-h pH test in 64/105 (61%) of patients (Table 8.7). The only statistically significant associations with abnormal acid exposure in the esophagus were male sex ($P = .038$) and the severity of chronic inflammation ($P = .024$). There was no significant association with acute inflammation in CM ($P = .81$).

TABLE 8.6 The Association of Gastritis With Clinical and Pathologic Characteristics in 141 Patients With Cardiac Mucosa (CM)

Characteristic	No of Patients	No of Patients With Gastritis	*P* Value
Total	141	30 (21%)	
Percent Time pH < 4			
4.5% or less	41	10 (24%)	.37
>4.5%	64	11 (17%)	
Unknown	36	9	
Acute Inflammation of CM			
No	115	16 (14%)	.001
Yes	26	14 (54%)	
Chronic Inflammation of CM			
Mild (1+)	34	3 (9%)	.042
Severe (2+ and 3+)	107	27 (25%)	
***Helicobacter pylori* Infection**			
No	121	10 (8%)	<.001
Yes	20	20 (100%)	

Selected data items are included. Excluded are age and gender.

TABLE 8.7 The Association Between an Abnormal 24-h pH Test (Percent Time the Esophageal Mucosa Was Exposed to a pH < 4 in a 24-h Period) in 105 Patients With a pH Test

Characteristic	No. of Patients	No. of Patients With Abnormal pH Test	Mean Time pH < 4% (95% CI)	*P* Value
Total	105	64 (61%)		
Age (Years)				
≤45	35	17 (49%)	4.8 (3.1, 7.0)	.09
46–59	40	27 (68%)	8.1 (5.9, 10.7)	
>60	30	29 (67%)	5.6 (3.6, 8.2)	
Gender				
Male	64	42 (86%)	7.4 (5.7, 9.4)	.038
Female	41	22 (54%)	4.7 (3.1, 6.6)	
Gastritis				
No	84	53 (63%)	6.6 (5.2, 8.2)	.24
Yes	21	11 (52%)	4.8 (2.6, 7.7)	
Acute Inflammation of CM				
No	86	54 (63%)	6.2 (4.8, 7.7)	.81
Yes	19	10 (53%)	6.6 (3.8, 10.1)	
Chronic Inflammation of CM				
Mild (1+)	25	10 (40%)	3.9 (2.1, 6.2)	.024
Severe (2+ and 3+)	80	54 (68%)	7.1 (5.6, 8.7)	
***Helicobacter pylori* Infection**				
No	89	55 (62%)	6.4 (5.1, 7.9)	.51
Yes	16	9 (56%)	5.3 (2.7, 8.8)	

CM, cardiac mucosa.

This study was on 141 consecutive patients referred to a foregut surgical unit. They likely had a constellation of esophageal pathology, probably dominated by GERD but likely including patients without GERD, such as diverticula, achalasia, other motility disorders, etc. We did not examine the patient's medical record, depending entirely on the 24-h pH test as a measure of reflux. At that time, we believed that the 24-h pH test was the most accurate and reliable criterion for the severity of GERD.

With our new understanding, this population is one that had a squamooxyntic gap of unknown length in a DDE. The biopsy taken at the GEJ was not adequate for measuring the squamooxyntic gap. In fact, the concepts of the squamooxyntic gap and the DDE were unknown to us during the study. It is likely that the range of abnormality in the pH test reflects a combination of symptomatic GERD and subclinical GERD. The range of the 95% confidence intervals for percent time pH<4 was 2.1% (in a patient with mild chronic carditis) to 10.7% in 46- to 59-year-old patients. This variation in the amount of esophageal acid exposure makes it reasonable to suggest that there was a mixture of symptomatic and subclinical GERD patients. In hindsight, it would have been wonderful if we had measured the squamooxyntic gap, but we could not do what we did not know then.

Other studies have confirmed several of the findings in our study.

In their study of asymptomatic volunteers, Robertson et al.[20] confirm that even small amounts (1–3 mm) of cardiac epithelium at the GEJ invariably show inflammation and reactive changes. They report: "In this group of healthy volunteers, cardiac mucosa was universally inflamed. As shown in Figure 2, inflammation defined by the density of neutrophils was significantly higher in cardiac mucosa than in the adjacent oxyntocardiac or distant gastric body and antral mucosa (all p values<0.001). Again, inflammation measured by the density of mononuclear cells was evidence in all 3 parts of the SCJ and was in its maximum density in cardiac mucosa (all p values<0.001 for cardia vs others). Reactive atypia was seen in all 3 types of SCJ mucosa, but the maximum activity was evident in the most distal squamous mucosa followed by cardiac mucosa. All biopsy specimens from the gastric body and antrum were virtually normal. Intestinal metaplasia at the cardia defined by the presence of goblet cells was seen in 4/62 volunteers."

This confirms the important finding in our study that cardiac epithelium is *always* inflamed even when the adjacent gastric epithelium was completely normal.

The study by Lembo et al.[33] followed a conference in Hamburg where I had presented my data suggesting that inflamed cardiac epithelium distal to the SCJ in endoscopically normal patients was a metaplastic esophageal epithelium and a diagnostic histologic feature of GERD rather than gastric mucosa. Dr. Fred Weinstein, the senior author of this paper from UCLA and a highly influential gastroenterologist at the time, was in the audience. He raised his hand after my talk and pronounced: "I will not be subjected to the histologic dictatorship of the University of Southern California. The mucosa distal to the SCJ in endoscopically normal patients is gastric, whatever its histology." The intense but friendly Los Angeles rivalry between the Trojans and Bruins extends beyond football!

Lembo et al.,[33] in this prospective study of 30 patients (21 men; mean age 52.7 years) with symptomatic GERD (heartburn, acid regurgitation, epigastric pain, or extraesophageal manifestation for a median duration of 7.3 years) who had a normal appearing esophagus at endoscopy, took biopsies in the SCJ sampling squamous and columnar epithelium, at 1–2 cm distal to the SCJ, and from the more distal stomach.

In the introduction, the authors state: "The gastric cardia occupies an area a few centimeters immediately below the squamocolumnar junction (Z-line). It contains mucous type glands similar to those of the pyloric antrum. In many instances it actually consists of mixed glands with both mucous glands and elements of oxyntic mucosa with parietal and chief cells. Inflammation of this area of the stomach had largely been ignored until recently when preliminary studies suggested that biopsy specimens of this area may be more reflective of GORD than those from the distal oesophagus." They cite the paper by Oberg et al.[29] from our group. The study was aimed "to examine the spectrum of mucosal injury in the gastric cardia in patients with minimally erosive GORD. We hypothesized that cardiac mucosa, if abnormal in GORD, should be maximally injured at the Z-line, the "battleground" rather than below it." The complete belief in the veracity of the endoscopic GEJ causes these highly experienced gastroenterologists to entertain the possibility that GERD can produce pathology in the stomach.

The authors did not classify the columnar epithelia into different histologic types; "carditis" was inflammation of whatever columnar epithelium was present distal to the endoscopic GEJ (=SCJ in these endoscopically normal patients) in the anatomic gastric cardia. 55/111 (50%) of the biopsy specimens at the Z-line had both squamous and columnar epithelia; 40/111 contained only columnar epithelium; 16/111 had only squamous epithelium.

The columnar mucosae in these biopsies were scored for inflammatory cells, epithelial cell abnormalities, intestinal metaplasia, and *H. pylori*. By their unique scoring system, 96% of the 55 biopsies that contained both squamous and columnar epithelium in continuity had a positive carditis score. 67% of the 40 Z-line biopsies that had only columnar epithelium had a positive carditis score. 17% of the biopsies taken 1–2 cm distal to the SCJ had a positive carditis score, significantly lower than the biopsies at the Z-line.

33% of the biopsies at the Z-line with squamous and columnar epithelium in the biopsies had goblet cells, significantly higher than in Z-line specimens with only columnar epithelium (17%) and those taken 1–2 cm distal to the SCJ (4%).

H. pylori gastritis was present in 4/30 (13%) of the patients. These four patients had carditis scores that were similar in biopsies of the Z-line that had squamous epithelium and columnar epithelium, in Z-line biopsies that had only columnar epithelium, and in biopsies taken 1–2 cm distal to the Z-line. The carditis scores were higher in these four patients with *H. pylori* gastritis than those without. The authors explain these data perfectly: "… the other main cause of carditis is *H. pylori*. Not surprisingly, *H. pylori* gastritis often extends to the cardia, which is consistent with the findings in our study that patients with *H. pylori* had a greater carditis score than did patients without *H. pylori*. However, in our GORD population and in some recent studies from others, the prevalence of *H. pylori* is one-third or less, suggesting that it is neither the exclusive nor the primary contributor to cardia injury."

This paper did not classify the histologic types of the columnar epithelia distal to the SCJ. Dr. Weinstein retained his right to his own classification of histology in the region! However, by performing biopsies at the SCJ and 1–2 cm distally, the study produced data that essentially confirmed that carditis and "intestinal metaplasia of the cardia" were limited to a very small extent immediately adjacent to the SCJ, rapidly declining within 1–2 cm distal to it. Unfortunately, they do not classify the histologic type of columnar metaplasia in which the inflammation was found. The pattern, though, suggests strongly that inflammation was limited to cardiac epithelium (with and without parietal and/or goblet cells) as I have defined in Chapter 4 and that the epithelium that was not inflamed in these patients (except in the 4 patients with *H. pylori* gastritis) was gastric oxyntic epithelium.

In their discussion, the authors seem to conclude: "Cardia injury in GORD is localised to the immediate vicinity of the squamocolumnar junction and that sampling the mucosa 1–2 cm below this junction can result in a notable underestimation of the severity and type of mucosal abnormalities."

However, the authors suggest a unique etiology for the cause of carditis: "We propose that carditis may be due to a number of different mechanisms: physiological reflux ("wear and tear"); gastro-oesophageal reflux; and *H. pylori*. The most common, in our opinion, is likely to be "wear and tear". By analogy, the squamous mucosa has been found in normal volunteers to exhibit the regenerative hyperplasia changes (basal cell hyperplasia and elongated dermal papillae) of reflux in the distal 2–3 cm of the oesophagus. These changes could represent the consequence of gastric contents "lapping at the shores of the oesophagus" in health. The same case can be made for carditis, perhaps even more so."

In his notoriously wonderful colorful manner, Dr. Weinstein is actually explaining the early pathogenesis of GERD that we have outlined in this chapter and well demonstrated in Ayazi et al.[18] and Robertson et al.[20] These show that gastric overdistension causes the distal esophageal squamous epithelium to descend to a point that is distal to the pH transition point. The gastric acid is indeed "lapping at the shores of the oesophagus," causing injury in the esophageal squamous epithelium and inducing cardiac metaplasia as is seen in this study. Dr. Weinstein's "wear and tear" and "physiologic reflux" are really the early stage of the pathogenesis of GERD. The insight of this great gastroenterologist who, incidentally, always read his own biopsies after the UCLA pathologists had read them is remarkable. It suggests that it takes a deep knowledge of both gastroenterology and pathology to interpret these biopsies adequately.

It is interesting to contemplate that the biopsies that were routinely performed distal to the SCJ in retroflex view in the USC Foregut Surgery Unit from 1990 did not attempt to sample the histologic SCJ. They were taken from within 1 cm of the SCJ and frequently contained only columnar epithelium. It is likely that the prevalence and severity of carditis was underestimated in these biopsies, but they were still adequate for recognition of the significance of carditis.

7. RELATIONSHIP OF CARDITIS TO *HELICOBACTER PYLORI*

In the study detailed above by Der et al.[32] from our unit, *H. pylori* infection was present in only 20/141 (14%) of the patients. 17 of these patients had a pangastritis; 15 of these patients had *H. pylori* in cardiac epithelium in the biopsy at the SCJ. The other three patients had *H. pylori* limited to the antrum.

Similarly, the *H. pylori* prevalence in Lembo et al.[33] (reviewed above) was 4/30 (13%), and the conclusion of the authors was that *H. pylori* was not the primary cause of carditis but represented extension of *H. pylori* gastritis into the cardia.

The association between carditis and *H. pylori* gastritis is that the vast majority of patients with carditis (defined histologically by the presence of inflammation in cardiac epithelium) have neither *H. pylori* infection nor gastritis. In these patients carditis is an isolated finding in the region of the GEJ. The conclusion from the data in Der et al.[32] is that carditis is not caused by *H. pylori*, but when carditis coexists with *H. pylori* gastritis, the infection spreads secondarily to involve cardiac and oxyntocardiac epithelium, aggravating inflammation therein.

This interpretation accounts for studies in the literature that reported an etiologic association of "carditis" with *H. pylori* infection. These studies originated in highly influential institutions in the period 1997–2003 and had a high impact at the time in negating my hypothesis that carditis was a specific manifestation of GERD. These studies had their desired impact.

The idea that carditis could be caused by either GERD or *H. pylori* became embedded in the minds of gastroenterologists and pathologists. The studies reporting *H. pylori* as a cause of carditis ceased around 2003 and no one today considers *H. pylori* to be a cause of carditis. However, no one still believes that carditis is caused *only* by GERD.

When the methods of these studies are evaluated carefully, some of the studies define "carditis" as the presence of inflammation in a biopsy taken from the anatomic gastric cardia distal to the endoscopic GEJ. As I have shown, the epithelium in this area can be cardiac or gastric oxyntic epithelium depending on the distance of the biopsy from the endoscopic GEJ. As such, carditis defined in this manner can be caused by both GERD (inflammation of cardiac epithelium) and *H. pylori* (inflammation of gastric oxyntic epithelium).

Goldblum et al.,[34] from the Cleveland Clinic, in a paper published in *Gastroenterology*, write in the method of endoscopic biopsy in 58 GERD patients without Barrett esophagus and 27 non-GERD controls: "Two biopsy specimens were obtained from each of four sites: distal esophagus, gastric antrum, fundus and cardia. Esophageal biopsies were taken 5 cm above the squamocolumnar junction away from areas of obvious esophagitis. Gastric cardia biopsy specimens were obtained with the endoscope in a retroflexed position within 5 mm from a normal appearing squamocolumnar junction." There was no attempt to classify the histologic type of the specimen from the gastric cardia. It is relevant that Dr. Goldblum was the senior author of the autopsy study of the GEJ published 2 years later that showed a mean length of cardiac epithelium of 1.8 (range 1–4) mm distal to the SCJ. These data are cited in the paper. It is therefore clear that some of these controls must have had either or both cardiac and oxyntic epithelium. The authors cite the data that have established the relationship between carditis and GERD. They then state: "An alternate hypothesis regarding the etiology of carditis is that it is secondary to *H. pylori* infection involving the cardia." They cite two studies where patients with *H. pylori* gastritis had a high incidence of *H. pylori* in the gastric cardia. This is correct but is a different perspective; it only shows that most patients with *H. pylori* gastritis have a pangastritis. There have been no reports of such a high prevalence of *H. pylori* infection in patients whose mucosal inflammation is limited to the proximal stomach.

In this study Goldblum et al.[34] report that the prevalence of carditis in the 58 GERD patients (40%) was nearly identical to that of controls of similar age (41%), whereas histologic evidence of esophagitis was more common (33%) in GERD patients than controls (7%). Of the 23 patients in the GERD group that had carditis, 22 patients had *H. pylori* infection. All 11 patients in the control group that had carditis had *H. pylori* infection. They conclude that their findings suggest a strong link between *H. pylori* infection and carditis.

Only 23/58 (40%) GERD patients had inflammation in their biopsy of the gastric cardia. Derakhshan et al.[25] showed that *all* asymptomatic volunteers with very small amounts of cardiac epithelium (histologically defined) distal to the SCJ had significant inflammation and reactive changes. The only explanation for the absence of "carditis" in 60% of GERD patients in the study by Goldblum et al.[34] was that the biopsies of the gastric cardia were taken at a significant distance from the SCJ and sampled gastric oxyntic rather than cardiac epithelium. Gastric oxyntic epithelium is frequently normal in patients who have GERD with inflammation in cardiac epithelium.[32,33]

In another study of 85 patients by Wieczorek et al.[35] with a diagnosis of "carditis" retrieved from the computerized files of Harvard Medical School hospitals published in 2003 in the *American Journal of Surgical Pathology,* the method of biopsy read: "The cardia was defined as the area of mucosa located immediately distal to (and within 3-5 mm of) the anatomic gastroesophageal junction. The anatomic junction was defined as the most proximal limit of gastric folds. Because patients with Barrett's esophagus were excluded, the gastroesophageal junction corresponded to the location of the squamocolumnar junction. All biopsies of the "cardia" contained the histologic squamocoloumnar junction." The patients were divided into three groups: probable GERD (30), probable *H. pylori* (25), and neither—the control group (30).

This requirement that the histologic SCJ was present in the biopsy removes any possibility that the biopsies were taken from a distance away from the SCJ and makes it very likely that the inflammation was in cardiac epithelium, and therefore true histologically defined carditis. This is confirmed by the fact that the histologic types of glandular epithelia found in all three groups were cardiac (mucous) or oxyntocardiac (mucous/oxyntic) with zero gastric oxyntic epithelium. There was no significant difference in the prevalence of the epithelial types in the three groups. Carditis (inflammation) was present in 76% of the *H. pylori* group, significantly higher than in the GERD group (50%; $P = .05$) and controls (30%; $P < .001$). Multilayered epithelium was seen significantly more frequently in the GERD group compared with the *H. pylori* group ($P = .01$) and controls ($P = .02$).

The study assumes that carditis is caused by both GERD and *H. pylori* and is focused on finding differences in the histologic features in the two groups. It concludes: "The histologic appearance of carditis may vary according to its

TABLE 8.8 Grade of Carditis in 116 Patients Correlated With the Presence of Heartburn, Endoscopic Esophagitis, Histologic Esophagitis, and *Helicobacter pylori* Infection

Grade of Carditis	0 (n=9)	1 (n=58)	2 (n=35)	3 (n=14)
Heartburn	43%	47%	38%	23%
Endoscopic esophagitis	38%	38%	38%	8%
Histologic esophagitis	50%	42%	51%	43%
H. pylori infection	0%	9% (n=5)	31% (n=11)	57% (n=8)

etiology. *H. pylori* carditis is characterized by greater overall inflammation (particularly plasmacytic and neutrophilic inflammation) in comparison with reflux carditis. Furthermore, we showed that multilayered epithelium not uncommonly occurs in the cardia of patients without Barrett's esophagus, and in this setting, is highly associated with GERD and reflux carditis."

The study is extremely well done with excellent data and its conclusions are accurate. What it does not do is provide data that permit evaluation of the impact of *H. pylori* in GERD patients and vice versa. This is the critical issue regarding the etiology of carditis. This study assumes that carditis is a manifestation of *H. pylori* infection. It is not. *H. pylori* causes gastritis, which can become a pangastritis involving the proximal stomach. Carditis is caused *only* by GERD. *H. pylori* secondarily involves GERD-induced cardiac epithelium when the infection spreads proximally. Because almost all people have cardiac epithelium resulting from GERD, patients with *H. pylori* pangastritis will have essentially a double-hit inflammation in cardiac epithelium.

The final study that I have selected for this discussion is by Spechler et al.[36] from the Harvard group published only in abstract form. This clearly shows how carditis is viewed. The definitions are perfect and the pathologic interpretations are totally reliable. The aim of the study was to evaluate the potential contribution of GERD and *H. pylori* infection in carditis (defined as inflammation in "gastric cardiac-type mucosa"). All adult patients having elective endoscopy over a 6-month period were invited to participate, irrespective of indication for endoscopy. Symptoms of GERD (heartburn one or more days per week), endoscopic esophagitis (grade 1 or more), histologic esophagitis, and histologic evidence of *H. pylori* infection were correlated with the grade of carditis on a scale of 0–3 using the updated Sydney system. Of the 156 patients in the study 116 (74%) had cardiac-type mucosa. The correlation between grade of carditis and other features evaluated were as follows (Table 8.8).

The only statistically significant correlation was that found between grade of carditis and the prevalence of *H. pylori* infection. Based on this, the authors reached the following conclusion: "Carditis was associated significantly with *H. pylori* infection, but not with symptoms or signs of GERD. These findings do not support an important role for GERD in the pathogenesis of inflammation of the gastric cardia" (p. A297).

With a perfect study design and a research group that I trust completely, how is it possible that they reach a conclusion that is diametrically opposite to our study of carditis? The data have similarities: (1) The prevalence of *H. pylori* infection in our study (20/141 or 14%) was not dissimilar to this study (24/116 or 21%), given the variation in the prevalence of *H. pylori* in different populations. The conclusion in this study that carditis is significantly associated with *H. pylori* is difficult to understand with 79% of the patients with carditis being negative. (2) 17/20 patients with *H. pylori* infection in our study had a pangastritis; there were no patients with *H. pylori* limited to the proximal stomach. No distal gastric biopsies were reported in this study. (3) In our study, the presence of acute inflammation and severity (grade 2 and 3) of chronic inflammation was associated with *H. pylori* infection. If we had used the Sydney grading system, our study would also have a strong association between *H. pylori* infection and carditis. This is not unexpected. The Sydney system was designed specifically for grading gastritis caused by *H. pylori*. It was not designed for grading carditis caused by GERD.

We showed that acute inflammation in carditis had no significant association in patients with GERD (abnormal acid exposure in a 24-h pH test). Carditis is diagnostic of GERD simply by its presence, and the severity of GERD is related to the length of cardiac epithelium, not the severity of inflammation. The severity of inflammation, particularly acute inflammation in cardiac epithelium, indicates secondary infection with *H. pylori*.

The correct interpretation is that cardiac epithelium distal to the endoscopic GEJ results from columnar metaplasia of the esophagus and is an early pathologic manifestation of GERD. In patients with no *H. pylori* gastritis, cardiac epithelium that is always inflamed is present as an isolated finding in this region. The amount or type of inflammation has no association with the severity of GERD. In the small number of patients who have associated *H. pylori* gastritis, the bacterium frequently causes a pangastritis with extension of infection to the cardiac epithelium. This results in increased acute and chronic inflammation in cardiac epithelium as a double-hit phenomenon. There is absolutely no evidence that *H. pylori* is responsible for generating metaplastic cardiac epithelium at the GEJ. This is highly unlikely because this would require squamous epithelial injury by *H. pylori*, which has never been reported.

The critical mind-set that is essential for reaching this conclusion and avoiding error is a recognition that cardiac epithelium never exists as a normal epithelium in the proximal stomach. Those who do not believe this will always fall into this trap.

8. SUMMARY AND CONCLUSIONS

In this chapter, I have described the mechanism whereby the distal esophageal squamous epithelium is exposed to gastric contents during periods of gastric distension. Despite significant resistance, this ultimately results in columnar metaplasia of the distal esophageal squamous epithelium.

This is the first step in the pathogenesis of LES damage where the normal state of no LES damage has been converted to the abnormal state of LES damage (Fig. 8.14), which, when it progresses to beyond its reserve capacity, causes LES incompetence and reflux. Damage to the abdominal LES is associated with metaplastic cardiac epithelium that is interposed between the true GEJ (proximal limit of gastric oxyntic epithelium) and the SCJ (Fig. 8.13). The first reliably recognizable change of GERD, i.e., the presence of *any* cardiac epithelium distal to the SCJ, has occurred. From this point of onset, increase in the length of cardiac epithelium in the DDE correlates better with progression of disease and cellular changes than any other criterion. The reason for this is that the length of cardiac epithelium distal to the endoscopic GEJ is a measure of LES damage, which is the cause of GERD.

REFERENCES

1. Chandrasoma PT. Histologic definition of gastro-esophageal reflux disease. *Curr Opin Gastroenterol* 2013;**29**:460–7.
2. Vakil N, van Zanten SV, Kahrilas P, Dent J, Jones B, The Global Consensus Group. The Montreal definition and classification of gastroesophageal reflux disease: a global evidence-based consensus. *Am J Gastroenterol* 2006;**101**:1900–20.
3. Rex DK, Cummings OW, Shaw M, Cumings MD, Wong RK, Vasudeva RS, Dunne D, Rahmani EY, Helper DJ. Screening for Barrett's esophagus in colonoscopy patients with and without heartburn. *Gastroenterology* 2003;**125**:1670–7.
4. Riddell RH. The biopsy diagnosis of gastroesophageal reflux disease, "carditis," and Barrett's esophagus, and sequelae of therapy. *Am J Surg Pathol* 1996;**20**(Suppl. 1):S31–51.
5. DeMeester TR, Johnson LF, Joseph GJ, Toscano MS, Hall AW, Skinner DB. Patterns of gastroesophageal reflux in health and disease. *Ann Surg* 1976;**184**:459–70.
6. Zaninotto G, DeMeester TR, Schwizer W, Johansson K-E, Cheng S-C. The lower esophageal sphincter in health and disease. *Am J Surg* 1988;**155**:104–11.
7. American Gastroenterological Association Medical Position Statement on the management of gastroesophageal reflux disease. *Gastroenterology* 2008;**135**:1383–91.
8. Chandrasoma PT, Wijetunge S, DeMeester SR, Hagen JA, DeMeester TR. The histologic squamo-oxyntic gap: an accurate and reproducible diagnostic marker of gastroesophageal reflux disease. *Am J Surg Pathol* 2010;**34**:1574–81.
9. Chandrasoma PT, Der R, Ma Y, et al. Histology of the gastroesophageal junction: an autopsy study. *Am J Surg Pathol* 2000;**24**:402–9.
10. Chandrasoma P, Wijetunge S, Ma Y, DeMeester S, Hagen J, DeMeester T. The dilated distal esophagus: a new entity that is the pathologic basis of early gastroesophageal reflux disease. *Am J Surg Pathol* 2011;**35**:1873–81.
11. Chandrasoma P, Makarewicz K, Wickramasinghe K, Ma YL, DeMeester TR. A proposal for a new validated histologic definition of the gastroesophageal junction. *Hum Pathol* 2006;**37**:40–7.
12. Curcic J, Roy S, Tech M, Schwizer A, Kaufman E, Forras-Kaufman Z, Menne D, Hebbard GS, Treier R, Boesiger P, Steingoetter A, Fried M, Schwizer W, Pal A, Fox M. Abnormal structure and function of the esophagogastric junction and proximal stomach in gastroesophageal reflux disease. *Am J Gastroenterol* 2014;**109**:658–67.
13. Korn O, Csendes A, Burdiles P, et al. Anatomic dilatation of the cardia and competence of the lower esophageal sphincter: a clinical and experimental study. *J Gastrointest Surg* 2000;**4**:398–406.
14. Chandrasoma P. Controversies of the cardiac mucosa and Barrett's esophagus. *Histopathology* 2005;**46**:361–73.
15. Chandrasoma PT, DeMeester TR. *GERD: from reflux to esophageal adenocarcinoma.* San Diego: Academic Press; 2006.
16. Odze RD. Unraveling the mystery of the gastroesophageal junction: a pathologist's perspective. *Am J Gastroenterol* 2005;**100**:1853–67.
17. Chandrasoma P. Pathophysiology of Barrett's esophagus. *Semin Thorac Cardiovasc Surg* 1997;**9**:270–8.
18. Ayazi S, Tamhankar A, DeMeester SR, et al. The impact of gastric distension on the lower esophageal sphincter and its exposure to acid gastric juice. *Ann Surg* 2010;**252**:57–62.
19. Tamhankar AP, DeMeester TR, Peters JH, et al. The effect of meal content, gastric emptying and gastric pH on the postprandial acid exposure of the lower esophageal sphincter (LES). *Gastroenterology* 2004;**126**(Suppl. 2):A-495.

20. Robertson EV, Derakhshan MH, Wirz AA, Lee YY, D=Seenan JP, Ballantyne SA, Hanvey SL, Kelman AW, Going JJ, McColl KE. Central obesity in asymptomatic volunteers is associated with increased intrasphincteric acid reflux and lengthening of the cardiac mucosa. *Gastroenterology* 2013;**145**:730–9.
21. Tobey NA, Carson JL, Alkiek RA, et al. Dilated intercellular spaces: a morphological feature of acid reflux-damaged human esophageal epithelium. *Gastroenterology* 1996;**111**:1200–5.
22. Tobey NA, Hosseini SS, Argore CM, Dobrucali AM, Awayda MS, Orlando RC. Dilated intercellular spaces and shunt permeability in non-erosive acid-damaged esophageal epithelium. *Am J Gastroenterol* 2004;**99**:13–22.
23. Souza RF, Huo X, Mittal V, Schuler CM, Carmack SW, Zhang HY, Zhang X, Yu C, Hormi-Carver K, Genta RM, Spechler SJ. Gastroesophageal reflux might cause esophagitis through a cytokine-mediated mechanism rather than caustic acid injury. *Gastroenterology* 2009;**137**:1776–84.
24. Dunbar KB, Agostori AT, Odze RD, Huo X, Pham TH, Cipher DJ, Castell DO, Genta RM, Souza RF, Spechler SJ. Association of acute gastroesophageal reflux disease with esophageal histologic changes. *JAMA* 2016;**315**:2104–12.
25. Derakhshan MH, Robertson EV, Lee YY, Harvey T, Ferner RK, Wirz AA, Orange C, Ballantyne SA, Hanvey SL, Going JJ, McColl KE. In healthy volunteers, immunohistochemistry supports squamous to columnar metaplasia as mechanism of expansion of cardia, aggravated by central obesity. *Gut* 2015;**64**(11):1705–14.
26. Milano F, van Baal JW, Buttar NS, Rygiel AM, de Kort F, DeMars CJ, Rosmolen WD, Bergman JJ, van Marle J, Wang KK, Peppelenbosch MP, Krishnadath KK. Bone morphogenesis protein 4 expressed in esophagitis induces a columnar phenotype in esophageal squamous cells. *Gastroenterology* 2007;**132**:2412–21.
27. Glickman JN, Fox V, Antonioli DA, Wang HH, Odze RD. Morphology of the cardia and significance of carditis in pediatric patients. *Am J Surg Pathol* 2002;**26**:1032–9.
28. Dresner SM, Griffin SM, Wayman J, Bennett MK, Hayes N, Raimes SA. Human model of duodenogastro-oesophageal reflux in the development of Barrett's metaplasia. *Br J Surg* 2003;**90**:1120–8.
29. Oberg S, Johansson J, Wenner J, Walther B. Metaplastic columnar mucosa in the cervical esophagus after esophagectomy. *Ann Surg* 2002;**235**:338–45.
30. Lord RV, Wickrmasinghe K, Johansson JJ, DeMeester SR, Brabender J, DeMeester TR. Cardiac mucosa in the remnant esophagus after esophagectomy is an acquired epithelium with Barrett's-like features. *Surgery* 2004;**136**:633–40.
31. Malfertheiner P, Nocon M, Vieth M, Stolte M, Jasperson D, Keolz HR, Labenz J, Leodolter A, Lind T, Richter K, Willich SN. Evolution of gastro-oesophageal reflux disease over 5 years under routine medical care – the ProGERD study. *Aliment Pharmacol Ther* 2012;**35**:154–64.
32. Der R, Tsao-Wei DD, DeMeester T, et al. Carditis: a manifestation of gastroesophageal reflux disease. *Am J Surg Pathol* 2001;**25**:245–52.
33. Lembo T, Ippoliti AF, Ramers C, Weinstein WM. Inflammation of the gastroesophageal junction (carditis) in patients with symptomatic gastroesophageal reflux disease; a prospective study. *Gut* 1999;**45**:484–8.
34. Goldblum JR, Vicari JJ, Falk GW, Rice TW, Peek RM, Easley K, Richter JE. Inflammation and intestinal metaplasia of the gastric cardia: the role of gastroesophageal reflux and *H. pylori* infection. *Gastroenterology* 1998;**114**:633–9.
35. Wieczorek TJ, Wang HH, Antonioli DA, Glickman JN, Odze RD. Pathologic features of reflux and *Helicobacter pylori*-associated carditis: a comparative study. *Am J Surg Pathol* 2003;**27**:960–8.
36. Spechler SJ, Wang HH, Chen YY, Zeroogian JM, Antonioli DA, Goyal RK. GERD vs *H. pylori* infections as potential causes of inflammation in the gastric cardia. *Gastroenterology* 1997;**112**:A297.

Chapter 9

Correlation of LES Damage and GERD

Gastroesophageal reflux disease (GERD) is well recognized as a disease that is caused by damage to the sphincter mechanism that exists in the distal esophagus. This lower esophageal sphincter (LES), when it is normal, is very effective in preventing reflux of gastric contents into the esophagus under virtually all circumstances. I have reviewed the normal LES and how it functions in Chapter 7. The normal LES has an overall length of around 50 mm and an abdominal length of ~35 mm.

It is remarkable that the present clinical diagnosis of GERD pays little or no attention to its causative agent: damage to the LES. This is largely because there is no test at present that can detect LES damage with accuracy.

The objective of this book is to change this status quo. By developing a method of assessment of LES damage, we can turn the symptom-based focus of diagnosis and management of GERD to one that concentrates on LES damage.

All present definitions of GERD require the presence of abnormal transsphincteric reflux into the thoracic esophagus across an LES that fails with sufficient frequency to produce abnormal acid exposure in the esophagus. Abnormal acid exposure is defined as a pH < 4 detected by an electrode situated 5 cm above the upper border of the LES for >4.5% of the period of the test.[1,2] This equates to 64 min/24 h of esophageal acid exposure.

This level of acid exposure that is deemed abnormal was selected to correlate with the occurrence of symptoms that define GERD.[1,2] At this level of acid exposure, the LES is defective. With an LES that is designed to prevent reflux (i.e., have zero acid exposure in the thoracic esophagus), this must represent a seriously damaged LES (Fig. 9.1).

The criteria for a defective LES that correlate with symptoms of GERD and abnormal acid exposure as defined by the pH test above are a mean LES pressure of <6 mmHg, a total LES length of <20 mm, and an abdominal LES length of <10 mm.[3]

The present diagnostic and management method of GERD is therefore to wait for sufficient LES damage and failure to cause symptoms that are sufficiently troublesome to satisfy the criterion of the Montreal definition.[4] This is necessarily late in the course of the pathogenesis of the disease, often delayed by several decades from the point of onset of the disease. Even after that, treatment is continuation of proton pump inhibitor (PPI) therapy, which does not reverse or prevent progression of LES damage or the severity of reflux.[5]

The new focus will depend on a study of how the normal LES progressively degrades from its normal state to the point of failure (Fig. 9.2). When this is done, a stage of disease prior to LES failure and prior to significant transsphincteric reflux into the thoracic esophagus can be defined. I will call this the phase of compensated LES damage (Fig. 9.3). In this phase, the LES shortens from normal to the point at which it begins to fail on rare occasion, after exhausting its entire reserve capacity. During this phase, the LES is damaged but has sufficient abdominal length to maintain competence.

LES damage begins in the distal end and first involves the abdominal segment of the LES. As LES damage progresses, it causes a shortening of the residual LES by an amount that is equivalent to the length of LES damage (Figs. 9.2 and 9.3). The shortening of the LES will be reflected in manometric studies, which defines the residual functional LES (Fig. 9.1). LES damage cannot be measured by any modality at present. In the new method, it can be measured histologically as the length of cardiac epithelium (with and without parietal and/or goblet cells) distal to the endoscopic gastroesophageal junction (GEJ), i.e., in the dilated distal esophagus (Fig. 9.4).

One of the critical determinants of competence of the LES depends is how much functional abdominal LES remains. When this decreases to <10 mm, LES failure begins.

All cellular changes and complications of GERD are highly correlated with the severity of transsphincteric reflux into the thoracic esophagus, which in turn is highly correlated with severity of LES damage. Controlling progression of LES damage is the obvious method of controlling reflux.

The new assessment of LES damage will permit early detection of LES damage. With the new test that measures LES damage at an early stage of the disease, I will develop a new algorithm that will accurately predict future progression of LES damage (see Chapter 19). This will allow identification of people at risk of progressing to severe LES damage during

GERD. http://dx.doi.org/10.1016/B978-0-12-809855-4.00009-9

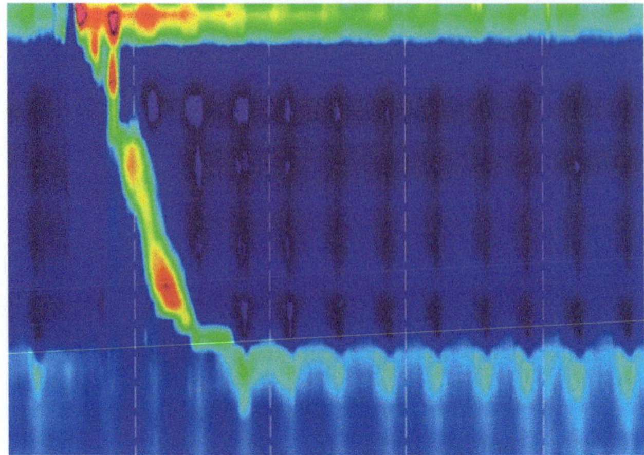

FIGURE 9.1 High-resolution manometry showing a markedly abnormal short lower esophageal sphincter (LES) with low pressure. Esophageal body motility and LES relaxation during swallowing is normal. The calculated total LES length is 1.1 cm, the abdominal LES length is 0 cm, and resting LES pressure is 9 mmHg *Study and interpretation by Chris Dengler, MD, Legato Medical Inc.*

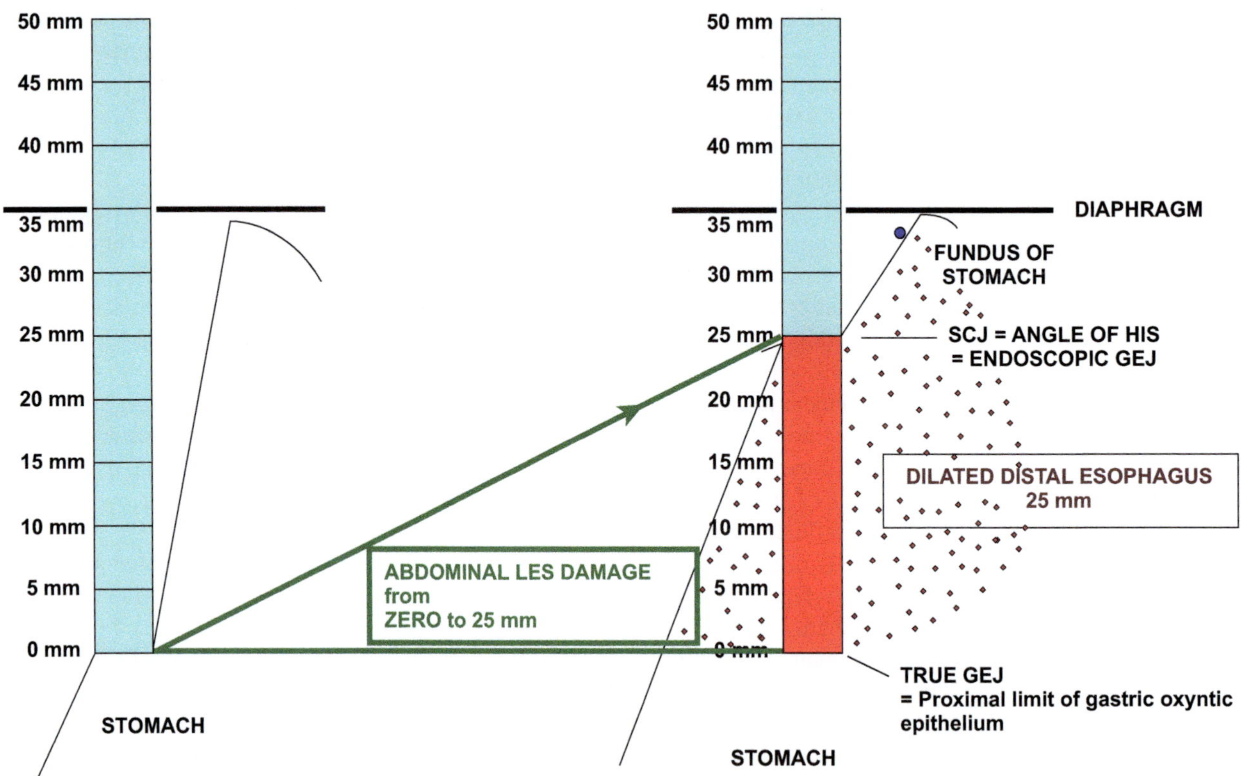

FIGURE 9.2 Progression of lower esophageal sphincter (LES) damage from the normal state (zero LES damage with an initial abdominal LES length of 35 mm) to the point of severe LES damage (25 mm LES damage with a residual abdominal LES length of 10 mm). Progression occurs at a different rate and only in people destined to develop gastroesophageal reflux disease as defined by present criteria for diagnosis. There is no present understanding of this progression. *GEJ*, gastroesophageal junction; *SCJ*, squamocolumnar junction. Damaged LES = red; dilated distal esophagus lined by cardiac epithelium = speckled area; peritoneal reflection = blue dot.

their lifetime. These are the patients at highest risk of progression to treatment failure, visible columnar-lined esophagus (CLE), Barrett esophagus, and adenocarcinoma. Identification of this minority group allows selective treatment that is designed to prevent or slow progression of LES damage with the objective of preventing severe GERD. If this can be done successfully, adenocarcinoma can be prevented.

Theoretically, if the assessment of LES damage is done at a very early stage, there can be an opportunity to prevent GERD altogether by preventing LES damage from exceeding 15 mm by appropriate interventions.

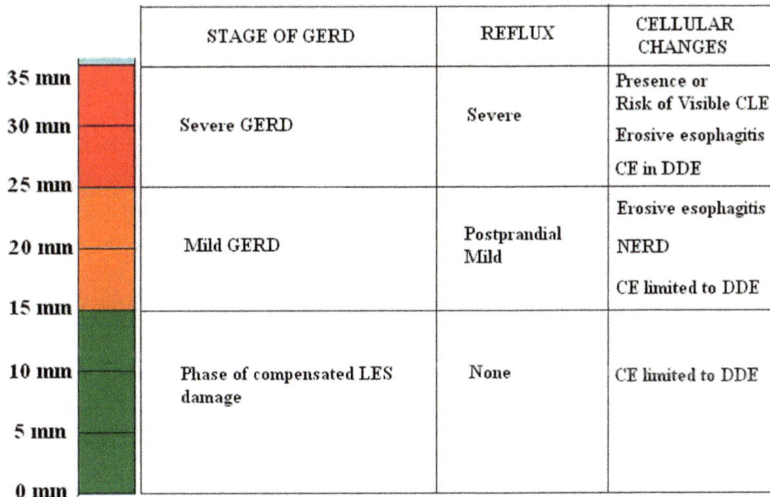

STAGE OF GERD	REFLUX	CELLULAR CHANGES
Severe GERD	Severe	Presence or Risk of Visible CLE Erosive esophagitis CE in DDE
Mild GERD	Postprandial Mild	Erosive esophagitis NERD CE limited to DDE
Phase of compensated LES damage	None	CE limited to DDE

FIGURE 9.3 Correlation between the amount of abdominal lower esophageal sphincter (LES) damage and status of the person from the standpoint of gastroesophageal reflux disease (GERD). In the phase of compensated LES damage, there is progression of LES damage from 0 to 15 mm, but the sphincter is competent. As the abdominal LES damage progresses from 15 to 25 mm, there is postprandial GERD of increasing severity. The final stage of severe GERD, at >25 mm of LES damage, is defined by the risk or presence of visible columnar-lined esophagus. Complete destruction of the abdominal LES (35 mm damage) is not uncommonly seen at manometry in GERD patients who have the test. *CE*, cardiac epithelium; *CLE*, columnar-lined esophagus; *DDE*, dilated distal esophagus.

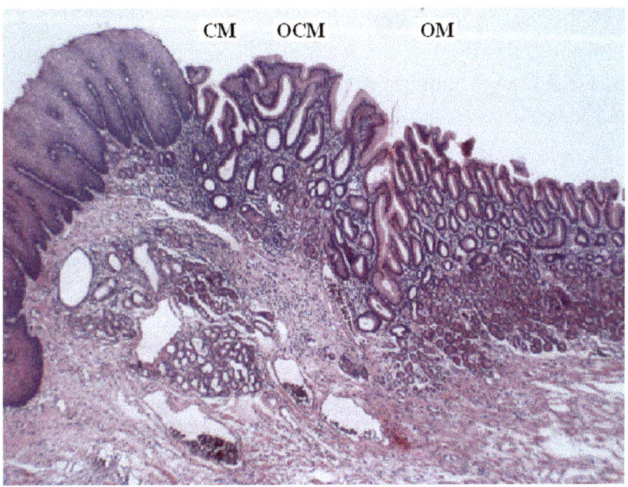

FIGURE 9.4 Sections across the squamocolumnar junction showing 2 mm of metaplastic cardiac (CM) and oxyntocardiac (OCM) mucosa between esophageal squamous epithelium (left) and gastric oxyntic mucosa (OM; on the right). The cardiac epithelium is inflamed and shows reactive hyperplasia of the foveolar region (reflux carditis). This is associated with 2 mm of abdominal lower esophageal sphincter damage.

1. LOWER ESOPHAGEAL SPHINCTER PARAMETERS THAT DEFINE LOWER ESOPHAGEAL SPHINCTER FAILURE

The availability of manometry in the early 1960s to define and study the LES was a stimulus to an attempt to correlate characteristics of the LES with GERD. Comparison of patients with symptoms of GERD and asymptomatic subjects showed that, as a group, the amplitude of the high-pressure zone was significantly lower in the GERD patients.[6] This suggested that LES pressure had a causal relationship with GERD.

However, the relationship between the magnitude of LES pressure and reflux was far from absolute. Some patients with a low basal LES pressure had a competent sphincter, whereas others with a high pressure had reflux.[6,7] Measuring basal LES pressure alone was therefore unreliable for the diagnosis of reflux.

At this time, technology that permitted quantitation of reflux from the stomach to the esophagus was also developed. This was achieved by measuring acid exposure by placing a pH-sensitive electrode in the thoracic esophagus. Monitoring the pH in this manner for 24 h or longer proved to be a reliable method of assessment of the amount of reflux.[1,2]

Indirectly, the 24-h pH test was an assessment of competency of the LES.

It also provided a means of correlating LES characteristics with objective evidence of reflux, which was a much superior indicator of reflux than the presence and severity of symptoms.

Thurer et al.,[7] in a study of 266 patients with symptoms suggestive of esophageal disease at the Department of Surgery, University of Chicago performed both manometric evaluation of the LES and 24-h pH monitoring of the distal esophagus. This showed that "the competency of the cardia is related to two manometric measurements, namely, the amplitude of the distal esophageal high pressure zone and the length of the abdominal esophagus.... The competency of the cardia improves as the amplitude of the distal esophageal high pressure zone increases... Only 17.6 per cent of those patients with a pressure of 30 mmHg or higher had an abnormal 24-hour pH reflux test compared with an 83.3 per cent incidence of an abnormal reflux test in patients with a pressure of 0 to 5 mmHg.... As the amount of distal esophagus exposed to the positive pressure environment of the abdomen increases, the incidence of an abnormal 24 hour pH reflux test decreases.... Only 38 per cent of patients with more than 3 cm of abdominal esophagus had an abnormal test, compared with an 81 per cent incidence in patients who had less than 1 cm of abdominal esophagus."

By this time (1974), an abnormal pH test had been defined as acid exposure (i.e., pH<4) in the distal esophagus 5 cm above the upper limit of the LES for >4.5% of the 24-h period.[1,2] This determination of abnormality was made by the fact that 95% of a volunteer population without clinical GERD had less than 4.5% acid exposure in the esophagus and patients with symptoms of GERD commonly had >4.5%.

It should be recognized that if the definition of an abnormal pH test was the presence of more than zero acid exposure, the percentage of patients with abnormal acid exposure in the study would have increased dramatically. This would have no meaning from a diagnostic standpoint if the patients did not have clinical evidence of GERD.

Significant numbers of asymptomatic volunteers had zero acid exposure suggesting that this was an alternative valid, albeit too sensitive, definition of abnormal esophageal acid exposure. It would indicate that an absolutely normal LES was capable of preventing reflux under all test circumstances.

The importance of the length of the abdominal segment of the LES is explained elegantly, and with incredible insight, as follows: "The in vitro study shows that the intraabdominal segment of the esophagus with normal intrinsic muscle tone has all the capabilities of serving as a valve to prevent the regurgitation of gastric contents into the esophagus. The important determinant for its function is the length exposed to the positive pressure environment of the abdomen. In this position, intraluminal pressure in the esophagus equals intraabdominal pressure plus the intrinsic esophageal muscle tone. As a consequence, based on the physical laws that govern the behavior of soft tubes, the intraabdominal segment of the esophagus is collapsed. In addition, the negative pressure in the thoracic esophagus results in its sucking the abdominal segment flat, and in doing so, helps to keep that segment closed. The force tending to open the abdominal segment of esophagus is the intragastric pressure, which equals intraabdominal pressure plus a small contribution from gastric muscle tone. The portion of intragastric pressure due to gastric muscle tone is therefore all that has to be controlled by the intrinsic esophageal muscle tone in order to maintain a collapsed abdominal esophagus, since all variations in intraabdominal pressure due to straining, coughing, and so on are applied simultaneously to both the stomach and the abdominal esophagus."

The only situation in which an adequate length of the abdominal esophagus cannot maintain competency is when gastric muscle tone exceeds esophageal muscle tone. According to Laplace's law, the pressure in a viscus is indirectly related to its radius. Since the esophageal radius is consistently smaller than gastric radius, it is difficult for gastric muscle tone to produce an intragastric pressure that will exceed the intraluminal pressure produced by the muscle tone of the abdominal esophagus. As a consequence, the cardia remains competent. An exception to this is the act of vomiting, when gastric hypermotility produces pressures that exceed esophageal muscle tone, and the stomach content is regurgitated into the esophagus and expelled out the mouth.

This study of the LES concentrates on the possible use of manometry as a test for reflux. In this regard, the authors conclude: "... the measurement of the high pressure zone or of the length of the abdominal esophagus in a single individual is not useful in determining the competency of the cardia... (and) is not of practical use in the management of patients.... An exception to this, however, is the patient who has a high pressure zone of less than 5 mmHg and an abdominal length of less than 1 cm in length. This person has a severe mechanical defect and incompetency of the cardia. In our experience, medical therapy has little to offer these individuals, and surgical repair should be performed."

The emphasis of the manometric assessment of the LES is to identify incompetency of the cardia of a sufficient degree to act as an indication for antireflux surgery in a patient with GERD.

The goal of a subsequent (1988) study from this group headed by Dr. DeMeester by Zaninotto et al.[3] is to evaluate the resistance imposed by the LES to "the flow of gastric juice from an environment above atmospheric pressure, the stomach, into an environment below atmospheric pressure, the esophagus. This resistance is due to the integrated mechanical effect of sphincter pressure, overall length and the length exposed to the positive environmental pressure of the abdomen. Each of these components can be measured by manometry."

TABLE 9.1 Normal Esophageal Parameters in 50 Healthy Volunteers

	Mean	SD	Median	Maximum	Minimum	Percentile			
						2.5	5	95	97.5
Lower Esophageal Sphincter Measurements									
Pressure (mmHg)	14.87	5.14	13.8	25.6	5.2	6.1	8		
Abdominal length (cm)	2.18	0.72	2.2	3.5[a]	0.8	0.89	1.1		
Overall length (cm)	3.65	0.68	3.6	5.5	2.4	2.4	2.6		
Esophageal Acid Exposure									
Total reflux time (%)	1.57	1.47	1.1	6	0			4.6	5.8
Reflux episodes	23.87	22.96	20	126	0			51	76

Only items in the table relevant for the discussion are reproduced. In the discussion in the text, I will concentrate on the abdominal length and total reflux time (% time pH < 4). These are highlighted in red.
[a]For abdominal length, one person with an abdominal length of >5 cm is excluded as an outlier.

The study consisted of 50 volunteers and 622 patients.

The 50 normal volunteers did not have a history of foregut symptoms, upper gastrointestinal disease, or previous esophageal, gastric, duodenal, or biliary tract surgery. All showed normal results on upper gastrointestinal barium studies, esophageal manometry, and 24-h pH monitoring. There were 19 men and 31 women with a mean age of 35.8 ± 10.3 years (median 33 years, range 23–71 years) and a mean body weight of 70.3 ± 14.7 kg (median 68 kg, range 47.6–131 kg).

The results of LES pressure, abdominal length, and overall length in the healthy subjects are summarized in Table 9.1.

The study also had a patient population consisting of 622 patients with foregut symptoms. Patients under 15 years and those who had undergone prior esophageal or gastric surgery were excluded. There were 295 men and 327 women with a mean age of 47.6 ± 19.6 years (median 50 years, range 15–82 years). All underwent manometry and 24-h esophageal pH monitoring.

Of the 622 patients, 324 (51%) had an increased esophageal acid exposure on 24-h pH monitoring. A sphincter pressure of 6 mmHg or less, an abdominal length of <1 cm, and an overall length of <2 cm had an association of 79%, 80%, and 79%, respectively, for increased esophageal acid exposure. This corresponded to the 2.5 percentile for pressure, 5 percentile for abdominal length, and the 2.5 percentile for overall length in the normal population. These values therefore provided the greatest discrimination for defining an incompetent LES.

Of the three defects of the LES that were associated with LES incompetence, an isolated low pressure was the most common, followed by the combination of a low pressure and short abdominal length, and an isolated short abdominal length. A low value in one of the three components could be compensated for by a higher value in the others, but low values in all three inevitably led to abnormal esophageal acid exposure.

Based on the data in this study, the criteria for an incompetent LES were defined as an LES pressure <6 mmHg, an abdominal length <1 cm, and an overall length <2 cm. These criteria were designed for maximum specificity because their primary use was for deciding which patients were most suited for antireflux surgery. These criteria were relatively insensitive for a diagnosis of GERD. As a result, their use was limited in the diagnosis of GERD.

When high-resolution manometry was introduced, it quickly replaced the traditional catheter-based manometry, largely because it was more convenient and comfortable for the patient. High-resolution manometry produced accurate measurements of the pressures in the esophagus and stomach including the LES. However, it produced measurements that were less accurate for measuring LES length than the best motorized pull-through catheter-based manometry it replaced. As a result, the use of esophageal manometry largely became a test for the diagnosis of esophageal motility disorders rather than a diagnostic test for GERD.

This was unfortunate. The diagnosis of GERD became almost completely divorced from its cause, which is damage to the LES.

2. CORRELATION BETWEEN LOWER ESOPHAGEAL SPHINCTER DAMAGE AND GASTROESOPHAGEAL REFLUX DISEASE

Measurement of the LES length has never been used as a diagnostic method for GERD. The use of manometry, right from the outset, has been mainly to identify patients whose LES damage is so severe that (1) medical therapy is unlikely to be effective and (2) antireflux surgery is likely to be the only method of providing relief from symptoms for these patients.

The data produced from this vantage point convincingly demonstrate a relationship between the presence of a defective sphincter (defined as pressure <6 mmHg, overall length <2 cm, or abdominal length <1 cm) and both abnormal reflux into the distal esophagus (defined as a pH<4 for >4.5% of the 24-h period) and the presence of mucosal damage in the distal esophagus.

The clinical section of another study from this group by Bonavina et al.[9] consisted of 45 asymptomatic subjects and 448 persons with symptoms suggestive of GERD. The asymptomatic controls had no history of upper gastrointestinal disease and normal findings on an upper gastrointestinal barium study. Their mean age was 35 years (range 18–63 years). The patient population had a mean age of 51 years (range 22–90 years). Both groups underwent esophageal manometry using the station pull-through technique and had 24-h esophageal pH monitoring.

Abnormal 24-h pH test was defined as a composite mathematical score that incorporated six parameters that exceeded by more than two standard deviations the mean score of the control subjects. A mechanically defective LES was defined by the presence of any of the following: a pressure <6 mmHg, abdominal length <1 cm, and overall length <2 cm.

All control subjects had a "normal" esophageal pH test result (pH<4 for <4.5% of the testing period). Manometry showed a mean LES pressure of 14.9 ± 4.6 mmHg, a mean abdominal length of 2.6 ± 0.8 cm, and a mean overall length of 4.2 ± 1.1 cm.

Manometry in the 448 patients with GERD symptoms showed a mechanically defective LES in only 191 (43%) patients; 108 of these patients had an LES pressure <6 mmHg, 111 had an abdominal LES length <1 cm, and 66 had an overall length <2 cm. None of the three manometric criteria used to define a defective LES were a sensitive method of identifying patients with GERD defined by the presence of symptoms.

In the symptomatic GERD patients the prevalence of an abnormal 24-h pH test decreased with increasing length of the abdominal LES length. With abdominal LES lengths of 0–1 cm, >1–2 cm, >2–3 cm, and >3 cm, the prevalence of an abnormal 24-h pH test was 83/111 (75%), 130/208 (63%), 48/102 (50%), and 11/27 (41%), respectively.

Manometry also had a relatively low correlation with an abnormal 24-h pH test as it was defined. The pH test was abnormal in 143 (75%) of the 191 patients who had a mechanically defective LES; 25% of the patients with a defective LES had a normal pH test.

Of the 257 patients who had a mechanically "normal" (i.e., no criteria of incompetence) LES, 24-h pH was abnormal in 50%. A "normal" manometry could not be used to predict the absence of abnormal reflux.

These findings show why manometry never became a diagnostic test for GERD. By the criteria used, the diagnostic sensitivity was very low. The only value of the test in practice was to identify those patients who would be most likely to be benefited by surgery to repair the LES.

The study by Stein et al.[10] also from the same group is aimed at defining the relationship between the mechanical characteristics of the LES and the prevalence and severity of esophageal mucosal injury, defined by the presence of esophagitis at endoscopy, stricture, or Barrett's esophagus in patients with GERD.

The study population consisted of 50 normal healthy volunteers and 205 consecutive patients with GERD documented by an increased esophageal exposure to gastric juice on 24-h esophageal pH monitoring. A third group of 67 consecutive patients had foregut symptoms from a cause other than GERD as shown by a normal esophageal pH test.

All subjects had standard manometry and 24-h pH monitoring. They also had upper gastrointestinal endoscopy. Of the 205 patients with GERD, macroscopic esophagitis was seen in 66 (grade 1 in 32, grade 2 in 20, and grade 3 in 14 patients), a stricture in 19, and Barrett's esophagus in 28 patients. The remaining 92 of the 205 GERD patients had no endoscopic evidence of mucosal injury.

Of the 67 patients with foregut symptoms and a normal pH test, mucosal erythema was present in 5 patients; none had a stricture or Barrett's esophagus. Barrett's esophagus was diagnosed by the histologic documentation of columnar epithelium lining the esophagus at least 3 cm above the endoscopic GEJ. There was no requirement for intestinal metaplasia on biopsy. At present, this would be classified as long-segment CLE with the likelihood of most, if not all, being positive for intestinal metaplasia.

A mechanically defective LES was present in 112 (55%) of the 205 patients with increased esophageal acid exposure, compared with 10/67 (15%) symptomatic patients with a normal pH test. In the GERD patients, the prevalence of a mechanically defective LES increased with the severity of mucosal injury and was significantly higher in patients with

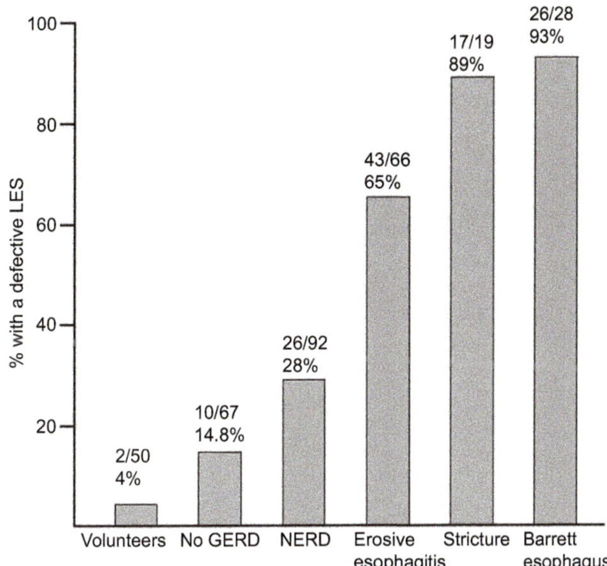

FIGURE 9.5 Percentage of people with a defective lower esophageal sphincter (LES) (one criterion of which is abdominal LES length <10mm). The percentage increases progressively from healthy volunteers to people with foregut symptoms and a normal 24-h pH test to persons with an abnormal 24-h pH test who are endoscopically normal (NERD), to those with erosive esophagitis, stricture, and Barrett esophagus. *GERD*, gastroesophageal reflux disease; *NERD*, nonerosive reflux disease.

esophagitis (65%), stricture (89%), and Barrett's esophagus (93%), as compared to patients with GERD without mucosal damage ($P<.001$, Fig. 9.5).

Numerous other studies have shown a direct correlation between the presence of abnormal acid exposure in a 24-h pH test and the presence of a mechanically defective LES and the prevalence and severity of mucosal abnormalities in the esophagus. They establish that decreasing lengths of the abdominal LES is associated with a progressive increase in the prevalence of abnormal acid exposure in the esophagus.

The data relating to a correlation between different lengths of the abdominal LES and esophageal mucosal abnormalities are sparse in the literature. These two studies from 1986 to 1992 suggest a wide range of abdominal LES length associated with abnormal reflux by a 24-h pH test. It is uncertain whether this lack of correlation is real or whether the measurements at the time were lacking in precision.

The 24-h pH test depends heavily on the correct placement of the electrode at 5cm above the upper border of the LES. Variation in placement has profound effects because there is a pH gradient in the esophagus during a reflux episode with the GEJ having the baseline gastric pH with gradual increase to neutral esophageal pH at the height of the column of refluxed gastric juice. Variations in placement of the pH electrode are likely to occur in practice.

3. LOWER ESOPHAGEAL SPHINCTER FAILURE

Intermittent failure of the LES, resulting from inappropriate or pathologic relaxation during times between swallows (excluding belching and vomiting), is the critical event that is responsible for transsphincteric reflux of gastric contents into the thoracic esophagus. It is widely recognized as the cause of GERD.

In patients with LES damage short of complete destruction of the abdominal LES, the main mechanism of LES failure is what is termed "transient lower esophageal sphincter relaxation (tLESR)." In such patients, the LES is sufficiently intact in the fasting state but fails increasingly with dynamic shortening that occurs during gastric overdistension produced in the postprandial state (Fig. 9.6).

An international group of workers[11] that included Dr. Dent was instrumental in producing much of the early research into LES relaxations that were not triggered by swallowing. In their introduction to this review, they write: "... traditional views held that gastroesophageal reflux was the consequence of a persistently weak LES. Measurement of esophageal motility at the time of reflux made in recumbent subjects, however, showed that most reflux events in both normal subjects and patients with reflux disease occurred during brief intermittent LES relaxations rather than because of persistently defective LES tone. This finding reorientated thinking about the major defect that underlies reflux disease: abnormally frequent reflux. The recognition of tLESRs provided an explanation for the occurrence of reflux in the majority of patients who have

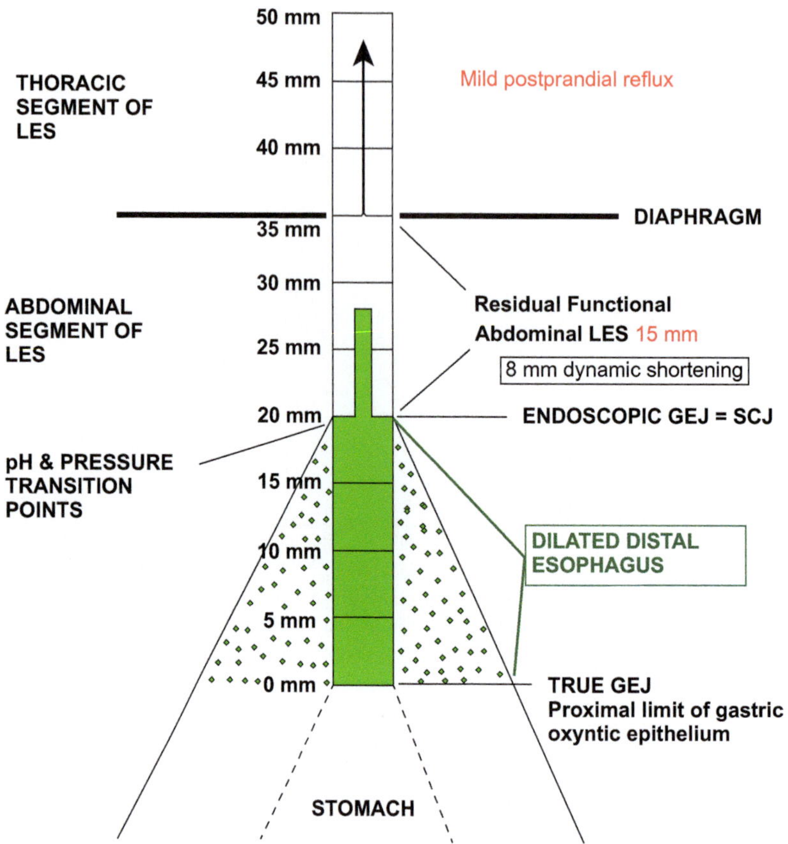

FIGURE 9.6 A person with 20 mm of abdominal lower esophageal sphincter (LES) damage (residual abdominal LES length of 15 mm). This person has a competent LES (abdominal length >10 mm) at baseline. However, during dynamic shortening associated with a heavy meal, the abdominal LES length becomes <10 mm (in this diagram, dynamic shortening of 8 mm causes the functional abdominal LES length to become 7 mm), resulting in the possibility of LES failure (transient LES relaxation).*GEJ*, gastroesophageal junction; *SCJ*, squamocolumnar junction.

normal LES resting pressure and changed the conceptual emphasis of LES dysfunction in reflux disease from LES strength to LES control."

Sudden LES failure (tLESR) is the single most common mechanism underlying gastroesophageal reflux. In normal subjects, the majority (70%–100%) of reflux episodes occur during tLESRs. The other postulated but less well-documented causes of reflux are swallow-induced LES relaxation associated with failed or incomplete primary peristalsis, persistently absent basal LES pressure, straining by deep inspiration, and increase in intraabdominal pressure.

In general, the proportion of reflux episodes that can be ascribed to tLESRs varies inversely with the severity of reflux disease, presumably because of the increasing likelihood of a defective LES as the severity of GERD increases. tLESRs are generally of longer duration (>10 s) than swallow-associated LES relaxations. They are characterized by an abrupt decrease in LES pressure to the level of intragastric pressure.

The authors attempt to explain tLESRs as being mediated by a neural mechanism: "The neural pathways that mediate tLESRs have not been explored in any depth, largely because they have been difficult to study… Current evidence favors that tLESR is a predominantly vagal reflex stimulated by receptors in the gastric fundus and pharynx (and possibly larynx)."

Because it was well known that gastric distension was a potent stimulus for tLESRs, the authors suggest that gastric distension triggers tLESRs through stimulation of tension receptors in the proximal stomach, particularly the gastric cardia. They suggest that afferent impulses that signal gastric distension project to the nucleus tractus solitarius and then to the dorsal motor nucleus of the vagus that contain the cell bodies of vagal efferents that project to the LES, causing relaxation.

The many studies cited in this review suggest that there is little in the way of hard evidence that proves the existence of a neural mechanism for causing tLESRs.

Most of the studies of motor events underlying gastroesophageal reflux had hitherto been done in resting, recumbent subjects. Schoeman et al.[12] studied reflux mechanisms in 10 healthy subjects (9 men, 1 women; median age 22 years, range 18–30 years) during different types of physicial activity. They recorded LES pressure; pressure in the stomach, esophagus, and pharynx; and pH 5 cm above the LES during 24 h that included moderate physical activity, periods of rest and sleep, standardized meals, and standardized exercise. All subjects had normal esophageal motor function; manometric mean basal LES pressure was 18.7 ± 4.0 mmHg.

Reflux occurred in all subjects, but the number of reflux episodes recorded in each subject ranged widely from 3 to 22. Most (81/123 or 66%) reflux episodes occurred in the 3-h postprandial period and the remainder (34%) in the fasting state. Reflux episodes were rare during exercise.

Esophageal acid exposure detected by the pH electrode located 5 cm above the LES, expressed as the percent time pH < 4 was in the normal range (0.71 ± 0.23 for the entire group). Percent time pH < 4 ranged from a low of 0.16 (±0.13) during exercise to a high of 1.65 ± 0.98 in the ambulatory phase. Reflux episodes per hour were all well within the normal range in these healthy subjects (0.57 ± 0.09 for the group), ranging from a low of 0.08 ± 0.06 in the supine posture to a high of 3.30 ± 1.69 in the ambulatory phase.

Basal LES pressure varied widely during the study, being significantly higher when the patients were supine and during the fasting state. The occurrence of reflux was unrelated to basal LES pressure. LES pressure was absent (<3 mm H_2O) at the onset of the reflux episode in 79%. tLESR was the mechanism of reflux in 82% of episodes; 13% were swallow-related LES relaxations and straining (2%) was not a major factor. The main effect of straining was that if it was present during a tLESR, it increased the likelihood of reflux. 2% of reflux episodes occurred while the basal LES pressure was >3 mm H_2O and one episode occurred during a period of sustained absence of LES pressure. tLESRs were significantly more common and of greater duration in the upright (vs. supine) state and in the postprandial (vs. fasting) state. tLESRs were accompanied by reflux in 197/584 (34%) events.

The authors concluded that tLESRs were the major mechanism of reflux in ambulatory and resting subjects. The occurrence of a tLESR does not necessarily mean that reflux occurs. This is related to the finding that a gastroesophageal common cavity occurred in only 66% of tLESR episodes. The LES can therefore relax to the pressure used to define a tLESR (3 mm H_2O) and still remain closed.

The authors do not explore the mechanism of tLESRs in this study. Their finding that there is a significant higher number and duration of tLESRs in the postprandial state compared to the fasting state seems to support that gastric distension is an important precipitant of tLESRs.

Also, their finding that tLESRs are unrelated to basal LES pressure confirms the lack of a good relationship between LES pressure and tLESRs. tLESRs appear to be an event resulting from a sudden decrease in LES pressure to <3 mm H_2O for no good reason in a manner that is not completely predictable.

This paper provides excellent data on the amount of reflux in "normal" young people. The overall acid exposure during the 24 h (expressed as percent time pH < 4) in this group of 10 healthy volunteers was 0.71 ± 0.23. "Normal" people reflux very little. I will show that "normal" and "healthy" that is based on the absence of GERD symptoms can include people with significant LES damage by the new pathologic assessment. This small amount of acid exposure in this group raises the question of whether a truly normal person with no LES damage can be defined as a person with zero acid exposure in the esophagus.

Kahrilas et al.[13] explore the occurrence of reflux as measured by a pH electrode placed 5 cm above the LES as they distend the stomach with air while measuring the length of the LES. This permits objective definition of the impact of changes in length of LES, gastric distension, and reflux.

In their introduction, the authors state their objectives: "Numerous manometric studies provide compelling evidence that tLESR is often the dominant mechanism of reflux. Similarly, it has been established that gastric distension is a potent stimulus for triggering tLESRs, probably through activation of tension receptors in the proximal stomach, particularly the gastric cardia."

It is important to note the clever juxtaposition of fact and dogma in the last sentence. There is excellent evidence that gastric distension is a potent stimulus for tLESRs. However, the fact that activation of tension receptors in the proximal stomach due to gastric distension triggers tLESRs is not supported by any hard evidence. As will be seen, it is not even supported by the data produced in this paper, as the authors themselves suggest in the discussion (see below).

The authors continue: "With hiatal hernia, the gastric cardia containing these tension receptors for eliciting tLESRs has presumably migrated proximally. Thus, it is reasonable to hypothesize that this anatomic aberration may alter the sensitivity of these tension receptors for eliciting tLESRs." The neural mechanism mediated by tension receptors in the cardia has seemingly become an established fact. "The aim of this study was to determine if and how hiatal hernia influences the vulnerability to gastroesophageal reflux and tLESRs in patients with GERD challenged with gastric air distension."

The study had three groups: (1) 8 healthy volunteers (5 men, 3 women; mean age 31 ± 2 years) free of gastrointestinal symptoms and without a history of upper gastrointestinal surgery; (2) 7 patients (4 men, 3 women; mean age 32 ± 2 years) with symptomatic reflux disease and absent Barrett esophagus at endoscopy who had no hiatal hernia; and (3) 8 patients (4 men, 4 women; mean age 37 ± 4 years) with symptomatic reflux disease and absent Barrett esophagus at endoscopy who had a hiatal hernia. The three groups represent increasing severity of GERD from "normal" to non–hernia-GERD to hernia-GERD.

Definition of the presence of a hiatal hernia included the placement of a steel clip at the squamocolumnar junction (SCJ) at endoscopy. Fluoroscopy was used to define the relationship of the clip at the SCJ to the center of the diaphragmatic hiatus. The requirement for a diagnosis of hiatal hernia was that the SCJ clip was at least 1 cm proximal to the hiatus. Eight patients with symptomatic GERD had a hiatal hernia by these criteria, seven did not and the eight normal subjects all had the SCJ clip at or distal to the hiatus.

In Chapter 10, I will discuss the significant error in the diagnosis of hiatal hernia using these criteria. The assumption made here that the SCJ is the true GEJ in the absence of Barrett esophagus is not correct; there is a very short dilated distal esophagus distal to the SCJ in most normal people that increases in length with increasing severity of GERD. Any question regarding the diagnosis of hiatal hernia, however, does not negate the findings in this study, as long as it is correct that GERD patients with hiatal hernia have more severe GERD than those without.

Manometry was performed with a 17-lumen catheter with 14 side-hole recording sites positioned 1 cm apart extending from the stomach across the LES into the esophagus. During the study, air was infused through the catheter into the stomach at 15 mL/min for 120 min. Swallowing was monitored with a submental electromyography recorder. Esophageal acid exposure was recorded with an electrode placed 5 cm above the proximal margin of the LES.

An esophageal reflux event was defined as either an abrupt decrease in esophageal pH to <4 for at least 5 s or if the pH was already <4 by a further abrupt decrease by at least 1 pH unit for at least 5 s. Manometric tracings were analyzed to characterize basal esophagogastric junction (EGJ) (LES) pressure, the length of the EGJ (LES) high-pressure zone, and the pressure activity associated with each reflux event. Motor events such as abdominal strain, slow drift of LES pressure, LES relaxation (swallow-induced or tLESR), and esophageal common cavity were recorded. Each reflux event was categorized as tLESR related, swallow related, strain related, or free on the basis of its temporal relationship to these motor events.

The three groups had no significant differences in the numbers of acid reflux events or percent time pH < 4 at baseline (Table 9.2). With air infusion, the number of acid reflux events and percent time pH < 4 increased significantly in all three groups compared with baseline. The increase in reflux episodes with air infusion was significantly greater in both GERD patient groups compared with normal controls. The increase in the non–hernia-GERD and hernia-GERD groups was not significantly different. It should be noted that the absence of significant differences may reflect the very small numbers in the three groups.

The mechanism of reflux varied insignificantly between the three groups at baseline. With air infusion, when the number of reflux episodes increased, the mechanism that increased significantly compared to baseline in all three groups was tLESRs. The corresponding increase in acid exposure during the period of air infusion was also significant only for reflux caused by tLESR.

The number of tLESRs during baseline recording was similar in all three groups. With gastric air infusion, normal controls and non–hernia-GERD patients had a median increase in tLESR frequency of 4.0 and 4.5 per hour ($P < .05$), respectively, and hernia patients had a median increase of 9.5 per hour ($P < .001$). The proportion of tLESRs associated with acid reflux events was not significantly different among groups. The data suggest "that these subject groups represent a continuum in the diminution of the threshold for the elicitation of tLESRs in response to gastric distension."

The peak basal EGJ (LES) pressure showed high intragroup variation and was not significantly different in the three groups. In contrast, the EGJ (LES) length in the baseline recording was significantly longer in the normal controls than in

TABLE 9.2 Number of Acid Reflux Events Per Hour and Percent Time With pH < 4 Among the Three Subject Groups

	Number of Acid Reflux Events/Hour		Percent Time With pH < 4	
	Baseline	Air infusion	Baseline	Air infusion
Normal controls	0 (0–2)	4 (1.8–4.3)[a]	0 (0–5.6)	5.4 (3.1–10.4)[a]
Non–hernia-GERD	0 (0–3)	6 (6.4–10.6)[a,b]	0 (0–4.9)	14.6 (5.6–16.6)[a,b]
Hernia-GERD	1 (0–7)	12.8 (6.5–26.3)[a,b,c]	1.8 (0–5.8)	22.7 (14–24.8)[a,b]

Data are presented as median (interquartile range).
GERD, gastroesophageal reflux disease.
[a]*P < .05 air infusion versus baseline.*
[b]*P < .05 versus normal controls.*
[c]*P = .07 versus non–hernia-GERD patients.*

the non–hernia-GERD and hernia-GERD patients ($P<.05$). Non–hernia-GERD patients also had a longer EGJ (LES) than hernia patients ($P<.05$). Intragastric distension with air infusion resulted in a gradual shortening of EGJ (LES) in all three groups, becoming significant 20–30 min after the beginning of air infusion (Fig. 9.7).

During recordings in which multiple side holes were positioned within the EGJ, it was often observed that the high pressure zone shortened before detecting a tLESR in the more proximal recording sites. This shortening was evidenced by a diminution in the amplitude of the basal pressure and of the crural diaphragm component of the recording, suggesting that the distal EGJ was opened as an early event in triggering a tLESR. In 50 instances where this pattern was observed, the distal EGJ "relaxation" preceded the tLESR by 4 +/- 0.3 seconds. This pattern was particularly evident among normal controls who had the longest EGJ (LES) high-pressure zones.

During intragastric air infusion, there was a gradually increased intragastric pressure, which was then restored to baseline after a tLESR. This is a phenomenon that is similar to what happens normally in belching. The three subject groups had a similar increase in intragastric pressure before tLESR: 4.5 ± 0.5, 3.8 ± 0.6, and 3.8 ± 0.9 mmHg in the control, non–hernia-GERD, and hernia-GERD groups, respectively.

The authors are actually describing in manometric terms what is now known to happen during gastric distension. As the stomach distends with either air in the study by Ayazi et al.[14] or a heavy meal in the study by Robertson et al.,[15] the functional (manometric) LES shortens because the gastric distension effaces the most distal part of the LES, which becomes taken up into the gastric contour (Fig. 9.8).

The study by Kahrilas et al.[13] shows clearly the two elements of critical change in the LES that are associated with the pathogenesis of GERD. Firstly, there is damage to the abdominal LES beginning distally that causes a progressive decrease of the baseline LES length (Fig. 9.7). This was seen in the three groups, which form a continuum of increasing GERD from "normal" to non–hernia-GERD to hernia-GERD. The baseline mean total LES length in the three groups (taken from Fig. 9.5 in the paper) is approximately 3.7, 2.9, and 2.2 cm in the "normal," non–hernia-GERD, and hernia-GERD groups.

Secondly, with gastric distension there is a further temporary effacement of the distal part of the residual functional LES. This dynamic shortening of the LES is superimposed on the permanent shortening of the LES. The dynamic additive shortening that occurs during air infusion in the three groups is similar, forming essentially parallel lines (Fig. 9.7). This indicates that the dynamic shortening that occurs with gastric distension is independent of the baseline residual LES length.

At any given time during this period, the total functional length of the LES is equal to baseline LES length by manometry minus the amount of shortening of the LES during gastric distension (Fig. 9.6). With gastric distension during air infusion, the functional residual length of the LES decreased from approximately 3.7 to 2.8 cm in the control group, 2.9 to 2.3 cm in the non–hernia-GERD group, and 2.2 to 1.4 cm in the hernia-GERD group (Fig. 9.7).

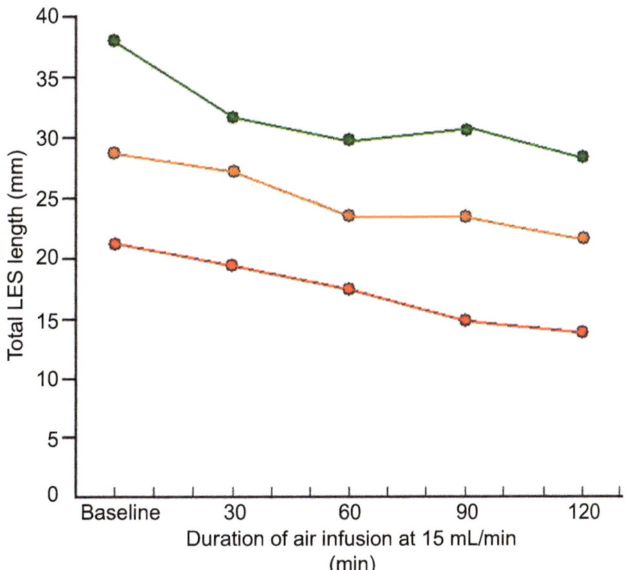

FIGURE 9.7 Three groups in the study by Kahrilas et al. showing progressively lower lengths of baseline total lower esophageal sphincter (LES) length. During air insufflation into the stomach at 15 mL/min, there is dynamic shortening of the LES that is similar in the three groups. It is in the patients with gastroesophageal reflux disease and hiatal hernia (red) that the functional total LES length falls well below 20 mm (the total length criterion for a defective LES). LES failure is significantly higher in this group than the normal person (green) and the non–hernia-GERD patient (orange).

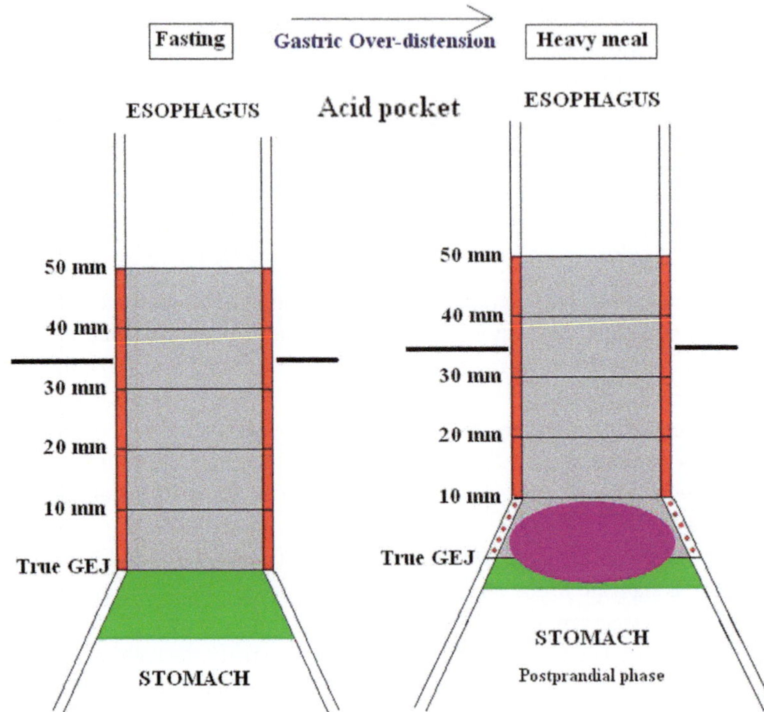

FIGURE 9.8 Changes associated with gastric overdistension caused by a heavy meal. The distal part of the abdominal lower esophageal sphincter (LES) becomes effaced (while wall with *red dots*) and moves down into the contour of the stomach. The manometric LES shortens by the amount of the effaced LES. The squamocolumnar junction moves distally and comes to lie below the pH transition point, and the distal squamous epithelium becomes exposed to the acid pocket (purple) that lies at the top of the food column in the distended stomach. *GEJ*, gastroesophageal junction.

With this decrease in the functional LES length when gastric distension is induced with air infusion, there is a significant incremental increase in tLESRs, acid reflux events, and esophageal acid exposure in the three groups. The increase was significantly more in the hernia-GERD group. This suggests that there is a close relationship between the occurrence of tLESR and *baseline LES length.* The similar amount of dynamic LES shortening that accompanied gastric distension has a greater impact in producing tLESRs in the hernia-GERD group than the others only because the baseline LES length is shorter at the initiation of gastric distension.

In their discussion, the authors suggest that "the tLESR frequency elicited by gastric distension in recumbent subjects was directly proportional to the degree of axial displacement of the SCJ relative to the midpoint of the diaphragmatic hiatus." They suggest that hiatal hernia may therefore predispose to GERD on the basis of increasing an individual's susceptibility to the dominant manometrically identified mechanism of reflux, tLESRs. They recognize that there is a considerable difficulty in defining hiatal hernia and suggest that this study has solved that problem by placing a clip at the SCJ at endoscopy. This is not necessarily true because the SCJ, when it has migrated cephalad in the patient with GERD, is not the distal end of the esophagus.

4. MECHANISM UNDERLYING LOWER ESOPHAGEAL SPHINCTER FAILURE (TRANSIENT LOWER ESOPHAGEAL SPHINCTER RELAXATION)

Kahrilas et al., in their study,[13] address the question of the mechanism of tLESRs. From their introduction (see above), it is likely that the authors' bias is that tLESRs are the result of a neural mechanism mediated by tension receptors in the cardia that recognize gastric distension and induce the tLESR.

The data in their study confuse them in this regard: "… the nearly identical ΔIGP (change in intragastric pressure) before tLESR among all subject groups in the present investigation suggests no alteration of the sensitivity of gastric tension receptors in GERD or with hiatal hernia. Rather, this finding suggests that the observed augmentation of tLESRs frequency during distension may result from anatomic distortion of the EGJ."

The authors are suggesting a mechanical alternative to the suggestion that a neural mechanism is involved with the production of tLESR.

This must be the case from their data. A similar ΔIGP in all three groups means that there is no significant difference in the amount of gastric pressure increase. As such, stimulation of any tension receptors in the proximal stomach will be the same in all three groups. Since the increase in the number of tLESRs was significantly greater from "normal" to non–hernia-GERD to hernia-GERD groups, a neural mechanism based on tension receptors in the stomach cannot be the mechanism for tLESR.

The amount of dynamic shortening of the LES during air infusion was also similar in the three groups and cannot therefore explain the incremental tLESR increase.

The only significant difference was in the baseline LES length, which decreased incrementally from the normal to the non–hernia-GERD to the hernia-GERD groups.

Gastric distension during air infusion causes the manometric LES length to decrease from 3.7 to 2.8 cm in the control group, 2.9 to 2.3 cm in the non–hernia-GERD group, and 2.2 to 1.4 cm in the hernia-GERD group (Fig. 9.7). At the baseline LES length before air infusion, the number of reflux events in all three groups is very small (0, 0, and 1). This means that the LES is largely competent in the fasting state even with a functional total LES length of 2.2–2.9 cm in the GERD patients. This is in keeping with the recognized total LES length associated with a defective LES, which is <2 cm.

As air infusion causes further LES shortening, acid reflux events increase significantly, but their number increases sharply in the hernia-GERD group. This is strong evidence that tLESRs are the result of a decline in the total functional residual LES length to a critical level of approximately <1.5–2.3 cm. It should be noted that these are total LES lengths that are reported. The finding is similar to that of studies of DeMeester et al.[8] and Zaninotto et al.[3] (reviewed above) where a decrease of the total LES length to <2 cm was a criterion of LES incompetence.

LES failure appears to be the result of shortening of the LES to a critical level. It is probable that this critical total length of the LES is variable within a small range around 2 cm. This critical length will first be reached during times of gastric distension when a temporary dynamic shortening of the LES occurs. When superimposed on a marginally adequate baseline LES length, this added shortening during gastric distension produces increased tLESRs and abnormal reflux (Fig. 9.6). This would represent postprandial GERD in response to a heavy meal.

With increasing LES damage, the baseline LES length shortens to a point at or close to the critical length. In this study, the baseline LES length of 2.2 cm in the hernia GERD group represents a marginally adequate LES much more than in the non–hernia-GERD group (2.9 cm) and the "normal" group (3.7 cm). In a patient with marginal baseline LES length, tLESR will be expected to occur with minimal gastric distension in contrast to the non–hernia-GERD group and "normal" group (Fig. 9.7).

From the data, it appears that LES pressure is maintained until the LES length reaches a critical point close to 2 cm at which time the LES pressure drops abruptly to a level close to zero, representing a tLESR. The exact mechanism whereby this happens is not known.

In patients with clinical GERD, the majority of tLESRs are associated with an open gastroesophageal cavity and reflux events where gastric contents flow along the pressure gradient through the open LES into the esophagus. This is the major mechanism for LES failure (tLESR) associated with GERD.

The study by Kahrilas et al.[13] shows that the frequency of LES failure (tLESR) is inversely related to the residual length of the LES as measured by manometry. According to our mathematical formula (residual LES length = initial LES length − LES damage), we can therefore conclude that frequency of LES failure is directly related to LES damage.

When we develop an accurate measure of LES damage, it is highly likely that this will become the best predictor of severity of GERD. The new pathologic assessment of the LES is limited to the abdominal segment of the LES. The length of the abdominal segment is approximately 1–1.5 cm less than the total LES length. This should be kept in mind when interpreting the lengths used in Kahrilas et al.,[13] which are for total LES length.

The mechanism of how the LES fails intermittently is critical to understanding reflux disease. There is a difference between a normal LES, a damaged LES that does not fail (i.e., is in the phase of compensated LES damage), and a defective LES that fails with increasing frequency as damage progresses. The best method of differentiating between these groups is by the amount of LES damage. A damaged LES that does not fail, i.e., an LES that is not damaged at all or one that is damaged within its reserve capacity, will not be associated with reflux. An LES that is critically damaged will first fail with dynamic shortening in the postprandial phase. As LES damage progressively increases, the frequency of LES failure and reflux episodes increases to a point where LES failure occurs at all times not necessarily related to meals.

Failure of the LES results in transsphincteric reflux where gastric content flows along the natural pressure gradient from the stomach to the esophagus. The amount of exposure of the thoracic esophagus to gastric contents is directly related to the severity of GERD and the occurrence of complications of GERD in the thoracic esophagus. These include symptoms of GERD, erosive esophagitis, CLE, intestinal metaplasia, dysplasia, and adenocarcinoma.

The severity of transesophageal reflux correlates with the frequency and duration of LES failure. This in turn is related to baseline functional LES length that is determined by the amount of LES damage. The severity of cellular changes in the thoracic esophagus can therefore be correlated with LES damage (Table 9.3).

The concept that I am trying to develop is that the abdominal LES undergoes progressive damage in a predictable sequence from normal (i.e., an initial length of ~35 mm with no damage) to a progressively decreasing length of the abdominal LES. The decrease in the functional abdominal LES length brings the patient ever closer to the critical <10 mm where the LES becomes incompetent in preventing reflux (Table 9.3).

LES failure is initially in the postprandial period associated with a heavy meal when gastric overdistention converts an LES that is competent in the fasting state (i.e., abdominal length >10 mm and total length >20 mm) to one that reaches the critical <10 mm abdominal and <20 mm total length because of superimposed dynamic shortening caused by the meal. As LES damage progresses toward 25 mm, LES failure with meals becomes more frequent and requires less gastric distension.

Between 35 mm (normal) and <10 mm (incompetent), the abdominal LES shortens by 25 mm (Fig. 9.2). During this progressive damage to the abdominal LES, the patient progresses from (1) having a competent LES with no reflux and (2) having mild postprandial GERD with reflux episodes that are not enough to cause severe erosive esophagitis or visible CLE. (3) When the LES damage >25 mm, the residual abdominal LES is <10 mm, LES failure and reflux into the thoracic esophagus become severe and difficult to control. This state of severe GERD is associated with a high risk of severe erosive esophagitis and Barrett esophagus.

The new pathologic assessment of abdominal LES damage provides a method of defining these categories with more accuracy than any other presently available method.

The actual progression of damage to the abdominal LES varies greatly in different people. This is dependent on the rate of shortening of the abdominal LES, which is in turn likely to be related to the patient's eating habits and inherent resistance of the LES to damage.

Most people in the population have such a slow rate of progression of LES damage that they never reach the stage of abnormality (>15 mm of abdominal LES damage) during their lifetime; they never get clinical GERD or any complication of GERD. At the other extreme, some people have a rapid rate of progression, reaching >25 mm of abdominal LES damage early in their lives. These people develop GERD early in their life, are difficult to control with PPIs, progress to treatment failure, progress to Barrett esophagus, and are at highest risk for adenocarcinoma. These

TABLE 9.3 Stages of Severity of Gastroesophageal Reflux Disease (GERD) as Assessed by Severity of Reflux Into the Thoracic Esophagus, Abdominal Lower Esophageal Sphincter (LES) Damage, Residual Abdominal LES Length, Mechanism of LES Failure, Probability of Cellular Changes in the Thoracic Esophagus, and Risk of Adenocarcinoma

	Abd-LES Damage	Residual Abd-LES Length	Severity of Reflux	Mechanisms of LES Failure	Cellular Changes	Risk of Cancer
Normal	0	35 mm	0	n/a	0	0
Compensated LES damage	>0–15 mm	<35–20 mm	0	n/a	0	0
Onset of GERD	>15–20 mm	<20–15 mm	>0 to normal pH test	tLESR	Minimal; no visible CLE	0
Mild clinical GERD	>20–25 mm	<15–10 mm	Abnormal pH test	tLESR	Esophagitis; no visible CLE	0
Severe GERD	**>25–35 mm[a]**	**<10 mm–0[a]**	**Abnormal pH test**	**tLESR + other mechanisms**	**Hiatus hernia; visible CLE**	**+**

CLE, columnar-lined esophagus; *tLESR*, transient lower esophageal sphincter relaxation.
[a]At this stage, LES damage extends from involving only the abdominal LES to involve the thoracic esophagus as well. The thoracic segment of the LES is usually 15–20 mm and acts as a residual barrier in patients with complete abdominal LES destruction.

cellular events correlate with amount of reflux into the thoracic esophagus, which is best predicted by the amount of LES damage.

The group at Northwestern University Medical School followed this with a study that concentrates on the mechanisms of LES failure in patients with the most severe LES damage characterized by the presence of a hiatus hernia.[16]

It also should be noted that the concept of the dilated distal esophagus had not been reported at the time of this study. To the authors, like all gastroenterologists, the esophagus ends at the endoscopic GEJ. As I will consider in Chapter 10, the concept of hiatus hernia will change significantly after the dilated distal esophagus is recognized. Some part of the distal dilated structure that these authors call hiatus hernia is likely the dilated distal esophagus. The hiatus hernia is that part of the viscus above the diaphragm that is within the herniated peritoneal sac and lined by gastric oxyntic epithelium, the location of which is not known to these investigators.

Their introduction: "The cardinal abnormality of gastroesophageal reflux disease (GERD) is incompetence of the esophagogastric junction (EGJ) to prevent reflux of gastric secretions. This functional compromise can be attributable to perturbation of a number of anatomical or physiological components of the EGJ, including the intrinsic lower esophageal sphincter (LES), the extrinsic compression of the distal esophagus at the diaphragmatic hiatus, the intra-abdominal location of the LES, or the physiology of transient LES relaxations (tLESRs). Physiological investigations suggest that the net results of these perturbations is an increased number of acid reflux events by three mechanisms: 1) tLESR, 2) strain-induced reflux in the setting of a hypothensive LES, or 3) refux during periods of low LES pressure or deglutitive relaxation."

Gaining acceptance as a cofactor in the pathogenesis of GERD is anatomical distortion of the EGJ inclusive of, but not limited to, hiatus hernia. When examining GERD in the context of hiatus hernia, certain distinctions become evident. Patients with more severe forms of GERD, such as erosive esophagitis and Barrett's metaplasia, almost invariably have a hiatus hernia, and GERD patients with hiatus hernia have increased acid exposure compared with patients without hiatus hernia. The mechanistic profile for reflux in hiatus hernia patients is also distinct. Whereas tLESRs can account for up to 90% of reflux events in normal subjects or in patients without hiatus hernia, patients with hiatus hernia have a more heterogeneous mechanistic profile with reflux episodes frequently occurring in the context of low LES pressure, straining, and swallow-associated LES relaxation.

In contemplating the occurrence of reflux in the setting of a relaxed or hypotensive sphincter, it is necessary to explore other mechanical attributes of the system that may account for a relaxed sphincter remaining closed in one case and physically open in another; one such attribute is compliance…

The aim of this study was to compare the distensibility of the EGJ both at rest and during relaxation of hiatus hernia patients with GERD with normal subjects using a combined barostatic/fluoroscopy technique that allowed for direct measurement of intraluminal EGJ diameter at predetermined intraluminal distension pressures.

The study uses complex methodology to arrive at the conclusions that are expressed in the abstract: "To quantify the effect of hiatus hernia on EGJ distensibility, 8 normal subjects and 9 GERD patients with hiatus hernia were studied with concurrent manometry, fluoroscopy, and stepwise controlled barostatic distension of the EGJ." During endoscopy, which was limited to the GERD patients, the SCJ was marked by a stainless steel clip. Hiatus hernia was defined as the presence of the SCJ clip between swallows was >1 cm above the hiatal impression.

The minimal barostatic pressure required to open the EGJ during the interswallow period was determined… The EGJ opening diameter was greater in hernia patients compared with normal subjects during deglutitive relaxation at all pressures, and EGJ length was 23% shorter. EGJ opening pressure among hernia patients was lower than normal subjects during the interswallow period. In conclusion, the EGJ of GERD patients with hiatus hernia was more distensible and shorter than in normal subjects. These findings partially explain why hiatus hernia patients are predisposed to reflux by mechanisms other than tLESRs, sustain greater volumes of reflux and have a reduced ability to discriminate gas from liquid reflux.

The study at least partly explains why reflux increases with extreme damage (near total damage of 35 mm to the abdominal segment of the LES and <6 mmHg mean LES pressure). Mechanisms of failure other than tLESRs begin to appear when the LES mean pressure decreases below the critical level where it approaches resting intragastric pressure. In such patients, the barrier provided by the LES to reflux essentially disappears and the residual compliance of the LES that keeps it closed is easily overcome by multiple inducing events other than gastric overdistension.

5. ONSET AND PROGRESSION OF ABDOMINAL LOWER ESOPHAGEAL SPHINCTER DAMAGE

There are two separate changes in the LES that influence its functional length at any given time. The first is permanent damage to the LES. This begins distally and progresses cephalad. Permanent LES damage is associated with cardiac metaplasia in the dilated distal esophagus. At manometry, it is represented by that part of the LES distal to the end of the high-pressure zone that cannot be distinguished from the stomach, i.e., it has lost its muscle tone and become gastricized.

The second change is a dynamic shortening that occurs when there is gastric overdistension (Figs. 9.6–9.8). Dynamic shortening causes the SCJ to move distally, reaching a level below the pH transition point (Fig. 9.8). During this period of dynamic effacement of the distal abdominal LES, the squamous epithelium becomes exposed to gastric contents (the acid pocket at the top of the food column). This is the mechanism whereby gastric overdistension causes squamous epithelium damage leading to cardiac metaplasia. In turn, cardiac metaplasia results in permanent abdominal LES damage.

5.1 Mechanism of Onset and Progression of Lower Esophageal Sphincter Damage

In the previous two chapters, I have considered the characteristics of the normal LES and described the mechanism whereby the normal distal abdominal LES can become exposed to gastric juice and undergo damage. During periods of gastric overdistension, the distal part of the LES is seen to shorten by manometric measurement. It is important to understand that the anatomic LES does not shorten during this process.

Anatomically, the distal part of the LES becomes effaced when the stomach overdistends, moving down into the gastric contour. This process of dynamic shortening can be demonstrated by inflating a balloon (Fig. 9.9). The balloon, like the stomach, is collapsed at rest. Its neck is its original length. As it is inflated, the balloon fills to capacity with minimal increase in pressure. This is similar to gastric filling with a normal meal. The neck of the balloon remains unchanged, as does the LES. If the balloon is overinflated, its neck shortens, the lower effaced part of the neck moves into the rounded contour of the balloon, and its interior is exposed to the interior of the balloon. This is exactly what happens when the stomach overdistends. With the balloon, the interior is exposed to harmless air; the squamous epithelium of the effaced esophagus encounters gastric contents.

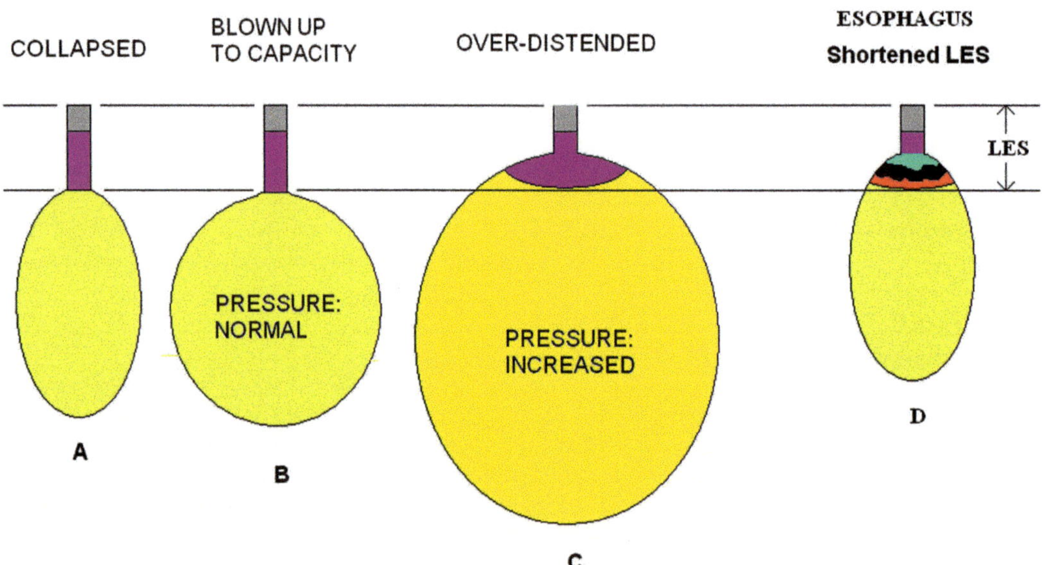

FIGURE 9.9 Analogy of blowing a balloon to explain effacement of the lower esophageal sphincter (LES) during dynamic shortening. The balloon, like the stomach is collapsed in the airless (fasting) state. As it is inflated (a meal enters the stomach) the balloon fills to its capacity (full stomach) without any change in its neck (LES). As it is inflated beyond capacity, the lower part of the neck (LES) moves into the contour of the balloon (stomach) and the neck (manometric residual LES) shortens. The interior of the effaced neck of the balloon (squamous epithelium of the effaced abdominal LES) is exposed to the air inside the balloon (gastric contents). Nothing further happens to the interior of the balloon because air does not cause damage. However, the squamous epithelium exposed to gastric juice undergoes (with time) columnar metaplasia, LES damage, and becomes permanently dilated. The dilated distal esophagus (blue, black, and red) is interposed between the shortened LES and the original gastroesophageal junction which is the proximal limit of gastric oxyntic epithelium. The length of the cardiac epithelium (blue, black, and red) lining the dilated distal esophagus is equal to the LES that has been damaged.

In effect, the distal squamous epithelium of the effaced LES comes to lie below the pH transition point and is exposed to the acid pocket that is present at the top of the food column and is damaged. This is a temporary change. With gastric emptying, distension subsides and the LES reverts to its baseline length.

The extent of dynamic LES shortening with gastric overdistension has been measured with air insufflation and after a heavy standardized meal. In the study by Robertson et al.[15] asymptomatic volunteers, the patients were fed a standardized meal of battered fish and chips with 150 mL of water over 20 min and instructed to eat until full. The shortening of the distal segment of the LES varied from 0.43 cm in upright position (from 2.12 to 1.69 cm) to 0.82 cm in the supine position (from 2.34 to 1.52 cm) in the subjects with central obesity. This shortening was sufficient to cause the SCJ (marked by clips) to move below the pH transition point where the squamous epithelium was exposed to gastric juice.

In the study by Ayazi et al.,[14] air boluses of 50 mL were injected every 30 s into the stomach of asymptomatic volunteers until a total of 750 mL was infused. There was a progressive decrease of overall LES length; the abdominal LES decreased by a median of 12 mm (from 26 to 14 mm). Again, the shortening of the LES was sufficient to cause the SCJ to move distal to the pH step-up point, indicating exposure of squamous epithelium to the acid pocket.

In the study of Kahrilas et al.,[13] air was infused at 15 mL/min for 120 min. The decrease in the total LES length ranged from ~0.6 cm in the non–hernia-GERD patient group to 0.9 cm in the asymptomatic patient group. During this period of gastric distension, the frequency of LES failure (tLESR), the number of reflux episodes, and the acid exposure of the thoracic esophagus increased. The amount of increase in reflux was directly proportional to the baseline length of the LES before air infusion was initiated.

Tamhankar et al.[17] demonstrated that the length of time the esophageal squamous epithelium was exposed to gastric juice was 150 min after ingestion of a large fatty meal.

This process of LES effacement can be demonstrated at endoscopy. Air insufflation into the stomach causes the SCJ to descend into the stomach and become part of the gastric contour. This recapitulates what happens during gastric overdistension. All of the squamous-lined esophagus that is visible at endoscopy when the stomach is distended by air insufflation will be exposed to gastric acid during a meal that has caused a similar amount of gastric overdistension.

5.2 Permanent Abdominal Lower Esophageal Sphincter Damage (Shortening)

The process of squamous epithelial exposure to gastric acid causes acute injury, leading to increased permeability of the epithelium. This permits entry of molecules in refluxed gastric juice into the epithelium, setting the stage for molecular events that cause columnar (cardiac) metaplasia of the distal squamous epithelium. This occurs very slowly, one cell at a time with a shortening that can be measured in millimeters per decade.

There may or may not be some symptoms associated with the initial squamous epithelial injury. This is likely to be transient postprandial epigastric discomfort, which is easily explained as a normal manifestation of an excessive meal. When cardiac metaplasia has occurred, this epithelium is not as pain sensitive. As such, the patient likely reverts to the asymptomatic condition.

Cardiac metaplasia causes loss of LES tone in the affected abdominal LES and the dilated distal esophagus, all of which are concordant in length.

5.3 The New Norm With a Damaged Abdominal Lower Esophageal Sphincter

When permanent abdominal LES damage occurs in a prior undamaged abdominal LES, the patient has developed a new norm. He/she has been converted from a normal person to one with permanent abdominal LES damage. This is the onset of the process of abdominal LES damage that ultimately results in GERD in those patients destined to develop the disease.

Let us begin with a normal person without any LES damage. The abdominal length of the LES is 35 mm, there is no cardiac epithelium (=no LES damage) between the esophageal squamous epithelum and gastric oxyntic epithelium (Fig. 9.10A). Let us assume that this person has gone on an eating binge over a 6-month period. Every time he/she distended the stomach during this period, the distal LES became effaced and the SCJ descended into the acid pocket below the pH transition point that was temporarily displaced cephalad (Fig. 9.8). Let us assume that this much of abuse of the LES with the eating binge caused 1 mm of the distal esophageal squamous epithelium to undergo cardiac metaplasia. This person now has 1 mm of abdominal LES damage (Fig. 9.10B). The status of this new norm can be defined by the following:

1. The normal zero gap between the distal squamous epithelium and gastric oxyntic epithelium has changed to a gap of 1 mm that is now composed of cardiac mucosa (with and without parietal cells; goblet cells are never seen with such a short squamo-oxyntic gap).

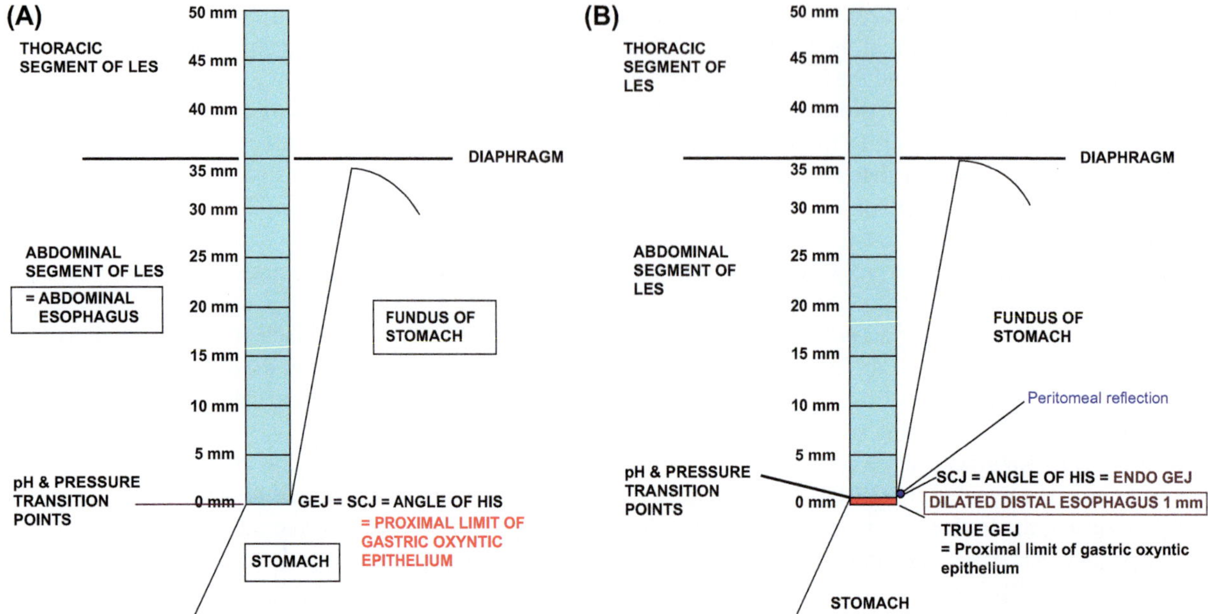

FIGURE 9.10 (A) The normal state with an undamaged (35 mm) abdominal lower esophageal sphincter (LES) and absence of cardiac epithelium between esophageal squamous epithelium (blue) and gastric oxyntic epithelium. (B) The distal 1 mm of the esophagus has undergone cardiac metaplasia (red), lost LES pressure and dilated. The abdominal LES is 1 mm shorter, but remains fully competent. The person has no reflux, no clinical features of gastroesophageal reflux disease (GERD), and manometry is not sensitive enough to detect the shortening. However, the patient has moved from the normal state to early GERD if GERD is defined by the presence of cardiac epithelium and/or LES damage. *GEJ*, gastroesophageal junction; *SCJ*, squamocolumnar junction.

2. The residual functional abdominal LES has shortened by 1 mm. This cannot be detected by manometry, which is not adequately sensitive.
3. There is no physiological consequence of this LES damage. The LES that has been shortened by 1 mm has a residual abdominal length of 34 mm. It remains competent. It does not fail (i.e., there are no tLESRs) with meals. There is no transsphincteric reflux into the thoracic esophagus. The patient has no GERD symptoms. There is no clinical impact. The patient will be considered "normal" or "asymptomatic."
4. The tubular shape of the distal 1 mm of the abdominal LES has been lost. That part of the abdominal esophagus that has lost LES tone becomes dilated, forming the dilated distal esophagus. This cannot be detected by any method other than histology.
5. Without the protection of intramural sphincter pressure, this 1 mm of the abdominal esophagus functionally behaves exactly like the stomach. It has become part of the reservoir and ceased to be part of the conduit that transmits food from the pharynx to the stomach. The distal abdominal esophagus that has lost LES pressure has become gastricized. Anatomically, it is no longer a tube. It dilates when the stomach fills and collapses with the stomach as it empties. Like any reservoir that dilates and contracts, it develops mucosal rugal folds.
6. At endoscopy and gross examination of resected or autopsy specimens, the 1 mm of the dilated distal esophagus that is lined by cardiac epithelium will be misinterpreted as proximal stomach. It is distal to the endoscopic GEJ (the proximal limit of rugal folds and end of the tubular esophagus).
7. The functional esophagus ends at the distal limit of the shortened LES. Instead of its original length of 40 cm, it is now 39.9 cm.
8. The pressure transition point from the distal functional LES to the stomach has moved 1 mm cephalad as has the SCJ. The pH transition point has also moved cephalad by 1 mm to coincide with the distal limit of the residual functional LES and the new SCJ.
9. The part of the distal esophagus that has lost LES pressure is no longer protected from gastric contents. The metaplastic cardiac epithelium that lines the dilated distal esophagus is continually exposed to gastric contents like the rest of the proximal stomach. Unlike gastric oxyntic epithelium, which is resistant, metaplastic cardiac epithelium is susceptible to damage by gastric juice. It shows inflammation.
10. The anatomic tubular esophagus in the abdomen has shortened by 1 mm.
11. The angle of His has become slightly less acute and moved slightly closer to the diaphragmatic hiatus.
12. The peritoneal reflection is now not at the angle of His but has moved 1 mm on to the gastric fundus. This can be seen externally with difficulty.

This new norm can be looked on as a compensatory mechanism to adjust to abdominal LES damage. The functional tubular esophagus has shortened; this has no significant physiologic change. The reservoir has added the dilated distal esophagus to its capacity; this is so small as to be insignificant.

5.4 Progression of Abdominal Lower Esophageal Sphincter Damage

In this new norm with an abdominal LES that has been shortened by 1 mm, the forces that operate when gastric overdistension occurs are similar to the previous norm, except that the capacity of the reservoir for the meal is slightly increased.

The functional residual LES becomes effaced and shortens temporarily as a result of gastric overdistension; the new SCJ moves distal to the pH transition point, becomes damaged, and undergoes cardiac metaplasia.

In this way, the length of cardiac epithelium (with and without parietal and goblet cells) increases progressively. Over time and very slowly but at different rates in different people, the dilated distal esophagus increases in length and the abdominal LES shortens, one cell and 0.05 mm at a time. Each episode wherein cardiac metaplasia occurs damages the abdominal LES permanently by a miniscule amount. Every time this happens, the process of cardiac metaplasia and LES damage is slowly eating into the reserve capacity of the LES.

There is a long way, however, before the abdominal LES is sufficiently damaged to result in LES failure (tLESR). This occurs when the abdominal LES reaches a residual functional length of <20 mm. When this state of LES damage is reached, a heavy meal that causes dynamic shortening of the LES by 10 mm brings the functional length of the abdominal LES to <10 mm, which is the length at which LES failure is precipitated (Fig. 9.6). This is the onset of reflux into the thoracic esophagus, which is initially postprandial.

When permanent LES damage progresses by another 10 mm, the residual abdominal LES length is 10 mm in the fasting state (Fig. 9.2). This person's LES is at failing threshold at rest. Reflux episodes and volume increase exponentially with further LES damage as it progresses to complete destruction of the abdominal LES. Over a long period, usually measured in decades, this person has progressed in the amount of LES damage from 0 to 25 mm (Fig. 9.2).

This concept of progression of pathologic change in GERD was beautifully described by Hayward in his paper in 1961[18]: "When the normal sphincteric… mechanism in the lower oesophagus… fails, reflux from the stomach occurs and acid and pepsin reach the squamous epithelium and begin to digest it. In quiet periods some healing occurs and … the destroyed squamous epithelium may re-form… or … cardiac epithelium … may replace it. Where this occurs the area is given considerable protection from future reflux because … cardiac epithelium resists acid-peptic digestion. Further reflux therefore attacks principally the squamous epithelium higher up. In the next remission it may be replaced by more … cardiac epithelium… With repetition over a long period the metaplastic … cardiac epithelium may creep higher and higher…"

The only error that Hayward makes is that he begins this story with the occurrence of reflux into the esophagus. This is the place that we recognize GERD to this day. We make the same error that Hayward made.

The real story begins much earlier when gastric overdistension causes exposure of the esophageal squamous epithelium at the distal end of the normal esophagus with an undamaged LES to the gastric acid pocket. This results in an asymptomatic phase of abdominal LES damage that degrades the LES length through its reserve capacity long before it reaches the level of damage that causes LES incompetence.

The cardiac metaplasia of esophageal squamous epithelium results at the beginning from gastric overdistension in the normal person, not by transesophageal reflux in the GERD patient. The story of progression of cardiac metaplasia of esophageal squamous epithelium is now understood from the beginning, not near the end of the book.

In our mathematical formula (assuming 35 mm as the initial abdominal LES length), the point where the abdominal LES reaches 10 mm represents a shortening of the abdominal LES by 25 mm.

At a rate of progression of abdominal LES damage of 0.1 mm per year, it would take 50 years for the abdominal LES to shorten 5 mm (Fig. 9.11). At this rate, the person will never develop GERD. In contrast, a person with LES damage progressing at a rate of 1 mm per year will take only 25 years to bring the abdominal LES to <10 mm from the initial 35 mm (Fig. 9.12). This patient will develop severe GERD during his/her lifetime.

The range of rate of progression of LES damage between 1 mm/decade and 1 mm/year represents the range of GERD (defined as the presence of cardiac epithelium distal to the endoscopic GEJ or amount of LES damage) in the population. 70% will never exhaust the reserve capacity of the LES; 20% will progress to GERD but will never develop visible CLE and be at risk of adenocarcinoma. 5%–10% will develop visible CLE and be at risk for adenocarcinoma. This last population is the one that needs intervention if the objective of management of GERD is to prevent adenocarcinoma.

The pathology of LES damage is the cardiac metaplasia that results in the dilated distal esophagus.[19] The pathology of progression of LES damage is the cardiac metaplasia of esophageal squamous epithelium that progressively increases in length in concordance with LES damage.[20]

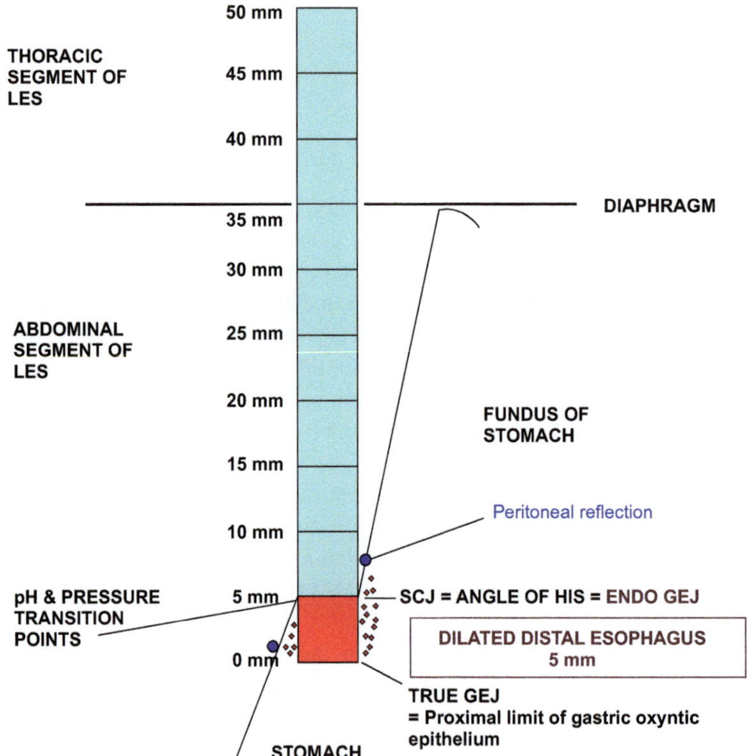

FIGURE 9.11 Progression of abdominal lower esophageal sphincter (LES) damage to 5 mm, as measured by the length of cardiac epithelium in the dilated distal esophagus (red, with red stippling). The time taken for the abdominal LES to shorten by this amount will depend on the rate of progression of abdominal LES damage in a given person. *GEJ*, gastroesophageal junction; *SCJ*, squamocolumnar junction.

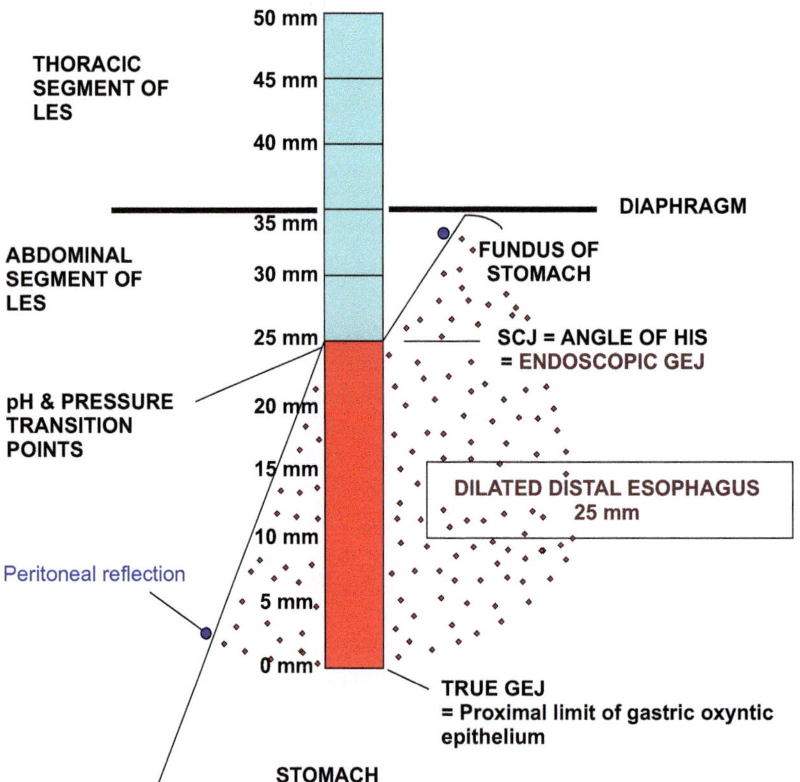

FIGURE 9.12 Progression of abdominal lower esophageal sphincter (LES) damage to 25 mm, as measured by the length of cardiac epithelium in the dilated distal esophagus (red, with red stippling). This person has reached the point of severe LES damage that is associated with troublesome symptoms with a risk of progression to treatment failure, visible columnar-lined esophagus, and a risk of adenocarcinoma. *GEJ*, gastroesophageal junction; *SCJ*, squamocolumnar junction.

All three elements under discussion are concordant, i.e., the length of the dilated distal esophagus, the length of the abdominal LES that has been damaged and lost its tonic pressure, and the length of cardiac epithelium (with and without parietal and/or goblet cells) distal to the endoscopic GEJ (end of tubular esophagus and/or proximal limit of rugal folds) up to the true GEJ (proximal limit of gastric oxyntic epithelium) are equal.[20]

5.5 Variables in the Production of Abdominal Lower Esophageal Sphincter Damage

The primary cause of abdominal LES damage is the result of a battle between the abdominal LES and an eating habit that overdistends the stomach repetitively and puts pressure on the LES from below.

There are a multitude of possible and little understood factors that govern the rate of LES damage in different people. These may include innate resistance of the LES, innate resistance of the squamous epithelium to acid and other molecules in gastric juice, innate resistance to develop columnar metaplasia, gastric size, and compliance that causes differences in changes of intragastric pressure during meals, and many others. These variables result in different rates of LES damage in different people with equivalent eating habits. This makes it impossible to predict the kind of eating that will cause GERD (Table 9.4). In general though, eating frequent high-volume meals with high-fat content is likely to be associated with GERD than more moderate volume meals.

The effect of unknown variables is seen in the relationship of GERD to obesity. Although GERD is more common in obese individuals, the relationship is far from absolute. Being thin does not provide immunity from developing GERD and being obese does not mean the patient will develop GERD.

This could be due to two things: (1) A diet producing obesity is one that is calorie rich; LES damage is caused by episodic eating to produce gastric overdistension. A person who ingests calories by snacking is less likely to damage the LES compared to a person who eats heavy meals regularly. (2) The variables of many factors in the LES and stomach may operate to make the LES more or less susceptible to damage with a given diet in different people. Which of these two factor categories is more important in the progression of LES damage is completely unknown and unstudied.

Being female offers some protection from progression of LES damage compared to being male. The reason for this is unknown but is likely to be related to innate differences in LES or mucosal resistance to either gastric overdistension or squamous epithelial damage when exposed to gastric contents.

Despite all the complicated variables that determine LES damage, the rate of progression of LES damage is likely to be constant over the long term for any given individual. LES damage is the end point of the battle between the LES and forces that are injurious. The variables are unlikely to change. Once a pattern of eating has been established, it is likely that the rate of progression of LES damage will remain linear over a long period, i.e., when measured in decades.

Certainly, change in eating habits from a GERD-unfriendly to a GERD-friendly diet is likely to be effective in changing the rate of progression of LES damage. This can become a practical method of controlling LES damage.

5.6 Trigger Foods as a Cause of Lower Esophageal Sphincter Failure

A significant number of patients with GERD complain that certain foods or drinks precipitate episodes of reflux. These foods vary in different patients but tend to remain constant in a given patient. The history is powerful in some patients and causes that

TABLE 9.4 Hypothetical Variation of Progression of Lower Esophageal Sphincter (LES) Damage in Persons With Different Levels of Resistance of the LES to be Damaged by a Diet That Puts High Pressure on the LES [Gastroesophageal Reflux Disease (GERD)-Unfriendly] or Low Pressure (GERD-Friendly)

	LES With High Resistance to Damage	LES With Average Resistance to Damage	LES With Low Resistance to Damage
GERD-friendly diet	Lowest rate of progression	Low rate of progression	Intermediate rate of progression
GERD-unfriendly diet	Low rate of progression	Intermediate rate of progression	Highest rate of progression

The exact rate of progression of LES damage will vary in different people, but is likely to remain constant with time in a given person.

person to avoid those foods religiously. The general belief in the population is that GERD may result from ingestion of many substances: alcoholic beverages, carbonated drinks, coffee, chocolate, spicy food, bananas, and many other specific items.

Theoretically, this would suggest that certain people have an innate mechanism whereby ingestion of a given food triggers LES failure in some way. No such mechanism has been demonstrated.

The question that must be addressed is whether a trigger food causes LES failure (tLESR) and reflux in a person with a normal or minimally damaged LES that is not otherwise susceptible to failure. This would mean that avoidance of that trigger factor would prevent reflux completely. It is possible that such people exist who have eliminated their reflux by avoiding trigger foods. They would have solved their problem and never reached the physician and never developed clinical GERD.

More commonly, trigger foods are specific dietary items that precipitate reflux in people who already have symptomatic GERD. The elimination of these trigger foods rarely cures their GERD; it is difficult to define whether avoidance of the trigger significantly diminishes the occurrence of LES failure and reflux. It is possible that these foods are recognized by the patient as triggers for a reflux episode that is precipitated by a meal in a person with a damaged LES where dynamic shortening causes reflux and the trigger is simply coincidental. Many patients with GERD will avoid trigger foods. There is no direct objective evidence as to how commonly such an action controls the amount of reflux in a patient.

6. RELATIONSHIP OF PROGRESSION OF LOWER ESOPHAGEAL SPHINCTER DAMAGE TO TIME

GERD is a chronic, progressive lifelong disease that is not curable by medical therapy. It is caused by progressive LES damage. The progression of LES damage is not decreased by acid suppression. The natural history of the disease is therefore not influenced by any medical therapy at present. It is influenced by factors that damage the LES (largely, dietary that cause gastric overdistension) and by procedures that attempt to augment and or repair the damaged LES.

It is important to understand the temporal course of LES damage. This will determine methods that are available to prevent its progression. To do this, I will use comparisons with the natural history of other chronic progressive disorders.

There is no evidence that LES damage results from one acute event. It is not like type I diabetes mellitus caused by (theoretically) an acute viral infection. The destruction of sufficient islet cells during this episode to cause diabetes leaves no alternative but to replace the insulin or develop some method of replacing lost beta cells in the islets.

There is no evidence that LES damage progresses by a series of relapses and remissions such as seen in chronic ulcerative colitis. LES damage does not remit spontaneously or by the use of any drugs, i.e., the LES damage does not reverse even partially once it has occurred.

There is also no evidence that LES damage progresses by multiple acute events that cause large increments of LES damage. In chronic pancreatitis associated with gallstones, multiple attacks of acute pancreatitis at wide intervals may result in progression of chronic pancreatitis. Removal of the cause of acute pancreatitis, i.e., addressing gallstone disease, can stop the progression to chronic pancreatitis.

Another mechanism of progression of a chronic disease is the development of a vicious cycle. A good example is a person who is infected with the human immunodeficiency virus. This attacks the immune cells, causing a reduction in T lymphocyte numbers. The lack of T cells decreases the body's ability to control the HIV virus, whose proliferation is facilitated, creating a vicious cycle leading to progressively increasing viral loads.

Such a vicious cycle does not occur in the case of LES damage. This is well demonstrated by Kahrilas et al.[13] In that study, there were three groups of patients defined as "normal," "non–hernia-GERD," and "hernia-GERD" representing three levels of increasing severity of GERD. These patients had progressively shorter baseline LES lengths, representing increasing permanent LES damage in the three groups. Air infusion to produce gastric dilatation resulted in a dynamic shortening of the LES from its baseline. The amount of shortening was identical in all three groups with the decrease in LES length associated with gastric distension forming essentially parallel lines. This shows that the effect of dynamic shortening associated with different levels of LES damage is the same. As such, the cause of permanent LES damage is the same irrespective of the status of the baseline residual LES. There is no vicious cycle effect. If there were, gastric distension would have resulted in a progressive increase in the amount of dynamic LES shortening as baseline LES length shortened.

All evidence points to a slowly progressive and inexorable LES damage measured in micrometers (one or few cells) at a time over many years and decades. Lee et al.[21] showed that the length of the abdominal LES decreases with age in GERD patients. This is associated with an increase in the esophageal acid exposure in the pH test. This provides support for a linear progression of abdominal LES damage with time.

This is similar to coronary atherosclerosis associated with a diet that causes changes in blood lipids that induce atherosclerosis. In both LES damage and coronary atherosclerosis, reductions and increases of the cause, i.e., different types of dietary changes, can alter the rate of progression. However, in both, the etiologic factor is unlikely to vary once it has been established by a habitual lifestyle.

Dietary modification to prevent LES damage in the population has a zero chance of gaining traction. Even with obesity and coronary artery disease that are well recognized in the population as serious health hazards, attempts to induce the population to restrict diets have not met with great success. Once LES damage has progressed to produce disease associated with troublesome symptoms, erosive esophagitis or visible CLE, dietary modification is relatively ineffective.

At an earlier stage in LES damage, if progression can be predicted in an individual, there may be a possibility of successfully inducing the person who has been identified as being at risk for progressive disease to change his/her diet from GERD-unfriendly to GERD-friendly by appropriate education.

The progression of both LES damage and coronary atherosclerosis is likely to vary from day to day and week to week, but over the long haul over several decades they are likely to have a relatively steady, linear rate of progression. Also, it is likely that the rate of progression of LES damage will remain the same into the future unless significant adjustment in diet takes place at an early stage.

This linear progression of LES damage permits the development of a simple algorithm if LES damage can be measured accurately. If one assumes that the LES is fully developed at age 15 years and that a person's lifestyle and dietary habit is established at that age, one can assume that LES damage will begin at that age and slowly progress at a linear rate that is different in different people because of their dietary habit and the effect it had on their LES.

The linearity of progression of LES damage is critically important in predicting future disease. In coronary atherosclerosis, narrowing of 70% indicates a sufficient risk of future ischemic heart disease to take action to correct the abnormality. Similarly, evaluation of the degree of LES damage at any given point in the life of the patient will predict future disease severity. I will explore this further in Chapter 18.

7. RATE OF PROGRESSION OF ABDOMINAL LOWER ESOPHAGEAL SPHINCTER DAMAGE

In practice, it is certain that the progression of LES damage occurs very slowly, cell by cell, millimeter by millimeter over a period of several decades. When one considers that the amount of shortening of the abdominal LES before the LES becomes incompetent is 25 mm (assuming an initial length of 35 mm and LES incompetence at a length of 10 mm), a rate of LES damage of 1 mm per year would result in severe GERD in 25 years from the onset of damage. Different rates of LES damage would result in widely different consequences in terms of the residual abdominal LES length and its ability to prevent reflux (Table 9.5).

By using the rate of LES damage and making some assumptions as I have done, a mathematical model of the progression of LES damage emerges. I will discuss in detail these assumptions, the probability that they are valid, and the ways in which they can be tested in the chapter on the pathologic assessment of LES damage (Chapter 16).

At this time, however, I would like to stress that a relatively minor change in the rate of progression of LES damage can have significant consequences in the future. If the sphincter becomes incompetent when its abdominal segment is <10 mm, as has been demonstrated, all patients in the Table 9.5 with <10 mm length (highlighted in red) can be expected to have severe GERD at rest at the age indicated in the diagram.

TABLE 9.5 Changes With Age of the Functional Residual Length of the Abdominal Lower Esophageal Sphincter (LES) Assuming That the Original Length at Maturity Is 35 mm, That LES Damage Begins at 15 Years and That LES Damage Has a Linear Progression Over the Long Term

Rate of LES Damage mm/decade	At 25 Years (mm)	At 35 Years (mm)	At 45 Years (mm)	At 55 Years (mm)	At 65 Years (mm)	At 75 Years (mm)
1	34	33	32	31	30	29
2	33	31	29	27	25	23
3	32	29	26	23	20	17
4	31	27	23	19	15	11
5	30	25	20	15	10	5
6	29	23	17	11	5	0
7	28	21	14	7	0	0
8	27	19	11	3	0	0
9	26	17	8	0	0	0
10	25	15	5	0	0	0

The abdominal lower esophageal sphincter (LES) lengths in green represent lengths at which the LES is likely to be competent. The lengths in orange represent an LES that is susceptible to failure with gastric distension (i.e., at risk of postprandial reflux). The lengths in red represent an LES that is below the length at which LES failure occurs at rest.

It is apparent from Table 9.5 that those people with a rate of LES damage of 3mm/decade or less will maintain a competent LES to age 65years. These represent the approximately 70% of the population who will never develop clinical GERD. I am one of these people. At age 56years, I persuaded my gastroenterologist to add an upper gastrointestinal endoscopy to my screening colonoscopy. I instructed him to take appropriate biopsies that permitted me to assess my abdominal LES damage. I have 4mm of LES damage. This means my rate to abdominal LES damage (assuming onset of damage at age 15years) 0.1mm/year. At that rate of LES damage, I will live my life with a sphincter that maintains competence. I will not develop GERD or esophageal adenocarcinoma. My LES will reach the critical 10 mm of abdominal length only if I live to be 250 years old!

At the other extreme, people whose LES shorten at the rate of 10mm/decade will be predicted to have a residual abdominal LES length of 15mm at age 35years. With dynamic shortening during a heavy meal, the functional abdominal LES length can dip below 10mm, the point of significant failure (tLESR). This patient is at risk for postprandial reflux. The rate of LES damage is so rapid, however, that this patient's functional abdominal LES length is 5mm at 45years. This would mean that this patient would develop significant LES failure around age 40years, progressing rapidly to severe GERD. This patient is at the highest risk for developing treatment failure, Barrett esophagus, and adenocarcinoma.

In Chapters 17–19, I will take this discussion to the next step where I will combine the pathologic assessment of LES damage with a method of predicting those people who are at highest risk for developing severe GERD defined by the severity of LES damage. The assessment of LES damage is highly accurate. The algorithm that is developed for prediction can be individualized to the patient, providing a prediction of future LES damage in that particular patient.

REFERENCES

1. DeMeester TR, Johnson LF, Joseph GJ, et al. Patterns of gastroesophageal reflux in health and disease. *Ann Surg* 1976;**184**:459–70.
2. Johnson LF, DeMeester TR. Twenty-four hour pH monitoring of the distal esophagus: a quantitative measure of gastroesophageal reflux. *Am J Gastroenterol* 1974;**63**:325–32.
3. Zaninotto G, DeMeester TR, Schwizer W, Johansson K-E, Cheng S-C. The lower esophageal sphincter in health and disease. *Am J Surg* 1988;**155**:104–11.
4. Vakil N, van Zanten SV, Kahrilas P, Dent J, Jones B, The Global Consensus Group. The Montreal definition and classification of gastroesophageal reflux disease: a global evidence-based consensus. *Am J Gastroenterol* 2006;**101**:1900–20.
5. Blonski W, Vela MF, Castell DO. Comparison of reflux frequency during prolonged multichannel intraluminal impedance and pH monitoring on and off acid suppression therapy. *J Clin Gastroenterol* 2009;**43**:816–20.
6. Haddad JK. Relation of gastroesophageal reflux to yield sphincter pressures. *Gastroenterology* 1970;**58**:175–84.
7. Thurer RL, DeMeester TR, Johnson LF. The distal esophageal sphincter and its relationship to gastroesophageal reflux. *J Surg Res* 1974;**16**:418–23.
8. DeMeester TR, Wernly JA, Bryant GH, Little AG, Skinner DB. Clinical and in vitro analysis of determinants of gastro-esophageal competence. A study of the principles of antireflux surgery. *Am J Surg* 1979;**137**:39–46.
9. Bonavina L, Evander A, DeMeester TR, Walther B, Cheng S-C, Palazzo L, Concannon JL. Length of the distal esophageal sphincter and competency of the cardia. *Am J Surg* 1986;**151**:25–34.
10. Stein HJ, Barlow AP, DeMeester TR, Hinder RA. Complications of gastroesophageal reflux disease. Role of the lower esophageal sphincter, esophageal acid and acid/alkaline exposure, and duodenogastric reflux. *Ann Surg* 1992;**216**:35–43.
11. Mittal RK, Holloway RH, Penagini R, Blackshaw LA, Dent J. Transient lower esophageal sphincter relaxation. *Gastroenterology* 1995;**109**:601–10.
12. Schoeman MN, Tippett MDM, Akkermans LMA, Dent J, Holloway RH. Mechanisms of gastroesophageal reflux in ambulatory healthy human subject. *Gastroenterology* 1995;**108**:83–91.
13. Kahrilas PJ, Shi G, Manka M, Joehl RJ. Increased frequency of transient lower esophageal sphincter relaxation induced by gastric distension in reflux patients with hiatal hernia. *Gastroenterology* 2000;**118**:688–95.
14. Ayazi S, Tamhankar A, DeMeester SR, et al. The impact of gastric distension on the lower esophageal sphincter and its exposure to acid gastric juice. *Ann Surg* 2010;**252**:57–62.
15. Robertson EV, Derakhshan MH, Wirz AA, Lee YY, Seenan JP, Ballantyne SA, Hanvey SL, Kelman AW, Going JJ, McColl KE. Central obesity in asymptomatic volunteers is associated with increased intrasphincteric acid reflux and lengthening of the cardiac mucosa. *Gastroenterology* 2013;**145**:730–9.
16. Pandolfino JE, Shi G, Curry J, Joehl RJ, Brasseur JG, Kahrilas P. Esophagogastric junction distensibility: a factor contributing to sphincter incompetence. *Am J Physiol Gastrointest Liver Physiol* 2002;**282**:G1052–8.
17. Tamhankar AP, DeMeester TR, Peters JH, et al. The effect of meal content, gastric emptying and gastric pH on the postprandial acid exposure of the lower esophageal sphincter (LES). *Gastroenterology* 2004;**126**(Suppl. 2):A-495.
18. Hayward J. The lower end of the oesophagus. *Thorax* 1961;**16**:36–41.
19. Chandrasoma P, Wijetunge S, Ma Y, DeMeester S, Hagen J, DeMeester T. The dilated distal esophagus: a new entity that is the pathologic basis of early gastroesophageal reflux disease. *Am J Surg Pathol* 2011;**35**:1873–81.
20. Chandrasoma P, Makarewicz K, Ma Y, DeMeester TR. A proposal for a new validated histologic definition of the gastroesophageal junction. *Hum Pathol* 2006;**37**:40–7.
21. Lee J, Anggiansah A, Anggiansah R, et al. Effects of age on the gastroesophageal junction, esophageal motility and reflux disease. *Clin Gastroenterol Hepatol* 2007;**5**:1392–8.

Chapter 10

The Effect of Damage to the Abdominal Segment of the LES: The Dilated Distal Esophagus

The dilated distal esophagus results as a consequence of loss of lower esophageal sphincter (LES) pressure (i.e., LES damage) in the abdominal esophagus.[1] Measurement of its length is therefore a method of assessing LES damage. The length of cardiac epithelium (with and without parietal and goblet cells) between the endoscopic gastroesophageal junction (GEJ) and true GEJ (proximal limit of gastric oxyntic epithelium) is equal to permanent LES shortening. Its length, measured by histology, is equal to LES damage with an accuracy that is within micrometers (Figs. 10.1 and 10.2).

The dilated distal esophagus is that part of the most distal esophagus that has lost the protection of the high pressure of the abdominal segment of the LES. LES damage begins distally and progresses in a cephalad direction.

At present, most people including experts in the field of esophageal disease do not recognize the dilated distal esophagus as an entity. Even worse, by using an incorrect opinion-based definition of the GEJ at endoscopy,[2,3] most of these experts believe that the dilated distal esophagus is the proximal part of the stomach that they call the gastric cardia.

The endoscopic GEJ, which is defined as the proximal limit of rugal folds and/or the end of the tubular esophagus, is located cephalad to the true GEJ (the proximal limit of gastric oxyntic epithelium) by the length of the dilated distal esophagus (Fig. 10.1).

The dilated distal esophagus is presently completely ignored at endoscopy. The current recommendation of all gastroenterology societies is to not take biopsies at or distal to the endoscopic GEJ in gastroesophageal reflux disease (GERD) patients who are endoscopically normal. All societies continue to accept the erroneous definition of the endoscopic GEJ, with full knowledge that it has no evidence base in support.[2,3]

It is well recognized that there is a variation in the length of the LES at manometry and that patients with GERD have a shortening of the LES compared with asymptomatic persons. Shortening of the LES is recognized as abnormal only when the total LES length is <20 mm and the abdominal length is <10 mm (Fig. 10.3).[4] When the LES has a total length > 20 mm and an abdominal length > 10 mm, it is generally regarded as "normal" (Fig. 10.4).

Patients with untreated achalasia typically have an LES that is usually of "normal" length but is abnormal in that it fails to relax during swallowing (Fig. 10.5). It is likely that patients with achalasia will have LES lengths that are closest to their initial length because, untreated, they have the least likelihood of anyone in the population of not damaging their LES. The reason for this is that failure of relaxation results in the esophagus acting as the reservoir for a meal, with slow emptying into the stomach. As a result, it is almost impossible for the patient with achalasia to fill his/her stomach to the point of overdistension that causes LES damage. Of course, LES damage can occur before the onset of achalasia and therefore the measured abdominal LES can be shortened depending on the age of the onset of achalasia (Fig. 10.4). If achalasia develops in a person who already has severe LES damage, GERD can be associated with achalasia.

In general, manometry is performed in GERD patients at a late stage of the disease, usually when treatment with proton pump inhibitors (PPIs), often of many years duration, has failed to control symptoms. In these patients, LES damage has progressed to an advanced stage with near-complete destruction of the abdominal segment (Fig. 10.3).

There is a marked variation in the residual length of the abdominal LES measured by manometry in asymptomatic patients (usually from 10 to 35 mm)[4] as well as in patients with GERD (usually from 0 to <10 mm). In GERD patients with an abdominal LES < 10 mm, it is certain that this length represents LES damage sufficient to cause a defective LES.

There is presently no clear understanding of the pathologic consequence of this LES damage in the distal abdominal esophagus. It is as if the damaged LES simply disappears into thin air without evoking a whisper of curiosity in the medical establishment. No one can really answer the question: "When the LES shortens, where does the part of the sphincter that disappeared in the manometric tracing go?"

GERD. http://dx.doi.org/10.1016/B978-0-12-809855-4.00010-5

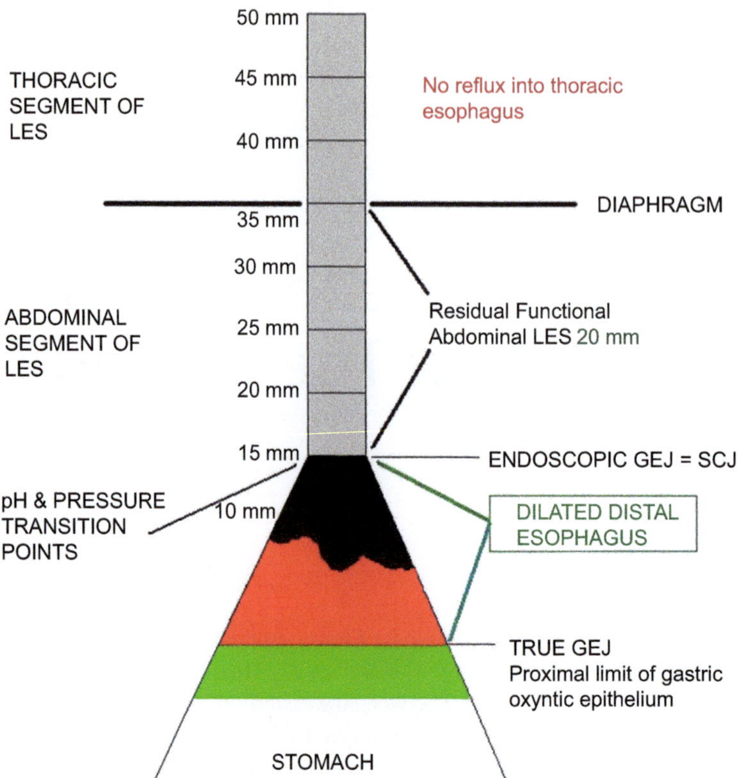

FIGURE 10.1 Diagrammatic representation of the status of the abdominal esophagus in a person with 15 mm of lower esophageal sphincter (LES) damage. This is the appearance in a resected specimen where the attachment of the peritoneal reflection to the diaphragm has been separated and the fundus collapses downward. The dilated distal esophagus measuring 15 mm is lined with cardiac (black) and oxyntocardiac (red) epithelium. The manometric length of the LES has shortened to 20 mm. The endoscopic gastroesophageal junction (GEJ) and squamocolumnar junction (SCJ) is at the end of the tubular esophagus, which corresponds to the distal limit of the LES at manometry. This is also the new pH and pressure transition point. The dilated distal esophagus is functionally part of the stomach; the metaplastic epithelia are exposed to gastric contents.

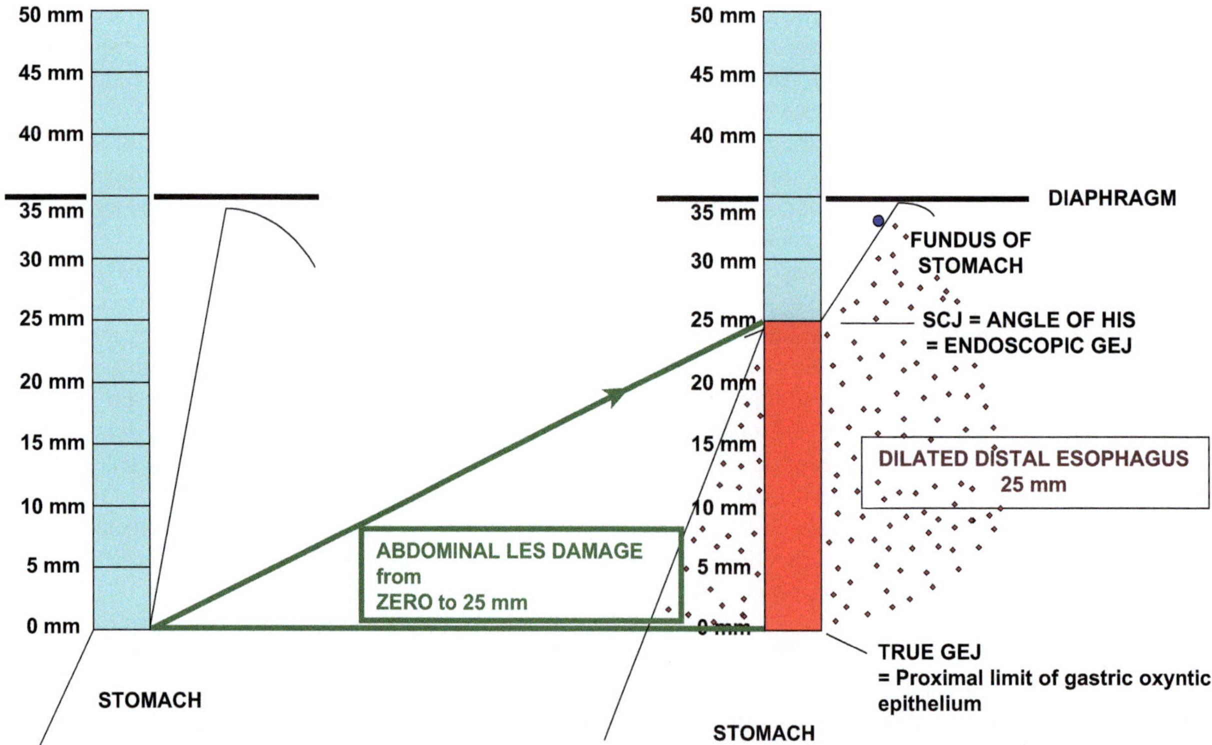

FIGURE 10.2 Conversion of the normal state (left) to the abnormal state with 25 mm of lower esophageal sphincter (LES) damage; diagram with the organs in place in the body. The end of the tubular stomach, squamocolumnar junction (SCJ), endoscopic gastroesophageal junction (GEJ), pH and pressure transition points are now 10 mm distal to the diaphragm, separated from the true GEJ by 25 mm. The peritoneal reflection (*blue dot*) is 25 mm up the contour of the gastric fundus; the dilated distal esophagus separates the angle of His, which has become more obtuse, from the peritoneal reflection. The *red stippled area* is the area of metaplastic cardiac epithelium in the dilated distal esophagus. The *solid red area* is cardiac epithelium lining the original abdominal esophagus before LES damage; this is the area that was originally lined by squamous epithelium. The true GEJ (proximal limit of gastric oxyntic epithelium) remains in its original place.

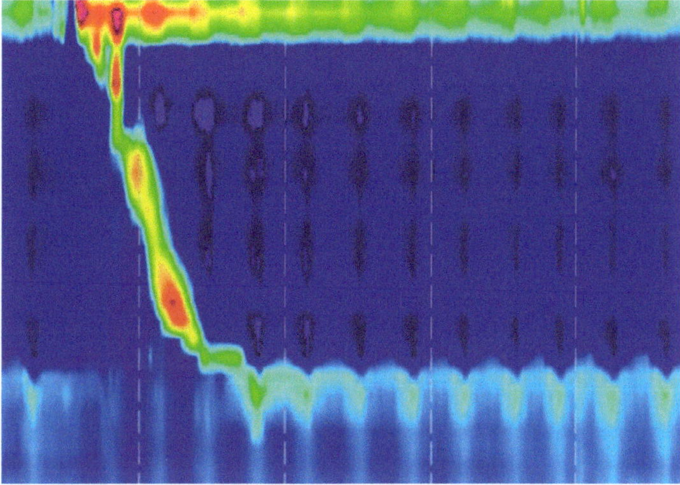

FIGURE 10.3 High-resolution manometry showing a markedly abnormal short lower esophageal sphincter (LES) with low pressure. Esophageal body motility and LES relaxation during swallowing is normal. The calculated total LES length is 1.1 cm, the abdominal LES length is 0 cm, and resting LES pressure is 9 mmHg. *Study and interpretation including calculated lengths is by Chris Dengler, MD, Legato Medical Inc.*

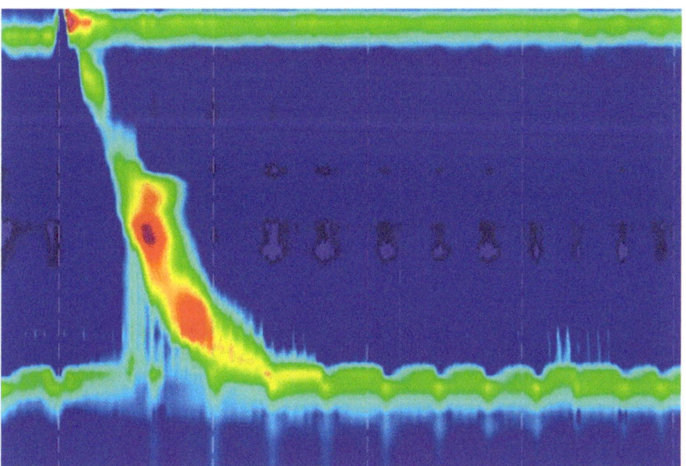

FIGURE 10.4 High-resolution manometry showing a "normal" lower esophageal sphincter (LES) with normal length and pressure of the LES, normal peristalsis, and normal LES relaxation with swallowing. The calculated LES total length is 3.3 cm; abdominal LES length is 2.2 cm; resting LES pressure is 34 mmHg. Note that Fig. 10.1, with 15 mm of abdominal LES damage, has a residual manometric LES length of 35 mm (total) and 20 mm (abdominal). *Study and interpretation including calculated lengths is by Chris Dengler, MD, Legato Medical Inc.*

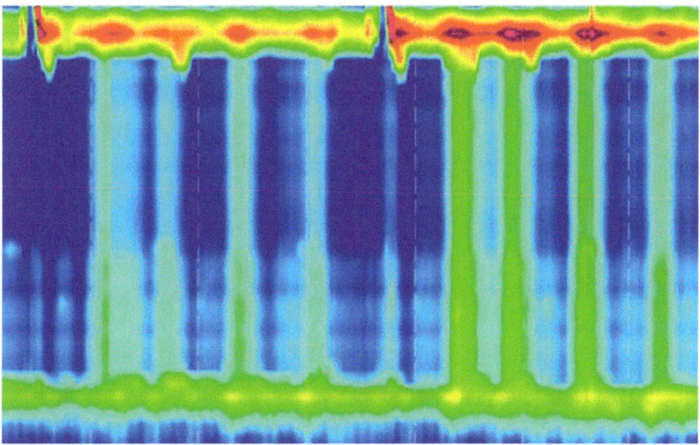

FIGURE 10.5 High-resolution manometry in a patient with type II achalasia at the first diagnostic assessment prior to treatment (i.e., no Botox, dilatation, or myotomy). There is no peristalsis and the lower esophageal sphincter (LES) does not relax during swallowing. The body of the esophagus is pressurized. The calculated total LES length is 3.6 cm, abdominal LES length is 2.5 cm, and resting LES pressure is 52 mmHg. According to the new method, this person has damaged 10 mm of the abdominal LES before developing achalasia, which protected the LES from further damage. *Study and interpretation by Chris Dengler, MD, Legato Medical Inc.*

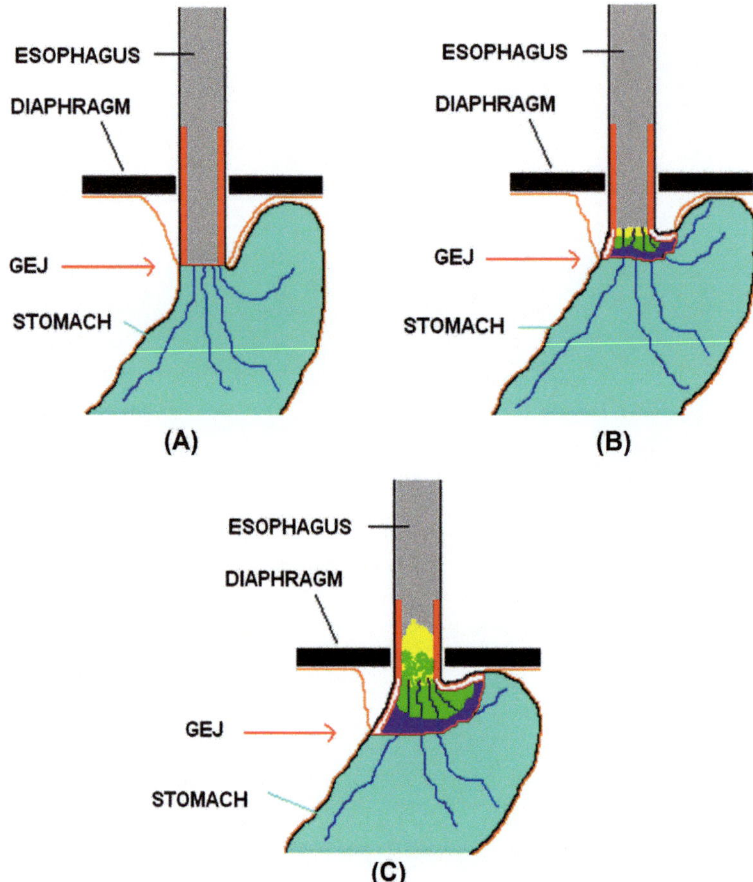

FIGURE 10.6 The three basic stages of pathology of gastroesophageal reflux disease: (A) The normal state without any cardiac epithelium or lower esophageal sphincter (LES) damage. (B) The person without visible columnar-lined esophagus (CLE), identified by the presence of a variable length of cardiac epithelium limited to the dilated distal esophagus, concordant with the length of LES damage. (C) The person with a visible CLE. This person has gastroesophageal reflux disease irrespective of symptoms. There is always a long associated dilated distal esophagus that indicates the severe LES damage required to cause visible CLE. *GEJ*, true gastroesophageal junction.

The answer to this question is that it goes nowhere. The part of the LES that is damaged becomes altered to the pathological anatomy of the dilated distal esophagus (Fig. 10.6A and B). With increasing shortening of the LES at manometry, the length of cardiac epithelium (with and without parietal and/or goblet cells), which is equal to LES damage, increases (Fig. 10.7).

The early pathologic events that lead to GERD and all its complications are misinterpreted as normal stomach. The crux of the early pathologic change in GERD is the conversion of the distal tubular abdominal esophagus into the dilated distal esophagus (Fig. 10.2). This is caused by cardiac metaplasia of squamous epithelium that is concordant with LES damage. At present, this sensitive and specific pathologic change of early GERD is mistaken for the normal proximal stomach (the gastric cardia).[1]

The error, if one looks at this area objectively, results in impossible contradictions:

1. One recognized cause of inflamed cardiac epithelium distal to a normal endoscopic GEJ ("carditis") is GERD. Carditis is commonly an isolated finding not associated with pathology in the remainder of the stomach and is associated with GERD.[5] *GERD cannot produce gastric pathology*. Cardiac epithelium distal to the endoscopic GEJ has shown to be similar morphologically, immunohistochemically, and at a molecular level to cardiac epithelium in visible columnar-lined esophagus (CLE).[6]
2. Intestinal metaplasia (IM) of the gastric cardia is both unassociated with gastric pathology in most patients and associated with GERD.[1] It is also similar to Barrett esophagus (BE) (i.e., visible CLE with IM) by histochemical, immunohistochemical, and molecular characteristics and different to gastric IM.[7] *GERD cannot produce IM in the stomach.*
3. Adenocarcinoma of the gastric cardia is associated with GERD.[8] *GERD cannot produce gastric adenocarcinoma.* Its epidemiology mimics esophageal adenocarcinoma and is different than gastric adenocarcinoma.[9] The seventh edition of the American Joint Commission on Cancer recommends that adenocarcinoma within 5 cm distal to the endoscopic GEJ be classified as esophageal rather than gastric in origin.[10] Adenocarcinoma of the region distal to the endoscopic GEJ, when evaluated by histology, is an adenocarcinoma of the dilated distal esophagus (Fig. 10.8).[11]

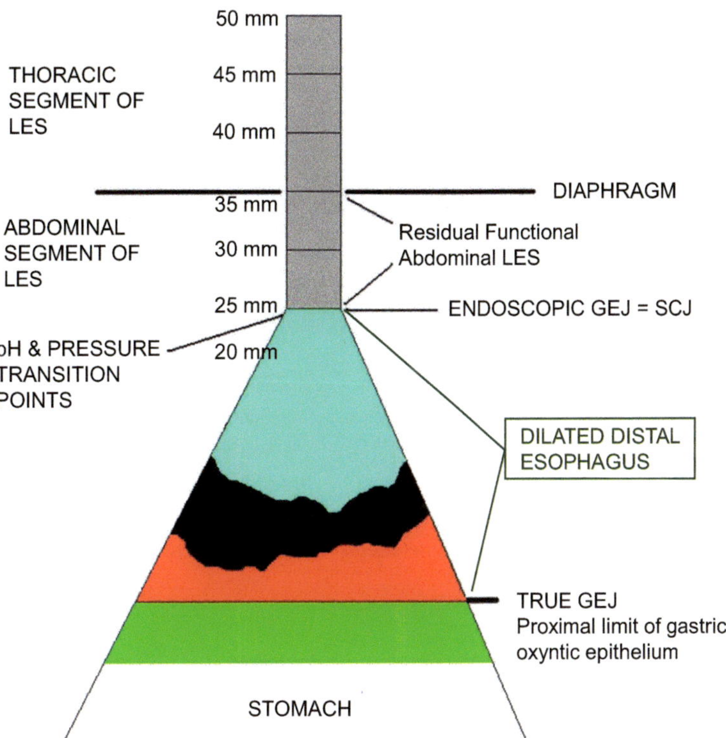

FIGURE 10.7 A person with lower esophageal sphincter (LES) damage of 25 cm. This is similar to the basic change in Fig. 10.1 except for the facts that the LES damage has progressed and the residual LES has shortened further to 10 mm. The dilated distal esophagus is shown to be lined by intestinal metaplasia (blue) in addition to the cardiac (black) and oxyntocardiac epithelia (red). The likelihood of intestinal metaplasia in the dilated distal esophagus likely increases with increasing length of metaplastic epithelium (LES damage). *GEJ*, gastroesophageal junction; *SCJ*, squamocolumnar junction.

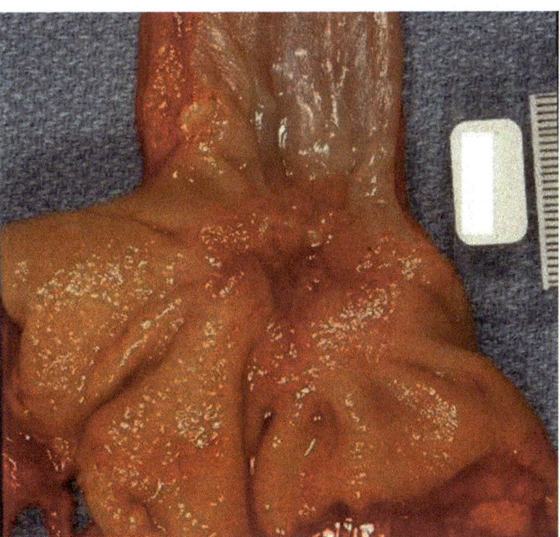

FIGURE 10.8 Ulcerated adenocarcinoma of the dilated distal esophagus. This has an epicenter distal to the end of the tubular esophagus. The rugal folds in the region have been obliterated by the tumor. The fact that this is an esophageal tumor is proved by the fact that the epithelium at the epicenter and distal edge of the tumor was metaplastic cardiac epithelium.

4. Increasing length of cardiac epithelium in the proximal stomach extending distally from the endoscopic GEJ is associated with increasing severity of GERD. The reason for this is that the perspective is wrong. Cardiac epithelium does not increase in length in a distal direction from the endoscopic GEJ. This would mean that cardiac epithelium replaces gastric oxyntic epithelium in patients with GERD. *This is impossible*. In reality, cardiac epithelium increases in length cephalad (proximally) from the true GEJ, which is the proximal limit of gastric oxyntic epithelium by columnar metaplasia of esophageal squamous epithelium (Fig. 10.9).

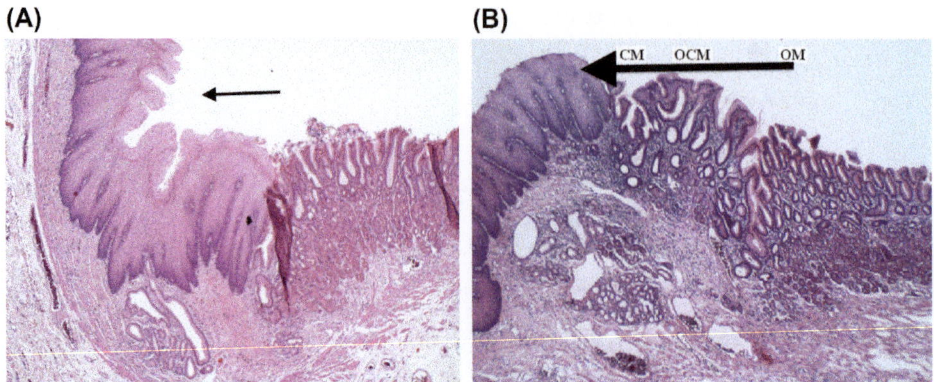

FIGURE 10.9 The histologic conversion of the normal state with no cardiac epithelium between squamous and gastric oxyntic epithelium (A) to the early abnormal state where the distal 2 mm of the esophageal squamous epithelium has been converted to metaplastic cardiac (CM) and oxyntocardiac (OCM) epithelium (B). This has caused separation of the squamocolumnar junction from the true gastroesophageal junction (GEJ). The cardiac metaplasia begins at the proximal limit of oxyntic epithelium and involves the squamous epithelium in a cephalad direction (*arrows*). Its direction of progression is blind to endoscopy until visible columnar-lined esophagus develops. Cardiac metaplasia *does not* begin at the endoscopic GEJ and extend distally, as is the present misperception. *CM*, cardiac mucosa; *OCM*, oxyntocardiac mucosa; *OM*, oxyntic mucosa.

We cannot manage an esophageal disease correctly with an incorrect definition of where the esophagus ends and the stomach begins. At present we believe that the esophagus ends at the proximal limit of the dilated distal esophagus (the endoscopic GEJ).

In reality, it ends at the distal end of the dilated distal esophagus (the proximal limit of gastric oxyntic epithelium, the true GEJ) (Fig. 10.9B), which has been reported to be 0–2.8 cm distal to the endoscopic GEJ.[12] This is the 2.8 cm that is the crux of the pathology of LES damage and early pathogenesis of GERD that is ignored.

It is not a surprise that GERD is a disease that is out of control with millions of people having symptoms that are inadequately controlled, hundreds of thousands of people developing BE, and more than 20,000 people in the United States developing esophageal adenocarcinoma annually with a mortality of 85%.

The crucial question is: "Can correction of this error result in a new management algorithm for GERD that will lead to earlier diagnosis and treatment and prevent severe GERD (treatment failure, BE, and adenocarcinoma)?"

This book provides a new practical method of early diagnosis of GERD by pathologic assessment of LES damage. This will open the possibility of preventing progression of LES damage with the prospect of effective prevention of the later stage of GERD including esophageal adenocarcinoma.

1. DEFINITION OF THE DILATED DISTAL ESOPHAGUS

The dilated distal esophagus is that part of the damaged abdominal esophagus that is between the endoscopic GEJ (end of the tubular esophagus and/or proximal limit of gastric rugal folds) and the true GEJ (proximal limit of gastric oxyntic epithelium defined histologically) (Fig. 10.1).[1]

It is lined by metaplastic cardiac epithelium (with and without parietal and/or goblet cells) that extends from the distal limit of esophageal squamous epithelium to the proximal limit of gastric oxyntic epithelium in a person who does not have a visible CLE at endoscopy (Figs. 10.1, 10.2, and 10.9B). The dilated distal esophagus is always abnormal. In the normal state, there is no dilated distal esophagus. The entire abdominal esophagus is a tube lined by squamous epithelium that transitions directly to gastric oxyntic epithelium at the true GEJ.

In a person with visible CLE at endoscopy, it is that area distal to the proximal limit of rugal folds (i.e., distal to the endoscopic GEJ) to the proximal limit of gastric oxyntic epithelium (the true GEJ) defined histologically.[1]

The dilated distal esophagus is also that part of the abdominal esophagus whose muscle wall has lost sphincter tone. The normal high pressure has disappeared and the pressure in this part of the esophagus is equal to gastric baseline pressure; that is, it has become manometrically "gastricized."

As a result, the dilated distal esophagus has ceased to function as part of the tube that transmits food from the pharynx to the stomach. It has become part of the reservoir that accommodates a meal. As the stomach fills and distends, the dilated distal esophagus also fills and distends. It is exposed to gastric contents without LES protection (Figs. 10.1 and 10.2). It, like any structure that has a reservoir function, develops mucosal rugal folds (Fig. 10.6).

In a person with early GERD (i.e., where LES damage and reflux into the esophagus is at a point where there is no visible CLE), three elements that constitute the dilated distal esophagus are concordant (Fig. 10.5B), i.e.,

the length of the dilated distal esophagus = length of cardiac epithelium (with and without parietal and/or goblet cells) between the endoscopic GEJ and gastric oxyntic epithelium = amount of shortening (damage) of the abdominal segment of the LES.

In practice and with presently available technology, measuring the length of cardiac epithelium is the only feasible method of defining the amount of LES damage. For the first time, there is an easy, accurate, and precise method to assess quantitative LES damage in any person. *This should be a game changer in the management of GERD.*

2. PATHOGENESIS OF THE DILATED DISTAL ESOPHAGUS

In the previous chapters, I have detailed how the normal state progresses through increasing LES damage to produce GERD. This is essentially the pathogenesis of the dilated distal esophagus. I will summarize those chapters here to introduce the anatomic consequences of loss or pressure in the abdominal LES.

Normally, the entire esophagus is a tubular organ. The entire normal abdominal part esophagus distal to the diaphragm is a high-pressure zone that represents the abdominal segment of the LES. The high pressure is maintained by tonic contraction of the smooth muscle of the esophagus.

The undamaged abdominal segment of the LES (and therefore the abdominal esophagus) has an unknown initial length in an individual person, but is likely to be ~35 mm (Figs. 10.1 and 10.2). This is the initial length of the abdominal LES. In a normal person, *the length of the abdominal esophagus = length of the abdominal segment of the LES.*

In a person without any LES damage (Figs. 10.6A and 10.9A), the tubular esophagus transitions to the normal saccular stomach with mucosal rugal folds at the acute angle of His. The entire normal esophagus is lined by squamous epithelium and the entire normal stomach is lined by gastric oxyntic epithelium. In the normal state, there is no cardiac epithelium (with and without parietal and/or goblet cells) between the esophageal squamous epithelium and gastric oxyntic epithelium, i.e., the squamooxyntic gap is zero (Fig. 10.9A).

Gastric overdistension associated with experimental air insufflation or a LES-unfriendly meal distorts the anatomy of this region, causing dynamic shortening of the LES such that the most distal part of the esophageal squamous epithelium becomes exposed to gastric juice (Fig. 10.10).[13,14] This causes injury that leads ultimately to columnar metaplasia of esophageal squamous epithelium, producing cardiac epithelium (Fig. 10.9B). Initially composed of only mucous cells, cardiac epithelium can secondarily develop parietal cells (becoming oxyntocardiac epithelium) or goblet cells (IM).

This process results in the interposition of cardiac epithelium between esophageal squamous epithelium (whose distal limit has moved cephalad because of columnar metaplasia) and the proximal limit of gastric oxyntic epithelium (the true GEJ, which is permanently static in its original anatomic location). The squamooxyntic gap has now developed (Figs. 10.6B and 10.9B).

The squamooxyntic gap, which is lined by cardiac epithelium (with and without parietal and/or goblet cells), is limited to the dilated distal esophagus in the early stage before visible CLE develops. When visible CLE develops, the metaplastic cardiac epithelium constituting the squamooxyntic gap extends above the diaphragm (i.e., above the dilated distal esophagus). Patients with visible CLE *always* have a dilated distal esophagus (Fig. 10.6C). In general, the length of the dilated distal esophagus is longer in patients with visible CLE than those without. This is because the presence of visible CLE correlates with a greater severity of LES damage.

Progression of LES damage in the abdominal esophagus occurs slowly and at varying rates in different people resulting in increasing amounts of cardiac epithelium. The cardiac epithelium always extends cephalad, i.e., proximally from the true GEJ (the proximal limit of gastric oxyntic epithelium), as increasing amounts of esophageal squamous epithelium are damaged by exposure to gastric juice during times of repeated gastric overdistension (Fig. 10.9).

3. VARIATION IN LENGTH OF THE DILATED DISTAL ESOPHAGUS

In Chapter 7, I suggested that the presence of cardiac epithelium (with and without parietal and/or goblet cells) was a perfect definition for GERD. This is correct in a theoretical sense only.[5,15]

The presence of cardiac epithelium (with and without parietal and/or goblet cells) in the squamooxyntic gap indeed separates the patient with GERD from the normal person with 100% sensitivity and specificity. It *always* represents esophageal squamous epithelial damage by exposure to gastric juice.

However, this definition is practically worthless because almost all adults in the world have cardiac epithelium (with and without parietal and/or goblet cells) and therefore have GERD if it is defined by the presence of cardiac epithelium.

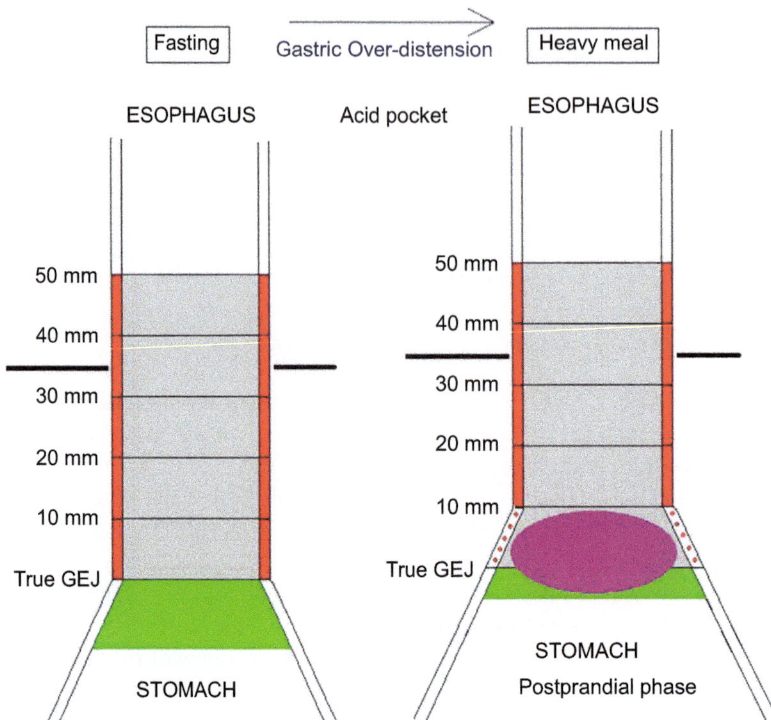

FIGURE 10.10 Dynamic shortening of the lower esophageal sphincter (LES) during the gastric overdistension of a heavy meal in a person with a normal LES (abdominal length 35 mm; no cardiac epithelium). The squamous epithelium has descended to below the pH transition point and is exposed to the acid pocket in the proximal stomach. This is the mechanism of initial LES damage. *GEJ*, gastroesophageal junction.

The practical significance of cardiac epithelium in the dilated distal esophagus is quantitative, not qualitative. *It is the length, not the mere presence, of cardiac epithelium that has critical value in diagnosis*. This is because the measured length of cardiac epithelium (with and without parietal and/or goblet cells) in the dilated distal esophagus is concordant with LES damage.

The theoretical variation of the length of the dilated distal esophagus is from greater than zero to the original length of the abdominal segment of the LES.

Zaninotto et al.[4] showed that the abdominal segment of the LES varies from 10 to 35 mm (excluding two outliers at either end) in 48 asymptomatic volunteers. This variation can mean one of the two things: (1) There is a marked variation in the normal initial length of the abdominal LES or (2) asymptomatic volunteers have varying degrees of damage to the abdominal segment of the LES at the time of the test. The probability that the second explanation is more likely is shown by the fact that a significant number of these volunteers had evidence of acid exposure in a pH test, some exceeding normal values. It is also supported by the fact that the majority of people in the population have variable lengths of cardiac epithelium. In our autopsy study, the length of cardiac and oxyntocardiac epithelium distal to the squamocolumnar junction (SCJ) was 0–15 mm.

The best evidence from the limited amounts of manometric and histologic data available suggests that the normal initial length of the abdominal LES is close to 35 mm with little individual variation.

The limited data available suggest that the theoretical limit of 35 mm for the extent of the dilated distal esophagus is reasonably accurate. There are two reasons for the present discrepancy of the theoretical (35 mm) and demonstrated (28 mm) limits:

1. The number of patients studied is inadequate for a true determination of the maximum extent of the dilated distal esophagus. This is very probable. In all, a total of less than 100 people have ever had this measurement.
2. There is an inherent difference in the manometric assessment of the LES in vivo and the measurement made on a resected specimen. It is well known that transecting the gastrointestinal tract can produce contraction of tissue in an unpredictable manner resulting from contraction of muscle. This may play a role in creating a discrepancy between an in vivo measurement with the organs in situ and that in the resected specimen. This uncertainty must be considered when measurements of the dilated distal esophagus are used in developing a treatment algorithm.

It is interesting to note that the historic extent of "normal" cardiac epithelium was 3 cm (30 mm). This measurement was accurate as a limit of the extent of cardiac epithelium that was present in this region. The measurement, which for decades defined the normal extent of cardiac epithelium in the proximal stomach, actually unwittingly delineated the approximate maximum extent of the dilated distal esophagus.

When autopsy studies[16,17] showed that the extent of cardiac epithelium was 0–8.75 mm in people without GERD, it should have called for an explanation of why some people had a much longer extent of cardiac epithelium. This did not happen. Expansion of cardiac epithelium distal to the endoscopic GEJ was conveniently ignored.

All the available evidence suggests that expansion of cardiac epithelium distal to the endoscopic GEJ is associated with increasing LES damage. The length of cardiac epithelium correlates with severity of GERD. This still remains to be accepted by the medical establishment.

4. MATHEMATICAL FORMULAE TO DEFINE THE NORMAL AND ABNORMAL STATES

The ability to define something by precise mathematical formulae is an indication of a high level understanding. This can now be achieved to describe the normal state of the abdominal esophagus, abdominal segment of the LES, and the histologic state:

1. The normal state where the abdominal esophagus is lined by squamous epithelium in its entirety changes in the abnormal state to:

   length of the abdominal esophagus = length of squamous epithelium + length of cardiac epithelium (with and without parietal and/or goblet cells) from the SCJ to the proximal limit of gastric oxyntic epithelium.

2. The normal state where mucosal rugal folds line only the stomach, which is lined by only gastric oxyntic epithelium, distal to the tubular esophagus changes in the abnormal state to:

   mucosal rugal folds = stomach lined by gastric oxyntic epithelium + dilated distal esophagus lined by cardiac epithelium (with and without parietal and/or goblet cells).

3. The normal state where the length of cardiac epithelium (with and without parietal and/or goblet cells) distal to the endoscopic GEJ in the person without visible CLE is zero changes in the abnormal state to:

   the length of cardiac epithelium = the dilated distal esophagus = amount of abdominal LES damage (i.e., shortening).

4. The normal state where the length of the initial abdominal LES is identical to the residual functional abdominal LES changes in the abnormal state to:

   length of the initial abdominal LES = length of the residual abdominal LES + the dilated distal esophagus (i.e., LES damage).

When the mathematical formulae in (3) and (4) are combined, we come to an important new understanding of the method to assess LES damage:

Initial length of abdominal LES = manometric length of residual (functional) abdominal LES + length of cardiac epithelium (with and without parietal and/or goblet cells).

There is now an explanation for all the contradictions that I have listed at the beginning of this chapter that the medical establishment has accepted so easily.

1. The presence of inflamed cardiac epithelium ("carditis") is not inflammation of the proximal stomach; it is pathology in the dilated distal esophagus resulting from cardiac metaplasia of esophageal squamous epithelium.[5]
2. IM of the gastric cardia is IM of the dilated distal esophagus. The reason why this epithelium mimics BE is that it is in every way equivalent to BE except that is invisible to endoscopy.
3. Adenocarcinoma of the gastric cardia and proximal stomach is adenocarcinoma of the dilated distal esophagus (Fig. 10.8).
4. The reason why increasing length of cardiac epithelium distal to the endoscopic GEJ is associated with GERD is because length of cardiac epithelium = amount of LES damage. As it lengthens, cardiac epithelium causes the endoscopic GEJ to move cephalad with the SCJ.

5. ANATOMICAL CHANGES ASSOCIATED WITH ABDOMINAL LOWER ESOPHAGEAL SPHINCTER DAMAGE

Abdominal LES damage is associated with numerous important anatomical changes in addition to the histologic change of cardiac metaplasia in the distal esophagus that is concordant in length.

5.1 Shortening of the Abdominal Esophagus

The length of the abdominal esophagus may vary in different individuals. The variation in the length of the abdominal esophagus is unknown because of the present difficulty in knowing what to measure. The proximal limit of the abdominal esophagus is the diaphragmatic hiatus. Its distal end is the end of the esophagus. Here, we face a problem because of our preconception that the entire abdominal esophagus is a tubular structure.

True shortening of the abdominal esophagus would require a longitudinal contraction of the abdominal esophagus. This could theoretically happen in GERD where fibrosis and subsequent contraction can result in shortening. However, it is also possible that "shortening" may result from misconception. Let us attempt to use a mathematical formula to illustrate this:

Length of the abdominal esophagus = length of the tubular abdominal esophagus + the length of the dilated distal esophagus.

The question that arises is whether a person who does not believe in the existence of a dilated distal esophagus (the majority of people) will correctly include it in the measurement of the abdominal esophagus or ignore it because it is perceived to be the proximal stomach. If it is largely ignored, the misperception of shortening of the abdominal esophagus will be extremely common.

There are no studies that I have encountered that have specifically measured the length of the abdominal esophagus. I have heard Dr. Jeffrey Peters, a surgical colleague for a long time, say during a lecture that the abdominal esophagus is very different in patients with GERD undergoing fundoplication and those with achalasia undergoing myotomy. The patient with GERD, according to Dr. Peters, has the end of the tubular abdominal esophagus much closer to the diaphragm than in the patient with achalasia. This is a telling statement. The patient with GERD of such severity as to require surgical sphincter repair is likely to have the longest dilated distal esophagus (i.e., the greatest LES damage). Is the "shortening" in the person with severe GERD a true longitudinal contraction or failure to include the dilated distal esophagus in measuring the abdominal esophagus?

Theoretically, the GEJ can be defined externally by the point at which the peritoneum reflects off the stomach onto the undersurface of the diaphragm (Fig. 10.2). This is the point of the true GEJ and close to the point of attachment of the lower leaf of the phrenoesophageal ligament. There is no evidence that the point of the peritoneal reflection changes when the LES is damaged and the dilated distal esophagus develops.

As such, there is a method to avoid the error of measuring the length of the abdominal esophagus at external examination. The abdominal esophagus is the distance from the diaphragmatic hiatus to the peritoneal reflection, irrespective of whether the esophagus is tubular or dilated.

I reviewed the historical paper by Bombeck et al. because it was cited in the paper by McClave et al.[18] as providing evidence that the proximal limit of rugal folds was the muscular junction between the esophagus and stomach. This claim was not true; Bombeck et al.[19] make no mention of gastric rugal folds. What I found instead was a very thoughtful paper with much valuable unbiased data.

The stated purpose of this autopsy study was to assess the role of the phrenoesophageal ligament in the function of the sphincter mechanism. A secondary goal was an attempt to delineate the muscular anatomy of the distal esophagus. The study was prompted by the emerging information at the time regarding the LES mechanism.

The study selected 56 cases out of 227 autopsies at Seattle King County Hospital. This included 8 cases with features of peptic esophagitis and/or hiatal hernia and 48 control patients who had no evidence of disease involving the esophageal hiatus or terminal esophagus. The esophagus was removed at autopsy with surrounding structures, fixed, and dissected with sections taken for histology.

The main objective was to define the attachment of the phrenoesophageal ligament to the esophagus. "The distance between the epithelial junction (this being the most constant and fixed reference point in the specimen) and the upper and lower insertions of the phrenoesophageal ligament were then measured."

This is a unique opportunity for complete study that permits correlation of external and mucosal landmarks. Similar autopsy studies are difficult to find in the literature. The results can be summarized as follows:

1. A total of 21 cases were examined for the location of the SCJ with reference to the point at which the esophagus flared out into the stomach. In 17 of these, it was found >1.0cm above this point and in the other four, it ranged from 0.5 to 1.0cm above. The average distance from the GEJ to the SCJ was 1.1cm.

   In 1966, the definition of the GEJ was the end of the tube where it flared into the saccular organ that was believed to be entirely stomach.[20] At that time, cardiac epithelium was regarded as normal in the distal 2cm of the tubular esophagus. This would suggest that, by the present belief that columnar epithelium in any part of the distal esophagus is abnormal, all 21 patients had visible CLE measuring a minimum of 0.5cm. The rugal folds were irrelevant at that time and are not mentioned.
2. The lower leaf of the phrenoesophageal ligament (which generally follows the peritoneal reflection, which is not mentioned in this paper) was found inserted into the esophagus at the SCJ in 13 cases and to a variable point below the junction in the other cases with the lowest point of insertion in 7 cases being 3.0cm below the junction *onto the fundus of the stomach.*

   The attachment of the lower leaf of the phrenoesophageal ligament approximates the peritoneal reflection and must certainly be proximal to the peritoneal reflection. If the true GEJ is defined by the peritoneal reflection, the above data indicate that the true GEJ in seven patients is *3cm below the SCJ at a point on the gastric fundus distal to the point of flaring of the esophagus to the stomach.* The SCJ was 5–20mm proximal to the end of the tubular esophagus in the 21 patients in whom it was measured. This means that up to 25mm of the esophagus was distal to the end of the tubular esophagus in seven patients; that is, it was dilated and perceived by these authors to be in the fundus of the stomach. The authors are providing excellent evidence for the dilated distal esophagus without knowing it.
3. The insertion of the upper leaf of the phrenoesophageal ligament in the normal patients was an average of 3.35cm above the SCJ in the 47 normal patients and 1.13cm in the 8 patients with peptic esophagitis. In three patients with hiatal hernia and peptic esophagitis, the insertion point was only 0.5cm above the SCJ.

These data are also very interesting. It shows that the insertion of the upper leaf of the phrenoesophageal ligament, which likely marks the most proximal point of the abdominal esophagus and the proximal limit of the abdominal LES, is at a variable distance proximal to the SCJ.

For purposes of argument, let us assume that the initial length of the abdominal segment of the LES is 35mm. The length of the abdominal LES consists of the part of the abdominal esophagus lined by squamous epithelium and the dilated distal esophagus lined by columnar epithelium distal to the SCJ. The length of columnar epithelium can therefore be calculated by the following complicated formula (only for discussion purposes):

Dilated distal esophagus = Initial length of the abdominal LES (assumed to be 35mm) − the measured distance between the insertion of the upper leaf of the phrenoesophageal ligament and the SCJ + the distance from the SCJ to the end of the tubular esophagus.

The calculation cannot be made because the lengths are not given for individual patients; I will use the average/minimum/maximum lengths for the groups with the understanding that inaccuracy is inevitable.

In the normal persons, the length of the dilated distal esophagus = 35mm (assumed initial length of the abdominal LES) − 33.5mm (measured average length from insertion of upper leaf to SCJ) + 5mm (the smallest gap between the SCJ and the end of the tubular esophagus) = a negative number for the length of dilated distal esophagus. With the likely inaccuracies, we can assume that the dilated distal esophagus is very small.

At the other extreme, in the person with GERD and hiatal hernia, the length of the dilated distal esophagus = 35mm (assumed initial length of the abdominal LES) − 5mm (minimum length from upper leaf to SCJ) + 5–20mm (the minimum/maximum gap between the SCJ and the end of the tubular esophagus) = 10–25mm dilated distal esophagus. This is, with all the inaccuracies, a positive number, likely demonstrating that the GERD patient with hiatal hernia has a much longer dilated distal esophagus than the normal person.

The stratification into these two groups is very similar to that in the two analogous groups ("normal" and "hernia-GERD") in the study of Kahrilas et al.[21] In this study, the baseline residual LES length decreased from "normal" to "hernia-negative GERD" to "hernia-positive GERD"; this means that the dilated distal esophagus increased in length.

This is remarkable because both studies were completely unrelated and done at a time when the dilated distal esophagus was completely unknown. The data in the two studies[19,21] show that the severity of GERD is directly related to the length of the dilated distal esophagus (=LES shortening) and inversely related to the manometric length of the residual LES.

The data also suggest that the shortening of the abdominal esophagus is more likely to be the result of misperception caused by an incorrect definition of the GEJ than a true longitudinal shortening by fibrosis. It is only when the variable length of the dilated distal esophagus is factored into the equation that we will learn the true reason for what is now called a "short esophagus." It would not surprise me if the data show no longitudinal displacement of the position of the peritoneal reflection or proximal limit of gastric epithelium in most patients with GERD. It is only when there is near-complete destruction of the abdominal LES resulting in a true sliding hiatal hernia that definite longitudinal displacement of the stomach occurs. In this situation, the peritoneal reflection moves into the thorax around the herniated proximal stomach.

5.2 Dilatation of the Distal Esophagus That Has Lost Lower Esophageal Sphincter Pressure

The maintenance of the tubular shape of the abdominal esophagus depends on the high pressure in the LES that is maintained by tonic contraction of the muscle wall. Unlike the thoracic esophagus, which has a negative pressure and maintains its tubular form without any intramural sphincter pressure, the abdominal esophagus has a baseline positive pressure that is identical to intragastric pressure (+5 mmHg). This is an intrinsic dilatory force that is resisted by the LES pressure.

When LES pressure is lost in the abdominal esophagus, the dilatory tendency causes the LES-less abdominal esophagus to dilate (Fig. 10.11). This tendency to dilatation is aggravated by increased gastric pressure during heavy meals where there is gastric overdistension.

This is the theoretical basis of the dilated distal esophagus. Although there are few data, there is evidence that the distal abdominal esophagus dilates when LES pressure is lost (Fig. 10.12).

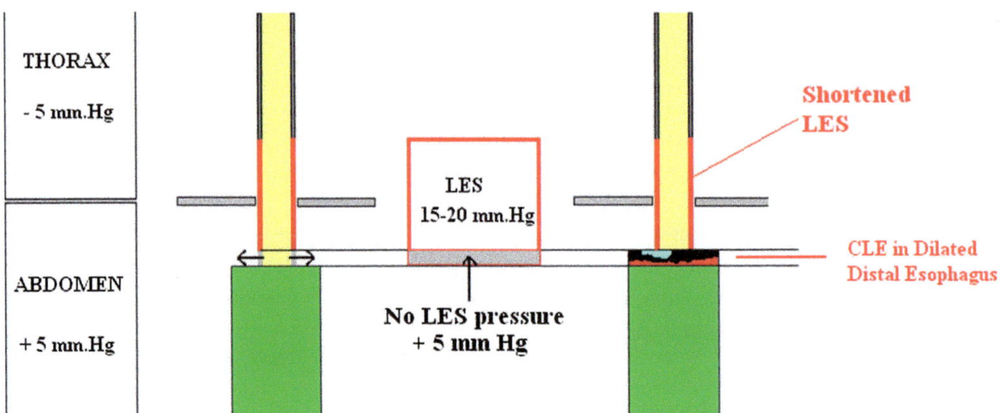

FIGURE 10.11 Pressure changes in a person who has lost 5 mm of the abdominal lower esophageal sphincter (LES), showing that the removal of LES tone in the abdominal esophagus causes that part of the esophagus to dilate. *CLE*, columnar-lined esophagus.

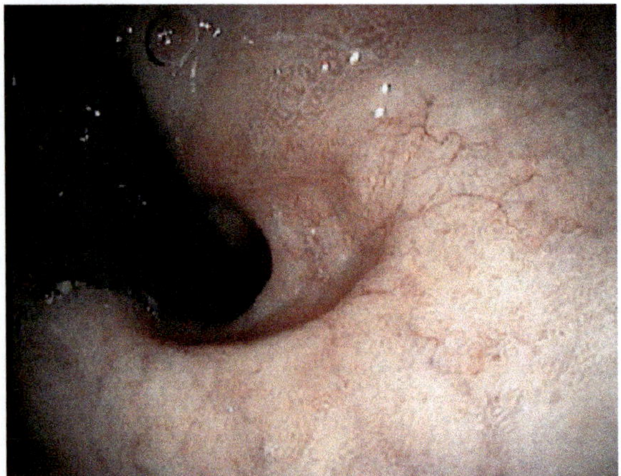

FIGURE 10.12 The contributor, Dr. Martin Riegler's description: "Retrograde endoscopic image into the dilated distal esophagus towards the lower aspect of the esophageal narrowing. Histology within the lower portion of the dilated esophagus showed nondysplastic Barrett esophagus, cardiac mucosa, and oxyntocardiac mucosa. Impedance manometry: absence of high-pressure zone in the lower end of the esophagus." *Photograph courtesy of Dr. Martin Riegler, Reflux Medical, Vienna, Austria.*

Korn et al.[22] measured the diameter of the cardia at the peritoneal reflection. The clinical part of the study had four patient groups: (1) The control group had 25 subjects (18 males; mean age 37.3 years, range 20–64 years) who were undergoing surgery for peptic ulcer disease or gallstones. None had symptoms of GERD, preoperative upper endoscopy was normal, and 24-h pH study was done in 20 patients, all of whom showed absence of pathologic acid reflux into the esophagus. (2) 45 patients with reflux esophagitis (20 males; mean age 45.2 years, range 19–75) who had been symptomatic for many years and dependent on ranitidine or omeprazole. Preoperative endoscopy was normal in 15 and showed erosive esophagitis in 30. (3) 17 patients with short-segment Barrett esophagus (SSBE) (8 males; mean age 49.5 years, range 26–69) with <3 cm visible CLE and IM on biopsy. (4) 15 patients with long-segment Barrett esophagus (LSBE) (mean age 53.8 years, range 26–68 years) with >3 cm visible CLE and IM on biopsy.

Manometry was done with a catheter and three characteristics of the sphincter measured: resting pressure, total length, and abdominal length. A 24-h pH monitoring was done with a catheter-based electrode placed 5 cm above the proximal edge of the LES.

Intraoperative measurements were made at the directly visualized esophagogastric junction (EGJ) or cardia, which was delineated by *the junction of the longitudinal esophageal muscular fibers and the gastric serosa after cleaning off the fat pad that surrounds the GEJ*. The exact perimeter of the cardia (i.e., at the point of the peritoneal reflection) was measured by wrapping a 00 size silk suture completely around it and then measuring its length in centimeters.

The average cardia perimeter was 6.3 cm in control subjects, 8.9 cm in GERD patients, and 13.8 cm in patients with BE. These values were significantly lower in control subjects than in patients with reflux esophagitis or BE ($P<0.001$). Patients with BE had significantly larger cardia perimeters than patients with reflux esophagitis ($P<0.001$), but values did not differ between patients within the SSBE and LSBE groups.

The abdominal sphincter length was 20±4 mm in the control group, significantly greater than the 10±6 mm in the reflux esophagitis group, which was significantly greater than the 3±3 mm in the SSBE group and the 4±4 mm in the LSBE group. These mean lengths of the abdominal LES suggest a competent LES in the control group, a borderline LES (10 mm mean abdominal length) in the reflux esophagitis group, and markedly compromised LES (all <10 mm) in the BE group where the length of the abdominal LES was zero in some patients.

This study shows by actual measurement the dilatation of the distal esophagus. The dilatation is measured at the original end of the LES where the esophageal muscle transitions to gastric muscle at the peritoneal reflection. The progressive increase in circumference of the esophagus at the point of the original GEJ from the control group (6.3 cm) to reflux esophagitis (8.9 cm) to BE (13.8 cm) groups proves that there is an increasing dilatation of the distal end of the esophagus.

This increasing dilatation is associated with the progressively greater shortening of the abdominal segment of the LES (from a mean of 20 mm in controls to 10 mm in reflux esophagitis to 3–4 mm in the BE group) (Table 10.1).

These findings provide strong evidence in support of the concept of the dilated distal esophagus and its relationship to shortening of the abdominal segment of the LES. In addition, it suggests that there is a quantitative relationship between shortening of the abdominal sphincter, circumference of dilated distal esophagus at the position of the peritoneal reflection (i.e., GEJ), and reflux-induced cellular changes in the esophageal body increasing progressively from normal→erosive esophagitis→BE. It is highly likely that the circumference of the distal esophagus at its distal end correlates with the increasing length of the dilated distal esophagus.

Two analogies for the conversion of the normal esophagus to the dilated distal esophagus have been told to me. The first, by Dr. Martin Riegler from Vienna, one of the great cradles of music in the world, is that the normal esophagus and GEJ is like a trumpet with a sharp angle of His that progressively transforms to a clarinet as the angle of His becomes more obtuse.

TABLE 10.1 Circumference at the Junction Between the Muscular Esophagus and Gastric Serosa (Peritoneal Reflection) at Surgery in Groups of Normal Controls and Patients With Gastroesophageal Reflux Disease and Barrett Esophagus. Circumference Was Measured by Wrapping a Silk Suture Around the Perimeter of the Peritoneal Reflection

	Circumference of Cardia (cm)	Abdominal Sphincter Length (mm)
Controls	6.3	20±4
Reflux esophagitis	8.9	10±6
Barrett esophagus[a]	13.8	3±3 (SSBE) and 4±4 (LSBE)

LSBE, long-segment Barrett esophagus; *SSBE*, short-segment Barrett esophagus.
[a]*There was no significant difference in these parameters between patients with short and long-segment Barrett esophagus.*

The second, by a radiologist who was an oenophile, is that the normal esophagus is like a bottle of Cabernet Sauvignon that transforms to a bottle of chardonnay. Both analogies are very expressive.

5.3 Changes in the Gastroesophageal (Hill) Valve

The Hill valve is a finding that many people use to assess the LES at endoscopy.[23] Not being an endoscopist, I am out of my depth here. However, I believe the assessment is valuable and will try to correlate changes in the appearance of the Hill valve to abdominal LES damage.

Oberg et al.[24] from the University of Southern California (USC) Foregut Surgery Unit, introduce the topic: "The primary components of the gastroesophageal barrier include not only the lower esophageal sphincter (LES) and the crural diaphragm but also the geometry of the gastroesophageal junction. Although the presence of an anatomic gastoesophageal valve has been discussed for more than a century, the major emphasis of the past several decades has been on the lower esophageal sphincter function. The significance of the geometry has received considerably less attention."

> *Recently, Hill et al have renewed the interest in the geometry of the gastroesophageal junction. They developed a grading system in which the geometry of the gastroesophageal valve was scored from I through IV based on its endoscopic appearance. They concluded that the endoscopic appearance of the valve was a better predictor of the presence or absence of reflux disease than LES pressure.*

A total of 268 consecutive patients (151 men; median age 51 years; range 16–89 years) with symptoms suggestive of GERD underwent endoscopy with biopsy, esophageal manometry, and 24-h pH monitoring. The geometry of the gastroesophageal valve was assessed with the endoscope in the retroflexed position as follows (Fig. 10.13):

Grade I: The presence of a prominent fold of tissue, closely approximated to the shaft of the endoscope and 3–4 cm along the lesser curve at the entrance of the esophagus into the stomach.
Grade II: A less prominent fold of tissue with occasional periods of opening and rapid closing around the endoscope with respiration.
Grade III: Absence of a prominent fold where the endoscope was not tightly gripped by the tissues.
Grade IV: Essentially no fold where the lumen of the esophagus gaped open, allowing the squamous epithelium to be viewed from below. This was accompanied by a hiatal hernia.

Three landmarks were recorded: (1) the position of the crural impression; (2) the GEJ (level where the gastric rugal folds end and the tubular esophagus begins); and (3) the SCJ. A hiatal hernia was diagnosed when the GEJ was 2 cm or more above the crural impression. The presence of erosive esophagitis and CLE (including length of proximal displacement of the SCJ from endoscopic GEJ) was recorded. Biopsies were taken according to the routine protocol of the Foregut Unit at USC. The diagnosis of BE required the presence of goblet cells in visible CLE.

Stationary manometry was performed with a water-perfused catheter with pressures of the LES measured with a stationary pull-through method (Table 10.2). A mechanically defective LES was defined by the presence of one of the following factors: resting pressure < 6 mmHg, overall length < 2 cm, and abdominal length < 1 cm. A 24-h pH monitoring was done with a pH electrode located 5 cm above the upper border of the LES. Abnormal esophageal acid exposure was defined as a pH < 4 for >4% of the measured time (95th percentile of healthy volunteers).

There was no difference in the median age of patients with different grades of valve. There was a significant male predominance in patients with grade II–IV valves compared with a grade I valve.

The relationship between the endoscopic appearance of the gastroesophageal valve and LES characteristics is shown in Table 10.2. LES resting pressure and length decreased with increasing grade of the valve.

Like many papers, the data unwittingly provide valuable information that is divorced from the authors' primary intent. These data are always unbiased and valuable. This is a population of patients with symptomatic GERD. In the new method of looking at GERD, this will represent a group whose LES damage has progressed over decades beyond the reserve capacity of the LES.

I have suggested that LES damage begins distally at the true GEJ and progresses cephalad, involving the abdominal segment of the LES with little initial involvement of the thoracic segment. The progression of these patients with increasing grades of the valve was largely associated with a decrease in the abdominal LES (from 14 to 12 to 8 to 4 mm in grades I through IV). The difference between abdominal and overall length remained essentially unchanged (from 16 to 14 to 14 to 14 mm in grades I through IV). The data show that LES shortening primarily involves the abdominal LES.

With the assumption that the initial length of the abdominal LES is 35 mm, I have suggested that there must be abdominal LES damage with a shortening of 15–20 mm to get to the point where the residual abdominal LES is 15–20 mm. At this level, LES failure results with dynamic shortening caused by heavy meals (i.e., postprandial reflux).

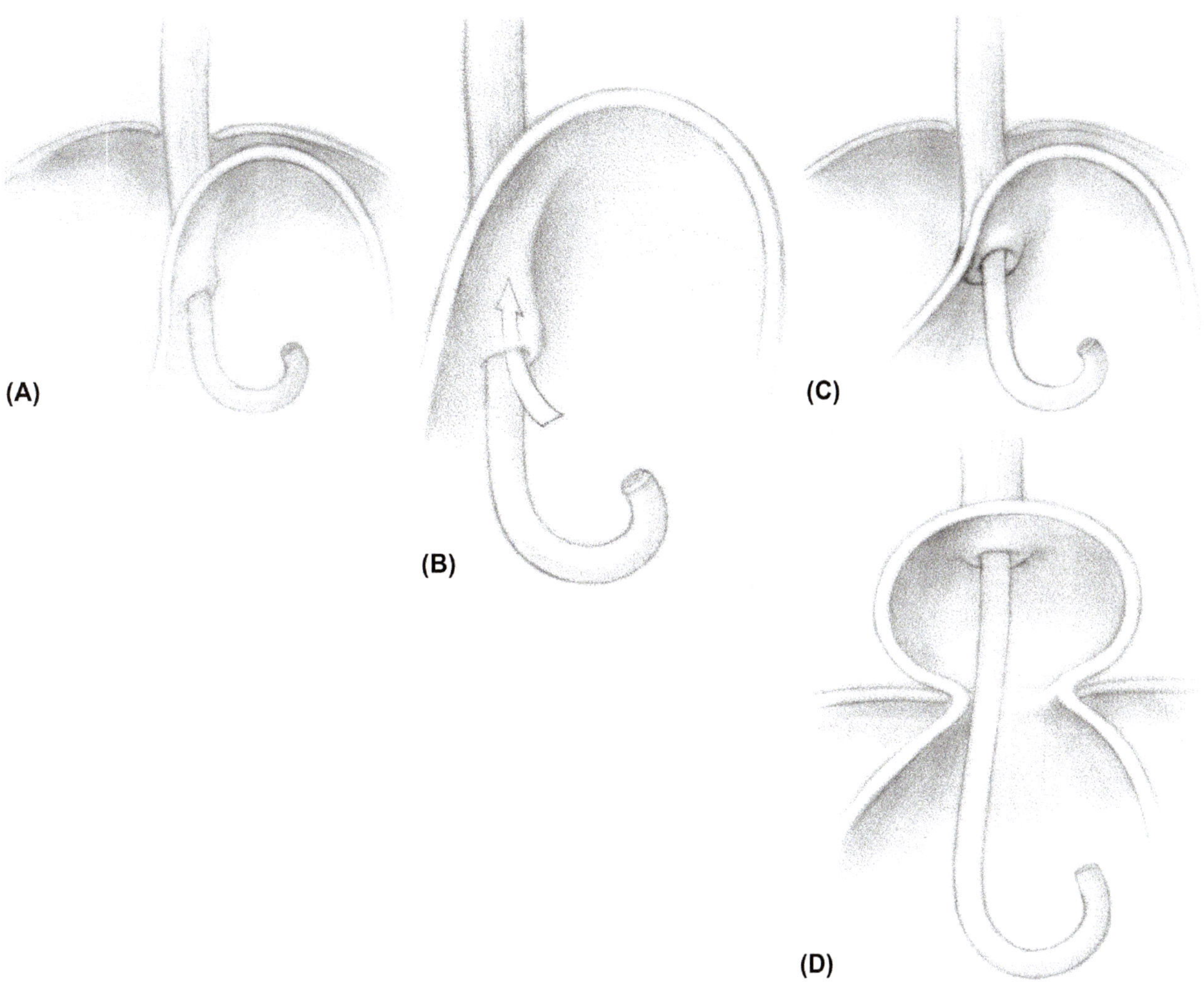

FIGURE 10.13 Increasing grades of Hill valve as seen in retroflex view at endoscopy. (A) Retroflexed view of a grade I valve, defined by the presence of a prominent fold of tissue closely approximated to the shaft of the endoscope and extending 3–4 cm along the lesser curve at the entrance of the esophagus into the stomach. (B) Endoscopic view of a grade II gastroesophageal valve. The fold of tissue is less prominent, and there are occasional periods of opening and rapid closing around the endoscope with respiration (*arrow*). (C) Grade III gastroesophageal valve. There is no prominent fold at the entrance of the esophagus into the stomach, and the endoscope is not tightly gripped by the tissues. (D) Grade IV gastroesophageal valve. The patient has a large hiatal hernia and essentially no fold where the lumen of the esophagus is gaping open, allowing the squamous epithelium to be viewed from below. *Reproduced with permission from Oberg S, Peters JH, DeMeester YR, Lor RV, Johansson J, Crookes PF, Bremner CG. Endoscopic grading of the gastroesophageal valve in patients with symptoms of gastroesophageal reflux disease.* Surg Endosc *1999;13:1184–88.*

TABLE 10.2 Lower Esophageal Sphincter (LES) Characteristics in Patients With Varying Grades of the Gastroesophageal Valve (Medians and Interquartile Range)

	Grade I	Grade II	Grade III	Grade IV
Number of patients	43	57	71	97
LES pressure (mmHg)	14.0 (9.0–21.2)	8.2 (3.5–13.5)[a]	8.8 (5.2–14.4)[a]	4.4 (2.1–7.6)[b]
Abdominal length (cm)	1.4 (1.0–1.8)	1.2 (0.6–1.6)[a]	0.8 (0.4–1.4)[a]	0.4 (0.2–0.8)
Overall length	3.0 (2.6–3.4)	2.6 (2.0–3.4)[a]	2.2 (1.4–3.2)[b]	1.8 (1.2–2.3)[b]
Defective LES	9/43 (20.9%)	32/57 (56.1%)[a]	43/71 (60.6%)[a]	85/97 (87.6%)[b]

[a] *$P<0.05$ versus grades I and IV.*
[b] *$P<0.05$ versus all other groups.*

All these patients had symptomatic GERD. They should, if what I have theorized is correct, have a residual abdominal LES of <20mm. The data in the study prove this is correct. In 43 patients with a grade I valve, the mean abdominal length was 14mm [interquartile range (IQR) 10–18mm]. The length of the abdominal LES progressively decreased from this point as the valve grade increased.

This suggests that symptoms develop when the abdominal LES is degraded through its reserve capacity of 15mm. The disease then progresses with the same rate of progression of LES damage.

Let us take the median age of patients with a grade I valve, which was 45 years. According to the new method of looking at the disease as an expression of varying rates of LES damage, the patient had an abdominal LES (assumed) of 35mm at age 15. In 30 years (at age 45 years), this patient had an abdominal length of 14mm (median abdominal length in patients with a grade I valve). Assuming a linear progression of abdominal LES damage, the rate of shortening is [(35 − 14)/30] 0.7mm/year. If the rate of LES degradation remains the same, the patient will reach an abdominal LES length of zero in (14/0.7) 20 years, i.e., at 65 years. There is a high likelihood that this patient will develop severe GERD in 20 years. Is there not an opportunity here to prevent this progression by an intervention that stops the progression of LES damage?

To contrast, let us take the IQR of the abdominal LES length in the 43 patients with a grade I valve and perform a similar calculation. The 25th percentile LES length was 10mm. The rate of progression of LES damage by our calculation is [(35 − 10)/35] 0.83mm/year. This patient will reach a zero abdominal LES in (10/0.83) 12 years, i.e., at age 57 years. The need to intervene is more urgent. The 75th percentile LES length was 18mm. The rate of progression of LES damage by our calculation is [(35 − 18)/30] 0.57mm/year. This patient will reach a zero abdominal LES in (18/0.57) 32 years, i.e., at age 77 years.

The impact of age is also dramatic on the prediction of future progression of LES damage. If one takes the 75th percentile age of the 43 patients with a grade I valve (78 years) and take the 75th percentile abdominal LES length of 18mm, using the same calculation, the rate of LES shortening is [(35 − 18)/63] 0.26mm/year. The time that this 78-year-old person takes to reach a zero abdominal LES is (18/0.26) 69 years i.e., at age 147 years; he/she will never reach a zero abdominal LES.

I have taken a zero abdominal LES as the endpoint for these calculations to make them more dramatic. The real endpoint that is critical for LES failure is 10mm. The calculations become different.

These calculations are for demonstrating a principle that can lead to a method of predicting future LES damage. They are highly unlikely to be accurate, mainly because of the inherent lack of precision of measurement of the residual LES by manometry and because the assumption that the initial length of the abdominal LES length is 35mm, which is uncertain. The calculations cannot be relied on for making clinical decisions because of this inaccuracy.

When, however, the new method of measuring abdominal LES becomes available, the ability to correlate severity of GERD with LES damage and predicting progression of LES damage into the future becomes much more feasible practically.

The data in this study show a progressive increase in acid exposure and a progressive increase in the prevalence of an abnormal pH test with increasing grades of the valve. This is expected because of the association with a progressive increase in LES damage in the different valve grades.

What is more interesting is the actual recorded acid exposure in the 43 patients with a grade I valve. The median acid exposure in this group was very low (around 2% of the 24 h period) with some patients having zero acid exposure. This provides evidence of the existence of a stage in the disease where the patient's LES is damaged within its reserve capacity where there is zero reflux. In these 43 patients (with an IQR of abdominal LES length of 10–18mm), there were patients who had zero acid exposure. This would suggest that a significant reduction in abdominal LES length from its initial 35mm is necessary for there to be *any* reflux.

If we set the normal value for acid exposure in the pH test as zero, it will discriminate between people whose LES damage is within their reserve capacity and those with a defective LES (in this study, between a subset of people with symptoms of GERD who have a grade I valve and all patients with grade II and above). The reason why this cannot be done with the pH test at present is that GERD is not recognized until symptoms develop, which is a relatively late stage of LES damage.

In the study, no patient with a grade I valve had erosive esophagitis or BE. Again, this is despite the fact that these 43 patients had an abdominal LES length of 1.4cm (IQR 1.0–1.8cm). This again suggests a correlation between a lesser degree of LES damage, grade I valve, a pH test within the normal range, and the absence of sufficient reflux to cause cellular changes in the thoracic esophagus.

The authors conclude that there is an association between the altered geometry of the gastroesophageal valve and the presence of reflux disease and its complications.

Koch et al.[25] studied 43 patients [15 women; mean age 49.9 ± 13.8 years; mean body mass index (BMI) 26.4 ± 3.4 kg/m^2] with proven GERD who underwent laparoscopic fundoplication. All patients had a long history of GERD prior to surgery.

The patients were selected from a total of 79 patients undergoing fundoplication by virtue of having had the following preoperative findings before surgery: (1) upper endoscopy with a documented Hill grade; (2) manometry; (3) ambulatory impedance monitoring; and (4) complete clinical evaluation including a quality-of-life questionnaire. The surface area of the diaphragmatic hiatus was measured using a standard calculation at the time of surgery.

Of the 43 patients, 8 (18.6%) had a grade I valve, 5 (11.6%) had a grade II valve, 12 (27.9%) had a grade III valve, and 18 (41.9%) had a grade IV valve. Significant positive correlations were present between the Hill valve grade and the DeMeester score for acid exposure, total number of reflux events, and hiatal surface area. There was no significant correlation with severity of GERD symptoms, quality-of-life measures, or mean LES pressure. No LES length measurements are given. The study provides evidence that increasing Hill valve grade is associated with increasing transsphincteric reflux.

What is the "gastroesophageal valve" that is being described by Hill and these authors? To the eye of someone who has never done endoscopy, it appears to be the abdominal esophagus that contains the endoscope as it enters the stomach. The opening is on the lesser curvature of the stomach and the abdominal esophagus appears to be a ridge of mucosa with a definable length as it is viewed from within the lumen of the stomach. This likely represents the apposition of the abdominal esophagus to the wall of the lesser curvature above its opening. As the valve grade increases, the length of the abdominal esophagus that is seen apposed to the lesser curvature appears to decrease, becoming essentially zero in the grade IV valve. The opening of the esophagus has moved to a point that appears more cephalad on the lesser curvature of the stomach.

This change is best explained by a progressive shortening of the abdominal LES that has resulted in a dilated distal esophagus. The part of the abdominal esophagus that has lost sphincter pressure has dilated and now appears to be part of the proximal stomach. The angle of His has become more obtuse as the tubular abdominal esophagus has become shorter.

The gap between the esophageal openings in a normal and grade IV valve likely represents the dilated distal esophagus. If true, this will correlate with increasing length of cardiac epithelium (with and without parietal and/or goblet cells) distal to the opening of the esophagus. This will be most easily measured along the lesser curvature where the mucosa is flatter than in the greater curvature. This area represents the optimal area for measurement of the dilated distal esophagus in the new test (Fig. 10.14).

5.4 Increase in the Acute Angle of His

During fetal development, the greater curvature of the stomach at and below the GEJ proliferates more rapidly than the lesser curvature, resulting in the protrusion of the fundus upward and outward (see Chapter 3). The upward movement of the fundus from the GEJ toward the diaphragm creates the angle of His.

In the normal state, the fundus of the stomach forms an acute angle as it passes sharply upward and to the left from the peritoneal reflection to reach the undersurface of the diaphragm (Fig. 10.2). When the distal esophagus loses sphincter pressure at its distal end, the tubular abdominal esophagus shortens and the damaged part of the esophagus dilates and is taken up onto the greater curve of the stomach. The shortened tubular esophagus ends closer to the diaphragmatic hiatus.

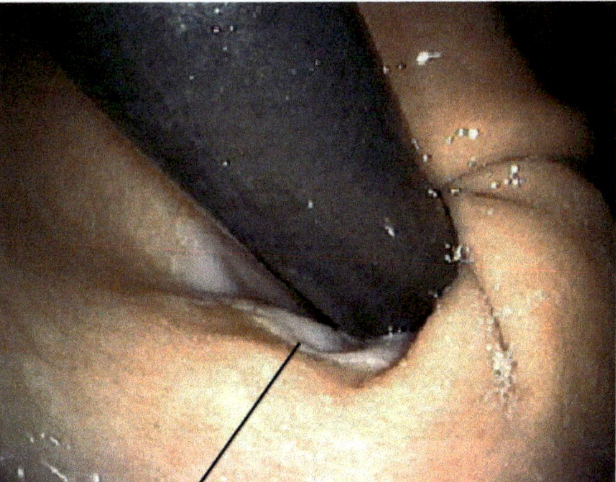

FIGURE 10.14 The contributor, Dr. Martin Riegler's description: "Retrograde endoscopic vision towards Hill III valve. Note the presence of the SCJ at the level of the diaphragm. Histology: cardiac mucosa, oxyntocardiac mucosa, squamous epithelium." The *vertical line* drawn by me shows the area that needs to be evaluated histologically to measure the length of metaplastic cardiac epithelium (= lower esophageal sphincter damage). *Photograph courtesy of Dr. Martin Riegler, Reflux Medical, Vienna, Austria.*

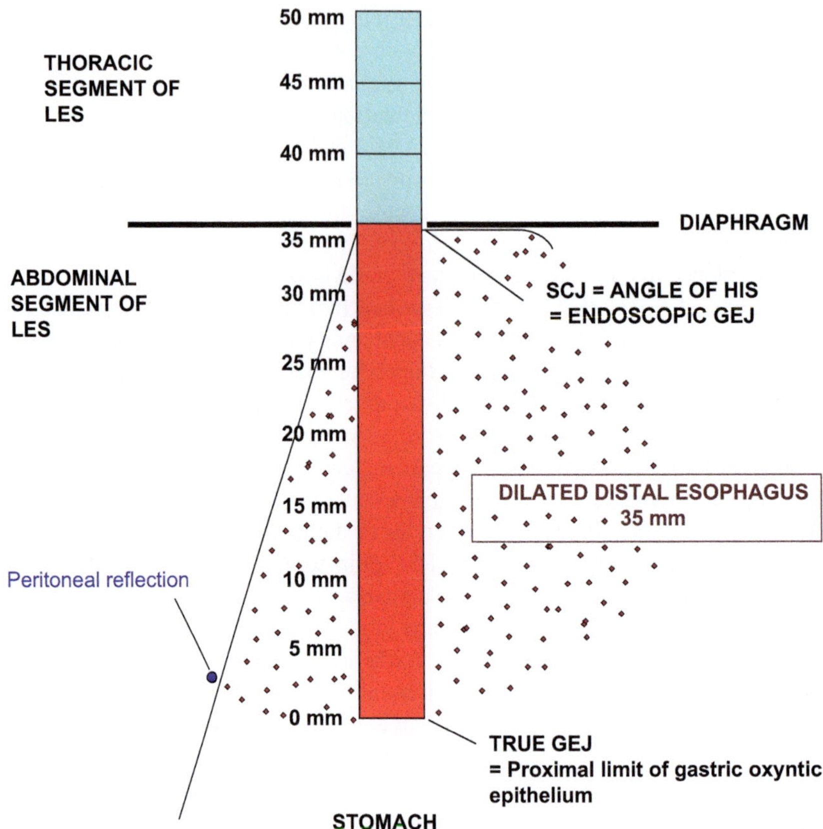

FIGURE 10.15 Diagrammatic representation of a person with complete destruction of the abdominal lower esophageal sphincter (LES). The tubular esophagus ends at the diaphragmatic hiatus, corresponding to a Hill IV valve; the angle of His is a right angle, flush with the diaphragm. It is highly likely that this person has severe reflux with severe erosive esophagitis and/or visible columnar-lined esophagus in the thoracic esophagus (not shown). The obliteration of the angle of His also sets the stage for a sliding hiatal hernia. *GEJ*, gastroesophageal junction; *SCJ*, squamocolumnar junction. Original abdominal esophagus, now lined with cardiac epithelium = red; dilated distal esophagus, 35 mm long = stippled area.

This causes the angle of His to become more obtuse. With complete destruction of the abdominal LES, the angle of His essentially becomes a right angle (Fig. 10.15).

Mathematically, taking up of the distal tubular esophagus into the contour of the stomach is the most likely if not the only method by which the angle of His can become more obtuse.

In a technologically advanced paper from Zurich, Curcic et al.[26] studied the structural differences at the GEJ in a group of 24 healthy volunteers and 24 patients with mild to moderate GERD using concurrent magnetic resonance imaging and high-resolution manometry.

In the introduction, the authors describe the limitations of animal studies, endoscopy, manometry, video fluoroscopy, and computed tomography in studying the functional anatomy of the junctional region. They used magnetic resonance imaging (MRI) images with high-resolution manometry to obtain a comprehensive assessment of three-dimensional functional anatomy of the GEJ in the fasting state and after a large test meal. This involved reconstruction of stacks of two-dimensional images to produce three-dimensional models of the distal esophagus and proximal stomach.

From these models, they could measure the insertion angle of the esophagus into the stomach, gastric morphology, and orientation within the abdomen. In addition, reflux events were imaged in real time by cine-MRI with concurrent manometry providing a continuous assessment of GEJ function including LES pressure and occurrence of transient LES relaxations (tLESRs).

The angle of His is the angle created by the tubal esophagus and the proximal stomach (defined more precisely in the study as the angle between the longitudinal axis of the esophagus and the tangent of the gastric fundus). The normally acute angle of His creates a powerful mechanical barrier to movement of the stomach into the thorax.

The study groups were 24 healthy volunteers (11 female; mean age 27 years; mean BMI 22.6 kg/m^2) and 24 GERD patients (10 female; mean age 39 years; mean BMI 25.2 kg/m^2). Healthy subjects had no symptoms suggestive of reflux, dyspepsia, or other digestive problems and were not taking any medication that could alter gastrointestinal function. The

GERD diagnosis was established by the presence of reflux esophagitis on endoscopy on the basis of Los Angeles (LA) classification and/or pathologic acid exposure on 24 h pH monitoring. Patients with a hiatus hernia>3 cm were excluded.

Patients came to the department after a minimum 4-h fasting period and a minimum 3 days off PPI treatment. After baseline manometry and MRI examination, patients ingested a standardized mixed solid and liquid meal (600 g, 939 kcal, 44% lipids). Half an hour later, subjects drank a liquid nutrient drink (0.5 kcal/mL) to maximum fullness. All liquids ingested were labeled with a paramagnetic contrast agent. Measurements were obtained in study phases of 4 h and 30 min: fasting, after meal, fullness, and emptying.

Details of methods involving MRI and manometry and image analysis are given. The definitions and calculations of the insertion angle between the esophageal axis and a tangent on the proximal fundus, shape of the proximal fundus above the GEJ, location at which the esophagus inserts into the proximal stomach, total gastric volume, and GEJ luminal diameter during reflux events are complicated and difficult for a mere mortal to comprehend.

This is expected from a paper originating from an institute of biomedical engineering. Reflux events and their duration were assessed in cine-MRI, and the LES pressure and length were measured for each study phase. Without the ability to evaluate their methods, I must trust these researchers implicitly as I evaluate their results.

All reflux events detected during cine-MRI were also detected on manometry recordings. The number of reflux events after the meal was similar in healthy volunteers and GERD patients (two vs. three reflux events) during a total observation time of 20 min (P=.092). Reflux duration was significantly longer in GERD patients (32.8 vs. 26.8 s; P=.002). The luminal opening diameter of the GEJ during reflux was larger in GERD patients (19.3 vs. 16.8 mm; P=.04).

The insertion angle of the esophagus into the proximal stomach, defined as the angle between the longitudinal axis of the esophagus and the tangent of the gastric fundus, was 55 ± 2.5 degrees in the healthy group after meal intake during expiration. It was wider (more obtuse) by 7 ± 3 degrees in the GERD patients (P=.03). Gastric filling did not alter this parameter. The angle was wider (more obtuse) in both groups during inspiration than during expiration (+3 ± 1 degree; P=.00001).

Proximal gastric curvature (i.e., distension) was similar in healthy subjects and GERD patients (P=.56). As expected, distension increased with gastric filling from the fasted to the postprandial phase and increased further after feeding to maximum fullness. Gastric emptying was faster in the healthy patients compared with the GERD patients (P=.009).

LES pressure after meal intake and during expiration was higher in healthy subjects than in GERD patients (20 ± 2 mmHg vs. 12 ± 2; P=.0001). Intraabdominal LES length after meal intake was 1.9 ± 0.2 cm in healthy subjects and significantly shorter in GERD patients (0.9 ± 0.3 cm; P=.0006). The length of the thoracic component of the sphincter was not different between the groups (P=.51).

In the discussion: "The primary outcome of this study was that the insertion angle of the esophagus into the stomach is wider (more obtuse) in GERD patients than in healthy subjects after ingestion of a large test meal. This anatomic relationship is referred to as the 'angle of His' in surgical studies.... An acute esophagogastric angle is essential for the proposed 'flap-valve' mechanism of reflux protection, as this increases the mechanical advantage by which the intra-abdominal esophagus is compressed by the proximal stomach as it fills after a meal. Such compression would limit GEJ opening during tLESRs and, as a result, reduce or prevent reflux. Consistent with the 'flap-valve' hypothesis, EGJ opening during reflux events was less in healthy volunteers than in GERD patients."

The authors make an interesting point regarding the luminal diameter of the GEJ during a reflux event. GEJ opening during reflux was significantly larger in GERD patients than in healthy subjects (19.3 vs. 16.8 mm). The authors comment: "... this is the first time that increased EGJ opening in GERD patients has been documented in vivo during reflux events by noninvasive imaging, According to Hagen-Poiseuille's law, this seemingly small (approximately +15%) difference in luminal diameter would result in a relatively large increase in reflux volume (approximately +75%) because flow is proportional to the radius to the fourth power. This is much greater than the 1:1 increase in flow that would be related to reduced LES nadir pressure, LES length, or increased abdominal pressure. Second, once reflux had occurred, esophageal volume clearance was significantly slower in GERD patients. Together, these observations provide a possible, mechanical explanation for the increase in 'volume reflux' and for prolonged reflux exposure that cause symptoms and mucosal damage in this condition."

All presently used measures of reflux, including pH testing and impedance, do not give any information regarding the volume of gastric fluid that enters the esophagus during a reflux episode. This is a critically important unknown value in the study of the pathology of transsphincteric GERD. The ability to measure the GEJ opening diameter during a reflux episode represents the first measure, albeit indirect, of reflux volume.

This paper shows that the angle of His increases from the baseline of 55 ± 2.5 degrees in the healthy subjects after meal intake to 62 ± 3 degrees in GERD patients. The angle of His becomes more obtuse. It should be noted that patients with the most severe GERD (i.e., those with hiatal hernia > 3 cm) were excluded from this study.

The extreme endpoint of this process whereby the angle of His becomes increasingly obtuse is complete destruction of the 2.5–3.5 cm of the abdominal sphincter. When this happens, the tubular esophagus ends immediately distal to the diaphragmatic hiatus and the angle of His becomes closer to a right angle (Fig. 10.15). In the study above, the angle of His moved from 55 degrees in the healthy non-GERD group who had a mean abdominal sphincter length of 1.9 cm after intake of a meal (i.e., dynamic LES shortening added to baseline permanent LES shortening) to 62 degrees in the GERD patients who had a mean abdominal sphincter length of 0.9 cm after a meal. A 1.0 cm decrease in the length of the abdominal sphincter resulted in an increase in the angle of His by 7 degrees. Assuming that the rate of increase of the angle remains the same with increasing abdominal LES shortening, if the entire 3 cm of the abdominal sphincter is lost, the angle of His will be expected to progress more toward a right angle.

The entire mechanism of increase in the angle of His in GERD can be related to the shortening of the abdominal LES. The greater the shortening, the more obtuse is the angle of His (Figs. 10.12A and 10.15).

5.5 Sliding Hiatal Hernia

A sliding hiatal hernia is a condition where there is herniation of the proximal part of the stomach through the diaphragmatic hiatus into the thorax, carrying with it a herniated peritoneal sac.

The herniation may be permanent. In this situation, the herniated stomach is present at all times above the diaphragm irrespective of the phase of respiration. Theoretically, a sliding hernia can result from two mechanisms: (1) A shortening of the esophagus has pulled the proximal stomach into the thorax or (2) increased intraabdominal pressure has pushed the stomach into the thorax. In the latter event, the esophagus is of normal length and should show redundancy or tortuosity in the thorax. This is not commonly seen in a sliding hiatal hernia.

A sliding hiatal hernia can be temporary or reducible; in this situation the stomach herniates into the thorax during inspiration and swallowing when there is upward vertical movement of the entire esophagus across the diaphragmatic hiatus. If the vertical movement is sufficient, the proximal stomach may herniate temporarily into the thorax, returning into the abdomen when the upward movement reverses.

The normal acute angle of His where the fundus of the stomach moves sharply upward to the undersurface of the diaphragm from the normal (35 mm) abdominal esophagus is a highly effective mechanism to prevent sliding of the stomach through the diaphragmatic hiatus.

When the abdominal LES is destroyed and the distal esophagus dilates to take up the contour of the stomach, the angle progressively becomes more obtuse. The anatomy that prevents sliding hiatal hernia is slowly degraded. Increasing LES damage, dilatation of the LES-less abdominal esophagus, and increase in the angle of His are the essential changes for the development of a sliding hiatal hernia (Fig. 10.15).

At surgery, there is no difficulty in diagnosing a hiatal hernia. There *must* be a herniated sac of peritoneum for a diagnosis of a hiatal hernia to be made. The stomach cannot herniate into the thorax without carrying with it the peritoneal lining that invests the entire stomach.

At endoscopy and radiology, where the diagnosis of sliding hiatal hernia is largely made at present, the diagnosis is more complicated. The peritoneal reflection cannot be seen at endoscopy or radiology. As such, the diagnosis must be made by the presence of what is believed to be stomach above the level of the diaphragm. This is complicated by two factors: (1) the vertical movement of the esophagus during respiration; and (2) the inaccuracy of the endoscopic definition of the GEJ. I will consider the impact of the latter factor in the section of diagnosis of a sliding hiatal hernia.

5.5.1 Historical Definition of Sliding Hiatal Hernia

The earliest descriptions and drawings of the different types of hiatal hernia, including sliding hiatal hernia made by Allison in 1948[27] drawings, are impeccable.

In 1948, peptic ulceration of the distal esophagus was believed to be caused by a congenital maldevelopment where the esophagus failed to reach its normal length. In the congenital short esophagus, the proximal stomach was in the thorax and the normal barriers to reflux were absent.

Allison writes, in this paper that refutes the concept of a congenital short esophagus: "Peptic ulcer of the oesophagus and congenitally short oesophagus became almost like the Beaver and the Butcher: 'You could never meet either alone!' Although such congenital abnormality cannot be denied, the burden of this report is that short oesophagus is usually an acquired condition due to defects in the diaphragm which allow a sliding hiatal hernia of the stomach."

This study consists of 74 patients with chronic peptic esophagitis with ulcers. 63 were adults (34 male), 13 with chronic esophagitis and recurrent acute ulcers and 50 with chronic ulcers; 4 who developed ulcers after surgery for cancers of the cardia; and 7 children between 6 months and 5 years.

This is the era before there were data on the presence or structure of the LES. The barrier to reflux was believed to be a combined effect of the diaphragm and a valvular mechanism in the GEJ region called the "cardia." Allison describes the mechanism of the cardia, as believed then, with an interesting experiment in presumably normal people at autopsy[27]: "If, soon after death, the stomach and oesophagus are removed, the pylorus sutured, and the organs laid on a flat bench, the stomach can be moderately distended with fluid with very little reflux into the oesophagus. As the fundus of the stomach distends, the angle between it and the oesophagus becomes more acute and a valve is produced at the cardia. The effectiveness of this valve may be increased by the elastic recoil of the oblique muscle fibres of the stomach. As these are stretched they will tend to narrow the cardia from before backwards, and the loop over the top of the cardia will accentuate the angulation between oesophagus and fundus much in the same way as the pubo-rectalis fibres of the levator ani do between the rectum and anus. During life the tonic action of these muscle fibres must be even more effective. When the organs are in the normal position in the body, the angle between the oesophagus and stomach is filled in by a thick wedge of muscle fibres of the diaphragmatic pinchcock. This also helps to maintain the angle, for the fundus of the stomach naturally passes upward under the dome of the diaphragm while the cardia is fixed at the hiatus."

After death, there is no tonic contraction of the LES. There is thus a purely mechanical component to the resistance to reflux that is based on what we now know to be the angle of His, the diaphragm, and possibly the sling fibers of the muscle wall of the stomach in this region. Allison strongly suggests that the way the abdominal esophagus inserts into the stomach creating an acute angle of His is crucial for maintaining the cardia (=valvular mechanism in the lower esophagus) and the stomach in its normal position.[27] In fact, with our new understanding of the pathogenesis of early LES damage, it is really the lack of LES damage that is crucial for maintaining the angle of His, the reverse of what Allison suggested.

In subsequent experiments, also in the Allison era, Nauta[28] showed that sectioning or removal of the sling fibers of the cardiac notch causes the angle of His to become more obtuse. Korn et al.[22] in an experimental model of the GEJ region of pigs that was modified with the use of elastic bands to increase the diameter of the cardia, showed that when dilatation of the cardia was simulated, the sphincter effect of the region decreased with decreased pressure, length, vector volume, and opening pressure. The opening pressure is the pressure at which increased gastric distension with fluid caused retrograde flow into the esophagus in this experimental model, i.e., it was a measure of sphincter competence. Again, the close relationship between the dilatation of the distal esophagus, increase in the angle of His, and competence of the anatomic sphincter mechanism is demonstrated. These are all the necessary anatomic changes that result ultimately in a sliding hiatal hernia.

Allison continues[27]: "During inspiration the negative suction force in the chest increases and the positive pressure in the abdomen increases. At this time, therefore, there is a tendency for the cardia to be drawn up into the mediastinum, or alternatively for the stomach contents to pass up into the oesophagus. But it is at this time that the crural fibres contract around the lower end of the oesophagus, closing its lumen and preventing herniation of the cardia."

Allison defines the sliding hernia by a shortened esophagus that draws up the GEJ into the thorax through the diaphragmatic hiatus. A peritoneal sac herniates into the thorax, surrounding the stomach. The esophagus is lined with squamous epithelium and the gastric mucosa consists of rugated columnar epithelium. According to Allison, the sliding hiatal hernia has caused the whole mechanism that prevents reflux to break down and is the cause of peptic ulceration of the distal esophagus.

It is obvious that I am an ardent admirer of Allison's work. His descriptions and definition of hiatal hernia, the mechanism of the cardia at a time when the LES was not defined, the definition of reflux esophagitis, and CLE are perfect and so radically different than what was believed universally then. He is responsible for eradicating widely held false beliefs such as congenitally short esophagus, Barrett's intrathoracic tubular stomach, that the GEJ was the SCJ, and congenital heterotopic gastric mucosa in the esophagus. His name should be remembered as the person in this field who drove the understanding of GERD in the correct direction in an unerringly accurate manner way ahead of his time. His words ring true to this day.

Sadly, Allison's greatness is inadequately honored. If I could suggest something with no hope of acceptance in this time where eponyms are not in favor, I would term visible CLE without IM as the Allison esophagus and limit BE to cardiac epithelium with IM.

5.5.2 Endoscopic and Radiologic Diagnosis of Sliding Hiatal Hernia

Sliding hiatal hernia is presently believed to be one of the most common anatomic abnormalities seen at endoscopy in a patient with GERD. It can also be seen, more rarely, in a person without symptoms of GERD. Allison, if he were alive today, may even say: "Gastroesophageal reflux disease and sliding hiatal hernia are almost like the Beaver and the Butcher: 'You could never meet either alone!'"

Whether the diagnosis of sliding hiatal hernia made by today's definitions is accurate is highly questionable. I believe there is a serious overdiagnosis of this entity. It is even possible that a sliding hiatal hernia is relatively uncommon and only accompanies severe GERD where the LES has undergone extreme damage.

This is not an unreasonable statement. It is more unreasonable to defend the present diagnosis of sliding hiatal hernia while using a definition of the GEJ that is incorrect and places the proximal limit of the stomach as much as 2 cm above the true GEJ. If one does not know where the stomach begins, how can one diagnose a sliding hiatal hernia? I would say: "Show me a herniated peritoneal sac in the thorax and I will believe that the patient has a sliding hiatal hernia."

Similar reservations about the diagnosis of esophageal shortening and sliding hiatal hernia have been expressed by other authors. Korn et al.[22] in a collaborative study in Dr. Attila Csendes' excellent unit in Chile and Hubert Stein's in Munich, studied the region of the lower esophagus at surgery in four groups of patients with reflux esophagitis (n=45), SSBE, LSBE (n=17 and 15 respectively), and controls without GERD (n=25). This study is reviewed previously in this chapter. After their careful study of the external anatomy of this region at surgery, they conclude: "A hiatal hernia, when present, is easily reduced. We have not found a true short esophagus in any of our patients, and in all of our patients the LES and esophagogastric junction were located in the abdomen." The study of this region at surgery provides two landmarks that are not seen at endoscopy: the peritoneal reflection, which accurately marks the true GEJ, and the peritoneal lining, which allows determination of whether there is a herniated peritoneal sac in the thorax.

The endoscopic diagnosis of sliding hiatal hernia appears superficially simple. If the endoscopic GEJ is above the diaphragmatic pinch, there is a sliding hiatal hernia. There is a recognized problem because of the vertical movement of the esophagus with respiration. Care is recommended and it is usual to diagnose a hiatal hernia when the endoscopic GEJ is >1 or 2 cm above the diaphragm to allow for this.

More experienced gastroenterologists recognize that the accurate diagnosis of hiatal hernia is not so simple. This is apparent in the study by Kahrilas et al.[21] that evaluated the influence of hiatal hernia on LES dysfunction (see Chapter 9). In this study there were three groups of patients, one with no symptomatic GERD and two with symptomatic reflux disease and absent BE at endoscopy.

Definition of the presence of a hiatal hernia included the placement of a steel clip at the SCJ at endoscopy. Fluoroscopy was used to define the relationship of the clip at the SCJ to the center of the diaphragmatic hiatus. The requirement for the diagnosis of hiatal hernia was that the SCJ clip was at least 1 cm proximal to the diaphragmatic hiatus at radiology. Eight patients with symptomatic GERD had a hiatal hernia by these criteria, seven did not, and the eight normal subjects in the study all had the SCJ clip at or distal to the hiatus.

The assumption is made that, in these people who have no visible CLE at endoscopy, i.e., when the proximal limit of rugal folds reach the SCJ, the SCJ is the true GEJ.

I have shown that there is a potentially serious error in this definition of the GEJ at endoscopy because the dilated distal esophagus is ignored. The error increases with increasing severity of GERD because the more severe the GERD, the greater is the LES damage, and longer the dilated distal esophagus.

One can get an idea of the length of the dilated distal esophagus distal to the SCJ in the three groups in the study by Kahrilas et al.[21] by the LES length (the study measured overall length, not abdominal length). The baseline mean LES length was approximately 3.7, 2.9, and 2.2 cm in the "normal," nonhernia-GERD, and hernia-GERD groups, respectively. This indicates that the patients with hiatal hernia had an LES that was shorter by 1.5 cm compared with the normal group; that is, the dilated distal esophagus distal to the SCJ was at least 1.5 cm distal. The true GEJ in the hernia group is therefore likely to be at least 1.5 cm distal to the SCJ. The 1 cm structure between the clips (at the SCJ) and diaphragm at radiology is the dilated distal esophagus, not the stomach and therefore not a sliding hernia. When the ignored dilated distal esophagus is included, the entire stomach would have been distal to the diaphragm. Even with the stringent criteria used in this study for the diagnosis of hiatal hernia, there is a strong probability of overdiagnosis of hiatal hernia. Endoscopy and radiology cannot demonstrate a peritoneal sac in the thorax.

In the absence of any histologic studies of sliding hiatal hernia, it is impossible to know whether the diagnosis is correct. It is only when the true GEJ (the proximal limit of gastric oxyntic epithelium) is defined and its relationship to the diaphragmatic hiatus determined that an accurate diagnosis of a sliding hiatal hernia is possible. Otherwise, it is possible that what is being called a sliding hiatal hernia is actually the dilated distal esophagus or even the thoracic esophagus that has dilated when there is severe LES damage below it. With total sphincter loss, abdominal pressure can theoretically be transmitted into the thorax and pressurize the thoracic esophagus, causing it to dilate. If this happens it can create the endoscopic and radiologic illusion of a hiatal hernia.

5.5.3 Manometric Diagnosis of Sliding Hiatal Hernia

A similar error exists in the manometric diagnosis of sliding hiatal hernia, which is made when the distal limit of the LES is at a level more proximal than the diaphragm (Fig. 10.16). Surely, this is not accurate. The LES is known to be shortened in GERD, sometimes >2 cm in patients with severe GERD. If this is true, the esophagus that has lost sphincter pressure is *distal* to the manometric level of distal end of the LES. Even a >2 cm hiatal hernia by manometry may have the true GEJ below the diaphragm resulting in an incorrect diagnosis.

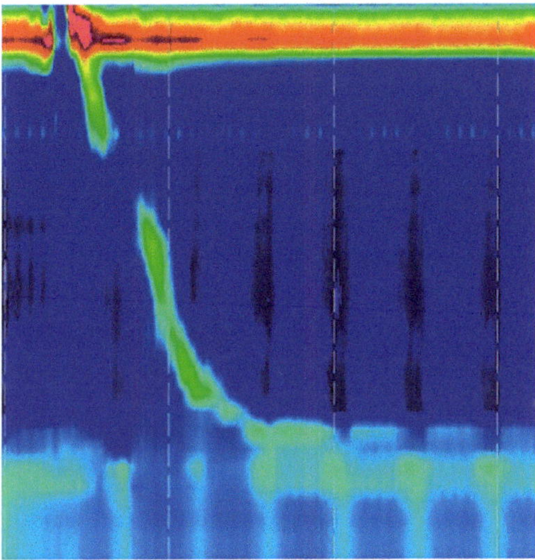

FIGURE 10.16 High-resolution manometry showing a markedly abnormal short lower esophageal sphincter (LES) with low pressure. There is a small hiatus hernia and weak motility in the esophageal body. The calculated total LES length is 1.4 cm, abdominal LES length is 0 cm, and resting LES pressure is 7 mmHg. *Study and interpretation including calculated lengths is by Chris Dengler, MD, Legato Medical Inc.*

5.5.4 Pathologic Diagnosis of a Sliding Hiatal Hernia

Accurate diagnosis of a sliding hiatal hernia is possible under all circumstances if the true anatomy and histology of the GERD-damaged esophagus is understood. Patients with the proximal limit of rugal folds located above the diaphragmatic hiatus are the candidates for this diagnostic protocol.

At endoscopy, the following anatomic points can be defined: (1) The endoscopic GEJ (the proximal limit of rugal folds). (2) The true GEJ, which is the level of the endoscopic GEJ + the distance from the endoscopic GEJ to the proximal limit of gastric oxyntic epithelium. Determination of the true GEJ requires biopsies distal to the endoscopic GEJ at 0.5–1 cm intervals until gastric oxyntic epithelium is reached. (3) The diaphragmatic hiatus, defined by the diaphragmatic pinch.

Using this diagnostic protocol permits a diagnosis of sliding hernia and short esophagus with accuracy using the following equation:

Level of endoscopic GEJ + dilated distal esophagus (measured length of cardiac epithelium from the endoscopic GEJ to the proximal limit of gastric oxyntic epithelium by histology) = Level of true GEJ.

If the level of the true GEJ as determined by this equation is above the diaphragmatic hiatus, the patient has a sliding hiatal hernia with a shortened esophagus. If the level of the true GEJ is below the diaphragmatic hiatus, the patient has no sliding hiatal hernia; the true GEJ and the entire stomach are in the abdomen.

We have taken biopsies from within an apparent endoscopic sliding hernia at unmeasured points distal to the proximal limit of rugal folds. These biopsies frequently show cardiac epithelium (with and without parietal and/or goblet cells) to a variable distance, proving this is dilated esophagus. However, we have not completed the biopsies until we reached the proximal limit of gastric oxyntic epithelium to make the required measurements. As such, I do not have the material to answer this question at this time.

The error in the diagnosis of sliding hiatal hernia is not of significant practical consequence. The presence of a sliding hiatal hernia is generally regarded as increasing the likelihood that GERD is severe. If it is being mistaken either partly or wholly for a dilated distal esophagus, the impact is the same. A large dilated distal esophagus correlates with severe LES damage. As such, it, like the hiatal hernia it is mistaken for, is indicative of severe GERD.

5.6 Enlargement of the Diaphragmatic Hiatus

I have listened to many surgeons describing enlargement of the diaphragmatic hiatus in patients with severe GERD. From what I can gather, this may influence the surgical technique of Nissen fundoplication by requiring "tightening" of the diaphragmatic hiatus by surgically approximating the crura. I have heard that this decreases the risk of the fundoplication herniating after surgery into the thorax.

I have also heard surgeons say they routinely measure the diaphragmatic hiatus to convert the subjective impression of an enlarged diaphragmatic hiatus to one that is more objective. Koch et al.[25] described the following method originally described by Granderath[29] in 2007 that provides a complicated formula with complete illustrations: "For measurement of the hiatus, the right and left crura and the crural commissure were dissected exactly. Then a ruler was introduced intra-abdominally. First, the length of the crus was measured in centimeters, beginning at the crural commissure and proceeding up to the edge where the pars flaccida begins (radius R). Afterward, the circuit between both crural edges were measured. Using these two values the hiatal surface area could be calculated by a formula." The complex method makes it no surprise that the hiatal surface area measurement is done so rarely.

Batirel et al.[30] did a prospective study of 28 patients (14 males; mean age 43.6 years) who underwent laparoscopic fundoplication for GERD. All patients underwent preoperative esophageal manometry and endoscopy. 22 patients had 24-h pH monitoring. 25 patients had hiatal hernia of varying size. 15 patients had erosive esophagitis of varying grades.

A standardized photographic image of the hiatus following hiatal dissection was used to evaluate hiatal size. The circumferential margin of the esophageal hiatus was drawn by the surgeon using a graphics program. The surface area of the hiatus was calculated by an author blind to surgical findings using a graphics program. The calculated hiatal area was divided by BMI to produce a second value, which was termed "hiatal index."

There was a significant negative relationship between hiatal area and mean LES pressure ($P=.005$). There was also a significant positive correlation between DeMeester scores on 24-h pH monitoring and hiatal area ($P=.035$) and hiatal index ($P=.01$). There was no significant correlation between BMI and hiatal area or index. There was a significant negative correlation between total LES length and hiatal area ($P=.013$) and hiatal index ($P=.038$).

This study demonstrates the difficulty in producing high-quality evidence regarding changes in the structure of the diaphragm in normal and GERD patients at surgery. It is difficult to find normal controls. Ethical issues restrict dissection of the region of the diaphragmatic hiatus required for photography and measurement in control patients without GERD not undergoing fundoplication. Many patients with GERD who undergo fundoplication have severe abnormalities including dilatation of the distal esophagus and hiatal hernia. Mild GERD does not commonly undergo surgical correction. The data in this paper only suggest that increase in hiatal area is associated with more severe GERD and more severe sphincter abnormality. It lends credence to the concept that the dilatation of the distal esophagus associated with LES damage results in secondary changes in the diaphragmatic hiatus. This would tend to aggravate LES failure by decreasing any effect that the diaphragm may have in preventing reflux.

5.7 Severe Changes in the Anatomy of the Abdominal Esophagus

The changes in the abdominal esophagus that are associated with LES damage are not easily defined. At surgery, the abdominal esophagus is not easily seen because it is separated from the peritoneal cavity by the conical collection of fibroadipose tissue between the esophagus and the peritoneum reflecting from the undersurface of the diaphragm to the GEJ.

At endoscopy, the area is extremely confusing because of the failure to accurately define the GEJ. I do not perform endoscopy, but some of the pictures that I have seen of this region in retroflex view are difficult to interpret in terms of the anatomy (Fig. 10.17). Similarly, in resected specimens, the area shows extreme distortion, making diagnosis of the exact location of the end of the esophagus impossible (Fig. 10.18).

These confusing situations can only be resolved when histology is used to accurately identify the location of the true GEJ by the proximal limit of gastric oxyntic epithelium.

6. CLINICAL SYNDROMES ASSOCIATED WITH THE DILATED DISTAL ESOPHAGUS

The dilated distal esophagus is a pathologic entity that is the result of LES damage. It is absent only in the normal state when there is neither LES damage nor metaplastic cardiac epithelium in the squamooxyntic gap. Its length increases with increasing LES damage. As a result, the length of the dilated distal esophagus correlates best with the occurrence of sphincter failure and the amount of reflux into the thoracic esophagus. In turn, the amount of reflux correlates best with the severity of cellular changes in the thoracic esophagus (Fig. 10.19).

6.1 Compensated Phase of Lower Esophageal Sphincter Damage

In the earliest stage of LES damage, the LES retains its competence and there is no reflux of gastric contents into the esophagus. This is the phase where LES damage is within its reserve capacity and is compensated by the fact that there is sufficient residual LES to prevent failure (tLESR). Theoretically, if the initial length of the abdominal LES in a given individual is

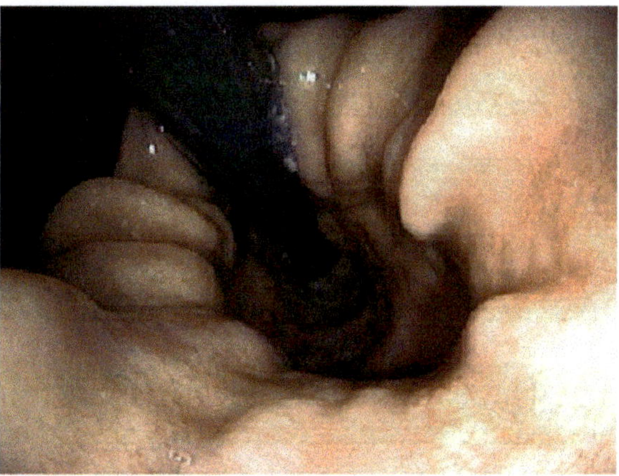

FIGURE 10.17 The contributor, Dr. Martin Riegler's description: "Retrograde endoscopic vision towards the Hill IV valve." This shows severe distortion of the anatomy distal to the diaphragm. The squamocolumnar junction (SCJ) is seen at the point of entry of the endoscope distal to the end of the tube. There is a large dilated area between the SCJ and the proximal limit of rugal folds that I cannot explain. Part of this is a dilated distal esophagus without rugal folds. The histologic localization of the true GEJ (proximal limit of gastric oxyntic epithelium) is the only way to correctly understand what has happened. *Photograph courtesy of Dr. Martin Riegler, Reflux Medical, Vienna, Austria.*

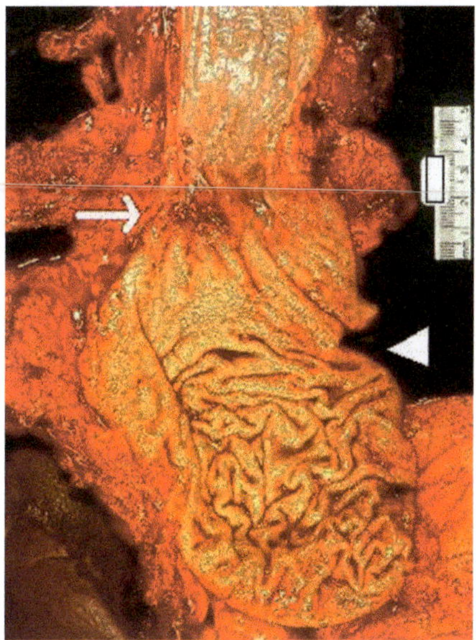

FIGURE 10.18 A resected esophagogastrectomy specimen showing severely distorted anatomy. There is an ulcerated adenocarcinoma at the end of the tubular esophagus. The proximal limit of well-defined rugal folds appears to be several centimeters distal to this. The dilated columnar-lined segment between the tumor and the proximal limit of rugal folds may represent a dilated distal esophagus, but histology is essential to define the anatomy accurately.

35 mm, the length at which failure occurs is 10 mm, and dynamic shortening during a heavy meal is 10 mm, the abdominal LES can shorten by 15 mm before it causes reflux. This is LES damage that has progressed to >40% of its initial length.

During this compensated phase of LES damage, the patient will maintain LES competence and there will be no reflux. The asymptomatic volunteers in Robertson et al. and Zaninotto et al.[4] will be among these patients. Many patients in these studies had zero reflux in the pH test. They have no symptoms suggesting GERD.

It is unknown whether the intermittent squamous epithelial damage that leads to columnar metaplasia during this phase results in transient vague symptoms such as epigastric pain and dyspepsia. These are so common and nonspecific that they will likely be ignored. The only abnormality these people will manifest is the dilated distal esophagus. Measuring this by the new test will permit stratification of these people by the length of the dilated distal esophagus (=LES damage). The test can be performed at any time in adult life; there is no requirement that symptoms of GERD be present.

The new algorithm to predict progression of LES damage into the future will potentially identify which of these people are at risk of developing GERD and progressing to severe GERD. The new test is therefore a powerful tool for evaluating

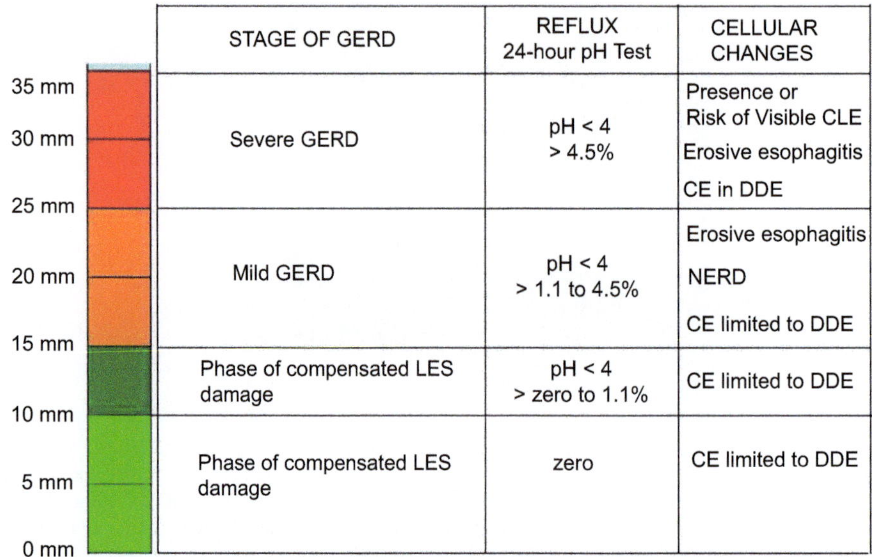

STAGE OF GERD	REFLUX 24-hour pH Test	CELLULAR CHANGES
Severe GERD	pH < 4 > 4.5%	Presence or Risk of Visible CLE; Erosive esophagitis; CE in DDE
Mild GERD	pH < 4 > 1.1 to 4.5%	Erosive esophagitis; NERD; CE limited to DDE
Phase of compensated LES damage	pH < 4 > zero to 1.1%	CE limited to DDE
Phase of compensated LES damage	zero	CE limited to DDE

FIGURE 10.19 Impact of lower esophageal sphincter (LES) damage showing correlation of LES damage with the phase of gastroesophageal reflux disease (GERD) as defined in the new method, severity of reflux (by a pH test), and probable cellular changes in the thoracic and dilated distal (abdominal) esophagus (DDE). *CE*, cardiac epithelium; *CLE*, columnar-lined esophagus; *NERD*, nonerosive reflux disease.

anyone in the population in terms of their future risk of progression to clinical GERD and severe GERD with BE and adenocarcinoma.

6.2 Symptomatic Gastroesophageal Reflux Disease Without Satisfying Montreal Definition

Further progression of LES damage (with increasing length of dilated distal esophagus) results in intermittent LES failure, causing reflux to an extent that is not adequate (i.e., pH test likely to be normal if it is done) to produce troublesome symptoms that satisfy the Montreal definition of GERD.[31]

These patients have mild episodic heartburn, usually after heavy meals, that does not trouble the person. Or they may have atypical symptoms such as epigastric pain and dyspepsia. These are likely to lead to a visit to the pharmacy and purchase of acid neutralizers for quick relief that is likely to be effective. Many will not need to even consult a physician.

In the absence of troublesome symptoms, they will not be diagnosed as GERD. These people have a dilated distal esophagus of varying lengths (15–20 mm of LES damage) (Fig. 10.19). Without endoscopy and biopsy and recognition of the concept of the dilated distal esophagus, this entity will not be recognized.

6.3 Clinical Gastroesophageal Reflux Disease Without Visible Columnar-Lined Esophagus

In the next phase of progression of LES damage, there is adequate reflux into the thoracic esophagus to produce an abnormal 24-h pH test and/or troublesome symptoms that fall within the Montreal definition of GERD.[31] The point of LES damage at which this happens is yet unknown but it is likely to be at the point where LES damage is 20–25 mm with a residual abdominal LES of 10–15 mm (Fig. 10.19). Further dynamic shortening during the postprandial period causes it to reach the critical <10 mm length where LES failure occurs with significant frequency.

This level of LES damage usually correlates with the appearance of typical symptoms of GERD, but this is not constant. Some patients with no symptoms have an abnormal pH test. This explains why people without symptoms of GERD can progress to BE and adenocarcinoma.

Cellular changes may be absent or show erosive esophagitis of increasing grade. The patient with symptoms of GERD with an abnormal pH test who is endoscopically normal falls within the definition of nonerosive reflux disease (NERD). This accounts for ~40% of patients with troublesome heartburn.

With further progression, changes of erosive esophagitis occur in the thoracic esophagus progressing from grade A to D in the LA classification that is based on the size and confluence of erosions. Erosive esophagitis is a reversible pathologic change. PPI therapy is effective in healing erosive esophagitis even when severe and preventing recurrence of erosions with maintenance therapy. The presence of severe erosive esophagitis had a 19.7% risk of developing visible CLE within 5 years with routine medical therapy.[32]

6.4 Visible Columnar-Lined Esophagus With or Without Symptomatic Gastroesophageal Reflux Disease

In the final phase of progression of LES damage, the dilated distal esophagus exceeds 25 mm and the residual baseline LES falls below the critical <10 mm level (Fig. 10.19). At this point LES failure becomes frequent, causing sufficient reflux into the thoracic esophagus to cause columnar metaplasia in the thoracic esophagus. With progression to complete destruction of the abdominal LES when the LES pressure in the entire 35 mm is lost (Figs. 10.15 and 10.16), the stage is set for the occurrence of a sliding hiatal hernia.

Unlike the very slow increase in the length of the dilated distal esophagus, the metaplastic change in the thoracic esophagus occurs much more rapidly. This is due to the high-volume intermittent reflux exploding into the thoracic esophagus as a result of LES failure. A large surface area of the squamous epithelium of the thoracic esophagus is exposed to gastric contents. This entire surface area is simultaneously affected and can undergo columnar metaplasia over a significant length in a short period.

In the Pro-GERD study,[32] 9.7% of 2721 patients who did not have visible CLE at the index endoscopy progressed to develop visible CLE (endoscopic BE) within 5 years. The occurrence of visible CLE correlated with male sex, the severity of erosive esophagitis at the index endoscopy, and regular use of PPIs. The maximum length of visible CLE was 5.5 cm. This is a much more rapid rate of columnar metaplasia than was seen in the generation of the dilated distal esophagus, where increasing length of cardiac epithelium was measured in millimeters per decade.

The reason for this difference in the rate of columnar metaplasia becomes clear when one considers the pathogenesis of the dilated distal esophagus and visible CLE. The dilated distal esophagus results from intermittent gastric overdistension that exposes a small area of squamous epithelium for a short time to gastric juice. This occurs in a patient with a competent LES. There is no transsphincteric reflux into the thoracic esophagus. The squamous epithelial injury is relatively minimal and cardiac metaplasia is very slowly progressive. In contrast, visible CLE in the thoracic esophagus occurs when the LES fails so frequently as to cause high-volume reflux into the esophagus, exposing a large surface area of the squamous epithelium.

The occurrence of visible CLE in the thoracic esophagus is a point of irreversibility. Short of ablation, significant lengths of columnar metaplasia in the esophagus cannot be reversed. It is the point of irreversibility for the patient with GERD. It is the point where progression to IM, dysplasia, and adenocarcinoma cannot be prevented be medical therapy.

Visible CLE is the point in the progression of GERD that I seek to prevent. If I succeed, esophageal cancer involving the thoracic esophagus will be prevented. There can be no adenocarcinoma in the thoracic esophagus without visible CLE.

I have provided theoretical correlates with highly specific lengths of LES damage to correlate with LES competence, reflux into the thoracic esophagus, and cellular changes. It must be understood that these are not based on data. There are no data. These numbers will change when the dilated distal esophagus is studied more extensively in the future.

7. HISTOLOGIC COMPOSITION OF THE DILATED DISTAL ESOPHAGUS

Based on the above clinical associations with the amount of LES damage, we can recognize two scenarios of the dilated distal esophagus: (1) The person who has a dilated distal esophagus as an isolated histologic finding without sufficient LES failure to produce visible CLE and (2) the patient whose dilated distal esophagus is associated with visible CLE. In general, the length of the dilated distal esophagus is greater in this second group.

7.1 The Isolated Dilated Distal Esophagus in the Patient With No Visible Columnar-Lined Esophagus

In this early stage of LES damage, the damaged sphincter functions within its reserve capacity. The baseline length of the sphincter is still adequate to prevent reflux even with the dynamic shortening that occurs during meals. It is presently misunderstood as representing "normal gastric cardia" and ignored, largely because of an incorrect endoscopic definition of the GEJ and belief that cardiac epithelium normally lines the proximal stomach.

The clinicopathologic features of an isolated dilated distal esophagus are as follows:

1. There is a histologic squamooxyntic gap between the SCJ and the proximal limit of gastric oxyntic epithelium in the endoscopically normal person. It can be recognized only by the presence of cardiac epithelium (with and without parietal and/or goblet cells). In patients without GERD, the length of the dilated distal esophagus is a maximum of 15 mm (Fig. 10.1). With increasing symptoms, the LES damage increases to 25 mm (Fig. 10.7).
2. Endoscopy shows no visible CLE. There is no visible separation of the proximal limit of rugal folds from the SCJ. Erosive esophagitis may be present. Presently, biopsy is not recommended in these patients, even when erosive esophagitis or typical GERD symptoms are present.

3. Symptoms vary from absent to severe; typical or atypical. Patients with symptoms have a length of dilated distal esophagus > 15 mm. This is the end of the compensated phase of LES damage. With >15 mm of LES damage, LES failure and reflux into the thoracic esophagus occurs, increasing progressively as LES damage increases to 35 mm.
4. Impedance and pH testing varies from zero to abnormal. The amount of acid exposure correlates with the length of the dilated distal esophagus.

The majority (approximately 60%–70%) of people in the population live their life with an isolated dilated distal esophagus without clinical GERD. They cannot be diagnosed by any present method. If the area is examined histologically, which is not usual because biopsies are not recommended in these people, the finding of cardiac epithelium will be regarded as normal proximal stomach by present histologic criteria.

Isolated dilated distal esophagus will be detected only by people who believe in the concept and take biopsies in this area in asymptomatic or GERD patients with normal endoscopy and at autopsy. To most of the medical world, the dilated distal esophagus does not exist. It is the proximal stomach. To such people, GERD begins when LES damage is close to 25 mm when symptoms become troublesome or visible CLE is seen at endoscopy.

Chandrasoma et al.[15,16] reported measured lengths of the dilated distal esophagus composed of cardiac and oxyntocardiac epithelium at autopsy (reviewed in Chapter 5). In the retrospective study of 61 persons where a single vertical section of the region was examined, the length of the dilated distal esophagus was 0 mm in 21, 0–5 mm in 25, 5–10 mm in 12, and 10–15 mm in 3. The maximum length was 15 mm. There were three patients with a history of "esophagitis" in the clinical summary; these three patients had a dilated distal esophagus measuring 3.7, 5.88, and 11 mm.

In the prospective part of the study, the entire circumference of the dilated distal esophagus was measured. Of the 18 patients, 10 (56%) had only oxyntocardiac epithelium without cardiac epithelium; in these persons the maximum length of oxyntocardiac epithelium was <2 mm. The eight persons who had both cardiac and oxyntocardiac epithelium had a dilated distal esophagus ranging in length from 0.475 to a maximum length 8.05 mm. In general, the dilated distal esophagus was shorter in patients with only oxyntocardiac epithelium than in patients with both cardiac and oxyntocardiac epithelium.

No IM was seen in any of the 90 persons in both retrospective and prospective parts of the study.

Pathology files of patients undergoing endoscopy and biopsy in the Foregut Surgery Unit of the Keck School of Medicine, USC, for any reason between 2004 and 2008 were reviewed in Chandrasoma et al.[1] We selected 714 patients satisfying the following criteria for inclusion: (1) there was no visible CLE, (2) retroflex biopsies were taken from within 1 cm distal to the end of the tubular esophagus from mucosa that had rugal folds (i.e., distal to the endoscopic GEJ); (3) squamous epithelium was present in a biopsy taken from the SCJ; (4) there was no dysplasia or adenocarcinoma; and (5) biopsies were taken from the distal antrum and/or body of stomach.

This study focuses on the epithelial composition of the dilated distal esophagus in patients whose squamooxyntic gap was limited to the dilated distal esophagus. The limitation to one set of biopsies taken within 1 cm distal to the end of the tubular esophagus meant that we could only divide these patients into those with <1 and 1 cm or more dilated distal esophagus.

The prevalence of the four columnar epithelial types in the retrograde biopsies in these patients is shown in Table 10.3.

TABLE 10.3 Prevalence of the Four Types of Columnar Epithelia in the Squamooxyntic Gap Limited to the Dilated Distal Esophagus in 714 Patients

Columnar Epithelial Types	Number of Patients	Gastric Pathology Present	Gastric Pathology Absent
GOM only	0	n/a	n/a
Cardiac epithelium with parietal cells (oxyntocardiac epithelium) only[a]	71 (9.9%)	See text	See text
Cardiac epithelium[b]	482 (67.5%)	See text	See text
Cardiac epithelium with goblet cells (intestinal metaplasia)[c]	161 (22.6%)	23 (14.3%)	138 (85.7%)
Total	714	71 (9.9%)	643 (0.1%)

GOM, gastric oxyntic epithelium.
[a]*Includes patients with gastric oxyntic epithelium.*
[b]*Includes patients with oxyntocardiac and gastric oxyntic epithelia.*
[c]*Includes patients with cardiac, oxyntocardiac, and gastric oxyntic epithelia.*

This study confirms the rarity of the normal state (i.e., a zero squamooxyntic gap) in a clinical endoscopy population. All patients had cardiac epithelium with and without parietal cells (oxyntocardiac epithelium) and/or goblet cells (IM) in their biopsies.

Of the 714 patients, 161 (22.6%) had IM. Of the 553 patients without IM, 71 (10%) had only oxyntocardiac epithelium; the remaining 482 (67.5%) had cardiac epithelium.

Of the 71 patients who had only oxyntocardiac epithelium, this was seen in the retroflex and antegrade biopsies in 38 patients, only in the retroflex biopsy in 20 patients (the antegrade biopsy had only squamous epithelium), and only in the antegrade biopsy in 13 patients (the retrograde biopsy had only gastric oxyntic epithelium).

Of the 482 patients with cardiac epithelium, 266 patients had only cardiac epithelium; the other 216 had oxyntocardiac epithelium. In 215 patients, cardiac epithelium was present in both retroflex and antegrade biopsies; in 146 patients, cardiac epithelium was seen only in the antegrade biopsy (the retroflex biopsy showed oxyntocardiac with or without gastric oxyntic epithelium); in 121 cardiac epithelium was seen in the retroflex biopsy only (76 of these had only squamous epithelium in the antegrade biopsy with the other 45 having squamous and oxyntocardiac epithelium). In all patients with cardiac epithelium where the histologic SCJ was present in one biopsy, cardiac epithelium transitioned directly to squamous epithelium.

Of the 161 patients with IM, 62 had it in both retroflex and antegrade biopsies, admixed in all cases with nonintestinalized cardiac epithelium. In 37 patients, IM was seen only in the antegrade biopsy, mixed with cardiac epithelium (the retroflex biopsy showed cardiac and oxyntocardiac epithelium in 30 patients and only oxyntocardiac epithelium in 7). In 62 patients, IM was seen only in the retroflex biopsy (the antegrade biopsy showed only squamous epithelium in all but one patient who had oxyntocardiac epithelium).

These findings indicate a uniform stratification of the four columnar epithelial cell types with only rare exceptions.[33] When all four columnar epithelial types are present, their distribution beginning at the SCJ is cardiac with IM, pure cardiac, oxyntocardiac, and gastric oxyntic epithelium. When IM is absent, it is cardiac, oxyntocardiac, and gastric oxyntic epithelium. In patients with oxyntocardiac epithelium only, this is interposed between squamous and gastric oxyntic epithelium.

The data from this study, when compared with the autopsy study,[16] show a difference in the prevalence of the three types of metaplastic columnar epithelium in the dilated distal esophagus. Clinical patients tend to have a much higher prevalence of IM than in the autopsy studies (22.6% vs. 0%) and a significantly lower percentage of persons with only oxyntocardiac epithelium (10% vs. 56%). In the autopsy study, the persons with cardiac and oxyntocardiac epithelium tended to have a longer dilated distal esophagus than those with oxyntocardiac epithelium alone.[16] This suggests strongly that the dilated distal esophagus in the clinical patients was longer than in the autopsy study.

These data also suggest that the prevalence of IM increases with increasing length of the dilated distal esophagus (Fig. 10.7).

Cardiac epithelium (with and without parietal and/or goblet cells) always showed inflammation and reactive epithelial changes associated with injury. We have shown previously that the Ki67 expression (indicative of proliferative activity) in the epithelial cells progressively increases from gastric oxyntic to oxyntocardiac to cardiac to intestinal metaplastic epithelia.[34]

Biopsies of the distal stomach (antrum and/or body) were normal (defined as absence of significant inflammation, *Helicobacter pylori* infection, atrophy, or IM) in 505/553 (91.3%) patients with nonintestinalized cardiac epithelium (with and without parietal cells). This was significantly higher than the 138/161 (85.7%) patients with IM ($P=<0.01$). The presence of cardiac epithelium (with and without parietal and/or goblet cells) in the dilated distal esophagus was therefore an isolated abnormality with normal gastric epithelium distally in 643/714 (90.1%) patients. Of the 714 patients, 71 (9.9%) patients who had pathologic lesions in the stomach, 46 had *H. pylori* gastritis, 30 had chronic atrophic gastritis with IM, and 4 had other pathologic lesions.

The finding that cardiac epithelium (with and without parietal and/or goblet cells) is isolated to the region immediately distal to the endoscopic GEJ without any abnormality in the stomach proper provides strong evidence that carditis is not a gastric disorder. Coupled with the data that these findings are associated with reflux, there is an essential certainty that this is an esophageal disorder associated with GERD. The conclusion that cardiac epithelium (with and without parietal and/or goblet cells) represents esophageal columnar metaplasia becomes inescapable. The fact that it is located distal to the endoscopic GEJ and that the area superficially looks more like stomach than esophagus does not make it stomach.

7.2 "Esophageal Markers" in the Dilated Distal Esophagus

There are well-recognized histologic markers of esophageal location that allow differentiation from gastric location. These include (1) absence of a peritoneal covering, (2) presence of submucosal glands, (3) islands of squamous epithelium within the columnar segment, (4) different layering of the muscular wall, (5) multilayered epithelium, and (6) immunohistochemical and molecular markers shared with esophageal CLE rather than gastric mucosa (see Chapter 5).

7.2.1 Submucosal Gland Ducts and Multilayered Epithelium

Submucosal glands develop during late fetal development after stratified squamous epithelium has replaced the fetal columnar epithelium of the esophagus[35] (see Chapter 3). In the human, stratified squamous epithelium is limited to the region proximal to the true GEJ; it never occurs in the stomach. As such, the presence of both submucosal glands (including the gland ducts that pass through the mucosa to open at the surface) and islands of squamous epithelium would be evidence of esophageal rather than gastric location of the dilated distal esophagus.

In Chapter 5, I reviewed two studies in resected specimens where full-thickness sections of the region distal to the end of the tubular esophagus and the proximal limit of rugal folds showed submucosal glands concordant with the extent of cardiac epithelium (with and without parietal and/or goblet cells).[12,36] Submucosal glands were never seen under gastric oxyntic epithelium.

Unfortunately, submucosal glands cannot be seen in mucosal biopsies. However, their ducts passing through the mucosa can be seen.[37] In a biopsy, the presence of submucosal gland ducts represents proof that the location of the biopsy is esophageal and not gastric. When the biopsy is taken distal to the endoscopic GEJ, it proves that the biopsy is taken from a dilated distal esophagus and not the proximal stomach.

Glickman et al. from the Harvard group, reported the presence of a multilayered epithelium as a feature of CLE.[38] Multilayered epithelium is likely a manifestation of the metaplastic process where squamous epithelium transforms into cardiac epithelium. It has no clinical significance but is useful as a marker of esophageal location of cardiac epithelium. Multilayered epithelium is not seen in pathologic states of gastric oxyntic epithelium.

The objective of the study from our group by Shi et al.[37]: "The ducts of esophageal glands and multilayered epithelium are two recognized histologic markers of esophageal mucosa. This study aims to document the prevalence and distribution of esophageal gland ducts and multilayered epithelium in the different glandular epithelial types in mucosal biopsies taken in the region of the gastroesophageal junction."

A total of 244 consecutive patients, irrespective of nature of symptoms, who had a retrograde biopsy taken at and distal to the proximal limit of rugal folds were selected. The biopsy set had a mean of 3.2 (range 2–4) specimens each resulting in 785 biopsy samples from this region. These biopsies were taken irrespective of the presence or absence of an endoscopic abnormality. When the patient had no endoscopic abnormality, the retroflex biopsies represented sampling of the normal SCJ. When the patient had a visible CLE, the biopsy represented sampling of the endoscopic GEJ at the proximal limit of rugal folds. In both groups, the biopsies under study were from the area of mucosa within 0.5 cm on either side of the endoscopic GEJ.

The glandular epithelia lining the surface were classified into gastric oxyntic, oxyntocardiac, cardiac, and cardiac with IM by criteria described in Chapter 4. Some biopsy samples contained more than one epithelial type; these were recorded separately.

Esophageal gland ducts were seen as rounded or elongated structures lined by a stratified basaloid squamous epithelium with or without a surface columnar cell layer that surrounds a lumen. Multilayered epithelium is seen in the surface or superficial foveolar region and consists of multiple layers of cells with basaloid squamous epithelium and overlying columnar cells. The epithelial type associated with gland ducts and multilayered epithelium was recorded.

The distribution of epithelial types in the 785 biopsies in the 244 patients is shown in Table 10.4. Esophageal gland ducts and multilayered epithelium were present only in cardiac epithelium (with and without parietal and/or goblet cells); they were not seen in gastric oxyntic epithelium. This is evidence that these biopsies were esophageal in location.

Gland ducts were present in 64/732 biopsies with glandular epithelium (8.7%; 13.6% when gastric oxyntic epithelium was excluded). Biopsies with cardiac epithelium with and without IM had a higher number of gland ducts (38/188; 20.2%) than oxyntocardiac epithelium (26/283; 9.2%). This is likely because oxyntocardiac epithelium is the most distal of the three types of cardiac epithelium in CLE and esophageal glands are less likely to be present in the most distal few millimeters of the esophagus.

Multilayered epithelium was also more commonly seen in cardiac epithelium with and without IM (43/188; 22.9%) than in oxyntocardiac epithelium (25/283; 8.8%). Multilayered epithelium has been suggested to represent an intermediate stage in the transformation of squamous to columnar epithelium.[38] If true, this would explain the higher frequency of multilayered epithelium in cardiac epithelium, which represents the metaplastic transitional zone compared to the more mature and stable oxyntocardiac epithelium.

The limitation of gland ducts and multilayered epithelium to cardiac epithelium (with and without parietal and/or goblet cells) is strong evidence that these epithelia are esophageal irrespective of whether the biopsy was taken at or distal to the endoscopic GEJ.

TABLE 10.4 Distribution of Epithelial Types and the Number of Biopsies With Esophageal Gland Ducts and Multilayered Epithelium in Each Epithelial Type

	Number of Biopsies	Esophageal Gland Ducts	Multilayered Epithelium
Gastric oxyntic epithelium	287	0	0
Oxyntocardiac epithelium	283	26 (9.2%)	25 (8.8%)
Cardiac epithelium	158	31 (19.6%)	34 (21.5%)
Cardiac epithelium with intestinal metaplasia	30	7 (23.3%)	9 (30%)
Squamous epithelium	53	n/a[b]	n/a
Total	785[a]	64 (8.2%)	68 (8.7%)
Cardiac epithelium (with and without parietal and/or goblet cells)	471	64 (13.6%)	68 (14.4%)

[a]The total number of biopsies is lower than the sum of biopsies containing different epithelial types because some biopsies contained more than one epithelial type.
[b]Biopsies of squamous epithelium rarely contained lamina propria to assess the presence of gland ducts.

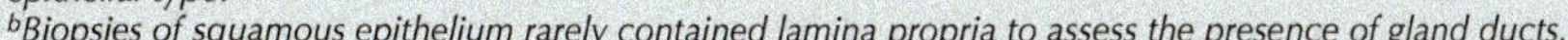

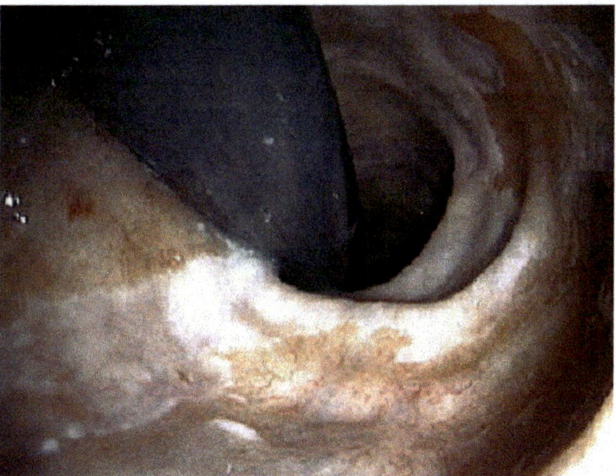

FIGURE 10.20 Retroflex view of the dilated distal esophagus showing irregular tongues of metaplastic columnar epithelium extending into the squamous epithelium, resulting in a serrated squamocolumnar junction distal to the endoscopic gastroesophageal junction (GEJ) but proximal to the true GEJ, which is distal to this area.

7.2.2 Squamous Epithelium

Fass et al.[39] in a study by two highly experienced gastroenterologists specialized in GERD over a 9-month period at the Tucson Veterans Affairs Medical Center, reported 16 patients with extension of squamous epithelium into the proximal stomach (Fig. 10.20). This was out of a total of 547 upper endoscopies done during that time, an incidence of squamous extension into the proximal stomach of 3%. The presence of squamous epithelium was determined by any extension of squamous epithelium beyond the endoscopic GEJ into the proximal stomach. Measurement of the length and width of the gastric squamous epithelium was assisted by a biopsy forceps with a 7 mm span.

Two biopsies were obtained from the squamous epithelium in the cardia during endoscopic retroflexion. Two additional biopsies were obtained from the cardia adjacent to the gastric squamous epithelium. The details of the study patients are shown in Table 10.5.

The indications for upper endoscopy were dysphagia (6), BE surveillance (5), failure of PPI therapy to control symptoms (4), and iron deficiency anemia (1). These 16 patients all had evidence of severe GERD; 6 had BE, all had hiatal hernias ranging in size from 3 to 8 cm, 4 had esophageal strictures, and 2 had erosive esophagitis. 12 patients reported heartburn, 8 reported regurgitation, and 9 had dysphagia.

TABLE 10.5 Endoscopic Findings in 16 Patients With Extension of Squamous Epithelium Into the Area Distal to the Endoscopic Gastroesophageal Junction

Age	Indications	Hiatal Hernia, cm	BE, cm	SE Length, cm	SE Width, cm	SE Tongue	Other Findings
70	Dysphagia	7.0	None	2.0	2.5	1	Stricture
65	BE surveillance	3.0	2.0	2.0	0.5	1	Schatski ring
65	BE surveillance	6.0	None	1.0	0.5	2	Stricture
50	Heartburn	3.0	1.0	1.5	2.0	1	
61	Heartburn	8.0	None	0.5	0.5	1	Stricture
65	BE surveillance	4.0	2.0	1.5	0.5	1	
46	Heartburn	3.0	None	3.0	1.5	1	
69	Dysphagia	6.0	None	3.0	1.0	1	
76	Dysphagia	4.0	None	3.0	1.5	2	
76	BE surveillance	3.0	4.0	0.5	2.5	1	
30	Heartburn	4.0	1.0	1.0	1.0	2[a]	
41	Dysphagia	3.0	None	0.5	0.5	2[a]	Stricture
65	BE surveillance	3.0	2.0	2.0	0.5	1	
58	Dysphagia	3.0	None	1.5	1.0	1	
61	Iron deficiency	3.0	None	1.0	0.5	1	
78	Dysphagia	4.0	None	0.5	0.5	1	

BE, Barrett esophagus; *SE*, squamous epithelium.
[a]*These patients had islands of squamous epithelium in addition to tongues.*

In their discussion, the authors' conviction that this is squamous epithelium in the proximal stomach is absolute. This reflects the conviction that most gastroenterologists have that the proximal limit of gastric folds is the true GEJ. These authors attempt to provide explanatory mechanisms for the finding: "Thus we hypothesize that an injurious process in the proximal stomach, within the hiatal hernia, may precipitate a regenerative downgrowth of esophageal squamous epithelium from the squamocolumnar junction into the cardia," and "Although acid reflux has been associated with mainly esophageal mucosal abnormalities, it has been recognized that mucosal damage may occur on the gastric side as well. Sentinel fold or polyp has been described at the proximal margin of a gastric fold within a hiatal hernia in patients with GERD. It is conceivable that acid may be one of the causes for injury of the proximal stomach."

The changes described here (including sentinel folds and polyps) are easy to understand when one recognizes that the esophagus extends distally beyond the endoscopic GEJ when it is lined by cardiac epithelium. In the patients in this study with severe GERD, characterized by hiatal hernia, BE, and strictures, the likelihood is that there is severe LES damage. As such, these patients will have a long dilated distal esophagus. The maximum length described in the literature is 28 mm. The theoretical limit is 35 mm (the initial length of the abdominal LES). The fact that the squamous epithelium was limited to 3 cm or less of the "proximal stomach" fits well with the concept that these changes are occurring in a dilated distal esophagus. What the authors are describing is not downward extension of esophageal squamous epithelium from the SCJ with intervening normal stomach. What they are describing is upward extension of tongues of cardiac epithelium (with and without parietal and/or goblet cells) into the esophagus from the true GEJ (the proximal limit of gastric oxyntic epithelium). The squamous epithelium the authors describe in the proximal stomach is actually residual squamous epithelium in the dilated distal esophagus (Fig. 10.20). The phenomenon is exactly how short segments of visible CLE are recognized above the endoscopic GEJ by serration of the SCJ, except that it is happening in the dilated distal esophagus at the true GEJ and therefore mistaken as a change in the proximal stomach.

The dilated distal esophagus has numerous pathologic changes that are seen in visible CLE in the thoracic esophagus. These include cardiac epithelium (with and without parietal and/or goblet cells) that is similar to the epithelium in visible

CLE morphologically, immunohistochemically, and at a molecular level; multilayered epithelium; submucosal glands and their ducts, and islands of residual squamous epithelium. All these changes are seen in the 3 cm distal to the endoscopic GEJ, a length that is close to the initial length of the abdominal segment of the LES whose destruction leads to the formation of the dilated distal esophagus.

7.3 Intestinal Metaplasia in the Dilated Distal Esophagus

In the study by Chandrasoma et al.[1] reviewed above, the prevalence of IM in the dilated distal esophagus was 161/714 patients (22.5%). This is similar to the study of Ringhofer et al.[40] that had a prevalence of IM of 8/35 (22.9%) in patients with GERD (see Chapter 5). This is a higher prevalence than in the asymptomatic volunteers reported by Robertson et al. (6.5%).[14]

The prevalence of IM in the dilated distal esophagus (i.e., in the retroflex biopsy taken distal to the endoscopic GEJ) is similar in the population with visible CLE as those with an isolated dilated distal esophagus.

7.3.1 Etiology and Diagnosis of Intestinal Metaplasia Immediately Distal to the Squamocolumnar Junction in Endoscopically Normal Patients

There are two completely different mechanisms by which IM develops immediately distal to the normal SCJ: (1) Progression to IM of GERD-induced cardiac metaplasia of squamous epithelium in the dilated distal esophagus, and (2) Chronic atrophic gastritis with IM involving the oxyntic epithelium of the proximal stomach. This is independent of GERD and can be the result of *H. pylori* gastritis or autoimmune gastritis. The two entities are completely different, but the histologic features may be similar, i.e., an inflamed epithelium consisting of mucous cells with IM. I have discussed this differential diagnosis in Chapter 4.

When a biopsy taken distal to the endoscopic GEJ shows IM, diagnosis is greatly facilitated by the availability of information regarding pathology in the esophagus above and the stomach below. In the esophagus, if there is visible CLE, biopsies will be taken according to the Seattle protocol or some variation thereof. Gastric pathology, in contrast, can be invisible to endoscopy. As such, the presence of biopsies from the distal body and antrum of the stomach is extremely important. Three diagnostic possibilities exist in the patient who has IM in a retroflex biopsy taken at or immediately distal to the endoscopic GEJ.

1. The patient has visible CLE with IM. The presence of IM in the biopsy distal to the endoscopic GEJ in these patients means that the IM is extending into the dilated distal esophagus. This is true irrespective of whether the patient has gastritis with or without IM. Patients with visible CLE *always* have a damaged LES and therefore a dilated distal esophagus;
2. In patients without visible CLE, the presence of IM immediately distal to the SCJ can represent IM in the dilated distal esophagus or extension of chronic atrophic pangastritis with IM to the most proximal stomach. The best way to differentiate between the two is to have a distal gastric biopsy. Our unit does this routinely. As such, we reported that 138/161 (85.7%) patients with IM in their retroflex biopsy had no gastric pathology whatsoever.[1] These patients certainly had isolated IM in the dilated distal esophagus.
3. Patients whose IM represents extension of chronic atrophic pangastritis into the proximal stomach. In these patients, the differentiation from coexistent IM in the dilated distal esophagus is more complicated but still possible in most cases (see Chapter 4).

In the study by Hirota et al.[41] from Walter Reed Army Medical Center, 889 patients were enrolled from January 1995 to September 1996. The following endoscopic landmarks were recorded: diaphragmatic hiatus, most proximal tip of the gastric fold, and SCJ. The authors use the term EGJ-SIM (specialized intestinal metaplasia at the EGJ), which is synonymous with IM in the columnar epithelium immediately distal to the SCJ. The diagnosis of EGJ-SIM was defined by the presence of a normal appearing SCJ, devoid of any tongues of pink CLE above the endoscopically defined EGJ, which was associated with SIM on antegrade biopsy specimens distal to the SCJ.

In their introduction, the authors are vehement that EGJ-SIM has nothing to do with reflux disease and should not be regarded as BE: "Were it not for the increasing incidence of adenocarcinoma of the esophagus and EGJ, Barrett's esophagus would have remained a curiosity resulting from the body's adaptation to severe reflux. However, when the SCJ is normal in appearance and SIM is found exclusively in the EGJ or in the cardia of the stomach, a descriptive term such as Barrett's esophagus, which implies an esophageal disorder, should not be used. More importantly, characterizing this as Barrett's would suggest a reflux-mediated disease…." One objective of the paper is to establish that EGJ-SIM is a disorder with different clinical characteristics than BE.

They write: "The increasing incidence of adenocarcinoma of the gastric cardia may be associated with this specific metaplasia in the EGJ or cardia." According to these highly renowned experts who author this paper, metaplastic intestinal epithelium is caused by something other than GERD. In our study of IM of this region, we showed that in 85.6% of patients, this pathology was isolated to this small area.[1] It is highly unlikely that any pathology other than GERD causes this change limited to a very small area distal to the endoscopic GEJ. The only reason why these authors are adamant is because of their absolute faith in the accuracy of the endoscopic definition of the GEJ. If the actual GEJ is distal to the endoscopic GEJ (as it is), this pathology is in the distal esophagus and caused by GERD.

Of the 889 patients, 151 had SIM; 40 had LSBE, 64 had SSBE, and 47 had EGJ-SIM. Selected clinical features of these patient groups are given in Table 10.6.

The main impact of this paper is that it established differences in the demographic groups in which BE and EGJ-SIM occurred. Both LSBE and SSBE were a disease of white males. In contrast, EGJ-SIM and the reference group without IM had a much smaller male preponderance and a much greater racial distribution. Patients with EGJ-SIM also tended to be older. There were less significant differences in tobacco and alcohol use in the groups.

H. pylori infection was more common in the EGJ-SIM group compared with the reference group (21% vs. 9%, $P = .05$). Two patients with EGJ-SIM had low-grade dysplasia and one had adenocarcinoma. Two patients with LSBE had dysplasia and two had adenocarcinoma. One SSBE patient had adenocarcinoma.

The prevalence of *H. pylori* in EGJ-SIM was 21%. In the 79% of patients without *H. pylori*, the occurrence of IM in cardiac epithelium distal to the endoscopic GEJ is very likely to be IM of the dilated distal esophagus and not proximal stomach.

7.3.2 Significance of Intestinal Metaplasia of the Dilated Distal Esophagus

Unfortunately, the recommendation that endoscopically normal patients with GERD should not have biopsies taken at and distal to the SCJ has left a huge vacuum in information relating to this condition. There are no long-term follow-up data relating to whether there is an increased risk of adenocarcinoma in patients with IM in the dilated distal esophagus.

There is one study with follow-up of patients with a diagnosis of IM in a biopsy from distal to the SCJ in endoscopically normal patients with GERD.[42] This is the Pro-GERD study in Europe (reviewed in Chapter 2). The biopsies were done at the initial endoscopy and the patients followed with repeat endoscopy at 2 and 5 years to evaluate disease progression.

Leodolter et al. introduce the study as follows: "Histological Barrett's esophagus (BE) is defined as the finding of unequivocal specialized intestinal metaplasia (SIM+) at histology in biopsies taken from esophageal z-line zones that have a normal endoscopic appearance. This is a frequent finding in patients presenting for routine upper gastrointestinal endoscopy as well as in patients with GERD, occurring in as many as 10-20% of biopsies. Although its clinical relevance is currently unknown, it has been speculated that it may be linked to the increasing incidence of cancers at the esophago-gastric junction (EGJ). Confirmed BE, or BE-SIM+ according to the Montreal classification, combines endoscopic suspicion of

TABLE 10.6 Selected Characteristics of the Patients in the Different Groups With Intestinal Metaplasia on Biopsy

	Total	SIM Negative	EGJ-SIM	SSBE	LSBE
No. of patients	889	738	47	64	40
Sex (M/F)		394/344	25/22	45/19	35/5
Age, years (median, range)		53 (19–85)	62 (34–86)	59 (21–86)	60 (37–88)
White %		485 (66%)	31 (66%)	55 (86%)	40 (100%)
Heartburn[a]	131 (15%)	113 (15%)	5 (11%)	9 (14%)	4 (10%)
GERD[a]	85 (10%)	71 (9.6%)	5 (11%)	8 (13%)	1 (2.5%)
BE screen/ surveillance[a]	56 (6.3%)	15 (2%)	0	14 (22%)	27 (66%)
Dysphagia[a]	147 (17%)	121 (16%)	9 (19%)	15 (23%)	2 (5%)
Abdominal pain[a]	163 (18%)	148 (20%)	12 (26%)	3 (4.7%)	0

BE, Barrett esophagus; *EGJ*, esophagogastric junction; *GERD*, gastroesophageal reflux disease; *LSBE*, long-segment Barrett esophagus; *SIM*, specialized intestinal metaplasia (synonymous with cardiac epithelium with intestinal metaplasia); *SSBE*, short-segment Barrett esophagus.
[a]*Symptom that was the primary indication for endoscopy.*

columnar epithelium with histological proof of SIM and is known to be a definite precancerous condition closely linked with GERD, although GERD without BE has also been associated with an increased risk of adenocarcinoma. Therefore, it is tempting to speculate that histological BE is the missing link between non-BE GERD and cancer development…. The aim of this sub-analysis was to evaluate progression of disease in the SIM+ sub-group of the Pro-GERD study."

There are some relatively radical viewpoints expressed in this introduction that fit perfectly with the new concept of the dilated distal esophagus. The presence of IM distal to the normal SCJ (i.e., distal to the endoscopic GEJ) is called "histologic BE" and suggested as being the lesion that is the basis of development of adenocarcinoma distal to the endoscopic GEJ in patients with "non-BE GERD." This is another way of saying that IM in cardiac epithelium distal to the endoscopic GEJ is IM of the dilated distal esophagus and is caused by GERD. In Chapter 12, I will provide evidence that IM in the dilated distal esophagus is the probable precursor lesion for adenocarcinoma of the dilated distal esophagus.

In the study, patients with GERD underwent an index endoscopy with biopsies taken 2 cm above the SCJ and from the endoscopically normal z-line with the aim of obtaining both squamous and columnar epithelium. Histologic diagnosis of SIM was made by specialized pathologists in the study. Patients were followed at 2 and 5 years for progression from SIM+ (histologic BE) to BE/SIM+ (endoscopically visible CLE with IM). Questionnaires that were filled by these patients during the follow-up period as part of the Pro-GERD study were analyzed to determine factors that were associated with progression.

Of the cohort of 6215 patients with GERD in the Pro-GERD study, the authors identified 171 patients with SIM+ ("histologic BE") at the baseline endoscopy. 38 of these patients had NERD while the remainder had erosive esophagitis of varying severity (the majority being LA grade A or B). When compared with 5540 non-SIM+ controls, these patients with SIM+ were significantly more likely to be male, higher age, current smokers, and positive for erosive esophagitis, with a higher Reflux Disease Questionnaire symptom score.

Of the 171 patients, 125 patients with SIM+ had a follow-up endoscopy at 2 years, and 68 patients had a follow-up endoscopy at 5 years. In all, 128 (75%) patients had studies at either 2 or 5 years; there was no follow-up in the other 43 patients.

At 2 years, 20 patients showed progression to BE/SIM+ (visible CLE with IM). All of these patients had erosive esophagitis at baseline (3 LA grade A, 10 grade B, 5 grade C, and 2 grade D). The length of visible CLE was 1 cm (9 patients), 2 cm (5 patients), 3 cm (2 patients), 5 cm (1 patient) and >5 cm (2 patients), and unknown length (1 patient).

After 2–5 years, a total of 33 patients (25.8%) had progressed from SIM+ to BE/SIM+. 31 of these had erosive esophagitis at the index endoscopy and 2 had NERD. 9 patients had endoscopies at both 2 and 5 years; all patients with BE/SIM+ at 2 years had it at 5 years as well.

By multivariate analysis, the characteristics at baseline significantly associated with progression to BE/SIM+ were smoking (P = .033), a >5 year history of GERD (P = .023), and the presence of LA grade C/D erosive esophagitis (P = .023). The presence of gastritis at baseline did not differ between those progressing to BE/SIM+ and not.

In their discussion, the authors state: "The results of this study indicate that SIM is a clinically relevant condition, as it is associated with endoscopically visible BE/SIM+ in approximately one-quarter of cases within 2–5 years." The authors are confident that the progression is real and not due to misclassification at the index endoscopy (i.e., BE/SIM+ being called SIM+). In addition to the training of the endoscopists before the study to specifically recognize BE, there was a much higher rate of progression of patients with SIM than in those without it (16% vs. 1.3% after 2 years and 25.8% vs. 2.4% after 5 years). This supports the accuracy of the index endoscopy in excluding visible CLE.

The authors analyzed the data for any association between progression to BE/SIM+ and the presence of *H. pylori* infection and distal gastritis. There was no association with *H. pylori* infection. There was a significantly lower severity of corpus gastritis in patients who progressed. This suggests that a higher grade of corpus gastritis, by reducing acid secretion in the stomach, may protect against the development of BE/SIM+.

They conclude their discussion: "Based on these findings and supporting data from the literature, the authors hypothesize that SIM+ at the cardia without overt BE could be the missing link explaining the risk of adenocarcinomas in the region of the EGJ in GERD patients without BE."

The results of this study are very interesting when viewed from the perspective that the region under discussion distal to the endoscopic GEJ is the dilated distal esophagus. If we assume that the occurrence of IM in the dilated distal esophagus increases with length of cardiac epithelium therein, as it does in visible CLE, the presence of SIM+ would be selecting those patients with the longest segments of dilated distal esophagus (Fig. 10.7).

This is equivalent to saying that the patients with SIM+ are those with the most severe LES damage at baseline. They are therefore those patients who have most exhausted the reserve capacity of the LES and are at the cusp of the <10 mm length of residual abdominal LES length where the sphincter fails and reflux into the thoracic esophagus becomes frequent. The high frequency of progression to BE/SIM+ in this study is likely to be the result of progression of LES damage from borderline to incompetent. This suggests that the actual measurement of abdominal LES damage by the new test would be a more direct predictor of imminent visible CLE than SIM+.

7.3.3 Relationship of Intestinal Metaplasia of the Dilated Distal Esophagus to Adenocarcinoma of the Dilated Distal Esophagus

In the absence of any significant long-term follow-up studies of patients with IM of the dilated distal esophagus, the question whether this change is a precursor lesion for adenocarcinoma in this region remains unanswered. The magnitude of risk, if any is present, is also unknown.

The dilated distal esophagus results from exposure of the esophageal squamous epithelium to gastric juice. The cardiac epithelium that results and its evolution into oxyntocardiac epithelium and IM is identical in its pathogenesis to visible CLE in the thoracic esophagus. It is logical, therefore, that the evolution of this epithelium to adenocarcinoma is also the same, i.e., by exposure of the cardiac epithelium to gastric juice leading sequentially to IM, dysplasia, and adenocarcinoma. I will show in Chapter 12 that IM in the dilated distal esophagus is the likely precursor lesion for adenocarcinoma of this region.

The only reason why this logic does not operate at present is that the dilated distal esophagus is mistaken for proximal stomach by the use of an incorrect definition of the GEJ. It is illogical that adenocarcinoma of the proximal stomach results from GERD. However, it is well accepted that adenocarcinoma of the "proximal stomach" is significantly associated with GERD. It has a similar epidemiology as adenocarcinoma of the thoracic esophagus. Its epidemiology is different than adenocarcinoma of the distal stomach.

However, there are differences in the type of exposure of the columnar epithelium in the dilated distal esophagus and that in the thoracic esophagus to gastric juice. These can influence the magnitude of risk of adenocarcinoma. I will discuss this further in Chapter 12.

8. RELATIONSHIP BETWEEN THE DILATED DISTAL ESOPHAGUS AND VISIBLE COLUMNAR-LINED ESOPHAGUS

The new understanding of the pathogenesis of GERD that I have presented removes all confusion regarding the association of the dilated distal esophagus and visible CLE. They are one and the same. They represent the full extent of the squamo-oxyntic gap. This begins at the true GEJ (proximal limit of gastric oxyntic epithelium) and progressively extends cephalad. In the first phase, cardiac epithelium in the gap increases slowly as the abdominal LES exhausts its reserve capacity. When this has happened and LES failure begins, the disease explodes into the thoracic esophagus. With increasing severity of LES damage, LES failure becomes more frequent, reflux increases quantitatively, ultimately resulting in visible CLE.

8.1 The Dilated Distal Esophagus in the Patient With Visible Columnar-Lined Esophagus

The dilated distal esophagus is an invariable precursor lesion to visible CLE in the thoracic esophagus (Fig. 10.6). Visible CLE results when severe reflux occurs into the thoracic esophagus in a patient with severe LES damage. This correlates with a long dilated distal esophagus lined by cardiac epithelium (with and without parietal and/or goblet cells). This may exceed 20 mm. In general, the dilated distal esophagus in patients with visible CLE is longer than in those without.

Chandrasoma et al.[36] showed that the length of the dilated distal esophagus was 10.3–20.5 mm in eight patients undergoing esophagectomy for GERD-related adenocarcinoma of the esophagus. This contrasted with lengths of 3.6 and 4.3 mm in two patients with squamous carcinoma (not GERD-related); this is similar to that seen in people without GERD.

The full extent of the CLE in any person varies in the following sequential manner (Fig. 10.6):

1. The normal person. This patient has no LES damage, zero cardiac epithelium (with and without parietal and/or goblet cells), i.e., zero CLE. This person's LES is fully competent. The patient has zero reflux.
2. The person with compensated LES damage. This patient has LES damage within the reserve capacity of the LES. Given an initial length of the abdominal LES of 35 mm and the point at which LES becomes incompetent, this is <15 mm of LES shortening. The patient has histologic cardiac epithelium limited to the dilated distal esophagus, which measures around 15 mm. The patient has no significant reflux.
3. The person with symptomatic GERD but no visible CLE. The LES has shortened further to around 10–20 mm (i.e., 15–25 mm of LES damage). Now, dynamic shortening of the LES can result in postprandial reflux. The pH test may become abnormal. There may be erosive esophagitis. However, the amount of reflux is inadequate to cause visible CLE in the thoracic esophagus. The microscopic cardiac epithelium is limited to the dilated distal esophagus and measures >20 mm.
   At this length of CLE in the dilated distal esophagus, IM develops in a small percentage of patients. These are the patients in the study of Leodolter et al.[42] who have a 25% incidence of BE (visible CLE + IM) within 5 years. In that study, SIM+ and erosive esophagitis were associated with progression to BE. Both of these are indicators of severe LES damage.

4. The patient with a visible CLE at endoscopy. This patient's full CLE includes the visible CLE+the histologic CLE in the dilated distal esophagus. This is the entire histologic squamooxyntic gap, extending from the SCJ to the true GEJ defined histologically by the proximal limit of gastric oxyntic epithelium.

There are few data on patients regarding the entire length of CLE. Because the GEJ is defined endoscopically as the proximal limit of rugal folds, biopsy of the visible CLE by the Seattle protocol usually fails to include the dilated distal esophagus. Even in the protocol used in our unit, only one biopsy is taken within 1 cm distal to the endoscopic GEJ. This frequently does not reach the true GEJ.

Available data where sampling of both the visible CLE and dilated distal esophagus has been done show that the distribution of the four columnar epithelial types is the same as with the isolated dilated distal esophagus. When all four columnar epithelia are present, the epithelial distribution from the SCJ distally is cardiac with IM, pure cardiac, oxyntocardiac, and gastric oxyntic epithelia. Gastric oxyntic epithelium is always distal to the endoscopic GEJ. The IM, when present, almost always begins at the SCJ and extends distally to a variable length of the squamooxyntic gap.[40]

8.2 Association of Adenocarcinoma of the Dilated Distal Esophagus With Barrett Esophagus

In the early 1990s, several centers reported on the presence of BE in patients with adenocarcinomas at the GEJ and "gastric cardia."

Clark et al.[43] showed that 13/31 (43%) of patients with cardiac adenocarcinomas were associated with BE. Cardiac tumors had shorter lengths of BE than those with adenocarcinomas in the thoracic esophagus (2.7 ± 1.8 vs. 7.4 ± 3.4 cm; $P = .01$). The finding suggests that approximately half of adenocarcinoma of the "cardia" arise in IM that extends down into the dilated distal esophagus from visible CLE.

Cameron et al.[44] studied resected esophagogastrectomy specimens with extensive histologic sampling around the tumor and in any visible CLE. BE was defined as the presence of IM above the GEJ (end of the tube); junctional carcinomas were defined as tumors centered 2 cm on either side of the GEJ. BE was found in 9 of 9 (100%) esophageal adenocarcinomas compared with 0 of 8 (0%) squamous carcinoma controls ($P < 0.001$). 10 of 24 (42%) junctional adenocarcinomas had a BE. In five specimens with junctional adenocarcinoma, the length of BE was <3 cm, and in five specimens it was ≥3 cm. The authors concluded that adenocarcinomas of the junctional region are associated with SSBE and LSBE.

Weston et al.[45] is an excellent study from 1997, a few years after SSBE was recognized. The study is aimed at providing data to understand the reasons for the increase in adenocarcinoma of the "gastric cardia" that paralleled the increase in esophageal adenocarcinoma since 1975.

The authors introduce their aims: "Barrett's esophagus, defined as the presence of specialized columnar epithelium anywhere within the tubular esophagus regardless of its extent, has been found in 42-64% of cases of adenocarcinoma of the cardia and esophagogastric junction.[43,44] The aim of the study was to prospectively evaluate the prevalence of cardia and non-cardia gastric intestinal metaplasia as well as the prevalence of cardia dysplasia in patients with Barrett's esophagus."

All consecutive patients undergoing elective upper endoscopy over a 15-month period found to have visible CLE with IM (divided into long and short segments based on whether the visible CLE was 2 cm or more or <2 cm) were selected. Patients who had visible CLE without IM in protocol four-quadrant biopsies were the control group.

Biopsies were taken from the cardia, which was defined as the 2 cm segment distal to the endoscopic GEJ (point where the tubular esophagus joined the stomach or, in patients with hiatal hernia, as the proximal limit of rugal folds). Four-quadrant cardia biopsies were taken 0.8–1.5 cm distal to the GEJ. Gastric biopsies were taken from the distal stomach to evaluate gastric IM. IM was diagnosed when definite goblet cells were seen in the hematoxylin and eosin section. Goblet cell density was graded, and grades of dysplasia and adenocarcinoma were recorded.

The final study groups were 59 patients with SSBE, 60 patients with LSBE, and 64 control patients who had visible CLE without IM. All three groups had a strong male predominance, but there were no age or sex differences between the three groups. BE was significantly more common in white patients than controls.

The cardia appeared endoscopically normal in all but one patient in whom the cardia appeared enlarged. Histologically, biopsies of the cardia (i.e., a location 8–15 mm distal to the endoscopic GEJ) showed typical cardiac mucosa in all study patients. There is no mention of parietal cells in this mucosa. The prevalence of IM, dysplasia and adenocarcinoma, and gastric IM in the three groups is shown in Table 10.7.

There was no statistical difference in the prevalence of noncardia gastric intestinal metaplasia among the three groups. There was a significantly greater prevalence of noncardia gastric intestinal metaplasia compared to cardia intestinal metaplasia for SSBE ($P = .02$), LSBE ($P = .02$), and controls ($P = .005$). The two LSBE patients who had cardia intestinal

TABLE 10.7 Demographic Features and Prevalence of Cardia and Gastric Intestinal Metaplasia and of Cardia and Barrett's Dysplasia and Adenocarcinoma in Patients With SSBE, LSBE, and Controls

	SSBE	LSBE	Control
Number	59	60	64
Age in years (mean, SD)	62.3 (13.3)	62.0 (12.1)	56.5 (13.5)
Gender (M/F)	58:1	60:0	62:2
Race (white:black:others)	53:6:0	60:0:0	55:8:1
Cardia intestinal metaplasia	6	2	0
Cardia dysplasia	1	0	0
Cardia adenocarcinoma	0	0	0
Barrett's LGD	5	14	n/a
Barrett's HGD	0	6	n/a
Barrett's adenocarcinoma	0	5	n/a
Gastric intestinal metaplasia	14	8	8

HGD, high-grade dysplasia; *LGD*, low-grade dysplasia; *LSBE*, long-segment Barrett esophagus; *SD*, standard deviation; *SSBE*, short-segment Barrett esophagus.

metaplasia did not have gastric intestinal metaplasia. Of the six SSBE patients who had cardia intestinal metaplasia two patients had gastric intestinal metaplasia.

In their discussion, the authors suggest that the known multiple-step model for progression of visible CLE to adenocarcinoma (IM→increasing dysplasia→adenocarcinoma) may also apply to cardia adenocarcinoma. They state: "Adenocarcinoma of the cardia may arise either de novo from gastric cardia epithelium through … cardia intestinal metaplasia, (or) from the downward extension (of) Barrett's intestinal metaplasia, or both. Our study does not provide the definitive answer. However, (we) speculate that intestinal metaplasia develops within the cardia and distal esophagus independently but as the result of a response to the same stimuli and risk factors. The fact that both adenocarcinoma of the esophagus and cardia have similar demographic patterns and have parallel, rising incidence rates provides indirect support for this."

They go on to suggest that IM of the cardia of Barrett's patients represents an early histologic marker for the development of cardiac adenocarcinoma.

The authors' analysis is perfectly accurate. However, their use of the false endoscopic GEJ results in a failure to recognize the obvious. The reason why adenocarcinoma of the esophagus and cardia are similar is because they are the same process, i.e., squamous epithelium→cardiac metaplasia→IM→increasing dysplasia→adenocarcinoma occurring in the esophagus as a result of exposure to gastric juice.

In adenocarcinoma of the thoracic esophagus, the process is caused by LES failure and severe reflux into the thoracic esophagus. In cardia adenocarcinoma, it is the identical cellular sequence occurring in the dilated distal esophagus where the first change is cardiac metaplasia of squamous epithelium resulting from exposure to gastric juice during gastric overdistension. The cardia in patients with BE is *always* the dilated distal esophagus and its epithelial evolution is the same as in the thoracic esophagus. It is not a change in native cardiac epithelium in the stomach because this does not exist. Recognition that the true GEJ is the proximal limit of gastric oxyntic epithelium and that cardiac epithelium is always a metaplastic esophageal epithelium results in absolute clarity.

Another important finding in the study of Weston et al.[45] relates to the control group. This control group consisted of patients with visible CLE who did not have IM. Visible CLE is a manifestation of abnormal reflux into the esophagus and is therefore *always* associated with a significantly damaged LES. As such, these patients *must* have a dilated distal esophagus lined by a significant length of cardiac epithelium. In this study, where the cardia biopsies were taken 8–15 mm distal to the endoscopic GEJ, *all* patients had typical cardiac epithelium. This indicates a segment of dilated distal esophagus of significant but unmeasured length. In contrast, in asymptomatic volunteers in the study by Robertson et al.[14] the length of cardiac epithelium was 1.75 and 2.5 mm in people without and with central obesity. If the same biopsies in the location of the study

TABLE 10.8 Relationship Between the Histology of the Dilated Distal Esophagus and the Thoracic Esophagus

	Histology Distal to Endoscopic GEJ	Histology Proximal to Endoscopic GEJ	Present Interpretation	Correct Interpretation
1. Normal endoscopy	Only GOM; No CE/IM	Squamous	Does not exist	Normal
2. Normal endoscopy	CE+ IM–	Squamous	Normal	DDE
3. Normal endoscopy	CE+ IM+	Squamous	IM of the gastric cardia[a]	IM of the DDE[a]
4. Visible CLE at endoscopy[b]	CE+ IM–	CE+ IM–	IM-visible CLE in the United States BE in the United Kingdom	IM-visible CLE + DDE
5. Visible CLE at endoscopy[b]	CE+ IM–	CE+ IM+	IM-visible CLE; IM in gastric cardia	IM-visible CLE + IM in DDE
6. Visible CLE at endoscopy[b]	CE+ IM+	CE+ IM–	BE	BE + DDE
7. Visible CLE at endoscopy[b]	CE+ IM+	CE+ IM+	BE+IM in gastric cardia	BE + IM in DDE

BE, Barrett esophagus; *CE*, cardiac epithelium; *CLE*, columnar-lined esophagus; *DDE*, dilated distal esophagus; *GEJ*, gastroesophageal junction; *GOM*, gastric oxyntic epithelium; *LES*, lower esophageal sphincter.
[a]*In Leodolter et al. 25% of these patients progressed to develop visible CLE with IM within 5 years.*
[b]*Not separated into short and long-segment Barrett esophagus.*

of Weston et al.[45] (i.e., 8–15 mm distal to the endoscopic GEJ) were taken in the volunteers of the study of Robertson et al.[14] the majority would have had no typical cardiac epithelium.

The data from Leodolter et al.[42] and Weston et al.[45] permit the development of a model for the relationship of the histology in the dilated distal esophagus in patients with visible CLE.

Table 10.8, which is based on the data in the two studies, is best interpreted in the following manner. In the normal person, there is no cardiac epithelium; the entire esophagus to the GEJ is lined by squamous epithelium and the entire proximal stomach distal to the GEJ is lined by gastric oxyntic epithelium (group 1 in the table).

Exposure of the squamous epithelium to gastric juice results in metaplasia to cardiac epithelium (group 2 in the table); this is associated with LES damage and the dilated distal esophagus, which develops rugal folds. The endoscopic GEJ separates from the true GEJ by the length of the dilated distal esophagus (=length of cardiac epithelium with and without parietal cells).

As LES damage progressively increases, the dilated distal esophagus elongates. At this point, one of two things can occur:

1. IM develops in the dilated distal esophagus (group 3 in Table 10.8) before visible CLE develops. In Leodolter et al.[42] 25% of these patients progress to develop visible CLE with IM within 5 years (group 7 in Table 10.8).
2. The LES damage precipitates LES failure and reflux into the thoracic esophagus resulting in visible CLE of varying lengths before IM develops in the dilated distal esophagus. In some of these patients, both the visible CLE and the dilated distal esophagus are lined by cardiac epithelium with no IM (control group in Weston et al.[42] and group 4 in the table). This is a controversial group at present for no real reason. These patients have severe GERD with a long dilated distal esophagus indicative of severe LES damage. They just do not have the milieu to induce IM in cardiac epithelium. In the United States, these patients are largely ignored because of a failure to understand the true meaning of the pathology. In the United Kingdom, they fall within the definition of BE. This is also not correct because in the absence of IM, there is no cancer risk and, unlike BE, the lesion is not yet premalignant.

Patients in group 4 of the table may progress to develop IM. When this happens, IM begins in the most proximal part of the visible CLE segment adjacent to the SCJ. This results in a patient who has IM limited to the thoracic esophagus (group 6 in the table).

Further evolution of the CLE can cause the progressive intestinalization of the cardiac epithelium distally in the visible CLE and then into the dilated distal esophagus. This usually occurs in a contiguous manner without skip areas. Evidence for this process is the progressive decline in the density of goblet cells from proximal to distal that is usual.[33]

8.3 Adenocarcinoma in the Dilated Distal Esophagus After Barrett Esophagus Ablation

The occurrence of IM in the dilated distal esophagus (groups 5 and 7 in the table) has a practical consequence when ablation is undertaken for BE. Radiofrequency ablation of BE generally ends distally at the endoscopic GEJ. In patients whose IM extends into the dilated distal esophagus, this area escapes ablation. This has resulted in the occurrence of adenocarcinoma in the dilated distal esophagus after ablation (reported as "adenocarcinoma of the gastric cardia following ablation of BE").

Sampliner et al.[46] in a 2006 study uses the following definitions of the GEJ and gastric cardia by endoscopy: "The EGJ was defined as the end of the tubular esophagus, the pinch at the end of the esophagus and/or the proximal margin of the gastric folds with minimal air insufflation. The gastric cardia is defined as the proximal 2 cm of the stomach distal to the EGJ."

The paper describes the early ablation of high-grade dysplasia in BE using photodynamic therapy. In 2006, ablation was an unproved alternative to esophagectomy for treating high-grade dysplasia. With the introduction of radiofrequency ablation and endoscopic mucosal resection, endoscopic treatment of early cancers has now become the standard of care. This study must therefore be taken in the spirit of the adventure that ablation represented then with these authors as pioneers.

The protocol for these patients were to take biopsies from any mucosal irregularities, followed by four-quadrant biopsies from the Barrett's epithelium (i.e., visible CLE) every 1–2 cm. They report three patients who had ablation of segments of BE measuring 5, 3, and 6 cm. Two patients had biopsies of the "gastric cardia" prior to ablation; both were "normal" without IM.

The three patients developed adenocarcinoma in the cardia at varying times after the ablation. In the first patient, the cardiac adenocarcinoma was found 3 years after ablation at 7 mm below the endoscopic GEJ; in the second patient, it was found after the final session of ablation at 2 mm into the gastric cardia; and in the third patient, it was found 1 year after ablation as a 1.5 cm mass in the cardia protruding from below the GEJ.

This report shows that using the incorrect endoscopic GEJ to demarcate the distal limit of ablation in the patient with LSBE results in the failure to address the microscopic CLE in the dilated distal esophagus. The occurrence of adenocarcinoma in this region proves that this area remains at risk (Fig. 10.21). Even if the initial biopsies show cardiac epithelium without IM, progression can occur with time from cardiac epithelium→IM→dysplasia→adenocarcinoma.

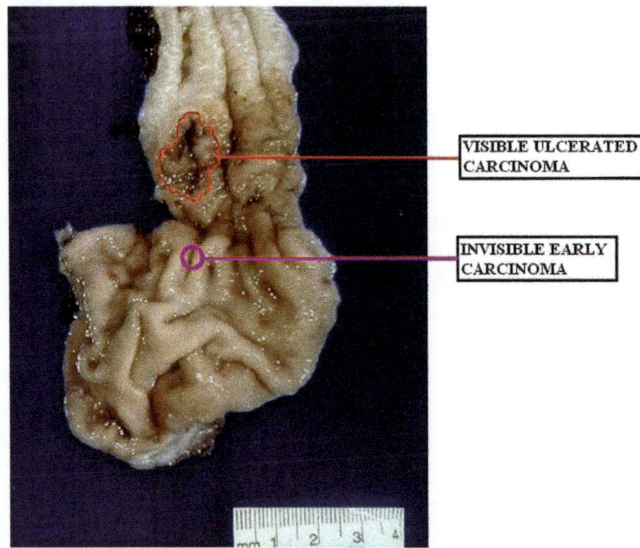

FIGURE 10.21 Patient with an adenocarcinoma of the thoracic esophagus arising in a long-segment of Barrett esophagus (*red circled area*). Histologic examination showed extension of intestinal metaplasia into the area distal to the end of the tubular esophagus into the area where there were mucosal rugal folds. A second early adenocarcinoma (*pink circle*) was present in this region which would have been regarded as "gastric cardia" because it is distal to the endoscopic GEJ. It is really a second tumor in the dilated distal esophagus.

This problem has been recognized by some authorities performing ablation for BE at present. They use the presence of IM at the distal edge of the neosquamous epithelium that results after ablation to indicate failure of complete ablation. Some extend initial radiofrequency ablation into the dilated distal esophagus while still believing it to be proximal stomach.

I will discuss adenocarcinoma of the dilated distal esophagus further in Chapter 12.

REFERENCES

1. Chandrasoma P, Wijetunge S, Ma Y, DeMeester S, Hagen J, DeMeester T. The dilated distal esophagus: a new entity that is the pathologic basis of early gastroesophageal reflux disease. *Am J Surg Pathol* 2011;**35**:1873–81.
2. Kahrilas P, Shaheen NJ, Vaezi MF. American gastroenterological association medical position statement on the management of gastroesophageal reflux disease. *Gastroenterology* 2008;**135**:1380–2.
3. Sharma P, McQuaid K, Dent J, Fennerty B, Sampliner R, Spechler S, Cameron A, Corley D, Falk G, Goldblum J, Hunter J, Jankowski J, Lundell L, Reid B, Shaheen N, Sonnenberg A, Wang K, Weinstein W. A critical review of the diagnosis and management of Barrett's esophagus: the AGA Chicago Workshop. *Gastroenterology* 2004;**127**:310–30.
4. Zaninotto G, DeMeester TR, Schwizer W, Johansson K-E, Cheng S-C. The lower esophageal sphincter in health and disease. *Am J Surg* 1988;**155**:104–11.
5. Der R, Tsao-Wei DD, DeMeester T, et al. Carditis: a manifestation of gastroesophageal reflux disease. *Am J Surg Pathol* 2001;**25**:245–52.
6. Derakhshan MH, Robertson EV, Lee YY, Harvey T, Ferner RK, Wirz AA, Orange C, Ballantyne SA, Hanvey SL, Going JJ, McColl KE. In healthy volunteers, immunohistochemistry supports squamous to columnar metaplasia as mechanism of expansion of cardia, aggravated by central obesity. *Gut* 2015;**64**:1705–14.
7. DeMeester SR. Cytokeratin and DAS-1 immunostaining reveal similarities among cardiac mucosa, CIM, and Barrett's esophagus. *Am J Gastroenterol* 2002;**97**:2514.
8. Lagergren J, Bergstrom R, Lindgren A, Nyren O. Symptomatic gastroesophageal reflux as a risk factor for esophageal adenocarcinoma. *N Engl J Med* 1999;**340**(11):825–31.
9. Devesa SS, Blot WJ, Fraumeni JF. Changing patterns in the incidence of esophageal and gastric carcinoma in the United States. *Cancer* 1998;**83**:2049–53.
10. American Joint Commission of Cancer Staging Manual. *Esophagus and esophagogastric junction*. 7th ed. 2010.
11. Chandrasoma PT, Wickramasinghe K, Ma Y, DeMeester TR. Adenocarcinomas of the distal esophagus and "gastric cardia" are predominantly esophageal adenocarcinomas. *Am J Surg Pathol* 2007;**31**:569–75.
12. Sarbia M, Donner A, Gabbert HE. Histopathology of the gastroesophageal junction. A study on 36 operation specimens. *Am J Surg Pathol* 2002;**26**:1207–12.
13. Ayazi S, Tamhankar A, DeMeester SR, et al. The impact of gastric distension on the lower esophageal sphincter and its exposure to acid gastric juice. *Ann Surg* 2010;**252**:57–62.
14. Robertson EV, Derakhshan MH, Wirz AA, Lee YY, Seenan JP, Ballantyne SA, Hanvey SL, Kelman AW, Going JJ, McColl KE. Central obesity in asymptomatic volunteers is associated with increased intrasphincteric acid reflux and lengthening of the cardiac mucosa. *Gastroenterology* 2013;**145**:730–9.
15. Chandrasoma PT. Histologic definition of gastro-esophageal reflux disease. *Curr Opin Gastroenterol* 2013;**29**:460–7.
16. Chandrasoma PT, Der R, Ma Y, et al. Histology of the gastroesophageal junction: an autopsy study. *Am J Surg Pathol* 2000;**24**:402–9.
17. Kilgore SP, Ormsby AH, Gramlich TL, et al. The gastric cardia: fact or fiction? *Am J Gastroenterol* 2000;**95**:921–4.
18. McClave SA, Boyce Jr HW, Gottfried MR. Early diagnosis of columnar lined esophagus: a new endoscopic diagnostic criterion. *Gastrointest Endosc* 1987;**33**:413–6.
19. Bombeck CT, Dillard DH, Nyhus LM. Muscular anatomy of the gastroesophageal junction and role of phrenoesophageal ligament. Autopsy study of sphincter mechanism. *Ann Surg* 1966;**164**:643–52.
20. Hayward J. The lower end of the oesophagus. *Thorax* 1961;**16**:36–41.
21. Kahrilas PJ, Shi G, Manka M, Joehl RJ. Increased frequency of transient lower esophageal sphincter relaxation induced by gastric distension in reflux patients with hiatal hernia. *Gastroenterology* 2000;**118**:688–95.
22. Korn O, Csendes A, Burdiles P, Braghetto I, Stein HJ. Anatomic dilatation of the cardia and competence of the lower esophageal sphincter: a clinical and experimental study. *J Gastrointest Surg* 2000;**4**:398–406.
23. Hill AD, Kozarek RA, Kraemer SJM, et al. The gastroesophageal flap valve: in vitro and in vivo observations. *Gastrointest Endosc* 1996;**44**:541–7.
24. Oberg S, Peters JH, DeMeester YR, Lor RV, Johansson J, Crookes PF, Bremner CG. Endoscopic grading of the gastroesophageal valve in patients with symptoms of gastroesophageal reflux disease. *Surg Endosc* 1999;**13**:1184–8.
25. Koch OO, Spaun G, Antoniou SA, et al. Endoscopic grading of the gastroesophageal flap valve is correlated with reflux activity and can predict the size of the esophageal hiatus in patients with gastroesophageal reflux disease. *Surg Endosc* 2013;**27**:4590–5.
26. Curcic J, Roy S, Tech M, Schwizer A, Kaufman E, Forras-Kaufman Z, Menne D, Hebbard GS, Treier R, Boesiger P, Steingoetter A, Fried M, Schwizer W, Pal A, Fox M. Abnormal structure and function of the esophagogastric junction and proximal stomach in gastroesophageal reflux disease. *Am J Gastroenterol* 2014;**109**:658–67.
27. Allison PR. Peptic ulcer of the oesophagus. *Thorax* 1948;**3**:20–42.
28. Nauta J. The closing mechanism between the oesophagus and the stomach. *Gastroenterol (Basel)* 1956;**86**:219–32.
29. Granderath FA. Measurement of the esophageal hiatus by calculation of hiatal surface area (HSA). Why, when and how? *Surg Endosc* 2007;**21**:2224–5.

30. Batirel HF, Uygur-Bayramicli O, Giral A, Ekici B, Bekiroglu N, Yildizeli B, Yuksel M. The size of the esophageal hiatus in gastroesophageal reflux pathophysiology: outcome of intraoperative measurements. *J Gastrointest Surg* 2010;**14**:38–44.
31. Vakil N, van Zanten SV, Kahrilas P, Dent J. Jones B and the global consensus group. The Montreal definition and classification of gastroesophageal reflux disease: a global evidence-based consensus. *Am J Gastroenterol* 2006;**101**:1900–20.
32. Malfertheiner P, Nocon M, Vieth M, Stolte M, Jasperson D, Keolz HR, Labenz J, Leodolter A, Lind T, Richter K, Willich SN. Evolution of gastro-oesophageal reflux disease over 5 years under routine medical care – the ProGERD study. *Aliment Pharmacol Ther* 2012;**35**:154–64.
33. Chandrasoma PT, Der R, Dalton P, Kobayashi G, Ma Y, Peters JH, DeMeester TR. Distribution and significance of epithelial types in columnar lined esophagus. *Am J Surg Pathol* 2001;**25**:1188–93.
34. Olvera M, Wickramasinghe K, Brynes R, Bu X, Ma Y, Chandrasoma P. Ki67 expression in different epithelial types in columnar lined esophagus indicates varying levels of expanded and aberrant proliferative patterns. *Histopathology* 2005;**47**:132–40.
35. Johns BAE. Developmental changes in the oesophageal epithelium in man. *J Anat* 1952;**86**:431–42.
36. Chandrasoma P, Makarewicz K, Wickramasinghe K, Ma YL, DeMeester TR. A proposal for a new validated histologic definition of the gastroesophageal junction. *Hum Pathol* 2006;**37**:40–7.
37. Shi L, Der R, Ma Y, Peters J, DeMeester T, Chandrasoma P. Gland ducts and multilayered epithelium in mucosal biopsies from the gastroesophageal-junction region are useful in characterizing esophageal location. *Dis Esophagus* 2005;**18**:87–92.
38. Glickman JN, Chen YY, Wang HH, Antonioli DA, Odze RD. Phenotypic characteristics of a distinctive multilayered epithelium suggests that it is a precursor in the development of Barrett's esophagus. *Am J Surg Pathol* 2001;**25**:569–78.
39. Fass R, Sampliner RE. Extension of squamous epithelium into the proximal stomach: a newly recognized mucosal abnormality. *Endoscopy* 2000;**32**:27–32.
40. Ringhofer C, Lenglinger J, Izay B, Kolarik K, Zacherl J, Fisler M, Wrba F, Chandrasoma PT, Cosentini EP, Prager G, Riegler M. Histopathology of the endoscopic esophagogastric junction in patients with gastroesophageal reflux disease. *Wien Klin Wochenschr* 2008;**120**:350–9.
41. Hirota WK, Loughney TM, Lazas DJ, Maydonovitch CL, Rholl V, Wong RKH. Specialized intestinal metaplasia, dysplasia and cancer of the esophagus and esophagogastric junction: prevalence and clinical data. *Gastroenterology* 1999;**116**:277–85.
42. Leodolter A, Nocon M, Vieth M, Lind T, Jaspersen D, Richter K, Willich S, Stolte M, Malfertheiner P, Labenz J. Progression of specialized intestinal metaplasia at the cardia to macroscopically evidence Barrett's esophagus: an entity of concern in the Pro-GERD study. *Scand J Gastroenterol* 2012;**47**:1429–35.
43. Clark GWB, Smyrk TC, Burdiles P, et al. Is Barrett's metaplasia the source of adenocarcinoma of the cardia? *Arch Surg* 1994;**129**:609–14.
44. Cameron AJ, Lomboy CT, Pera HA, et al. Adenocarcinoma of the esophagogastric junction and Barrett's esophagus. *Gastroenterology* 1995;**109**:1541–6.
45. Weston AP, Krmpotich PT, Cherian R, Dixon A, Topalovski M. Prospective evaluation of intestinal metaplasia and dysplasia within the cardia of patients with Barrett's esophagus. *Dig Dis Sci* 1997;**42**:597–602.
46. Sampliner RE, Camargo E, Prasad AR. Association of ablation of Barrett's esophagus with high grade dysplasia and adenocarcinoma of the gastric cardia. *Dis Esophagus* 2006;**19**:277–9.

Chapter 11

Columnar-Lined Esophagus (Microscopic and Visible) and Barrett Esophagus

Successful total outcome of management of a disease can be assessed in many different ways. Of all measures of outcome, the one that is most important for any disease is mortality rate. If this number is large and is increasing without any prospect of control, the disease represents a serious problem. By this viewpoint, gastroesophageal reflux disease (GERD) is a serious problem. Death from adenocarcinoma increased sevenfold from 1975 to 2006 and continues to increase.[1]

Defining GERD by symptoms and having symptom control be the objective of management does not have any chance of preventing adenocarcinoma.[2] The reason for this is that while there is a good correlation between the presence or severity of symptoms and adenocarcinoma, there are a significant number of people who do not have significant GERD symptoms that develop cancer.

Without a reliable diagnostic test for the diagnosis of GERD, the treatment is at an empiric level with the sole aim of controlling the secretion of gastric acid or neutralizing secreted acid.[2] This controls symptoms of GERD and improves quality of life, which is the objective of treatment. The success in achieving this objective with 70% of patients being satisfied with their treatment allows justification of this management algorithm.

However, acid-reducing therapy has no effect on lower esophageal sphincter (LES) damage, which is the cause of GERD. Treatment therefore does not reduce reflux.[3] There is evidence that the disease progresses at a cellular level ultimately resulting in visible columnar-lined esophagus (CLE) that progresses to Barrett esophagus (BE) and then to adenocarcinoma.[4,5] This disease progression is most likely the result of progression of LES damage. Without a reliable test to assess progression of LES damage, there are few options to alter this disease course.

Unfortunately control of symptoms with proton pump inhibitor (PPI) therapy does not prevent progression in the GERD→adenocarcinoma sequence. Nason et al.[6] showed that the prevalence of BE is greater in patients who are well controlled with PPI therapy than those who are not controlled. Malfertheimer et al.[4] showed that treatment of GERD in the community setting resulted in a 9.7% overall incidence of visible CLE within 5 years in patients who had no visible CLE at the index endoscopy. The risk of progression to BE was associated in that study with male sex, the severity of erosive esophagitis at the index endoscopy, and regular PPI use.

Only about 15% of GERD patients reach a point of failure of symptom control to satisfy the main present indication for endoscopy. Those in this group who have a diagnosis of BE at that endoscopy enter a surveillance program aimed at early detection of dysplasia and cancer. Early neoplasia in these patients can now be treated effectively with minimally invasive methods not requiring esophagectomy.

In the other 85% of patients, progression through BE and increasing dysplasia to adenocarcinoma, often over decades, goes undetected. They have never had an indication for endoscopy by present guidelines. These patients present when their advanced-stage adenocarcinoma obstructs the esophagus and causes dysphagia. They have a dismal survival and require highly morbid and ineffective treatment methods such as radical esophagectomy, radiation, and chemotherapy, which are often only palliative.

The usual method of following progression of a chronic disease from onset to death is at a cellular level with changes defined by gross and microscopic pathologic changes. In GERD, this does not happen. The normal state and early stages of disease are undefined and seriously misunderstood.[7] When patients reach the definitional point of GERD where their symptoms have become "troublesome," they have LES damage that has been permitted to progress to an advanced state where they already have evidence or impending risk of severe disease manifested by symptoms that are difficult to control, erosive esophagitis, and visible CLE. These are the present hallmarks of diagnosis of GERD. In reality they are manifestations of the end stage of LES failure.

GERD. http://dx.doi.org/10.1016/B978-0-12-809855-4.00011-7

If we are to make an effort to prevent esophageal adenocarcinoma, we must have a new primary objective. It cannot be symptom control and improvement of quality of life. The proven result of management with this primary objective has had an unintended consequence: deaths of thousands of people in the United States from adenocarcinoma each year.[1]

The new histologic assessment of LES damage that I am proposing will provide a method of early diagnosis of GERD based on its causation. It will also allow prediction of future LES damage. This will provide a reliable identification of risk of progression in the future while the patient still has a relatively intact LES. This will provide a new opportunity for early intervention with new objectives.

I will suggest two new objectives for the treatment of GERD:

1. Prevent LES damage reaching a critical point where the abdominal segment of the LES reaches a defined amount of shortening at which the patient is at risk of developing visible CLE. For the present, I will assume this is 25 mm of LES damage, which leaves a residual abdominal LES of 10 mm. This is the abdominal length that presently defines a defective LES.[8] Prevention of the abdominal LES reaching this level of damage is the primary objective (Fig. 11.1).
2. Prevent the occurrence of visible CLE (Figs. 11.2 and 11.3). This is the secondary objective, which is achieved when the primary objective is achieved. It is irrelevant whether the visible CLE is short segment (Fig. 11.2) or long segment (Fig. 11.3).

At present, there are no studies that tell us the exact amount of abdominal LES damage that correlates with the presence of visible CLE. When data accumulate, the abdominal LES length at which visible CLE occurs will be known and the LES damage objective will likely change. Until then, I will use 10 mm of residual abdominal LES length to define this objective.

The primary reason for the lack of data is that this entire concept of the dilated distal esophagus that I present is either not understood or not accepted. Acceptance of this concept requires revision of the presently used incorrect endoscopic definition of the gastroesophageal junction (GEJ) and the recognition that the damaged abdominal LES is the dilated distal esophagus between the endoscopic and true GEJ (proximal limit of gastric oxyntic epithelium)[7,9] (Fig. 11.4). When the concept is accepted, there is a need for a new biopsy device that will provide a suitable mucosal biopsy sample to make the measurement of LES damage with accuracy (see Chapter 17).

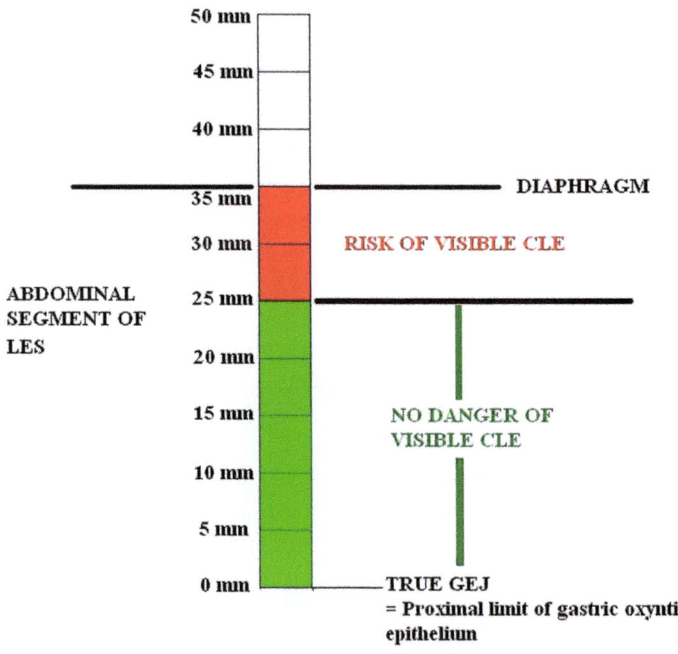

FIGURE 11.1 Different levels of lower esophageal sphincter (LES) damage correlated with risk of adenocarcinoma of the thoracic esophagus. It is assumed at this time that visible columnar-lined esophagus (CLE) develops with abdominal LES damage >25 mm. If this is true, it means that if a person can be maintained within abdominal LES damage <25 mm for his/her lifetime, he/she will never develop a risk of adenocarcinoma. Failure of prevention of adenocarcinoma is defined as permitting a person to progress to >25 mm abdominal LES damage. This happens all the time with present management of gastroesophageal reflux disease in patients who develop troublesome symptoms that define the disease. *GEJ*, gastroesophageal junction.

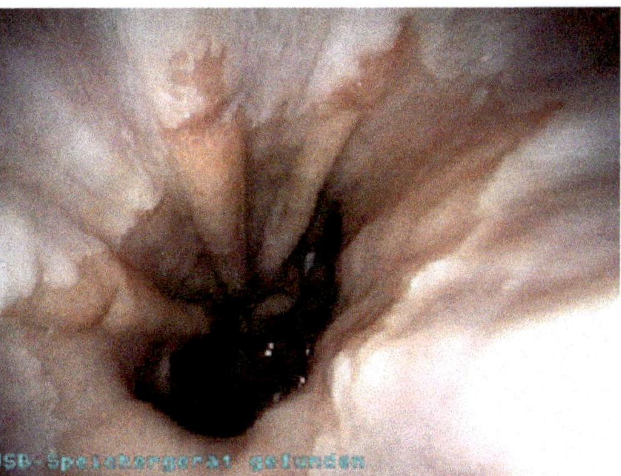

FIGURE 11.2 Short-segment visible columnar-lined esophagus showing 1-cm separation of the squamocolumnar junction (SCJ) and proximal limit of rugal folds and serration of the SCJ. Note that the significant segment of cardiac metaplasia in the dilated distal esophagus that is distal to the proximal limit of rugal folds is misinterpreted as normal stomach. Histology in this person was negative for intestinal metaplasia.

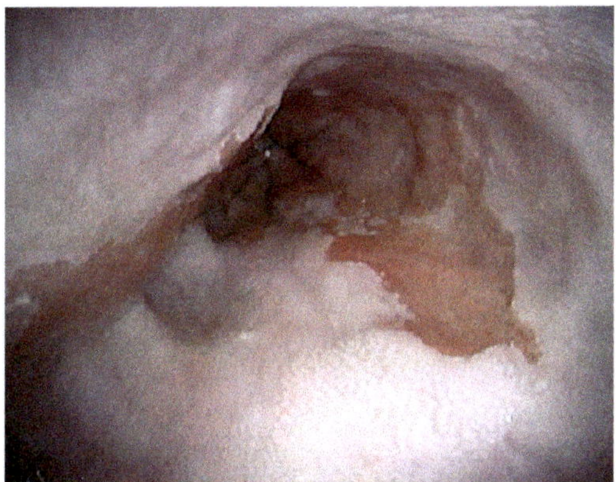

FIGURE 11.3 Visible long-segment columnar-lined esophagus, with 10-cm separation of the squamocolumnar junction and proximal limit of rugal folds by metaplastic cardiac epithelium. Biopsy was positive for intestinal metaplasia.

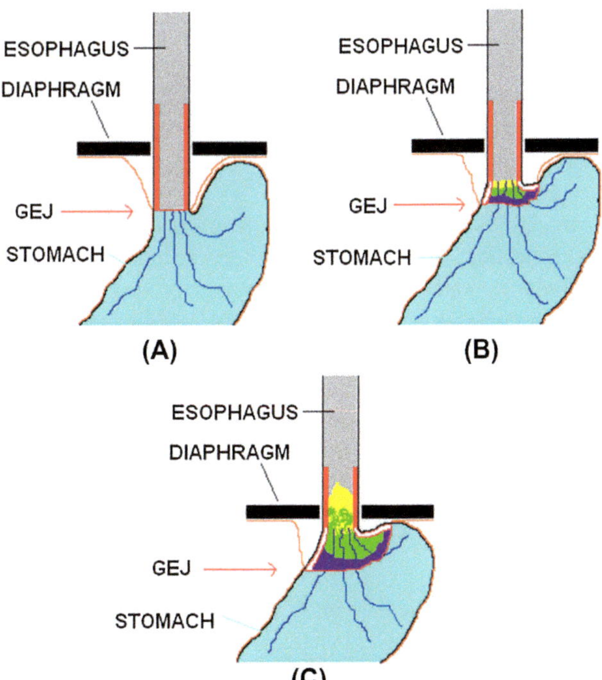

FIGURE 11.4 Only the rare person with no cardiac epithelium will have no columnar-lined esophagus (CLE) (A). In patients that are in the compensated phase of lower esophageal sphincter (LES) damage (i.e., no reflux) and with clinical gastroesophageal reflux disease without visible CLE, there will be metaplastic cardiac epithelium limited to the dilated distal esophagus (B). The occurrence of visible CLE (C) is associated with the longest segments of cardiac epithelium in the dilated distal esophagus, correlating with severe LES damage. *GEJ*, gastroesophageal junction.

The objective of the new proposed management of GERD is to measure LES damage at an early stage, identify those at high risk for progression to visible CLE in the future, and develop a method of preventing or slowing down the progression of LES damage such that it never reaches the critical point at which visible CLE occurs. Present methods exist for repairing a damaged LES. While these are likely to be effective, the needed method is likely to be feasible with less invasive procedures.

If visible CLE is prevented, adenocarcinoma of the thoracic esophagus will be prevented. The recognition of the true nature of the dilated distal esophagus, which is a measure of LES damage, leads to a new method of assessing GERD that has a theoretical basis of achieving this objective.

1. THE PAST: HISTORICAL EVOLUTION OF COLUMNAR-LINED ESOPHAGUS

In the first half of the 20th century, patients who had long segments of columnar epithelium in the tubular esophagus were seen at both endoscopy and in resected specimens, often associated with peptic ulcers in the esophagus. Back then, this was simply a medical curiosity that had no clinical significance if they were not associated with ulcers. This was reasonable because esophageal adenocarcinoma was a rare entity; the first case was not reported until 1952.

1.1 The Perfect Beginning: 1953–57

Isolated reports of what now can be interpreted as CLE existed in the literature of the first half of the 20th century. These were generally regarded as a congenital anomaly in which the tubular esophagus contained heterotopic gastric mucosa. In the four years between 1953 and 1957, the medical field transformed from creating total confusion to near complete understanding of CLE in some of the most thoughtful papers ever published.

In 1950, Barrett[10] defined the esophagus as "that part of the foregut distal to the cricopharyngeal sphincter that was lined by squamous epithelium." All columnar epithelium in the foregut proximal to the duodenum was therefore gastric because the squamocolumnar junction (SCJ) defined the GEJ at endoscopy. When the SCJ was displaced cephalad, this definition caused the entire CLE to be misinterpreted as a tubular intrathoracic stomach associated with a congenitally short esophagus.

In 1953, Allison and Johnstone refuted Barrett's definition of the GEJ. They brought a new insight into the problem that was largely spawned by a more accurate definition of the GEJ by the position of the peritoneal reflection. This was based on their examination of resection specimens of the distal esophagus and stomach. The external surface of the stomach is covered in its entire extent by peritoneum. The peritoneum reflects from the stomach at the GEJ to the undersurface of the diaphragm. The abdominal esophagus is separated from the peritoneum by adventitial tissue consisting of adipose and fibrovascular tissue, including the phrenoesophageal ligaments.

The paper of Allison and Johnstone[11] teaches us the value of accurate definition of the GEJ and the fact that the use of an incorrect endoscopic definition of the GEJ results in serious error of interpretation. The difference between Barrett's definition[10] and that of Allison and Johnstone[11] could be over 10 cm in patients with a long CLE.

Allison and Johnstone write[11]: "All would agree that the muscular tube extending from the pharynx downwards and lined with squamous epithelium may be correctly referred to as the oesophagus or gullet. The dilated sac covered with peritoneum and lined with gastric mucous membrane is obviously stomach whether it lies in the abdomen or is herniated into the mediastinum."

This paper addresses the controversial situation where there is a viscus not covered with peritoneum, which was esophagus because it was above the peritoneal reflection, but was lined by columnar mucosa on the inside. Disagreeing with Barrett's suggestion of calling this entity a tubular stomach, Allison and Johnstone provided strong evidence that this was actually the esophagus.[11]

Allison and Johnstone's evidence that the columnar-lined structure in the thorax was esophagus and not stomach[11] (Fig. 11.5): (1) It was not covered by peritoneum; (2) It had a musculature that was two-layered like the normal esophagus rather than the more complex three-layered muscle wall of the stomach; (3) It was associated with islands of squamous epithelium within the columnar mucosa; (4) It was devoid of oxyntic (parietal) cells in the mucosa; (5) It was associated with typical esophageal mucous glands in both mucosa and submucosa. All these criteria are accurate except that subsequent reports showed that oxyntic (parietal) cells can be seen in the distal part of this viscus in what we now call oxyntocardiac epithelium.[12]

Allison and Johnstone's[11] series consisted of 115 patients with stenosis from esophageal ulcer (i.e., ulcer entirely within squamous epithelium, presumably caused by reflux) and 10 with an ulcer with "gastric" mucosa below it. Among the 115 patients with esophageal ulcers, there were 11 patients with "gastric" mucosa between the area of stricture and the sliding hiatal hernia that was lined by peritoneum on the outside. In a total of 125 cases, 21 patients had evidence of "gastric" mucosa lining the esophagus between the SCJ and the GEJ, which was defined by the peritoneal reflection.

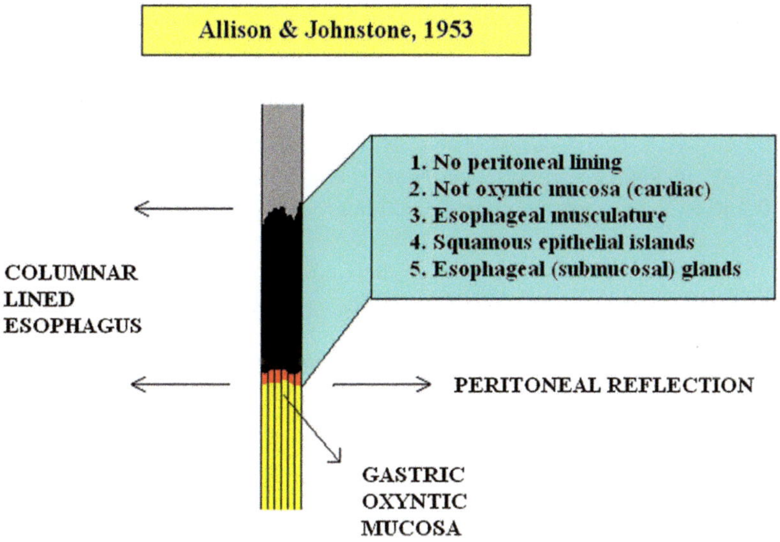

Allison & Johnstone, 1953

1. No peritoneal lining
2. Not oxyntic mucosa (cardiac)
3. Esophageal musculature
4. Squamous epithelial islands
5. Esophageal (submucosal) glands

COLUMNAR LINED ESOPHAGUS

PERITONEAL REFLECTION

GASTRIC OXYNTIC MUCOSA

FIGURE 11.5 The criteria used by Allison and Johnstone to differentiate esophagus from stomach and convince Norman Barrett that his tubular intrathoracic stomach distal to the squamocolumnar junction was actually a columnar-lined esophagus.

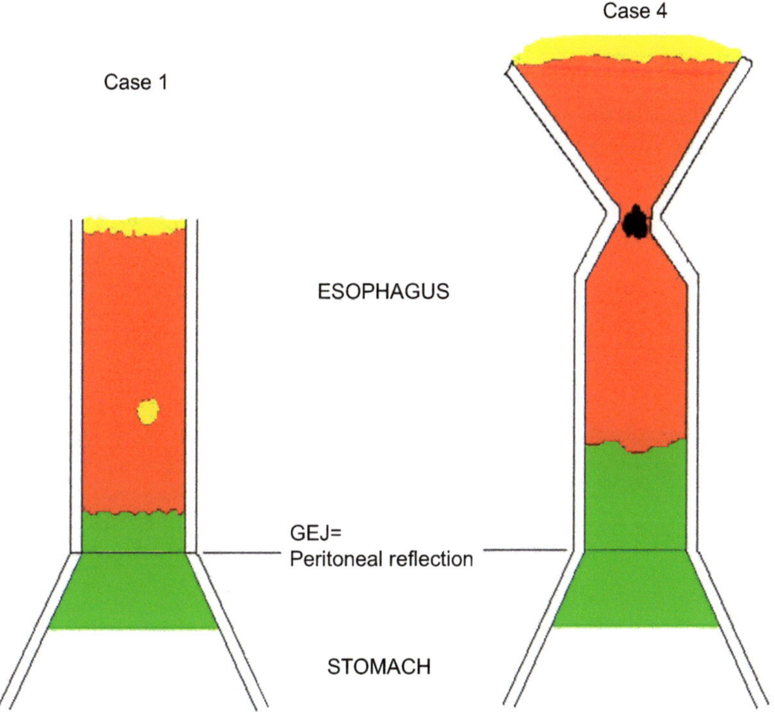

FIGURE 11.6 Diagrammatic illustration of Allison and Johnstone's Cases 1 and 4. This shows a long segment of visible columnar-lined esophagus lined entirely by cardiac (red) and columnar epithelium with parietal cells (green). The peritoneal reflection was (correctly) used to define the true gastroesophageal junction. Note the squamous island (yellow) described in Case 1.

Allison and Johnstone[11] gave detailed reports of clinical features, radiology, and pathology of seven patients who had resections of the distal esophagus and proximal stomach. These seven cases that are described in detail are summarized here because they give excellent insight into the entity of CLE and the histologic features of the various normal and abnormal areas. Data such as these are very rare and provide insight into GERD before the time of effective acid suppression. The pathologic examination of the resected specimen by Dr. Collins in Cases 1 and 4 deserves particular attention (Fig. 11.6).

Case 1 was a 66-year-old woman who had recurrent attacks of violent pain in the epigastrium and lower chest behind the sternum. Radiologically the patient had a mixed type of hiatal hernia, partly sliding and partly paraesophageal. Endoscopically, there was superficial esophagitis and leucoplakia at 23 cm from the incisor teeth and "gastric" mucosa encountered at 24 cm. There was no stenosis. As the esophagoscope traversed the "gastric" mucosa, some bleeding occurred from the exuberant folds of mucosa.

The pathologic description by Dr. D.H. Collins: "The resected specimen comprises a substantial portion of the body and cardia of the stomach, measuring about 13 cm horizontally and 7 cm vertically, together with 5 cm of esophagus... There is no sharp demarcation of mucosa at the anatomical cardia, although the mucous lining appears to become ironed out in the esophagus. The esophagus was separated with a knife from the stomach along the line of peritoneal reflection. A vertical slice was then made through the center of the reconstituted specimen, i.e. up the posterior wall, and three vertically continuous blocks were prepared, and a horizontal block then made across the extreme upper limit of the esophageal mucosa."

> *Histology: The stomach below the anatomical junction with the esophagus is lined by gastric mucosa of fundal type.... Cardiac glands and cardiac gastric mucosa do not appear until 0.6 cm up the anatomical oesophagus, and esophageal glands appear at 2 cm up the oesophagus underlying columnar cell mucous membrane. The first islet of squamous epithelium appears at 3 cm up the anatomical esophagus, but predominantly gastric mucosa continues to the upper limit of the specimen. The transverse sections across the uppermost part of the removed mucosa shows alternations of squamous and gastric mucosa. The rather villous type of cardiac mucosa, its lack of depth, and a diffuse fibrosis of the submucosa in the zone for 2 to 4 cm above the stomach orifice, suggest healing of previous shallow erosions. Goblet cells were not mentioned.*

"The clinical diagnosis of reflux oesophagitis, lower oesophagus lined with gastric mucous membrane, and mixed sliding and para-oesophageal hernia was, therefore, confirmed." The pathologic diagnosis is remarkable for its clarity. The entire CLE from the SCJ to the true GEJ (peritoneal reflection) is within this viscus without a peritoneal covering. There is no cardiac epithelium in the stomach distal to the peritoneal reflection.

Case 2 showed gastric mucous membrane immediately below a stricture at 28 cm. Biopsy at 31 cm showed "gastric" mucosa with goblet cells.

Case 3 showed a tight stricture with ulceration at 21 cm with gastric mucosa below the stenosis. Biopsies showed squamous epithelium with simple ulceration at 21 cm, inflamed cardiac mucosa at 26 cm, and gastric type mucosa at 31 cm. Goblet cells were not mentioned.

Case 4 showed a tight stricture at 30 cm. Resection was performed for postendoscopy perforation. The specimen showed a tubular esophagus measuring 5 cm with a 7-mm ulcer in the middle. The proximal 5 mm was lined by squamous epithelium; the remainder of the esophagus is lined by gastric mucosa of which the proximal two-thirds are of atrophic cardiac type without oxyntic cells. The lower third of the specimen, which is called a sliding hiatal hernia, is lined by fundal-type gastric mucosa with oxyntic cells. Goblet cells were not mentioned.

Case 5 was a resected specimen consisting of 6 cm of distal esophagus and proximal stomach. "The stomach, up to its anatomical junction with the esophagus (presumably defined by the peritoneal reflection as in Case 1), is lined by gastric mucosa with glands of fundal type. Cardiac glands only are seen in the immediate neighborhood of the GEJ, and the cardiac type of gastric mucosa extends up the esophagus for 3 cm before joining with the squamous epithelium. Small islands of squamous epithelium are present in the cardiac mucosa." Goblet cells were not mentioned.

Case 6 showed esophagitis from 20 to 25 cm, a deep ulcer with constriction at 25 cm, with the mucosa distal to the ulcer being gastric. This was resected. The resection specimen showed an upper segment 1.5 cm long lined by squamous epithelium, below which was a segment lined by gastric cardiac–type epithelium with goblet cells and two deep ulcers. Esophageal submucosal glands were seen beneath the cardiac mucosa. The stomach distal to the ulcerated zone was lined by fundal-type mucosa with oxyntic cells.

Case 7 showed squamous lining down to 30 cm where there was a mass that proved to be an adenocarcinoma. The resected specimen included 9 cm of esophagus with the proximal stomach. The tumor arose 1 cm above the cardiac orifice and extended upward for 6 cm before giving way to squamous mucosa. The epithelium below the tumor is gastric in type with cardiac glands. Goblet cells were not mentioned.

An analysis of this study in the light of present knowledge would suggest that all these patients had severe GERD with ulcers and strictures and that 21/125 patients had long segments of CLE. Of the seven patients with whom the histology was described in detail, the mucosa (called "gastric") was cardiac in type in all with goblet cells indicative of intestinal metaplasia in two and an adenocarcinoma in the third.

In all cases where information is given, the mucosa at and below the GEJ (defined by the peritoneal reflection in the resected specimens) was gastric fundal in type with oxyntic cells.

Allison and Johnstone's use of the peritoneal reflection results in a perfect description of the CLE. If we stop at the understanding that these authors had in 1953, we will recognize that they clearly identify a proximal esophagus lined by squamous epithelium, the normal stomach below the peritoneal reflection lined by gastric fundal (=oxyntic) epithelium, and an intervening CLE lined by cardiac epithelium with goblet cells in two patients and complicated by adenocarcinoma in the third. Scattered parietal cells are described in the cardiac epithelium (=oxyntocardiac epithelium) in the distal part of the columnar-lined segment.

They are describing the entire squamooxyntic gap from the true GEJ (gastric-fundic/oxyntic epithelium at the peritoneal reflection) to the SCJ. This consists of cardiac and oxyntocardiac epithelium in all cases and intestinal metaplasia in two. They do not concern themselves with the appearance of this segment (i.e., tubular or not) or the presence of rugal folds.

By using an accurate definition of the GEJ (the peritoneal reflection), they reached the truth. They did not fall into the trap of using the end of the tubular structure as esophagus. Also they did not fall into the trap of defining the stomach by the presence of rugal folds. These incorrect definitions of the GEJ were correctly not used at the time.

In reviewing any literature, there is usually one paper that appears to be the definitive end to a problem that has previously caused confusion.[13] This is the paper by Norman Barrett in 1957. As Barrett writes in his introduction, which is entitled "Definitions": "This paper concerns a condition whose existence is denied by some, misunderstood by others, and ignored by the majority of surgeons. It has been called a variety of names which have confused the story because they have suggested incorrect etiologic explanations: congenital short esophagus, ectopic gastric mucosa, short esophagus, and the lower esophagus lined by gastric epithelium are but a few. At the present time the most accurate description is that it is a state in which the lower end of the esophagus is lined by columnar epithelium. This does not commit us to ideas which could be wrong, but it carries certain implications which must be clarified."

As an aside, I can write a similar statement today regarding the dilated distal esophagus: "The dilated distal esophagus concerns a condition whose existence is denied by some, misunderstood by others, and ignored by the majority of physicians. It has been called a variety of names which have confused the story because they have suggested incorrect etiologic explanations: normal proximal stomach, normal gastric cardia, and carditis are but a few. At the present time the most accurate description is that it is a state in which the lower end of the esophagus is lined by columnar epithelium. This does not commit us to ideas which could be wrong, but it carries certain implications which must be clarified."

Barrett is reversing the histologic definition that he proposed in 1950 that the esophagus was that part of the foregut, which is lined by squamous epithelium.[10] He is essentially agreeing with Allison and Johnstone[11] that the esophagus can be lined by columnar epithelium. Importantly, he is changing Allison and Johnstone's term "esophagus lined by gastric mucous membrane" to the more noncommittal and accurate term "columnar-lined esophagus."

Barrett[13] goes on to attempt precise definition of several terms:

1. "The esophagus (=gullet) is the narrow passage through which food and drink pass from the pharynx to the stomach." With this new definition, there is no longer even an attempt to define the lower end of the esophagus. It is interesting that the only histologic definition of the GEJ that was ever attempted was Barrett's assertion that it was the distal limit of squamous epithelium.[10] When he changed his mind in 1957,[13] the only histologic definition of the GEJ that was ever accepted disappeared. This was appropriate because the definition was wrong. However, no one has since defined the distal limit of the esophagus in histologic terms until 2007 when we proposed that it was the proximal limit of gastric oxyntic epithelium.[9] This new histologic definition is yet to be accepted or recognized. We continue to use the incorrect endoscopic definitions of the GEJ.[2,14]
2. "The 'cardia' describes the mechanism (flap valve or whatever it be) which prevents reflux from the stomach into the esophagus. The word does not refer to the site at which the gullet widens out into the stomach nor to a hypothetical muscle sphincter which some believe separates the two."

I have not yet figured out how the term "cardia" changed from this definition by Barrett to becoming synonymous with the proximal stomach.

Barrett clearly differentiates CLE from a sliding hiatal hernia: "The latter is an acquired deformity, which may be present at birth, in which part of the stomach has migrated from below the esophageal hiatus into the mediastinum.... Allison has said that sliding hiatal hernia and a lower esophagus lined by columnar epithelium often exist together in one and the same patient. In such a person the proximal alimentary canal would be constituted in this way: pharynx, esophagus lined by squamous epithelium, esophagus lined by columnar epithelium, sliding hiatal hernia, and stomach."

Barrett's understanding of the normal histology of the region is superb (not uncommon in surgeons of that era): "It is principally in the carnivora that the epithelium changes abruptly from squamous to columnar precisely at the level of the 'cardiac valve'. And in this group the esophagus widens into the stomach at the same place as that which the epithelium changes."

In 1957, Barrett understands that the entire normal esophagus is lined by squamous epithelium. He goes on to differentiate CLE from gastric mucosa histologically: "The thing which importantly distinguishes the stomach of a man from the esophagus is its mucosa. The mucosa of the stomach is formed of special tall columnar cells which are prolonged deeply into simple or branched tubules whose components have undergone further differentiation into cardiac, fundal, and pyloric glands. These glands secrete hydrochloric acid, mucus, pepsin, and secretin. Functionally the columnar mucous membrane in these esophageal cases secretes little if any acid, pepsin, or secretin. Histologically, the upper part of the esophagus is normal and there is an obvious and abrupt transition between this and the unusual segment at the bottom. Just below this boundary, whose margin is corrugated, the columnar cells are flat and arranged in shallow, tubular glands among which lie mucus secreting units. There are no oxyntic cells and the structure of these glands resembles the normal deep esophageal glands. Lower in the esophagus the simple tubular crypts give place to more typical gastric mucous membrane. Scattered oxyntic cells appear. At any level in this abnormal segment, perfectly formed patches of squamous epithelium occur. These findings, which are similar to those described by Allison and Johnstone, suggest that the abnormal epithelium, despite its looks, does not function exactly as stomach and probably secretes little digestive juice."

His description of histopathologic findings is impeccable: "Surgeons who have studied the histology of specimens removed at operation have found that the greater part of the unusual epithelium consists of simple tubular glands which secrete mucus but which include few true glandular elements. There are some oxyntic cells at the lower end of most specimens and these may be situated proximal to the junction between the gullet and the stomach.... The junction between the squamous and columnar epithelia in these cases is generally situated at about 20 to 25 cm from the incisor teeth and not just above the diaphragm as it is in most sliding hiatal hernias."

The histologic descriptions are unbelievably accurate. Barrett is describing perfectly the proximal segment of the CLE lined by cardiac epithelium and oxyntocardiac epithelium with scattered parietal cells in the more distal region. He does not even fall into the obvious trap that the presence of parietal cells indicates gastric epithelium.

Despite the fact that Allison and Johnstone described the presence of goblet cells in the cardiac epithelium,[11] Barrett makes no mention of these. It must be recognized that Barrett is describing very long segments of CLE. If the SCJ is at 20–25 cm, the segments must be well in excess of 10 cm. To describe the histology as cardiac without mention of goblet cells suggests that intestinal metaplasia was uncommon and when present, focal, in these long segments of CLE. This is a dramatic difference from the present condition where intestinal metaplasia is present in nearly all segments of CLE greater than 3 cm in length and is seen in much more of the columnar segment.

Barrett then speculates on the sphincter mechanism: "Is there, in these patients, a 'cardiac valve' mechanism, and if so where is it? My observations suggest that two types of cases occur. The first is the patient who has the lower part of the gullet lined by columnar epithelium and no pathologic lesion at the oesophagogastric junction. On a barium swallow and when examined radiologically, this individual has what appears to be a normal gullet. The 'cardiac valve' is in its usual place and the mechanism does not permit reflux (Note: he would have been surprised if he had access to manometry and a test to measure reflux—he is describing patients with long-segment Barrett esophagus (LSBE) of today without a hiatal hernia who usually have a severely abnormal sphincter and severe reflux); that is, the cardiac valve is a long way below the level at which the change-over in the epithelium occurs, and between the valve and the change-over there is a segment of gullet lined by abnormal epithelium. The fact that a 'cardiac valve' can behave normally in such cases indicates that the esophagogastric angle is more important to this function than the exact histology of the epithelium at the level of the valve. The second is that described by Allison and Johnstone who reported that in most of their patients the lesion was complicated by the presence of a typical sliding hiatal hernia below the abnormal mucous membrane. In these there was no 'cardiac valve' and free reflux of gastric contents was usual."

His speculation of the pathogenesis comes very close to the truth: "… if the cardiac valve of a normal person were to become incompetent and if the lower esophagus were, as a result, to be bathed for a long time by digestive gastric juice, the squamous epithelium could be eaten away and totally replaced by more quickly growing columnar cells. This concept might explain the site of the deformity, the fact that many cases occur in patients who have an incompetent cardia due to a sliding hiatal hernia, and the fact that many patients are elderly and have a history of heartburn dating back many years."

This is the first mention in the literature of columnar metaplasia of the squamous lining of the esophagus although Barrett does not use the term metaplasia.

Barrett is confused by findings in animal experiments, and his explanation of the etiology of CLE fails at the end: "But defects, produced experimentally in the squamous epithelium of the esophagus of the dogs, heal by squamous regeneration." He handles his uncertainty well: "None of these suggestions smooth away all difficulties and the etiology remains open to speculation." However, in his summation, he goes back to the probability of a congenital abnormality: "It is suggested that the lesion under discussion should be called the lower esophagus lined by columnar epithelium, and that it is probably the result of a failure of the embryonic lining of the gullet to achieve normal maturity."

Barrett's paper in 1957[13] established and described with clarity what we now recognize as the CLE.

One effect of Barrett's paper was that there was no longer an endoscopic definition of the GEJ (= distal limit of CLE) in a patient with CLE. The only definition of the junction mentioned by Barrett was the junction between the peritoneum-lined stomach with or without a sliding hiatal hernia and the esophagus lined with columnar epithelium, a criterion that could be applied accurately only to resected specimens and was not seen at endoscopy.

Unfortunately, both Allison and Barrett did not have the knowledge that what they were describing as CLE was caused by GERD. They regarded this as a separate entity from reflux esophagitis with an undetermined cause, probably congenital. They failed to recognize that CLE in most cases that Allison described were associated with ulcers and strictures caused by reflux esophagitis.[11]

Their histologic descriptions show that they could differentiate the epithelium of the CLE from the stomach even when parietal cells were present in the CLE. They had the data to translate their external definition of the GEJ (the junction between the peritoneum-lined stomach and the CLE) to a histologic definition. They had shown that the mucosa at and below the peritoneal reflection was gastric oxyntic mucosa and the CLE above it was cardiac and oxyntocardiac epithelia. They just did not define the distal end of the esophagus histologically, probably because their belief that the CLE was a congenital gastric heterotopia made that unnecessary.

1.2 The Period of Confusion: 1957 to Present

Much of the confusion from the time of Allison and Johnstone's and Barrett's papers in 1953–57[11,13] to the present regarding CLE is based on the following:

1. CLE required to be defined endoscopically. The criteria used by Allison and Barrett to define the entity were based on the definition of the GEJ as the peritoneal reflection and the fact that full thickness sections were available to permit use of the submucosal glands, peritoneal lining, and muscle wall to differentiate esophagus from stomach.[11,13]
2. The control for the definition of the disease shifted from surgeons who were very conversant with anatomy and histology of resected specimens to internists (gastroenterologists) whose training and expertise were more limited to endoscopy with less emphasis on anatomy and histology.
3. The GEJ, which was precisely defined by the peritoneal reflection, could no longer be identified with accuracy at endoscopy. The endoscopic definitions of the GEJ that developed were and continue to be inaccurate. In reality, the GEJ cannot be defined by endoscopic examination because there is no difference in the endoscopic appearance of CLE and gastric oxyntic epithelium. Only histology can differentiate these.

While it is fitting that Barrett's name be applied to the CLE he named, the term Barrett esophagus is a good example of why eponymous names should be avoided. BE in 1960 is different than BE in 1970 (a purely endoscopic diagnosis), in 1985 (limited to those with intestinal metaplasia), and today (inclusive of short-segment disease). BE to a person in England today is not BE to one in Europe and the United States; the Englishman does not require intestinal metaplasia whereas the American does. As such, it is often impossible to be certain as to what any user of the term Barrett esophagus means and what the listener understands. Confusion reigns.

Barrett himself recommended that the term "columnar-lined esophagus" replace Allison and Johnstone's term "esophagus lined by gastric mucous membrane" because it was purely descriptive without any connotation of etiology. This simple change was probably responsible for the rapid recognition that the epithelial lining in the CLE was not gastric (either congenital or acquired), but rather a metaplastic esophageal columnar epithelium caused by GERD.

In this discussion, I will use the term CLE, as Barrett suggested, for this endoscopic abnormality that he defined. I will use the term Barrett esophagus only to show how the definition of the term changed and why confusion resulted from this change and how the confusion remains to this day. In all other circumstances I will use BE synonymously with visible CLE composed of cardiac epithelium with intestinal metaplasia.

1.3 The Evolution of Confusion in the Definition of the Gastroesophageal Junction

With Barrett's paper of 1957, the entity of the CLE became established in the literature. Over the next few years, it became clear that CLE was not a congenital abnormality but rather caused by GERD. It therefore became a significant clinical entity that required to be defined accurately.

There was no issue with defining the upper limit of CLE; this was the SCJ. The problematic issue was the definition of the distal limit of the CLE. When the GEJ was defined externally by the peritoneal reflection, the distal limit of CLE was clear. However, attempts to replace this external definition at endoscopy created enormous confusion and significant error.

The reason why endoscopy was doomed from the outset in its ability to define the GEJ was that it has no ability to differentiate metaplastic columnar epithelium lining the esophagus from gastric mucosa. Failure to recognize this resulted in serial false definitions of the GEJ. Endoscopists fell into a common trap in medicine: they were stretching the capability of their instrument to do what was impossible. The definitions that were developed ascribed various endoscopic landmarks to represent the GEJ. These were accepted without any evidence that they accurately defined the GEJ. The conviction in these opinion-based definitions led to serious errors that exist to this day.

The lack of evidence for definition of the GEJ led to an unusual viewpoint expressed in the American Gastroenterological Association (AGA) Institute technical review of GERD.[2] In their statement of the definition of the GEJ, they state: "Regardless of how many citations are identified by literature review, there can be no criterion standard definition for the GEJ… Hence these questions can only be answered by opinion, and presumably the best opinion upon which to base the answers is that of experts."

This is a statement that defies belief. The appropriate reaction to the inability to define something is to say that it cannot be defined, not to produce an opinion-based incorrect definition. The reason for this is that when an expert or, worse, a consensus of experts, introduces an opinion-based definition, the entire world follows as if that was based on evidence. Critical scrutiny and independent thinking against expert-defined guidelines is not a strength of most practicing physicians. This is clearly the case with the endoscopic definitions of the GEJ. In papers on the subject, the opinion-based incorrect definitions of the GEJ at endoscopy are followed implicitly, resulting in a body of literature over many decades that is flawed.

The entirety of the confusion that exists about the endoscopic and histologic assessment of CLE is derived from the false opinion-based endoscopic definition of the GEJ (see Chapter 6). This leads to the failure to place cardiac epithelium in its correct organ, the esophagus. Instead, it is regarded incorrectly as being in the proximal stomach.

As we emphasized in the previous section, the papers of Allison and Johnston[11] and Barrett[13] that are based on the evidence-based definition of the GEJ by the position of the peritoneal reflection are clear on these two things: (1) The GEJ is where CLE meets the gastric oxyntic epithelium of the stomach and (2) cardiac epithelium (with and without parietal and/or goblet cells) is only found *proximal* to the GEJ in the esophagus. The epithelium at and distal to the junction is gastric oxyntic epithelium. *The histologic gastroesophageal junction is the junction between metaplastic columnar epithelium (= cardiac and oxyntocardiac) in CLE and normal gastric epithelium (= gastric oxyntic epithelium).*

All endoscopic definitions of the GEJ that have been developed from 1961 to this day are false and responsible for all the confusion.

1.3.1 The Gastroesophageal Junction Is the End of the Tube

In 1961, Hayward[15] described the lower esophagus as a region where the pathology, the physiology, and even the anatomy were not quite clear. His definitions were largely functional and his main conclusion was: "The oesophagus is defined as the tube conducting food from the throat to the stomach and all of this tube is regarded as oesophagus, irrespective of its lining."

Hayward's reasoning is logical and raises important fundamental questions about the anatomy and physiology of this region. The esophagus is in fact the tube that transmits food from the mouth to the stomach. The stomach is the sac that acts as the reservoir that accommodates a meal.

The problem with this reasoning is that, while true in the normal patient, it is untrue in the patient with GERD. In Chapter 10, I demonstrated that when the distal part of the esophagus loses sphincter pressure, i.e., when the sphincter is shortened beginning at its distal end, it dilates (the dilated distal esophagus). The functioning tubular esophagus that transmits food becomes shorter. The GERD-damaged dilated distal esophagus becomes part of the reservoir. While it is esophagus, it has ceased functioning like esophagus and functions like the stomach. It becomes "gastricized."

The extent of the dilated distal esophagus is equal to the amount of shortening of the abdominal LES and equal to the amount of cardiac epithelium distal to the end of the tube. This variable extent of dilated distal esophagus is mistaken as proximal stomach by every definition of the endoscopic GEJ that is accepted today.

Hayward, in his paper, attempts to define the GEJ and the cardia.[15] With his definition of the esophagus as the tube in mind, he writes: "From the outside it is easy to see where the tube ends and the pouch begins. As well as the abrupt change in shape there is the peritoneal reflection from the stomach to diaphragm to mark the junction. It is therefore a definite, easily identified, anatomical region and will be referred to in this paper as the oesophago-gastric junction. From the inside the only line of demarcation apparent is the wavy line of junction of stratified squamous with columnar epithelium; but it is not at the level of the oesophago-gastric junction as defined above. It is 1 or 2 cm higher… the columnar epithelium also passes without change across the junction into the stomach for a centimeter or two before it gradually alters to gastric epithelium of fundal type."

Hayward states the problem: "This has put us in the ridiculous position of having the oesophago-gastric junction at a different level according to whether the tube is viewed from the inside or outside. I firmly believe that this anomaly should be corrected and that the only way to correct it is to go back to the simple definition given above. The oesophagus is a tube and all of this tube is oesophagus, regardless of its lining. The true junction is where the conducting tube changes to the digesting pouch. This means that the oesophagus is not lined exclusively by squamous epithelium. Its lower 1 to 2 cm is lined by columnar (cardiac) epithelium, which also extends a little way into the stomach."

Hayward does not give a reason as to why he believes that his observation of 1–2 cm of columnar epithelium in the distal esophagus is normal. In 1961, endoscopy was an uncommon procedure that was limited to the most severely symptomatic GERD patients. It is not unreasonable that many of these patients were abnormal and had CLE. It is unreasonable to use this population to define the normal state.

For three decades after Hayward, endoscopists believed that the presence of 2 (or 3) cm of columnar epithelium in the distal esophagus was normal. A diagnosis of CLE was made only when there was >2 (or 3) cm of columnar epithelium at endoscopy. Patients were not subject to biopsy when they had <2 (or 3) cm. What we call short-segment Barrett esophagus (SSBE) today was completely ignored for three decades.

Hayward describes and illustrates the location of cardiac epithelium based on his definition of the esophagus as a tube. He describes the columnar epithelium in the distal esophagus: "This columnar epithelium forms simple tubular glands, confined to the mucosa. It appears to produce a mucous secretion containing no digestive juices, but it obviously shares with fundal and pyloric epithelium the ability to resist acid-peptic digestion, because part of it is always in the stomach."

Hayward's illustration of this region shows cardiac epithelium lining the distal 2 cm of the tube and the proximal 3 cm of the sac (Fig. 11.7). He suggests that the epithelium be called "junctional epithelium" rather than "cardiac epithelium" because of the confusion that existed regarding the term "cardia."

In terms of where this epithelium lies: "I suggest that junctional epithelium should be regarded as esophageal. Thus the esophageal lining should be described as having mainly stratified squamous epithelium with glands in the submucosa, but partly columnar epithelium with glands in the mucosa… in the lower one to two centimetres. The stomach should be described as lined by two sorts of epithelium, fundal and pyloric, except for a small area round the esophageal opening where oesophageal junctional epithelium protrudes into it."

He is repeating Allison and Johnstone's[11] and Barrett's[13] assertion that the stomach is lined by fundal (= gastric oxyntic) epithelium and that cardiac epithelium is esophageal. But his illustration shows cardiac epithelium well into the proximal stomach (Fig. 11.7).

In fact, Hayward correctly suggests that cancer arising in the "cardia" should be called "junctional carcinoma" and be regarded as esophageal and not gastric. In 1961, Hayward suggested a designation that the American Joint Commission on Cancer accepted 50 years later in their seventh edition.[16] For five decades, adenocarcinoma of the cardia has been classified erroneously as a gastric cancer.

Hayward places the blame for the confusion as to the anatomic location of cardiac epithelium on histopathologists: "Under the microscope the oesophago-gastric junction cannot be identified, so it is not surprising that histologists adopted

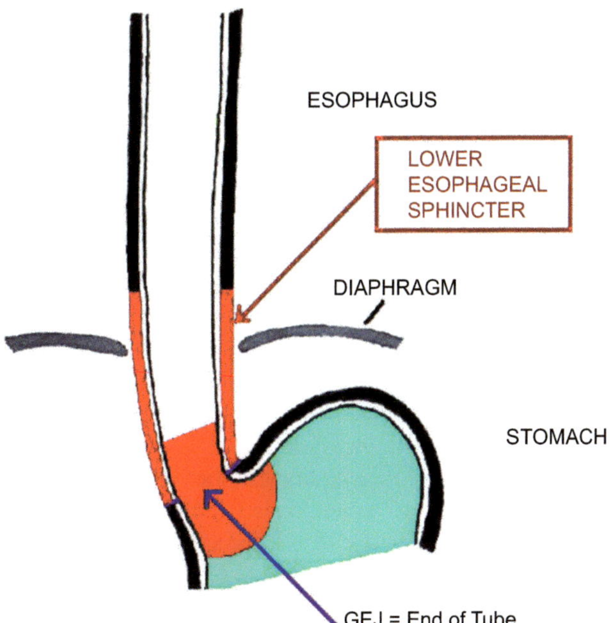

FIGURE 11.7 Diagrammatic representation of Hayward's illustration that places cardiac epithelium normally lining the distal 2 cm of the esophagus and proximal 2–3 cm of the stomach. *GEJ*, gastroesophageal junction.

the only line of demarcation available to them, namely the sudden change from squamous to columnar epithelium, and called the squamous epithelium oesophageal and columnar epithelium gastric."

There is truth in this statement. Some pathologists even today call cardiac epithelium distal to the SCJ "gastric cardiac epithelium" even when the biopsy is labeled "distal esophagus."

Hayward's paper had many accurate conclusions, but these were largely ignored. What had tremendous impact was his definition of the esophagus as a tube because he was giving the endoscopists a practical method of evaluating CLE. It rapidly became the standard and resulted in the following opinion-based dogmas:

1. The endoscopic GEJ was the end of the tube at the point where it flared into the saccular stomach.
2. The presence of 2 cm (which later became 3 cm) of columnar epithelium in the distal esophagus was normal; its presence was not an indication for biopsy.
3. CLE, which he recognized as being a metaplastic process caused by GERD, could be diagnosed only when there was columnar epithelium at a point greater than 2 cm above the end of the tube (i.e., the GEJ).
4. Histology was not essential because CLE was an endoscopic diagnosis.
5. Histology could not differentiate distal esophagus from the proximal stomach because identical cardiac mucosa was present normally for 2–3 cm on both sides of the end of the tube. It therefore had no value. This dogma was based on Hayward's diagram, which inaccurately showed cardiac epithelium in the proximal stomach (Fig. 11.7). His own writing clearly indicated that cardiac epithelium was esophageal and that the stomach was lined by fundic (oxyntic) and pyloric antral epithelia only.

Hayward's dogma regarding the extent and anatomic location of cardiac epithelium had a profound and long-standing impact on the pathology literature. In the 1997 edition of the widely used textbook *Histology for Pathologists*, DeNardi and Riddell,[17] in describing the histology of the esophagus, writes: "The mucosal gastroesophageal junction… normally lies within the lower esophageal sphincter, found usually within 2 cm of the muscular junction as defined by the proximal limit of rugal folds; thus, the distal 2 cm of the esophagus is lined by columnar epithelium that is identical to that found in the gastric cardia." The assumption is that the muscular (true gastroesophageal) junction is equivalent to the proximal limit of rugal folds. There is no evidence for this.

In the same textbook, in the chapter on the histology of the stomach, Owen[18] writes: "Extending distally from the gastroesophageal junction for approximately 1 to 2 cm is the cardiac mucosa… At the lower end of the esophagus, there is a change from nonkeratinizing squamous epithelium to columnar epithelium, which is abrupt, both grossly and microscopically. The position of this squamocolumnar junction is variable and does not coincide with the strict anatomic esophago-gastric junction, that is, the point where the tubular esophagus becomes the saccular stomach. The mucosal junction commonly is located 0.5 to 2.5 cm proximal to the anatomic junction… The lower portion of the esophagus, below the Z-line, is therefore covered by cardiac-type gastric mucosa."

1.3.2 Histologic Classification of Columnar-Lined Esophagus and Importance of Intestinal Metaplasia

The next significant advance after Hayward's paper was a histologic classification of CLE published in the 1976 New England Journal of Medicine by Paull et al.[12] from the Harvard group of gastroenterologists who were leaders in the field (see Chapter 4 for review). Prior to the study, histology had no value in the diagnosis of CLE, which was made by endoscopic criteria. By this time, the medical establishment had honored Norman Barrett by calling CLE "Barrett esophagus." The title of Paull et al.'s paper[12] uses Barrett esophagus synonymously with CLE.

In a detailed histologic study of CLE above the proximal limit of the LES (defined by manometry), they classified CLE into three histologic subtypes: junctional (=cardiac), specialized (=intestinal), and gastric-fundic type (=oxyntocardiac). These three became recognized as three histologic subtypes of columnar-lined (at the time = Barrett) esophagus (Fig. 11.8).

It was known that all esophageal columnar epithelia arose by a process of metaplasia when esophageal squamous epithelium was exposed to a sufficient amount of gastric juice delivered to the thoracic esophagus by an incompetent LES. The presence of endoscopically visible CLE was therefore a highly specific indicator for GERD-induced esophageal damage.

In the early 1980s, the group at the University of Washington led by Haggitt[19,20] produced strong evidence that only CLE with intestinal metaplasia was associated with an increased risk for adenocarcinoma. Patients with only nonintestinalized cardiac and oxyntocardiac epithelial types had no demonstrable risk. This suggested that carcinogens in gastric contents could only express its carcinogenicity if the visible CLE in the thoracic esophagus had intestinal metaplasia (Fig. 11.9).

At this time, the increasing incidence of esophageal adenocarcinoma was being recognized. Haggitt, in 1992,[21] declared esophageal adenocarcinoma, which was still in the early stages of the beginning of the rising trend of cancer, "an epidemic." Two decades later, with cancer rates having increased six- to sevenfold, Haggitt's "epidemic" of esophageal adenocarcinoma was described in a Wall Street Journal article as a "hidden killer."

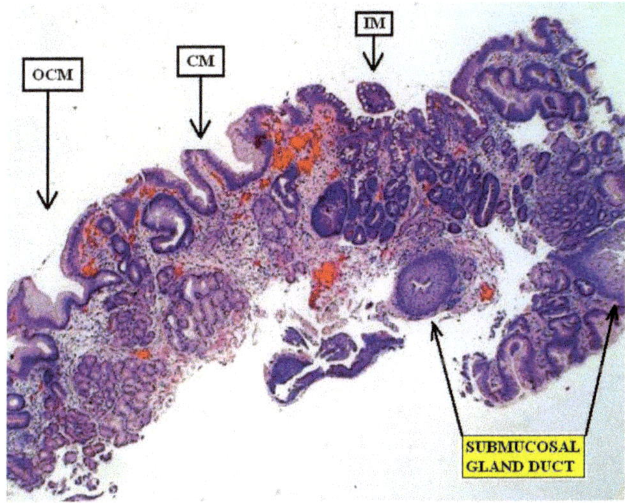

FIGURE 11.8 The three types of epithelium constituting metaplastic columnar epithelium in the esophagus. When all the three epithelia are present in columnar-lined esophagus, their distribution from cephalad distally is cardiac epithelium with intestinal metaplasia (to the right), cardiac epithelium (center), and oxyntocardiac epithelium (to the left). Note the squamous-lined ducts of submucosal glands that identify the location of this biopsy as esophageal. *CM*, cardiac mucosa; *IM*, intestinal metaplasia, *OCM*, oxyntocardiac mucosa.

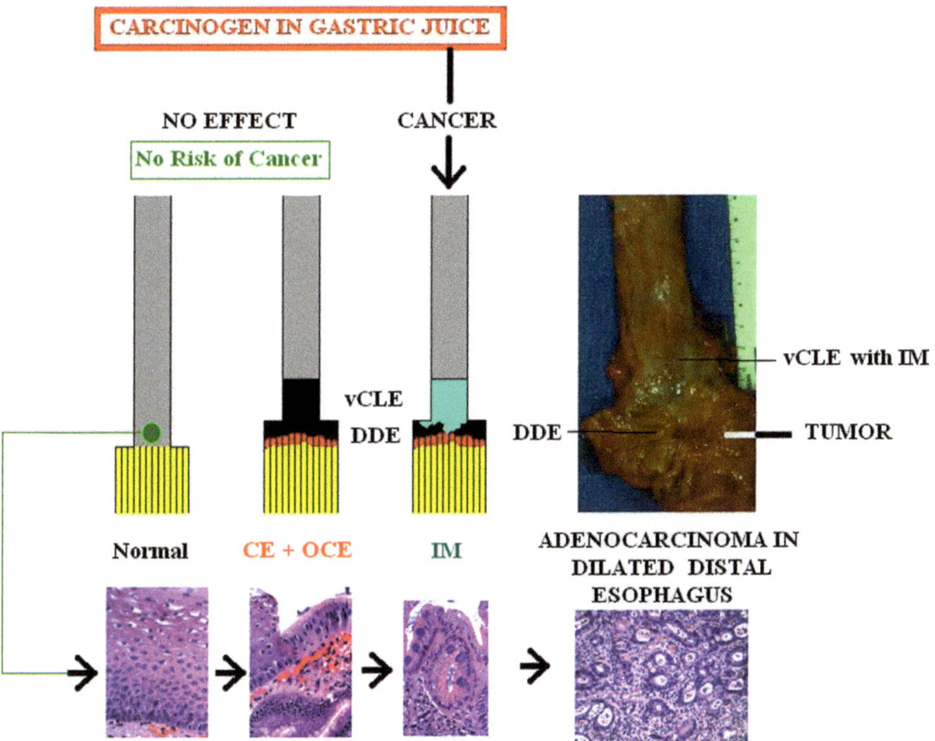

FIGURE 11.9 The target epithelium for carcinogenesis is marked by the presence of intestinal metaplasia (i.e., goblet cells) in cardiac epithelium. All other epithelia in the esophagus are not at proximate risk for carcinogenesis. They can develop a risk in the future if intestinal metaplasia develops. *CE*, cardiac epithelium; *DDE*, dilated distal esophagus; *IM*, intestinal metaplasia; *OCE*, oxyntocardiac epithelium; *vCLE*, visible columnar-lined esophagus.

The medical community had done an excellent job of not making known to the public one of the most dramatic increases in a cancer type in the history of medicine.

Unfortunately, in one of the greatest tragedies to medicine in the 1990s, Rodger Haggitt, one of the most astute gastrointestinal pathologists that ever lived, died tragically before his time. If he was alive, I am convinced by conversations I had with him before his death that he would have led the charge to control the increasing incidence of GERD-induced adenocarcinoma based on sound principles based on cellular pathology.

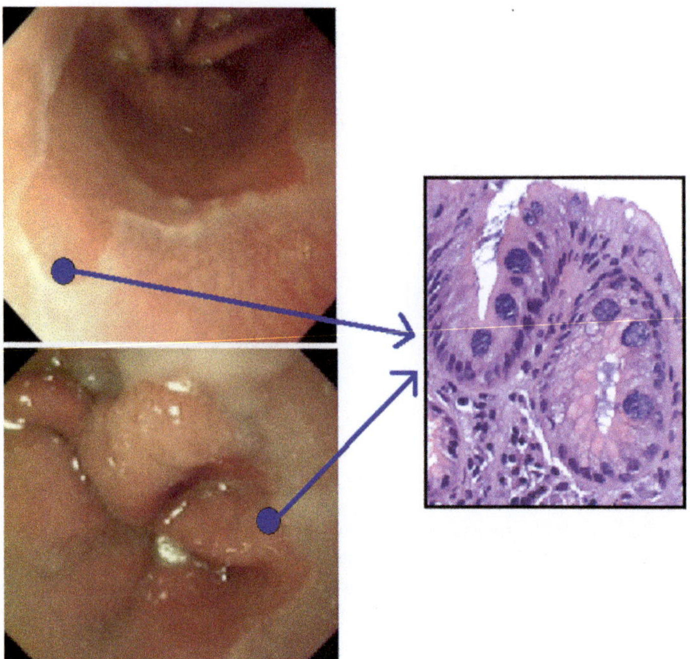

FIGURE 11.10 The definition of Barrett esophagus in the United States requires the presence of intestinal metaplasia in a biopsy taken from visible columnar-lined esophagus.

With the recognition that cancer risk was limited to the intestinal type of metaplasia, the definition of BE changed and became limited to the intestinal type of CLE (Fig. 11.10). The newly defined Barrett esophagus was now not only an entity that was caused by GERD-induced columnar metaplasia but also an important premalignant disease that indicated an increased risk of adenocarcinoma (Fig. 11.9).

If the eponym BE did not exist, there would have been no problem. CLE would be classified into its three histologic types, and the intestinal type would be the only type that was a marker for increased cancer risk. All three types would be diagnostic of GERD.

What should have happened is one of the two things: (1) BE should have continued to be used to be synonymous with visible CLE and classified as to the presence and absence of intestinal metaplasia or (2) the term visible CLE should have been retained and the term BE limited to CLE with intestinal metaplasia. The first would then indicate severe GERD and the latter a premalignant state with a risk of future adenocarcinoma.

Incredibly, neither of these happened. Endoscopists confused the issue by using the alternative terms at will. They kept using the term Barrett esophagus for endoscopic CLE despite the fact that the term now required histologic evidence of intestinal metaplasia. Many gastroenterologists still do this. When questioned, they change it to "endoscopic Barrett esophagus" to distinguish it from true BE with proven intestinal metaplasia. It becomes worse; In England and Japan, endoscopic CLE is equivalent to BE even in the absence of intestinal metaplasia (Table 11.1).

Even the Prague system[22] of measurement states it is a method of assessing BE, creating the illusion that BE is an endoscopic diagnosis. It is not. The Prague measurement is for visible CLE. It is not a measure of the amount of intestinal metaplasia within the columnar segment. Mixing the two terms results in profound misconceptions and prevents using the extent of intestinal metaplasia as a possible differential risk indicator in BE.

The typical reports of the endoscopist and pathologist faced with an endoscopically visible CLE are as follows:

Endoscopic Report: There was a visible segment of Barrett esophagus characterized by tongues of salmon pink mucosa with a total length of 1.5 cm and a circumferential involvement of 0.5 cm proximal to the proximal limit of rugal folds. Biopsies were taken.

Pathologic Report: Gastric cardiac–type epithelium with inflammation. No intestinal metaplasia identified.

Interpretation: The patient does not have Barrett esophagus (in the United States and Europe) or the patient has Barrett esophagus (in England and Japan).

"Not Barrett esophagus" is not a diagnosis. In the United States, no further consideration is given to the endoscopic abnormality of visible CLE when biopsies are negative for intestinal metaplasia. The patient is not placed under surveillance. In England, he/she is placed under surveillance.

TABLE 11.1 Classification of Columnar-Lined Esophagus by Histologic Types With Known Associated Cancer Risk With Variations in Relation to Historical Times, Present, and Future

Endoscopy	Histologic Types	Etiology	Risk of Adenocarcinoma
Past			
CLE = tubular stomach distal to GEJ (= SCJ)—before 1953	Gastric cardiac epithelium; IM±	Congenital gastric heterotopia	0; cancer not reported
CLE—after 1953 (Allison and Johnstone)	Gastric cardiac epithelium; IM±	Congenital gastric heterotopia	0; cancer not recognized
CLE—after 1957 (Barrett)	Cardiac epithelium; IM±	Metaplasia (?)	0; cancer not recognized
CLE < 2 cm—after 1961 (Hayward)	Biopsy not indicated; cardiac epithelium	Normal	Not recognized
CLE > 2 cm—after 1961 (Hayward)	Biopsy not indicated; cardiac epithelium	GERD	Not recognized
CLE > 2 cm—before 1976	Biopsy not necessary	GERD	Beginning to be recognized
CLE > 2 cm—after 1976 (Paull et al.)	IM+ (BE = LSBE)	GERD	Beginning to be recognized
CLE > 2 cm—after 1976 (Paull et al.)	Cardiac epithelium; IM–	GERD	0
CLE < 2 cm—after 1994 (Spechler et al.)	IM+ (SSBE)	GERD	Present
CLE < 2 cm—after 1994 (Spechler et al.)	Cardiac epithelium; IM–	GERD	0 (recently questioned)
Present			
No CLE at endoscopy	Biopsy not indicated; unknown	N/A; normal	Not known; no evidence
No CLE at endoscopy	If biopsy is done; IM+	IM of gastric cardia	Not known; no evidence
No CLE at endoscopy	If biopsy is done; IM–	Normal gastric cardia	0
CLE < 2 cm, at endoscopy	IM+	SSBE	Present
CLE < 2 cm, at endoscopy	IM–	Endoscopic SSBE (?); normal (?)	Present (in the United Kingdom) zero (in the United States);
CLE > 2 cm, at endoscopy	IM+	LSBE	Present
CLE > 2 cm, at endoscopy	IM–	Endoscopic LSBE	Zero (in the United States); present (in the United Kingdom)
Future			
No CLE at endoscopy	No cardiac epithelium	Normal	0
No CLE at endoscopy	Cardiac epithelium; IM+	IM in dilated distal esophagus	Probable; low; no evidence
No CLE at endoscopy	Cardiac epithelium; IM–	GERD/dilated distal esophagus	0
CLE at endoscopy	IM+	BE	Present
CLE at endoscopy	IM–	Endoscopic BE	Zero (in the United States) present (in the United Kingdom)

BE, barrett esophagus; *CLE*, columnar-lined esophagus; *GEJ*, gastroesophageal junction; *GERD*, gastroesophageal reflux disease; *IM*, intestinal metaplasia; *LSBE*, long-segment Barrett esophagus; *SCJ*, squamocolumnar junction; *SSBE*, short-segment Barrett esophagus.

In the new method of looking at GERD, the presence of visible CLE signifies failure of GERD treatment and a point at which the patient has developed a cellular change that does not reverse with and can progress to cancer despite medical therapy. The presence or absence of intestinal metaplasia *only* means that the patient has developed the essential target epithelium that is at risk for carcinogenesis.

A simple change in the precision of terminology (actually recommended by Barrett himself) may lead to an increase in diagnostic accuracy of both the endoscopic and pathology report of the same patient described earlier:

Endoscopic Report: There was a visible segment of columnar-lined esophagus characterized by tongues of salmon pink mucosa with a total length of 1.5 cm and a circumferential involvement of 0.5 cm between the proximal limit of rugal folds. Biopsies were taken.
Pathologic Report: Cardiac epithelium with inflammation. No intestinal metaplasia identified.
Interpretation: The patient has columnar-lined esophagus of cardiac type without intestinal metaplasia (all over the world).

This patient has visible CLE that is diagnostic for GERD. In fact, it indicates severe damage to the squamous epithelium of the body of the esophagus associated with severe reflux into the thoracic esophagus due to frequent sphincter failure [transient lower esophageal sphincter relaxation (tLESR)]. Whether it has intestinal metaplasia or not, visible CLE is the result of severe LES damage. It is always associated with a dilated distal esophagus, often >25 mm in length.

GERD is a progressive chronic disease. The lack of intestinal metaplasia at the time of the biopsy (even if one assumes there is no false negative diagnosis due to inadequate sampling) does not mean that intestinal metaplasia will not develop in the future. The biopsy result "negative for intestinal metaplasia" only means there is no risk of adenocarcinoma at that point in time.

Ignoring future cancer risk in such a patient is irrational. Some type of surveillance is necessary not for detection of cancer but for possible evolution to intestinal metaplasia. When this is done, as it is in England, intestinal metaplasia frequently becomes apparent in subsequent endoscopic biopsies.[23]

1.3.3 The Gastroesophageal Junction Is the Proximal Limit of "Gastric" Rugal Folds

Hayward's definition of the GEJ as the end of the tube (i.e., the point where the tube flares into the sac distal to it) is still used to define the GEJ endoscopically. It has issues, however, in the patient with a sliding hiatal hernia. In such cases, it is difficult to define the end of the tube.

To find a more accurate method of defining the GEJ in these patients, McClave et al.[14] (reviewed in Chapter 8) suggested a new endoscopic definition of the GEJ in 1987 as the proximal limit of "gastric" rugal folds. This was a reproducible endoscopic landmark. The study by McClave et al. had 18 patients with CLE diagnosed elsewhere and 4 controls without evidence of symptomatic GERD. The 18 patients had the SCJ displaced >2 cm from the proximal limit of rugal folds (12 had >3 cm displacement). Interestingly, the four controls had their SCJ separated from the proximal limit of rugal folds but by <2 cm (<1 cm in 3 controls). Three control patients had a hiatal hernia and one had mild esophagitis. Back then, the distal 2 cm of the esophagus was believed to be lined normally by columnar epithelium. This has now changed; any columnar epithelium above the proximal limit of rugal folds is now visible CLE. By today's criteria, therefore, all controls of McClave et al. had visible CLE.

In the study of McClave et al.[14] biopsies from the proximal limit of rugal folds contained cardiac and oxyntocardiac epithelium (junctional and fundic-type, following the terminology of Paull et al.[12]). These represent the proximal region of the dilated distal esophagus. The absence of gastric oxyntic epithelium at this level indicates that this point is proximal to the true GEJ by an unknown length.

McClave et al.[14] presented no valid evidence that the proximal limit of rugal folds had an accurate correlation with the true GEJ. They stated that it corresponded to the muscular esophagogastric junction, citing a prior autopsy report by Bombeck et al.[24] That citation (reviewed in Chapter 10) provided no such correlation; gastric mucosal folds were not even mentioned in that study.

This endoscopic landmark, despite the complete lack of evidence that it was the true GEJ, was rapidly accepted as a new and better definition of the GEJ at endoscopy, superseding the end of the tube as a definition. It could be used in every patient, irrespective of the presence of a sliding hiatal hernia.

The use of the proximal limit of gastric rugal folds to define the GEJ at endoscopy resulted in a new method of measuring what is believed to be the entire extent of CLE. By the Prague system, BE is measured from the most proximal point in the SCJ to the proximal limit of gastric rugal folds.

Errors have a habit of repeating themselves. Norman Barrett's incorrect definition of the GEJ (the distal limit of squamous epithelium[13]) resulted in the entire CLE being regarded as a tubular intrathoracic stomach. The present incorrect definition of the GEJ (the proximal limit of rugal folds[14]) causes the visible CLE to be recognized but results in the distal region of the CLE to be incorrectly regarded as the proximal stomach (Fig. 11.4).

I have provided evidence in Chapter 6 that this endoscopic definition of the GEJ is incorrect except in normal people who do not have a dilated distal esophagus. The proximal limit of rugal folds lies proximal to the dilated distal esophagus, which develops rugal folds. The true distal limit of CLE is the distal limit of the dilated distal esophagus where metaplastic oxyntocardiac epithelium transitions into gastric oxyntic epithelium. When this is accepted, the entire length of the CLE will be recognized (Fig. 11.4).

As LES damage increases and the length of the dilated distal esophagus increases, the endoscopic GEJ separates further from the true GEJ (the proximal limit of gastric oxyntic epithelium). The critical elongation of the dilated distal esophagus that signifies increasing LES damage comes to be interpreted as normal gastric cardia. Incredibly, by virtue of this error, many expert gastroenterologists hold the opinion that the gastric cardia expands with increasing GERD. GERD cannot cause pathology in the stomach.

1.4 The Birth of Short-Segment Barrett Esophagus

In the 1961–94 period, BE was defined as the presence of intestinal metaplasia in a biopsy taken from a patient with a CLE exceeding 2 (or 3) cm in length. The presence of columnar epithelium in the distal 2 (or 3) cm of the esophagus was regarded as "normal" and therefore not subjected to biopsy. This changed in 1994.

In a paper published in Lancet in 1994, Spechler et al.[25] summarized the state-of-the-art endoscopy that existed at that time: "Columnar epithelium in the oesophagus has a characteristic red color and velvet-like texture that contrasts sharply with the pale, glossy appearance of adjacent squamous epithelium. Although endoscopic examination can usually distinguish columnar epithelium from squamous epithelium in the oesophagus, the several types of columnar epithelia cannot be differentiated on endoscopic appearance alone. The distinction between specialized columnar epithelium (SCE) and gastric columnar epithelium can be made only by histologic examination of biopsy specimens. Gastric columnar epithelium may normally line a short segment of the distal esophagus. Therefore endoscopists usually diagnose Barrett's oesophagus only when they see columnar epithelium extending well above the gastro-oesophageal junction… short segments of SCE in the distal oesophagus are not recognized as abnormal… Endoscopists usually do not obtain biopsy specimens from columnar epithelium that involves only a short segment of the distal oesophagus, and therefore the characteristics of patients who have short segments of SCE at the gastro-oesophageal junction have not been well defined" (p. 1533).

The study of Spechler et al.[25] included all adults scheduled for elective endoscopy. For patients without a hiatal hernia, the GEJ was defined as the point where the tubular esophagus flared to become a saclike structure. For patients with hiatal hernia, the GEJ was defined as the proximal margin of rugal folds in the hernia pouch. Two biopsy specimens were obtained from the columnar epithelium at the SCJ to seek specialized columnar epithelium (SCE) (i.e., cardiac epithelium with intestinal metaplasia). One biopsy specimen was obtained from the squamous epithelium to seek esophagitis.

This was a population where a majority of the subjects had clinical evidence of GERD. 86% of patients with visible CLE (i.e., CLE>3 cm) and ~60% of those without visible CLE (i.e., CLE<3 cm) had a symptom of GERD (Table 11.2).

The patients were divided into two groups by their endoscopic findings:

1. 16 patients with endoscopically apparent Barrett esophagus who had columnar epithelium extending >3 cm from the GEJ. All these 16 patients had SCE (i.e., cardiac epithelium with intestinal metaplasia).

TABLE 11.2 Clinical Correlates of the Patients in the Study of Spechler et al.[8]

	Endoscopic Barrett's (i.e., >3 cm)	No Endoscopic Barrett's (i.e., <3 cm)	No Endoscopic Barrett's (i.e., <3 cm)
	SCE on biopsy (16/16)	SCE on biopsy (26/140)	No SCE on biopsy (114/140)
M:F	10/6 (1.7:1)	17/9 (1.9:1)	49/65 (0.8:1)
White race	16 (100%)	26 (100%)	99 (87%)
Heartburn	11 (69%)	14 (54%)	62 (54%)
Any GERD symptom	12 (86%)	15 (58%)	72 (63%)
Esophagitis (endoscopic/histologic)	8 (50%)	19 (73%)	30 (44%)

GERD, gastroesophageal reflux disease; *SCE*, specialized columnar epithelium.

2. 140 patients without endoscopically apparent Barrett esophagus who had <3 cm of CLE (Fig. 11.2). 26/140 of this group had SCE (i.e., cardiac epithelium with intestinal metaplasia); the other 114 did not. By default, these patients must have had cardiac epithelium without intestinal metaplasia. The introduction of Spechler et al. suggests that this was interpreted as "gastric cardiac epithelium" despite the fact that it was above the GEJ, and according to the belief back then, normal in the distal esophagus.

The 26 patients in group (2) with intestinal metaplasia are described in detail: "in 9 (35%) the epithelial squamocolumnar junction and the gastro-oesophageal junction were at the same level so that the squamocolumnar junction appeared as a relatively straight line at the very end of the oesophagus. In 17 patients (65%), columnar epithelium extended up to 3 cm into the distal esophagus. In these cases the squamocolumnar junction appeared as a wavy or zigzag line, occasionally with eccentric tongues of columnar epithelium projecting as far as 3 cm up the oesophagus" (p. 1534).

This paper was responsible for the ultimate recognition of SSBE. The data that 26/140 (18.6%) of patients with <3 cm of CLE had intestinal metaplasia were considered important because of the understanding that SCE (cardiac epithelium with intestinal metaplasia) was an abnormal finding. This resulted in a change in the guidelines for biopsy at endoscopy. Biopsy was recommended when any columnar epithelium was seen at endoscopy rather than the prior requirement of >3 cm CLE.

The recognition of SSBE changed the definition of endoscopic normalcy as an esophagus lined *entirely* by squamous epithelium, at least to precipitate a biopsy. The presence of SCE (i.e., cardiac epithelium with intestinal metaplasia) in a biopsy from any endoscopically visible CLE was diagnostic of BE with those patients who had <3 cm being designated as "SSBE" to distinguish this lesion from LSBE, which was diagnosed when >3 cm of visible CLE was present.

Over the next decade, SSBE was shown to be an indicator for increased cancer risk. It is interesting that failure to biopsy CLE segments <2 cm between 1980 and 1994 must have resulted in the failure to identify cancer risk in up to 20% of patients with visible CLE < 3 cm at endoscopy.

2. THE PRESENT STATE OF THE ART

The present state of the art is one of confusion between the terms "columnar-lined esophagus," which is used variably and interchangeably with "Barrett esophagus" by different gastroenterologists at different times. Terms such as "endoscopic Barrett esophagus," "ultrashort-segment Barrett esophagus," "histologic Barrett esophagus," and "Barrett esophagus, confirmed," confuse the issue further.

2.1 Present Definition of Columnar-Lined Esophagus

CLE is defined now as the presence at endoscopy of columnar epithelium between the SCJ and the proximal limit of rugal folds. Unfortunately, it is uncommon for endoscopists to use the accurate term "visible CLE" for this finding. Most endoscopy reports will describe this as BE.

Visible CLE may be circumferential (Fig. 11.3) or seen as tongues of salmon-pink columnar epithelium extending upward into the squamous epithelium causing serration of the SCJ or Z-line (Fig. 11.2). The difficulty in recognizing very small amounts on noncircumferential columnar epithelium has made some authorities require that a minimum of 5 mm of CLE be seen before the diagnosis is made. The guidelines of the British Society of Gastroenterology requires visible CLE to be >1 cm.[26] Other authorities do not have this minimum requirement; any visible CLE is acceptable. This results in a variation of data between papers that cannot be reconciled.

The exact position of the proximal limit of rugal folds tends to vary with insufflation of air into the stomach during endoscopy. Some papers make it a point of reporting the state of insufflation in their method of evaluating the position of the endoscopic GEJ. Most do not. It is theoretically possible to create an artificially visible CLE by producing gastric overdistension with air at endoscopy.

The person whose SCJ is not separated from the endoscopic GEJ (proximal limit of rugal folds) does not have CLE according to present thinking. This is an error created by the false definition of the GEJ. Columnar metaplasia extends distal to the endoscopic GEJ into the dilated distal esophagus. It is appropriate to use the lack of visible CLE to define the normal endoscopic appearance.

However, it must be recognized that this normal endoscopy does not define the normal state. The true GEJ is distal to the endoscopic GEJ by the length of the dilated distal esophagus defined by histology (Fig. 11.4).

When a visible CLE is identified or suspected at endoscopy, measurement of its extent is recommended but seen only in sporadic endoscopy reports. This method of measurement is standardized as the Prague classification.[22]

To quantify visible CLE, the following steps are required: (1) Identify the endoscopic GEJ by the proximal limit of rugal folds. The use of the proximal limit of rugal folds to define the GEJ is essential in patients with a sliding hiatal hernia.

(2) Measure to the nearest centimeter the distance of the GEJ from the incisor teeth. (3) Measure to the nearest centimeter the distance from the incisor teeth of the most proximal circumferential extent of suspected columnar metaplasia. (4) Measure to the nearest centimeter the distance from the incisor teeth the maximum extent of suspected columnar metaplasia (tops of tongue-like extensions). (5) Subtract measurement of #3 and #4 from #2.

For example, if the endoscopic GEJ is at 38 cm, the circumferential segment extends to 36 cm and the maximum tongue is at 33 cm, the Prague designation for this endoscopic CLE is **C2 M5**.

If a patient with GERD has no visible CLE at endoscopy, biopsy is not recommended in both the British and American Gastroenterology Association guidelines.

2.2 Present Definition(s) of Barrett Esophagus

The objective of defining BE is not clearly stated. Two objectives should be satisfied: (1) to define an entity by criteria that are proven to indicate an increased risk of future adenocarcinoma and (2) to define criteria without which adenocarcinoma cannot arise.

The objective should not be to simply define future increased risk of cancer. All patients with GERD, irrespective of the presence of symptoms or any cellular change at the time of examination, are at risk for future adenocarcinoma. For example, a symptomatic GERD patient has a proven risk of future adenocarcinoma in the thoracic esophagus even if that patient has no visible CLE. The cellular change must progress from endoscopically normal squamous epithelium in the thoracic esophagus that is exposed to reflux→visible CLE→intestinal metaplasia→increasing dysplasia→adenocarcinoma. Each of these states has a proven risk of future adenocarcinoma because of progression along this sequence with time.

The term Barrett esophagus should define the point in this sequence that the cellular change has made the epithelium susceptible to carcinogenesis and without which cellular changes leading to cancer will not occur.

Certainly squamous epithelium exposed to reflux without visible CLE cannot transform directly to adenocarcinoma. There is controversy as to whether cardiac epithelium in CLE can transform directly to adenocarcinoma without passing through intestinal metaplasia. The answer to that controversy is a definite no. I will address this issue in Chapter 12.

The practical objective of defining BE is to determine the cellular criterion that will define the need for endoscopic surveillance. This is not a matter of cancer risk; it is a matter of resources. Theoretically, all patients with GERD will benefit from endoscopic surveillance. This is not feasible. With resources available at this point in time, endoscopic surveillance is recommended only for patients who have endoscopy (indicated in the United States only for patients who develop treatment failure or alarm symptoms like dysphagia) when visible CLE with intestinal metaplasia is found at histology. Even in this group, questions are being raised regarding the cost-effectiveness of surveillance.

Even this method of selecting patients for endoscopic surveillance in GERD is not completely rational. Nason et al.[6] showed that GERD patients whose symptoms were well controlled with PPI therapy had a higher prevalence of visible CLE with intestinal metaplasia than those patients who were not well controlled. The authors conclude that their study "highlights potential causes for the ineffectiveness of the current cancer-screening guidelines that recommend screening for BE only in patients with poorly controlled GERD."

The simple definition of BE used by most gastroenterologists the world over is: "The presence of intestinal metaplasia in a biopsy taken from an endoscopically visible segment of columnar lined esophagus."

The definition recommended by the British Society of Gastroenterologists is[26]: "Barrett's oesophagus is defined as an oesophagus in which any portion of the normal distal squamous epithelial lining has been replaced by metaplastic columnar epithelium, which is clearly visible endoscopically (≥1 cm) above the GOJ and confirmed histopathologically from oesophageal biopsies." The biopsies are to confirm CLE of any histologic type; there is no requirement for intestinal metaplasia.

The most recent definition of BE by the AGA is[27]: "The condition in which any extent of metaplastic columnar epithelium that predisposes to cancer development replaces the stratified squamous epithelium that normally lines the distal esophagus. At present, intestinal metaplasia is required for the diagnosis of Barrett's esophagus because intestinal metaplasia is the only type of esophageal columnar epithelium that clearly predisposes to malignancy."

In the discussion, however, the expert panel indicates that although there is no evidence to support nonintestinalized epithelium being at risk, it is a matter worthy of consideration.

This new American definition was designed to accommodate the opinion expressed by experts in pathology who express the view that CLE without intestinal metaplasia is an at-risk epithelium.[28] This is not an issue. Obviously cardiac epithelium has a risk of adenocarcinoma in the future, only because it can progress with time to develop intestinal metaplasia. I will provide evidence in Chapter 12 that cardiac epithelium never develops adenocarcinoma directly without the intermediate step of intestinal metaplasia.

The more interesting element of the new American definition is that it does not require the presence of visible CLE for a diagnosis of BE. I am not certain whether this was omitted intentionally or by oversight. Whatever the reason, the new definition opens the door to recognizing intestinal metaplasia of the dilated distal esophagus as a premalignant epithelium. Although not accepted, the dilated distal esophagus is indeed the distal esophagus and falls within the criteria of the new American definition. Although the magnitude of risk of cancer of intestinal metaplasia in the dilated distal esophagus is not known, there is strong evidence that it is the precursor of adenocarcinoma of the dilated distal esophagus.

Ultimately, when the dust settles the definition of BE will be simple: "Barrett esophagus is columnar lined esophagus with intestinal metaplasia." This is the only epithelium in the esophagus that is susceptible to carcinogenesis. Adenocarcinoma will not arise in a person who does not have intestinal metaplasia in the esophagus. The limitation of a requirement for visible CLE will disappear when more evidence accumulates that intestinal metaplasia in the dilated distal esophagus is the precursor lesion for adenocarcinoma of the dilated distal esophagus, at present misunderstood as proximal cardia by many.

2.3 Resolution of the Confusion Between Columnar-Lined Esophagus and Barrett Esophagus

The terminology of columnar-lined and Barrett esophagus is extremely confusing at present. The resolution of this confusion is simple. The abnormality where a columnar segment is seen between the endoscopic GEJ (proximal limit of rugal folds) and the SCJ is visible CLE. Visible CLE is subject to biopsy. Depending on the biopsy result, visible CLE is divided into visible CLE with and without intestinal metaplasia.

If the term Barrett esophagus is to be used, a global consensus must be developed as to what this term means: the present majority global viewpoint is that BE is limited to visible CLE with intestinal metaplasia. This is the only epithelial type that is proven to be susceptible to carcinogens that lead to adenocarcinoma.

In the Montreal document,[29] the consensus resolution of this conflict in terminology by the experts is the following statement #33: "The term Barrett's esophagus is variably interpreted at the present time and lacks the clarity needed for clinical and scientific communication about columnar metaplasia of the esophageal mucosa."

This statement, though reaching consensus, had significant disagreement among the experts. In the discussion: "The term 'Barrett's esophagus' is currently confusing and ambiguous because the spectrum of what is currently referred to as 'Barrett's esophagus' ranges from some clinicians making this diagnosis solely on the basis of endoscopic appearance of any extent, to the requirement that intestinal-type esophageal columnar epithelium be proven histologically before this diagnosis is made."

The disagreement between experts is significant. In the AGA Chicago workshop, reported in 2004,[30] only 72% of 18 physicians reviewing the literature on BE agreed that intestinal metaplasia documented by histology was a prerequisite for the diagnosis of BE; 16% had major reservations with this requirement and 12% rejected it.

The simple solution of dropping the eponym is deemed impossible. Statement #38 of the Montreal document[29] concluded as follows: "… it was decided that the eponymous term 'Barrett' should be retained in any definition because it would be futile and counterproductive to try to remove such an embedded use from general use."

Surely, this is not true. The entire population of the United Kingdom changed from pounds-shillings-pence to a decimal system, the European Union (except Britain) unified to adopt the Euro, the entire world (except the United States) changed seamlessly from tons-pounds-ounces and feet-inches to the decimal system. Why not Barrett's esophagus where change is necessary only among highly trained physicians?

The discussion within statement #38[29] itself has one rational basis for a solution to the terminology that is similar to the one I have proposed: "Columnar lined esophagus should be called Barrett's esophagus and the presence or absence of intestinal-type metaplasia specified." Unfortunately, it is still uncommon for any gastroenterologist or pathologist to follow this part of the recommendation. The term Barrett esophagus continues to be used in a manner that is ambiguous.

I suggest that we go back to what Norman Barrett said about this entity when he recommended the term "columnar-lined esophagus:" "This paper concerns a condition whose existence is denied by some, misunderstood by others, and ignored by the majority of surgeons… At the present time, the most accurate description is that it is a state in which the lower end of the esophagus is lined by columnar epithelium. This does not commit us to ideas which could be wrong, but it carries certain implications which must be clarified."

The terminology from Allsion and Johnstone's "esophagus lined by gastric mucous membrane" to "Barrett's columnar-lined esophagus" changed on a dime. I believe that Mr. Barrett would have preferred that we follow his recommendation rather than introduce confusion by applying his name for the entity.

In this book, I have avoided this confusion in a simple manner. I refer to the presence of CLE seen at endoscopy as "visible CLE" and qualify whether it has intestinal metaplasia or not based on the result of biopsy. I have always qualified the term Barrett esophagus when used as "visible CLE with intestinal metaplasia." The findings of visible CLE and intestinal metaplasia have independent value. The use of the term Barrett esophagus in the way it is done at present causes the significance of the endoscopic finding of visible CLE without intestinal metaplasia to diminish.

2.4 What Is Visible Columnar-Lined Esophagus Without Intestinal Metaplasia?

The presence of visible CLE without intestinal metaplasia is a critical event in the new understanding of GERD that I present in this book. It is a manifestation of GERD that indicates both severe reflux into the thoracic esophagus and the point of irreversibility of the cellular change in the adenocarcinoma sequence. Visible CLE does not reverse with PPI therapy. It progresses in successively diminishing numbers of patients to intestinal metaplasia, increasing dysplasia, and adenocarcinoma.

The interpretation of the results from the study of Spechler et al.[25] reviewed above is a good example of how gastroenterologists presently view visible CLE when the biopsies are negative for intestinal metaplasia. The study simply ignores this group from any further analysis or discussion. Does that patient simply become "not SSBE?"

The visible CLE<3 cm in these patients will consist of cardiac epithelium without intestinal metaplasia. I am not certain whether gastroenterologists regard these patients as having CLE (by endoscopy) with metaplastic cardiac epithelium or whether they regard these patients as normal (i.e., normal cardiac epithelium in the distal esophagus). In either event, these patients with endoscopic CLE who have cardiac epithelium are basically ignored. The finding has no diagnostic connotation; pathologists often will report these biopsies as "cardiac epithelium with inflammation" or "gastric cardiac epithelium." Gastroenterologists are only interested in whether the pathologist reports intestinal metaplasia. If not, the patient simply goes back into the pool of patients being treated with PPIs and are not subject to any kind of surveillance in the United States.

In most papers relating to BE, including Spechler et al.,[25] the number of patients with visible CLE is stated and those who do not have intestinal metaplasia are completely ignored in the analysis as if they have no consequence. In Westhoff et al.[31] and Gerson et al.,[32] visible CLE without intestinal metaplasia is not of enough significance to even be enumerated; only the number of patients with BE (i.e., visible CLE + intestinal metaplasia) is mentioned. In the Pro-GERD study,[4] visible CLE is given importance and listed as "Barrett esophagus" in contrast to visible CLE with intestinal metaplasia, which is termed "Barrett esophagus, confirmed." In studies from England, because the definition of BE does not need intestinal metaplasia, visible CLE is equivalent to BE.

In reality, the patient with visible CLE composed of cardiac epithelium has histologic evidence of reflux into the thoracic esophagus of sufficient severity to cause cardiac metaplasia of esophageal squamous epithelium. This is 100% specific for GERD. However, this obvious histologic criterion for GERD is ignored. Except in our unit, where the finding of cardiac epithelium in a biopsy from anywhere in this region receives a diagnosis of "reflux carditis" and has been a recognized criterion for GERD for the past 20+ years.[33]

2.5 Intestinal Metaplasia at the Squamocolumnar Junction in the Person Without Visible Columnar-Lined Esophagus

In the study of Spechler et al.[25] 9 of the 26 patients with SCE (cardiac epithelium with intestinal metaplasia) had normal endoscopy (i.e., no visible CLE with the SCJ extending to the proximal limit of rugal folds at the end of the esophagus). This is also abnormal by virtue of the presence of SCE (i.e., cardiac epithelium with intestinal metaplasia), exactly the same as the presence of SCE in patients with visible CLE<3 cm.

The significance or lack thereof of intestinal metaplasia in a biopsy taken at the endoscopically normal SCJ in a person, with or without symptoms of GERD, is a conundrum that exists to the present day. AGA guidelines for taking biopsies recommend that biopsies be not taken in a person without visible CLE.[2] The rationale for this is that the SCJ is the GEJ in a person who is endoscopically normal. As such, any columnar epithelium distal to the SCJ is "gastric." The presence of any type of epithelium and any pathologic abnormality (other than dysplasia or adenocarcinoma) has no known significance.

This is a circular argument that has no merit. The reason why the significance of this obviously pathologic lesion is unknown is because biopsies have never been taken as a routine in these patients. As a result, the significance of cardiac epithelium (with and without intestinal metaplasia) remains unknown.

When it is recognized that the present endoscopic definition of the GEJ is incorrect, the interpretation of the finding of intestinal metaplasia at the SCJ in the endoscopically normal patient becomes clear. This finding represents intestinal metaplasia in the dilated distal esophagus (see Chapter 10). There are very few studies with long-term follow-up in this group of patients, largely because of the failure to biopsy endoscopically normal people. If biopsies were recommended in this group in 1994 when the data of Spechler et al.[25] were published, the information regarding cancer risk in this group that would have accumulated would have been invaluable.

Leodolter et al.[34] showed that patients with intestinal metaplasia in biopsies taken at or immediately distal to the proximal limit of rugal folds had a 25.8% rate of progressing to visible CLE within 5 years. This finding indicates (1) that there is a clear link between intestinal metaplasia distal to the endoscopic GEJ and visible CLE. Leodolter et al.[34] call this "histologic Barrett esophagus" suggesting the link. (2) That there is a good reason to take biopsies from endoscopically normal patients with GERD. Despite these powerful data, the AGA recommendation continues to be to not take biopsies from the endoscopically normal GEJ.[2]

3. THE FUTURE: THE COMPLETE UNDERSTANDING OF COLUMNAR-LINED ESOPHAGUS

The eradication of the dogma of the endoscopic GEJ being the proximal limit of rugal folds and cardiac epithelium being gastric in location results is the definition of the dilated distal esophagus. This is lined by the same epithelial types (cardiac epithelium with and without intestinal metaplasia and oxyntocardiac epithelium) that have been recognized in visible CLE in the thoracic esophagus from the earliest time.

3.1 The Origin of Columnar-Lined Esophagus From the Normal State

CLE is defined histologically as cardiac epithelium with and without parietal and/or goblet cells (i.e., cardiac epithelium with or without intestinal metaplasia and oxyntocardiac epithelium). The beginning CLE *always* begins at the normal state where the entire esophagus is lined by squamous epithelium and the entire proximal stomach by gastric oxyntic epithelium (i.e., the squamooxyntic gap is 0). There is normally no cardiac epithelium.[7]

Once it has begun at the true GEJ (proximal limit of gastric oxyntic epithelium) CLE *always* progresses in a cephalad direction (Fig. 11.11). Measurement of cardiac mucosal length must *always* begin at the proximal limit of gastric oxyntic epithelium. When this is done, it is an extremely sensitive indicator of abdominal LES damage. At present, cardiac epithelial length is measured from the incorrect endoscopic GEJ in a distal direction. This results in complete chaos. It gives the totally erroneous impression that the gastric cardia lengthens as the severity of GERD increases.

Distal esophageal columnar metaplasia of esophageal squamous epithelium occurs as a result of exposure to gastric juice.[35,36] The proximal limit of gastric oxyntic epithelium remains as a constant marker of the true GEJ because it is unaffected by gastric juice. I have described the mechanism whereby distal squamous epithelium is exposed to gastric contents during times of gastric overdistension in Chapters 7 and 9.

Starting from the true GEJ, the CLE lengthens in two sequential steps that are completely different in their pathogenesis: (1) Progressive increase in abdominal LES damage with CLE limited to the dilated distal esophagus[37] and (2) the occurrence of visible CLE in the thoracic esophagus (Table 11.3).

3.2 Increase in Length of Microscopic Columnar-Lined Esophagus in the Dilated Distal Esophagus Within the Reserve Capacity of the Lower Esophageal Sphincter

From its onset, abdominal LES damage slowly increases at a variable rate measured at 1–10 mm/decade. Concordant with LES damage, the columnar metaplastic segment in the dilated distal esophagus increases in length. Theoretically, 15 mm of dilated distal esophagus represents the reserve capacity of the LES. Damage causing up to 15 mm of LES shortening leaves a residual abdominal LES of >20 mm (assumed initial length of 35 mm) (Fig. 11.12). At this length the LES is capable of maintaining competence at rest and can withstand a 10 mm further dynamic shortening of the LES associated with gastric overdistension of a heavy meal without reaching the <10 mm critical point of abdominal LES length that is associated with LES failure (tLESR). The LES remains competent and no significant reflux occurs into the thoracic esophagus. The squamous epithelium above the endoscopic GEJ is protected.

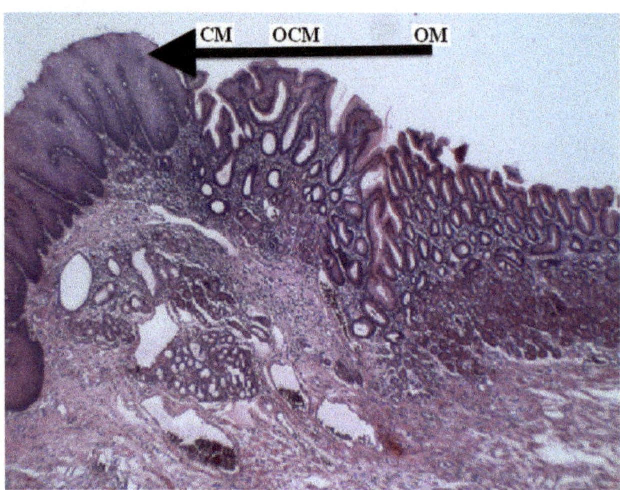

FIGURE 11.11 The direction of progression of cardiac metaplasia of squamous epithelium is from the true gastroesophageal junction (the proximal limit of gastric oxyntic epithelium—OM) in a cephalad direction (*arrow*) to involve increasing amounts of squamous epithelium. This shows very early (2 mm) columnar-lined esophagus with cardiac mucosa (CM) and oxyntocardiac mucosa (OCM) epithelia that have replaced the distal 2 mm of esophageal squamous epithelium. *OM*, oxyntic mucosa.

TABLE 11.3 Stages in the Sequential Progression of Columnar-Lined Esophagus (CLE), Beginning at the Normal State Where CLE Is Absent

	Manometric Abd-LES Length	Length of CLE in DDE	LES Failure/Reflux Into Thoracic Esophagus	Symptoms of GERD	Visible CLE Above Endoscopic GEJ
Normal	35 mm	0	0	0	0
Compensated LES damage	<35–20 mm	>0–15 mm	0	0	0
Postprandial LES incompetence	<20–10 mm	>15–25 mm	Rare/mildly abnormal pH test	Postprandial heartburn	0/at minimal risk
Fasting LES incompetence	<10–0	>25–35 mm	Frequent/severely abnormal pH test	Severe GERD symptoms	Present/at high risk

Abd-LES, abdominal segment of lower esophageal sphincter; *DDE*, dilated distal esophagus; *GERD*, gastroesophageal reflux disease.

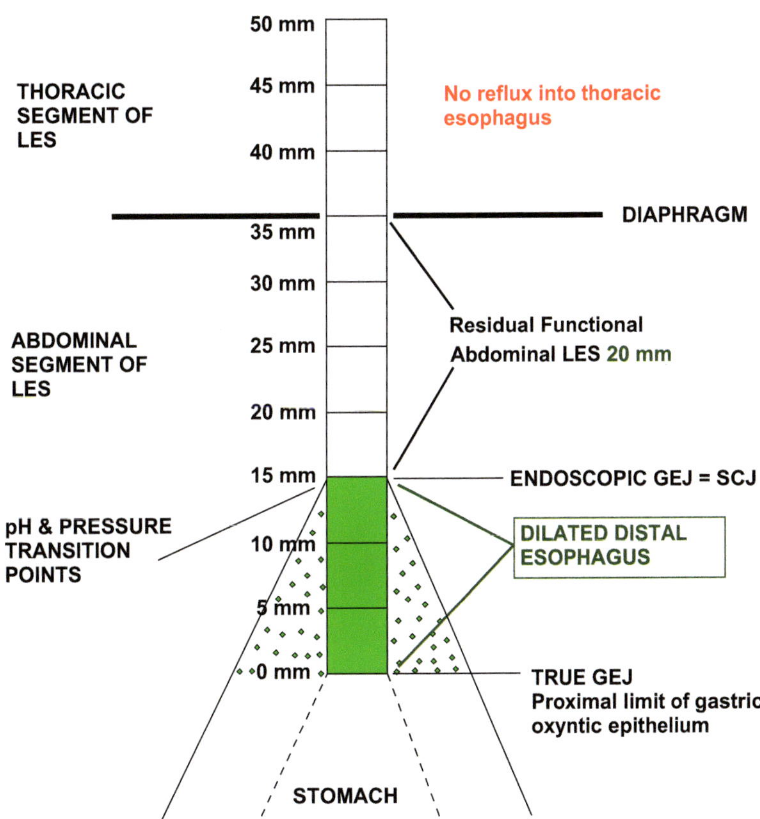

FIGURE 11.12 Progression of cardiac metaplasia to a length of 15 mm in the dilated distal esophagus, associated with 15 mm of lower esophageal sphincter (LES) damage. This is the end point of the reserve capacity of LES damage where the LES still retains competence. The person has no reflux. *GEJ*, gastroesophageal junction; *SCJ*, squamocolumnar junction.

This is the phase where LES damage is present but compensated by the reserve capacity of the abdominal LES. The only abnormality in the patient is a dilated distal esophagus lined by >0–15 mm of cardiac epithelium (with and without parietal and/or goblet cells).

3.3 Clinical Gastroesophageal Reflux Disease (Without Visible Columnar-Lined Esophagus)

The next stage represents further increase in the length of CLE in the dilated distal esophagus between >15 and 25 mm. The abdominal LES has exhausted its reserve capacity. The residual abdominal LES length is <20–10 mm (Fig. 11.13).

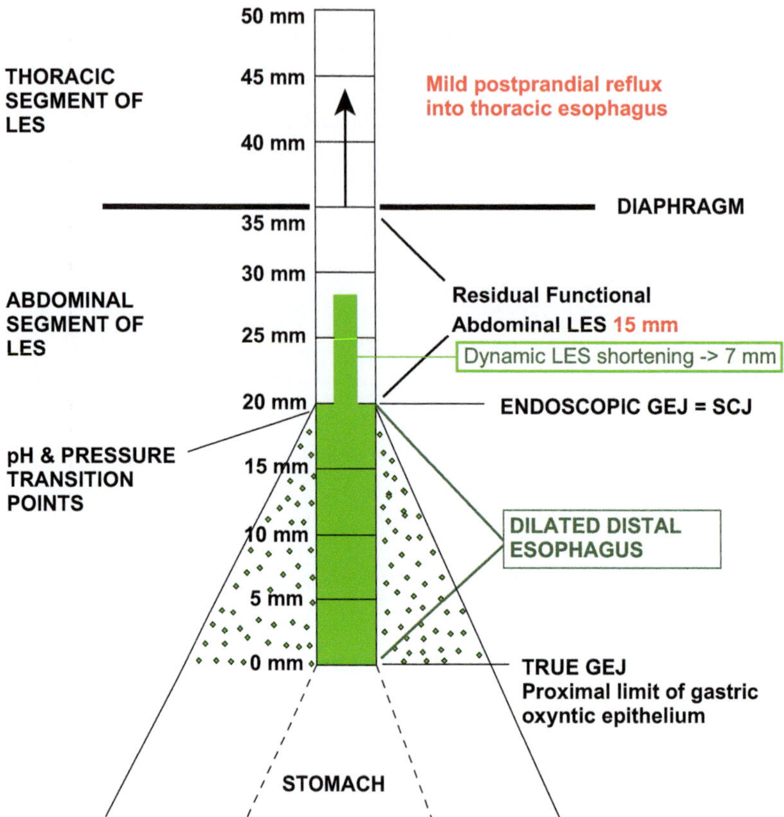

FIGURE 11.13 Progression of cardiac metaplasia to a length of 20 mm in the dilated distal esophagus. The residual functioning abdominal lower esophageal sphincter (LES) is now 15 mm. With dynamic shortening of 8 mm of the LES during a heavy meal, the postprandial LES length has become 7 mm, below the <10 mm threshold for LES failure. Postprandial reflux is now possible. *GEJ*, gastroesophageal junction; *SCJ*, squamocolumnar junction.

With gastric overdistension of a very heavy meal that causes dynamic shortening of 10 mm or more, the functional abdominal LES length decreases to <10 mm where the LES fails and reflux begins. Rare postprandial reflux episodes occur into the thoracic esophagus with the heaviest of meals at the beginning of this stage. The patient can compensate for this damage voluntarily by limiting meal size and content, prompted by the knowledge of the type of meal that precipitates GERD symptoms. Most patients in this group will either have no symptoms or symptoms that do not meet the troublesome requirement that is necessary for the diagnosis of clinical GERD.

As the CLE in the dilated distal esophagus (and abdominal LES damage) progresses from >15 to 25 mm, the residual abdominal LES length decreases from <20 to 10 mm. The LES retains competence during the fasting phase but becomes increasingly susceptible to failure with increasingly smaller meals. Voluntary control of meal size and content may control LES damage but this becomes increasingly difficult. Intermittent LES failure (tLESR) begins to occur increasingly frequently with meals as the LES damage reaches 25 mm.

At this point within this stage, the patient usually develops typical GERD symptoms and acid exposure progressively increases from near zero in the beginning of this stage to become abnormal (pH<4 for >4.5% of the period).

LES failure (tLESR) causes reflux into the thoracic esophagus along the pressure gradient that naturally exists from the stomach (+5 mmHg) to midesophageal level (nadir of negative pressure in the thoracic esophagus). The column of gastric contents refluxed reaches the midesophagus and higher. This results in simultaneous exposure of a large area of thoracic esophageal squamous epithelium, resulting in damage. Initially, this damage is microscopic, progressing to increasing grades of erosive esophagitis. There is a time period in the course of the disease where the patient is symptomatic but has no visible CLE. This is the "window of opportunity" to prevent visible CLE that is wasted by the present management algorithm.

By present management guidelines, the patient is placed on empiric acid-reducing therapy with the objective of controlling symptoms. The following things can happen to this patient, unbeknownst to the physician:

1. The patient remains in control of symptoms for the rest of his/her life without progressing to visible CLE.
2. The patient remains in control of symptoms for the rest of his/her life but develops visible CLE that does not progress to adenocarcinoma.

3. The patient remains in control for the rest of his/her life but develops visible CLE that progresses to adenocarcinoma, usually advanced, without ever developing an indication for endoscopy.
4. The patient's symptoms fail to be controlled by PPI therapy. Endoscopy is indicated only in these patients.

There is absolutely no ability or attempt to stratify these patients into these categories at the time of beginning treatment. The stages are recognized only when they reach their end point.

The ability to manage advanced adenocarcinoma and failure of PPIs to control symptoms is highly unsatisfactory. Patients with adenocarcinoma have a high mortality; those with treatment failure have their quality of life disrupted. The inability of recognizing these groups at the beginning of treatment causes the window of opportunity to be lost.

The reason for the existence of these different patient groups is that their LES damage progresses unimpeded by PPI therapy and their end point is determined by the rate of progression of their LES damage. Let us assume that the onset of symptoms begins with abdominal LES shortening of 15 mm and that treatment failure occurs at 25 mm damage. This represents an incremental shortening of 10 mm of abdominal LES damage. Let us assume the patient is 40 years old at the onset of symptoms. If the rate of abdominal LES shortening is 2 mm/decade, the patient will be 90 years old when he/she reaches the critical point of abdominal LES damage (<10 mm). If, however, the 40-year-old patient has a rate of abdominal LES shortening of 5 mm/decade, he/she will reach the critical level of LES damage by 60 years. If damage is 10 mm/decade, he/she will get into trouble by 50 years.

I will show that the new method of assessment of LES damage permits identification of the likely outcomes of patients *at or even before the onset of treatment*. This creates a window of opportunity. The minority of patients who are predicted to progress to an LES damage severity that has a high risk of developing treatment failure and visible CLE can enter a treatment protocol that is aimed at preventing progression of LES damage. The majority of patients who are predicted to maintain LES competence in the long term can be treated with PPIs with the objective of symptom control without risk of treatment failure, visible CLE, or adenocarcinoma.

3.4 Formation of Visible Columnar-Lined Esophagus

The final stage of progression of CLE is the development of visible CLE. This results only when there is sufficient LES failure (tLESR) to severely damage the squamous epithelium. Columnar metaplasia is the irreversible end point of squamous epithelial damage. The development of visible CLE in the thoracic esophagus results from the exposure of a large area of the thoracic esophagus simultaneously to gastric contents that reflux along a pressure gradient when the LES fails (tLESR). The column or refluxed gastric juice commonly reaches the midesophagus. As a result, unlike CLE in the dilated distal esophagus that is associated with slow degradation due to intermittent gastric overdistension, visible CLE appears rapidly in the thoracic esophagus. In the Pro-GERD study,[4] 9.7% of GERD patients who had no visible CLE at the index endoscopy developed visible CLE within 5 years; the length of visible CLE reached or exceeded 5 cm (50 mm) in two patients.

In the new method of understanding, visible CLE is a disastrous end point of GERD that represents failure of management and irreversibility. Prevention of visible CLE is the goal of the new management. With continued PPI therapy, the progression to adenocarcinoma is inevitable in visible CLE in the small minority of patients destined to progress to cancer.

At present, gastroenterologists regard visible CLE as something not worth screening for in patients GERD unless they fail treatment, and to be ignored if it is seen at endoscopy but biopsies do not show intestinal metaplasia.

In the new perspective, visible CLE indicates damage to the abdominal LES to have produced sufficient LES failure and reflux into the thoracic esophagus to cause columnar metaplasia in squamous epithelium. At this stage, there is likely to be abdominal LES damage >25 mm where the residual abdominal LES is <10 mm (assuming a normal abdominal LES length of 35 mm) (Fig. 11.14) and the pH test is almost always abnormal (pH<4>4.5% of the time).

Visible CLE at endoscopy is presently the only type of CLE that is recognized by gastroenterologists. It is a late consequence of LES damage. The initial 25 mm of LES damage that is manifested as CLE in the dilated distal esophagus is distal to the endoscopic GEJ and mistaken for normal proximal stomach. The invariable prelude to visible CLE, which is CLE in the dilated distal esophagus, has been missed as it increased in length from 0 to 25 mm over many decades of a patient's life.

4. PREVALENCE OF COLUMNAR-LINED ESOPHAGUS

4.1 Prevalence of Columnar-Lined Esophagus in the Dilated Distal Esophagus

With rare exception, all adults in the population of the United States and Europe will have CLE in the dilated distal esophagus. This means that all adults have evidence of LES damage. This is not unusual in chronic progressive diseases. Nearly all adults in the United States will have evidence of atherosclerosis in the abdominal aorta.

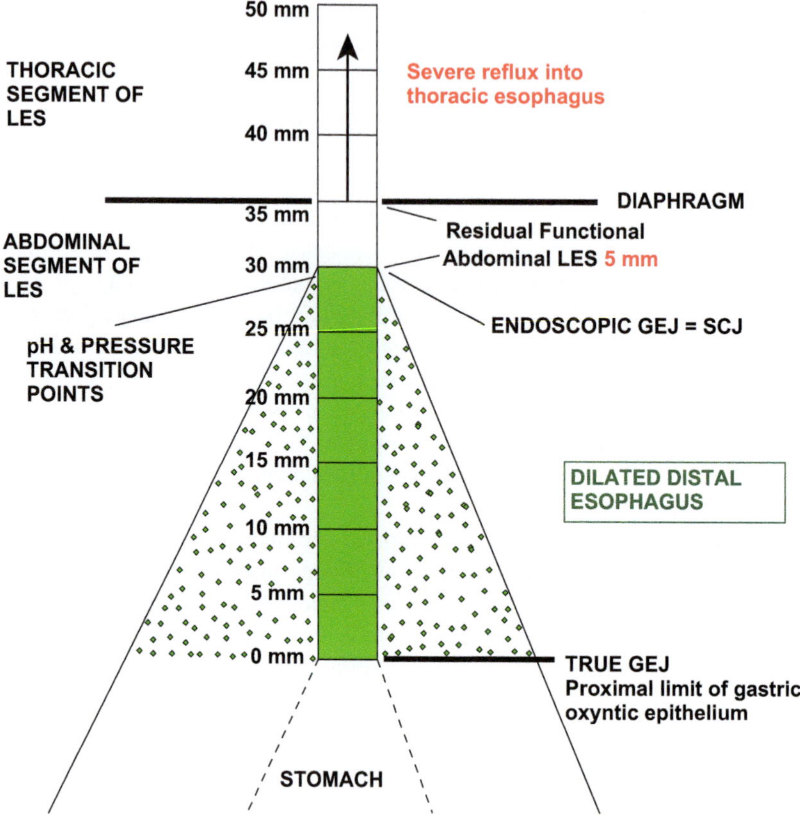

FIGURE 11.14 Progression of cardiac metaplasia to a length of 30 mm in the dilated distal esophagus. The residual functioning abdominal lower esophageal sphincter (LES) is now 5 mm which means that it is a defective LES at rest and associated with frequent LES failure and severe reflux. Such patients are at high risk for visible CLE in the thoracic esophagus (not shown). *GEJ*, gastroesophageal junction; *SCJ*, squamocolumnar junction.

There are two critical criteria that determine clinically significant dilated distal esophagus:

1. The occurrence of intestinal metaplasia in CLE within the dilated distal esophagus. This defines the presence of a risk of adenocarcinoma. Intestinal metaplasia in CLE limited to the dilated distal esophagus is likely the precursor lesion for adenocarcinoma of the dilated distal esophagus. This may occur in patients with LES damage that is within the compensated stage (<15 mm). These patients are commonly asymptomatic and may not have LES failure with reflux into the thoracic esophagus.
2. LES damage that exceeds the reserve capacity of the sphincter and results in LES failure (tLESR) and significant reflux into the esophagus. This begins at LES damage >15 mm. This is the onset of postprandial reflux. As LES damage progressively increases from this point, it causes increasing reflux leading to increase in symptomatic GERD, and progression to erosive esophagitis, visible CLE, and adenocarcinoma.

There are very few studies in the literature that report the prevalence of intestinal metaplasia in the dilated distal esophagus. The reasons for this are that the dilated distal esophagus is mistaken for proximal stomach by the presently used incorrect definition of the endoscopic GEJ and management guidelines of all gastroenterology societies recommend that biopsies should not be taken in the patient who is endoscopically normal.[2]

In our study of 714 patients with CLE limited to the dilated distal esophagus, 161 (22.6%) had intestinal metaplasia.[37] This percentage is likely to be higher than in the general population because our study is based on patients referred to a surgical unit specializing in the management of complicated GERD. In Ronkainen et al. (see below), the prevalence of intestinal metaplasia in biopsies of people without visible CLE was 4.5%.

There is no study in the literature that has accurately measured the length of the dilated distal esophagus at endoscopy, including our studies. Ringhofer et al.[38] (see Chapter 5 for review) showed a variation in the length of CLE in the dilated distal esophagus with multilevel biopsies performed at the endoscopic GEJ and 5 and 10 mm distal to it.

4.2 Prevalence of Visible Columnar-Lined Esophagus in the General Population

Endoscopic screening for visible CLE is not recommended in any person with GERD at present. There are relatively few studies with screening endoscopy in a large group of people with either no symptoms of GERD or without consideration of the presence of symptoms of GERD. These have shown significant variation in the prevalence of both visible CLE and intestinal metaplasia within visible CLE.

In a study by Ronkainen et al. a randomly selected sample was drawn from the entire population of two communities in Sweden with a total population of 21,610 people.[39] By a series of random computer maneuvers, a representative sample of 1000 people was selected and agreed to an upper endoscopy to evaluate the prevalence of BE. The study group completed the validated abdominal symptom questionnaire. Gastroesophageal reflux symptoms were defined as troublesome heartburn and/or acid regurgitation over the past 3 months.

At endoscopy, visible CLE was identified by the presence of salmon-pink mucosa in either a circumferential upward shift of the SCJ or in adjacent mucosal tongues or islands. The endoscopic GEJ was defined as the junction of the proximal gastric folds and the tubular esophagus. BE was defined by the presence of intestinal metaplasia in biopsies from the visible CLE. It was defined as LSBE if the visible CLE was 2 cm or more and SSBE if it was <2 cm.

Suspected CLE was seen at endoscopy in 103/1000 (10.3%) patients of mean age 55.7 years with 60.2% men. 12 (1.2%) had a long segment of visible CLE.

BE (i.e., visible CLE with intestinal metaplasia on biopsy) was present in 16 (1.6%) patients with a mean age of 57 years (9 men). LSBE occurred in 5 (0.5%) individuals; 11 (1.1%) had SSBE. The prevalence of intestinal metaplasia in the esophageal biopsies was higher if the visible CLE was longer; 41.7% in visible CLE of 2 cm or more, 12.1% in segments <2 cm, and 4.5% in those without visible CLE who had intestinal metaplasia.

The prevalence of BE (visible CLE with intestinal metaplasia) was 2.3% in people with reflux symptoms, not significantly higher than the 1.2% in those without ($P = .18$). This suggests that the use of symptoms of reflux is not an effective criterion to select people for screening endoscopy to find BE.

The study concluded that the prevalence of BE was 1.6% in this random sample of the population. The prevalence of visible CLE was 10.3%.[39]

In Rex et al. the authors identify the problem of unidentified BE (visible CLE with intestinal metaplasia; BE) in the population. They write: "… most patients with BE go unrecognized in clinical practice. More than 90% of patients with esophageal adenocarcinomas never enter screening programs because they are not recognized to be at risk prior to presenting with cancer-related dysphagia. Although the cost-effectiveness of screening for BE is not established, screening for BE has been recommended for persons at high risk. Identified risk factors include chronic heartburn,… older age, male gender, white race, family history of BE or esophageal adenocarcinoma, obesity, and a large hiatal hernia." The study is designed to establish the prevalence of BE.

The study group consisted of 961 outpatients older than 40 years who were scheduled for screening colonoscopy and accepted an invitation to have a screening upper endoscopy as an add-on procedure. They had no clinical indication for and had not had a previous upper endoscopy. Patients completed two validated heartburn questionnaires prior to their procedure.

At endoscopy, the following landmarks were identified: (1) the SCJ; (2) the endoscopic GEJ, defined as the proximal end of the gastric rugal folds where the tubular esophagus meets the stomach; and (3) the diaphragmatic hiatus. LSBE was defined as columnar epithelium with specialized intestinal metaplasia (SIM) equal to or greater than 3 cm proximal to the GEJ; SSBE was defined as columnar epithelium with SIM <3 cm above the GEJ. Only patients with >5 mm of columnar epithelium above the GEJ were considered abnormal.

Four-quadrant biopsy samples were taken from the tubular esophagus at least every 2 cm in the case of circumferential segments and at least one sample from each tongue in the case of columnar-epithelial tongues extending into the squamous epithelium. Two samples were taken from the gastric cardia (defined as the proximal edge of the gastric folds, just distal to the end of the tubular esophagus.). Beginning with the 150th patient, biopsy samples were taken from the gastric antrum and body.

The results of the study can be summarized as follows:

1. Prevalence of heartburn: 59.1% of patients had never had heartburn; 25.1% reported heartburn < once per week; 6.4% had heartburn once per week; 6.5% several times per week; and 2.7% had heartburn daily.
2. Visible CLE with and without intestinal metaplasia: A total of 176/961 patients (18.3%) had visible CLE > 5 mm from the endoscopic GEJ. 12/961 (1.2%) of patients had long segment visible CLE; all had intestinal metaplasia. 164/961 (17.1%) patients had visible CLE measuring 5–29 mm. Of these, 53 had intestinal metaplasia; 111 did not; the histology in those without intestinal metaplasia was not given but can be assumed to be cardiac epithelium (with and without parietal cells). The total number of patients with visible CLE + SIM was 65/961 (6.8%).

3. Association of BE with symptoms: Among 556 subjects who had never had heartburn, the prevalence of BE and LSBE was 5.6% and 0.3%, respectively. Among 384 subjects with a history of any heartburn, the prevalence of BE and LSBE was 8.3% and 2.6%, respectively. No data are provided from the relationship between symptoms and the prevalence of visible CLE without intestinal metaplasia.
4. Erosive reflux esophagitis was present in 155/961 (16.1%) patients. It was more common in patients with BE (31%) than in those without (15%; P = .005).
5. Of the 12 patients with LSBE, one each had low-grade dysplasia and "indefinite for dysplasia." One patient with SSBE was "indefinite for dysplasia." No patient had high-grade dysplasia or adenocarcinoma.

The critical item of data in this study is that, in this unselected population, 176/961 (18.3%) had visible CLE exceeding 0.5 cm above the endoscopic GEJ. None of these patients had an indication for endoscopy. Many did not have any symptom suggesting GERD. These patients, by the significance for visible CLE that I have suggested, have severe LES damage and severe GERD with an irreversible cellular change. This is an extremely high number in the general population that has severe and irreversible GERD.

The number of patients with BE (i.e., visible CLE + SIM) was 65/961 (6.8%).

In a study from Stanford, Gerson et al.[32] write in their introduction: "Approximately 10–14% of subjects who undergo endoscopy for GERD evaluation will be found to have BE, a condition that is most common in white men older than 50 years of age… Because the increase in esophageal adenocarcinoma does not appear to be solely attributable to a rise in the prevalence of GERD, it is possible that many cases of BE occur in asymptomatic adults who do not experience symptoms of GERD or do not seek medical attention for GERD symptoms… The aim of this study was to assess the prevalence of BE in asymptomatic adults undergoing routine sigmoidoscopy for colorectal cancer screening."

727 subjects scheduled for screening sigmoidoscopy were invited to have an additional upper gastrointestinal endoscopy. 408 subjects responded and were interviewed; 240 were excluded because of GERD symptoms (heartburn more frequently than once a month), use of antisecretory drugs, or a history of prior upper endoscopy; 58 refused to participate.

This left an ultimate study group of 110 subjects of mean age 61 (SD 9.1, range 50–80) years, 80 (73%) were white, 101 were male, and all but 6 were veterans. 51% of the subjects never experienced heartburn; 17% and 11% had it once or twice per year, respectively; the other 21% had mild postprandial heartburn relieved by antacid use up to 3–6 times per year. The mean reflux symptom index score was 0.1 ± 0.4 (range 0–2) as expected in an asymptomatic population without GERD symptoms.

At endoscopy, the GEJ was defined as the junction of the proximal gastric folds and tubular esophagus. Subjects were classified as normal (no visible CLE between GEJ and SCJ), LSBE (>3 cm), and SSBE (<3 cm). BE was defined as visible CLE with intestinal metaplasia. Endoscopically normal subjects with intestinal metaplasia at the normal GEJ were designated SIM–esophagogastric junction (EGJ). The number of subjects with visible CLE who were negative for intestinal metaplasia was not reported.

BE (visible CLE with intestinal metaplasia) was found in 27 (25%) subjects, 8 with LSBE and 19 with SSBE. 17 (16%) of the subjects had SIM–EGJ, 47% of whom were *Helicobacter pylori* positive.

The rate of visible CLE in the general population was 10.3% in Ronkainen et al., 18.3% in Rex et al., and not reported in Gerson et al. (but has to be greater than 25%, the number of patients with visible CLE + intestinal metaplasia). The purest study group and the lowest prevalence of visible CLE is the study of Ronkainen et al. I will take this as the most conservative estimate of prevalence of visible CLE in the asymptomatic population.

This means that 10% of people in the general population with and without symptoms of GERD have sufficient reflux into the thoracic esophagus to cause visible CLE. 10% of the ~200 million adults in the United States is 20 million people.

The three studies show that the present recommendation to not screen patients who are not symptomatic for GERD leaves a significant population of people in the general population with undetected visible CLE. I will show that these patients represent the population that, by virtue of the presence of visible CLE, progresses to intestinal metaplasia → increasing dysplasia → adenocarcinoma without developing troublesome symptoms of GERD.

It is no surprise that many people who develop esophageal adenocarcinoma present for the first time with dysphagia caused by late stage disease. Only 15% of patients presenting with esophageal adenocarcinoma have ever had an endoscopy before their cancer diagnosis.

The data also show the serious disconnect between the severity of symptoms of GERD and the success of PPI therapy in controlling GERD symptoms with severity of LES damage. Ultimately, progression of GERD depends on the rate of progression of LES damage, not symptoms. As long as we define GERD by symptoms and severity of GERD by quality of life, we will never change outcomes, particularly progression to adenocarcinoma.

4.3 Prevalence of Visible Columnar-Lined Esophagus in Patients With Gastroesophageal Reflux Disease of Any Duration

Endoscopy is limited to those patients who fail PPI therapy. There are relatively few studies with screening endoscopy in a large group of people with symptoms of GERD where endoscopy was performed without the requirement of failure of control of symptoms with PPI therapy.

The Pro-GERD study sponsored by Astra-Zeneca in Europe recruited 6215 patients with their main symptom suggestive of GERD such as heartburn without restriction of prior duration of GERD from 1253 centers in Germany (1185), Austria (33), and Switzerland (35).[41]

All patients underwent endoscopy in central locations by endoscopists who were trained specifically for this study to grade erosive esophagitis, recognize visible CLE, and take biopsies according to protocol. The endoscopy was performed without an indication, i.e., if not for the study, these patients would have been treated with acid suppression by their primary care physician without endoscopy unless treatment failure or dysphagia occurred. As such, the results represent findings that will be present if GERD patients who were well controlled were subject to endoscopy.

At endoscopy, the patients were classified into three groups: nonerosive reflux disease (NERD), erosive reflux disease (ERD), and BE. BE was defined as "patients with endoscopic suspicion (i.e., indications of any columnar-lined epithelium in the esophagus) or histologic proof of Barrett's mucosa, as assessed in the Pro-GERD core pathology laboratory." The definition of BE is therefore equivalent in this study to visible CLE without separation into patients who were positive and negative for intestinal metaplasia.

The study excluded patients found to have any gastrointestinal malignancy. Five patients with esophageal adenocarcinoma were found at endoscopy; these were not included.

The number of patients in each of the three groups at the baseline endoscopy is shown in Table 11.4.

In patients with GERD as defined by this study, 11.3% of patients had already developed visible CLE at the time of endoscopy. Because the endoscopies were done for the study and would have been otherwise not recommended, this represents the probable percentage of people under treatment by physicians for symptoms of GERD who harbor visible CLE.

This was a GERD population with more than mild symptomatic GERD; 46% of patients in the study reported daily symptoms and an additional 44% had symptoms at least 2 days per week. Two-thirds reported moderate or severe symptoms off medication during the week before endoscopy.

There was no difference in the GERD score (a measure of GERD severity developed for the study) among NERD, ERD, and visible CLE patients. The factors that were significantly associated with visible CLE versus NERD were increasing age, male gender, duration of GERD symptoms, body mass index, smoking, high alcohol consumption, and previous intake of PPI. Patients with visible CLE were 1.5 and 2 times more likely to have had a long duration of GERD than ERD and NERD groups, respectively.

This first report of the Pro-GERD study reported visible CLE only without mention of the presence of the percentage of intestinal metaplasia. The prevalence of visible CLE in this study of patients with significant symptomatic GERD population under good symptom control with PPI therapy (11.3%) is comparable to that in the Ronkainen et al.[39] population-based study (10.3%).

In the study done by Westhoff et al.[31] BE is defined as the presence of intestinal metaplasia in a biopsy taken from visible CLE in the tubular esophagus. It is defined into long and short segment by the endoscopic length being more or less than 3 cm. The authors report only BE; visible CLE without intestinal metaplasia is not mentioned.

The authors' reason for the study is stated: "The prevalence of BE varies from as much as 2% to 25% in prior studies and depends on the indication for endoscopy, the nature of the population studied, and the definition of Barrett esophagus… The aims of the present study were to determine the frequency of LSBE (>3 cm) and SSBE (<3 cm) in consecutive patients presenting for a first endoscopic evaluation… with the indication of GERD and to determine whether there are differences between patients with LSBE and SSBE."

TABLE 11.4 Patient Numbers in the Three Groups at the Index Endoscopy Done at the Time of Entry Into the Pro-GERD Study

	Total Cohort	NERD	ERD	Barrett Esophagus
Number	6215	2853 (45.9%)	2660 (42.8%)	702 (11.3%)

ERD, erosive esophagitis; *GERD*, gastroesophageal reflux disease; *NERD*, nonerosive reflux disease.

The study group consisted of 378 patients (94% men, 86% white, median age 56 years with a range of 27–93 years) presenting for a first upper endoscopy at the Veterans Affairs Medical Center in Kansas, Missouri, with the indication of GERD (heartburn and/or regurgitation at least twice per week). Patients with alarm symptoms (dysphagia, weight loss, anemia, GI bleeding) were excluded.

At endoscopy, the GEJ was defined as the "pinch" at the distal end of the esophagus, coinciding with the proximal margin of the gastric rugal folds. Biopsy specimens were obtained only if columnar-appearing mucosa was detected in the distal esophagus at a point that was clearly proximal to the GEJ. Intestinal metaplasia was a requirement for the diagnosis as BE.

The patients reported having heartburn for a median of 5 years (range 1–7 years) and/or regurgitation for a median of 4 years (range 1–7 years).

BE was present in 50/378 (13.2%) patients, 94% male, and 98% white with a median age of 62 years (range 29–81 years). 66% of the patients had GERD for >5 years. There was no significant difference between LSBE and SSBE for male gender, median age, median duration of heartburn or regurgitation, tobacco use, family history of GERD, hiatal hernia, or body mass index.

In the discussion of the Pro-GERD study,[4] the authors state: "in clinical care it is hardly possible to predict erosive GERD or BE based on reflux symptoms alone… To optimize patient care, the knowledge of other factors may help to distinguish patients with an increased risk of endoscopic lesions of the esophagus and, as a result, a higher risk of developing physical complications of GERD. This is of particular importance to the occurrence of BE. It would be preferable to predict BE because it is regarded as a risk factor of the esophageal adenocarcinoma."

The crux of the problem is our present inability to identify the minority of patients at high risk for adenocarcinoma before they reach that state. There is therefore no opportunity to prevent adenocarcinoma. Even in a population with symptomatic GERD, there is presently no method (symptom-based or demographic) to identify a category of patient more likely to have visible CLE with an accuracy that justifies screening endoscopy. Even if endoscopy is done, there is no criterion short of visible CLE with intestinal metaplasia that justifies surveillance. This results in the unsatisfactory end result that only 15% of all patients who develop adenocarcinoma have had an endoscopy and a diagnosis of BE prior to presenting with cancer.

I will show that pathologic assessment of LES damage by the new method proposed in this book has the potential to provide the ability to predict those patients who are most likely to progress from the asymptomatic state and early GERD to visible CLE. By declaring visible CLE as the point in the disease that represents failure, the new method will strive to develop a new treatment algorithm that is aimed at preventing patients from progressing to this point. If successful, esophageal adenocarcinoma will be prevented.

5. CLINICAL MANIFESTATIONS OF COLUMNAR-LINED ESOPHAGUS

There is very little data on the clinical manifestations of patients with CLE limited to the dilated distal esophagus. This is not surprising because the entity of the dilated distal esophagus is not recognized, misinterpreted as proximal stomach, and not studied. I have discussed the possible clinical syndromes associated with CLE limited to the dilated distal esophagus in Chapter 10. These syndromes all precede the troublesome symptoms that define GERD at present. Most are not recognized clinically as GERD.

The clinical manifestations in the patient with a visible CLE are essentially those of GERD. There are demographic and clinical features that indicate a higher likelihood of visible CLE in patients with symptomatic GERD (Table 11.5). However, no feature or combination of features is sufficiently predictive of visible CLE to constitute a reason for endoscopy.

Without any ability to recognize patients with GERD who are at increased risk for visible CLE, the only indication for endoscopy remains treatment failure and the occurrence of dysphagia.

As a result, visible CLE is found in two groups: (1) patients with treatment failure in GERD, which is the main indication for endoscopy in GERD; and (2) patients undergoing endoscopy for reasons other than GERD who are incidentally found to have visible CLE.

6. HISTOLOGY OF COLUMNAR-LINED ESOPHAGUS

The histology of CLE is simple, consistent, and nonrandom. Throughout its length, it is composed of one, two, or three variants of cardiac epithelium arranged in a distribution that rarely varies. Histologic criteria are also seriously underutilized in the management and risk assessment of the patient with GERD. There is much more to the histology of CLE than whether or not it is seen at endoscopy and whether or not intestinal metaplasia is present.

Columnar metaplasia of the esophagus that results in CLE is *always* the result of chronic and frequently repeated exposure of esophageal squamous epithelium to gastric contents. In the dilated distal esophagus, the exposure to gastric contents

TABLE 11.5 Demographic and Clinical Features of Gastroesophageal Reflux Disease That Are Associated With a Higher Prevalence of Visible Columnar-Lined Esophagus at Endoscopy

Feature	Criterion	Comment
Age	>50 years	Wide age range
Sex	Male	Highly significant
Ethnicity	Caucasian	Low prevalence in other races
Family history of BE or adenocarcinoma	Yes	Rare
Country of domicile	Western Europe, United States	Low prevalence in Asia
Socioeconomic status	Affluent	
Body habitus	Increased BMI, central obesity	Can occur in lean persons
Severity and duration of symptoms	More severe, longer duration	Can occur in asymptomatic people
Good control of symptoms with PPI	Yes	
Smoking	Yes	
Alcohol	Yes	
Regular PPI use	Yes	May represent indication bias

BE, Barrett esophagus; *BMI*, body mass index; *PPI*, proton pump inhibitors.

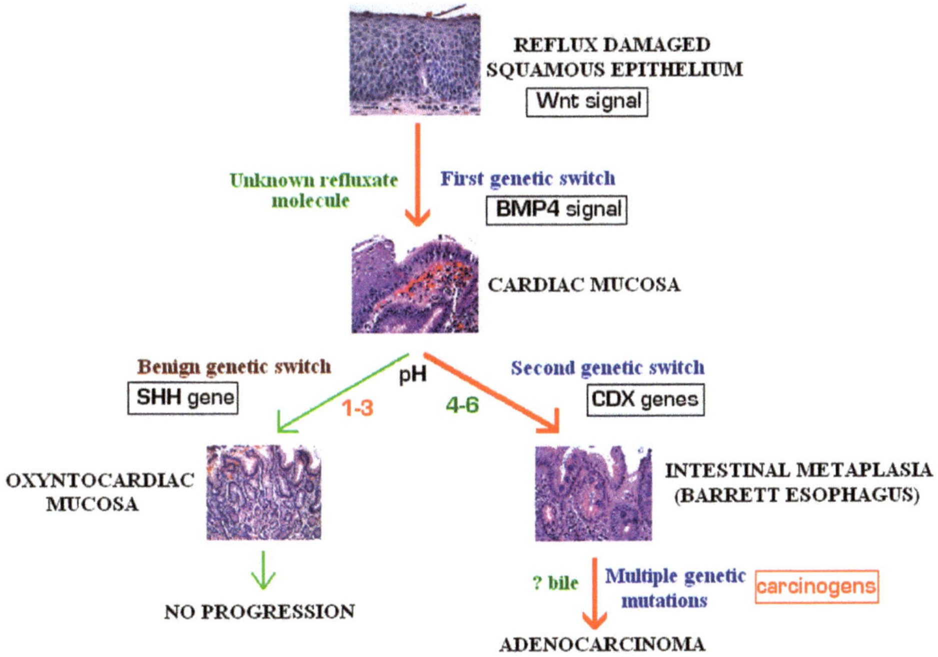

FIGURE 11.15 Evolution of cardiac epithelium in columnar-lined esophagus, both in the thoracic and abdominal (dilated) esophagus. Cardiac epithelium is the first epithelium that occurs when squamous epithelium undergoes columnar metaplasia. Cardiac epithelium can differentiate in one of the two directions: (A) oxyntocardiac epithelium by the occurrence of parietal cells; this is a stable epithelium that does not progress to intestinal metaplasia or adenocarcinoma and (B) intestinal metaplasia by the occurrence of goblet cells; this is the target epithelium for carcinogenesis.

occurs during times of gastric overdistension. In the thoracic esophagus the exposure is the result of LES failure (tLESR) that causes reflux of gastric contents into the thoracic esophagus.

The first epithelium produced by columnar metaplasia of esophageal squamous epithelium is cardiac epithelium. Cardiac epithelium is the only epithelium that results directly from metaplasia of squamous epithelium. The two variants of cardiac epithelium arise when a second metaplastic event results in the development of specialized cells (Fig. 11.15).

6.1 Types of Columnar Epithelia in Columnar-Lined Esophagus

Cardiac epithelium that is exposed to gastric contents may undergo two subsequent metaplasias characterized by the development of specialized cells: (1) When it develops goblet cells, it is called cardiac epithelium with intestinal metaplasia; (2) When it develops parietal cells, it is called oxyntocardiac epithelium. These secondary metaplasias can occur in both the dilated distal esophagus as well as in visible CLE.

Cardiac epithelium in the dilated distal esophagus and visible CLE is the same epithelium, morphologically and by expression of immunohistochemical and molecular markers.[42] It is different than any epithelium found in the stomach. Cardiac epithelium is not present as a normal epithelium in the proximal stomach.

Other specialized cells such as pancreatic cells, Paneth cells, chief cells, neuroendocrine cells can also develop in cardiac epithelium, but they have no known clinical significance at this time and are ignored for definition purposes.

Cardiac epithelium, which consists of only mucous cells, oxyntocardiac epithelium, which has, in addition, parietal cells, and cardiac epithelium with intestinal metaplasia, characterized by goblet cells are the three epithelial types that, in varying amounts, constitute the entire CLE. No other epithelial types occur between the distal end of residual esophageal squamous epithelium and the true GEJ defined as the proximal limit of gastric oxyntic epithelium.

The classification of epithelial types is made for each foveolar unit. Adjacent foveolar units can have different epithelial types, i.e., cardiac with and without intestinal metaplasia, and oxyntocardiac (Fig. 11.8). It is extremely unusual for both parietal and goblet cells to be present in one foveolar unit.

I have defined these epithelial types and discussed the way they are diagnosed in Chapter 4. The classification of the three types is simple and, with very little training, can be mastered by any pathologist. The interobserver variation is very low. To have an accurate and reproducible diagnostic criterion has great value. In this book, I have used the synonymous phrases "cardiac epithelium with and without parietal and/or goblet cells" and "cardiac epithelium with or without intestinal metaplasia and oxyntocardiac epithelium" to represent all metaplastic esophageal cardiac epithelium with both its evolutionary variants.

6.2 Histology of Columnar-Lined Esophagus Limited to the Dilated Distal Esophagus

Most people in the population without symptoms of GERD have microscopic CLE limited to the dilated distal esophagus. This, being distal to the endoscopic GEJ (proximal limit of rugal folds) and macroscopic pathologic GEJ (end of the tubular esophagus), is presently misinterpreted as proximal stomach (see Chapter 10).

I have described the histology of the dilated distal esophagus in people without visible CLE in Chapter 10. The measurement of the dilated distal esophagus is limited to study of autopsy and resection specimens. In our unit, a retroflex biopsy is routinely taken from the area distal to the endoscopic GEJ. The findings in this biopsy set in over 15,000 patients over 20 years in our unit are the basis of this book. It is a unique experience. Even in our unit, though, the biopsy protocol distal to the endoscopic GEJ does not permit measurement of the dilated distal esophagus. This is presently a huge vacuum in what I consider to be critical data.

At autopsy, patients without clinical GERD during life have a short (<15 mm) dilated distal esophagus. This correlates with abdominal LES damage <15 mm, which is within the reserve capacity of the abdominal LES and therefore associated with normal LES competence and the absence of reflux.

Normal patients have zero metaplastic columnar epithelium. Some people without GERD symptoms with very short (<5 mm) have only oxyntocardiac epithelium at autopsy. Others with cardiac and oxyntocardiac epithelium have lengths of CLE that are usually less than 15 mm.

Patients with early and mild symptomatic GERD also commonly have CLE limited to the dilated distal esophagus. Present guidelines for biopsy at endoscopy recommend that endoscopically normal patients with GERD should not have biopsies taken. This has greatly limited the data available regarding the histology of the dilated distal esophagus.

Marsman et al.[43] selected patients undergoing endoscopy for any reason who had a normal endoscopic GEJ for biopsy. They analyzed only those patients who had the actual histologic SCJ in one biopsy sample. 24/63 (38%) had a direct transition from squamous epithelium to oxyntocardiac epithelium without any intervening pure cardiac epithelium without parietal cells. The other 39 (62%) of patients had cardiac epithelium adjacent to the squamous epithelium. They did not report any patient with intestinal metaplasia. Even more interesting is their report of those people in their study who were excluded because they did not have the histologic squamous epithelium in one biopsy piece. The prevalence of cardiac epithelium in this group was 2%. Marsman et al.[43] were reporting the histology of CLE limited to the dilated distal esophagus, although their conclusion was that all people had cardiac epithelium in the proximal stomach.

Ringhofer et al.[38] by performing multilevel biopsies at the endoscopic GEJ and 5 and 10 mm distal to it, showed that the three epithelial types are stratified in a nonrandom manner within the segment. Intestinal metaplasia, when present, occupies the most proximal part of the segment next to the SCJ whereas oxyntocardiac epithelium is most distal, adjacent to gastric oxyntic epithelium at the true GEJ. Cardiac epithelium, when present is proximal to oxyntocardiac epithelium. When intestinal metaplasia is present, cardiac epithelium is intermediate between intestinal and oxyntocardiac epithelia.

These data suggest that the presence of intestinal metaplasia in the dilated distal esophagus is predictive of the longest segments of CLE in the dilated distal esophagus. This would mean that intestinal metaplasia in the dilated distal esophagus is a marker for more severe abdominal LES damage. There is some support for this.

Leodolter et al.[34] in the Pro-GERD study, showed that patients who had intestinal metaplasia at the normal endoscopic GEJ (= intestinal metaplasia in the dilated distal esophagus) had a 25.8% incidence of visible CLE within the next 5 years. This is best explained by a correlation between intestinal metaplasia at the normal endoscopic GEJ and greater severity of LES damage. Minimal progression of LES damage from that point results in severe reflux and visible CLE.

The presence of goblet cells in cardiac epithelium distal to an endoscopically normal GEJ is not recognized presently as being an abnormality in the esophagus, although all the evidence points to that. It is variably termed "intestinal metaplasia of the gastric cardia," "histologic Barrett esophagus," "ultra-SSBE," or "specialized intestinal metaplasia of the esophagogastric junction (SIM-EGJ)" creating impossible confusion where none is necessary. All of these terms mean one thing: "intestinal metaplasia of the dilated distal esophagus."

6.3 Histology of Visible Columnar-Lined Esophagus

Visible CLE is the end result of reflux resulting from severe LES damage beyond its reserve capacity. It is therefore *always* associated with and preceded by CLE in the dilated distal esophagus. The length of the associated dilated distal esophagus is usually >15 mm and often longer when there is a visible CLE. As LES damage (= length of dilated distal esophagus) progresses from 15 cm upward, the probability of visible CLE increases.

The visible CLE is *always* continuous with the CLE of the dilated distal esophagus. The epithelial composition of CLE in the two locations is identical morphologically, the amount and nature of inflammation, and by the presence of numerous common immunohistochemical and molecular markers (see Chapter 5).[42]

When reading the literature, the plethora of terms used gives the false impression that the histology of visible CLE is extremely complicated. One reason for this is the incorrect division of the CLE into a visible CLE proximal to the endoscopic GEJ and gastric cardia distal to it. It falsely converts one continuous CLE (beginning at the proximal limit of gastric oxyntic epithelium and progressively lengthening, first limited to the dilated distal esophagus distal to the endoscopic GEJ and then involving the thoracic esophagus proximal to the endoscopic GEJ) into two completely different entities (visible CLE and normal stomach).

The majority of gastroenterologists and some pathologists will avoid using the term "cardiac mucosa/epithelium" to visible CLE because of their belief that cardiac epithelium is gastric. Visible CLE with intestinal metaplasia is rarely called "cardiac epithelium with intestinal metaplasia" probably because of its similarity to "intestinal metaplasia of the cardia," which is incorrectly believed to be gastric. Instead, terms such as "specialized columnar epithelium," "SIM," and "BE, confirmed" are used. There is nothing specialized about this epithelium. It is cardiac epithelium with intestinal metaplasia. It is the same as intestinal metaplasia in CLE distal to the endoscopic GEJ.

The reality is simple. The only normal epithelia in the esophagus and proximal stomach are squamous and gastric oxyntic epithelium. Columnar metaplasia of the squamous epithelium results in cardiac epithelium, which is always esophageal and never normal or gastric.

Cardiac epithelium begins at the true GEJ (proximal limit of gastric oxyntic epithelium), slowly increases in length with increasing metaplasia of esophageal squamous epithelium within the dilated distal esophagus. This is accompanied by damage to the abdominal LES of equal length. When the LES becomes incompetent and causes uncontrolled reflux, CLE develops in the thoracic esophagus.

In all cases, the entire CLE segment from the SCJ to the true GEJ (= proximal limit of gastric epithelium) is one unit with a stratification of the epithelial types into oxyntocardiac distally adjacent to gastric oxyntic epithelium, cardiac epithelium proximal to this and, when present, intestinal metaplasia most proximal and adjacent to the squamous epithelium. Histologically, the single segment of CLE straddles the endoscopic GEJ.

7. THE DETECTION OF INTESTINAL METAPLASIA IN COLUMNAR-LINED ESOPHAGUS BY ENDOSCOPIC BIOPSY

There is no attempt at present to detect intestinal metaplasia in the dilated distal esophagus. All gastroenterology associations recommend that biopsy should not be done in the endoscopically normal person, even in the presence of symptoms of GERD.[27]

Intestinal metaplasia in the dilated distal esophagus is important for two reasons: (1) As shown in Leodolter et al.[34] (see Chapter 10), the presence of intestinal metaplasia in a biopsy taken from the SCJ in the endoscopically normal person predicts a 25.8% incidence of visible CLE within 5 years; (2) It is the precursor lesion for adenocarcinoma of the dilated distal esophagus. Although adenocarcinomas arising within 5 cm distal to the endoscopic GEJ are now required to be classified as esophageal by the American Joint Commission on Cancer (seventh edition), the failure to biopsy this region has resulted in the complete failure to define precursor lesions by prospective studies.

When the world decides to biopsy the normal endoscopic GEJ, it is critically important to ensure that the biopsies straddle the histologic SCJ. This is not always easy and requires expertise; the SCJ is a moving target in the breathing person. Intestinal metaplasia is commonly limited to a few foveolar units immediately adjacent to the SCJ. If the actual junction is missed, there is a high risk of false negative diagnosis.

For most gastroenterologists, the only relevant issue in visible CLE is whether or not intestinal metaplasia is present. In most of the world (except England and Japan), only patients with intestinal metaplasia will enter a BE endoscopic surveillance program.

There should be intense interest in the most efficient and accurate endoscopic biopsy protocol to achieve the objective of finding intestinal metaplasia. Unfortunately, this has not happened. There is no standardized biopsy protocol. Endoscopic biopsy of CLE has a wide variation in clinical practice, with a tendency to be inadequate in more cases than not. As a result, there can be little confidence in the negative diagnosis of intestinal metaplasia.

Gastroenterologists know that intestinal metaplasia is not present throughout the segment of visible CLE. They also know that there is a significant risk of sampling error that may result in a false negative diagnosis of BE.

Statement #36 of the Montreal document[29] addressed the issue of biopsy sampling as follows: "Multiple, closely spaced biopsies are necessary to characterize endoscopically suspected esophageal metaplasia." In the discussion: "This current biopsy practice must sample all areas of metaplastic mucosa as thoroughly as possible. The best researched biopsy protocol is four-quadrant biopsies every 1 cm for circumferential metaplastic segments, which is substantially more sensitive than sampling at 2-cm intervals. This approach has been variably modified to include biopsies at the top of tongue-like metaplastic extensions. These onerous and usually expensive protocols are generally not accepted as best practice."

The last sentence of this statement drives the reality of general practice. The Seattle protocol (see below), is rarely followed. Biopsy of the CLE is random and often inadequate. The reason for this is that most endoscopists taking the biopsies have no understanding of the distribution of intestinal metaplasia within the CLE. Most have never sat down at a microscope to look at the histology of biopsies they have taken.

Endoscopic biopsy protocols for the index endoscopy of a patient found to have visible CLE should have the objective of finding intestinal metaplasia.

7.1 Presently Recommended and Used Biopsy Protocols

1. The Seattle biopsy protocol: This requires four-quadrant biopsies at 1 or 2 cm intervals. As stated in the Montreal document, this is time consuming and expensive. It is also not necessary to achieve the objective of finding intestinal metaplasia at the first endoscopy. The Seattle protocol is ideal for surveillance of patients with known BE where the objective is to find dysplasia.
2. Random sampling: It is not uncommon for me to receive one specimen bottle with 2–6 biopsy pieces from a 5-cm segment of visible CLE taken randomly from various areas in the visible CLE.
3. Modified Seattle protocol: In the best case scenario in practice, the biopsies are more numerous and submitted in multiple bottles that are labeled with regard to distance from the incisor teeth. Rarely, these will have biopsies from the SCJ that straddle the histologic junction.

There is a common misconception that the three types of epithelia in CLE are distributed in a mosaic pattern. This is not true. The three types of epithelia are layered within the CLE segment, with intestinal metaplasia being proximal, cardiac epithelium intermediate, and oxyntocardiac epithelium distal. Biopsy protocols to detect intestinal metaplasia on the basis of a mosaic rather than layered distribution of intestinal metaplasia will have a higher false negative rate for an equal number of biopsies taken.

Recognizing that random biopsies from the segment of visible CLE was the rule in practice, Harrison et al.[44] studied the effect of variations in the number of biopsies taken on the detection rate of intestinal metaplasia. This study is primarily centered in England with international input.

In the introduction: "The clinical importance of Barrett's esophagus is derived from its premalignant nature, with intestinal metaplasia alone being premalignant." They discuss the difference between the American definition of BE, which requires histologically proven intestinal metaplasia, and the British definition that does not. They continue: "… the

American definition is a useful one since it… selects out those patients at risk of developing adenocarcinoma (intestinal metaplasia positive) and rejects those not at risk (intestinal metaplasia negative). If this U.S. definition is to be applied worldwide, it is clear that the histological assessment must be highly accurate and detect all cases of intestinal metaplasia that are present."

Harrison et al. cite a study by Ishaq et al.[45] of 228 UK specialists in gastroenterology that indicated that the average number of biopsies taken from visible CLE was four in total. "Logistical constraints" was the reason given for the low number of biopsies.

Harrison et al.[44] is a prospective study of 125 patients identified as having CLE at endoscopy performed between 1998 and 2002 at the University Hospital, Birmingham. The patients had a visible CLE 1 cm or more in length (maximum 11 cm; minimum 1 cm; mean 4.9 cm). 74% were men, mean age 65 years (range: 41–85 years), and 56% were being treated with PPI. 64% of patients underwent subsequent surveillance endoscopies yielding a total of 296 sets of endoscopic biopsies in these 125 patients.

The number and site of biopsies taken were left to the discretion of the individual endoscopist. The clinician simply continued their normal practice and thus the number of biopsies was influenced by time available, length of CLE, and the presence of stricture or suspicious lesions. Dr. Jankowski, a renowned expert, performed 66% of the procedures, the remainder by other trained members of the team.

The patients were arbitrarily allocated to one of two arms: 1–4 biopsies (N = 150) and >4 biopsies (N = 149). The number of biopsies taken in relation to CLE length was calculated. In 82%, at least 4 biopsies were taken every 1–2 cm of CLE, fulfilling the minimum requirement of the Seattle biopsy protocol (average 5.56 biopsies taken, average length of CLE 3.1 cm).

Intestinal metaplasia was diagnosed in the routine hematoxylin and eosin stain by the presence of goblet cells, irrespective of their number. Endoscopies with six biopsies or more were chosen for evaluation by alcian blue/periodic acid Schiff staining. This group had intestinal metaplasia on H&E stain in 60% of cases, making this a suitable group in which to determine if the additional staining enhanced the detection of intestinal metaplasia.

A total of 1646 individual biopsies from 296 endoscopies performed on 125 patients were examined. Intestinal metaplasia was demonstrated in 80/125 (64%) patients and 150/296 (51%) endoscopies (Table 11.6). Only 557/1646 (34%) individual biopsies contained intestinal metaplasia. Of 296 endoscopies, 166 (56%) contained cardiac mucosa, 158 (53%) fundic mucosa (probably oxyntocardiac rather than gastric oxyntic based on the fact that the biopsies were from visible CLE), 64 (22%) glandular mucosa of no special type (this is a designation that has not been defined anywhere), and 111 (38%) islands of squamous mucosa. It is uncertain whether these nonintestinal epithelia were seen in the entire 296 endoscopies or only in those without intestinal metaplasia.

The detection of intestinal metaplasia increased with increasing numbers of biopsies per endoscopy. Patients were stratified into those who had 1–4, 5–8, 9–12, and 13–16 biopsies per endoscopy (6 patients who had >16 biopsies per endoscopy were excluded because they had mucosal irregularity with dysplasia). There was a significant increase in frequency of

TABLE 11.6 Number and Distribution of Endoscopies (EGD) and Biopsies for the Study Group

# EGD/ Patient	Number of Patients	Total Number of EGDs	Number of Individual Bxs	% of Patients With IM	% of Biopsies With IM	Average Number Bxs per EGD (Range)	Length of CLE Mean (Range)
1	32	32	162	65.6	31.5	5.06 (1–26)	5.3 cm (2–10)
2	54	108	505	57.4	25.6	4.68 (1–23)	3.5 cm (1–10)
3	23	69	398	56.5	29.1	5.77 (1–29)	4.4 cm (1–10)
4 or more (average: 6.5)	16	87	581	87.5	46.3	5.59 (1–34)	6 cm (2–11)
Total	125	296	1646	64	34	5.56 (1–34)	4.9 cm (1–11)

EGD, Esophagogastroduodenoscopy; *IM*, intestinal metaplasia.

intestinal metaplasia in patients with five to eight biopsies compared with one to four biopsies ($P<.001$) per endoscopy. No significant difference was seen with higher biopsy numbers per endoscopy.

The number of biopsies and the length of CLE were associated with an increased rate of detection of intestinal metaplasia, but only the biopsy numbers was statistically significant. There was an increase in the detection rate of intestinal metaplasia in the proximal area of the esophagus, but this was only significant in older patients. Males (99/178; 56%) had a significantly higher prevalence of intestinal metaplasia than women (42/98; 42%) ($P=.04$).

Alcian blue/periodic acid Schiff stain was done in 92 cases with >6 biopsies. This produced no change in 51 (55%) of cases; in the other 41 cases, it identified goblet cells in biopsies that were negative on H&E. However, in 36 of these, other biopsies were positive on H&E. The stain contributed to an overall 5.4% gain in the detection of intestinal metaplasia in five cases. They also state in the discussion that cytokeratin 7/20 staining did not add significantly to the detection of intestinal metaplasia.

In their discussion, these authors from the highest circles of gastroenterology in the United Kingdom, state: "We would concur with the guidelines from the American College of Gastroenterology and suggest that the diagnosis of Barrett's esophagus requires the histological demonstration of intestinal metaplasia within a columnar-lined esophagus." I have encountered other eminent gastroenterologists in the United Kingdom that have the same opinion, which is contrary to the British Society of Gastroenterologists definition of BE.

In summary, they state: "We would recommend that a minimum of 8 random biopsies be taken at index endoscopy to confidently exclude intestinal metaplasia, and assessed using H&E staining alone… Absence of intestinal metaplasia in a patient who has been subjected to less biopsies than this does not imply absence of intestinal metaplasia but may instead simply reflect an inadequate test especially in longer segments."

7.2 The Optimum Biopsy Protocol for Detecting Intestinal Metaplasia in Columnar-Lined Esophagus

The objective of an index endoscopy when visible CLE is seen is to find intestinal metaplasia if it is present and confidently exclude intestinal metaplasia if the biopsy is negative. The optimum biopsy protocol to achieve this is elusive. While the Seattle protocol is recommended, it is rarely followed in practice. The suggestion of Harrison et al.[44] that a minimum of eight random biopsies will achieve the objective is questionable. Their finding of intestinal metaplasia in 150/296 (51%) of endoscopies in patients with a mean CLE length of 4.9 cm (range 1–11 cm) raises the inevitable question as to whether there was a significant rate of false negative diagnoses. Their conclusion that the failure of eight random biopsies to find intestinal metaplasia "confidently excluded" the possibility of a false negative diagnosis is based only on the fact that taking additional random biopsies did not increase the detection rate.

In my experience, the most likely cause of a false negative diagnosis of intestinal metaplasia is the failure to take biopsies that straddle the SCJ. There are many cases with small numbers of goblet cells, and in these the goblet cells are limited to the foveolar complexes immediately adjacent to the SCJ (see below).

I would suggest that it is important to take biopsies directed at the most proximal region of the CLE immediately adjacent to the SCJ. This is the region of CLE where goblet cells will first occur and, if the number of goblet cells is limited, will be most likely to be found in a biopsy. If a four-quadrant biopsy straddling the SCJ does not have goblet cells, it is highly unlikely that they will be present more distally. If there are tongues of columnar epithelium, the highest point of the tongue has the highest probability of having goblet cells.

In a letter to the editor reacting to Harrison et al.[44] I suggested that four directed biopsies taken at the CLE immediately distal to the SCJ would be more effective than any number of random biopsies. It is clear that the biopsies in Harrison et al.[44] did not straddle the SCJ; they reported that only 111 (38%) of endoscopies had "islands of squamous epithelium." Their study establishes that an increase in the number of random biopsies over eight did not increase yield of intestinal metaplasia. I would venture to guess that adding biopsies that straddled the SCJ would have increased the yield significantly. We will never know.

The optimum biopsy protocol, in my opinion, is to take four-quadrant biopsies straddling the SCJ. This requires fewer biopsies than either the Seattle protocol or Harrison's eight random biopsies and will likely have the lowest false negative rate for detection of intestinal metaplasia in visible CLE. The only situation where other biopsies are necessary is when a lesion suspicious for dysplasia or malignancy is seen.

The Montreal document,[29] in the discussion of biopsy protocol, does make reference to our study (see below): "It has been found that intestinal-type metaplasia is most prevalent at the most proximal end of metaplasia." However, this led to no recommended biopsy protocol to replace the Seattle protocol. Failure to use the constant distribution of intestinal metaplasia adjacent to the SCJ in CLE is as responsible for false negative diagnosis of BE as the number of biopsies taken.

7.3 Assessing Effectiveness of Biopsy Protocols for Detecting Intestinal Metaplasia in Visible Columnar-Lined Esophagus

This is a critical issue that has a powerful impact on patient management. The finding of intestinal metaplasia in visible CLE leads to surveillance and early detection of adenocarcinoma, minimizing treatment and improving the survival of GERD patients who progress to adenocarcinoma. Unfortunately, this issue has not been addressed.

Similar issues where sampling is critical have been addressed in other areas. For example, it has been shown that adequate lymph node dissection is essential for accurate staging. By consensus, it has been decided that a minimum of 15 lymph nodes are necessary in a colon resection to establish node negative status. If a colon resection has less than 15 nodes, the probability of a false negative nodal status is assumed and the patient treated as if the nodes were positive. This has resulted in a tremendous improvement in the node dissections and pathologic examination in colon resections for cancer.

A similar minimal requirement for biopsy sampling in visible CLE should be established that establishes the threshold biopsy protocol where a negative finding of intestinal metaplasia in visible CLE defines a true negative. If the sampling is below that level, the patient should be placed under surveillance at least until the next endoscopy that has an adequate sample. This will ensure that gastroenterologists are forced to reach that standard because otherwise they subject the patient to another endoscopy. It will cause poor biopsy protocols that are prevalent today to disappear.

The critical question, when assessing a biopsy protocol for adequacy, is how many biopsies were taken for a given length of visible CLE? The variation in biopsy protocol is likely largely responsible for the large differences in the prevalence of intestinal metaplasia in different series. For example, in two studies of the prevalence of intestinal visible CLE with intestinal metaplasia in populations without GERD, Ronkainen et al.[39] reported a 1.6% prevalence contrasting with a prevalence of 25% in Gerson et al.[32]

The total number of individual biopsies taken at endoscopy per centimeter length of CLE is an excellent frame of reference for adequacy of a biopsy protocol. In studies, data for individual patients are almost never given. As such, the number of biopsies must be related to the mean CLE length for comparison.

When this is done (Table 11.7) it is seen that there is a wide range of sampling in the different studies. In Harrison et al.[44] a total of 1646 biopsies were taken from 296 endoscopies (mean of 5.6 biopsies per endoscopy) with a mean length of CLE in the group being 4.9 cm. This calculates out to 1.1 biopsies per centimeter of CLE. In contrast, in Chandrasoma et al.[46] (reviewed below), 424 biopsies were taken from 32 endoscopies with a mean CLE length of 5.7 cm. This sampling can be expressed as 13.3 biopsies per endoscopy and 2.3 biopsies per centimeter of CLE. In Gatenby et al.[23] (see Chapter 12), the total number of biopsies was 3568 from 1751 endoscopies with visible CLE (mean length 4.93 cm). The biopsies per patient were 2.0 and biopsies per centimeter of CLE were 0.4. The type of sampling was random in Harrison et al.[44] and Gatenby et al.[23] and directed in Chandrasoma et al.[46] where biopsies were taken at 1–2 cm intervals with an effort made to take biopsies straddling the SCJ.

These three biopsy protocols were the usual for the practices conducting the studies. Two studies were retrospective; Harrison et al.[44] were prospective but the taking of the biopsies was left to the discretion of the endoscopist. This variation is likely to be representative of the variation in the biopsy protocol adopted in the community for evaluating visible CLE. The one proviso is that the two studies with the lower sampling were in England[23,44] and Chandrasoma et al.[47] from a center in the United States with a very strict and unusually extensive biopsy protocol. There is little data on biopsy protocols used in community practice in the United States. Somewhere in the middle of these three biopsy protocols is the minimum acceptable protocol.

TABLE 11.7 Biopsy Protocol Comparison in Three Studies With Detailed Methodology That Permits Assessment of Biopsy Protocol

Study	# per EGD	# Bxs	Bxs/EGD	Random Versus Levels	Mean CLE Length (cm)	Bx/cm of CLE	SCJ Targeted	IM Prevalence
Chandrasoma et al.	32	424	13.3	Levels	5.7	2.3	Yes	N/A[a]
Harrison et al.	296	1646	5.6	Random	4.9	1.1	No	150 (51%)
Gatenby et al.	1751	3568	2.0	Random	4.93	0.4	No	Not stated[b]

Unfortunately, a comparison of the detection rate between these studies is not possible. The table shows, however, the marked variation in the biopsy protocol in terms of numbers of biopsies per endoscopy and number of biopsies per cm length of CLE.
Bx, biopsy; *CLE*, columnar-lined esophagus; *EGD*, esophagogastroduodenoscopy; *IM*, intestinal metaplasia; *SCJ*, squamocolumnar junction.
[a]*IM prevalence was 100%; the study was designed not to test detection rate of IM in CLE; it was designed to test distribution of goblet cells in visible CLE with IM.*
[b]*Does not report prevalence of IM in the total patient population; study is designed to show results of follow-up of IM negative patients.*

8. THE DISTRIBUTION OF EPITHELIAL TYPES IN VISIBLE COLUMNAR-LINED ESOPHAGUS

The distribution of the three variants of cardiac epithelium follows a very constant pattern. Recognizing the nonrandom distribution provides valuable insight as to the causation of the changes in cellular differentiation.

The paper from the Harvard group led by Dr. Raj Goyal is one of the seminal papers on the histology of CLE. It is reviewed in Chapter 4. After classifying the three epithelial types in visible CLE (at the time synonymous with BE), this paper by Paull et al.[12] went on to describe the distribution of the 3 epithelial types in their 11 patients. The authors used a biopsy protocol that was controlled by manometry. The LES was identified and biopsies were taken at measured levels from within the LES and proximal to the LES at intervals of 1–3 cm from the upper border of the LES to a point proximal to the SCJ above. The total number of biopsies in the 11 patients was 122, representing generous sampling that allowed mapping the different types of epithelium within the segment of CLE.

9/11 patients with CLE above the LES had SCE (cardiac epithelium with intestinal metaplasia). Of these nine patients, the intestinal epithelium with goblet cells extended over a length of 4.0–14.5 cm in seven patients, but was present in only one biopsy in the other 2 patients. In five of the nine patients with intestinal metaplasia, either gastric-fundic type (oxyntocardiac) or junctional type (cardiac) epithelium was interposed between the intestinal epithelium and LES. The interposed gastric-fundic type (oxyntocardiac) epithelium extended over at least 2, 6, and 10 cm above the LES in three of these patients. In the two patients with junctional (cardiac) epithelium, this extended at least 1 and 4 cm above the LES. The two patients who did not have intestinal metaplasia had either a mixture of junctional (cardiac) and gastric-fundic (oxyntocardiac) epithelia or gastric-fundic type (oxyntocardiac) epithelium alone. This showed "marked atrophic changes characterized by sparse and shortened glands containing a variable number of parietal cells."

The authors provided a detailed map of the various epithelial types in four of their patients (Fig. 11.16). They stated: "When present, specialized columnar-type (intestinal) epithelium was always the most proximally located, and the gastric-fundic type (oxyntocardiac) always the most distally located columnar epithelium. Junctional-type (cardiac) epithelium, when present, was interposed between the gastric fundic type (oxyntocardiac) and the specialized columnar type (intestinal) or squamous epithelium."

In their figure, intestinal metaplasia when present, extends contiguously down from the SCJ without skip areas. In one case, it extends down into the LES zone with no other epithelial type shown. In a second, it transitions to cardiac epithelium. In the third it transitions to oxyntocardiac with no intervening cardiac epithelium. The fourth case in their figure shows a

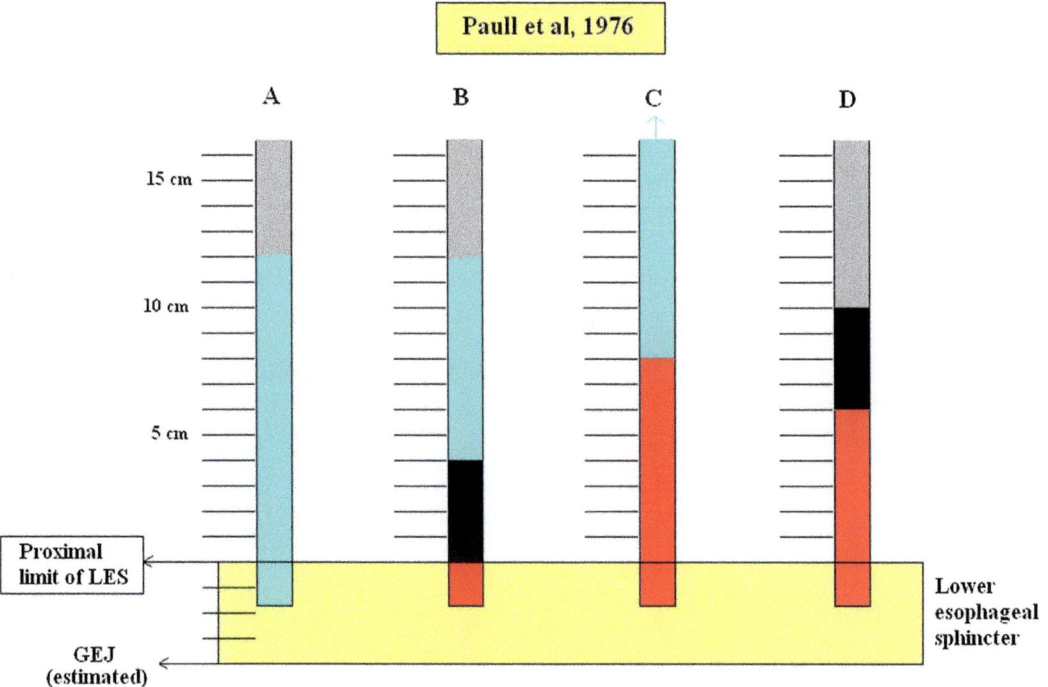

FIGURE 11.16 Prevalence and distribution of the three epithelial types in four patients detailed in Paull et al. in their 1976 paper. When all the three epithelia are present, the distribution is cardiac epithelium with intestinal metaplasia (blue), cardiac epithelium (black), and oxyntocardiac epithelium (red). One patient with a 10-cm segment of visible columnar-lined esophagus above the upper border of the lower esophageal sphincter (LES) had no intestinal metaplasia. *GEJ*, gastroesophageal junction.

small area of cardiac epithelium adjacent to the SCJ followed by a 6–7 segment of oxyntocardaic epithelium. Except for the one case with all intestinal metaplasia, the LES zone contains oxyntocardiac epithelium in three cases.

In Chandrasoma et al.[46] we evaluated the distribution of the different epithelial types in 32 patients with visible CLE measuring a minimum of 2 cm (range 2–16 cm; mean 5.7 cm; median 5.0 cm) who had intestinal metaplasia in at least one biopsy (i.e., with criteria for BE). This study differs from that of Paull et al. because the segment of CLE extends to the endoscopic GEJ, which includes the entire functional LES.

These patients had 2–5 biopsies per level at 1–2 cm measured intervals within the columnar-lined segment. The number of biopsies containing metaplastic columnar epithelium (cardiac epithelium with and without parietal and/or goblet cells) in the 32 patients was 424 (range per patient: 5–37; mean 13.3). When a biopsy contained intestinal metaplasia, the density of goblet cells in each biopsy was graded using the following criteria: grade 0: no goblet cells; grade 1: <one-third of the foveolar-gland units contain goblet cells; grade 2: one to two-thirds of the foveolar-gland units contain goblet cells; grade 3: more than two-thirds of the glands contain goblet cells. An overall intestinal metaplasia grade was then assigned to each level based on the mean grade for all biopsies at that level.

This permitted mapping of the three epithelial types and the density of goblet cells in the columnar-lined segment. The distribution of intestinal metaplasia in the 424 biopsies from CLE in these 32 patients is shown in Table 11.8.

This shows that intestinal metaplasia favored the most proximal level in the segment of CLE. If biopsies were taken only at the most proximal level (i.e., immediately distal to the SCJ with the biopsy straddling the histologic junction) of a patient with visible CLE measuring 2–16 cm, intestinal metaplasia would have been detected in all patients and in 64/68 (94%) of the biopsies.

The mean density of goblet cells in the biopsies at the most proximal and most distal level, as determined by our grading method, is shown in Table 11.9. Because the diagnosis of intestinal metaplasia is made when there is a single definite goblet cell in cardiac epithelium, the likelihood of a false negative diagnosis is the least in biopsies at the most proximal level.

Cardiac epithelium was present at all levels in all patients. The distribution of oxyntocardiac epithelium was the exact reverse of intestinal metaplasia. The number of parietal cells was maximal in the distal part of the CLE and progressively decreased in the more proximal region. This is identical to the study of Paull et al.[12] and the original descriptions of histology of the columnar-lined segment by Allison and Johnstone[11] and Barrett.[13]

In Harrison et al.[44] the authors comment in their discussion: "Chandrasoma et al, also demonstrated a gradient of intestinal metaplasia within columnar-lined esophagus, with 94% of biopsies from the proximal edge of columnar-lined esophagus being positive for intestinal metaplasia compared with 39% of biopsies at the distal edge, and that within intestinal metaplasia at the proximal edge there was an increased frequency of goblet cells compared to the more distal biopsies containing IM; we were unable to confirm this in this study."

The distribution of epithelial types within the segment of CLE cannot be defined with random sampling where the biopsies are not separated out by measured location. Careful biopsy technique is required to recognize the distribution of epithelial types in CLE.

TABLE 11.8 Distribution of Intestinal Metaplasia Within the Segment of Columnar-Lined Esophagus

	# Patients With IM (%)	# Biopsies With IM (%)	# Biopsies Without IM (%)
Total	32	311/424 (73%)	113/424 (27%)
Most proximal level	32 (100%)	64/68 (94%)	4/68 (6%)
Most distal level	22 (69%)	40/102 (39%)	62/102 (61%)

IM, intestinal metaplasia.

TABLE 11.9 Density of Goblet Cells at Most Proximal and Distal Biopsy Levels

	Grade 0	Grade 1	Grade 2	Grade 3
Most proximal level	0	5	6	21
Most distal level	10	8	13	1

In examining thousands of cases with biopsies taken by Dr. DeMeester's unit in a standard manner over 20 years, I can state that the proximal predominance of intestinal epithelium in a segment of visible CLE is true with only rare exception. Anecdotal statements such as this are dangerous, but 15,000 patients over 20 years coupled with multiple studies that provide a sample that proves the point justifies even anecdotal statements.

9. DIFFERENCE IN THE PREVALENCE OF INTESTINAL METAPLASIA WITH DIFFERENT LENGTH OF VISIBLE COLUMNAR-LINED ESOPHAGUS

The variation of the prevalence of intestinal metaplasia in visible CLE of different lengths is also well established and recognized. In general, the longer the segment of visible CLE, the higher the prevalence of intestinal metaplasia.

Spechler et al.[25] in his seminal paper in 1994 showed that 100% of 16 patients with visible CLE > 3 cm had intestinal metaplasia ("specialized columnar epithelium" was the term used in that paper). In patients with <3 cm of visible CLE, only 26/140 (18.6%) patients had intestinal metaplasia. The <3 cm group included patients who were endoscopically normal; though the total number of such patients is unknown, it was reported that 9/26 patients with intestinal metaplasia were endoscopically normal.

Chandrasoma et al. is a study of 959 patients in whom endoscopic biopsies were taken according to the standard protocol used in the Foregut Surgery unit of the University of Southern California.[47] In endoscopically normal people, irrespective of symptoms, antegrade biopsies were taken straddling the SCJ, retroflex biopsies from the area within 1 cm distal to the SCJ, and distal gastric biopsies from antrum and body. In patients with visible CLE, additional four-quadrant biopsies were taken at 1–2 cm intervals between the endoscopic GEJ (proximal limit of rugal folds) and the cephalad-displaced SCJ. The epithelial types at each level were classified according to our standard definitions into squamous, cardiac with intestinal metaplasia, cardiac, oxyntocardiac, and gastric oxyntic epithelia (see Chapter 4).

The definition of CLE was the distance between the most proximal point of the SCJ and gastric oxyntic epithelium (i.e., histologic CLE). The biopsy protocol was inadequate to sample the entire CLE as shown by the absence of gastric oxyntic epithelium in a significant number of retroflex biopsies. This study predated our realization that the dilated distal esophagus could extend well beyond 1 cm distal to the SCJ.

The length of histologic CLE was determined as follows:

Less than 1 cm CLE: These patients did not have a visible CLE and had cardiac epithelium (with and without parietal and/or goblet cells) in the measured biopsy from the SCJ and/or the unmeasured retroflex biopsy within 1 cm distal to the SCJ. This was CLE limited to the dilated distal esophagus.

1–2 cm CLE: Cardiac epithelium (with and without parietal and/or goblet cells) present in two measured biopsies either 1 or 2 cm apart, including the retroflex biopsy which was within 1 cm distal to the endoscopic GEJ.

3–4 cm CLE: Cardiac epithelium (with and without parietal and/or goblet cells) present in two or greater measured biopsies either 3 or 4 cm apart.

5+ cm CLE: Cardiac epithelium (with and without parietal and/or goblet cells) present in two or more measured biopsies 5 cm or more apart.

The distribution and prevalence of the different epithelial types in the four groups is shown in Table 11.10 (Fig. 11.17).

This population is not comparable to Spechler et al.[25] and other studies. It is from a surgical unit specializing in management of complex patients with a possible need for surgery. There is a bias toward a larger proportion of patients with severe GERD and visible CLE with intestinal metaplasia.

Despite this, the study has the advantage of large numbers, standardized biopsy protocol, and standardized histology. The prevalence of intestinal metaplasia increases from 14.8% in endoscopically normal patients to 100% in patients with visible CLE > 5 cm. Of the 148 patients with visible CLE in the study, 128 (86.5%) had intestinal metaplasia. This is likely due to the fact that patients with BE were selected for referral to this surgical unit.

20% of endoscopically normal patients had only squamous and gastric oxyntic epithelium. Though classified as "normal" it is unknown whether they had unsampled cardiac epithelium (with or without parietal and/or goblet cells). In many of these cases, the actual histologic SCJ was not present in any biopsy sample (no effort was made during this time to take biopsies straddling the SCJ). As such, the possibility that small amounts of cardiac epithelium (with or without parietal and/or goblet cells) were missed cannot be excluded.

Patients with only oxyntocardiac epithelium as their metaplastic epithelium are largely limited to those who are endoscopically normal (with the exception of 1 patient in group 2). Cardiac epithelium with and without intestinal metaplasia is almost always present in patients with visible CLE, with intestinal metaplasia increasing as the length of CLE increases.

I will discuss the significance of the distribution and variation in the prevalence of intestinal metaplasia in CLE in the chapter on esophageal adenocarcinoma.

TABLE 11.10 Distribution of Epithelial Types in the 959 Patients Divided Into the Four Groups Defined by the Length of Histologic Columnar-Lined Esophagus (Including the Retroflex Biopsy Distal to the Endoscopic Gastroesophageal Junction)

Group	Definition	Significance[a]	Number (%)
1	**CLE<1 cm**		**811**
1a	Only squamous and GOE	Normal	161 (19.9%)
1b	OCE+ CE– IM–	Compensated GERD	158 (19.4%)
1c	CE+ IM–	Mild GERD	372 (45.9%)
1d	IM+	Mild GERD with IM	120 (14.8%)
2	**CLE 1–2 cm**		**54**
2a	Only squamous and GOE	Normal	0 (0%)
2b	OCE+ CE– IM–	Compensated GERD	1 (3.8%)
2c	CE+ IM–	Severe GERD	15 (27.8%)
2d	IM+	Severe GERD with IM	38 (70.4%)
3	**CLE 3–4 cm**		**38**
3a	Only squamous and GOE	Normal	0 (0%)
3b	OCE+ CE– IM–	Compensated GERD	0 (0%)
3c	CE+ IM–	Severe GERD	4 (10.5%)
3d	IM+	Severe GERD with IM	34 (89.5%)
4	**CLE 5 cm or more**		**56**
4a	Only squamous and GOE	Normal	0 (0%)
4b	OCE+ CE– IM–	Compensated GERD	0(0%)
4c	CE+ IM–	Severe GERD	0(0%)
4d	IM+	Severe GERD with IM	56 (100%)

CE, cardiac epithelium; *GERD*, gastroesophageal reflux disease; *GOE*, gastric oxyntic epithelium; *IM*, intestinal metaplasia; *OCE*, oxyntocardiac epithelium. Each defined group is divided into a: only oxyntic and squamous epithelium; b: oxyntocardiac epithelium also present; c: cardiac and oxyntocardiac also present; d: intestinal metaplasia present.
[a]*GERD is divided into mild and severe in this table based on the presence or absence of visible CLE at endoscopy.*

10. (FUTURE) DEFINITIONS BASED ON THE NEW CONCEPTS

The CLE can be defined in its entirety only by a combination of endoscopy and histology. The proximal limit of CLE is the SCJ. The distal limit of the CLE is the junction between cardiac epithelium (with and without parietal and/or goblet cells) and gastric oxyntic epithelium. This can be defined only by histologic sampling of the area distal to the endoscopic GEJ until the junction is reached.

The true simplified definitions of CLE are given in Table 11.11. I have used precise terms without using the term BE. This provides much more clarity with regard to the risk of adenocarcinoma in the thoracic and dilated distal esophagus. In particular, there is a clear recognition of a large population of people who are not at risk for adenocarcinoma in both thoracic and dilated distal esophagus. This is presently not possible. With the present symptom-based management method that is used universally today, no one in the population has a guarantee that they are free of risk of adenocarcinoma. Today, even a person who has never had symptoms can present for the first time with advanced adenocarcinoma.

This classification follows a theme that I am pursuing in this book of classifying severity of GERD into mild and severe based on the presence of visible CLE at endoscopy. This recognizes that visible CLE is a critical event in the progression of GERD. It is the point where LES damage is so severe that it has resulted in frequent LES failure (tLESR) and maximum cellular abnormality in the thoracic esophagus short of neoplasia. It is also the point where cellular changes in the thoracic

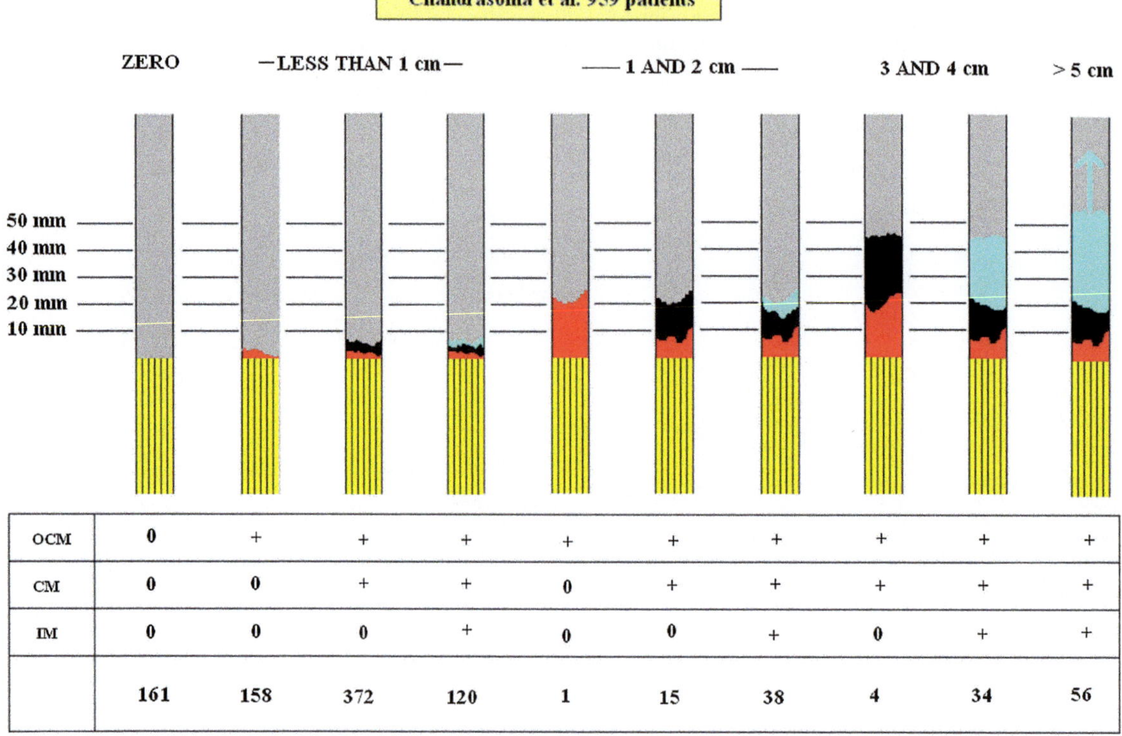

OCM	0	+	+	+	+	+	+	+	+	+
CM	0	0	+	+	0	+	+	+	+	+
IM	0	0	0	+	0	0	+	0	+	+
	161	158	372	120	1	15	38	4	34	56

FIGURE 11.17 Prevalence and distribution of the three epithelial types in 959 patients in the paper from our unit in 2003. 100% of patients with columnar-lined esophagus>5cm had intestinal metaplasia. *CM*, cardiac mucosa; *IM*, intestinal metaplasia; *OCM*, oxyntocardiac mucosa.

TABLE 11.11 Entire Spectrum of Columnar-Lined Esophagus and Their Diagnostic and Prognostic Considerations

CLE in Dilated Distal Esophagus Distal to Endoscopic GEJ	Visible CLE Proximal to Endoscopic GEJ	Severity of GERD	Cancer Risk
0	0	No GERD	0
Present; <15 mm; IM negative	0	Compensated LES damage; no GERD	0
Present; >15 mm; IM negative	0	Mild GERD	0
Present; any length; IM present	0	Mild GERD	Zero in thoracic esophagus; present in DDE (?)
Present; IM negative	Present; IM negative	Severe GERD	0
Present; IM negative	Present; IM positive	Severe GERD	Present in thoracic esophagus; zero in DDE
Present; IM positive	Present; IM positive	Severe GERD	Present in both thoracic and dilated distal esophagus

Note that the prognosis and risk of cancer is liable to change with progression of the disease with time in all patients. The interpretation and cancer risk is for the moment in time at which these findings exist.
CLE, columnar-lined esophagus; *DDE*, dilated distal esophagus; *GEJ*, gastroesophageal junction; *GERD*, gastroesophageal reflux disease; *IM*, intestinal metaplasia.

esophagus become irreversible with PPI therapy. Progression of visible CLE to intestinal metaplasia, increasing dysplasia, and adenocarcinoma is out of the control of PPI therapy.

While the proximal limit of rugal folds is not accurate in defining the true GEJ, it is nevertheless an important endoscopic landmark. It reliably identifies the proximal limit of the dilated distal esophagus.

11. RELATIONSHIP OF COLUMNAR-LINED ESOPHAGUS TO ADENOCARCINOMA

All CLE can progress in the sequence of cardiac epithelium→intestinal metaplasia→increasing dysplasia→adenocarcinoma, irrespective of whether it is recognized as visible CLE or in the dilated distal esophagus. At present, the viewpoint of the medical establishment is that the only proven risk indicator for esophageal adenocarcinoma is visible CLE in the thoracic esophagus with intestinal metaplasia.

The lack of evidence that CLE in the dilated distal esophagus progresses to adenocarcinoma is the result of a failure to take biopsies in the patient who is endoscopically normal. The evidence supporting the progression of CLE in the dilated distal esophagus to adenocarcinoma is that adenocarcinoma of the dilated distal esophagus has shown a parallel increase incidence to adenocarcinoma of the thoracic esophagus in the past 35 years. The proof is that adenocarcinoma of the dilated distal esophagus is associated with GERD.

REFERENCES

1. Pohl H, Sirovich B, Welch HG. Esophageal adenocarcinoma incidence: are we reaching the peak? *Cancer Epidemiol Biomark Prev* 2010;**19**:1468–70.
2. Kahrilas PJ, Shaheen NJ, Vaezi MF. American Gastroenterological Association Medical Position Statement on the Management of Gastroesophageal Reflux Disease. *Gastroenterology* 2008;**135**:1380–2.
3. Blonski W, Vela MF, Castell DO. Comparison of reflux frequency during prolonged multichannel intraluminal impedance and pH monitoring on and off acid suppression therapy. *J Clin Gastroenterol* 2009;**43**:816–20.
4. Malfertheiner P, Nocon M, Vieth M, Stolte M, Jasperson D, Koelz HR, Labenz J, Leodolter A, Lind T, Richter K, Willich SN. Evolution of gastro-oesophageal reflux disease over 5 years under routine medical care – the Pro-GERD study. *Aliment Pharmacol Ther* 2012;**35**:154–64.
5. Falkenback D, Oberg S, Johnsson F, Johansson J. Is the course of gastroesophageal reflux disease progressive? A 21 year follow-up. *Scand J Gastroenterol* 2009;**44**:1277–87.
6. Nason KS, Wichienkuer PP, Awais O, Schuchert MJ, Luketich JD, O'Rourke RW, Hunter JG, Morris CD, Jobe BA. Gastroesophageal reflux disease symptom severity, proton pump inhibitor use, and esophageal carcinogenesis. *Arch Surg* 2011;**146**:851–8.
7. Chandrasoma PT. Histologic definition of gastro-esophageal reflux disease. *Curr Opin Gastroenterol* 2013;**29**:460–7.
8. Zaninotto G, DeMeester TR, Schwizer W, Johansson KE, Cheng SC. The lower esophageal sphincter in health and disease. *Am J Surg* 1988;**155**:104–11.
9. Chandrasoma P, Makarewicz K, Wickramasinghe K, Ma YL, DeMeester TR. A proposal for a new validated histologic definition of the gastroesophageal junction. *Hum Pathol* 2006;**37**:40–7.
10. Barrett NR. Chronic peptic ulcer of the oesophagus and 'oesophagitis'. *Br J Surg* 1950;**38**:175–82.
11. Allison PR, Johnstone AS. The oesophagus lined with gastric mucous membrane. *Thorax* 1953;**8**:87–101.
12. Paull A, Trier JS, Dalton MD, Camp RC, Loeb P, Goyal RK. The histologic spectrum of Barrett's esophagus. *N Eng J Med* 1976;**295**:476–80.
13. Barrett NR. The lower esophagus lined by columnar epithelium. *Surgery* 1957;**41**:881–94.
14. McClave SA, Boyce Jr HW, Gottfried MR. Early diagnosis of columnar lined esophagus: a new endoscopic diagnostic criterion. *Gastrointest Endosc* 1987;**33**:413–6.
15. Hayward J. The lower end of the oesophagus. *Thorax* 1961;**16**:36–41.
16. American Joint Commission of Cancer Staging Manual. *Esophagus and esophagogastric junction*. 7th ed. 2010.
17. DiNardi FG, Riddell RH. Esophagus. In: Sternberg SS, editor. *Histology for pathologists*. 2nd ed. Philadelphia: Lippincott-Raven Publishers; 1997. p. 461–80.
18. Owen DA. Stomach. In: Sternberg SS, editor. *Histology of for pathologists*. 2nd ed. Philadelphia: Lippincott-Raven Publishers; 1997. p. 481–93.
19. Haggitt RC, Tryzelaar J, Ellis FH, et al. Adenocarcinoma complicating columnar epithelium-lined (Barrett's) esophagus. *Am J Clin Pathol* 1978;**70**:1–5.
20. Haggitt RC, Dean PJ. Adenocarcinoma in Barrett's epithelium. In: Spechler SJ, Goyal RK, editors. *Barrett's esophagus: pathophysiology, diagnosis and management*. New York: Elsevier Science Publishing; 1985. p. 153–66.
21. Haggitt RC. Adenocarcinoma in Barrett's esophagus: a new epidemic? *Hum Pathol* 1992;**23**:475–6.
22. Sharma P, Dent J, Armstrong D, et al. The development and validation of an endoscopic grading system for Barrett's esophagus: the Prague C & M criteria. *Gastroenterology* 2006;**131**:1392–9.
23. Gatenby PAC, Ramus JR, Caygill CPJ, Shepherd NA, Watson A. Relevance of the detection of intestinal metaplasia in non-dysplastic columnar epithelium. *Scand J Gastroenterol* 2008;**43**:524–30.
24. Bombeck CT, Dillard DH, Nyhus LM. Muscular anatomy of the gastroesophageal junction and role of phreno-esophageal ligament. Autopsy study of sphincter mechanism. *Ann Surg* 1966;**164**:643–54.
25. Spechler SJ, Zeroogian JM, Antonioli DA, Wang HH, Goyal RK. Prevalence of metaplasia at the gastroesophageal junction. *Lancet* 1994;**344**:1533–6.
26. Fitzgerald RC, di Pietro M, Ragunath K. British Society of Gastroenterology guidelines on the diagnosis and management of Barrett's oesophagus. *Gut* 2014;**63**:7–42.
27. Shaheen NJ, Falk GW, Iyer PG, Gerson L. ACG clinical guideline: diagnosis and management of Barrett's esophagus. *Am J Gastroenterol* November 2015;**3**. http://dx.doi.org/10.1038/ajg.2015.322. Advance online publication.
28. Riddell RH, Odze RD. Definition of Barrett's esophagus: time for a rethink – is intestinal metaplasia dead? *Am J Gastroenterol* 2009;**104**:2588–94.
29. Vakil N, van Zanten SV, Kahrilas P, Dent J. Jones B and the global consensus group. The Montreal definition and classification of gastroesophageal reflux disease: a global evidence-based consensus. *Am J Gastroenterol* 2006;**101**:1900–20.

30. Sharma P, McQuaid K, Dent J, Fennerty B, Sampliner R, Spechler S, Cameron A, Corley D, Falk G, Goldblum J, Hunter J, Jankowski J, Lundell L, Reid B, Shaheen N, Sonnenberg A, Wang K, Weinstein W. A critical review of the diagnosis and management of Barrett's esophagus: the AGA Chicago Workshop. *Gastroenterology* 2004;**127**:310–30.
31. Westhoff B, Brotze S, Weston A, McElhinney C, Cherian R, Mayo M, Smith HJ, Sharma P. The frequency of Barrett's esophagus in high risk patients with chronic GERD. *Gastrointest Endosc* 2005;**61**:226–31.
32. Gerson LB, Shetler K, Triadafilopoulos G. Prevalence of Barrett's esophagus in asymptomatic individuals. *Gastroenterology* 2002;**123**:461–7.
33. Der R, Tsao-Wei DD, DeMeester T, et al. Carditis: a manifestation of gastroesophageal reflux disease. *Am J Surg Pathol* 2001;**25**:245–52.
34. Leodolter A, Nocon M, Vieth M, Lind T, Jaspersen D, Richter K, Willich S, Stolte M, Malfertheiner P, Labenz J. Progression of specialized intestinal metaplasia at the cardia to macroscopically evidence Barrett's esophagus: an entity of concern in the Pro-GERD study. *Scand J Gastroenterol* 2012;**47**:1429–35.
35. Ayazi S, Tamhankar A, DeMeester SR, Zehetner J, Wu C, Lipham JC, Hagen JA, DeMeester TR. The impact of gastric distension on the lower esophageal sphincter and its exposure to acid gastric juice. *Ann Surg* 2010;**252**:57–62.
36. Robertson EV, Derakhshan MH, Wirz AA, Lee YY, Seenan JP, Ballantyne SA, Hanvey SL, Kelman AW, Going JJ, McColl KE. Central obesity in asymptomatic volunteers is associated with increased intrasphincteric acid reflux and lengthening of the cardiac mucosa. *Gastroenterology* 2013;**145**:730–9.
37. Chandrasoma P, Wijetunge S, Ma Y, DeMeester S, Hagen J, DeMeester T. The dilated distal esophagus: a new entity that is the pathologic basis of early gastroesophageal reflux disease. *Am J Surg Pathol* 2011;**35**:1873–81.
38. Ringhofer C, Lenglinger J, Izay B, Kolarik K, Zacherl J, Fisler M, Wrba F, Chandrasoma PT, Cosentini EP, Prager G, Riegler M. Histopathology of the endoscopic esophagogastric junction in patients with gastroesophageal reflux disease. *Wien Klin Wochenschr* 2008;**120**:350–9.
39. Ronkainen J, Aro P, Storskrubb T, Johansson S-E, Lind T, Bolling-Sternevald E, Vieth M, Stolte M, Talley NJ, Agreus L. Prevalence of Barrett's esophagus in the general population: an endoscopic study. *Gastroenterology* 2005;**129**:1825–31.
40. Deleted in review.
41. Kulig M, Nocon M, Vieth M, Leodolter A, Jaspersen D, Labenz J, Meyer-Sabellek W, Stolte M, Lind T, Malfertheimer P, Willich SN. Risk factors of gastroesophageal reflux disease: methodology and first epidemiological results of the ProGERD study. *J Clin Invest* 2004;**57**:580–9.
42. Derakhshan MH, Robertson EV, Lee YY, Harvey T, Ferner RK, Wirz AA, Orange C, Ballantyne SA, Hanvey SL, Going JJ, McColl KE. In healthy volunteers, immunohistochemistry supports squamous to columnar metaplasia as mechanism of expansion of cardia, aggravated by central obesity. *Gut* 2015;**64**(11):1705–14.
43. Marsman WA, van Sandyck JW, Tytgat GNJ, ten Kate FJW, van Lanschot JJB. The presence and mucin histochemistry of cardiac type mucosa at the esophagogastric junction. *Am J Gastroenterol* 2004;**99**:212–7.
44. Harrison R, Perry I, Haddadin W, McDonald S, Bryan R, Abrams K, Sampliner R, Talley NJ, Moayyedi P, Jankowski JA. Detection of Intestinal Metaplasia in Barrett's esophagus: an observational comparator study suggests the need for a minimum of eight biopsies. *Am J Gastroenterol* 2007;**102**:1154–61.
45. Ishaq S, Harper E, Brown J, et al. Survey of current clinical practice in the diagnosis, management and surveillance of Barrett's metaplasia. A UK national survey. *Gut* 2003;**53**:A32.
46. Chandrasoma PT, Der R, Dalton P, Kobayashi G, Ma Y, Peters J, DeMeester T. Distribution and significance of epithelial types in columnar lined esophagus. *Am J Surg Pathol* 2001;**25**:1188–93.
47. Chandrasoma PT, Der R, Ma Y, Peters J, DeMeester T. Histologic classification of patients based on mapping biopsies of the gastroesophageal junction. *Am J Surg Pathol* 2003;**27**:929–36.

Chapter 12

Esophageal Adenocarcinoma

Esophageal adenocarcinoma is the most lethal endpoint of gastroesophageal reflux disease (GERD). GERD is, from a practical standpoint, the only cause of esophageal adenocarcinoma.

Fortunately, the risk of esophageal adenocarcinoma is miniscule when considered as a percentage risk in a person who has symptoms of GERD under treatment. Patients with GERD are either never informed about the cancer risk or reassured when they inquire about the possibility of cancer.

Unless there is full and documented disclosure about the risk of future carcinoma to patients, there are potential future medicolegal implications of such reassurance. The few patients who develop adenocarcinoma after many years of effective symptom control with proton pump inhibitors (PPIs) are often surprised to learn that cancer was related to their GERD. It is probable that no one had advised them about cancer risk or told them that PPI treatment does not prevent cancer. PPI therapy now has a 25-year record of use in GERD during which period esophageal adenocarcinoma has increased in a steep upward incline.[1]

The failure to inform the patient of the clearly recognized link between GERD and cancer[2] results in the worrisome statistic that only 10%–15% of patients who develop adenocarcinoma have ever had an endoscopy. This is because guidelines of GERD management recommend that endoscopy is not indicated unless treatment fails to control symptoms.[3]

A patient who develops adenocarcinoma without ever having been told that endoscopy may have led to a diagnosis of Barrett esophagus can take exception to not having been fully informed of the potential that existed to save their lives. This is because Barrett surveillance is known to lead to early diagnosis of dysplasia and adenocarcinoma that can be effectively treated with endotherapy without the need for esophagectomy and with a significantly better survival probability.[4]

Nason et al.[5] showed that the incidence of "adenocarcinogenesis" (i.e., Barrett esophagus, dysplasia, and adenocarcinoma) is significantly higher in GERD patients who are well controlled with PPI therapy.

While the percentage risk in the individual patient is small, esophageal adenocarcinoma is a serious problem. Because the number of patients with GERD is so high, even the very small percentage risk translates to a high total incidence. An estimated 20,000 people will develop esophageal adenocarcinoma (including adenocarcinoma of the dilated distal esophagus) in 2017.

Even the patient who is diagnosed with Barrett esophagus has a low risk of cancer, estimated now to be 0.2%–0.3% per year.[6] This translates to a 4%–6% risk in 20 years. Again, with the high prevalence of Barrett esophagus in the population, this seemingly low risk translates to a high overall incidence of adenocarcinoma.

1. DOES PRESENT MEDICAL THERAPY PROMOTE ESOPHAGEAL ADENOCARCINOMA?

It is appropriate to emphasize rather than minimize the risk of adenocarcinoma in the management of the GERD patient. It is the most dreaded consequence of GERD. The present management does nothing to prevent progression to adenocarcinoma. PPI therapy does not prevent the occurrence of any of the well-recognized steps in the GERD to adenocarcinoma sequence: squamous epithelium → visible CLE → intestinal metaplasia → dysplasia → cancer.

An important question to address is whether there is any possibility that the present treatment of GERD in any way promotes cancer. There is vehement denial by gastroenterologists of this possibility when the question is raised. However, there is no randomized trial that proves the lack of promotion of cancer by present medical therapy. In fact, if evaluated objectively without emotion and bias, the balance of evidence favors a role for acid-reducing therapy in promoting the incidence of cancer in patients with GERD rather than the reverse.

GERD. http://dx.doi.org/10.1016/B978-0-12-809855-4.00012-9

Acid-reducing drug therapy, mainly with PPIs, is the only treatment of GERD, used in all but a few patients who opt for procedures to repair the lower esophageal sphincter (LES). While these drugs have a positive effect on the squamous epithelium and symptom control, they do not prevent progression of LES damage or reduce the amount of reflux. The idea that controlling symptoms with treatment must reduce cancer risk has no substance.

By alkalinizing baseline gastric juice as their method of action, acid-reducing drugs change the pH milieu of the esophagus in patients with GERD. Between reflux episodes, the esophageal pH is near neutral. During reflux episodes, medical therapy converts the pH milieu of the esophagus from strong-acid reflux without PPIs to weak-acid reflux with PPI therapy. No one has addressed the question: "Does increasing the pH of the esophagus during reflux have an effect in promoting any cellular change that represents progression of GERD to adenocarcinoma?"

There is a temporal association between our increasing ability to increase the pH of gastric juice that is episodically refluxed into the esophagus when the LES fails in the GERD patient and the increase in the prevalence of Barrett esophagus and esophageal adenocarcinoma.[1] This began with acid neutralizers in the 1950s to H_2 receptor antagonists in the 1960s to PPIs in the 1980s.

This epidemiologic fact makes it imperative that we ensure that our treatment of GERD that effectively controls symptoms and heals erosive esophagitis does not at the same time promote Barrett esophagus and adenocarcinoma. These logical reasons, not disproved by evidence, to believe that this, in fact, is what happens.

Carcinogenesis is a multistep process: GERD→columnar-lined esophagus (CLE) in the dilated distal esophagus→CLE in the thoracic esophagus→intestinal metaplasia in CLE (Barrett esophagus)→adenocarcinoma. The increasing incidence of adenocarcinoma can arise if there is an increase in any one or more of these steps (Fig. 12.1).

1.1 Increased Incidence and Severity of Gastroesophageal Reflux Disease

The primary cause of adenocarcinoma in the thoracic esophagus is the reflux of gastric contents into the esophagus. The cellular changes resulting from reflux are caused by luminal exposure of the esophageal epithelium to gastric contents. Lagergren et al.[2] showed a strong correlation between the severity and duration of heartburn and/or regurgitation with both adenocarcinoma of the thoracic esophagus and dilated distal esophagus ("gastric cardia").

However, in that study,[2] the association with symptoms was not absolute. 40% of patients developing adenocarcinoma of the thoracic esophagus, and 71% developing adenocarcinoma of the dilated distal esophagus had no symptoms as defined in the study (heartburn and/or regurgitation at least once per week).

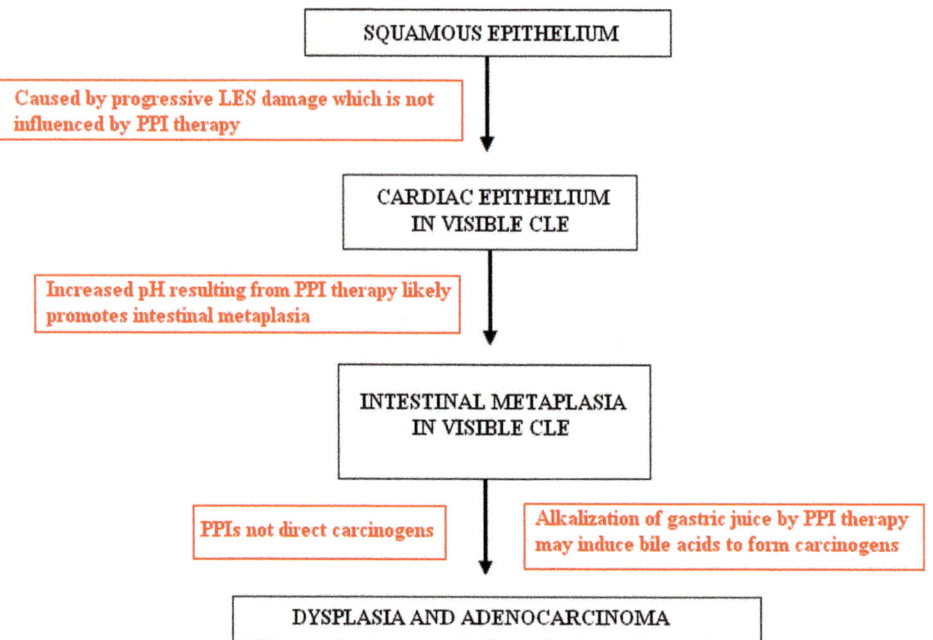

FIGURE 12.1 The gastroesophageal reflux disease to adenocarcinoma sequence in the thoracic esophagus involves cardiac metaplasia of squamous epithelium, intestinal metaplasia in cardiac epithelium, and carcinogenesis in intestinal metaplasia. Proton pump inhibitor (PPI) therapy does not prevent visible columnar-lined esophagus (CLE) because lower esophageal sphincter (LES) damage progresses despite PPI therapy. PPI therapy, by increasing the pH in gastric juice, likely promotes intestinal metaplasia in cardiac epithelium in the esophagus and may act as an indirect carcinogen by converting bile acids to carcinogenic derivatives.

This suggests that symptoms as well as their control with PPI therapy correlate poorly with cellular changes leading to adenocarcinoma. All studies of the incidence and prevalence of GERD in the population have been based on assessing symptoms of GERD. These show an increase in the prevalence of GERD in the population between 2005 and 2014.[7,8] Even a small increase in the prevalence of GERD may, by increasing the prevalence and length of CLE in the esophagus, be responsible for the rise in cancer. However, data of prevalence of GERD measured by symptoms are unreliable at best.

I have shown that the best criterion to assess severity of GERD is by assessment of abdominal LES damage, increasing severity of which leads to increasing length of cardiac epithelium in the dilated distal esophagus, ultimately reaching the point where the LES becomes incompetent, causing reflux (Fig. 12.2). The occurrence of visible CLE is indicative of severe reflux into the thoracic esophagus. In the new method of correlating severity of GERD with LES damage (see Chapter 16), it is likely that visible CLE is limited to patients with >25 mm LES damage (i.e., a residual abdominal LES length of <10 mm) (Fig. 12.3).

To date, the absence of any study of severity of LES damage makes it impossible to determine whether the prevalence of GERD has actually increased. There is a theoretical basis for an increase in LES damage in the population. Overeating, which is primarily responsible for LES damage, has increased in the past three decades, concurrent with the obesity epidemic. However, it does not date back to the 1950s when the epidemic of esophageal adenocarcinoma began, given the expected lag phase of development of carcinoma.

Increase in CLE length in the dilated distal esophagus has not been studied to any extent because most experts presently regard it as the "normal proximal stomach." The area distal to the endoscopic gastroesophageal junction (GEJ) is not biopsied, per guidelines of management.[3]

In summary, an increase in abdominal LES damage and the incidence and severity of GERD may contribute to the increased cancer risk, but it is unlikely to be the entire reason. This question is difficult to answer at present because of our inability to accurately assess the incidence and severity of GERD historically because of a lack of a standard definition for GERD. When data accumulate with the new histologic test for LES damage, a clearer picture will emerge.

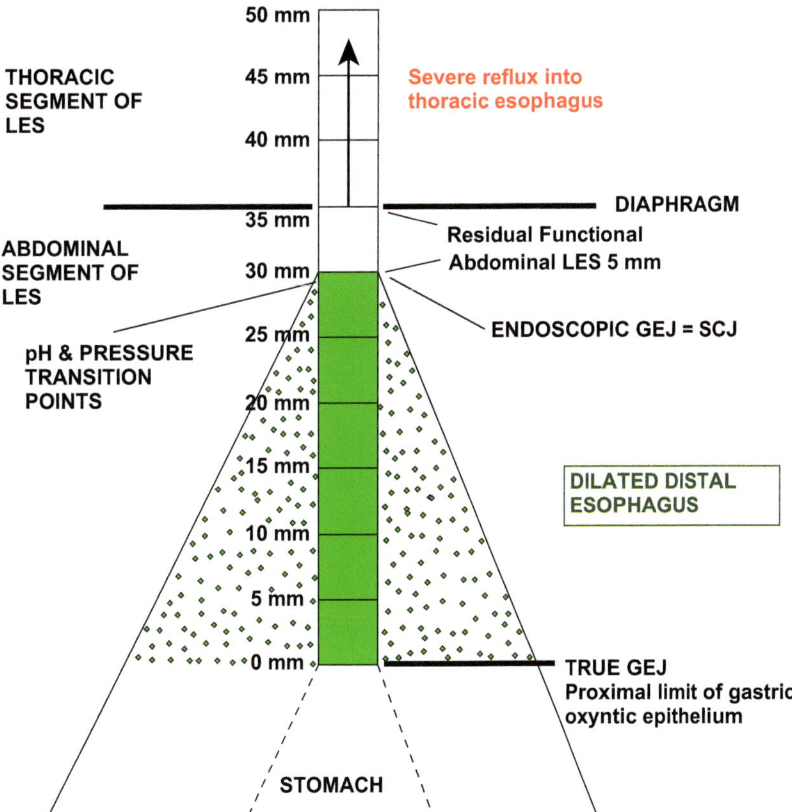

FIGURE 12.2 A patient with 30 mm of abdominal lower esophageal sphincter (LES) damage. The residual abdominal LES length is 5 mm, below the 10 mm length that is associated with LES failure. The defective LES is associated with severe reflux into the thoracic esophagus (*arrow*). This person is at risk for development of severe erosive esophagitis and visible columnar-lined esophagus (not shown). *GEJ*, gastroesophageal junction; *SCJ*, squamocolumnar junction.

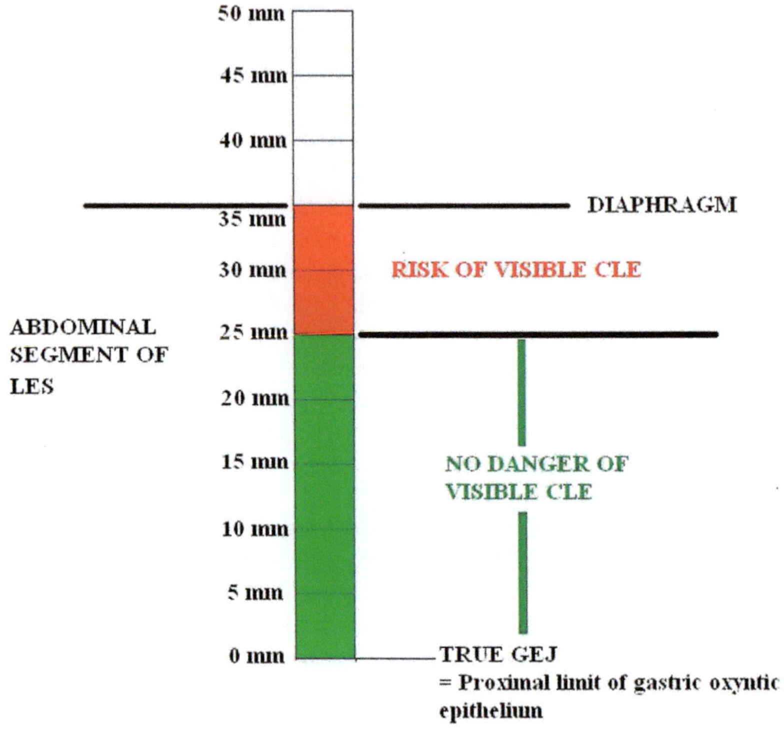

FIGURE 12.3 The abdominal lower esophageal sphincter (LES) is 35 mm in the undamaged state. It can undergo damage to a theoretical level of 25 mm before reflux into the thoracic esophagus become sufficient to cause visible columnar-lined esophagus (CLE). Persons with <25 mm abdominal LES damage are not at risk for visible CLE (*green zone*); those with >25 mm (*red zone*) may develop visible CLE. If progression of abdominal LES damage can be kept below 25 mm for the duration of a person's life, adenocarcinoma can be prevented. *GEJ*, gastroesophageal junction.

1.2 Increase in pH of Gastric Contents as a Promoter of Adenocarcinoma in Patients With Gastroesophageal Reflux Disease

There is no randomized prospective clinical trial that has shown whether the use of acid-reducing drugs in GERD affects the incidence of esophageal adenocarcinoma positively or negatively.

Most prospective studies that evaluate the association of cancer with GERD do not begin with GERD. They begin when Barrett esophagus (visible CLE with intestinal metaplasia) has been diagnosed. To most experts, this is when the cancer risk begins. This is true. Despite recent controversy,[9] there is no cancer risk before intestinal metaplasia occurs in CLE.

However, factors in the phase of GERD→Barrett esophagus may actually be more important in explaining the increasing incidence of adenocarcinoma (Fig. 12.1). Anything that increases the number of persons entering the pool of people with Barrett esophagus will, by increasing the number of people at risk, increase the incidence of adenocarcinoma. This will happen even if the conversion from Barrett esophagus→adenocarcinoma remains constant.

If acid-reducing drugs increase the risk for adenocarcinoma by promoting the GERD→Barrett esophagus step, that association will only emerge when the incidence of cancer in patients without an endoscopic diagnosis of Barrett esophagus who are treated with PPIs is studied. Even then, the interpretation of data can be difficult because of indication bias. Indication bias means that the intake of acid-reducing drugs will vary with the severity of GERD. It is therefore important to correct for symptom severity when evaluating medical therapy. In general, studies divide acid-reducing drug use by the type (PPIs vs. H_2 receptor antagonists vs. acid neutralizers) and frequency of use (never users vs. intermittent users vs. regular users).

In Lagergren et al.[2] which studied the risk of cancer in patients with symptomatic GERD, there is a statement within the body of the paper that suggests that medication increases risk of adenocarcinoma even when corrected for symptom severity: "We compared the risk of esophageal adenocarcinoma among persons who used medication for symptoms of reflux at least five years before the interview with that among symptomatic persons who did not use such medication. The odds ratio was 3.0 (95 percent confidence interval, 2.0 to 4.6) without adjustment for severity of symptoms and 2.9 (95 percent confidence interval, 1.9 to 4.6) with this adjustment" (pp. 827–829).

In an epidemiologic study done at the University of Southern California of 220 patients with esophageal adenocarcinoma matched against 1356 control patients, Duan et al.[10] reported an odds ratio of 6.32 (95% CI 3.14–12.6; $P<.01$) in patients who had used nonprescription acid neutralizers compared with controls who had never used such medications. This study is reviewed in detail later in the chapter.

1.3 Increase in Visible Columnar-Lined Esophagus

Any increase in visible CLE will potentially cause an increase in the incidence of adenocarcinoma in the thoracic esophagus. In fact, it is an absolute certainty that adenocarcinoma will not arise in the thoracic esophagus without preexisting visible CLE.

There is excellent recent evidence that there is a very significant rate of progression from no visible CLE to visible CLE in patients being treated empirically for GERD. The Pro-GERD study[11] has shown, in 2721 patients being treated for GERD routinely by community physicians, a conversion from no endoscopically visible CLE at an index endoscopy to the presence of endoscopic CLE at a rate of 9.7% within 5 years.

Even more troubling, the study showed a significantly increased association between the incidence of visible CLE and the regular use of PPIs compared to no use and on-demand use of acid-suppressive drugs. Multivariate analysis did not show a significant relationship to symptom severity, making indication bias unlikely. While this is not necessarily the proof that PPIs promote the occurrence of visible CLE, it, nevertheless, proves that treatment with regular PPIs is associated with and does not prevent this conversion.

A conversion rate in patients with GERD from no visible CLE (=no cancer risk) to visible CLE (cancer risk of 0.2%–0.3% per year) of 10% every 5 years may explain a large part of the sevenfold increase in the incidence of esophageal adenocarcinoma that has occurred in the past 40 years.

There are no data as to whether this 10% increase in incidence of visible CLE is recent, although its association with PPIs and no other acid-reducing agent suggests that the risk has increased as effectiveness of acid reduction by drug therapy has improved.

The present management guidelines that emphasize empiric PPI therapy for GERD without endoscopy until the point of treatment failure ensure that any increase in CLE will remain hidden. The Pro-GERD study was designed specifically to test progression in patients with GERD being treated with acid-reducing therapy who did not have an indication for endoscopy.[11]

Long segments of visible CLE existed in the 1950s.[12,13] However, endoscopy was so rarely performed at the time that there is no idea as to its prevalence in GERD patients than compared with the present time. Also, endoscopic criteria that define visible CLE have changed and lacked consistency over time.[14] Even today, there is lack of agreement as to whether visible CLE means any visible CLE or only counts when it is >1 cm.[15] Certainly, there has never been any effort to measure CLE in the dilated distal esophagus, which, because it correlates with LES damage,[16,17] may actually be the most important determinant of cancer incidence.

In summary, there are no data relating to changes in the prevalence and length of CLE historically. However, there is excellent evidence that there is a significant (around 10% per 5 years) conversion of no visible CLE to visible CLE in patients being treated with PPIs for GERD.

1.4 Increase in Intestinal Metaplasia in Columnar-Lined Esophagus

Historically, Allison et al.'s[12] and Barrett's[13] papers had excellent descriptions of CLE. They described long segments of columnar esophagus with a relative paucity of intestinal metaplasia limited to the proximal regions of the CLE segment (Fig. 12.4).

The dramatic change since the 1950s seems to have been the increase in prevalence and extent of intestinal metaplasia in CLE. It is difficult to find historical studies with detailed histologic mapping of segments of CLE, making it very difficult to get a clear picture.

In Chapter 11, I reviewed the data on the distribution of intestinal metaplasia in patients with visible CLE. Although the information is scanty, these data suggest that there has been an increase in the prevalence and extent of intestinal metaplasia within a segment of CLE of a given length from Paull et al.[18] in 1976 (Fig. 12.5) to Chandrasoma et al.[19,20] in the early 2000s (Fig. 12.6).

More certain, however, is the distribution of intestinal metaplasia within a columnar-lined segment. At all times in history, intestinal metaplasia has occurred preferentially at the most proximal end of the segment of CLE (Fig. 12.6).

In addition, at all times in history, the prevalence of intestinal metaplasia has been directly related to the length of the segment of CLE. In Chandrasoma et al.[20] the prevalence of intestinal metaplasia in CLE lengths of <1, 1–2, 3–4, and 5 cm and more was 15%, 70%, 90%, and 100% (Fig. 12.6).

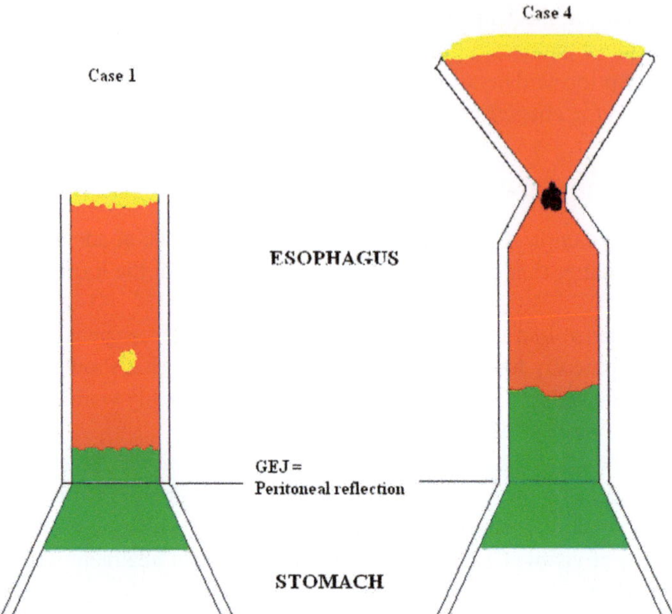

FIGURE 12.4 Two described cases (1 and 4) in Allison and Johnstone's paper of 1953 showing long segments (5 and 7.5 cm) of columnar-lined esophagus above the peritoneal reflection [true gastroesophageal junction (GEJ)]. There is no description of intestinal metaplasia.

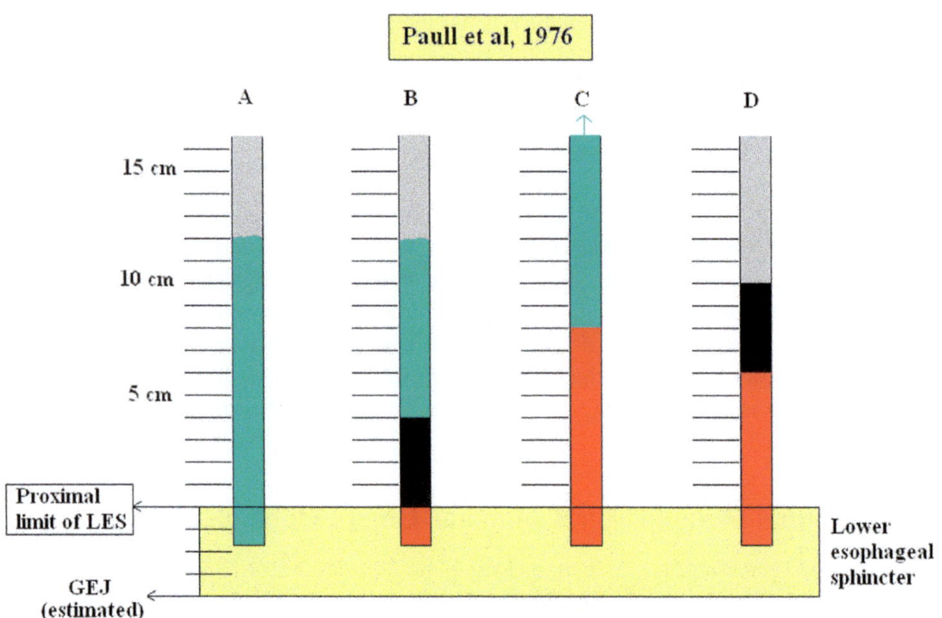

FIGURE 12.5 Four histologically mapped cases in Paull et al. in 1976. The prevalence of intestinal metaplasia (blue) appears to have increased compared with Fig. 12.4. There is still one patient with 10 cm visible columnar-lined esophagus above the upper border of the lower esophageal sphincter (LES) without intestinal metaplasia. *GEJ*, gastroesophageal junction. Cardiac epithelium = black; oxyntocardiac epithelium = red.

The two findings indicate that the probability of cardiac epithelium in CLE developing intestinal metaplasia increases with increasing distance from the GEJ. This is remarkable. All gastric molecules that reflux into the esophagus have their highest number at the GEJ, progressively decreasing to zero at the height of the column of refluxate. As a result, pathologic events in reflux tend to be maximal at the GEJ. The only thing that is greater at the top of the column of refluxate than at the GEJ is pH (i.e., decreasing acidity).

The indisputable fact that intestinal metaplasia increases in prevalence with increasing distance from the GEJ must mean that it is caused by a higher than lower pH. If that is the case, a higher pH of gastric contents resulting from the use of acid-reducing agents will promote intestinal metaplasia in CLE. With increasing effectiveness in controlling gastric acid,

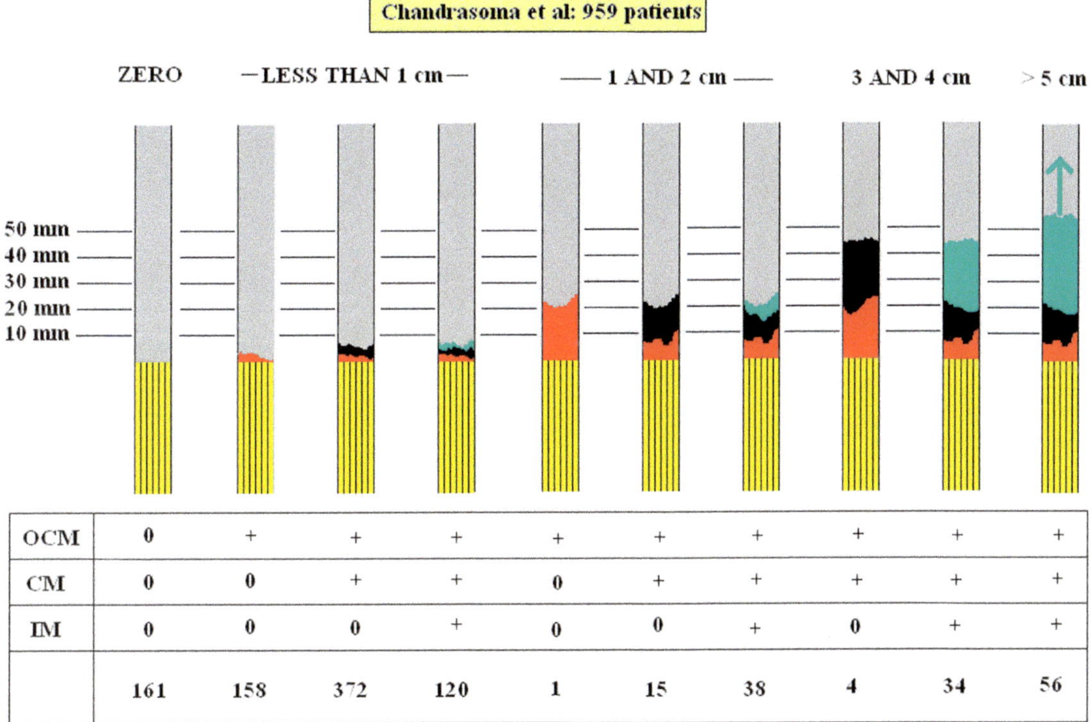

OCM	0	+	+	+	+	+	+	+	+	+
CM	0	0	+	+	0	+	+	+	+	+
IM	0	0	0	+	0	0	+	0	+	+
	161	158	372	120	1	15	38	4	34	56

FIGURE 12.6 Mapping study of 959 patients in 2003, showing 100% of patients with columnar-lined esophagus (CLE) >5 cm with intestinal metaplasia (blue). CLE is measured from 1 cm distal to the endoscopic gastroesophageal junction (GEJ) by the DeMeester biopsy protocol. *CM,* cardiac mucosa; *IM*, intestinal metaplasia; *OCM*, oxyntocardiac mucosa. Cardiac epithelium = black; oxyntocardiac epithelium = red.

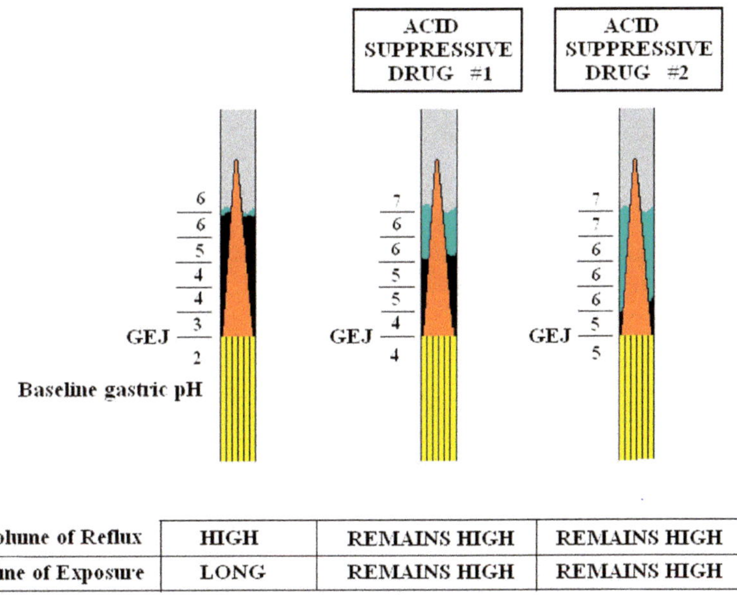

Volume of Reflux	HIGH	REMAINS HIGH	REMAINS HIGH
Time of Exposure	LONG	REMAINS HIGH	REMAINS HIGH

FIGURE 12.7 The pH gradient that is created in the esophagus during a reflux episode in three patients with different baseline gastric pH. The esophagus shows an increase in the pH throughout with two acid-reducing agents that increase baseline gastric pH from 2 to 4 and 5. The volume and height of reflux is the same in the three persons. *GEJ*, gastroesophageal junction. For illustration, IM is shown to occur at a pH of 6 in the esophagus. This is theoretical.

the esophageal pH during reflux has increased steadily over the years (Fig. 12.7). It is therefore logical that it will be associated with an increase in the prevalence and extent of intestinal metaplasia in CLE.

In a later section of this chapter, the effect of alkalinization of gastric contents on bile acid derivatives is discussed. There is evidence that the bile acids produced in a higher pH milieu promote the activation of CDX2 in esophageal cell lines. CDX2 is highly overexpressed in IM and is the probable molecular event that induces intestinal metaplasia in cardiac epithelium (see Chapter 14).

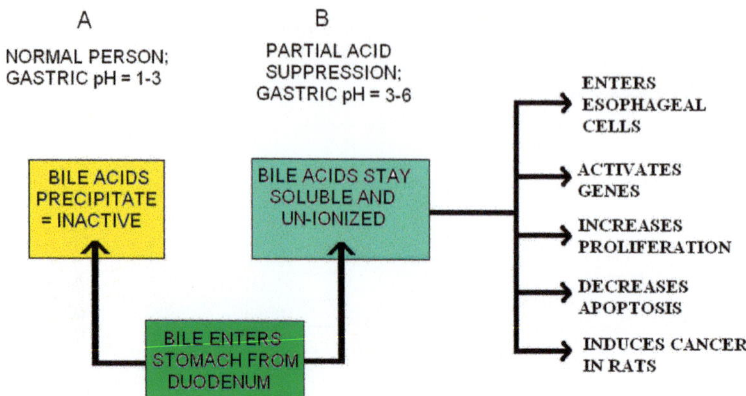

FIGURE 12.8 Effect of increase in pH of gastric juice by proton pump inhibitors on bile acids refluxed into the stomach from the duodenum. In the higher pH milieu, bile acids are converted into un-ionized derivatives that have been shown to enter esophageal cells and lead to molecular abnormalities.

When analyzing all the probable reasons for an increase in the esophageal adenocarcinoma over the past decade, the most likely reason is that the creation of a progressively increased pH milieu in the esophagus during reflux episodes has resulted in promoting the development of intestinal metaplasia in CLE. This, by itself, can explain the increasing incidence of esophageal adenocarcinoma.

1.5 Do Acid-Reducing Drugs Have a Direct Carcinogenic Effect?

The answer to this question is a definite "no." There is no evidence that any drug used to reduce gastric acid is a direct carcinogen. The beginning of the increase in the incidence of adenocarcinoma predated the introduction of PPIs by two decades. Given the lag time required for carcinogenesis, it is impossible to implicate either H_2 receptor antagonists or PPIs as direct carcinogenic agents.

It is also highly unlikely that the pH increase in the esophagus resulting from use of acid-reducing drugs has a direct carcinogenic effect on the epithelium.

1.6 Do Acid-Reducing Agents Have an Indirect Carcinogenic Effect?

The exact agent or agents in gastric contents that are involved in inducing the multiple mutations in the target epithelium in the esophagus (intestinalized cardiac epithelium) to progress to adenocarcinoma are not known. There is suspicion that the likely agent resides in bile that commonly refluxes into the stomach from the duodenum.

There is some evidence that an alkaline milieu created in the stomach by acid-suppressive agents may act on bile components (bile acids, bile salts) to produce agents that have an increased carcinogenic potential (Fig. 12.8). I will discuss this further in Section 9 of this chapter.

2. SHOULD ADENOCARCINOMA BE INCLUDED IN THE DEFINITION OF GASTROESOPHAGEAL REFLUX DISEASE?

The present focus of definition and management of GERD is the relatively easy task of improving quality of life directed only at patients who develop symptoms they consider troublesome. The occurrence of adenocarcinoma is covered but cleverly hidden in the Montreal definition[21] under the generic reference to "complications." It may be time to consider changing the very definition of GERD to openly acknowledge the cancer risk if we are to mount a concerted effort to control the increasing incidence of adenocarcinoma and mortality rate.

The present Montreal definition is[21]: "GERD is a condition which develops when the reflux of stomach contents causes troublesome symptoms **and/or complications**." If the last phrase (highlighted) of the definition is changed to "**and/or rarely but increasingly causes adenocarcinoma**." The problem is bared for the world to see.

This change in the definition of GERD simply emphasizes rather than minimizing the universally accepted fact that GERD-induced adenocarcinoma is lethal and is a cause of significant and increasing mortality. Highlighting the fact increases the stress placed on the medical establishment to find a solution at a time when no solution is even in the horizon. It forces the ostrich to take its head out of the sand and face reality.

3. THE CHANGING DEFINITION OF ESOPHAGEAL ADENOCARCINOMA

The definition of esophageal adenocarcinoma should be simple: Esophageal adenocarcinoma is an adenocarcinoma that arises in the esophagus.

Esophageal adenocarcinoma has two theoretical sources:

1. Metaplastic columnar epithelium that replaces the normal squamous epithelium when it is exposed to gastric juice, irrespective of whether or not the patient has symptoms of GERD. This is responsible for virtually all esophageal adenocarcinomas.
2. Mucous glands that are normally present in the mucosa and submucosa of the esophagus are very rarely the source of adenocarcinoma. In all my >30 year experience as a pathologist, I have encountered only one convincing case. I had finished a lecture in Riyadh, Saudi Arabia, many years ago when a pathologist approached me. He wanted me to see an unusual case he had encountered. This was a high-grade adenocarcinoma in the midesophagus with transmural invasion that was predominantly subepithelial with a normal overlying squamous epithelium. At endoscopy, the esophagus all the way to its distal limit consisted of unremarkable squamous epithelium without visible CLE. There was a small amount of unintestinalized cardiac epithelium distal to the squamocolumnar junction (SCJ). This case is the only exception to my rule that all esophageal adenocarcinoma arise in metaplastic columnar epithelium.

The definition of esophageal adenocarcinoma is not simple as it can be; it is complicated by the failure to accurately define the esophagus. If one does not know where the esophagus ends, it is impossible to avoid error in classification. This is presently the case, but the situation is slowly improving.

There is no problem in defining a cancer as esophageal when it is limited to the tubular thoracic esophagus (Fig. 12.9). The problem arises in cancers that are in and distal to the region of the endoscopic GEJ (end of tubular esophagus and/or proximal limit or rugal folds) (Figs. 12.10 and 12.11).

The general method of assigning a tumor in the region of the endoscopic GEJ to esophageal or gastric origin is to define the epicenter of the tumor and see its relationship to the GEJ. If the epicenter is on the esophageal side, the tumor is esophageal; if it is on the gastric side, it is gastric. The accuracy of this method depends on whether the tumor growth is uniform in all directions. This is reasonable though not proven for tumors that straddle the GEJ. It is the best we can do. Although it will create error when the growth favors one direction, this is likely to be insignificant.

The present generally accepted definition of the esophagus is that it is a tube extending from the cricopharyngeal sphincter to the proximal limit of rugal folds, which is the endoscopic GEJ. The proximal limit of rugal folds is at the end of the tube unless there is a hiatal hernia in which case the proximal limit of rugal folds takes precedence over the end of the tube

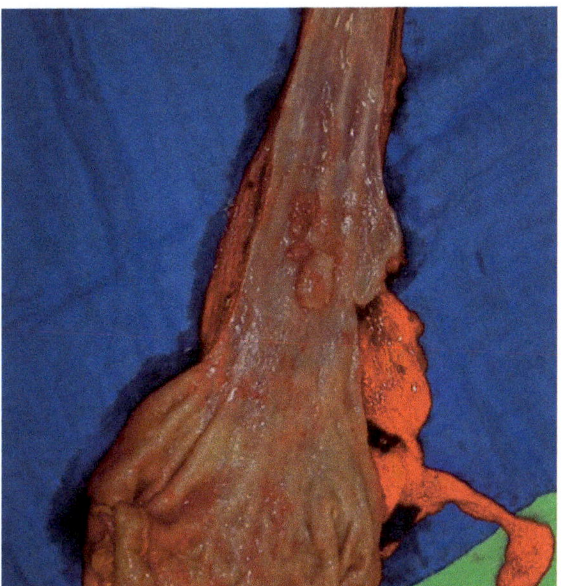

FIGURE 12.9 Adenocarcinoma of the tubular thoracic esophagus, seen as a polypoid mass above the proximal limit of rugal folds. *Photograph courtesy of Dr. Martin Riegler, Reflux Medical, Vienna, Austria.*

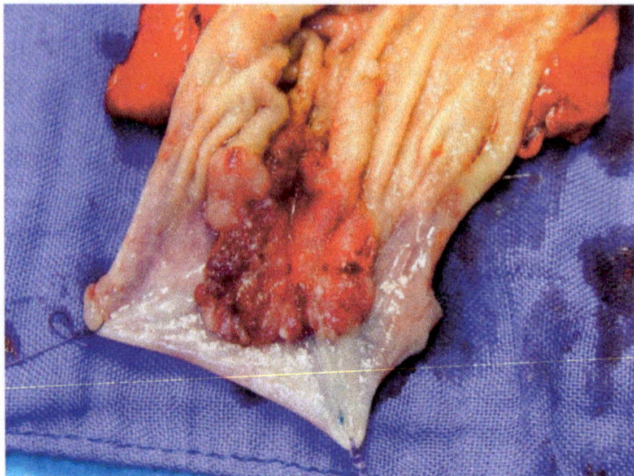

FIGURE 12.10 Adenocarcinoma of the region of the gastroesophageal junction. The tumor has its epicenter above the proximal limit of the rugal folds and appears to be arising from a short segment of visible columnar-lined esophagus seen as a flat salmon-pink columnar epithelium between the squamo-columnar junction and rugal folds. The tumor extends distally into the area distal to the rugal folds (into the dilated distal esophagus defined by histology as being above the proximal limit of gastric oxyntic epithelium. *Photograph courtesy of Dr. Martin Riegler, Reflux Medical, Vienna, Austria.*

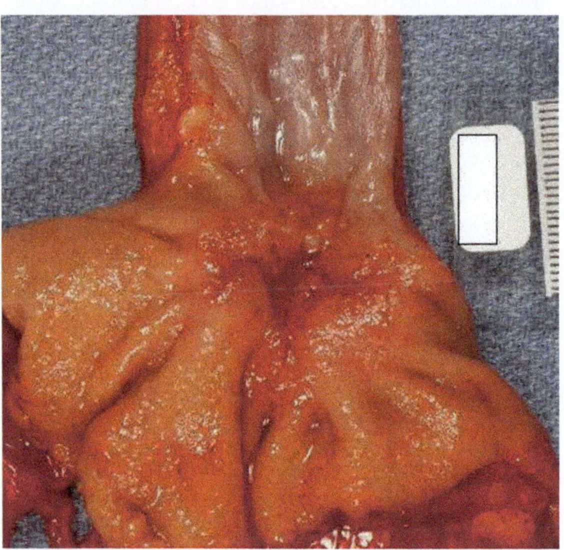

FIGURE 12.11 Adenocarcinoma of the dilated distal esophagus, seen as an ulcerated mass distal to the end of the tubular esophagus in a person without visible columnar-lined esophagus. Histology showed cardiac epithelium at the distal edge of the tumor, indicating that the tumor was entirely within the dilated distal esophagus.

to define the GEJ. There is no evidence base of any kind that these endoscopic definitions of the GEJ correctly represent the true GEJ.[21]

I have provided evidence in the preceding chapters that these endoscopic definitions of the GEJ are incorrect.[16] The damaged distal region of the esophagus in the patient with GERD (which is the dilated distal esophagus) is mistaken for proximal stomach by the use of these incorrect endoscopic definitions of the GEJ. The use of this definition of the GEJ must necessarily result in confusion and error regarding whether adenocarcinomas that occur in and around the endoscopic GEJ are esophageal or proximal gastric cancers. I will show that when the correct definition of the GEJ (histologically defined as the proximal limit of gastric oxyntic epithelium) is used, the confusion disappears in the vast majority of cases.

Histologic appearance of the tumor cannot differentiate between adenocarcinoma of the esophagus or stomach. There is considerable variation in the appearance in patterns of growth and cell type in esophageal and gastric adenocarcinomas with an overlap between the two. Intestinal-type adenocarcinomas of varying differentiation are most common (Fig. 12.12). However, papillary, signet ring cell, mucinous, and many other types occur in both organs. Similarly, special staining and the use of immunohistochemical and molecular markers cannot accurately differentiate esophageal from gastric adenocarcinoma.

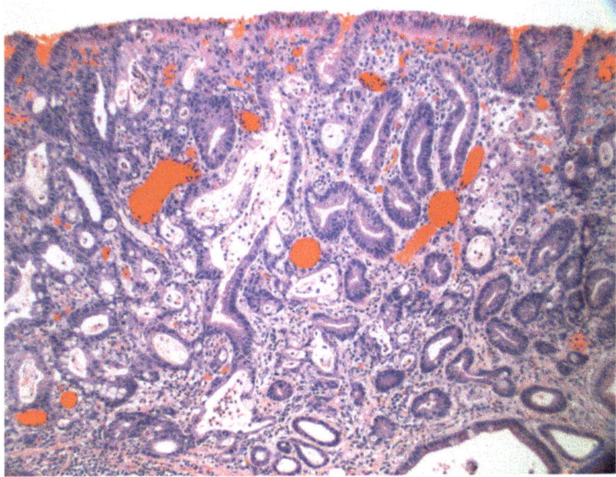

FIGURE 12.12 Adenocarcinoma of the esophagus, intestinal type, well differentiated.

I will first describe the presently used methods used for defining adenocarcinomas of the GEJ region. Unfortunately, the application of these methods is not uniform throughout the world, resulting in difficulty in comparing studies. The Siewert classification is widely used internationally.[22] In the United States, the guidelines of the American Joint Commission on Cancer (AJCC) are followed.[23] The AJCC guidelines regarding classification of esophageal and proximal gastric adenocarcinoma changed suddenly between the sixth and seventh editions in 2010. Finally, I will propose a new method of classification based on the new histologic definition of the GEJ that I have proposed.

3.1 The Siewert Classification of Adenocarcinoma of the Gastroesophageal Junctional Region

This is a widely used classification of adenocarcinoma in the GEJ region. It is based on a consensus conference held in 1998 under the auspices of the International Gastric Cancer Association and the International Society for Diseases of the Esophagus. This was published in the following summary form as a leading article in the *British Journal of Surgery*.[22]

The authors introduce the need for a classification: "In the Western world there has been an alarming rise in the incidence and prevalence of adenocarcinoma at the oesophagogastric junction. However, much discrepancy exists in the current literature about the classification and optimal management of this tumour. While some consider and treat all tumours arising in the vicinity of the oesophagogastric junction as oesophageal cancer, others regard them as gastric cancer, or even as an entity separate from both oesophageal and gastric cancer. This confusion is due largely to the borderline location between the oesophagus and stomach, the ambiguous use of the term 'cardia carcinoma', and an inability of the current Union Internacional Contra la Cancrum system to classify and stage these lesions. Consequently the data reported in the literature regarding the epidemiology, diagnosis and management are not comparable; it is hardly surprising that the outcome and benefit of surgical resection for adenocarcinoma of the oesophagogastric junction are matters of some controversy."

At a recent consensus conference.... all participating experts agreed that a clear definition and classification of tumours arising in the proximity of the oesophagogastric junction is essential to clarify these inconsistencies. We have defined and described adenocarcinomas of the oesophagogastric junction as tumours that have their centre within 5 cm proximal and distal of the anatomical cardia and have differentiated the following three distinct tumor entities within this area.

Type I tumour: Adenocarcinoma of the distal oesophagus which usually arises from an area with specialized intestinal metaplasia of the oesophagus (i.e. Barrett's oesophagus) and which may infiltrate the oesophagogastric junction from above.

Type II tumour: True carcinoma of the cardia arising from the cardiac epithelium or short segments with intestinal metaplasia at the oesophagogastric junction; this entity is also often referred to as 'junctional carcinoma'.

Type III tumour: Subcardial gastric carcinoma which infiltrates the oeosphagogastric junction and distal oesophagus from below.

The assignment of a lesion into one of these types is purely morphological and based on the anatomical location of the tumour centre or, in patients with an advanced tumour, the location of the tumour mass. Classification ***can be performed easily*** *based on a combination of contrast radiography, endoscopy with orthograde and retroflexed view of the oesophagogastric junction, computed tomography and the intraoperative appearance.*

Although all adenocarcinomas arising in the vicinity of the oesophagogastric junction share a number of common epidemiological and morphological features, a series of observations in the recent literature provides justification for such a differentiation.

These observations that justify separation of these adenocarcinomas into three groups are as follows:

1. A preponderance of male sex is significantly more marked in patients with type I tumors than type II and III tumors.
2. Patients with type I tumor are more likely to have a hiatal hernia and a long history of GERD compared with patients with type II and III tumors.
3. Specialized intestinal metaplasia in the distal esophagus (Barrett esophagus) is found in 81% of patients with type I tumors, 11% in type II, and 2% in type III tumors. Intestinal metaplasia at or below the gastric cardia is present in 75% of type I and less frequently in type II (32%) and type III (9%) tumors.
4. Undifferentiated tumors account for 51% type I, 56% type II, and 71% type III tumors. Tumors with a "nonintestinal" growth pattern account for 24%, 34%, and 45% of types I, II, and III tumors.
5. Primary direction of lymphatic spread is both proximal and distal in type I, more distal than proximal in type II and distal in type III.

The authors concluded that these differences suggest a possible heterogeneity in the pathogenesis and biological behavior of adenocarcinomas in the junctional region and called for further study. "All experts at the consensus conference agreed that the classification outlined above should form the basis for defining and reporting treatment of adenocarcinoma arising in the vicinity of the oesophagogastric junction."

The Siewert classification is notable for its lack of defined criteria to define the GEJ with which to compare the tumor center. It says that classification is easily performed based on a combination of modalities (radiology, endoscopy, computed tomography, and intraoperative appearance) without defining specific criteria for any of the modalities. The assumption is that it is an easy matter to define the GEJ and the problem of classification exists because, as stated in the introduction, "of the borderline location (of the tumor) between the oesophagus and stomach."

If, by some chance, the "easy" definition of the GEJ is incorrect, then the classification and terminology cease to have any meaning. The fact that widespread acceptance and application of this classification has not removed controversy around the issue regarding adenocarcinomas of this region in the nearly 18 years since its publication suggests that its impact in solving this problem has been minimal.

The most likely reason for this is that the definition of the GEJ is not "easy." In fact, we have shown that none of the criteria that were used permitted recognition of the dilated distal esophagus and the true GEJ.

In a more recent study from our unit, Leers et al.[24] examines whether the use of the Siewert classification to guide the selection of the surgical approach to the three types is justified. They state: "Type I tumors, which are considered esophageal in origin, are typically treated by means of an esophagectomy with abdominal and lower mediastinal node dissection. Type III tumors, which are considered to represent true gastric cancers, are typically treated with gastrectomy and resection of the regional abdominal lymph nodes. The management of type II (or GEJ) tumors remains less clear, with the debate centered on whether these are best treated like proximal gastric cancer or distal esophageal cancer. In addition, the location of potentially involved lymph nodes and the extent of lymphadenectomy necessary to adequately treat GEJ tumors remain disputed. Therefore the purpose of this study was to compare the prevalence and distribution of lymph node metastases, the prevalence and type of recurrence, and survival in patients with adenocarcinoma of the GEJ versus adenocarcinoma of the distal esophagus with the goal of determining whether an effort to distinguish between these tumors is warranted in clinical practice."

The study retrospectively reviewed 613 patients who underwent resection for adenocarcinoma within 5 cm of the GEJ from 1987 to 2007. The GEJ was defined as the site where the proximal gastric folds met the tubular esophagus with the stomach decompressed. By examining the endoscopy report, tumors were classified into 301 type I distal esophageal tumors (epicenter of the tumor 2–5 cm above the top of the rugal folds), 208 type II GEJ tumors (epicenter of the tumor within 2 cm above or below the proximal gastric folds), and 8 type III proximal gastric tumors (epicenter 2–5 cm below the proximal gastric folds). The 8 proximal gastric tumors and 96 tumors that were too large for accurate placement into types I or II were excluded.

There were no differences in age, sex, or body mass index (BMI) between type I and II tumors. Patients with type I tumors were significantly more likely to have reflux symptoms and Barrett esophagus and be detected within a surveillance program than type II. Intramucosal carcinoma was more prevalent in type I and transmural tumors in type II tumors, a probable effect of the greater likelihood of surveillance in patients with type I tumors.

There was no difference in the prevalence of lymph node involvement based on the depth of tumor invasion in type I and II tumors. There was no difference in the prevalence of systemic disease.

The distribution of involved lymph nodes was similar in type I and II tumors. More than 40% of node-positive patients in both groups had involved nodes in the mediastinum. In 9% of type I and 8% of type II tumors, a positive mediastinal node was the only site of lymph node involvement (all resected abdominal nodes were negative).

The median follow-up and overall and disease-specific survival were similar in both groups. The 5- and 10-year survival was 45% and 25% for patients with distal esophageal tumors compared with 38% and 31% for GEJ tumors. Recurrent cancer was found in 41% of type I and 39% of type II tumors. The type of recurrence (i.e., locoregional, distant nodes, and systemic) was also similar with systemic recurrence being the majority in both groups.

In the discussion, the authors make the point of the importance of mediastinal lymph node dissection in patients with GEJ tumors because of the high prevalence of mediastinal node involvement and the fact that when one node was positive it was mediastinal rather than abdominal.

The authors conclude: "There are inherent inaccuracies in trying to determine the precise location of the GEJ and the relationship of the epicenter of a cancer to the GEJ. Given these difficulties and the lack of a significant difference in the biologic behavior between adenocarcinoma of the distal esophagus and GEJ, we suggest that efforts to determine the precise origin of the tumor are not necessary and that an esophagectomy, preferably an en bloc resection, is appropriate surgical therapy for adenocarcinoma in either location."

This conclusion has the merit of recognizing the impossible and adjusting the basis of management on things that can be proven. Rather than attempt to define the GEJ endoscopically, which is impossible, they suggest that management should be based on the fact that these two tumors are biologically similar in their pattern of nodal metastasis, recurrence rates, and survival after a similar surgical procedure.

The main difference between type I and type II tumors is that the former has a higher association with symptomatic GERD and Barrett esophagus than the latter. This difference can be reasonably explained by the fact that type I tumors occur in conventional long-segment Barrett esophagus, which is associated with severe transsphincteric reflux with frequent hiatal hernia, as was found. In contrast, it is likely that a significant number of type II tumors were associated with less transsphincteric reflux with a lower incidence of symptomatic GERD and hiatal hernia, and either a shorter-segment Barrett esophagus or intestinal metaplasia in the dilated distal esophagus.

If one assumes that type II tumors that were within 2 cm distal to the GEJ and type III (subcardiac) tumors were largely adenocarcinomas in the dilated distal esophagus, many would have arisen with intestinal metaplasia limited to the dilated distal esophagus. The low incidence of symptomatic GERD, conventional Barrett esophagus, and hiatal hernia is typical of GERD limited to the dilated distal esophagus (See Chapter 11).

Robertson et al.[25] showed that 6.5% of asymptomatic patients without a visible CLE had intestinal metaplasia in the dilated distal esophagus. This is the probable source of adenocarcinoma of the dilated distal esophagus.

Unless proven otherwise, the likelihood is that type I tumors involve the tubal esophagus, type II tumors involve the junction between the tubal esophagus and the dilated distal esophagus, and type III tumors are limited to the dilated distal esophagus.

The possibility now arises that true proximal gastric adenocarcinomas are extremely uncommon and most so classified in the past are actually esophageal. This is the only explanation for the epidemiology of tumors previously called proximal gastric cancer being similar to esophageal adenocarcinoma and different than distal gastric adenocarcinoma.

3.2 American Joint Commission of Cancer Staging Manual: Esophagus and Esophagogastric Junction, Seventh Edition (2010)

In the AJCC staging recommendations,[23] adenocarcinoma of the esophagus is divided into cervical, upper thoracic, middle thoracic, and lower thoracic esophagus/esophagogastric junction.

The lower thoracic esophagus is bordered superiorly by the inferior pulmonary veins and inferiorly by the stomach. Because it is the end of the esophagus, it includes the esophagogastric junction…. It normally passes through the diaphragm to reach the stomach, but there is a variable intra-abdominal portion, and because of hiatal hernia, the portion may be absent.

The arbitrary 10-cm segment encompassing the distal 5 cm of the esophagus and proximal 5 cm of the stomach, with the EGJ in the middle, is an area of contention. Cancers arising in this segment have been variably staged as esophageal or gastric tumors, depending on orientation of the treating physician. ***In this edition, cancers whose epicenter is in the lower thoracic esophagus, EGJ or within the proximal 5 cm of the stomach (cardia) that extend into the EGJ or esophagus (Siewert III) are stage grouped similar to adenocarcinoma of the esophagus.*** *Although Siewert and colleagues subtype EGJ cancers (types I, II and III), not only do their data support a single stage-grouping scheme across this area, but also they demonstrate that prognosis depends on cancer classification (T,N,M,G) and not Siewert type. All other cancers with an epicenter in the stomach greater than 5 cm distal to the EGJ, or those within 5 cm of the EGJ but not extending into the EGJ or esophagus are stage-grouped using the gastric (non-EGJ) cancer staging system.*

The seventh edition is a radical change from the sixth edition where tumors 5 cm on either side of the GEJ were stage-grouped as gastric cancers. In the discussion the AJCC states: "Cancers arising in this segment have been variably staged as esophageal or gastric tumors, depending on orientation of the treating physician." This is an admission of a remarkable situation that borders on chaos. How is it reasonable for two treating physicians to look at the same tumor and, depending on their "orientation," one call it esophageal and the other gastric?

In essence, though, what the AJCC is doing is switching from the side that says these tumors are gastric in the sixth edition to the side that says they are esophageal in the seventh. They really do not give any good reason for this change.

The strongly positive effect of the seventh edition stage-grouping is that most of these tumors in the junctional region are, in fact, esophageal carcinomas involving the distal tubal esophagus and the dilated distal esophagus. By changing the designation from gastric in the sixth edition to esophageal in the seventh, the AJCC is moving closer to more accurate classification. These tumors are well known to have epidemiologic characteristics that are much more like esophageal than gastric carcinoma, so the change brings these tumors in line with epidemiology.

The impact of the change is likely to be a sudden sharp increase in the incidence of esophageal adenocarcinoma in the United States once the seventh edition classification goes into effect. Tumors previously classified as gastric carcinoma will suddenly be included in the statistics of esophageal cancers. This number is around 8000 per year, resulting in a significant shift. This shift will bring the incidence of esophageal carcinoma closer to the real number and prevent the masking of the problem of GERD-induced cancer by classifying adenocarcinoma in the dilated distal esophagus as a gastric cancer unrelated to GERD.

3.3 The Proposed New Definition of Adenocarcinoma of the Esophagus Based on the New Histologic Definition of the Gastroesophageal Junction

After our paper in 2007 on the proposal for a new definition of the GEJ as the proximal limit of gastric oxyntic epithelium based on histology,[16] we examined esophagectomy specimens of this region to determine the effect of this new proposed histologic definition of the GEJ on the classification of tumors in this region.[26]

This paper[26] was written at the time the classification of tumors that involved the GEJ was based on the AJCC sixth edition, where they were to be classified as gastric tumors.

Somewhat surprisingly, epidemiologic studies have shown that there is a correlation between adenocarcinoma of the gastric cardia and symptomatic gastro-esophageal reflux, although this correlation is weaker than that for adenocarcinoma of the distal esophagus.

Classifying tumors of this region into distal esophageal and gastric cardiac is based on endoscopic and radiologic criteria or pathologic examination of resected specimens. Endoscopically, the gastroesophageal junction is defined as the proximal limit of rugal folds. In 2000, the Association of Directors of Anatomic and Surgical Pathology (ADASP) made recommendations for standardizing the classification of these tumors in resected specimens. They advocated the use of the relationship of the epicenter of the tumor to the grossly defined gastro-esophageal junction. This group of experts defined the gastroesophageal junction as the horizontal line drawn across the end of the tubular esophagus. When classified by the present definitions of the gastro-esophageal junction, there is evidence that gastric cardiac adenocarcinomas commonly arise in intestinal metaplasia of the gastric cardia, and that they bear similarities to distal esophageal adenocarcinoma.

We have suggested that the gastro-esophageal junction is most accurately defined by histologic determination of the proximal limit of gastric oxyntic mucosa. The area proximal to this, which is esophagus, is lined by squamous epithelium in normal people and a variable extent of metaplastic esophageal columnar epithelium (oxynto-cardiac, cardiac, and intestinal) in patients with reflux damage. In our recent study of 10 esophagectomy specimens, we showed that there was a region between the end of the tubular esophagus (the present definition of the gastro-esophageal junction) and the proximal limit of gastric oxyntic mucosa (the true gastro-esophageal junction) that was identified as esophagus rather than stomach by the presence of submucosal esophageal glands. We designated this area as the "dilated end-stage esophagus." The use of the end of the tubular esophagus resulted in a discrepancy of 0.31 to 2.05 cm between what is presently called the gastro-esophageal junction and what we have shown is the true gastro-esophageal junction. Sarbia et al, in a similar study of esophagectomy specimens, showed that cardiac and oxynto-cardiac mucosa can extend to 2.8 cm beyond the end of the tubular esophagus. This error would have the potential of distal esophageal adenocarcinomas being incorrectly designated as gastric cardiac adenocarcinomas.

We undertook this study to compare how adenocarcinomas of this region would be classified on the basis of the presently accepted gross definition of the gastro-esophageal junction for use in resected specimens and the new histologic definition of the gastro-esophageal junction that we have proposed. We were unable to test these tumors against the presently accepted endoscopic definition of the gastro-esophageal junction; in many of these cases, the rugal folds are greatly distorted in this region by the presence of the tumor.

Seventy-four patients who underwent esophagectomy for adenocarcinoma of the distal esophagus and gastric cardia during the years 1997–2000 were selected for study. The relationship of the epicenter of the tumor to the grossly defined GEJ was used to classify these as distal esophageal and gastric cardiac tumors. The GEJ was defined as a horizontal line drawn between the end of the tubular esophagus and the saccular stomach.[27]

The specimens were sectioned extensively in a manner that permitted evaluation of the mucosal types present immediately proximal to the tumor, immediately distal to the tumor, and at the lateral edge of the epicenter of the tumor (Fig. 12.13). The last determination was not possible in 10 cases where the tumor involved the full circumference of the esophagus.

The epithelia were classified by criteria defined in Chapter 4 into squamous, cardiac, oxyntocardiac, intestinal, and gastric oxyntic types. The epithelial type at the proximal, distal, and central lateral edge of each tumor was recorded. When more than one epithelial type was present, priority was given as follows: intestinal > cardiac > oxyntocardiac > squamous and gastric oxyntic. This was based on our belief that the first three represents the gradation of these epithelial types in this region from most to least abnormal of reflux-induced columnar epithelia of the esophagus regarding cancer risk, and the last two are normal epithelia of the esophagus and stomach, respectively.

There were 58 (78%) men with a M:F ratio of 3.6:1. By the present criterion of the GEJ (a line drawn across the end of the tubular esophagus[27]), the epicenter of 38 (51%) tumors was in the distal esophagus and that of 36 (49%) was in the gastric cardia. Of the 38 distal esophageal cancers, 30 (79%) were men; of the 36 gastric cardiac cancers, 28 (75%) were men. The mean age was 64.1 years (median: 66, range: 31–86 years). The gross tumor size ranged from grossly invisible to >15 cm. Histologically, they were all pure adenocarcinomas and ranged from well differentiated to poorly differentiated and showed many different morphologic subtypes including tubular, mucinous, papillary, signet ring cell, solid, microcystic, and clear cell adenocarcinoma. The depth of invasion of the tumors was intramucosal in 14 (18.9%), submucosal in 8 (10.8%), intramural in 4 (5.4%), and transmural in 48 (64.9%) patients. *Helicobacter pylori* was present in gastric biopsies in 10 (14%) patients, 5 each with distal esophageal and gastric cardiac tumors. Lymph node involvement was present in 1/14 (7%) intramucosal, 2/8 (25%) submucosal, 3/4 (75%) intramural, and 43/48 (90%) transmural tumors.

Table 12.1 shows the epithelial type at the proximal edge, lateral epicenter, and distal edge of the tumors. The epithelium at the lateral edge of the epicenter of the tumor in 34 patients with noncircumferential tumors classified as being in the distal esophagus was squamous in 2 (5%), intestinal in 25 (66%), cardiac in 5 (13%), and oxyntocardiac in 2 (5%), with none having gastric oxyntic epithelium.

In the 30 patients with noncircumferential tumors in the gastric cardia, the lateral edge at the epicenter showed squamous in 3 (8%), intestinal in 9 (25%), cardiac in 16 (44%), oxyntocardiac in 2 (6%), and gastric oxyntic in 0.

None of the 64 patients with noncircumferential tumor had gastric oxyntic epithelium at the lateral edge of the epicenter of the tumor. If one uses the histologic definition of the GEJ (proximal limit of gastric epithelium), the lateral edge of the epicenter of these 64 tumors was all proximal to the true GEJ, i.e., in the distal esophagus. They could be divided into 34 tumors in the tubular esophagus and 30 tumors in thc dilated distal esophagus.

In the 10 patients with circumferential tumors, 4 were classified as distal esophageal and 6 as gastric cardiac by their relation to the end of the tubular esophagus. The epithelium at the epicenter of the tumor could not be determined. The epithelium at the distal edge of these tumors showed cardiac epithelium in three cases, oxyntocardiac epithelium in four

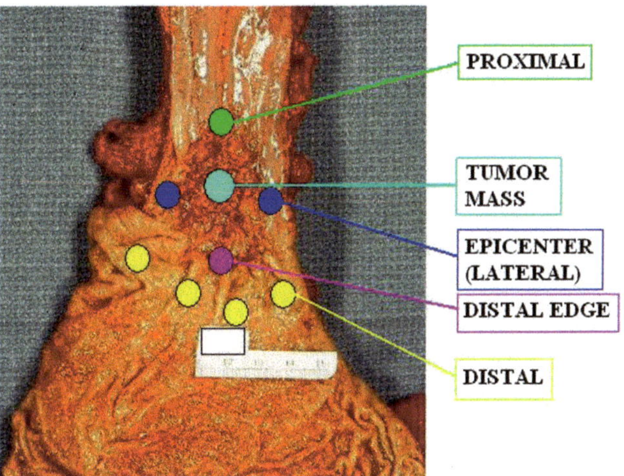

FIGURE 12.13 Adenocarcinoma in the region of the end of the tubular esophagus. This shows the method of taking sections from the proximal, lateral, and distal margins of the tumor that allows accurate histologic definition of whether the carcinoma is esophageal or gastric in origin.

TABLE 12.1 Epithelial Types Found at Proximal Edge, Distal Edge, and Lateral Edge at the Epicenter of Distal Esophageal and Gastric Cardiac Adenocarcinomas Classified by Its Relationship to the End of the Tubular Esophagus

Epithelial Type	Distal Esophageal Tumors (n = 38)	"Gastric Cardiac" Tumors (n = 38)
At the Proximal Edge of Tumor		
Squamous	16 (42%)	27 (75%)
Intestinal	20 (53%)	5 (14%)
Cardiac	2 (5%)	4 (11%)
Oxyntocardiac	0	0
Gastric oxyntic	0	0
At the Lateral Edge of Tumor Epicenter		
Squamous	2 (5%)	3 (8%)
Intestinal	25 (66%)	9 (25%)
Cardiac	5 (13%)	16 (44%)
Oxyntocardiac	2 (5%)	2 (6%)
Gastric oxyntic	0	0
Unknown (circumferential tumor)	4 (11%)	6 (17%)
At the Distal Edge of Tumor		
Squamous	0	0
Intestinal	12 (32%)	4 (11%)
Cardiac	13 (34%)	14 (39%)
Oxyntocardiac	7 (18%)	8 (22%)
Gastric oxyntic	8 (16%)	10 (28%)

cases, and gastric oxyntic epithelium in three cases. This indicates that the entire tumor was proximal to gastric oxyntic epithelium (i.e., the true GEJ) in seven patients, making them adenocarcinomas of the distal esophagus that had not extended distal to the GEJ.

In the three patients with circumferential tumors that had gastric oxyntic epithelium at the distal edge, it was impossible to determine whether these arose in the esophagus and extended into the stomach or whether they were primary gastric tumors that extended into the esophagus. These three tumors were large, measuring 4, 6.5, and 11 cm in greatest diameter with transmural invasion, positive lymph nodes (17 of 62, 8 of 26, and 9 of 36 nodes were involved), lymphovascular invasion, undermining of the squamous epithelium at the proximal edge, and showing satellite nodules in the deep mucosa away from the main tumor mass. They were advanced tumors where accurate definition of primary site is known to be difficult.

None of the 74 tumors had either oxyntocardiac or gastric oxyntic epithelium at the proximal edge of the tumor. 43 had squamous epithelium, 25 had intestinal metaplasia, and 6 had cardiac epithelium. The 25 patients with intestinal metaplasia in the proximal edge were 20 with tumors classified as distal esophageal and 5 as gastric cardiac. The finding of intestinal metaplasia at the proximal edge in these tumors suggests that adenocarcinoma arose in the distal region of intestinal metaplasia within the columnar-lined segment. 14 patients (12 distal esophageal and 2 gastric cardiac) had intestinal metaplasia at the distal end of the tumor, indicating that adenocarcinoma is sometimes surrounded by intestinal metaplasia within the columnar-lined segment.

Of the 74 tumors, only 16 (6 classified as distal esophageal and 10 gastric cardiac) had gastric oxyntic epithelium at the distal edge of the tumor. The other 58 tumors were therefore entirely in the esophagus and without any doubt primary in the esophagus. The 16 patients with gastric oxyntic epithelium at the distal edge had tumors that straddled the true GEJ. These were large transmural tumors (mean size 6 cm, median size 5.5 cm, range 2–11 cm). Thirteen of these patients had nongastric oxyntic epithelia at the lateral edge of the epicenter and were therefore very likely to be esophageal primaries extending across the GEJ into the stomach. Three were circumferential where the epithelium at the lateral edge could not be evaluated. Only these three tumors were questionable as to origin in the esophagus or stomach.

This study shows that when these 74 tumors of the GEJ region are classified by present criteria using the end of the tubular esophagus as the GEJ, 38 were distal esophageal, and 36 were gastric cardiac. We reclassified these 74 tumors using the relationship of the epicenter of the tumor to the proximal limit of gastric oxyntic epithelium defined histologically. This was validated in a previous study as the true GEJ.[16] When this was done, all 64 noncircumferential tumors had intestinal, cardiac, or oxyntocardiac epithelium at the epicenter and were reclassified as esophageal adenocarcinoma. Seven of the ten circumferential tumors had cardiac and oxyntocardiac epithelium at the distal end and were therefore proximal to the GEJ and entirely esophageal. This meant that reclassification by histology resulted in 71 tumors being esophageal. The other three (circumferential tumors with gastric oxyntic epithelium at the distal edge) could not be defined because they straddled the GEJ.

In addition, 58 of the tumors had intestinal, cardiac, or oxyntocardiac epithelium at their distal edge, indicating that the entire tumor was proximal to the GEJ, i.e., esophageal without crossing the GEJ. While the relationship of the tumor epicenter to the GEJ is a reasonable criterion for deciding the origin of the tumor, this assumes equal growth in all directions. The most specific criterion is when the distal edge of the tumor is proximal to the GEJ as in these 58 cases.

Of the 16 tumors, 13 that had gastric oxyntic epithelium at the distal edge were esophageal (epicenter with intestinal, cardiac, or oxyntocardiac epithelium) crossing the GEJ in their distal region.

When the new validated histologic criterion of the GEJ is used, all tumors that are presently classified as distal esophageal are in the distal part of the tubular esophagus. Most tumors that are presently classified as gastric cardiac have their origin in the dilated distal esophagus between the end of the tubular esophagus and the proximal limit of gastric oxyntic epithelium defined by histology.

3.4 Method of Classifying Adenocarcinoma as Esophageal or Gastric

The accuracy of the histologic definition of the GEJ lends precision to the classification of tumors in this region. There is no need for observer interpretation. Defining the origin of a tumor only requires a method to accurately locate the proximal limit of gastric oxyntic epithelium. This can only be done by histology for the following reasons:

1. CLE is identical to gastric oxyntic epithelium by all modalities other than histology. This includes endoscopy and radiology.
2. The distal abdominal esophagus, when it loses the protection of the resting sphincter pressure, dilates and becomes part of the contour of the stomach. The dilated distal esophagus is interposed between the end of the tubular esophagus and the true GEJ to the extent that the sphincter is shortened. This leads to the illusion at endoscopy, radiology, surgery, and gross pathologic examination of resected specimens that the dilated distal esophagus is part of the stomach.

Radiologists are fooled into thinking that the esophagus ends where the tube ends. Physiologists studying manometric tracings are fooled into thinking that the esophagus ends at the distal limit of the sphincter; they do not take into account the part of the sphincter that has shortened, which has an identical manometric footprint as the stomach. Endoscopists are fooled into thinking that the stomach begins at the point of the proximal limit of rugal folds. Rugal folds are not limited to the stomach. When the distal esophagus dilates, it becomes part of the reservoir and, like any reservoir, develops mucosal folds.

The proximal limit of rugal folds is the junction between the tubular esophagus and the dilated distal esophagus, proximal to the true GEJ by zero to a maximum reported length of 2.8 cm, and a theoretical maximum of 3.5 cm, depending on how much of the LES has shortened.

The new definition of the GEJ by histology converts a 10 cm gastroesophageal zone to an imaginary line, causing the "zone" to disappear. The definition of esophageal and proximal gastric adenocarcinoma increases in precision by an amazing 10 cm. This is surely an advance in understanding.

Correlation between the new classification we have proposed and Siewert's classification[22] now makes some sense. Type III tumors in the Siewert classification—subcardiac tumors—are adenocarcinomas of the dilated distal esophagus. 79% of these had no hiatal hernia, 71% had no history of GERD, and 98% did not have Barrett esophagus (i.e., visible CLE in the thoracic esophagus with intestinal metaplasia). These tumors arise in asymptomatic patients with CLE limited to the dilated distal esophagus. Their source is the 6.5% of asymptomatic patients in the study by Robertson et al.[25] who had intestinal metaplasia distal to a normal endoscopic GEJ.

The AJCC seventh edition[23] guidelines for staging cancers of this region reach an endpoint that is similar to the new classification that we have proposed. By edict, the AJCC defines tumors that are 5 cm on either side of the GEJ (end of the tube and/or proximal limit of rugal folds) as esophageal. They are actually correct in doing this, but they should recognize that the reason why they are correct is that they are adjusting to the fact that the true GEJ is distal to the endoscopic GEJ.

The anomaly between classification of tumors in this region and epidemiology now disappears. There is no longer a need to explain why "adenocarcinoma of the gastric cardia or proximal stomach" is associated with symptomatic GERD and has an epidemiology that is similar to adenocarcinoma of the esophagus rather than the distal stomach. The reason for this is simple: they are esophageal cancers, not gastric.

The only reliably identifiable landmarks at endoscopy and on gross examination of a resected specimen that includes the junctional region are (1) the SCJ; (2) the end of the tubular esophagus; and (3) the proximal limit of rugal folds. In patients without a CLE these three points are at the same level. In patients with a visible CLE in the tubular esophagus, the SCJ is proximal to the end of the tube and the proximal limit of rugal folds. When a tumor is present in this area, this will frequently have caused varying distortion of the mucosal landmarks.

3.4.1 Biopsy Protocol to Define Origin of Adenocarcinoma at Endoscopy

At endoscopy, the relationship of the tumor to the endoscopic GEJ is noted. In addition to sampling the tumor for diagnosis, biopsies are taken at the interface of the tumor edge with the mucosa at the proximal and distal ends and at the epicenter, i.e., the lateral edges of the middle of the tumor. The origin can be defined by the epithelial type present in each of these biopsies (Table 12.2). The system can also define the origin of tumors that have crossed the GEJ and involves both the esophagus and

TABLE 12.2 Origin and Location of Tumors in the Junctional Region as Defined by the Relationship of the Epicenter of the Tumor at the Lateral Edge to the True Gastroesophageal Junction (GEJ) (Proximal Limit of Gastric Oxyntic Epithelium). The Epithelial Type at the Proximal and Distal Edges of the Tumor Define Whether the Tumor Is Entirely in the Esophagus, Entirely in the Stomach, or Straddling the True GEJ

Origin of Tumor	Endoscopic Appearance	Proximal Edge	Lateral Edge/Epicenter	Distal Edge
Thoracic esophagus	Visible CLE; tumor above endo-GEJ	Squamous/CLE	Squamous/CLE above endo-GEJ	CLE above endo-GEJ
Thoracic esophagus→DDE	Tumor epicenter above endo-GEJ; extending <5 cm below endo-GEJ	Squamous/CLE	Squamous/CLE above endo-GEJ	CLE below endo-GEJ
Thoracic esophagus→DDE→stomach	Tumor epicenter above endo-GEJ; extending >5 cm distal to endo-GEJ	Squamous/CLE	CLE above endo-GEJ	GOE
DDE in patient with BE	Visible CLE; tumor <5 cm below endo-GEJ	CLE below endo-GEJ	CLE below endo-GEJ	CLE below endo-GEJ
DDE in patient with BE→thoracic esophagus	Visible CLE; tumor epicenter below endo-GEJ; extending above endo-GEJ	Squamous/ CLE above endo-GEJ	CLE below endo-GEJ	CLE below endo-GEJ
DDE in patient without BE	Tumor <5 cm below endo-GEJ	CLE below endo-GEJ	CLE below endo-GEJ	CLE below endo-GEJ
DDE in patient without BE→thoracic esophagus	No visible CLE; tumor epicenter below endo-GEJ; extending above endo-GEJ	Squamous	CLE below endo-GEJ	CLE below endo-GEJ
DDE in patient without BE→stomach	Tumor <5 cm below endo-GEJ	CLE below endo-GEJ	CLE below endo-GEJ	GOE
Gastric tumor	Tumor >5 cm below endo-GEJ	GOE	GOE	GOE
Gastric tumor→DDE	Tumor >5 cm below endo-GEJ	CLE below endo-GEJ	GOE	GOE

BE, Barrett esophagus; *CLE*, columnar-lined esophagus (above endo-GEJ in thoracic esophagus and distal to endo-GEJ in DDE); *DDE*, dilated distal esophagus; *Endo-GEJ*, endoscopic gastroesophageal junction; *GOE*, gastric oxyntic epithelium. *Arrow* indicates extension of tumor from site of origin (thoracic esophagus/DDE/stomach) up or down to adjacent compartment.

stomach. In circumferential tumors, there is no lateral edge and the epicenter cannot be defined. In these cases, which are usually advanced tumors, the only possible biopsies are at the proximal and distal edges. The origin remains undefinable.

This biopsy protocol is largely academic and only recommended as a method that is available to the endoscopist if there is any question about whether a tumor is esophageal or gastric in origin. Biopsy will be unnecessary if it is accepted that almost all these tumors are esophageal. As shown in Leers et al.[24] the clinical features and surgical management of these tumors are similar.

3.4.2 Gross Dissection of an Esophagogastrectomy Specimen

In the cases of gross examination of a resected specimen with an adenocarcinoma, sectioning to define the origin and extent of a carcinoma should be mandatory. The localization of the tumor to the esophagus and stomach is an essential part of the pathology report and is valuable in epidemiologic studies.

In noncircumferential tumors, sections are taken perpendicular to the edge of the tumor at the proximal and distal edges and at both lateral edges of the tumor in its middle. With circumferential tumors where there is no lateral edge to define the origin accurately, sections are taken from the proximal and distal tumor interfaces with surrounding mucosa.

3.4.3 Interpretation of Histologic Findings

Depending on the type of epithelium present at these tumor interfaces with surrounding epithelium, the following conclusions can be drawn (Table 12.2): (1) The origin of the tumor is defined by the epithelial type at the epicenter of the tumor (section taken at the lateral edge of the middle of the tumor). If this is either squamous or CLE (i.e., cardiac epithelium with and without goblet and/or parietal cells), the tumor is esophageal. This can be subdivided into tumors arising in the thoracic esophagus or dilated distal esophagus based on whether the epicenter is CLE above or below the endoscopic GEJ. (2) Extension of the tumor from its site of origin into adjacent compartment is defined by the epithelial types at the proximal and distal edges of the tumor.

An adenocarcinoma is esophageal without question if the entire tumor is proximal to true GEJ; that is, the epithelium at distal interface of the tumor is cardiac (with and without parietal and/or goblet cells). If the epicenter of the tumor is cardiac epithelium (with and without parietal and/or goblet cells) and the distal edge of the tumor has gastric oxyntic epithelium, the tumor is esophageal with extension into the stomach.

An adenocarcinoma is gastric without question if the proximal edge of the tumor consists of gastric oxyntic epithelium. If the proximal edge of the tumor is cardiac epithelium (with and without parietal and/or goblet cells), the tumor is assumed to be gastric with extension into the esophagus across the GEJ if the lateral edge of the epicenter of the tumor consists of gastric oxyntic epithelium.

4. EPIDEMIOLOGY OF ESOPHAGEAL ADENOCARCINOMA

As early as 1992, Rodger Haggitt referred to the increasing incidence of esophageal adenocarcinoma as an epidemic.[28] From 1975 to 2005, esophageal adenocarcinoma has been by far the most rapidly increasing cancer type in the United States and Western Europe[29] (Fig. 12.14).

In 1998, Devesa et al.[30] assessing data from the Surveillance, Epidemiology, and End-Results (SEER) program of the National Cancer Institute reported that the annual incidence among white males rose more than 350% from the 1974–76 period to the 1992–94 period. Devesa et al.[30] also reported that there was a parallel increase in the incidence of adenocarcinoma of the proximal stomach during that same time span.

A study by Pohl and Welch in 2005[29] showed that the incidence continued to increase (Fig. 12.15). They reported that the incidence rose more than sixfold from 1975 to 2001. Esophageal adenocarcinoma, which had been one-sixth that of esophageal squamous carcinoma in 1975, overtook the incidence of squamous carcinoma in the mid-1990s (Fig. 12.15). Esophageal adenocarcinoma is a lethal cancer type; the mortality from esophageal adenocarcinoma has closely paralleled the increasing incidence (Fig. 12.16).

Pohl and Welch[29] also reported that adenocarcinoma of the proximal stomach had an epidemiology that was similar to esophageal adenocarcinoma (Fig. 12.17) and very different to that of distal gastric cancer, which had declined in incidence during this same period. In 1999, Lagergren et al.[2] showed that both esophageal and proximal gastric adenocarcinoma were associated with GERD.

In 2010, Pohl et al.[1] updated the epidemiologic data to 2006. In this update, they limited the study to adenocarcinoma of the esophagus, excluding proximal gastric cancer. They used data from the SEER database of the National Cancer Institute. SEER 9 has collected information on all newly diagnosed malignancies since 1973 among residents of the same SEER geographic areas, representing ~10% of the population of the United States.

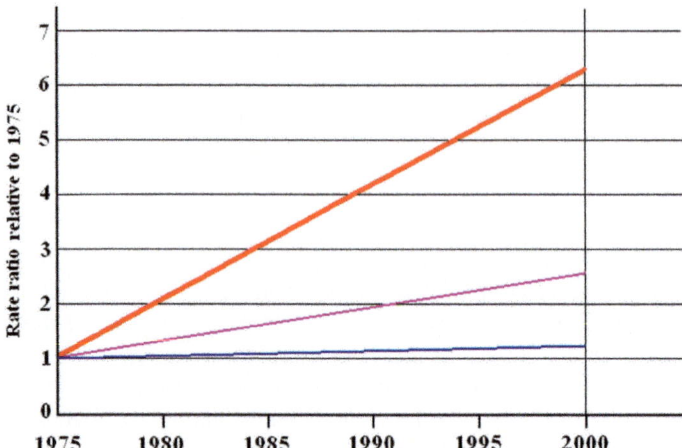

FIGURE 12.14 The change in incidence of adenocarcinoma of the esophagus (*red line*), adenocarcinoma of the breast (*pink line*), and colorectal adenocarcinoma (*black line*) from 1975 to 2000. The graph is modified from Pohl et al. drawing a straight line between the two endpoints of time in the study, removing annual fluctuations.

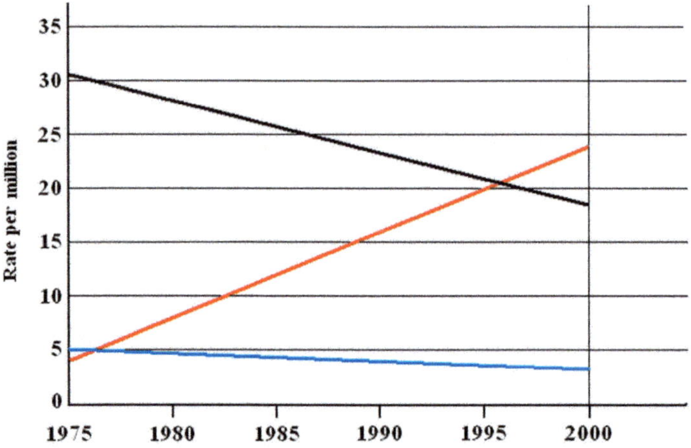

FIGURE 12.15 Change in the incidence of squamous carcinoma (*black line*) and adenocarcinoma (*red line*) of the esophagus, showing the marked differences in the two cancer types between 1973 and 2000. The *blue line* denotes cancers that were unclassified. Adenocarcinoma, which overtook squamous carcinoma incidence in 1995, continues to increase in relation to squamous carcinoma.

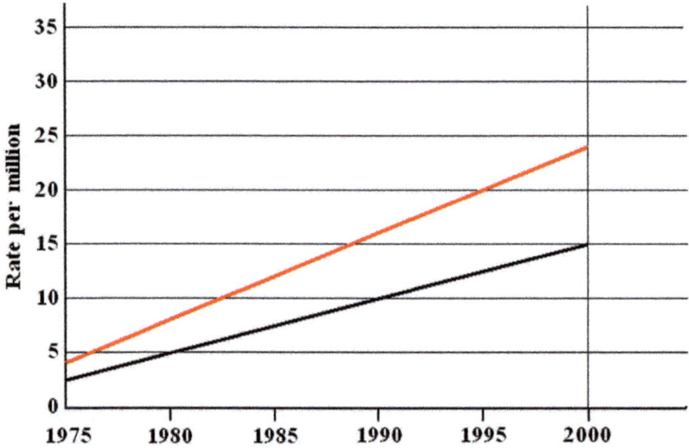

FIGURE 12.16 The mortality rate (*black line*) of esophageal adenocarcinoma mirrors the incidence rate (*red line*), a feature of cancers with a high mortality rate.

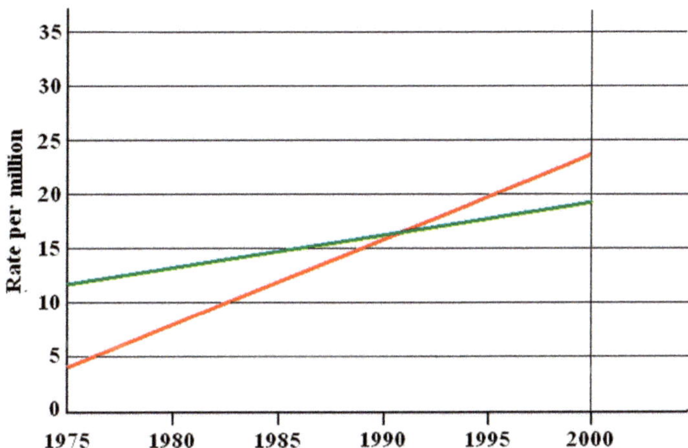

FIGURE 12.17 The change in the incidence of gastric cardiac (proximal gastric) adenocarcinoma (*green line*) shows a greater similarity to esophageal adenocarcinoma (*red line*) than distal gastric adenocarcinoma (not shown). Distal gastric cancer decreased in incidence during this same period.

They first examined the change in the age-adjusted overall incidence of esophageal adenocarcinoma, defined anatomically as located in the esophagus (ICD-O-3 code 150–159) and histologically as adenocarcinoma.

It should be noted that the AJCC classification change of esophageal adenocarcinoma occurred after this study. This classification now potentially includes many cases previously diagnosed as proximal gastric cancers as esophageal. As such, the incidence rates of esophageal adenocarcinoma would be expected to show an increase after 2010 because proximal gastric adenocarcinoma would be added.

They also examined the stage-specific distribution using the SEER historic stage variable, which defines localized (early) stage as a cancer limited to the esophagus, regional stage as a cancer extending beyond the esophagus into adjacent tissues or regional lymph nodes, and distant stage as the presence of metastatic disease. SEER does not use the T(umor) N(ode)M(etastasis) staging system for esophageal adenocarcinoma. To evaluate time trends, they used the Joinpoint regression analysis developed by the NCI to test the significance of changes in trend.

Their results are as follows:

> "(a) Overall incidence trends: Figure 1 shows the dramatic overall increase in incidence of esophageal adenocarcinoma between 1973 (3.6 cases per million) and 2006 (25.6 cases per million), a 7-fold increase (Fig. 12.18). Figure 1 also suggests a change in the slope of the rising incidence curve in the late 1990s or early 2000s. Using Joinpoint regression analysis, the inflection point is estimated as occurring in 1996. From 1973 to 1996, the incidence of esophageal adenocarcinoma increased from 3.6 to 21.9 per million, an annual increase of 8.2% (95% confidence interval, 7.7-8.8%). From 1996 to 2006, incidence increased only to 25.6 per million, reflecting an annual increase of 1.3% (95% confidence interval, 1.9-4.7%), a significant decline in the slope of the incident curve."
>
> "(b) Stage-specific incidence trends: Figure 2 shows that among all stage-specific incidence trends, it is the early stage trend that has changed the most over the last decade. Although there is no significant change in trend for regional or distant stage disease during this time period, Figure 3 shows a significant trend change for early stage disease. After a steep annual increase of 10.0% (95% confidence interval, 8.8-11.3%) prior to 1999, incidence subsequently declined by 1.6% (95% confidence interval, 9.5-7.0%) per year. The incidence of unstaged disease has declined over the same period."

The authors subsequently corrected an error in the confidence interval of early stage cancer after 1999 to −9.5% to 7.0%.

In their discussion, the authors try to find possible explanations for the fact that the incidence of early stage cancers has plateaued in 1999 without any significant change in the rate of increase in regional and advanced cancers. Their reasons are as follows:

First, the observed incidence pattern may be due to chance. And simply reflect annual fluctuations. We acknowledge that it may be too early to draw firm conclusions about the changes in incidence.

... Second, improved staging techniques (such as endoscopic ultrasound and computed tomography) might have led to greater detection of affected lymph nodes or metastatic lesions. This may, in turn, have led to stage migration, in which patients may have become more likely to be diagnosed as having regional or distant metastasis and less likely to be diagnosed with early stage disease. Given that we do not see a concomitant acceleration in later stage incidence, we think this is unlikely.

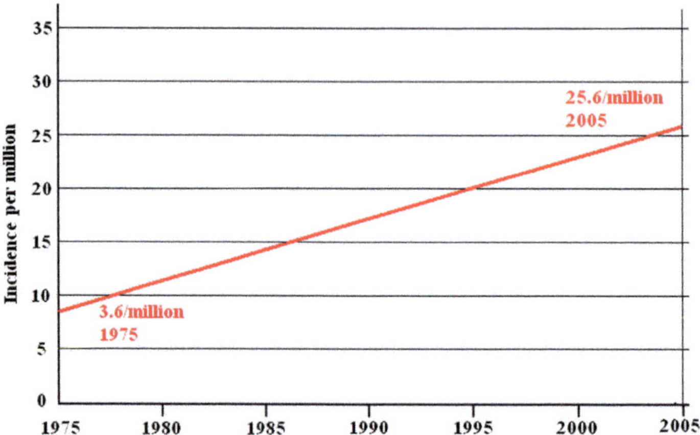

FIGURE 12.18 The incidence of esophageal adenocarcinoma continued to increase from 2000 (the previous study by the same authors) to 2005.

Third, it may be that the change in early stage disease incidence is real, but there has been insufficient time for a decreased incidence of later stage disease to appear. We can think of three possible explanations why early disease stage could truly be decreasing:

(1) Exposure to initiating factors, such as reflux disease and increased BMI, might have plateaued. (2) The transition from Barrett esophagus to cancer might have slowed. Aspirin and PPIs have been reported to decrease the progression of Barrett esophagus to cancer. (3) Barrett screening and surveillance may be effective. Increased detection of dysplastic lesions with endoscopic or surgical ablation may prevent progression to cancer.

There is another possible explanation for the decline in incidence. The authors include only ICD code 150–159 for inclusion in this study. All adenocarcinomas in the GEJ region were classified at this time as gastric cancers. It is well known that the location of adenocarcinoma of the esophagus has moved progressively distally in the past four decades. If that is true, then some tumors that were esophageal could have site-migrated from codes 150–159 to code 160 and not be included in this study. This results from the error of classification wherein code 160 was classified as gastric.

With the change in the definition of tumors in this region by the AJCC in its seventh edition (2010),[23] the real incidence of esophageal adenocarcinoma, which now includes cases previously classified as gastric, will become apparent in epidemiologic studies of cases after 2010.

5. PROPOSED NEW TERMINOLOGY FOR ESOPHAGEAL ADENOCARCINOMA

Presently, the designation of "esophageal adenocarcinoma" is only applied universally and uniformly to adenocarcinomas that arise >5 cm above the endoscopic GEJ.

There is a wide range of terms that are applied for adenocarcinomas that are 5 cm on either side of the endoscopic GEJ. Many people will simply call all such tumors "junctional adenocarcinoma" or "adenocarcinoma of the GE junctional zone." Tumors do not arise from junctions; junctions are not zones. They must arise either from the esophagus or stomach. They may extend into one from the other by growth and infiltration, but they arise in one or the other. The use of these terms is an expression of the inability to define the true GEJ.

Esophageal and gastric adenocarcinomas have a different etiology (GERD vs. *H. pylori*, respectively) and epidemiology (the former is increasing and the latter decreasing in incidence). The epidemiology of "junctional adenocarcinoma" within 5 cm *distal to the endoscopic GEJ* has an epidemiology similar to esophageal than gastric adenocarcinoma: this suggests that what falls under the definition of the proximal stomach is actually esophagus.

Worldwide, most people probably use the Siewert terminology[22] for "junctional adenocarcinoma": type I (distal esophageal), type II (junctional), and type III (subcardiac). Some will use the term "adenocarcinoma of the gastric cardia or proximal stomach" for tumors largely distal to the endoscopic GEJ. Theoretically, pathologists are supposed to follow the AJCC classification, but whether this happens universally is not known.

This could mean that a tumor in the area distal to the endoscopic GEJ can be called a "junctional," "subcardiac," "Siewert type III," "adenocarcinoma of the gastric cardia or proximal stomach," or "esophageal adenocarcinoma." The potential confusion is incredible, but surprisingly well accepted.

The use of the proposed new histologic classification and new method of using histology to classify adenocarcinomas in the esophagus and stomach will provide clarity that does not presently exist. Only two terms are used: esophageal adenocarcinoma and gastric adenocarcinoma.

An esophageal adenocarcinoma is a tumor whose epicenter (lateral edge at midpoint) abuts cardiac epithelium (with and without parietal and/or goblet cells). If the distal edge of the tumor also abuts cardiac epithelium (with and without parietal and/or goblet cells), the tumor is entirely within the esophagus. If the distal edge of an esophageal cancer abuts gastric oxyntic epithelium, it is an esophageal cancer that has extended across the true GEJ into the stomach.

In the new method of using histology for classification, there are two different clinicopathological types of esophageal adenocarcinoma: (1) adenocarcinoma arising in patients with CLE limited to the dilated distal esophagus distal to the endoscopic GEJ without visible CLE above the endoscopic GEJ (Figs. 12.11 and 12.19A) and (2) adenocarcinoma arising in patients with visible CLE with intestinal metaplasia (Barrett esophagus). These tumors may occur in the thoracic esophagus above the endoscopic GEJ (Figs. 12.9 and 12.19B) or in the dilated distal esophagus between the endoscopic GEJ and the true GEJ (Figs. 12.19C and 12.20).

A gastric adenocarcinoma is a tumor whose epicenter (lateral edge at midpoint) abuts gastric oxyntic epithelium. If the proximal edge of a gastric cancer also abuts gastric oxyntic epithelium, the tumor is entirely in the stomach. If the proximal edge of a gastric cancer abuts squamous epithelium or cardiac epithelium (with and without parietal and/or goblet cells), the tumor has extended into the esophagus.

It may not be possible to classify advanced tumors with circumferential involvement by the criterion of the epithelium at their epicenter because the tumor has no interface with normal epithelium at a lateral edge. In these cases, a tumor is definitely esophageal if the distal edge abuts cardiac epithelium (with and without parietal and/or goblet cells) and definitely gastric if the proximal edge abuts gastric oxyntic epithelium. All other circumferential tumors should be designated as esophageal unless their epicenter is >5 cm distal to the perceived (likely inaccurate because of distortion) GEJ. Chandrasoma et al.[26] showed that most adenocarcinomas of this region are esophageal in origin.

5.1 Adenocarcinoma of the Thoracic Esophagus in Patients With Visible Columnar-Lined Esophagus (Figs. 12.9, 12.10, and 12.19B)

Barrett esophagus (visible CLE with intestinal metaplasia) is measured at endoscopy by the Prague C+M method. This is limited to visible CLE above the endoscopic GEJ.

I have suggested that visible CLE above the endoscopic GEJ represents the point of irreversibility for the cellular changes in the thoracic esophagus. Its presence is the result of severe reflux, as indicated by an abnormal (usually markedly abnormal) pH test. Severe reflux in turn is the result of severe LES damage with a residual LES length that is <10 mm for the abdominal segment and <20 mm for the total LES (Fig. 12.2). The abdominal LES is commonly shortened by >25 mm from its initial length of 35 mm.

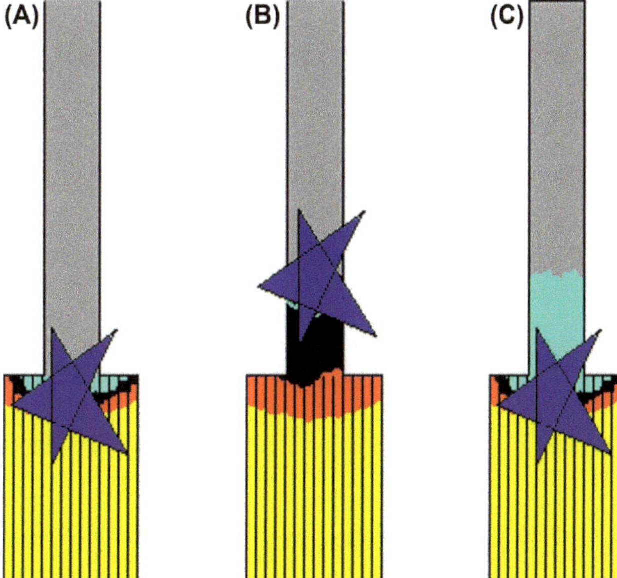

FIGURE 12.19 Three types of esophageal adenocarcinoma: (A) adenocarcinoma of the dilated distal esophagus in a patient without visible columnar-lined esophagus (CLE); (B) adenocarcinoma of the thoracic esophagus in a patient with visible CLE; (C) adenocarcinoma of the dilated distal esophagus in a patient with visible CLE.

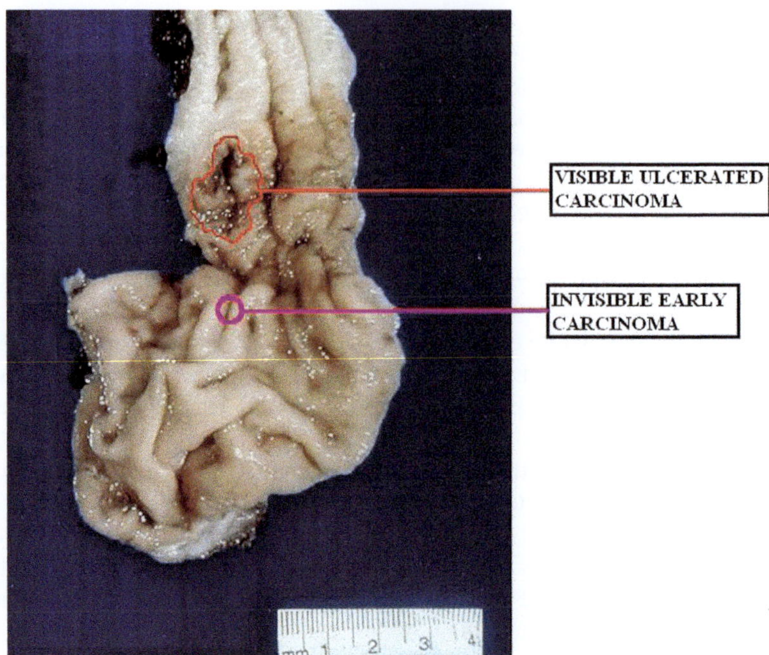

FIGURE 12.20 Two independent adenocarcinomas arising in a long segment of Barrett esophagus. The first is a large ulcerated mass in the thoracic esophagus immediately below the squamocolumnar junction. The second is an incidental, microscopic, early intramucosal adenocarcinoma in the columnar-lined esophagus in the dilated distal esophagus distal to the end of the tubular esophagus and the proximal limit of rugal folds. This tumor was above the proximal limit of gastric oxyntic epithelium.

Presently undetectable, LES damage can be measured exactly in the new method by the length of cardiac epithelium (with and without parietal and/or goblet cells) distal to the endoscopic GEJ.

Patients with visible CLE who develop thoracic esophageal adenocarcinomas are much more commonly males than females and have severe reflux (as shown by an abnormal 24-h pH test) and severe LES damage (as shown by a dilated distal esophagus that is usually >20 mm and a residual abdominal LES that is <10 mm at manometry). Most have symptoms of GERD, often of long duration and often severe and difficult to control. Symptoms, however, may be absent, minimal, or well controlled with PPI therapy. The angle of His has become obtuse with LES damage; sliding hiatal hernias are commonly present.

In Lagergren et al.[2] these are the patients designated as esophageal adenocarcinoma. Sixty percent had heartburn and/or regurgitation more than once per week. Their odds ratio for adenocarcinoma progressively increased to 43.5 (95% CI 18.3–103.5) for patients with the highest symptom severity and duration of symptoms. Forty percent were "asymptomatic" within the definition used.

These patients are those most likely to have had a diagnosis of Barrett esophagus made by endoscopy and be on surveillance. They have a chance of early detection of their cancer and minimally invasive endotherapy that has a high rate of cure. Unfortunately, that is a minority of people in this group.

5.2 Adenocarcinoma of the Dilated Distal Esophagus in a Patient With Visible Columnar-Lined Esophagus (Figs. 12.19C and 12.20)

These patients are clinically similar to patients with adenocarcinoma of the thoracic esophagus except for the fact that their tumor is centered distal to the endoscopic GEJ. The CLE is visible at endoscopy and may be short or long segment and appears normal. The tumor arises distal to the endoscopic GEJ.

This situation is easy to understand in the new method where the CLE extends unbroken from above the endoscopic GEJ into the dilated distal esophagus. The segment of CLE extends from the cephalad-displaced SCJ to the true GEJ (proximal limit of gastric oxyntic epithelium) involving the thoracic esophagus above and the dilated distal esophagus below the endoscopic GEJ. The tumor has arisen in the distal region of the CLE segment in the dilated distal esophagus distal to the endoscopic GEJ.

The demographic profile, symptoms and severity of reflux, and the association with hiatal hernia are identical to patients with Barrett esophagus whose adenocarcinoma is in the thoracic esophagus. In Lagergren et al.[2] this group will be included in "adenocarcinoma of the gastric cardia" and will be the majority of patients among the 29% who were symptomatic within this group.

They are also patients who may have a diagnosis of Barrett esophagus and be on surveillance. I am uncertain how gastroenterologists react when they encounter a patient who has an adenocarcinoma distal to the endoscopic GEJ during surveillance endoscopy. Does the endoscopist call this an adenocarcinoma of the proximal stomach complicating Barrett esophagus? Or, does he/she call it an esophageal adenocarcinoma that is found distal to the GEJ? Both choices are irrational. There is no better proof that the true GEJ is distal to the incorrectly defined endoscopic GEJ than this scenario.

5.3 Adenocarcinoma Arising in the Dilated Distal Esophagus in Patients With No Visible Columnar-Lined Esophagus Above the Endoscopic Gastroesophageal Junction (Figs. 12.11 and 12.19A)

This group is defined by an adenocarcinoma that arises distal to the endoscopic GEJ in a patient with no visible CLE at endoscopy. They are esophageal in the new method of classification because the epithelium at the epicenter (lateral edge at midpoint) is cardiac (with and without parietal and/or goblet cells). In many cases, the tumor is entirely esophageal as the epithelium at the distal edge is also cardiac (with and without parietal and/or goblet cells). They may extend proximally into the region above the endoscopic GEJ (lined by squamous epithelium) or distally into the stomach (lined by gastric oxyntic epithelium).

These tumors are presently called "subcardiac" (Siewert type III) or "junctional" (Siewert type II) tumors by those who use that classification and adenocarcinomas of the gastric cardia/proximal stomach by many others.[22] Siewert type II differs from type III only by the fact that the tumor centered distal to the endoscopic GEJ has extended proximally to straddle the endoscopic GEJ.

In Siewert et al.[22] (reviewed above), junctional (type II) and subcardiac (type III) adenocarcinomas are associated with Barrett esophagus (i.e., visible CLE with intestinal metaplasia above the endoscopic GEJ) in only 11% and 2% of cases. In the classification that is proposed here, these would be adenocarcinomas arising in patients with Barrett esophagus. The majority of type II (89%) and type III (98%) adenocarcinomas (excepting the 11% and 2% with Barrett esophagus) would be classified in the new system as adenocarcinomas of the dilated distal esophagus in patients without Barrett esophagus.

5.4 Differences Between Esophageal Adenocarcinomas of the Dilated Distal Esophagus With and Without Visible Columnar-Lined Esophagus

In Siewert et al.[22] subcardiac (type III) and junctional tumors (type II) differed from distal esophageal (type I) adenocarcinomas in the following ways: (1) lower preponderance of male sex; (2) less likely to have a hiatal hernia; (3) less likely to have a long history of GERD; (4) less likely to have intestinal metaplasia in the distal esophagus (2% and 11% with Barrett esophagus in types III and II compared with 81% in type I); (5) less likely to have intestinal metaplasia at or below the endoscopic GEJ (9% and 32% in types III and II compared with 75% in type I); (6) more likely to have undifferentiated tumors (51% type I, 56% type II, and 71% type III); and (7) more likely to have a "nonintestinal" growth pattern (24%, 34%, and 45% of types I, II, and III tumors).

Patients who develop adenocarcinoma in CLE limited to the dilated distal esophagus are those that I have classified in the new system as being in the phase of compensated LES damage (with LES damage<15 mm) or having mild GERD (with LES damage >15–25 mm), defined by the absence of visible CLE. This correlates with a severity of reflux that is less than in patients with visible CLE. These patients have less severe LES damage, usually in the 15–25 mm range, as measured by the length of cardiac epithelium (with and without parietal and/or goblet cells) in the dilated distal esophagus.

In Lagergren et al.[2] 187/262 (71%) patients with "adenocarcinoma of the gastric cardia" (defined as any tumor with its epicenter within 2 cm proximal to 3 cm distal to the endoscopic GEJ) did not have symptoms of reflux (defined as heartburn and/or regurgitation at least once per week). This was significantly higher than the 76/189 (40%) with esophageal adenocarcinoma (i.e., tumors with their epicenter >2 cm proximal to the endoscopic GEJ) who did not have reflux symptoms. When one excludes patients with Barrett esophagus that were included with this group in Lagergren et al.[21] (11% type II and 2% type III), the probability of lack of symptoms in this group increases further.

5.5 Gastric Adenocarcinoma

Gastric adenocarcinoma is defined as a tumor with its epicenter (lateral edge at the midpoint of the tumor) abutting gastric oxyntic epithelium, irrespective of its distance from the perceived GEJ.

Gastric adenocarcinoma has no association with GERD, is associated with *H. pylori* gastritis, and has decreased in incidence in the United States and Western Europe over the past four decades.[30] There is an inverse relationship between the incidence of esophageal and gastric adenocarcinoma in the population of any given nation. Gastric adenocarcinoma tends to have the highest incidence in Far East Asia where the incidence of GERD[8] and esophageal adenocarcinoma is the lowest.

There is no longer any need to classify gastric adenocarcinomas as "proximal" and "distal." This terminology should be abandoned because the historical association of "proximal gastric adenocarcinoma" with GERD creates confusion. This should be recognized as esophageal adenocarcinoma in the dilated distal esophagus. Gastric carcinoma *cannot* be the result of GERD.

The paper by Hamilton et al.[31] which I consider a classic in the literature of esophageal adenocarcinoma, shows the different types of adenocarcinoma in this region in a more detailed manner than in any other paper. It seems to clearly differentiate adenocarcinomas into esophageal and gastric, with all tumors of the GEJ being considered along with esophageal adenocarcinoma. Their "gastric adenocarcinomas" are limited to tumors in the body and antrum. It is remarkable that their figures seem to show the GEJ distal to the end of the tubular esophagus and they have no trouble in assigning a designation of Barrett adenocarcinoma to tumors that are distal to the GEJ.

In their discussion they state: "Because a substantial portion of the adenocarcinomas in the gastric cardiac region consists of Barrett adenocarcinomas, the relative importance of Barrett carcinoma as a cause of upper gastrointestinal tract cancer appears likely to increase as the decline in gastric cancer incidence continues." This is a prophecy made in 1988 that became recognized as true in the next two decades.

The paper was written at a time before the definition of the GEJ by McClave et al.[32] as the proximal limit of rugal folds had gained acceptance and before short segment Barrett esophagus had been recognized.[33] Then, 3 cm of columnar epithelium in the distal esophagus was regarded as a normal finding and the normal gastric cardia occupied the proximal 3 cm of the stomach.[14]

Sixty-one consecutively accessioned esophagogastrectomy specimens between 1980 and 1986 with the tumor mass centered in the esophagus or esophagogastric junction (including gastric cardia) were identified from the surgical pathology index of The Johns Hopkins Hospital in Baltimore.

The specimens were opened along the wall opposite the tumor, pinned on a wax block, fixed overnight, photographed, and the entire columnar-lined segment including the GEJ region was submitted for histology. The sections were diagrammed on the photograph, permitting histologic mapping. The epithelial types in Barrett esophagus were classified as specialized with intestinal metaplasia, cardiac type (junctional), and fundic type according to the criteria of Paull et al.[18] that I have followed.[17]

The esophago-gastric junction was identified on the basis of at least two of four possible gross and histological landmarks: the most distal location of esophageal submucosal glands, the transition from the tubular configuration of the esophagus into the stomach, the most distal location of squamous epithelium, and the change in orientation of the layers of the muscularis propria in longitudinal oriented sections.

In the specimens in which Barrett mucosa was histopathologically identified, the extent of the Barrett esophagus above the esophagogastric junction was determined morphometrically from the ruler included in the photograph and the gross description. The upper extent of the Barrett esophagus was defined by the most proximal Barrett mucosa (Fig 1a) or ulcerated tumor (Fig 1b) in the specimen. This latter criterion was applied because of the possibility that pre-existing Barrett mucosa had been overgrown by the tumor.

The study population of 61 patients was compared with three other populations identified from the surgical pathology files in the same period: (1) 69 patients who underwent gastrectomy for adenocarcinoma of the gastric body or antrum, (2) 32 patients with biopsy-proven adenocarcinoma of the esophagus or esophagogastric region who did not undergo resection, and (3) 149 patients with Barrett esophagus diagnosed by endoscopic biopsy showing intestinal metaplasia.

The prevalence of Barrett esophagus (i.e., intestinal metaplasia) was 39/61 (64%) patients. These were designated as Barrett adenocarcinoma. These had a striking preponderance of white men (34/39; 87%). All Barrett carcinomas were located in the lower third of the esophagus. The extent of Barrett esophagus in each specimen, along with the location and size of the associated adenocarcinomas, are shown in Fig. 2. The median length of Barrett esophagus in the 39 cases was 5 cm (range 1–12 cm).

The Barrett adenocarcinomas were generally centered within a few centimeters of the esophagogastric junction (mean 2 ± 0.3 cm). Four large Barrett-associated tumors appeared to be centered distal to the junction.

The characteristics of the 39 Barrett adenocarcinomas are compared with the 22 cases in the original 61 cases of the study population not demonstrated to have Barrett epithelium histologically and with the 69 cases of adenocarcinoma of the gastric body and antrum.

TABLE 12.3 Comparison of Cases of Adenocarcinoma of the Esophagus and Stomach Resected Between 1980 and 1986

	Adenocarcinoma of Esophagus or Esophagogastric Junction		
Characteristic	**BE Identified**	**No BE Identified**	**Gastric Adenocarcinoma**
Prevalence of white men	34/39 (87%)	16/22 (73%)	21/69 (30%)
Mean age (years ± SE)	64 ± 1	56 ± 4	65 ± 1
Age range (years)	46–83	24–85	36–86
Mean length from epicenter to GEJ (cm)	2 ± 0.3	0 ± 0.5	n/a
Mean tumor extent (cm)	5 ± 0.4	5 ± 0.4	Not examined
Percentage (%) of cases completely studied by histology	33/39 (85%)	14/22 (64%)	Not examined

BE, Barrett esophagus; *GEJ*, gastroesophageal junction; *SE*, standard error.

The 61 study group tumors had a predominance of white males irrespective of whether they had demonstrated Barrett esophagus or not (Table 12.3). The two tumors were also similar for tumor size but differed significantly in their relationship to the GEJ; those without Barrett esophagus had their tumor epicenter at or within 0.5 cm on either side of the GEJ compared to Barrett carcinomas, which were a mean of 2 cm above the GEJ. There were no significant differences between the 61 patients with resected adenocarcinomas and the 32 cases of proven adenocarcinoma that were not resected.

In their discussion, the authors address the possibility that adenocarcinomas arising in Barrett esophagus are underestimated. 1980–86 was early in the epidemic of esophageal adenocarcinoma that resulted in a sevenfold rise in incidence from 1973 to 2006. The rising incidence had already been recognized by 1988. They write: "Barrett esophagus appears to be under-recognized in routinely sampled esophagogastrectomy specimens with adenocarcinoma, as well as at endoscopy."

At the time Barrett adenocarcinoma was a small percentage of esophageal cancer, which was dominated by squamous carcinoma. "Based on an estimate of 10% and the projected 9700 new cases of esophageal cancer in the United States in 1987, fewer than 1000 cases of Barrett adenocarcinoma would be expected nationwide during the year." What a difference three decades has made when these numbers are contrasted with the data of Pohl et al.[1] in 2010!

6. THE WINDOW OF OPPORTUNITY FOR PREVENTING ESOPHAGEAL ADENOCARCINOMA

In the new method that I am proposing, the primary goal of management of GERD is to develop a rational method of preventing esophageal adenocarcinoma.

This is not to minimize the importance of symptom control and improvement of quality of life. It is to attack GERD at its root cause by recognizing that all complications of GERD result from LES damage, not from acid. Success in preventing esophageal adenocarcinoma will likely have the "side effect" of better symptom control with less treatment failure.

Presently, esophageal adenocarcinoma arises in a manner that is practically unpredictable. No one in the population is immune. There are risk indicators (see below), but these do not provide actionable information that precipitates any type of screening method. The only points of intervention in the present guidelines that essentially treat all symptomatic patients with PPIs with the aim of symptom control are (1) failure to control symptoms with PPI therapy and (2) dysphagia that raises the specter of cancer. Eighty-five percent of people who develop adenocarcinoma have never had an endoscopy.

This management algorithm results in the detection of Barrett esophagus with entry into a surveillance protocol that detects early cancer in 15% of all patients who develop cancer. Even the detection of Barrett esophagus does not predict who among these patients will develop cancer. Most of these patients continue to be treated with PPIs with the objective of best symptom control possible. Antireflux surgery is not regarded as a procedure that is justified to prevent cancer. The objective is not preventing cancer, it is finding cancer that develops at an earlier stage, thereby improving prognosis.

The new pathologic assessment of LES damage and prediction of future severe GERD can be used in all people as a screening method. It is not feasible in practice to screen all people at this time because the assessment requires endoscopy. Resources are not available for population screening by endoscopy and it will never be cost-effective. In the future, if methodology develops to measure LES damage by methods that do not need standard endoscopy, wider screening may become possible (see Chapter 20).

The window of opportunity for preventing cancer is the period between identifying a point in the disease that can precipitate screening for LES damage and the occurrence of irreversibility in the adenocarcinoma sequence. This window is different for adenocarcinoma arising in patients with and without Barrett esophagus (visible CLE with intestinal metaplasia).

6.1 The Window of Opportunity for Preventing Adenocarcinoma in Patients With Visible Columnar-Lined Esophagus

The occurrence of visible CLE is the point at which the sequential cellular changes leading to adenocarcinoma in the thoracic esophagus become irreversible. Visible CLE→intestinal metaplasia→increasing dysplasia→adenocarcinoma cannot be prevented by medical therapy. Everything before this point, including severe erosive esophagitis is reversible with PPI therapy.

The objective is to find a point in the disease that can be correlated with the possibility of visible CLE before this has occurred. The new measurement of LES damage by histology will identify this point. Let us say that a large study measuring LES damage has shown that LES damage of 25 mm is the smallest amount of LES damage at which visible CLE occurs. The objective of management will then become the prevention of LES damage reaching 25 mm.

The new test uses an algorithm to convert the measured length of LES damage at any point in time into a prediction of the extent of future LES damage (see Chapter 19). By the test, therefore, a patient who has 15 mm of LES damage at age x years can be predicted to reach 25 mm of LES damage at age y years. There is a window of opportunity to prevent progression of LES damage to the critical level of y − x years. In practice, the y − x interval is likely to be several decades.

If the new test becomes simple, safe, and cheap without needing endoscopy, there is no problem. The entire population can be screened for LES damage at a defined age (35 years and/or the onset of any symptom that may represent symptomatic GERD).

The test will identify the extent of LES damage at the time of the test and divide the population into two groups: (1) people who will not reach LES damage of 25 mm until age 80 years; (2) people who will develop LES damage of 25 mm at a variable time in the future before age 80 years and be at high risk for visible CLE. The test will allow the first group to be ignored; they will not develop visible CLE by an age that allows progression to cancer, assuming a lag period of at least 10 years. The second group is focused on as requiring some intervention to prevent or slow the rate of progression of LES damage such that 25 mm of damage will never be reached. This will prevent the development of visible CLE, which means that esophageal adenocarcinoma will be prevented.

At present, though, the need for endoscopy to perform the test for LES damage makes it necessary to define criteria to screen a limited number of people within available resources. Ideally, this point will provide the maximum time before the patient progresses to visible CLE. The longer the window of opportunity, the more likely it is that any intervention will succeed.

I have suggested that visible CLE is caused by a severity of reflux that results from LES damage >25 mm. At this point of LES damage, the residual LES<10 mm and below the level at which it is incompetent at rest and LES failure (transient lower esophageal sphincter relaxation) is frequent. The patient is at high risk of visible CLE if this has not already developed.

Before a person reaches abdominal LES damage >25 mm, he/she has progressed over many decades from zero abdominal LES damage (with an initial length of 35 mm), exhausting the reserve capacity of the LES (abdominal LES damage >0–15 mm with a residual length of 20 mm) slowly over many decades, and developed symptomatic GERD (15–25 mm of abdominal LES damage), initially postprandial and progressively getting worse, but controlled by PPI therapy. PPI therapy masks LES progression until treatment failure occurs.

A recognizable window of opportunity exists between the time of onset of symptoms (15 mm LES damage) and the development of severe GERD (>25 mm LES damage where visible CLE becomes likely). During this period, the LES damage progresses by 10 mm. The duration of this sweet spot can vary from 25 years in a person with a rate of progressive LES damage of 4 mm/decade to 10 years in a person with a rate of LES damage of 10 mm/decade.

At present, the treating physician has no option other than to treat the patient with PPIs, with hope and prayer that the patient will not progress to severe GERD, treatment failure, and visible CLE. The physician is blind to the critical progression of abdominal LES damage from 15 to 25 mm. The unfortunate minority of patients who progress to the irreversible points of treatment failure and visible CLE is out of luck.

Control of symptoms with PPI therapy has no bearing on progression of LES damage. Nason et al.[5] showed that patients who are well controlled have a higher risk of progression to visible CLE.

The new diagnostic test that measures LES damage and predicts who among these patients are at highest risk to progress to severe GERD provides an alternative to hope and prayer. Within the relatively longer duration of the sweet spot between onset of symptoms and severe GERD, there is now an opportunity of aggressive intervention in that minority of patients. Intervention has the goal of slowing or stopping the inexorable progression of LES damage. This is possible because of the sudden new ability by the new test to identify those patients at risk. Because the percentage of people progressing to visible CLE is ~10%, the numbers are manageable.

Kulig et al.[34] in the first report of the Pro-GERD study confirmed the existence of the window of opportunity that I have identified. Of the 6215 patients with symptomatic GERD enrolled in the study, the index endoscopy showed nonerosive reflux disease in 2853 (46%), erosive reflux disease in 2660 (43%), and visible CLE in only 702 (11%). This cohort of GERD patients was well beyond the onset of symptoms; 46% had daily symptoms, 44% had symptoms at least 2 days/week, and the duration of symptoms was variable over a wide range.

Even in this group of fairly advanced symptomatic GERD, visible CLE and irreversibility had been reached only in 11% of patients. The 89% without visible CLE are still open to preventive measures. The prevalence of visible CLE would likely have been much less at the onset of symptoms, providing a significant window of opportunity for prevention of visible CLE. In the ProGERD study, 9.7% of persons without visible CLE at the index endoscopy developed it within 5 years.[11] The new test would have been able to predict these patients, permitting intervention and prevention.

Kulig et al.[34] reported that visible CLE was associated with increasing age, male gender, duration of GERD symptoms, obesity, smoking, alcohol consumption, and PPI use. These are among other valuable factors that may add to the equation to stratify risk in GERD patients.

For example, the indication for endoscopy in GERD can change from the present guideline of failure of PPI therapy to control symptoms to something like: "endoscopy is recommended for any patient with GERD symptoms that have been on daily PPI therapy for a period of 1 month or more irrespective of symptom control." The availability of the new test with predictive value of disease progression will make that a no-brainer and impossible to deny.

Another indication for endoscopic screening for LES damage that will be difficult to deny will be the point of presentation with symptoms of GERD that precipitates a trial of empiric PPI therapy. It is known that this empiric test results in an unknown but possibly large number of people being placed on long-term PPI therapy unnecessarily. The high cost and recent recognition of complications of long-term PPI therapy are well-recognized problems. The presence of the new test that can provide a result that will exclude GERD (i.e., LES damage <15 mm) makes endoscopy before empiric PPI treatment attractive (see Chapter 21).

The window of opportunity for identifying patients at risk for progression to severe GERD, visible CLE, and treatment failure is real, large, and of sufficient duration for effective preventive measures. If we are successful in preventing progression to visible CLE, adenocarcinoma of the thoracic esophagus is prevented. There can be no cancer without visible CLE; squamous epithelium does not progress to adenocarcinoma.

6.2 The Window of Opportunity for Preventing Adenocarcinoma of the Dilated Distal Esophagus in Patients Without Visible Columnar-Lined Esophagus

Adenocarcinoma may arise in the dilated distal esophagus in patients without visible CLE. In Siewert et al.[22] Barrett esophagus was present only in 11% of patients with type II (junctional) and in 2% of type III (subcardiac) adenocarcinomas.

There are almost no data in this group of people who have CLE limited to a dilated distal esophagus. This is largely because almost all physicians who perform endoscopy firmly believe in the accuracy of the endoscopic GEJ. To them the dilated distal esophagus, which is distal to the endoscopic GEJ, is "proximal stomach" or "gastric cardia" and therefore has no relevance in the diagnosis of GERD. Biopsies in this region are not recommended and almost never performed at endoscopy.

The little information that exists in the literature indicates that the length of the dilated distal esophagus is a reported >0–28 mm. It has been measured only in rare studies of autopsy patients and those undergoing esophagogastrectomy.[35,36]

Patients with a dilated distal esophagus can be divided into (1) people with compensated abdominal LES damage (<15 mm) who have a competent LES with damage that is within the reserve capacity of the LES and have no reflux and no typical GERD symptoms and (2) patients with mild GERD who have LES damage from 15 to 25 mm. These patients progress from intermittent and rare LES failure in the postprandial period at 15 mm LES damage to frequent LES failure at 25 mm LES damage.

The point of irreversibility for adenocarcinoma in the thoracic esophagus was visible CLE. I suggested that the window of opportunity for preventing adenocarcinoma of the thoracic esophagus is somewhere between the onset of symptomatic GERD (15 mm abdominal LES damage) and the occurrence of visible CLE (high risk at 25 mm of abdominal LES damage).

The dilated distal esophagus is different. All patients with a dilated distal esophagus have CLE; the very definition of the entity is the presence of CLE distal to the endoscopic GEJ. As such, the presence of CLE in the dilated distal esophagus cannot be used as the point of irreversibility in a practical sense.

The CLE in the dilated distal esophagus is regularly exposed to acidic gastric juice during heavy meals. If carcinogens exist in gastric juice, the dilated distal esophagus is exposed. The fact that adenocarcinoma of the dilated distal esophagus is so uncommon in a population with a virtually 100% prevalence of CLE must mean that there is a second specific cellular element within CLE that is required. This is very likely to be intestinal metaplasia in the dilated distal esophagus.

If true, the point at which cancer risk begins is the development of intestinal metaplasia in CLE of the dilated distal esophagus. Patients without intestinal metaplasia have no cancer risk.

There are little or no data regarding the factors associated with the development of intestinal metaplasia in CLE in the dilated distal esophagus. Marsman et al.[37] in a study of 102 persons with normal endoscopy, reported cardiac and oxyntocardiac epithelium but no intestinal metaplasia. In the detailed prospective part of our autopsy study,[35] none of the persons with a CLE length <10 mm had intestinal metaplasia.

Taking an optimistic viewpoint based on very little data, it can be postulated that intestinal metaplasia increases in prevalence with increasing length of CLE in the dilated distal esophagus. Let us assume, as the little available data suggest, that the risk of intestinal metaplasia in the dilated distal esophagus becomes significant when LES damage progresses to 25 mm. If that assumption is true, the same window of opportunity that we identified for preventing adenocarcinoma in the thoracic esophagus will also operate for preventing adenocarcinoma of the dilated distal esophagus.

If, however, intestinal metaplasia occurs in the dilated distal esophagus <15 mm in the asymptomatic person, and this progresses to cancer, there is no rational method for preventing adenocarcinoma in the dilated distal esophagus short of early-age population screening. Only data from future studies will answer this question.

7. TARGET EPITHELIUM FOR CARCINOGENESIS: INTESTINAL METAPLASIA

There are three basic factors required for carcinogenesis in the esophagus: (1) the presence of an epithelial cell type that is susceptible to a carcinogen in gastric juice; (2) an adequate concentration of carcinogen in gastric juice; and (3) exposure of the susceptible cell in the esophagus to an adequate dose of the carcinogen.

Although the carcinogen for esophageal adenocarcinoma is yet not known, we can reasonably assume that the concentration of that carcinogen in gastric juice has a variation in the population that is independent of the presence of GERD.

Patients with a zero carcinogen level are not at risk even if they have the most severe reflux. Patients with a high carcinogen level will be at higher risk than those with lower levels, but only when the other two conditions are satisfied. Even the patient with the highest possible carcinogen level in gastric juice is protected from esophageal carcinogenesis if there is no susceptible cell and/or no reflux adequate to deliver the carcinogen to the anatomic location in the esophagus where the susceptible cell resides within the esophagus.

This means two things with the present assumption that intestinal metaplasia is the *only* target cell for carcinogenesis: (1) If intestinal metaplasia is not present in the area of effective carcinogen delivery, cancer will not occur; and (2) if delivery of carcinogen can be reliably curtailed so that it does not reach the target cell, cancer will not occur. This of course has the proviso that reflux must be stopped before the necessary mutations for cancer have occurred and the cancer cell is in its lag phase.

I have defined the point of irreversibility in the reflux to adenocarcinoma sequence in the thoracic esophagus as the occurrence of visible CLE. The change from squamous to cardiac epithelium in the thoracic esophagus begins a progression that is not reversed or slowed by acid-reducing therapy for GERD. This progression is cardiac epithelium → intestinal metaplasia → increasing dysplasia → adenocarcinoma.

While cardiac epithelium is the point of irreversibility, the only target epithelium that is universally recognized as being susceptible to carcinogenesis in the esophagus is cardiac epithelium with intestinal metaplasia. There is a time lapse from the occurrence of cardiac metaplasia to the development of intestinal metaplasia. This varies from months to decades. This gap is an additional opportunity in the patient with visible CLE to intervene in the prevention of esophageal adenocarcinoma.

If carcinogenesis can occur to any significant extent in cardiac epithelium without intestinal metaplasia or oxyntocardiac epithelium, the window of opportunity disappears and prevention of esophageal adenocarcinoma becomes much more difficult.

7.1 The Present Belief in the United States: Visible Columnar-Lined Esophagus Is Susceptible to Carcinogenesis Only When It Has Intestinal Metaplasia

This is the one most important thing that the medical establishment understood correctly from the outset. In short order after Paull et al.[18] classified the histologic types present in visible CLE above the LES into specialized (=cardiac epithelium with intestinal metaplasia), junctional (=cardiac), and fundic-gland type (=oxyntocardiac type) in 1976, Rodger Haggitt recognized that only the intestinal type of CLE was at risk for esophageal adenocarcinoma.[38] This was rapidly accepted in the United States and most of the world. It resulted in the change of the definition of Barrett esophagus to "visible CLE with intestinal metaplasia."

The Montreal experts who were responsible for the present definition of GERD[21] clearly reached consensus at their meeting in 2006 that it was critical to differentiate "endoscopically suspected esophageal metaplasia" that was positive and negative for intestinal metaplasia.

They further stated, in statement 38: "A literature search failed to reveal any systematic review or meta-analysis of the risk of esophageal adenocarcinoma of definite esophageal columnar metaplasia in which intestinal metaplasia had not been shown to be present, despite careful biopsy sampling."

The present definition of Barrett esophagus in the American Gastroenterological Association medical position statement on Barrett esophagus[39]: "Barrett's esophagus is the condition in which any extent of metaplastic columnar epithelium that predisposes to cancer development replaces the stratified squamous epithelium that normally lines the distal esophagus. Presently, intestinal metaplasia is required for the diagnosis of Barrett's esophagus because intestinal metaplasia is the only type of esophageal columnar epithelium that clearly predisposes to malignancy."

In the discussion of the definition: "The diagnosis of Barrett's esophagus can be suspected when, during endoscopic examination, columnar epithelium is observed to extend above the gastroesophageal junction (GEJ) into the tubular esophagus.... If Barrett's esophagus is to be considered a medical condition rather than merely an anatomic curiosity, then it should have clinical importance. The columnar-lined esophagus has clinical importance primarily because it predisposes to esophageal cancer. Therefore, Barrett's esophagus can be defined conceptually as the condition in which any extent of metaplastic columnar epithelium that predisposes to cancer development replaces the stratified squamous epithelium that normally lines the distal esophagus."

Intestinal-type epithelium in the esophagus is clearly abnormal and metaplastic and predisposes to cancer development. Therefore, an esophageal biopsy specimen that shows intestinal-type epithelium above the GEJ establishes the diagnosis of Barrett's esophagus.

The matter appears to be settled. Evidence was powerful that columnar epithelia in the esophagus were not susceptible to carcinogenesis if there was no intestinal metaplasia. In the United States and most of the world, endoscopic surveillance was limited to patients with visible CLE with intestinal metaplasia.

In 2012, we undertook a study to examine this issue.[40] The objectives of this retrospective study were to determine the target epithelium for progression to adenocarcinoma in visible CLE. Stated in a simpler way, was the change cardiac epithelium→dysplasia and adenocarcinoma possible or did the sequence cardiac epithelium→intestinal metaplasia→dysplasia and adenocarcinoma operate in all cases?

The study consisted of reviewing 2586 reports in the pathology database of patients undergoing biopsy in the Foregut Surgery Unit of the Keck School of Medicine of the University of Southern California between 2004 and 2008. All patients were biopsied by a faculty person according to our established protocol. We selected two groups of patients: Group A: 214 patients who had visible CLE with systematic four-quadrant multilevel biopsies at 1–2 cm intervals from the entire visible CLE, irrespective of the presence of any lesions. For inclusion, at least two biopsy levels from the tubular esophagus had to show metaplastic columnar epithelium and the most proximal biopsy set had to have at least one biopsy that straddled the histologic SCJ. Group B: 109 patients who had visible neoplastic lesions; in these cases, the objective of the biopsies was to establish a histologic diagnosis of cancer before treatment. In our retrospective review, it was clear that once a neoplasm was identified, the level of interest of the endoscopist in taking biopsies in the surrounding mucosa decreased dramatically.

We excluded 468 patients who had a previous endoscopy with biopsy and 1795 patients who did not have a visible CLE or had less than two biopsy levels in the tubular esophagus with metaplastic columnar epithelium.

Intestinal metaplasia was diagnosed when definite goblet cells were identified in routine stain. Alcian blue stain was not used. Mucin distended pseudogoblet cells in the cardiac epithelium, multilayered epithelium, and gland ducts were not diagnosed as intestinal metaplasia. Low- and high-grade dysplasia and adenocarcinoma were diagnosed by standard criteria.

In the 214 patients with systematic four-quadrant, multilevel biopsies taken at 1–2 cm intervals from the entire visible CLE (Group A), including biopsies that straddled the SCJ, intestinal metaplasia was present in 187 (87.4%) patients. The other 27 patients with visible CLE had cardiac and oxyntocardiac epithelium with an observed transition of squamous epithelium to cardiac epithelium without intestinal metaplasia at the SCJ.

All 80 patients with a columnar metaplastic length>5 cm had intestinal metaplasia. The prevalence of intestinal metaplasia at a length of 1–5 cm was <100%, increasing from 19/34 (55.9%) in patients with 1 cm CLE to 31/38 (81.6%) at 2 cm and >85% at 3, 4, and 5 cm.

Dysplasia or adenocarcinoma was present in 55/214 (25.7%) patients. 17 had low-grade dysplasia, 17 had high-grade dysplasia, and 21 had invasive adenocarcinoma. The high prevalence of dysplasia and adenocarcinoma at the index endoscopy reflects the referral pattern in this specialized surgical unit. All 55 patients with dysplasia or adenocarcinoma were in the group of 187 patients who had intestinal metaplasia within the columnar-lined segment, irrespective of length. None of the 27 patients with nonintestinalized columnar metaplasia had dysplasia or adenocarcinoma, a finding that was significantly lower than in patients with intestinal metaplasia ($P=.01$).

The 109 patients (Group B) with dysplasia and/or adenocarcinoma in their tubular esophagus who underwent random, nonsystematic biopsies mostly had obvious visible lesions. None had low-grade dysplasia, 2 had high-grade dysplasia, and 107 had invasive adenocarcinoma. In the 49/109 patients, the biopsies contained only cancerous tissue. In the 60 patients who had columnar epithelium in addition to tumor, intestinal metaplasia was present in both patients who had high-grade dysplasia and 32/58 (55.2%) patients with invasive adenocarcinoma. The total amount of nondysplastic columnar epithelium present in these limited biopsies was generally small.

In our discussion, we consider the question of false-negative diagnosis of intestinal metaplasia because of inadequate sampling. "An important question is whether there exists any biopsy protocol that would remove the risk of a false-negative diagnosis."

Our study shows that when the four-quadrant biopsy protocol is adhered to, and the most proximal level includes a biopsy that straddles the SCJ, there is a separation of patients with and without intestinal metaplasia. That this is a real difference is shown by two facts: (1) There is a significant difference in the prevalence of intestinal metaplasia at different lengths of CLE, increasingly progressive from 55% in patients with 1 cm length to 100% when the length is >5 cm. (2) Dysplasia and adenocarcinoma occurred only in patients who were positive for intestinal metaplasia (55/187); in contrast, 0/27 patient without intestinal metaplasia had dysplasia and adenocarcinoma ($P=.01$).

Our findings showed that the biopsy protocol that we used separated patients with and without intestinal metaplasia and that only patients with intestinal metaplasia progressed to dysplasia and adenocarcinoma. The sequence for esophageal adenocarcinoma was therefore cardiac epithelium→intestinal metaplasia→dysplasia and adenocarcinoma. Cardiac epithelium without intestinal metaplasia did not directly progress to dysplasia and adenocarcinoma.

7.2 An Alternate Recent Viewpoint: Nongoblet Columnar Epithelium Is Susceptible to Carcinogenesis Without Intestinal Metaplasia

There has been a recent trend in the United States to question the requirement of intestinal metaplasia in the diagnosis of Barrett esophagus based on several review papers in the literature.

In the first of these, Riddell and Odze,[9] two highly influential gastrointestinal pathologists in North America with a special interest in esophageal pathology, suggest that the American definition of Barrett esophagus should change to fall in line with the British definition where Barrett esophagus does not require intestinal metaplasia in visible CLE.[41]

Their reasons for this conclusion include (1) evidence that nongoblet epithelium may be at risk for cancer; (2) evidence that nongoblet columnar epithelium may be "intestinalized" at a molecular level; and (3) the absence of residual intestinal metaplasia in the mucosa around esophageal adenocarcinoma. Let us examine these reasons.

7.2.1 Cancer Risk in Nongoblet Columnar Epithelium

Riddell and Odze state[9]: "There is little doubt that dysplasia and carcinoma may develop within areas of intestinal metaplasia, generally defined by the presence of goblet cells. The main clinical issue is whether non-goblet epithelium is also at risk for neoplasia, and if so, to what degree. The data presented here suggests overwhelmingly that nongoblet columnar epithelium in Barrett esophagus is abnormal, and likely has a propensity for neoplastic transformation that is similar to that seen in mucosa with goblet cells."

They cite two clinical studies from England as evidence. Gatenby et al.[42] followed 322 patients without and 612 patients with intestinal metaplasia at the index endoscopy and showed that there was no significant difference in the rate of development of dysplasia or adenocarcinoma between the two groups; 19.8% of those with and 15.2% of those without intestinal metaplasia progressed to neoplasia in the 10-year follow-up period. In a similar long-term follow-up study of 379 patients by Kelty et al.[43] also from England, the rate of development of adenocarcinoma was similar in patients with (4.5%) and those without intestinal metaplasia (3.6%) at the index endoscopy.

However, these two studies have serious issues relating to biopsy sampling that make this conclusion highly questionable. I will review this in the next section. Chandrasoma et al.[40] in the study reviewed above show convincing evidence that, with adequate sampling, dysplasia and adenocarcinoma arise only in patients whose CLE has intestinalized cardiac epithelium. Cardiac epithelium without intestinal metaplasia has no proven potential to directly progress to cancer.

7.2.2 Nongoblet Columnar Epithelium May Be "Intestinalized" at a Molecular Level

Liu et al.[44] in 2009 from the Harvard laboratory of Dr. Odze studied mucosal biopsies from 68 patients with columnar metaplasia, 22 without goblet cells and 46 with goblet cells, and 14 gastric controls. Four molecular parameters of DNA content (peak DNA index, DNA content heterogeneity index, percentage of cells with DNA > 5N, and aneuploidy rate) were studied using image cytometry and analysis of high-fidelity DNA histograms. The authors report that patients with and without goblet cells showed similar DNA content abnormalities.

The authors also point to the presence of similar expression of a variety of molecular markers in nongoblet and goblet cell esophageal columnar epithelia. These include CDX2, Heppar-1, Villin, and DAS-1. They cite a study by Hahn et al.[45] where MUC-2, DAS-1, Villin, and CDX2 were expressed similarly in nongoblet and goblet cell positive epithelium from patients without and with intestinal metaplasia in their CLE. Liu et al.[44] suggest that this represents "intestinalization" of nongoblet epithelium at a molecular level.

These data are interesting and worthy of future studies. However, they certainly do not warrant the replacement of goblet cells with any of these molecular markers as the means of diagnosing intestinal metaplasia. These molecular changes are not evidence of "intestinalization" in the sense that they indicate future cancer risk. Until a study is done that shows that DNA content changes, CDX2, Villin, or DAS-1 or some combination of these is associated with increased risk of future adenocarcinoma, they simply remain interesting changes that require a lot more study. As of now, it is the presence of goblet cells and not some concept of "molecular intestinalization" that has a proven cancer risk.

7.2.3 Absence of Residual Intestinal Metaplasia Around Esophageal Adenocarcinoma in Endoscopic Mucosal Resections

Riddell and Odze[9] cite papers that show the absence of goblet cells in mucosa around esophageal adenocarcinoma in endoscopic mucosal resections (EMRs) as evidence that carcinoma arises in nongoblet epithelium.

They point to the pathologic study by Takubo et al.[46] who, in a study of EMRs in Germany, showed that in 70% of patients, the mucosa adjacent to the cancers was cardiac/fundic type rather than intestinal. In addition, only 43.4% of patients had intestinal metaplasia in the entire EMR specimen. This is similar to the finding of intestinal metaplasia in 31.2% (34/109) patients in Group B of Chandrasoma et al.[40] reviewed above where a visible neoplastic lesion was present. In both situations, the objective of the procedure is diagnosis or staging of a visible tumor, not detection of residual intestinal metaplasia. EMR is essentially a large biopsy with greater sampling than multiple biopsies taken from a visible tumor at endoscopy. The great value of EMR over biopsy is its depth and the ability to define depth of invasion. Depth does not impact the finding of residual intestinal metaplasia.

This group of patients reported by Takubo et al.[46] is from a large cohort of patients with Barrett esophagus under surveillance in a specialized unit in Germany. It is only in this setting that minute adenocarcinomas amenable to endoscopic resection are found with any frequency. In Germany, the diagnosis of Barrett esophagus, and therefore entry into the surveillance program, requires the presence of intestinal metaplasia. Therefore, it is highly likely that all these patients had a prior biopsy diagnosis of intestinal metaplasia away from the tumor. These are not patients with nonintestinalized esophageal columnar metaplasia. They are patients with intestinal metaplasia that is not present in the limited EMR specimen.

In a recent study,[47] we evaluated EMRs for adenocarcinoma in our unit.

In this study, we reviewed 27 EMR specimens in 21 patients with the specific objective of defining the histologic type of the epithelia adjacent to and surrounding the neoplasm. We looked at the composition of the surface of the EMR specimen for percentage of dysplastic and cancerous tissue and types of nondysplastic epithelia, the epithelial types immediately adjacent to the neoplasm, the surface composition of the entire specimen, and the presence of goblet cells in the neoplastic epithelium. Goblet cells were diagnosed when they were definite in the routine hematoxylin and eosin section. No special stains were used.

After examination of the EMR specimen, we retrieved previous biopsy slides in our unit for all patients. In the cases of patients who did not have intestinal metaplasia in the EMR or prior biopsy specimens, we reviewed the clinical history and slides from an esophagectomy specimen that may have followed the EMR. We evaluated these for evidence of intestinal metaplasia.

We recorded the presence of goblet cells in dysplastic and cancerous epithelia in the EMR specimens. However, we did not consider this as representing residual intestinal metaplasia. Our definition of residual intestinal metaplasia required goblet cells in nondysplastic epithelium.

Of the 27 EMR specimens, 1 had high-grade dysplasia only, 17 had intramucosal adenocarcinoma, and 9 had adenocarcinoma invading into the submucosa. The tumors were low-grade intestinal type in 24, high-grade intestinal type in 2, and diffuse signet ring cell carcinoma in 1.

Of the 27 EMRs, 17 (63%) had intestinal metaplasia in nondysplastic CLE adjacent to the cancer (14), not adjacent to but in surrounding mucosa (1), under squamous epithelium in buried glands (1), and under neoplastic tissue (1). Ten (37%) had no intestinal metaplasia in nondysplastic CLE anywhere in the specimen.

Of the 10 EMRs without intestinal metaplasia, 3 were in two patients who had intestinal metaplasia in another EMR specimen at the same endoscopy. Five of the other patients had a single EMR specimen and the sixth had two EMRs, both without intestinal metaplasia. Of the 21 patients in the study, 15 (71%) had intestinal metaplasia in at least 1 EMR specimen; 6 (29%) did not.

Of the seven EMR specimens in six patients without intestinal metaplasia, the percentage of nondysplastic columnar epithelium occupying the surface area was 0% in three EMRs, 5% in one, 10% in two, and 55% in one. The paucity of nondysplastic CLE in six of these seven EMRs is a reasonable explanation for the absence of intestinal metaplasia.

Further study in the seven EMR specimens from six patients that had no intestinal metaplasia showed the following: (1) four EMR specimens in three patients had intestinal metaplasia in a previous biopsy specimen; (2) two patients had intestinal metaplasia in the esophagectomy performed 11 and 29 days after the EMR that was negative for intestinal metaplasia.

The one remaining patient had no intestinal metaplasia in the EMR, prior biopsy or post-EMR esophagectomy specimen. She was a 72-year-old woman who had a four-decade history of heartburn with a diagnosis of Barrett esophagus with surveillance for two decades. She had low-grade dysplasia that progressed to high grade and was treated with radiofrequency ablation 3 years prior to being seen in our unit. Material from prior material was not available for review. She was referred to our unit with a mass just below the Z-line that proved to be adenocarcinoma on biopsy. There was no nondysplastic columnar epithelium in the biopsies. The EMR specimen had a surface area of 80% neoplastic tissue, 10% squamous epithelium, and 10% nondysplastic CLE without intestinal metaplasia. The subsequent esophagectomy specimen showed a low-grade adenocarcinoma in short segment CLE of pathologic stage pT1bN0. There were no goblet cells in nondysplastic CLE. There were goblet cells in the dysplastic epithelium in both the biopsy and sections of the esophagectomy.

Goblet cells were present in dysplastic and cancerous epithelium in 22 EMRs in nine patients. If goblet cells in dysplastic and cancerous epithelia were included in the definition of residual intestinal metaplasia, all patients in this study would have had intestinal metaplasia.

In the discussion, we suggest the reasons why goblet cells may not be found in a resection specimen for esophageal adenocarcinoma: (1) sampling error; (2) overgrowth of goblet cells by the growing tumor; and (3) loss of goblet cells in the dysplastic process. Theisen et al.[48] reported that neoadjuvant chemotherapy can unmask intestinal metaplasia in patients without residual intestinal metaplasia preoperatively.

In our study, no reasons were necessary to explain absence of goblet cells. Careful study of clinical history and other specimens before and after the EMR resulted in the finding of intestinal metaplasia in all cases. From a practical standpoint, it is impossible to define the target cell from which a cancer arises. It is only possible to say that only patients who have intestinal metaplasia somewhere in their visible CLE are at risk for cancer.

7.2.4 Absence of Residual Intestinal Metaplasia Around Esophageal Adenocarcinoma in Esophagectomy Specimens

Riddell and Odze[9] also cite studies that report the absence of residual goblet cells in a significant percentage of resected esophageal adenocarcinoma specimens. They cite our study where intestinal metaplasia was observed in 87% of cancers that developed in the distal esophagus above the level of the GEJ, but in only 40% and 45% of GEJ and cardia cancers, respectively.[26] They cite data from numerous other large studies that show a significant percentage of patients with absence of residual intestinal metaplasia in esophagectomies performed for adenocarcinoma. Most of this data show that residual intestinal metaplasia is less common in tumors arising in short segments of columnar metaplasia.

When we evaluate esophagectomy specimens, our primary objectives are to satisfy staging requirements as required in synoptic reporting. The number of lymph nodes that we find in our esophagectomy specimens is among the largest in any series (usually over 40 lymph nodes). We do not have the objective of demonstrating intestinal metaplasia, largely because of our belief that all adenocarcinomas arise in intestinal metaplasia. The finding of residual intestinal metaplasia is purely academic; it has no practical significance. This viewpoint has been expressed previously by Hamilton et al.[31]

With the recognition that our sampling of the esophageal mucosa around the tumor was suboptimal and that false-negative diagnosis of residual intestinal metaplasia was possible, we carefully assessed the literature. It was likely that pathologists existed who were more compulsive than our group about sampling the mucosa around tumors. We found six studies from highly reputable centers that reported a total of 163 esophagectomies for adenocarcinoma in which 161 (98.8%) had residual intestinal metaplasia (Table 12.4).

This was evidence that the main reason why residual intestinal metaplasia is reported as being absent in a significant number of cases in most studies is sampling error. This includes our study. The alternative explanations of loss of goblet cells in dysplasia and overgrowth of intestinal metaplasia by the tumor to explain the absence of intestinal metaplasia around cancers are not necessary.

7.2.5 Summary

The evidence presented above should remove all doubts that the invariable progression of change leading to adenocarcinoma requires the intermediate step of intestinal metaplasia in CLE. The goblet cell, which is a terminally differentiated postmitotic cell, is not the target cell for carcinogens. The target cell for carcinogens is the proliferative cell in the deep region of the foveolar pit that has the molecular transformation that has resulted in its ability to differentiate into a goblet cell. The goblet cell is the marker that indicates the area of CLE that is susceptible to the action of carcinogens.

CLE without intestinal metaplasia is not susceptible to carcinogenesis.

However, things are never that simple. Riddell and Odze[9] are highly influential pathologists in North America. Their review paper resulted in a subtle change in the definition of Barrett esophagus in the AGA guidelines. From the prior definitive statement that Barrett esophagus was the presence of intestinal metaplasia in visible CLE, the wording became more convoluted.

In the discussion under the definition of Barrett esophagus: "Recent data suggest that cardia-type epithelium in the esophagus is also abnormal and may predispose to malignancy. The key unanswered clinical question for patients who have cardia-type epithelium in the distal esophagus is this: What is the risk of developing esophageal cancer? A great majority of studies on the risk of cancer in Barrett's esophagus have included patients with esophageal intestinal metaplasia either primarily or exclusively. Although cardia-type epithelium might be a risk factor for malignancy, the magnitude of that risk remains unclear."

> *Presently, there are insufficient data to make meaningful recommendations regarding management of patients who have solely cardia-type epithelium in the esophagus, and we do not recommend use of the term "Barrett's esophagus" for those patients. Based on this lack of data, it is justified not to perform endoscopic surveillance for patients solely with cardia-type epithelium in the esophagus.*

The new AGA definition of Barrett esophagus ("the condition in which any extent of metaplastic columnar epithelium that predisposes to cancer development replaces the stratified squamous epithelium that normally lines the distal esophagus") that is intended to open the door to the cancer risk of nonintestinalized cardiac epithelium in visible CLE may have

TABLE 12.4 Selected Esophagectomy Studies With the Most Complete Sampling to Demonstrate the Prevalence of Residual Intestinal Metaplasia (IM) in Patients With Esophageal Adenocarcinoma (EAC)

References	Place of Origin	No. of Patients With EAC	Number (%) With Residual IM
Ruol A et al.	Padova, Italy	26	25 (96.2%)
Skinner DB et al.	Chicago, IL	20	20 (100%)
Cameron AJ et al.	Rochester, MN	9	9 (100%)
Rosenberg JC et al.	Detroit, MI	9	9 (100%)
Paraf F et al.	Paris, France	67	66 (98.5%)
Van Sandick JW et al.	Amsterdam, Holland	32	32 (100%)

an unintended positive consequence. When the dilated distal esophagus becomes accepted as being the esophagus rather than the stomach, intestinal metaplasia therein will fit into the definition of Barrett esophagus because it is very likely at risk for progressing to adenocarcinoma.

8. INCREASED RISK FOR CARCINOGENESIS: DECISION FOR SURVEILLANCE

There is significant confusion about the presence of cancer risk and susceptibility of an epithelium to carcinogenesis.

8.1 Difference Between Increased Risk of Cancer and Susceptibility to Cancer

A patient with symptomatic GERD has an increased risk of adenocarcinoma irrespective of the epithelial lining of the esophagus. This risk increases with increasing severity and duration of reflux.[2]

The only patient with a zero risk of adenocarcinoma in the thoracic esophagus is the person who has zero reflux. As soon as the thoracic squamous epithelium is exposed to refluxed gastric juice, molecular changes occur that begins the long road to adenocarcinoma (see Chapter 15).

The totality of the sequence of cellular changes and the risk associated with future adenocarcinoma is as follows: normal squamous epithelium that has not been exposed to reflux (no risk)→squamous epithelium exposed to reflux (risk level 1)→cardiac epithelium (risk level 2)→intestinal metaplasia (risk level 3)→low-grade dysplasia (risk level 4)→high-grade dysplasia (risk level 5)→adenocarcinoma.

The increased risk for adenocarcinoma in the thoracic esophagus exists even if the patient with evidence of reflux does not have visible CLE. Certainly, the squamous epithelium that lines the esophagus in such a patient is not a susceptible cell type for carcinogens in gastric juice that are delivered to the esophagus by reflux. Squamous carcinoma has no association with GERD. Adenocarcinoma cannot arise in squamous epithelium; it must be preceded by columnar metaplasia.

Similarly, cardiac epithelium is also an epithelium at increased risk for future cancer because GERD causes cardiac epithelium to evolve with time into intestinal metaplasia in a significant number of patients. The risk is higher than for squamous epithelium because only one step is involved: visible CLE→intestinal metaplasia. However, the presence of an increased risk of future cancer in cardiac epithelium without intestinal metaplasia does not mean that the cells in cardiac epithelium are susceptible to carcinogenesis.

The question that needs to be answered is: “Can cardiac epithelium progress directly to adenocarcinoma without intestinal metaplasia as a necessary intermediate step?” In the previous section, I presented what I hope is strong evidence that the answer to this question is a vehement “no.”

8.2 Definition of Barrett Esophagus in Great Britain and Japan

The definition of Barrett esophagus in the guidelines of the British Society of Gastroenterologists does not require the presence of intestinal metaplasia in visible CLE.[41] In Japan, where the diagnosis of Barrett esophagus is made at endoscopy alone, patients with very short (<0.5 cm) segments of endoscopic Barrett esophagus are excluded if there are no other factors associated with typical Barrett esophagus.[49]

This is not necessarily a contradiction with the American position. Of course, nonintestinalized cardiac epithelium is at increased risk of future cancer. It is perfectly appropriate to recommend surveillance for all visible CLE. Progression of cellular change despite present treatment with PPIs will result in intestinal metaplasia in a variable percentage of patients depending on length of visible CLE. Visible CLE in our new method is the point or irreversibility in the GERD→adenocarcinoma sequence.

It is highly likely that any expansion of endoscopic surveillance to visible CLE will result in earlier diagnosis and improvement of survival in esophageal cancer. The British position of surveillance for all visible CLE will produce better results than the more restrictive American requirement of intestinal metaplasia as long as the indication and frequency of endoscopy for GERD is the same. Conversations I have had with British physicians involved in the management of GERD suggest, at an anecdotal level, that the frequency of endoscopy for GERD in the United Kingdom is much lower than in the United States.

The question is merely whether the expansion of surveillance to patients without intestinal metaplasia is cost-effective and within the present availability of resources. This it not likely because there will likely be a subset of patients whose visible CLE may not progress to intestinal metaplasia. If intestinal metaplasia is the only epithelium susceptible to carcinogenesis, those patients would have been subject to the cost and discomfort of surveillance without reason.

8.3 The British Rationale

The rationale for the British definition is presented by Gatenby et al.[42]

The stated aim of this study: "In the USA, detection of intestinal metaplasia is a requirement for enrolment in surveillance programmes for dysplasia and adenocarcinoma in columnar-lined oesophagus. In the UK, it is believed that the failure to detect intestinal metaplasia at index endoscopy does not imply its absence within the columnarized segment or that the tissue is not at risk of neoplastic transformation. The aim of this study was to investigate the factors predicting the probability of detection of intestinal metaplasia in the columnarized segment."

The opinion in the United Kingdom is that if sufficient biopsies are taken over an adequate period, then intestinal metaplasia will effectively always be demonstrated in visible CLE. This is possibly close to the truth for visible CLE >5 or even 3 cm, but certainly not true for short-segment visible CLE. Only 12% of UK clinicians undertaking surveillance in patients with visible CLE based this on the presence or absence of intestinal metaplasia even before the British Society of Gastroenterologists made this recommendation.[50]

The study by Gatenby et al.[42] was a retrospective observational cohort study of the natural history of CLE. Patients diagnosed with nondysplastic CLE at seven centers throughout the United Kingdom represented the study group. Researchers visited the centers and extracted information from the medical records. No interventions in patient care were undertaken as part of the study. The centers involved practiced their own surveillance endoscopy and biopsy protocols.

The finding of intestinal metaplasia was analyzed in 3568 biopsies from 1751 patients. Of these 2347 (65.8%) demonstrated intestinal metaplasia and 1221 (34.2%) did not. At logistic regression, intestinal metaplasia was found to be significantly correlated with male gender ($P<.001$), age at which the biopsy was taken, number of tissue samples, and first recorded CLE segment length. The mean length for CLE with and without intestinal metaplasia was 5.75 and 4.93 cm, respectively.

The study examined the development of intestinal metaplasia in CLE in 322 patients followed up for a median of 3.5 years. 54.8% of the 322 patients with CLE without intestinal metaplasia at the index endoscopy progressed to intestinal metaplasia or dysplasia at 3 years and 90.8% at 10 years (Table 12.5).

The rate of development of intestinal metaplasia, low-grade dysplasia, high-grade dysplasia, or adenocarcinoma is shown in Table 12.6. The hazard ratios (HRs) comparing visible CLE with and without intestinal metaplasia were not statistically different between the two groups for all combinations of progression to low-grade dysplasia, high-grade dysplasia, and adenocarcinoma.

TABLE 12.5 Survival Analysis of Detection of Intestinal Metaplasia in 322 Patients Initially Diagnosed With Columnar-Lined Esophagus Without Intestinal Metaplasia (IM)

Follow-up Period	6 Months	1 Year	2 Years	3 Years	5 Years	10 Years
Proportion developed IM	12.7%	22.6%	41.0%	54.8%	72.0%	90.8%
Number still under follow-up	313	265	173	123	57	7

TABLE 12.6 Number of the Patients in Cohort Who Developed Dysplasia and Adenocarcinoma (AC)

Initial Histology (Number)	Target Histology	Number (%) Developing Change
CLE IM− (322)	CLE IM+, LGD, HGD, or AC	217 (67.4%)
	LGD, HGD, or AC	49 (15.2%)
	HGD or AC	12 (3.7%)
	AC	10 (3.1%)
CLE IM+ (612)	LGD, HGD or AC	121 (19.8%)
	HGD or AC	29 (4.7%)
	AC	20 (3.2%)

CLE, columnar-lined esophagus; *HGD*, high-grade dysplasia; *IM*, intestinal metaplasia; *LGD*, low-grade dysplasia.

The authors state: "Of the 322 patients who did not have intestinal metaplasia on their initial histologic examinations, 10 developed adenocarcinoma (and a further 2 patients developed HGD). These patients would have been excluded from surveillance at the time of CLO diagnosis if the American College of Gastroenterology guidelines (which require that IM is demonstrated on biopsies to make a diagnosis of Barrett's esophagus and to allow enrolment in a surveillance program) were adhered to." This is correct.

They continue: "Consequently, these adenocarcinoma patients would likely have presented clinically with a poorer prognosis than would have been the case if they had been detected within the surveillance programmes." This is also true.

The British rationale is based on the assumption that there is no sampling protocol that can remove the possibility of significant false negative diagnosis of intestinal metaplasia. Gatenby et al.[42] state as such in their discussion: "If sufficient biopsy samples were taken over time, the vast majority of patients without IM would subsequently demonstrate IM." They also suggest that if intestinal metaplasia is a necessary criterion for surveillance, many patients with a false-negative diagnosis of intestinal metaplasia at the time of index endoscopy will be inappropriately denied surveillance. The last phrase in their paper is key: "… especially if the local practice is to take fewer biopsies at that index endoscopy."

The British rationale is valid in two circumstances: (1) if there is no biopsy protocol that can minimize the false-negative diagnosis of intestinal metaplasia in visible CLE and (2) if the biopsy protocol that is performed at the index endoscopy is inadequate.

In the previous section, I showed that there is a biopsy protocol that can effectively separate patients with and without intestinal metaplasia such that only those with intestinal metaplasia develop cancer. If this biopsy protocol is followed, it can identify a higher risk group (visible CLE with intestinal metaplasia) that will improve the cost-effectiveness of surveillance. This is the rationale of the American guidelines for including intestinal metaplasia as a requirement for surveillance.

However, if the biopsy protocol is inadequate in the patient with visible CLE, the probability of a false-negative diagnosis makes the British surveillance recommendation superior to the American.

8.4 The Adequacy of Biopsy Sampling in the United States

It is easy for me to state that the American position on surveillance is better than the British. I can point to the difference in the biopsy sampling in Gatenby et al.[42] and our paper[26] and convince the reader that British endoscopists do not biopsy their patients with visible CLE adequately. In Gatenby et al.[42] the number of biopsies taken was 3568 biopsies from 1751 patients. This was 2.04 per patient. The mean length for CLE with and without intestinal metaplasia was 5.75 and 4.93 cm, respectively. Our biopsy protocol would take 16–20 biopsies from a 5-cm segment of visible CLE. In our study, 85% of patients with 3, 4, and 5 cm CLE segments had intestinal metaplasia and all patients with CLE>5 cm had intestinal metaplasia at the index endoscopy. This suggests that the final rate of 90.8% of intestinal metaplasia in Gatenby et al.[42] after 10 years was likely close to the true rate of intestinal metaplasia at the index endoscopy. False-negative diagnosis was responsible for much of the "progression" from no intestinal metaplasia to intestinal metaplasia. It is easy to say that they should have biopsied more thoroughly.

One of the problems with the biopsy protocol in Gatenby et al.[42] is that it is impossible to detect true progression from cardiac epithelium to intestinal metaplasia. In England, it will be simply assumed that the second biopsy at a subsequent endoscopy showing intestinal metaplasia indicates that the first biopsy was a false-negative diagnosis.

The expert group that developed the Montreal document reached the following consensus on the matter of biopsy sampling in Statement 36[21]: "Multiple, closely spaced biopsies are necessary to characterize endoscopically suspected esophageal metaplasia." The evidence grade for this statement was high, but there was less than complete agreement among experts (four disagreed with the statement). This is not unexpected as intestinal metaplasia is not required for diagnosis of Barrett esophagus in England. If this was a panel of experts from England alone, disagreement would likely have been greater.

In their discussion of this statement, they state[21]: "Effective management of the risk for esophageal adenocarcinoma requires sensitive detection of intestinal-type metaplasia and high grade dysplasia…. The best researched biopsy protocol is 4-quadrant biopsies every 1 cm for circumferential metaplastic segments, which is substantially more sensitive than sampling at 2 cm intervals. This approach has been variably modified to include biopsies at the top of tongue-like metaplastic extensions. These onerous and usually expensive protocols are generally not accepted as best practice."

The last sentence drives practical biopsy methodology of visible CLE in the United States. With experts discouraging the effective biopsy protocol as "onerous, expensive, and not best practice" without providing a replacement protocol, community gastroenterologists have no standard. I see biopsy protocols that vary from two to three random biopsies to those that closely follow the Seattle protocol.

In the United States, a GERD patient who does not have intestinal metaplasia in biopsies taken from visible CLE is declared as "not having Barrett esophagus" and no surveillance or any follow-up endoscopy is recommended. This makes less sense than placing all these people under Barrett esophagus surveillance, as is the case in Great Britain.

Even with a perfect biopsy protocol, visible CLE without intestinal metaplasia should not be ignored. The indisputable fact is that a significant number of such patients will progress with time to intestinal metaplasia. If there is no follow-up endoscopy, these patients are at risk to progress through intestinal metaplasia that is not detected to advanced stage adenocarcinoma.

A reasonable compromise between the practice in Great Britain and the United States would be best. All patients should have a biopsy protocol designed to minimize false-negative diagnosis of intestinal metaplasia. This will permit the most cost-effective surveillance of those patients with intestinal metaplasia who are at immediate risk. Adequate biopsy is more cost-effective than commitment to long-term surveillance.

The patient with visible CLE without intestinal metaplasia should not enter a regular Barrett esophagus surveillance protocol, but there should be a recommendation for a repeat endoscopy at some point in the future, possibly in 5 years to see if they have progressed to intestinal metaplasia.

There is also the possibility of stratifying biopsy protocol and surveillance by the length of the visible CLE segment. In patients with a segment length of >3 cm, the patient has a >85% probability of intestinal metaplasia.[20] They can enter surveillance without the need for proving intestinal metaplasia. Where the segment length is <3 cm, an extensive biopsy protocol requires fewer total biopsies. The decision for surveillance can be made in those patients by the presence of intestinal metaplasia.

9. RISK INDICATORS FOR ESOPHAGEAL ADENOCARCINOMA

Esophageal adenocarcinoma is unusual because it has a sequence of multiple cellular changes leading to adenocarcinoma: squamous epithelium → cardiac metaplasia → intestinal metaplasia → increasing dysplasia and adenocarcinoma.

As the patient progresses from one stage to the next, the risk of future cancer increases. The point at which the esophageal epithelium becomes susceptible to carcinogenesis is when intestinal metaplasia occurs in cardiac epithelium. Within a segment of CLE of any length, the probability is that carcinogenesis is limited to that area of CLE marked by the presence of intestinal metaplasia. The areas lined by squamous, cardiac, and oxyntocardiac epithelia are probably not susceptible to carcinogenesis.

The only patient proven to be at risk for carcinogenesis at the time of the assessment is the patient with intestinal metaplasia in visible CLE. A patient who has intestinal metaplasia limited to the dilated distal esophagus has an unknown risk at the present time, largely because it has not been studied. However, it is likely that intestinal metaplasia is the precursor lesion for adenocarcinoma of the dilated distal esophagus.

The only metaplastic columnar epithelium that is not at risk for progression to adenocarcinoma is oxyntocardiac epithelium. The reason is that the conversion of cardiac epithelium to oxyntocardiac epithelium represents an endpoint of metaplastic change. Oxyntocardiac epithelium does not progress to intestinal metaplasia. The areas of CLE lined by oxyntocardiac epithelium are not at risk for cancer.

Each stage of the GERD → adenocarcinoma sequence can be correlated with increasing LES damage. With the new ability to measure LES damage histologically, a new avenue of study of this sequence becomes possible (Table 12.7). This is likely to correlate better with cancer risk than any other factor because LES damage is the cause of reflux and the risk of cancer is likely to be most closely associated with the severity of reflux.

The demographic and clinical factors recognized as being associated with esophageal adenocarcinoma are increasing age; male gender; Caucasian race; the presence, severity, and duration of GERD symptoms; genetic factors; obesity (measured as BMI or waist circumference); smoking; alcohol; yet unknown carcinogens; and Barrett esophagus. The association with male gender, race, and severity of GERD symptoms is less pronounced in adenocarcinoma of the dilated distal esophagus in patients with no visible CLE.

Even the thought that acid-suppressive drugs used in the treatment of GERD can promote adenocarcinoma is taboo.

Of all these associated factors, the only one that influences management at the present time is the diagnosis of Barrett esophagus. None of the other factors represent an indication for endoscopy. All patients with GERD are treated empirically with PPIs with the goal of symptom control. The main indication for endoscopy is failure of PPI therapy to control symptoms. Nason et al.[5] showed that there is a negative correlation between control of symptoms with PPIs and the prevalence of Barrett esophagus; that is, patients who are well controlled with PPI therapy have a higher prevalence of Barrett esophagus. The criterion of treatment failure as an indication for endoscopy in GERD patients is not designed to maximize the probability of detection of Barrett esophagus.

TABLE 12.7 Steps in the Progression of Changes From Normal to Esophageal Adenocarcinoma; Cancer Risk Is for the Moment in Time That Assessment Is Made. Progression From One Stage to the Next Can Alter Cancer Risk

Category	LES Damage	Squamous Epithelium	CLE Without IM	IM	Reflux	Cancer Risk
Normal	0	Normal	0	0	0	0
Compensated LES damage	>0–15 mm	Damage in abdLES	Limited to DDE	0	0	0
Mild GERD	15–25 mm	Damage in abdLES	Limited to DDE	0	Mild	0
Mild GERD	15–25 mm	Damage in abdLES	Limited to DDE	In DDE	Mild	Low (only in DDE)
Severe GERD	>25 mm	Erosive esophagitis	In DDE and above GEJ	0	Severe	0
Severe GERD	>25 mm	Erosive esophagitis	In DDE and above GEJ	Above GEJ only	Severe	+ (only in thoracic esophagus)
Severe GERD	>25 mm	Erosive esophagitis	In DDE and above GEJ	Above GEJ and in DDE	Severe	+ (in thoracic esophagus and DDE)

abdLES, abdominal lower esophageal sphincter; *CLE*, columnar-lined esophagus; *DDE*, dilated distal esophagus; *GEJ*, gastroesophageal junction; *GERD*, gastroesophageal reflux disease; *IM*, intestinal metaplasia; *LES*, lower esophageal sphincter.

The factors that increase the risk of adenocarcinoma can operate at any stage of GERD. Anything that promotes LES damage to the point of producing reflux to an amount that induces the development of visible CLE will increase the risk of cancer. Anything that promotes the development of intestinal metaplasia in both the thoracic esophagus and the dilated distal esophagus will further increase the risk of cancer. Identifying the precise point in the sequence that a risk factor operates is helpful in both understanding the progression to cancer and developing preventive methods (Table 12.8).

When considering factors associated with esophageal adenocarcinoma, it is difficult to know when an associated factor actually acts. For example, acid-reducing therapy will reduce squamous epithelial damage, but likely promotes intestinal metaplasia in cardiac epithelium. The impact of acid-reducing therapy will therefore depend on the point in the disease where acid-reducing therapy begins.

A few years ago, I had come to the audience question part of a long lecture on reflux disease at the International Academy of Pathology in Hong Kong. An extraordinarily smart pathologist in the audience asked: "Dr. Chandrasoma, if what you are saying is correct, will it be possible to prevent esophageal adenocarcinoma if we put PPIs in tap water so that everyone was acid suppressed from birth?" I answered: "Yes, it is. In fact, in Hong Kong, the incidence of esophageal adenocarcinoma is low because it is very likely that early *H. pylori* gastritis in the population actually does exactly what you suggest."

9.1 Age

In GERD patients, critical events occur at given ages dependent largely on the status of the LES. If the LES is competent, there is no reflux. As the LES becomes increasingly incompetent, reflux gets progressively worse and cellular changes progress. It is the variation of the rate of progression of LES damage that results in differences in the ages at which GERD events occur in different people. This is a critical element in the new method of managing GERD that is based on the new ability to assess LES damage by pathology.

When viewed as an inexorably progressive disease from the point of view of LES damage, it becomes apparent that the disease begins early in life and passes through a phase of compensated LES damage during which the dilated distal esophagus increases in length from 0–15 mm before the onset of symptoms.

In 70% of the population, LES damage never passes 15 mm. The LES remains competent, reflux does not occur, and the patient never develops cellular changes of GERD in the thoracic esophagus.

The rate of progression of LES damage will determine onset of reflux and symptoms. The age of onset of symptoms of GERD varies greatly, but is most common around age 40 years.

Barrett esophagus (visible CLE with intestinal metaplasia), which is an event associated with further LES damage, is commonly diagnosed over age 50 years, and adenocarcinoma is most common after age 60 years.

TABLE 12.8 Point of Maximum Impact of Factors Associated With the Causation of Esophageal Adenocarcinoma

Category	Normal Squamous→RE	RE→CE	CE→IM	IM→Carcinogenesis
Age	+	+	+	+
Gender (male)	++	++	0	0
Race (Caucasian)	++	++	0	0
Genetic/familial	0?	0?	+	+
Smoking/alcohol	+	+	0	+?
Obesity	++	++	0?	0
GERD	++	++	+	+
Helicobacter pylori	Decrease	Decrease	0	0
Acid	++	+	Decrease?	0
Bile	+	+?	+?	++
Acid-reducing therapy	Decrease	Decrease	++	+?
Carcinogen	0	0	0	++

CE, cardiac epithelium; *GERD*, gastroesophageal reflux disease; *IM*, intestinal metaplasia; *RE*, reflux esophagitis; *?*, questionable.

9.2 Gender

There is a marked male predominance in the incidence of adenocarcinoma of the thoracic esophagus. This is paralleled by a male predominance of symptomatic GERD and Barrett esophagus.

The strong male predominance is much less pronounced in adenocarcinoma of the dilated distal esophagus that is not associated with visible CLE.

The reasons for this are poorly understood. There is some way in which the female LES maintains competence better than the male or some way in which the squamous epithelium of the thoracic esophagus resists damage when exposed to reflux.

9.3 Race

GERD is predominantly a disease of white persons. Again, the reasons are poorly understood. It is probable that it has something to do with dietary habits. Because of this, GERD has a marked variation in geographic prevalence; the incidence and prevalence are much higher in the United Kingdom, Western Europe, the United States, and Australia than in Asia.

Part of the reason for this may be the population prevalence of *H. pylori* infection. There is an inverse correlation between *H. pylori* prevalence and esophageal adenocarcinoma. There is also an inverse correlation between gastric adenocarcinoma prevalence (which is related to *H. pylori* prevalence) and that of esophageal adenocarcinoma.

9.4 Genetic Factors

Inherited genetic factors play a very limited role in esophageal adenocarcinoma. A family history of Barrett esophagus and/or esophageal adenocarcinoma is regarded with some concern, but this is not sufficient to recommend screening. No genetic abnormality that can be tested for is available for screening patients for Barrett esophagus.

9.5 Smoking and Alcohol

Smoking and alcohol are much greater risk factors for squamous carcinoma of the esophagus. In many studies, a statistical association has been found between smoking and alcohol consumption and adenocarcinoma. Patients with GERD are commonly advised not to smoke and consume alcohol, but this is more because they may act as reflux triggers than to prevent adenocarcinoma.

9.6 Obesity

One of the most dramatic epidemics of the past 50 years has been an increase in obesity in the entire world population. This began in the Western world but is spreading to Asian countries. This epidemic has been primarily the result of greater availability of food. Food has also become cheaper, and relationship between obesity and opulence has become less. Poverty, particularly in the Western world, is as likely to be associated with obesity as undernutrition.

Fast-food restaurants have proliferated throughout the world, dispensing super-sized meals with high calorie and fat content. The large volume causes gastric distension; the high calorie intake causes obesity. The high fat content causes obesity and decreases the rate of gastric emptying, compounding the distending effect of high volume.

The millennial generation has also developed a new obsession with food. "Foodies," or people who chase new creations in restaurants in a manner similar to oenophiles who chase good wine, abound. Restaurants cater to this new market with evermore delicious food. Food intake is now driven more by taste than the primeval need to reduce hunger. It is becoming the new drug of high society. The food addiction has resulted in the worldwide obesity epidemic.

Obesity has an association with GERD, Barrett esophagus, and esophageal adenocarcinoma. The maximum point of impact of obesity is most likely to be in promoting LES damage. The repetitive overeating that causes obesity is likely to put increasing stress on the LES.

A review of the literature shows evidence that the risk of adenocarcinomas of the esophagus and gastric cardia increases in a dose-dependent manner with increasing BMI.[51,52] This relationship is enhanced when BMI 20 years earlier[52] or at age 40 years[51] is considered. Locke et al.[53] reported that BMI>30 kg/m^2 was independently associated with frequent (at least weekly) symptoms of GERD such as heartburn and regurgitation.

A recent study from the Norfolk and Norwich University Hospital in England[54] attempted to address the timing of risk factors in the development of Barrett esophagus and esophageal adenocarcinoma. They followed a cohort of 24,068 people aged 39–79 years recruited between 1993 and 1997 into the prospective EPIC-Norfolk Study until December 2008.

104 participants were diagnosed with Barrett esophagus and 66 with esophageal adenocarcinoma during the follow-up period.

For a diagnosis of Barrett esophagus, the endoscopy report required the endoscopist to document characteristic appearance of CLE and the pathologist to report columnar metaplasia of intestinal, gastric, or mosaic types. As is the case in England, the definition of Barrett esophagus does not require intestinal metaplasia. (I will use the more accurate term visible CLE to avoid confusion.)

The authors acknowledge that there is likely to be an underdiagnosis of visible CLE because the people did not undergo routine endoscopy. The median age at diagnosis of visible CLE in the 104 patients was 67.0 years (range 41.3–84.4 years) and 79.8% were men and the diagnosis was made after a median follow-up time of 6.2 years (range 1.1–13.3 years). The median length of visible CLE was 5 cm (range 1–10 cm). The columnar epithelium was classified as intestinal in 70%, gastric in 9%, mosaic in 16%, and not reported in 5%. Seventy-seven percent of patients had reflux symptoms. Five of the incident visible CLE cases progressed to adenocarcinoma during follow-up.

Sixty-six patients developed esophageal adenocarcinoma (four others were excluded for incomplete data). The median age at diagnosis was 73 years (range 52–86 years) and the median time of follow-up before diagnosis was 6.2 years (range 0.6–11.8 years). Eighty-seven percent of the patients in whom the information was available had the tumor involving the GEJ.

A BMI >23 kg/m^2 was associated with a greater risk of visible CLE (HR 3.73; 95% CI 1.37–10.16 compared to BMI 18.5–23). Neither smoking nor alcohol consumption had a significant association. For esophageal adenocarcinoma, risk was significantly associated with only the highest BMI category (>35 vs. 18.5–23: HR 4.95, 95% CI 1.11–22.17).

The authors conclude: "Obesity may be involved early in carcinogenesis." The term "carcinogenesis" is used here in the broad sense. In reality, what the study shows is that obesity is associated with the cellular changes that result in the generation of the susceptible cell (visible CLE with intestinal metaplasia in 70%) in the thoracic esophagus. The study suggests that obesity is not associated with the actual carcinogenetic mechanism wherein adenocarcinoma is induced in the intestinal epithelium.

If it is true that obesity is involved in the early steps of carcinogenesis, it can be best investigated by evaluating the relationship of BMI with the types of epithelia seen in the esophagus. Robertson et al.[25] showed that asymptomatic volunteers with central obesity had longer segments of cardiac epithelium than those with no central obesity.

In a study from our unit, we studied the association of BMI with the presence of different epithelial types in the esophagus.[55]

Among patients who presented to the USC Foregut Surgery service and had systematic biopsies of the distal esophagus, GEJ region, and the stomach between 1998 and 2000, we identified 174 Barrett esophagus patients (visible CLE with intestinal metaplasia), 333 patients with cardiac epithelium and no intestinal metaplasia, and 274 controls who had neither intestinal or cardiac epithelium. The controls had squamous epithelium, oxyntocardiac epithelium, or gastric oxyntic epithelium in their biopsies. All patients with Barrett esophagus had cardiac epithelium without intestinal metaplasia in some of the biopsy specimens.

BMI was calculated on weight and height abstracted from the patient's chart; these were measured at the time of endoscopy.

We determined the length of metaplastic epithelium by the interval between the most distal and proximal biopsy levels that contained cardiac and intestinal epithelia. Biopsies containing metaplastic epithelium at a minimum of two measured levels were required for determination of the length of CLE. The length of CLE was measured in 103 patients.

Age and gender distribution of the patients with Barrett esophagus and cardiac metaplasia and controls are shown in Table 12.9.

Barrett esophagus patients tended to be older and males. Controls and those with cardiac epithelium had equal gender distribution contrasting with the male predominance in Barrett esophagus patients.

A strong dose–response relationship between BMI and Barrett esophagus was observed (P=.0004). The multivariate adjusted odds ratio for Barrett esophagus comparing obese (BMI>30 kg/m^2) to lean (BMI<22 kg/m^2) individuals was 3.3 (95% CI 1.6–6.7).

A substantial, although more modest, dose–dependent relationship was observed for BMI and risk of cardiac epithelium being present (P=.03). The multivariate adjusted odds ratio for obese (BMI>30 kg/m^2) versus lean (BMI<22 kg/m^2) individuals was 1.8 (95% CI 1.04–3.1).

Linear regression analysis showed that the length of CLE increased with increasing BMI (P=.04) in the 103 cases where this measurement was possible (86 Barrett esophagus and 17 cardiac epithelium).

In the discussions, Bu et al.[55] state: "Recently, evidence has been presented that cardiac metaplasia represents an intermediate stage between reflux and Barrett esophagus. The finding in this study that all patients with Barrett esophagus had cardiac epithelium and cardiac epithelium existed without Barrett esophagus strongly suggests that this is true. The complete reflux→adenocarcinoma sequence would then be as follows: reflux→cardiac metaplasia of squamous epithelium→intestinal metaplasia of cardiac epithelium or Barrett esophagus→increasing dysplasia→adenocarcinoma."

The data in the present study suggest that BMI not only predicts increased risk of adenocarcinoma but also increased risk of the precursor stages of GERD→cardiac metaplasia and Barrett esophagus. The data also showed a relationship

TABLE 12.9 Age and Gender Distribution of Patients With Barrett Esophagus (BE) [Cardiac Epithelium (CE) With Intestinal Metaplasia], CE, and Controls (With Squamous, Oxyntocardiac, and/or Gastric Oxyntic Epithelia)

Variables	BE (n = 174)	CE (n = 333)	Controls (n = 274)
Age (Years)			
18–34	10 (5.7%)	37 (11.1%)	34 (12.4%)
35–44	22 (12.6%)	63 (18.9%)	53 (19.3%)
45–54	42 (24.1%)	84 (25.2%)	77 (28.1%)
55–64	49 (28.2%)	69 (20.7%)	48 (17.5%)
65–74	34 (19.5%)	55 (16.5%)	34 (12.4%)
>75	17 (9.8%)	25 (7.5%)	28 (10.2%)
<55	74 (42.5%)	184 (55.3%)	164 (59.9%)
>55	100 (57.5%)	149 (44.7%)	110 (40.1%)
Gender			
Men	123 (70.7%)	169 (50.8%)	141 (51.5%)
Women	51 (29.3%)	164 (49.3%)	133 (48.5%)

between BMI and the length of CLE, which is also known to correlate with the prevalence of Barrett esophagus. This suggests that the primary effect of higher BMI is to increase the amount of reflux, resulting in visible CLE. The steps in the transformation of cardiac epithelium→Barrett esophagus→adenocarcinoma may have some other dominant etiologic factor.

The conclusion of the study is that "high BMI or obesity is a significant risk factor for development of cardiac metaplasia and Barrett esophagus. CLE segment length increases with increasing BMI. These results fit with BMI acting at an earlier time point in the reflux→adenocarcinoma sequence and enhance our understanding about the progression of this disease sequence."

9.7 Symptomatic Gastroesophageal Reflux Disease

The 1999 epidemiologic study in the *New England Journal of Medicine* by Lagergren et al.[2] provided powerful and undeniable evidence of the relationship between symptomatic GERD and adenocarcinoma of the esophagus.

In this nationwide, population-based, case–control study in Sweden, the authors interviewed 189 patients with esophageal adenocarcinoma and 252 patients with adenocarcinoma of the cardia (now classified by the AJCC as esophageal and not gastric) for information on the subjects' history of gastroesophageal reflux. Controls included 802 subjects from the general population and 167 patients with squamous carcinoma. The odds ratios were calculated by logistic regression.

Symptomatic gastroesophageal reflux was defined as the presence of heartburn, regurgitation, or both occurring at a frequency of at least once per week. To avoid reverse causality (i.e., to avoid collecting data on reflux caused by adenocarcinoma), the authors disregarded symptoms that had occurred less than 5 years before the interview. To further evaluate the effect of severity of symptoms, the symptoms were graded with scores for heartburn only (1 point), regurgitation only (1 point), heartburn and regurgitation combined (1.5 points), nightly symptoms (yes=2; no=0 points), frequency of symptoms (0 points=once per week, 1 point=2–6 times/week, 2 points=7–15 times/week, 3 points=more than 15 times/week). The location of the tumor and histologic type were verified by endoscopic data and histologic examination. The definition of adenocarcinoma of the cardia: "For a case to be classified as a cancer of the cardia, the tumor had to have its center within 2 cm proximal, or 3 cm distal, to the gastroesophageal junction." Esophageal adenocarcinomas had their epicenter >2 cm above the GEJ. This definition applied only to adenocarcinoma: "Squamous cell carcinomas were classified as esophageal even if the location was the gastric cardia" (Table 12.10).

Among patients with symptomatic reflux, compared with patients without symptoms, the odds ratios were 7.7 (95% confidence interval, 5.3–11.4) for esophageal adenocarcinomas. Persons with severe reflux (4.5 points or higher) had a risk

TABLE 12.10 Reflux Symptom Score and Duration of Reflux Symptoms Related to Odds Ratios for Adenocarcinoma of the Esophagus and Gastric Cardia

	Esophageal Adenocarcinoma (Number: 189)		Adenocarcinoma of Gastric Cardia (Number: 262)	
	Number (%)	Odds Ratio (95% CI)	Number (%)	Odds Ratio (95% CI)
Symptom Severity (Score)				
No symptoms	76 (40%)	1.0	187 (71%)	1.0
Score 1–2 points	10 (5%)	1.4 (0.7, 3.0)	30 (11%)	1.7 (1.0, 2.9)
Score 2.5–4 points	39 (21%)	8.1 (4.7, 16.1)	27 (10%)	1.8 (1.0, 3.2)
Score 4.5–6.5 points	64 (34%)	20.0 (11.6–34.6)	18 (7%)	2.8 (1.6, 5.0)
Symptom Duration				
No symptoms	76 (40%)	1.0	187 (71%)	1.0
<12 years	31 (16%)	7.5 (4.3, 13.5)	19 (7%)	1.6 (0.9, 2.9)
12–20 years	42 (22%)	5.2 (3.1, 8.6)	34 (13%)	1.8 (1.1, 2.9)
>20 years	40 (21%)	16.4 (8.3, 28.4)	22 (8%)	3.3 (1.8, 6.3)

CI, confidence interval.

of esophageal adenocarcinoma that was 20 times higher than asymptomatic persons. Among persons with reflux duration of more than 20 years and severe symptoms, the odds ratio was 43.5 (95% confidence interval, 18.3–103.5) as compared to asymptomatic persons.

Symptomatic reflux was also associated with a smaller increased risk of adenocarcinoma of the cardia; the odds ratio was 2.0 (95% confidence interval, 1.4–2.9). The risk increased with the increasing severity of symptoms and the duration of symptoms but not to the same extent as esophageal adenocarcinoma. Persons who reported both severe symptoms and long duration of symptoms had an odds ratio of 4.4 (95% confidence interval, 1.7–11.0) of developing adenocarcinoma of the cardia compared with asymptomatic persons.

The authors concluded that there was a strong and probably causal relationship between gastroesophageal reflux and esophageal adenocarcinoma and a weak relationship with adenocarcinoma of the cardia.

The study's conclusions regarding adenocarcinoma of the gastric cardia is based on the classification at the time that required tumors centered on either side of the junction be called gastric cardia (proximal gastric) cancers. It makes no sense that GERD can be causally associated with gastric cancer.

If one examines the definition used by Lagergren et al.[2] for adenocarcinoma of the gastric cardia (within 2 cm proximal and 3 cm distal to the GEJ), it is almost exactly concordant with the LES. The maximum extent of dilated distal esophagus reported is 28 mm distal to the endoscopic GEJ, and the thoracic segment of the GEJ, which is above the endoscopic GEJ, is estimated to be 15 mm in length. The adenocarcinomas of the "gastric cardia" in this study are actually adenocarcinomas arising in short-segment Barrett esophagus (within 2 cm proximal to the endoscopic GEJ) and in the dilated distal esophagus. The association with symptomatic GERD now becomes clear. Also concordant is that the data show that the presence and severity of symptoms is greater in the patients with adenocarcinoma of the esophagus compared to adenocarcinoma of the "cardia."

The other very significant finding in the study is that 76/189 (40%) patients with adenocarcinoma of the esophagus and 187/262 (71%) patients with adenocarcinoma of the cardia were asymptomatic by the study definition of symptomatic GERD. The presence of heartburn and/or regurgitation for less than once a week is likely to correlate with symptomatic GERD that is not "troublesome." If that is true, 263/451 (58%) patients with adenocarcinoma in both locations would have developed their cancer without ever falling within the Montreal definition of GERD.[21] They would not have reached the attention of a gastroenterologist with the capability of performing endoscopy. They had no chance because their progression from squamous epithelial damage→cardiac metaplasia→intestinal metaplasia→increasing dysplasia→adenocarcinoma would have gone undetected. The duration of progression from symptomatic GERD to adenocarcinoma is also interesting. 82/113 (73%) and 56/75 (75%) patients with symptomatic GERD who developed esophageal and "cardiac" adenocarcinoma had symptoms for more than 12 years. There is a long period between onset of symptoms of GERD and adenocarcinoma in patients who are symptomatic. This is the period of opportunity to prevent adenocarcinoma at least in the symptomatic population.

As clearly shown in the Lagergren et al.[2] study, the absence of GERD symptoms is no guarantee of immunity from esophageal adenocarcinoma. There is absolutely no way to predict which of the 10% of asymptomatic persons in the general population harbors Barrett esophagus and is at risk to develop adenocarcinoma in the future.

9.8 Acid-Reducing Therapy for Gastroesophageal Reflux Disease

At the start of this chapter, I explored the possibility that converting the esophageal milieu during a reflux episode from strong acid to weak acid is responsible for the induction of intestinal metaplasia in visible CLE. I will show in Chapter 15 that activation of CDX2 that likely induces intestinal metaplasia is promoted in a more alkaline pH milieu.

9.9 Carcinogens

An important unknown in the genesis of adenocarcinoma in GERD is the carcinogenic agent that is responsible for esophageal adenocarcinoma. There is evidence that the carcinogen is derived from bile that reaches the stomach by bile reflux. While bile is not believed to be implicated in gastric carcinogenesis (which is primarily related to *H. pylori*), many authorities believe that bile components are likely carcinogens for esophageal adenocarcinogenesis.[56] *H. pylori* infection has not been associated with esophageal adenocarcinoma.

Kauer et al.[57] reported that 60% of patients with GERD have duodenal admixture of the gastric contents that reflux into the esophagus. The severity of mucosal damage was increased in patients who refluxed both gastric and duodenal juice compared with patients who refluxed gastric juice alone.[57]

This 1997 study was undertaken to determine the concentration and type of bile acids refluxed into the esophagus. They used a portable, battery-powered pump that allowed ambulatory aspiration of esophageal contents. The aspiration tube was placed 5 cm above the upper border of the manometrically defined LES. Aspiration was measured in the upright (7 h), postprandial (2 h after a meal), and in the supine (8 h) position.

A total of 43 healthy volunteers (29 males; median age 32 years, range 19–48 years) and 37 patients (30 males; median age 52 years, range 21–67 years; GERD confirmed by abnormal esophageal acid exposure in a 24-h pH test) represented the study population.

32/37 (86%) patients with GERD showed at least trace amounts of bile during one or more collection periods compared to 25/43 (58%) of volunteers. The amount of bile aspirated during the total 17-h collection period was significantly greater in patients. The mean bile acid reflux rate (Table 12.11) during the three aspiration periods was significantly higher in patients.

On a molar basis, 60% of the total bile acids consisted of glycocholic acid, 16% glyco-deoxycholic acid, and 15% glycol-cheno-deoxycholic acid. Other bile acids constituted the remaining 9%.

The source of bile acids in the refluxate is the entry of duodenal contents into the stomach (duodenogastric reflux). This occurs in the majority of patients. Duodenogastric reflux is increased in patients who have had cholecystectomy.[58] Lagergren et al.[59] in a population-based cohort study of 345,251 patients followed up for a mean of 15 years reported an incidence of 126 new cases of esophageal adenocarcinomain patients with cholecystectomy, greater than expected (Standardized Incidence Ratio 1.29, 1.07–1.53). There was no increase in squamous carcinoma in this cohort and an unoperated cohort of patients with gallstones had no increased risk of adenocarcinoma.

There are little data on the effect of specific bile acids in their conjugated and unconjugated forms on esophageal squamous and metaplastic columnar epithelia. It is known that these acids can cause damage by a direct toxic effect on the cells. It has also been shown that bile acids can be concentrated in the cytoplasm of esophageal cells, thereby opening the possibility that they may be involved in cell interactions that can result in genetic changes.

There are marked differences in the behavior of bile acids depending on the pH of the solution in which they reside. Bile acids enter cells when they are in their nonionized lipophilic form, which occurs at a pH close to their pKa. After entering the cell, the bile acid is ionized because of the relatively high intracellular pH. This results in the bile acids becoming concentrated in the cell.

Under normal circumstances, deconjugated bile acids are rarely present in the stomach because the bacteria normally needed for deconjugation are not present in the acid environment. In patients with high gastric pH due to atrophic gastritis or acid-suppressive therapy, the removal of the acid environment may cause changes in the chemistry of bile acids. The possibility that PPI therapy changes bile acid chemistry in a way that might adversely impact its effects on esophageal epithelia in patients with duodenogastroesophageal reflux should be investigated.

Hu et al.[60] studied the molecular changes in cells exposed to bile acid derivatives. They state their objectives: "The presence of bile acids in the refluxed material has been consistently observed in patients with Barrett's esophagus, strongly suggesting that they are important in its pathogenesis. The exact molecular mechanisms underlying this intestinal metaplastic and/or differentiation process remains largely unknown."

Caudal-related homeobox 2 (CDX2) is a homeobox factor that plays an important role in the early differentiation and maintenance of intestinal epithelium. Immunohistochemical staining studies have recently confirmed that CDX2 protein is overexpressed in human Barrett's epithelium… The aim of this study was to investigate whether bile acid exposure can affect CDX2 mRNA expression in a variety of human esophageal cell lines.

Four human esophageal cell lines were used: (1) Het-1A, a human esophageal squamous epithelial cell line immortalized by the SV40 transfection; (2) SEG-1, a Barrett's esophageal adenocarcinoma cell line; (3) HKESC-1 and HKESC-2, esophageal squamous carcinoma cell lines.

TABLE 12.11 Amount of Bile Acid Reflux Rate (μmol/h) for Volunteers and Patients With Gastroesophageal Reflux During Postprandial, Upright, and Supine Periods. Values Are Reported as Mean ± SEM

	Postprandial	Upright	Supine	*P* Value
Volunteers (n = 43)	0.1 ± 0.04	0.02 ± 0.01	0.03 ± 0.01	<0.01[a]
Patients (n = 37)	3.7 ± 2.9	0.08 ± 0.03	0.4 ± 0.3	<0.04[b]
P value	<0.002	<0.007	<0.004	

[a]*Postprandial versus upright and supine.*
[b]*Postprandial versus upright.*

The four esophageal cell lines were exposed to 100, 300, and 1000 μM deoxycholic acid (DCA), chenodeoxycholic (CDC) acid, and glycocholic(GC) acid in serum-free medium for 1, 2, 4, 8, or 24 h, respectively. The cells were harvested immediately before bile acid exposure (t_0) and at each time point (t_1, t_2, t_4, t_8, t_{24}) and total RNA extracted. This was reverse-transcribed to cDNA, which was used for the polymerase chain reaction for real-time quantitative mRNA expression.

CDX2 mRNA was minimal to absent before bile acid exposure in all four cell lines. In the normal squamous epithelial cell line Hwt-1A, CDX2 expression was highly upregulated by 1000 μM DCA treatment for 8 h. In Barrett esophagus adenocarcinoma cells (SEG-1), CDX2 was highly upregulated in a dose- and time-dependent fashion. In esophageal squamous carcinoma cell lines (HKESC-1 and 2), CDX2 expression was highly upregulated by 1000 μM DCA exposure for 1 h and the upregulation maintained for the 2–24 h time points. CDX2 upregulation was similar after exposure to CDC acid in all the cell lines, although to a lesser extent than DCA. GC acid exposure had no measurable effect on CDX2 expression in any of the cell lines. Quantitation of CDX2 mRNA expression by real-time PCR confirmed the pattern of upregulation. The maximum induction of CDX2 expression (1973-fold increase) was seen in SEG-1 adenocarcinoma cells after 1000 μM DCA exposure for 24 h.

The results show that bile acid exposure can induce CDX2 expression in normal human esophageal and cancer cell lines. CDX2 mRNA expression increased dramatically with unconjugated bile acids DCA and CDC, but not with the conjugated bile acid GC.

The authors conclude that the findings in this study "support the role of bile acids in the pathogenesis of Barrett's esophagus and link the clinical evidence of a high prevalence of intraluminal bile acids in Barrett's esophagus to expression of the gene thought to be responsible for the phenotypic appearance of intestinal metaplasia."

These findings are interesting in that they show that bile acids induce CDX2 in human esophageal cells. Unfortunately, the experiment lacks the critical cell type that is involved in intestinal metaplasia in the reflux to adenocarcinoma sequence. Cardiac epithelium is not available for study because cell lines of cardiac epithelium have not been developed. The reason for this is that cardiac epithelium is believed to be a normal gastric epithelium and not recognized as being a metaplastic esophageal epithelium caused by GERD.

REFERENCES

1. Pohl H, Sirovich B, Welch G. Esophageal adenocarcinoma incidence: are we reaching the peak? *Cancer Epidemiol Biomark Prev* 2010;**19**:1468–70.
2. Lagergren J, Bergstrom R, Lindgren A, Nyren O. Symptomatic gastroesophageal reflux as a risk factor for esophageal adenocarcinoma. *N Engl J Med* 1999;**340**:825–31.
3. Kahrilas PJ, Shaheen NJ, Vaezi MF. American Gastroenterological Association medical position statement on the management of gastroesophageal reflux disease. *Gastroenterology* 2008;**135**:1380–2.
4. Peters JH, Clark GWB, Ireland AP, et al. Outcome of adenocarcinoma arising in Barrett's esophagus in endoscopically surveyed and nonsurveyed patients. *J Thorac Cardiovasc Surg* 1994;**108**:813–21.
5. Nason KS, Wichienkuer PP, Awais O, Schuchert MJ, Luketich JD, O'Rourke RW, Hunter JG, Morris CD, Jobe BA. Gastroesophageal reflux disease symptom severity, proton pump inhibitor use, and esophageal carcinogenesis. *Arch Surg* 2011;**146**:851–8.
6. Hvid-Jensen F, Pedersen L, Drewes AM, et al. Incidence of adenocarcinoma among patients with Barrett's esophagus. *N Engl J Med* 2011;**365**:1375–83.
7. Dent J, El-Serag HB, Wallander MA, et al. Epidemiology of gastro-oesophageal reflux disease: a systematic review. *Gut* 2005;**54**:710–7.
8. El-Serag HB, Sweet S, Winchester CC, et al. Update on the epidemiology of gastro-oesophageal reflux disease: a systematic review. *Gut* 2014;**63**:871–80.
9. Riddell RH, Odze RD. Definition of Barrett's esophagus: time for a rethink – is intestinal metaplasia dead? *Am J Gastroenterol* 2009;**104**:2588–94.
10. Duan L, Wu AH, Sullivan-Halley J, Bernstein L. Antacid drug use and risk of esophageal and gastric adenocarcinoma in Los Angeles County. *Cancer Epidemiol Biomark Prev* 2009;**18**:526–33.
11. Malfertheiner P, Nocon M, Vieth M, Stolte M, Jasperson D, Koelz HR, Labenz J, Leodolter A, Lind T, Richter K, Willich SN. Evolution of gastro-oesophageal reflux disease over 5 years under routine medical care – the Pro-GERD study. *Aliment Pharmacol Ther* 2012;**35**:154–64.
12. Allison PR, Johnstone AS, Royce GB. Short esophagus with simple peptic ulceration. *J Thorac Surg* 1943;**12**:432–57.
13. Barrett NR. The lower esophagus lined by columnar epithelium. *Surgery* 1957;**41**:881–94.
14. Hayward J. The lower end of the oesophagus. *Thorax* 1961;**16**:36–41.
15. Shaheen NJ, Falk GW, Iyer PG, Gerson L. ACG clinical guideline: diagnosis and management of Barrett's esophagus. *Am J Gastroenterol* November 3, 2015. http://dx.doi.org/10.1038/ajg.2015.322. Advance online publication.
16. Chandrasoma P, Makarewicz K, Ma Y, DeMeester TR. A proposal for a new validated histologic definition of the gastroesophageal junction. *Hum Pathol* 2006;**37**:40–7.
17. Chandrasoma PT. Histologic definition of gastro-esophageal reflux disease. *Curr Opin Gastroenterol* 2013;**29**:460–7.
18. Paull A, Trier JS, Dalton MD, Camp RC, Loeb P, Goyal RK. The histologic spectrum of Barrett's esophagus. *N Engl J Med* 1976;**295**:476–80.
19. Chandrasoma PT, Der R, Dalton P, Kobayashi G, Ma Y, Peters J, DeMeester T. Distribution and significance of epithelial types in columnar-lined esophagus. *Am J Surg Pathol* 2001;**25**:1188–93.

20. Chandrasoma PT, Der R, Ma Y, Peters J, DeMeester T. Histologic classification of patients based on mapping biopsies of the gastroesophageal junction. *Am J Surg Pathol* 2003;**27**:929–36.
21. Sharma P, McQuaid K, Dent J, Fennerty B, Sampliner R, Spechler S, Cameron A, Corley D, Falk G, Goldblum J, Hunter J, Jankowski J, Lundell L, Reid B, Shaheen N, Sonnenberg A, Wang K, Weinstein W. A critical review of the diagnosis and management of Barrett's esophagus: the AGA Chicago Workshop. *Gastroenterology* 2004;**127**:310–30.
22. Siewert JR, Stein HJ. Classification of adenocarcinoma of the oesophagogastric junction. *Br J Surg* 1998;**85**:1457–9.
23. American Joint Commission of Cancer Staging Manual. *Esophagus and esophagogastric junction*. 7th ed. 2010.
24. Leers JM, DeMeester SR, Chan N, Ayazi S, Oezcelik A, Abate E, Banki F, Lipham JC, Hagen JA, DeMeester TR. Clinical characteristics, biologic behavior, and survival after esophagectomy are similar for adenocarcinoma of the gastroesophageal junction and the distal esophagus. *J Thorac Cardiovasc Surg* 2009;**138**:594–602.
25. Robertson EV, Derakhshan MH, Wirz AA, Lee YY, Seenan JP, Ballantyne SA, Hanvey SL, Kelman AW, Going JJ, McColl KE. Central obesity in asymptomatic volunteers is associated with increased intrasphincteric acid reflux and lengthening of the cardiac mucosa. *Gastroenterology* 2013;**145**:730–9.
26. Chandrasoma P, Wickramasinghe K, Ma Y, DeMeester T. Adenocarcinomas of the distal esophagus and "gastric cardia" are predominantly esophageal carcinomas. *Am J Surg Pathol* 2007;**31**:569–75.
27. Association of Directors of Anatomic and Surgical Pathology. Recommendations for reporting of resected esophageal adenocarcinomas. *Am J Surg Pathol* 2000;**31**:1188–90.
28. Haggitt RC. Adenocarcinoma in Barrett's esophagus: a new epidemic? *Hum Pathol* 1992;**23**:475–6.
29. Pohl H, Welch G. The role of overdiagnosis and reclassification in the marked increase of esophageal adenocarcinoma incidence. *J Nat Cancer Inst* 2005;**97**:142–6.
30. Devesa SS, Blot WJ, Fraumeni JF. Changing patterns in the incidence of esophageal and gastric carcinoma in the United States. *Cancer* 1998;**83**:2049–53.
31. Hamilton SR, Smith RRL, Cameron JL. Prevalence and characteristics of Barrett esophagus in patients with adenocarcinoma of the esophagus or esophagogastric junction. *Hum Pathol* 1988;**19**:942–8.
32. McClave SA, Boyce Jr HW, Gottfried MR. Early diagnosis of columnar lined esophagus: a new endoscopic diagnostic criterion. *Gastrointest Endosc* 1987;**33**:413–6.
33. Spechler SJ, Zeroogian JM, Antonioli DA, Wang HH, Goyal RK. Prevalence of metaplasia at the gastroesophageal junction. *Lancet* 1994;**344**:1533–6.
34. Kulig M, Nocon M, Vieth M, Leodolter A, Jaspersen D, Labenz J, Meyer-Sabellek W, Stolte M, Lind T, Malfertheimer P, Willich SN. Risk factors of gastroesophageal reflux disease: methodology and first epidemiological results of the ProGERD study. *J Clin Invest* 2004;**57**:580–9.
35. Chandrasoma PT, Der R, Ma Y, Dalton P, Taira M. Histology of the gastroesophageal junction. An autopsy study. *Am J Surg Pathol* 2000;**24**:402.
36. Sarbia M, Donner A, Gabbert HE. Histopathology of the gastroesophageal junction. A study on 36 operation specimens. *Am J Surg Pathol* 2002;**26**:1207–12.
37. Marsman WA, van Sandyck JW, Tytgat GNJ, ten Kate FJW, van Lanschot JJB. The presence and mucin histochemistry of cardiac type mucosa at the esophagogastric junction. *Am J Gastroenterol* 2004;**99**:212–7.
38. Haggitt RC, Tryzelaar J, Ellis FH, et al. Adenocarcinoma complicating columnar epithelium-lined (Barrett's) esophagus. *Am J Clin Pathol* 1978;**70**:1–5.
39. Spechler SJ, Sharma P, Souza RF, Inadomi JM, Shaheen NJ. American Gastroenterological Association medical position statement on the management of Barrett's esophagus. *Gastroenterology* 2011;**140**:1084–91.
40. Chandrasoma P, Wijetunge S, DeMeester S, Ma Y, Hagen J, Zamis L, DeMeester T. Columnar-lined esophagus without intestinal metaplasia has no proven risk of adenocarcinoma. *Am J Surg Pathol* 2012;**36**:1–7.
41. Fitzgerald RC, di Pietro M, Ragunath K. British Society of Gastroenterology guidelines on the diagnosis and management of Barrett's oesophagus. *Gut* 2014;**63**:7–42.
42. Gatenby PAC, Ramus JR, Caygill CPJ, Shepherd NA, Watson A. Relevance of the detection of intestinal metaplasia in non-dysplastic columnar epithelium. *Scand J Gatreoenterol* 2008;**43**:524–30.
43. Kelty CJ, Gough MD, van Wyk G, et al. Barrett's oesophagus: intestinal metaplasia is not essential for cancer risk. *Scand J Gastroenterol* 2007;**42**:1271–4.
44. Liu W. Metaplastic esophageal columnar epithelium without goblet cells shows DNA content abnormalities similar to goblet cell containing epithelium. *Am J Gastroenterol* 2009;**104**:816–24.
45. Hahn H, Blount PL, Ayub K, et al. Intestinal differentiation in metaplastic, non-goblet columnar epithelium in the esophagus. *Am J Surg Pathol* 2009;**33**:1006–15.
46. Takubo K, Aida J, Naomoto Y, et al. Cardiac rather than intestinal-type background in endoscopic resection specimens of minute Barrett adenocarcinoma. *Hum Pathol* 2009;**40**:65–74.
47. Smith J, Garcia A, Zhang R, DeMeester SR, Vallone J, Chandrasoma P. Intestinal metaplasia is present in most if not all patients who have undergone endoscopic mucosal resection for esophageal adenocarcinoma. *Am J Surg Pathol* 2016;**40**:537–43.
48. Theisen J, Stein HJ, Dittler HJ, Feith M, Moebius C, Kauer WKH, Werner M, Siewert JR. Preoperative chemotherapy unmasks underlying Barrett's mucosa in patients with adenocarcinoma of the distal esophagus. *Surg Endosc* 2002;**16**:671–3.
49. Ogiya K, Kawano T, Ito E, et al. Lower esophageal palisade vessels and the definition of Barrett's esophagus. *Dis Esophagus* 2008;**21**:645–9.
50. Smith MA, Maxwell-Armstrong CA, Welch NT, Schofield JH. Surveillance for Barrett's oesophagus in the UK. *Br J Surg* 1999;**86**:276–80.

51. Wu AH, Wan P, Bernstein L. A multiethnic population-based study of smoking, alcohol and body size and risk of adenocarcinoma of the stomach and esophagus. *Cancer Causes Control* 2001;**12**:721–32.
52. Lagergren J, Bergstrom R, Nynen O. Association between body mass index and adenocarcinoma of the esophagus and gastric cardia. *Ann Intern Med* 1999;**130**:883–90.
53. Locke III GR, Tally NJ, Fett SL, Zinsmeister AR, Melton LJ. Risk factors associated with symptoms of gastroesophageal reflux. *Am J Med* 1999;**106**:642–9.
54. Yates M, Cheung E, Luben R, Igali L, Fitzgerald R, Khaw K-T, Hart A. Body mass index, smoking, and alcohol and risks of Barrett's esophagus and esophageal adenocarcinoma: a UK prospective cohort study. *Dig Dis Sci* 2014;**59**:1552–9.
55. Bu X, Ma Y, Der R, DeMeester T, Bernstein L, Chandrasoma PT. Body mass index is associated with Barrett esophagus and cardiac mucosal metaplasia. *Dig Dis Sci* 2006;**51**:1589–94.
56. Kauer WKH, Peters JH, DeMeester TR, Feussner H, Ireland AP, Stein HJ, Siewert RJ. Composition and concentration of bile acid reflux into the esophagus in patients with gastroesophageal reflux disease. *Surgery* 1997;**122**:874–81.
57. Kauer WKH, Peters JH, DeMeester TR, Ireland AP, Bremner CG, Hagen JA. Mixed reflux of gastric juice is more harmful to the esophagus than gastric juice alone: the need for surgical therapy re-emphasized. *Ann Surg* 1995;**222**:525–33.
58. Kunsch S, Neesse A, Huth H, Steinkamp M, Klaus J, Adler G, Gress TM, Ellenrieder V. Increased duodeno-gastro-esophageal reflux (DGER) in symptomatic GERD patients with a history of cholecystectomy. *Z Gastroenterol* 2009;**47**:744–8.
59. Lagergren J, Mattsson F. Colecystectomy as a risk factor for esophageal adenocarcinoma. *Br J Surg* 2011;**98**:1133–7.
60. Hu Y, Williams V, Gellersen O, Jones C, Watson TJ, Peters JH. The pathogenesis of Barrett's esophagus: secondary bile acids upregulate intestinal differentiation factor CDX2 expression in esophageal cells. *J Gastrointest Surg* 2007;**11**:827–34.

Chapter 13

Progression of GERD at the Clinical Level

Gastroesophageal reflux disease (GERD) is a chronic progressive disease with two faces: (1) a face that is manifested by squamous epithelial injury that is largely responsible for pain and reversed by acid suppression (Fig. 13.1) and (2) a more lethal columnar epithelial face that is not prevented by acid suppression and manifests as Barrett esophagus (Fig. 13.2) and adenocarcinoma (Fig. 13.3). Beginning in the 1950s, GERD has slowly transformed itself from the first to the second face (Table 13.1).[1]

The medical establishment has not adjusted to this change. It is still treating GERD like it was a disease caused by acid. It addresses only one face of GERD while the other has exploded out of control before our eyes.

It is time to put all our effort into changing the objectives of management. It is no longer sufficient to have the sole objective be symptom control and maintenance of quality of life. Total concentration toward achieving this objective leads to failure, with up to 30% of GERD patients becoming incompletely controlled and their quality of life disrupted, another significant number being treated long term with proton pump inhibitors (PPIs) without a diagnosis of GERD, and 20,000 people developing adenocarcinoma annually in the United States.

GERD must be addressed at its root cause, which is lower esophageal sphincter (LES) damage. LES damage is responsible for both iterations of GERD; its control will prevent progression of both severe squamous epithelial injury, and the occurrence and complications of columnar metaplasia.

In the 1950s, GERD progressed in the absence of effective acid-suppressive agents. This was manifested largely by uncontrollable squamous epithelial damage causing reflux esophagitis→increasing erosive esophagitis→severe ulceration and strictures.[2] This pathway created untold misery but was not lethal.

Acid-reducing drug treatment has reversed the progression of squamous epithelial damage. Symptoms that caused untold misery in the 1950s have been largely controlled; the quality of life for GERD patients has greatly improved.

It is the columnar metaplasia iteration of GERD that has raised its ugly head over this period. From being nonexistent (the first reported case was by Morson in 1952[3]), esophageal adenocarcinoma has undergone a sevenfold increase from 1973 to 2006.[1] In the 1950s CLE was a medical curiosity believed to be a harmless congenital heterotopia of gastric epithelium in the esophagus. Now it is a killer disease.

Acid-reducing agents used in the treatment of GERD exert their effect by alkalinizing gastric juice.[4] Because the present goal of treatment is control of symptoms, PPIs are considered to be highly successful in managing patients with GERD. Acid-reducing drugs decrease severity of heartburn to near zero in a majority and provide some measure of control in most patients.

Seemingly, GERD is "cured" because cure is defined as improvement in the quality of life. GERD is commonly advertised by the medical community and perceived by the population at large as being caused by acid and cured by adequate suppression of acid secretion.

This is a dangerous and fallacious viewpoint. It permits the illusion of cure while the critical sequence of cellular changes that lead to adenocarcinoma progress unbeknownst to the patient and physician. Unlike in the 1950s, the acid-suppressed patient's probability of progressing to columnar metaplasia→intestinal metaplasia→dysplasia→adenocarcinoma has increased.

The majority (70%) of GERD patients do not develop either failure of PPIs to control symptoms adequately or adenocarcinoma during their lifetime. However, a significant minority does. 30% of patients will have varying degrees of failure of symptom control and become dissatisfied with their quality of life. This has been called "disruptive GERD" and has a high cost (see Chapter 2).

Approximately 10% of the population develops Barrett esophagus. Annually, ~20,000 patients in the United States will develop adenocarcinoma of the thoracic and dilated distal esophagus, a lethal cancer.

Progression to treatment failure, Barrett esophagus, and adenocarcinoma occurs while the patient is being treated with PPIs. There is no attempt to prevent these complications. A cure that does nothing to prevent the major lethal complication

GERD. http://dx.doi.org/10.1016/B978-0-12-809855-4.00013-0

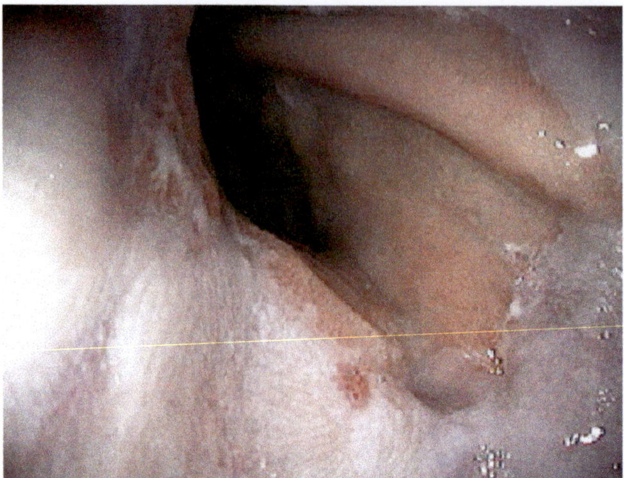

FIGURE 13.1 Erosive esophagitis. Dr. Martin Riegler's description: "Antegrade endoscopic image of the SCJ with esophagitis, LA grade B, in the 6 and 9 o'clock position with fibrin covering the surface of the SCJ." Histology: squamous epithelium and cardiac epithelium at the SCJ (squamocolumnar junction); oxyntocardiac epithelium distal to this. *Photograph courtesy of Dr. Martin Riegler, Vienna, Austria.*

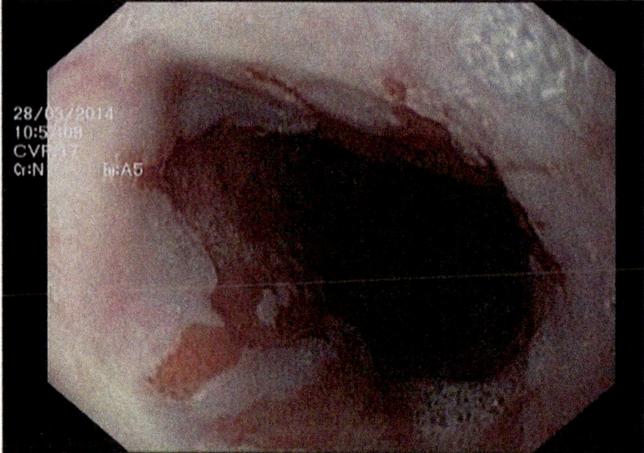

FIGURE 13.2 Long segment of visible columnar-lined epithelium in the esophagus showing replacement of squamous epithelium with tongues of pink columnar epithelium. Biopsy showed intestinal metaplasia. *Photograph courtesy of Dr. Martin Riegler, Vienna, Austria.*

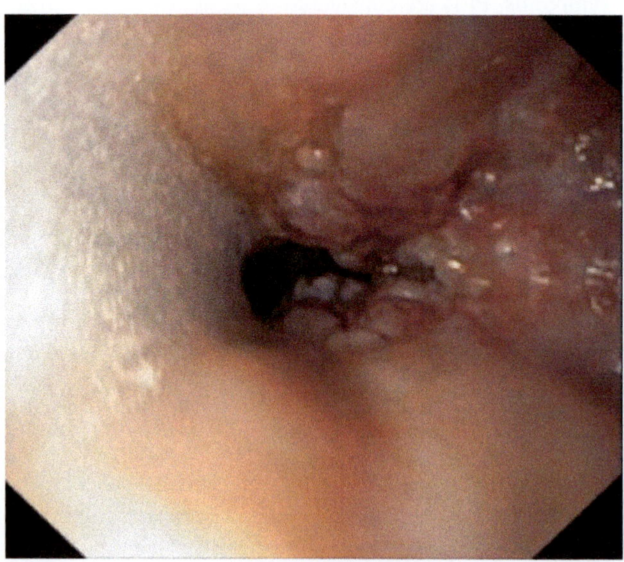

FIGURE 13.3 Endoscopic view showing mucosal nodularity in a patient undergoing endoscopy for refractory gastroesophageal reflux disease. Biopsy showed adenocarcinoma. *Photograph courtesy of Dr. Martin Riegler, Vienna, Austria.*

TABLE 13.1 The Two Faces of Gastroesophageal Reflux Disease (GERD) and Its Change From the 1950s to the Present

Manifestation of GERD	1950s	Present
Incidence of GERD	Unknown, but not rare	Increasing; 30% of population
1. Squamous Injury		
Heartburn	Uncontrollable	Fully controlled in 56%
Quality of life	Miserable	Satisfactory in 70%; Less than satisfactory in 30%
Erosive esophagitis	Severe; out of control	Improved; healable in 85%
Deep ulcers with hemorrhage	Common	Rare
Strictures with dysphagia	Common	Rare
Squamous carcinoma	No	No
2. Columnar Metaplasia		
Visible CLE	Present; prevalence unknown	Present; common
Length of visible CLE	Longer	Shorter
Intestinal metaplasia	Rare	Common; increasing
Adenocarcinoma	Almost nonexistent First reported case in 1952	Increased sevenfold from 1973 to 2006; 20,000/year in the United States
Location of adenocarcinoma	More proximal	More distal
Mortality From GERD	Very low	Increased sevenfold since 1973
3. Extra Esophageal GERD	Not recognized	Increasingly recognized

CLE, columnar-lined esophagus.

of GERD is no cure at all. Acid reduction does not decrease the rate of progression of LES damage. It does not stop or reduce reflux.[5] The natural progression of LES damage is expressed fully.

GERD is a disease caused by progressive mechanical degradation of the LES. It is treated like a disease that is caused by acid. By the time we recognize GERD as a disease, the LES damage has reached an advanced stage. Nothing can restore the mechanically damaged LES. The horse has left the barn.

1. CATEGORIZATION OF GASTROESOPHAGEAL REFLUX DISEASE

The present understanding and the management algorithm with heavy dependence on empiric PPI therapy without any diagnostic testing divide patients with GERD into the following practical categorical groups:

1. Patients who never develop symptoms that define GERD.
2. Patients who have GERD and remain under control for the entirety of their lives with long-term acid-reducing therapy without developing any problems.
3. Patients, either with or without GERD symptoms, either well controlled or not, who develop dysphagia resulting from advanced stage adenocarcinoma. They have never had an endoscopy. The cellular progression from normal to advanced adenocarcinoma over a period of decades has been missed with no attempt to find it (Fig. 13.4). This group represents 85% of people who develop esophageal adenocarcinoma.
4. Patients whose symptoms are not completely controlled by PPI therapy. These patients suffer a decline in quality of life, with symptoms such as nocturnal reflux, interference with normal eating and sleeping, chronic cough, and hoarseness. They frequently require dose escalation and ultimately reach a point where a physician declares treatment failure. It is at this point that endoscopy is indicated.[6] Endoscopy may show no abnormality (nonerosive reflux disease, NERD), erosive esophagitis (Fig. 13.1), visible CLE, Barrett esophagus (Fig. 13.2), or, rarely, adenocarcinoma (Fig. 13.3).

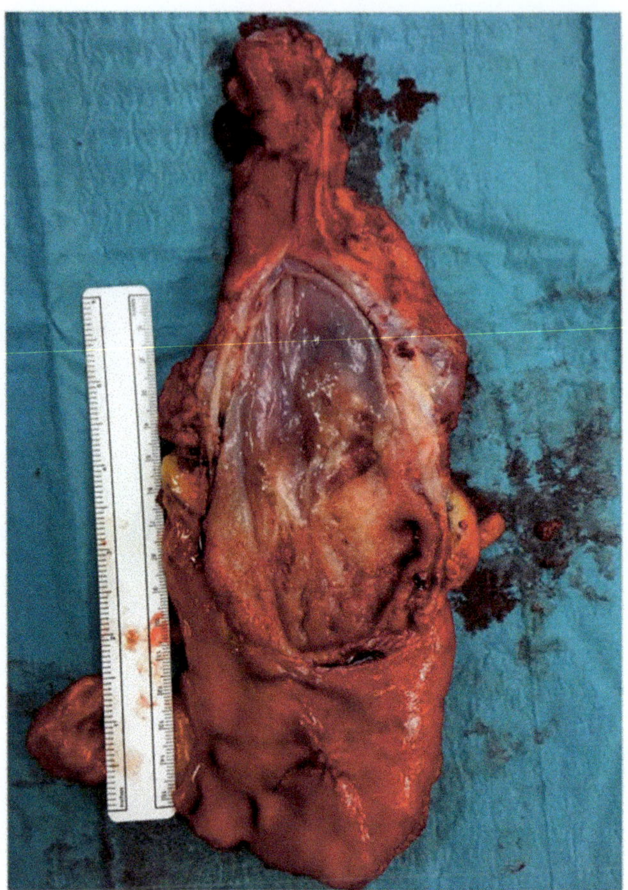

FIGURE 13.4 Esophagectomy specimen showing an advanced adenocarcinoma in the distal esophagus. This is seen as a large bulky mass obstruction in the esophagus. The patient presented with dysphagia. *Photograph courtesy of Dr. Martin Riegler, Vienna, Austria.*

5. Patients with Barrett esophagus diagnosed at endoscopy and biopsy. These patients are placed on an endoscopic surveillance program to detect dysplasia and adenocarcinoma. 15% of all esophageal adenocarcinomas are detected in this surveillance program. These are early cancers (Fig. 13.5) that are effectively treated presently with endotherapy and have a much better survival than overall esophageal adenocarcinoma.
6. Patients who opt for endoscopic or surgical procedures to augment or repair their LES. Nissen fundoplication is effective in decreasing reflux and there is evidence that it may decrease cancer risk, but it does not completely prevent adenocarcinoma. The percentage of GERD patients undergoing LES augmenting procedures is very small: in the United States ~20,000 patients per year.

This categorization of GERD patients determines the level of care that is delivered to these patients. The majority of patients with easily controlled GERD are treated in a pharmacy or in the primary care setting. Patients who progress to treatment failure are referred to a gastroenterologist. If they develop dysphagia due to adenocarcinoma, they quickly migrate to the world of oncologic surgeons and physicians who deliver chemotherapy and radiation.

The only patients who benefit from the present management algorithm of GERD from the point of view of mortality are the few patients who are detected with early cancer within a Barrett esophagus surveillance program.

2. IS GASTROESOPHAGEAL REFLUX DISEASE A CATEGORICAL OR PROGRESSIVE DISEASE?

The above six groups (those who have no GERD, those who are well controlled with drugs, poorly controlled with drugs, have Barrett esophagus, are treated by LES augmenting procedures, and who develop cancer) appear categorical in the sense that each group tends to be separate problems that only rarely move from one group to the next. Much of the literature reflects this categorization (Table 13.2).

In the American Gastroenterological Association Institute Technical Review on the Management of GERD,[6] the natural history of GERD was debated: "Two potential paradigms for viewing the natural history of GERD exist. In the first, GERD

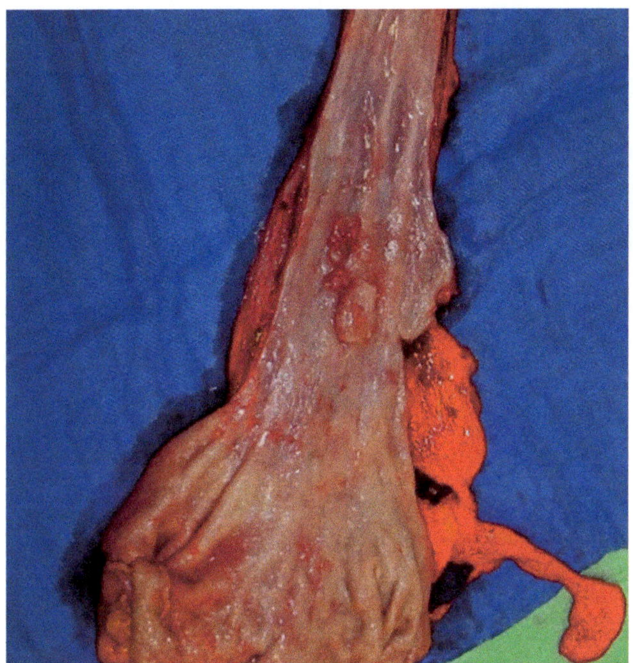

FIGURE 13.5 Early adenocarcinoma of the distal esophagus detected during surveillance for Barrett esophagus. *Photograph courtesy of Dr. Martin Riegler, Vienna, Austria.*

TABLE 13.2 Categories of Gastroesophageal Reflux Disease (GERD) in the General Population Based on How They Are Managed

	Self Medicate	Family Practitioner	Internist	Gastroente rologist	GERD Specialist	GERD Surgeon
OTC PPIs	+	+	+	+	+	+
Prescription PPIs	–	→ +	+	+	+	+
Endoscopy	–	–	–	→ +	+	+
pH/impedance, manometry	–	–	–	+	+	+
EMR/ ablation	–	–	–	–	→ +	+
TIF, Stretta	–	–	–	–	+	+
Linx/ antireflux surgery	–	–	–	–	–	→ +

The *arrows* point to the progression of GERD. However, this progression is masked in the eye of the patient and physician. Patients who self-medicate with acid-reducing drugs purchased over the counter (OTC) at the pharmacy (number determinable only by estimation of the use of OTC acid-suppressive drugs) likely move to the care of a physician infrequently. Most stay self-diagnosed and medicated. Patients who move out of the next three categories of increasing severity represent severe GERD treated by gastroenterologists specialized in the disease that involves only about 10% of patients with the disease. This is the GERD that receives most of the academic attention and literature.
EMR, *endoscopic mucosal resection; PPI*, proton pump inhibitor; *TIF*, transoral incisionless fundoplication.

is viewed as a progressive disease such that, in the absence of effective intervention, today's patient with nonerosive disease becomes tomorrow's patient with erosive disease, who then becomes a candidate for the development of Barrett's esophagus. This 'spectrum of disease' approach has been contrasted with the view that GERD may be a disease with phenotypically discrete 'categories', such as nonerosive disease, erosive esophagitis, and Barrett's esophagus. In this phenotypically preordained view, conversion from one disease manifestation to another is distinctly unusual, and subjects generally stay in their initial category."

When asked the question whether GERD is a categorical disease or a progressive disease, most expert gastroenterologists will agree that it is a chronic progressive disease. However, a significant number of expert gastroenterologists and probably a majority of primary care physicians will opine that it is categorical.

The above categorization of GERD is largely caused by the lack of availability of a method of reliable objective of assessment of GERD. The dependence on symptoms for diagnosis and assessment of severity and control of GERD by quality of life questionnaires masks true progression. Diagnostic tests used at present such as endoscopy, pH and impedance testing, and manometry are highly specific but not adequately sensitive for early diagnosis.

Endoscopy is delayed till the time of treatment failure. Without endoscopy, the only categories of GERD are "controlled with drugs" and "refractory to drugs to a variable extent." Primary care physicians treat the first category; gastroenterologists treat the second.

It is when endoscopy is performed in the second category that entities such as erosive esophagitis, Barrett esophagus, and adenocarcinoma appear. Gastroenterology journals are replete with studies on these patients. All these studies generally begin with the results of endoscopy. These categories exist only in the peripheral vision of primary care physicians who treat the majority of GERD patients.

3. GASTROESOPHAGEAL REFLUX DISEASE IS ALWAYS A PROGRESSIVE DISEASE FROM THE PERSPECTIVE OF LOWER ESOPHAGEAL SPHINCTER DAMAGE

It is the definition of GERD by symptoms and its treatment with acid suppression that creates the above categories of GERD. It is an artificial categorization. It divides the responsibility of management into its categories and prevents management of the disease in its entirety.

Progression of GERD as defined by its cause, LES damage, is inexorable and not affected in any positive manner by present medical therapy. The progression of LES damage is allowed to express its natural history without any attempt at prevention except in the rare patient who has a sphincter augmentation procedure.

The conflict as to whether GERD is categorical or progressive resolves when the progression of GERD is appreciated as a progression of LES damage at varying rates in different individuals. In the scenario of LES damage, progression of the disease is inexorably progressive in all people. The endpoint of the disease is based on the rate of progression, the ensuing LES damage, and the life span of the individual (Table 13.3). I will explore LES damage and its progression more fully in Chapter 16.

In Table 13.3, the progressive shortening of the abdominal LES is shown at rates varying from 1 to 10 mm/decade of life, beginning with an assumed abdominal LES length of 35 mm.

The areas in green represent a functional abdominal LES length of 20 mm or more, which predicts a competent LES without reflux (patient with no GERD). At this length, the LES can withstand a dynamic shortening of the LES of 10 mm with a heavy meal without reaching the critical point where the LES is likely to fail (i.e., <10 mm).

The orange zone is a residual abdominal LES length of 10–20 mm, which predicts intermittent LES failure, usually postprandial, with mild GERD (easily controlled with PPI therapy). The red zone is an abdominal

TABLE 13.3 Changes With Age of the Functional Residual Length of the Abdominal Lower Esophageal Sphincter (LES) Assuming That the Original Length at Maturity Is 35 mm, That LES Damage Begins at Age 15 Years, and That LES Damage Has a Linear Progression Over the Long Term

Rate of LES Damage (mm/decade)	At 25 Years (in mm)	At 35 Years (in mm)	At 45 Years (in mm)	At 55 Years (in mm)	At 65 Years (in mm)
1	34	33	32	31	30
2	33	31	29	27	25
3	32	29	26	23	20
4	31	27	23	19	15
5	30	25	20	15	10
6	29	23	17	11	5
7	28	21	14	7	0
8	27	19	11	3	0
9	26	17	8	0	0
10	25	15	5	0	0

The abdominal LES lengths in green represent lengths at which the LES is likely to be competent. The lengths in orange represent an LES that is susceptible to failure with gastric distension (i.e., at risk of postprandial reflux). The lengths in red represent an LES that is below the length at which LES failure occurs; reflux occurs at rest and with minimum provocation.

LES <10 mm, which predicts frequent LES incompetence and severe reflux (treatment failure, Barrett esophagus, and adenocarcinoma).

The artificial separation of GERD into clinical categories is entirely dependent on different rates of progression of LES damage. The severity of untreated GERD from both squamous and columnar epithelial standpoint correlates with the amount of reflux into the esophagus, which in turn depends on the amount of LES damage. PPI therapy reverses squamous epithelial changes but not progression of LES damage, the amount of reflux, or the cellular progression to adenocarcinoma.

Medical therapy is doomed to continued failure if the objective of treatment is preventing adenocarcinoma. The present management algorithm would be perfect if not for the occurrence of adenocarcinoma. We have treated it as perfect despite the occurrence of adenocarcinoma, simply watching as its numbers have increased each year over the past 50 years.

4. IMPACT OF OVERWHELMING NUMBERS IN GASTROESOPHAGEAL REFLUX DISEASE

GERD exists as an uncontrollable problem largely because of the massive numbers of people who have the disease (Fig. 13.6). In terms of prevalence, GERD dwarfs all other chronic diseases. Any attempt at controlling major problems of GERD will fail unless these numbers are taken into account.

Categorization of GERD has largely evolved to accommodate GERD patients at different levels of care in a manner that does not overwhelm resources that are available in the best way possible. GERD patients are kept away from the expensive and limited resources of gastroenterologists and endoscopy centers until the disease has progressed to significant failure. Management reacts to complications rather than being proactive in preventing complications. This is a prescription to the failed state of GERD management that exists today. It must change.

A large number of people (30% of the adult population of the United States, which is approximately 75–80 million) who develop heartburn are induced by the highly effective commercials of the pharmaceutical companies to go to the nearby pharmacy for relief. Others will access the Internet where the information from even the American Gastroenterological Association site will advise them that no diagnostic testing is indicated and that the treatment is to control symptoms with acid-reducing drugs.[6]

At the pharmacy, the person with heartburn finds that a plethora of acid-reducing agents from simple acid neutralizers to the most powerful PPIs can be purchased over the counter (Figs. 13.7 and 13.8). Many of these people control their symptoms effectively by self-medication. Their belief that heartburn is a nuisance that can interfere with their eating and is easily controlled with a pill is confirmed. With a little luck, they never need a consultation with the medical profession.

If the patient consults a general practitioner, an internist, or a gastroenterologist, either because self-medication has failed or he/she desires physician control, the result is similar. The recommended treatment algorithm is to use empiric PPI therapy. If the symptoms resolve, the patient is placed on maintenance acid-reducing drugs as needed to prevent recurrence. The majority of patients with GERD are well controlled in the long term by such empiric therapy.

To general practitioners and internists, GERD is a very common disease easily controlled with the miraculous acid-suppressive drugs. This is the only category of GERD they see. This is 70% or more of the GERD population (56 million in the United States).

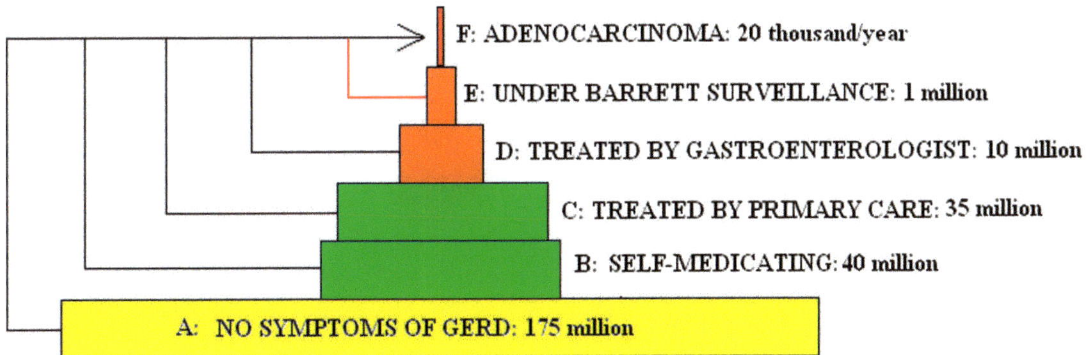

FIGURE 13.6 Approximate numbers of people with different categories of gastroesophageal reflux disease (GERD) in the United States. 70% of the adult population (~175 million) have no symptoms of GERD. The other 30% with GERD are estimated at 40 million persons self-medicating with over-the-counter acid-reducing agents as needed, 35 million under treatment for GERD in the primary care setting with prescribed proton pump inhibitors. Gastroenterologists, who manage the more severe GERD patients, take care of 10 million with ~1 million under surveillance for Barrett esophagus. All these categories of people can progress to adenocarcinoma. While the percentage of people developing cancer decreases from top to bottom, only 10%–15% of people developing adenocarcinoma are detected during Barrett surveillance (*red line*).

FIGURE 13.7 Shelf at the local pharmacy showing the prominent place given to the vast array of acid-reducing agents available for over-the-counter purchase. This photograph was taken in 2006. The most powerful drug at the time was Prilosec (omeprazole).

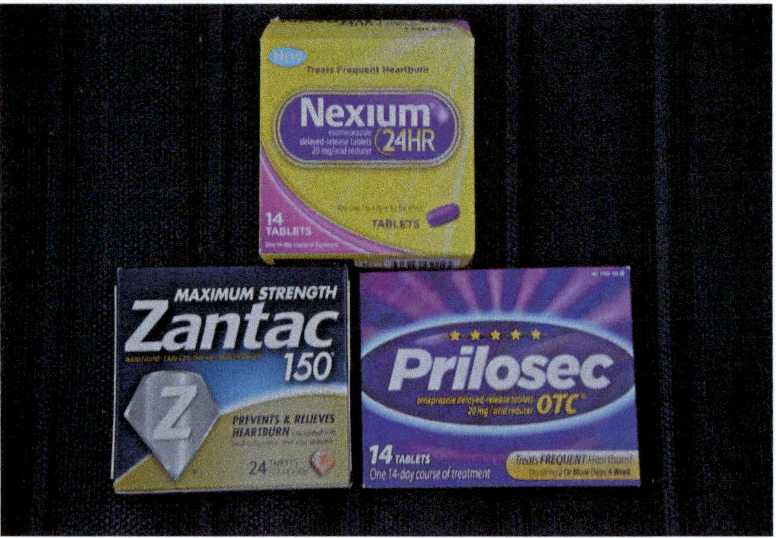

FIGURE 13.8 The three drugs given pride of place in 2016 include Nexium, Prilosec, and Zantac. The largest packages available over the counter in the United States for all three drugs have over 40 pills. There are no warnings on the label regarding the need to see a physician or that heartburn is associated with cancer.

Approximately 30% of patients (about 24 million in the United States) will experience symptoms that are not controlled to their satisfaction by PPI therapy. These patients will seek or be referred to a gastroenterologist who has a higher level of expertise in the treatment of GERD. The gastroenterologist will treat the patient with different PPIs at higher doses and more frequently. In many cases, adequate albeit incomplete symptom control is achieved.

TABLE 13.4 Endoscopic Categories of Gastroesophageal Reflux Disease (GERD)

	Biopsy	Biopsy Objective	Biopsy Result	Follow-Up Therapy	Endoscopic Surveillance
Normal endoscopy	No	n/a	n/a	Continue PPI therapy	No
Erosive esophagitis	No/yes	To exclude CLE	No CLE; reflux esophagitis	Continue PPI therapy	No
Erosive esophagitis	No/yes	To exclude CLE	CLE+ IM−	Continue PPI therapy	No (United States) Yes (England)
Erosive esophagitis	No/yes	To exclude CLE	CLE+ IM+	Continue PPI therapy	Yes
Visible CLE	Yes	To detect IM	IM−	Continue PPI therapy	No (United States) Yes (England)
Visible CLE	Yes	To detect IM	IM+	Continue PPI therapy	Yes

Endoscopic categorization of GERD patients is limited to the small percentage of patients with GERD whose symptoms fail to be controlled with self-medication or PPI therapy prescribed by primary care physicians. This is, according to guidelines of management of GERD, the main indication for endoscopy in GERD.
CLE, columnar-lined esophagus; *IM*, intenstinal metaplasia; *PPI*, proton pump inhibitor.

Failure of medical treatment is the indication for diagnostic testing. Let us assume that 30% of the patients referred to gastroenterologists (10 million) fail to be controlled adequately. The common first test is an endoscopy (Table 13.4). This will divide the patient into nonerosive disease, erosive esophagitis, or visible CLE. Biopsies will be taken only in patients with visible CLE; if intestinal metaplasia is found, they will be diagnosed with Barrett esophagus (Fig. 13.2).

Erosive esophagitis, an abnormal pH test (i.e., a pH<4 for >4.5% of the testing period), or a damaged LES (mean pressure <6 mmHg, total length <20 mm, or abdominal length <10 mm) confirms the diagnosis of GERD. However, normal endoscopy, a normal pH test, and normal manometry do not negate the diagnosis. The net result of this is that all patients continue to be treated with PPIs at whatever dose is needed within the maximum.

Many of these patients continue to have significant decline in the quality of life because of their symptoms. However, they have only two options: (1) continue PPI therapy and tolerate with dissatisfaction the lowered quality of life and (2) opt for sphincter augmentation procedures. The majority of gastroenterologists have little enthusiasm regarding sphincter repair procedures. As a result, these procedures are done relatively rarely. As numbers of these surgeries decline, the surgical expertise available declines. A vicious cycle has developed wherein antireflux surgery has an increasing probability of failure and complications.

Present management treats all symptomatic GERD patients with a one-size-fits-all algorithm that we know will result in some level of treatment failure in ~30% (24 million in the United States), an estimated 1–2.5 million in endoscopic surveillance programs for Barrett esophagus, and adenocarcinoma in 20,000 people per year in the United States.

5. WHO DEVELOPS ESOPHAGEAL ADENOCARCINOMA?

From a practical standpoint, only patients with GERD develop adenocarcinoma. GERD is the only disease that can cause the squamous epithelium→cardiac metaplasia→intestinal metaplasia→increasing dysplasia→adenocarcinoma sequence.

The 1999 epidemiologic study in the *New England Journal of Medicine* by Lagergren et al.[7] provided powerful and undeniable evidence of the relationship between symptomatic GERD and adenocarcinoma of the esophagus. This paper is reviewed in Chapter 12.

Symptomatic gastroesophageal reflux was defined in Lagergren et al.[7] as the presence of heartburn, regurgitation, or both occurring at a frequency of at least once per week. Symptoms were graded by severity and duration. Adenocarcinomas were divided into esophageal [epicenter >2 cm proximal to the endoscopic gastroesophageal junction (GEJ)] and cardiac (epicenter within 2 cm proximal to 3 cm distal to the GEJ). The category of adenocarcinoma of the cardia includes esophageal adenocarcinoma arising in the distal thoracic esophagus and the dilated distal esophagus.

Among patients with symptomatic reflux, compared with patients without symptoms, the odds ratios were 7.7 (95% confidence interval, 5.3–11.4) for esophageal adenocarcinomas. Persons with severe reflux (4.5 points or higher) had a risk of esophageal adenocarcinoma that was 20 times higher than asymptomatic persons. Among persons with reflux duration of more than 20 years and severe symptoms, the odds ratio was 43.5 (95% confidence interval, 18.3–103.5) as compared to asymptomatic persons.

Symptomatic reflux was also associated with a smaller increased risk of adenocarcinoma of the cardia; the odds ratio was 2.0 (95% confidence interval, 1.4–2.9). The risk increased with the increasing severity of symptoms and the duration of symptoms but not to the same extent as esophageal adenocarcinoma. Persons who reported both severe symptoms and long duration of symptoms had an odds ratio of 4.4 (95% confidence interval, 1.7–11.0) of developing adenocarcinoma of the cardia compared with asymptomatic persons.

The authors concluded that there was a strong and probably causal relationship between gastroesophageal reflux and esophageal adenocarcinoma and a weak relationship with adenocarcinoma of the cardia.

The other very significant finding in the study is that 76/189 (40%) of patients with adenocarcinoma of the esophagus and 187/262 (71%) patients with adenocarcinoma of the cardia were asymptomatic by the study definition of symptomatic GERD. The presence of heartburn and/or regurgitation for less than once a week is likely to correlate with symptomatic GERD that is not "troublesome."

If that is true, 263/451 (58%) of patients with adenocarcinoma in both locations would have developed their cancer without ever falling within the Montreal definition of GERD. They would not have reached the attention of a gastroenterologist with the capability of performing endoscopy. They had no chance because their progression from squamous epithelial damage→cardiac metaplasia→intestinal metaplasia→increasing dysplasia→adenocarcinoma would have gone undetected.

The duration of progression from symptomatic GERD to adenocarcinoma is also interesting. 82/113 (73%) and 56/75 (75%) of patients with symptomatic GERD who developed esophageal and "cardiac" adenocarcinoma had symptoms for more than 12 years. There is a long period between onset of symptoms of GERD and adenocarcinoma in patients who are symptomatic. This is the period of opportunity to prevent adenocarcinoma at least in the symptomatic population.

As clearly shown in the study by Lagergren et al.,[7] the absence of GERD symptoms is no guarantee of immunity from esophageal adenocarcinoma. There is absolutely no way to identify the ~10% of asymptomatic persons in the general population who harbor Barrett esophagus and are at risk to develop adenocarcinoma in the future.

Patients whose symptoms are well controlled with PPI therapy are also not protected from progression to Barrett esophagus and, therefore, adenocarcinoma. The Progression of GERD (Pro-GERD) study[8] showed that 25% of GERD patients being treated for symptom control converted from no visible CLE to visible CLE within 5 years and that this was significantly associated with regular PPI therapy. Nason et al.[9] showed that people who had better symptom control with PPIs had a higher risk of Barrett esophagus than those who were not well controlled.

Both asymptomatic persons and GERD patients whose symptoms are well controlled with drugs have no indication for endoscopy. The vast majority of these people have no problem during their life. The adenocarcinoma risk in the population that has never had endoscopy is miniscule. However, this population that has never had an endoscopy accounts for 85% of the 20,000 patients who develop esophageal adenocarcinoma annually in the United States. They present with advanced stage disease that has a high probability of a lethal outcome (Fig. 13.4). Because of the large numbers involved, a "miniscule" percentage risk means ~17,000 people annually in the United States (Table 13.5).

The entire establishment treating GERD minimizes the risk of adenocarcinoma in patients with GERD. In fact, the experts who produced the consensus definition of GERD in the Genval Workshop in 1999[10] introduced the phrase "after adequate reassurance of the benign nature of their symptoms." How does an odds ratio that reaches 43.5 translate into "benign"?

The first time the risk of adenocarcinoma enters the mind of a treating physician is when endoscopy and biopsy has resulted in a diagnosis of Barrett esophagus. Then, the patient enters an endoscopic surveillance program. The presently cited risk of cancer in a patient with Barrett esophagus is 0.2% per year, a decrease from the earlier number of 0.5% per year.[11]

The goal of surveillance is to detect dysplasia (Figs. 13.9 and 13.10) and early adenocarcinoma. It has been shown that, despite some practical limitations, patients who develop adenocarcinoma while in a surveillance program for Barrett esophagus have earlier stage cancers that are more frequently amenable to endotherapy without the need for esophagectomy and have a better overall survival rate.[12]

Unfortunately, as the enthusiasm for surveillance in Barrett esophagus declines, the intervals for endoscopy increase, and a significant number of patients who develop adenocarcinoma do so with tumors that, although "early" in that they do not cause dysphagia, require esophagectomy because of submucosal invasion (Fig. 13.5).

TABLE 13.5 Risk of Esophageal Adenocarcinoma in the Different Clinical Categories of Gastroesophageal Reflux Disease Expressed in Percentage Risk and Absolute Numbers

Category	% in Population	Number in Population	% Risk of Cancer	Number of People Getting Cancer	Cancer Stage
Asymptomatic[a]	70%	175 million	Mininscule	8000	Advanced
Symptomatic	30%	75 million	Extremely low	12000	3000 early; rest advanced
Symptomatic; self medicating	50%	40 million	Extremely low	4000	Advanced
Symptomatic; physician-managed No endoscopy	50%	35 million	Very low	4000	Advanced
Symptomatic; endoscopy = No BE	75%–90% of endoscopies	7–9 million	Low	1000	Advanced
Symptomatic endoscopy = BE	10%–25% of endoscopies	1–2.5 million	0.2%– 0.5% per year	3000	Early

[a]Definition of asymptomatic (Lagergren et al.): heartburn and/or regurgitation at least once in a week. Orange: symptomatic patients with no indication for surveillance; Green: symptomatic patients under surveillance. Percentage risk, which is commonly used to assess risk of adenocarcinoma, tends to underestimate risk because it ignores the denominator, which is the population at risk. Number of adults in the United States=250 million. Note: Numbers are all approximate. It is assumed that the total number of esophageal cancers per year is 20,000 and 85% occurs in people who have never had an endoscopy. *BE*, Barrett esophagus.

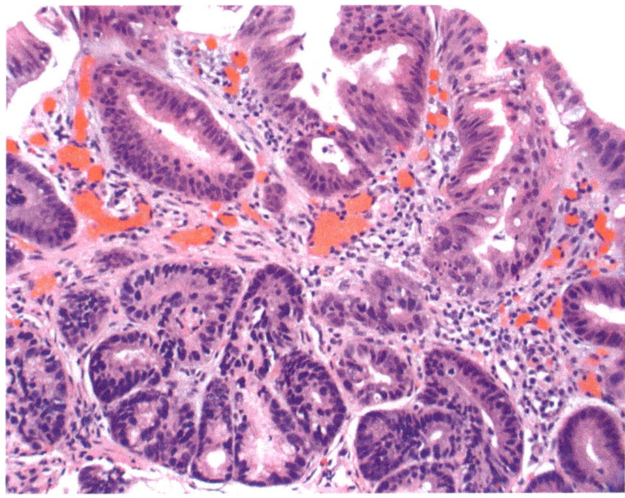

FIGURE 13.9 Low- and high-grade dysplasia in Barrett esophagus.

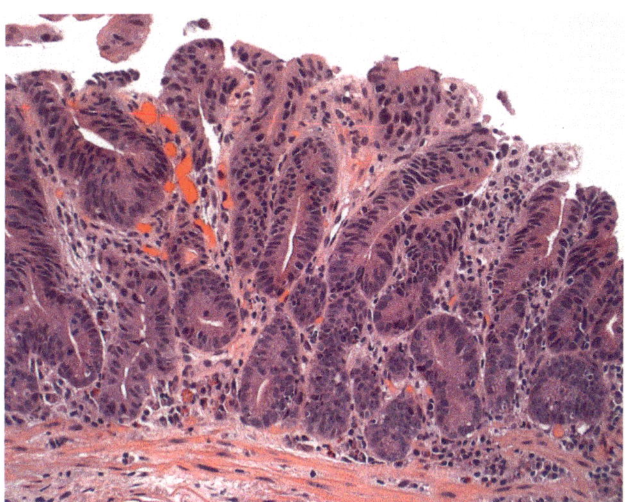

FIGURE 13.10 High-grade dysplasia in Barrett esophagus.

There is no known method to prevent the occurrence of adenocarcinoma in patients with Barrett esophagus. PPI therapy has been shown to both decrease and increase the risk of cancer in different studies.[13,14] Antireflux surgery is not regarded at present as a cancer-preventive operation in a patient with Barrett esophagus.[15]

Ablation of Barrett esophagus with either radio frequency or cryotherapy is effective, when combined with acid suppression, in replacing the columnar-lined esophagus (CLE) with new squamous epithelium. This procedure, however, is only approved at present for dysplastic Barrett esophagus and early cancer. It is not cost-effective for nondysplastic Barrett esophagus. It also has a significant failure rate with recurrence of Barrett esophagus on long-term follow-up.

In practice, most general gastroenterologists who have a small number of patients with Barrett esophagus under their care rarely if ever encounter a patient with high-grade dysplasia or adenocarcinoma. It is easy for them to believe that cancer is a rare event in Barrett esophagus.

The approximate number of patients who will develop esophageal adenocarcinoma annually in the United States in 2016 is 20,000. This number has increased every year for the past 40 years and continues to increase.[1] In addition to its high overall mortality of 85%, treatment of late-stage esophageal adenocarcinoma with esophagectomy, chemotherapy, and radiation represents untold misery.

For any category of physician to do nothing to prevent cancer in these 20,000 patients is not a satisfactory situation. Unfortunately, that is a reality that everyone must accept today. There is nothing that can be done. There is no method of early diagnosis of GERD, there is no method of slowing or preventing progression of LES damage, no method of preventing the occurrence of Barrett esophagus, and no method of preventing adenocarcinoma.

Patients diagnosed with dysplasia and adenocarcinoma represent a small minority of the original population that is diagnosed with GERD. It is generally treated by highly specialized gastroenterologists and surgeons capable of performing endoscopic mucosal resection, radio-frequency ablation, and esophagectomy. It never enters the practice of the primary care physician except for the patient who presents with dysphagia who is immediately referred for endoscopy to a gastroenterologist.

As a pathologist, my experience has been the total opposite of the primary care physician. Most pathologists do not encounter the early stages of GERD because these patients do not have an indication for endoscopy or biopsy. The pathologist enters the scene at an advanced stage of the disease to identify intestinal metaplasia, dysplasia, and adenocarcinoma in biopsies from visible CLE.

When I began my pathology practice in 1982, I rarely saw an esophageal adenocarcinoma. There were 10 squamous carcinomas for one adenocarcinoma. Now, I encounter 10 adenocarcinomas for every squamous carcinoma. The increase in the incidence of GERD-induced cancer has been the most dramatic medical phenomenon that I and most physicians in the field have experienced.

6. COST OF TREATING GASTROESOPHAGEAL REFLUX DISEASE WITH THE PRESENT MANAGEMENT ALGORITHM

As I contemplate a possible change in the management of GERD by a new test based on assessment of LES damage, it is important to ensure that the overall cost of treating GERD does not increase unreasonably. If the new method proves to be effective in preventing severe GERD that diminishes quality of life in a large number of patients and/or turns downward the curve of increasing incidence of esophageal adenocarcinoma, some cost increase may be acceptable. The goal, however, is to decrease the overall cost while producing better control of GERD and decreasing cancer.

Cost must be divided into a financial cost and a human cost that is dependent on patient suffering. As physicians, we will always emphasize the human cost; this is the reason for our existence. However, more and more, financial cost drives treatment. It is therefore important to assess and control cost increases associated with any new management. Otherwise, its implementation is likely to fail at the outset, preventing even testing of the new method.

Finally, though, if the new method is shown to be effective in reversing the five-decade trend of increasing GERD-induced adenocarcinoma that seems to have no end, the pressures exerted by physicians, patients, and the medicolegal system in the United States will force change even at a significantly increased cost.

Each category of GERD that we defined in the previous section has its own cost structure. The costs involved for each category vary (Table 13.6). The following expenditures exist in the present management structure:

1. Cost of over-the-counter sales of acid-reducing agents (acid neutralizers, H_2 receptor antagonists, and PPIs) to people who believe they have GERD.
2. Outpatient visits to primary care physicians and cost of prescriptions for acid-suppressive agents (H_2 receptor antagonists and PPIs). PPIs are among the most widely prescribed drugs in the United States and represent significant cost centers for medicare and health insurance companies.

TABLE 13.6 Present Expenditures Incurred by Present Management of Gastroesophageal Reflux Disease (GERD) in Each of the Categories of GERD

Category	OTC Drugs	Physician Visits + Long-Term PPI	Initial Endoscopy + Testing	Barrett Surveillance	Treatment of Cancer	Cost of Decreased QOL Due to GERD
Asymptomatic patient	Zero	Zero	Zero	Zero	High	Zero
Self-medicated	Billions	Zero	Zero	Zero	High	Zero
Primary care physician; well controlled	Low	Billions	Zero	Zero	Very high	Zero
Gastroenterologist; poorly controlled	Low	Billions	High	Zero	High	20 billion
Barrett esophagus	Low	Varies with GERD category	High	Very high	High; early cancer	Varies with GERD category
Adenocarcinoma	Low	Varies with GERD category	Varies with GERD category	Varies with GERD category	Very high; advanced cancer	Varies with GERD category

Red highlights represent major causes of expenditure in the present management of GERD. The exact amount of expenditure is unknown. Broad estimates are given to demonstrate the massive costs associated with the disease. Refer Fig. 13.6 for estimated numbers of each category of patient in the population.
OTC, over the counter; *PPI*, proton pump inhibitor; *QOL*, quality of life.

3. Cost of significant overuse of these drugs, largely due to oversensitivity in the diagnosis of GERD for virtually any symptom in the upper aerodigestive tract. This leads to an empiric trial of PPI therapy and, if the symptoms respond, the drugs are used in the long term.
4. Cost of endoscopy and biopsy. Endoscopy requires an endoscopy suite (outpatient or hospital), a gastroenterologist, conscious sedation, preoperative assessment, and postoperative care. It has a low but not zero risk of complications.
5. Cost of long-term surveillance for patients with Barrett esophagus.
6. Cost of treating cancer. This may be endotherapy for early cancer, usually detected during surveillance for Barrett esophagus. Advanced cancer management requires clinical staging procedures (endoscopic ultrasound, CT, MRI, and/or PET scan), treatment (esophagectomy, radiation, and chemotherapy), palliation, and end-of-life care.
7. There is an increasing risk of class action and individual lawsuits resulting from complications of PPI therapy. At present, these are limited to minor complications such as hip fractures and chronic renal disease. If, in the future, PPIs are shown to be implicated in the promotion of Barrett esophagus and adenocarcinoma, the ensuing litigation has the potential to erase all profits made from the sales of these drugs.

The major expenditures in the present management of GERD (highlighted in red in Table 13.6) are as follows:

1. Massive use of lifelong acid-reducing therapy in 30% of the population in the United States (75 million) amounting to over $10 billion per year at the front end of the disease. This has a significant positive impact on quality of life in the majority of patients with GERD and is easily justified. However, what cannot be justified is the fact that an estimated 30% of these people are taking these drugs with a false-positive diagnosis of GERD because of empiric use of the drugs. This is not only a waste of dollars but also exposes these patients unnecessarily to the complications of PPI therapy.
2. Massive expenditure resulting from complications of GERD includes the economic cost of poorly controlled GERD that has been estimated at $20 billion per year in the United States (see Chapter 2), cost of endoscopy in poorly controlled patients, cost of Barrett surveillance, and cost of treating cancer. All these complications are the result of GERD that has been allowed to progress from the onset to severe GERD (associated with severe reflux of gastric contents into the thoracic esophagus). The benefit of PPI therapy for the 70% of patients with GERD cannot compensate for the misery and expenditure in the 30% of patients who fail to be controlled and develop Barrett esophagus and cancer.

There is no alternative to this cost structure with the present understanding of the disease and the way the guidelines recommend management. The reasons for this are lack of (1) a reliable method of early diagnosis of GERD; (2) a method to accurately assess severity of GERD until treatment failure caused by severe GERD occurs; (3) a method to predict and identify those patients in the population, with or without GERD, who will progress to severe GERD in the future.

We are at a management impasse in a common disease that is costing the population untold billions of dollars while watching a cancer incidence that has exploded in the past five decades and continues to increase.

The solution will not come from tinkering with GERD in its present categories. The only improvement to date from a cancer standpoint has been the detection of dysplasia and early cancer in patients with Barrett esophagus. These employ expensive technologies that allow minimally invasive treatment with high rates of cure. But, in the big scheme of things, this impacts only 15% of all patients who develop esophageal adenocarcinoma.

The problem can only be solved by addressing GERD from the bottom up in a comprehensive manner. This will necessitate the development of a method to identify persons in the early stages of GERD, including asymptomatic GERD, who are at highest risk of progressing to severe GERD. If this is coupled with interventions that can prevent progression to severe GERD, all of the very expensive complications of GERD (poor quality of life associated with treatment failure, Barrett esophagus, and cancer) will be eradicated.

The new method that is based on the pathologic assessment of LES damage has the potential to achieve this. In Chapter 21, I will evaluate the change in the cost structure associated with the new proposed management.

7. CLINICAL PREDICTION OF FUTURE DISEASE PROGRESSION

In any chronic disease, the ability to predict which patients with early stage disease are likely to progress to severe complicated disease is essential for effective management of the disease. Early diagnosis and identification of patients at high risk for progression to severe, life-threatening disease permits early intervention directed selectively at these patients.

There is presently no attempt or ability to accurately predict the stepwise progression of GERD to adenocarcinoma. Known cellular progression (squamous epithelium→visible CLE→Barrett esophagus→adenocarcinoma) is simply allowed to occur without any attempt at predicting or preventing them. Complications are managed as they arise and in the small minority of patients in whom they are detected before the development of adenocarcinoma. The management of GERD is totally reactive, not proactive.

There is no present screening to detect Barrett esophagus, which is present in significant numbers in the general population and in patients with symptomatic GERD even if their symptoms are controlled. The rare patient in this group who progresses to cancer is out of luck. The cancer progresses to an advanced stage before it produces dysphagia. The treatment of these patients is associated with high morbidity and an overall poor survival. Unfortunately, they are 85% of patients who develop GERD-induced cancer.

Such a fatalistic management algorithm that permits progression and discovers complications when they occur without any attempt at screening for them or preventing outcomes (except in the 15% of patients who develop cancer while under surveillance for Barrett esophagus) is a nihilistic and futile approach to the treatment of a disease.

Some clinical features in the GERD patients are predictive of progression to complications. These include the need for dose escalation to control symptoms, the occurrence of nocturnal symptoms that wake up the patient from his/her sleep, severe regurgitation, and chronic cough. Although recognized as predictors for future complications, they are not by themselves necessarily indications for endoscopy or any change in management.

Presently, there is no way, before endoscopy is performed, to predict which patients have or will progress to develop Barrett esophagus. While there are defined risk factors for Barrett esophagus such as age >50 years, Caucasian race, male gender, central obesity, smoking, and alcohol, their presence does not lead to practical recommendations. There is no defined high-risk group that represents an indication for screening.

There is no way, at any stage of the disease, to predict which patients have the highest probability of developing adenocarcinoma in the future. Even in patients with Barrett esophagus who are under surveillance, treatment is to use PPIs to control symptoms and wait for development of dysplasia or adenocarcinoma before anything different is done.

The lack of ability to identify the minority of GERD patients at high risk of progression to severe GERD (defined as inability to control symptoms with PPIs, the occurrence of Barrett esophagus or adenocarcinoma) precludes managing these patients differently than those who will be controlled satisfactorily.

There is evidence for two endoscopic criteria that are predictive of future Barrett esophagus in the GERD patient. In the Pro-GERD study,[8] the presence of severe erosive esophagitis [Los Angeles (LA) grade C/D] was associated with a 19.7% incidence of visible CLE in the next 5 years, compared with 5.4% of patients who had no erosive esophagitis. This increased incidence was significantly greater in patients treated with regular PPIs than in those with no PPIs. PPI therapy was highly effective in healing erosive esophagitis in these patients. The disease progressed to Barrett esophagus even as PPIs effectively controlled erosive esophagitis.

The second criterion, also discovered in the Pro-GERD study,[16] was the presence, in the endoscopically normal patient with GERD, of intestinal metaplasia at the normal squamocolumnar junction (SCJ). This finding was associated with a 25.8% risk of progression to Barrett esophagus within 5 years.

Despite these data, no action is taken by gastroenterologists to prevent progression to Barrett esophagus. The fact that intestinal metaplasia at the normal SCJ has a 25.8% risk of Barrett esophagus in the near future has not resulted in a recommendation to even take biopsies in the endoscopically normal GERD patient. Severe erosive esophagitis is treated with PPIs despite the evidence that regular PPI use is associated with progression to Barrett esophagus. An opportunity to test whether a sphincter augmentation procedure will prevent the progression to Barrett esophagus in these recognized situations is missed. It is missed largely because most gastroenterologists do not adequately appreciate the serious nature of Barrett esophagus.

The new method of measuring the damage (shortening) of the abdominal segment of the sphincter, which is the basis of this book, permits accurate assessment of severity of GERD and will be able to predict the future progression of GERD to treatment failure and Barrett esophagus. This can be done in people at all stages of the disease from the earliest asymptomatic stage of LES damage. The hope is that the test will permit separating the minority that is at high risk of progression from those that will remain well controlled by PPI therapy. This will permit an aggressive approach of treatment directed only at those patients who need it to prevent complicated GERD.

8. GASTROESOPHAGEAL REFLUX DISEASE IS A PROGRESSIVE DISEASE

Because of the categorization of GERD among different groups of physicians, there are relatively few studies regarding the progression of GERD to Barrett esophagus or adenocarcinoma. This is because the early stages of GERD are either completely ignored or treated empirically by primary care patients. In general, most patients reach a gastroenterologist for specialized treatment only when treatment has failed and endoscopy is indicated.

As a result, prospective research studies are rarely done in patients with early-stage disease who are being treated by family practitioners and internists with adequate follow-up to define progression toward adenocarcinoma. To these primary care physicians, GERD is a simple and settled disease that is highly successfully treated by the recommended guidelines. GERD is only a small part of the practices of each individual primary care physician, and focus on GERD is rare. Unless there is a concerted central effort to collect these patients from multiple sources, the number of patients is inadequate for any study to have adequate statistical power.

Gastroenterologists, being more academic, produce most of the literature on GERD. This is the reason why the literature on GERD is dominated by the management of refractory GERD, Barrett esophagus, and adenocarcinoma.

Studies that show the progression of GERD require an index endoscopy at an early stage of GERD in a large cohort of patients, a long follow-up period, and a second endoscopy to evaluate what elements of the disease have remained under control and what elements have progressed.

The best example of such a study is the Pro-GERD study.[8] This study was conducted in Europe (mainly Germany) and was sponsored by Astra-Zeneca, manufacturers of omeprazole and esomeprazole, the most widely prescribed PPIs.

The Pro-GERD study is a prospective follow-up study of a cohort of 6215 patients diagnosed with GERD. The study was sponsored by Astra-Zeneca, Europe. Patients had an index endoscopy at entry into the study at designated centers by endoscopists who were specially trained to conform to study criteria. In the clinical setting, endoscopy would not have been indicated at this early stage of the disease by a physician following the management guidelines for GERD.

The patients were then sent back to their primary physicians for treatment. Of the 6215 patients originally enrolled in the study, 2721 patients attended a follow-up at 5 years for a second or third (many had a second endoscopy at 2 years) endoscopy done at the same designated centers.

In the intervening period, the patients were required to respond to questionnaires relating to their symptoms, quality of life, and the use of acid-reducing drugs. No guidance was given to the primary care physicians regarding how these patients should be treated.

The information in this study provides a powerful insight into what happens at an endoscopic level to patients with a diagnosis of GERD over a 5-year period of follow-up during which they received treatment at the discretion of their primary physician. This study is therefore designed to see what happens to GERD patients in real life.

The 2721 patients were divided at the index endoscopy into four groups: NERD, LA grade A/B, LA grade C/D, and visible CLE (=endoscopic Barrett esophagus). The mean age was similar (55/56 years ± 12/13 years) and the GERD symptom score was similar in all the four groups, i.e., by the criterion of symptom score, all groups were identical.

1224 patients had NERD, i.e., had no erosive esophagitis or CLE; 1044 had LA grade A/B; 213 had LA grade C/D; and 240 had visible CLE. The 240 (8.8%) patients with visible CLE at the index endoscopy were excluded. This is unfortunate

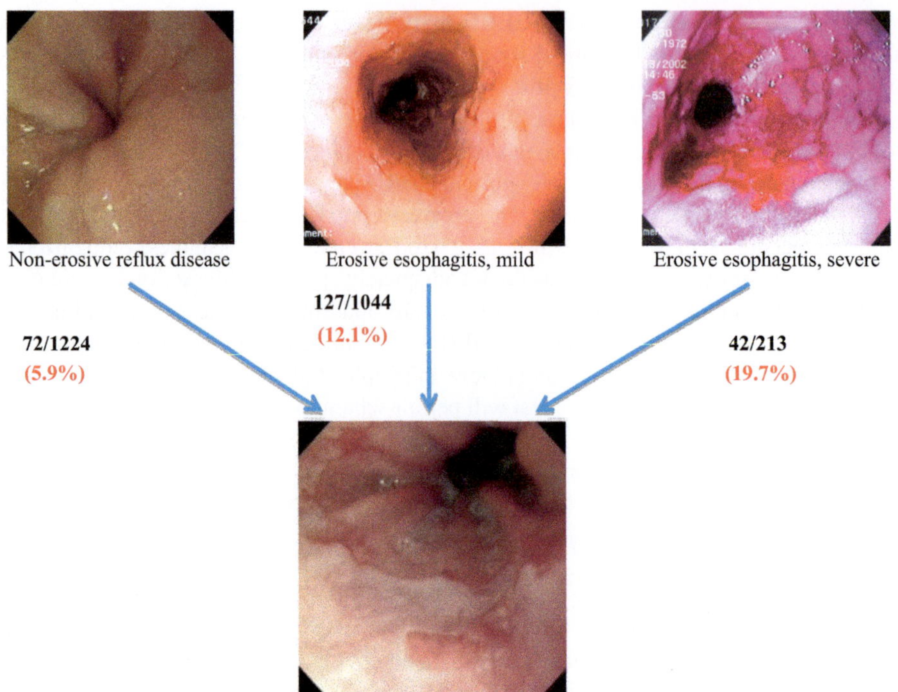

FIGURE 13.11 Progression of gastroesophageal reflux disease without visible columnar-lined epithelium (CLE) to visible CLE during a 5 year period of treatment in the primary care setting. A total of 9.7% progressed to visible CLE, the risk being significantly associated with severity of erosive esophagitis at the index endoscopy.

because it would have provided a 5-year follow-up on the progression of patients who were negative for intestinal metaplasia to intestinal metaplasia as well as progression to dysplasia and cancer.

In Chapter 2, the part of the study that evaluated the effect of treatment on erosive esophagitis was reviewed in detail. This showed that PPIs are highly effective drugs in controlling acid-induced damage of the squamous epithelium. Routine treatment of GERD by primary care physicians over a 5-year period resulted in a reduction in the number of patients with severe (LA C/D) erosive disease from 188 at baseline to 29 at the end of 5 years. Of these 11 stayed at LA C/D throughout the 5 years, 9 each progressed from nonerosive disease and LA A/B disease.

The fact that progression was significantly less common in patients on regular PPI use suggests that some of the failure to heal and prevent severe erosive esophagitis was the result of inadequate acid suppression. This is not unexpected because medical therapy is designed to control symptoms, and the correlation between symptoms and erosive esophagitis is not perfect.

Patients with Los Angeles grades C/D were the most difficult to control. However, even these patients with severe squamous epithelial injury had significant healing of erosive esophagitis. This attests to the power of PPI therapy. No patient developed serious squamous epithelial complications such as deep ulcers and complicated strictures that were common before the era of effective acid suppression.

Many treating physicians will use these data to suggest that PPIs stabilize the progression of GERD and actually cause it to regress based on the fact that symptoms, which define GERD, become less troublesome and erosive esophagitis reverses in the majority of patients. By these criteria, which are used almost exclusively in studies to define the efficacy of treatment of GERD, PPI therapy is wonderful.

The story changes completely when one considers progression of GERD to Barrett esophagus. At 5 years, 9.7% of the overall 2721 patients had progressed from no CLE to visible CLE. There was a strong positive correlation between the severity of erosive esophagitis and the progression to visible CLE, which occurred in 72/1224 (5.9%) of NERD, 127/1044 (12.1%) in LA grade A/B erosive esophagitis, and 42/213 (19.7%) in LA grade C/D erosive esophagitis patients (Fig. 13.11).

The factors significantly associated with progression to visible CLE at 5 years were (1) a negative association with female gender (P = .041); (2) alcohol intake (P = .033); (3) erosive esophagitis compared with NERD (P < .001); for LA grade A/B, the odds ratio was 2.04 (95% CI 1.47–2.82); for LA grade C/D, the odds ratio was higher at 3.31 (95% CI 2.10–5.22); and (4) regular PPI use (P = .019).

The value of the Pro-GERD study is that it reports progression using a criterion other than symptoms and healing of erosive esophagitis. It shows that a significant number of the same patients whose erosive esophagitis heals progress from not having Barrett esophagus at the index biopsy to developing it within the 5-year period of follow-up.

This evidence confirms that continuing weak-acid reflux into the esophageal body in patients treated with PPIs to control symptoms and heal erosive esophagitis simultaneously promotes Barrett esophagus. The amount of reflux is sufficient to induce columnar metaplasia of the squamous epithelium of the esophageal body. Columnar metaplasia occurs at a higher rate in patients who have been on regular PPIs than in those with no PPIs.

These findings in the Pro-GERD study were confirmed by Ronkainen et al.[17] in a community-based endoscopic study. Patients with an initial endoscopy that had no evidence of visible CLE (i.e., either normal or with erosive esophagitis) were invited for a follow-up endoscopy 5 years later. 284 patients agreed. The overall incidence of Barrett esophagus was 9.9/1000 patient years. Of 118 patients with NERD at the index endoscopy, 11 progressed to erosive esophagitis and 2 to Barrett esophagus. Of 90 patients with erosive esophagitis at the index endoscopy, 8 progressed to Barrett esophagus. The presence of erosive esophagitis at the index endoscopy was independently associated with a relative risk of 5.2 (95% confidence interval 1.2–22.9) of having developed Barrett esophagus 5 years later.

Many gastroenterologists largely ignore or discount the data in the Pro-GERD study. They explain the finding that regular PPI use is associated significantly with a higher incidence of Barrett esophagus as indication bias. According to this, the reason why regular PPI use is associated with the increased incidence of Barrett esophagus is that regular PPI use is more common in patients with more severe GERD. It is the severity of GERD symptoms that is responsible for the increased incidence of Barrett esophagus, not regular PPI use. This is unlikely. In the Pro-GERD study,[8] there was no difference in symptom scores among the various endoscopic categories (NERD, LA A/B, and LA C/D erosive esophagitis). Also multivariate analysis of all factors associated with the incidence of Barrett esophagus at 5 years found no association between symptom severity and Barrett esophagus.

While there may be argument about whether PPIs actually promote Barrett esophagus or the association is the result of indication bias, there is no doubt about the reality. Patients with GERD treated with regular PPI convert from no visible CLE to short- and long-segment CLE at a rate of nearly 10% within 5 years. The reasons are arguable. The fact is not. The fact is what is critically important.

I will show in Chapter 14 that the distribution of intestinal metaplasia in CLE provides strong evidence that increase in pH of gastric juice is responsible for intestinal metaplasia in CLE. Intestinal metaplasia is not a direct effect of PPIs. It is an effect of increasing the pH in the esophagus during reflux episodes. This began with acid neutralizers in the 1950s becoming more effective with H_2 receptor antagonists in the 1960s and PPIs in the 1980s. The culprit is not the drug, it is the success the pharmaceutical industry has had in producing drugs with increasing ability to alkalinize gastric juice.

The Pro-GERD study shows clearly that symptom control and healing of erosive esophagitis comes with a price. In the same cohort, regular treatment of GERD causes 10% of the patients to progress to Barrett esophagus in 5 years and the likelihood of Barrett esophagus is higher with more effective alkalinization of gastric juice with medical therapy.

This is the likely basis of how GERD has transformed from being a nonlethal debilitating chronic disease resulting from squamous epithelial disease to a less debilitating but more lethal disease resulting from columnar metaplasia since the 1970s. The conversion of a population of patients with GERD who have no CLE at endoscopy and therefore at no risk for cancer to those with Barrett esophagus and a 0.2% per year risk for cancer at the rate of 10% per 5 years can explain much of the increasing incidence of adenocarcinoma in GERD.

The Pro-GERD study is highly credible. That regular PPI use promoted progression to Barrett esophagus in a cohort of patients being treated for GERD was demonstrated in a study sponsored by Astra-Zeneca with two of the authors of the paper being employees of Astra-Zeneca. The study methods state that the authors had freedom to report the data as they saw fit without interference from the sponsors of the study. The fact that this actually happened is a miracle of integrity in medicine.

The study data also demonstrate that present guidelines for treating GERD actively prevent the detection of endoscopic progression of GERD to Barrett esophagus. The progression of 10% of GERD patients without CLE to CLE in 5 years would remain hidden because most if not all the patients in the study cohort would have had adequate symptom control and therefore no indication for endoscopy according to the guidelines for management of GERD.

Incredibly, the guidelines for management of GERD do nothing to prevent the progression of a patient from being at no risk for cancer to one that is at a significant risk. The guidelines are also responsible for the fact that the majority of patients in the population with Barrett esophagus remain undetected until they develop advanced cancer.

Another excellent follow-up study by Falkenback et al.[18] reviewed patients admitted to the esophageal laboratory at the Department of Surgery, Lund University Hospital in Sweden between August 1984 and December 1988 with the main complaint suggesting typical GERD. Inclusion into the study required that there was no previous surgery of the esophagus or stomach, that the patient had undergone upper gastrointestinal endoscopy, a complete esophageal manometry study not suggesting a named motility disorder other than findings in accordance with GERD, and a pathologic 24-h ambulatory esophageal pH study with a percentage time pH<4 of 3.8%–10%. Of the 126 patients who had complete manometry and pH studies that met study criteria, 52 did not meet other inclusion criteria (34 had surgical treatment, 6 patients could not be found, and 12 were deceased). The 74 eligible patients were approached by telephone and mail with an invitation to undergo a new set of studies. Of these, 34 declined to participate in the study, leaving 40 evaluable eligible patients.

The baseline studies done in 1984–88 on the 40 patients were retrieved. Follow-up studies were done in 2007–08, a mean of 20.7 (range 18.8–23.5) years after the initial studies. Studies included upper gastrointestinal endoscopy, esophageal manometry, ambulatory 24-h pH-metry, *Helicobacter pylori* assessment, and a clinical evaluation using multiple structured questionnaires.

Endoscopic assessment included evaluation for hiatal hernia, Hill valve grade, the presence and severity of erosive esophagitis, and the presence of Barrett esophagus. The diagnosis of Barrett esophagus was made when a tongue of metaplastic columnar epithelium >0.5 cm was seen and confirmed by histology (Table 13.7).

Comparison between baseline and follow-up 18.8–23.5 years later showed that the use of acid suppressants had increased significantly. Frequent (>2–3 times per week) use of acid suppressants had increased from a baseline of 16 (40%) to 28 (70%) at the follow-up evaluation (P=.007).

There was also significant progression of endoscopic abnormality. At baseline, 24 patients had NERD, 16 had erosive esophagitis, and no patient had endoscopic Barrett esophagus. At the follow-up examination, the number of patients with erosive esophagitis had increased from 16 (40%) to 29 (72.5%) (P=.001), and the number of patients with endoscopic Barrett esophagus had increased from 0 to 18 (45%) (P<.001) and those with positive histology to 10 (25%) (P=.002).

In the 24 patients who had NERD at baseline, the follow-up endoscopy showed erosive esophagitis in 16 (67%), endoscopic Barrett esophagus in 10 (42%), and histologically confirmed Barrett esophagus in 4 (17%). Progression to Barrett

TABLE 13.7 Characteristics of the 40 Patients at Baseline and Follow-Up

Characteristic	Baseline	Follow-Up	*P* Value
Follow-up; years (mean; range)	20.7 (18.8–23.5)		
Age at follow-up; years (mean; range)		64.8 (53–77)	
Males:females	27:13		
Mean BMI (kg/m^2)	26.0	28.8	<.001
Smokers; number (%)	14 (35%)	3 (7.5%)	.002
Acid suppressants >2–3 times/week; n (%)	16 (40%)	28 (70%)	.007
Hiatal hernia present; n (%)	26 (65%)	24 (60%)	.45
Hiatal hernia mean size	2.2 cm	3.1 cm	.28
Esophagitis present; n (%)	16 (40%)	29 (72.5%)	.001
Endoscopic Barrett esophagus; n (%)	0	18 (45%)	<.001
Endoscopic Barrett esophagus with IM; n (%)	0	10 (25%)	.002
Total percent time pH<4	6.5	7.1	.44
Number of reflux episodes	44.9	97.6	<.001
Number of reflux episodes >5 min	3.8	3.9	.76
Longest reflux episode; minutes	23.8	13.8	.001
Mean LES pressure (mmHg)	11.0	12.4	.99
Total LES length (cm)	3.6	3.4	.58
Abdominal LES length (cm)	1.7	1.9	.30

Blue: significant differences between the two evaluations.
IM, intenstinal metaplasia; *LES*, lower esophageal sphincter.

esophagus occurred in 8/16 (50%) patients who had erosive esophagitis at baseline; 7 of these 8 (88%) patients had Barrett esophagus confirmed by histology.

No predictors for esophagitis were identified at baseline. Predictors of severe esophagitis (Savary–Miller grade 3/4) were pathologic pH-metrics and intraabdominal LES length. Surprisingly, there was no significant difference in mean LES pressure, total LES length, and abdominal LES length between the baseline and follow-up measurement. This cannot be attributed to technical reasons; accurate manometry was available in 1984.

The results of this study with a longer period of follow-up of GERD patients show progression from NERD→erosive esophagitis→endoscopic Barrett esophagus (i.e., visible CLE)→histologic Barrett esophagus (i.e., visible CLE with intestinal metaplasia). The progression appears to occur at different rates in different patients. Many patients do not progress, and the ability to predict progression at the baseline examination is poor at best. The present method of treatment of GERD seems to leave no alternative but to treat GERD with acid-reducing therapy and simply wait for the inevitable progression in a minority of patients.

The Pro-GERD study shows a much better control of erosive esophagitis compared with this study suggesting that the increased experience with the range of PPI drugs has resulted in better control of acid-induced changes. However, the progression to Barrett esophagus has remained unchanged or is worsening.

9. DEFINITION OF IRREVERSIBILITY OF GASTROESOPHAGEAL REFLUX DISEASE

In any chronic disease, it is critical to establish the point in time where pathologic changes evolve from being reversible to irreversible. At that point, if the physician has no ability to prevent progression to those pathologic changes, all control is lost. Recognition of the presence of such a point in a disease should result in a concerted effort to prevent the patient from reaching that point or, if that point is reached, developing the ability to reverse the change or control further progression. Otherwise, treatment is doomed to failure and adverse outcomes will become inevitable.

I have used the example of ischemic heart disease to demonstrate this principle. The point of significant coronary artery narrowing is around 70%. At that point, the vessel fails to deliver adequate blood during exercise causing angina on exertion. If nothing is done to prevent or treat this level of coronary artery narrowing, the patient is at risk for clinical ischemic heart disease and its progression to angina at rest, myocardial infarction, and death.

Damage to the LES is to GERD what coronary artery narrowing is to ischemic heart disease. Both are irreversible from the outset. However, both have a large reserve capacity that permits it to maintain competent function. In coronary artery disease, critical narrowing is 70% of the luminal area.

Defining the point of irreversibility at which the amount of LES damage results in a sufficient degree of failure of its sphincter function to result in visible CLE is critical to the management of GERD. If LES damage can be contained below this threshold point, visible CLE in the thoracic esophagus can be prevented. If that is achieved, Barrett esophagus and adenocarcinoma will be prevented.

The progression of GERD from its onset, from the point of view of LES damage, and how it correlates with the amount of reflux into the thoracic esophagus, symptoms, and cellular changes in the thoracic esophagus is shown in Table 13.8. I will use, based on best evidence available, an original length of the abdominal segment of the LES as 35 mm, the point at which LES failure occurs as 10 mm, with dynamic shortening of the LES during a heavy meal as 10 mm.

This shows that as LES damage causes the abdominal LES to shorten, there is an early stage where the damage is within the reserve capacity of the LES where there is no significant reflux into the thoracic esophagus and no cellular changes are seen therein. In this stage, the disease is manifested only by cardiac epithelium (with and without parietal and/or goblet cells) in the dilated distal esophagus, whose length is concordant with the amount of shortening of the abdominal LES.

The end of the reserve capacity is an abdominal LES shortening of 15 mm, when the residual abdominal LES length is 20 mm (Fig. 13.12). Up to this point the LES is competent and there is no reflux. With a 20 mm long abdominal LES, ingestion of a heavy meal that causes a dynamic shortening of the abdominal LES of 10 mm causes the abdominal LES to reach the critical 10 mm, but it does not dip <10 mm level at which LES failure (transient lower esophageal relaxation, tLESR) and reflux begins. Further damage from this point results in LES failure (tLESR) and reflux.

At 25 mm of abdominal LES damage, the residual abdominal LES length is 10 mm and is in danger of failure (tLESR) without gastric distension (Fig. 13.13). The period during which the abdominal LES shortens from 15 to 25 mm represents the period of transition from mild to severe GERD, the latter defined as treatment failure and/or visible CLE. With a rate of progression of LES damage of 5 mm/decade, it will take two decades for LES damage to progress from 15 to 25 mm. The time frame varies with the rate of progression of LES damage.

In a significant number of GERD patients, diagnostic testing is delayed to a point of extreme damage of the LES, often with a hiatal hernia (Figs. 13.14 and 13.15).

TABLE 13.8 Progression of Gastroesophageal Reflux Disease From the Point of View of Lower Esophageal Sphincter (LES) Damage, Amount of Reflux Into the Thoracic Esophagus, Cellular Change in the Thoracic Esophagus, and Symptoms

Abdominal LES Length—Fasting (mm)	Abdominal LES Length—Postprandial (mm)	Amount of Reflux Above LES	Cellular Changes Above eGEJ	Symptoms
35	25	0	0	0
20	10	0–<4.5%	0	0
15	5	0–<4.5%	NERD—mild erosive esophagitis	0/+/controlled
10	0	>4.5%	NERD— severe erosive esophagitis	0/+/controlled
<10–0	0	>>4.5%	Visible CLE	0/+/controlled
<10–0	0	>>4.5%	Barrett esophagus	0/+/controlled
<10–0	0	>>4.5%	Dysplasia	0/+/controlled
<10–0	0	>>4.5%	Adenocarcinoma	0/+/controlled

Irreversibility of cellular change (red) is correlated with severe LES damage and abnormal reflux by the pH test, but not by symptoms (presence, severity or control with PPI).

The *arrow* on the left shows inexorable disease progression when seen from the viewpoint of LES damage (shortening); the *arrow* on the right shows the disease progression when viewed by symptoms in patients treated with acid reduction. PPIs impact the disease positively only to control symptoms and heal esophagitis; it does not impact progression of visible CLE.

CLE, columnar-lined esophagus; *eGEJ*, endoscopic gastroesophageal junction; *LES*; lower esophageal sphincter; *NERD*, nonerosive reflux disease; *PPI*, proton pump inhibitor.

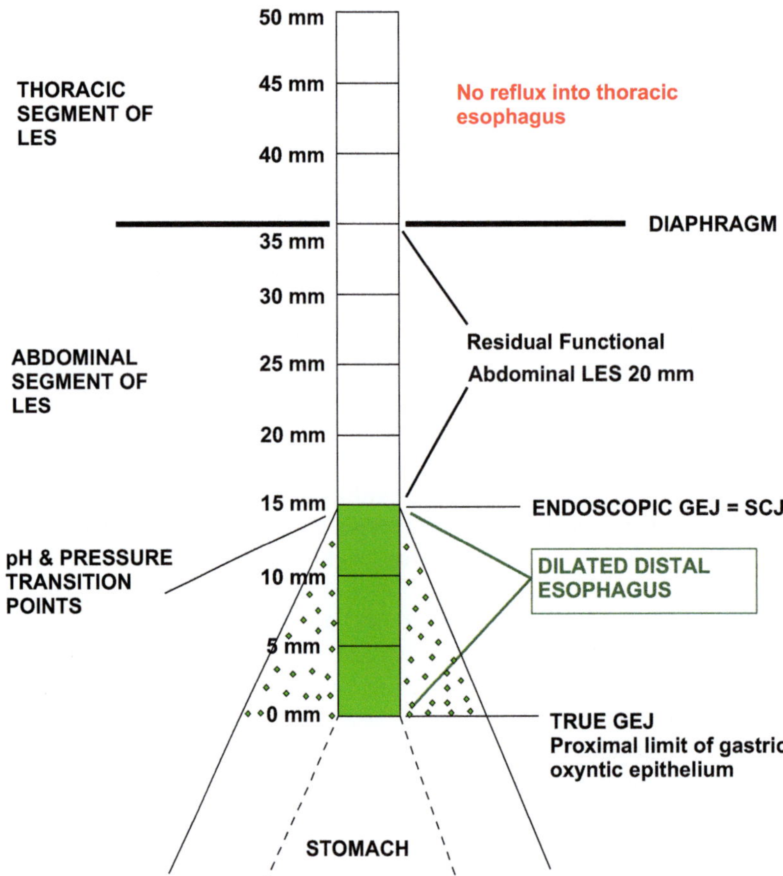

FIGURE 13.12 Abdominal lower esophageal sphincter (LES) damage of 15 mm. This is the endpoint of the phase of compensated LES damage. Although damaged, the residual length of the abdominal LES is 20 mm (assuming an initial length of 35 mm). This means that even with dynamic shortening of 10 mm with a heavy meal, the residual abdominal LES does not reach the critical <10 mm length that is associated with LES failure. *GEJ*, gastroesophageal junction; *SCJ*, squamocolumnar junction.

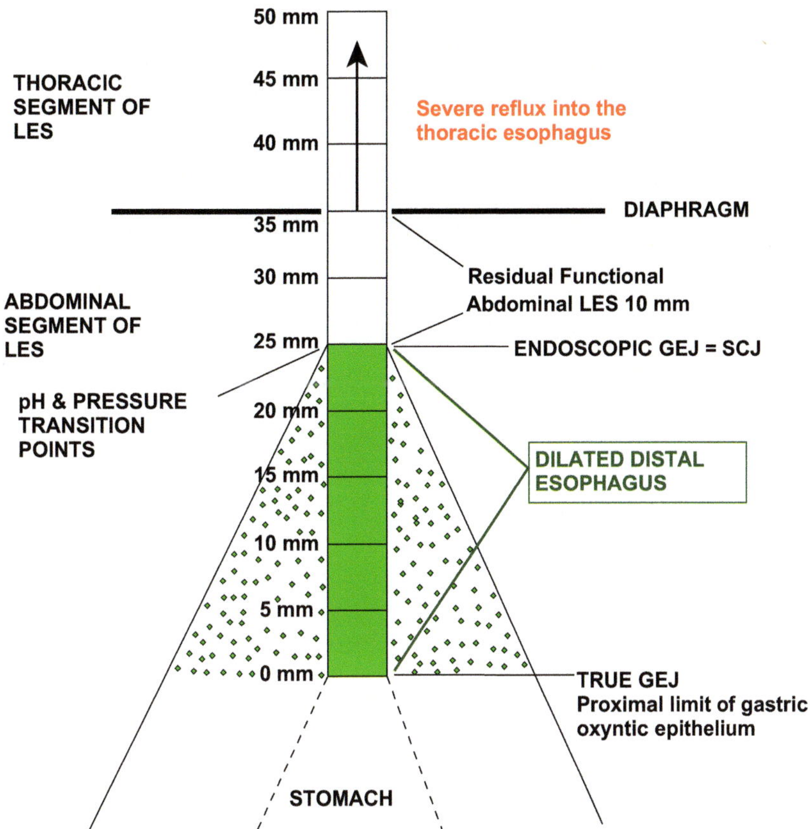

FIGURE 13.13 Abdominal lower esophageal sphincter (LES) damage of 25 mm. The residual length of the abdominal LES is 10 mm (assuming an initial length of 35 mm) in the fasting state. This means that any stress on the LES causes the functional abdominal LES length to reach the critical <10 mm length. The patient has frequent LES failure and severe reflux into the thoracic esophagus. This patient is at risk for visible columnar-lined epithelium. It should be recognized that this is the point at which gastroesophageal reflux disease reaches criteria of definition by present manometric criteria for a defective LES. *GEJ*, gastroesophageal junction; *SCJ*, squamocolumnar junction.

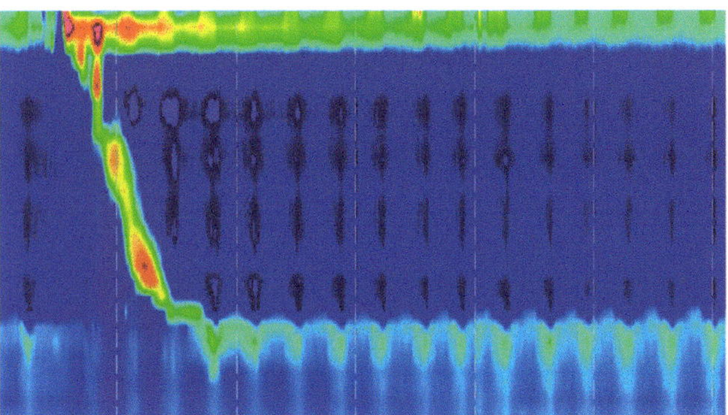

Short, low pressure LES, no hiatus hernia, normal motility

FIGURE 13.14 High-resolution manometry showing a markedly abnormal short lower esophageal sphincter (LES) with low pressure. The calculated total LES length is 1.1 cm, the abdominal LES length is 0 cm, and resting LES pressure is 9 mmHg. *Study and interpretation including calculated lengths is by Chris Dengler, MD, Legato Medical Inc.*

LES damage is irreversible from its earliest time. However, like coronary artery narrowing, damage becomes significant from the clinical point of view only when it has exhausted its reserve capacity and begins to fail, causing reflux into the thoracic esophagus. It is at this point that cellular changes begin to occur in the esophagus.

The progression of cellular change correlates well with severity of reflux, which in turn correlates well with severity of LES damage, as defined by LES shortening.

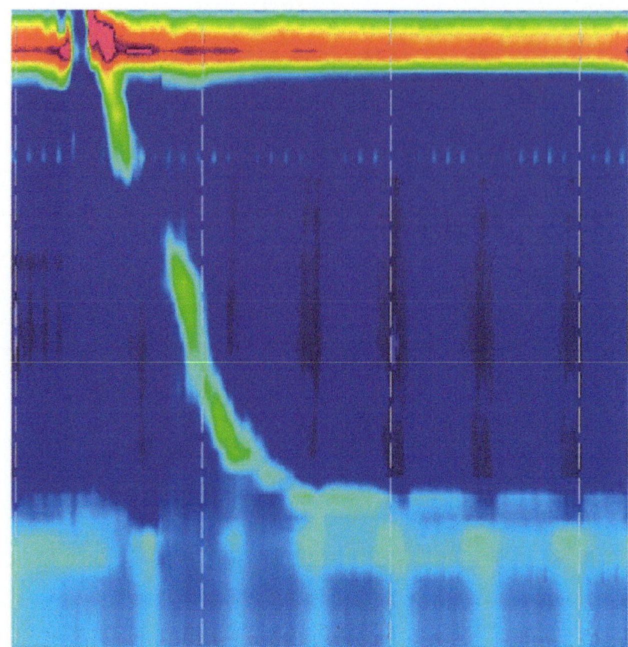

Short, low pressure LES, small hiatus hernia, weak motility

FIGURE 13.15 High-resolution manometry showing a markedly abnormal short lower esophageal sphincter (LES) with low pressure. There is a small hiatus hernia and weak motility in the esophageal body. The calculated total LES length is 1.4 cm, abdominal LES length is 0 cm, and resting LES pressure is 7 mmHg. *Study and interpretation including calculated lengths is by Chris Dengler, MD, Legato Medical Inc.*

The sequence of cellular change in the esophagus above the endoscopic GEJ is as follows, in the order of association with increasing LES damage:

1. No visible microscopic, electron microscopic, and molecular changes in the squamous epithelium of the thoracic esophagus. This is likely limited to the normal state and in those people whose LES is damaged within its reserve capacity. The LES is competent and there is no reflux into the esophageal body (i.e., the pH test is zero).
2. No visible endoscopic or microscopic evidence of squamous epithelial injury. However, evidence of injury is present at ultrastructural (dilated intercellular spaces) and molecular (low-level expression of bone morphogenetic protein-4 and trefoil factor family-3) level. These patients have LES damage sufficient to cause mild reflux into the thoracic esophagus. The patient may be asymptomatic or have mild symptoms of GERD.
3. No visible endoscopic evidence of squamous injury but with microscopic features of reflux disease (basal cell hyperplasia, papillary elongation, intraepithelial eosinophils).
4. Erosive esophagitis of increasing grade seen at endoscopy (Fig. 13.1).
   All of the above changes are reversible with PPI therapy.
   Reversal of these squamous epithelial changes by PPI therapy comes with a price. The patients progress to develop columnar metaplasia in the thoracic esophagus:
5. Visible CLE composed of cardiac and oxyntocardiac epithelium without intestinal metaplasia (Fig. 13.16).
6. Visible CLE with intestinal metaplasia (Barrett esophagus) (Fig. 13.2).
7. Increasing dysplasia in Barrett esophagus (Figs. 13.9 and 13.10).
8. Adenocarcinoma (Fig. 13.17).

The occurrence of visible CLE represents the point of irreversibility for the patient with GERD. Visible CLE does not reverse with medical therapy. It progresses to intestinal metaplasia, increasing dysplasia and adenocarcinoma despite PPI therapy that controls symptoms and erosive esophagitis.

In the Pro-GERD study, 19.7% of patients with LA grade C/D erosive esophagitis developed visible CLE within 5 years. The data in Falkenbeck et al.[18] suggest that the progression to Barrett esophagus in these patients increases with time up to 20 years.

I will show that all the evidence suggests that a significant reason for the increase in adenocarcinoma in the past 60 years is the increase in the conversion of patients with GERD to visible CLE and Barrett esophagus. Attempting to prevent cancer

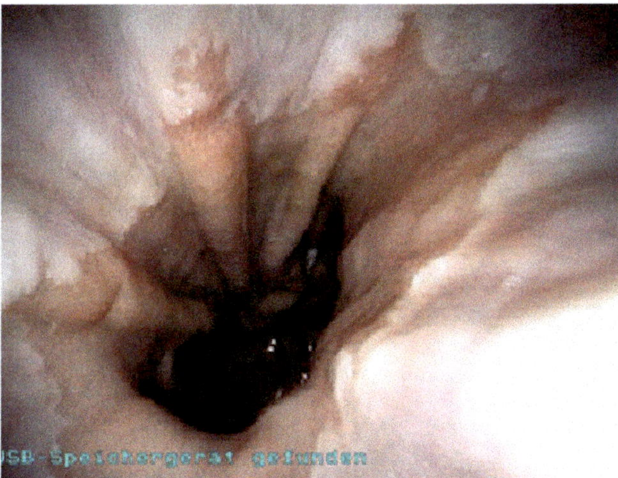

FIGURE 13.16 Antegrade endoscopic view of a 1-cm segment of visible columnar-lined epithelium. This is seen as a salmon pink flat columnar epithelium between the proximal limit of rugal folds and the irregular squamocolumnar junction. Biopsy showed cardiac epithelium without intestinal metaplasia. *Photograph courtesy of Dr. Martin Riegler, Vienna, Austria.*

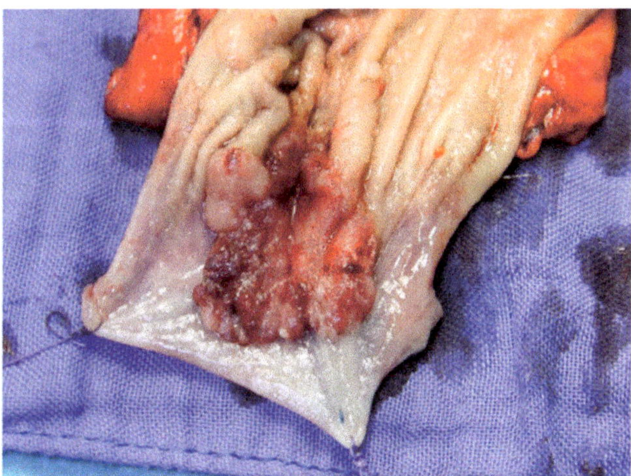

FIGURE 13.17 Adenocarcinoma of the esophagus arising in a short segment of flat salmon pink columnar epithelium between the squamocolumnar junction and the proximal limit of rugal folds. The tumor extends below the proximal limit if rugal folds into the dilated distal esophagus.

in patients with Barrett esophagus will not decrease overall cancer incidence if the primary reason for the increase in cancer incidence is that present treatment continually adds to the number at risk by the fact that it promotes the development of Barrett esophagus.

Another reasonable definition of irreversibility in GERD is the failure of the maximum PPI therapy to control symptoms adequately. This situation correlates with severe LES damage and severe reflux. Even the most potent drug therapy cannot control symptoms. This could be due to either breakthrough acid secretion (at best, PPI therapy maintains intragastric pH at <4 for 20h) or molecules other than acid being responsible for symptoms. The only alternative to suffering the reduced quality of life with maximum PPI therapy for these patients is a procedure, endoscopic or surgical, to repair the LES. These are performed only in a small number of these patients at this time.

10. PREVENTION OF THE IRREVERSIBLE STATE

I have defined irreversibility in GERD as failure of acid-reducing drugs to control symptoms adequately and/or the occurrence of visible CLE. The new goal of treatment that we must try to achieve is the prevention of irreversibility. If we achieve this objective, adenocarcinoma of the esophagus will be prevented. Adenocarcinoma of the thoracic esophagus does not arise without visible CLE.

These two criteria of irreversibility result from cellular events in two different epithelial types in the esophagus. Failure to control symptoms results primarily from continued squamous epithelial injury caused by reflux that overwhelms the ability of the protective effect of PPIs. Visible CLE results from columnar metaplasia of squamous epithelium, also caused by severe reflux. Visible CLE sets the stage for progression to adenocarcinoma. Once CLE undergoes intestinal metaplasia, this progression is unpredictable. It depends on the activity of an unknown carcinogenic mechanism that attacks CLE with intestinal metaplasia.

10.1 Failure of Present Gastroesophageal Reflux Disease Management to Prevent Visible Columnar-Lined Epithelium

Present management focuses entirely on reducing squamous epithelial injury by alkalinizing the refluxate. This succeeds in controlling symptoms until the frequency and volume of reflux episodes increases to a point where even maximum PPIs cannot control symptoms in those who reach irreversible treatment failure.

PPI therapy actually works at a cross purpose, as shown clearly in the Pro-GERD study. At the same time that it is effective in reducing squamous epithelial injury and symptoms, Malfertheimer et al.[8] showed that PPI therapy induces visible CLE at a rate of 10% of the GERD population every 5 years. While it tries valiantly and often successfully to achieve symptom control, PPI therapy unquestionably precipitates the second criterion of irreversibility, visible CLE. Nason et al. showed that the more effective the medical therapy is in controlling symptoms, the higher the progression to Barrett esophagus. Lagergren et al.[7] showed that use of medication had an odds ratio, after adjustment for severity of symptoms, of 2.9 (95% confidence interval, 1.9–4.6) causing esophageal adenocarcinoma compared to persons not using medications.

The above effects of acid-reducing therapy have been confirmed by the overall direction of GERD outcomes over the past five decades. While quality of life has improved greatly for millions of GERD patients, the incidence of Barrett esophagus and adenocarcinoma has increased out of control.

10.2 New Management Objectives

To be effective in its entirety, the management of GERD must change from one that works at a cross purpose to one that has a singular objective that will prevent both criteria of irreversibility.

This should be easy. Both these criteria of irreversibility result from severe reflux, which in turn results from severe LES damage. This is the root cause of reflux, which is necessary for both treatment failure and visible CLE.

The point of attack must be aimed at preventing severe LES damage in those patients destined to progress to that state. If we set as a new management goal the prevention of the critical amount of LES damage that results in LES failure and reflux sufficient to produce either treatment failure or visible CLE, the problem will be solved.

According to the new model that correlates LES damage with reflux, most patients will begin developing symptoms in the postprandial period when the abdominal LES is around 20 mm (=abdominal LES damage of 15 mm) (Fig. 13.12). A very heavy meal with gastric overdistension can cause dynamic shortening of the LES by a further 10 mm and bring the patient's functional abdominal LES to 10 mm, the point of LES failure and reflux. According to Table 13.8, severe GERD with a high risk of treatment failure and/or visible CLE occurs at an abdominal LES length of around 10 mm (25 mm damage) (Fig. 13.13). The exact amount of LES damage that heralds the risk of visible CLE will emerge when data are produced in future clinical studies using the new pathologic test for LES damage.

If the patient has the new assessment of LES damage described in Chapter 17 at the very earliest point in the onset of symptoms, it is likely that the patient has a gap of 10 mm of abdominal LES length between the LES damage at the time of symptom onset and the critical 10 mm residual abdominal LES length that precipitates LES failure and severe enough reflux to produce our criteria of irreversibility—treatment failure and visible CLE.

This is the new window of opportunity. With a 5 mm/decade progression of LES damage, 10 mm of LES damage will take 20 years. In the study by Falkenbeck et al.[18] 45% of patients progressed from GERD with no visible CLE to visible CLE during a follow-up period of 20 years. It will take longer or shorter with slower or faster progression of the abdominal LES damage.

At present, all patients who develop symptoms of GERD are treated empirically with acid-reducing drugs. It does not matter whether the patient purchases OTC drugs or receives them from a physician by prescription. This heals squamous epithelial damage and symptoms. However, progression of LES damage continues unchecked. It is completely unaffected by acid-reducing therapy. The medical establishment is simply permitting the abdominal LES to shorten from 20 mm at onset of symptoms to 10 mm. At that point in the disease, physicians get serious about GERD. It is too late. The window of opportunity to prevent LES damage to its critical 10 mm point has been lost.

In Chapter 18, I will show that the progression of LES damage can be predicted with the new histologic assessment. For example, if a person has his/her LES damage measured by the new test, an algorithm can be developed that will predict when in the future the patient will reach the critical point of 25 mm abdominal LES damage that corresponds to a residual abdominal LES length of 10 mm. This will happen only in a minority of patients with GERD. This is the minority that will progress to treatment failure and visible CLE.

By having a method that separates the large majority of patients who will never progress to abdominal LES damage of 10 mm from the minority that will so progress, we can stratify management of GERD. The majority who will not reach an abdominal LES length of 10 mm can be confidently maintained on acid-reducing drugs with full knowledge that they will not reach irreversibility defined by treatment failure and visible CLE. The minority who are predicted by the algorithm to progress to an abdominal LES length of 10 mm in the future must be aggressively treated to prevent that outcome. If that can be achieved, all patients with GERD will remain well controlled and not at risk for treatment failure or adenocarcinoma.

Even the idea that there is a theoretical method of preventing esophageal adenocarcinoma is a marked improvement from the present nihilism regarding esophageal adenocarcinoma.

REFERENCES

1. Pohl H, Sirovich B, Welch HG. Esophageal adenocarcinoma incidence: are we reaching the peak? *Cancer Epidemiol Biomarkers Prev* 2010;**19**:1468–70.
2. Allison PR. Peptic ulcer of the oesophagus. *Thorax* 1948;**3**:20–42.
3. Morson BC, Belcher BR. Adenocarcinoma of the oesophagus and ectopic gastric mucosa. *Br J Cancer* 1952;**6**:127–30.
4. Katz PO. Intragastric acid suppression. *Aliment Pharmacol Ther* 2004;**20**:399.
5. Blonski W. Comparison of reflux frequency. *J Clin Gastroenterol* 2009;**43**:816.
6. Kahrilas P, Shaheen NJ, Vaezi MF. American gastroenterological association medical position statement on the management of gastroesophageal reflux disease. *Gastroenterology* 2008;**135**:1380–2.
7. Lagergren J, Bergstrom R, Lindgren A, Nyren O. Symptomatic gastroesophageal reflux as a risk factor for esophageal adenocarcinoma. *N Engl J Med* 1999;**340**:825–31.
8. Malfertheiner P, Nocon M, Vieth M, Stolte M, Jasperson D, Koelz HR, Labenz J, Leodolter A, Lind T, Richter K, Willich SN. Evolution of gastro-oesophageal reflux disease over 5 years under routine medical care – the Pro-GERD study. *Aliment Pharmacol Ther* 2012;**35**:154–64.
9. Nason KS, Wichienkuer PP, Awais O, Schuchert MJ, Luketich JD, O'Rourke RW, Hunter JG, Morris CD, Jobe BA. Gastroesophageal reflux disease symptom severity, proton pump inhibitor use, and esophageal carcinogenesis. *Arch Surg* 2011;**146**:851–8.
10. An evidence-based appraisal of reflux disease management – the Genval Workshop Report. *Gut* 1999;**44**(Suppl. 2):S1–16.
11. Hvid-Jensen F, Pedersen L, Drewes AM, et al. Incidence of adenocarcinoma among patients with Barrett's esophagus. *N Engl J Med* 2011;**365**:1375–83.
12. Peters JH, Clark GWB, Ireland AP, et al. Outcome of adenocarcinoma arising in Barrett's esophagus in endoscopically surveyed and nonsurveyed patients. *J Thorac Cardiovasc Surg* 1994;**108**:813–21.
13. Kastelein F. Proton pump inhibitors reduce the risk of neoplastic progression in patients with Barrett's esophagus. *Clin Gastroenterol Hepatol* 2013;**11**:382.
14. Hvid-Jensen F. Proton pump inhibitor use may not prevent oesophageal adenocarcinoma in Barrett's oesophagus. *Aliment Pharmacol Ther* 2014;**39**:984.
15. Chang EY, Morris CD, Seltman AK, et al. The effect of antireflux surgery on esophageal carcinogenesis in patients with Barrett esophagus. A systematic review. *Ann Surg* 2007;**246**:11–21.
16. Leodolter A, Nocon M, Vieth M, Lind T, Jasperson D, Richter K, Willich S, Stolte M, Malfertheiner P, Labenz J. Progression of specialized intestinal metaplasia at the cardia to macroscopically evident Barrett's esophagus: an entity of concern in the Pro-GERD study. *Scand J Gastroenterol* 2012;**47**:1429–35.
17. Ronkainen J, Talley N, Storsknubb T, et al. Erosive esophagitis is a risk factor for Barrett's esophagus: a community-based endoscopic follow-up study. *Am J Gastroenterol* 2011;**106**:1946–52.
18. Falkenback D, Oberg S, Johnsson F, Johansson J. Is the course of gastroesophageal reflux disease progressive? A 21 year follow-up. *Scand J Gastroenterol* 2009;**44**:1277–87.

Chapter 14

Progression of GERD at a Pathological Level

The progression of gastroesophageal reflux disease (GERD) at a pathological level is not relevant to the study or clinical management of GERD at the present time. The critical early stages of GERD where the squamous epithelium of the most distal abdominal esophagus undergoes columnar metaplasia, loses sphincter tone, and forms the dilated distal esophagus are not yet recognized.

The reasons for this failure are (1) the false belief that cardiac epithelium is normally found in the proximal stomach[1,2]; (2) the false belief that the endoscopic definition of the gastroesophageal junction (GEJ) (proximal limit of rugal folds) is correct[3]; (3) the systemic failure to biopsy the region distal to the endoscopic GEJ in the endoscopically normal person, with or without GERD symptoms[4]; (4) the definition of GERD that requires "troublesome symptoms."[5]

The fallacy of these dogmas and recommendations will be recognized sometime in the near future. The evidence is overwhelming and cannot be ignored for much longer.

The primary result of this failure is that there is no present test that has practical value in the diagnosis of GERD. The disease is recognized only at the point of severity of lower esophageal sphincter (LES) damage where LES failure (tLESR) is so frequent as to produce sufficient reflux to cause troublesome symptoms or, worse, adenocarcinoma without preexisting troublesome symptoms.

When a patient is diagnosed with GERD in this manner, the disease is already at an advanced pathologic stage. Pathology is only valuable at the present time in the assessment of progression from advanced disease defined by the presence of visible columnar-lined esophagus (CLE) at endoscopy→Barrett esophagus→increasing dysplasia→adenocarcinoma.

GERD, as defined at present, is only recognized either at the cusp of irreversibility characterized by treatment failure and visible CLE or beyond the point of irreversibility. Treatment directed at this group has little chance of success. This is painfully apparent with 30% patients with GERD whose lives have been disrupted in a significant manner and an uncontrolled increase in adenocarcinoma.[6]

In this chapter I will present pathology in a manner that is designed to understand the pathogenesis of GERD at a cellular level defined by pathologic changes in the entire extent of the esophagus, including the dilated distal esophagus.

This will begin with the definition of the normal state (see Chapter 5), the changes that are associated with early LES damage as it progresses to exhaust its reserve capacity to the point where it starts failing to cause reflux. The new understanding will result in a practical method of early diagnosis and, hopefully, prevention of the devastating complications of GERD such as disruption of quality of life because of treatment failure and adenocarcinoma.

1. HISTOLOGIC DEFINITION OF GASTROESOPHAGEAL REFLUX DISEASE

In 2013, I received an unexpected invitation from Stuart Spechler, one of the most authoritative gastroenterologists in the areas of GERD and Barrett esophagus. He was editing an issue of the *Current Opinion in Gastroenterology* and had thought that my opinion was sufficiently valuable for inclusion. It was the first positive expression of interest in my new ideas from any gastroenterologist. I had previous invitations from surgeons, largely because of Dr. Tom DeMeester's reputation, and pathologists. People who believe these new concepts were limited to these specialties; with few exceptions, gastroenterologists have not been receptive.

This was not the first invited review I have written. From the initial suggestion that cardiac epithelium was metaplastic esophageal rather than normal gastric,[7,8] the concept has solidified and expanded with time to defining the true GEJ as opposed to the false endoscopic GEJ,[3] and the recognition of the dilated distal esophagus.[9] This has led to a new histologic method of measuring LES damage, which is developed to its fullest in this book.

GERD. http://dx.doi.org/10.1016/B978-0-12-809855-4.00014-2

In the introduction of my chapter in Current Opinion in Gastroenterology[10]: "The definition of a disease is a statement of the level of scientific understanding of the disease. At present, GERD is practically defined merely by the presence of its main symptom, heartburn, without a standardized frequency, or duration… Histologic definition only comes into play in the diagnosis of GERD complications like Barrett esophagus, dysplasia and adenocarcinoma. Whereas other histologic changes exist in GERD, these presently have no definitional value."

"The natural scientific evolution of most diseases progresses from clinical to histologic definitions. This has not yet happened in GERD. In this article, I will explore the possibility of defining GERD by histology." After discussing epithelial types seen in the distal esophagus and stomach, I proposed: "… a histologic definition of the normal state as the absence of any epithelia other than esophageal squamous epithelium and gastric oxyntic mucosa, which lines the proximal stomach… The normal person has only two epithelial types: squamous epithelium in the esophagus, which can be damaged by exposure to gastric contents; and gastric oxyntic mucosa, which resists damage by gastric contents… The presence of any oxyntocardiac, cardiac, or intestinal epithelia between esophageal squamous and gastric oxyntic mucosa… is a histologic finding that indicates cellular damage caused by GERD. I contend that this definition is more specific, sensitive, and scientific than the clinical definition of troublesome symptoms."

From this definition of normal, I attempted to classify the progression of GERD by a combination of histologic and endoscopic findings (Table 14.1).

The above definitions are accurate, based on evidence, and valuable from an academic standpoint. However, they have no relevance in terms of practical value in the diagnosis of GERD. If one were to define GERD by the presence of metaplastic cardiac epithelium (with and without parietal and/or goblet cells) in the squamooxyntic gap, nearly everyone in the population would have GERD.[2,11] This is correct, but hardly of any practical value.

The main value of these definitions was to show that the presently used definition of the GEJ at endoscopy was fatally flawed. To expect that to cause a change in the way endoscopy was interpreted or change the management of GERD was unreasonable at best.

TABLE 14.1 New Definitions of the Normal State and Gastroesophageal Reflux Disease (GERD) Based on Histologic Criteria

	Histologic Criteria
Normal	Squamous epithelium lining the entire esophagus, transitioning at the GEJ to gastric oxyntic mucosa. Squamooxyntic gap = zero.
Squamooxyntic gap	Gap between distal limit of squamous epithelium and proximal limit of gastric oxyntic mucosa, composed of oxyntocardiac ± cardiac ± intestinal epithelia.
Dilated distal esophagus	The area distal to the tubal esophagus, containing rugal folds, that is composed of oxyntocardiac ± cardiac ± intestinal epithelia. Mistaken by present endoscopic definition of the gastroesophageal junction (GEJ) as gastric cardia.
Columnar-lined esophagus (CLE)	Presence of a squamooxyntic gap of any length (=endoscopically visible CLE in tubal esophagus + dilated distal esophagus).
Absence of chronic GERD	Absence of a squamooxyntic gap (=absence of histologically defined CLE)
GERD	Presence of a squamooxyntic gap of any length (=presence of histologically defined CLE)
GEJ	Proximal limit of gastric oxyntic mucosa. In normal: junction between gastric oxyntic mucosa and esophageal squamous epithelium. In GERD patients: junction between gastric oxyntic mucosa and CLE.
Severity of GERD	
Mild GERD	Endoscopy: no visible CLE; biopsy: oxyntocardiac ± cardiac epithelia (=CLE) in the squamooxyntic gap limited to the dilated distal esophagus.
Moderate GERD[a]	Endoscopy: visible CLE <2 cm; biopsy: squamooxyntic gap with CLE extent = visible CLE + CLE in dilated distal esophagus.
Severe GERD[a]	Endoscopy: visible CLE >2 cm; biopsy: squamooxyntic gap with CLE extent = visible CLE + CLE in dilated distal esophagus.

[a]*In this book, I have combined these into one category of severe GERD based on the presence of visible CLE. This recognizes the lack of need to separate short- and long-segment visible CLE.*

The main value of this review paper to me in retrospect is to demonstrate that nothing that I wrote had any practical value. I was trying to push concepts that were best kept in the ivory towers of academia. They had no interest to the practicing physician.

Who cares where the esophagus ends? Who cares what the GEJ is? Who cares that cardiac epithelium is esophageal and not gastric? What value is there in a histologic definition of GERD if it means that everyone has GERD? To reiterate one of the classic political statements of the past election year: "What difference does it make?"

In that review, I described the pathogenesis and the variation of the length of the dilated distal esophagus: "Chandrasoma et al. showed that oxyntocardiac, cardiac and intestinal epithelia extended into the region with rugal folds distal to the end of the tubal esophagus. This extension varied in length from 0.36 to 2.06 cm. The presence of these epithelia in this area was concordant with the presence of submucosal glands… This area, which is distal to the point of flaring of the tubal esophagus and contains rugal folds, is the dilated distal esophagus… The presence of a dilated distal esophagus correlates with shortening of the lower esophageal sphincter, which correlates with progressive GERD… With destruction of the sphincter and dilatation of the end of the tubal esophagus, the tubal part of the abdominal esophagus shortens and the angle of His becomes more obtuse, increasing the susceptibility to sliding hiatal hernia."

It was after this review that I realized the importance of the concordance of the dilated distal esophagus and abdominal LES damage that is the crux of this book. This new understanding changed what had been an academic exercise to a method of changing the way GERD is diagnosed and managed in a fundamental way.

The practical value of these data is not the *presence* of CLE distal to the endoscopic GEJ in the dilated distal esophagus, but that *the length of CLE distal to the endoscopic GEJ in the dilated distal esophagus is an exact measure of the amount of damage (shortening) of the abdominal segment of the LES.*

This is the basis of a new critical diagnostic test—the histologic assessment of LES damage—that can measure severity of the etiology of GERD long before clinical GERD rears its ugly head.

The new ability to measure LS damage should drive change with a sense of urgency. It changes the present attitude of cancer prevention in GERD which can be described in real terms as: "Diagnose GERD as late as possible; delay endoscopy as much as possible; never biopsy anything other than visible CLE; give PPIs to everyone to control symptoms; hope that the patient does not belong in the 30% who will develop treatment failure; hope and pray that the patient will not develop cancer with the certainty that there will be 20,000 times this year that this hope and prayer will not be answered in the USA."

The new method will be: "Diagnose GERD by assessment of abdominal LES damage as early as practically possible but before the development of visible CLE; Identify patients who will progress to severe GERD characterized by treatment failure and visible CLE; treat only those patients who are not at risk of developing visible CLE with PPIs, confident that they will remain well controlled and not develop cancer; Intervene early and aggressively by methods to slow or prevent progression of LES damage in those patients at high risk of progression to severe GERD (=visible CLE); Decrease the incidence of treatment failure and adenocarcinoma."

Time will tell whether and how quickly this change occurs.

2. VERTICAL AND HORIZONTAL PATHOLOGIC PROGRESSION OF GASTROESOPHAGEAL REFLUX DISEASE

The natural progression of GERD at a pathological level progresses in vertical and horizon directions.

Vertical progression depends on the extent of esophageal squamous epithelium that undergoes damage by exposure to gastric contents, beginning at the true GEJ (proximal limit of gastric oxyntic epithelium) and progressing cephalad.

It increases in the following manner: normal state→columnar metaplasia limited to the dilated distal esophagus→reflux esophagitis in the thoracic esophagus→short- and long-segment visible CLE (Fig. 14.1).

Vertical progression reflects progression of severity of GERD. The greater the vertical progression, the greater is the severity of LES damage, greater is the amount of reflux into the thoracic esophagus, and greater is the vertical extent of cellular damage in the esophagus. Vertical progression that involves the thoracic esophagus causes failure of proton pump inhibitor (PPI) therapy to control symptoms completely. This is a state where the amount of reflux into the thoracic esophagus overwhelms the capacity of PPIs. When reflux reaches the point at which visible CLE occurs, adenocarcinoma cannot be prevented with any medical therapy in the people destined for that end point.

The objective of the new management that I am trying to develop is to prevent visible CLE and/or treatment failure, which represent irreversible cellular change in the thoracic esophagus. Anything short of that is reversible.

Horizontal progression is the progression of the epithelial type at each level of the esophagus as it is exposed to gastric contents. This can be expressed as: normal→reflux-induced damage to squamous epithelium→cardiac metaplasia→bidirectional

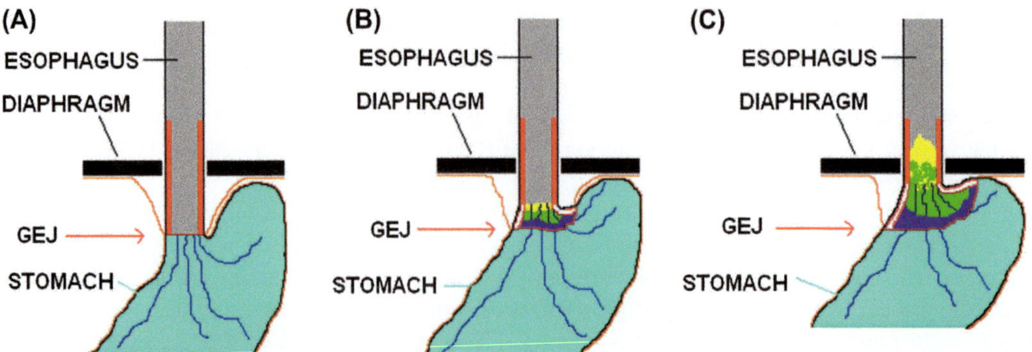

FIGURE 14.1 Vertical progression of the pathology of GERD. This begins with the normal state, characterized by an esophagus lined by squamous epithelium (gray) and a stomach lined by gastric oxyntic epithelium (blue) (A). Cardiac metaplasia of the squamous epithelium begins at the proximal limit of gastric oxyntic epithelium (the true GEJ) and progresses cephalad. In the first stage, cardiac epithelium is limited to the abdominal esophagus and is associated with LES damage (B). When LES damage reaches a critical point, reflux begins and causes cellular changes in the thoracic esophagus culminating in visible CLE (C). (Yellow = cardiac epithelium with intestinal metaplasia; green = cardiac epithelium; purple = oxyntocardiac epithelium.) *CLE*, columnar-lined esophagus; *GEJ*, gastroesophageal junction; *GERD*, gastroesophageal reflux disease; *LES*, lower esophageal sphincter.

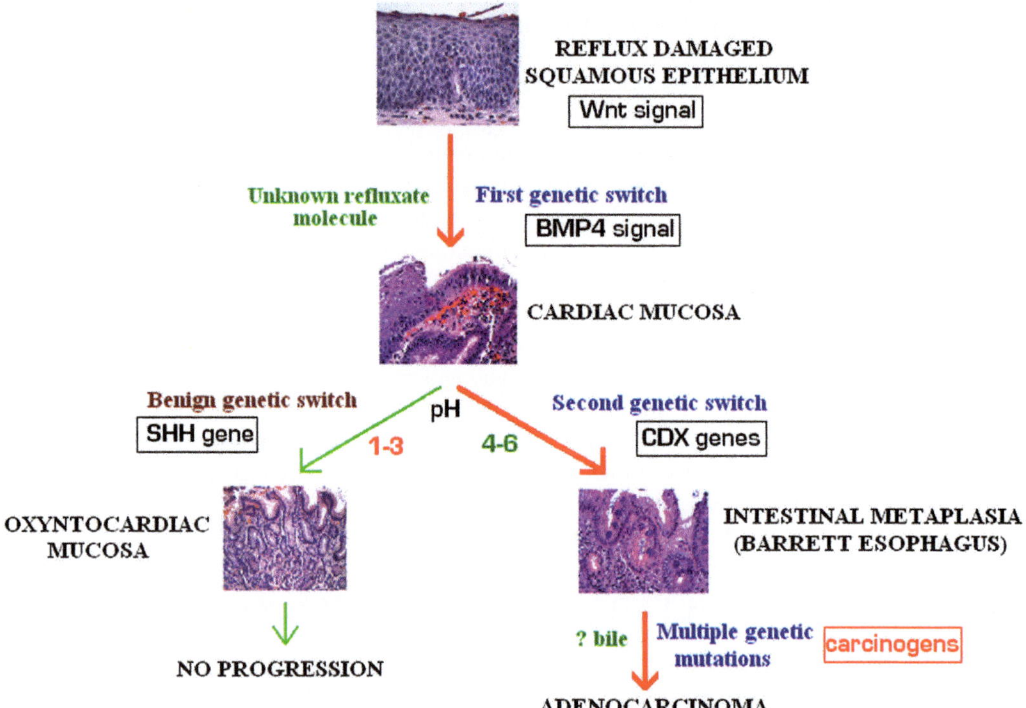

FIGURE 14.2 Horizontal progression of cellular changes of GERD. Each horizontal level of the esophagus is lined by squamous epithelium or metaplastic columnar epithelium. The progression of change at each level depends on exposure to gastric contents. The steps in the progression are squamous→cardiac epithelium; cardiac epithelium to either oxyntocardiac or intestinal epithelium; intestinalized cardiac epithelium to adenocarcinoma. The end point of change varies at different horizontal levels. *GERD*, gastroesophageal reflux disease.

evolution to oxyntocardiac and intestinal epithelium→progression of intestinal metaplasia→increasing dysplasia and adenocarcinoma (Fig. 14.2). In a given patient, the exact change that occurs will vary at each horizontal level of the esophagus depending on the milieu at that level.

The exposure of esophageal epithelium to gastric contents can be continuous or intermittent, and when intermittent, its frequency and duration can vary. The composition, volume, and concentration of the constituents of the gastric contents and the pH milieu will vary with circumstance and treatment.

In general, horizontal progression at any given level will be a reflection of the change in the epithelium resulting from that exposure. There will be changes that increase progression such as increasing severity of erosive esophagitis and intestinal metaplasia of cardiac epithelium leading to increased risk of adenocarcinoma. Conversely, changes may result

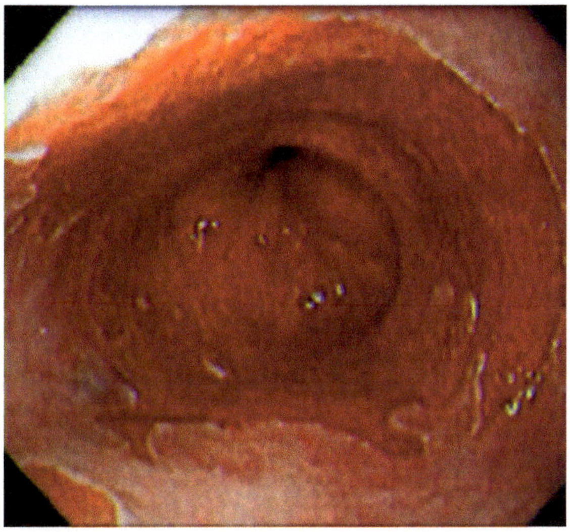

FIGURE 14.3 Endoscopy showing visible columnar-lined esophagus.

in healing of erosions and the development of oxyntocardiac from cardiac epithelium that represents a decreased risk of adenocarcinoma at that level.

Each horizontal level in the esophagus is its own microenvironment and different to its neighboring levels. The milieu at any horizontal level is not the same during every reflux episode. Every reflux episode is potentially different, based on the pressure gradients and other factors that determine the volume and composition of the refluxate. The cellular changes at each horizontal level are therefore an expression of cumulative damage over a long period.

Many epithelial changes in the esophagus in GERD are, with exceptions because of the variations, layered rather than random because of the predictable differences in the milieu at each level. Finding explanations for horizontal changes will shed light on pathogenesis of the various manifestation of GERD in a systematic manner.

All horizontal changes in the esophagus are reversible except columnar metaplasia (Fig. 14.3). The only type of columnar metaplastic epithelium that is at risk of progression to adenocarcinoma is cardiac epithelium with intestinal metaplasia (see Chapter 12) (Fig. 14.4).

Even where the objective of preventing visible CLE fails, preventing the progression of cardiac epithelium to intestinal metaplasia can prevent adenocarcinoma. However, this is much less practical because it is not predictable and the window of opportunity for prevention is very short. The safe point of irreversibility with a reasonably long window of opportunity for prevention is visible CLE.

3. FACTORS CAUSING VERTICAL AND HORIZONTAL PROGRESSION

The esophagus and stomach in a normal person is uncomplicated. Except during swallowing, when the LES is open and there is a common cavity between the esophagus and stomach, the two organs are separate.

During swallowing, the peristaltic wave in the esophagus creates a pressure gradient that passes from the esophagus to the stomach, preventing reflux. The entire esophagus, lined by squamous epithelium, is sequestered at all other times from the stomach contents by a competent LES. Never the twain shall meet; nothing in the stomach shall touch the esophageal squamous epithelium (Fig. 14.5).

A competent LES sharply separates the low pH (1–2) environment of the stomach distal to the distal end of the functional LES from the higher pH above. This point is called the pH transition point or pH step-up point (Fig. 14.5). It is at the true GEJ in the normal person without any LES damage (i.e., the junction between the SCJ and gastric oxyntic epithelium). In a person who has abdominal LES damage, the pH transition point moves cephalad with the SCJ, remaining at the proximal limit of the dilated distal esophagus. In both of these people, the LES maintains the separation during a meal as long as the stomach accommodates the meal without causing gastric overdistension (Fig. 14.6). With such a meal, the pH or composition of gastric contents is irrelevant.

This blissful state is altered by gastric overdistension, usually by a heavy meal. This results in dynamic shortening of the abdominal LES and the movement of the pH transition point to a point above the SCJ (Fig. 14.7). This results in exposure of the distal squamous epithelium to gastric juice. This dynamic change with gastric distension occurs to the same amount in the person with or without abdominal LES damage.

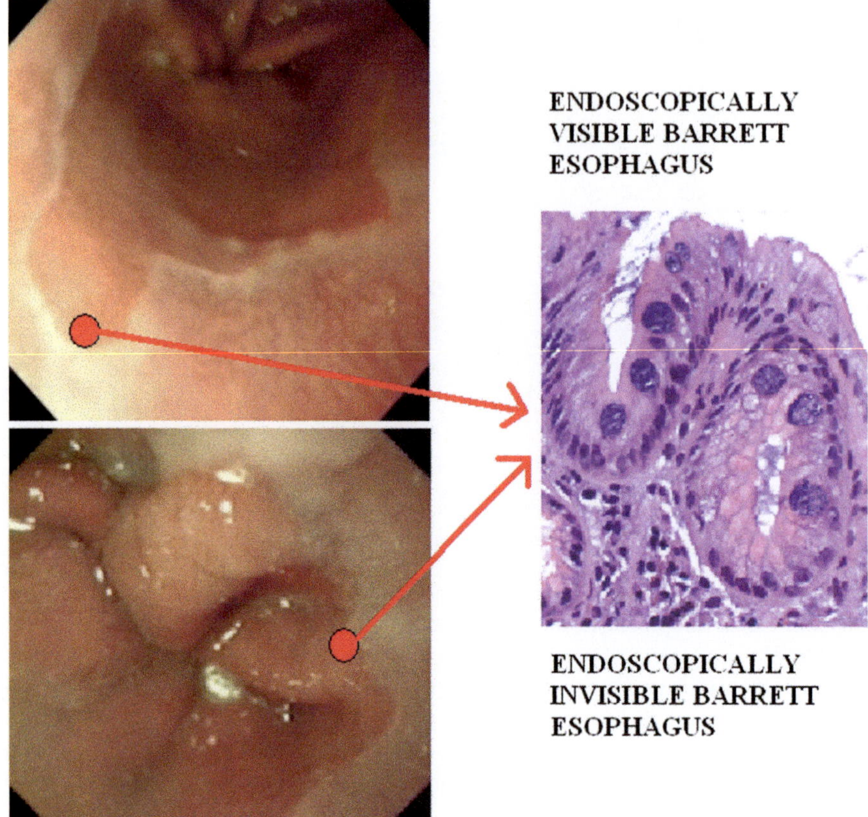

FIGURE 14.4 Endoscopy showing two lengths of columnar-lined esophagus with intestinal metaplasia in biopsies taken immediately distal to the squamocolumnar junction.

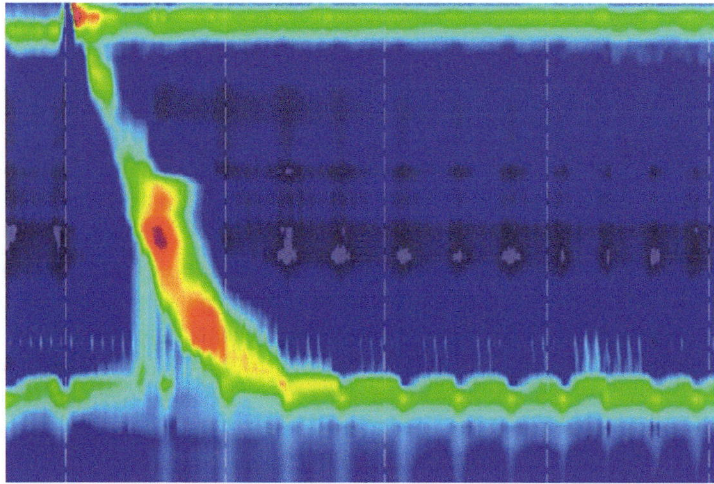

FIGURE 14.5 High-resolution manometry in a person without clinical GERD. The lower esophageal sphincter has a total length of 3.3 cm with an abdominal length of 2.2 cm. Peristalsis is normal and the LES relaxes normally. Between swallows, the LES separates the low pressure (darker blue) and higher pH in the esophagus from the higher pressure (lighter blue) and low pH in the stomach. The transition points of pressure and pH is at the distal end of the LES, which, in a person with normal endoscopy, is the location of the SCJ. *LES*, lower esophageal sphincter; *SCJ*, squamocolumnar junction. *Photograph and calculation of LES length: courtesy of Dr. Chris Dengler, Legato Medical Inc.*

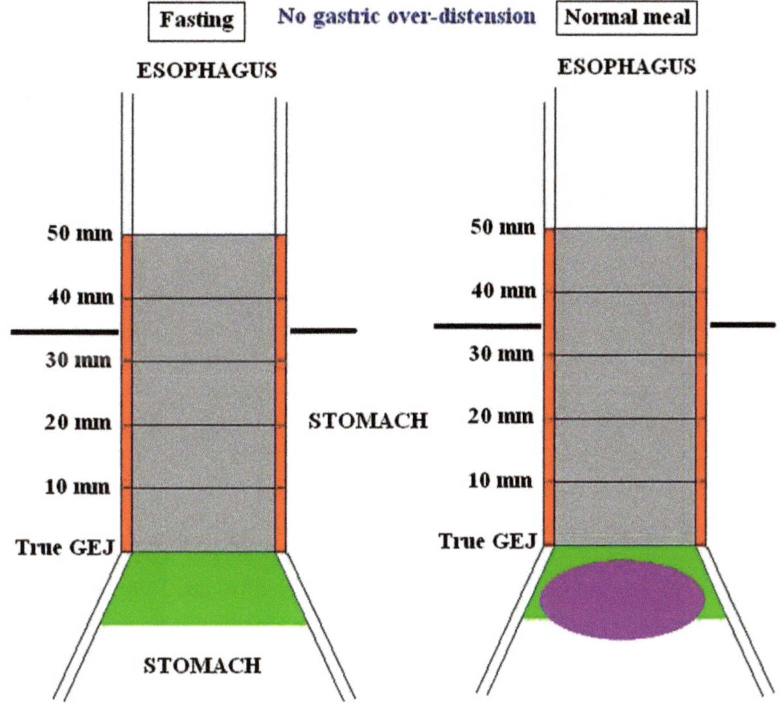

FIGURE 14.6 Physiology of a normal meal that is accommodated by the stomach without overdistension in a person who has no LES damage. The LES remains unchanged as the stomach fills to capacity. An acid pocket develops immediately below the GEJ (=SCJ) due to acid secretion during the meal. This acid pocket is kept separate from the esophageal squamous epithelium (gray) by the LES. *GEJ*, gastroesophageal junction; *LES*, lower esophageal sphincter; *SCJ*, squamocolumnar junction.

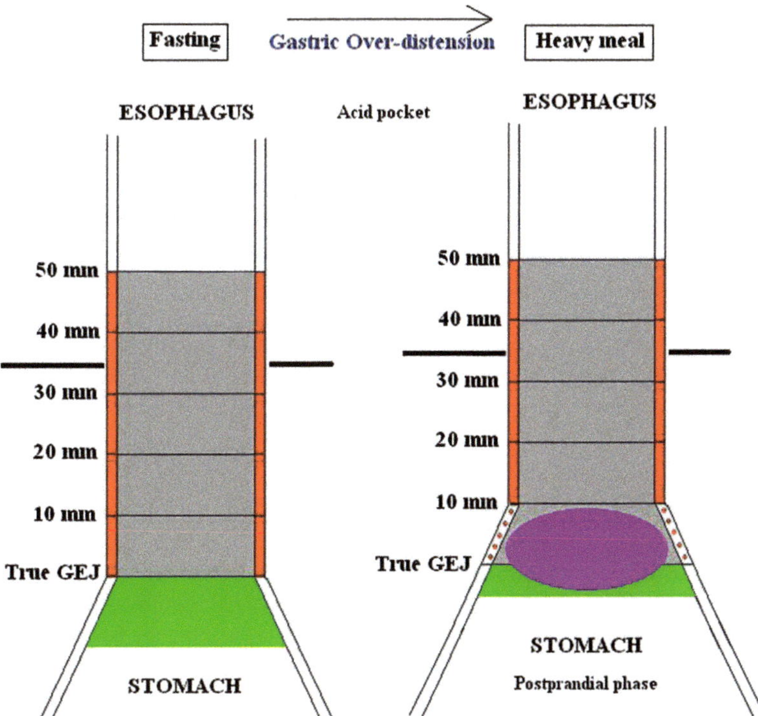

FIGURE 14.7 Physiology of a heavy meal that causes gastric overdistension. The LES becomes effaced (dynamic "shortening" shown here as being 10 mm) causing the SCJ to descend below the pH transition point. This causes the squamous epithelium to become exposed to the acid pocket for the duration of dynamic effacement. This is the fundamental cause of GERD in that it is the mechanism by which cardiac metaplasia occurs and the LES undergoes damage. *GERD*, gastroesophageal reflux disease; *LES*, lower esophageal sphincter; *SCJ*, squamocolumnar junction.

This is the cause of GERD at its most fundamental level. To understand what happens during this interaction at all levels of the esophagus that become exposed, it is important to define the variation in the composition of gastric contents, what happens when the stomach fills and overfills, and what happens when reflux occurs into the thoracic esophagus.

3.1 Composition of Gastric Contents

In the fasting state, the stomach is empty except for the small volume of basal gastric secretion, which is strongly acidic (pH 1–2). The gastric secretion contains, in addition to H^+ ions, the enzyme pepsin that is produced by the chief cells and the mucous secretion of the surface and foveolar cells. The mucin forms a thin layer on the surface of the epithelium that protects the cells from autodigestion by luminal strong acid.

Gastric contents can have many other elements: (1) food; (2) saliva; (3) intrinsic factor; (4) bile, in people who regurgitate bile from the duodenum to the stomach. Bile reflux into the stomach is believed to be extremely common. Many people will use the term duodenogastroesophageal reflux in preference to gastroesophageal reflux. There is no present interest in testing patients with GERD for bile reflux although a methodology (the Bilitec probe, which measures bilirubin) is available. Bile reflux causes mild damage to the distal gastric mucosa, causing reactive changes in the epithelium, edema, and mild inflammation. These features are not specific and, while attempts have been made, the pathologic diagnosis of bile reflux gastropathy is not practically feasible because of lack of specificity.

Pathologic states can alter the normal gastric juice composition. Chronic atrophic gastritis, both autoimmune and secondary to *Helicobacter pylori* infection, is associated with hypochlorhydria and increased resting pH. Zollinger Ellison syndrome is associated with increased acid secretion in the fasting state. Bile reflux into the stomach is increased after cholecystectomy.

H. pylori is believed to have a profound effect on the prevalence and progression of GERD. Populations, largely in East Asia, that have a high prevalence of *H. pylori* infection tend to have a lower prevalence of GERD, lower incidence of esophageal adenocarcinoma, and higher incidence of distal gastric adenocarcinoma. This has been at least partially ascribed to the hypochlorhydric state that is produced by *H. pylori* infection. When infection occurs in early life, the patient is naturally acid suppressed and this decreases initial squamous epithelial injury when gastric overdistension exposes the squamous epithelium to gastric contents. As a result, the rate of cardiac metaplasia and LES damage is slowed. Even a slight impact on the rate of LES damage can cause a significant decrease in the prevalence and severity of GERD in such populations.

Treatment with acid-reducing agents has a powerful impact on gastric contents. The treatment of symptomatic GERD is based on maintaining intragastric pH at a level that prevents acid-induced damage of thoracic esophageal epithelium during reflux episodes. It has been shown that when acid suppression maintains an intragastric pH>4 for >12h/day, the likelihood of healing erosive esophagitis and preventing recurrence of erosive esophagitis are high. Acid-reducing therapy therefore converts strong acid reflux to weak acid reflux exposure of the thoracic esophageal epithelium. Unlike *H. pylori* infection, PPI therapy is started after the onset of established reflux-induced changes in the esophagus. As such, they do not prevent GERD.

3.2 Gastric Filling and the Acid Pocket

In a normal person, the stomach fills with ingestion of a meal. The alkaline food mixes with the small volume of the resting phase gastric juice that has a pH of 1–2. The acid is buffered and the pH increases as food collects. This stimulates gastrin secretion by the G cells of the antrum, resulting in the stimulation of acid secretion by the parietal cells in the fundus and body of the stomach.

In 2001, Fletcher et al.[12] showed that an acid pocket accumulated at the top of the food column in the "cardia." Using dual pH electrode pull-through studies they showed that 15min after a meal, the pH at a point immediately below the GEJ was 1.6, lower than the intragastric pH of 4.4 more distally.

The formation of the acid pocket is believed to be due to acid secretion by the parietal cells during the gastric phase of acid secretion. The secreted acid tends to layer in the area between the intraluminal food column and the mucosa, mixing with the buffered food relatively slowly. This is maximal in the proximal stomach, which is the last area to fill during a meal. The lesser muscle activity in the proximal stomach exaggerates the paucity of mixing between the secreted acid and the food. The secreted acid in the "cardia" thereby escapes the buffering action of the food, remaining as an acid pocket (Figs. 14.6 and 14.7).

As the stomach empties, the food column moves distally and with full emptying that usually takes approximately 2h, the fasting state is restored. In the fasting state with the stomach empty, the area distal to the SCJ has very little gastric acid.

Fletcher et al.[12] measured the length of the acid pocket to be 2cm in healthy volunteers. Pandolfino et al.,[13] using a more sensitive method, measured the acid pocket in healthy volunteers to be 1cm.

Clarke et al.,[14] using a high-resolution pH catheter with 12 sensors attached to the esophageal wall with clips, showed that the sensor 0.5 cm below the SCJ ("the cardia") had only minimal acidity during the fasting state, markedly increasing in the postprandial period and reaching maximum acidity 60–70 min after the meal.

The length of the acid pocket is longer in GERD patients than in volunteers, ranging from 3 to 6.5 cm in length.[13–15] The length is even greater in patients with hiatal hernia. Beaumont et al.[15] reported a mean length of the acid pocket of 5.0 cm and a maximum length of 6.1 cm in patients with large hiatal hernias.

In all of these studies, the acid pocket is measured in the "cardia" defined as the area distal to the SCJ. The data suggest that in the more normal person, there is a resistance to expansion of the acid pocket proximally that keeps the pocket small. As the LES is damaged and the dilated distal esophagus increases in length with the cephalad displacement of the SCJ, the acid pocket elongates. Increasing length of the acid pocket in sequential increments in normal, nonhernia GERD to patients with hiatal hernia is identical to the vertical progression of change in the dilated distal esophagus. The data suggest, to me at least, that the acid pocket in the proximal stomach finds it increasingly easier to extend upward into the dilated distal esophagus as it elongates than it does into the normal esophagus that is protected by LES pressure.

Kwiatek et al.[16] showed that there is a progressive increase in the distensibility of the GEJ region as disease progressed from the asymptomatic person to the person with GERD. Again, the possibility arises that this is related to the loss of LES tone in the distal abdominal LES resulting in a dilated distal esophagus.

In an excellent review by Mitchell et al., the authors discuss GERD treatments that affect the acid pocket.[17]

1. PPI therapy: In a study of healthy volunteers, rabeprazole reduced the number and size of acid pockets, but did not abolish them; acid pockets were detected in 19% of the treated group compared with 37% in the placebo group.[18] In GERD patients taking PPIs, the acid pocket was smaller and more likely to be below the diaphragm. As a result, the number of acidic reflux events was decreased although the total number of reflux events was similar.[19] The authors suggest that the persistence of the acid pocket with PPI therapy could partly explain why some patients continue to experience symptoms despite PPI therapy. They also suggest that persistence of nonacid reflux can continue to damage the esophagus because of its pepsin and bile content.
2. Antacid medications have been used for treating GERD for over 30 years. It is only recently that alginate–antacid medications have been studied to assess their effect on the acid pocket. Alginates form a buffering layer or raft above the food column with no effect on the pH, therefore acting purely as a physical barrier.[20] Antacids, on the other hand can neutralize the acid pocket. Neutralizing capability is dependent on the distribution of the antacid in the stomach. Gaviscon Advance is the most studied acid neutralizer. This has been shown to accumulate in the proximal stomach near the GEJ compared to other antacid products that move quickly toward the distal stomach, losing its efficacy.[21] The authors conclude that alginate–antacid medications may have a use alongside PPIs in the management of refractory GERD and as maintenance acid reducing therapy in patients who prefer to not continue PPI therapy long term after initial treatment has produced control.

It is important to note that the first impulse of the gastroenterologist when they encounter the existence of the acid pocket is to find ways in which it can be neutralized. There is never a consideration that acid neutralization may have a negative effect. To the gastroenterologist, acid is always bad and a higher pH is always good. They never consider the possibility that the negative effects that have occurred in GERD like Barrett esophagus and adenocarcinoma may be a side effect of that mindset.

I was peripherally involved with a study conducted by Dr. Leslie Bernstein, an eminent and highly respected epidemiologist at the University of Southern California, to assess the association between esophageal adenocarcinoma and acid-reducing drugs. This was largely before PPIs came into the market. The main acid reducers were H_2 receptor antagonists and antacids. After a wonderfully thorough and well-controlled study was completed, Dr. Bernstein called me and indicated that although there was a highly significant odds ratio for an association between esophageal adenocarcinoma and acid-reducing drugs, she thought the study had produced a negative result. Her reason was that the odds ratio, although significant, was lower for H_2 receptor antagonists than for acid neutralizers. Since it was well known that H_2 receptor antagonists were superior to acid neutralizers in terms of acid suppression, the result had to be wrong. The odds ratio for adenocarcinoma patients using acid neutralizers compared to controls who had never used antacids was 6.32 (95% CI 3.14–12.6; $P<.01$).[22] (See Chapter 2 for a more complete review of this paper.)

At the time, the existence of the acid pocket was not known. In our discussions, I suggested to her that the result that acid neutralizers were more likely to be associated with adenocarcinoma than H_2 receptor antagonists may be valid. Acid neutralizers are used for temporary rapid relief during an episode of heartburn or to prevent heartburn when the patient knows that heartburn is likely. They alkalinize the refluxate better than H_2 receptor antagonists because their use is at the time of reflux. "Efficacy" of acid-reducing drugs is presently measured by how they control intragastric pH, not pH of the refluxate. Dr. Bernstein agreed to place this explanation in the discussion of that paper. With the discovery of the acid pocket, the explanation of how drugs that neutralize acid can be associated with esophageal adenocarcinoma becomes even more persuasive.

3.3 "Short-Segment Reflux" or "Intrasphincteric Reflux"

Normally, with the LES closed, there is a sharp demarcation of intragastric and intrasphincteric (esophageal) pressure and pH at the lower border of the LES. When the LES fails and a reflux episode occurs across an open cavity between the stomach and esophagus, a pH gradient is created from intragastric pH (usually 1–2) to intraesophageal pH (usually 7) at the top of the column of refluxed material.

The 24-h (or longer duration with the Bravo capsule) pH test is the gold standard for measuring reflux into the thoracic esophagus. Careful examination of the test methodology suggests that the pH test is not an adequately sensitive assessment of reflux. At best, it is a good method of establishing the presence of reflux of considerable severity. It is designed to be highly specific for reflux. The original studies were done at the University of Chicago by a group of foregut surgeons.[23] Their primary objective was to identify patients who were clearly abnormal and in whom antireflux surgery would have a high probability of producing a good outcome. It is not designed to be sensitive enough for the diagnosis of GERD.

The abnormal amount of reflux was defined by an abnormal acid exposure (pH < 4 for >4.5% of the measuring period) at a point 5 cm above the upper border of the LES where the measuring device was placed. By this test, a patient was required to have a pH < 4 at that point in the thoracic esophagus for 64 min per 24 h is to be considered as having abnormal acid exposure. Superficially, this seems like a very high level of acid exposure.

The point of placement of the pH electrode was determined to avoid the problem of the electrode slipping into the stomach during swallowing when the esophagus could shorten by as much as 3 cm.

The >4.5% time requirement for acid exposure was determined by comparing a group of asymptomatic volunteers with patients with symptoms of GERD and determining the amount of acid exposure that best distinguished between the two groups. An abnormal pH test was therefore designed to differentiate between patients with GERD symptoms and those without. This was appropriate because GERD was defined at the time by the presence of symptoms and the objective of treatment (including surgery) was the control of symptoms. Adenocarcinoma was not yet the large problem it is today.

The methodology of the pH test has never been adjusted from the criteria established at its inception. Its value has therefore remained the same, i.e., to provide a test that would correlate with symptoms with sufficient specificity to predict a positive outcome with antireflux surgery.

The generation of a reflux episode depends on LES failure (i.e., tLESR) where the LES pressure drops close to zero, the development of an open cavity between the stomach and the esophagus, and the presence of a pressure gradient between the stomach and esophagus.

In general, the number of molecules within the refluxed column of gastric contents will be highest at the GEJ where the column of refluxate has the largest volume, and decrease proximally, becoming zero at the top of the refluxate column. The number of all molecules in the refluxate at any given level of the esophagus will be directly proportional to the volume and height of the column of reflux as well as the concentration of that molecule in the refluxate.

For example, a carcinogenic molecule will have a maximum effect in the distal esophagus, making that area of the esophagus most susceptible. Similarly, H^+ ions will have their maximum effect immediately above the GEJ. The pH at different levels of the esophagus will vary with the baseline pH in the gastric contents or the acid pocket at the time of reflux.

There is only one thing in the refluxate that is more active in the more proximal than distal part of the esophagus. That is alkalinity (increasing pH) which increases as the number of H^+ ions decrease. A reflux episode creates a pH gradient that is equal to baseline gastric pH at the distal end of the esophagus to near neutral at the height of the column of refluxate (Fig. 14.8). Any cellular change that favors the more proximal esophagus is likely related to a higher pH. The most dramatic cellular change in GERD that occurs selectively in the more proximal esophagus is intestinal metaplasia in columnar epithelium.

Variations in the height of the column of refluxate are likely to be significant both in a given patient with different reflux episodes as well as between patients. The fact that there is a marked variation in the length of CLE in different patients and that short-segment Barrett esophagus is much more common than long-segment suggests that such a variation is likely.

What is measured in the routine 24-h pH test is the acid exposure (defined by pH < 4) at one point 5 cm above the upper border of the LES. If the electrode is placed 10 cm above the upper border of the LES,[24] the recorded acid exposure (i.e., % time pH < 4) is significantly reduced compared to an electrode located 5 cm above the LES.

There have been a few measurements of esophageal acid exposure in the area between GEJ and 5 cm above the upper border of the LES. In 2004, Fletcher et al.[25] studied the acid exposure of the most distal esophagus by placing a pH electrode 5 mm above the SCJ in addition to the more proximal electrode.

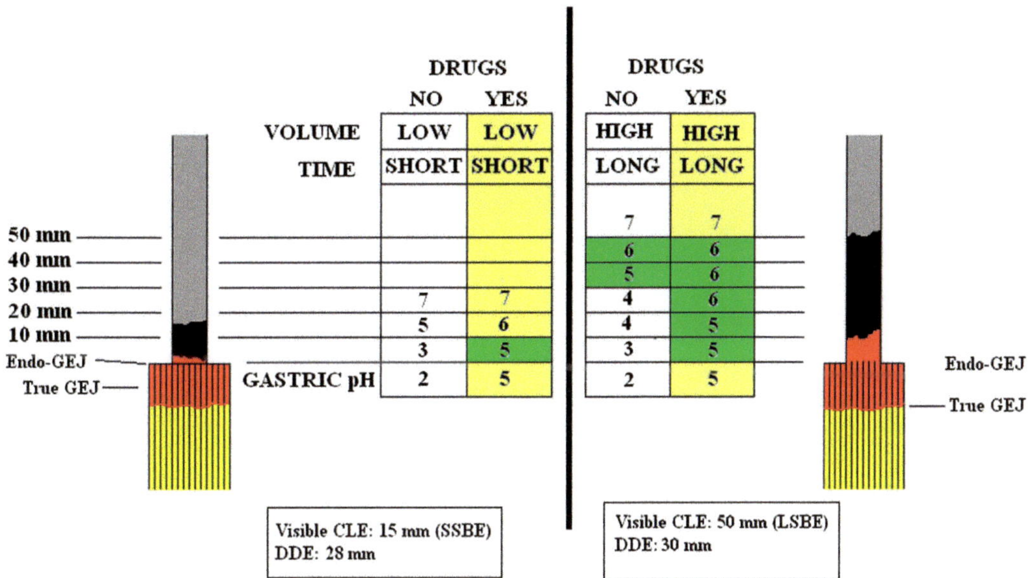

FIGURE 14.8 Reflux into the esophagus occurs when the LES fails and an open cavity is produced between the stomach and esophagus. Gastric contents flow up into the thoracic esophagus along the pressure gradient during the reflux episode. The height reached by the refluxate varies with the pressures. The pH of the esophagus at varying horizontal levels of the esophagus varies from lowest at the GEJ to highest at the top of the column of refluxate. Cellular changes result from the action of all these variables. Two scenarios with short- and long-segment visible CLE are shown in persons with and without acid suppressive drug therapy. Note that visible CLE is always preceded by significant LES damage as shown by a dilated distal esophagus measuring >25 cm in both scenarios. *CLE*, columnar-lined esophagus; *GEJ*, gastroesophageal junction; *LES*, lower esophageal sphincter; *LSBE*, long-segment Barrett esophagus; *SSBE*, short-segment Barrett esophagus. Cardiac epithelium = black; oxyntocardiac epithelium = red; pH < 5 or >6 in tubular esophagus = solid yellow; pH 5–6 = green.

In their introduction: "We recently reported that acid in the most proximal cardia region of the stomach escapes the buffering effects of food and remains highly acidic during the postpartum period. In addition, we observed that this unbuffered pocket of acid may traverse the squamocolumnar junction and extend 1–2 cm into the distal esophagus. This suggested the existence of short segment reflux with acid reaching the most distal intrasphincteric segment of the oesophagus but without traversing the sphincter."

"In studies of gastro-oesophageal reflux, the oesophageal pH electrode is traditionally positioned 5 cm above the proximal limit of the LOS. This convention was adopted early on to avoid the electrode slipping into the stomach during swallowing when the oesophagus shortens by 2–3 cm. As a consequence, a conventionally placed pH electrode will only detect acid refluxing into the distal oesophagus if it reaches this point 5 cm above the LOS."

The authors suggest that shorter columns of refluxate may not reach the conventional position of the pH electrode and underestimate acid reflux. They point to the fact that short-segment Barrett esophagus and intestinal metaplasia of the GEJ are 3.5 times more prevalent than long-segment Barrett's.

The study enrolled 14 patients with chronic dyspepsia (described as retrosternal discomfort, epigastric discomfort, and bloating) with normal upper endoscopy, no evidence of esophagitis or hiatal hernia, absent *H. pylori* gastritis, and a normal 24-h pH test (pH<4; mean 2.7%; range 0.6–4.8). 11 patients completed the study; the mean age was 43 years (range 28–56); 6 of the 11 patients were men.

pH monitoring was undertaken with a two-channel pH catheter with two pH electrodes located 5 cm apart. The catheter was positioned so that the two pH electrodes were clipped 0.5 and 5.5 cm proximal to the SCJ. Manometry was performed by station pull-through technique with measurement of the length of the LES and pressure inversion point. The patients were monitored for 24 h, during which they were given a standard lunch (soup, sandwich, and rice pudding snack) and dinner (battered cod and french fries).

Three patients could not complete the study, leaving eleven patients for data analysis. The proximal pH electrode was positioned 5.5 cm above the SCJ, which was a mean of 4.4 cm (range 3–6) above the upper border of the LES. The distal pH electrode was positioned 0.5 cm above the SCJ and confirmed by manometry to be a mean of 0.6 cm below the upper border of the LES.

The recorded acid exposure was significantly higher in the pH electrode 0.5 cm above the SCJ than at 5.5 cm above the SCJ over the 24 h and in both upright and supine position (Table 14.2) and in the preprandial and postprandial states (defined as 3 h before and after the evening meal) (Table 14.3).

TABLE 14.2 Esophageal Acid Exposure and DeMeester Score Measured 55 mm (Conventional Reflux) and 5 mm (Short-Segment Reflux) Above the Squamocolumnar Junction

	% Time pH < 4 (Total)	% Time pH < 4 (Upright)	% Time pH < 4 (Supine)	DeMeester Score
Conventional reflux	1.8 (0.8–4)	2.3 (0.9–4.3)	1.3 (0–6.1)	8 (4/19)
Short-segment reflux	11.7 (2.4–36.3)	12.7 (2.3–32.1)	10.5 (1.2–42.9)	45 (12–131)
P value	<.001	<.001	<.001	<.001

TABLE 14.3 Preprandial and Postprandial Esophageal Acid Exposure Measured 55 mm (Conventional Reflux) and 5 mm (Short-Segment Reflux) Above the Squamocolumnar Junction

	% Time pH < 4 (Preprandial)	% Time pH < 4 (Postprandial)
Conventional reflux	1.6 (0–6.7)	2.8 (0–9.3)
Short-segment reflux	14.2 (1.5–57.6)	21.8 (1–55.5)
P value	<.001	<.001

The number of recorded individual reflux episodes (defined as starting when the pH fell below 4 and ending when the pH rose above 4) was also significantly greater in the distal electrode (168; range 51–350 vs. 33; range 14–53, $P<.001$). The pH nadir (median value of the lowest pH recorded during each reflux episode) was 2.0 in the distal electrode and 2.9 in the proximal ($P<.01$). The acid exposure (pH<4) in the electrode 0.5 cm above the SCJ was significantly ($P<.05$) greater in the postprandial period (21.8%; range 1.0–55.3) than in the preprandial period (14.2%; range 1.5–57.6).

There was no significant difference in the mean duration of reflux episodes in the two electrodes. The pH tracing was similar in both recording positions. There was no correlation between the character of the patients' symptoms and the extent of acid exposure in either electrode.

The authors' interpretation of data is interesting. It is based on the assumption that these are "normal" persons, i.e., without symptoms that fall under the definition of GERD, with normal endoscopy and a 24-h pH test that is within normal limits.

The authors state: "Our findings indicate that acid may reflux onto the most distal squamous oesophageal mucosa without extending more proximally. We would suggest that an appropriate term to describe this phenomenon is 'short segment reflux'."

This is a questionable conclusion. The authors clearly show that when a "reflux episode" is detected in the lower pH electrode, it is also recorded in the proximal electrode. There is no difference in the pH tracing of the two electrodes. This would suggest that these are true reflux episodes caused by LES failure (tLESR) that extend from the stomach to the electrode 5.5 cm above the SCJ, traversing the lower pH electrode as it must.

The difference in acid exposure between the two electrodes may be explained by the fact that a pH gradient is created during an episode of reflux where the pH is lowest (=acid exposure highest), closer to the GEJ in the lower electrode than in the upper electrode.

Another possible reason for the higher acid exposure over the 24-h period in the distal electrode is acid exposure of this region that is unrelated to "reflux episodes." In a subsequent paper by this amazing Glasgow group headed by Professor McColl, Robertson et al.[26] clearly demonstrate the mechanism whereby gastric distension associated with a meal causes the SCJ to come to a point that is distal to the pH transition point (Fig. 14.7). This is the mechanism that causes cardiac metaplasia and LES damage before the onset of clinical GERD. This cephalad displacement of the pH transition point in the postprandial phase is likely to be recorded in the lower electrode, but not in the upper electrode.

The patients in this study are "normal" only in the absence of typical GERD symptoms, visible endoscopic abnormality, and a "normal" 24-h pH test as defined at present. They are likely not "normal." They have symptoms that do not fall within the definition of GERD and a pH test where the acid exposure (pH<4) in the upper electrode is between 0% and 4.5%. They also have evidence of significant exposure of the squamous epithelium in the distal 0.5 cm esophagus to acid, likely caused by dynamic shortening of the LES in the postprandial phase.

I believe that what Fletcher et al.[25] are describing in clinical and physiological terms is the subclinical phase of GERD that presently does not fall within the definition of GERD. The most interesting data in the paper to my mind is the range

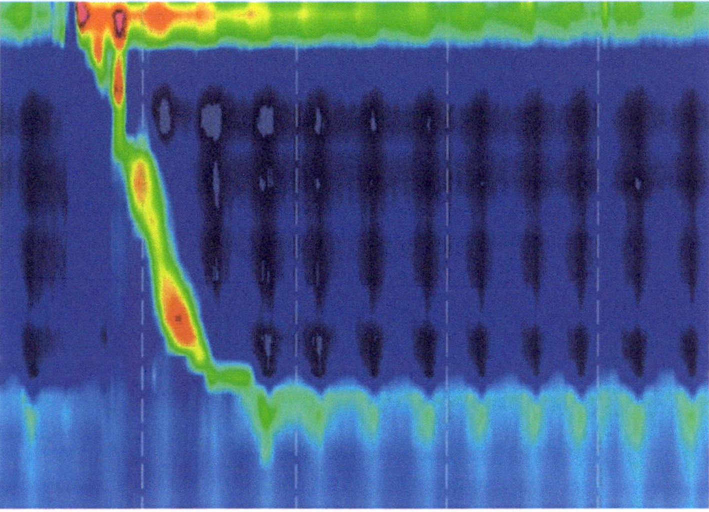

FIGURE 14.9 High-resolution manometry showing severe LES damage. The total LES length is 1.1 cm with zero abdominal LES length. *LES*, lower esophageal sphincter. *Photograph and calculation of LES length: courtesy of Dr. Chris Dengler, Legato Medical Inc.*

of abnormality reported in these 11 patients: (1) total acid exposure (pH<4) in the upper electrode ranged from 0.8% to 4% compared with 2.4%–36.3% in the lower electrode; (2) preprandial acid exposure in the upper electrode ranged from 0% to 6.7% compared with 1.5%–57.6% in the lower electrode; (3) postprandial acid exposure in the upper electrode ranged from 0% to 9.3% compared with 1%–55.5% in the lower electrode; (4) number of reflux episodes in the upper electrode ranged from 14 to 53 compared with 51–350 in the lower electrode; (5) the DeMeester score in the upper electrode ranged from 4 to 19 compared with 12–131 in the lower electrode.

This is a homogeneous population only when defined by the absence of present criteria of diagnosis of GERD. However, they are not a homogeneous population by any stretch of the imagination in relation to many other characteristics. At the low (and more "normal") extreme of the group, total acid exposure in the conventional pH test (pH<4) was 0.8%, pre- and postprandial acid exposure was 0 and the number of reflux episodes was 14 with a DeMeester score of 4. At the high (and more abnormal) extreme of the group, total acid exposure in the conventional pH test (pH<4) was 4%, pre- and postprandial acid exposure was 6.7 and 9.3, respectively, and the number of reflux episodes was 53 with a DeMeester score of 19.

The best explanation for this diversity is that these patients have increasing amounts of LES damage in the 15–25 mm range. Those patients with abdominal LES damage close to 15 mm will be at the low end of this group and those closer to 25 mm at the high end. This is a group of persons who have exhausted their reserve capacity of abdominal LES damage. They are in the early stage of GERD that can only be accurately diagnosed by the fact that they have a dilated distal esophagus >15 mm in length. At 15 mm of LES damage, they begin to get postprandial reflux with dynamic shortening of 10 mm with a very heavy meal. When they reach the critical 25 mm of abdominal LES damage, their residual abdominal LES is 10 mm and there is very little stress on the LES that is required to produce reflux.

If persons in the 15 mm LES damage end of this group self-medicate or are treated with PPI therapy and their LES damage is allowed to progress, some will progress to severe GERD and some will not. At this point in time, they are in the window of opportunity for identifying LES damage by the new test and applying the algorithm that predicts future progression with the goal of preventing severe GERD in the future.

At the present time, LES damage is simply allowed to progress until an advanced stage where severe GERD develops, manifested by treatment failure and a significant risk of visible CLE. Manometry done in these patients frequently shows complete destruction of the abdominal LES (Fig. 14.9).

3.4 Reflux of Gastric Contents Into the Esophagus

Reflux of gastric contents into the thoracic esophagus is the cause of clinical GERD. It requires failure of the LES (tLESR). Reflux does not occur to any significant extent in people who have a normal LES or one where the LES damage is compensated by virtue of the fact that the amount of damage is within the reserve capacity of the LES. I have suggested that the point where the reserve capacity is exhausted is abdominal LES damage up to 15 mm.

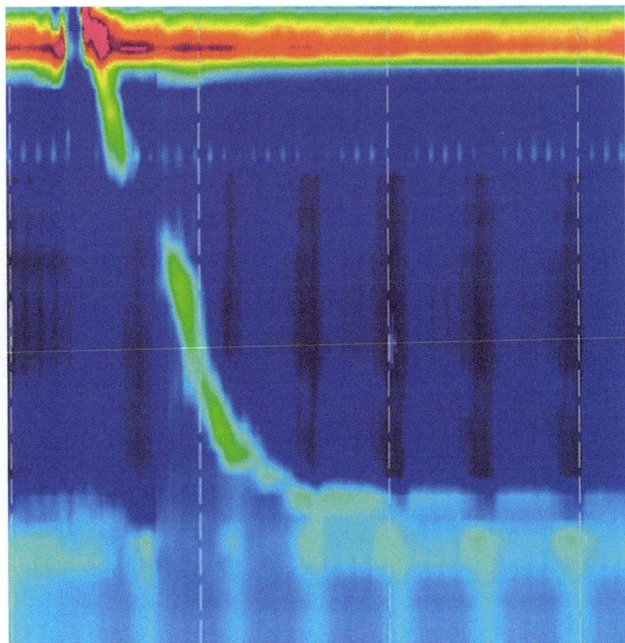

FIGURE 14.10 High-resolution manometry showing severe LES damage. The total LES length is 1.4 cm with zero abdominal LES length. The peristaltic wave generates an abnormally low propulsive pressure and there is a small hiatal hernia, both features that are common in patients with severe GERD. *GERD*, gastroesophageal reflux disease; *LES*, lower esophageal sphincter. *Photograph and calculation of LES length: courtesy of Dr. Chris Dengler, Legato Medical Inc.*

Reflux varies in volume, frequency, duration, and the height to which the refluxate rises in the esophagus. The height to which the refluxate reaches depends on the pressure gradient that exists at the time when the LES fails.

It is the frequency of LES failure (tLESR) that is largely determined by LES damage. The duration is largely dependent on how quickly the refluxed gastric contents are cleared from the esophagus. This is likely a function of esophageal body motility. With complete destruction of the abdominal LES, the angle of His is obliterated and sliding hiatal hernia occurs (Fig. 14.10). Severe reflux-induced damage of the thoracic esophagus resulting from complete abdominal LES destruction causes abnormal body motility and ineffective peristalsis (Fig. 14.10). This aggravates thoracic esophageal epithelial damage due to decreased clearing, creating a vicious cycle effect of cellular damage.

During a reflux episode, failure of the LES (tLESR) creates a common cavity, which causes the gastric contents to move rapidly along the natural pressure gradient into the esophagus.

The idea that the column of refluxate is like a water fountain that operates intermittently came to me while watching the dancing fountains in front of the Bellagio in Las Vegas. There is a valve at the tip of the nozzle and a pressurized water source below it. When the valve is opened, a column of water shoots up into the air. The column has a conical shape with maximal volume at the bottom and zero volume at the top. By changing the pressure below the valve, the operators of the fountain can vary the height of the water column (Fig. 14.8).

Similarly, LES failure (tLESR) produces a column of refluxate that has a maximal volume at the functional GEJ and slowly decreases, reaching zero at the top of the column in the esophagus. The height of the column within the esophagus probably varies with the intragastric pressure at the time of LES failure.

The number of molecules at any level of the esophagus is the sum of the concentration of that molecule and the volume of gastric contents at that level. As such, the number of any given molecule in gastric contents is maximal at the functional GEJ and decreases to zero at the height of the column.

This includes H^+ ions. The pH gradient varies from the functional GEJ where the pH is at the baseline gastric pH (normally 1–2), progressively increasing to the neutral pH 7 normally found in the esophagus at the height of the column of refluxate (Fig. 14.8).

The excellent report by Tharalson et al.[27] from the University of Arizona in Tucson studied 17 patients with varying lengths of Barrett esophagus with 24-h esophageal pH monitoring using a pH probe with four sensors located 5 cm apart. Barrett esophagus was defined by the presence of intestinal metaplasia on biopsy. Barrett's length was measured from the proximal margin of continuous Barrett's epithelium to the end of the tubular esophagus or the proximal margin of hiatal hernia folds. No data are provided regarding the distribution of the three epithelial types in this visible CLE.

A reflux episode was defined as a decrease in pH to <4 and reflux time as the interval until the pH is >4. The data were reported for each of the four sensors as the percent time that the pH was <4 during the 24h of the test. The values from mean acid exposure expressed as mean percentage time pH<4 at each sensor height (expressed as cm above the lower esophageal sphincter) are shown in Table 14.4.

The authors used a complicated mathematical calculation to show that there was a statistically significant relationship between the rate of change in acid exposure and the length of Barrett esophagus.

The data in the study are more interesting than the authors' conclusion. They show that the acid exposure (and, similarly, exposure to all other molecules in the refluxate that accompanies the acid) progressively decreases from the endoscopic GEJ to the proximal esophagus 16cm above the endoscopic GEJ when the total reflux over a 24-h period is measured. This study essentially confirms that an episode of reflux creates a pH gradient in the esophagus from a low pH in the distal region to a pH approaching neutral in the more proximal part. The top of the column appears to be in the mid-esophagus which is the nadir of pressure in the thoracic esophagus.

The results of this study can be compared in part to those of Fletcher et al.[25] (see above) who placed two pH electrodes at 0.5 and 5.5cm above the SCJ in endoscopically normal persons without GERD symptoms, visible CLE, or an abnormal pH test. This population was clearly at an earlier stage in the progression of GERD than Tharalson et al.'s[27] patients with Barrett esophagus. The contrast in the acid exposures in the two studies is interesting (Table 14.5). The total acid exposure at both levels is much greater in the patients with Barrett esophagus in Tharalson et al.[27] compared with the "no-GERD dyspeptic" patients in Fletcher et al.[25] This is evidence of profound LES damage in patients with Barrett esophagus compared with lesser LES damage in the "no-GERD/dyspepsia" group. In our new method of understanding, LES damage is >25mm in the presence of visible CLE and an estimated 15–20mm in Fletcher et al.'s[25] patient group (see above). What a difference 10mm of LES damage makes after the reserve capacity of the LES has been exhausted!

If a similar study is performed in a person who is on PPI therapy in sufficient dosage to maintain gastric baseline pH>4, it becomes obvious that the 24-h pH test will show a pH<4 in any of the sensors only when an episode of reflux coincides with a time when the PPI therapy fails in its objective to keep baseline gastric pH>4.

TABLE 14.4 The Acid Exposure Expressed as Percentage Time pH<4 in Four Different Sensors Located at 1, 6, 11, and 16cm Above the Endoscopic Gastroesophageal Junction (GEJ) in 17 Patients With Barrett Esophagus

Sensor Height (cm) Above Endo-GEJ	Total Study Time (SD)	Upright Position (SD)	Supine Position (SD)
1	**26.4%** (14.6)	**24.2%** (11.4)	**39.8%** (25.5)
6	**23.2%** (15.8)	**21.4%** (15.8)	**25.8%** (20.8)
11	**12.6%** (8.0)	**10.2%** (8.0)	**15.8%** (13.2)
16	**6.7%** (6.9)	**5.3%** (4.5)	**8.4%** (12.0)

TABLE 14.5 Comparison of Acid Exposure at Two Electrodes Placed at or Near the Conventional Position and More Distally Near the Endoscopic GEJ in Two Studies

	Tharalson et al.	Fletcher et al.
Patient characteristics	Barrett esophagus	Dyspepsia—no GERD symptoms, normal pH test, normal endoscopy
Position of upper pH electrode	6cm above endoscopic GEJ	5.5cm above endoscopic GEJ (=SCJ)
Position of lower pH electrode	1cm above endoscopic GEJ	0.5cm above endoscopic GEJ (=SCJ)
Total acid exposure (pH<4) in upper electrode	**23.2%** (SD=15.8)	**1.8%** (range: 0.8–4)
Total acid exposure (pH<4) in lower electrode	**26.4%** (SD=14.6)	**11.7%** (range 2.4–36.3)

GEJ, gastroesophageal junction; *GERD*, gastroesophageal reflux disease; *SCJ*, squamocolumnar junction.

In the patient who is on PPI therapy, the goal of therapy is to maintain gastric pH > 4 for as much of the day as necessary to prevent squamous epithelial damage when reflux occurs. In an acid suppressed person, the pH at the functional GEJ is higher (around 4), increasing progressively in the more cephalad esophagus.

In effect, the esophagus has become much more alkalinized in the patient who is on acid suppression. While this is the reason why it heals and prevents squamous epithelial damage and reduces heartburn, it is also the likely reason why these people were shown in the Pro-GERD study to progress to Barrett esophagus.[28] In Fitzgerald et al.[29] (see below), a pH of 3.5 was the pH where intestinal differentiation was maximal; at a pH < 2.5 there was complete inhibition of intestinal differentiation. Does this mean that alkalinization of the esophagus with PPIs creates the very milieu that promotes Barrett esophagus?

3.5 Types of Exposure of Esophageal Epithelium to Gastric Contents

There are two distinct types of exposure of esophageal epithelia to gastric contents: (1) In the compensated phase of LES damage, there is >0–15 mm of metaplastic cardiac epithelium in the dilated distal esophagus. This is exposed to gastric contents without sphincter protection. The exposure varies with the amount of gastric filling. (2) In people who progress beyond the phase of compensated abdominal LES damage, intermittent LES failure results in episodic exposure of a large area of the thoracic esophagus due to transsphincteric reflux. In the early stages (abdominal LES damage 15–25 mm), the esophagus is lined by squamous epithelium. In severe GERD (abdominal LES damage >25 mm) with visible columnar epithelium, even more frequent episodic exposure of large amounts of the thoracic esophagus, lined now by both native squamous and metaplastic columnar epithelium, occurs.

The possible types of exposure of the esophageal epithelium are therefore relatively continuous exposure of columnar epithelia in the dilated distal esophagus and episodic exposure of both squamous and columnar epithelia in the thoracic esophagus. A beautiful study by Fitzgerald et al.[29] in the Stanford laboratory of Dr. George Triadofilopoulos showed the different effects on cellular proliferation resulting from intermittent versus continuous exposure of esophageal epithelia to gastric contents.

From the introduction: "Barrett's esophagus is a commonly noted condition in humans, whereby normal squamous esophageal epithelium becomes replaced by metaplastic intestine-like epithelium containing goblet cells. This specialized intestinal metaplasia is noted in approximately 10% of patients suffering from GERD and is associated with nearly a 30-fold increased risk for the development of adenocarcinoma of the esophagus and gastric cardia."

"For any given patient, Barrett esophagus is heterogeneous on many levels: histopathologically in terms of the degree of dysplasia and adenocarcinoma, on a cellular level in terms of cell differentiation, and at a molecular level… and DNA content… The pattern of reflux disease, and hence acid exposure of Barrett esophagus, is also very heterogeneous. For example, reflux events are influenced by posture, have a circadian pattern and the amount of reflux that occurs in any one individual varies from day to day."

"To address the effect of acid in Barrett esophagus we first investigated whether exposure of Barrett esophagus to acid directly affects cell proliferation and differentiation. Secondly, we investigated whether there is a relationship between the pattern of acid exposure and the heterogeneity of this complex epithelium. These questions were addressed using an ex-vivo approach in which endoscopic biopsies were cultured in an organ tissue culture system. The pH of the media was acidified either continuously, or for a 1-hour pulse followed by 24-hour culture at pH 7.4. At specified time points, cell differentiation was assessed ultrastructurally and by measuring villin expression; cell proliferation was examined using tritiated thymidine incorporation or proliferating cell nuclear antigen (PCNA) expression."

The study results showed that villin expression in Barrett esophagus increased after culture in continuous acid. Villin correlates with the presence of microvilli in Barrett epithelium and is a marker of intestinal cell differentiation. Villin expression was maximal at a pH 3.5, increasing from a baseline of 25%–50% and 82% of Barrett esophagus samples after 6 and 24 h culture (contrasting with no significant increase in specimens cultured at pH 7.4). Ultrastructural examination confirmed significantly increased microvilli in the samples cultured at 3.5 compared with pH 7.4 (12.9 vs. 5.25 microvilli; $P < .001$), indicating greater maturation of the brush border.

Cell proliferation, as assessed by tritiated thymidine uptake and PCNA expression in response to all pH exposures was greatest in the Barrett esophagus samples compared with esophageal squamous epithelium and duodenum. Continuous acid exposure reduced cell proliferation in all tissue types. In contrast, after a pulse acid exposure of 1-h followed by culture at neutral pH for 24 h, Barrett esophagus tissue proliferated six times faster than normal esophagus or duodenum ($P < .01$) as well as Barrett esophagus tissue cultured continuously at pH 7.4 for the 24 h period ($P < .05$). A marked cell proliferation difference was noted in PCNA expression between continuous acid versus acid-pulsed Barrett esophagus culture. PCNA staining was seen in a mean of gland cells at pH 7.4 (control) which decreased to 4% of gland cells after continuous acid exposure. In contrast, PCNA expression increased dramatically to 35% ($P<.05$) after a short acid pulse.

The authors interpret their results to be evidence that acid affects cell proliferation and differentiation in Barrett esophagus and that the change depends on the pattern of acid exposure. Continuous acid exposure results in increased differentiation (villin production) and reduced cell proliferation suggesting a differentiated phenotype. In contrast, a short acid pulse increases cell proliferation dramatically without an increase in differentiation (villin production) suggesting a relatively undifferentiated phenotype.

They state: "These findings may be significant in terms of the heterogeneity and the variable risk of neoplastic progression within an area of Barrett esophagus… We chose a short acid pulse in order to mimic the physiologic effect of short reflux episodes… Extrapolating our data to the situation in vivo, one model that can be envisioned is that variable patterns of acid exposure in Barrett esophagus may contribute to the complex, heterogeneous epithelium seen in Barrett esophagus with a consequent variable risk of neoplastic progression. As such, Barrett esophagus cells pulsed with acid would proliferate preferentially and may in turn have a higher risk of developing dysplasia. In contrast, cells which have prolonged acid exposure would proliferate slowly, become more differentiated, and hence would be predicted to have a lower likelihood of progressing towards dysplasia."

The authors run into trouble because of the difficulty in explaining how acid exposure can vary in Barrett esophagus: "This model assumes that cells at the same level of the esophagus do indeed have variable degrees of acid exposure as a result of the reflux of gastroduodenal contents."

To the authors, at the time of publication of the paper in 1996, Barrett esophagus was limited to visible CLE and always caused by reflux of gastric contents into the thoracic esophagus as a result of LES failure. Therefore, all exposure of esophageal epithelium was episodic. The only nonepisodic exposure was in the dilated distal esophagus, which is still not recognized as esophagus.

Dr. Triadofilopoulos is well known to me and he has always been intensely interested in the concept of the dilated distal esophagus that I was describing. In fact, he was the person who suggested that I change the designation of this entity from dilated end-stage esophagus to dilated distal esophagus.

This study, where he is the senior author, explains beautifully why columnar metaplasia in the dilated distal esophagus can be expected to have a much lower risk of adenocarcinoma than adenocarcinoma arising in visible CLE in the thoracic esophagus.

Because it is not protected by sphincter pressure, the columnar epithelium in the dilated distal esophagus is subject to continuous acid exposure of the acid pocket at the top of the food column when the stomach is full. In contrast, columnar epithelium in the thoracic esophagus is always exposed to acid pulses during the short duration reflux episodes associated with intermittent LES failure (tLESR).

Another important point in this study: the maximum differentiation of columnar epithelium, as shown by increased villin expression, occurred at a pH of 3.5. Villin expression was not detected in Barrett esophagus explants grown at pH<2.5. Strong acid protects against intestinal differentiation; weak acid promotes it.

For the past 20 years, I have stated at many conferences: "Acid is bad and acid neutralization is good for symptom control in GERD. However, acid is good and acid neutralization is bad for cancer in GERD." Neutralizing the acid pocket may help in symptom control, but it drives CLE toward intestinal metaplasia; it creates a premalignant epithelium where none existed. The Fitzgerald et al.'s[29] study provides the cellular basis for this statement.

4. VERTICAL PATHOLOGICAL PROGRESSION OF GASTROESOPHAGEAL REFLUX DISEASE

Vertical progression involves pathological damage to increasing lengths of esophageal lining epithelium. This begins at the true GEJ (the junction between esophageal squamous epithelium and gastric oxyntic epithelium), and progresses cephalad.

The amount of esophagus involved shows variation from >0 to the entire length of the esophagus. In some cases, the reflux-induced abnormalities can extend proximal to the esophagus to cause pathology in the oropharynx, larynx, and airways. This causes extraesophageal pathology in GERD and many syndromes that are defined in the Montreal definition of GERD.[5] Although well recognized, I will not consider extraesophageal reflux disease in this book.

The terminology of the pathology must be simple but precise. I recommend that we retain the term GERD for the disease as it is presently defined by the Montreal definition,[5] i.e., when sufficient reflux occurs into the thoracic esophagus to cause troublesome symptoms and/or complications. The change that I propose is a subclassification of what is presently called "not GERD" into normal, compensated LES damage, and subclinical (i.e., with no symptoms at all or symptoms that do not qualify for the Montreal definition) GERD. These are all presently regarded as irrelevant. Pathology associated with these entities is not recognized and/or misinterpreted as changes in the proximal stomach.

Vertical progression occurs in two distinct phases (Fig. 14.1):

1. Columnar metaplasia associated with slow progression of LES damage in the region of the abdominal esophagus resulting in formation of the dilated distal esophagus.
2. Squamous epithelial changes and visible CLE in the thoracic esophagus as a result of LES failure and reflux of gastric contents into the thoracic esophagus.

The first phase of vertical progression of GERD involving the dilated distal esophagus occurs in almost all people at varying rates. The existence of the normal state is very rare in the adult population in the United States.

4.1 The Normal State

In the normal state (see Chapter 5), the entire esophagus is tubular and lined by squamous epithelium. It transitions at the true GEJ to gastric oxyntic epithelium. The GEJ in this normal state is defined by all of the following: (1) the distal end of the tubular esophagus, (2) the distal limit of squamous epithelium, (3) the proximal limit of saccular organ that is lined gastric oxyntic epithelium, and which has mucosal rugal folds, (4) the distal limit of the LES, and (5) the peritoneal reflection. There is no cardiac epithelium (with and without parietal and/or goblet cells) (Fig. 14.11).

4.2 Phase 1 of Vertical Progression: Compensated Lower Esophageal Sphincter Damage

The normal state is converted to the pathologic state from the mechanical effect of gastric overdistension that causes dynamic effacement of the distal end of the LES (see Chapters 7 and 9). When this happens, the squamous epithelium of the esophagus descends into the acid pocket that exists below the pH transition point (Fig. 14.7). Squamous epithelial damage occurs and, with repeated exposure, ultimately results in columnar metaplasia and an associated loss of LES pressure in the area that undergoes columnar metaplasia (Fig. 14.12).

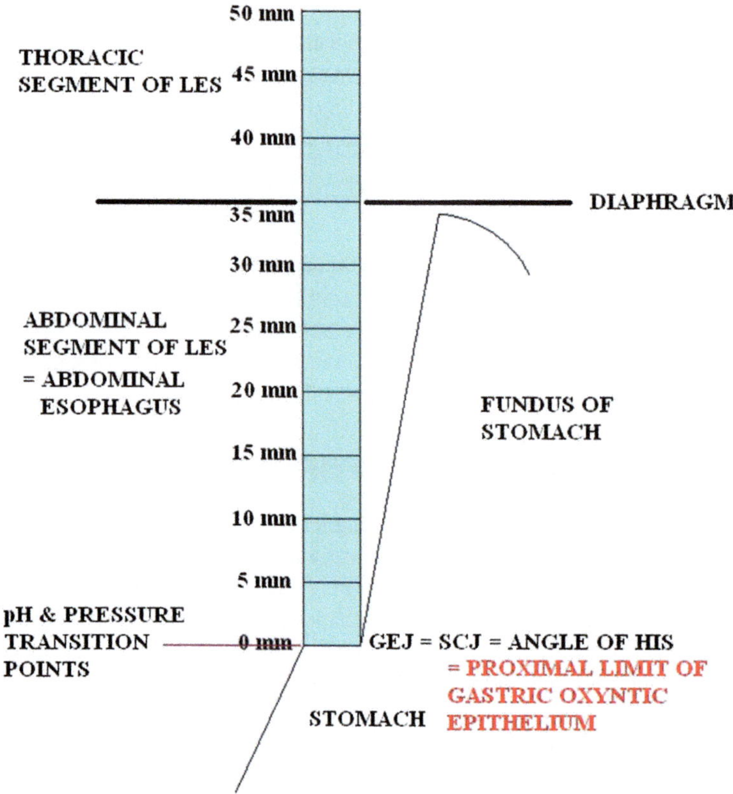

FIGURE 14.11 The normal state. The total LES length is 50 mm with 15 mm thoracic and 35 mm representing the normal length of the abdominal segment. Note that the entire esophagus is lined by squamous epithelium (blue) and the stomach by gastric oxyntic epithelium (white). The angle of His is acute. *GEJ*, gastroesophageal junction; *LES*, lower esophageal sphincter; *SCJ*, squamocolumnar junction.

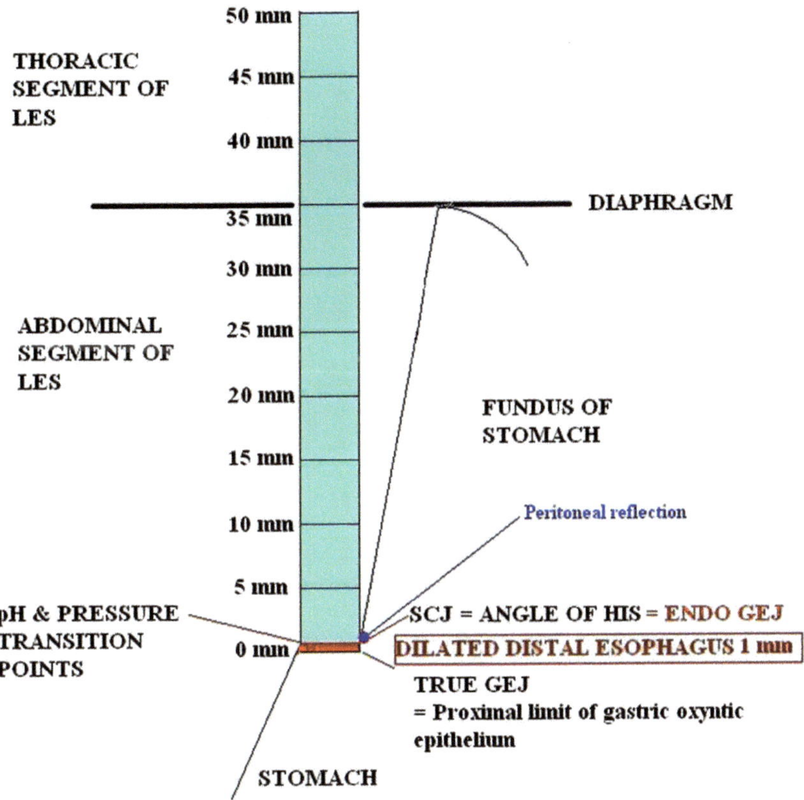

FIGURE 14.12 Very early stage of minimal LES damage with loss of 1 mm of the abdominal LES. The 1-mm cephalad movement of the SCJ due to cardiac metaplasia measuring 1 mm (red) is recognizable only by histology. Note that the functional LES is 1 mm shorter and separated by 1 mm of the dilated distal esophagus from the true GEJ. The angle of His is slightly less acute. *GEJ*, gastroesophageal junction; *LES*, lower esophageal sphincter; *SCJ*, squamocolumnar junction.

Under the influence of the positive intragastric pressure, the abdominal esophagus that loses sphincter tone dilates to form the dilated distal esophagus. Without the protection of the LES, the dilated distal esophagus distends with gastric filling and is not protected from the luminal gastric contents.

The rate of vertical progression of columnar metaplasia is a function of the continuous battle between overeating that causes gastric overdistension and the resistance of the LES. With an established eating habit, LES damage very likely increases in a linear manner over time. The length of metaplastic columnar epithelium progressively increases (Fig. 14.13) always in concordance with dilatation of the abdominal esophagus and shortening (damage) of the abdominal LES (Figs. 14.1B, 14.11, 14.12, and 14.14).

The initial phase of increasing length of the metaplastic columnar epithelium in the dilated distal esophagus is very slow. However, it is critically important because it represents the slow degradation of the reserve capacity of the LES. The abdominal segment of the LES is precious real estate. The best evidence suggests that it has an original length of 35 mm. When it reaches 10 mm, it is associated with significant LES failure (tLESR) and reflux of gastric contents into the thoracic esophagus.

A person therefore has a reserve capacity of the LES of only about 15 mm (Fig. 14.14). When 15 mm of the abdominal LES is destroyed, the residual LES is 20 mm. When this person eats a heavy meal that produces dynamic LES shortening of <10 mm, the postprandial length of the functional abdominal LES becomes >10 mm.

The phase of LES shortening from 35 to 20 mm represents the compensated phase of LES damage. The LES remains competent even during a heavy meal that causes 10 mm of dynamic shortening. There is no LES failure (tLESR) and therefore no reflux into the thoracic esophagus unless a meal is ingested that causes dynamic shortening >10 mm.

The only pathologic abnormality in this compensated stage of LES damage is the presence of cardiac epithelium distal to the endoscopic GEJ in the dilated distal esophagus to a length of >0–15 mm (Figs. 14.1B and 14.13). Histologically, people with a length of columnar metaplasia of 5 mm or less commonly have only oxyntocardiac epithelium. Those with 5–15 mm have cardiac and oxyntocardiac epithelium. Intestinal metaplasia in the dilated distal esophagus is likely to be uncommon with lengths <15 mm.

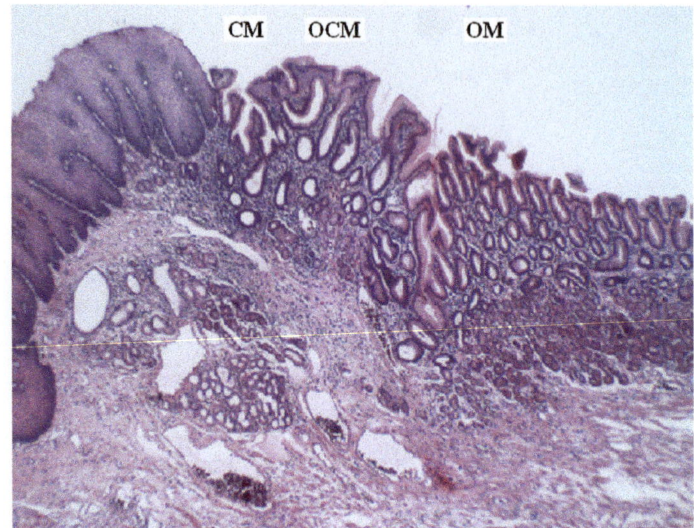

FIGURE 14.13 Histologic section showing cardiac (CM) and oxyntocardiac (OCM) epithelium of 2mm separating the squamous epithelium (on left) from gastric oxyntic epithelium (OM; the true gastroesophageal junction). This corresponds to 2mm of abdominal lower esophageal sphincter damage. *CM*, cardiac mucosa; *OCM*, oxyntocardiac mucosa; *OM*, oxyntic mucosa.

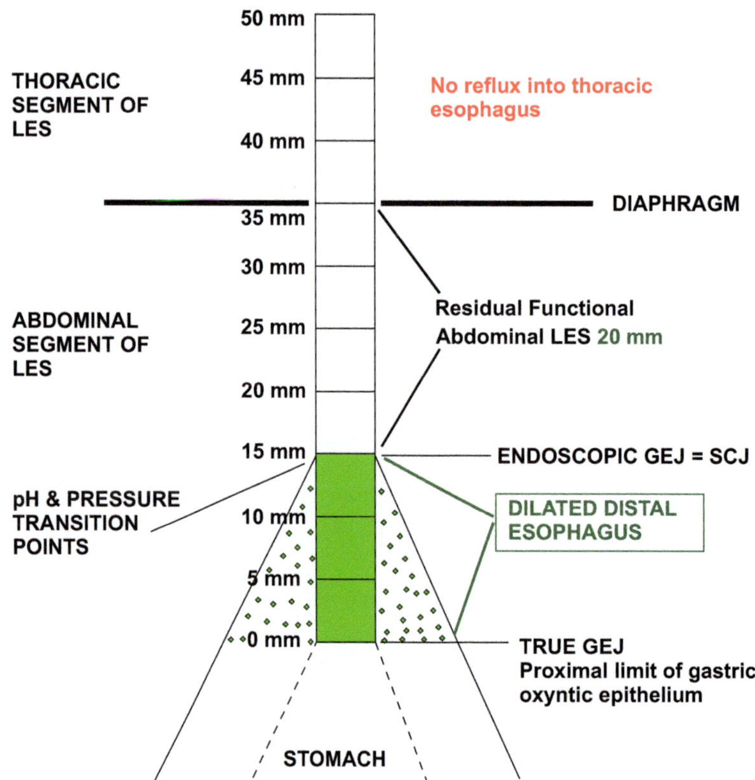

FIGURE 14.14 Progression of LES damage to 15mm as shown by a 15mm length of cardiac epithelium (green) between the esophageal squamous epithelium and gastric oxyntic epithelium, the proximal limit of which marks the true GEJ. The 15mm of cardiac epithelium is associated with LES damage of 15mm and a concordant dilated distal esophagus. The SCJ, endoscopic GEJ, end of the tubular esophagus, the pH and pressure transition points have moved 15mm proximal to the true GEJ. The angle of His is more obtuse. *GEJ*, gastroesophageal junction; *LES*, lower esophageal sphincter; *SCJ*, squamocolumnar junction.

4.3 Phase 2 of Vertical Progression: Subclinical Gastroesophageal Reflux Disease

As LES damage passes the 15-mm point (=residual abdominal LES length decreasing below 20mm), the reserve capacity of the abdominal LES has become exhausted. With 10mm of dynamic shortening during a heavy meal, LES failure (tLESR) may be precipitated and postprandial reflux results (Fig. 14.15).

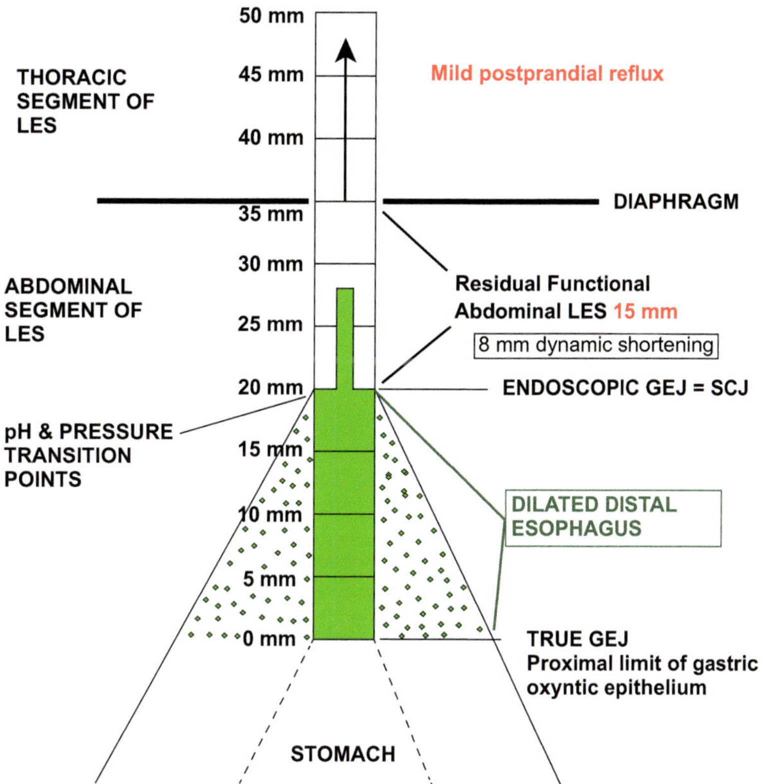

FIGURE 14.15 Further progression of LES damage to 20mm. The residual functioning LES is 15mm and now susceptible to postprandial reflux if dynamic shortening of the LES >5mm occurs with a meal. This brings the functional abdominal LES to <10mm where LES failure occurs. In the figure, dynamic LES shortening is shown as 8 mm, making the functional abdominal LES 7 mm. *GEJ*, gastroesophageal junction; *LES*, lower esophageal sphincter; *SCJ*, squamocolumnar junction.

LES failure and reflux are very uncommon with an abdominal LES that is damaged a little above 15mm. Most of these patients will be asymptomatic or have mild occasional symptoms that can be ignored.

As the patient's abdominal LES damage progresses from 15 to 20mm, the ability to eat normally decreases. Increasingly less heavy meals cause LES failure (tLESR) and symptoms. However, many of these patients will be asymptomatic or have mild dyspepsia. When typical symptoms are present, they are still not severe or frequent, and they are easily controlled with self-medication or abstention from heavy meals. The patient often does not consider the symptoms troublesome enough to consult a physician; therefore, the patient still does not fall within the Montreal definition of GERD.

In most studies of GERD, patients in this group will fall into the "control group" when GERD is defined by the presence of symptoms once a week or daily. These patients, frequently obese middle-aged white males, are the targets for advertising for over the counter acid-reducing agents. These advertisements are most prevalent in the most-watched sporting events. The advertising is truthful. For most of these patients the acid-reducing drugs produce excellent symptom relief and allow them to eat a large fatty meal without developing heartburn.

Endoscopy in this phase of the disease is usually normal or will show minimal erosive esophagitis. The pH test is greater than zero but has often not reached the accepted abnormal value of a pH<4 for 4.5% of the testing period.

The squamous epithelium of the esophagus may show mild histologic reflux esophagitis without erosion. The patients will have CLE to a length of 15–20mm limited to the dilated distal esophagus.

4.4 Phase 3 of Vertical Progression: Clinical Gastroesophageal Reflux Disease

As LES damage passes the 20-mm point (=residual abdominal LES length decreasing below 15mm) and progresses toward 25mm, the amount of LES failure (tLESR) and reflux into the thoracic esophagus progressively gets worse (Fig. 14.16). Symptoms increase in severity and the limitation to the postprandial phase decreases.

Acid exposure in the thoracic esophagus, as assessed in the pH test, becomes more likely to be abnormal, i.e., pH<4 for >4.5% of the testing period. Control of symptoms with PPI therapy is still adequate. Endoscopy, if done, will likely show no change or mild erosive esophagitis. Visible CLE is very unlikely.

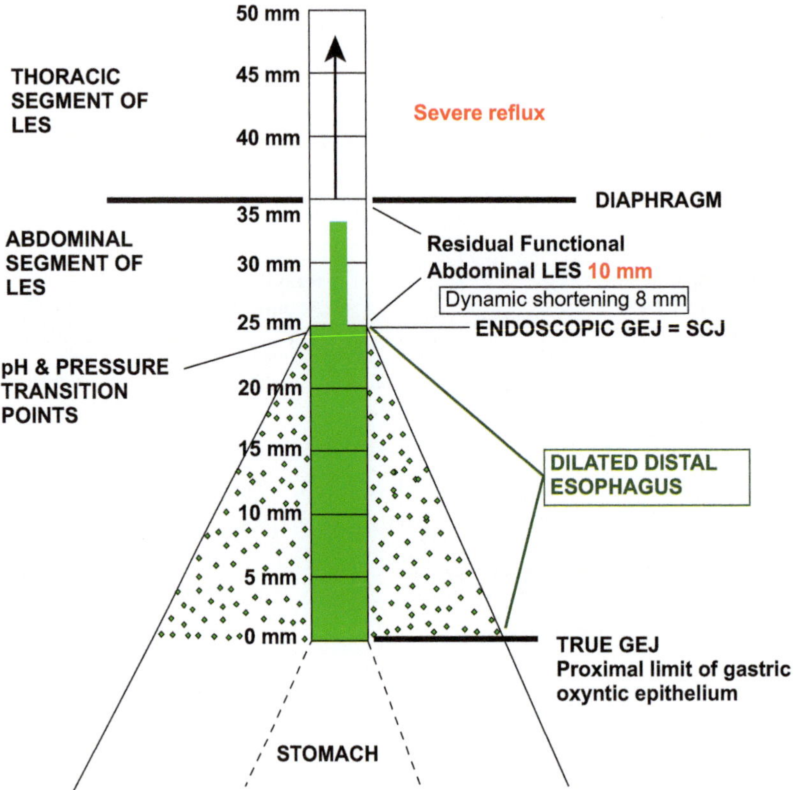

FIGURE 14.16 Further progression of LES damage to 25 mm. The residual functional LES is 10 mm. Reflux is increased further with minimal stresses on the LES that cause any dynamic effacement of the LES. *GEJ*, gastroesophageal junction; *LES*, lower esophageal sphincter; *SCJ*, squamocolumnar junction.

The population of patients in phases 2 and 3 of vertical progression (i.e., abdominal LES damage of 15–25 mm) represents the primary target for the new test of LES damage. They are at a point that is after the onset of symptoms (15 mm LES damage) and the development of severe GERD (>25 mm LES damage).

Kulig et al.,[30] in the Pro-GERD study, reported the endoscopic features of 6215 patients being treated in the community for GERD. These patients had significant symptomatic GERD but no indication for endoscopy. 2853 (46%) had nonerosive reflux disease, 2660 (43%) had erosive reflux disease, and 702 (11%) had visible CLE. By our criteria, 89% of these patients had reversible disease. For 11%, endoscopy was already too late if the objective was the prevention of adenocarcinoma.

In general, the closer the patient is to LES damage of 15 mm, the less likely is there to be visible CLE. The closer the patient is to LES damage of 25 cm, the more likely is visible CLE, although the probability of visible CLE is low with LES damage <25 mm. With the availability of an effective test, it is highly likely that the indication for endoscopy in GERD will move to an earlier stage of the disease. This is particularly true because effectiveness of intervention to prevent progression of LES damage is likely to be greater with a lesser degree of LES damage.

4.5 Phase 4 of Vertical Progression: Severe Gastroesophageal Reflux Disease

Patients with severe GERD are defined by the failure of PPI therapy to control symptoms and/or the presence of visible CLE at endoscopy (Fig. 14.3). These are the criteria of irreversibility in the management of GERD. They are the points of the disease that the new management method aims to prevent.

Visible CLE begins as nonintestinalized cardiac epithelium. There is a lag phase between the onset of visible CLE and intestinal metaplasia. If it is possible to prevent intestinal metaplasia in cardiac epithelium, cancer can be prevented. Acid-reducing drug therapy cannot prevent intestinal metaplasia in cardiac epithelium. Antireflux surgery and endoscopic LES augmenting procedures may have the capability of preventing intestinal metaplasia.

Severe GERD is associated with abdominal LES damage >25 mm where the residual abdominal LES is <10 mm (Fig. 14.17). LRS failure (tLESR) is common. As LES damage progresses to complete destruction of the abdominal and then the thoracic segment of the LES, LES failure becomes frequent and severe (a pH test with the pH<4 usually much >4.5%) and the probability of severe erosive esophagitis, failure of PPI therapy to control symptoms, and visible CLE become increasingly greater.

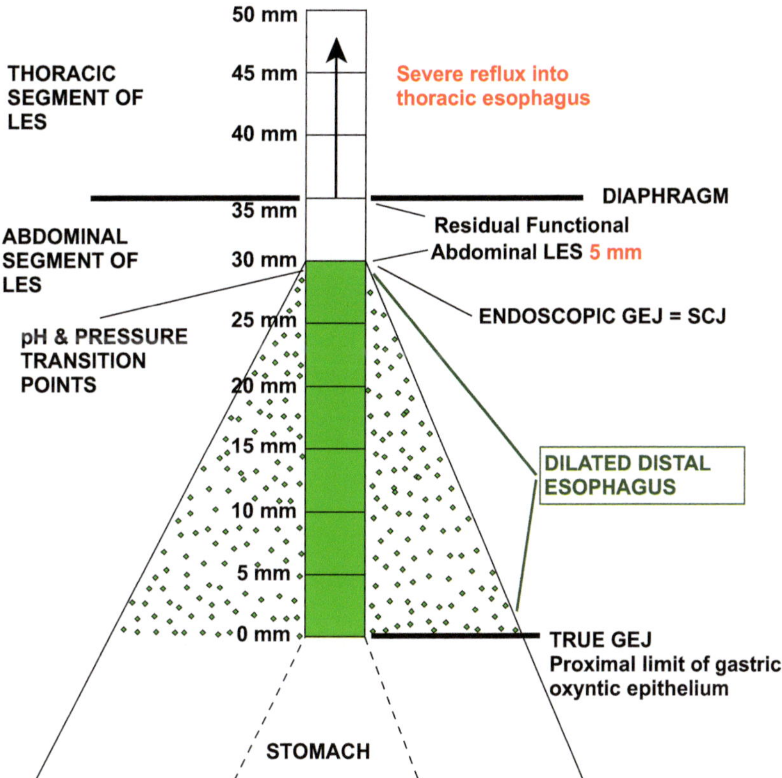

FIGURE 14.17 Progression of LES damage to 30mm and into the severe stage with uncontrollable reflux because the residual LES at rest is less than 10mm. It should be noted that this stage of LES damage is less than that shown in the manometric studies that showed a zero length of the abdominal LES. This is the stage at which patients with GERD commonly reach the care of a gastroenterologist for management of treatment failure. *GEJ*, gastroesophageal junction; *LES*, lower esophageal sphincter; *SCJ*, squamocolumnar junction.

The presence of visible CLE is *always* preceded by CLE in the dilated distal esophagus distal to the endoscopic GEJ. The length of CLE in the dilated distal esophagus is in general greater in patients with visible CLE than in those without.

The rate of development of visible CLE in the thoracic esophagus is much more rapid than the progressive increase of CLE in the dilated distal esophagus. The reason is that LES failure causes the reflux of a large volume of reflux that frequently reaches the mid-esophagus. A large area of the thoracic esophageal squamous epithelium is therefore simultaneously exposed to gastric contents.

In the Pro-GERD study,[28] some of the patients who progressed from having no visible CLE at the index endoscopy to having visible CLE 5years later, had lengths of CLE that exceeded 50mm. Compared to the millimetric increments over years of CLE in the dilated distal esophagus, visible CLE can be termed an explosive occurrence.

There is relatively little difference between short-segment and long-segment visible CLE and Barrett esophagus in terms of associated severity of reflux in the 24-h pH test and severity of LES damage. This should not be surprising. All lengths of visible CLE are the result of LES failure that creates an open cavity between the stomach and esophagus, causing gastric contents to reflux along the pressure gradient. The distance to which the reflux column rises depends more on the pressure gradient than severity of LES damage.

4.6 Vertical Stage Migration

Vertical progression is not static. LES damage is always progressive, but at varying rates in different people. Stage migration is therefore always possible. However, most patients will remain within their stage because of the relatively large range of LES damage that defines each stage.

People in Phase 1 vertical progression (i.e., the phase of compensated LES damage) at any point in their lives can be divided into the following groups, based on the projected rate of progression of LES damage:

1. People whose rate of LES degradation (either because they eat well or their LES has a high resistance or both) is so slow that they do not reach the threshold 15mm LES damage at any point in their lives. This is the 70% of the population who will never develop clinical GERD and are not at risk for esophageal adenocarcinoma.

TABLE 14.6 Phases of Vertical Progression of Pathologic Changes in the Esophagus

	AbdLES Damage (mm)	Residual AbdLES (mm)	Reflux 24 h pH Test	Control With PPI	Cellular Changes
Normal	Zero	35	Zero	n/a	Zero
Phase 1: compensated LES damage	>0–15	<35–20	Zero to pH<4 for 1.1%	n/a	CLE in DDE 15 mm or less
Phase 2: subclinical GERD	>15–20	<20–15	>0 to pH<4 for <4.5%	Yes	CLE in DDE >15–20 mm NERD Mild EE
Phase 3: clinical GERD	>20–25	<15–10	pH<4 for < or >4.5%	Yes	CLE in DDE >20–25 mm NERD More severe EE
Phase 4: severe GERD	>25–35	<10–0	pH<4 for >4.5%	Poor	CLE in DDE >25–35 mm Visible CLE

CLE, columnar-lined esophagus; *DDE*, dilated distal esophagus; *EE*, erosive esophagitis; *GERD*, gastroesophageal reflux disease; *LES*, lower esophageal sphincter; *NERD*, nonerosive reflux disease; *PPI*, proton pump inhibitor.

2. People with a rate of LES degradation that causes them to reach the >15 mm during middle life but never progress to LES damage >25 mm. The majority of such people will retain sufficient abdominal LES length to never develop treatment failure or visible CLE. They remain well controlled with drug therapy and will never develop esophageal adenocarcinoma. They are 75% of the 30% of the population that develops clinical GERD.
3. The last group of people degrade their abdominal LES rapidly enough to run out of reserve capacity early, develop significant LES failure (tLESR) at a relatively young age, and progress to near complete destruction of the abdominal LES during their lifetime. These patients have frequent LES failure (tLESR) causing a markedly increased number of reflux episodes. The severity of reflux makes control with PPIs difficult. The severe reflux also induces severe pathologic changes in the thoracic esophagus including severe erosive esophagitis and visible CLE, Barrett esophagus, and adenocarcinoma. This is the group that needs to be identified early and treated aggressively to prevent irreversible disease by some form of sphincter augmentation *before* the sphincter has been damaged to the point of failure.

People in the higher stages of vertical progression at the time of the test can similarly be grouped into (1) those who will remain controlled with PPI therapy for their lifetime without a risk of visible CLE or adenocarcinoma and (2) those who will progress to >25 mm of LES damage and develop severe GERD. By the new test and prediction of future LES damage, all patients can be stratified into different treatment strategies, all aimed at ensuring they do not progress in their lifetime to LES damage >25 mm as long as the test is done before severe LES damage has occurred (Table 14.6).

5. HORIZONTAL PATHOLOGIC PROGRESSION OF GASTROESOPHAGEAL REFLUX DISEASE

Horizontal progression of GERD defines the pathologic change that takes place at each level of the esophagus that has been damaged by exposure to gastric contents. This progression of histologic change is different at various levels and different in the squamous epithelium and metaplastic columnar epithelium that may be present at each level.

The pressures and pH in the esophagus vary in different parts in the resting state, during swallowing, in the postprandial state, and during a reflux episode. The composition and concentration of various components of gastric juice also varies during fasting, in the postprandial state and during a reflux episode. These factors are responsible for the variation in changes that occur in the different areas of the esophagus.

5.1 The Normal State

In the normal state, an undamaged and totally competent LES prevents any part of the esophagus from coming into contact with gastric contents. The most distal squamous epithelial cell is above the pH transition point (i.e., at the normal esophageal pH that is near neutral) at all times. There is no acid exposure and therefore the normal state of the zero squamooxyntic gap is maintained (Figs. 14.1A, 14.11, and 14.18).

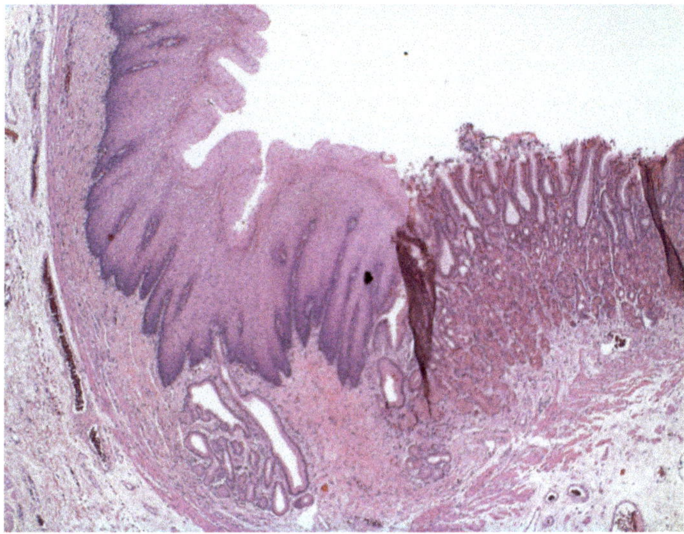

FIGURE 14.18 The normal state at histology, characterized by direct transition of esophageal squamous epithelium to gastric oxyntic epithelium. There is no cardiac epithelium and therefore no lower esophageal sphincter damage. This was from a resection for squamous carcinoma in a 67-year-old patient.

This normal state is rare. Almost everyone in the population will overdistend their stomach to expose their squamous epithelium to gastric contents sufficiently frequently to produce microscopic amounts of cardiac epithelium in their most distal esophagus.

5.2 Horizontal Cellular Changes in the Phase of Compensated Lower Esophageal Sphincter Damage

LES damage, which is the first evidence of conversion of normal to the abnormal state begins when the squamous epithelium in the most distal esophagus is exposed to gastric contents. This results from gastric overdistension that causes dynamic effacement of the distal LES such that the SCJ descends below the pH transition point into the acid pocket (Fig. 14.6).

The squamous epithelium undergoes repeated episodes of acid damage (with or without a contribution from pepsin and bile). The resistance of the squamous epithelium to undergo columnar metaplasia is high. In most instances, the squamous epithelium that is exposed transiently to gastric contents during a meal undergoes mild damage that reverses.

It is only when exposures are repeated frequently enough to prevent complete healing between exposures that cumulative damage occurs. This progressively increases permeability of the squamous epithelium sufficiently to induce columnar metaplasia. In Chapter 15, I will provide evidence that this cumulative damage is accompanied by increasing molecular changes in the squamous epithelium that culminates in metaplasia to cardiac epithelium.

The vertical progression is extremely slow, causing increase in the length of the CLE by a few millimeters per decade (Fig. 14.13). The rate of associated progressive shortening of the abdominal LES varies in concordance with increasing length of cardiac metaplasia.

In patients who stay within Phase 1 of vertical progression, the length of CLE in the dilated distal esophagus is <15 mm for the entirety of their lives (Fig. 14.9). The patient has no LES failure, no reflux into the thoracic esophagus, no symptoms of GERD, and no exposure to acid-reducing drugs. Therefore, there are no cellular changes in the thoracic esophageal squamous epithelium, i.e., no visible CLE or adenocarcinoma.

However, this <15 mm of CLE in the dilated distal esophagus is exposed to the acid pocket during meals because the epithelium therein is no longer protected by the LES.

It is likely that the acid pocket has a conical shape with its volume decreasing in a cephalad direction from the top of the food column to the endoscopic GEJ. The dilated distal esophagus is likely collapsed in the empty stomach. However, when the stomach fills it becomes exposed to every molecule in the acid pocket.

The exposure is probably maximal at the interface with the food column (at the distal end of the dilated distal esophagus) and decreases proximally within the dilated distal esophagus to the endoscopic GEJ. This gradient includes H^+ ions, creating a pH gradient from bottom (most acid) to the top (least acid). Acid exposure is therefore likely to be greatest at the distal end of the dilated distal esophagus adjacent to the true GEJ and least immediately distal to the SCJ.

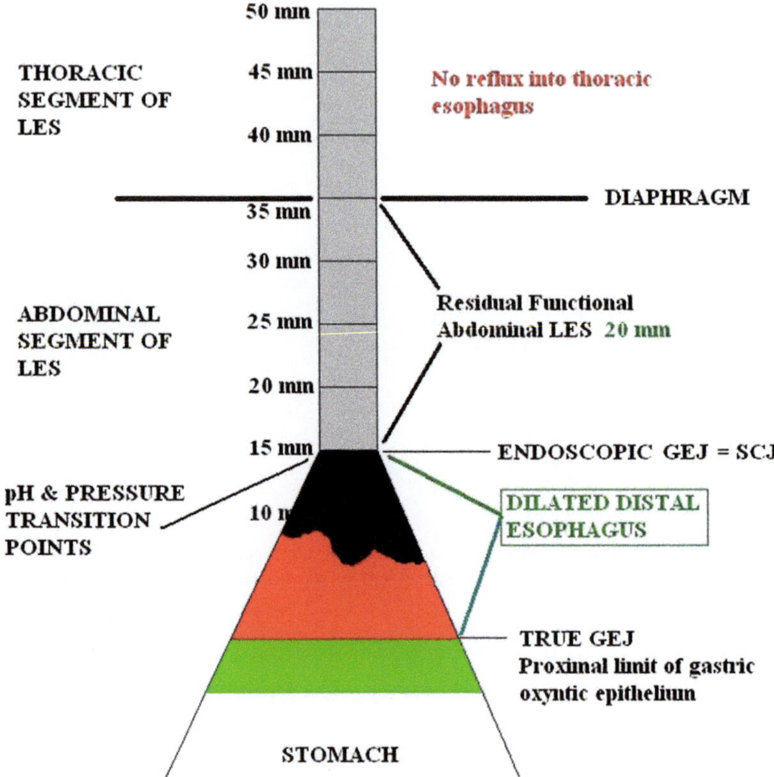

FIGURE 14.19 Patient with 15 mm of LES damage showing the typical distribution of epithelial types in the dilated distal esophagus. The residual squamous epithelium of the esophagus transitions to cardiac epithelium (black) in the dilated distal esophagus at the endoscopic GEJ. The distal part of the dilated distal esophagus is composed of oxyntocardiac epithelium (red) which transitions to gastric oxyntic epithelium at the true GEJ. *GEJ*, gastroesophageal junction; *LES*, lower esophageal sphincter; *SCJ*, squamocolumnar junction.

The type of exposure of the columnar epithelia in the dilated distal esophagus is more continuous than intermittent. The acid pocket likely moves into the dilated distal esophagus during every meal and remains in place until the stomach empties sufficiently. As such the <15 mm of cardiac epithelium in the dilated distal esophagus is likely exposed to the pH 1–2 in the acid pocket. In Fitzgerald et al.,[29] continuous exposure promoted intestinal differentiation at a pH of 3.5 while a pH of <2.5 inhibited intestinal differentiation.

The impact of exposure to the strong acid in the dilated distal esophagus is reflected in the distribution of its epithelia. In the autopsy study of Chandrasoma et al.,[2] the length of dilated distal esophagus varied from 0 to 15 mm with 56% of patients showing only oxyntocardiac epithelium and the other 44 showing cardiac and oxyntocardiac epithelium (Fig. 14.19).

Patients with only oxyntocardiac epithelium usually had a dilated distal esophagus <5 mm. When present, the cardiac epithelium was always proximal, adjacent to the SCJ and oxyntocardiac distal, adjacent to gastric oxyntic epithelium.

This distribution of the two epithelia suggests that cardiac epithelium converts to oxyntocardiac epithelium in the area of strongest acid exposure (i.e., lowest pH in the distal part of the acid pocket).

Marsman et al.,[11] in a study of people with normal endoscopy showed that when biopsies are taken straddling the squamocolumnar junction, there was a direct transition from squamous to oxyntocardiac epithelium without cardiac epithelium in 32% of people. The others had cardiac and oxyntocardiac epithelium.

People with <15 mm of LES damage will be seen predominantly at autopsy,[2,31] resected specimens in people without GERD[32] and in endoscopic biopsies taken distal to the normal SCJ in asymptomatic volunteers.[26]

5.3 Intestinal Metaplasia of the Dilated Distal Esophagus

The scanty data available suggest that intestinal metaplasia in the dilated distal esophagus is rare in the patient with compensated LES damage (i.e., <15 mm). It was not seen in the autopsy studies of people without GERD.[2,31] This is important. The absence of intestinal metaplasia means that these patients do not have a risk of progressing to adenocarcinoma of the dilated distal esophagus.

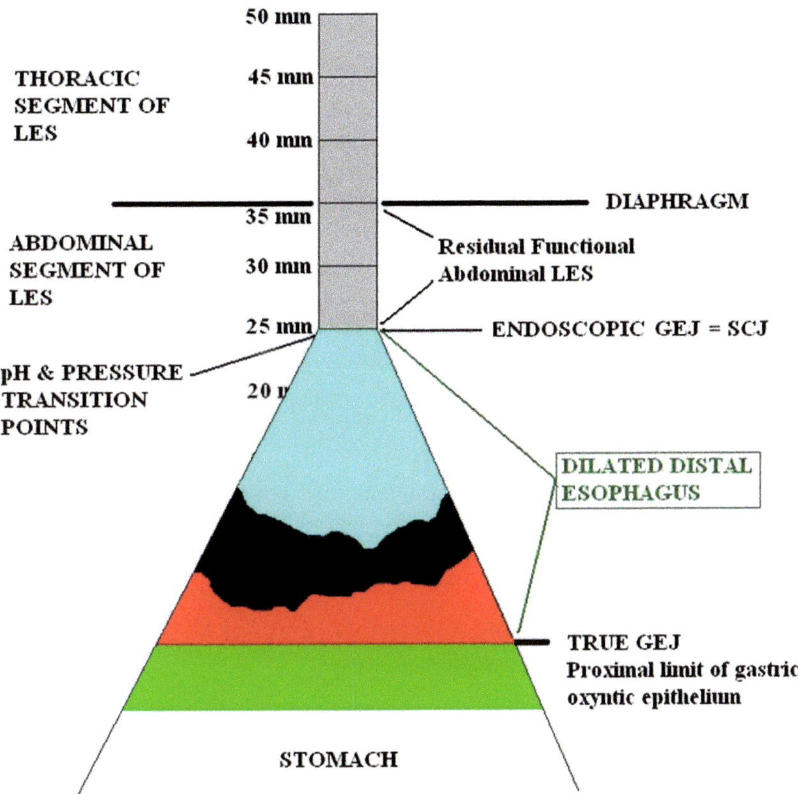

FIGURE 14.20 Patients with 25 mm of LES damage without visible CLE. This patient shows intestinal metaplasia (blue) in the proximal region of the dilated distal esophagus adjacent to the SCJ. *GEJ*, gastroesophageal junction; *LES*, lower esophageal sphincter; *SCJ*, squamocolumnar junction.

The likelihood of intestinal metaplasia in the dilated distal esophagus increases progressively when LES damage progresses from 15 to 35 mm. (i.e., from the onset of GERD, through increasing severity of GERD to complete abdominal LES destruction). At some point in this progression, it is likely that the patient develops LES incompetence and starts taking acid-reducing drugs.

The minimum length of dilated distal esophagus associated with intestinal metaplasia is unknown, but likely to be closer to 35 mm than 15 mm.

In the Pro-GERD study,[33] 25% of GERD patients who had intestinal metaplasia at the normal SCJ progressed to visible CLE within 5 years. This suggests that progression of LES damage during the 5 years moved the patient from the stage of clinical GERD (>20–25 mm) to severe GERD (>25 mm of LES damage). It therefore suggests that intestinal metaplasia of the dilated distal esophagus is an indicator of LES damage and dilated distal esophagus that is closer to 25 mm than 15 mm in patients with clinical GERD without visible CLE (Fig. 14.20).

When found, intestinal metaplasia in the dilated distal esophagus as well as goblet cell density has the highest prevalence in the most proximal region adjacent to the SCJ. It commonly occupies a relatively short length immediately distal to the SCJ followed by cardiac epithelium and oxyntocardiac epithelium.

This distribution of epithelial types in the dilated distal esophagus provides insight to the factors responsible for evolution of cardiac epithelium in its two separate directions: intestinal metaplasia and oxyntocardiac epithelium. The conversion of cardiac epithelium, the initial product of columnar metaplasia, to oxyntocardiac epithelium and intestinal metaplasia, respectively, result from two completely different genetic switches (see Chapter 15).

Oxyntocardiac epithelium, which is seen in the more acidic distal region of the dilated distal esophagus, results from activation of Sonic Hedgehog gene. This genetic switch is likely stimulated by the low pH (strong acid) environment that is present distally.

In contrast, intestinal metaplasia, which is seen in the weaker acid proximal region of the dilated distal esophagus, results from activation of Cdx-2 gene. This genetic switch is likely stimulated by continuous exposure to a higher pH (less acid) environment seen proximally. If this is true, it would mean that alkalinization of gastric juice with acid-reducing therapy will promote intestinal metaplasia in the dilated distal esophagus.

5.4 Horizontal Cellular Changes in Gastroesophageal Reflux Disease in the Thoracic Esophagus Without Visible Columnar-Lined Esophagus

In patients with abdominal LES damage of >15mm, LES failure (tLESR) becomes possible in the postprandial state. The squamous epithelium in the thoracic esophagus is exposed to reflux and undergoes damage.

Reflux into the thoracic esophagus causes damage in the squamous epithelium that is primarily the result of acid with pepsin and bile contributing. Damage will therefore be greatest at the SCJ and progressively decrease in the more proximal epithelium, which is at a higher pH during a reflux episode.

Damage in the squamous epithelium progresses endoscopically through no abnormality (nonerosive reflux disease)→increasing grades of erosive esophagitis→deep ulcers and strictures. The last stage has become uncommon with the availability of acid-suppressive therapy.

At a cellular level, squamous epithelial damage progresses from dilated intercellular spaces (ultrastructural and then microscopic)→microscopic reflux esophagitis without erosive esophagitis (Fig. 14.21)→increasing grades of erosive esophagitis.

The progression of squamous epithelial changes varies greatly depending on whether or not the patient is on acid-suppressive therapy. Most of these patients will be symptomatic and on acid-reducing therapy. Most studies reporting endoscopic changes of GERD require the study patients to stop taking PPIs for at least 2weeks before the test to produce some semblance of the nontreated condition in the esophagus.

5.5 Horizontal Cellular Changes in Severe Gastroesophageal Reflux Disease in the Thoracic Esophagus With Visible Columnar-Lined Esophagus

There is recent epidemiologic evidence that the prevalence of GERD (defined by heartburn and/or regurgitation more than once per week, the presence of troublesome symptoms as defined by the Montreal definition, and GERD diagnosed by a clinician) has increased significantly from 1991 to the present time.[34]

If it is correct that clinical GERD has increased, it means that the rate of LES damage and reflux into the thoracic esophagus has increased in this time frame. An increase in reflux results in increased damage to squamous epithelium with increased permeability. It also means increased exposure of the squamous epithelium to all molecules in gastric contents, one of which must cause the genetic switch that induces cardiac metaplasia. This translates into a likely increase in the incidence of visible CLE in the past 30years.

The occurrence of CLE above the endoscopic GEJ in the thoracic esophagus is the definition of severe GERD that I have suggested in this book (Fig. 14.3). It is the point of irreversibility in the carcinoma sequence. Visible CLE does not reverse with PPI therapy.

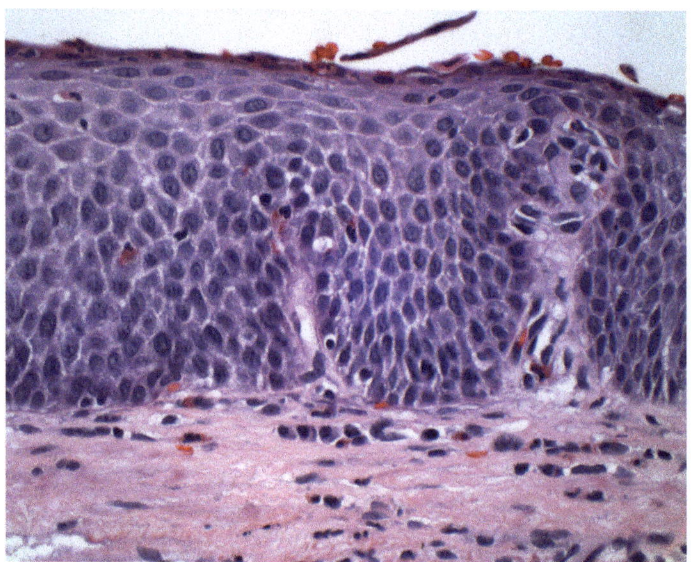

FIGURE 14.21 Histological appearance of reflux esophagitis showing basal cell hyperplasia, papillary elongation, and scattered intraepithelial eosinophils.

5.5.1 *Gastroesophageal Reflux Disease With Squamous Epithelium in Thoracic Esophagus→Visible Columnar-Lined Esophagus in the Pre–Acid Suppression Era*

The best method to evaluate the natural history of progression of squamous epithelial change is to go back to historical papers and compare the findings in that era with the present. Beautiful descriptions of endoscopic, radiologic, and histologic changes in the esophagus are to be found in the papers of Allison and Johnstone[35] and Barrett[36] in the 1950s (see Chapters 1, 5, and 11 where these papers are reviewed in detail).

Without acid suppressive drug therapy in the 1950s, those patients who progressed to severe GERD had extremely severe squamous epithelial changes. The incidence and prevalence of these changes are impossible to define because endoscopy was limited to the most severely affected patients. They had suffered in relative silence the decreasing quality of life as they progressed from microscopic change to erosive esophagitis of increasing grade. Endoscopy was indicated largely when they developed bleeding ulcers or dysphagia resulting from strictures.[35]

In the 1950s and before, patients with long segments of visible CLE were seen and reported.[37] Allison and Johnstone,[35] in their seminal paper describing CLE in 1953, reported visible CLE in 21/125 patients: These 125 patients in the study all had strictures in the esophagus with 10 patients having stenosis associated with ulcers. In a population of 125 patients with severe reflux damage in the esophagus, the prevalence of visible CLE in 21 seems low.

Barrett in 1957[36] reported that in most cases of CLE, the SCJ was 25–30 cm from the incisors, representing CLE of at least 10–15 cm. He made no mention of the presence of intestinal metaplasia in his detailed histologic descriptions.

These historical papers show that squamous epithelium underwent metaplasia to cardiac epithelium with severe reflux at a time when there was no significant ability to reduce gastric acid. The development of intestinal metaplasia in cardiac epithelium in visible CLE in this era seems to have been much less than it is today (Fig. 14.22).

There is a suggestion in the literature and a belief in many gastroenterologists that the length of visible CLE has shortened during the time frame of availability of acid-suppressive drugs, i.e., from the 1970s when H_2 receptor antagonists were introduced. This is likely to be correct. Increasing the pH of the refluxate would tend to limit initial squamous epithelial injury to a shorter length of the thoracic esophagus and thereby reduces visible CLE length.

5.5.2 *Gastroesophageal Reflux Disease With Squamous Epithelium→Visible Columnar-Lined Esophagus in the Present Time*

Today's situation is superficially improved with regard to squamous epithelial changes in the thoracic esophagus in patients with severe GERD. With adequate PPI therapy, the squamous epithelial changes reverse with complete healing at the endoscopic level in 85% of patients. Healing is least likely to be complete with the most severe grades of erosive esophagitis.

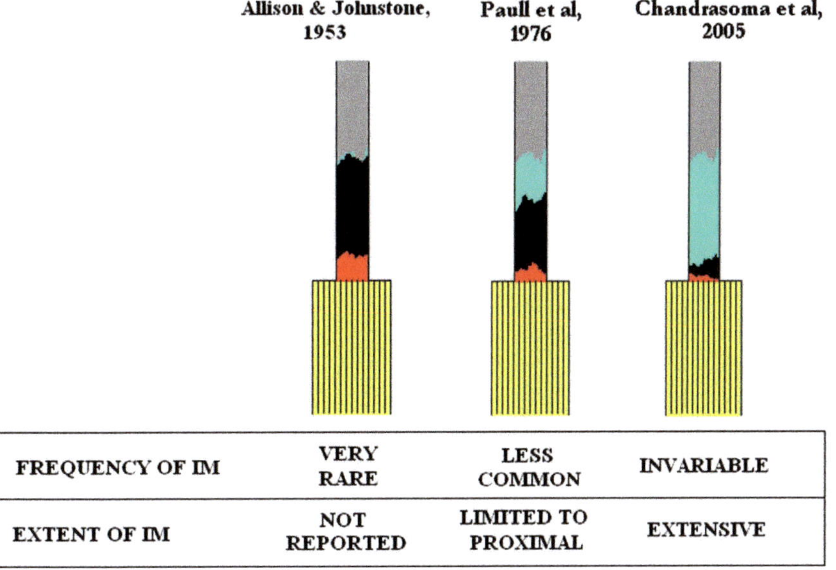

Allison & Johnstone, 1953
Paull et al, 1976
Chandrasoma et al, 2005
FREQUENCY OF IM
VERY RARE
LESS COMMON
INVARIABLE
EXTENT OF IM
NOT REPORTED
LIMITED TO PROXIMAL
EXTENSIVE

FIGURE 14.22 Historical change in the prevalence and extent of intestinal metaplasia in a segment of visible columnar-lined esophagus of comparable length. The prevalence and extent of intestinal metaplasia has increased from 1953 to 1976 to 2003. *IM*, intestinal metaplasia.

The angry ulcers and complex strictures of yesteryear have been replaced by a higher prevalence of visible CLE. The Pro-GERD study clearly shows this: while patients with GERD are being treated with PPIs, the erosive esophagitis reverses, and their visible CLE appears.[28] The appearance of visible CLE may superficially be more comfortable to the endoscopist because the epithelium is flat and appears benign. It may seem easier to manage than deep ulcers and complex strictures. This is an illusion. Visible CLE is the harbinger of adenocarcinoma, which is rarer but much more lethal than ulcers and strictures.

The prevalence of visible CLE varies in different screening studies. Ronkainen et al.,[38] in a population-based study in Finland, randomly selected 1000 patients for endoscopy and found visible CLE in 103 (10.3%; mean age 55.9 years; 60% male). Rex et al.,[39] who offered addition of an upper GI endoscopic screening for patients presenting for colonoscopy screening, found visible CLE in 176/961 (18.3%) patients. Gerson et al.,[40] in a study that screened asymptomatic persons in California, reported that 27/110 patients (25%; mean age 61 years; 92% male) had visible CLE with intestinal metaplasia (the number without intestinal metaplasia was not stated). The prevalence of visible CLE at screening is therefore 10%–25% based on differences in demographic and geographic variation.

5.5.3 Development of Intestinal Metaplasia in Visible Columnar-Lined Esophagus

Intestinal metaplasia must be present in cardiac epithelium for carcinogenesis; goblet cells mark the epithelium that is susceptible to carcinogenic activity (Fig. 14.4).

The sequential progression from cardiac epithelium→intestinal metaplasia is not impeded by PPI therapy. In fact, the alkalinization of the esophagus during reflux caused by PPI therapy may promote this step.

5.5.4 Cause of Intestinal Metaplasia in Visible Columnar-Lined Esophagus

There is rarely any attempt to describe or map the extent of intestinal metaplasia within the segment of visible CLE. The literature appears to suggest that if a single goblet cell is found in visible CLE, the entire segment of visible CLE is intestinalized. This is far from the truth (Fig. 14.23). To any pathologist that encounters Seattle protocol biopsies, the variation in the presence and density of goblet cells in biopsies from different areas of the CLE is infinite (Fig. 14.24).

In Chapter 11, I reviewed the literature that clearly shows the following undisputable and generally accepted facts about the prevalence and distribution of intestinal metaplasia in a segment of visible CLE: (1) Goblet cells, if they appear, first do so at the most proximal region of the visible CLE, adjacent to the SCJ; (2) The prevalence of intestinal metaplasia in a visible CLE segment increases with increasing length of visible CLE; (3) The distribution of intestinal metaplasia usually extends distally from its point of origin at the SCJ to a variable length of the visible CLE without skip areas. (4) Nonintestinalized cardiac epithelium is admixed with intestinal metaplasia; the density of goblet cells within this mixed epithelium is greatest at the proximal region of the segment of visible CLE and decreases distally. (5) Oxyntocardiac epithelium is limited to the most distal region of the CLE, usually in the dilated distal esophagus.

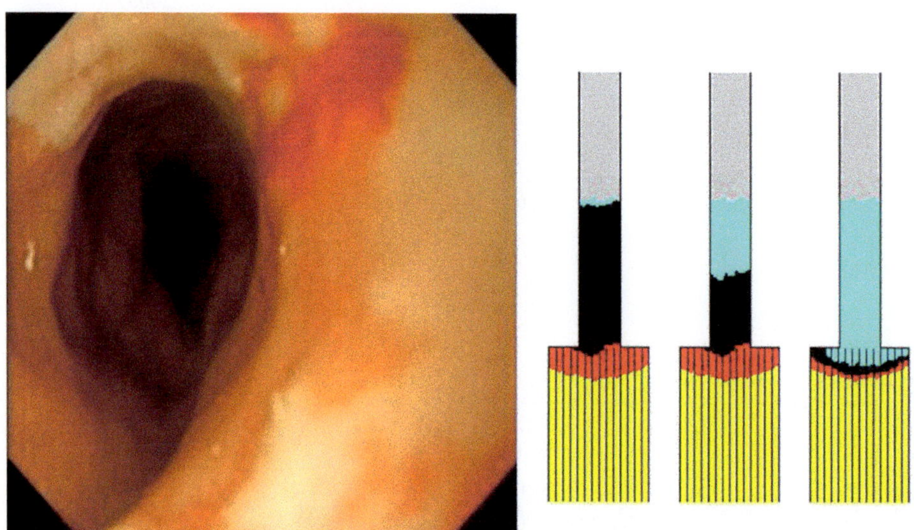

FIGURE 14.23 Visible columnar-lined esophagus (CLE) with intestinal metaplasia (i.e., Barrett esophagus), long segment at endoscopy. The presence and extent of intestinal metaplasia within this CLE segment is very variable and can be established only by mapping biopsies taken at 1 cm horizontal levels within the visible CLE. Intestinal metaplasia = blue; cardiac epithelium = black; oxyntocardiac epithelium = red.

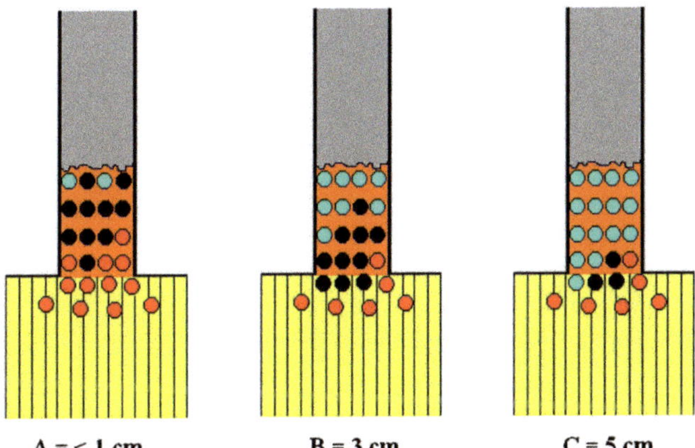

FIGURE 14.24 Histologic findings in a 5-cm segment of visible columnar-lined esophagus (CLE) with four-quadrant biopsies taken at 1-cm intervals. The distribution of intestinal metaplasia, cardiac epithelium, and oxyntocardiac epithelium, when mapped accurately shows a significant variation in different people. This is rarely done in practice and the significance of the extent of intestinal metaplasia in visible CLE has never been tested as a risk indicator for future adenocarcinoma.

Careful evaluation of the relatively scanty histologic descriptions in historical papers suggests the following: (1) The prevalence of intestinal metaplasia has increased from 1953 to the present; (2) The extent of intestinal metaplasia has increased. In the papers before 1960, intestinal metaplasia was limited to a very small region of the most proximal part of long segments of visible CLE. Today, it frequently extends into the dilated distal esophagus, involving the entire visible CLE segment[3]; (3) Adenocarcinoma has moved from its preferred location in the mid-esophagus in the 1950s to a more distal region of the visible CLE around the endoscopic GEJ and in the dilated distal esophagus.[41]

Paull et al.,[42] in their classic 1976 paper, reported that 9 out of 11 patients with long segments of visible CLE proximal to the upper border of the LES had intestinal metaplasia. Of these 9 patients, the intestinal epithelium with goblet cells extended over a length of 4.0–14.5 cm in 7 patients, but was present in only one biopsy in the other 2. The two patients who did not have intestinal metaplasia had either a mixture of junctional (cardiac) and gastric–fundic (oxyntocardiac) epithelia or gastric-fundic type (oxyntocardiac) epithelium alone. The authors provided a detailed map of the various epithelial types in four of their patients. When present, specialized columnar-type (intestinal) epithelium was always the most proximally located and the gastric-fundic type (oxyntocardiac) always the most distally located columnar epithelium. Junctional-type (cardiac) epithelium, when present, was interposed between the gastric-fundic type (oxyntocardiac) and the specialized columnar-type (intestinal) or squamous epithelium.

In their figure, intestinal metaplasia, when present, extended contiguously down from the SCJ without skip areas. In one case, it extended down into the LES zone with no other epithelial type shown. In the second, it transitioned to cardiac epithelium. In the third, it transitioned to oxyntocardiac with no intervening cardiac epithelium. The fourth case in their figure shows a small area of cardiac epithelium adjacent to the SCJ followed by 6–7 segments of oxyntocardiac epithelium. Except for the one case with all intestinal metaplasia, the LES zone contains oxyntocardiac epithelium in three cases.

When I compared the findings in Paull et al.[42] with what I routinely see in patients with visible CLE who have biopsies taken according to the Seattle protocol in our unit, the following differences are apparent: (1) The distribution of intestinal metaplasia has the same proximal predominance. (2) The extent of intestinal metaplasia within the CLE segment has increased, i.e., the intestinal metaplasia now extends further distally. (3) The prevalence of oxyntocardiac epithelium in the CLE above the LES has decreased.

These patterns of distribution and their changes over the years provide powerful clues as to the cause of intestinal metaplasia. The segment of visible CLE is a series of horizontal levels in the esophagus exposed to intermittent reflux episodes. The reflux of gastric contents during LES failure (tLESR) results in a gradient of every component within the gastric juice from the endoscopic GEJ to the top of the column of refluxate. The volume is greatest at the GEJ, decreasing in conical shape to zero at the height of the column. Every horizontal level in the esophagus that is in the path of the refluxate column is exposed to a different number (=concentration × volume) of every molecule in gastric juice. This is maximal at the GEJ and least at the height of the column of refluxate. Therefore, if one observes the location of any cellular change and that change is constant, it is possible to postulate the cause for that change.

For example, if the entire thoracic esophagus is lined by squamous epithelium and erosive esophagitis is seen maximally in the distal esophagus immediately above the endoscopic GEJ, it is likely that this change is caused by strong acid milieu. This is true that erosive esophagitis is most prevalent in the distal esophagus and is caused by acid.

It is also true that visible CLE is seen maximally in the most distal esophagus. Short-segment (<3 cm) visible CLE is three to four times more prevalent than long-segment visible CLE.[43,44] This would also suggest that the squamous epithelium→cardiac metaplasia step is favored in a stronger acid environment. It is also possible that other molecules in gastric juice (e.g., bile salt derivatives) that induce the genetic switch are involved. These will also be maximally present in the distal region of the CLE.

If, however, a cellular change occurs away from the point of maximum exposure to H^+ ions, i.e., at the most proximal limit of the visible CLE adjacent to the SCJ, it must follow that the change is favored by a weaker acid milieu in the esophagus and inhibited by a stronger acid milieu. This is true for the cardiac epithelium→intestinal metaplasia step. The likelihood of intestinal metaplasia increases whenever cardiac epithelium is present in areas of the esophagus exposed to the higher pH environment (pH 4–6) required for activation of the Cdx-2 gene pathway that results in intestinal metaplasia. This occurs in two situations:

1. When cardiac epithelium is present in the more proximal region of the thoracic esophagus. The longer the segment of visible CLE, the greater the prevalence of intestinal metaplasia (Table 14.7). At the present time, intestinal metaplasia is present in >95% of patients with visible CLE, which is >3 cm long.[43] The prevalence of intestinal metaplasia is significantly less as CLE length decreases from 3 cm. In contrast, the few studies from historic times show a much lower prevalence of intestinal metaplasia in segments of visible CLE of comparable lengths (Table 14.7).
2. When the baseline gastric pH is more alkaline than the normal resting gastric pH of 1–2. This is true in people with chronic atrophic gastritis (autoimmune or *H. pylori* induced) where parietal cell destruction causes hypochlorhydria. It is also true in patients treated with adequate acid-suppressive drugs where the objective is to maintain gastric pH>4. When reflux occurs in these patients the entire esophagus will be exposed to the higher pH environment. Also in these patients who cannot secrete acid normally, the acid pocket will likely be at a higher pH.

As the power to increase gastric pH with ever more powerful acid-suppressive drugs has increased, the pH in the esophagus during reflux has increased, favoring intestinal metaplasia in cardiac epithelium.[45]

A study from our unit by Theodorou et al.[45] examined the possibility whether there was a correlation between goblet cell density in visible CLE and the pH gradient that is created by reflux into the body of the esophagus.

The study consisted of a group of patients with a very long segment of Barrett esophagus group and a control group. The Barrett esophagus (defined as visible CLE with intestinal metaplasia) group consisted of six prospectively recruited patients (three males; median age 48 years; range 37–62 years). They all had heartburn for a duration that ranged from 6 to 20 years. The median length of the columnar-lined segment was 8 cm with a range of 6–10 cm. All had abnormal 24-h pH tests (median time pH<4 was 28.4%; Interquartile range 21.6%–56.2%). The control group consisted of five asymptomatic persons (four males) with normal 24-h pH tests.

TABLE 14.7 Prevalence of Intestinal Metaplasia in Historical Papers That Provide Reliable Data Regarding the Presence of Intestinal Metaplasia in Visible CLE

References	No. of Patients With CLE	Mean Length of Visible CLE	Prevalence of IM
Allison and Johnstone[41a]	7[a]	Long, unstated	3/7 (43%)[b]
Paull et al.[42]	11	>10 cm	9/11 (81.8%)
Spechler et al.[42a]	16	>3 cm	16 (100%)
	140	<3 cm[c]	26/140 (18.6%)
Rex et al.[39]	12	>3 cm	12/12 (100%)
	164	<3 cm	53/164 (32.3%)
Chandrasoma et al.[43]	94	>3 cm	90/94 (95.8%)
	54	<3 cm visible CLE	38 (70.4%)
Chandrasoma et al.[9]	811	No visible CLE	120 (14.8%)

CLE, columnar-lined esophagus; *IM*, intestinal metaplasia.
[a]*Only patients with detailed histologic descriptions are used.*
[b]*Includes two patients with goblet cells and one with adenocarcinoma.*
[c]*Unlike other studies, this includes people with normal endoscopy.*

All patients and normal volunteers underwent a separate multilevel 24-h measurement of esophageal pH. Endoscopy was performed only on the Barrett patients. The institutional review board that approved the study stipulated that endoscopy would not be performed on normal subjects.

The multilevel pH test was done over a 24-h period using a custom-designed catheter containing six pH sensors. Acid-suppressive medication was discontinued before the test. The catheter was inserted transnasally and placed such that most distal sensor was in the stomach 6 cm below the manometrically defined upper border of the lower esophageal sphincter. The remaining sensors were located 1 cm below and 1, 3, 5, and 8 cm above the upper border of the lower esophageal sphincter. The pH values recorded by each sensor were used to calculate the median 24-h luminal pH exposure at each level.

(Note: The patients with Barrett esophagus have severe reflux. If this correlates with near complete destruction of the abdominal LES, the gastric sensor is still in the true proximal stomach, but the sensor located 1 cm below the upper border of the LES may be in the dilated distal esophagus. This contrasts with the no-GERD people where abdominal LES damage is <15 mm, the residual abdominal LES length is >20 mm and the thoracic LES is intact. The sensor 1 cm below the upper border of the LES is therefore in the intact LES.)

Endoscopy was done in the Barrett esophagus patients. The length of CLE was measured from the squamocolumnar junction to the proximal limit of rugal folds. Four-quadrant biopsies were taken every 2 cm in the CLE. Goblet cells were identified in routine hematoxylin and eosin–stained sections and graded on a 0–3 scale based on the percentage of foveolar complexes containing goblet cells (0 = zero; 1 = less than one-third; 2 = one- to two-thirds; 3 = more than two-thirds). I read all the slides specifically for the study; the routine pathology report done by me does not report on goblet cell density. I was blinded to the results of the pH test.

All patients with Barrett esophagus had a mechanically defective sphincter by manometry while the normal persons had a normal sphincter. The median gastric pH was similar in both groups (pH 1.9 in the Barrett group). The median pH in the sensor placed 1 cm below the upper limit of the LES was significantly lower in the Barrett esophagus group compared to normal (2.8 vs. 6.1; $P = .001$).

(Note: The significantly lower pH in the Barrett group indicates that the point 1 cm below the upper border of the LES is likely to be exposed to gastric pH, i.e., below the pH transition point in the Barrett group, and protected by the intact LES in the control group.)

In the normal subjects, there was no difference in the median pH recorded by the sensor within the sphincter and the four sensors in the esophagus. In patients with Barrett esophagus, there was a significant difference in the median pH recorded by the sensor in the sphincter (pH 2.8), increasing to pH 4.7 at 8 cm above the sphincter.

(Note: This shows that the entire region extending from 1 cm below to 8 cm above the upper border of the LES is protected by the competent LES in the control group. The control group has no reflux. In contrast, the Barrett group shows a pH gradient from pH 1.9 in the stomach to pH 2.8 at a point 1 cm below to pH 4.7 at a point 8 cm above the upper border of the LES. This part of the esophagus therefore has a varying pH milieu that is 2.8 in the dilated distal esophagus, changing progressively to 4.7 at a point 8 cm above the LES.)

We divided the total length of the columnar-lined segment into three horizontal segments for the patients with Barrett esophagus to permit comparison of goblet cell density. In patients with Barrett esophagus, the median length of the proximal, middle, and distal third was 2.8, 2.5, and 2.7 cm, respectively. The median pH in these compartments was 4.4 in the proximal third, 2.4 in the middle third, and 2.2 in the distal third. The median goblet cell density was grade 3 in the proximal third, grade 2 in the middle third, and grade 1 in the distal third. This suggests that a pH > 4 is the optimum milieu for the cardiac epithelium → intestinal metaplasia step in a patient with visible CLE.

It must be recognized that the study was performed in a pH environment that was created artificially by discontinuing PPI therapy before the test. The Barrett patients had been on PPI therapy for a long time. As such, the chronic pH in the esophagus in the Barrett group is likely to have been higher than those recorded in the test. The goblet cell density seen is the result of the chronic exposure to reflux that has been alkalinized by PPI therapy. The pH recorded at the time of the test are likely to be significantly lower than the reality of exposure of these patient's esophagi during years of PPI therapy.

This study provides support for the explanation that the prevalence of intestinal metaplasia in the proximal part of the CLE is that this region has a higher pH luminal environment than the more distal thoracic esophagus and intrasphincteric region.

It also suggests that alkalinization promotes intestinal metaplasia. This means that PPI therapy promotes the cardiac epithelium → intestinal metaplasia step. They convert the patient who is not at risk for adenocarcinoma to Barrett esophagus which is the at-risk epithelium. Acid suppression promotes the precancerous epithelium and by doing so increases the incidence of adenocarcinoma. The increasing success in the ability of drugs to alkalinize the refluxate is the most reasonable explanation for the increase in the incidence of adenocarcinoma over this period of time.

There is sufficient logic and data to ask an important question: "Does increasing the pH of the refluxate with acid-reducing drugs promote intestinal metaplasia to an extent that this is the reason for the increased and increasing incidence of esophageal adenocarcinoma?" It is becoming difficult to answer no to this question with any confidence.

6. ESOPHAGEAL CARCINOGENESIS

There is no question that esophageal adenocarcinoma results from reflux of gastric contents into the esophagus. There is also no question that the risk of adenocarcinoma increases with increasing reflux. Lagergren et al.[46] showed convincingly that increasing severity and duration of symptomatic reflux causes a progressive increase in the risk of both adenocarcinoma of the thoracic esophagus and the dilated distal esophagus ("gastric cardia").

There is also no doubt that Barrett esophagus (defined as visible CLE with intestinal metaplasia) confers a significantly increased risk of esophageal adenocarcinoma. The magnitude of this risk has recently been downgraded from the previously accepted 0.5% per year to 0.2%–0.3% per year.[47]

There is much argument and controversy regarding the impact of PPI therapy on the incidence of esophageal adenocarcinoma in patients with Barrett esophagus. Kasterlein et al.[48] suggested that PPI therapy reduces the incidence of esophageal adenocarcinoma in patients with Barrett esophagus by 75%. Hvid-Jensen et al.,[49] attempting to duplicate this finding, came to the opposite conclusion, although the authors invoked the possibility that indication bias could have been responsible for their finding that esophageal adenocarcinoma had a significantly positive association with PPI use.

The question and argument as to whether PPI therapy influences the progression to adenocarcinoma is without merit in the practical sense. It just causes an unnecessary diversion.

The reality is certain. Present treatment algorithms for GERD increase the prevalence of Barrett esophagus. In the Pro-GERD study,[28] regular treatment of GERD patients in the primary care setting converted 9.7% of the overall population of 2741 patients from endoscopically negative for visible CLE to positive. Many had intestinal metaplasia. There was a significant association between conversion to visible CLE and regular PPI use.

It does not matter whether PPIs caused the change. The change occurred under the watch of PPI therapy. With no alternative management recommended for GERD, there is an infusion of cancer risk in GERD patients at nearly 10% per 5 years. PPI therapy has had its shot for 25 years during which cancer rates have continued to rise. Whatever data is produced in studies, that reality is certain. If these problems are not addressed, nothing will reliably stop the increasing incidence of adenocarcinoma.

6.1 The Mechanism of Carcinogenesis

Carcinogenesis is ultimately the result of the exposure of the target epithelium in the esophagus to a carcinogenic agent. The carcinogenic molecule is almost certainly in gastric juice and delivered to the target cells by reflux of gastric contents into the esophagus. The highest risk for carcinogenesis is when the target cell is in the thoracic esophagus and reflux occurs as a result of LES damage and failure. The risk of carcinoma of CLE limited to the dilated distal esophagus is unknown, but probably present and very small.

Increasing dysplasia and adenocarcinoma occur when there is a carcinogen that is delivered in adequate dosage to a specific target epithelium in the esophagus. This target cell is marked by columnar epithelium in CLE that has intestinal metaplasia.[50]

The mechanism of carcinogenesis is best understood by looking at the esophagus in terms of a series of horizontal levels, each with its own microenvironment during an episode of reflux. The refluxate causes a gradient of volume and number of every molecule in the refluxate from the endoscopic GEJ to the height of the column. Clearly, anything that increases the frequency and volume of reflux will tend to increase carcinogenesis.

The two elements required for carcinogenesis, intestinal metaplasia and adequate carcinogen dose, are maximal at the two opposite ends of the esophagus. Intestinal metaplasia is maximal at the proximal limit of CLE adjacent to the SCJ. Carcinogen dose is maximal at the most distal region of the thoracic esophagus, immediately above the GEJ. Carcinogenesis occurs when there is a perfect storm with the intestinal epithelium extending sufficiently distal to reach a point where it meets an adequate carcinogen dose (Fig. 14.25).

The concept of different cellular changes and milieu at different horizontal levels in the esophagus makes it extremely easy to understand carcinogenesis. At any level in the esophagus, there is an epithelial lining. If this lining is anything other than intestinal metaplasia, there is no risk of cancer at that level. The risk at any level can be defined because the target epithelium is known.

Similarly, at any level of the esophagus, there is a carcinogen dose during an episode of reflux unless the patient has no carcinogen in the stomach. The concentration of carcinogen in the stomach is totally unpredictable because the carcinogen

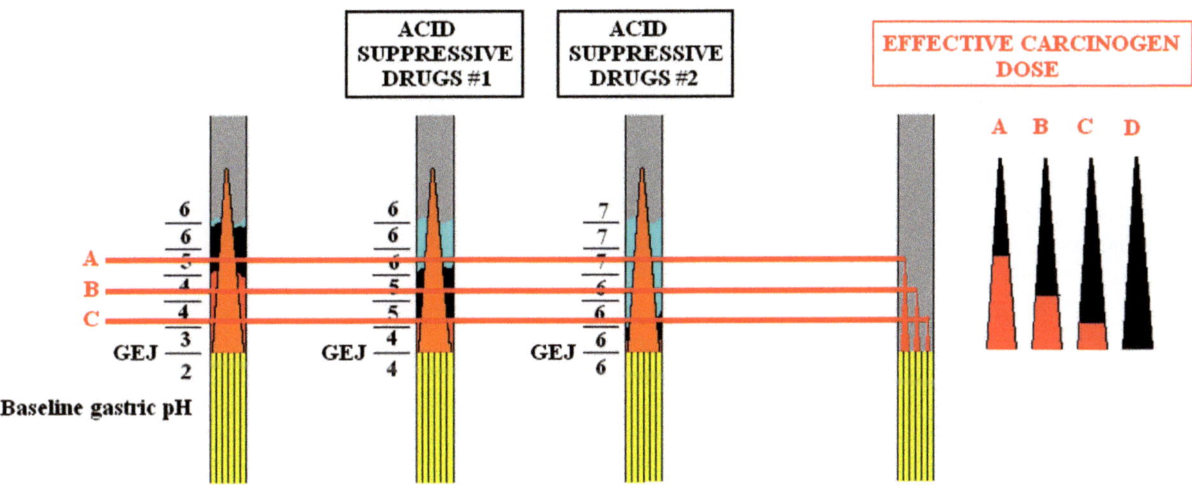

Volume of Reflux	HIGH	REMAINS HIGH	REMAINS HIGH
Time of Exposure	LONG	REMAINS HIGH	REMAINS HIGH
Amount of IM	SMALL	INCREASED	MORE INCREASED
Amount of OCM	MODERATE	LESS	STILL LESS
CANCER RISK	ZERO	LOW	HIGH

FIGURE 14.25 The theoretical relationship of extent of intestinal metaplasia within a segment of visible columnar-lined esophagus, pH milieu of the esophagus on patients with and without acid suppression of varying effectiveness, and carcinogen dose. *GEJ*, gastroesophageal junction; *IM*, intestinal metaplasia; *OCM*, oxyntocardiac mucosa.

is unknown and it likely varies greatly in different people. Two things are certain: (1) the carcinogen dose is always maximal immediately above the GEJ and (2) the carcinogen dose will increase at all levels as the severity (volume and frequency) of reflux increases.

Carcinogenesis will occur only at the horizontal levels in the esophagus where a sufficient carcinogenic dose is delivered by reflux episodes to intestinal metaplastic epithelium (Fig. 14.25).

Theoretically, carcinogenesis will be promoted by the following changes in the esophageal lining and milieu during reflux:

1. Distal migration of intestinal metaplasia within the CLE from its point of origin at the SCJ. The closer the intestinal metaplasia is to the GEJ, the more likely it is that the carcinogen dose will be adequate. I have shown that one of the effects of alkalinizing the refluxate may be to increase the extent of intestinal metaplasia within a segment of CLE.
2. Increased carcinogen dose. Without definite knowledge of the carcinogens involved, this remains an unknown quantity in carcinogenesis.
3. Increased reflux. Ultimately, luminal carcinogenesis is dependent on delivery of carcinogen to the esophagus. The greater the LES damage, the greater is the volume and frequency of reflux episodes, and the greater the risk of carcinoma. The risk of carcinoma is associated with severity of reflux as shown by an abnormal pH test and a mechanically defective LES.

6.2 Relationship Between Length of Columnar-Lined Esophagus and Risk of Adenocarcinoma

Present methods of assessing CLE histologically are so inadequate that there is no rational method of understanding the relationship between the length of Barrett esophagus and risk of adenocarcinoma. The reason for this is that there is no attempt whatsoever to define the extent of intestinal metaplasia within a segment of CLE.

The quantitation of "Barrett esophagus" is done at the time of endoscopy by the Prague C+M method.[51] This is entirely endoscopic. As such, it is quantitating CLE, not CLE with intestinal metaplasia, which defines Barrett esophagus. When biopsies come back that indicate the presence of *any* intestinal metaplasia, the entire segment quantitated by the Prague method becomes Barrett esophagus. This is not accurate. Barrett esophagus requires intestinal metaplasia, it is not equivalent to CLE.

A 2-cm segment of visible CLE with intestinal metaplasia throughout has more target cells for carcinogenesis than a 10-cm segment of CLE with intestinal metaplasia limited to the proximal 1 cm of the CLE. This 2-cm short-segment Barrett esophagus is at high risk because it has more target epithelium closer to the GEJ than the long 10-cm segment with intestinal metaplasia limited to an area 9 cm proximal to the GEJ. The 2-cm short segment of Barrett esophagus with intestinal metaplasia over its entire extent is typical of today's Barrett esophagus; the 10-cm long segment with intestinal limited to the proximal region was typical of Barrett esophagus in the 1950s.

It is therefore not surprising that the data relating the risk of cancer with length of CLE is confusing and without a definite conclusion. As a result, the recommended surveillance for both long- and short-segment Barrett esophagus is similar.

There is some conflict in the literature regarding the risk of malignancy in short-segment Barrett esophagus. In Weston et al.,[52] 10/108 patients developed multifocal high-grade dysplasia or adenocarcinoma. This broke down into 0 of 51 patients with a Barrett length of <2 cm, 1/28 patients with a length of 2–6 cm, and 9/39 patients with a length >6 cm. Although the numbers are small, they suggest that the risk of cancer increases with increasing length of CLE.

In 1993, Menke-Pluymers et al.,[53] in a retrospective case control study compared 96 patients with benign Barrett esophagus and 62 patients with adenocarcinoma referred to the Rotterdam Esophageal Tumor Study Group between 1978 and 1985. They reported that the length of CLE correlated with cancer risk. Doubling the length of CLE increased risk by a factor of 1.7. This study predated the recognition of short-segment Barrett esophagus. In 2002, Avidan et al.[54] also reported that the risk of high-grade dysplasia and adenocarcinoma increased with increasing length of Barrett esophagus.

These studies are partially contradicted by the data in the largest and best-controlled study by Rudolph et al.[55] This was a prospective cohort study from the University of Washington that consisted of 309 patients with Barrett esophagus. The patients were monitored for progression to adenocarcinoma by repeated endoscopy and biopsy for an average of 3.8 years. When all baseline histologic diagnoses were included, the incidence of cancer was 2.5 per 100 person-years for a Barrett length of <3 cm, 2.8 for a 3–6 cm length, 3.2 for a 7–10 cm length, and 7.0 for a length >10 cm. After adjusting for histologic diagnosis at study entry, the authors found that segment length was not related to cancer risk for the entire cohort ($P>.2$). A 5-cm difference in segment length had a 1.7-fold (95% CI: 0.8- to 3.8-fold) increased risk.

Much more important than the question of whether there was a difference in the risk of cancer with increasing segment length was the finding that 7/83 patients with a Barrett length of <3 cm developed cancer. Based on this, the authors suggested: "Until further data are available, the frequency of endoscopic surveillance should be selected without regard to segment length."

6.3 Preferred Location for Dysplasia and Adenocarcinoma in the Esophagus

If the target for the carcinogen is only the epithelium that has intestinal metaplasia and the carcinogen is delivered from below by reflux of gastric contents into the esophagus, the preferred location for the occurrence of dysplasia and adenocarcinoma can theoretically be predicted to be the junction between intestinalized and nonintestinalized epithelium within the segment of CLE. This is the place where the carcinogen meets the target epithelium.

This relationship is likely to be true only in a broad sense with significant individual variation. Carcinogenic mutations are random events whose likelihood is increased by (1) the number of target cells exposed to the carcinogen; (2) the frequency of exposure; and (3) the proliferative rate of the target cells. A mutation occurs when an adequate dose of carcinogen interacts with a target cell that is in the mitotic cycle and therefore most susceptible to the mutational change.

Historically, in the 1950s, esophageal adenocarcinoma tended to favor the mid-esophagus in the proximal region of the segment of CLE.[35,56] The first and second reported case in 1952 and 1953 were both in the thoracic esophagus close to the SCJ and well above the end of the tubular esophagus, i.e., in the proximal region of CLE.

With the passing of decades, the preferred location esophageal adenocarcinoma has progressively moved more distally in the thoracic esophagus. Today, tumors are commonly seen around and distal to the endoscopic GEJ in the dilated distal esophagus.

Theisen et al.[41] studied 213 patients with histologically proven esophageal adenocarcinoma in the distal esophagus (Siewert type I). Junctional and gastric cardiac tumors (types II and III) were excluded. This was a selected group comprising 134 patients with early (T1) cancer and 79 patients with locally advanced (T3) cancers receiving neoadjuvant chemotherapy. Barrett esophagus was diagnosed by the presence of a visible columnar segment that showed intestinal metaplasia with goblet cells on biopsy (Table 14.8).

The patients were 15 women and 198 men with a mean age of 54 years (range 19–74 years). In the early cancer group 112/134 (83.5%) patients had intestinal metaplasia on biopsy; in the other 22 extensive biopsies failed to demonstrate intestinal metaplasia. In the group of patients with advanced cancers, intestinal metaplasia was present in 59/79 (75%) patients. After chemotherapy, repeat endoscopy and biopsies were done. These showed that 18 of the 20 patients who did not have

TABLE 14.8 Prevalence of the Location and Distribution of the Cancer Development in Patients With Early and Advanced Esophageal Adenocarcinoma

Location of Tumor Within Barrett Segment	Early Adenocarcinoma	Advanced Adenocarcinoma
Distal third	92/112 (82%)	65/77 (84%)
Proximal and middle third	20/112 (18%)	12/77 (16%)
P value	<.05	<.05

TABLE 14.9 Epithelial Types Found at Proximal Edge, Distal Edge, and Lateral Edge at the Epicenter of Distal Esophageal and Gastric Cardiac Adenocarcinomas Classified by Its Relationship to the End of the Tubular Esophagus

	Number of Cases	Proximal Edge (Number)	Lateral Epicenter (Number)	Distal Edge (Number)
Distal esophageal	38	Sq (16), IM (20), CE (2)	Sq (2), IM (25), CE (5), OCE (2)	IM (12), CE (13), OCE (7), GOE (8)
Gastric cardiac	36	Sq (27), IM (5), CE(4)[a]	Sq (3), IM (9), CE (16), OCE (2)[b]	IM (4), CE (14), OCE (8), GOE (10)

Gastric cardiac tumors would now be carcinomas of the dilated distal esophagus.
CE, cardiac epithelium; *GOE*, gastric oxyntic epithelium; *IM*, intestinal metaplasia; *OCE*, oxyntocardiac epithelium; *Sq*, squamous epithelium.
[a]*Four tumors were circumferential.*
[b]*Six tumors were circumferential.*

intestinal metaplasia prior to neoadjuvant therapy now had biopsies showing intestinal metaplasia. The overall prevalence of intestinal metaplasia in this group was 77/79 (97.4%). In the entire study group, intestinal metaplasia was present in 189/213 (88.7%) patients.

The median length of the Barrett esophagus segments was 3.6cm (range 2–15cm). In the 112 patients with early lesions who had intestinal metaplasia, thc tumors were located in the distal third of the Barrett segment in 92/112 (82%) of patients. In the 77 patients with advanced lesions who had intestinal metaplasia, the remaining tumor after neoadjuvant therapy was also located in the distal third of the Barrett segment in 85%. Tumors were located in the proximal or middle third of the Barrett segment in 18% of early tumors and 15% of advanced tumors.

In their discussion, the authors try to explain the preferred origin of adenocarcinoma in the distal third of the segment of Barrett esophagus: "A reason why these tumors occur predominantly in the distal portion of a segment of intestinal metaplasia is found in the so-called second line of defense, the luminal clearance capacity of the esophagus. Esophageal clearance is influenced by gravity, peristalsis, and submucosal glands secreting bicarbonates. Some studies suggest that the severity of esophageal epithelium damage is related to peristaltic dysfunction. It has been shown that peristaltic dysfunction may impair esophageal emptying of a refluxed bolus and result in prolonged acid clearing time. This, taken together with the fact that bile acids in a moderate acidic environment such as pH 3-5 have the most harmful potential for the columnar epithelium, might explain the predominant location in the distal third of the Barrett's segment for the development of adenocarcinoma within a segment of specialized intestinal metaplasia."

The authors are suggesting a mechanism wherein the distal region of the esophagus is exposed to carcinogen for a longer time because the refluxate does not get cleared out of the esophagus. This is reasonable. It would be interesting to study esophageal motility in patients with Barrett esophagus and see whether disordered motility is associated with an increased risk of cancer.

An alternate explanation is that the region of the distal third of the Barrett segment is the region of maximum interaction between carcinogen and target epithelium.

In our study of 74 esophageal adenocarcinomas[57] (38 distal esophageal and 36 distal to the end of the tubular esophagus) (reviewed in Chapter 12), the epithelium at the lateral edge of the epicenter of the tumor in 34 patients with noncircumferential tumors in the distal esophagus was intestinal or cardiac in 30/34 (88%) (see Table 14.9). In the 30 patients with noncircumferential tumors in the dilated distal esophagus, the lateral edge at the epicenter showed intestinal or cardiac epithelium in 25/30 (83%). The epithelium at the proximal edge of the majority of both groups was squamous and intestinal and the distal edge was predominantly nonintestinalized CLE and gastric oxyntic epithelium. This suggests that the tumor origin was most commonly at the interface of intestinal and cardiac epithelium within the CLE segment. This is the point where the carcinogen and susceptible cell meet in the esophagus.

The evidence for the preferred location for adenocarcinoma being the distal limit of intestinal epithelium is strong. This is also strong evidence that cardiac epithelium without intestinal metaplasia is not at any significant risk. Otherwise, the preferred location for carcinogenesis would have been distal to the junction of intestinal and cardiac epithelium where the carcinogen dose is greater.

7. PRESENT CONCEPT OF PATHOLOGIC PROGRESSION: A STARK CONTRAST

At the present time, the only progression that is followed in a practical sense by histopathologic criteria based on endoscopic biopsy is visible CLE with intestinal metaplasia→increasing dysplasia→adenocarcinoma. Everything before CLE with intestinal metaplasia is completely ignored.

GERD does not exist from a pathologic diagnostic standpoint until empiric PPI therapy fails to control symptoms. This is the usual indication for endoscopy.

When endoscopy is performed, and is not abnormal, biopsies are not done. Pathology has no role. The dilated distal esophagus and its relationship to LES damage is not understood and/or accepted.

The presence of esophagitis is based on endoscopic criteria. If the patient has erosive esophagitis, it is graded by the Los Angeles classification. Biopsies are only performed to confirm that what the endoscopist believes is erosive esophagitis is not masking CLE. The recognized criteria for the diagnosis of reflux esophagitis have no practical relevance. Biopsies are never taken in patients without erosive esophagitis to see if microscopic evidence of reflux esophagitis is present. At present, biopsies are frequently done to exclude idiopathic eosinophilic esophagitis.

Biopsies are done when visible CLE is found with the objective of finding intestinal metaplasia. When the biopsies are negative for intestinal metaplasia, the patient does not have a diagnosis except "not Barrett esophagus" if you happen to be in the United States but not in England. Although the Seattle protocol is recommended in the patient with visible CLE, it is declared to be "too expensive, time-consuming, and not best practice." Biopsies are done with no standard and often randomly. With these inadequate biopsy protocols, there is a high probability of a false negative diagnosis of intestinal metaplasia.

This is not of great concern because the enthusiasm for surveillance of Barrett esophagus is not high and decreasing as new statistics show that the risk of cancer in patients with Barrett esophagus is 0.2%–0.3% rather than 0.5% per year.[47] The decline in the cancer risk raises valid questions about the cost-effectiveness of Barrett esophagus surveillance.

The only role of the pathologist is in the diagnosis of intestinal metaplasia in biopsies from visible CLE and in the diagnosis and grading of dysplasia and adenocarcinoma in surveillance biopsies in patients with Barrett esophagus.

The pathology literature is deafeningly silent. There is no ability to assess LES damage, no ability to assess progression of GERD, no pathologic criteria defining cancer risk in Barrett esophagus. This silence is not the fault of the pathologists. They cannot write about what they do not see. With the present recommendations for biopsy at endoscopy, gastroenterologists do not produce specimens that permit the assessment of early stages of GERD where LES damage causes cardiac metaplasia in the dilated distal esophagus.

Pathologists can assess progression in the neoplastic process in Barrett surveillance, but only if they are given adequate samples. Pathologists do poorly in the diagnosis of low-grade dysplasia, better in the diagnosis of high-grade dysplasia, and are excellent in diagnosing cancer in biopsies and excellent in staging cancer in endoscopic mucosal resection and esophagogastrectomy specimens.

With the new ability of pathology to measure abdominal LES damage accurately, it is probable that the role of the pathologist in the management of GERD will increase substantially in the future.

REFERENCES

1. Chandrasoma P. Controversies of the cardiac mucosa and Barrett's esophagus. *Histopathology* 2005;**46**:361–73.
2. Chandrasoma PT, Der R, Ma Y, et al. Histology of the gastroesophageal junction: an autopsy study. *Am J Surg Pathol* 2000;**24**:402–9.
3. Chandrasoma P, Makarewicz K, Wickramasinghe K, Ma YL, DeMeester TR. A proposal for a new validated histologic definition of the gastroesophageal junction. *Hum Pathol* 2006;**37**:40–7.
4. Kahrilas PJ, Shaheen NJ, Vaezi MF. American Gastroenterological Association Institute technical review on the management of gastroesophageal reflux disease. *Gastroenterology* 2008;**135**:1392–413.
5. Vakil N, van Zanten SV, Kahrilas P, Dent J, Jones B, The Global Consensus Group. The Montreal definition and classification of gastroesophageal reflux disease: a global evidence-based consensus. *Am J Gastroenterol* 2006;**101**:1900–20.
6. Pohl H, Sirovich B, Welch HG. Esophageal adenocarcinoma incidence: are we reaching the peak? *Cancer Epidemiol Biomark Prev* 2010;**19**:1468–70.

7. Clark GWB, Ireland AP, Chandrasoma P, DeMeester TR, Peters JH, Bremner CG. Inflammation and metaplasia in the transitional mucosa of the epithelium of the gastroesophageal junction: a new marker for gastroesophageal reflux disease. *Gastroenterology* 1994;**106**:A63.
8. Chandrasoma P. Pathophysiology of Barrett's esophagus. *Semin Thorac Cardiovasc Surg* 1997;**9**:270–8.
9. Chandrasoma P, Wijetunge S, Ma Y, DeMeester S, Hagen J, DeMeester T. The dilated distal esophagus: a new entity that is the pathologic basis of early gastroesophageal reflux disease. *Am J Surg Pathol* 2011;**35**:1873–81.
10. Chandrasoma PT. Histologic definition of gastro-esophageal reflux disease. *Curr Opin Gastroenterol* 2013;**29**:460–7.
11. Marsman WA, van Sandyck JW, Tytgat GN, ten Kate FJ, van Lanschot JJ. The presence and mucin histochemistry of cardiac type mucosa at the esophagogastric junction. *Am J Gastroenterol* 2004;**99**:212–7.
12. Fletcher J, Wirz A, Young J, et al. Unbuffered highly acidic gastric juice exists at the gastroesophageal junction after a meal. *Gastroenterology* 2001;**121**:775–83.
13. Pandolfino JE, Zhang Q, Ghosh SK, et al. Acidity surrounding the squamocolumnar junction in GERD patients: "acid pocket" versus "acid film". *Am J Gastroenterol* 2007;**102**:2633–41.
14. Clarke AT, Wirz AA, Seenan JP, et al. Paradox of gastric cardia: it becomes more acidic following meals while the rest of stomach becomes less acidic. *Gut* 2009;**58**:904–9.
15. Beaumont H, Bennink RJ, de Jong J, et al. The position of the acid pocket as a major risk factor for acidic reflux in healthy subjects and patients with GORD. *Gut* 2010;**59**:441–51.
16. Kwiatek MA, Pandolfino JE, Hirano I, Kahrilas PJ. Esophagogastric junction distensibility assessed with an endoscopic functional luminal imaging probe (EndoFlip). *Gastrointest Endosc* 2010;**72**:272–8.
17. Mitchell DR, Derakhshan MH, Robertson EV, McColl KE. The role of the acid pocket in gastroesophageal reflux disease. *J Clin Gastroenterol* 2016;**50**:111–9.
18. Vo L, Simonian HP, Doma S, et al. The effect of rabeprazole on regional gastric acidity and in the postprandial cardia/gastro-oesophageal junction acid layer in normal subjects: a randomized, double-blind placebo-controlled study. *Aliment Pharmacol Ther* 2005;**21**:1321–30.
19. Rohoff WO, Bennink RJ, Boeckxstaens GE. Proton pump inhibitors reduce the size and acidity of the acid pocket in the stomach. *Clin Gastroenterol Hepatol* 2014;**12**:1101–7.
20. Kwiatek MA, Roman S, Fareeduddin A, et al. An alginate-antacid formulation (Gaviscon Double Action Liquid) can eliminate or displace the post-prandial "acid pocket" in symptomatic GERD patients. *Aliment Pharmacol Ther* 2011;**34**:59–66.
21. Sweis R, Kaufman E, Anggiansah A, et al. Post-prandial reflux suppression by a raft-forming alginate (Gaviscon Advance) compared to a simple antacid documented by magnetic resonance imaging and pH impedance monitoring: mechanistic assessment in healthy volunteers and randomized, controlled, double-blind study in reflux patients. *Aliment Pharmacol Ther* 2013;**37**:1093–102.
22. Duan L, Wu AH, Sullivan-Halley J, Bernstein L. Antacid drug use and risk of esophageal and gastric adenocarcinoma in Los Angeles County. *Cancer Epidemiol Biomark Prev* 2009;**18**:526–33.
23. DeMeester TR, Johnson LF. The evaluation of objective measurement of gastroesophageal reflux and their contribution to patient management. *Surg Clin North Am* 1976;**56**:39–53.
24. Anggiansah A, Sumboonnanonda K, Wang J, et al. Significantly reduced acid detection at 10 centimeters compared to 5 centimeters above lower esophageal sphincter in patients with acid reflux. *Am J Gastroenterol* 1993;**88**:842–6.
25. Fletcher J, Wirz A, Henry E, McColl KE. Studies of acid exposure immediately above the gastro-oesophageal squamocolumnar junction: evidence of short segment reflux. *Gut* 2004;**53**:168–73.
26. Robertson EV, Derakhshan MH, Wirz AA, Lee YY, Seenan JP, Ballantyne SA, Hanvey SL, Kelman AW, Going JJ, McColl KE. Central obesity in asymptomatic volunteers is associated with increased intrasphincteric acid reflux and lengthening of the cardiac mucosa. *Gastroenterology* 2013;**145**:730–9.
27. Tharalson EF, Martinez SD, Garewal HS, Sampliner RE, Cui H, Pulliam G, Fass R. Relationship between rate of change in acid exposure along the esophagus and length of Barrett's epithelium. *Am J Gastroenterol* 2002;**97**:851–6.
28. Malfertheiner P, Nocon M, Vieth M, Stolte M, Jasperson D, Keolz HR, Labenz J, Leodolter A, Lind T, Richter K, Willich SN. Evolution of gastro-oesophageal reflux disease over 5 years under routine medical care – the ProGERD study. *Aliment Pharmacol Ther* 2012;**35**:154–64.
29. Fitzgerald RC, Omary MB, Triadofilopoulos G. Dynamic effects of acid on Barrett's esophagus. An ex vivo proliferation and differentiation model. *J Clin Invest* 1996;**98**:2120–8.
30. Kulig M, Nocon M, Vieth M, Leodolter A, Jaspersen D, Labenz J, Meyer-Sabellek W, Stolte M, Lind T, Malfertheimer P, Willich SN. Risk factors of gastroesophageal reflux disease: methodology and first epidemiological results of the ProGERD study. *J Clin Invest* 2004;**57**:580–9.
31. Kilgore SP, Ormsby AH, Gramlich TL, et al. The gastric cardia: fact or fiction? *Am J Gastroenterol* 2000;**95**:921–4.
32. Sarbia M, Donner A, Gabbert HE. Histopathology of the gastroesophageal junction. A study on 36 operation specimens. *Am J Surg Pathol* 2002;**26**:1207–12.
33. Leodolter A, Nocon M, Vieth M, Lind T, Jaspersen D, Richter K, Willich S, Stolte M, Malfertheiner P, Labenz J. Progression of specialized intestinal metaplasia at the cardia to macroscopically evidence Barrett's esophagus: an entity of concern in the Pro-GERD study. *Scand J Gastroenterol* 2012;**47**:1429–35.
34. El-Serag HB, Sweet S, Winchester CC, et al. Update on the epidemiology of gastro-oesophageal reflux disease: a systematic review. *Gut* 2014;**63**:871–80.
35. Allison PR, Johnstone AS, Royce GB. Short esophagus with simple peptic ulceration. *J Thorac Surg* 1943;**12**:432–57.
36. Barrett NR. The lower esophagus lined by columnar epithelium. *Surgery* 1957;**41**:881–94.

37. Lyall A. Chronic peptic ulcer of the oesophagus: a report of eight cases. *Br J Surg* 1937;**24**:534–47.
38. Ronkainen J, Aro P, Storskrubb T, Johansson S-E, Lind T, Bolling-Sternevald E, Vieth M, Stolte M, Talley NJ, Agreus L. Prevalence of Barrett's esophagus in the general population: an endoscopic study. *Gastroenterology* 2005;**129**:1825–31.
39. Rex DK, Cummings OW, Shaw M, Cumings MD, Wong RK, Vasudeva RS, Dunne D, Rahmani EY, Helper DJ. Screening for Barrett's esophagus in colonoscopy patients with and without heartburn. *Gastroenterology* 2003;**125**:1670–7.
40. Gerson LB, Shetler K, Triadafilopoulos G. Prevalence of Barrett's esophagus in asymptomatic individuals. *Gastroenterology* 2002;**123**:461–7.
41. Theisen J, Stein HJ, Feith M, Kauer WKH, Dittler HJ, Pirchi D, Siewert JR. Preferred location for the development of esophageal adenocarcinoma within a segment of intestinal metaplasia. *Surg Endosc* 2006;**20**:235–8.
41a. Allison PR, Johnstone AS. The oesophagus lined with gastric mucous membrane. *Thorax* 1953;**8**:87–101.
42. Paull A, Trier JS, Dalton MD, Camp RC, Loeb P, Goyal RK. The histologic spectrum of Barrett's esophagus. *N Engl J Med* 1976;**295**:476–80.
42a. Spechler SJ, Zeroogian JM, Antonioli DA, et al. Prevalence of metaplasia at the gastroesophageal junction. *Lancet* 1994;**344**:1533–6.
43. Chandrasoma PT, Der R, Ma Y, Peters J, DeMeester T. Histologic classification of patients based on mapping biopsies of the gastroesophageal junction. *Am J Surg Pathol* 2003;**27**:929–36.
44. Chandrasoma PT, Wijetunge S, DeMeester SR, Hagen JA, DeMeester TR. The histologic squamo-oxyntic gap: an accurate and reproducible diagnostic marker of gastroesophageal reflux disease. *Am J Surg Pathol* 2010;**34**:1574–81.
45. Theodorou D, Ayazi S, DeMeester SR, Zehetner J, Peyre CG, Grant KS, Augustin F, Oh DS, Lipham JC, Chandrasoma PT, Hagen JA, DeMeester TR. Intraluminal pH and goblet cell density in Barrett's esophagus. *J Gastrointest Surg* 2012;**16**:469–74.
46. Lagergren J, Bergstrom R, Lindgren A, Nyren O. Symptomatic gastroesophageal reflux as a risk factor for esophageal adenocarcinoma. *N Engl J Med* 1999;**340**:825–31.
47. Hvid-Jensen F, Pedersen L, Drewes AM, et al. Incidence of adenocarcinoma among patients with Barrett's esophagus. *N Engl J Med* 2011;**365**: 1375–83.
48. Kastelein F, Spaander MC, Steyerberg EW, Bierman K, Valkhoff VE, Kuipers EJ, Bruno MJ. Proton pump inhibitors reduce the risk of neoplastic progression in patients with Barrett's esophagus. *Clin Gastroenterol Hepatol* 2013;**11**:382–8.
49. Hvid-Jensen F, Pedersen L, Funch-Jensen P, Drewes AM. Proton pump inhibitor use may not prevent high grade dysplasia and oesophageal adenocarcinoma in Barrett's oesophagus. *Aliment Pharmacol Ther* 2014;**39**:984–91.
50. Chandrasoma P, Wijetunge S, DeMeester S, Ma Y, Hagen J, Zamis L, DeMeester T. Columnar lined esophagus without intestinal metaplasia has no proven risk of adenocarcinoma. *Am J Surg Pathol* 2012;**36**:1–7.
51. Sharma P, Dent J, Armstrong D, et al. The development and validation of an endoscopic grading system for Barrett's esophagus: the Prague C & M criteria. *Gastroenterology* 2006;**131**:1392–9.
52. Weston AP, Badr AS, Hassanein RS. Prospective multivariate analysis of clinical, endoscopic, and histological factors predictive of development of Barrett's multifocal high grade dysplasia or adenocarcinoma. *Am J Gastroenterol* 1999;**94**:3413–9.
53. Menke-Pluymers MB, Hop WC, Dees J, et al. Risk factors for the development of an adenocarcinoma in columnar-lined (Barrett's) esophagus: The Rotterdam Esophageal Tumor Study Group. *Cancer* 1993;**72**:1155–8.
54. Avidan B, Sonnenberg A, Schnell TG, et al. Hiatal hernia size, Barrett's length, and severity of acid reflux are all risk factors for esophageal adenocarcinoma. *Am J Gastroenterol* 2002;**97**:1930–6.
55. Rudolph RE, Vaughan TL, Storer BE, et al. Effect of segment length on risk for neoplastic progression in patients with Barrett esophagus. *Ann Intern Med* 2000;**132**:612–20.
56. Morson BC, Belcher BR. Adenocarcinoma of the oesophagus and ectopic gastric mucosa. *Br J Cancer* 1952;**6**:127–30.
57. Chandrasoma PT, Wickramasinghe K, Ma Y, DeMeester TR. Adenocarcinomas of the distal esophagus and "gastric cardia" are predominantly esophageal adenocarcinomas. *Am J Surg Pathol* 2007;**31**:569–75.

Chapter 15

Molecular Evolution of Esophageal Epithelial Metaplasia

There are two largely interdependent mechanisms that operate in the pathogenesis of the cellular changes within the esophagus in a patient with gastroesophageal reflux disease (GERD):

1. Cellular injury and the reaction to it that is largely the result of toxic components in gastric juice such as H^+ ions (acid), pepsin, and toxic molecules in bile. These are responsible for increased rate of cell loss at the surface of the epithelium, cell death, release of cytokines from the damaged and inflammatory cells,[1] regenerative activity with increased rates of cellular proliferation, and inflammation resulting in edema,[2] leukocyte infiltration, smooth muscle proliferation, vascular changes, and fibrosis. These direct and indirect injury-related changes are maximally seen in squamous epithelium. Metaplastic columnar epithelium in general is more resistant to these injurious agents in gastric contents. Within the three types of cardiac epithelium, there are also differences in the amount of injury, proliferative activity, and inflammation.[3,4]
2. Molecular changes that occur by activation and suppression of genes in the epithelium and possibly mesenchymal cells in the lamina propria under the basement membrane. These involve two types of change:
   a. Genetic "switches," which are potentially reversible changes in cells resulting from suppression and activation of normal genes involved in signaling. These appear to be caused by changes in the milieu around the cells. Theoretically there is an opportunity to control these genetic switches by changing the milieu. Genetic switches in the genes that signal the direction of differentiation are likely responsible for different types of metaplasia. These genetic switches are predictable, constant, and relatively easy to detect and understand.
   b. More permanent genetic changes resulting from gene deletion, transcription error, or mutation. These abnormal genetic changes include innocent mutations that have no impact and those that drive the neoplastic process. The mutations that drive neoplasia are unpredictable and likely to be highly complex with the possibility of multiple pathways to neoplasia. When an abnormal gene is found, it is difficult to know whether it is a critical change involved in neoplasia or one that represents background clutter and has no real practical role in carcinogenesis.

The neoplastic progression in the esophagus, driven by genetic changes: squamous epithelium (not at risk)→cardiac epithelium (not at risk)→cardiac epithelium with intestinal metaplasia (at risk)→increasing dysplasia (increasing risk) →adenocarcinoma (Fig. 15.1).

I will consider the first three steps of this pathway in this chapter. These result from genetic switches and prime the squamous epithelium of the esophagus to cardiac epithelium with intestinal metaplasia that is premalignant.

The transformation of the at-risk target epithelium, which is cardiac epithelium with intestinal metaplasia, to increasing dysplasia and adenocarcinoma is not clearly understood. It results from mutations that are unpredictable and not completely worked out at this time. It is largely outside the scope of this book.

1. FETAL DEVELOPMENT IN THE ESOPHAGUS AND STOMACH (SEE CHAPTER 3)

The fetal epithelial lining of the esophagus is a complex entity.[5] The early primitive endodermal stratified columnar epithelium of the esophagus goes through multiple phenotypic expressions before it becomes lined entirely by stratified squamous epithelium. Intermediate fetal epithelia include a pseudostratified ciliated epithelium that dominates the second trimester esophagus until squamous epithelium develops in the 22nd week of gestation. Squamous epithelium is seen first in the mid-esophagus and progressively extends both cephalad and caudad. Small amounts of nonciliated fetal columnar epithelium are present between the distal limit of squamous epithelium and the angle of His in the third trimester and,

GERD. http://dx.doi.org/10.1016/B978-0-12-809855-4.00015-4

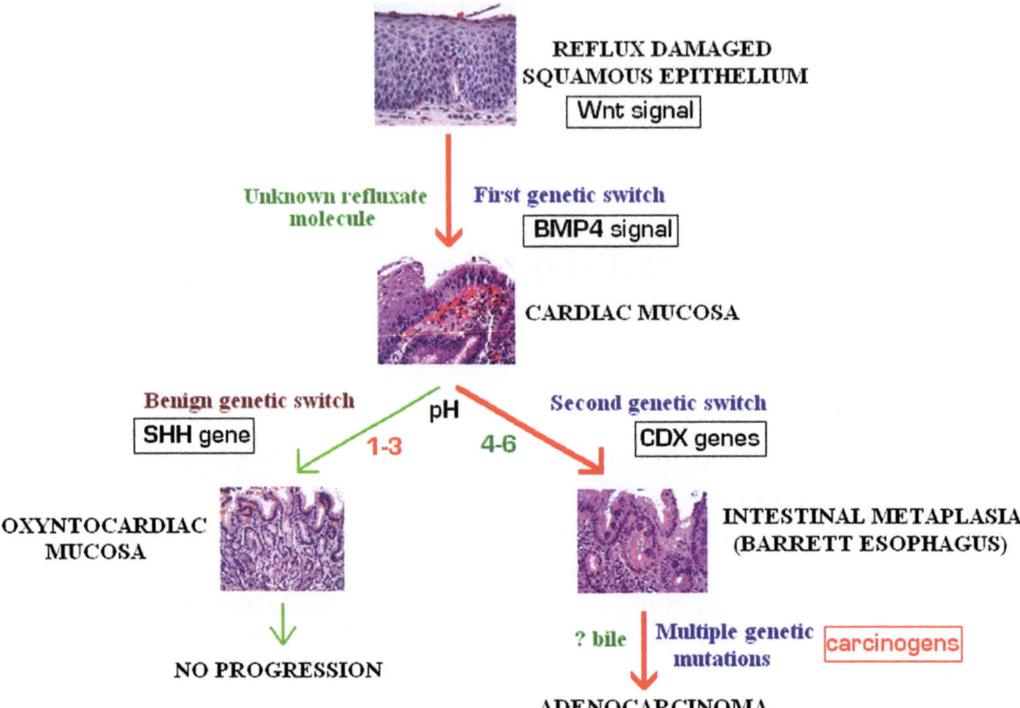

FIGURE 15.1 Sequence of cellular progression from squamous epithelium to adenocarcinoma. All changes are the result of exposure of the target epithelium for each step to gastric contents.

sometimes, after birth during infancy.[6–8] When development is complete, all fetal columnar epithelium in the esophagus becomes replaced by stratified squamous epithelium.

The stomach, in contrast, develops in a much less chaotic manner.[9] The primitive endodermal stratified columnar epithelium of the stomach transforms in early gestation into a flat columnar epithelium. This invaginates to form a foveolar gland complex in the 13th week of gestation. The stem cells migrate from the surface to the deep region of the foveolar pit, where they remain. When the stem cells proliferate, they provide two types of daughter cells: (1) cells that migrate upward in the foveolar pit to the surface where they replenish cells at the surface and (2) cells that migrate downward into the gastric gland where they differentiate into chief and parietal cells.

From the outset, the fetal epithelium in the stomach resembles gastric oxyntic epithelium with a surface layer and foveolar pit consisting of mucous cells and a gland below the foveolar pit in which parietal cells can be seen as early as the 17th week. The epithelium at the angle of His [the fetal gastroesophageal junction (GEJ)] and distal to this is gastric oxyntic epithelium from the early second trimester.[6] After this time, the gastric oxyntic epithelium grows to its adult thickness, maintaining its general architecture. When development is complete, all fetal columnar epithelium in the proximal stomach becomes replaced by gastric oxyntic epithelium.

2. DYNAMIC STRUCTURE OF ADULT ESOPHAGEAL AND GASTRIC EPITHELIA

Once development is completed, stratified squamous epithelium lines the entire esophagus and gastric oxyntic epithelium lines the entire proximal stomach. The dynamics of epithelial proliferation and differentiation into different cell types within the epithelium is important if one is to understand the mechanisms behind metaplastic change.[10]

The esophageal squamous epithelium is a stratified epithelium of numerous cell layers that can be divided into three functional zones: a basal layer (stratum germinativum) that contains the stem cells, an intermediate zone that contains prickle cells (stratum spinosum) that are in transit, moving from the basal region to the surface where the functional mature squamous cells are found. The fully mature functional cells at the surface are fully keratinized and nonmitotic. Surface mature cells are continually shed, being replenished from below.

The stem cells in the basal layer are found mostly in the interpapillary region. When a stem cell divides, it may produce two identical basal cells, one basal and one intermediate (prickle) cell, or two prickle cells. Those daughter cells that retain their original characteristics remain in the basal layer. The daughter cells that have characteristics of intermediate (prickle)

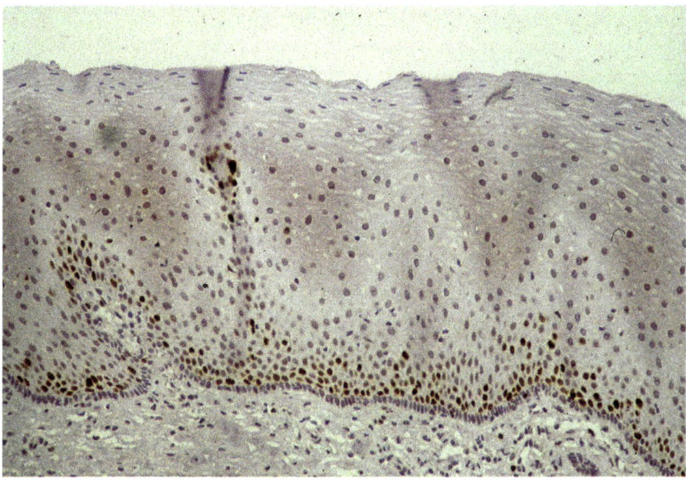

FIGURE 15.2 Normal stratified squamous epithelium of the esophagus, stained with Ki67, which shows nuclear staining of cells in the mitotic cycle. The proliferative zone is limited to 2–3 cells above the basal layer.

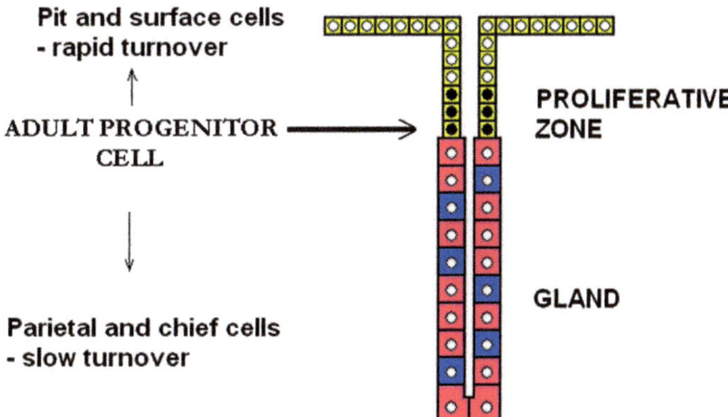

FIGURE 15.3 The bidirectional flow of cells from the division of the stem cells located in the deep foveolar region. The majority of the cells are "pit cells" that move up into the foveolar pit and to the surface to replace cells that continuously shed at the surface. A minority of stem cell divisions differentiate into precursors of zymogenic (blue) and oxyntic (red) cells that move downward into the glands.

cells migrate upward to replenish the cells lost at the surface. The prickle cells in the intermediate zone are proliferative in the suprabasal region and progressively increase their keratinization and decrease their ability to undergo mitotic division as they move upward (Fig. 15.2). The average turnover time of the whole epithelium is about 7.5 days in humans.

The normal structure of the squamous epithelium as seen in a biopsy is a snapshot of this dynamic process. If, for any reason, the rate at which mature surface cells are shed at the surface increases, the basal cells and proliferative cells in the suprabasal region increase their proliferation rate to compensate. This is expressed phenotypically as basal cell hyperplasia and elongation of papillae, two cellular features that are seen when squamous epithelium is exposed to gastric acid.[11] They are not specific for reflux because these changes can be seen in any condition associated with increased rate of cell loss from the surface.

In contrast to the unidirectional flow of cells from the basal region to the surface in esophageal squamous epithelium, gastric oxyntic epithelium has a bidirectional flow of cells (Fig. 15.3). The stem cells are located in the isthmus of the gland below the base of the foveolar pit. All cells lining the foveolar pit gland unit of the oxyntic epithelium originate from the isthmal stem cells. Stem cell divisions give rise to three main cell lineages:

1. Pit cell lineage: The pit cells migrate upward in the foveolar pit, reaching the surface in approximately 5 days. Pit cells in the deep foveolar region at the border of the isthmus retain their ability to divide. Those in the mid and superficial foveolar region and surface are postmitotic terminally differentiated cells.
2. Zymogenic cell lineage: These cells form the mucous neck cells and move progressively downward in the tubular gland, populating the lower part of the gland. As they migrate downward, their phenotype changes from mucous to serous. They ultimately become the pepsinogen-secreting chief cells. These zymogenic cells are postmitotic cells with a long turnover time of approximately 6 months.

3. Parietal cell lineage: Parietal cells develop from the isthmal stem cells and migrate both upward into the base of the foveolar pit as well as downward to populate the superficial half of the tubular gastric gland below the isthmus. They are also long lived with a turnover time of around 2 months.

Special stains for mucins allow differentiation of the different cell lineages in the gastric glands. Periodic acid Schiff stain positively stain the pit cells a dark purple, prezymogenic mucous neck cells a pink color, and the zymogenic chief cells a light blue color, with the parietal cells being completely negative. Gastric mucins that are an important component of mucus can be used as genetic markers for cells of different lineage. MUC5AC stains gastric pit cells whereas gastric gland cells produce MUC6.

The majority of stem cell divisions produce cells that become pit cells migrating to the surface. In normal oxyntic epithelium, the foveolar pit is short with a small number of proliferative pit cells in the basal region. By the mid level of the foveolar pit, the cells have stopped proliferating. In pathologic states characterized by increased cell turnover at the surface, the foveolar pit elongates and the proliferating pit cells increase in number and can be seen in the mid region of the pit in addition to the base.

3. MOLECULAR SIGNALING IN THE FETAL ESOPHAGUS AND STOMACH

These phenotypic expressions of epithelial development in the esophagus and stomach must represent sequential activation and suppression of genes that signal epithelial differentiation. These signals are responsible for converting the primitive endodermal stratified columnar epithelium to all of the phenotypic iterations seen during fetal life.

In writing this chapter, I must give credit to one group that has done much of the research into molecular pathways involved in the metaplastic events in the esophagus. This is the Department of Gastroenterology and Hepatology at the Academic Medical Center in Amsterdam. I will cite the work of this unit extensively in this chapter. Much of my understanding of molecular changes is derived from their work.

There is limited data regarding the signaling mechanisms that operate in the fetal endoderm of the esophagus. From an excellent review on the topic by Krishnadath[12]: “It is known that protein sonic hedgehog (Shh) and bone morphogenetic proteins (BMPs) are highly expressed throughout embryonic development of several organs. It has been shown that for normal endoderm development of the mid- and hindgut, there is an important interaction between the notochord (mesenchyme) and the overlying endoderm. In this interaction, Shh and BMP-4 are key players in transforming the primordial stratified epithelium of the endoderm into a simple columnar epithelial lining. Shh and BMP proteins, together with Notch and Wnts, subsequently take part in the further differentiation of the intestinal mucosa into crypts and villi, and in the adult small and large intestine these factors are critical in the homeostasis and crypt renewal.”

> *In mouse embryos it was confirmed that both Shh and BMP-4 are highly expressed in the notochord of the anterior foregut. Importantly, the action of BMP-4 on the development of the foregut seems to be closely regulated by noggin, the natural antagonist of BMP proteins… Shh and BMP-4 are presumably highly expressed in the mesenchyme of the esophagus during early embryogenesis, but it seems that their action is closely balanced through simultaneous expression of the inhibitory noggin. Thus, it seems that the inhibitory effect of noggin on BMP-4 prevents development of the primary endoderm, a stratified type of epithelium, into a columnar type of mucosa as is normally seen in the trachea and in the mid- and hindgut. In the adult esophagus, Shh and BMP-4 are not expressed and the BMP (downstream) pathway seems to be dormant.*

The various genetic pathways used by Shh and BMP-4 are highly complicated. However, in a highly simplified (and maybe oversimplified) summary, it appears that these are highly active in the early embryogenesis of the esophagus but become suppressed after development is completed.

Ellison et al.[13] reported a case of a 23-week fetus that had a single focus of intestinal epithelium with goblet cells in the esophagus proximal to the GEJ. This is beautifully and convincingly illustrated. It is reported to contain acid mucin by positive staining with Alcian blue at pH 2.5. Exceptionally, therefore, there can be transient expression of aberrant epithelial types in the fetal esophagus including goblet cells.

Signaling in the fetal gastric epithelium is less well studied. However, the fact that the organization and development of the basic cell lineages in fetal gastric epithelium resembles that of adult oxyntic epithelium, it seems likely that the fetal stomach establishes a signaling mechanism that resembles that of the adult stomach early in gestation. Sonic hedgehog is essential in gastric organogenesis. Shh null mice do not develop gastric epithelium.[14] The stomachs of these mice contain intestinal epithelium.

4. MOLECULAR SIGNALING IN THE ADULT ESOPHAGUS AND STOMACH

Ali et al.’s study evaluates expression of the Wnt pathway components in the homeostasis of the normal human esophagus.[15] The authors introduce the subject: “The molecular genetic mechanisms governing the biology and pathobiology of the squamous mucosa of the esophagus have received little attention compared with other areas of the digestive system. Available studies have mostly focused on malignant tissues and Barrett metaplasia. Consequently the mechanisms involved in homeostasis of the esophageal squamous tissue remain for the most part unexplored… Earlier genome wide expression study of normal looking esophageal squamous mucosa has shown differential expression of the Wnt (Wingless-type) modulators, DDK (Dickkopf) homologs among healthy individuals and patients with reflux esophagitis and Barrett metaplasia, suggesting involvement of the Wnt signaling in the biology of the esophageal squamous mucosa. "

Full-thickness human esophageal tissues obtained from organ donors were confirmed to be normal by gross and microscopic examination. Cells were dissected from the lamina propria and the basal, intermediate, and surface epithelial layers were tested by incredibly sophisticated molecular methods for expression of Wnt ligands (1, 2b, 3, 3a, 5a, 5b), receptors, modulating proteins (Dkk 1,3,4), and intracellular components that form part of the Wnt system in all three layers and the lamina propria. The results of the tests are detailed in terms of the expression of these various markers in different cells.

The expression of these components of the Wnt system is given in detail. In the authors' discussion: "The study findings indicate that within the layers of the human esophageal squamous mucosa various components of the Wnt signaling pathway are distributed in a location-dependent manner."

The Wnt pathway consists of the canonical and noncanonical pathways. The activation of the canonical pathway leads to cytosolic stabilization and nuclear localization of β–catenin, which induces Wnt-related gene expression that leads to cell proliferation. The noncanonical pathway does not use the β–catenin–mediated gene expression and is believed to be nonproliferative and stimulates cell differentiation.

Our findings of greater expression of canonical Wnt 1, 2b, and 3a in the basal cell layer compared with the other layers is consistent with the proposed role of these molecules in stimulating cell proliferation as has been observed in other cells and organs… In contrast to the preponderance of expression of the canonical Wnt ligands primarily in the basal cell layer and lamina propria of the esophagus, the expression of non-canonical ligands is distributed differently throughout various layers. In that Wnt 5a is expressed mostly in the differentiated intermediate cell layer but the greatest level of expression of Wnt 5b is observed in the basal cell layer and lamina propria… The non-canonical ligand may perform various functions in the squamous mucosa including stimulation of differentiation…

The study provides evidence that the Wnt family is involved in the differentiation and maintenance of the squamous phenotype, and control of cell proliferation in the normal human squamous epithelium. In the absence of other candidate genes for this function at this time, these data suggest that continued differentiation of esophageal epithelium in the squamous direction is dependent on the normal activity of the Wnt gene family.

In Van den Brink et al.'s study of the genetic signaling involved in the stomach, the authors summarize the organization of the gastric oxyntic epithelium[16]: "… the epithelial cells of the gastric mucosa are organized in vertical tubular units. These consist of an apical pit region, and an isthmus just below the pit, whereas the actual gland region forms the lower part of the vertical unit… The progenitor cell of the gastric unit is located somewhere in the region of the isthmus, in the middle of the vertical tubular unit, and gives rise to all gastric epithelial cells that migrate either up or down from this point."

In this detailed and complicated molecular study with technology at the highest level of sophistication, the authors show that Sonic hedgehog (Shh) is expressed in the epithelium of the adult human and murine stomach and that Shh controls gastric epithelial proliferation in a compartmentalized fashion in the mouse.

The authors describe the Shh protein: "Sonic hedgehog is produced as a 47-49 kilodalton precursor protein. After the covalent attachment of a cholesterol moiety to the precursor protein, a 19-kilodalton signaling protein is cleaved from the precursor by autoproteolysis. Whereas the remaining 29-41 kilodalton carboxy terminal fragment can freely diffuse from the cell, the signaling peptide remains tethered to the cell membrane by virtue of its cholesterol modification. Additional levels of control seem to determine whether Shh acts short range or is released from the cell and acts more distally."

The results of their study can be summarized as follows:

1. Both Shh precursor protein and large amounts of the cleaved aminoterminal protein are present in the murine stomach when assessed by immunoblot. In contrast, it was not found in the duodenum. By immunohistochemistry, heavy staining was found in the fundic glands of both murine and human stomach, whereas no Shh was detected in the duodenum of either.
2. By the use of differential mucin staining with periodic acid Schiff, MUC5AC and MUC6 immunostaining, it was shown that Shh expression is restricted to the gland compartment in both murine and human gastric glands. In the human, Shh expression is greatest at the pit-gland transition with staining intensity decreasing toward the base of the gland. By double staining with H^+,K^+-ATPase (proton pump), it was shown that Shh was largely limited to the parietal cells in humans.

3. The expression of the Shh target receptor "patched" (Ptc) indicates Shh receptiveness as well as active Shh signaling. Parietal cells in both human and mouse expressed Ptc, whereas the pit cells were negative. The results for expression of putative Shh target in the TGF-β family and the transcription factor HNF3β showed expression of the latter in parietal cells. BMP-4 was not expressed, being limited to interstitial mesenchymal cells. The expression of both HNF3β and BMP-4 were decreased in mice treated with cyclopamine, a Shh inhibitor, confirming that these were indeed downstream targets of Shh.

The study concludes that Shh may play a role both in the organogenesis of gastric epithelium as well as remaining active in maintaining the normal dynamic function and phenotype of the gastric epithelium in adult life. They are careful in their conclusions: "... because Shh seems necessary for gastric epithelial differentiation during organogenesis, our finding that Shh is expressed and functional in adult gastric epithelium is interesting, and although our data offer no direct evidence for such a role, we speculate that Shh may play a role in epithelial differentiation in the adult."

In a follow-up to their earlier study, Van den Brink et al.,[17] in collaboration with Massachusetts General Hospital and Brigham and Women's Hospital, studied Shh expression along the normal adult human and rodent gastrointestinal tract as well as in intestinal metaplasia of the stomach, "gastric and intestinal metaplasia of the esophagus" (i.e., cardiac epithelium with and without parietal and/or goblet cells), and gastric heterotopia in Meckel's diverticulum.

The study tested normal tissues from at least 10 different patients from each compartment of the gastrointestinal tract, 16 patients with intestinal metaplasia of the stomach, 13 resection specimens of Meckel's diverticulum, and 6 resection specimens of patients with Barrett esophagus for expression of Shh. They used in situ hybridization and immunohistochemistry, the latter with simultaneous staining for multiple epitopes (Shh, MUC5AC, MUC2) to evaluate cells expressing Shh.

Shh was found to be expressed in the following locations and situations:

1. In the normal human and rodent gastrointestinal tract, Shh (using an antibody that recognizes the Shh precursor protein) was exclusively detected in the fundic glands of the stomach. No Shh staining was observed in the normal esophagus or intestine.
2. Shh expression was completely lost in areas of intestinal metaplasia of the stomach.
3. Of the 13 Meckel's diverticula, all 8 specimens that had heterotopic gastric epithelium with fundic glands showed Shh expression in the parietal cells. The five specimens that did not contain gastric fundic epithelium were negative for Shh.
4. "We examined oesophageal resection specimens of six patients with Barrett's esophagus for expression of both the H^+K^+-ATPase expression and Shh. We found one resection specimen with areas of gastric metaplasia of fundic type glands. Complete overlap of H^+K^+-ATPase expression and Shh expression was found in this specimen whereas all oesophageal (including the submucosal glands) and intestinal tissue in the resection specimens were negative for Shh. This indicates that the switch in differentiation from squamous to gastric epithelial tissue with fundic glands is accompanied by induction of Shh expression."

This paper provides compelling evidence that the phenotypic expression of parietal cells in the gastrointestinal tract in both normal and abnormal states is closely associated with and possibly induced by Shh activation. This includes oxyntocardiac epithelium in the esophagus as the only epithelium therein to express Shh.

5. STRUCTURE AND MOLECULAR SIGNALING IN THE FETAL AND ADULT INTESTINE

The relevance of studying intestinal development and signaling is that there is a possibility that the same mechanisms that direct and maintain phenotypic expression of intestinal epithelium during organogenesis and in adult life may be similar to those involved in inducing intestinal metaplasia in the esophagus in patients with GERD-induced columnar-lined esophagus (CLE) as well as intestinal metaplasia of the stomach in chronic atrophic gastritis.

The dominant signaling mechanism for both development and maintenance of the normal phenotype of the colon and small intestine involves a complex interaction of Cdx1 and Cdx2. These two genes are expressed in the developing intestine from the earliest stage through fetal development into adult life.

Cdx1 and Cdx2 are members of the caudal-related homeobox gene family. They are expressed in a large number of early embryonic tissues, but very early in fetal development, expression of Cdx1 and Cdx2 is lost in all but the developing small intestine and colon. This pattern of expression continues to adulthood.

Silberg et al.[18] show a complicated relationship between the expression of Cdx1 and Cdx2 which vary depending on proximal and distal location in the intestine. Cdx1 is expressed distally and Cdx2 in the proximal intestine with an overlap in the midgut region. These differences may be responsible for the development of the crypt + villous architecture of the small intestine and the purely crypt architecture of the colon.

In the discussion, the authors summarize their findings: "The results help elucidate the expression of these 2 related genes in 4 aspects of intestinal development, including (1) early patterning of intestinal endoderm along the horizontal axis, (2) patterning of cellular compartments along the crypt-villus or vertical axis, (3) gene expression at the suckling-weaning transition, and (4) maintenance of the adult epithelial phenotype."

The authors point to the aberrant development of the intestine in heterozygous Cdx2 null mice, which develop polyp-like lesions in the colon with loss of expression of Cdx2 within these lesions. The polyps contain areas of keratinizing stratified squamous epithelium, as found in the esophagus and gastric and small intestinal-type mucosa.

The development of the intestine is directed from early in gestation to adult life by the expression of Cdx1 and Cdx2. These genes remain expressed in the small intestine and colon in adult life. The two genes are suppressed in the remainder of the gastrointestinal tract in the normal state. Expression of Cdx2 occurs in the stomach in chronic atrophic gastritis with intestinal metaplasia and in the esophagus in patients with Barrett esophagus, which have cardiac epithelium with intestinal metaplasia.

6. SUMMARY OF MOLECULAR SIGNALING IN THE GASTROINTESTINAL TRACT

I run the serious risk in this summary of oversimplifying the genetic signaling that is involved in organogenesis of the various compartment of the gastrointestinal tract and those involved in maintaining the adult phenotype of the esophagus, stomach, and intestine. However, this book is meant for clinicians involved in the treatment of patients with GERD. I apologize to any molecular researcher who is reading this summary.

It appears that there are a relatively small number of signaling genes that are involved in transforming the primitive stratified columnar epithelium of the fetal endodermal tube into the esophagus, stomach, and intestine. These genes are those in the Wnt family, TGF-β family (the BMPs), Shh, and the Cdx genes of the homeobox family.

In the early stages, these genes are involved in basic things such as converting the stratified fetal epithelium of the endodermal tube into a single columnar cell layer. As development proceeds, all but one of these genes become dominant in a given compartment. The others are suppressed.

1. In the esophagus, active genes of the Wnt family appear in the 22nd week of gestation when the ciliated columnar epithelium of the early esophagus becomes replaced by stratified squamous epithelium. BMP-4, Shh, and Cdx1 and Cdx2 are suppressed in the fully developed esophagus.
2. In thc stomach, which is lined entirely by gastric oxyntic epithelium (except for the pyloric antrum), Shh is active, inducing a glandular epithelium with parietal cells. Wnt, BMP-4, and Cdx genes are suppressed. Shh null mice do not develop gastric oxyntic epithelium; their stomach contains intestinal epithelium.
3. In the intestine, Cdx1 and Cdx2 are active, inducing the small intestine with its crypt-villus architecture and the colon with its crypt-only architecture. Wnt, BMP-4, and Shh are suppressed. Heterozygous Cdx2 null mice develop colonic polyps with aberrant stratified squamous and gastric epithelium.

With this background model of genetic control of normal epithelial differentiation in the gastrointestinal tract, it becomes relatively easy to understand the process of metaplasia as changes in differentiation induced by switches in the expression of signaling genes.

There is very little data as to what maintains these signaling genes in their active state in the various parts of the gastrointestinal tract during adult life. I will propose that genetic switches that result in metaplasia result from changes in the milieu in which the epithelium resides.

In this regard, it is interesting that Wnt expression in the esophagus is associated with a near neutral luminal pH, Shh expression in the normal stomach is associated with a strong acid environment and Cdx1 and Cdx2 expression in the intestine with an alkaline environment.

In the stomach, atrophy or destruction of parietal cells with an increase in gastric pH appears to result in suppression on Shh and activation of Cdx2, resulting in intestinal metaplasia.

In the esophagus, exposure of the squamous epithelium to gastric acid appears to induce BMP-4 resulting in metaplasia to cardiac epithelium. Cardiac epithelium in the esophagus evolves in two directions based on the pH to which the epithelium is exposed.

During a reflux episode, a pH gradient is created by the retrograde flow of gastric juice into the esophagus. This gradient is strongly acidic at the GEJ and becomes near neutral at the height of the refluxate column (Fig. 15.4). Cdx2 activation occurs in the area farthest from the GEJ that has the highest pH, with the maximum prevalence seen at the squamocolumnar junction (SCJ) in the longest segments of visible CLE. In contrast, Shh activation occurs in the strong acid milieu of the region of the GEJ where oxyntocardiac epithelium is seen.

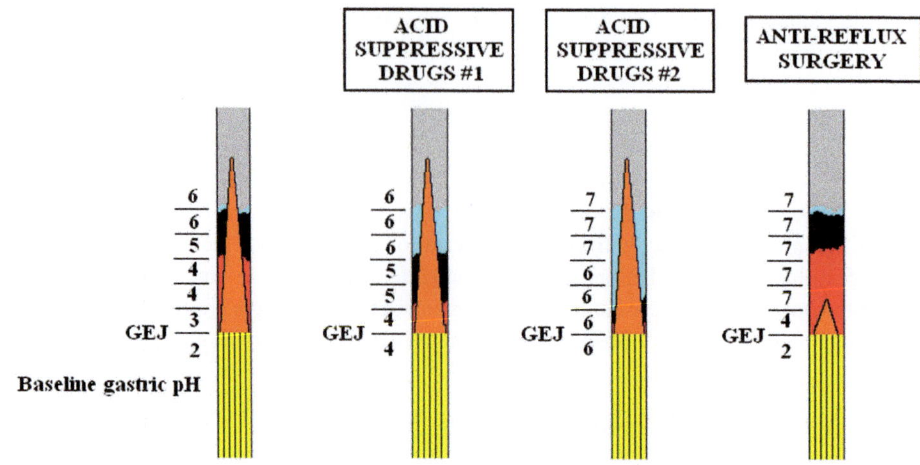

Volume of Reflux	HIGH	REMAINS HIGH	REMAINS HIGH	DECREASED
Time of Exposure	LONG	REMAINS HIGH	REMAINS HIGH	DECREASED
Amount of IM	SMALL	INCREASED	MORE INCREASED	SAME OR LESS
Amount of OCM	MODERATE	LESS	STILL LESS	INCREASED

FIGURE 15.4 The pH gradient that is created in the esophagus during a reflux episode. This varies from the lowest pH at the GEJ to the highest pH at the top of the column of refluxate. Treatment with acid-suppressive drugs dramatically converts the pH milieu at all levels in the esophagus to a higher (less acidic) level. *GEJ*, gastroesophageal junction; *IM*, intestinal metaplasia; *OCM*, oxyntocardiac mucosa.

7. TRANSDIFFERENTIATION AND TRANSCOMMITMENT

The exact mechanism whereby the squamous epithelium of the esophagus is converted to columnar epithelium has been a matter of debate for many decades. The original theory was that injury to the squamous epithelium caused the migration of gastric columnar epithelial cells upward into the esophagus to replace the denuded squamous epithelium. This concept lost favor two decades ago but has reemerged in some circles as a possible mechanism.

The current belief is that erosion of the squamous epithelium is not necessary for metaplasia to occur. In the absence of erosion, there is no reason or possibility for upward migration of columnar cells from the stomach. The change from squamous to columnar must occur in intact epithelium. There are two possible ways in which this can happen; these are not mutually exclusive.

The first is a process of transdifferentiation. In this mechanism, the differentiated squamous epithelial cells that have arisen from the stem cells undergo a change in the signaling that causes them to change their differentiation from squamous to a columnar phenotype. This change can therefore occur at any level of the squamous epithelium but is more likely in the proliferative suprabasal prickle cell zone than in the mature surface layer. In the superficial cells, the molecular change must, in addition to changing the direction of signaling, reactivate the ability of the mature cells to undergo mitotic division to express the signaling change. In the proliferative cells, it is only the change in the direction of signaling that is required.

Phenotypically, histologic evaluation of the metaplastic columnar epithelium in the esophagus commonly shows the presence of a multilayered epithelium.[19] This usually has several layers of basal and suprabasal cells with squamous characteristic and a columnar epithelial layer on the surface. These cells express both squamous and columnar cytokeratins, suggesting a change from squamous to columnar in the mid region of the squamous epithelium. Multilayered epithelium likely reorganizes into a single-layered columnar epithelium with the passage of time. It has been designated a precursor epithelium of CLE.

The second mechanism is called transcommitment. In this mechanism, the stem cells in the esophageal epithelium, by virtue of a change in the differentiating gene signaling, commit to differentiate into a columnar epithelial type rather than the normal squamous epithelium. This would result in a fundamental alteration of the attachment mechanism of stratified squamous epithelium, leading to a sloughing of the squamous epithelium above the newly differentiated columnar cell at the base of the epithelium. There would be no multilayered epithelium.

The stem cells in the normal esophagus reside in the basal layer of the epithelium. These would be the cells undergoing the metaplastic change in the transcommitment model. However, stem cells also exist in the gland ducts of the submusosal

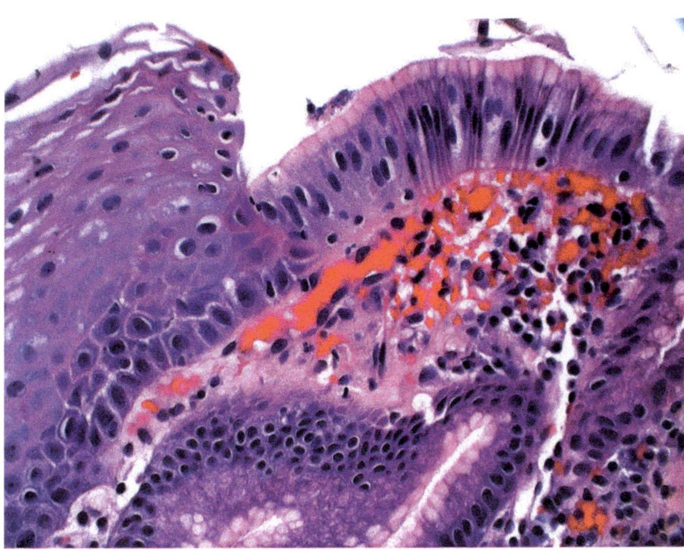

FIGURE 15.5 Cardiac metaplasia of squamous epithelium. The change in the differentiating signal in the stratified squamous epithelium alters cell attachments, causing the squamous epithelium above the new columnar cell to slough off, leaving the basal region cells to express the new differentiating signal that induces columnar differentiation.

and mucosal glands of the esophagus. These stem cells could be a source of repopulation of a denuded mucosal surface with stem cells in injury. They would then migrate to cover the denuded epithelium and differentiate according to the genetic signal that drives differentiation. They are believed by some authorities to be responsible for squamous islands that are seen in areas of CLE. That would represent stem cells with a squamous signal that have migrated to the surface around the opening of the gland ducts to produce an area at the surface that is lined by squamous epithelium.

Another source of stem cells for the esophagus is circulating stem cells in the blood stream. Theoretically, these cells can repopulate the denuded esophagus and develop signals that are site specific and differentiate in a manner directed by those signals. Communication with signaling genes in the mesenchymal cells of the lamina propria of the esophageal mucosa would explain how these circulating stem cells learn that they are in the esophagus.

By whatever mechanism squamous epithelium undergoes columnar metaplasia, the first epithelium that is produced will be a flat columnar epithelium composed of mucous cells without any other type of differentiated cells such as a goblet cell or a parietal cell. This is cardiac epithelium.[20] No evidence exists for the simultaneous expression of multiple signaling genes that would allow direct metaplasia of squamous epithelium to cardiac epithelium with intestinal metaplasia or to oxyntocardiac epithelium. All the evidence points to a two-step process: (1) squamous → cardiac epithelium; (2) cardiac to either oxyntocardiac or intestinalized cardiac epithelium.

8. HISTOLOGIC PATHWAYS OF EVOLUTION OF CARDIAC EPITHELIUM

The metaplasia of squamous epithelium results in the separation of the SCJ, which moves cephalad, from the true GEJ. Metaplastic cardiac epithelium in the distal esophagus becomes interposed between squamous epithelium and gastric oxyntic epithelium. This is initially limited to the dilated distal esophagus during the phase of compensated lower esophageal sphincter (LES) damage.[21] With increasing LES damage, abnormal reflux occurs into the thoracic esophagus, ultimately resulting in visible CLE.

The pathogenesis of cardiac metaplasia of squamous epithelium, either in the dilated distal esophagus or in the production of visible CLE, is identical. Squamous epithelium is damaged by exposure to gastric juice, which induces molecular events that result in metaplasia to cardiac epithelium. In the dilated distal esophagus, the squamous epithelium is exposed to gastric contents during times of gastric distension. In the thoracic esophagus, it is exposed when reflux occurs secondary to LES failure.

The change in signaling from squamous to columnar causes a dramatic change in the way the cells are attached in the epithelium. The squamous cell is intimately connected to its neighbors by tight junctions and desmosomes. When the basal region squamous cell changes its differentiation signal to columnar, it results in a cell that has no ability to adhere to the overlying squamous cell (Fig. 15.5). A cleavage of the epithelium results and the superficial part of the squamous epithelium sloughs off, leaving either a columnar cell on the basement membrane or a multilayered epithelium, which likely slowly reverts to a single layer of columnar epithelium.

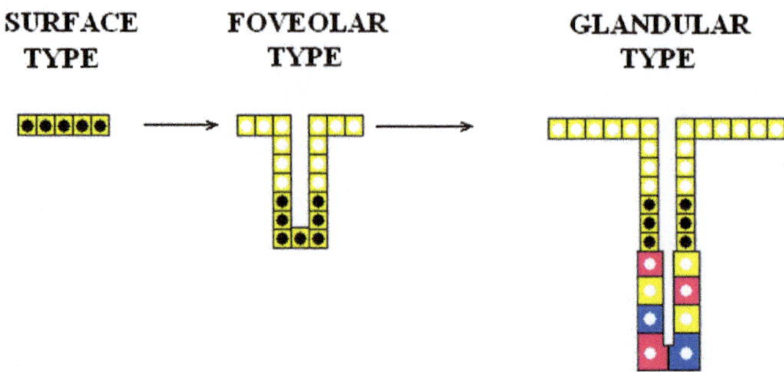

FIGURE 15.6 The stages of maturation of cardiac epithelium. Metaplasia from squamous epithelium produces a flat epithelium consisting of mucous cells, which include the stem cells. These quickly invaginate into a foveolar pit where the stem cells move into the deep foveolar region. The final development of cardiac epithelium is the formation of a gland below the foveolar pit. In cardiac epithelium, this consists only of mucous cells. Shown here is oxyntocardiac epithelium that has, in addition to the mucous cells (yellow), parietal (red) and chief (blue) cells in the gland.

The first epithelium that results from metaplasia is a flat columnar epithelium composed of undifferentiated mucous cells that now line the esophageal mucosa (Fig. 15.6). This is the "surface-type" of cardiac epithelium (see Chapter 4). It is rarely seen. It consists of stem cells, proliferative cells (Ki67$^+$), and differentiated mucous cells.

The surface type of cardiac is a unique epithelium that only occurs during the process of columnar metaplasia of esophageal squamous epithelium. It results from expression of the genetic signal that replaces the Wnt signal that determined squamous differentiation. This is a new genetic signal not expressed anywhere in the adult gastrointestinal tract. There is evidence that the signal is related to BMP-4 (see later).

Under the influence of this new differentiating signal, the epithelium develops a foveolar pit that recapitulates the developmental stages of the fetal epithelia of the stomach and intestine (Fig. 15.6). However, it does not initially produce either parietal cells or goblet cells. It produces a foveolar pit that is lined entirely by mucous cells. This is the foveolar type of cardiac epithelium. It is a common epithelial type in CLE. The foveolar type of cardiac epithelium has the stem cells in the deep part of the foveolar pit like all gastrointestinal epithelia. It has a proliferative zone in the deep foveolar region and terminally differentiated mucous cells lining the surface. It is as if the epithelium is designed to protect the stem cells by moving them from the surface to a more sequestered and protected region in the deep foveolar pit.

The final phenotypic event that occurs in cardiac epithelium is the development of a glandular unit below the foveolar pit. Again, the glandular unit that is formed is unique. It is composed of only mucous cells initially. It is composed of long-lived cells that are regenerated from stem cells in the deep part of the foveolar pit that move downward away from the surface. This is a similar bidirectional pattern of growth typically seen in the tubular glands of gastric epithelium. However, the lack of Shh causes the cardiac glands to be lobulated and composed entirely of mucous cells. They can be identified as glands by the fact that they are seen below the foveolar pit, which shows the proliferative zone of the foveolar-gland unit.

The fully developed cardiac epithelium consists of a focal point in the deep foveolar region that contains the stem cells that replenish all the cells of the epithelium. Stem cell division produces daughter cells, most of which move upward toward the surface to replenish surface cells that are continually shed. A minority of the cells move downward to replenish the glandular cells that are lost by apoptosis. In the absence of a sufficient Shh or Cdx signal, parietal cells and goblet cells are not seen.

It is important to recognize that cardiac epithelium is an epithelial type that is unique to the esophagus. It is the result of columnar metaplasia caused by exposure of esophageal squamous epithelium to gastric juice. It is the result of a new signaling gene, BMP-4 and its downstream targets, unique to this pathology that has arisen in the esophagus. There is no other BMP-4 driven epithelium in the adult gastrointestinal tract.

Cardiac epithelium is *always* the first epithelium that results when squamous epithelium undergoes columnar metaplasia. Cardiac epithelium with intestinal metaplasia and oxyntocardiac epithelium arise from cardiac epithelium in a second step by another change in the signaling mechanism of the epithelium (see below). All people with CLE have cardiac epithelium at some point in time. Almost all people with cardiac epithelium develop oxyntocardiac epithelium at the distal end of the segment of CLE. A significant number of people with very small (<1 cm) length of CLE may convert all their cardiac epithelium to oxyntocardiac epithelium. A minority, usually patients with more severe LES damage and longer segments of CLE, develop intestinal metaplasia. Intestinal metaplasia occurs in the proximal region of the CLE adjacent to the SCJ. I will discuss in detail the events associated with the evolution of cardiac epithelium to intestinal metaplasia and oxyntocardiac epithelium below.

9. SQUAMOUS EPITHELIUM→METAPLASIA TO CARDIAC EPITHELIUM→INTESTINAL METAPLASIA

The entire spectrum of cellular changes in GERD is caused by exposure of the esophageal epithelium to gastric juice. Without exposure of the squamous epithelium to gastric juice, there will be no columnar metaplasia or LES damage, and GERD will not occur.

The earliest damage involves the normal squamous epithelium in the most distal esophagus immediately above the true GEJ where it transitions to gastric oxyntic epithelium that lines the entire proximal stomach.

The normal LES extends to the end of the esophagus and separates the squamous epithelium from the contents in the gastric lumen. Expressed in another way, the distal limit of the LES is a point where the luminal pH changes from the acidic milieu of the stomach distally to the neutral (or near neutral) milieu of the esophageal lumen. This is called the pH transition point (or pH step-up point). This point is also the point at which manometry shows a step down of pressure from the resting LES pressure (>2 mmHg above gastric baseline at its distal limit) to baseline intragastric pressure (+5 mmHg).

In Chapters 6, 7, and 9, I have detailed the normal LES and how it is progressively damaged by repetitive gastric overdistension and how this ultimately results in LES failure and reflux into the thoracic esophagus.

9.1 Interaction Between Proliferative Cells in the Squamous Epithelium and Molecules That Induce Metaplasia

The normal squamous epithelium is designed to facilitate the rapid transit of the food bolus from the pharynx to the stomach. Its surface is lubricated by mucin secreted by mucus glands in the mucosa and submucosa. It is a stratified epithelium, allowing for natural removal of the surface layers by the abrasive action of the rapidly moving food bolus.

The epithelium is also impermeable to the entry of luminal molecules. In a normal person, this is not important because the swallowed food bolus and fluids are propelled quickly by peristalsis into the stomach. When the squamous epithelium is exposed to gastric juice, however, it becomes exposed to a myriad of molecules in gastric juice derived from food, gastric secretion, and in patients who have duodenogastric reflux, duodenal contents. These can remain in contact with the squamous epithelium during both gastric overdistension and reflux for significant periods of time. The inability of the molecules to enter the epithelium prevents them from gaining access to the proliferative cells in the basal and intermediate cell layers.

Tobey et al.[2] showed that the squamous epithelium is damaged by acid, causing separation of epithelial cells due to damage of tight junctions. He also showed that with increasing acid-induced damage, the epithelium permitted entry of molecules of increasing size to an increasing depth in the epithelium.[22]

Exposure to gastric juice causes acid-induced increase in permeability of the squamous epithelium allowing the entry of a variety of molecules in gastric juice. An unknown but very commonly present molecule in gastric juice must interact with the squamous epithelial cells to produce the genetic switch that causes columnar metaplasia.

The phenotype of the normal adult stratified squamous epithelium is maintained by activity of the signaling genes in the Wnt family.[15] BMP-4, Shh, and Cdx genes are suppressed. The Wnt genes are expressed in all the cells including the stem cells in the basal layer, the proliferative cells in the prickle cell layer, and the mature cells in the surface layer. The mature cells near the surface are postmitotic cells incapable of cell division. Genetic changes in postmitotic cells do not have an impact because cells must divide to cause a change in differentiation.

The increased permeability of the squamous epithelium when exposed to gastric juice must therefore be of sufficient severity to allow the critical unknown large molecule in gastric juice to penetrate the epithelium to a depth where it can interact with proliferative cells in the prickle cell or basal layer. This is the basis whereby the genetic switch occurs that overrides the Wnt signal and results in columnar metaplasia.

9.2 Temporal Sequence of Appearance of Metaplastic Columnar Epithelia in the Esophagus

There should be no doubt that the metaplastic process in the esophagus begins with the transformation of squamous epithelium to cardiac epithelium. Pure cardiac epithelium *always* precedes both intestinal metaplasia and oxyntocardiac epithelium. I am not certain that all physicians involved in the management of GERD recognize this fact. Sometimes, I wonder whether even some expert gastroenterologists believe that metaplasia of squamous epithelium can result directly in cardiac epithelium with intestinal metaplasia. *There is no evidence that this ever happens at either a histologic or molecular level.*

The sequence of molecular genetic switches that I will describe in this chapter will clearly demonstrate this fact. The sequence of molecular change with its associated phenotypic expression in the epithelium of the esophagus is as follows:

Step 1: Squamous epithelium to cardiac epithelium. This results from the active Wnt signal being replaced by a newly activated genetic signal that induces cardiac metaplasia (Fig. 15.7). There is evidence that the BMP-4 pathway is activated

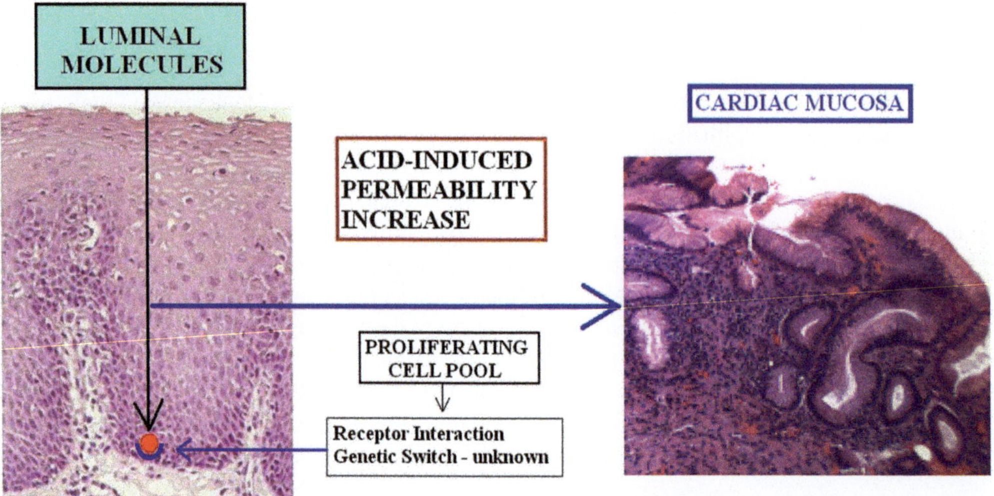

FIGURE 15.7 Cardiac epithelium is the first epithelium that develops from columnar metaplasia of squamous epithelium. When acid-induced damage of the squamous epithelium causes increased permeability, luminal molecules enter the epithelium. Interaction of these molecules with proliferating epithelial cells results in a change in the differentiating signal to one that directs the cells toward columnar differentiation. This new signal is not yet certain but is likely to be related to BMP-4.

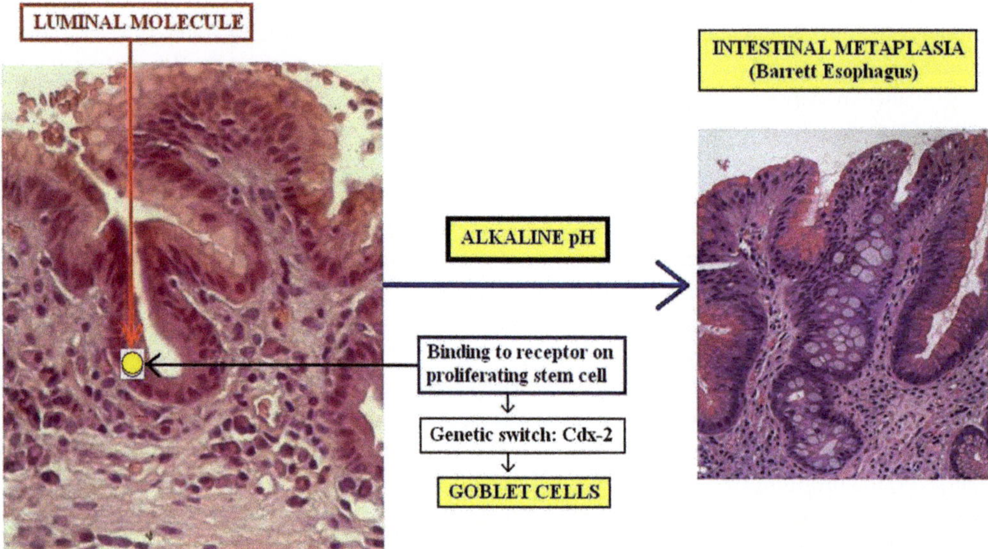

FIGURE 15.8 Cardiac epithelium represents a pivotal epithelium that can differentiate in one of two directions. Expression of Cdx2 causes intestinal metaplasia with goblet cells. The conversion of cardiac epithelum to intestinal metaplasia represents progression in the gastroesophageal reflux disease→adenocarcinoma pathway.

in this process, but there is still uncertainty whether this is the actual signal involved or a passenger genetic change that is associated with an yet undetermined differentiating signal.

Step 2A: Cardiac epithelium to oxyntocardiac epithelium. This results in the distal region of the CLE when Shh is activated and induces the appearance of parietal cells in cardiac epithelium (Fig. 15.8).

Step 2B: Cardiac epithelium to intestinal metaplasia. This results in the proximal region of the CLE when Cdx2 is activated and induces the appearance of goblet cells in cardiac epithelium (Fig. 15.9).

Steps 2A and 2B are mutually exclusive (Fig. 15.10). It is extremely uncommon to see both parietal cells and goblet cells in one foveolar-gland unit. Another way of saying this is that Shh and Cdx2 cannot be expressed in the stem cell in the same foveolar unit. This is important. Oxyntocardiac epithelium is a stable epithelium, which is a positive end point in the metaplastic process. As it does not progress to intestinal metaplasia, it does not progress to adenocarcinoma.

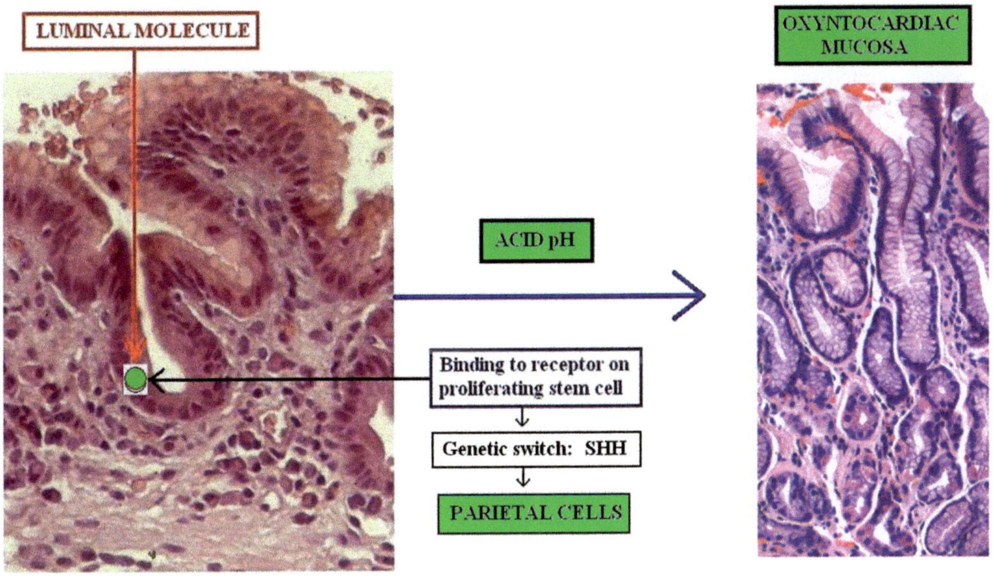

FIGURE 15.9 Expression of Sonic Hedgehog (SHH) gene causes the epithelium to form oxyntocardiac epithelium with parietal cells. Oxyntocardiac epithelium is a benign, stable epithelium with no cancer risk. It does not develop intestinal metaplasia.

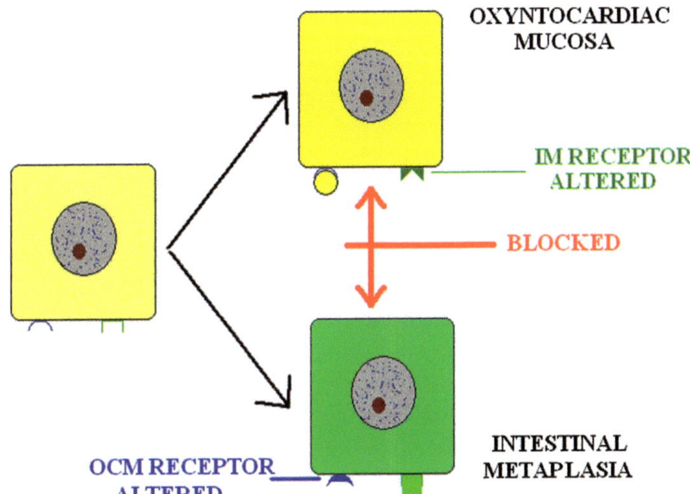

FIGURE 15.10 Theoretical depiction of how the expression of Sonic Hedgehog and Cdx2 genes in the cardiac epithelium appears to be mutually exclusive. It is extremely uncommon to find both parietal and goblet cells in one foveolar complex of metaplastic columnar epithelium.

There are two clinical situations that establish the fact that the transformation of esophageal epithelium to intestinal metaplasia is a two-step process: the histology of CLE in children and the metaplastic change that occurs in the esophagus above the anastomotic line after esophagogastrectomy.

9.2.1 Columnar-Lined Esophagus in Children

Reflux-induced visible CLE in children consists predominantly of cardiac epithelium without intestinal metaplasia.

Hassall, in an excellent review of the subject, found that the diagnosis of "Barrett esophagus" is often made in children based on the presence of endoscopically visible CLE without the requirement of intestinal metaplasia.[23] He reviewed 14 reports from the literature from 1984 to 1996 in which 119 children were reported to have Barrett esophagus. Only 43 of these had intestinal metaplasia.

In the review, Hassall reported his unpublished experience: "Over a 12-year period (1985–1996), only 7 children were newly diagnosed as having Barrett esophagus, (i.e. with specialized intestinal metaplasia), for a prevalence of 0.02% of all pediatric upper gastrointestinal endoscopies during that period. Their ages were 8–17 years (mean 14 years). In this unit, it is routine practice at endoscopy to document esophagogastric landmarks and take biopsy specimens from the Z-line and tubular esophagus; use of this protocol makes it highly unlikely that Barrett esophagus would be missed" (p536).

Hassall also commented on his previously reported study from 1985 of 11 patients with "Barrett esophagus." Only 5 of these 11 patients had specialized mucosa with goblet cells. The other 6 patients had "biopsy specimens containing cardiac mucosa as proximal as 4–15 cm above the lower esophageal sphincter." (p536). Hassall suggested that it is likely that Barrett esophagus was missed because of sampling error of the biopsies. He considered less likely the possibility that a 10- or 15-cm CLE in a child does not contain goblet cells. The youngest patient reported in the literature with intestinal metaplasia was 5 years old.[24]

Even when intestinal metaplasia is present in children with visible CLE, the density of goblet cells is less than in adults. Qualman et al.[24] reported that pediatric patients had 25 or fewer goblet cells per square millimeter of visible CLE compared with a mean of 57/mm^2 of CLE in 41- to 80-year-old patients.

Most children with visible CLE had long segments of CLE with a mean of 4, 7, and 8.5 cm where the lengths were reported. 60%–70% of children with visible CLE have significant comorbid diseases such as mental retardation, cerebral palsy, cystic fibrosis, repaired esophageal atresia, and chemotherapy for malignancies.

Hassall explains the rarity of intestinal metaplasia in children even when long segments of visible CLE are present: "Perhaps Barrett's esophagus evolves in some children by the development of goblet cell metaplasia in cardiac mucosa. In other words, perhaps cardiac mucosa in the tubular esophagus is a precursor of Barrett's esophagus in children."

9.2.2 Metaplastic Change in Esophagus After Esophagogastrectomy

Patients who undergo esophagogastrectomy with an anastomosis created between the distal margin of the resected esophagus and the gastric body represent a human experiment to evaluate the progression of epithelial changes associated with reflux. Because the surgery removes the LES, these patients almost invariably have gastroesophageal reflux. Because the distal esophagus and proximal stomach have been resected, there is no "normal gastric cardiac epithelium." The squamous epithelium of the esophagus transitions directly to gastric oxyntic epithelium at the line of anastomosis. Sequential biopsy above the anastomotic line will show the progression of cellular changes.

An excellent study by Dresner et al. from the Northern Esophagogastric Cancer Unit at the Royal Victoria Infirmary in Newcastle-upon-Tyne studied the outcome in 40 patients who had an intrathoracic esophagogastrostomy (i.e., gastric pull up) after subtotal esophagectomy for Barrett adenocarcinoma and high-grade dysplasia (26 patients) or squamous carcinoma (14 patients).[25] The patients experienced severe duodenogastroesophageal reflux, verified in 30 patients by combined 24-h ambulatory pH and bilirubin monitoring.

Serial endoscopic assessment and systematic biopsy at the esophagogastric anastomosis was done over a 36-month period. They had a total of 130 (median 3, range 1–8) examinations. The authors defined Barrett esophagus as follows: "The definition of Barrett's mucosa included both specialized and non-specialized columnar epithelium from the oesophageal remnant. The former epithelium was identified by the presence of goblet cells and is referred to as 'intestinal metaplasia-type' Barrett's mucosa. Non-specialized epithelium, which has no goblet cells and has features similar to those of the mucosa of the gastric cardia or fundus, is referred to as 'cardiac-type' or 'fundic-type' Barrett's mucosa respectively." This is a detailed and completely unambiguous definition of the histology.

At the end of the study, 7 patients had normal squamous epithelium, 14 had reflux esophagitis of varying grades, and 19 patients had esophageal columnar epithelium. Biopsies from the columnar mucosa showed cardiac-type epithelium in 10 patients and intestinal metaplasia in 9 patients. Endoscopically, the columnar epithelium was seen as single tongues (6), multiple tongues (2), noncircumferential patches (5), and circumferential segments (6). The measured extent of columnar epithelium ranged from 0.5 to 3.5 cm (median 1.9 cm). Coexistent reflux esophagitis in the squamous epithelium was present in 14 of the 19 patients with columnar metaplasia.

At the index endoscopy in this group, only 9 patients showed columnar metaplasia. In the other 10 patients, metaplastic progression was observed over 6–36 months. In these patients, there was evidence of squamous esophagitis before progression to columnar metaplasia.

The initial detection of columnar epithelium in the esophagus was made a median of 14 (range 3–118) months after surgery. Of the nine patients with columnar mucosa at the index endoscopy, seven had cardiac-type epithelium and two had intestinal metaplasia. All 10 patients who developed columnar mucosa during follow-up initially had cardiac-type epithelium; intestinal metaplasia was identified later in 7 patients. The median time to the development of cardiac-type epithelium was 14 (range 3–22) months, whereas intestinal metaplasia was first detected significantly later at a median of 27 (range 11–118) months.

The gastric mucosa below the anastomosis demonstrated "gastric body epithelium in all cases, with various degrees of quiescent and active gastritis."

The occurrence of Barrett's metaplasia was similar irrespective of the histologic subtype of the resected tumor. Patients with esophageal columnar epithelium had significantly higher acid ($P=.016$) and bilirubin ($P=.011$) reflux compared with patients who had only squamous epithelium at the end of the study period.

The authors concluded that severe duodenogastroesophageal reflux occurring after subtotal esophagectomy causes a sequence of changes in the esophageal squamous epithelium, which shows temporal progression from reflux esophagitis to columnar metaplasia of cardiac type without goblet cells and eventually to intestinal metaplasia.

It is also relevant that the epithelium below the anastomotic line showed no change during the follow-up period. There was no development of cardiac epithelium in the stomach. The reflux-induced changes are strictly limited to the mucosa above the anastomotic line. This should lead one to the conclusion that in the patient with an intact esophagus and stomach, all pathologic changes in the region of the GEJ must be esophageal. There can be no "expansion" of the gastric cardia resulting from GERD.

10. MOLECULAR BASIS OF SQUAMOUS EPITHELIUM→CARDIAC EPITHELIUM

The molecular change that induces cardiac metaplasia in squamous epithelium in response to exposure to gastric juice is very likely to be bone morphogenetic protein-4 (BMP-4). This is a protein belonging to the transforming growth factor (TGF)-β family. Members of the TGF-β family are involved in controlling cellular differentiation, migration, and proliferation. BMPs are 30–35 kilodalton proteins. They were originally shown to play a role in bone formation but were also found to be essential during embryonic development (see Section 1). BMPs induce the formation of a heterodimeric complex of the BMP receptor type I and type II. This receptor complex signals downstream by phosphorylating specific BMP receptor–regulated Smads (1, 5, and 8). The P-Smad 1/5/8 forms a complex with Smad 4, which then translocates into the nucleus, where certain target genes such as ID2 can be transcribed. Studies have shown that BMPs are activated during inflammation and injury.

In a study by Milano et al.,[26] the authors aim to provide evidence that reflux-induced metaplastic transformation of inflamed squamous epithelium to columnar-type epithelium is mediated by BMP-4. For the study, they used the following. (1) Biopsy samples from 28 patients with nondysplastic Barrett esophagus (median length 5.4 cm; range 2–13 cm; median age 67 years; range 39–88 years). Although not specifically stated, this likely represents visible CLE with intestinal metaplasia because the patients were under surveillance. Biopsies were taken from the visible CLE 2 cm above the endoscopic GEJ, normal squamous epithelium 2 cm above the visible CLE, and from squamous epithelium with esophagitis (at least LA grade B) in 6 patients. (2) Primary cell cultures prepared from 15 patients who underwent biopsy. (3) A rat model of Barrett esophagus. They used highly sophisticated molecular techniques to assess the expression of BMP-4, P-Smad 1/5/8, ID2, and Smad-4 in the various cell types.

In their discussion, the authors present their hypothesis: "… we recently found that BMP-4 is exclusively expressed in Barrett esophagus and absent in normal squamous epithelium. Because Barrett esophagus is associated with chronic inflammation as a result of reflux of gastric contents damaging the esophageal mucosa, we hypothesized that these inflammatory changes could induce the production of BMP-4, which subsequently triggers trans-differentiation of squamous epithelial cells into a columnar cell type, resembling metaplastic Barrett mucosal cells, that replaces the normal squamous epithelium."

Using Western blot analysis and immunohistochemistry, the authors showed that BMP-4 and its downstream targets, P-Smad 1/5/8 and ID2, are present in esophagitis but not in normal squamous epithelium. This indicates that BMP-4 (normally absent in squamous epithelium) is upregulated in esophagitis and its downstream pathway is activated.

Tests performed on the primary cell cultures of normal squamous epithelium showed that treatment of squamous cells with recombinant human BMP-4 showed rapid expression of P-Smad 1/5/8 that increased with time, while the control squamous cell cultures did not show such activation of Smad proteins. Blocking the pathway with the BMP antagonist noggin showed inhibition of Smad phosphorylation.

To test the hypothesis that activation of the BMP-4 pathway in esophagitis resulted in columnar metaplasia, the authors looked for changes in the cytokeratin profile of the squamous epithelial cell primary cultures after incubation with BMP-4 for 5 h. The cytokeratin profile of squamous epithelial cells ($CK7^{-}$ $CK20^{-}$ $CK10/13^{+}$) is different than that of columnar epithelium ($CK7^{+}$ $CK20^{+}$ $CK10/13^{-}$). After incubation, Western blot analysis showed that there was upregulation of CK7 and CK20 while CK10/13 was still present, but decreased, in these cells. CK7 and CK20 could also be demonstrated by immunohistochemistry. This indicated that treatment of squamous cells with BMP-4 induces a shift in the cytokeratin expression from squamous to columnar.

The authors conclude: "Our findings suggest that BMP-4 is involved in initiating the process of transformation of normal esophageal squamous mucosa into a columnar type of cells."

In all these studies, cardiac epithelium was above the endoscopic GEJ in visible CLE. There are no studies of BMP-4, and PSmad expression in the cardiac epithelium found distal to the endoscopic GEJ. The reason for this is that the area distal to the endoscopic GEJ is incorrectly believed to be proximal stomach rather than dilated distal esophagus.

Derakhshan et al.,[27] in a study of asymptomatic volunteers, demonstrated that cardiac epithelium distal to the normal endoscopic GEJ had features similar to cardiac epithelium in visible CLE and different to gastric epithelium. It would be interesting if cardiac epithelium distal to the endoscopic GEJ showed expression of the BMP-4 and PSmad pathway because all studies of adult gastric oxyntic epithelium have never shown expression BMP-4. Normal gastric oxyntic epithelium expresses Sonic Hedgehog gene and, in chronic atrophic gastritis with intestinal metaplasia, Cdx2 is activated.

11. MOLECULAR BASIS OF INTESTINAL METAPLASIA IN CARDIAC EPITHELIUM

In 2012, the Amsterdam group associated with another excellent group based in Spain to do a follow-up prospective study on the remnant esophagus of patients who had undergone esophagectomy.[28]

In the introduction, Castillo et al.[28] present the viewpoint held by many gastroenterologists: "Barrett's esophagus is characterized by a phenotypic switch in the epithelial cells of the distal esophagus from the normal stratified squamous mucosa to an intestinal columnar cell type." They continue: "However, a novel concept suggests that a non-specialized columnar type of metaplasia, also known as cardiac-type epithelium, characterized by monotonous mucous cells may rather be an intermediate metaplastic stage between inflamed squamous epithelium and intestinal-type metaplasia." It is amazing that this is regarded as a novel concept. There has never been any suggestion at any time in the literature that intestinal metaplasia arises directly from squamous epithelium.

Altered expression of key tissue-type-regulatory genes, also referred to as master switch genes, induced by reflux stimulation seems to drive the metaplastic conversion of keratinocytes into a columnar cell type. Recent studies establish Hedgehog and BMP-4 signaling as important interconnected regulatory pathways that contribute to the early stage of this phenotypic shift.

> *A recent experimental study provided evidence that BMP-4 is not only involved in the transition of squamous epithelium into non-specialized columnar epithelium but also crucial for the development of intestinal metaplasia through direct interaction with Cdx2, a homeotic gene of the para-homeobox gene family, to induce transcription of intestine specific genes such as MUC2.*
>
> *The aim of this study was to investigate the incidence of columnar metaplasia that appears in the esophageal remnant at specific times following subtotal esophagectomy and, specifically, to identify the sequence of molecular events, BMP-4, Cdx2, Cdx1, that parallel this phenotypic change.*

The products of the Cdx genes, members of the caudal homeobox gene family, are important in the early fetal differentiation and maintenance in the adult of normal intestinal epithelium.[18] The Cdx genes are expressed throughout the adult small and large intestine distal to the gastroduodenal junction. It is not expressed in the adult stomach or esophagus. However, Cdx2 is expressed when intestinal metaplasia occurs in the stomach. This study evaluated expression of Cdx genes and proteins in the remnant esophagus in patients following esophagectomy.

18 patients (16 males; mean age 61.9 years; range 49–77 years) who had a subtotal esophagectomy with gastric pull-up for adenocarcinoma ($n=13$) or squamous carcinoma ($n=5$) were included. A total of 72 endoscopic examinations were done over the study period. The number of patients and endoscopic findings at each time period are shown in Table 15.1. More than 50% of the patients had erosive esophagitis in the remnant esophagus at each follow-up time and at least 33% had an area of visible CLE.

TABLE 15.1 Endoscopic Findings at Each Follow-Up Time

Time (Months)	Number	Erosive Esophagitis	Columnar Mucosa
6	18	9 (50%)	6 (33%)
12	17	9 (53%)	8 (47%)
18	15	8 (53%)	7 (47%)
24	14	7 (50%)	6 (43%)
36	8	5 (63%)	4 (50%)

Of the 18 patients, 10 (56%) had columnar epithelium in the remnant esophagus. The prevalence of columnar epithelium increased along the time period. The CLE were first detected between 6 and 18 months postesophagectomy in 8/10 patients. The mean length of CLE was 15.5 mm (range 5–30 mm). Of the 10 patients, 8 who developed CLE had an esophagectomy for adenocarcinoma and the other 2 for squamous carcinoma. Histologically, 76 biopsies of the metaplastic columnar epithelium showed cardiac epithelium (nonspecialized columnar epithelium) in all 10 patients. Focal intestinal metaplasia was identified in cardiac epithelium in four biopsies from two patients; in one patient the intestinal metaplasia was seen at 18 and 24 months and in the other at 36 months.

Immunohistoehcmistry for Cdx1 and Cdx2 was negative in the squamous epithelium. Nuclear expression of Cdx2 was seen in 48/76 (63%) biopsies with cardiac epithelium in 6/10 patients with CLE. "In some cases a pattern of expression with up to 3 solitary Cdx2 positive cells in the crypt region of the glands was observed (Fig 3a) while in other cases groups of at least 3 positive nuclei per gland were detected (Fig 3b). 2 patients showed few isolated glands fully expressing Cdx2 at 18 and 24 months in the first case… and at 36 months in the second case (Fig 3c, d)… A few goblet cells were identified within all these Cdx2 positive glands. Glands fully expressing Cdx2 also co-expressed Cdx1… MUC2 showed very focal staining only in those glands fully expressing Cdx2 and harboring isolated goblet cells."

There was positive nuclear expression of P-Smad 1/5/8, a downstream target of BMP-4 in the squamous epithelium. Expression of PSmad 1/5/8 was higher in the cardiac epithelium.

Cdx2 mRNA was expressed in eight biopsy samples of squamous epithelium, five of which were in patients who also had cardiac epithelium. Cdx2 mRNA expression was significantly higher in cardiac epithelium than in squamous epithelium. Similarly, BMP-4 mRNA expression was highest in cardiac epithelium, followed by squamous epithelium adjacent to the new SCJ.

The results of this study provide evidence supporting an important role for BMP-4 in the process of conversion of inflamed squamous epithelium to cardiac epithelium. "Development of nonspecialized columnar phenotype might depend on reaching a certain threshold of BMP-4 expression in the stroma of squamous epithelium at some specific stage." The authors express caution in the interpretation of the data and warn the reader that it is possible that BMP-4 activation may be a passenger phenomenon than the true cause of columnar metaplasia.

Castillo et al.'s[28] study also shows that Cdx2 expression is associated with the metaplastic process. Cdx2 mRNA is expressed at low levels in the squamous epithelium adjacent to cardiac epithelium without immunohistochemical evidence of Cdx2 positivity. In some biopsies of cardiac epithelium, Cdx2 is expressed very focally by immunohistochemistry. It is when Cdx2 expression involves entire glands in cardiac epithelium that goblet cells appear along with expression of the intestinal markers MUC2 and villin. Again, it appears that phenotypic intestinal metaplasia occurs when a significant threshold of Cdx2 expression is reached within cardiac epithelium.

Castillo et al.'s[28] study shows the molecular changes associated with the early phenotypic expression of intestinal metaplasia in cardiac epithelium. Cdx2 expression in intestinal metaplasia within CLE has been well documented in the literature, occurring at a higher level when goblet cells are numerous.

Phillips et al. studied, using immunohistochemistry, Cdx2 expression in metaplastic columnar epithelium of the esophagus, including cardiac (junctional-type), intestinal metaplasia, increasing dysplasia, and adenocarcinoma.[29] Their hypothesis: "… Cdx2 gene expression occurs in the development of intestinal metaplasia of the esophagus and its encoded protein might be a marker for the histopathologic diagnosis of Barrett's esophagus."

The study material consisted of 134 paraffin blocks from 116 esophageal biopsies or resection specimens retrieved from the Pathology Department of the University of Virginia Medical Center. These consisted of 62 blocks from 62 esophageal biopsies showing cardiac (junctional) epithelium only (34 from patients without and 28 from patients with a biopsy-confirmed diagnosis of Barrett esophagus), 34 blocks from 33 cases of Barrett esophagus without dysplasia, and 38 blocks from 21 cases of Barrett esophagus with low-grade dysplasia (13), high-grade dysplasia (19), or adenocarcinoma (6).

A PAS-Alcian blue stain was evaluated for the presence of acid mucin containing nongoblet columnar cells and goblet cells. Immunohistochemical stains for Cdx2 were evaluated for the presence of nuclear staining in goblet cells, nongoblet columnar epithelial cells, as well as in dysplastic and invasive malignant glandular epithelium. Any nuclear staining was considered positive.

The authors classified the 62 patients with only cardiac epithelium into (1) 13 cases that were histologically suspicious for intestinal metaplasia (i.e., probably with pseudogoblet cells); these were all positive for acid mucin on the PAS-Alcian blue stain (i.e., columnar blue cells). 10 of these 13 cases with pseudogoblet cells exhibited focal Cdx2 positivity. The staining was confined to nuclei in small areas of glandular epithelium, and the intensity of staining tended to be less than that in the unequivocal cases of Barrett esophagus. (2) 49 cases without any histologic suspicion for intestinal metaplasia (i.e., no pseudo-goblet cells); 35 of these 49 cases contained columnar blue cells in the PAS-Alcian blue stain. 10 of these 49 cases were focally positive for Cdx2.

The total number of cases that had cells that were positive for PAS-Alcian blue was 48/62 (77%); these included 17/20 cases that were Cdx2 positive and 31/42 that were Cdx2 negative. The authors concluded: "… our results suggest that the finding of acid mucin in columnar epithelium lacking goblet cell morphology may not be a reliable indicator of cells committed to a pathway of intestinal differentiation."

The total number of cases that were positive for Cdx2 in cases of cardiac epithelium only was 20/62 (30%). All these showed only focal positivity. Each of the 34 cases from 33 patients with Barrett esophagus without dysplasia showed goblet cell and non-goblet cell columnar cell positivity for Cdx2. Nuclei of most of the goblet cells were positive.

The author's conclusion regarding the utility of Cdx2 in the diagnosis of intestinal metaplasia: "… Cdx2 protein is a sensitive marker of intestinal metaplasia… and may be useful in detecting histologically equivocal cases of Barrett's esophagus."

This conclusion is hardly justified because 30% of patients with cardiac epithelium only were Cdx2 positive. It would mean that the use of Cdx2 for diagnosis of intestinal metaplasia would include those cases without phenotypic goblet cells. All the evidence for Barrett esophagus being a premalignant condition has been based on goblet cells being present. Long-term prospective studies will be needed to show that Cdx2 expression has a premalignant potential before this can be used for the diagnosis of Barrett esophagus.

Cdx2 was expressed in all 38 blocks of 21 patients with dysplasia and adenocarcinoma. The Cdx2 positivity progressively decreased from nondysplastic Barrett esophagus to low-grade dysplasia to high-grade dysplasia to well-differentiated adenocarcinoma and poorly differentiated adenocarcinoma. This suggests that Cdx2 is not a significant agent in the process that transforms Barrett esophagus to increasing dysplasia and adenocarcinoma.

The study of Vallbohmer et al.[30] from our unit at the University of Southern California is designed to expand on Phillips et al.'s[29] finding that Cdx2 is expressed in a minority of patients with cardiac epithelium but in all patients with intestinal metaplasia. "…frequency of expression does not give quantitative data regarding the level of expression in a given individual. The advantage of measuring quantitative gene expression using real-time polymerase chain reaction (PCR) is the generation of numerical values allowing precise assessment of Cdx2 at each step in the development of intestinal metaplasia."

Tissue biopsies obtained at endoscopy from 107 patients with symptoms of GERD were snap frozen in liquid nitrogen. In 25 patients without visible CLE, biopsies were taken from squamous epithelium and duodenal mucosa; patients with visible CLE had biopsies taken from this segment and the Cdx2 expression values were averaged for each patient. The frozen samples were embedded in OCT medium, sections cut, examined to select areas of different epithelial types [squamous ($n=62$), cardiac ($n=19$), oxyntocardiac ($n=14$), intestinal metaplasia ($n=15$), and duodenal ($n=26$) epithelium] for microdissection. The dissected flakes of tissue were processed for quantification of Cdx2 mRNA expression by real-time PCR.

There were 107 patients (63 men) with a mean age of 51 years (range 18–80 years). Ambulatory 24-h pH monitoring was done in 98/107 (92%) patients; the median composite pH score was 27.3 (normal <14.7). There were statistically significant differences in Cdx2 gene expression among all five histological groups ($P<.001$) (Table 15.2).

TABLE 15.2 Cdx2 mRNA Expression Levels in the Different Histological Groups

Group	Number of Patients	Cdx2 x 100β-actin mRNA Expression, Median (25th–75th Percentile)
Normal squamous epithelium	62	0.01 (0.01–0.05)
Cardiac epithelium	19	0.4 (0.3–0.71)
Oxyntocardiac epithelium	14	0.76 (0.28–1.14)
Intestinal metaplasia	15	6.72 (3.97–8.08)
Duodenal epithelium	26	39.64 (25.98–55.29)

Cdx2 expression was lowest in squamous epithelium and highest in duodenal mucosa control tissue. Cdx2 gene expression levels were similar in cardiac and oxyntocardiac epithelia and both were significantly higher than squamous epithelium. There was an even greater increase in Cdx2 expression in intestinalized cardiac epithelium.

12. SUMMARY OF THE SQUAMOUS→CARDIAC→INTESTINAL METAPLASTIC PROCESS

The association of the genetic changes described above does not necessarily mean that these genes are the ones that actually drive the processes of metaplasia. They may simply be passenger genes that are associated with the process. However, the fact that BMP-4, Shh, and Cdx genes are active as differentiating signals during embryogenesis of the gastrointestinal tract make them logical drivers of the metaplastic process.

The data suggest that there are two sequential steps in the metaplastic process wherein squamous epithelium transforms into cardiac epithelium and then, often after a significant lapse of time and in a minority of patients, intestinal metaplasia develops in cardiac epithelium. Both these metaplastic steps appear to be dependent not on simple expression of the genes involved, but a cumulative increase in the amount of the gene expressed. As such, it would be a serious error to use the presence of a gene or gene product in an epithelium to indicate that the metaplastic event has indeed occurred or that it is imminent in the cell involved.

For example, BMP-4 and Cdx2 are expressed in minute amounts in squamous epithelium and this is clearly not indicative of either columnar metaplasia or even a commitment of the squamous epithelium to undergo columnar metaplasia in the future.

The activation of the BMP-4 pathway that is suggested as being responsible for the metaplasia of squamous to cardiac epithelium appears to occur in a quantitatively stepwise manner. BMP-4 expression in normal squamous epithelium well above the SCJ is very low, increasing in injured squamous epithelium immediately above the SCJ, and becoming highest in cardiac epithelium. The exact level at which the metaplastic change is effected is unknown but has to be intermediate between the level of expression in injured squamous epithelium and cardiac epithelium. It is the phenotypic expression of the metaplasia that permits diagnosis of columnar metaplasia, not the presence of the driving gene. In the future, it is possible that we will know the exact quantity of gene expression that is needed to produce the phenotypic change.

The same is true of Cdx2, the gene that has been suggested as driving the occurrence of intestinal metaplasia in cardiac epithelium. Cdx2 gene activation is seen at low levels in squamous epithelium increasing progressively in cardiac epithelium to intestinal metaplasia. The protein product of the gene is more specific, seen in small amounts in a minority of cardiac epithelial cells and increasing when goblet cells appear. Again, it is the phenotypic appearance of goblet cells that defines intestinal metaplasia. Mere expression of either the Cdx2 gene or its protein has no diagnostic relevance at the present time.

13. CARDIAC EPITHELIUM TO OXYNTOCARDIAC EPITHELIUM

There is very little data on the process by which cardiac epithelium develops parietal cells to transform into oxyntocardiac epithelium. There are many reasons for this:

1. Oxyntocardiac epithelium occurs as the only metaplastic esophageal epithelium in people with very short (<1 cm) segments of columnar metaplasia limited to the dilated distal esophagus. These are patients who are either asymptomatic or have mild symptoms and whose endoscopy is invariably normal (i.e., without visible CLE). They rarely undergo endoscopic biopsy. In Marsman et al.,[31] a study of endoscopically normal patients, irrespective of indication for endoscopy, in whom the histologic SCJ was present, 38% had a direct transition from squamous to oxyntocardiac epithelium with no pure cardiac epithelium. In our autopsy study 56% of 18 subjects who had a complete circumferential evaluation of the SCJ had only oxyntocardiac epithelium between the squamous and gastric oxyntic epithelia.
2. Oxyntocardiac epithelium is often lumped with cardiac epithelium when it is seen in biopsies taken from visible CLE.[31]
3. When found in biopsies distal to the endoscopic GEJ, they are considered to be a normal epithelium in the proximal stomach. Oxyntocardiac epithelium is, by virtue of the terminology used by Paull et al.,[32] called "gastric fundic-type epithelium." When numerous parietal cells are present, it is often misclassified as gastric oxyntic epithelium.

Oxyntocardiac epithelium is an important epithelial type in CLE. The reason for this is that it represents a dead end in the progression of columnar epithelium that does not lead to intestinal metaplasia, dysplasia, and adenocarcinoma (Fig. 15.1). With rare exception, a foveolar gland complex that has differentiated to develop parietal cells, does not develop goblet cells. This means that persons who convert all their metaplastic cardiac epithelium to oxyntocardiac epithelium have no risk of progression to adenocarcinoma. It is therefore the most benign and stable epithelium in the esophagus. While benign, squamous and cardiac epithelia can be transformed by exposure to gastric juice into cardiac and intestinal metaplastic epithelia, respectively. All evidence suggests that oxyntocardiac epithelium, once formed, does not change.

The development of parietal cells in metaplastic cardiac epithelium is reminiscent of the early development of fetal gastric oxyntic epithelium, which is most likely influenced by Sonic Hedgehog gene. Van den Brink et al.[17] reported that Shh is expressed in concert with H^+K^+ ATP-ase in areas of visible CLE with gastric fundic type (i.e., oxyntocardiac) metaplasia. This suggested that Shh may be involved in the transformation of cardiac to oxyntocardiac epithelium in CLE.

Despite the uniform endoscopic appearance of long-segment Barrett esophagus, the histology is variable. Within the segment of visible CLE, intestinal metaplasia is most prevalent proximally, oxyntocardiac epithelium is most prevalent distally, and pure cardiac epithelium is seen in between. Oh et al, from our unit, correlated the distribution of gene expression with histologic types found in CLE. "We hypothesized that within long segment Barrett esophagus, expression of Cdx2, a gene involved in intestinal differentiation, would be maximal proximally, whereas Sonic hedgehog (Shh), a gene responsible for maintaining gastric mucosa, would be maximal distally."

This study examined biopsies taken every 2 cm from 13 patients with long-segment Barrett esophagus (at least three levels with columnar epithelium). The samples were grouped by location as proximal, middle, and distal. After microdissection, RNA isolation, and reverse transcription, mRNA expression was measured with quantitative real-time polymerase chain reaction.

The median visible CLE length was 8 cm (range 4–17 cm). Cdx2 expression was highest proximally with decreasing expression distally. In contrast Shh expression was highest distally.

We concluded that "The expression of Cdx2 and Shh parallel the typical distribution of intestinal and cardiac (and oxyntocardiac) mucosa, respectively. This suggests that the expression of these genes may drive the phenotypic appearance of the columnar mucosa within long segment Barrett esophagus."

14. EXPLANATION OF DISTRIBUTION OF EPITHELIA IN COLUMNAR-LINED ESOPHAGUS

In CLE, whether it is visible above the endoscopic GEJ or microscopic in the dilated distal esophagus, there is a nonrandom distribution of the three epithelial types. When all three types of metaplastic columnar epithelia are present, intestinal metaplasia favors the proximal region adjacent to the SCJ, oxyntocardiac epithelium favors the distal region adjacent to gastric oxyntic epithelium, and cardiac epithelium is found in between. When two epithelial types are present, it is always cardiac and oxyntocardiac epithelium with the latter favoring the distal region. When one epithelial type is present, it is always only oxyntocardiac epithelium.

Oh et al.[33] showed that this phenotypic distribution of the epithelial types is matched by the pattern of molecular gene expression. When all three epithelia are present, Cdx-2 is expressed in the proximal region where intestinal metaplasia is seen and Shh is expressed in the distal region where parietal cells are present. BMP-4 and PSmad are expressed throughout the CLE segment. When cardiac and oxyntocardiac epithelia are present, there is expression of BMP-4 throughout and Shh in the distal areas with parietal cells. Cdx-2 expression is present but at a much lower level than when goblet cells are present. When only oxyntocardiac epithelium is present, BMP-4, Shh, and very low levels of Cdx-2 are present.

When the distribution of epithelia follows a nonrandom distribution such as is seen here, it usually means that goblet cells and parietal cells arise in cardiac epithelium due to an environmental factor rather than chance. It is easy to understand why a change occurs distally because this is the area where the concentration of every molecule in gastric juice is the highest. It is more difficult to understand why intestinal metaplasia occurs proximally and why its prevalence increases with the length of visible CLE, i.e., why it is more likely to occur as the distance of cardiac epithelium from the GEJ increases.

One factor that could explain this is pH. During an episode of reflux, a column of gastric juice flows into the thoracic esophagus along the pressure gradient. The normal pH of the region is immediately changed. Normally, the competent LES separates the low pH gastric lumen from the near neutral pH in the lumen of the esophagus. The pH transition point corresponds to the distal limit of the functional LES. When the LES fails (i.e., a tLESR occurs) an open cavity is formed and the flow of gastric juice creates a pH gradient from gastric baseline at the GEJ to near neutral at the top of the column of refluxate. As a result, the esophagus in the proximal region of the column of refluxate is at a pH around 5–7. The fact that intestinal metaplasia is favored in this region suggests that intestinal metaplasia is favored by a less acid (i.e., pH 5–7) milieu. Similarly, the fact that parietal cells occur distally suggests that differentiation into parietal cells is favored by a strong acid (pH 1–2) milieu.

15. PRACTICAL VALUE OF UNDERSTANDING MOLECULAR PATHWAYS

The utility of understanding molecular pathways of metaplasia in the esophagus in GERD is the vision of being able to control the processes whereby one cell type transforms to another.

Cardiac metaplasia of esophageal squamous epithelium occurs very early in the dilated distal esophagus in the patient without LES failure, significant reflux, or symptoms of GERD.

However, cardiac metaplasia in the squamous epithelium of the thoracic esophagus occurs at a point in the disease that is later than the occurrence of symptoms of GERD. In the Pro-GERD study,[34] 702/6215 (11.3%) of patients in the initial cohort had visible CLE at the index endoscopy, which was performed some time, often several years, after the onset of symptoms.

More than 90% of patients with sufficient reflux into the thoracic esophagus to cause symptomatic GERD will have a thoracic esophagus lined by squamous epithelium. In this group, BMP-4 expression progressively increases in the squamous epithelium as it is increasingly damaged. If a molecule can be identified that can practically inhibit the expression of BMP-4, it is possible that cardiac metaplasia of the thoracic esophagus can be prevented. In the absence of columnar metaplasia, adenocarcinoma cannot develop in the thoracic esophagus. GERD-induced cancer in the thoracic esophagus can therefore be prevented by administering this molecule to patients with symptomatic GERD who do not have visible CLE.

Similarly, in patients with visible CLE without intestinal metaplasia, Cdx-2 expression is at low levels. Development of a molecule that will prevent the level of Cdx-2 expression necessary to cause intestinal metaplasia would theoretically prevent Barrett esophagus (traditional type with intestinal metaplasia). Because intestinal metaplasia is a necessary precursor for adenocarcinoma, this would also prevent the occurrence of adenocarcinoma. Reversal of Cdx-2 expression can achieve a similar result if such reversal would reverse intestinal metaplasia.

Many years ago, I fantasized about a molecule that would transform cardiac epithelium to oxyntocardiac epithelium. This transformation can be done either by developing a molecule that will stimulate Shh activation in cardiac epithelium or by changing the milieu of the cardiac epithelial cell to be strongly acidic, which may result in the natural activation of Shh in cardiac epithelium.

My dream molecule was a H^+ ion that was attached to an inert ligand. The size of the H^+ ion/ligand would be such that it was too large to enter the squamous epithelium but small enough to reach the proliferative cells in cardiac epithelium through the opening of the foveolar pit. When it reached the bottom of the foveolar pit, the ligand would magically dissociate (by a mechanism I could never even visualize) into the H^+ ion and the inert ligand. The H^+ ion would then react with the cardiac epithelial cell to activate Shh and induce parietal cells. If one converted all the cardiac epithelium into oxyntocardiac epithelium, adenocarcinoma would be prevented completely because oxyntocardiac epithelium was stable and did not progress to intestinal metaplasia. If the H^+ ion/ligand was not absorbed into the system and harmless, it could be taken like regular vitamin pill, continuously converting any cardiac epithelium into oxyntocardiac epithelium. Theoretically it could, by preventing intestinal metaplasia, convert CLE into a harmless entity that was more resistant to reflux than squamous or other metaplastic columnar epithelia.

The practical utility of molecular mechanisms of controlling the metaplastic pathway is still many years away, if they ever come to pass.

Metaplastic events in esophageal epithelia are not the fundamental cause of adenocarcinoma in the esophagus. These events require exposure of the esophageal squamous epithelium to gastric juice. The LES is designed to prevent that exposure. LES damage is therefore the primary cause of reflux.

GERD is a mechanical abnormality wherein the LES undergoes progressive damage. It should be treated as such with effort directed at controlling and preventing LES damage. Preventing LES damage will result in ensuring that the LES performs its function of protecting esophageal squamous epithelium from gastric contents. Without exposure of the squamous epithelium to gastric contents, no molecular or histologic changes leading to adenocarcinoma of the esophagus will take place. LES damage is the root cause of adenocarcinoma.

The problem with the present management and research into GERD is that it is aimed at GERD as a chemical and molecular disease. Acid does not cause GERD; the use of acid-reducing drugs will never solve the GERD problem because it does not prevent progression of LES damage. No molecular manipulation is necessary if there is no LES failure that permits metaplasia to occur.

Both acid suppression and molecular manipulation are aimed at managing the effects of LES failure. We should be concentrating on preventing the degree of LES damage that is sufficient to result in LES failure.

REFERENCES

1. Souza RF, Huo X, Mittal V, Schuler CM, Carmack SW, Zhang HY, Zhang X, Yu C, Hormi-Carver K, Genta RM, Spechler SJ. Gastroesophageal reflux might cause esophagitis through a cytokine-mediated mechanism rather than caustic acid injury. *Gastroenterology* 2009;**137**:1776–84.
2. Tobey NA, Carson JL, Alkiek RA, et al. Dilated intercellular spaces: a morphological feature of acid reflux-damaged human esophageal epithelium. *Gastroenterology* 1996;**111**:1200–5.
3. Der R, Tsao-Wei DD, DeMeester T, et al. Carditis: a manifestation of gastroesophageal reflux disease. *Am J Surg Pathol* 2001;**25**:245–52.
4. Olvera M, Wickramasinghe K, Brynes R, Bu X, Ma Y, Chandrasoma P. Ki67 expression in different epithelial types in columnar lined esophagus indicates varying levels of expanded and aberrant proliferative patterns. *Histopathology* 2005;**47**:132–40.

5. Johns BAE. Developmental changes in the oesophageal epithelium in man. *J Anat* 1952;**86**:431–42.
6. De Hertogh G, Van Eyken P, Ectors N, Tack J, Geboes K. On the existence and location of cardiac mucosa; an autopsy study in embryos, fetuses, and infants. *Gut* 2003;**52**:791–6.
7. Park YS, Park HJ, Kang GH, Kim CJ, Chi JG. Histology of gastroesophageal junction in fetal and pediatric autopsy. *Arch Pathol Lab Med* 2003;**127**:451–5.
8. Derdoy JJ, Bergwerk A, Cohen H, Kline M, Monforte HL, Thomas DW. The gastric cardia: to be or not to be? *Am J Surg Pathol* 2003;**27**:499–504.
9. Menard D, Arsenault P. Cell proliferation in developing human stomach. *Anat Embryol* 1990;**182**:509–16.
10. Karam SM. Lineage commitment and maturation of epithelial cells in the gut. *Front Biol* 1999;**4**:286–98.
11. Riddell RH. The biopsy diagnosis of gastroesophageal reflux disease, "carditis," and Barrett's esophagus, and sequelae of therapy. *Am J Surg Pathol* 1996;**20**(Suppl. 1):S31–51.
12. Krishnadath KK. Novel findings in the pathogenesis of esophageal columnar metaplasia or Barrett's esophagus. *Curr Opin Gastroenterol* 2007;**23**:440–5.
13. Ellison E, Hassall E, Dimmick JE. Mucin histochemistry of the developing gastroesophageal junction. *Pediatr Pathol Lab Med* 1996;**16**:195–206.
14. Ramalho-Santos M, Melton DA, McMahon AP. Hedgehog signals regulate multiple aspects of gastrointestinal development. *Development* 2000;**127**:2763–72.
15. Ali I, Rafiee P, Zheng Y, Johnson C, Banerjee B, Haasler G, Jacob H, Shaker R. Intramucosal distribution of WNT signaling components in human esophagus. *J Clin Gastroenterol* 2009;**43**:327–37.
16. Van den Brink GR, Hardwick JCH, Tytgat GNJ, Brink MA, ten Kate FJ, van Deventer JH, Peppelenbosch MP. Sonic Hedgehog regulates gastric gland morphogenesis in man and mouse. *Gastroenterology* 2001;**121**:317–28.
17. Van den Brink GR, Hardwick JCH, Nielsen C, Xu C, ten Kate FJ, Glickman J, van Deventer SJH, Roberts DJ, Peppelenbosch MP. Sonic Hedgehog expression correlates with fundic gland differentiation in the adult gastrointestinal tract. *Gut* 2002;**51**:628–33.
18. Silberg DG, Swain GP, Suh ER, Traber PG. Cdx1 and Cdx2 expression during intestinal development. *Gastroenterology* 2000;**110**:961–71.
19. Glickman JN, Chen Y-Y, Wang HH, Antonioli DA, Odze RD. Phenotypic characteristics of a distinctive multilayered epithelium suggests that it is a precursor in the development of Barrett's esophagus. *Am J Surg Pathol* 2001;**25**:569–78.
20. Chandrasoma P. Controversies of the cardiac mucosa and Barrett's esophagus. *Histopathology* 2005;**46**:361–73.
21. Chandrasoma P, Wijetunge S, Ma Y, DeMeester S, Hagen J, DeMeester T. The dilated distal esophagus: a new entity that is the pathologic basis of early gastroesophageal reflux disease. *Am J Surg Pathol* 2011;**35**:1873–81.
22. Tobey NA, Hosseini SS, Argore CM, Dobrucali AM, Awayda MS, Orlando RC. Dilated intercellular spaces and shunt permeability in non-erosive acid-damaged esophageal epithelium. *Am J Gastroenterol* 2004;**99**:13–22.
23. Hassall E. Columnar lined esophagus in children. *Gastroenterol Clin North Am* 1997;**26**:533–48.
24. Qualman SJ, Murray RD, McClung HJ, Lucas J. Intestinal metaplasia is age related in Barrett's esophagus. *Arch Pathol Lab Med* 1990;**114**:1236–40.
25. Dresner SM, Griffin SM, Wayman J, Bennett MK, Hayes N, Raimes SA. Human model of duodenogastro-esophageal reflux in the development of Barrett's metaplasia. *Br J Surg* 2003;**90**:1120–8.
26. Milano F, van Baal JWPM, Buttar NS, Rygiel AM, de Kort F, Demars CJ, Rosmolen WD, Bergman JJGHM, van Marle J, Wang KK, Peppelenbosch MP, Krishnadath KK. Bone morphogenetic protein 4 expressed in esophagitis induces a columnar phenotype in esophageal squamous cells. *Gastroenterology* 2007;**132**:2412–21.
27. Derakhshan MH, Robertson EV, Lee YY, Harvey T, Ferner RK, Wirz AA, Orange C, Ballantyne SA, Hanvey SL, Going JJ, McColl KE. In healthy volunteers, immunohistochemistry supports squamous to columnar metaplasia as mechanism of expansion of cardia, aggravated by central obesity. *Gut* 2015;**64**:1705–14.
28. Castillo D, Puig S, Iglesias M, Seoane A, de Bolos C, Munitiz V, Parrilla P, Comerma L, Poulsom R, Krishnadath KK, Grande L, Pera M. Activation of the BMP-4 pathway and early expression of CDX2 characterize non-specialized columnar metaplasia in a human model of Barrett's esophagus. *J Gastrointest Surg* 2012;**16**:227–37.
29. Phillips RW, Frierson Jr HF, Moskaluk CA. Cdx2 as a marker of epithelial intestinal differentiation in the esophagus. *Am J Surg Pathol* 2003;**27**:1442–7.
30. Vallbohmer D, DeMeester SR, Peters JH, Oh DS, Kuramochi H, Shimizu D, Hagen JA, Danenberg KD, Danenberg PV, DeMeester TR, Chandrasoma PT. Cdx-2 expression in squamous and metaplastic columnar epithelia of the esophagus. *Dis Esophagus* 2006;**19**:260–6.
31. Marsman WA, van Sandyck JW, Tytgat GNJ, ten Kate FJW, van Lanschot JJB. The presence and mucin histochemistry of cardiac type mucosa at the esophagogastric junction. *Am J Gastroenterol* 2004;**99**:212–7.
32. Paull A, Trier JS, Dalton MD, Camp RC, Loeb P, Goyal RK. The histologic spectrum of Barrett's esophagus. *N Engl J Med* 1976;**295**:476–80.
33. Oh DS, DeMeester SR, Mori R, Kuramochi H, Tanaka K, Hagen JA, Danenberg PV, Danenberg KD, Chandrasoma P, DeMeester TR. Cdx-2 and Sonic Hedgehog gene expression in the proximal, middle, and distal regions of columnar mucosa in long segment Barrett's esophagus. *Gastroenterology* 2008;**134**(4):A442–3. Suppl. 1.
34. Kulig M, Nocon M, Vieth M, Leodolter A, Jaspersen D, Labenz J, Meyer-Sabellek W, Stolte M, Lind T, Malfertheimer P, Willich SN. Risk factors of gastroesophageal reflux disease: methodology and first epidemiological results of the ProGERD study. *J Clin Invest* 2004;**57**:580–9.

Chapter 16

Progression of GERD From the Perspective of LES Damage

Gastroesophageal reflux disease (GERD) results when the lower esophageal sphincter (LES) fails in its function of sequestering the esophageal squamous epithelium from exposure to gastric contents.

It would seem that, at least from a theoretical standpoint, the severity and progression of GERD in terms of abnormal acid exposure in the pH test and cellular changes that occur in the thoracic esophagus would most accurately correlate with the degree of LES damage (Fig. 16.1; Table 16.1). Correlation of LES damage with symptoms is less reliable in that even patients with severe reflux can sometimes have minimal symptoms.

It is known that symptoms do not correlate well with cellular changes in the esophagus. The present diagnosis of GERD, which is almost totally dependent on the presence and severity of symptoms, is not predictive of present and future cellular changes. This makes management of GERD with proton pump inhibitors (PPIs) with the objective of controlling symptoms illogical and results in significant failure resulting from progression of undetected cellular changes in a substantial number of patients.

Nason et al. reported that a patient who has good control of symptoms with PPI therapy is at increased risk for Barrett esophagus than one who is uncontrolled.[1] Patients who do not have troublesome symptoms that satisfy the definition of GERD are not immune from the risk of adenocarcinoma.[2] Future progression to severe GERD characterized by inability to control symptoms and a decline in quality of life is not predictable. These events are simply allowed to happen without any ability or attempt at prevention.

That endoscopy, which is the only method of detecting cellular changes, is delayed until treatment has failed to control symptoms results in the unfortunate reality that 85% of all patients presenting with adenocarcinoma have never had an endoscopy. The patients often have advanced stage cancers with the need for radical therapy that has a high morbidity and a very low probability of survival.

The new method of measuring LES damage by histology provides an opportunity to understand the disease from its outset without depending on thc presence of symptoms. When this is done, it will be recognized that present criteria for definition and diagnosis of GERD identify the disease at an advanced stage. A phase of slowly progressive LES damage over several decades exists before GERD is recognized by present criteria.

A new opportunity for early diagnosis by the severity of LES damage and early intervention by methods to prevent or slow down the progression of LES damage in selected cases now emerges.

In Fig. 16.1 and Table 16.1, the disease progression from the normal state to the most severe GERD is represented in terms of degree of abdominal LES damage. It becomes apparent in this method of understanding that the Montreal definition of GERD that requires "troublesome" symptoms occurs with 50%–70% of degradation of the LES. The new test for LES damage allows recognition of the disease long before it satisfies the Montreal definition.

The entire population can now be divided into the following categories by the new histologic test for the abdominal LES damage done on a person without GERD symptoms at age 35 years. The test, which measures the length of dilated distal esophagus in a 20 mm biopsy sample taken distal to the endoscopic gastroesophageal junction (GEJ), has an accuracy to a micrometer (Table 16.2).

(1) 70% of people stay in the green zone without developing any significant reflux during their lifetime (person A in Table 16.2 with a rate of LES damage of 2.85 mm/decade calculated by the result of the test);
(2) 20% of people stay in the orange zone with reflux sufficient to cause symptoms but easily controlled with PPI therapy and without risk of visible columnar-lined esophagus (CLE) or adenocarcinoma (person B in Table 16.2 with a rate of LES damage of 3.57 mm/decade calculated by the result of the test);
(3) 10% of people who enter the red zone with a high risk of treatment failure, severe erosive esophagitis, intestinal metaplasia of the dilated distal esophagus, visible CLE, and adenocarcinoma. This is the population that progresses to adenocarcinoma (person C in Table 16.2 with a rate of LES damage of 5.00 mm/decade calculated by the result of the test).

GERD. http://dx.doi.org/10.1016/B978-0-12-809855-4.00016-6

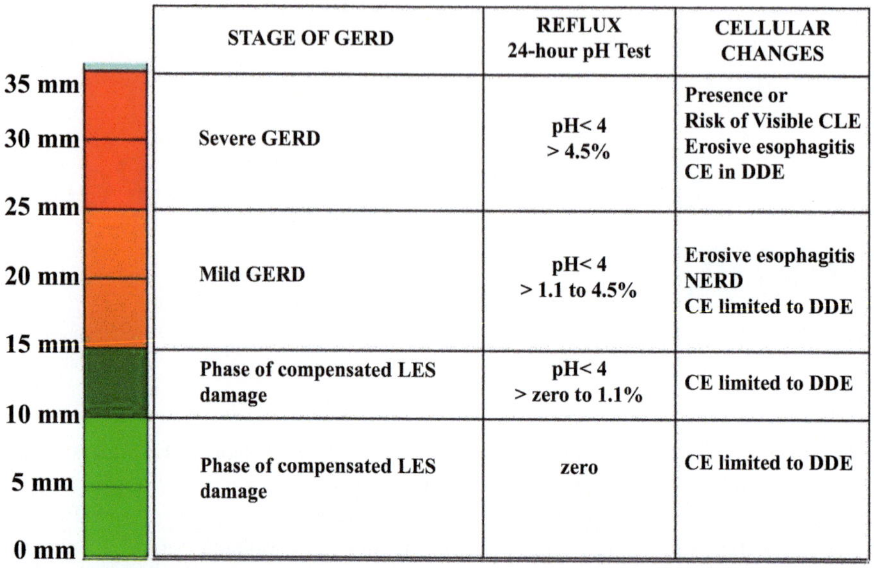

	STAGE OF GERD	REFLUX 24-hour pH Test	CELLULAR CHANGES
35 mm 30 mm 25 mm	Severe GERD	pH< 4 > 4.5%	Presence or Risk of Visible CLE Erosive esophagitis CE in DDE
20 mm 15 mm	Mild GERD	pH< 4 > 1.1 to 4.5%	Erosive esophagitis NERD CE limited to DDE
10 mm	Phase of compensated LES damage	pH< 4 > zero to 1.1%	CE limited to DDE
5 mm 0 mm	Phase of compensated LES damage	zero	CE limited to DDE

FIGURE 16.1 Diagrammatic representation of the correlation between severity of abdominal lower esophageal sphincter (LES) damage measured by the new pathologic test, severity of reflux (pH test), and cellular changes in the esophagus. *CE*, cardiac epithelium; *CLE*, columnar-lined esophagus; *DDE*, dilated distal esophagus; *GERD*, gastroesophageal reflux disease.

TABLE 16.1 Correlation of Lower Esophageal Sphincter (LES) Damage With Reflux Into the Thoracic Esophagus, Cellular Changes in the Thoracic Esophagus, and Symptoms

Abdomen LES Damage (mm)[a]	Residual Abdomen LES Length (mm)[b]	Reflux Into Thoracic Esophagus[c,d]	Cellular Changes in Thoracic Esophagus[e]	Symptoms[e]
0	35	0	0	0
5	30	0	0	0
10	25	0	0	0
15	20	pH test >0%–<4.5%	Likely zero	Zero to postprandial symptoms
20	15	pH test normal/abnormal	Zero to possible erosive esophagitis	Zero to postprandial symptoms
25	10	pH test >4.5%	Zero to probable erosive esophagitis	Zero to probable Troublesome symptoms
30	5	pH test >4.5%	Severe erosive esophagitis, visible CLE	Zero to uncontrollable symptoms
35	0	pH test >4.5%	Severe erosive esophagitis, visible CLE	Zero to uncontrollable symptoms

This assumes that the initial length of the abdominal segment of the LES is 35 mm and that LES failure correlates with a functional abdominal LES length of <10 mm. Also assumed is that dynamic shortening of the LES with a heavy meal is a maximum of 10 mm. Evidence supports all these assumptions.
[a]Abdominal LES damage is measured by the length of cardiac epithelium (with and without parietal and/or goblet cells) from the squamocolumnar junction/endoscopic gastroesophageal junction to the proximal limit of gastric oxyntic epithelium.
[b]The residual length is calculated as 35 mm minus measured LES damage.
[c]Reflux into the thoracic esophagus is assumed to occur when LES failure (transient lower esophageal relaxation) occurs; data indicate that this occurs when the functional abdominal LES length is <10 mm.
[d]Reflux is postprandial when the functional length of the abdominal LES is <10 mm after dynamic shortening of 10 mm.
[e]Cellular changes in the thoracic esophagus and typical GERD symptoms are assumed to be the result of reflux into the thoracic esophagus.
CLE, columnar-lined esophagus.

TABLE 16.2 Classification of the Population Based on Future Lower Esophageal Sphincter (LES) Damage

	Rate of Progression of a-LES Damage (mm/decade)[a]	a-LES at Age 15 (mm)	Measured Damage at Age 35 Years (mm)	a-LES at Age 35 Years (mm)	LES Damage at Age 85 (mm)
Person A	2.85	35	5.7	29.3	15
Person B	3.57	35	7.1	27.9	25
Person C	5.00	35	10.0	25.0	35

This is based on an initial abdominal LES length of 35 mm, onset of LES damage at age 15 years, linear progression of LES damage, and projected future abdominal LES damage at age 85 years. It is assumed that the new histologic test is performed at age 35 years (test result highlighted in red).
[a]Rate of progression of abdominal LES damage calculated by the measured abdominal LES damage (=length of dilated distal esophagus) by the test done at age 35 years divided by 20 (the difference between ages 15 and 35 years).
a-LES, abdominal LES.

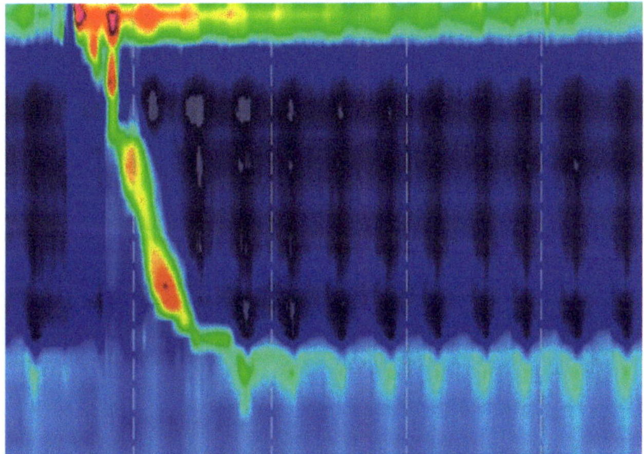

FIGURE 16.2 High-resolution manometry showing a defective lower esophageal sphincter (LES). The total length is 1.1 cm and the abdominal length is zero. This represents complete destruction of the abdominal segment of the LES. *Tracing and calculation of data from tracing by Dr. Chris Dengler, Legato Medical, Inc.*

The method of predicting future LES damage status can predict the future GERD status of these patients (Chapter 19). Patient A can be assured he/she will never develop GERD. Patient B can be told that he/she will likely develop GERD but will be well controlled with PPI therapy and has no risk of future adenocarcinoma. Patient C requires to be managed aggressively with an objective of slowing down or preventing progression of LES damage. This aggressive management starts at age 35 years when the test is done. At this time, the person has no symptoms and an abdominal LES length of 25 mm. If the rate of progression of abdominal LES damage can be slowed down from the present 5 to 1 mm/decade, the projected abdominal LES length at age 85 years is 15 mm; the person remains in the green zone and GERD will have been prevented. If it can be slowed to 2 mm/decade, the abdominal LES length at age 85 years is 25 mm; the person will develop GERD but never have reflux severe to produce visible CLE; adenocarcinoma is prevented.

The problem at present is the inability to accurately assess LES damage. Manometry only permits definition of a defective LES (mean pressure<6 mmHg, total length<20 mm; abdominal length<10 mm).[3] Manometry does not permit recognition of early LES damage before the LES becomes defective (Fig. 16.2). At manometry, the damaged distal LES has a pressure that is equal to the stomach. This makes it impossible to use LES damage as a practical diagnostic test for GERD. It also means that there is no possibility of intervention to prevent progression of LES damage; the LES damage is already beyond control when manometry declares it to be defective.

In the next chapters, I will describe the method of assessment of LES damage by pathology (Chapter 17). An algorithm developed from this measurement permits prediction of progression of LES damage into the future (Chapter 18). That diagnostic system will provide the new ability to test any person at any time after 30 years of age and identify who is at risk of entering the red zone with severe GERD and the risk of visible CLE at some point in their future.

This early diagnosis will allow intervention limited to those people at risk. If we can develop an intervention that will slow down or stop progression of LES damage, the red zone can be eradicated. There will no longer be people whose lives are disrupted by uncontrollable GERD symptoms. Barrett esophagus and adenocarcinoma of the esophagus will be a thing of the past, like complex ulceration and strictures that plagued patients with GERD in the 1950s.

In this chapter, I will define the pathologic stages of vertical progression of GERD in terms of abdominal LES damage. Any category of LES damage that I present in this book will be far less accurate than the correlations derived from actual measurement of abdominal LES damage by the new test in clinical trials once the appropriate biopsy device has been made available for use. The new measurement of LES damage by histology will have an accuracy that is in the range of micromillimeters. These are individualized to each person at that level.

1. THE NORMAL STATE (ZERO LOWER ESOPHAGEAL SPHINCTER DAMAGE)

A normal LES is not designed to be impenetrable under all circumstances (Fig. 16.3). It is designed to be impenetrable under normal circumstances. To understand what this means, it is necessary to define both the normal LES and circumstances.

A normal LES straddles the diaphragm, extending distally to the entire length of the abdominal segment of the esophagus and proximally to ~15mm of the distal thoracic esophagus. For purposes of clarity, I will be dogmatic and provide measurements for the LES that are likely to be close to the median value and ignore individual variation, which is largely unknown. This variation will be determined only after actual study when the technical requirements of taking the appropriate biopsy sample has been met and adequate numbers of patients have been subjected to the measurement of LES damage.

The total LES length is 50mm, with 35mm around the abdominal esophagus. The mucosal lining of the entire LES is stratified squamous like the rest of the esophagus. It transitions at the GEJ to the stomach, which is lined entirely by gastric oxyntic epithelium (Fig. 16.3).

The LES is the distal 50mm of the esophagus. It is not a separate organ between the esophagus and the stomach. There is a tendency in the literature to use the terms "GEJ," "GEJ zone," "cardia" for the 50mm of the LES. The reason for these imprecise and vague terms is that the pathology of LES damage is not understood. The normal LES has a tubular shape and is part of the tube that transmits food into the stomach. The maintenance of the tubular shape of the LES is

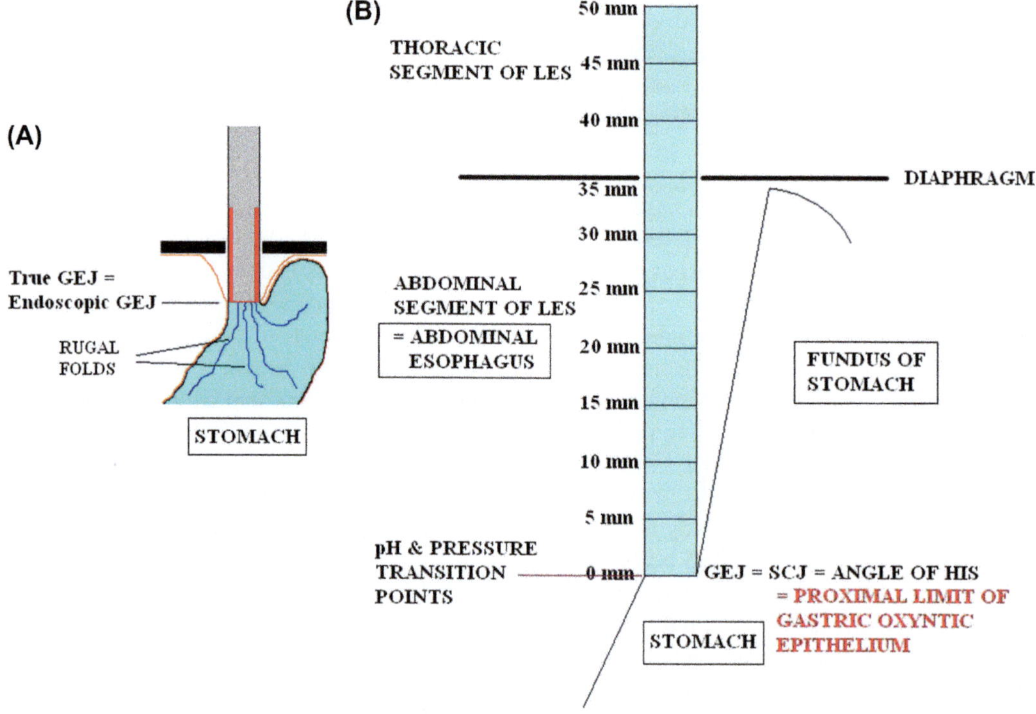

FIGURE 16.3 The normal state. (A) The entire esophagus is lined by squamous epithelium (gray), which transitions to gastric oxyntic epithelium (blue) at the true gastroesophageal junction (GEJ). There is no cardiac epithelium. The lower esophageal sphincter (LES) (*shown as a red wall*) straddles the diaphragm. Rugal folds reach the squamocolumnar junction (SCJ). (B) Magnified diagrammatic representation of the normal LES, with a total length of 50mm of which the abdominal segment is 35mm. The entire esophagus to the distal limit of the LES is lined by squamous epithelium (blue). Gastric oxyntic epithelium is white. The pH and pressure transition point are distal to the SCJ. The angle of His is acute.

dependent on the tonic contraction of the LES muscle. The high pressure of the LES prevents dilatation of the abdominal esophagus, which has an innate positive luminal pressure.

In its normal state, the LES is a high pressure zone with a mean pressure of around 15–30 mmHg. The LES pressure is higher than intragastric pressure in the fasting state and acts as a barrier that prevents esophageal squamous epithelium from being exposed to gastric juice. The LES separates the low pH environment of the stomach from the near-neutral pH of the esophagus. The squamocolumnar junction (SCJ) is at or above the point where gastric pH transitions to esophageal pH (the pH transition or "step-up" point) (Fig. 16.3B).

"Normal circumstance" is defined as a state where the dynamic changes that occur with ingestion of a meal do not result in any distortion of the LES that causes the squamous epithelium becoming exposed to gastric contents. This includes the fasting state when the stomach is empty, swallowing when the LES is relaxed but peristalsis creates a pressure gradient from the esophagus to the stomach, and during a "normal" meal that does not overdistend the stomach. With a normal meal, the intragastric pressure does not rise sufficiently to produce dynamic shortening of the LES and the pH transition point remains below the SCJ (Fig. 16.4).

Under these normal circumstances, there is no exposure of the esophageal squamous epithelium to gastric contents. The squamous epithelium that is never exposed to gastric acid continues to line the entire esophagus and transitions to gastric oxyntic epithelium at the GEJ. There is no possibility of damage or cardiac metaplasia. The squamooxyntic gap remains at zero (Fig. 16.5). The LES maintains its normal length of 35 mm during the meal and normal pressure. The LES does not fail. There is no reflux into the thoracic esophagus.

It is extremely uncommon to encounter this normal state except at autopsy in young people without any evidence of GERD during their lifetime.[4]

The normal state can be summarized in the following terms:

Initial abdominal LES length: (assumed with best evidence) 35 mm;
LES damage: (measured by squamooxyntic gap in dilated distal esophagus) zero;
Residual abdominal LES length by manometry: 35 mm;
Abdominal LES length after a heavy meal: (assuming dynamic shortening of 10 mm due to gastric overdistension associated with a heavy meal) 25 mm;
Probability of LES failure (transient lower esophageal relaxation, tLESR): (probability of abdominal LES reaching the critical failure level of <10 mm) zero;

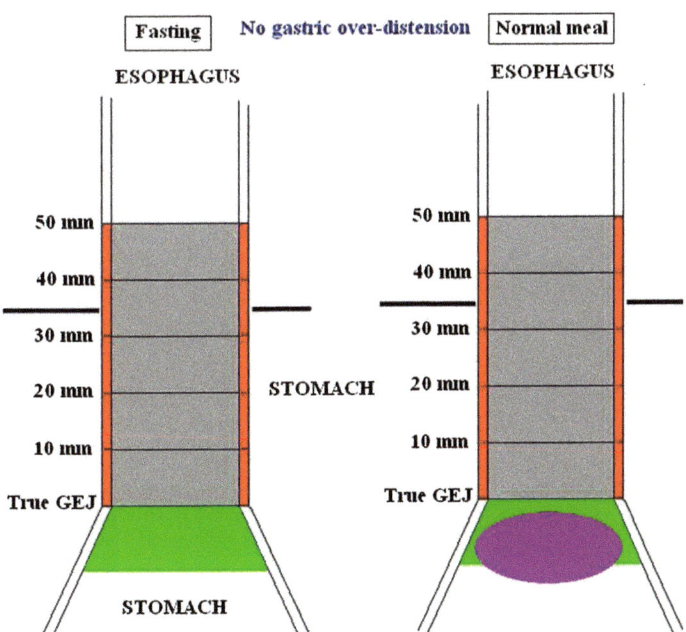

FIGURE 16.4 Changes in the esophagus and stomach with ingestion of a normal meal, i.e., one that is within gastric capacity and does not cause overdistension. The lower esophageal sphincter (LES) is unchanged. The full stomach contains an acid pocket (pink) below the pH transition point. The squamous epithelium (gray) is protected from this by the LES. The acid pocket is in the stomach lined by gastric oxyntic epithelium (green). *GEJ*, gastroesophageal junction.

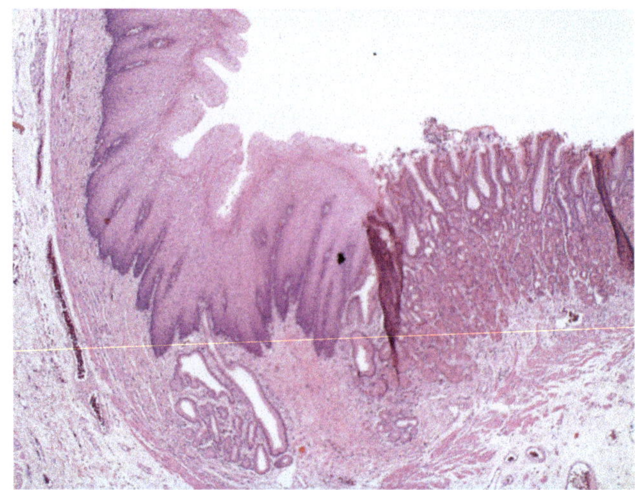

FIGURE 16.5 The normal state in a histologic section across the gastroesophageal junction (=squamocolumnar junction) in the normal state. The squamous epithelium transitions directly to gastric oxyntic epithelium. This was in a 67-year-old man who had a resection for squamous carcinoma of the esophagus.

24-h pH test: 0% time pH<4;
Cellular changes in thoracic esophagus above endoscopic GEJ: zero;
Symptoms: zero;
Risk of adenocarcinoma: zero at this point in time; needs three steps in the future: squamous epithelium→visible CLE→intestinal metaplasia→adenocarcinoma.

2. PHASE OF COMPENSATED LOWER ESOPHAGEAL SPHINCTER DAMAGE (>0–15 MM LOWER ESOPHAGEAL SPHINCTER DAMAGE)

The "abnormal circumstance" as far as the LES is concerned is a meal (Fig. 16.6) that is heavy enough to produce gastric overdistension and a rise in intragastric pressure resulting in dynamic effacement of the LES. The manometric length shortens. This causes a caudal movement of the SCJ such that the esophageal squamous epithelium reaches a point below the pH transition point and becomes exposed to the acid pocket at the top of the food column.[5,6]

Repeated exposure of the most distal squamous epithelium ultimately results in cardiac metaplasia (Fig. 16.7). Cardiac metaplasia of the distal abdominal esophagus leads to loss of LES tone, causing permanent LES shortening (Fig. 16.8).

LES damage changes the pressures in the area of damage. The basal positive (+5 mmHg) intraluminal pressure of the abdominal esophagus is not counteracted by LES pressure. As such the abdominal esophagus that has lost LES tone dilates (Fig. 16.8). It ceases to function as the esophagus and functions like the stomach. It is "gastricized" at an anatomic (dilatation with rugal folds) (Fig. 16.9), manometric (equal to gastric pressure) (Fig. 16.10), and functional level where the dilated distal esophagus becomes a part of the reservoir that accepts a meal rather than the tube that transmits the food.

At this phase of compensated LES damage, the only recognizable abnormality is the presence of cardiac epithelium between the cephalad-displaced SCJ and the true GEJ (proximal limit of rugal folds) (Fig. 16.7). To pathologists who believe that cardiac epithelium is a normal epithelium in the proximal stomach, the dilated distal esophagus has completed its deception. To them, the esophagus ends at the end of the tube at the proximal limit of rugal folds. The cardiac epithelium distal to that is proximal stomach. There is one obvious question that highlights the flaw in this perception. *There is no argument that patients with severe GERD have a greatly shortened abdominal LES; where did that shortened LES go?*

It is only those who recognize that LES shortening results from cardiac metaplasia of esophageal squamous epithelium that will understand that the pathology of LES damage is the dilated distal esophagus. The area distal to the end of the tubular esophagus is a dilated segment of esophagus that has lost LES tone. It can be recognized histologically by cardiac epithelium (with and without parietal and/or goblet cells) between the SCJ and the true GEJ, which is the proximal limit of gastric oxyntic epithelium.

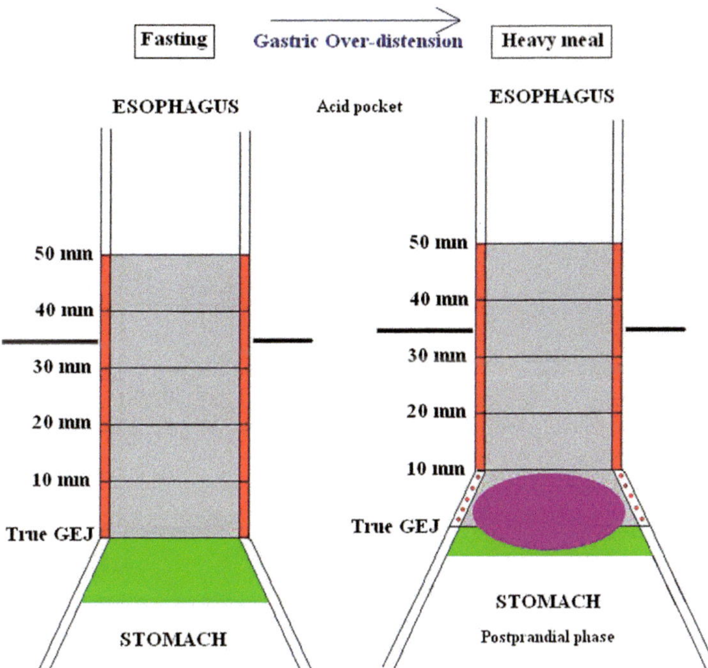

FIGURE 16.6 Changes in the lower esophageal sphincter (LES) resulting from gastric overdistension caused by a heavy meal. The lower 10 mm of the normal LES has been effaced (*red stippling* in the wall) causing that part of the esophagus to dilate. The squamocolumnar junction has descended to a point below the pH transition point. The squamous epithelium (gray) is exposed to the acid pocket (pink). This is the fundamental cause of gastroesophageal reflux disease. *GEJ*, gastroesophageal junction.

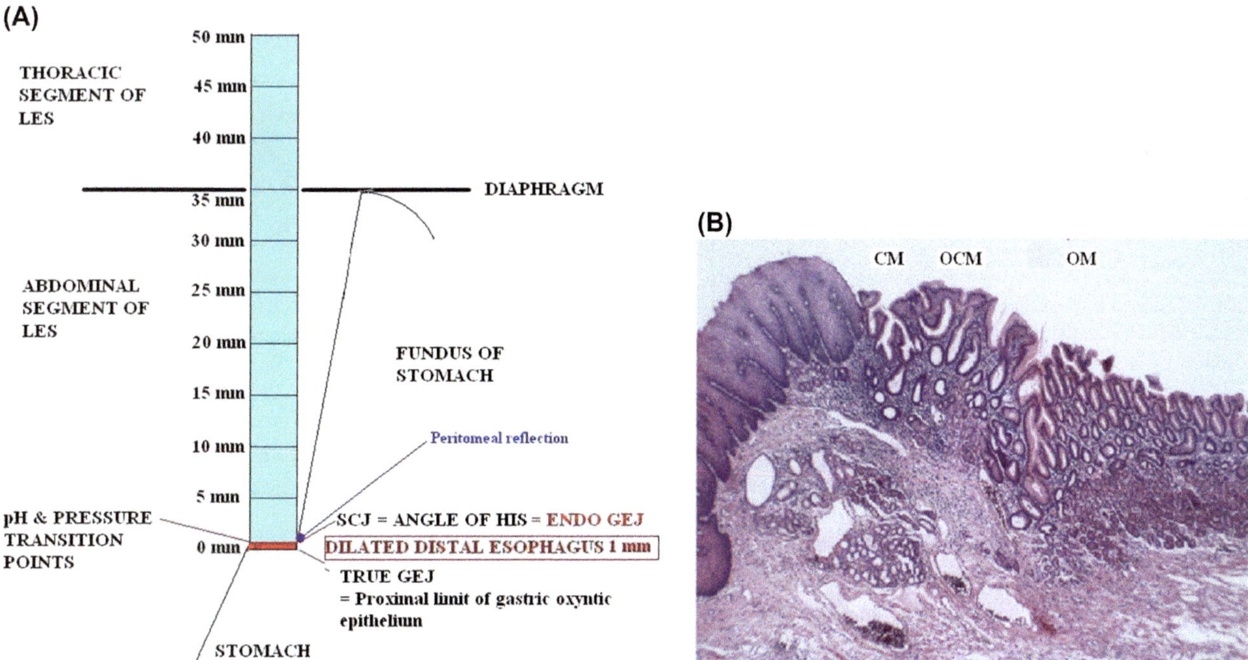

FIGURE 16.7 Very early stage of gastroesophageal reflux disease soon after the onset of lower esophageal sphincter (LES) damage. (A) Diagram showing 1 mm of cardiac epithelium (red) between the squamous epithelium (blue) and gastric oxyntic epithelium (white). This patient will have a competent LES that is 34.9 mm long. (B) The only recognizable abnormality is the presence of a 1 mm segment of cardiac (CM) and oxyntocardiac (OCM) epithelium in a histologic section across the squamocolumnar junction (SCJ). *GEJ*, gastroesophageal junction; *OM*, oxyntic mucosa.

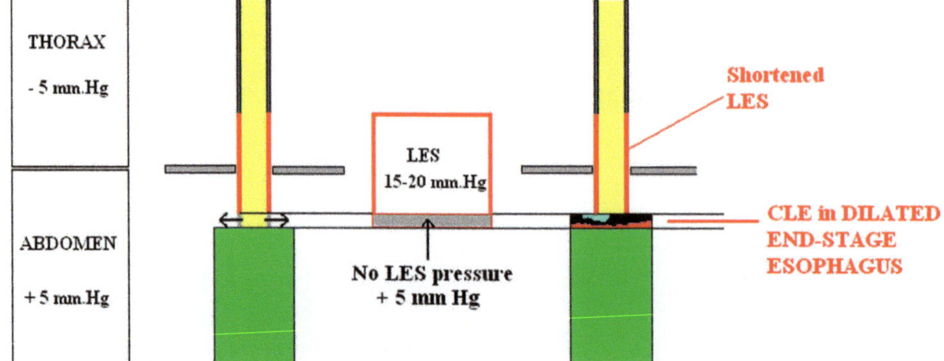

FIGURE 16.8 Cardiac metaplasia is associated with loss of lower esophageal sphincter (LES) pressure, possibly by interruption of a local neuromuscular reflex arc resulting from loss of afferent innervation when squamous epithelium is replaced by cardiac epithelium. Loss of pressure in the abdominal LES causes expression of the positive intraabdominal pressure in this part of the esophagus and causes dilatation. *CLE*, columnar-lined esophagus.

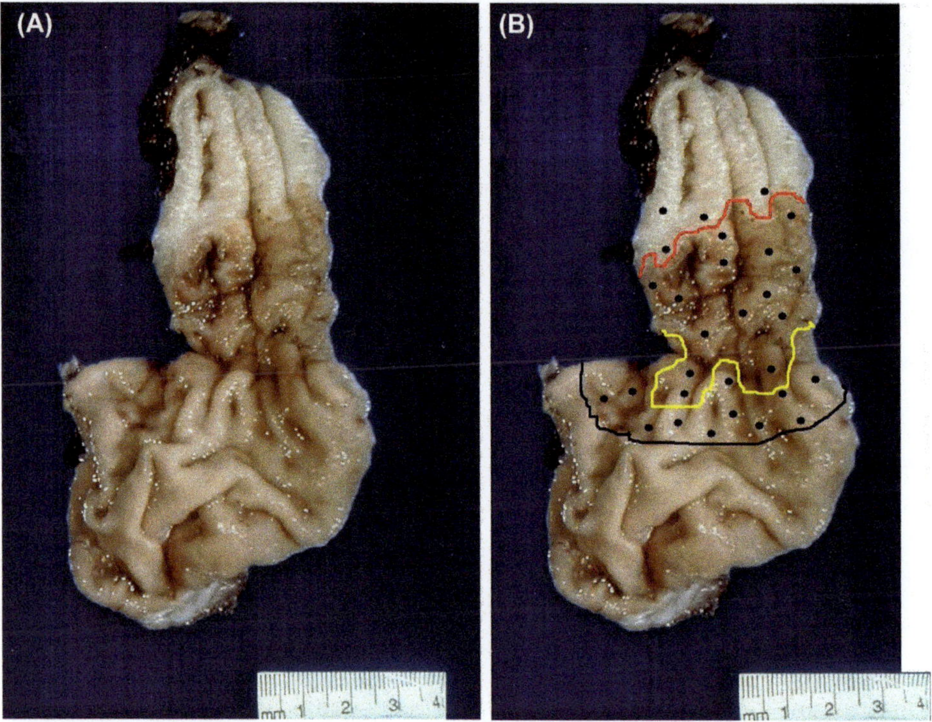

FIGURE 16.9 (A) The dilated distal esophagus exactly mimics the stomach. It is the proximal part of the sac distal to the end of the tubular esophagus and has rugal folds. (B) Histologic examination of this esophagogastrectomy shows extension of cardiac epithelium for 2.05 cm distal to the tubular esophagus (*black line*). That this is the esophagus is proven by the presence of submucosal glands (*black dots*).

Over time and very slowly in a progression measured by millimetric increments per decade, the dilated distal esophagus increases in length (measured by the length of cardiac epithelium with and without parietal and/or goblet cells) and concordant with the degree of abdominal LES damage (=shortening).

The severity of LES damage depends entirely on the rate of progression of squamous epithelial damage. Each individual can be defined by the exact length of LES damage at any point in life. This condition changes with time, always increasing because LES damage is irreversible and inexorably progressive.

In the first phase of progression of LES damage, the LES retains its competence because it has a reserve capacity. If the original length of the abdominal LES is 35 mm, it can shorten by 15 mm and still be competent, i.e., not associated with LES failure and reflux into the thoracic esophagus.

The length of the abdominal LES that correlates with LES failure sufficient to cause symptomatic GERD is 10 mm. With 15 mm damage, the residual length of the abdominal LES is 20 mm. With a heavy meal that distends the stomach, there

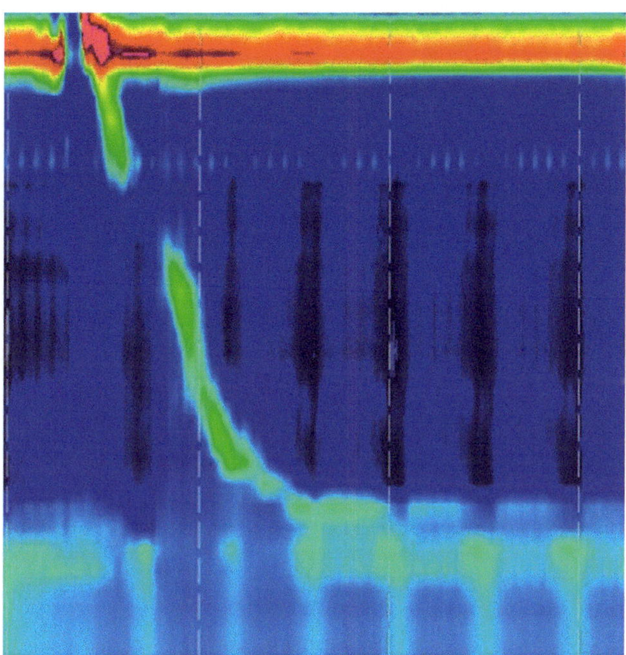

FIGURE 16.10 High-resolution manometry showing a severely damaged lower esophageal sphincter (LES) (total length 1.4cm, abdominal length 0). The entire residual LES is above the diaphragmatic impression. This is the present diagnostic criterion of a hiatal hernia. *Tracing and calculation of data from tracing by Dr. Chris Dengler, Legato Medical, Inc.*

can be dynamic shortening of the LES by 10mm, bringing the functional residual abdominal LES to the critical 10mm. Postprandial reflux therefore can be expected to begin with abdominal LES damage of >15mm.

The LES shortens from the distal end very slowly, with metaplasia occurring cell by cell over decades. I will consider the status of the patient at three points within the phase of compensated LES damage. This will show that the only change that will occur during this phase is an elongation of cardiac epithelium (with and without parietal cells); intestinal metaplasia is highly unlikely in the dilated distal esophagus of >0–15mm in length.

The age of the patient at which these points of LES damage are reached will depend entirely on the rate of progression of LES damage. In the 70% of the population that never develops GERD, LES damage remains <15mm during their lifetime. I am one of these people. At age 56years, my LES damage was 4mm. I will show that there is a certainty that I will not reach 15mm of LES damage even if I live up to 150years.

The relationship of age to abdominal LES damage was studied by Lee et al. at the laboratory of Dr. Mark Fox at Guy's and St. Thomas NHS Foundation Trust in London.[7]

The aim of this study was to assess "(1) whether progressive physiologic degradation of the reflux barrier and esophageal dysfunction are associated with increasing age and (2) whether these changes explain the increased severity of reflux disease reported in the elderly population."

The study group consisted of 1307 patients with typical reflux symptoms who were studied for severity of symptoms by a standard questionnaire; they also had esophageal manometry and 24-h ambulatory pH monitoring. Of these, 74 failed manometry. The mean age of the study group was 49±14years (range: 15–92years) with a significant male preponderance (58%). The age distribution was parametric, with 99 aged <30years, 255 aged 30–39, 299 aged 40–49, 313 aged 50–59, 205 aged 60–69, and 96 aged >70years.

By univariate analysis, there was a highly significant correlation between esophageal acid exposure and both the LES pressure ($P<.0001$) and the abdominal LES length ($P<.0001$) (Fig. 16.11). Esophageal acid exposure was more impacted by structural degradation of the reflux barrier (i.e., decreasing abdominal LES length) and was more pronounced than that of functional weakness (i.e., decreasing LES pressure).

The abdominal LES lengths and the corresponding esophageal acid exposure are shown in Table 16.3 with numbers for abdominal LES length taken from Fig. 16.1.

By multivariate analysis, there were only two physiologic variables with independent effects on 24-h esophageal acid exposure: LES pressure ($P<.0001$) and abdominal LES length ($P<.0004$). Acid exposure increased 1.0% for every 10mm decrease in abdominal LES length (95% CI 0.6%–1.4%).

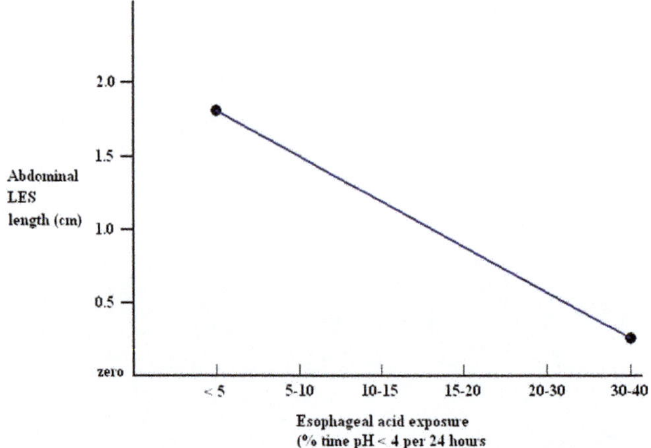

FIGURE 16.11 Correlation between abdominal lower esophageal sphincter (LES) length and reflux showing increasing acid exposure as the LES shortens in gastroesophageal reflux disease (GERD) patients. Note that the measured residual length of the abdominal LES in these GERD patients ranges from 3 to 18 mm (i.e., 17–32 mm LES damage), all above the 15 mm of LES damage, which is the theoretical end of the phase of compensation.*GEJ Data from Lee J, Anggiansah A, Anggiansah R, et al. Effects of age on the gastroesophageal junction, esophageal motility and reflux disease.* Clin Gastroenterol Hepatol *2007;5:1392–8.*

TABLE 16.3 Abdominal Lower Esophageal Sphincter (LES) Length (in millimeters) With the Corresponding Acid Exposure in the 24-h Ambulatory pH Study (Expressed as % Time pH < 4/24 h)

Esophageal Acid Exposure % time pH < 4/24 h	Length of Abdominal LES (mm)
<5	18
5–10	14
10–15	10
15–20	12
20–30	7
30–40	3

Note that in this population of patients with gastroesophageal reflux disease (GERD), the maximum manometric length of the abdominal LES is 19 mm. All these patients with symptomatic GERD are beyond the phase of compensated LES damage.

A separate analysis was performed to assess whether these factors were responsible for any change in reflux severity with aging. Increasing age was associated with increasing percentage 24-h esophageal acid exposure ($P < .0001$) with an increase of 1.1% time esophageal pH < 4/24 h for every additional decade. On multivariate regression, increasing age was associated with decreasing abdominal LES length ($P < .001$) (Fig. 16.12). The decrease in the length of the abdominal LES shows a linear decline with lengths of 15, 13, 13, 12, 9, and 7 mm for <30, 30–40, 40–49, 50–59, 60–69, and >70 years, respectively. The difference in the mean abdominal LES length in the 30–40 year group and that in the 60–69 year group (i.e., a three-decade interval) is 4 mm. These measurements are mean measurements for the group at one point in time. It would be very interesting to see data for change in abdominal LES lengths in a longitudinal time study of individual GERD patients.

This confirms that the progression of abdominal LES damage is extremely slow and within the range that I have suggested is required for the pathogenesis of GERD.

2.1 Abdominal Lower Esophageal Sphincter Damage of 5 mm

Initial abdominal LES length: (assumed with best evidence) 35 mm (Fig. 16.13);
LES damage: (measured by squamooxyntic gap in dilated distal esophagus) 5 mm;
Residual abdominal LES length by manometry: 30 mm;

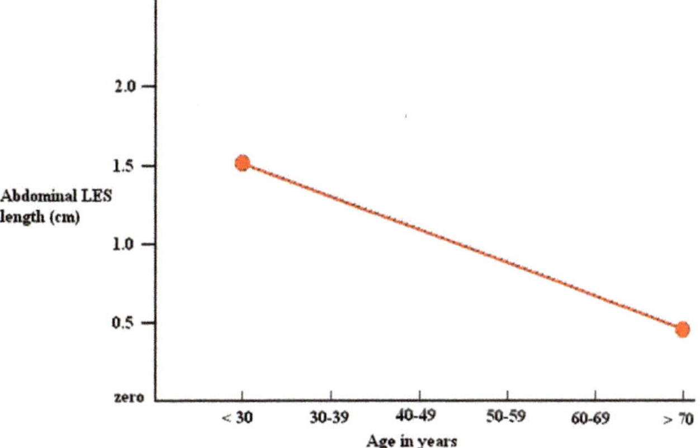

FIGURE 16.12 The length of the abdominal lower esophageal sphincter (LES) decreased in an almost perfect linear slope in this population with increasing age. While this is in the group, it suggests that the concept of a linear progression of LES damage over a long term is likely correct. *Data from Lee J, Anggiansah A, Anggiansah R, et al. Effects of age on the gastroesophageal junction, esophageal motility and reflux disease.* Clin Gastroenterol Hepatol *2007;5:1392–8.*

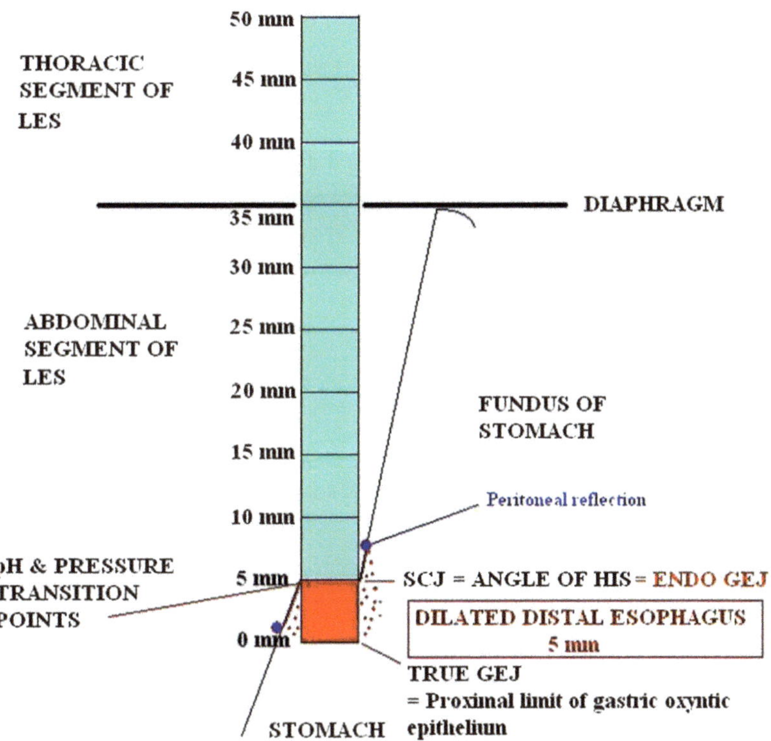

FIGURE 16.13 Diagrammatic representation of 5 mm of lower esophageal sphincter (LES) damage, concordant with 5 mm of a dilated distal esophagus lined by cardiac epithelium (red). *GEJ*, gastroesophageal junction; *SCJ*, squamocolumnar junction. *Data from Lee J, Anggiansah A, Anggiansah R, et al. Effects of age on the gastroesophageal junction, esophageal motility and reflux disease.* Clin Gastroenterol Hepatol *2007;5:1392–8.*

Abdominal LES length after a heavy meal: (assuming dynamic shortening of 10mm due to gastric overdistension associated with a heavy meal) 20mm;
Probability of LES failure (tLESR): (probability of abdominal LES reaching the critical failure level of 10mm) zero;
24-h pH test: 0% time pH<4
Cellular changes in thoracic esophagus above endoscopic GEJ: zero;
Symptoms: zero.
Risk of adenocarcinoma: zero at this point in time; needs three steps: squamous→visible CLE→intestinal metaplasia →adenocarcinoma

The person with 5 mm of LES shortening is very similar to the person without any LES damage. In the postprandial state after dynamic LES shortening by 10 mm, the abdominal LES has a functional length of 20 mm, well above the 10 mm length at which LES failure (transient lower esophageal relaxation, tLESR) occurs. There is no reflux into the thoracic esophagus.

The only difference is that this person has a gap between the distal end of the squamous epithelium and the proximal limit of gastric oxyntic epithelium of 5 mm. This is composed of metaplastic cardiac epithelium (with and without parietal cells) lining the dilated distal esophagus. The abdominal LES is shortened by 5 mm. This is within the reserve capacity of the LES. It is still totally competent in performing its function of preventing reflux. The probability of intestinal metaplasia is very low.

2.2 Abdominal Lower Esophageal Sphincter Damage of 10 mm

Initial abdominal LES length: (assumed with best evidence) 35 mm (Fig. 16.14);
LES damage: (measured by squamooxyntic gap in dilated distal esophagus) 10 mm;
Residual abdominal LES length by manometry: 25 mm;
Abdominal LES length after a heavy meal: (assuming dynamic shortening of 10 mm due to gastric overdistension associated with a heavy meal) 15 mm;
Probability of LES failure (tLESR): (probability of abdominal LES reaching the critical failure level of 10 mm) zero;
24-h pH test: 0% time pH<4
Cellular changes in thoracic esophagus above endoscopic GEJ: zero;
Symptoms: zero;
Risk of adenocarcinoma: zero at this point in time; needs three steps: squamous→visible CLE→intestinal metaplasia→adenocarcinoma.

This person is essentially identical to the prior person with the abdominal LES shortening of 5 mm. The only difference is that the dilated distal esophagus (length of cardiac epithelium with and without parietal cells) is 10 mm long. The functional abdominal LES length is 15 mm in the postprandial period, still greater than the critical 10 mm length. The LES

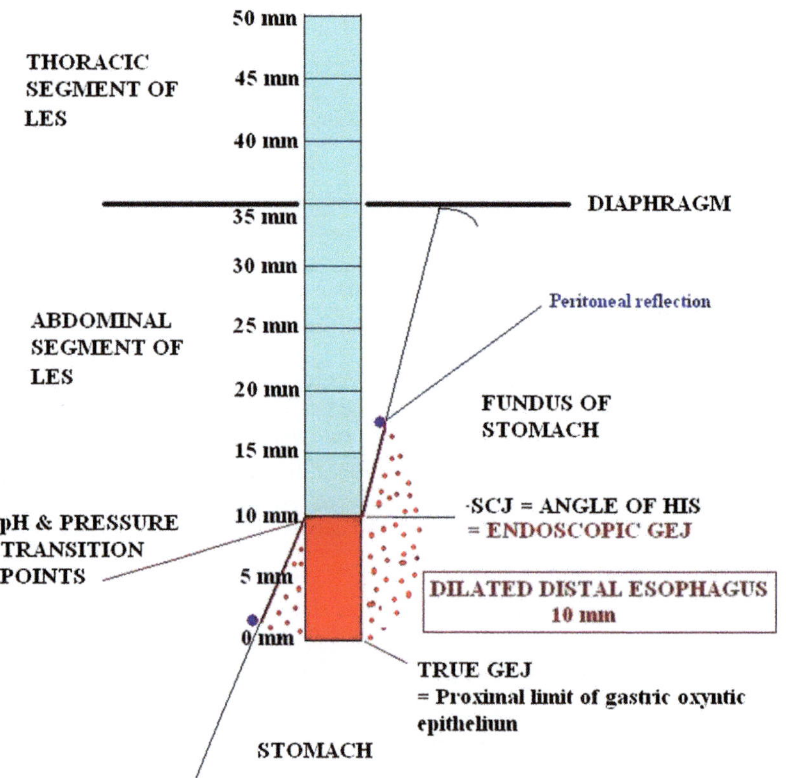

FIGURE 16.14 Diagrammatic representation of 10 mm of lower esophageal sphincter (LES) damage, concordant with 10 mm of a dilated distal esophagus lined by cardiac epithelium (red). *GEJ*, gastroesophageal junction; *SCJ*, squamocolumnar junction.

is competent. The patient has no reflux. The probability of intestinal metaplasia is low, but higher than in the person with 5 mm of LES damage.

There is, however, one significant difference between patients with abdominal LES damage of 5 versus 10 mm. The latter has used up more of the reserve capacity of the abdominal LES and is therefore in greater danger of progressing to clinical GERD, assuming a similar rate of LES damage progression.

2.3 Abdominal Lower Esophageal Sphincter Damage of 15 mm

Initial abdominal LES length: (assumed with best evidence) 35 mm (Fig. 16.15);
LES damage: (measured by squamooxyntic gap in dilated distal esophagus) 15 mm;
Residual abdominal LES length by manometry: 20 mm;
Abdominal LES length after a heavy meal: (assuming dynamic shortening of 10 mm due to gastric overdistension associated with a heavy meal) 10 mm;
Probability of LES failure (tLESR): (probability of abdominal LES reaching the critical failure level of 10 mm) zero;
24-h pH test: 0% time pH<4
Cellular changes in thoracic esophagus above endoscopic GEJ: zero
Symptoms: zero;
Risk of adenocarcinoma: zero at this point in time; needs three steps: squamous→visible CLE→intestinal metaplasia to adenocarcinoma.

The three points of LES damage ranging from >0 to 15 mm represent the "prereflux stage" of GERD. It results from squamous epithelial damage of the distal 15 mm of the esophagus. The damage is so slow and so intermittent that it is not symptomatic, except possibly for minor discomfort that commonly accompanies binge eating and regarded as normal. This is the phase of compensated LES damage where the reserve capacity of the abdominal LES is being slowly degraded.

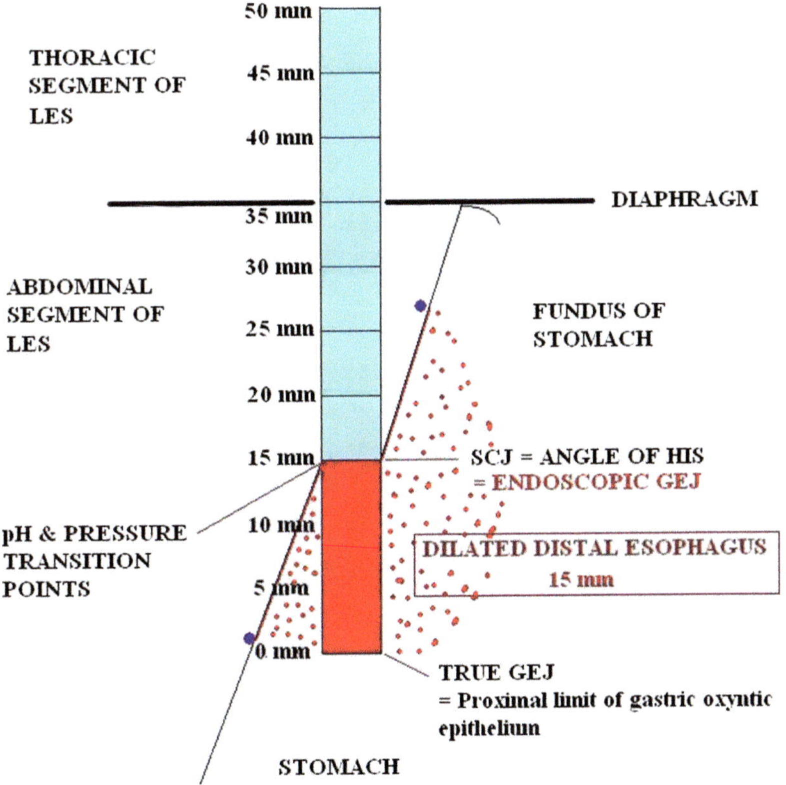

FIGURE 16.15 Diagrammatic representation of 15 mm of lower esophageal sphincter (LES) damage, concordant with 15 mm of a dilated distal esophagus lined by cardiac epithelium (red). This is the end of the phase of compensated LES damage. *GEJ*, gastroesophageal junction; *SCJ*, squamocolumnar junction.

Although the LES remains competent, the mechanism whereby it is damaged is the same as that is responsible for GERD; exposure of esophageal squamous epithelium to gastric contents.

This suggests that the very term GERD is a possible misnomer for the disease. GERD, if defined as esophageal squamous epithelial damage caused by exposure to gastric contents, occurs in this phase of compensated LES damage without reflux into the thoracic esophagus. Therefore, the terms "preclinical GERD," "asymptomatic GERD," and "subclinical GERD" are all appropriate to describe this phase of the disease.

The only manifestation of this phase of compensated LES damage are (1) manometric shortening of the LES that has not reached the point where it is recognized as abnormal (Fig. 16.16), and (2) the histologic presence of cardiac epithelium (with and without parietal cells) up to 15 mm in length in the dilated distal esophagus.

To those who believe that cardiac epithelium distal to the endoscopic GEJ is normal proximal gastric epithelium, these patients will be regarded as normal. To those who believe in the concept of the dilated distal esophagus, this is a patient who has progressived in the degree of abdominal LES damage from 0 to 15 mm.

There is a highly significant difference between patients with 15 mm of LES damage and those with lesser damage. People with abdominal LES damage of 15 mm are at the end of the reserve capacity of the LES and are at the cusp of onset of clinical GERD.

3. CLINICAL GASTROESOPHAGEAL REFLUX DISEASE (>15–35 MM LOWER ESOPHAGEAL SPHINCTER DAMAGE)

LES shortening >15 mm means a residual abdominal LES length of 20 mm. Dynamic LES shortening of the LES of 10 mm associated with gastric overdistension will result in a functional abdominal LES length of <10 mm in the postprandial period. LES failure (tLESR) is now possible, causing reflux into the thoracic esophagus. This can cause symptoms and cellular damage in the squamous epithelium above the endoscopic GEJ.

Progression of LES damage from >15 mm to complete destruction of the abdominal LES (35 mm damage) results in progression of cellular changes from electron and light microscopic reflux esophagitis (nonerosive reflux disease)→increasing grades of erosive esophagitis (including ulcers and strictures)→visible CLE→intestinal metaplasia (Barrett esophagus)→increasing dysplasia→adenocarcinoma.

There is a high likelihood of a strong correlation between LES damage and the sequence of changes up to visible CLE in patients without acid-reducing therapy (as was seen historically in the 1950s). However, with effective acid-reducing therapy, squamous epithelial damage has decreased in severity and the progression to columnar epithelial changes has increased.

The only reason for this is that progression of LES damage continues while the patients are being treated with PPIs. The severity of reflux continues to increase as LES damage progresses. The progression of cellular change correlates with severity of reflux. It is not altered positively by converting strong acid reflux to weak acid reflux[8]; the progression of cardiac epithelium→intestinal metaplasia may actually be promoted by PPI therapy.

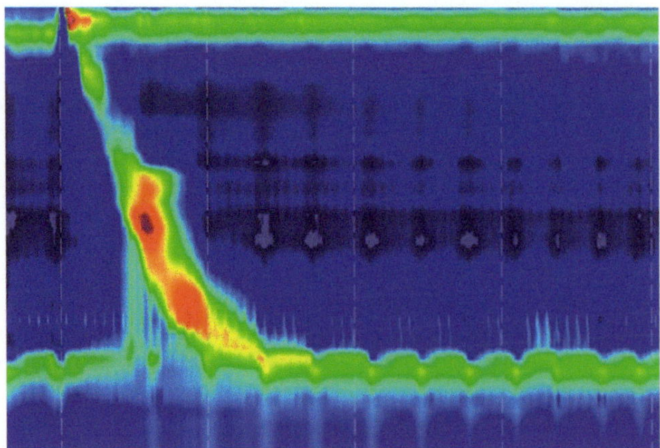

FIGURE 16.16 High-resolution manometry that is considered normal. The lower esophageal sphincter (LES) has a high resting pressure, relaxes normally with swallows, and has a total length of 34 mm and an abdominal length of 22 mm. In the new method of understanding, this is an LES with 13 mm of damage, toward the end of the phase of compensated damage corresponding to the status shown in Fig. 16.15.

The exact level of LES damage associated with each of these changes is uncertain. I have assigned the point of onset of LES failure is reached when damage exceeds 15 mm. This is the point of onset of the earliest reflux-induced changes in the thoracic esophagus. Whether it is accurate will be determined by future studies that correlate measured LES damage with the 24-h pH test.

Similarly, assigning 25 mm of LES damage is a reasonable point for the GERD-damaged squamous epithelium→visible CLE step. Future studies that provide more accurate data correlating LES damage severity with the occurrence of visible CLE are necessary to establish this point as the lowest level of damage associated with visible CLE. This is crucial if the new management objective is the prevention of visible CLE.

3.1 Abdominal Lower Esophageal Sphincter Damage of 20 mm

Initial abdominal LES length: (assumed with best evidence) 35 mm (Fig. 16.17);
LES damage: (measured by squamooxyntic gap in dilated distal esophagus) 20 mm;
Residual abdominal LES length by manometry: 15 mm;
Abdominal LES length after a heavy meal: (assuming dynamic shortening of 10 mm due to gastric overdistension associated with a heavy meal) 5 mm;
Probability of LES failure (tLESR): (probability of abdominal LES reaching the critical failure level of 10 mm) significant in postprandial state;
24-h pH test: >0 and <4.5% time pH<4 (closer to 4.5%);
Cellular changes in thoracic esophagus above endoscopic GEJ: microscopic reflux esophagitis (dilated intercellular spaces) or mild erosive esophagitis;
Symptoms: postprandial heartburn;

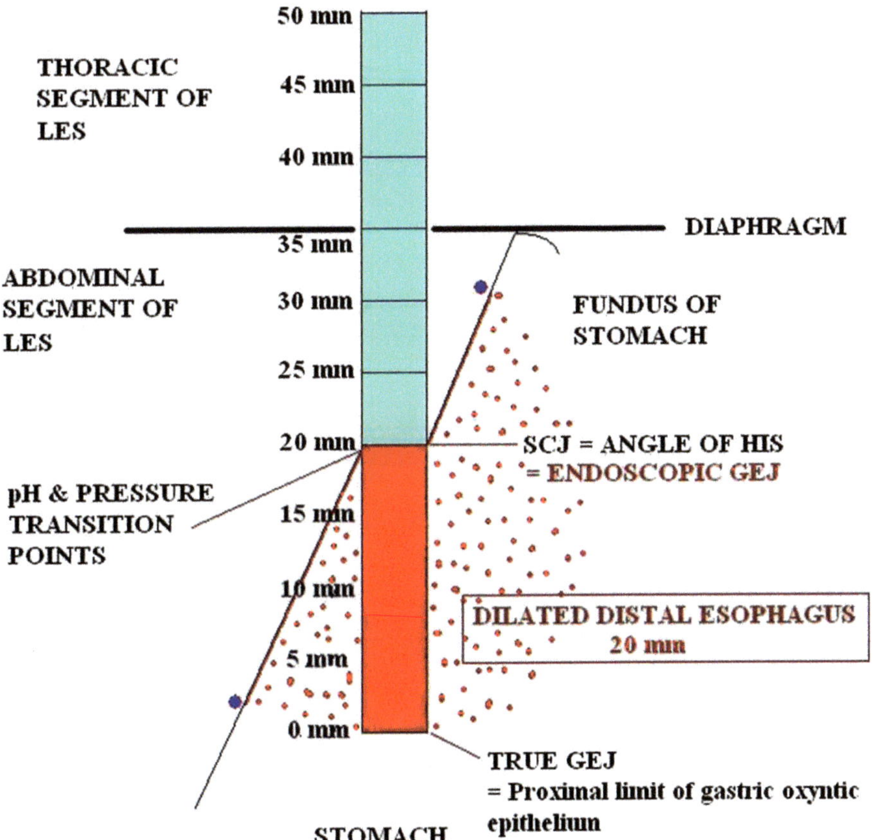

FIGURE 16.17 Diagrammatic representation of 20 mm of lower esophageal sphincter (LES) damage, concordant with 20 mm of a dilated distal esophagus lined by cardiac epithelium (red). Patients have more frequent LES failure and more severe gastroesophageal reflux disease with a greater probability of erosive esophagitis in the thoracic esophagus. *GEJ*, gastroesophageal junction; *SCJ*, squamocolumnar junction. Fig. 16.16

Risk of adenocarcinoma: zero at this point in time; needs three steps: squamous→visible CLE→intestinal metaplasia to adenocarcinoma.

This group has sufficient LES damage to have a significant risk of postprandial heartburn with any meal of significant volume. Their threshold of dynamic shortening due to gastric overdistension before the functional length of the abdominal LES reaches the critical 10mm has decreased to 5mm. Postprandial LES failure (tLESR) and reflux occur infrequently but limit meals to smaller size and quality. Patients tend to recognize "trigger foods" and avoid them. Alcohol and possibly smoking have an adverse impact, particularly when taken with meals.

These people are likely to be able to control their symptoms by adjusting their diet. Alternatively, they have learned to take acid-reducing drugs before partaking in a meal they know will likely cause heartburn. They have largely learned to do this from the incessant advertising with a strong bias to be shown at televised sporting event. They are directed at the middle-aged white male who habitually eats a GERD-unfriendly diet of a double cheeseburger with fries and a shake or pizza or nachos with mounds of cheese while watching the equivalent of a super bowl with their friends. Some well-known celebrity appears and tells them not to worry, but only if they have prepared themselves with a Nexium. Or, someone else will advise them that if they have forgotten to take their Nexium, they can still be all right with a more quickly acting Zantac or Tums. Take a pill (to neutralize the acid pocket), eat to your heart's content, and be free of heartburn!

The pills simply prevent the pain that controls their pain by neutralizing acid. The pills, however, are enablers. They permit the patient with >15–20mm of LES damage to eat excessively and continue to damage their LES. The natural control of LES damage that is heartburn is removed. Acid neutralizers for these people with early GERD are the equivalent of using painkilling drugs to enable running on an injured knee. They allow aggravating the cause of the problem while controlling its result.

This will be the group of patients who are self-medicating themselves successfully with over-the-counter acid reducers or under the care of primary care physicians. They are still not likely to fall within the definition of GERD because, in most of them, they do not perceive their symptoms as "troublesome." Advertising has convinced them that their heartburn is just a nuisance that can be avoided with a miracle pill. It is not a disease. Physicians sometimes use the term "episodic heartburn" to describe these patients, suggesting that this is still not GERD.

There is a universal effort to minimize this stage of early GERD at the onset of symptoms. This is reasonable. The number of people with this mild, often unrecognized GERD is in the millions and can overwhelm the system if they access the health-care delivery system at the onset of symptoms.

This is the group of people who will be included in "control" or "asymptomatic" or "normal" arms of GERD studies where the definition of GERD is "troublesome (Montreal definition)"[9] or "heartburn and/or regurgitation at least once a week (Lagergren et al)"[2] or "heartburn less than once a month (Gerson et al)"[10] or "heartburn equal to or greater than three times per week controlled by acid suppression (Kahrilas et al)".[11]

In the new system of understanding, this is GERD where LES damage has exhausted the reserve capacity of the LES. The diagnosis is made by LES damage >15 mm, not symptoms. Every 1mm increment in LES damage from this point onward has an exponentially greater impact than a similar progression during the phase of compensated LES damage. Now, every millimeter of further damage increases the likelihood and frequency of LES failure and reflux into the thoracic esophagus.

Ultimately, a significant number of this group of people will progress to develop symptoms that are "troublesome," be defined as having GERD, and be taken seriously by the medical establishment, being deemed to be "worthy" of the attention of gastroenterologists.

The reason why this management algorithm is acceptable is that only a relatively small percentage of people with "episodic" or "nontroublesome" GERD reach the amount of LES damage necessary to reach 25mm LES damage during their lifetime. They stay controlled with self-medication and acid-reducing therapy delivered cheaply by the pharmacy or their primary care physician.

The failure to discriminate between people who remain with <25mm LES damage in their lifetime and those progressing to this level of severe GERD is the crux of the GERD problem. Using symptom severity and symptom control is an ineffective method of recognizing those patients who will progress to severe GERD. We just treat them empirically and let them express themselves at the point of progression beyond 25mm of LES damage.

Unfortunately, this expression of severe GERD is sometimes and increasingly dysphagia resulting from adenocarcinoma. LES damage has been allowed to progress to this level of severity even as the patient is well controlled with PPIs. They have never reached the point of treatment failure where endoscopy is presently indicated.

The objective of the new method of management is early diagnosis of LES damage by the new test and prediction of future LES damage by the new algorithm. This will permit identification of the high-risk group long before they progress to the point of severe GERD defined by visible CLE. Aggressive intervention to prevent progression of LES damage while their LES is relatively intact will provide a window of opportunity to prevent adenocarcinoma. This window of opportunity can be many decades.

3.2 Abdominal Lower Esophageal Sphincter Damage of 25 mm

Initial abdominal LES length: (assumed with best evidence) 35 mm (Fig. 16.18);
LES damage: (measured by squamooxyntic gap in dilated distal esophagus) 25 mm;
Residual abdominal LES length by manometry: 10 mm;
Abdominal LES length after a heavy meal: (assuming dynamic shortening of >0–10 mm due to gastric overdistension associated with a heavy meal) 0–<10 mm;
Probability of LES failure (tLESR): (probability of abdominal LES reaching the critical failure level of 10 mm) very high in postprandial state; significant at rest;
24-h pH test: >4.5% time pH<4;
Cellular changes in thoracic esophagus above endoscopic GEJ: significant risk of severe erosive esophagitis and/or visible CLE;
Symptoms: asymptomatic (rare) to troublesome symptoms; probably well controlled with a good regimen of PPI therapy under gastroenterologist care; significant risk of failure to control symptoms completely with PPIs;
Risk of adenocarcinoma: zero at this point in time; needs three steps: squamous→visible CLE→intestinal metaplasia to adenocarcinoma.

Most people in the population never reach 25 mm of LES damage in their lifetime. This is the point of LES damage that I believe, from a purely theoretical standpoint, is the cusp of risk for the development of visible CLE. This is therefore the point in time that is the endpoint of reliable freedom from cancer risk for the patient with GERD. This is also the point that is the target for preventing LES damage from reaching if the objective of preventing adenocarcinoma is to be achieved.

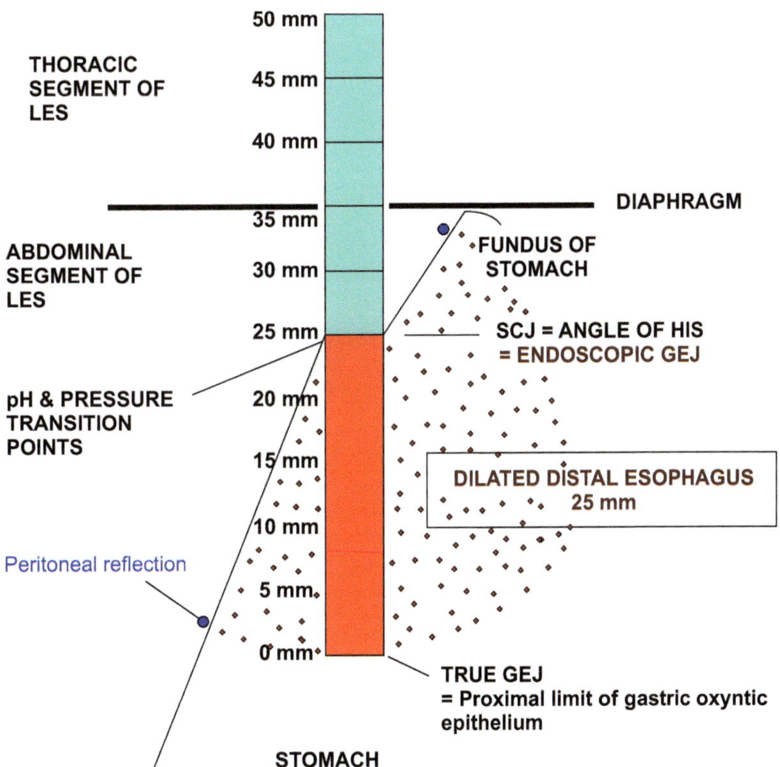

FIGURE 16.18 Diagrammatic representation of 25 mm of LES damage, concordant with 25 mm of a dilated distal esophagus lined by cardiac epithelium (red). Patients have a residual abdominal LES of 10 mm and likely to have severe gastroesophageal reflux disease with difficulty in controlling symptoms with proton pump inhibitor therapy. *GEJ*, gastroesophageal junction; *SCJ*, squamocolumnar junction.

Again, it is important to emphasize that the amount of LES damage at which visible CLE develops will be determined by actual testing for LES damage when this test becomes available. The 25 mm LES damage is yet a theoretical number.

At 25 mm of LES damage, patients have a residual abdominal LES length of 10 mm. Any pressure on the LES by eating, abdominal strain, and other stresses can induce LES failure. Pandolfino et al.[12] reported that with increasing LES damage (defined in their study as the presence of a hiatal hernia), the mechanism of LES failure progressively shifts from tLESRs (which result from dynamic shortening of the LES to the critical length <10 mm) to other mechanisms.

These patients are likely to have troublesome symptoms with incomplete or increasing difficulty in controlling their symptoms with prescription PPI therapy. Their care has likely shifted from a primary care physician to a gastroenterologist. They still have quality-of-life measurements that are tolerable. Many, however, have lives that have been disrupted by their disease to some extent with increased absences from work, increased doctor visits, decreased productivity, and general dissatisfaction with life.

These patients are the most studied in the gastroenterology literature. They are also more likely than all previous groups to either have or soon develop treatment failure and severe erosive esophagitis. They do not have visible CLE, but only by our present assumption that visible CLE occurs at >25 mm LES damage. They are either at or on the verge of developing visible CLE, which is the criterion of irreversibility that I have defined.

Still, the only management in these patients at the present time is continuation of empiric PPI therapy, often with dose escalation and the need for maximum and most frequent dosage. They still may not fall into the definition of "treatment failure" that is the present recommended indication for endoscopy. There is no sense of urgency even at this point. Visible CLE is not recognized as a serious problem. If endoscopy is done and there is visible CLE with no intestinal metaplasia, these patients simply go back to being treated with PPIs.

Without visible CLE, these patients are at no immediate risk of adenocarcinoma. With only squamous epithelium in the thoracic esophagus, they still need three steps (squamous→visible CLE→intestinal metaplasia→adenocarcinoma) in the GERD→adenocarcinoma sequence.

Patients in the 20–25 mm range of LES damage can be recognized by present clinical methods by the appearance of two findings: (1) the presence of severe erosive esophagitis (LA grade C/D) and (2) intestinal metaplasia at the endoscopically normal GEJ, i.e., immediately distal to the SCJ at the proximal limit of the dilated distal esophagus. In the Pro-GERD study, both these findings predicted the progression to visible CLE within the next 5 years; 19.7% of patients with severe erosive esophagitis,[13] and 25.8% of patients with intestinal metaplasia at the SCJ[14] progressed to visible CLE within 5 years. The reason for this is that these two features indicate abdominal LES damage close to the 25 mm required for visible CLE to occur. They are indicators of LES damage at the cusp of the >25 mm required for severe reflux and risk of visible CLE.

3.3 Abdominal Lower Esophageal Sphincter Damage of >25–35 mm

Initial abdominal LES length: (assumed with best evidence) 35 mm (Figs. 16.19 and 16.20);
LES damage: (measured by squamooxyntic gap in dilated distal esophagus) 30–35 mm;
Residual abdominal LES length by manometry: 0–5 mm;
Abdominal LES length after a heavy meal: (assuming dynamic shortening of any length due to gastric overdistension associated with a heavy meal) 0–<5 mm;
Probability of LES failure (tLESR): (probability of abdominal LES reaching the critical failure level of 10 mm) already present at rest;
24-h pH test: >4.5% time pH<4;
Cellular changes in thoracic esophagus: (above endoscopic GEJ) high probability of visible CLE with or without intestinal metaplasia (Barrett esophagus)
Symptoms: asymptomatic to troublesome symptoms; likely to be not satisfactorily controlled with maximum PPI therapy (refractory GERD);
Risk of adenocarcinoma: Present; needs only one or two steps if there is visible CLE with and without intestinal metaplasia, respectively;→intestinal metaplasia (Barrett esophagus)→adenocarcinoma.

This patient group represents the endpoint of abdominal LES damage characterized by near complete or complete destruction (35 mm). This is defined by the presence of visible CLE at endoscopy, irrespective of amount of LES damage. It is the point of irreversibility. Patients who reach this point will develop adenocarcinoma if they are so destined. Medical therapy does not prevent the incidence of cancer.

At >25 mm of LES damage, the patient does not have a competent LES even at rest. LES failure (tLESR) and reflux into the esophagus occurs with minimal stress placed on the LES, either by eating or by other things that influence abdominal

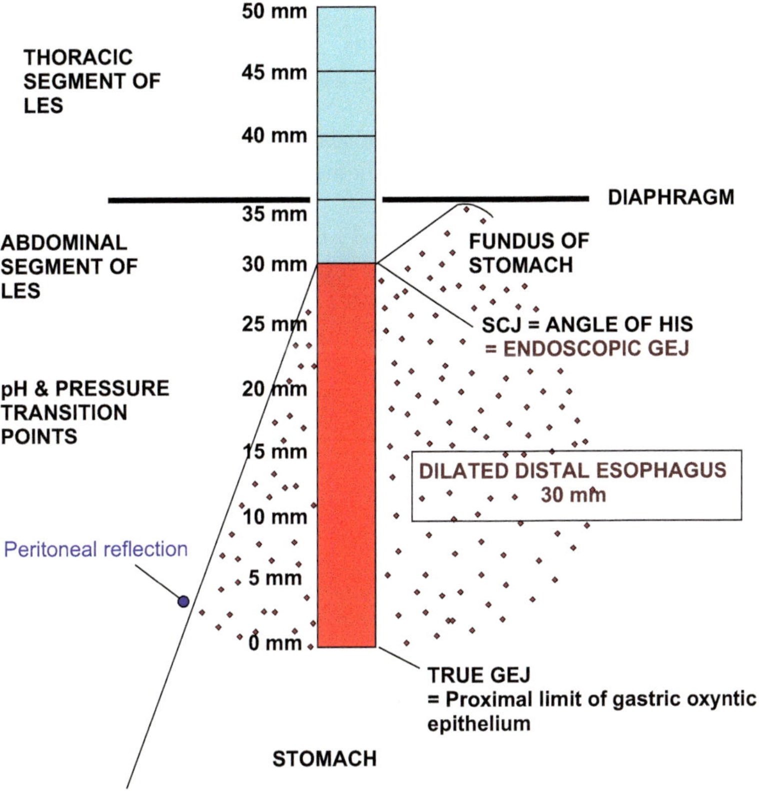

FIGURE 16.19 Diagrammatic representation of 30 mm of lower esophageal sphincter (LES) damage, concordant with 30 mm of a dilated distal esophagus lined by cardiac epithelium (red). The abdominal LES length is 5 mm; reflux is severe and the probability of visible columnar-lined esophagus and the onset of cancer risk are higher. *GEJ*, gastroesophageal junction; *SCJ*, squamocolumnar junction.

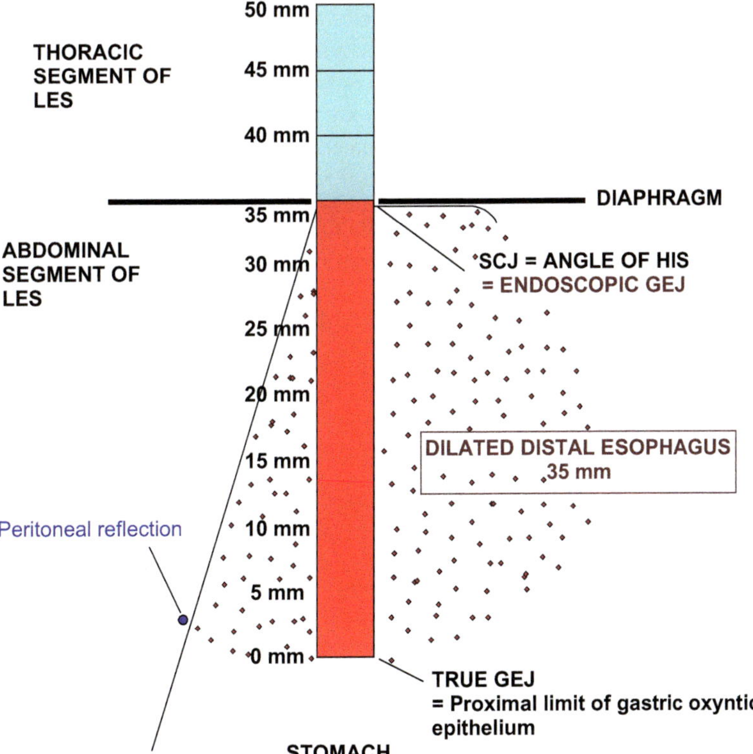

FIGURE 16.20 Diagrammatic representation of lower esophageal sphincter (LES) damage that has completely destroyed the abdominal LES, concordant with 35 mm of a dilated distal esophagus lined by cardiac epithelium (red). Patients have severe gastroesophageal reflux disease with a high risk of visible columnar-lined esophagus and adenocarcinoma. Note that the angle of His is now nearly a right angle. Hiatal hernia becomes increasingly common (see Fig. 16.10). *GEJ*, gastroesophageal junction; *SCJ*, squamocolumnar junction.

and thoracic pressure. This group will have the highest prevalence of treatment failure, severe erosive esophagitis, visible CLE, Barrett esophagus, and the highest risk of adenocarcinoma.

As the LES is damaged >25 mm and progresses to complete loss of the entire 35 mm, the tubular abdominal esophagus shortens with increasing length of the dilated distal esophagus. The angle of His becomes increasingly obtuse. At 35 mm abdominal LES damage, the entire abdominal esophagus has dilated and the tubular esophagus is flush with the diaphragm. The angle of His is close to a right angle. The tendency to a sliding hiatal hernia progressively increases during this phase of the disease (Fig. 16.10).

In such patients who have lost the entirety of the abdominal LES, the thoracic segment of the LES represents the only resistance to reflux. When a hiatal hernia develops, the thoracic segment of the LES moves further cephalad into the thorax, proximal to the dilated distal esophagus and the herniated stomach. In such patients, the diaphragmatic pinch at the base of the hiatal hernia may provide an additional resistance to reflux.

When looked at progression from the point of view of measured LES damage, this end stage (abdominal LES damage of >25–35 mm with a residual length of 0–10 mm) appears very far from the onset of disease (abdominal LES length 35 mm). The LES damage has taken many decades to progress from 0 to 30–35 mm and the abdominal LES to shorten from 35 to 0–5 mm.

The present diagnostic and management algorithms have ignored LES destruction from 0 to 15 mm where the LES has progressively exhausted its reserve capacity while remaining competent. During this phase, the patient has no reflux and no symptoms resulting from reflux. This phase can only be detected by screening asymptomatic people.

The present management algorithm has further ignored abdominal LES damage in the range of 15–20 mm where the patient is symptomatic but not to the extent of reaching the "troublesome" requirement of usual frequencies used to define GERD. They may have "dyspepsia," "episodic heartburn," or atypical and nonspecific symptoms not recognized presently as being caused by GERD.

The onset of clinically recognizable GERD occurs when the LES has been damaged to the point of significant failure (20–25 mm), initially postprandial where dynamic shortening of the LES during a heavy meal pushes a marginally competent LES below the failure (tLESR) threshold at <10 mm of abdominal LES length. Initially, this can be controlled by changing the diet to be more "GERD-friendly" or by taking over-the-counter acid reducers. Dietary restriction largely limits the occurrence of tLESRs that are associated with dynamic shortening.

It is around 25 mm when troublesome symptoms, abnormal pH tests, and manometric criteria for LES failure occur. At this point of LES damage, LES failure becomes less related to meals because non-tLESR mechanisms increasingly become the mechanism of LES failure. Dietary restrictions at this stage of the disease are ineffective in preventing reflux.

Symptoms and the ability of acid-reducing drugs to control symptoms correlate poorly with severity of reflux and cellular changes in the thoracic esophagus. There is, however, a subset of patients whose LES damage will never progress >25 mm because of a slow rate of progression. These patients are the majority of patients successfully managed over the long term with acid-reducing therapy. They may need dose escalation but remain in control.

The present management with acid reduction fails for the following reasons:

(1) In the minority of patients who progress to 25–35 mm abdominal LES damage, PPI therapy ultimately fails to control symptoms. There is good evidence that patients with refractory GERD have a higher probability of a defective LES (abdominal LES <10 mm).
(2) The nature of cellular progression is altered by PPI therapy, resulting in excellent control of squamous epithelial changes, but permitting cardiac metaplasia to occur and possibly promoting visible CLE.
(3) Treatment failure and visible CLE, which are my definitions of failure, occur primarily because of progression to abdominal LES damage into the 25–35 mm range. Present therapy fails because it does not in any way decrease the rate of progression of LES damage or does anything to prevent progression.

In the 1950s, before effective acid reduction was possible, the cellular changes in the thoracic esophagus progressed to extreme squamous epithelium injury and visible CLE, but rarely progressed to intestinal metaplasia and adenocarcinoma. The major impact of acid-reducing therapy has likely been the induction of visible CLE and promoting intestinal metaplasia in visible CLE.

The preamble to significant LES failure (tLESR) at >20–25 mm LES damage, which is slowly progressive LES damage over many decades is not even recognized as an abnormality. LES damage begins when poor dietary habits begin early in life. We have assumed that onset of damage to be at age 15 years. From this point to the point where the patient develops troublesome symptoms and is diagnosed with GERD, there are two to three decades of slowly progressive LES degradation within its reserve capacity. This is manifested *only* by the increasing length of cardiac and/or oxyntocardiac epithelium in the dilated distal esophagus.[15]

The increase in length of cardiac epithelium (with or without parietal and/or goblet cells) from 0 to 35 mm is a measure of progression of LES damage. It is largely unseen because of the practice of not taking biopsies distal to the endoscopic GEJ both in the endoscopically normal patient and in the patient with visible CLE.

Even when biopsies are done, cardiac epithelium (with or without parietal and/or goblet cells) distal to the endoscopic GEJ is misinterpreted as "normal proximal stomach/gastric cardia" or "intestinal metaplasia of the gastric cardia." With the powerful evidence base that I have presented in this book that this interpretation is absolutely false, there is no excuse to ignore the dilated distal esophagus by failing to evaluate the region distal to the endoscopic GEJ by histology.

4. COMPARISON BETWEEN ASSESSMENT OF GASTROESOPHAGEAL REFLUX DISEASE PROGRESSION BY CLINICAL, PATHOLOGICAL MOLECULAR CRITERIA AND LOWER ESOPHAGEAL SPHINCTER DAMAGE

In the introduction of Chapter 14, I suggested that progression of GERD as a pathologically defined disease progressed inexorably at a constant rate that varies with the rate of progression of LES damage.

In this chapter, I showed the probable clinical and pathological changes associated with each 5 mm increment of abdominal LES damage. From a logical standpoint, supported by a relative scanty evidence base I have shown that there is likely to be excellent correlation between the vertical progression of cellular changes and LES damage. The cellular changes progress through normal (no cellular changes), through a phase of compensated LES damage (dilated distal esophagus), subclinical GERD (microscopic cellular changes in the squamous epithelium of the thoracic esophagus), clinically recognized GERD (erosive esophagitis), to severe GERD where the patient develops severe erosive esophagitis and visible CLE. The endpoint of the disease in any given patient is based on the rate of progression of LES damage and the life span of the individual.

In Table 16.4, the progressive shortening of the abdominal LES is shown at rates varying from 1 mm to 10 mm/decade of life. The areas in green represent a functional abdominal LES length of 20 mm or more, which predicts a competent LES without reflux (patient with no clinical GERD), 10–20 mm, which predicts intermittent LES failure with mild GERD (easily controlled with PPI therapy), and <10 mm, which predicts severe LES incompetence and severe reflux (treatment failure, visible CLE, and adenocarcinoma).

By this viewpoint of the disease, the present understanding and management of GERD using symptoms and control of symptoms with PPI therapy as the defining criteria is a poor method of assessment of the severity and progression of GERD.

The reasons for this are as follows:

(1) Symptoms are not specific for GERD. Combined with the fact that there is no sensitive diagnostic test that can exclude GERD as the cause of symptoms, an empiric PPI test is used as the standard method of diagnosis of GERD. This may result in a false-positive test resulting in a significant number of people being treated, often long term, for GERD with no certainty that they have GERD.
(2) Symptom severity does not correlate well with cellular changes. 40% of patients who develop adenocarcinoma of the thoracic esophagus and 71% of those who develop adenocarcinoma of the dilated distal esophagus have heartburn and/or regurgitation less than once a week.[2]
(3) Symptoms are easily controlled in the early stages with acid-reducing drugs, masking LES damage and cellular progression.
(4) Symptom control does not mean lack of disease progression of LES damage. PPI therapy does not have any positive impact on the LES or the severity of reflux. It simply converts strong acid reflux to weak acid reflux of equal severity.[8]
(5) Symptom control does not mean that progression of cellular changes does not occur. Squamous epithelial damage in the thoracic esophagus reverses at the same time columnar metaplasia increases. Nason et al.[1] reported that good symptom control with PPIs is associated with a higher prevalence of Barrett esophagus.

This lack of correlation between symptoms and cellular change/LES damage results in GERD being managed as a categorical disease. A patient with early GERD is first characterized as "non-GERD" by the Montreal definition that requires the patient to have "troublesome" symptoms. Advertising suggests that this is a nuisance and not a serious disease and pushes patients to self-medication. A person with troublesome symptoms that fit the definition of GERD is initially treated as if he/she has disease that is easily curable with PPI therapy. They are treated by primary care physicians and gastroenterologists without any diagnostic testing.

The disease is only taken seriously when control becomes difficult. At this point they reach a gastroenterologist. Even here, endoscopy is commonly delayed until a decision is made that treatment has failed. This is the indication for endoscopy,

TABLE 16.4 Changes With Age of the Functional Residual Length of the Abdominal Lower Esophageal Sphincter (LES) Assuming That the Original Length at Maturity is 35 mm, That LES Damage Begins at Age 15 Years, and That LES Damage Has a Linear Progression Over the Long Term

Rate of LES Damage (mm/decade)	At 25 Years (mm)	At 35 Years (mm)	At 45 Years (mm)	At 55 Years (mm)	At 65 Years (mm)	At 75 Years (mm)
1	34	33	32	31	30	29
2	33	31	29	27	25	23
3	32	29	26	23	20	17
4	31	27	23	19	15	11
5	30	25	20	15	10	5
6	29	23	17	11	5	0
7	28	21	14	7	0	0
8	27	19	11	3	0	0
9	26	17	8	0	0	0
10	25	15	5	0	0	0

The abdominal LES lengths in green represent lengths at which the LES is likely to be competent. The lengths in orange represent an LES that is susceptible to failure with gastric distension (i.e., at risk of postprandial reflux). The lengths in red represent an LES that is below the length at which LES failure occurs at rest.

unless dysphagia has occurred earlier and raised the possibility of cancer. This is the point of diagnostic testing. In most patients, this is LES damage of ~25 mm.

Diagnostic testing is by endoscopy. This aims to define severity of erosive esophagitis or visible CLE. There are only three options for the patient at this stage:

(1) Continuation with PPI therapy despite its failure to control symptoms;
(2) Antireflux procedures, which are not favored by gastroenterologists in the United States; and
(3) Barrett esophagus surveillance if the patient has this diagnosis by endoscopy and biopsy.

All these are poor options for the patient.

From the point of view of LES damage, GERD has been allowed to reach the orange and red stages in Table 16.2 without any attempt at early diagnosis or any attempt at controlling progression of cellular changes or LES damage. By the time gastroenterologists start becoming serious, the horse has left the barn. Their ability and success in managing red zone disease are dismal.

What needs to happen in GERD is diagnosis at an early stage (green or early orange stage in Tables 16.1 and 16.4), and the recognition of which minority of patients in the green and orange stage will progress to the red stage in the future.

This is possible only when a method of assessment of LES damage becomes available. When this happens, the patients predicted to remain in the green and orange stages for the rest of their lives can be treated with acid-reducing drugs as needed by primary care physicians. The minority of patients predicted to progress to the red stage *must* have an intervention that will slow or stop progression of LES damage.

I will discuss the method of LES assessment by pathology, the method of predicting progression of LES damage into the future, and the methods available to slow down and stop progression of LES damage in the remaining chapters.

REFERENCES

1. Nason KS, Wichienkuer PP, Awais O, Schuchert MJ, Luketich JD, O'Rourke RW, Hunter JG, Morris CD, Jobe BA. Gastroesophageal reflux disease symptom severity, proton pump inhibitor use, and esophageal carcinogenesis. *Arch Surg* 2011;**146**:851–8.
2. Lagergren J, Bergstrom R, Lindgren A, Nyren O. Symptomatic gastroesophageal reflux as a risk factor for esophageal adenocarcinoma. *N Engl J Med* 1999;**340**:825–31.
3. Zaninotto G, DeMeester TR, Schwizer W, Johansson K-E, Cheng S-C. The lower esophageal sphincter in health and disease. *Am J Surg* 1988;**155**:104–11.
4. Chandrasoma PT, Der R, Ma Y, et al. Histology of the gastroesophageal junction: an autopsy study. *Am J Surg Pathol* 2000;**24**:402–9.

5. Ayazi S, Tamhankar A, DeMeester SR, Zehetner J, Wu C, Lipham JC, Hagen JA, DeMeester TR. The impact of gastric distension on the lower esophageal sphincter and its exposure to acid gastric juice. *Ann Surg* 2010;**252**:57–62.
6. Robertson EV, Derakhshan MH, Wirz AA, Lee YY, Seenan JP, Ballantyne SA, Hanvey SL, Kelman AW, Going JJ, McColl KE. Central obesity in asymptomatic volunteers is associated with increased intrasphincteric acid reflux and lengthening of the cardiac mucosa. *Gastroenterology* 2013;**145**:730–9.
7. Lee J, Anggiansah A, Anggiansah R, et al. Effects of age on the gastroesophageal junction, esophageal motility and reflux disease. *Clin Gastroenterol Hepatol* 2007;**5**:1392–8.
8. Blonski W, Vela MF, Castell DO. Comparison of reflux frequency during prolonged multichannel intraluminal impedance and pH monitoring on and off acid suppression therapy. *J Clin Gastroenterol* 2009;**43**:816–20.
9. Vakil N, van Zanten SV, Kahrilas P, Dent J, Jones B, The Global Consensus Group. The Montreal definition and classification of gastroesophageal reflux disease: a global evidence-based consensus. *Am J Gastroenterol* 2006;**101**:1900–20.
10. Gerson LB, Shetler K, Triadafilopoulos G. Prevalence of Barrett's esophagus in asymptomatic individuals. *Gastroenterology* 2002;**123**:461–7.
11. Kahrilas PJ, Shi G, Manka M, Joehl RJ. Increased frequency of transient lower esophageal sphincter relaxation induced by gastric distension in reflux patients with hiatal hernia. *Gastroenterology* 2000;**118**:688–95.
12. Pandolfino JE, Shi G, Curry J, Joehl RJ, Brasseur JG, Kahrilas P. Esophagogastric junction distensibility: a factor contributing to sphincter incompetence. *Am J Physiol Gastrointest Liver Physiol* 2002;**282**:G1052–8.
13. Malfertheiner P, Nocon M, Vieth M, Stolte M, Jasperson D, Keolz HR, Labenz J, Leodolter A, Lind T, Richter K, Willich SN. Evolution of gastro-oesophageal reflux disease over 5 years under routine medical care – the ProGERD study. *Aliment Pharmacol Ther* 2012;**35**:154–64.
14. Leodolter A, Nocon M, Vieth M, Lind T, Jasperson D, Richter K, Willich S, Stolte M, Malfertheiner P, Labenz J. Progression of specialized intestinal metaplasia at the cardia to macroscopically evident Barrett's esophagus: an entity of concern in the Pro-GERD study. *Scand J Gastroenterol* 2012;**47**:1429–35.
15. Chandrasoma P, Wijetunge S, Ma Y, DeMeester S, Hagen J, DeMeester T. The dilated distal esophagus: a new entity that is the pathologic basis of early gastroesophageal reflux disease. *Am J Surg Pathol* 2011;**35**:1873–81.

Chapter 17

New Pathologic Test of LES Damage

Assessment of lower esophageal sphincter (LES) damage, which is universally recognized as the cause of gastroesophageal reflux disease (GERD), does not enter into the equation at any point in the present diagnostic algorithm of the patient with GERD. This is largely because there is no method at present to measure LES damage. In this chapter, a new pathologic assessment of LES damage is proposed.

At present, LES assessment of any kind is not a diagnostic method of the study or management of GERD. Endoscopy is blind to the location of the LES. Gastroenterologists rarely use manometry as a diagnostic test for GERD. Most gastroenterologists have little or no interest in the LES as they manage their GERD patients at a level that is based largely on the presence of symptoms and the objective of symptom control.[1]

1. FEASIBILITY OF THE NEW TEST FOR LOWER ESOPHAGEAL SPHINCTER DAMAGE

The structure of the LES and the way in which it is damaged was detailed in Chapters 7–9. I recommend a quick review of those chapters. I will present here a more detailed analysis of the theory and method of assessing LES damage by biopsy.

At the outset, it must be recognized that this is an entirely new diagnostic test that is being proposed. As such, it is to be expected that the presently available technology will not be adequately designed to provide the required accuracy of measurement. Also, by necessity, there is little direct evidence to support the method. All the evidence is indirect from data in papers about the LES.

Presently available biopsy techniques provide only a theoretical basis for a measurement to show its potential value. However, this method is likely to be poorly accepted by gastroenterologists because of the need for multiple biopsies at endoscopy, increasing the time and effort of endoscopy. It is also very dependent on operator ability, care, and technique. Even in the hands of the most expert gastroenterologist using extreme care, there is a high likelihood of significant error in measurement of the dilated distal esophagus (LES damage) with presently available endoscopic biopsy devices. These inevitable technical errors will make it appear that the new method of assessment of LES damage is incorrect. The method should therefore not be tried with available biopsy technology.

The real assessment of LES damage will require a new biopsy device that will permit an accurate and precise measurement of the dilated distal esophagus. This will require removing a mucosal biopsy specimen that measures 20–25 mm long, 2 mm wide, and 1 mm deep. This will be equivalent to a specimen taken from an autopsy or resection specimen. It is my belief that the production of this biopsy device will be relatively easy for medical device engineers. Accurate measurement is the key to both demonstrating proof of concept and success of the new test.

2. ASSESSMENT OF LOWER ESOPHAGEAL SPHINCTER DAMAGE WITH PRESENT TECHNOLOGY

Three elements of the LES determine its function: mean LES pressure, abdominal length, and total length (including the thoracic segment).[2] In Chapter 9, the relative importance of these factors and criteria that predict LES failure was described.

Early belief, still present in the minds of many physicians, was that mean LES pressure was the critical determinant of LES competence.[3] However, it was soon recognized that there was no good correlation between LES pressure and GERD until the mean LES pressure decreased to a very low level (<6 mmHg) from its normal range of 15–40 mmHg.[4]

Evidence was produced that the length of the abdominal segment of the LES was a critical determinant of LES competence, at least as important as mean LES pressure.[5] It correlated well with the occurrence of LES failure [transient LES relaxation (tLESR)] and reflux episodes.

GERD. http://dx.doi.org/10.1016/B978-0-12-809855-4.00017-8

2.1 Importance of the Abdominal Segment of the Lower Esophageal Sphincter

LES damage begins at its distal end, which is the true gastroesophageal junction (GEJ) [proximal limit of gastric oxyntic epithelium (GOE)], and progresses cephalad. As such, the first part of the LES to undergo damage is the abdominal segment of the LES. Damage to the thoracic segment of the LES occurs at a relatively late stage of the disease, after substantial or total damage of the abdominal LES.

It has also been shown that the length of the abdominal segment of the LES is a critical element for the maintenance of LES competence.[5] As such, this chapter will largely concentrate on assessment of the abdominal segment of the LES. Another reason for concentrating on the abdominal LES is that the new pathologic assessment is based entirely on damage to the abdominal LES.

The abdominal segment of the LES can be defined comprehensively at all times by the following mathematical formula:

Abdominal LES damage = Initial abdominal LES length − Residual abdominal LES length (manometric)

This formula applies to the resting LES. During a meal when the stomach distends, there is a temporary dynamic shortening of the abdominal LES that is additive to the permanent (baseline) shortening.

The functional length of the abdominal LES during the postprandial phase = Residual abdominal LES length (manometric) − Dynamic shortening of the abdominal LES during a meal.

With a heavy meal, the dynamic shortening of the abdominal LES can be as much as 10 mm.[6] With gastric emptying, the dynamic shortening reverses and the abdominal LES length reverts to the residual (baseline) length. In a person with early GERD, the postprandial phase is the time of maximum susceptibility of the LES to failure because of the dynamic shortening. With a severely damaged LES, reflux is less predictably related to meals.

At present, these formulae cannot be used in any practical manner because (1) the initial abdominal length of any person is unknown; (2) the residual length of the abdominal LES cannot be measured with adequate accuracy by manometry to have value in the diagnosis of early GERD; and (3) abdominal LES damage cannot be measured by any modality.

The formula will resolve if two of the three elements are known. Having a new method that permits accurate measurement of abdominal LES damage provides an accurate assessment of one of the elements that does not exist at present. It opens the door to a potential new method of comprehensive assessment of the LES if one of the other two elements is known with accuracy.

The determination of the other two elements has problems at present, but these limitations exist because of present lack of demand. When a demand arises, it is not difficult to believe that more accurate data on the initial length of the abdominal LES will become available and the accuracy of its manometric measurement will improve with new technology. Also, multiple measurements of LES damage may provide a method of defining both initial LES length and the rate of progression of LES damage.

If the present limitations of one of these two measurements can be removed and measurement of LES damage by the new test is accurate, there will be a comprehensive LES assessment by the above formula. This has the potential to become the primary diagnostic method for GERD that has maximum sensitivity and specificity of diagnosis. It can become a stand-alone diagnostic test that can be used at any stage of the disease, including the earliest phase before clinical GERD develops.

This diagnostic assessment of the LES has the potential to replace all other diagnostic tests for GERD, particularly the flawed empiric PPI test that is now used. It has the potential to improve the diagnosis and management of GERD dramatically.

2.2 Initial Length of the Abdominal Lower Esophageal Sphincter

The initial length of the abdominal LES in any person is unknown. It is also unknown whether there is an individual variation in the length of the abdominal LES. This is important. Theoretically, people with a longer initial abdominal LES will be more resistant to develop LES failure because they have a greater reserve capacity. Except for male gender, no susceptibility to GERD has been convincingly demonstrated in any group to suggest that differences in innate LES characteristics exist. Known population differences in GERD prevalence can be explained by reasons other than variation in initial LES length, for example, *H. pylori* infection in Asia.

The variation in the initial length of the abdominal LES will only be known when it is measured in a large number of people asymptomatic for GERD. The only such data available at present are in the 50 volunteers in the study of Zaninotto et al.[2] In this group, the distribution of abdominal LES length measured by manometry (i.e., functional length) is shown in Table 17.1. This data has an almost perfect Gaussian distribution.

The upper limit of the length of the abdominal LES has a sharp demarcation at 3.5 cm (if the single person >5 cm is excluded on the reasonable basis that this is an outlier). There are, if this data can be extrapolated to the population, no people with an abdominal LES length >35 mm. The lower end also has a demarcation, but this disappears when patients with GERD are considered. Such patients have a length of the abdominal LES that decreases from 1 to 0 cm as the severity of GERD increases. As such, the range of length of the entire population, with and without evidence of GERD, is 0–35 mm.

TABLE 17.1 Length of the Abdominal Lower Esophageal Sphincter (LES) in the 48 Volunteers (Excluding One Patient With >5 cm and One <1 cm) Divided Into Quintiles

Abdominal LES Length in Volunteers (= Functional Length)	Number of Volunteers	Decrease From Assumed Initial Length of 3.5 cm (= LES Damage)[a]
1.0–1.5 cm	6	2.0–2.5 cm
1.5–2.0 cm	10	1.5–2.0 cm
2.0–2.5 cm	17	1.0–1.5 cm
2.5–3.0 cm	11	0.5–1.0 cm
3.0–3.5 cm	5	0–0.5 cm

[a]If one assumes that initial length of the abdominal LES is 35 mm in all people, these measurements indicate a decrease (by LES damage) of the length of the abdominal LES.

There are two ways to interpret this distribution. The first is that the best estimate of the initial length of the abdominal LES is 35 mm with little or no individual variation. This is the highest initial length from which reduction has taken place in these 50 "normal" subjects.

The second is that there is a significant individual variation of initial length in this population. In general, most experts today believe that the criterion of a defective LES is an abdominal length <10 mm and everything else is "normal" which equates to individual variation.

There is also the possibility that the truth is a combination of the two. This is most likely. However, if the individual variation is small (e.g., 2 mm or less), the assumption of an initial value of 35 mm can still be valuable and can potentially be used in the formula.

If one assumes that the initial length of the abdominal LES is 35 mm in all people, the variation of the length of the abdominal LES in the volunteers of Zaninotto et al.[2] represents a shortening of the LES length from its initial length of 35 mm. This causes the formula to simplify to the following:

LES damage = 35 mm – Functional (residual) length

Theoretically, by using this formula, and based on the measurement of the functional length by manometry, we can determine the severity of damage (= shortening) of the abdominal LES (Table 17.1).

This calculation assumes two facts: (1) The initial length in all people is 35 mm and that there is no individual variation. (2) The manometric measurement of the abdominal LES is accurate.

At present, the belief is that the only value of the manometric measurement of the LES is the detection of a defective LES that correlates with severe GERD sufficient to cause symptoms and an abnormal 24-h pH test. Every other measurement of the abdominal LES is "normal" and therefore ignored. The variation in length that exists largely in the asymptomatic population is ignored and ascribed to an innate "individual variation." There has never been any attempt to use the manometric measurement to define LES damage short of the <10 mm that defines a defective LES.

There is value to undertaking the exercise of attempting to calculate LES damage by using the formula LES damage (i.e., shortening) = initial abdominal LES length – functional abdominal LES length (by manometry).

Chandrasoma et al.,[7] in a study of esophagectomy specimens, showed that there was a concordance between the length of cardiac epithelium distal to the endoscopic GEJ and the dilated distal esophagus. The cause of dilated distal esophagus is LES damage. This means that the length of cardiac epithelium (with and without parietal and/or goblet cells) distal to the endoscopic GEJ is a measure of abdominal LES damage.

Very few people in the population have zero cardiac epithelium (which is the normal state—see Chapter 5). In autopsies of persons without GERD, the length of cardiac epithelium was 0–15 mm.[8] Sarbia et al.,[9] in a study of esophagectomy specimens of patients with squamous carcinoma (unrelated to GERD), reported a median length of cardiac an oxyntocardiac epithelium distal to the end of the tubular esophagus of 11 mm (range: 1–28 mm). This indicates that most persons without GERD have a range of LES damage from 0 to 15 mm with some having up to 28 mm of LES damage in esophagectomies done for squamous carcinoma.

When the data in the study of Zaninotto et al.[2] are looked at from this point of view, 5/49 (excluding one outlier with >5 cm) "normal" subjects had a functional abdominal LES length measured by manometry of >3 cm, 33/49 had a length >2 cm, and 16/49 have a length <2 cm. All these people were asymptomatic. Surely, an abdominal LES of >3 cm cannot be the same level of competence in preventing reflux as one of <2 cm. Evidence relating to acid exposure of this group of asymptomatic persons in Zaninotto et al.[2] strongly suggests this is not the case.

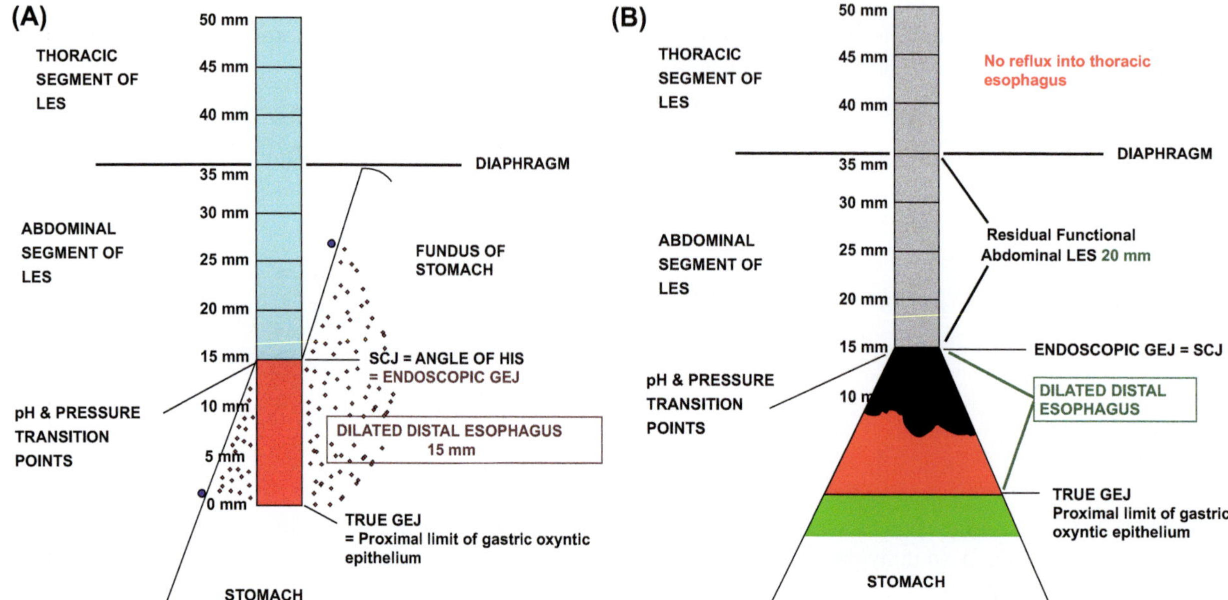

FIGURE 17.1 (A) Diagrammatic representation of the status of the lower esophageal sphincter (LES) in an asymptomatic person with abdominal LES damage of 15mm measured by the new histologic test. This area of the esophagus without LES pressure has dilated. The distal 15mm of squamous epithelium (gray) has undergone cardiac metaplasia (red); the cardiac epithelium now lines the dilated 15mm of the esophagus (*red stippled*). The squamocolumnar junction (SCJ), pH, and pressure transition points have moved cephalad by 15mm. The angle of His is more obtuse. The peritoneal reflection (*blue dot*) is at the true GEJ at the distal end of the dilated distal esophagus. The residual abdominal LES length has decreased from its initial 35–20mm. (B) Same person showing the dilated distal esophagus in its complete form lined by cardiac (black) and oxyntocardiac (red) epithelia between the SCJ and true gastroesophageal junction (GEJ).

If one assumes that the initial abdominal LES length is 35 mm, most of these subjects have LES damage (Table 17.1) varying from 0 to 25mm, amazingly similar to the data of LES damage in the autopsy[8] and esophagectomy[9] studies above.

The decline in functional (manometric) length has resulted from LES damage beginning at the distal end, causing the distal end of the functional LES to move cephalad. Though these subjects are asymptomatic, they are not "normal." Most do not have significant LES failure, abnormal reflux by present criteria, symptoms, or complications of GERD. But they do have a damaged LES. Most of them are in the phase of compensated LES damage.

The new test for LES damage is based on the histologic measurement of the dilated distal esophagus [i.e., cardiac epithelium with and without parietal and goblet cells between the squamocolumnar junction (SCJ) and proximal limit of GOE]. This is an accurate measure of LES damage to micrometer level. The formula for the residual length of the abdominal LES now becomes:

Functional (residual) length of abdominal LES = 35 mm − LES damage (measured by new histologic test)

For example, if the new test shows LES damage of 15mm, the residual functional abdominal LES length will be 35–15mm = 20mm (Fig. 17.1). The pathologic measure that defines LES damage is the length of cardiac epithelium (with and without parietal and/or goblet cells) between the endoscopic GEJ [= SCJ in persons without visible columnar-lined esophagus (CLE)] and the true GEJ (the proximal limit of GOE) (Fig. 17.1A and B).

The pH test in the volunteers in Zaninotto et al. showed a range of esophageal acid exposure (i.e., pH < 4) from 0% to 6% during the 24h (Table 17.2). The percent time pH < 4 was >4.6% in 5% of these volunteers and >5.8% in 2.5%. This was the rationale in establishing the normal value for the 24-h pH test as <4.5% acid exposure (i.e., pH < 4). This was the point at which the test could discriminate between "normal" defined as the lack of symptoms and patients with symptomatic GERD, 95% of whom had an acid exposure >4.5%.[2] A "normal" acid exposure expressed as pH < 4 for 4.5% of the time in the thoracic esophagus is not normal; it just correlates with symptomatic GERD.

It is reasonable to suggest that the normal state is the absence of any reflux, as was seen in those volunteers in this study with zero reflux. In the study of asymptomatic volunteers by Robertson et al.,[6] zero reflux was common. 51 asymptomatic volunteers were tested for 30min after a 12-h fast. The acid exposure (i.e., pH < 4) at a point 5cm above the upper border of the LES was 0%–0.2%. When these subjects were given a full meal and retested in the postprandial period, the acid exposure was 0%–1.7%.

TABLE 17.2 Length of the Abdominal Segment of the Lower Esophageal Sphincter (LES) and Esophageal Acid Exposure in the 24 h pH Test in the 50 Volunteers in Zaninotto et al.

	Length of Abdominal Segment of the LES (cm)	Acid Exposure in 24-h pH Test (% Time pH < 4)
Minimum	0.8	0
Maximum (excluding 5 cm outlier)	3.5	6
Mean	2.18	1.57
Standard deviation	0.72	1.47
Median	2.2	1.1
95 percentile	>1.1	<4.6
97.5 percentile	>0.89	<5.8

Blue, highlighted means and medians for abdominal LES length and acid exposure.

TABLE 17.3 Sequence of Progression of Damage to the Abdominal Segment of the Lower Esophageal Sphincter (LES) From No Damage to Damage Sufficient to Cause LES Failure

LES Status	Functional Abdominal LES Length (Manometry)	LES Damage (New Pathologic Test)	Esophageal Acid Exposure (% Time)
Normal	Initial (35 mm)	0	0
Compensated	35–20 mm	>0–15 mm	0%–1.1%
Subclinical LES damage	20–15 mm	15–20 mm	>0–4.5
Abnormal (clinical GERD)	15–10 mm	20–25 mm	>4.5
Severely abnormal	<10 mm	>25 mm	Much >4.5

GERD, gastroesophageal reflux disease.

If we take zero reflux as normal, then the range of acid exposure seen in the study of Zaninotto et al.[2] would represent increasing amounts of abnormal reflux even though this increased amount did not reach the defined abnormal level or cause symptoms of GERD.

In contrast to the "normal" volunteers, 80% of the 622 patients in the study with an abdominal LES length of <1 cm had an increased esophageal exposure to gastric acid. This contrasted with the finding that only 5% of the 50 volunteers in the study had >4.6% time pH<4 in the esophagus in the 24-h pH study.

Though the study data do not permit correlation between the length of the abdominal LES and pH study results in individual patients, it would be logical that the 5% of patients with an abdominal LES length 0.8–1.1 would have been those most likely to have abnormal amounts of reflux in the pH test.

If true, the pH test can be a method of quantitating LES failure beginning with 0% acid exposure through a stage where the acid exposure is in the range considered normal at present (i.e., >0%–4.5%) to abnormal (>4.5%). Abnormal in turn can be quantitated in increasing amounts of acid exposure (Table 17.3).

The most interesting new concept that emerges in this classification based on LES damage is the possibility that a limited amount of LES damage may occur without any thoracic esophageal acid exposure in the pH test. In Table 17.3, I have hypothesized that amount to be <15 mm of LES shortening. This is based on the assumption that the initial length of the abdominal LES is 35 mm and <15 mm LES damage still leaves a residual functioning abdominal LES of >20 mm. The data for dynamic shortening of the LES during a heavy meal suggest a maximum of 10 mm. This would mean that these patients will retain a functional abdominal LES of 10 mm even with the gastric distension of the heavy meal. This is above the <10 mm length that is recognized to be associated with LES failure.

Of course, such hypotheses have a possibility of being false when put to the test. The assumption of a uniform initial length of abdominal LES in the population is very likely to have some limited and unknown individual variation. With the new ability to measure LES damage histologically, the initial length can be defined with greater accuracy in the future.

2.3 Manometric Measurement of Residual Length of the Abdominal Lower Esophageal Sphincter

The only available method to assess the LES at present is by manometry. Pathology has no method of identifying the intact LES. The initial manometric tests were done with a slow motorized pull through of a pressure sensor.[10] This provides a tracing of the pressures in the full length of the LES and permits accurate measurement of the total length, respiratory inversion point, mean pressure, and abdominal length.

This older method has been largely replaced by high-resolution manometry. This uses multiple pressure sensors located at intervals of 10 mm on the catheter. This provides a pressure tracing of the entire esophagus and stomach over a time period that permits accurate definition of resting pressure in the esophagus, LES and stomach, and the nature of peristalsis. The main present use of high-resolution manometry is the detailed classification of motility disorders of the esophagus. It has little or no role in the diagnosis of early GERD. It can identify a severely defective LES.

The ability of high-resolution manometry to measure the length of the LES accurately is believed to be limited in its accuracy. However, it is possible that the lack of clinical demand of manometry as a practical method of diagnosis of GERD may be responsible for underutilization of the technology. It is possible that if there is a newly recognized clinical significance of abdominal LES length between 10 and 35 cm (presently called "normal"), a concerted effort may result in greater accuracy of measurement.

The inherent inaccuracy of high-resolution manometry is based on the spacing of its pressure sensors at 10 mm intervals. The abdominal LES can be considered to be precious real estate in the prevention of reflux; very small changes in abdominal LES length are highly significant. Attempting to measure small changes within a 25 mm length with pressure sensors placed at 10 mm intervals is optimistic at best.

There is an inbuilt computer program that applies a correction to the measurement of the LES length, in essence averaging the readings between two sensors, but this does not solve the problem adequately.

It is probable that measurement of the abdominal LES by high-resolution manometry is not adequately accurate to be applied with confidence in the assessment of LES damage. It is adequate to detect a defective LES where the abdominal LES length is <10 mm.

When the new test for LES damaged becomes available, the actual level of accuracy of assessing the residual abdominal LES length can be verified by using the formula: *Residual abdominal LES length by manometry = 35 mm (assumed initial length of abdominal LES) − measured length of LES damage by the new histologic test.* If this test shows that manometry is not adequately accurate, new research can be directed to improving accuracy, e.g., by increasing the number of pressure sensors across the region of the LES.

2.4 Comprehensive Lower Esophageal Sphincter Assessment

The complete assessment of the LES is the sum of the residual LES as measured by manometry + LES damage measured by the new histologic test. The two methods of assessment are complementary; pathology cannot measure the intact LES, and manometry cannot measure the damaged LES.

Manometry measures only the LES that remains functional. LES damage at its distal end cannot be measured or detected by manometry. When damaged, the intramural pressure in the area of the damaged LES disappears and the intraluminal pressure only is expressed. The intraluminal pressure of the abdominal esophagus that has lost LES pressure is identical to intragastric pressure. The damaged LES is manometrically identical to the stomach.

At manometry, the distal end of the LES is defined as the point where the high pressure of the LES decreases to intragastric pressure. This is the distal end of the functional (residual) LES. The damaged LES distal to the functional LES is identical to, and usually misinterpreted as, the stomach at manometry (Figs. 17.2 and 17.3).

There is a gap between the distal end of the LES at manometry and the proximal end of the stomach; this is the dilated distal esophagus where the LES has been damaged. Manometry cannot define the true GEJ (proximal limit of GOE) where the original abdominal esophagus and LES ended.

If one assumes that the initial length of the abdominal LES is 35 mm, then one can use the manometric measurement to calculate LES damage. For example, in Fig. 17.1A, which is a normal LES by manometry, the measured abdominal length is 22 mm. While this is >10 mm that defines an LES that is not defective, using the formula would indicate that there is LES damage of 35–22 mm = 13 mm. This abdominal LES damage is within the range of the phase of compensated LES damage. The abdominal LES is sufficiently long to be competent, but is damaged within its reserve capacity. In contrast, in Fig. 17.1B, the sphincter is defective with a residual abdominal length at manometry being zero. This would indicate abdominal LES damage of 35 mm, well into the severe GERD stage in the new method.

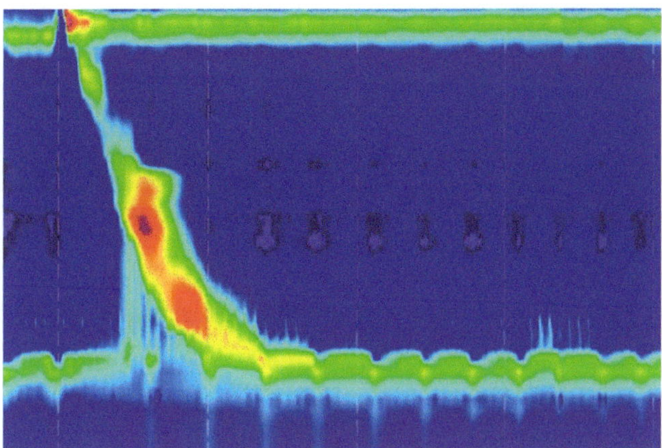

FIGURE 17.2 High-resolution manometry showing a "normal" lower esophageal spincter. This had a total length of 3.4 cm and an abdominal length of 2.2 cm. In the new method, the distal end of the manometric lower esophageal sphincter (LES) is proximal to the true gastroesophageal junction (proximal limit of rugal folds) by the length of the dilated distal esophagus. Assuming an initial length of abdominal LES as 35 mm and the manometric measurement is accurate, there is a dilated distal esophagus 13 mm long distal to the manometric end of the residual LES. This manometric finding is similar to the person depicted in Fig. 17.1.

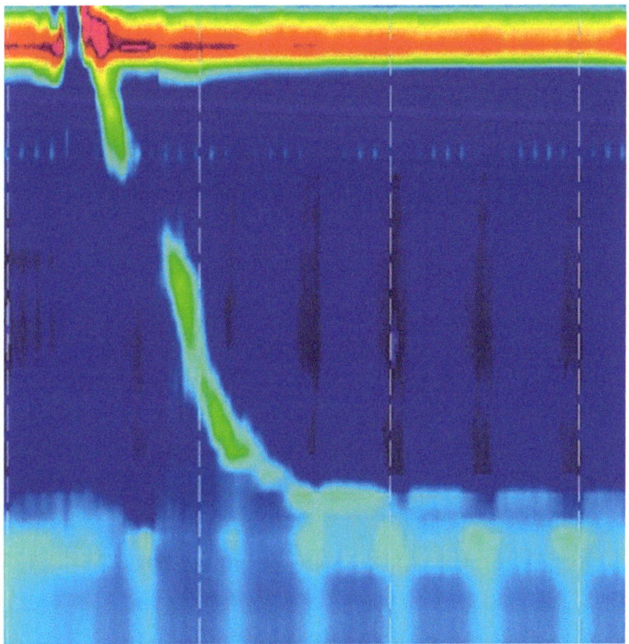

FIGURE 17.3 High-resolution manometry showing a markedly abnormal lower esophageal sphincter (LES) with a total length of 1.4 cm and an abdominal length of 0. The fact that the distal end of the manometric LES is above the diaphragm is taken as evidence of the presence of a hiatal hernia. This is false. An abdominal LES length of zero means there is a dilated distal esophagus (i.e., LES damage) 35 mm in length, bringing the true gastroesophageal junction below the diaphragm.

With the individual variation of the initial length of the abdominal LES being unknown and the manometric assessment of the residual abdominal LES being classified as "normal" versus "defective," there is no attempt at the present time to utilize manometry in the assessment of GERD.

This changes significantly with the new ability to measure LES damage. The two measurements together will provide a complete assessment of the LES. The value of this is potentially immense in the diagnosis and management of GERD. The accuracy of the new measurement of LES damage by pathology will, with testing of large numbers of people, allow accurate determination of (1) the amount of individual variation in the initial length of the abdominal LES and (2) accuracy of the measurement of the abdominal LES by high-resolution manometry.

2.5 Mechanism of Lower Esophageal Sphincter Damage

In Chapter 9, I described the mechanism of LES damage, which is exposure of the esophageal squamous epithelium to gastric acid during periods of gastric overdistension,[6,11] eventually resulting in cardiac metaplasia.

When cardiac epithelium replaces squamous epithelium, the mechanism that maintains LES tone fails. The exact manner in which this happens is unknown. I have suggested that columnar metaplasia interrupts the afferent limb of a local neural reflex that maintains muscle tone because of the loss of normal intraepithelial nerve endings in the squamous epithelium.

Loss of LES tone, which is concordant with cardiac metaplasia, results in the dilated distal esophagus and migration of the SCJ cephalad (Fig. 17.1). The endoscopic GEJ also moves cephalad as the dilated distal esophagus develops rugal folds (see Chapter 16).

In this person with compensated LES damage, the anatomic and histologic elements from proximal to distal are squamous epithelium in the entire thoracic esophagus, the diaphragm, squamous epithelium in the abdominal esophagus up to the endoscopic GEJ, the endoscopic GEJ (SCJ, end of the tubular esophagus and proximal limit of rugal folds), the dilated distal esophagus (lined by cardiac and oxyntocardiac epithelium; length > 0–15 mm), and the true GEJ (proximal limit of GOE) (Fig. 17.1). The present errors in interpretation of this anatomic reality are the incorrect definition of the endoscopic GEJ and the belief that the cardiac epithelium lining the dilated distal esophagus is the normal proximal stomach.

As LES damage exceeds 15 mm and progressively increases to 35 mm (complete damage of the abdominal LES), LES failure begins and reflux episodes progressively increase in frequency and very probably volume (see Chapter 16). The dilated distal esophagus lengthens concordant with LES damage.[7]

The following additional changes occur with LES damage >15 mm: (1) reflux induced damage in the thoracic esophagus (microscopic changes followed by erosive esophagitis); (2) intestinal metaplasia (IM) in the dilated distal esophagus, resulting from elongation of the dilated distal esophagus to >15 mm; (3) sliding hiatal hernia; and finally (4) visible CLE in the thoracic esophagus, progressing to intestinal metaplasia, dysplasia to adenocarcinoma. These changes represent progression of disease at the cellular level. Refractory GERD, where proton pump inhibitor (PPI) therapy does not control symptoms, visible CLE, Barrett esophagus and adenocarcinoma are the irreversible end points of this stage.

2.6 Pathologic Assessment of the Lower Esophageal Sphincter

Unfortunately, the functional and damaged LES cannot be defined by its muscular structure. There is no gross thickening or anything distinctive at histology of the muscle that is different than the non-LES esophageal muscle. Complex arrangements of the muscle fibers that traverse the GEJ have been described that suggest there is a structural basis for the sphincter action.[12] However, even if true, this is not something that is ever done in routine pathology practice.

The pathologist is blind to the position of the LES at routine examination of esophagogastrectomy specimens and at autopsy. There is no method at the present time that is capable of measuring any component of the LES by endoscopic biopsies.

3. PATHOLOGIC MEASUREMENT OF THE DILATED DISTAL ESOPHAGUS

The new method of pathologic assessment depends on the fact that LES damage is associated with pathologic changes that can be recognized. While the pathologist cannot recognize the normal LES that is lined by squamous epithelium, he/she can measure the cardiac epithelium (with and without parietal and/or goblet cells) that lines the dilated distal esophagus. This measurement is equal to the length of abdominal LES damage (= shortening of the abdominal LES).[7]

Histologic assessment of the length of cardiac epithelium (with and without parietal and/or goblet cells) distal to the endoscopic GEJ is the only way that the damaged LES can be measured.

The damaged LES cannot be measured by manometry, endoscopy, or gross examination of a resected specimen or at autopsy. With all of these modalities, the dilated distal esophagus is indistinguishable from the stomach. The dilated distal esophagus is distal to the distal end of the manometric LES and has a resting pressure equal to the stomach; it is dilated and has taken the contour of the stomach, lies distal to the endosopic GEJ, distal to the end of the tubular esophagus, and has rugal folds. It is presently mistaken as being the proximal stomach by everyone except those pathologists who know how to interpret the histology of this region. The number of such pathologists in the world is very small at this time.

3.1 Measurement of the Dilated Distal Esophagus at Autopsy

The measurement of the dilated distal esophagus can be made at autopsy.[8] This requires taking sections from the full circumference of the end of the endoscopic GEJ (the SCJ in persons without visible CLE) extending distally for a length of

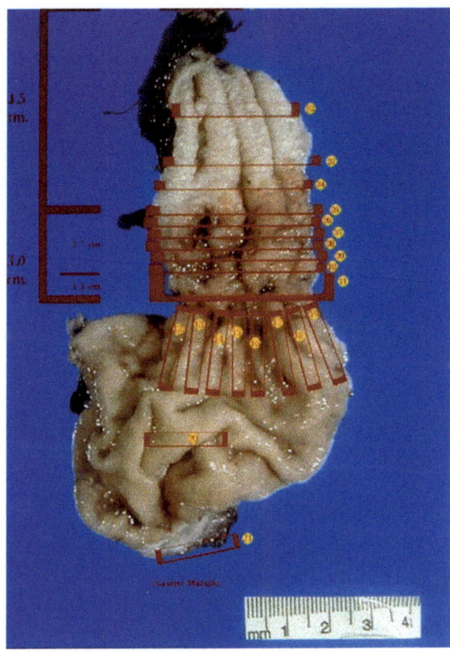

FIGURE 17.4 Method of sectioning of a resected esophagogastrectomy specimen. The specimen is cut horizontally at the end of the tube (the gross definition of the gastroesophageal junction). The entire circumference of the area distal to this is cut vertically into 2 mm sections for microscopic evaluation that allows measurement of the dilated distal esophagus (= abdominal lower esophageal sphincter damage).

35 mm. This specimen will reach the proximal limit of GOE. Inking the specimen for orientation is unnecessary because the histology will permit recognition of proximal from distal ends. Ideally, the specimen is pinned out on a cork board and fixed overnight in formalin. After fixation, vertical sections are taken, numbered, and separately submitted, permitting mapping the epithelia of the entire area.

In a person with no visible CLE, the proximal limit of this section will be proximal to the SCJ containing the distal squamous epithelium, extending distally in a vertical orientation. In such a person, the proximal limit of the measurement will be the distal edge of the squamous epithelium. The distance from this point to the proximal limit of gastric oxyntic mucosa is the dilated distal esophagus (Fig. 17.1).[13]

The dilated distal esophagus can be measured accurately using an ocular micrometer. It is lined by cardiac epithelium (with and without parietal and/or goblet cells). This can be differentiated from squamous epithelium proximally and GOE distally. Differentiating cardiac epithelium (with and without parietal and/or goblet cells) from squamous epithelium at the proximal end is easy. Differentiating oxyntocardiac epithelium, which occurs in the distal region of the squamooxyntic gap, from GOE requires a little training, aided in difficult cases by the use of a digested periodic acid Schiff (PAS) stain (see Chapter 5).

In a person with a visible CLE, a horizontal line is drawn across the end of the tubular esophagus, which is the gross pathology definition of the GEJ. In these cases, the point where the dilated distal esophagus begins is the proximal edge of the section. The distal end is the proximal limit of GOE.

The specimen is usually taken after the body has been refrigerated for a variable time after death. The changes in length of this region with refrigeration of the body, time after autopsy, and dissection are unknown.

3.2 Measurement of the Dilated Distal Esophagus in Resected Specimens

The required sections in resected esophagogastrectomy specimens are identical to those taken at autopsy. The specimen is separated at the end of the tubular esophagus. Vertical full thickness sections are taken from the entire circumference of the 35 mm distal to the end of the tube (Fig. 17.4). The length of cardiac epithelium (with and without parietal and/or goblet cells) between the distal limit of squamous epithelium (in the person without visible CLE) and the GOE is measured (Fig. 17.1). In persons with visible CLE, the end of the tubular esophagus and/or the proximal limit of rugal folds is used as the proximal limit of the dilated distal esophagus. The distal limit is always the proximal limit of GOE.

In both autopsy and resection specimens, the sections are transmural, including mucosa, submucosa, the muscle wall, and adventitia. These sections permit identifying and mapping the epithelial types at the surface and the presence of submucosal glands. The presence of submucosal glands and their ducts that traverse the mucosa is absolute evidence of esophageal location of the overlying epithelial type. Submucosal glands do not occur in the human stomach.

In our autopsy study done in 2000,[8] I did not have the knowledge or wisdom to search for submucosal glands and document their presence. However, in our study of resection specimens in 2007,[7] the presence of submucosal glands was documented and found to be present to a length concordant with the length of the dilated distal esophagus. This proves that the dilated distal esophagus is not the stomach as is believed at present. In a similar study, Sarbia et al.[9] also reported the presence of submucosal glands in the dilated distal esophagus.

The resected specimen is different than the autopsy specimen and these differences can potentially lead to differences in the measured length. The resection requires separation of the esophagus and stomach at the two ends in the live patient. The contraction of muscle can lead to changes in mucosal length. The existence and magnitude of such contraction is presently unknown. They can result in an unknown but potentially significant difference between the measurements of the dilated distal esophagus at autopsy compared with resected specimens. Both measurements can in turn be different than measurements made in biopsies taken in vivo at endoscopy.

3.3 Measurement of the Dilated Distal Esophagus in Endoscopic Specimens

The devices available for biopsy at present are not designed to obtain the optimal sample required for accurate measurement of the dilated distal esophagus at endoscopy. While a variety of devices are available for sampling the dilated distal esophagus, none are suitable.

Endoscopic submucosal dissection, where the endoscopist accesses the submucosa and operates in that plane, can dissect out a large, full thickness, mucosal specimen. However, it has several limitations for this purpose: (1) It is a complicated and difficult procedure that is presently limited to a relatively few academic centers; (2) It is a time-consuming procedure that requires a considerable amount of training and expertise; (3) It has a higher complication rate than biopsies; (4) It does not allow accurate orientation; and (5) It is unnecessarily too deep. As such, it is highly unlikely to be used to obtain a mucosal specimen for a diagnostic test for GERD. At present, endoscopic submucosal dissection is used to treat early esophageal cancer.

Endoscopic mucosal resection is similarly used to obtain a circular specimen of mucosal and submucosa measuring up to 20 mm in size. This procedure essentially lifts the mucosa and submucosa into a cup by suction and cuts the tissue out in the plane of the deep submucosa. Endoscopic resection is also limited to the treatment of early cancers of the esophagus. Like endoscopic submucosal dissection, the procedure is limited to academic centers, requires expertise, takes a longer time, and has a higher complication rate than biopsies. For these reasons it is unlikely to be used for a simple diagnostic test.

Biopsy forceps are too small to measure the entire extent of the dilated distal esophagus except in people with a very short segment of cardiac epithelium (with and without parietal and/or goblet cells). The regular biopsy forceps takes a 4 mm specimen.

Robertson et al.,[6] using a jumbo (8 mm) forceps, measured cardiac epithelial length in asymptomatic volunteers. Persons with and without central obesity had 2.50 and 1.75 mm of cardiac epithelium distal to the SCJ. This is not the entire dilated distal esophagus, which includes both cardiac and oxyntocardiac epithelium. In Robertson et al.,[6] oxyntocardiac epithelium was not included in the measurement. It is likely that the 8-mm biopsy forceps used did not reach the proximal limit of GOE in many of these patients.

4. A MULTILEVEL BIOPSY PROTOCOL TO MEASURE THE DILATED DISTAL ESOPHAGUS

The only way the dilated distal esophagus can be measured with present biopsy forceps is to employ a multilevel biopsy protocol distal to the endoscopic GEJ. I will outline this below. It has serious limitations. A multilevel biopsy protocol distal to the endoscopic GEJ has been used to study patients on three occasions, two of which are published and one personal: (1) Jain et al.[14] took biopsies at the endoscopic GEJ, 1 and 2 cm distal to the endoscopic GEJ; (2) Ringhofer et al.[15] took biopsies at the endoscopic GEJ, 0.5 and 1 mm distal to the endoscopic GEJ.

In both studies, the objective of the biopsy protocol was to see how many patients had intestinal, cardiac, oxyntocardiac (in Ringhofer et al.[15]), and GOE at the multiple different levels distal to the endoscopic GEJ. They did not attempt to measure the length of cardiac epithelium (with and without parietal and/or goblet cells) in the squamooxyntic gap: (3) at my own endoscopy (I requested that the gastroenterologist add an upper gastrointestinal endoscopy to my screening colonoscopy), I instructed my gastroenterologist to take multilevel biopsies according to the protocol outlined below. This was successful in providing a measure of the dilated distal esophagus only because I had a dilated distal esophagus <5 mm that was captured entirely in the first biopsy level.

Apart from my biopsy series, no one has yet used a biopsy protocol at endoscopy to measure the length of cardiac epithelium (with and without parietal and/or goblet cells) distal to the endoscopic GEJ. This is solely because no one knows

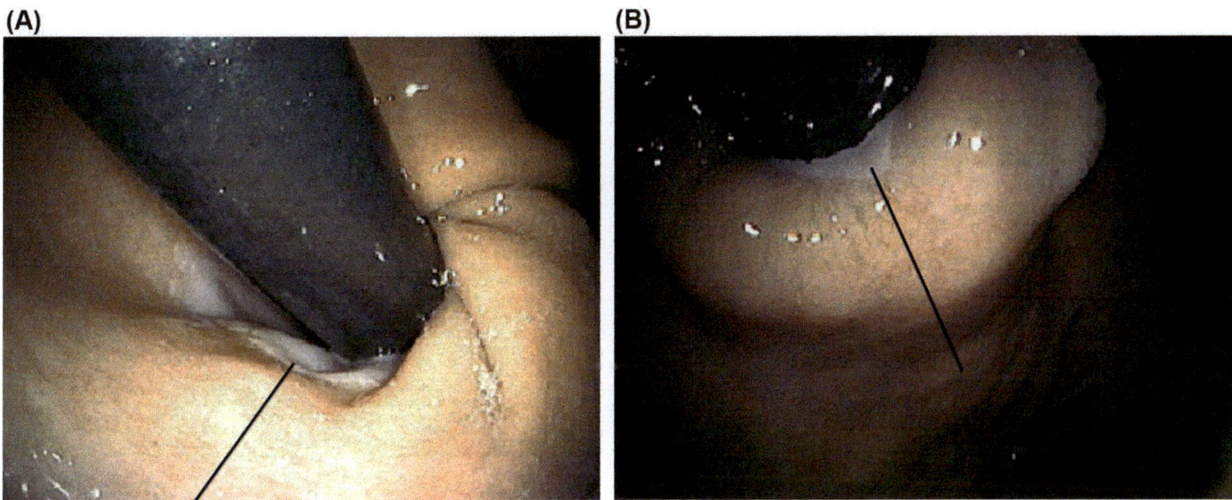

FIGURE 17.5 (A) The junctional region seen on retroflex view at endoscopy. The squamocolumnar junction (SCJ) and the area distal to the SCJ represent the target area for biopsies to evaluate the length of the dilated distal esophagus. (B) The SCJ on retroflex view is on a mucosal fold. *Photographs: Courtesy of Dr. Martin Riegler, Reflux Medical, Vienna, Austria.*

the importance of this measurement at present. Whether it is possible to develop a biopsy method or not will depend largely on the ingenuity of endoscopists. I will describe the method I think will best achieve this, but I am not an endoscopist and therefore not the best person to develop a protocol.

In developing a protocol, I will assume the use of a standard endoscope, a standard 4-mm biopsy forceps, and standard processing of the samples with well-established histologic methods of diagnosis of histologic types. Larger (6–8 mm) biopsy forceps are available and may decrease the number of biopsies needed, but are less familiar to most endoscopists.

All biopsy forceps sample only the mucosa. The depth of the biopsy is commonly partial thickness of the mucosa with some biopsies reaching muscularis mucosae. A reasonable partial thickness of the mucosa as seen in most mucosal biopsies at present is an adequate sample to classify the epithelial type into squamous, cardiac with IM, cardiac, oxyntocardiac, and gastric oxyntic.

4.1 The New Biopsy Protocol

The new biopsy protocol is to be used primarily for those patients without a visible CLE at endoscopy. Patients who have a visible CLE at endoscopy have already reached the irreversible state as defined. The presence of visible CLE is a better criterion of severe LES damage than any histologic measurement.

The level of each biopsy level will be defined in relation to the SCJ, which is a highly reliable landmark (Fig. 17.5). In patients without visible CLE, the SCJ is at the level of the proximal limit of the rugal folds (the endoscopic GEJ). In patients without a hiatal hernia, this is also the end of the tubular esophagus. The biopsy method is probably not necessary in people with hiatal hernia, another sign of severe LES damage without the need for measurement.

The area distal to the SCJ is visualized in retroflex position with sufficient air insufflation to make the SCJ visible in this view (Fig. 17.5A). The SCJ is reference point zero. If there is folding of the mucosa, this should be leveled out as much as possible with insufflation (Fig. 17.5B). Using the open 4-mm biopsy forceps as a guide, the 25 mm length distal to the SCJ is divided into five 5-mm horizontal cylindrical segments (0–5 mm, 5–10 mm, 10–15 mm, 15–20 mm, and 20–25 mm) (Fig. 17.6). It is likely that the 20–25 mm level is unnecessary because this amount of abdominal LES damage is already too advanced for application of this method. The new management method has, as its objective, preventing advancement of abdominal LES damage reaching 25 mm. At 25 mm, the disease is either irreversible or close to it.

Biopsy sets are taken separately within each of these four segments. I believe that one vertical row of biopsies will be adequate (Fig. 17.6). However, in an initial testing phase, 4-quadrant biopsies may be necessary to define the position in the circumference of the SCJ that provides the most accurate measurement. The following specimens will be taken, appropriately labeled, and placed separately in a standard, formalin-filled specimen bottle for transport to the pathology department.

Specimen A: The first biopsy set is taken from the first cylindrical segment with the proximal limit of the forceps just straddling the SCJ and the distal limit of the forceps within 5 mm distal to the SCJ. The method of taking this biopsy is described well in the methods section of Robertson et al.[6]: "Junctional biopsy specimens were taken perpendicular

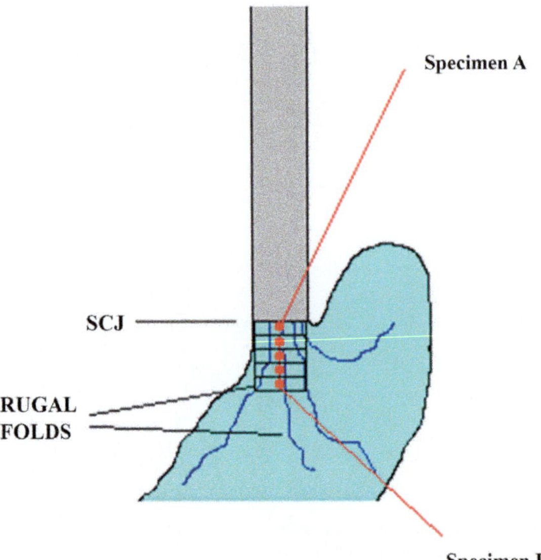

FIGURE 17.6 Diagrammatic representation of the multilevel biopsy protocol designed to assess the length of the dilated distal esophagus. The segment distal to the squamocolumnar junction (SCJ) is divided into 5-mm horizontal segments. Biopsies are taken in one vertical row along the lesser curvature within these 5-mm segments. These are submitted separately as specimens A through E.

to the squamocolumnar junction and targeted to include just enough squamous mucosa at the proximal end to confirm positioning."

Specimen B: The second biopsy set is taken from the second (5–10 mm) cylindrical segment with the proximal limit of the forceps immediately distal to the first biopsy and the distal limit of the forceps just within 10 mm distal to the SCJ as assessed by the open biopsy forceps.

Specimen C: The third biopsy set is taken from the third (10–15 mm) cylindrical segment with the proximal limit of the forceps immediately distal to the second biopsy and the distal limit of the forceps just within 15 mm distal to the SCJ.

Specimen D: The fourth biopsy set is taken from the fourth (15–20 mm) cylindrical segment with the proximal limit of the forceps immediately distal to the third biopsy and the distal limit of the forceps just within 2.0 cm distal to the SCJ.

I will limit the discussion to this protocol. A Specimen E can be taken from the fifth cylindrical segment distal to the SCJ (20–25 mm), but this is practically not likely to be necessary.

While not required for the assessment of the sphincter, it is recommended that an additional sample be taken from the distal stomach, two biopsies each from body and antrum. This will allow the detection of gastritis and *Helicobacter pylori* infection and help in the interpretation of the protocol biopsies in patients who have pathology in the stomach.I recognize that taking biopsies from the region of the dilated distal esophagus presents significant practical difficulties. For example, bleeding obscures the view as soon as the first biopsy is taken. With care, however, the error in the biopsy can be minimized to a point where the information may be sufficiently accurate for assessment.

However, there is no doubt that this biopsy protocol will not be acceptable to gastroenterologists at large because it is so labor intensive, time-consuming and expensive, generating five pathology specimens per patient. It also produces five mucosal specimens that are unoriented to each other, precluding accurate measurement. This creates an inherent error in the measurement of at least 2.5 mm. The following discussion is more for illustrative theoretical understanding of the method rather than suggesting a practical method.

4.2 Pathologic Examination of the Biopsies

The biopsies are placed in formalin containers and processed routinely for histology. The biopsy set from each level (A–D) is processed as a separate specimen. Three levels are cut from each biopsy set and stained with hematoxylin and eosin (routine) stain. It is recommended that a PAS after digestion with amylase (Di-PAS) be used as a special stain. This facilitates differentiation of cardiac, oxyntocardiac, and gastric oxyntic epithelia (see Chapter 4). The Di-PAS stain is a very commonly used stain in any pathology laboratory anywhere in the world.

The requirements of the pathologist in his/her examination of the specimens are explained in the following sections.

4.2.1 Define Adequacy and Accuracy of the Biopsy Method

The adequacy of mucosal biopsies in depth and quantity of mucosa present is not likely to be a problem. It is unusual for mucosal biopsies to not have the depth to permit accurate classification of mucosal types.

The question of accuracy with which the operator follows the biopsy protocol is entirely another matter. The intent is to procure biopsies that are separate in their vertical orientation, i.e., the entire extent of one biopsy is proximal or distal from the entire extent of the immediately adjacent biopsy without overlap. If this is true, then the biopsies should permit identification of the location of the true GEJ (the junction between metaplastic oxyntocardiac epithelium, which is *always* the most distal of the three metaplastic columnar epithelia, and GOE).

In most cases, this will mean that one level (specimen) will contain both cardiac (with and without parietal and/or goblet cells) and GOE. Biopsies proximal to this level will contain only cardiac epithelium (with and without parietal and/or goblet cells), and biopsies distal to it will contain only GOE.

In a less common situation, the true GEJ will not be present in one specimen; the transition will be between two specimens. In this event, all biopsies proximal to that point will have cardiac epithelium (with and without parietal and/or goblet cells), and all biopsies distal to that point will have only GOE.

If either of these distributions is present in the protocol biopsies, it can be assumed that the biopsies have been taken according to protocol with accuracy.

If a mixture of cardiac epithelium (with and without parietal and/or goblet cells) and GOE is present in more than one level, it will indicate that the biopsies have been taken in such a way that they overlap. This seriously complicates interpretation. The probability of overlap increases when multiple biopsies are taken from each level at different points in the circumference. It may be more effective to take biopsies in one vertical line. However, bleeding associated with such a protocol may create a significant error.

Even as I write this, I am led to despair that this biopsy protocol will not be followed accurately or consistently, especially in the larger endoscopy community at large. The probability of variation in the care and time taken to follow the biopsy protocol is likely to be high. This will limit practical application of the new method and initial studies of its usefulness will likely fail, not because the diagnostic method is flawed, but because of operator error in the way the biopsies are taken. I would discourage any attempt to do this until appropriate biopsy technology is developed.

4.2.2 Define the Most Proximal Biopsy Set That Contains Gastric Oxyntic Epithelium

The following interpretation should be limited to those biopsy sets that have satisfied the criteria for accuracy that I have defined above. Any series of biopsies with cardiac epithelium (with and without parietal and/or goblet cells) and GOE at more than one level must be rejected. Otherwise it will contaminate data and make the interpretation of the method confusing and unintelligible.

The first step for the pathologist is to accurately differentiate metaplastic oxyntocardiac epithelium and GOE. Both have parietal cells. Oxyntocardiac epithelium is thinner, is less organized with lobular architecture of the glands, has more lamina propria and chronic inflammatory cells between the glands; mucous cells are interspersed with parietal cells in the glands, seen easily in the PAS stain. GOE contains straight tubular glands composed of parietal and chief cells without mucous cells (PAS stain is positive only in the surface and foveolar pit cells) and has minimal lamina propria and inflammatory cells.

With directed attention to the problem, some experience, and the use of the Di-PAS stain, differentiation of metaplastic oxyntocardiac and normal GOE has a high level of accuracy. Again, sloppy pathology can be a problem. Mistaking oxyntocardiac epithelium for GOE will underestimate LES damage. Small variations in measurement can produce significant error in the method. Compulsive pathology is critical. At least in the initial stage of testing the method in clinical trials to evaluate utility, the biopsies must be read in a centralized place, preferably by me or one of many pathologists who have worked with me in applying this method. Later, after the method has been validated, it will move into the community with some requirement of training for pathologists. This should not be difficult; the method is easy to learn as long as the pathologist is committed.

In patients with chronic atrophic gastritis, distinction between oxyntocardiac and atrophic GOE can be difficult. The presence of biopsies from the distal stomach is very useful for this differential diagnosis (see Chapter 5).

The biopsy set that contains a mixture of cardiac epithelium (with and without parietal and/or goblet cells) and GOE is now identified. In the rare event where cardiac epithelium (with and without parietal and/or goblet cells) and GOE are completely separated in all biopsy sets, the biopsy set that consists entirely of cardiac epithelium (with and without parietal and/or goblet cells) is noted.

4.2.3 Estimate the Percentage of Gastric Oxyntic Epithelium in the First Biopsy Set in Which It Is Encountered

The accuracy of measurement by the biopsy protocol is limited by the lack of vertical orientation and 5-mm intervals of the biopsies. Any grade (see below) will therefore have a range of 5 mm. For example, grade 2 is abdominal LES damage of 5–10 mm. In the evaluation of the patient's GERD severity and prediction of future progression of sphincter damage, this 5-mm difference is highly significant and will create considerable uncertainty.

With this biopsy protocol, a correction can be applied to the measurement provided by the biopsy interval by estimation of the percentage of GOE in the most proximal biopsy set that has GOE. This will provide information as to whether the patient's true GEJ is nearer the upper or lower limit of the range for the grade.

It is suggested that the pathologist defines the percentage of GOE in the first biopsy that it appears. The percentage will be divided into <33%, 33%–66%, >66% or 100% GOE. This will permit subdivision of each grade into three subgrades. For example, grade 2 (which, on the whole represents 5–10 mm of LES damage) will be divided into grade 2-minimum (5 mm LES damage) when the entire specimen consists of GOE; 2-low (6.25 mm LES damage) if GOE is >66%, grade 2-mean (7.5 mm LES damage) if GOE is 33%–66%, grade 2-high (8.75 mm of LES damage) if GOE is <33%.

If the percentage of GOE in the critical biopsy is not provided in the pathology report, the default value will be the mean value. For example, grade 2 will be shortening of 5–10 mm (mean −7.5 mm). Present high-resolution manometry measures LES length by the mean between two sensors separated by 10 mm. The error is similar to or greater than this biopsy protocol with this correction if the percentage of GOE is not used.

Any time complicated calculations need to be made to correct an obvious problem in measurement, there is a probability of significant error. However, with care in the performance of biopsies according to this protocol, I believe that, even with the major limitations associated with this method, the differences in the measured LES damage will provide a much clearer picture of the patient's GERD status and future progression than anything that is available at present. Its potential for precision is much greater with a better biopsy technology.

4.3 The Pathology Report

The pathology report on these biopsies would be succinct and precise. I am reproducing the biopsy report from my own set of biopsies as assessed by me based on the interpretation of findings in my biopsies:

1. **Protocol biopsy accuracy**: confirmed. (The accuracy of the biopsy protocol is shown by the fact that the only GOE was found distal to the first biopsy showing GOE.)
2. **Pathologic findings**
   a. Biopsy at segment 0–5 mm distal to SCJ:
      i. Squamous epithelium showing focal basal cell hyperplasia and dilated intercellular spaces;
      ii. Columnar epithelium consisting of cardiac epithelium + oxyntocardiac epithelium;
      iii. No intestinal metaplasia;
      iv. GOE present and is 20% of total columnar epithelium in biopsy.
   b. Biopsy at segment 5–10 mm distal to SCJ: GOE, normal.
   c. Biopsy at segment 10–15 mm distal to SCJ: GOE, normal.
   d. Biopsy at segment 15–20 cm distal to SCJ: GOE, normal.
3. **Comment**:
   a. Most proximal biopsy level with GOE: Specimen A (0–5 mm).
   b. Percentage of GOE at first biopsy level it is found: 20%.
   c. LES damage grade: 1-high (3.75 mm).
4. **Synoptic report (optional)**
   It will be easy to develop a synoptic report that requires the pathologist to choose between specified options as follows:
   a. Biopsy protocol accurate: **Yes**/No;
   b. Most proximal biopsy level at which GOE is seen (select one): **A (SCJ to 5 mm)**; B (5–10 mm); C (10–15 mm); D (15–20 mm); none of the above;
   c. Percentage of GOE at most proximal level it is seen (select one): **<33%**; 33%–66%; >66%; 100%; no GOE seen at any biopsy level;
   d. IM (select one): Present/**Absent**.

   The advantage of synoptic reporting is that the pathologist will not be able to fail to place critical information in the pathology report.

TABLE 17.4 Grading System for Lower Esophageal Sphincter (LES) Damage Based on Findings in the Four Biopsies at 0–5 mm (A), 5–10 mm (B), 10–15 mm (C), and 15–20 mm (D)

Grade and Amount of LES Damage	Biopsy A (0–5 mm)	Biopsy B (5–10 mm)	Biopsy C (10–15 mm)	Biopsy D (15–20 mm)
0	Sq + GOE	GOE	GOE	GOE
1 (>0–5 mm)	Sq + CE + GOE	GOE	GOE	GOE
2 (5–10 mm)	Sq + CE	CE + GOE	GOE	GOE
3 (10–15 mm)	Sq + CE	CE	CE + GOE	GOE
4 (15–20 mm)	Sq + CE	CE	CE	CE + GOE
5 (>20 mm)	Sq + CE	CE	CE	CE

CE, cardiac epithelium (with and without parietal and/or goblet cells); *GOE*, gastric oxyntic epithelium; *Sq*, squamous epithelium.

4.4 Grading of Severity of Lower Esophageal Sphincter Damage

The severity of LES damage by the above biopsy protocol will be based on the most proximal biopsy showing GOE and the percentage of GOE in that biopsy (Table 17.4).

The above grading system identifies six grades (0–5) with each grade having a range of 5 mm. This results in a potential error of up to 5 mm that is highly significant. The critical range of LES damage in the patient with clinical GERD is from 15 mm (<15 mm LES damage is the phase of compensated LES damage) to 25 mm (>25 is the point of severe LES damage associated with a high risk of irreversible cellular changes, i.e., visible CLE).

Attempting to measure a 10 mm range with a 5 mm error is optimistic at best and foolish at worst. For this reason, I have suggested applying a correction by assessing the amount of GOE at the first level it is found. This assumes that if the true GEJ is closer to the low end of the range of a given grade, the biopsy will show a higher percentage of GOE than if it is closer to the high end of the range.

High-resolution manometry, which attempts to measure the LES with pressure sensors situated 10 mm apart, has a similar potential error. Manometry attempts to minimize this error by providing a computer-based "correction" that calculates the mean between two sensors. That correction does not make the measurement of the LES at high-resolution manometry adequately sufficient to be used in practice as a diagnostic test for GERD. A similar fate probably awaits the suggested "correction" in this biopsy protocol, even though this has more logical sense.

The person without any LES damage will have a direct transition from squamous epithelium to GOE (Fig. 17.7A). The person without any LES damage will show no cardiac epithelium in any of the biopsies. Specimen A will show a transition of squamous epithelium to GOE. All other biopsies will show only GOE (Fig. 17.7B).

In persons with LES damage, the findings in the biopsies allow grading the LES damage (shortening) based on the length of cardiac epithelium (with and without parietal and/or goblet cells) in the following manner (Fig. 17.8):

1. **Grade 0**:
   Definitional criteria: No cardiac epithelium (with and without parietal and/or goblet cells); the SCJ shows direct transition from squamous to GOE. All levels have 100% GOE [except squamous epithelium (Fig. 17.7B)].
   Interpretation: No LES damage. This is extremely rare in adults.
2. **Grade 1**: (Fig. 17.8A)
   Definitional criteria: Cardiac epithelium (with and without parietal and/or goblet cells) present; the most proximal biopsy containing GOE is in Specimen A within 5 mm distal to SCJ.
   Interpretation: LES damage is >0–5 mm.
   Correction (subgrade):
   Grade 1-low (GOE >66% in Specimen A): LES damage is 1.25 mm.
   Grade 1-mean (GOE 33%–66% in Specimen A): LES damage is 2.5 mm.
   Grade 1-high (GOE <33% in Specimen A): LES damage is 3.75 mm.
3. **Grade 2**: (Fig. 17.8B)
   Definitional criteria: Cardiac epithelium (with and without parietal and/or goblet cells) present; the most proximal biopsy containing GOE is in Specimen B within 5–10 mm distal to SCJ.

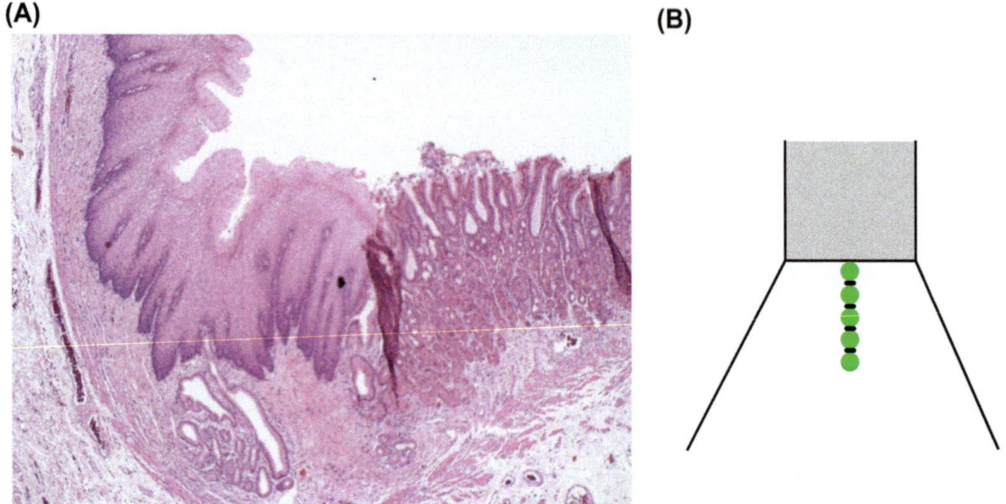

FIGURE 17.7 The normal state: (A) Histologically shown by the absence of cardiac epithelium between squamous epithelium and gastric oxyntic epithelium (GOE); (B) The biopsy specimen A will show a direct transition of squamous epithelium to GOE. All biopsies distal to the squamocolumnar junction (specimens A–E) will be entirely composed of GOE (green). There is no lower esophageal sphincter damage.

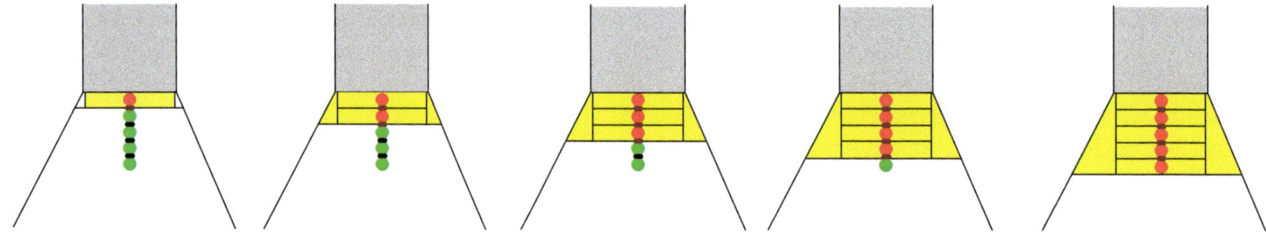

FIGURE 17.8 The presence of cardiac epithelium (red in biopsies; yellow in the dilated distal esophagus) in the biopsies permits assessment of the length of the dilated distal esophagus (=lower esophageal sphincter damage) with significant inaccuracy. (A) Cardiac epithelium present only in specimen A=cardiac epithelial length >0–5 mm. (B) Cardiac epithelium present in specimens A and B=cardiac epithelial length >5–10 mm. (C) Cardiac epithelium present in specimens A, B, and C=cardiac epithelial length >10–15 mm. (D) Cardiac epithelium present in specimens A, B, C, and D=cardiac epithelial length >15–20 mm. (E) Cardiac epithelium present in all specimens A through E=cardiac epithelial length >20 mm.

Interpretation: LES damage is 5–10 mm

Correction (subgrade):

Grade 2-minimum (GOE 100% in Specimen B): LES damage is 5 mm.
Grade 2-low (GOE >66% in Specimen B): LES damage is 6.25 mm.
Grade 2-mean (GOE 33%–66% in Specimen B): LES damage is 7.5 mm.
Grade 2-high (GOE <33% in Specimen B): LES damage is 8.75 mm.

4. **Grade 3**: (Fig. 17.8C)

Definitional criteria: Cardiac epithelium (with and without parietal and/or goblet cells) present; the most proximal biopsy containing GOE is in Specimen C within 10–15 mm distal to SCJ.

Interpretation: LES damage is 10–15 mm.

Correction (subgrade):

Grade 3-minimum (GOE 100% in Specimen C): LES damage is 10 mm.
Grade 3-low (GOE >66% in Specimen C): LES damage is 11.25 mm.
Grade 3-mean (GOE 33%–66% in Specimen C): LES damage is 12.5 mm.
Grade 3-high (GOE <33% in Specimen C): LES damage is 13.75 mm.

5. **Grade 4**: (Fig. 17.8D)

Definitional criteria: Cardiac epithelium (with and without parietal and/or goblet cells) present; the most proximal biopsy containing GOE is in Specimen D within 15–20 mm distal to SCJ.

Interpretation: The abdominal sphincter shortening is 15–20 mm

Correction (subgrade):

Grade 4-minimum (GOE 100% in Specimen D): LES damage is 15 mm.
Grade 4-low (GOE >66% in Specimen D): LES damage is 16.25 mm.
Grade 4-mean (GOE 33%–66% in Specimen D): LES damage is 17.5 mm.
Grade 4-high (GOE <33% in Specimen D): LES damage is 18.75 mm.

6. **Grade 5**: (Fig. 17.8E)
   Definitional criteria: Cardiac epithelium (with and without parietal and/or goblet cells) present; no GOE is found in any biopsy including specimen D.
   Interpretation: LES damage is >20 mm.

The biopsy protocol therefore provides a grading of LES damage from 0 to >20 mm at increments of 1.25 mm. This is a measurement that has never been conceived previously. At present, there is no method of measuring LES damage. The area in which the measurements are taken is *distal* to the endoscopic GEJ, which most physicians, including gastroenterologists and pathologist specialized in GERD, believe is proximal stomach.

The present manometric diagnostic criterion for the abdominal LES that indicates a defective LES is a length <10 mm. The new test with the grading that is developed therefore aims to measure abdominal LES damage *before* it reaches the length at which it is presently recognized as abnormal. In the present understanding, the abdominal LES is considered normal/not abnormal till it reaches a residual manometric length of <10 mm. In the new understanding, the only LES that is normal is one that has retained its original length of 35 mm. The abdominal LES is progressively degraded from that point, slowly shortening from 35 mm to the 20 mm point (the end of the phase of compensated LES damage) where it begins to fail in the postprandial state and then to 25 mm where its residual length is 10 mm, the point where it is presently recognized as a criterion for a defective LES.

The problem in the present understanding of GERD is that the disease is defined at the point of LES damage that signifies LES failure. The entire period of slow degradation of the abdominal LES is regarded as "normal/no GERD." All the criteria for the present diagnosis of GERD (symptoms, manometry, pH/impedance testing) are designed to detect failure of the LES, not LES damage before it fails.

In my own endoscopic biopsies, LES damage is grade 1-high based on the fact that GOE was found in Specimen A and constituted 20% of the volume of columnar epithelium in that biopsy. This is LES damage of 3.75 mm. I have a residual abdominal LES (assuming an initial length of 35 mm) of >30 mm. My LES is perfectly competent. However, I do not have a normal abdominal LES. My LES damage is simply well within the reserve capacity of the LES.

No one has ever measured the dilated distal esophagus in the manner suggested above except the gastroenterologist who did my endoscopy and took the biopsies as I instructed him. No one in the population, except me, knows the status of his/her LES damage.

There is only one study in the literature by Ringhofer et al.[15] where multiple measured biopsies were taken from distal to the endoscopic GEJ and the pathologic examination evaluated cardiac epithelium (with and without parietal and/or goblet cells) as I have suggested above.

Dr. Martin Riegler was an attendee at the USC Foregut Surgery Conference in Hawaii for several years through the turn of the millennium. He was a surgeon in charge of the esophageal surgery unit in the Medical University of Vienna. He became extremely interested in the cardiac epithelium concept and sat down and talked with me for many hours on many beaches in the various islands of Hawaii and informed himself on every aspect of the concept.

Dr. Riegler was one of the decreasing number of surgeons who do their own endoscopy. He was not dependent on gastroenterologists and, like some adventurous surgeons, willing to defy the recommendation for endoscopic biopsy in the guidelines of the American Gastroenterology Association. On his return to Vienna he developed his own biopsy protocol, expanding the one that was used at USC.

He initially had trouble finding a pathologist willing to read the biopsies in the manner he had learned at the conference and he had gleaned by reading the pathology literature. Finally, after much effort, he persuaded a senior pathologist in the University to read the biopsies. This was Dr. Fritz Wrba.

Dr. Wrba, who I got to know later as a superb surgical pathologist, was initially skeptical but was sufficiently open-minded to test the concept and became quickly convinced when he started looking at the biopsies. He recognized their value and became an enthusiastic ally in the project. The Riegler–Wrba partnership was the second surgeon/pathologist duo to be involved in a systematic routine examination of biopsies from the GEJ region after the DeMeester–Chandrasoma partnership. Other surgeon–pathologist partnerships may exist but I do not know of any gastroenterologist–pathologist partnership that works under the principle that cardiac epithelium distal to the endoscopic GEJ is always metaplastic.

My authorship in this paper is based only on the fact that I was invited to Vienna after the data had been collected and analyzed to provide confirmation of the pathologic classification of the epithelial types that had been made by Dr. Wrba following the definitions that we had published—Dr. Riegler called this histologic diagnostic method the "Paull–Chandrasoma classification."

The interobserver concordance between Dr. Wrba and me was extremely high. In a second blind look at the 1998 biopsy samples from 102 patients, there was minor disagreement in only four cases: one diagnosis of IM was changed to cardiac epithelium with pseudogoblet cells (i.e., no IM) and three cases called GOE were changed to oxyntocardiac epithelium. This low rate of interobserver variation demonstrated that the pathologic method was sound and easily applicable.

The aim of the study was to compare the findings at endoscopy and histopathology of the GEJ in patients with GERD using defined multilevel biopsies.[15] The introduction of the paper essentially reproduces what has been written under this section and shows complete understanding and assimilation of the new concept into patient management.

102 consecutive patients, over a 7-month period in 2006, with clinical symptoms of GERD (heartburn and regurgitation) presenting at the University Clinic of Surgery, Vienna, prospectively underwent video esophagogastroduodenoscopy and biopsy sampling of the GEJ region. All patients had heartburn and regurgitation and a clinical diagnosis of GERD (typical esophageal syndrome); pH testing was not done. Patients with known Barrett esophagus, history of dysphagia, prior thoracic and abdominal surgery except appendectomy and cholecystectomy were excluded. All patients were on treatment with a PPI (2×40 mg omeprazole daily) for at least 2 weeks prior to endoscopy. No non-GERD controls were included.

At endoscopy, after assessment of the duodenum and the stomach, which included four biopsies from the gastric antrum and body, the GEJ region was examined using retrograde and antegrade views. The esophagogastric valve was assessed during retroflex viewing according to Hill classification (grades I–IV). The presence of esophagitis was recorded. Endoscopic classification of patients was as follows:

1. Normal junction: absence of endoscopically visible columnar-lined esophagus, i.e., squamous epithelium ends at the proximal limit of rugal folds.
2. Abnormal junction was defined as the presence of any visible CLE between the proximally displaced Z-line and the proximal margin of rugal folds. When present, the length of CLE was noted.
3. Hiatal hernia was diagnosed if the proximal limit of rugal folds was 2 cm or more proximal to the crural impression with the stomach decompressed.

Multilevel biopsies were done at the endoscopic GEJ and on both sides of it at intervals of 5 mm. The level of the proximal limit of rugal folds was designated 0 and biopsies proximal to this were labeled +5 mm and those distal to it were designated −5, −10 mm. In endoscopically normal patients, the biopsy protocol included a minimum of 1–4 biopsies from four levels, starting distal to proximal at −10, −5, 0, +5 mm. The last of these were from squamous-lined mucosa 5 mm above the endoscopic GEJ. In patients with a visible columnar-lined esophagus, biopsies were taken in 5 mm steps until the level 5 mm above the level of the SCJ.

Luminal content, despite precautions to prevent blood from obscuring the procedure, precluded completing the biopsy protocol in 14 patients who each had one less biopsy level than required. Histologic examination was based on diagnostic criteria for columnar epithelia that are exactly as described in Chapter 5.

The authors' viewpoint of histology: "Cardiac mucosa, with and without intestinal metaplasia, and oxyntocardiac mucosa were considered to be of esophageal origin, irrespective of endoscopic appearance." (The authors cite two previous papers by them that validate this histologic method.) They call these collectively as histologic CLE. "If biopsies obtained at one level contain more than one type of CLE, the worst histopathology is listed first, according to the following order: IM > cardiac > oxyntocardiac mucosa. Squamous epithelium and oxyntic mucosa were considered to be normal lining of the esophagus and stomach, respectively."

The authors assess the length of histopathologic CLE: "By definition, CLE is interposed between esophageal squamous epithelium and gastric oxyntic mucosa. Transition from squamous epithelium to CLE and CLE to oxyntic mucosa was defined as the proximal and distal limits of CLE, respectively. The length of histopathologic CLE was measured as the distance between biopsy levels positive for proximal and distal CLE limits. If CLE was obtained at only one biopsy level, CLE length was categorized as being <5 mm. For comparison with histopathologic CLE, 5 and 10 mm lengths of endoscopically visible CLE equal 2 and 3 biopsy levels, respectively." (Note: I have taken the liberty to convert the measurement units from centimeters in the paper to millimeters to be concordant with this chapter).

This study is the only one in the literature where the methodology permits assessment of the presence of cardiac epithelium (with and without parietal/goblet cells—their histologic CLE) at varying lengths distal to the endoscopic GEJ.

The biopsy protocol is designed to test for the presence of cardiac epithelium (with and without parietal/goblet cells) at the various levels rather than to measure its length. This study has a more detailed biopsy protocol to study this region than the standard biopsy protocol used by the USC group which takes a single biopsy distal to the endoscopic GEJ. It represents the best biopsy study method of this region reported to date.

The patient population had a typical esophageal GERD syndrome with symptom relief on PPI treatment. 60% had endoscopically visible CLE (defined as *any* visible CLE, including short tongues resulting in serration of the Z-line) and 40% a normal appearing endoscopic GEJ. Endoscopic esophagitis was absent in 76%; when present esophagitis was equally distributed between grades A and B with no patient having grades C and D. It should be noted that this is a population that is on omeprazole 40 mg twice daily for at least 2 weeks. Hiatal hernia was present in 25%. The gastroesophageal flap valve was Hill grades I and II in patients with a normal appearing endoscopic GEJ in 85.7% whereas Hill grades III and IV were seen at significantly (P = .001) higher frequency in patients with endoscopically visible CLE (43.4%).

A total of 1998 biopsies were obtained from the 102 patients (mean 20.3, SD 6.3 per patient; range 4–37). Note: In these studies, it is important to pay attention to this piece of information. In Glickman et al.,[16] the mean number of biopsies per patient was 1.2 with a range of 1–4. In Marsman et al.,[17] the biopsy protocol was two biopsies each 2 cm above the SCJ and two biopsies just below the SCJ. The number of biopsies is 10–20 times greater in this study. Considering the importance of sampling demonstrated in Marsman et al.,[17] the data in this study should have much greater accuracy than all other studies in the literature, including our own. Just this fact makes this study superior to those with lesser sampling.

All patients had histologically defined CLE irrespective of the presence or absence of endoscopically visible CLE. In no patient did squamous mucosa transition directly to gastric oxyntic mucosa. CLE consisted of oxyntocardiac epithelium only in 12 (11.8%) patients, cardiac and oxyntocardiac epithelium in 71 (69.6%) patients, and cardiac epithelium with IM in 19 (18.6%) patients. In the patients who had a mixture of cardiac and oxyntocardiac epithelia, the former was always located proximally and the latter distally within a CLE segment.

Biopsy level–dependent distribution of epithelial types is shown in Table 17.5.

Squamous epithelium was limited to biopsy levels at (0 mm) and above (5, 10, and >10 mm) the level of the GEJ. GOE was limited to biopsy levels at (0 mm) and below (−5 and −10 mm) the level of the GEJ. Cardiac epithelium (with and without parietal and/or goblet cells) straddled the level of the endoscopic GEJ. The CLE above the GEJ was CLE that was visible at endoscopy and is not controversial. The CLE below the GEJ is the area of controversy. Based on the present definition of the endoscopic GEJ, this area will be classified as "proximal stomach." This is the area that we hypothesize represent dilated distal esophagus lined by metaplastic esophageal cardiac epithelium (with and without parietal/goblet cells).

The presence of histologic CLE was maximal at level 0 (level of the proximal limit of rugal folds) and decreased in biopsies obtained from above and below level 0. Decrease of CLE-positive biopsies was paralleled by an increase of biopsies containing gastric oxyntic mucosa at distal biopsy levels and those including squamous epithelium at proximal biopsy levels.

98 (97%) patients had CLE at level 0, 78 (81%) at level −5 mm, and 28 (28%) at level −10 mm. Only three patients had no CLE at level 0. One of these patients had oxyntic mucosa at level 0 and CLE at levels +5 and 10 mm; the other two patients had squamous epithelium at level 0, oxyntocardiac mucosa at level 5 mm, and gastric oxyntic mucosa at level 10 mm.

IM was present in 19 (18.6%) patients, always in cardiac mucosa. In 16 of these patients, IM was found in the most proximal segment of CLE at the SCJ, irrespective of CLE and IM length. In the remaining 3 patients, IM was found in the

TABLE 17.5 Biopsy Level–Dependent Distribution of Mucosal Types (n = 102)

Level	Patients n (%)	OM n (%)	CLE—All Types n (%)	OCM n (%)	CM n (%)	IM n (%)	Squamous
>+10 mm	7	0	4 (58%)	0 (0)	2 (29)	2 (29)	3 (42)
+10 mm	34	0	6 (18%)	1 (3%)	3 (9%)	2 (6%)	28 (82%)
+5 mm	97	0	39 (40%)	9 (9%)	25 (26%)	5 (5%)	58 (60%)
0.0	101	1 (1%)	98 (97%)	21 (21%)	65 (64%)	12 (12%)	2 (2%)
–5 mm	96	18 (19%)	78 (81%)	57 (59%)	12 (13%)	9 (9%)	0
–10 mm	100	72 (72%)	28 (28%)	16 (16%)	8 (8%)	4 (4%)	0

CLE, columnar-lined esophagus; *CM*, cardiac mucosa; *IM*, intestinal metaplasia; *OCM*, oxyntocardiac mucosa; *OM*, oxyntic mucosa.

biopsy 5 mm distal to the SCJ with cardiac epithelium at the junction. The mapping of the entire squamooxyntic gap suggests a constant distribution of the three epithelia within the entire squamooxyntic gap. Intestinalized cardiac epithelium is seen proximally adjacent to squamous epithelium, and oxyntocardiac epithelium distally adjacent to gastric oxyntic epithelium. Irrespective of the length of the squamooxyntic gap, the distribution of the three epithelia shows that there is one gap that extends from the squamous epithelium to GOE.

In the 19 patients with IM, the entire length of CLE was present (defined by the presence of GOE in the level −10 mm biopsy) in 11 (57.9%) patients. In the other eight patients, the level −10 mm biopsies showed IM (four patients), cardiac epithelium (three patients), and oxyntocardiac epithelium in one patient. This indicates that the dilated distal esophagus in these eight patients >10 mm.

In their discussion, the authors conclude: "… the data of the present study and other recent studies indicate that the endoscopically visible esophagogastric junction does not coincide with the histopathologic junction in the majority of patients with GERD and that the presence of CLE cannot be excluded by endoscopy. Therefore, the assumption that the level of the rise of endoscopically visible 'gastric folds' is the esophagogastric junction proves to be incorrect in such patients. This is in keeping with a novel concept that has recently been introduced by Chandrasoma and DeMeester: CLE within endoscopically visible 'gastric folds' in fact represents 'dilated end-stage esophagus', where loss of function of the lower esophageal sphincter favors gastric-type folding of CLE." (Note: We have replaced the term "dilated end-stage esophagus" with "dilated distal esophagus" since that time).

This paper shows the prevalence and extent of cardiac epithelium in a population of symptomatic GERD patients. The length of cardiac epithelium (with and without parietal/goblet cells) distal to the endoscopic GEJ can be partially measured from the data. It was >10 mm in 28 (28%) patients who had CLE in the biopsies at −10 mm; it was 5–10 mm in 50 patients who had CLE at −5 mm but not at −10 mm; and 0–5 mm in 20 patients who had CLE at 0 mm but not at −5 mm.

At the time this paper was published, I had not yet realized that the length of metaplastic cardiac epithelium (with and without parietal and/or goblet cells) distal to the endoscopic GEJ represented LES damage. There was no attempt to measure the length of the dilated distal esophagus. The data show, however, that the length of cardiac epithelium (with and without parietal and/or goblet cells) distal to the endoscopic GEJ varies considerably in this population with GERD. This indicates that different patients with symptomatic GERD have different amounts of LES damage. Whether this correlates with severity of GERD (as determined by pH testing, which was not done) is unknown in this patient group.

The overall length of cardiac epithelium (with and without parietal and/or goblet cells) found distal to the endoscopic GEJ in this study is, in general, lower than would be expected in a population of patients with symptomatic GERD. This has two explanations: (1) the theoretical calculations I have used based on the assumption of an initial abdominal LES length of 35 mm are not correct; and (2) the biopsy protocol differed from that I have described and was not intended to accurately measure the dilated distal esophagus completely.

These biopsies are at −5 and −10 mm distal to the endoscopic GEJ. It is likely that the center of the biopsy is at −5 and −10 mm, making the sampling not equivalent to the 0–5 and 5–10 mm segments as in the biopsy protocol. This has the potential of making the measurements different. However, it points to the possibility that adjustment may need to be made in the calculations as the new method is tested.

Also possible is the inherent error in the placement of the biopsy forceps when multilevel biopsies are preformed, even when care is taken. The discrepancy seen in this study with the theoretical measurements makes a powerful case for a better biopsy technique with a better biopsy device.

5. THE NEED FOR A NEW BIOPSY DEVICE

The biopsy protocol that is described above is cumbersome, time-consuming, and highly dependent on operator skill and attentiveness to detail. As such, it is very unlikely to gain wide acceptance or reach a high level of accuracy in practice. With the specter of inaccuracy associated with failure of adherence to a challenging biopsy protocol, this method is unlikely to have success in the practical sense.

The main value of the description of the above protocol was to show the theoretical feasibility of dividing the abdominal LES into segments of damage by increments of 1.25 mm.

The resistance from the gastroenterology community to implementing the new method will be powerful, and rightly so. GERD is such a common disease that even if this biopsy protocol was recommended because of its potential value, gastroenterologists will balk at the incursion of this laborious biopsy protocol into their schedules with what will very likely be no significant increase in reimbursement.

For this reason, I believe that testing and application of the new method should await the development of a new biopsy device that will allow taking of a single specimen 20- to 25-mm long and will permit exact measurement of LES damage

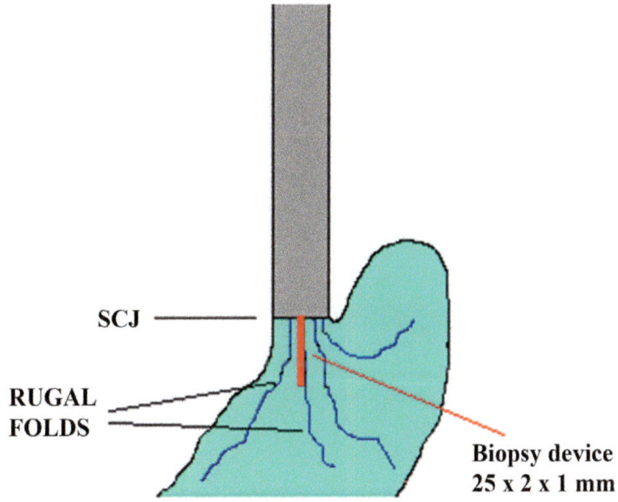

SCJ
RUGAL FOLDS
Biopsy device 25 x 2 x 1 mm

FIGURE 17.9 Biopsy protocol of the region of the dilated distal esophagus distal to the squamocolumnar junction (SCJ) using the new biopsy device measuring 25×2×1 mm. The device is placed such that its proximal end is proximal to the SCJ and oriented vertically along the lesser curvature. This allows exact measurement of the dilated distal esophagus.

(Fig. 17.9) without all the complex calculations that will have to be applied to this multilevel biopsy protocol with its inherent errors. The biopsy needs to be introduced by endoscopy and placed, in retroflex view, vertically such that its proximal end is immediately proximal to the SCJ (Fig. 17.5A and B). It must have adequate flexibility to be placed flush against the surface of the mucosa such that a continuous mucosal sample measuring 20–25 mm can be obtained intact without fragmentation.

This sets the stage for an incredible opportunity for entrepreneurial technology-oriented companies to develop a new type of biopsy device that will permit the taking of a single 20- to 25-mm long mucosa 2 mm in width and 1 mm deep.

If placed accurately such that it is vertically oriented with its proximal end immediately proximal to the SCJ, the sample will provide an *exact* measurement of the dilated distal esophagus with an accuracy that is measured in micrometers. The need for the complex pathology report and the complex grading system that I have proposed above will disappear. The 5 mm range in the biopsy protocol that has an uncertain correction will disappear. The need to apply a correction that will divide each 5-mm segment into four segments of 1.25 mm will disappear.

All that will be needed is to introduce the new device through the biopsy channel of the endoscope, position the 20- to 25-mm biopsy device vertically along the lesser curvature from the SCJ at the esophageal opening extending distally. Methods can be devised to facilitate the correct positioning of the biopsy forceps. The time taken will, with experience, not greatly exceed that of a single biopsy with a standard biopsy forceps.

The cylindrical biopsy device can be made to double as a transport device. Injection of formalin into the device with the specimen inside will convert the device into a container that can be placed in a standard formalin-filled specimen bottle to transport the specimen to the pathology laboratory. This will minimize handling of the specimen and allow it to fix in situ before it is removed from the biopsy container, thereby increasing the likelihood that the specimen will not fragment. The objective is to have a single 2-mm wide, 1-mm deep mucosal biopsy that has a length of 20–25 mm. No orientation is necessary for the specimen because (1) the squamous epithelium will represent the proximal end of the specimen or (2) if squamous epithelium is not present (if the patient has visible CLE), the proximal end will be intestinal or cardiac with oxyntocardiac and GOE marking the distal end.

The specimen is processed and examined as a single pathology specimen that will provide an exact measurement of cardiac epithelium (with and without parietal and/or goblet cells) distal to the SCJ. This requires one routine stain with the option of a second Di-PAS stain that is available in every laboratory. The cost of the diagnostic test is the cost of any single biopsy sample stained with the routine hematoxylin and eosin stain.

The measurement requires a standard microscope fitted with an ocular micrometer, again a readily available and relatively inexpensive piece of equipment in modern microscopes.

The single continuous mucosal specimen will provide a precise measurement from the distal limit of squamous epithelium to GOE (Fig. 17.10A–C). It is clean, hopefully easy to procure, accurate without question, and needs no manipulation of measurements. The three illustrated examples show a measured length of cardiac epithelium of zero (the normal state; Fig. 17.10A), 9.89 mm (within the phase of compensated LES damage; Fig. 17.10B), and 19.84 mm (patient with clinical GERD; Fig. 17.10C).

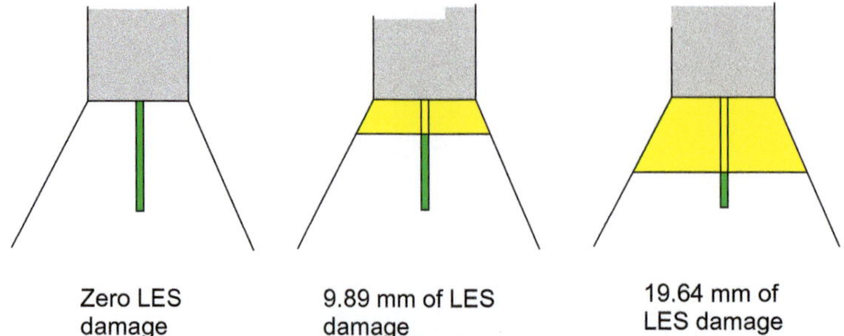

FIGURE 17.10 Results obtained with the new biopsy device provide an accurate measurement of the dilated distal esophagus (yellow=cardiac epithelium with and without parietal and/or goblet cells; green=gastric oxyntic epithelium). (A) No lower esophageal sphincter (LES) damage (no cardiac epithelium). (B) Measured LES damage of 9.89 mm. The patient is asymptomatic and within the range of the compensated phase of LES damage (up to 15 mm). (C) Measured LES damage of 19.64 mm. Exact measurement by histology provides a powerful tool for the diagnosis of gastroesophageal reflux disease.

The pathologist needs minimal training to recognize the histologic types with precision and only the time taken for any other biopsy sample to make the required measurement. This measurement will provide the exact LES damage in the patient, accurate to within micrometers (Fig. 17.10).

There are potential errors in the measurement that are unavoidable and unpredictable. These result from possible contraction of the biopsy specimen after it has been removed. This is unlikely because the biopsy is not aimed to access muscularis mucosae and therefore no muscle contraction is involved. Shrinkage is also possible with formalin fixation.

The best that can be done is to standardize the biopsy procedure and minimize handling of the specimen until several hours after allowing fixation within the device. If there is shrinkage, the measurements will require an adjustment to the calculations that I have used to define various stages of the process. However, if standardization in the way the specimen is taken, fixed, transported, and processed is achieved, the artifacts created will be uniform. The measurements of the lengths of the abdominal LES that define the phase of compensated LES damage, subclinical GERD, clinical GERD, and severe GERD as obtained by the new method with its standardized artifacts will become apparent as data accumulate.

The elaborate and cumbersome grading system that I have suggested above with the multilevel biopsy protocol is replaced by an individualized assessment of LES damage in each patient that will be at micrometer increments from 0 to 25 mm.

The pathology report will be simple:

1. Squamooxyntic gap composed of (select one): zero gap/oxyntocardiac epithelium only/cardiac and oxyntocardiac epithelium/cardiac epithelium with IM, cardiac and oxyntocardiac epithelia;
2. Length of the squamooxyntic gap: x mm.

Suddenly, a new and critical assessment becomes available for clinical use. Suddenly, GERD becomes a disease that can be defined by its etiology, LES damage, from its earliest point to the occurrence of LES failure and total destruction.

This opens the door to developing the ability to potentially prevent severe GERD in which treatment failure, visible CLE, Barrett esophagus, and esophageal adenocarcinoma occur. If the device can be adapted to procure a specimen easily, safely, and cheaply in a physician's office without need for endoscopy, this test can become a screening test with the potential to eradicate GERD itself as a human disease.

The potential commercial value of this device should be obvious and should create a push to develop such a new biopsy device. The only conceivable reason why such a simple device does not exist at the present time is that it has never been conceptualized.

I have developed and submitted for patent a prototype for such a biopsy device that can be passed through the biopsy channel of a standard endoscope as well as applied without the need for endoscopy. This will, if made, permit a 20- to 30-mm-long biopsy to be taken that will allow for exact LES damage assessment. It is difficult to imagine a reason why, assuming that measurement proves to be valuable, the test fails to become a critical test for early diagnosis in a patient suspected of having GERD (Chapter 21).

REFERENCES

1. Kahrilas PJ, Shaheen NJ, Vaezi MF. American gastroenterological association medical position statement on the management of gastroesophageal reflux disease. *Gastroenterology* 2008;**135**:1380–2.
2. Zaninotto G, DeMeester TR, Schwizer W, Johansson K-E, Cheng S-C. The lower esophageal sphincter in health and disease. *Am J Surg* 1988;**155**:104–11.

3. Haddad JK. Relation of gastroesophageal reflux to yield sphincter pressures. *Gastroenterology* 1970;**58**:175–84.
4. Thurer RL, DeMeester TR, Johnson LF. The distal esophageal sphincter and its relationship to gastroesophageal reflux. *J Surg Res* 1974;**16**:418–23.
5. DeMeester TR, Wernly JA, Bryant GH, Little AG, Skinner DB. Clinical and in vitro analysis of determinants of gastroesophageal competence. A study of the principles of antireflux surgery. *Am J Surg* 1979;**137**:39–46.
6. Robertson EV, Derakhshan MH, Wirz AA, Lee YY, Seenan JP, Ballantyne SA, Hanvey SL, Kelman AW, Going JJ, McColl KE. Central obesity in asymptomatic volunteers is associated with increased intrasphincteric acid reflux and lengthening of the cardiac mucosa. *Gastroenterology* 2013;**145**:730–9.
7. Chandrasoma P, Makarewicz K, Wickramasinghe K, Ma YL, DeMeester TR. A proposal for a new validated histologic definition of the gastroesophageal junction. *Hum Pathol* 2006;**37**:40–7.
8. Chandrasoma PT, Der R, Ma Y, et al. Histology of the gastroesophageal junction: an autopsy study. *Am J Surg Pathol* 2000;**24**:402–9.
9. Sarbia M, Donner A, Gabbert HE. Histopathology of the gastroesophageal junction. A study on 36 operation specimens. *Am J Surg Pathol* 2002;**26**:1207–12.
10. Campos GM, Oberg S, Gastal O, et al. Manometry of the lower esophageal sphincter: inter- and intraindividual variability of slow motorized pull-through versus station pull-through manometry. *Dig Dis Sci* 2003;**48**:1057–61.
11. Ayazi S, Tamhankar A, DeMeester SR, Zehetner J, Wu C, Lipham JC, Hagen JA, DeMeester TR. The impact of gastric distension on the lower esophageal sphincter and its exposure to acid gastric juice. *Ann Surg* 2010;**252**:57–62.
12. Stein HJ, Liebermann-Meffert D, DeMeester TR, Siewert JR. Three-dimensional pressure image and muscular structure of the human lower esophageal sphincter. *Surgery* 1995;**117**:692–8.
13. Chandrasoma P, Wijetunge S, Ma Y, DeMeester S, Hagen J, DeMeester T. The dilated distal esophagus: a new entity that is the pathologic basis of early gastroesophageal reflux disease. *Am J Surg Pathol* 2011;**35**:1873–81.
14. Jain R, Aquino D, Harford WV, Lee E, Spechler SJ. Cardiac epithelium is found infrequently in the gastric cardia. *Gastroenterology* 1998;**114**:A160. [Abstract].
15. Ringhofer C, Lenglinger J, Izay B, Kolarik K, Zacherl J, Fisler M, Wrba F, Chandrasoma PT, Cosentini EP, Prager G, Riegler M. Histopathology of the endoscopic esophagogastric junction in patients with gastroesophageal reflux disease. *Wien Klin Wochenschr* 2008;**120**:350–9.
16. Glickman JN, Fox V, Antonioli DA, Wang HH, Odze RD. Morphology of the cardia and significance of carditis in pediatric patients. *Am J Surg Pathol* 2002;**26**:1032–9.
17. Marsman WA, van Sandyck JW, Tytgat GNJ, ten Kate FJW, van Lanschot JJB. The presence and mucin histochemistry of cardiac type mucosa at the esophagogastric junction. *Am J Gastroenterol* 2004;**99**:212–7.

Chapter 18

New Method of Functional Assessment of the LES

The ability to measure abdominal lower esophageal sphincter (LES) damage opens new ways of correlating the function of the LES at various levels of progression of LES damage. The function of the LES is to prevent reflux of gastric contents into the thoracic esophagus across the pressure gradient that exists between the stomach and esophagus.

1. PRESENT ASSESSMENT OF REFLUX

Reflux is commonly measured by either the pH test or by impedance technology. I will limit this discussion to the pH test, which is more widely used. The main value of impedance over the pH test is that it has the ability to measure reflux in a patient who is on proton pump inhibitor (PPI) therapy.

The pH test uses acid in the gastric contents as the marker of reflux. It places a pH sensor at a point 5 cm above the upper limit of the LES to detect acidification of the esophagus, defined as a drop in the normal near neutral pH of the esophagus at this location to <4. This defines a reflux episode. The number and duration of reflux episodes in the supine and upright position are recorded over a period of continuous monitoring.

At present, pH is most often measured by implanting a capsule with the pH sensor ("Bravo capsule") in the mucosa of the esophagus at the designated point. Wireless technology is used to monitor the pH recorded by the electrode. The Bravo capsule stays in place for 48 h and sometimes up to 96 h, permitting a long monitoring period. It is then shed into the lumen.

The placement of the pH sensor in the esophagus at the designated 5 cm above the upper border of the LES is important. All the normal values that have been established are based on this location of the pH sensor. During a reflux episode, there is a pH gradient in the esophagus where the refluxate is most acidic at the gastroesophageal junction (GEJ) with a progressive increase in pH at higher regions of the column of refluxing gastric contents, reaching neutrality at the top of the column. As such, any variation in the position of the sensor can create error.

The patient is advised to conduct his normal life during the period of pH monitoring. Meals and sleep times are recorded. The patient is also given a method of recording the occurrence of symptoms, allowing association of symptoms with episodes of reflux. The result of the pH test is reported as percent time the pH is <4 during the 24-h period, or by a more complex analysis of multiple parameters that produces a DeMeester score.

In a patient with gastroesophageal reflux disease (GERD) who is being treated with acid suppressive drugs, the patient must be taken off the drugs for a period of at least 2 weeks to permit the test to quantitate reflux. Impedance testing, which senses the entry of air and liquid into the esophagus, can detect reflux episodes in the patient who is on acid suppressive drugs.

The pH test was initially developed and studied by the same group of surgeons who did the early research on manometry, headed by Dr. Tom DeMeester.[1] Their objective was to establish a gold standard for the diagnosis of GERD. Their primary interest at the time was in identifying a level of reflux into the body of the esophagus that was definitely abnormal at a clinical level. The presence of an abnormal pH test had a high probability at the time that medical therapy was unlikely to be able to control symptoms effectively. When PPIs entered the market, this changed because of the greater efficacy of PPIs in suppressing acid and controlling symptoms.

The amount of acid exposure in the pH test that best discriminated between patients with GERD symptoms and asymptomatic people was used to define the cut-off point between normal and abnormal. This amount was a pH<4 for >4.5% of the testing period.[1] At the time, this level of reflux correlated with symptoms too severe to be controlled by medical therapy and indicated that the patient was likely to benefit from antireflux surgery. It also correlated with a defective LES defined by manometry as mean pressure<6 mmHg, total length<20 mm, and abdominal length<10 mm. In contrast with patients with GERD 95% of asymptomatic volunteers had a pH < 4 for <4.5% of the time and an abdominal LES length>10 mm.[2]

GERD. http://dx.doi.org/10.1016/B978-0-12-809855-4.00018-X

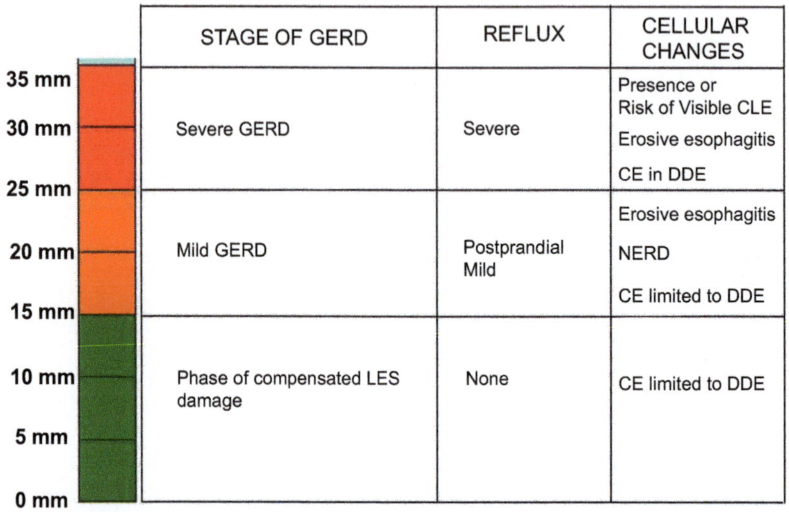

	STAGE OF GERD	REFLUX	CELLULAR CHANGES
35 mm 30 mm 25 mm	Severe GERD	Severe	Presence or Risk of Visible CLE Erosive esophagitis CE in DDE
20 mm 15 mm	Mild GERD	Postprandial Mild	Erosive esophagitis NERD CE limited to DDE
10 mm 5 mm 0 mm	Phase of compensated LES damage	None	CE limited to DDE

FIGURE 18.1 Correlation between abdominal LES damage, reflux, and cellular changes in the thoracic and dilated distal esophagus: normal state (zero LES damage), the phase of compensated LES damage (abdominal LES damage of >0–15 mm), mild clinical GERD (15–25 mm), and severe GERD (25–35 mm). Only the last stage is clearly recognized presently as GERD. *CE*, cardiac epithelium; *CLE*, columnar-lined esophagus; *DDE*, dilated distal esophagus; *GERD*, gastroesophageal reflux disease; *LES*, lower esophageal sphincter; *NERD*, nonerosive reflux disease.

The objective of the test was high specificity of diagnosis. One reason for this was to develop an objective criterion that would justify the performance of antireflux surgery. An abnormal pH test is, or should be, a requirement for performing antireflux surgery in the quest to cure the GERD patient at a symptomatic level. Again, with the advent of PPI therapy which has an efficacy similar to antireflux surgery in most patients, the enthusiasm for antireflux surgery in the treatment of GERD has declined. Surgery is effective, but too often creates side effects that are as much of a problem as the symptoms it controls. Surgery is now limited to patients who fail PPI therapy, not those with an abnormal pH test.

If the LES is designed to prevent reflux, the expectation is that the pH test should have zero reflux. There is therefore a gap of 64 min between the expectation of zero reflux and the 4.5% of the day (i.e., 64 min) that is presently defined as the duration of abnormal acid exposure in the esophagus.

All the present diagnostic criteria for GERD have a similar gap. This includes endoscopic and histologic diagnosis, interpretation of manometry, and interpretation of the pH test. The criteria for abnormality are set to define GERD at an advanced stage.

I have shown that recognition of the gap between a defective abdominal LES (defined as <10 mm in length) and "normal" resolves into progressive shortening from an initial length of 35 to 10 mm. There is a 25-mm gap between a normal abdominal LES (i.e., zero damage with a length of 35 mm) and a defective abdominal LES. Definition of this gap by the new pathologic measurement of the dilated distal esophagus results in the recognition of stages of disease at a cellular level corresponding to different degrees of LES damage from the very onset of the process. This allows developing new criteria for correlating LES damage levels with reflux and cellular changes in the esophagus (Fig. 18.1).

Similarly recognition of these gaps for other diagnostic tests for GERD can result in new criteria of definition of abnormality at endoscopy, histology, and the interpretation of the pH test and manometry that permit recognition of the disease at a much earlier stage and greatly increase the potential value of these tests (Table 18.1).

2. CORRELATION OF ABDOMINAL LOWER ESOPHAGEAL SPHINCTER DAMAGE AND REFLUX

In the previous chapters, I have established a new method of defining the structure of the LES on the basis of a new pathologic measurement of LES damage. Based on this, the normal LES has a length of 35 mm [with possible minor individual variation (Fig. 18.2)]. LES damage initially passes through a compensated phase of where it exhausts its reserve capacity. During this phase, the LES is damaged but remains competent. I have suggested that this phase is 15 mm of abdominal LES damage (Fig. 18.3). At this level of damage, the LES will have a residual functional length of 20 mm. It can withstand 10 mm of dynamic shortening during gastric overdistension caused by a heavy meal and still remain above the <10 mm abdominal length that is associated with LES failure.

I have suggested that there is no LES failure or reflux during this phase of compensated LES damage, i.e., the pH test will be zero. It is probable, though, that the pH test will be zero at no LES damage and slowly rise as it is damaged to 15 mm.

TABLE 18.1 Present and Future Criteria for Diagnosis of GERD. The Present Criteria of These Diagnostic Tests Are Set to Define the Disease With High Specificity at an Advanced Stage of the Disease. The New Criteria Define the Disease With High Sensitivity at a Much Earlier Stage

Diagnostic Test	Present Criterion of Abnormality	Proposed Future Criterion of Abnormality	What Is Missed (Error)
Identification of the GEJ	Proximal limit of rugal folds, defined at endoscopy	Proximal limit of gastric oxyntic epithelium, defined by histology	Up to 35 mm of dilated distal esophagus
Endoscopy	Erosive esophagitis and visible CLE above endoscopic GEJ	Cardiac epithelium distal to the endoscopic GEJ, by biopsy	Up to 35 mm of dilated distal esophagus
Length of cardiac epithelium distal to endoscopic GEJ	None; normal is believed to be 1–4 mm	>0–35 mm	Up to 35 mm of dilated distal esophagus
GERD symptoms	Troublesome	Irrelevant; early disease is asymptomatic	Entire early stage of disease
Empiric PPI test	Response of symptoms to a trial of PPI therapy = GERD	Irrelevant; obsolete	Up to 30% false positive diagnosis
Abdominal LES	<10 mm in length	<35 mm in length	25 mm of a nondefective abdominal LES
24-h pH test	pH < 4 for >4.5% (>64 min/24 h)	pH < 4 for 0 min	64 min/day of acid exposure

CLE, columnar-lined esophagus; *GEJ*, gastroesophageal junction; *GERD*, gastroesophageal reflux disease; *LES*, lower esophageal sphincter; *PPI*, proton pump inhibitors.

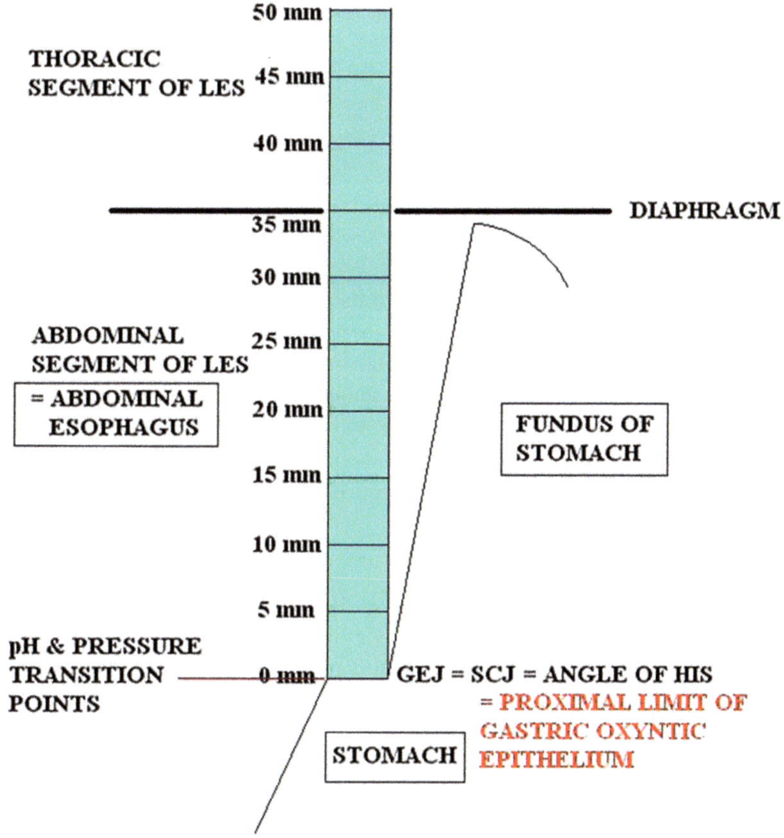

FIGURE 18.2 The normal state, defined manometrically as an abdominal LES with no damage and its full length of 35 mm. This correlates with zero cardiac epithelium between the SCJ and the proximal limit of gastric epithelium. *GEJ*, gastroesophageal junction; *LES*, lower esophageal sphincter; *SCJ*, squamocolumnar junction.

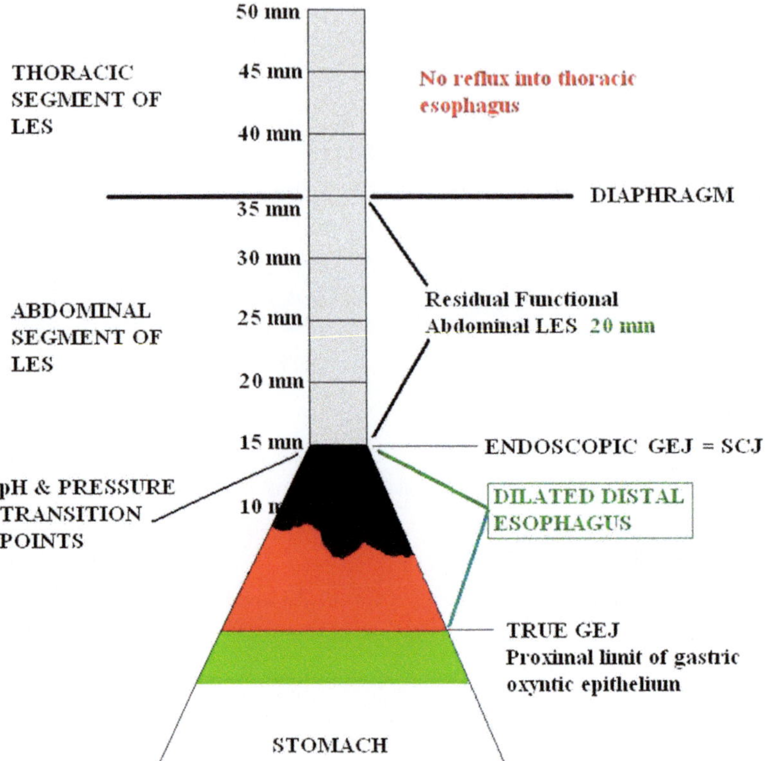

FIGURE 18.3 The end of the compensated phase of LES damage (abdominal LES damage >0–15 mm). The LES damaged has reached the end-point of its reserve capacity. The LES is still competent and there is no reflux. There is 15 mm of a dilated distal esophagus lined by cardiac (black) and oxynto-cardiac epithelium. *GEJ*, gastroesophageal junction; *LES*, lower esophageal sphincter; *SCJ*, squamocolumnar junction.

The next stage of LES damage extends from >15 to 25 mm. During this stage, postprandial dynamic shortening of the LES causes the functional length to become <10 mm, resulting in postprandial reflux. This progressively increases in severity until the patient develops significant reflux at 25 mm of LES damage where the fasting residual abdominal LES is 10 mm, on the cusp of being permanently <10 mm (Fig. 18.4). It is likely that the amount of reflux in the pH test progressively increases as it progresses from 15 to 25 mm from a very low level at 15 mm of damage to close to the abnormal level of pH<4 for >4.5% of the time.

The final stage of abdominal LES damage is the stage of severe LES damage (>25 to complete destruction at 35 mm). The 24-h pH is commonly abnormal (i.e., pH<4 for >4.5% of the 24-h period). This is the stage of disease that presently has a positive 24-h pH test. The earlier stages of LES damage are not recognized by this established criterion of the pH test.

I am trying to suggest that, like the gradation of LES damage in micrometer increments detected by the new pathologic test for LES damage, a similar gradation of increasing acid exposure of the esophagus exists. It is highly likely that there will be a correlation between the increments of LES damage and acid exposure by the pH test. The correlation is likely to be nonlinear with acid reflux increasing slowly as the LES shortens from 0 to 15 mm and very rapidly beyond 25 mm damage. There is some evidence that this is the case. The question is whether new norms for the pH test can be identified that correlate with earlier stages of the disease.

3. RESIDUAL (FUNCTIONAL) ABDOMINAL LOWER ESOPHAGEAL SPHINCTER LENGTH AND REFLUX

From the standpoint of its function, it is the residual length of the LES that primarily determines LES competence. There is good evidence that the length of abdominal LES that best discriminates between the presence of sufficient reflux to cause symptoms and the asymptomatic state is 10 mm.[2]

Dynamic (temporary) shortening of the abdominal LES occurs during gastric overdistension. This can be induced by air insufflation or by a heavy meal.[3–5] The dynamic shortening with a heavy meal ranges from 5 to 10 mm.[5]

When the LES is competent, it does not fail under any circumstance and there is no "significant" reflux into the thoracic esophagus. I would like to believe that "significant" is equal to zero. However, it may be more practical to assume that even a normal LES can occasionally be associated with some reflux. Some authorities recognize the presence of "physiologic reflux."

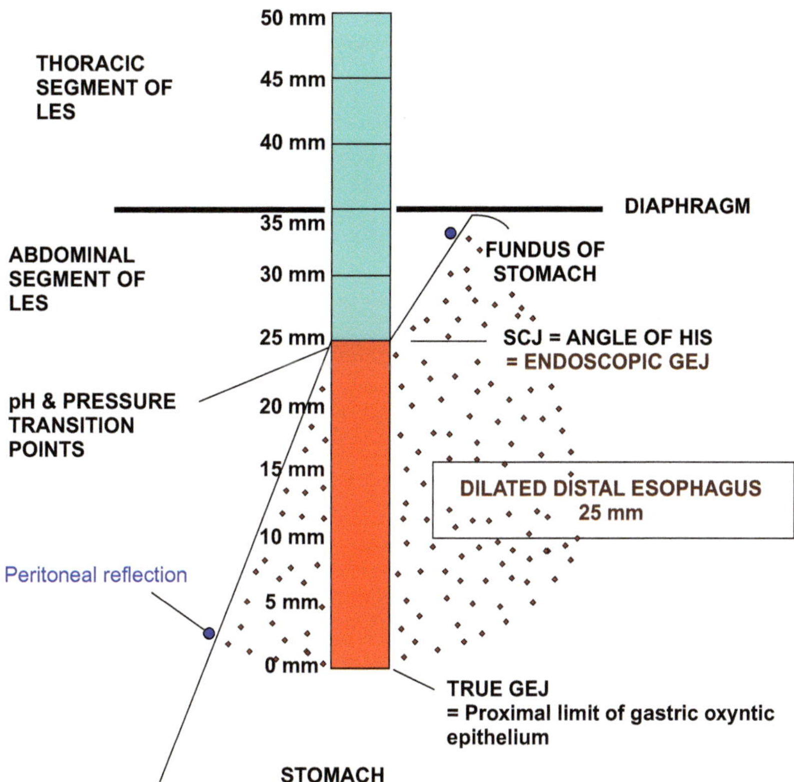

FIGURE 18.4 New definition of the onset of severe gastroesophageal reflux disease is an abdominal LES damage of 25 mm and a residual (manometric) length of 10 mm. The dilated distal esophagus is shown in red in the original tubular esophagus, and stippled in its dilated state. The angle of His is more obtuse. The LES is incompetent with frequent reflux episodes resulting from minimal stresses. *GEJ*, gastroesophageal junction; *LES*, lower esophageal sphincter; *SCJ*, squamocolumnar junction.

While this is possible, it is safer to believe that all reflux results from LES failure. The concept of "physiologic reflux" is impossible to prove and tends to act as an excuse to avoid the possibility of unexplainable reflux. There is no evidence whatsoever that reflux of gastric contents into the thoracic esophagus serves any useful function. People with zero reflux in a 24-h pH test exist and are likely to be more normal than people with any "physiologic" reflux.

LES failure by any mechanism results in an open cavity between the stomach and esophagus. This causes a flow of gastric contents into the thoracic esophagus along the pressure gradient that exists between the abdominal stomach (+5 mmHg) and the thoracic esophagus (−5 mmHg). The nadir of intraluminal pressure is in the midesophagus near the arch of the aorta. The column of refluxate can be expected to reach this level whenever the LES fails.

LES failure is initially related to meals and associated with heavy meals because of this dynamic shortening. As a result, postprandial reflux can be expected to begin when the residual abdominal LES in the fasting state is 15–20 mm. At this residual abdominal LES length, a dynamic shortening of 10 mm will result in a defective (<10 mm) LES at the height of gastric overdistension.

3.1 New Norms for the pH Test Correlated With Manometry

An abnormal pH test is the objective gold standard of diagnosis of GERD. If a person has a positive test, it is diagnostic of GERD as defined by the presence of significant reflux. It has a high correlation with the presence of symptoms, which are often troublesome.

A normal pH test, however, does not mean that the patient does not have reflux. The definition of normalcy in the test is presently designed to have high specificity at the expense of sensitivity. For that reason, it should not be surprising that the pH test is not widely used for the diagnosis of GERD except in patients who are being considered for antireflux surgery.

An abnormal pH test, as defined at present, correlates with symptomatic GERD and a severely damaged LES (often <10 mm abdominal length). An abnormal pH test means that there is acidification (pH < 4) of a point in the esophageal body 5 cm above the upper border of the LES for 64 min of a 24-h period.

Surely, reflux that causes acidification of the esophagus for a lesser duration is not "normal." Again, we face the uncomfortable position that a person who has reflux for 60 min a day is "normal" while another person with 64 min of acid

TABLE 18.2 Length of the Abdominal Segment of the Lower Esophageal Sphincter (LES) and Esophageal Acid Exposure in the 24-h pH Test in the 50 Volunteers in Zaninotto et al.

	Length of Abdominal Segment of the LES (mm)	Acid Exposure in 24-h pH Test (% Time pH < 4)
Minimum	8	0
Maximum (excluding 5 cm outlier)	35	6
Mean	21.8	1.57
Standard deviation	7.2	1.47
Median	22	1.1
95 percentile	>11	<4.6
97.5 percentile	>8.9	<5.8

The red numbers and blue numbers are correlative, i.e., the minimum length of abdominal LES (8) correlates with maximum pH (6) and the maximum length of abdominal LES (35) correlates with the minimum pH value (0).

exposure is "abnormal." Even worse, the same person can have these "normal" and "abnormal" values on two separate days. Again, this makes no sense.

The important question arises as to how much reflux will be present with lesser amounts of LES damage as the abdominal LES damage progresses from 0 to 25 mm of damage at which time the residual fasting abdominal LES reaches the critical 10 mm and the LES becomes defective.

In Zaninotto et al.[2] (reviewed in Chapter 9), the distribution of abdominal LES length in 49 asymptomatic volunteers (excluding one outlier with 50 mm) was 30–35 mm (n=5); 25–30 mm (n=11); 20–25 mm (n=17); 15–20 mm (n=10); and 10–15 mm (n=6). This is a near perfect Gaussian distribution with the mean length (21.8 mm) almost equaling the median length (22 mm).

The pH test in these asymptomatic volunteers (as a group) had a minimum of 0%, a maximum of 6% (86 min/24 h), a mean of 1.57% (23 min/24 h) with a standard deviation of 1.47, and a median of 1.1% (16 min/24 h). The acid exposure of this group does not have a Gaussian distribution; the mean (1.57%) is higher than the median (1.1%). The curve has a positive bias, i.e., there are significantly more persons with a pH exposure less than the mean of 1.57% than more. 50% of the persons had a pH of <1.1% (16 min/24 h) (Table 18.2).

If one assumes that the abdominal LES length is important to maintaining LES competence, it would suggest that the LES maintains its competence while it becomes shorter during that phase in which it loses its reserve capacity. When it reaches a critical length, further shortening causes a significant and progressive increase in LES failure [transient LES relaxation (tLESR)], producing a rapid increase in reflux and acid exposure of the thoracic esophagus.

The present norms for the 24-h pH test and abdominal LES length at manometry were developed to correlate with symptoms, which reflect LES failure and reflux sufficient to cause symptoms. These values were the 95th percentile of these tests in the asymptomatic group.

In my theoretical argument, I have suggested that the abdominal LES can shorten from its initial length of 35 to 20 mm before reflux would start occurring in the postprandial phase. This 15 mm of abdominal LES damage is the compensated phase where the damaged LES is still competent and reflux is unlikely. In Zaninotto et al.'s[2] 50 volunteers, the median length of the abdominal LES was 22 mm (Fig. 18.5). With the assumption that the initial length was 35 mm, this would mean that the median LES shortening in this group was 13 mm. The median acid exposure for this group in the pH test was 1.1% of the 24-h period. If these median values are used to define the norm, it would be reasonable to suggest that persons with LES damage within the compensated phase of LES damage would have a pH<4 for <1.1% of the 24-h period (Fig. 18.6).

Actual studies in asymptomatic volunteers can be designed to identify the critical point of abdominal LES length at which the sharp upward rise in acid exposure begins. If this point of abdominal LES length is defined, both the pH test and manometry can have new normal values that are designed to make the tests highly sensitive for the diagnosis GERD.

Present criteria for defining an abnormal pH test are primarily based on correlations with troublesome symptoms that define GERD. These become relevant only when reflux is severe, i.e., pH<4 for >4.5% of the 24-h period. An abnormal pH is seen in progressively higher percentage of patients with increasing cellular damage in the thoracic esophagus from nonerosive reflux disease to erosive esophagitis to Barrett esophagus.[6]

The new values will provide a sensitive diagnosis of GERD at or before the onset of GERD symptoms during the phase of compensated LES damage. The present normal values can remain as evidence of proof of severe GERD sufficient to

24-hr pH
6.0
5.0
4.0
3.0
2.0
1.0
0
6
4.6
1.1
0
5
8
10
11
15
20
22
25
30
35
Abdominal LES
length (mm)

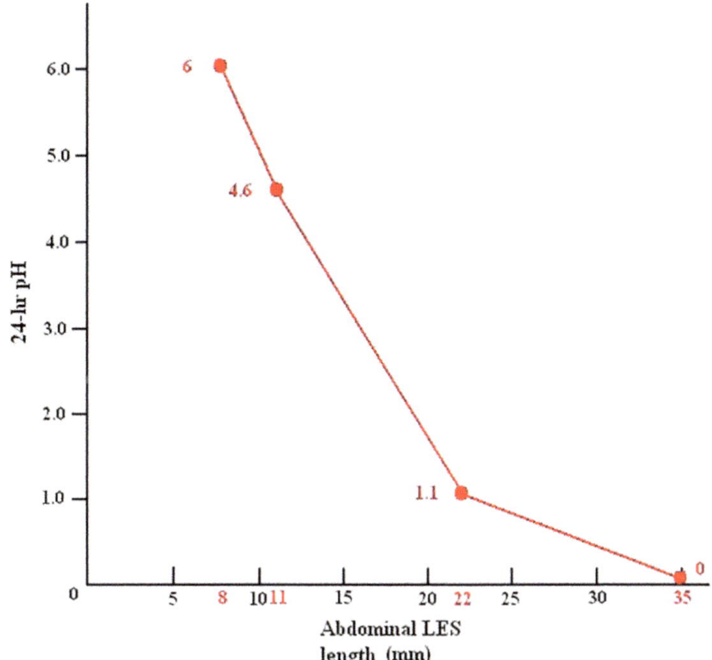

FIGURE 18.5 Data in Zaninotto et al.[2] correlating residual lower esophageal sphincter (LES) length as measured by manometry with acid exposure by the 24-h pH test. This shows that the initial phase of LES damage to the median LES length of 22 mm is associated with 0 to pH<4 for 1.1% of the 24 h. As LES length decreases from the median, there is a sharper rise in acid exposure per unit of LES shortening. At the 95th percentile of abdominal LES length in these asymptomatic volunteers, the 11 mm length correlates with a pH<4 for 4.6% of the 24-h period.

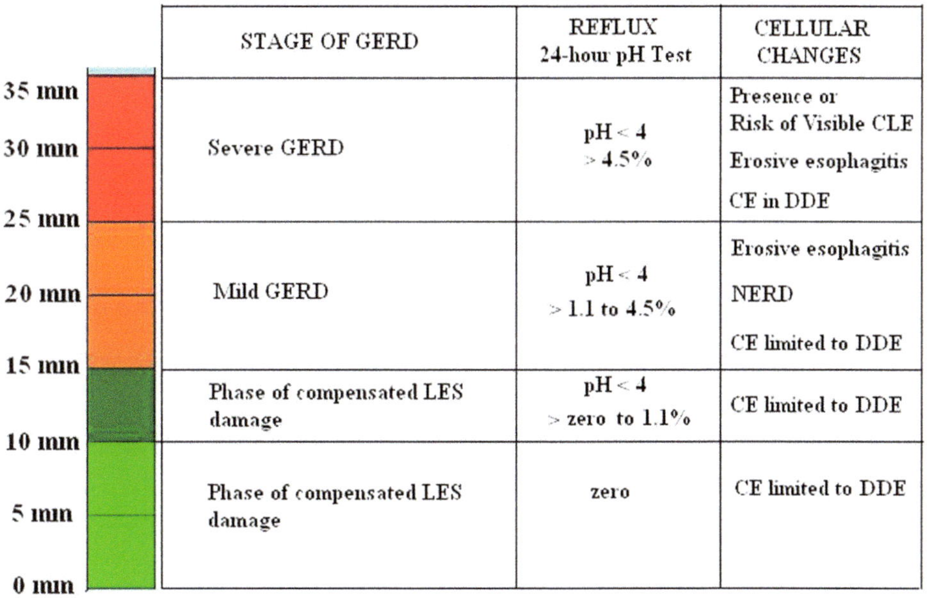

35 mm
30 mm
25 mm
20 mm
15 mm
10 mm
5 mm
0 mm
STAGE OF GERD
REFLUX 24-hour pH Test
CELLULAR CHANGES
Severe GERD
pH < 4 > 4.5%
Presence or Risk of Visible CLE
Erosive esophagitis
CE in DDE
Mild GERD
pH < 4 > 1.1 to 4.5%
Erosive esophagitis
NERD
CE limited to DDE
Phase of compensated LES damage
pH < 4 > zero to 1.1%
CE limited to DDE
Phase of compensated LES damage
zero
CE limited to DDE

FIGURE 18.6 Correlation of the three stages of LES damage with severity of reflux and cellular changes in the thoracic esophagus. This is similar to Fig. 18.3 except that the severity of reflux has been changed from descriptive to objective criteria based on the 24-h pH test. *CE*, cardiac epithelium; *CLE*, columnar-lined esophagus; *DDE*, dilated distal esophagus; *GERD*, gastroesophageal reflux disease; *LES*, lower esophageal sphincter; *NERD*, nonerosive reflux disease.

consider surgical intervention. Until these studies are done, I would suggest that a 24-h pH test result of pH<4 for 1.1% of the period and an abdominal LES length of 22 mm are excellent numbers that define the end of the compensated phase of LES damage and the onset of reflux (Table 18.3). These are based on the data in Zaninotto et al.,[2] which was the study that defined the existing criteria of abnormality of the pH test (Fig. 18.5).

TABLE 18.3 New Proposed Values for the Definition of Normal in the 24-H pH Test and Manometry (Only Abdominal Length Is Used) Aimed at Making the Tests Highly Sensitive for the Diagnosis of Gastroesophageal Reflux Disease. These Are Contrasted With the Normal Values in the Test That Are Presently Used

Test	Present Definition of Normal	New Definition of Normal	Rationale for New Definition
24-h pH test	pH<4 for <4.5% of 24h (<64min/24h)	pH<4 for <1.1% of 24h (<16min/24h)*	Median acid exposure in Zaninotto et al's median LES damage was 1.1%
AbdLES length by manometry	<10mm	>22mm**	Median AbdLES length in Zaninotto et al. was 22mm

AbdLES, abdominal segment of the LES; *, median pH in asymptomatic volunteers in Zaninotto et al.[2]; **, median length of abdominal LES in asymptomatic volunteers in Zaninotto et al.[2]

As the LES damage progresses from a residual manometric length of 22mm, resulting in further shortening, there will be a steeper climb in the 24-h pH test. In Zaninotto et al.,[2] the 95th percentile of the abdominal LES in asymptomatic volunteers measured 11mm (Fig. 18.5). The corresponding acid exposure at the 95th percentile was 4.6%. The 11mm shortening from 22mm (median) to 11mm (95th percentile) had increased acid exposure in the thoracic esophagus fourfold.

In further analyzing the data in this study, again assuming a perfect inverse correlation between abdominal LES length and acid exposure, the following can be deduced: (1) The patient with the longest abdominal LES (35mm) had an acid exposure of zero. (2) The shortening of the abdominal LES by 13mm from 35 to 22mm resulted in an acid exposure increase from 0% to 1.1%. (3) The patient with the shortest abdominal LES (8mm) had an acid exposure of 6%; the shortening from the 11mm at the 95th percentile to 8mm (shortest) caused the acid exposure to increase from 4.6% to 6%.

These new values are concordant with the theoretical concept of the LES that I have developed in this book. I have suggested that the reserve capacity of the abdominal LES is 15mm. With 15mm abdominal LES damage, the residual (manometric) length of the abdominal LES will be 20mm (initial length of 35mm−LES damage of 15mm). At 20mm residual abdominal LES length, dynamic shortening of the LES during meals (by 10mm) can cause the functional length of the abdominal LES to become <10mm, causing infrequent LES failure (tLESR) and reflux limited to the postprandial phase. This will be seen in a 24-h pH test.

These values are also concordant with the findings at autopsy in persons without symptoms of GERD.[7,8] The length of the dilated distal esophagus (=abdominal LES damage), as measured by the length of cardiac epithelium between the squamocolumnar junction (SCJ) and the proximal limit of gastric oxyntic epithelium, was <15mm. This corresponds with a manometric length of the residual abdominal LES in asymptomatic people with pH<4 of >22mm that defines the phase of compensated LES damage.

By this analysis, I can suggest the following two sets of normal values in the 24-h pH test and manometry, designed for two different purposes:

1. Normal values to assess the end of the compensated phase of LES damage and the onset of LES failure
   a. Manometric residual baseline length of the abdominal LES of 22mm or more;
   b. Abdominal LES damage by histologic measurement of 13mm or less;
   c. The acid exposure (pH<4) in the thoracic esophagus in the 24-h pH test of 1.1% or less.
2. Normal values to assess the onset of severe LES damage (present criteria)
   a. Manometric residual baseline length of the abdominal LES of <10mm;
   b. Abdominal LES damage by histologic measurement of >25mm;
   c. The acid exposure (pH<4) in the thoracic esophagus in the 24-h pH test of >4.5%.

Between these two is the present nebulous stage of the disease where the patient may or may not have symptoms. If they have symptoms, they are commonly episodic ("episodic heartburn") and often not "troublesome" enough to garner the serious attention of physicians. The symptoms, if present, are almost always easily controlled with acid-reducing pills, obtained either over the counter or from a primary care physician. There is no thought in these patients of performing endoscopy or any diagnostic testing other than an empiric PPI test. While this management seems reasonable, a small percentage but large absolute number of these patients progress to refractory GERD with uncontrolled symptoms, and adenocarcinoma or presenting with dysphagia resulting from advanced cancer.

In the new method of understanding that I am trying to develop, the present point at which GERD is defined and diagnosed clinically is the point that represents a high risk of impending treatment failure and visible columnar-lined esophagus (CLE). The objective is to diagnose LES damage well before this point and prevent LES damage reaching 25 mm (or residual manometric length of 10 mm) at which severe reflux (pH < 4 for 64 min of the day) occurs.

With these new norms, there is a new value to both manometry and pH testing as tests that will complement the new histologic assessment of LES damage and provide a comprehensive overview of the structural and functional status of the LES in any person, irrespective of the presence of symptoms of GERD. This would be a powerful tool in the early diagnosis of GERD that can be used to define, for the first time, the basis of GERD before the patient develops GERD by the criteria of the present Montreal definition.[9]

3.2 A New Gastroesophageal Reflux Disease Stress Test

In the fasting state, the manometrically defined length of the abdominal LES can be expressed by the following formula:

Baseline (fasting) abdominal LES length (measured by manometry) = original total sphincter length (35 mm, assumed by best evidence) − length of abdominal LES damage (measured by histology).

It has been well established that the abdominal LES undergoes significant dynamic shortening during meals associated with gastric overdistension. As intragastric pressure increases, it overcomes the LES pressure in the distal region of the LES. There is a dynamic shortening of the distal LES that is temporary; the LES slowly returns to its fasting length as the stomach empties. The formula for LES status in the postprandial state is as follows:

Postprandial abdominal LES length = original length of the abdominal LES (35 mm, assumed by best evidence) − length of abdominal LES damage (measured by histology) − amount of dynamic shortening of the abdominal LES because of gastric distension (measured by manometry)

The functional abdominal LES length at any point in time is the critical measure of LES competence. When the length of the abdominal sphincter is <10 mm, either in the fasting or postprandial state, sphincter failure (tLESr) will occur with sufficient frequency to cause clinical symptoms and cellular changes in the thoracic esophagus.[6]

Dynamic shortening of the LES in the postprandial state becomes relevant in people whose abdominal LES has a fasting length that is within the range in which the added shortening during a meal results in the abdominal LES length reaching the critical <10 mm point of failure. If this happens, LES failure (tLESR) and reflux may occur during the postprandial phase.

The actual amount of dynamic shortening of the LES varies with the volume of the meal, rate of gastric emptying, the LES pressure, and many other factors. It is difficult to predict. In a study of asymptomatic volunteers, Ayazi et al.[4] showed that the distal end of the LES shortened from 44.5 cm from the nares to 42.8 cm when the stomach was distended with air. This was a dynamic shortening of 17 mm.

Kahrilas et al.,[3] showed that the amount of shortening resulting from slow controlled air insufflation was similar and between 5 and 10 mm in asymptomatic (normal) people and those with symptomatic GERD with and without a hiatal hernia.

Robertson et al.,[5] from the McColl group in Glasgow, produced data on acid exposure of the thoracic esophagus in a pH test correlated with abdominal LES length (see Chapter 7). The asymptomatic volunteers in that study were divided into those with and without central obesity. The volunteers with central obesity had greater lengths of cardiac epithelium and shorter distal LES lengths. In the volunteers with central obesity, the fasting upright distal (abdominal) LES length was a mean of 21.2 mm (interquartile range, 16.3, 26.1 mm). During the testing period, the pH in this group at a point 5 cm above the SCJ was 0%–0.2% in the fasting state. It increased from 0% to 1.7% in the postprandial state when the distal LES length decreased to a mean of 16.9 mm (interquartile range, 12.5–21.3). The shortest distal LES length reported in this study in the fasting state was 16.3 mm. This (assuming an initial length of 35 mm) means that abdominal LES damage was a maximum of 18.7 mm in this group. The pH in the fasting state was 0%–0.2%, indicating that the abdominal LES could withstand shortening by 18.7 mm without becoming incompetent. In Robertson et al.,[5] the decrease in abdominal LES length was 4–5 mm after a heavy meal where the persons were asked to ingest fish and chips until they were full.

An effective method of introducing an assessment of dynamic shortening of the abdominal LES is to devise a stress test. This will provide information as to whether gastric distension will induce LES failure by measuring the effect of such failure, i.e., precipitating increased episodes of reflux and increased acid exposure during the postprandial period.

The GERD stress test consists of a heavy meal that has the greatest probability of inducing gastric overdistension with maximum dynamic shortening of the LES. Some patients will know the type of meal that is most likely to produce symptoms; this is the best choice for those patients. For people without a known precipitating diet, something of the nature of a double cheeseburger with a chocolate milk shake and as much french fries as the patient needs to feel a sense of fullness. In England, fried codfish and chips with a milk shake are generally preferred. Other menus can be designed that will give the patient a choice and make the test appealing rather than an intrusion. The patient is asked to ingest the meal as quickly as possible and record times and volumes.

A stress test has most value in a person who is asymptomatic or minimally symptomatic with GERD (abdominal LES damage of 15–20mm). The stress test is a simple addition to the new comprehensive battery of tests in the early diagnosis of GERD (see later). It is similar in concept to a stress test for ischemic heart disease, except that the stress of a heavy meal for testing LES function is likely to be more pleasurable than exercise on a treadmill for testing coronary artery function.

The exact amount of increases in reflux episodes and acid exposure that define a positive test will have to be determined by actual data from performing the test.

4. COMBINED MANOMETRIC AND PH TESTING FOR GASTROESOPHAGEAL REFLUX DISEASE

It is interesting to predict how the new histologic test of LES damage may impact the present diagnosis and management of GERD.

4.1 Present Endoscopy in the Patient Who Has Failed Proton Pump Inhibitor Therapy

In the present guidelines, endoscopy is indicated in patients whose symptoms have failed to be controlled with maximum PPI therapy.[10] Present requirements of endoscopy include a careful visual examination of the esophagus and stomach, sometimes aided by special techniques, most commonly narrow band imaging. More rarely, and generally in academic settings, confocal microscopy and chromoendoscopy may be utilized.

If there is no abnormality seen, no biopsies are taken. If erosive esophagitis is seen, biopsies may or may not be taken. If taken, the objective is to exclude metaplastic columnar epithelium above the endoscopic GEJ that may be masked by the erosive esophagitis. If visible CLE is seen, biopsies are taken to diagnose intestinal metaplasia, which leads to a diagnosis of Barrett esophagus. If there is any visible pathology, biopsies are taken of that pathology.

The following endoscopic findings result in a change in the management of the patient: (1) If Barrett esophagus (defined as visible CLE with intestinal metaplasia in the United States) is present, the patient enters an endoscopic surveillance program because of the risk of future cancer. (2) If other specific pathology is found, that is addressed according to the diagnosis.

All patients, irrespective of endoscopic findings, are maintained on acid-reducing drug therapy as required to control symptoms. Rarely, control is so poor with optimum PPI therapy as to consider an endoscopic or surgical antireflux procedure.

There is no method at present to detect those patients who are being treated without proven GERD. These patients are being given PPIs long term, unnecessarily if GERD is not the cause of their symptoms. This is recognized as a significant problem with the recent recognition of adverse effects associated with chronic PPI use.

In a minority of patients, a Bravo capsule is placed 5cm above the LES for a pH test. In such cases, the patient is usually kept off PPI therapy for 2 weeks prior to the endoscopy. A positive pH test (defined as a pH<4 for 4.5% of the 24h and/or an abnormal DeMeester score) confirms the diagnosis of GERD. A positive test changes very little; the patient is maintained on PPIs as needed to control treatment with the option of antireflux surgery.

A negative pH test can be interpreted in one of two ways: (1) The patient has no GERD. Some expert gastroenterologists will use a negative test as evidence that the symptoms are not caused by GERD. This results in diagnoses of "functional heartburn" and "hypersensitive esophagus." There is a large literature on how these should be treated. (2) The negative pH test represents a false negative test, i.e., its sensitivity is not adequate to exclude GERD. These patients are maintained on PPI therapy by many gastroenterologists who subscribe to this viewpoint.

Analysis of this endoscopy protocol suggests that there is little value to endoscopy in this patient population except for the finding of Barrett esophagus, which represents 10%–30% of patients, or the rare discovery of other pathology, e.g., adenocarcinoma.

Even the finding of Barrett esophagus is of questionable value according to some experts because the risk of cancer is so low that it raises questions about the cost-effectiveness of endoscopic surveillance. There is good evidence that Barrett surveillance is proven to save lives by early detection and treatment of dysplasia and cancer.[11]

The present enthusiasm of primary care physicians, who manage the vast majority of patients with GERD, to refer their patients to a gastroenterologist for endoscopy is low because of its limited value. Apart from endoscopy, the primary care physician has the same tools for treating the GERD patient as a gastroenterologist.

There is evidence that the following two endoscopic findings have predictive value for the development of visible CLE within the next 5years of treatment with acid-reducing drugs:

1. Severe erosive esophagitis, LA grade C/D; 19.7% of these patients will develop visible CLE within 5years even as their erosive esophagitis is controlled with PPI therapy.[12]
2. In patients who are endoscopically normal, the presence of intestinal metaplasia at the SCJ (i.e., intestinal metaplasia of the dilated distal esophagus); 25.8% of these patients will develop visible CLE in the next 5years while being treated

with PPIs.[13] This second group is not detected at present because of the recommendation that biopsy should not be performed in patients without endoscopic abnormality.[5]

4.2 A New Comprehensive Gastroesophageal Reflux Disease Assessment

The availability of a histologic measurement of LES damage adds a new dimension to diagnostic testing for GERD. This new comprehensive GERD test will provide valuable information at any point in the adult life of a person (i.e., after age 30 unless GERD symptoms develop earlier). This is true irrespective of whether the person has clinical evidence of GERD or not.

Initially, the test will be limited to be an adjunct to endoscopy. If the test provides valuable information and becomes the basis for change that significantly impacts the progression of GERD and prevents refractory GERD and reduces the incidence of adenocarcinoma, expansion will become necessary. This will likely require new instrumentation to permit measurement without the need for endoscopy. At present, I will limit the test to require endoscopy.

I will also assume that a new biopsy device to procure a 25-mm long biopsy of the dilated distal esophagus (i.e., distal to the SCJ in the endoscopically normal person) has been developed successfully.

The new testing is limited to patients who do not have visible CLE. The presence of visible CLE is by itself a proof of severe LES damage and irreversible GERD. Management in these patients is designed to manage visible CLE with the assumption of severe LES damage. The same is probably true of the patient with a hiatal hernia, who can also be deemed to have severe LES damage.

The patient will undergo manometry immediately prior to endoscopy. This can be done by high-resolution manometry, a slow motorized pull-through, or a new device that has been developed to provide a higher level of accuracy of measurement of the LES than at present.

At endoscopy, which is done after PPI therapy has been withdrawn for 2 weeks, the following new protocol is followed:

1. In retroflex view, the new biopsy device will be positioned vertically along the lesser curvature with its proximal end 1 mm cephalad to the SCJ (visualized by insufflation as necessary). The device will be activated to procure the 25 mm × 2 mm × 1 mm vertical mucosal sample along the lesser curvature. I have chosen the lesser curvature for two reasons: it is the most flat part of the circumference of this area, and it is the part of the circumference that is reported to have the most adenocarcinoma, which suggests that it is most likely to be representative of the maximum LES damage.
2. A Bravo pH device will be fixed to the esophageal squamous epithelium 5 cm above the upper border of the LES, guided by the results of manometry. This provides a pH monitor for the next 48 h. The patient is instructed to record the times of occurrence of symptoms, supine position (in bed), and ingestion of meals.
3. On the day after endoscopy, the patient will be given a stress test with a preplanned standardized meal. This can be done either at home (if the patient is reliable and has support) or in the doctor's office. The patient is instructed to record the meal volume, content and time of ingestion of the meal.

The additional time taken at endoscopy to perform the new biopsy and fix the Bravo capsule is likely to be readily acceptable to gastroenterologists, as long as the new data prove to be clinically useful in patient care. The cost of the procedure will remain equal to present endoscopy with biopsy and fixing a Bravo capsule. From a patient standpoint, the endoscopy will be no different than the standard endoscopy. It will not take a significant amount of increased time and the complication rate will not increase. The GERD stress test has potential to be attractive and a potential incentive if catered to the patient's dietary desires.

After the biopsy is taken, fixative is added to the specimen in biopsy device. The specimen within the device is then transported to the pathology laboratory for processing, microscopic assessment, and measurement of the length of the dilated distal esophagus.

The complete GERD assessment will provide the following new information:

1. Abdominal segment LES damage in millimeters to two decimal points by histologic measurement of the length of cardiac epithelium (with and without parietal and/or goblet cells) using standard microscopy with an ocular micrometer.
2. Length of the residual (baseline, fasting) abdominal LES as measured by manometry, hopefully at a higher level of accuracy with newly developed technology.
3. 48-h acid exposure in the pH test, including symptom association and response to normal meals, and upright and supine positions.
4. Response to dynamic shortening of the LES induced by the stress test as indicated by a change in the number of reflux episodes (=LES failure = tLESR) and acid exposure in the 2-h postprandial period following the ingestion of the stress test meal on the second day.

The information from these tests will provide unique data that comprehensively assess the condition of the abdominal segment of the LES (residual length by manometry and LES damage by the histologic measurement). This will provide accurate information regarding the initial abdominal LES length (=manometric length+pathologic measurement of LES damage).

The Bravo pH test will also provide data over a prolonged time period on the level of competence of the LES in preventing reflux as well as how the LES responds to the GERD stress test.

4.3 New Normal Values for Manometry and pH Test

The test will be interpreted according to the following normal values that I have developed earlier:

1. Abdominal LES damage (as measured by histology): <13 mm = phase of compensated LES damage; reflux highly unlikely to be the cause of patient's symptoms, if any exist.
2. Residual abdominal LES length (manometric): >22 mm = phase of compensated LES damage; reflux highly unlikely to be the cause of patient's symptoms, if any exist.
3. 24-h acid exposure (pH test, excluding postprandial period after stress test): <1.1% pH < 4 for the testing period = phase of compensated LES damage; reflux highly unlikely to be the cause of patient's symptoms, if any exist.
4. Acid exposure in 2-h postprandial period after GERD stress test: acid exposure < 2% of time pH < 4 = phase of compensated LES damage; reflux highly unlikely to be the cause of patient's symptoms, if any exist.

These values are designed for maximal sensitivity for the detection of significant abdominal LES damage. They are based on the best available evidence but the data are sparse and these criteria will need to be adjusted as new data accumulate.

A patient who is normal in all four of these values *can be defined as not having GERD*. It is theoretically highly unlikely that a patient with LES damage < 15 mm, a residual LES > 20 mm, acid exposure of <1% over 24 h, and an acid exposure of <2% following the GERD stress test has sufficient LES failure and reflux into the thoracic esophagus to produce symptoms of GERD. If such a patient is being treated for GERD with PPIs, the treatment should be stopped and an alternate cause for symptoms found.

The test will therefore provide a method of preventing the present problem where an empiric PPI test has led to the diagnosis of GERD and the patient is on long-term PPI therapy without the ability to confirm the diagnosis in a reliable manner. These tests will provide that reliable answer (see Chapter 20).

If the patient is abnormal in all four parameters, the diagnosis of GERD is confirmed, irrespective of the presence of symptoms. Most patients with symptomatic GERD will have LES damage > 15 mm, a residual abdominal LES length < 20 mm level, and an acid exposure (pH < 4) >1% but less than the present value of 4.5% which defines an abnormal pH test at present.

The patient being abnormal with one, two, or three of these parameters will need to be evaluated when data accumulate. Presently available data in the literature do not permit predicting the sensitivity and specificity of these situations for the diagnosis of GERD.

There is a possibility of interesting new diagnoses with this new testing. For example, many patients with GERD believe that certain foods trigger LES failure and reflux episodes. It is probable that the relationship between trigger foods and reflux is coincidental, i.e., the person who has LES damage at the cusp of LES failure will relate the initial failure of the LES that occurs with meals to the foods associated with the meal. These patients will have abnormalities in the criteria enumerated above in the new assessment of GERD. If, however, a patient has LES damage < 15 mm, a residual manometric abdominal LES length > 20 mm, a 24-h pH of <1% and has an abnormal stress test that correlates with a specific food, a definitive diagnosis of a trigger food can be made. This implies that the food in some manner causes the failure of the LES by a mechanism that is different than the mechanical failure that I have assumed is true for all LES failure.

The presently used norms for manometry and the pH test have a different meaning with this new method of assessment:

1. A pH test that shows acid exposure (pH < 4) for >4.5% of the testing period indicates severe LES damage. It is highly specific for the diagnosis of GERD and is a reason for consideration of antireflux procedures because this degree of reflux is unlikely to be controlled in the long term by medical therapy.
2. An abdominal LES length of <10 mm, along with a mean LES pressure < 6 mmHg, and a total LES length of <20 mm indicate a defective LES. This is again highly specific for GERD and represents a reason for considering an antireflux procedure.

These pH and manometric criteria have a significant risk of visible CLE either at the time of assessment or in the near future. They are therefore the criteria that indicate advanced disease that the new management attempts to prevent.

In all people who undergo the test, the critical determination is the length of LES damage. This is the only parameter tested that can be used to predict future progression of GERD (see Chapter 19).

It is probable that this comprehensive testing is unnecessary for the future management of GERD. If the single histologic measurement of LES damage provides sufficient information with regard to the present severity and future projection of LES damage, it may suffice as a stand-alone single test for GERD that will guide the entire management of the patient. However, in the phase of clinical testing the new test of LES damage, comprehensive testing as described will be necessary to establish the norms for the new test and establish its correlations with cellular progression of GERD.

REFERENCES

1. Jamieson JR, Stein HJ, DeMeester TR, et al. Ambulatory 24-hour esophageal pH monitoring: normal values, optimal thresholds, specificity, sensitivity, and reproducibility. *Am J Gastroenterol* 1992;**87**:1102–11.
2. (a) Zaninotto G, DeMeester TR, Schwizer W, Johansson K-E, Cheng S-C. The lower esophageal sphincter in health and disease. *Am J Surg* 1988;**155**:104–11.
   (b) Chandrasoma P, Wijetunge S, Ma Y, DeMeester S, Hagen J, DeMeester T. The dilated distal esophagus: a new entity that is the pathologic basis of early gastroesophageal reflux disease. *Am J Surg Pathol* 2011;**35**:1873–81.
3. Kahrilas PJ, Shi G, Manka M, Joehl RJ. Increased frequency of transient lower esophageal sphincter relaxation induced by gastric distension in reflux patients with hiatal hernia. *Gastroenterology* 2000;**118**:688–95.
4. Ayazi S, Tamhankar A, DeMeester SR, et al. The impact of gastric distension on the lower esophageal sphincter and its exposure to acid gastric juice. *Ann Surg* 2010;**252**:57–62.
5. Robertson EV, Derakhshan MH, Wirz AA, Lee YY, Seenan JP, Ballantyne SA, Hanvey SL, Kelman AW, Going JJ, McColl KE. Central obesity in asymptomatic volunteers is associated with increased intrasphincteric acid reflux and lengthening of the cardiac mucosa. *Gastroenterology* 2013;**145**:730–9.
6. Stein HJ, Barlow AP, DeMeester TR, Hinder RA. Complications of gastroesophageal reflux disease. Role of the lower esophageal sphincter, esophageal acid and acid/alkaline exposure, and duodenogastric reflux. *Ann Surg* 1992;**216**:35–43.
7. Chandrasoma PT, Der R, Ma Y, et al. Histology of the gastroesophageal junction: an autopsy study. *Am J Surg Pathol* 2000;**24**:402–9.
8. Kilgore SP, Ormsby AH, Gramlich TL, et al. The gastric cardia: fact or fiction? *Am J Gastroenterol* 2000;**95**:921–4.
9. Vakil N, van Zanten SV, Kahrilas P, Dent J, Jones B, and the global consensus group. The Montreal definition and classification of gastroesophageal reflux disease: a global evidence-based consensus. Am J Gastroenterol 2006;**101**:1900–20.
10. Pj K, Shaheen NJ, Vaezi MF. American gastroenterological association medical position statement on the management of gastroesophageal reflux disease. *Gastroenterology* 2008;**135**:1380–2.
11. Peters JH, Clark GW, Ireland AP, et al. Outcome of adenocarcinoma arising in Barrett's esophagus in endoscopically surveyed and nonsurveyed patients. *J Thorac Cardiovasc Surg* 1994;**108**:813–21.
12. Malfertheiner P, Nocon M, Vieth M, Stolte M, Jasperson D, Keolz HR, Labenz J, Leodolter A, Lind T, Richter K, Willich SN. Evolution of gastro-oesophageal reflux disease over 5 years under routine medical care – the ProGERD study. *Aliment Pharmacol Ther* 2012;**35**:154–64.
13. Leodolter A, Nocon M, Vieth M, Lind T, Jasperson D, Richter K, Willich S, Stolte M, Malfertheiner P, Labenz J. Progression of specialized intestinal metaplasia at the cardia to macroscopically evident Barrett's esophagus: an entity of concern in the Pro-GERD study. *Scand J Gastroenterol* 2012;**47**:1429–35.

Chapter 19

Prediction of Future Progression of LES Damage

I have presented gastroesophageal reflux disease (GERD) as a chronic progressive disease resulting from damage to the lower esophageal sphincter (LES). I have suggested that the entire disease spectrum can be understood by what happens to the 35 mm of the abdominal segment of the LES that represents the normal LES at full development.

The progression of LES damage is orderly and likely to be highly predictable. This means that in a given person, once a rate of progression of LES damage is established, it will progress at basically the same rate throughout life. Everyone begins with a normal LES, defined as an abdominal length of 35 mm and degrades the LES at differing rates. Based on the rate of shortening of the abdominal LES, the patient will reach a definable end point at the end of life. That end point determines at what stage of GERD the person reaches, based entirely on how much of the abdominal LES is damaged. The entire progression of the disease can be followed from 35 mm to that end point of abdominal LES damage at the end of the projected life span.

In this method, GERD has a subclinical phase of LES damage within its reserve capacity and three levels of LES failure leading to increasing reflux with increasing severity of cellular change (Table 19.1). The amount of abdominal LES damage has a reliable correlation with the amount of reflux and cellular changes in the esophagus (Fig. 19.1). It is far simpler to understand GERD by this method than trying to define it by symptoms of dizzying number, severity, and varying specificity for GERD as is done by the present Montreal definition.[1]

The normal state where the LES is undamaged, which correlates with the absence of cardiac epithelium (with and without parietal and/or goblet cells) between esophageal squamous and gastric oxyntic epithelia, is extremely rare. Nearly everyone has some cardiac or oxyntocardiac epithelium distal to the endoscopic gastroesophageal junction (GEJ).[2–4]

LES damage results when metaplastic cardiac epithelium replaces esophageal squamous epithelium. This begins at the true GEJ (the proximal limit of rugal folds) and extends cephalad, separating the squamocolumnar junction (SCJ) from the true GEJ. The length of cardiac epithelium between squamous and oxyntic epithelia represents the dilated distal esophagus in the person without visible columnar-lined esophagus (CLE). The length of the dilated distal esophagus, which can be measured by the length of cardiac epithelium (with and without parietal and/or goblet cells), is concordant with the degree of LES damage.[5]

I have correlated the degree of abdominal LES shortening with the status of the patient in terms of the various clinicopathological stages of GERD (Chapter 16). These stages are defined by the amount of reflux into the thoracic esophagus by the pH test and the presence of cellular changes in the thoracic esophagus. Symptoms, while they have a broad correlation with LES damage, can be absent in patients with severe GERD, including adenocarcinoma.

1. When the abdominal LES damage is <15 mm (residual abdominal LES length >20 mm, assuming an initial length of 35 mm), the LES is competent, rarely fails, and is associated with no or minimal reflux into the thoracic esophagus. This is the phase of compensated LES damage (Fig. 19.1). Endoscopy is normal.
2. Between 15 and 20 mm LES damage (residual length of 15–20 mm assuming an initial length of 35 mm), the person develops subclinical GERD that usually does not satisfy the present definition of GERD; the person is either asymptomatic or has mild symptoms that are not “troublesome” (Fig. 19.1).[1] Endoscopy is normal or shows mild erosive esophagitis. Histology may show microscopic changes in the squamous epithelium.
3. Between 20 and 25 mm of LES damage (residual length of 10–15 mm assuming an initial length of 35 mm), the person has mild GERD that may satisfy the Montreal definition[1] without visible CLE. Endoscopy and histology may show increasing grades of erosive esophagitis.
4. With >25 mm LES damage (residual length of <10 mm assuming an initial length of 35 mm), the patient has severe GERD with failure of proton pump inhibitor (PPI) therapy to control symptoms satisfactorily either have or be at imminent risk for the development of visible CLE (Fig. 19.1). Visible CLE defines the stage of severe GERD where the

GERD. http://dx.doi.org/10.1016/B978-0-12-809855-4.00019-1

TABLE 19.1 Severity of Gastroesophageal Reflux Disease (GERD) as Defined by the Extent of Abdominal Lower Esophageal Sphincter (LES) Damage[a]

Severity of GERD	Residual Abdominal LES Length (Manometry) (mm)	Abdominal LES Damage (Histology) (mm)	pH Test (% Time pH < 4)
Normal	35	0	0
Phase of compensated LES damage	<35–20	>0–15	1.1% or less
Mild GERD (without visible CLE)	<20–10	>15–25	>1.1%–4.5%
Severe GERD (visible CLE present or imminent)	<10	>25	>4.5%

[a]*This is measured histologically as the length of cardiac epithelium (with or without parietal and/or goblet cells) between esophageal squamous epithelium and gastric oxyntic epithelium in an endoscopically normal person and between the endoscopic GEJ and gastric oxyntic epithelium in a patient with visible CLE.*

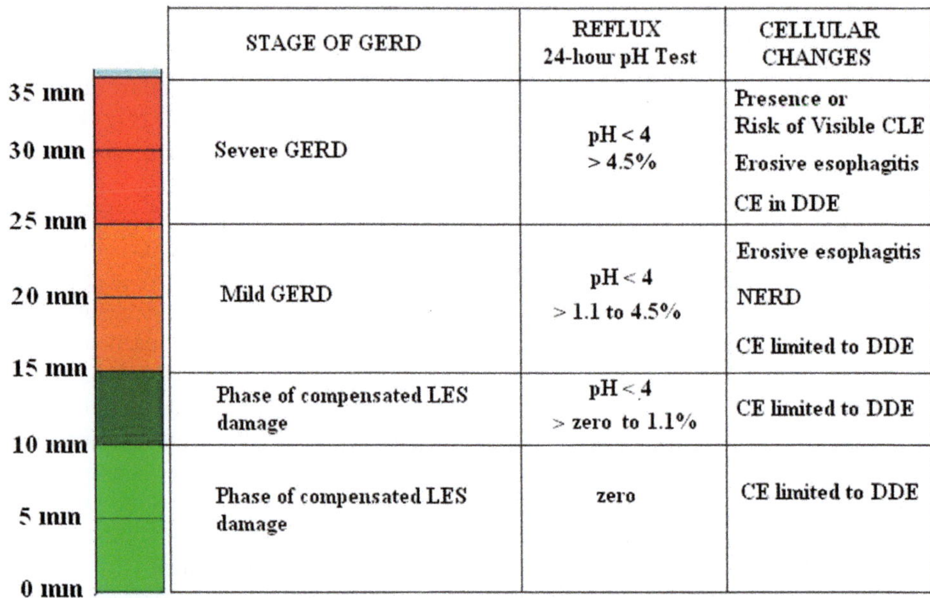

FIGURE 19.1 Stages of gastroesophageal reflux disease (GERD) based on the measured length of abdominal lower esophageal sphincter (LES) damage. This permits division of the early stages of the disease into normal (zero LES damage), the phase of compensated LES damage (>0–15 mm), mild clinical GERD (15–25 mm), and severe GERD (>25 mm). *CE*, cardiac epithelium; *CLE*, columnar lined esophagus; *DDE*, dilated distal esophagus.

cellular change is irreversible. The >25 mm abdominal LES damage is the assumed earliest point where visible CLE may occur. The exact abdominal LES damage associated with visible CLE will be determined by future studies.

These correlations do not work when GERD is viewed from the point of view of symptoms and their control, as is the case at present. A person without symptoms typical of GERD can be in any of the above stages of LES damage including severe damage with visible CLE and even adenocarcinoma.

I have previously used the analogy of GERD to ischemic heart disease. They are both chronic progressive diseases. They are both caused by eating disorders. GERD is the result of meals that induce repeated gastric overdistension that causes LES damage from below; ischemic heart disease is caused by a diet that induces coronary atherosclerosis.

TABLE 19.2 Severity of Ischemic Heart Disease (IHD) as Defined by the Extent of Narrowing by Coronary Arteries[a]

IHD	Coronary Artery Luminal Area (Angiography)	Coronary Artery Narrowing (Angiography)	Impact to Myocradium	Stress Test
No IHD	100%	0	0	0
Phase of compensated coronary artery narrowing	<100%–50%	>0%–50%	0	0
Subclinical IHD	<50%–30%	>50%–70%	0	EKG/echo abnormalities on stress test
Mild IHD	<30%–20%	>70%–80%	Ischemia on exertion	Not performed
Severe IHD	<20%–0%	>80%–100%	Ischemia at rest; MI, death	Not performed

For purposes of discussion, this is measured by coronary angiography as the totality of arterial supply of the left ventricle, which varies with the predominance of right and left coronary arteries and collateral circulation.
MI, myocardial infarction. Ischemic heart disease differs from GERD in that myocardial infarction and death can result from coronary artery thrombosis or hemorrhage into a plaque that can result in occlusion with relatively mild coronary artery disease. LES damage in GERD has no such acute events, but is complicated by a lethal chronic event, adenocarcinoma.
[a]*The values given in the table are not real values backed by evidence that I have researched for this table. They are meant to be illustrative, to show the similarity of progression of ischemic heart disease to GERD.*

The new understanding of GERD that I have proposed based on LES damage is essentially identical to the present understanding of ischemic heart disease (Table 19.2). In both conditions, there is an unpredictable lethal event that complicates severe disease. In coronary atherosclerosis, it is occlusion by thrombosis. In GERD, it is neoplasia in Barrett esophagus.

When I attended medical school in the 1970s, ischemic heart disease was defined and understood by its symptoms—angina of effort, angina at rest, crescendo angina, myocardial infarction, and sudden death (let us suggest this was the "Montreal definition of ischemic heart disease"). At the time, angina was treated with vasodilators (let us suggest these were the PPIs of ischemic heart disease). Like PPIs in GERD, these were ineffective at preventing progression of disease. We simply waited for myocardial infarction and death that was inevitable in a minority of patients (let us suggest this is equivalent to adenocarcinoma in GERD). The incidence of myocardial infarction and death was rising, similar to adenocarcinoma of the esophagus.

Ischemic heart disease changed its course when its cause, coronary artery narrowing by atherosclerosis, was addressed. Coronary angiography and stress testing led to the ability to predict which patients were the high risk groups. Intervention by coronary artery bypass, balloon dilatation, and coronary arterial stents transformed the disease. The incidence of myocardial infarction and mortality from ischemic heart disease decreased.

My hope is that addressing LES damage with new diagnostic testing that I have proposed will lead to predicting which patients will progress to severe GERD. This allows intervention to prevent or slow progression of LES damage by dietary modification with GERD-friendly diets, surgical (LINX, fundoplication) and endoscopic [EndoStim, Transoral Incisionless Fundoplication (TIF), Stretta] methods of controlling LES damage. Like balloons and stents, which were unknown in the 1970s, my hope is that the medical device industry will create yet unimagined new and highly effective and minimally invasive methods of preventing or slowing progression of LES damage in the future. Preventing progression of damage of a partially damaged LES surely must be easier than trying to repair a severely damaged LES.

The primary objective of the new management is the reversal of the rising curve of esophageal adenocarcinoma toward zero.

1. THE POINT OF ONSET OF LOWER ESOPHAGEAL SPHINCTER DAMAGE FROM INITIAL LENGTH

Everyone starts with a normal LES that has no damage. It is not easy to define the anatomic structure of a normal LES. There is little or no information of LES growth during childhood or at the age of full maturation when the LES reaches its adult length.

In Chapter 17, I discussed methods that will become available to define the initial LES length when the new proposed test for measuring LES damage is in place. Until then, based on the evidence that I have presented in Chapter 17, I will assume that the initial abdominal LES length is 35 mm with little individual variation and that full maturation to this length is reached at age 15 years. This can be adjusted upward if the person grows significantly after age 15 years and/or data from future studies indicate otherwise.

The other critical point that needs to be known is the time of onset of LES damage. This can be defined in an individual by two measurements of LES damage that are separated by a significant time interval in the two-measurement protocol (see below). Until then, I will assume that the onset of LES damage has an onset at age 15 years. This is based on the fact that the cause of LES damage is an eating habit that puts pressure on the LES from below resulting in exposure of esophageal squamous epithelium to gastric contents.

It is reasonable that the onset of LES damage is around the age at which the adult dietary habit of a child is established. 15 years seems a reasonable age, but any age around that time can be chosen. If there is a history of significant childhood obesity, this age can be moved to 10 years. Ultimately, though, study data from measuring LES damage in the future will provide more accurate data regarding the onset of LES damage that can be individualized.

2. POINT OF LOWER ESOPHAGEAL SPHINCTER DAMAGE THAT RESULTS IN LOWER ESOPHAGEAL SPHINCTER FAILURE

Ultimately, if one looks at GERD from the point of view of abdominal LES damage, the single most important variable determinant of LES competence is its residual length. While this can be measured by manometry, the present accuracy of this measurement has not been sufficient to permit it to be practically useful in the diagnosis of GERD or correlation with cellular changes. While it is known that shortening of the LES is associated with GERD, the only practical diagnostic criteria available at present are criteria that define a defective LES, which are mean LES pressure <6 mmHg, overall length <20 mm, and abdominal length <10 mm.[6]

LES pressure has been shown to correlate poorly with reflux. It is possible that improvement of manometric technology will increase the accuracy of the measurement of LES length. With the introduction of the new histologic test for abdominal LES damage, the residual length of the abdominal can be calculated by the formula:

Residual abdominal LES length = 35 mm (assumed initial length) − measured abdominal LES damage.

The reason why the residual length of the abdominal LES is the best determinant of LES incompetence is that the shortest functional length of the abdominal LES is at the point in time when the stomach is maximally distended, usually at maximum gastric filling with a heavy meal. The residual baseline length of the abdominal LES shortens by the amount of dynamic shortening associated with a heavy meal. A heavy meal is therefore the equivalent of a stress test for the LES. The above equation for the residual length of the abdominal LES during the postprandial phase becomes:

Residual abdominal LES length = 35 mm (assumed initial length) − measured abdominal LES damage − dynamic shortening of abdominal LES in the postprandial phase.

Robertson et al.[4] showed that a heavy meal can cause a dynamic shortening of the abdominal LES by 5–10 mm. I have assumed that a maximally heavy meal in the United States (which is reasonably assumed to at the high end of one in the United Kingdom!) will cause a dynamic shortening of 10 mm.

A heavy meal that causes a dynamic shortening of 10 mm can therefore cause an LES that is competent in the fasting state to fail in the postprandial state. Postprandial reflux is the first evidence of GERD. Unfortunately, postprandial reflux is not necessarily associated with postprandial symptoms and can remain hidden for significant periods of time.

By my theoretical assessment, a residual abdominal LES of 20 mm or greater is competent because it can accommodate a 10 mm dynamic shortening during a heavy meal without reaching the <10 mm length that is known to be associated with significant LES failure [transient lower esophageal relaxation (tLESR)].[7] A residual abdominal LES length of 20 mm, by the first formula above, is the result of 15 mm of abdominal LES damage. 15 mm of abdominal LES damage is the end point of the compensated phase of LES damage where the LES remains competent even with the dynamic shortening associated with a heavy meal.

A person with an LES length that is within the compensated phase of LES damage is significantly less likely to develop LES failure when subject to the dynamic shortening of gastric distension. Kahrilas et al.[7] showed that LES failure (tLESR) and acid exposure with gastric distension produced by air insufflation were directly proportional to the length of the baseline residual LES.

3. IMPACT OF FACTORS OTHER THAN LOWER ESOPHAGEAL SPHINCTER DAMAGE IN GASTROESOPHAGEAL REFLUX DISEASE

It is probable that attempting to explain LES competence by the single factor of abdominal length of the LES is an oversimplification. There are likely to be many other factors that contribute to LES competence, e.g., diaphragmatic crural action, the degree of distensibility of the LES to gastric distension, the compliance of the stomach that determines overdistension and intragastric pressure changes during meals, the exact LES pressure across its length, the total length of the LES, and many other factors. However, there are no reliable and widely available measures of any of these factors.

The new test for LES damage can provide, for the first time, an accurate measure of an element that is of primary importance in maintenance of LES competence. The test requires only the development of a new biopsy device to obtain the required piece of mucosa. After that is achieved, the test uses universally available and inexpensive standard methodology. This is the same histologic processing and microscopic analysis that is used for every endoscopic biopsy. It is presently governed by the same 88305 CPT code that is used for every endoscopic biopsy. The cost of producing the pathologic result of the measured abdominal LES damage is equivalent to any other endoscopic biopsy. The enormous value of the new test is its extremely low cost and lack of any need for capital expenditure for new equipment.

It is reasonable to say that GERD results primarily from progressive LES damage, which begins in the distal region and causes shortening of the abdominal LES. It is absolutely true that this progressive abdominal LES damage causes shortening of the LES. It is not true that any other factor associated with LES competence has a similar absolute relationship. It is therefore incumbent on the medical community to evaluate the impact of the ability of measuring LES damage by the new histologic test.

It is also true that all present physiologic evaluations of GERD are done at the point of the present diagnosis of GERD by the Montreal definition of troublesome symptoms. Endoscopy is delayed even further to the point of treatment failure. This is the point at which gastroenterologists commonly become involved in the management of GERD. Much of the gastroenterology literature deals with the management of patients with LES failure that has caused sufficiently severe reflux into the thoracic esophagus to produce troublesome symptoms.

In Table 19.1 the likelihood is that at this stage of the disease, there is an abdominal LES that has shortened from 35 to 10 mm. At this level of LES destruction, it is very likely that other compensatory mechanisms come into play to prop up the defective LES. The probability that LES damage is the sole pathologic abnormality and the sole or main reason for reflux is much higher in the earlier stages of preclinical GERD that this new method is attempting to detect and manage.

4. PROGRESSION OF LOWER ESOPHAGEAL SPHINCTER DAMAGE WITH INCREASING AGE

Lee et al.,[8] in a study of 1307 patients with GERD with a mean age of 49 ± 14 years and a parametric age distribution ranging from 15 to 92 years, showed, by univariate analysis, there was a highly significant correlation between esophageal acid exposure and both LES pressure ($P<.0001$) and abdominal LES length ($P<.0001$). Esophageal acid exposure was more impacted by structural degradation of the reflux barrier (i.e., decreasing abdominal LES length) and was more pronounced than that of functional weakness (i.e., decreasing LES pressure).

On multivariate regression, increasing age was associated with decreasing abdominal LES length ($P<.001$) (Fig. 19.2). The decrease in the length of the abdominal LES showed a linear decline with lengths of 15, 13, 13, 12, 9, and 7 mm for <30, 30–40, 40–49, 50–59, 60–69, and >70 years, respectively. The difference in the mean abdominal LES length in the 30–40 year group and that in the 60–69 year group (i.e., a three-decade interval) was 4 mm.

Table 19.3 shows the status of the abdominal LES during the lifetime of an individual based on an initial abdominal LES length of 35 mm and varying rates of progression of LES damage, which is assumed to begin at age 15 years. The table also assumes linear progression of LES damage over the long term (see below). The red zones indicate a residual abdominal LES length of <10 mm (point of severe GERD), and the orange zones indicate a residual abdominal LES length of <20–10 mm, which is the stage of clinical GERD of increasing severity short of severe GERD (see Table 19.1).

The above table is obviously impacted by the assumption of the initial length of the abdominal LES being 35 mm. Tables 19.4 and 19.5 show the progression of LES damage with age with assumed initial lengths of 32 and 29 mm, respectively, with the different rates of progression.

It will be seen that there is a significant but not massive shift of the points of onset of clinical GERD (orange) and severe GERD (red) with the shorter initial abdominal LES lengths. A shorter initial length of the abdominal LES will result in a shorter phase of compensated LES damage with LES failure and reflux of increasing severity occurring at a younger age for an identical rate of LES damage progression. However, if the test for LES damage is done at an early age, the time available for intervention is still substantial even with the error.

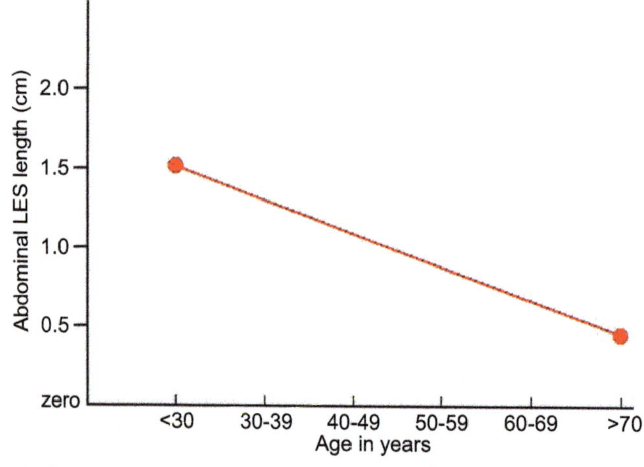

FIGURE 19.2 Correlation of the length of the abdominal lower esophageal sphincter (LES) with age in a large population of patients with gastroesophageal reflux disease in the study by Lee et al.,[8] showing that there is a decrease in the mean length with increasing age.

TABLE 19.3 Changes With Age of the Functional Residual Length of the Abdominal Lower Esophageal Sphincter (LES) Assuming That the Original Length at Maturity Is 35 mm, LES Damage Begins at Age 15 years, and LES Damage Has a Linear Progression Over the Long Term

Rate of LES Damage mm/decade	At 25 Years (mm)	At 35 Years (mm)	At 45 Years (mm)	At 55 Years (mm)	At 65 Years (mm)	At 75 Years (mm)
1	34	33	32	31	30	29
2	33	31	29	27	25	23
3	32	29	26	23	20	17
4	31	27	23	19	15	11
5	30	25	20	15	10	5
6	29	23	17	11	5	0
7	28	21	14	7	0	0
8	27	19	11	3	0	0
9	26	17	8	0	0	0
10	25	15	5	0	0	0

The abdominal LES lengths in green represent lengths at which the LES is likely to be competent. The lengths in orange represent an LES that is susceptible to failure with gastric distension and dynamic shortening (i.e., at risk of postprandial reflux). The lengths in red represent an LES that is below the length at which LES failure occurs at rest.

TABLE 19.4 Changes With Age of the Functional Residual Length of the Abdominal Lower Esophageal Sphincter (LES) Assuming That the Original Length at Maturity Is 32 mm, LES Damage Begins at Age 15 years, and LES Damage Has a Linear Progression Over the Long Term

Rate of LES Damage mm/decade	At 25 Years (mm)	At 35 Years (mm)	At 45 Years (mm)	At 55 Years (mm)	At 65 Years (mm)	At 75 Years (mm)
1	31	30	29	28	27	26
2	30	28	26	24	22	20
3	29	26	23	20	17	14
4	28	24	20	16	12	8
5	27	22	17	12	7	2
6	26	20	14	8	2	0
7	25	18	11	4	0	0
8	24	16	8	0	0	0
9	23	14	5	0	0	0
10	22	12	2	0	0	0

The abdominal LES lengths in green represent lengths at which the LES is likely to be competent. The lengths in orange represent an LES that is susceptible to failure with gastric distension (i.e., at risk of postprandial reflux). The lengths in red represent an LES that is below the length at which LES failure occurs at rest.

TABLE 19.5 Changes With Age of the Functional Residual Length of the Abdominal Lower Esophageal Sphincter (LES) Assuming That the Original Length at Maturity Is 29 mm, LES Damage Begins at Age 15 years, and LES Damage Has a Linear Progression Over the Long Term

Rate of LES Damage mm/decade	At 25 Years (mm)	At 35 Years (mm)	At 45 Years (mm)	At 55 Years (mm)	At 65 Years (mm)	At 75 Years (mm)
1	28	27	26	25	24	23
2	27	25	23	21	19	17
3	26	23	20	17	14	11
4	25	21	17	13	9	5
5	24	19	14	9	4	0
6	23	17	11	5	0	0
7	22	15	8	1	0	0
8	21	13	5	0	0	0
9	20	11	3	0	0	0
10	19	9	0	0	0	0

The abdominal LES lengths in green represent lengths at which the LES is likely to be competent. The lengths in orange represent an LES that is susceptible to failure with gastric distension (i.e., at risk of postprandial reflux). The lengths in red represent an LES that is below the length at which LES failure occurs at rest.

These tables are only for illustration of the new method in this present state of darkness. When new testing for LES damage provides data, there will be no need for any tables. The initial length of any patient's abdominal LES will be known within the level of accuracy of manometric and histologic measurement.

The data relating to the age of onset of clinical GERD and severe GERD in the population should provide an idea as to which of these tables are most likely to be true. However, the inability to accurately define both the onset of clinical GERD and severe GERD by present methods makes it impossible to gather these data from the literature.

In most large series of GERD, the median age of the patient population is around 45–50 years with an interquartile range that is usually around 30–75 years.[8,9] This is not the age of onset of GERD; it is the point at which physicians take GERD seriously by virtue of symptoms that have become troublesome and caused disruption of the quality of life. The correlation of that point of GERD with LES damage is not reliable.

It may be clearer to gastroenterologists experienced with treating large numbers of GERD patients with varying disease severity to look at these graphs and come to some conclusion at an anecdotal opinion-based level that is more accurate than I can derive from the literature.

5. TYPE OF PROGRESSION OF LOWER ESOPHAGEAL SPHINCTER DAMAGE: LINEAR OR NOT?

In any chronic disease, the method of progression is critical in designing an appropriate management method.

For example, chronic ulcerative colitis progresses by intermittent relapses that occur at unpredictable intervals during which the disease is in remission. If the cause of relapse is prevented or treated when it happens, i.e., by immunosuppressive therapy, the progression of chronic ulcerative colitis can be slowed.

In contrast, chronic alcoholic liver disease and pancreatitis progress in a more linear manner as long as the alcohol habit is similar over the long term. Alcoholic intake can be acutely increased during binge episodes, producing attacks of acute liver injury or pancreatitis. However, the overall progression of the chronic disease tends to be linear over the long term. This means that while an alcoholic binge may cause sudden acute disease, this is balanced out in the progression to chronic liver and pancreatic disease over a long time frame of many decades. This is true as long as the basic cause of injury, which is alcohol ingestion, remains unchanged over this period. At any point, if the cause is removed, progression can be slowed as long as irreversibly progressive disease has not already been established.

GERD is more like alcoholic liver disease than ulcerative colitis. Once a person's eating habit has been established, the impact of repetitive injurious effect of gastric overdistension on the LES, which results in LES damage, is likely to progress

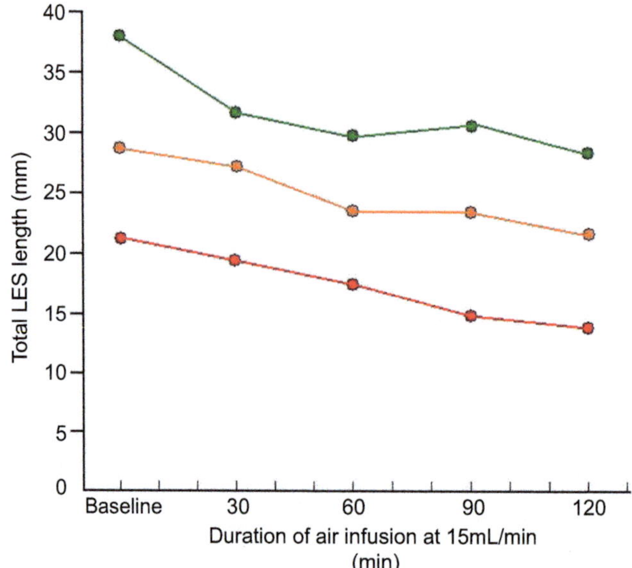

FIGURE 19.3 Decrease in baseline total lower esophageal sphincter (LES) length with gastric distension induced by insufflation of air into the stomach at 15 mL/min in the study of Kahrilas et al.[7] The dynamic shortening of the LES is similar in the normal (green), gastroesophageal reflux disease (GERD) without hiatal hernia (orange), and severe GERD with hiatal hernia. The similar decrease in the three groups indicates that amount of dynamic shortening of the LES does not change with decrease in baseline LES length.

in a linear manner over the long term. A binge of overeating during the Christmas season likely produces more damage than a period of diminished intake during lent. However, over the long term, progression of LES damage will likely balance out. Like alcohol, if the eating habit changes, progression can slow, but will not be effective if the cellular changes have reached irreversibility. Like alcohol, the probability that a poor eating habit will change voluntarily is low. Even progression to obesity does not commonly cause a significant change in the eating habit of many patients.

There is some evidence in the literature that supports a linear progression of LES damage. In Kahrilas et al.,[7] three patient groups were defined: "normal," which is really no symptomatic GERD; GERD without hiatal hernia, which is GERD of milder severity than GERD with hiatal hernia. The baseline total LES length showed a stepwise decrease among the groups.

The study produced gastric distension with air insufflation into the stomach at a constant rate. The dynamic shortening of the LES was identical in the three groups, decreasing in essentially parallel lines (Fig. 19.3). This shows that the baseline residual length of the LES does not influence the amount of gastric distension and dynamic LES shortening caused by air insufflation. This shows that the cause of LES damage (gastric overdistension) has no vicious cycle effect. It affects the LES in the same way at all levels of baseline LES length. This suggests that the exposure of esophageal squamous epithelium to gastric acid during gastric overdistension is likely to be the same irrespective of the amount of abdominal LES damage. As a result, the rate of cardiac metaplasia and progression of LES damage is likely to be linear.

I have assumed the linearity of the rate of LES progression on the basis of scanty data that are presently available in the literature (Fig. 19.4). The new test to measure LES damage will provide data that will more definitively show how LES damage progresses with time.

6. ASSESSING THE RATE OF PROGRESSION OF LOWER ESOPHAGEAL SPHINCTER DAMAGE

Assessing the rate of progression of LES damage is theoretically simple. All that is needed is two accurate measurements of LES damage separated by a reasonable time interval.

6.1 The Two-Measure Method

The two-measure method for determining progression of LES damage is theoretically the most accurate for assessing the rate of progression of LES damage as long as the two points of measurement are separated by a sufficient time interval to ensure linearity of progression of LES damage.

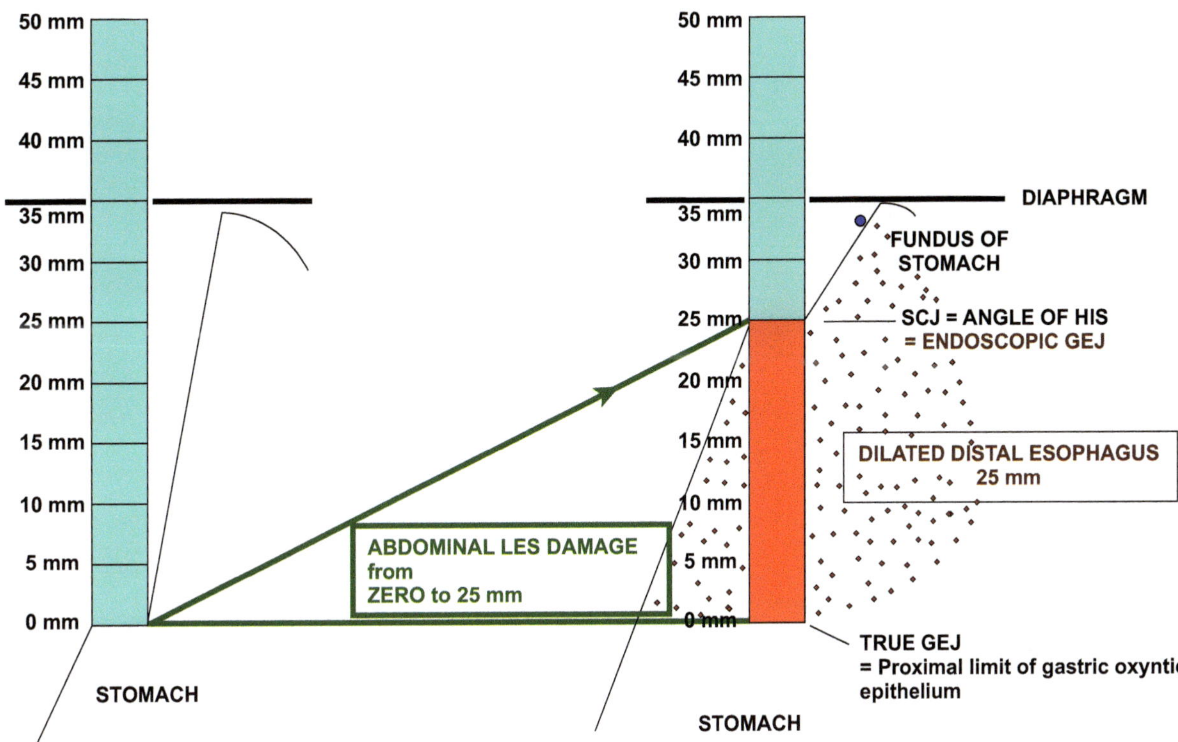

FIGURE 19.4 Progression of abdominal lower esophageal sphincter (LES) damage from 0 to 25 mm (severe gastroesophageal reflux disease). The progression of abdominal LES damage is shown as a *straight line*, indicating linear progression. *GEJ*, gastroesophageal junction; *SCJ*, squamocolumnar junction.

With two such measurements of abdominal LES damage, the rate of progression of abdominal LES damage is calculated by the formula:

$$\text{Rate of progression of LES damage} = \frac{\text{Measurement \#2} - \text{Measurement \#1}}{\text{the time interval between the two tests}}$$

This cannot be applied by present methodology because there is no way to measure LES damage. With the new histologic test for LES damage (=length of cardiac epithelium with and without parietal and/or goblet cells distal to the endoscopic GEJ), the measurement becomes not only possible, but also extremely accurate (Fig. 19.5).

As an example, if one measures the length of abdominal LES damage at ages 20 and 50 years in any individual, one can easily calculate the rate of progression of abdominal LES damage during this 30 years of life. Let us take two scenarios:

1. If LES damage at ages 20 and 50 years is 2 and 8 mm, respectively, the rate of progression of LES damage is 2 mm/decade. This person with 8 mm of LES damage (residual abdominal LES length of 27 mm) will not have GERD at age 50 years; he/she is in the phase of compensated LES damage where the LES is competent (i.e., an abdominal length >20 mm) (Fig. 19.5).
2. If LES damage at ages 20 and 50 years is 4 and 19 mm, respectively, the rate of progression of LES damage is 5 mm/decade. This person with 19 mm of LES damage at age 50 years (16 mm residual abdominal LES) is likely to have mild GERD at age 50 years. While the LES is competent at rest, dynamic shortening of 7 mm during a heavy meal can cause the abdominal LES length to become <10 mm, triggering episodes of LES failure (tLESR), reflux into the thoracic esophagus causing mild abnormality in the pH test and possible postprandial symptoms (orange zone) (Fig. 19.5).

If one assumes that the rate of progression of LES damage is constant over the long term, i.e., a person who damages the LES at a certain rate will continue to damage it at the same rate going forward, the calculated rate can be projected to predict the length of the abdominal LES into the future.

The measurements at the two points of life can be graphed with the patient's age at the time of each measurement, allowing a complete extrapolation of abdominal LES length into both the past to define the onset of disease and into the future to define the projected length of the abdominal LES during the remainder of the patient's life (Fig. 19.5).

In the examples above, the person "(a)" with a rate of progression of LES damage of 2 mm/decade will have a further LES shortening of 6 mm at age 80 years. This will leave him/her with a residual abdominal LES length of 21 mm, which is

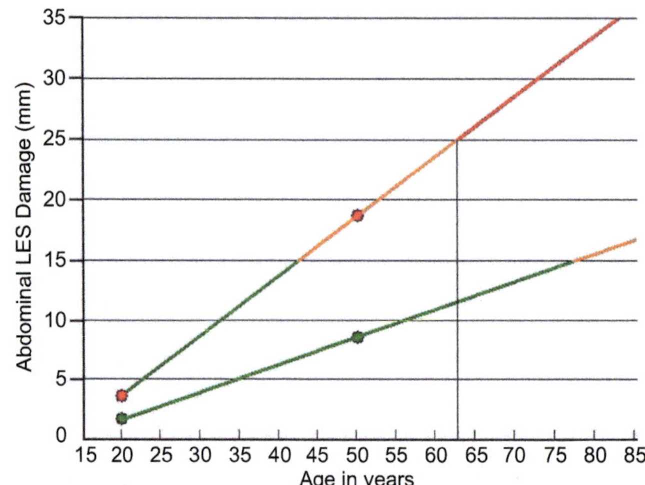

FIGURE 19.5 The method of predicting future lower esophageal sphincter (LES) damage to the end of expected life span. Two measurements of abdominal LES damage at ages 20 and 50 years permit extrapolation of LES damage into the future, based on the assumption that progression of LES damage will be linear. The person with lower LES damage in the tests will not enter the critical point of LES damage (25 mm) within his/her life span. The person with higher LES damage will reach 25 mm of LES damage at age 62 years.

yet within the phase of compensated LES damage (Fig. 19.5). The LES is competent even with dynamic shortening with a heavy meal. This person has lived his/her life with an eating habit that has kept him/her within the reserve capacity of the LES. He/she is one of the 70% of the population who never develops GERD. Even if this person progresses into early GERD with infrequent LES failure and mild reflux (orange zone) after age 80 years, it is highly likely that he/she will never reach the stage of >25 mm of LES damage that results in visible CLE and adenocarcinoma. Lee et al.[8] showed that even with decreasing abdominal LES length and increasing acid exposure, elderly persons tend to remain asymptomatic.

In contrast, the person "(b)" with a progression rate of 5 mm/decade will show the following progressive increase in abdominal LES damage as follows: at age 50 years = 19 mm; at age 60 = 24 mm; at age 70 = 29 mm. This person can be predicted at the time of the measurement (at any age between 20 and 50 years, where he/she may be asymptomatic or at the onset of GERD) to enter the red zone of severe GERD at age 63 years (Fig. 19.5).

This prediction allows for an intervention to control the progression of LES damage at sometime between the ages of 20 and 50 years when the test is performed (when he/she is asymptomatic or at the early onset of postprandial symptoms) and age 62 years before severe GERD (abdominal LES length <10 mm) develops. If this intervention is achieved and LES damage is prevented from reaching 25 mm, visible CLE is prevented and adenocarcinoma never occurs. The management objective therefore converts a person who will enter the risk pool for adenocarcinoma in the future to one who will never be at risk.

If one examines the progression of abdominal LES damage in person "b" in Fig. 19.5, it will be seen that the age at which he/she exhausts the reserve capacity of the abdominal LES (=15 mm of abdominal LES damage) is 42 years. This is, by the new method of understanding GERD, the onset of mild postprandial symptoms. If the test is performed at the onset of symptoms (i.e., age 42 years in this person), it will be predicted that he/she will reach the 10 mm length of severe GERD at age 62 years.

If an intervention is done at age 42 years (Fig. 19.6), all that needs to be achieved is to slow the progression of damage of a relatively intact LES. To keep this person from getting to the critical point of 25 mm LES damage before age 80 years, the rate of LES damage needs to be slowed from 5 to 2.6 mm/decade (Fig. 19.6). If progression of LES damage is completely stopped, the LES damage remains at 15 mm and GERD does not progress at all. There is no necessity in this person of any procedure to repair the sphincter. This will only be needed if nothing is done and abdominal LES damage is permitted to reach 25 mm of damage at age 62 years. This is the person's fate with the present management protocol for GERD.

Presently recommended management consists of giving this patient PPIs to control symptoms and just waiting for him/her to reach the point of severe GERD at age 60–70 years. By this time, the patient is at risk of having progressed to severe LES damage, severe reflux (abnormal pH test by present definition of norms), possible treatment failure, and possible visible CLE.

In the example mentioned in Fig. 19.5, I showed the two measurements being done at ages 20 and 50 years in a person's life. It can be similarly applied to any person at any age, preferably above age 30 years with the second measurement being made any time >10 years later. There is no need for the person to have symptoms.

The fact that the new diagnostic test to measure abdominal LES damage by histology provides a theoretical method of predicting patients who are at high risk to progress to GERD and severe LES damage 10–40 years in the future is extremely important.

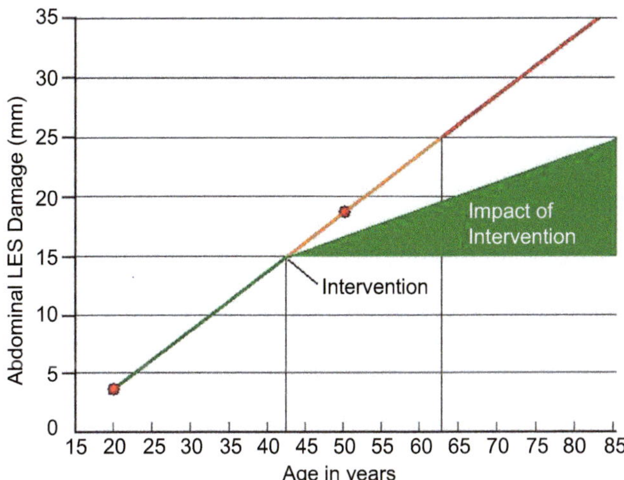

FIGURE 19.6 The person with the higher lower esophageal sphincter (LES) damage in Fig. 19.5 is shown having an intervention to prevent LES damage from reaching 25 mm done at age 42 years. This ranges from arresting progression when the LES damage will stay at the level of the test (15 mm) or slowing progression such that the slope of LES damage in the future is altered to achieve the objective of the person not reaching 25 mm of abdominal LES damage.

It permits identification of those patients destined to develop severe GERD (defined by treatment failure and/or visible CLE) at a point where they have no symptoms or are at the onset of symptoms and their LES is relatively intact.

At present, this two-measure method is impractical solely because the measurement of abdominal LES damage is not possible without endoscopy. In the future, if a simple test to measure abdominal LES damage by a method that does not need endoscopy becomes available, this theoretical two-measure method can become a reality. It only needs the ingenuity of the medical device industry to develop a device that can be passed to the area distal to the SCJ and either directly measures (e.g., by a photograph of the area whose image can be analyzed to make the measurement) or by obtaining a 25 × 2 × 1 mm mucosal sample by a suitable biopsy device delivered to this point in some manner (e.g., a guidewire with a miniaturized camera that permits visualization of the SCJ). If this test can be done quickly and safely with low cost in a doctor's office without sedation, the two-measure method becomes feasible as a screening methodology.

If the two-measure method of assessment of the rate of progression of LES damage fulfills the promise it has of being able to predict LES damage into the future, it can become a stand-alone test that can completely remove the need for all other diagnostic testing for GERD.

6.2 The One-Measure Method

As long as the measurement of abdominal LES damage requires endoscopy, the two-measure method of assessing the rate of progression of abdominal LES damage is not likely to be acceptable or feasible from the standpoint of cost and availability of resources.

The prediction of future LES damage in any person can be made with a single measurement of abdominal LES damage, but it is more complicated and requires certain assumptions that may compromise the accuracy of the prediction by a yet unknown amount. The feasibility of this one-measure method will become apparent when data from extensive testing establish the reality.

Using one measurement of abdominal LES damage, the rate of progression of LES damage in an individual can be defined by assuming the age on onset of LES damage, i.e., when LES damage was zero.

The assumed age of onset of LES damage (i.e., point zero of LES damage) artificially replaces the second measure needed for calculating the rate of progression of LES damage. The rate of progression can be calculated with one measurement of LES damage by the formula:

$$\text{Rate of progression of abdominal LES damage} = \frac{\text{Measured LES damage (by histologic test) at age x years}}{\text{Time difference between test and assumed age of onset}}$$

For example, if it is known (or assumed) that the age of onset of LES damage was age 15 years, then the measured LES damage by histology at (for example) age 50 would provide the rate of progression of LES damage: *Measured abdominal LES damage/35 years = rate of LES damage/year.*

6.3 The Age of Onset of Abdominal Lower Esophageal Sphincter Damage: The Second Measure in the One-Measure Method

Development of a slope of future LES damage is impossible with one measurement of LES damage. If there is no LES damage at the time of measurement, the onset may be in the future. If there is any LES damage, the onset was at some unknown point in the past.

The closer the squamo-oxyntic gap gets to zero, the more likely it will be close to the onset of LES damage. However, because the rate of progression of LES damage is so variable, it is impossible to know when damage began. In Chandrasoma et al.,[10] there was a 61-year-old person with no cardiac epithelium in any section, no cardiac and oxyntocardiac epithelium in 2/6 sections, and a maximum length of oxyntocardiac epithelium of 1.00 mm. This finding may mean one of the two things: (1) LES damage began when he was old or (2) LES damage began when he was young but progressed extremely slowly from 0 to 1 mm in 61 years. The rarity of children who have no cardiac and oxyntocardiac epithelium at autopsy[11] makes it highly likely that the second scenario is more likely.

In Section 1 above, I suggested that the onset of LES damage is likely to be related to the age at which a person develops an adult eating habit. In normal people, this is likely around 15–20 years. It may be earlier, at 10–15 years, in some people. There may be a relationship between the presence of childhood obesity and the onset of LES damage. From a logical standpoint, it seems reasonable to assume that the onset of LES damage is at age 15 years with a correction made to an onset at age 10 years if there is a history of childhood obesity.

The onset of abdominal LES damage represents the second measure of abdominal LES damage (i.e., zero at age 15 or 10 years if there is a history of childhood obesity) to produce the slope of future LES damage in conjunction with the one actual measurement of LES damage made at a later point in the person's life.

The assumptions of the age of onset of LES damage are a reasonable starting point of discussion. They will become more definite once data accumulate with testing for abdominal LES damage with the new pathologic test. In particular, accurate definition of the onset of LES damage will be available in people who have two measures of LES damage at different ages (see above).

An error in the assumption of the age of onset of LES damage produces a significant but small change in the slope of future LES damage (Fig. 19.7). For example, if the test for LES damage done at age 40 shows LES damage of 7 mm, the patient can be predicted to not reach the critical 10 mm residual abdominal LES length at age 85 years irrespective of an onset between 10 and 20 years. If the test at age 40 years shows 14 mm of LES damage, the slope of future LES damage shows the person reaching 10 mm of residual abdominal LES length at age 59 years if the onset of LES damage was at 10 years and 68 years if onset was at 20 years (Fig. 19.7). The only change is that the time lapse from the test to the point at which critical damage is reached is less with an earlier onset than assumed. However, there is still a window of 19–28 years to intervene to slow progression of LES damage sufficiently. The extremely long lag phase between the time of the test and LES failure decreases the implication of any reasonable error in the age of onset. In marginal cases, a second test can be done, providing a two-measure slope.

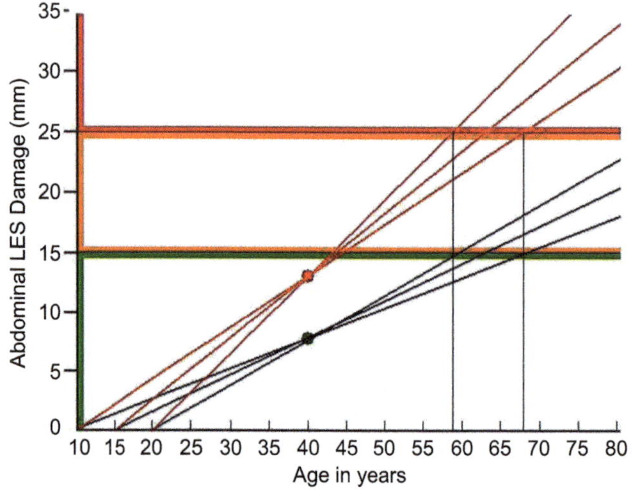

FIGURE 19.7 For predicting future lower esophageal sphincter (LES) damage by one measurement of LES damage, the onset of LES damage is assumed to be at 15 years of age. The variation in the future progression of LES damage changes if the actual onset is different. The variation in the slope is shown for two persons with an onset of LES damage at 10 and 20 years. The slope changes significantly. However, if the test is performed early in life, there is only a small shortening of the window for preventing progression.

LES damage is present in most children under age 15 years.[10] This suggests that if there is an error in the age of onset of LES damage, it will be an onset younger than the assumed age 15 years. With an age on onset before 15 years, the point in the future at which the LES fails will be later in life.

The most accurate data to attempt to define onset of LES damage with one measurement exist in autopsy studies of children who died from causes unrelated to esophageal disease. In Chandrasoma et al.,[10] there were five persons who were at age 15 years or less (Table 19.6).

The cardiac + oxyntocardiac mucosal measurement is exactly the measurement that is used to define LES damage in the new test. This shows that using the age 15 to define the age of onset of LES damage has an error based on this severely limited data ranging from 0.475 to 4.000 mm with a median of 2.875 mm.

In contrast, 7/13 (54%) persons in this autopsy study who were 15 years or more had a squamo-oxyntic gap of >4 mm (Table 19.7).

TABLE 19.6 Age, Cause of Death, and Maximum Length of Cardiac + Oxyntocardiac Epithelium Between the Distal Limit of Squamous Epithelium and Gastric Oxyntic Epithelium in Persons 15 years and Under

Age (Years)	Cause of Death	CM + OCM (Maximum Length; mm)
3	Drowning	2.875
11	Stab injury	0.475
12	Gunshot injury	1.375
14	Drug overdose	4.000
15	Trauma	3.200

Measurements made on sections of the entire circumference at autopsy.
CM, cardiac mucosa; *OCM*, oxyntocardiac mucosa.

TABLE 19.7 Age, Cause of Death, and Mean and Maximum Length of Cardiac + Oxyntocardiac Epithelium Between the Distal Limit of Squamous Epithelium and Gastric Oxyntic Epithelium in Persons Over 15 years of Age

Age (Years)	Cause of Death	CM + OCM (Maximum Length; mm)
17	Gunshot injury	3.000
18	Gunshot injury	4.250
20	Electrocution	1.500
20	Gunshot injury	7.250
21	Blunt trauma	4.950
21	Gunshot injury	2.375
22	Gunshot injury	3.925
25	Gunshot injury	5.375
26	Head trauma	2.550
29	Drug overdose	8.050
41	Head trauma	4.800
49	Drug overdose	7.500
61	Blunt trauma	1.000

Measurements made on sections of the entire circumference at autopsy.
CM, cardiac mucosa; *OCM*, oxyntocardiac mucosa.

TABLE 19.8 Calculated Rate of Progression of Lower Esophageal Sphincter (LES) Damage Accounting for a Possible Error of 2 and 4mm by Assuming the Age of Onset to be 15years

Assumed LES Damage at Age 15years	Measured LES Damage at Age 40years (mm)	Rate of Progression of LES Damage (mm/decade)
0	14	14/25=5.6
Error of 2 mm	14	12/25=4.8
Error of 4 mm	14	10/25=4.0

For this example, I will assume that LES damage in a person at age 40years has been measured at 14mm.

These data indicate that the onset of LES damage is very early in life, shown by the fact that no person in this study had zero cardiac and/or oxyntocardiac epithelium when the entire circumference was examined. Nine (50%) of these patients had a zero squamo-oxyntic gap in some part of the circumference of the SCJ.

Until other data become available, it is reasonable to suggest that using the age of 10–15years as the assumed age of onset of LES damage will have an error that is likely to be around 2mm with a maximum of 4mm. Additionally, it is more likely that an individual establishes an adult eating habit around age 15 years than earlier in childhood, making this age the beginning of a more linear progression of LES damage.

When one uses the age of onset of LES damage to calculate the rate of progression of LES damage, this will be an error that has significance and compromises accuracy (Table 19.8). However, when one considers its impact in the projection of future LES damage, it still can provide information that can lead to effective prevention of progression in a subgroup of people projected to develop severe LES damage in the future.

The potential for error, however, suggests that the one-measure test is best regarded as a screening test rather than a one-time definitive test of LES damage. It will identify the majority of people who have no risk of progressing to severe LES damage and a minority of people who should be kept under observation with a repeated measurement of LES damage in the future. A repeated measurement will convert the evaluation into the much more accurate two-measure method (see above) where the age of onset of LES damage become irrelevant. The feasibility of a second test will dramatically increase if the measurement can be made safely and easily with a nonendoscopic method of obtaining the required measurement of abdominal LES damage.

6.4 Algorithms to Establish the End Point of Lower Esophageal Sphincter Damage

Projections for future LES damage can easily be made with either the two-measure method or one-measure method of assessing the rate of progression of LES damage. Simple computer- or application-based programs for smartphones can be developed to provide projections into the future.

1. In the two-measure method, the entry into a simple program of the following data points allows an exact projection of LES damage backward into the past to determine age of onset of LES damage (i.e., when LES damage crosses the x-axis or zero) and into the future: (1) patient's date of birth; (2) patient's first measured length of LES damage and date of measurement; and (3) patient's second measured length of LES damage and date of measurement. This will provide an individualized slope of LES damage that will show the complete progression of LES damage from zero to any point in the future. This slope can be divided into a green zone (LES damage 0 to 15mm), an orange zone (LES damage >15–25mm), and a red zone (LES damage >25mm). Fig. 19.5 shows two representative slopes for the two patients discussed above with two measurements of abdominal LES damage at ages 20 and 50years of 2 and 8mm and 4 and 19mm, respectively.
2. In the one-measure method, the entry into a simple program of the following data points allows an exact projection of LES damage into the future within the margin of errors discussed above: (1) patient's date of birth; (2) patient's measured length of LES damage; (3) date of measurement; and (4) presence or absence of childhood obesity.

The program will add 15years to the date of birth for the onset of LES damage (i.e., zero) at the assumed age of 15years unless there is a history of childhood obesity when it will add 10years for earlier onset of LES damage. This will provide an individualized slope of LES damage that will show the complete progression of LES damage from zero to any point in the future. This slope can be divided into a green zone (LES damage 0 to 15mm), an orange zone (LES damage >15–25mm), and a red zone (LES damage >25mm).

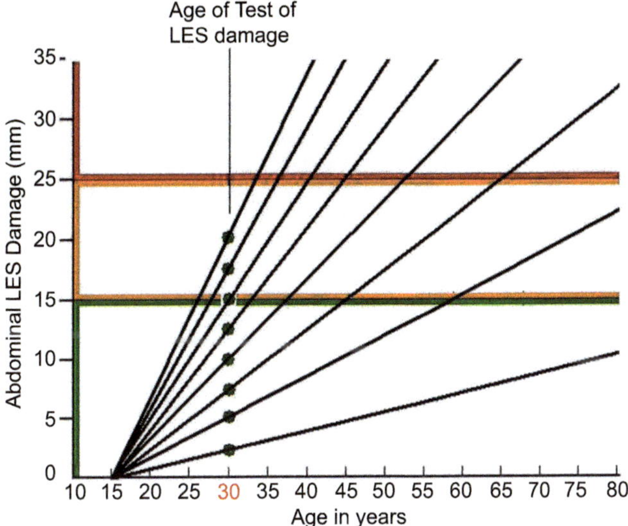

FIGURE 19.8 Diagrammatic prediction of future lower esophageal sphincter (LES) damage with the histologic test for LES damage done in a 30-year-old person. The variation in the slope with a test result for LES damage ranging from 2.5 to 20 mm in 2.5 mm increments is shown. The ages at which the person reaches the various levels of LES damage that define stages of GERD are seen.

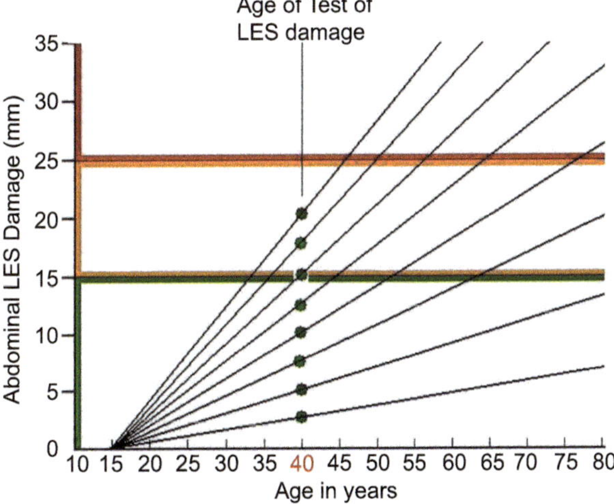

FIGURE 19.9 Diagrammatic prediction of future lower esophageal sphincter (LES) damage with the histologic test for LES damage done in a 40-year-old person. The slope is flatter for each level of LES damage at this age than at age 30 years.

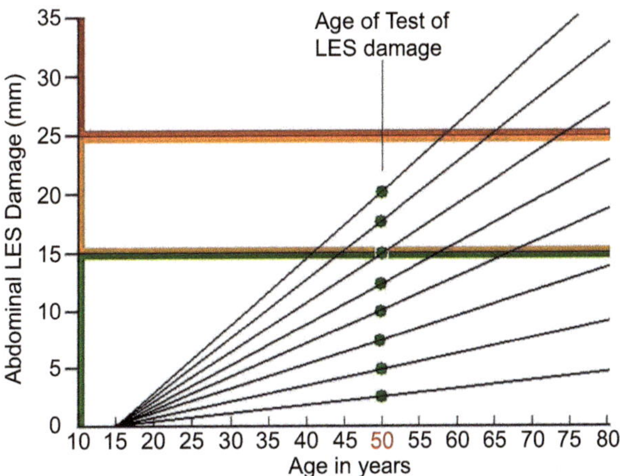

FIGURE 19.10 Diagrammatic prediction of future lower esophageal sphincter (LES) damage with the histologic test for LES damage done in a 50-year-old person. The slope is flatter for each level of LES damage at this age than at age 30 and 40 years.

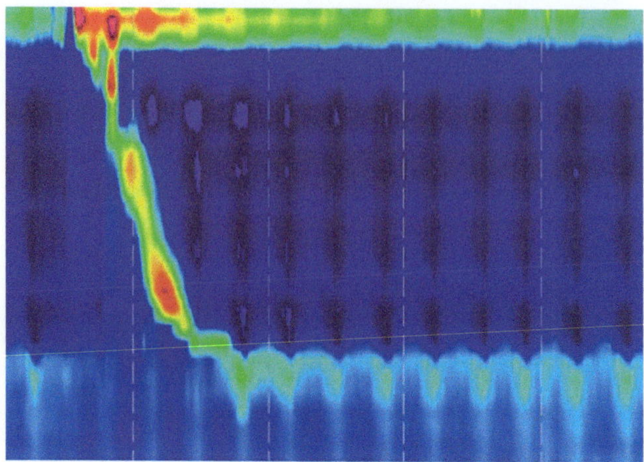

FIGURE 19.11 High-resolution manometry in a patient with gastroesophageal reflux disease showing the typical findings in patients who undergo testing at present when they have treatment failure. In this tracing, the abdominal lower esophageal sphincter (LES) is 0 with 1.1 cm of the thoracic segment remaining. Intervention is not possible; the only possible means of addressing the destroyed LES is an antireflux procedure. *Tracing and calculations of length by Dr. Chris Dengler, Legato Medical Inc.*

Figs. 19.8–19.10 show the projections for selected ages of measurement and measured abdominal LES damage of 2.5–20 mm in increments of 2.5 mm. These show a range of residual abdominal LES length of 32.5–15 mm in persons whose age is 30 years (Fig. 19.8), 40 years (Fig. 19.9), and 50 years (Fig. 19.10).

These slopes show the projected LES damage in the future up to age 80 years. The time it takes for a given person to reach the critical 10 mm level of length of the abdominal LES that signifies severe GERD varies with the age of the person and the amount of LES damage on the date of the test.

These are examples for illustration of the impact of the algorithm. In reality, with use of the algorithm, the exact individualized slope for the patient that is controlled by his/her date of birth and actual measured LES damage in millimeters to two decimal points will be instantly calculated.

Showing a patient his/her personalized progression of LES damage into the future on a computer screen, smartphone, or printed graph with the slope of LES damage in backgrounds of green, orange, and red zones that correlate with no GERD, mild GERD, and severe GERD at different ages is likely to be a powerful tool for communicating to him/her the exact future course of their disease.

Figs. 19.8–19.10 provide some examples of the kind of slope one gets from the application of the algorithm. The actual slope for each person is uniquely individual because of the infinite variation of date of birth, measured length of abdominal LES damage by histology to within a micrometer, and the date on which the test is performed.

The number of people predicted to reach the critical 10 mm level will vary with the timing of the test. If the test is done at the time of troublesome symptoms that defines GERD at the present time or later, when treatment fails to control symptoms, the majority of patients will already be at or below the 10 mm critical level for the residual abdominal LES as measured by manometry (Fig. 19.11). The test has no point in such patients; there is no opportunity to do anything other than antireflux surgery to repair the LES, which is not in favor.

The earlier in the life of a person the test is done, the longer the window of opportunity for prevention. The objective is to make certain that LES damage does not reach the 25 mm level where the person is at risk for visible CLE. If that is achieved, adenocarcinoma of the esophagus will be effectively prevented.

7. MY VISION FOR THE FUTURE OF MANAGEMENT OF GASTROESOPHAGEAL REFLUX DISEASE

My vision of the future will be the potential eradication of GERD as a human disease by preventing LES damage from progressing to the point where it exceeds the 15 mm of damage that defines the compensated phase of LES damage within its reserve capacity.

This vision of the future will require the development of a simplified, cheap, safe, and nonendoscopic screening test for LES damage in a doctor's office at age 30 years. The measured abdominal LES length will identify, by a simple algorithm for a one-measure method, the significant minority of people at high risk for developing GERD and severe GERD in the future with the projected age at which these are likely. These selected few can then be managed initially with a trial of aggressive dietary control to decrease progression of LES damage. This can be assessed in 3–5 years with a second measure to see if the progression curve has changed. This second measure will provide an updated and more accurate prediction of future LES damage.

If it is not practically feasible to prevent GERD because of sheer numbers and availability of resources, the secondary objective is to prevent GERD from progressing to severe GERD. This means preventing progression of LES damage to the 10mm point at which LES failure becomes severe. This level of LES damage is necessary for treatment failure, visible CLE, and adenocarcinoma. This lesser objective is eminently feasible even with the limitation of endoscopy being necessary for obtaining the appropriate biopsy sample.

If it appears that LES damage is progressing to a point of danger, intervention by a simple new and yet undiscovered technique to prevent progression of damage to the LES can be performed. This will be focused on those patients predicted to enter the red zone.

There are interventions that are presently used to repair the LES that will likely achieve the lesser objective of preventing progression of LES damage rather than its present goal of repairing a defective LES. Fundoplication, LINX, TIF, Stretta, and EndoStim all may be found to have the capability of preventing progression of LES damage at a higher rate of success with fewer complications than in the case when they are used as now to repair the severely damaged LES.

One can define the point of danger as:

1. Prediction that reflux will begin within 5 years, i.e., prediction of an abdominal LES length of 15 or 20mm. Intervention at this point can prevent the onset of GERD. The disease will be eradicated. If preferred, the patient can opt to develop GERD if the projection is that severe GERD will never occur, i.e., abdominal LES length will never reach 10mm in his/her lifetime. In such patients, GERD can be controlled with medical therapy for the long term with no risk of future treatment failure, visible CLE, or adenocarcinoma. This will, however, subject the person to complications of long-term PPI use.
2. Prediction that severe GERD will occur within 5 years, i.e., prediction of an abdominal LES length of 10mm within 5 years. Successful intervention at the point at which the measurement is made will maintain the patient with mild GERD for the rest of his/her life. If symptoms persist, they can be controlled with PPIs. The patient will never progress to develop severe GERD. Treatment failure, visible CLE, and adenocarcinoma will not occur. Failure to undergo screening with the new test or failure to intervene at this point will be the only reasons why a person progresses to severe GERD.

The new choice for the potential GERD patient with an abnormal test that predicts future mild GERD is whether they want to have the intervention early to prevent GERD or whether they want to delay the intervention and live their life taking PPIs to control GERD effectively without a risk of severe GERD, visible CLE, and cancer.

The choice would really depend on whether industry has developed an easy, effective method of both performing the screening test for LES damage and effectively intervening to prevent progression of LES damage when this is indicated.

What a different vision to the present mess of GERD!

REFERENCES

1. Vakil N, van Zanten SV, Kahrilas P, Dent J, Jones B, Global Consensus Group. The Montreal definition and classification of gastroesophageal reflux disease: a global evidence-based consensus. *Am J Gastroenterol* 2006;**101**:1900–20.
2. Ringhofer C, Lenglinger J, Izay B, Kolarik K, Zacherl J, Fisler M, Wrba F, Chandrasoma PT, Cosentini EP, Prager G, Riegler M. Histopathology of the endoscopic esophagogastric junction in patients with gastroesophageal reflux disease. *Wien Klin Wochenschr* 2008;**120**:350–9.
3. Marsman WA, van Sandyck JW, Tytgat GNJ, ten Kate FJW, van Lanschot JJB. The presence and mucin histochemistry of cardiac type mucosa at the esophagogastric junction. *Am J Gastroenterol* 2004;**99**:212–7.
4. Robertson EV, Derakhshan MH, Wirz AA, Lee YY, Seenan JP, Ballantyne SA, Hanvey SL, Kelman AW, Going JJ, McColl KE. Central obesity in asymptomatic volunteers is associated with increased intrasphincteric acid reflux and lengthening of the cardiac mucosa. *Gastroenterology* 2013;**145**:730–9.
5. Chandrasoma P, Makarewicz K, Wickramasinghe K, Ma YL, DeMeester TR. A proposal for a new validated histologic definition of the gastroesophageal junction. *Hum Pathol* 2006;**37**:40–7.
6. Zaninotto G, DeMeester TR, Schwizer W, Johansson KE, Cheng SC. The lower esophageal sphincter in health and disease. *Am J Surg* 1988;**155**:104–11.
7. Kahrilas PJ, Shi G, Manka M, Joehl RJ. Increased frequency of transient lower esophageal sphincter relaxation induced by gastric distension in reflux patients with hiatal hernia. *Gastroenterology* 2000;**118**:688–95.
8. Lee J, Anggiansah A, Anggiansah R, et al. Effects of age on the gastroesophageal junction, esophageal motility and reflux disease. *Clin Gastroenterol Hepatol* 2007;**5**:1392–8.
9. Kulig M, Nocon M, Vieth M, Leodolter A, Jaspersen D, Labenz J, Meyer-Sabellek W, Stolte M, Lind T, Malfertheimer P, Willich SN. Risk factors of gastroesophageal reflux disease: methodology and first epidemiological results of the ProGERD study. *J Clin Invest* 2004;**57**:580–9.
10. Chandrasoma PT, Der R, Ma Y, et al. Histology of the gastroesophageal junction: an autopsy study. *Am J Surg Pathol* 2000;**24**:402–9.
11. Kilgore SP, Ormsby AH, Gramlich TL, et al. The gastric cardia: fact or fiction? *Am J Gastroenterol* 2000;**95**:921–4.

Chapter 20

Proof of Concept of the New Diagnostic Method

The objective of this book is to present a new method of understanding the pathophysiology and pathology of gastroesophageal reflux disease (GERD) that is based on abdominal lower esophageal sphincter (LES) damage. After I realized that a new test was available, I took this new method to what my imagination believed was its ultimate endpoint in terms of its potential value.

The goal is to drive the medical establishment to look at the disease in a completely different way.

It is to drive GERD research in a completely new direction rather than the present research that is largely aimed at more effective acid suppression.

It is to provide a new practical diagnostic method on the basis of an objective method based on pathology rather than on the subjectivity of symptoms.

It is to change management goals with a new objective of preventing severe abdominal LES damage and visible columnar-lined esophagus (CLE) rather than symptom control.

The potential that this new concept offers includes the possibility of preventing GERD completely; of preventing disruption of the quality of life by the inadequacy of present acid-reducing therapy; of preventing people without GERD being treated with proton pump inhibitors (PPIs) when they do not have GERD; of preventing Barrett esophagus; and, most importantly, of preventing esophageal adenocarcinoma.

These are lofty and ambitious goals. Even the fact that these goals are suggested as a result of a novel method of assessment of GERD that has not been even contemplated previously is noteworthy. The method that can theoretically achieve all these objectives is based on a single diagnostic test; this suggests that it must address the root cause of the disease. This is true; it addresses LES damage, the known fundamental cause of GERD.

The entire disease is defined from the normal state of the abdominal LES to its endpoint of complete destruction by a new test that can measure abdominal LES damage by histology. Depending on one relatively simple need that must be satisfied for its implementation, the test is easy, very accurate, and cheap. It requires no new expensive technology. It is based on pathology, the time-honored diagnostic method for most gastrointestinal diseases.

The abdominal LES has an initial length in the normal state of ~35 mm. When destroyed completely, it has a zero length. The new method permits recognizing the entire transition from the normal state to severe GERD where the abdominal LES is completely destroyed in terms of measured progression of LES damage from 0 to 35 mm with micrometer accuracy (Fig. 20.1).

I have shown that there is evidence that everyone damages the abdominal LES, but to a variable amount that is determined by a variable rate of progression of LES damage.

If a person ends his/her life with abdominal LES damage 15 mm or less (i.e., with a residual abdominal LES length of 20 mm or more), he/she would never have had GERD (Fig. 20.1). This is 70% of the population. The person has lived his/her life causing abdominal LES damage within its reserve capacity. At 20 mm, the abdominal LES has a baseline residual length that is sufficient to maintain competence. The reason for this is that LES failure (transient LES relaxation) occurs when the abdominal LES has a functional length of <10 mm. When the person with a residual abdominal LES length of 20 mm eats a heavy meal that overdistends the stomach and causes dynamic shortening by an additional 10 mm, the critical length of 10 mm where LES failure occurs is reached but not exceeded in the postprandial phase. 15 mm abdominal LES damage is the endpoint of the phase of compensated LES damage where no reflux occurs.

A patient who destroys >15–25 mm during his/her entire life span (i.e., with a residual abdominal LES length of <20–10 mm) will develop reflux with severity increasing with progressive shortening of the LES within this range (Fig. 20.1). This is 70% of the population who develop GERD. At 15–20 mm of LES damage, they have subclinical or mild GERD that is often ignored or effectively treated with over-the-counter acid-reducing agents. At 20–25 mm of

GERD. http://dx.doi.org/10.1016/B978-0-12-809855-4.00020-8

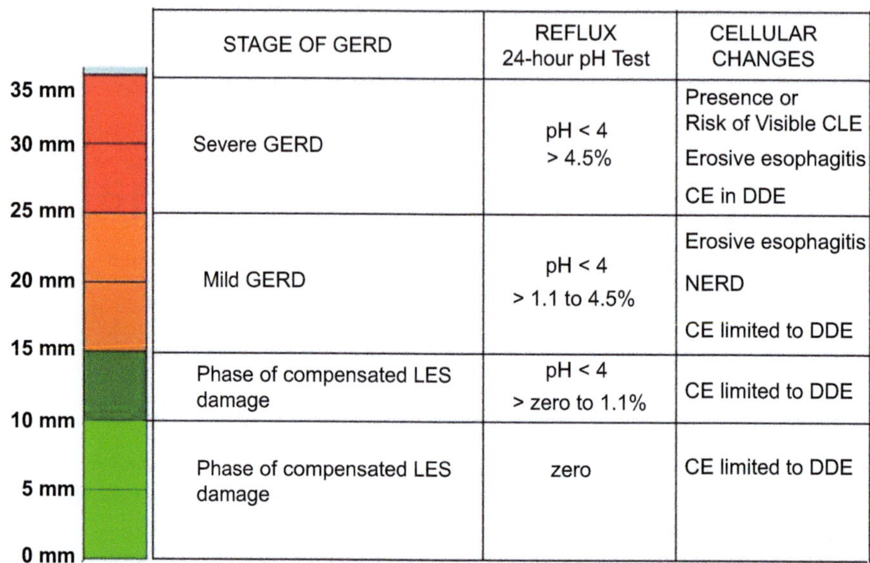

STAGE OF GERD	REFLUX 24-hour pH Test	CELLULAR CHANGES
Severe GERD	pH < 4 > 4.5%	Presence or Risk of Visible CLE; Erosive esophagitis; CE in DDE
Mild GERD	pH < 4 > 1.1 to 4.5%	Erosive esophagitis; NERD; CE limited to DDE
Phase of compensated LES damage	pH < 4 > zero to 1.1%	CE limited to DDE
Phase of compensated LES damage	zero	CE limited to DDE

FIGURE 20.1 Correlation between lower esophageal sphincter (LES) damage and cellular changes in the thoracic esophagus as a consequence of reflux. In the compensated phase of LES damage (*green zone*), the abdominal LES damage progresses from 0 to 15 mm. Although it has lost 40% of its length, the LES remains competent. There is no or minimal reflux and therefore no cellular changes in the thoracic esophagus. Postprandial reflux begins as the LES damage >15 mm and increases until it reaches 25 mm (*orange zone*). At >25 mm of damage (*red zone*), the residual abdominal LES is <10 mm and reflux becomes severe.

LES damage, they have well-established but controllable GERD with no risk of visible CLE and esophageal adenocarcinoma. It is important to recognize that it is not the LES damage at the time of the test that is important; it is the predicted LES damage in the future. Management must be based on the endpoint of LES damage in the future. Otherwise, PPI therapy is doomed to failure in a patient who is under control at the time of the test but will progress to >25 mm of LES damage in the future.

If the person destroys >25 mm (i.e., with a residual abdominal LES of <10 mm) of the abdominal LES to complete destruction (Fig. 20.1), the person has uncontrollable LES failure, incessant reflux with minimal provocation, and severe GERD with a high risk of visible CLE. These are the people who develop adenocarcinoma.

This simple assessment of GERD is not presently possible because LES damage cannot be measured. With the new test that allows accurate measurement of LES damage by histology, the door opens to a new understanding based on the root cause of GERD. The little information available suggests that this criterion correlates much better with severity of reflux into the thoracic esophagus and cellular changes therein than the presence of symptoms or any other diagnostic test that is presently used to guide management of GERD.

The potential of the test for LES damage is so high that it can become the stand-alone sole test for GERD that will direct management. The need for endoscopy and all other diagnostic testing can potentially become obsolete in the management.

1. THE NEED FOR PROOF OF CONCEPT

In this section, I will summarize the strong evidence base that supports the theory of the new method. The method is new and revolutionary. The method requires complete reevaluation of many steps in the progression of GERD from this new vantage point. The evidence base for many of these steps exists at present time but is not believed, ignored, or misinterpreted.

For some steps, however, the data are scanty albeit always confirmatory and never contradictory of the new method. However, there is a need for proof of concept with further clinical testing.

For those elements that have powerful support in evidence, it is only adherence to false dogma that prevents acceptance. This is true for the false dogma that cardiac epithelium is a normal lining epithelium of the stomach and that the end of the tubular esophagus and proximal limit of rugal folds is the true GEJ. These two errors result in the failure to recognize the pathology of early LES damage, the dilated distal esophagus, which is mistaken by most of the world as the normal proximal stomach.

Think of it: we fail to manage GERD appropriately at the present time because we believe implicitly and with no evidence that the initial phase of LES damage that results in the dilated distal esophagus represents normal stomach! Is it a surprise that we fail?

2. CARDIAC EPITHELIUM

2.1 The Truth

Cardiac epithelium is never normal and never gastric; it is always esophageal and results from columnar metaplasia of esophageal squamous epithelium.[1,2] There is only one etiology for cardiac metaplasia of the squamous epithelium of the esophagus: exposure to gastric contents. When cardiac epithelium (with and without parietal and/or goblet cells) is found in a biopsy taken at and around any definition of the GEJ, it is absolute evidence that the location of the biopsy is esophageal and not gastric. It is proximal to the true GEJ, which is defined by the proximal limit of gastric oxyntic epithelium.[3]

Cardiac metaplasia results from two completely different methods of exposure of squamous epithelium to gastric contents:

1. In a person with a competent LES and no reflux, the squamous epithelium is exposed during periods of gastric overdistension when dynamic shortening of the abdominal LES causes the squamocolumnar junction (SCJ) to move distally beyond the pH transition point (Fig. 20.2).[4,5] This is a transient exposure limited to the postprandial phase when the stomach is full and overdistended. Damage to the squamous epithelium during a single exposure is mild and reversible. It requires repeated cumulative damage to cause cardiac metaplasia. This is the mechanism of development of the dilated distal esophagus and LES damage.[6] This cardiac metaplasia of the squamous epithelium of the abdominal esophagus is slow and is measured in millimeters per decade (Fig. 20.3).
2. In a person whose LES is incompetent, reflux of gastric contents into the thoracic esophagus results in exposure of a large area of squamous epithelium to gastric contents (Fig. 20.4). This, when reflux is sufficiently severe, causes cardiac metaplasia in the thoracic esophagus (i.e., visible CLE) (Fig. 20.5). A large surface area of the esophagus can undergo metaplasia quickly, measured by centimeters per year,[7] in contrast to metaplasia in the abdominal esophagus.

2.2 The Present Evidence Base

It was in 1994 that I first proposed the concept that cardiac epithelium found distal to what everyone believes is the GEJ (defined by the end of the tubular esophagus and/or the proximal limit of rugal folds) was always esophageal in its origin and not the stomach, and always abnormal, resulting from metaplasia of esophageal squamous epithelium.[8–10]

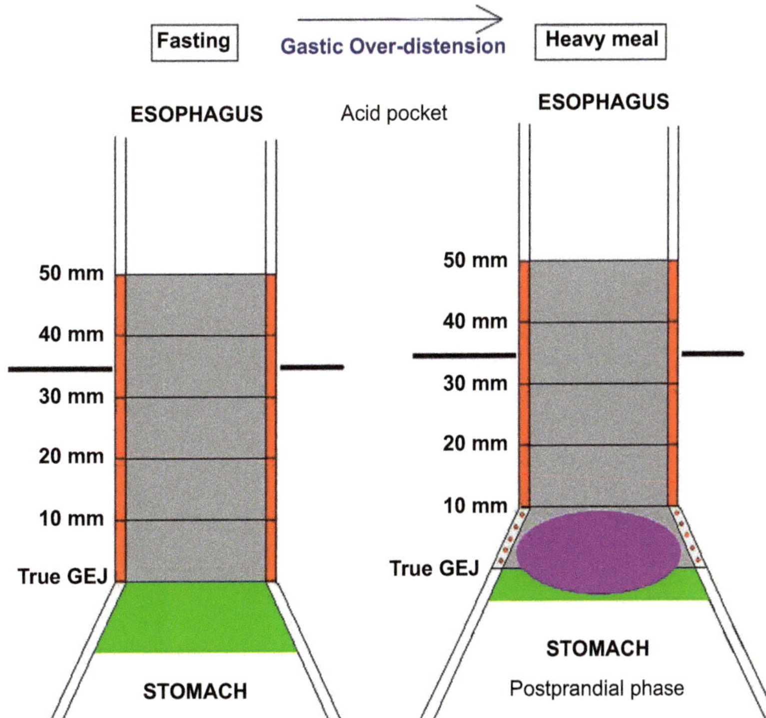

FIGURE 20.2 The mechanism of lower esophageal sphincter (LES) damage. Gastric overdistension causes effacement of the distal LES and causes the distal squamous epithelium lining the effaced LES to become exposed to the acid pocket below the pH transition point. This ultimately leads to cardiac metaplasia, which results in permanent LES damage and the dilated distal esophagus.

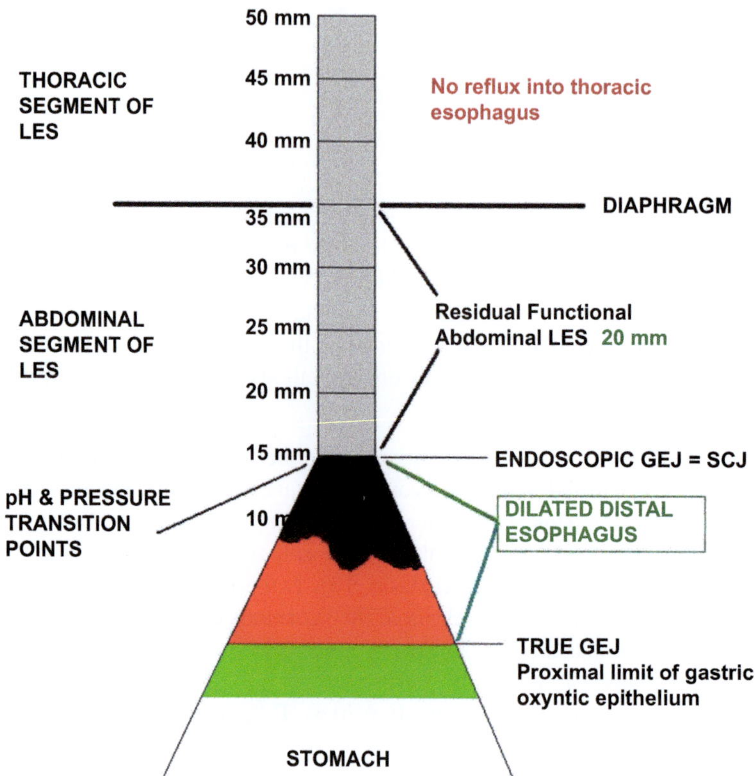

FIGURE 20.3 Diagrammatic representation of the person with 15 mm abdominal lower esophageal sphincter (LES) damage. The LES damage can be measured by the length of cardiac epithelium (black and red) between the squamocolumnar junction (SCJ) (=endoscopic gastroesophageal junction (GEJ) in the person without visible columnar-lined esophagus) and gastric oxyntic epithelium. This area of the abdominal LES that has no sphincter protection is dilated. The pH transition point has moved up along with the SCJ (=GEJ).

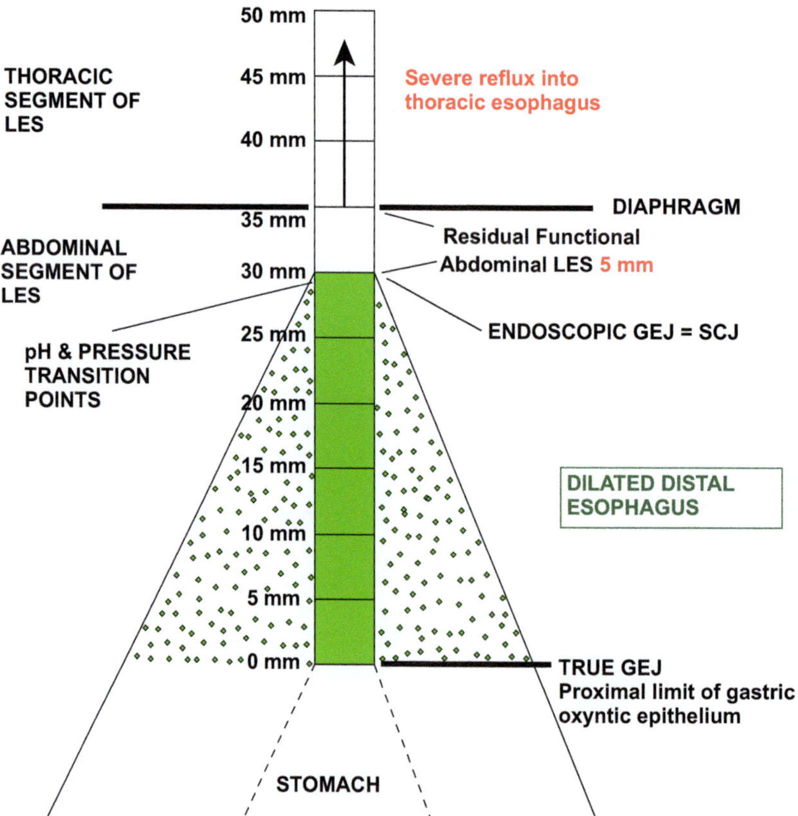

FIGURE 20.4 Lower esophageal sphincter (LES) in a patient with severe gastroesophageal reflux disease. Abdominal LES damage is 30 mm with a residual abdominal LES of 5 mm. There is frequent reflux because the abdominal LES is <10 mm, which correlates with a defective LES. There is a 30 mm dilated distal esophagus between the endoscopic gastroesophageal junction (GEJ) and the true GEJ lined by cardiac epithelium with and without parietal and/or goblet cells (green).

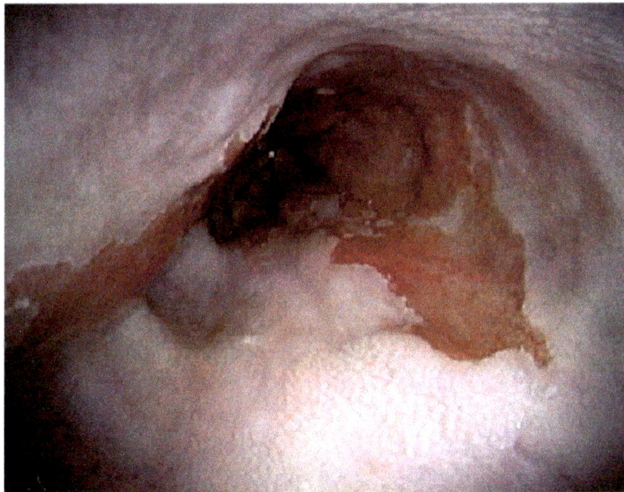

FIGURE 20.5 Visible columnar-lined esophagus showing clear demarcation between the squamous epithelium (white) and columnar epithelium (pink). *Photograph courtesy of Dr. Martin Riegler, Reflux Medical, Vienna, Austria.*

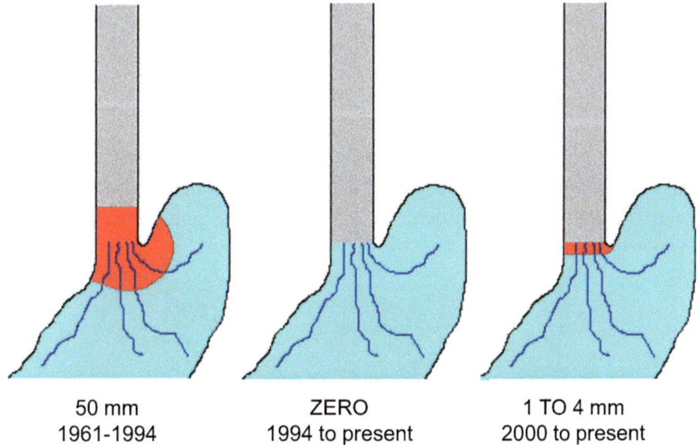

FIGURE 20.6 Historical changes in the accepted "normal" extent of cardiac epithelium. In 1961, Hayward opined that the presence of cardiac epithelium (red) in the distal 2 cm of the esophagus and proximal 3 cm of the stomach was normal (*left*). In 1994, I proposed that cardiac epithelium was normally absent (*center*). This is presently accepted by a minority of physicians. The majority of physicians still accept the alternate incorrect viewpoint that 1–4 mm of cardiac epithelium is normally present distal to the endoscopic gastroesophageal junction in the proximal stomach (*right*).

Back then, it was believed (without any evidence) that cardiac epithelium normally lined the distal 2–3 cm of the esophagus and up to 3 cm of the proximal stomach.[11] This was believed to be a buffer epithelium necessary to prevent "digestion" of esophageal squamous epithelium by gastric acid. It was a concept dating back to the era before the LES was accurately defined. The concept had never adjusted to the fact that the LES was responsible for protecting esophageal squamous epithelium from exposure to gastric contents. Dogmas linger; they are harder to kill than cancer cells.

For the next decade, I was preoccupied with trying to convince people I was right. The idea that cardiac epithelium was not a gastric epithelium met with intense resistance that has still not disappeared.[12] In a 2012 lecture, I pointed out that I had succeeded in decreasing the accepted normal extent of cardiac epithelium in the proximal stomach from 3 cm in 1994 to 0.1 cm (Fig. 20.6). This 0.1 cm is still regarded as normal "stomach" by too many people and prevents forward progress of understanding of GERD.[12]

2.3 The Need for Proof of Concept

I no longer feel the desire to convince people of the fact that cardiac epithelium is always esophageal. I have presented the evidence in Chapter 5 that cardiac epithelium is never gastric and always the result of exposure of esophageal squamous epithelium to gastric contents. The evidence is indisputable although there will always be people who will dispute it.

I believe it is reasonable to say: "Cardiac epithelium is never normal and never gastric; it is always esophageal and results from columnar metaplasia of esophageal squamous epithelium."

A sufficient number of people with open minds have now recognized this truth. A few have adapted their endoscopy biopsy protocol to make use of this fact.

The rest will follow at their own pace. I think I can move forward without waiting for them.

3. THE DEFINITION OF THE TRUE GASTROESOPHAGEAL JUNCTION

3.1 The Truth

The true GEJ is defined histologically as the proximal limit of gastric oxyntic epithelium. It is the junction of the metaplastic cardiac epithelium (i.e., the distal limit of the dilated distal esophagus) with the proximal limit of gastric oxyntic epithelium (i.e., the stomach) (Fig. 20.3).[3]

The true GEJ is not the end of the tubular esophagus or proximal limit of rugal folds, criteria that are presently used universally to define the GEJ at endoscopy and gross examination at autopsy and in resected specimens.[13–15] Neither endoscopy nor gross examination can define the true GEJ at this time because there is no method other than histology that allows differentiation between cardiac epithelium (with and without parietal and/or goblet cells) from gastric oxyntic epithelium. Histology permits differentiation with accuracy and precision (Fig. 20.7).

In a person with no visible CLE at endoscopy, the present definitions of the GEJ represent the proximal limit of the dilated distal esophagus. As such, the gap between the endoscopic and gross definitions of the GEJ and the true GEJ is misinterpreted as "proximal stomach" by the presently accepted definitions of the GEJ. This gap represents the dilated distal esophagus, which is equal to LES damage, and can be identified by measuring the length of cardiac epithelium between the SCJ and the proximal limit of gastric oxyntic epithelium (Fig. 20.3).[6]

3.2 The Present Evidence Base

In 2007, I realized that the presence of cardiac epithelium distal to the endoscopic GEJ meant that the endoscopic GEJ was incorrect and that what was being called the proximal stomach must include part of the damaged distal esophagus that had dilated and developed rugal folds, essentially mimicking the stomach anatomically and endoscopically. The study of

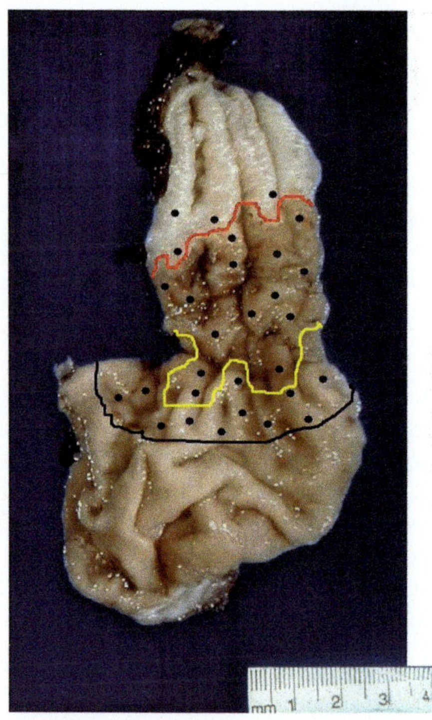

FIGURE 20.7 A esophagogastrectomy specimen showing a long segment of visible columnar-lined esophagus in the thoracic esophagus with an adenocarcinoma. There is a well-demarcated junction between the tube and sac; the rugal folds extend to the end of the tube. The present definitions of the gastroesophageal junction (GEJ) are the end of the tube and the proximal limit of rugal folds. Histologic examination shows that the esophagus extends distal to the endoscopic and gross GEJ to the true GEJ (proximal limit of gastric oxyntic epithelium; *black line*). The area between the end of the tube and the true GEJ is dilated, has rugal folds, and is lined by cardiac epithelium (with parietal and goblet cells). This is the dilated distal esophagus, which results from abdominal lower esophageal sphincter damage. *Red line,* squamocolumnar junction; *yellow line,* distal limit of intestinal metaplasia.

esophagectomy specimens proved that the area that looked like the proximal stomach distal to the end of the tubular esophagus and the proximal limit of rugal folds contained submucosal glands, proving it was the esophagus (Fig. 20.7).[6] The dilated distal esophagus could only be identified by histology; it was that area that was lined by cardiac epithelium (with and without parietal and/or goblet cells). This was the same evidence that Allison and Johnstone used to convince Norman Barrett in 1953 to change his mind that his "tubular intrathoracic stomach" was actually the CLE.[16–18]

The gastroenterology community completely ignores the proof of the true GEJ that belies the universally accepted present definition of the GEJ. They continue to recommend the use of the proximal limit of rugal folds as the definition of the GEJ.[19] They do this while openly admitting that this definition is a purely opinion-based definition not based on any evidence.[15]

3.3 The Need for Proof of Concept

In Chapter 8, I have presented the indisputable evidence that the present definitions of the GEJ place the end of the esophagus at a point that is zero (rarely, in persons without any LES damage) up to 35mm (in people with severe GERD) proximal to the true GEJ. I believe it is reasonable to say: *The true GEJ is defined histologically as the proximal limit of gastric oxyntic epithelium.*

Anyone who believes that the GEJ can be defined accurately at endoscopy is mistaken. The error has serious repercussions because those physicians fail to recognize the underlying pathology in their patients, losing the opportunity for early diagnosis and potential prevention of progression of LES damage to the end stages of the disease.

4. RELATIONSHIP BETWEEN LOWER ESOPHAGEAL SPHINCTER DAMAGE AND CARDIAC METAPLASIA

4.1 The Truth

The esophageal squamous epithelium at the distal end of the functional abdominal LES is exposed to gastric contents during gastric overdistension. With repeated exposure, cumulative damage to the squamous epithelium results in metaplasia to cardiac epithelium.

Cardiac metaplasia in the abdominal esophagus results in loss of LES pressure and causes dilatation of the abdominal esophagus. There is concordance between the length of cardiac epithelium and abdominal LES damage as well as dilatation of the abdominal esophagus that loses LES tone.

LES damage begins, as does cardiac metaplasia, at the distal limit of the abdominal LES. Its progression results in a progressive shortening of the abdominal LES that is equal to the length of cardiac epithelium (with and without parietal and/or goblet cells) between the endoscopic GEJ (which is the SCJ in a person without visible CLE) and the proximal limit of gastric oxyntic epithelium.

4.2 The Present Evidence Base

In Chapter 10, I presented elegant studies that show how gastric overdistension changes the anatomy at the end of the functional LES. 5–10mm of the residual LES is effaced, causing the SCJ to descend below the pH transition point into the acid pocket at the top of the food column during the overdistended postprandial period.[4,5] Squamous epithelial damage results by this mechanism (Fig. 20.2).

There is evidence at a molecular level that exposure to gastric contents causes changes in the differentiating genetic signals in squamous epithelium that result in sequential changes to cardiac epithelium and intestinal metaplasia (see Chapter 15).

While there is good evidence that cardiac metaplasia results in loss of LES pressure (i.e., LES damage), there are few data regarding the mechanism for this relationship. I have suggested in Chapter 7 that alteration in the afferent innervation of the epithelium resulting from the conversion of the richly innervated squamous epithelium by cardiac epithelium that is devoid of innervation may interrupt a local reflex arc that is responsible for maintaining LES tone. This is conjectural. Other mechanisms, such as cytokines released when squamous epithelial damage occurs, may explain loss of tone but cannot explain the concordance between cardiac metaplasia and LES damage.

4.3 The Need for Proof of Concept

This basis of squamous epithelial damage resulting from exposure to gastric contents during gastric overdistension has been replicated in multiple studies. In the absence of any evidence to the contrary, it is reasonable to state that this is proved.

There is strong, albeit scanty, evidence without contradiction that there is a close relationship between the length of cardiac epithelium (with and without parietal and/or goblet cells) distal to the endoscopic GEJ and both LES damage and the dilatation of the abdominal esophagus that has lost LES tone.

However, this is an area that has been ignored at endoscopy because of the misconception that the esophagus ends at the endoscopic GEJ, i.e., at the proximal limit of the dilated distal esophagus. With the availability of the new test, concentrated study of this area becomes possible for the first time. New data will be generated quickly that will refine the relationships between the length of cardiac epithelium, length of the dilated distal esophagus, and manometric length of the undamaged residual abdominal LES.

There is no explanation as to why cardiac metaplasia of the abdominal esophagus is so intimately associated with loss of LES function. Studies regarding how the LES normally maintains its tone and what things cause it to lose muscle tone that controls LES pressure do not exist.

There is much that needs to be done.

5. THE NEW PATHOLOGIC ASSESSMENT OF LOWER ESOPHAGEAL SPHINCTER DAMAGE

5.1 The Truth

The length of the dilated distal esophagus measured histologically as the gap between the endoscopic GEJ (=SCJ in a person without visible CLE) and the proximal limit of gastric oxyntic epithelium is equal to LES damage. This gap is composed of cardiac epithelium (with and without parietal and/or goblet cells) and can be measured with a high level of accuracy.

5.2 The Present Evidence Base

At present, the proposed new pathologic assessment is based on the evidence that has correlated the length of cardiac epithelium distal to the endoscopic GEJ with dilatation of that part of the abdominal esophagus lined by cardiac epithelium.[3] The intraluminal pressure in the abdominal esophagus that is +5 mmHg must be counteracted by a higher LES pressure to maintain the tubular shape of the abdominal esophagus. When LES tone is lost, the abdominal esophagus must dilate. As the length of LES damage increases, the circumference of the abdominal esophagus at the true GEJ increases. This has been shown to be true by measurement.[20]

5.3 The Need for Proof of Concept

While logic and available evidence suggests that this pathologic assessment will accurately measure LES damage, there are very few data in the literature. The reason for the paucity of data is the present misunderstanding of the pathophysiology of LES damage that has resulted in the dilated distal esophagus being regarded as proximal stomach. This has prevented pathologic study of this region.

There is opportunity for those who embrace this concept to study this area and increase the database that shows a direct correlation between LES damage and the length of the dilated distal esophagus.

Additional studies using the ability of the new test to measure LES damage open the door to a complete characterization of the LES in all people. The present limitation of manometry that limits definition to criteria that define a defective LES will be replaced by the ability to define the initial length of the LES in a person (=manomteric length + measured LES damage).

6. PROCUREMENT OF TISSUE TO MEASURE THE LENGTH OF CARDIAC EPITHELIUM

6.1 The Truth

The entire new method is based on the measurement of cardiac epithelium (with and without parietal and/or goblet cells) distal to the endoscopic GEJ (the SCJ in persons without visible CLE). This requires identification of the junction between squamous and cardiac epithelium at the proximal end, which is easy, and the junction between oxyntocardiac and gastric oxyntic epithelia at the distal limit. The latter is invisible at endoscopy and is located at a point that varies from 0 to 35 mm distal to the endoscopic GEJ. To be certain to reach the true GEJ, a vertical biopsy that obtains a single continuous piece of mucosal tissue 35 mm long from the SCJ needs to be taken (Fig. 20.8). If the objective is early diagnosis, a biopsy length of 20 mm is probably adequate.

This has never been done. There is no biopsy device at present that can do this. Endoscopic mucosal resections have effectively removed early neoplastic lesions up to 20 mm large in the dilated distal esophagus, showing that it is probably feasible to obtain a biopsy of this length.

6.2 The Present Evidence Base

The present limitation of biopsy size to 8 mm with a jumbo forceps has precluded measurement beyond this length. An 8 mm biopsy of the SCJ is valuable in the diagnosis of GERD for the following reasons:

1. If it shows the entire length of cardiac epithelium, i.e., if it has squamous epithelium at one end and gastric oxyntic epithelium at the other, it can provide an accurate measurement of the length of cardiac epithelium (Fig. 20.9).

   In Robertson et al.[5] all asymptomatic volunteers had a pure cardiac epithelial length of <8 mm with a mean of 1.75 and 2.50 mm in those with and without central obesity. This measurement did not include oxyntocardiac epithelium; as such, it was not a measure of the dilated distal esophagus.

   It is uncommon in present clinical practice to find very short segments of cardiac epithelium entirely present in an 8 mm biopsy. The reason for this is that endoscopy and biopsy are presently performed at the point of severe GERD, usually with treatment failure. At this stage of the disease, LES damage is severe and the length of the dilated distal esophagus is almost always >8 mm.

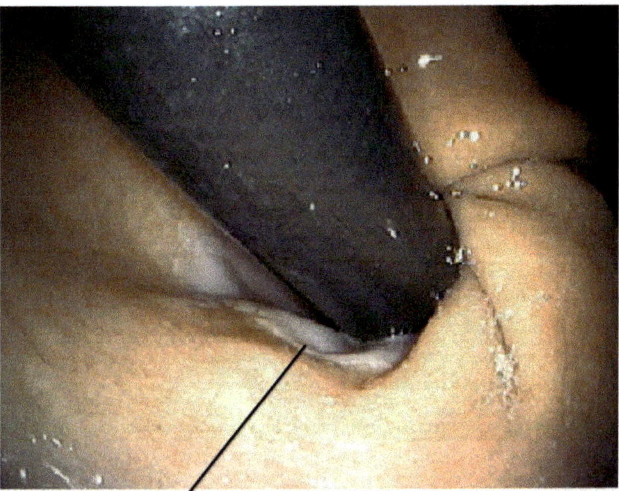

FIGURE 20.8 Retroflex view of the region that requires to be sampled by the new biopsy device. The squamocolumnar junction is seen just below the opening of the tubular esophagus (occupied by the endoscope). The biopsy device must procure a sample of mucosa 20–25 mm long extending vertically down from the distal limit of squamous epithelium (*black line*). *Photograph courtesy of Dr. Martin Riegler, Reflux Medical, Vienna, Austria.*

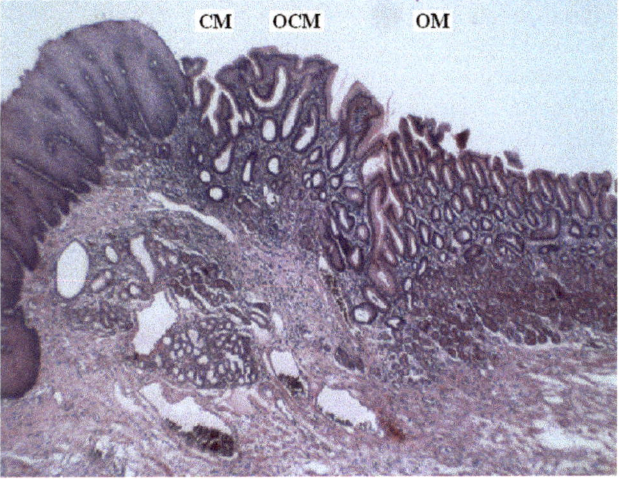

FIGURE 20.9 Biopsy from the squamocolumnar junction showing a 2 mm length of cardiac (CM) and oxyntocardiac (OCM) epithelium with squamous epithelium on the left and gastric oxyntic epithelium on the right. The entire length of the dilated distal esophagus is present in this biopsy, permitting accurate measurement.

However, if the pathologic test for LES damage is performed early, the 8 mm biopsy may be adequate for measurement. The importance of this is that if the full length of cardiac epithelium is <8 mm, the person under study does not have GERD. If the patient has symptoms, this will prove that the symptoms are not caused by GERD. PPI therapy can therefore be avoided.

2. If the biopsy shows intestinal metaplasia, it is predictive of the occurrence of visible CLE within the next 5 years.[21] This is independent of the length of the dilated distal esophagus. As such, this finding is valuable irrespective of the size of the biopsy. It is completely beyond my understanding why biopsies are not recommended in GERD patients who are endoscopically normal. This critical piece of information is easily available but not found simply because of the failure to take a biopsy.

All the measurements of the length of cardiac epithelium in the dilated distal esophagus that have been reported to date have been made either on material removed at autopsy or on sections taken in resection specimens. In these sections, the measurement of cardiac epithelium has been possible to a micrometer level. However, there is no certainty that these measured lengths are the actual in vivo length of cardiac epithelium.

Autopsy causes postmortem changes that can uniquely impact the measurement of cardiac epithelial length in unknown ways. Our measurement of the length of cardiac + oxyntocardiac epithelium ranging from 0 to 8.05 mm in the 18 completely studied persons is accurate. However, it is uncertain whether the upper limit accurately reflects the length of the dilated distal esophagus while the person was living. There is a certainty that the lower limit of zero is accurate; cardiac epithelium cannot disappear as a result of autopsy change without complete autolysis, which was not present in these cases.

Similarly, the act of transecting the esophagus and stomach during esophagectomy is very likely to produce contraction of the muscle of both muscularis externa and muscularis mucosae in a manner that is variable and unpredictable.

As a result, when we reported that the two esophagectomy specimens in patients with squamous carcinoma had cardiac epithelial lengths of 3.1 and 4.3 mm, and the eight patients with adenocarcinoma had lengths ranging from 10.3 to 20.5 mm, the accuracy of the measurements is certain.[3] However, whether they are the lengths that were present in vivo before the esophagus was resected cannot be guaranteed.

It is likely, based on the logic that changes resulting from transecting the muscle are likely to affect all specimens in a similar way, that the conclusion that cardiac epithelial length is greater in adenocarcinoma than in squamous carcinoma is correct. However, there is much less confidence that these lengths will be the same as the actual lengths in vivo.

I have suggested that a multilevel biopsy protocol, if done with precision, can provide a reasonably accurate measurement of the length of cardiac epithelium (see Chapter 17). This is optimistic at best and completely worthless at worst. It has an error rate of at least 2.5 mm, which is significant when the LES damage range that separates onset of GERD from severe GERD is 10 mm.

6.3 The Need for Proof of Concept

Ultimately, an accurate measurement of the cardiac epithelial length will depend on the availability of a new biopsy device that can remove a single longitudinal piece of mucosa extending from the SCJ (in the endoscopically normal person) to a vertical point 20–25 mm distal to the SCJ. This specimen may also shrink after it is removed, but its shrinkage can be standardized by a method of fixation and processing that is uniform. In that way, the artifact that is created will be uniform across the population, making the measurements usable.

There is also the question whether the biopsy device can procure an unbroken, continuous piece of mucosa measuring 20 mm. The area distal to the SCJ in retroflex view is not totally flat even along the lesser curvature where I have suggested the biopsy must be taken. Some flexibility of the biopsy device and a method to flatten the mucosa adequately will be necessary. The rugal folds are largely in a vertical direction and should not greatly impede procurement of the sample. However, since the test is being planned for an early stage of GERD, the likelihood of relatively normal anatomy will make the test more feasible than perceived at present. Endoscopists contemplating this should imagine the anatomy encountered in the non-GERD patient undergoing endoscopy for another reason. This is the anatomy that will likely be encountered when performing the test early or in the course of GERD or before its onset.

There is a tremendous commercial incentive to developing such a new biopsy device. Once it is developed, there should be a similar rush to provide data regarding the length of cardiac epithelium (with and without parietal and/or goblet cells) and how the length correlates with manometric abdominal LES length, reflux as measured by acid exposure in the pH test, impedance tests, symptoms, treatment failure, and cellular changes in the thoracic esophagus (erosive esophagitis, visible CLE, Barrett esophagus, adenocarcinoma).

I believe there are a sufficient number of endoscopists who can be persuaded to spearhead this exploration into the unknown with a spirit of adventure. The huge advantage is that all people have some degree of abdominal LES damage and 30% of the population has symptomatic GERD. To accrue 10,000 people for a study requires 50 endoscopists to recruit 200 subjects each. It is not optimistic to believe that such a large study can be done relatively quickly. The Pro-GERD study recruited 6215 people in relatively short order. I personally know 50 people who do endoscopy in GERD patients who will enthusiastically participate. A large clinical trial is feasible as soon as the new biopsy device is produced and shown in animal studies to be capable of obtaining an adequate mucosal specimen.

7. CORRELATION BETWEEN LOWER ESOPHAGEAL SPHINCTER DAMAGE AND REFLUX

7.1 The Truth

Reflux results from the failure of the LES. When the LES fails, transient relaxation of the LES occurs, leading to reflux of gastric contents along the pressure gradient from the stomach and midesophagus. The abdominal length that has been shown to correlate with LES failure is <10 mm.

When the residual abdominal LES is <10 mm, LES failure and reflux episodes are frequent, occurring with little stress. When the residual abdominal LES length is <20 mm, the LES is susceptible to failure after meals that cause a temporary dynamic abdominal LES shortening of 10 mm. This brings the functional abdominal LES length to <10 mm during the height of gastric overdistension (Fig. 20.10). Postprandial reflux occurs, increasing in frequency as the abdominal LES shortens from 20 mm (the onset of symptoms of GERD) to 10 mm.

The initial abdominal LES length is ~35 mm. This means that with LES damage between 0 and 15 mm (i.e., a residual abdominal LES length of 20 mm or more), the sphincter is competent. It does not fail unless dynamic shortening >10 mm occurs. This is the phase of compensated LES damage (Fig. 20.1).

At present, this phase of compensated LES damage is not recognized. GERD is defined by the presence of "troublesome" symptoms, which corresponds, in this new method of understanding, to a residual abdominal LES length of 10–15 mm (i.e., LES damage of 20–25 mm).

The present management of GERD can therefore be seen to be the management of a person whose LES damage has reached an advanced stage (Fig. 20.4). We are managing patients with a complete failure of the LES.

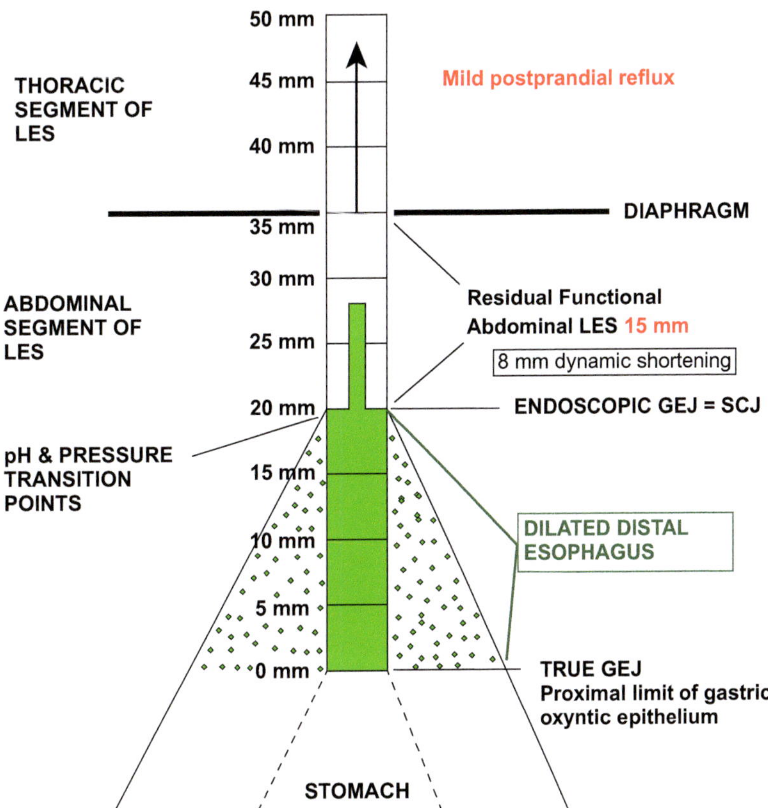

FIGURE 20.10 Patient with 20 mm of abdominal lower esophageal sphincter (LES) damage (residual abdominal LES of 15 mm). The LES is competent at rest but is susceptible to failure in the postprandial phase if a heavy meal causes gastric overdistension. This results in dynamic shortening of the abdominal LES (shown here as 8 mm), which causes the functional LES length to decrease to 7 mm. The LES becomes incompetent in the postprandial period.

The new method is designed to detect LES damage earlier by the new test and provide an opportunity to limit progression of LES damage to 20mm. The new objective is to prevent LES failure before it happens, not to manage it after it happens.

7.2 The Present Evidence Base

The correlations that exist between LES damage as measured by the length of the dilated distal esophagus and severity of reflux are largely theoretical. It is based on data derived largely from studies in autopsy and esophageal specimens from our unit that exactly measured the gap between the SCJ and the proximal limit of gastric oxyntic epithelium. These showed a variation in the length of the dilated distal esophagus (Table 20.1).

In the retrospective part of our autopsy study of 61 patients where there was a single section from the region distal to the SCJ for examination, 21 patients had zero cardiac + oxyntocardiac epithelium, 25 had 0–5 mm, 12 had 5–10 mm, and 3 had 11, 11, and 15 mm.

The limited data show that the length of CLE in asymptomatic and autopsy populations as well as most patients with esophagectomy for squamous carcinoma (a non-GERD-related cancer) was <15 mm. In contrast, eight patients with esophageal adenocarcinoma (a GERD-related cancer) had a length of cardiac and oxyntocardiac epithelium of >10 mm with a maximum length of 20.5 mm.

There are only two studies[22,23] where multilevel biopsies were done distal to the endoscopic GEJ. These studies do not permit accurate measurement of the dilated distal esophagus. Instead, they provide data regarding the prevalence of the different epithelial types at different points distal to the endoscopic GEJ.

Jain et al.[22] with four-quadrant biopsies taken at the Z-line, at GEJ, 1 cm below and 2 cm below the GEJ in 31 patients, reported that cardiac epithelium (they did not include oxyntocardiac epithelium) was present at the GEJ in 35%, 1 cm below the GEJ in 14%, and 2 cm below the GEJ in 3%. The failure to include oxyntocardiac precludes conclusions regarding the length of the dilated distal esophagus.

Ringhofer et al.[23] with four-quadrant biopsies taken at the GEJ and 5 and 10 mm distal to the GEJ in 102 patients with GERD, reported that cardiac epithelium (with and without parietal and/or goblet cells) decreased progressively in prevalence distal to the GEJ with 28% being positive at 10 mm distal to the GEJ. These data permit the conclusion that 28% of GERD patients had a dilated distal esophagus >10 mm long. Beyond this, no conclusions are possible.

These are the only studies that have sampled the area distal to the endoscopic GEJ by any biopsy technique that allows any possibility of measurement. The reasons for the paucity of data are (1) that there is no perceived need to take biopsies distal to the endoscopic GEJ. After all, the area distal to the endoscopic GEJ is incorrectly believed to be stomach and this cannot be involved in GERD. As one says this, there is the contradictory view that adenocarcinoma distal to the endoscopic GEJ (the "gastric cardia") is associated with GERD. (2) Even when biopsies are performed, the only way measurement is

TABLE 20.1 Evidence From Studies on Measurement of the Gap Between the Endoscopic Gastroesophageal Junction (GEJ) (End of Tubular Esophagus and Proximal Limit of Rugal Folds) and the True GEJ (Proximal Limit of Gastric Oxyntic Epithelium) Measured by Histology in Autopsy, Endoscopic Biopsy, and in Resection Specimens

Study	Study Population	Number of Patients	Length of CE	Length of CE + OCE
Chandrasoma et al.; retrospective	Autopsy; single section	61	0 to unknown	0–15 mm
Chandrasoma el al.; prospective	Autopsy; circumferential	18	0–2.75 mm	0.475–8.05 mm
Kilgore et al.	Pediatric autopsy	30	1–4 mm	Not reported
Chandrasoma et al.	SCC; resection specimen	2	Not reported	3.1, 4.3 cm
Chandrasoma et al.	Adeno-CA; resection specimen	8	Not reported	10.3–20.5 mm
Sarbia et al.	Resection specimen; SCC	36	0–15 mm	1–28 mm (median: 11 mm)
Robertson et al.; central obesity+	Endoscopy with jumbo biopsies	24	Median: 2.50 mm	Not reported
Robertson et al.; central obesity−	Endoscopy with jumbo biopsies	27	Median: 1.75 mm	Not reported

CE, cardiac epithelium; *OCE*, oxyntocardiac epithelium; *SCC*, squamous cell carcinoma.

possible is with a large jumbo biopsy forceps, which is limited to 8 mm distal to the SCJ or by taking multilevel biopsies. Presently, there is no biopsy instrument that has the ability to take a 20–25 mm vertical mucosal specimen extending vertically distal to the endoscopic GEJ (=SCJ in the endoscopically normal person). Accurate measurements in the literature are limited to resected and autopsy specimens where a long vertical sections can be taken.

It is important to note that the entire concept presented in this book resulted not from measured biopsies of the dilated distal esophagus; rather, it arose out of examination of the unmeasured retroflex biopsies taken within 10 mm distal to the endoscopic GEJ. Logical interpretation and studies in resected and autopsy specimens were all that was available to develop this concept.

7.3 The Need for Proof of Concept

This is a new concept and a new test that has a strong theoretical basis but few data in support. Development of a new biopsy device that can take a measurable length of mucosa distal to the endoscopic GEJ will open the door to clinical trials designed to establish accurate data that will establish correlations between measured LES damage and reflux (as measured in a pH test) and cellular changes in the esophagus.

The most critical determination will be the minimum amount of LES damage that can be associated with visible CLE. If the goal of the new management method is to prevent adenocarcinoma, its primary objective must be to prevent visible CLE. If this is achieved, adenocarcinoma of the thoracic esophagus will be prevented; without visible CLE, there can be no adenocarcinoma. The way to prevent visible CLE is to keep LES damage from reaching the point where the risk of visible CLE begins.

I have theorized that LES damage of 25 mm is the minimum level for visible CLE to occur. If studies show that visible CLE occurs at lower levels, the objective of management must change to prevent LES damage from reaching that defined amount if adenocarcinoma is to be prevented.

8. PREDICTION OF FUTURE LOWER ESOPHAGEAL SPHINCTER DAMAGE

8.1 The Truth

Progression of GERD is likely to be best assessed by progression of LES damage with time. LES damage is likely to be inexorably progressive. Being a disorder associated with an eating habit that causes gastric overdistension, it is likely that the progression of LES damage has a linear progression over the long term.

This means that if a person has LES damage of 5 mm over a period of 10 years, it is reasonable to assume that a similar 5 mm of damage will occur in the next 10 years.

Progression of LES damage may be impacted by two events:

1. A permanent and significant change in eating habit that puts less pressure on the LES can slow the rate of progression. However, without reason, it is uncommon for eating habits to change.
2. The development of postprandial heartburn. Pain is the body's natural defense. When a person develops heartburn in response to a meal, there is a probability that the fear and knowledge of pain results in a change in eating habit. Watching celebrities on television advertise acid-reducing drugs with the suggestion that one can eat anything without heartburn after swallowing a pill should make one wonder whether acid suppression removes the natural defense that is designed to slow LES progression.

The concept of linear progression has a logical basis, but it is still only a hypothesis. It is supported by all present evidence. It is also supported by the fact that this is the method of progression of many other slowly progressive chronic diseases such as coronary atherosclerosis and chronic glomerulonephritis. In both these diseases, the movement from inexact clinical diagnostic criteria (angina of effort, angina at rest, myocardial infarction; nephrotic syndrome, uremia) to objective criteria of assessment of the cause of the clinical syndrome (coronary artery narrowing; loss of glomeruli measured by GFR) resulted in better understanding, the ability to predict future risk, and the ability to take measures to control progression.

In both these entities, the assumption is that progression of the primary cause of the disease is linear. It is not unreasonable to assume that progression of LES damage will also be linear over the long term.

8.2 The Present Evidence Base

Among the clinical predictors of disease progression in GERD are male sex, obesity, the need for dose escalation to control symptoms, supine reflux, nocturnal reflux, regurgitation, chronic cough, or hoarseness. None of these predictors are sufficiently accurate to be used in practice to induce a change in management.

Endoscopically, the presence of severe erosive esophagitis is known to predict the occurrence of visible CLE in 19.7% of patients within the next 5 years. Pathologically, the finding of intestinal metaplasia at a normal SCJ has a 25.8% risk of progression to visible CLE in the next 5 years. Neither of these endoscopic and histologic predictors of visible CLE has had any practical impact.

As a result, GERD is presently managed in a purely reactive manner, treating all people alike until they fail treatment or develop dysphagia, which are the indications for endoscopy.

All of these clinical indicators of progression of GERD have the common factor of being associated with increasing acid exposure of the esophagus when tested by a 24-h pH test. Male patients with GERD have higher acid exposure levels than females.[24] Failure of PPIs to control symptoms is associated with a higher frequency of an abnormal pH test than well-controlled GERD. Similarly, increasing severity of erosive esophagitis has a higher acid exposure than a GERD patient with either nonerosive disease or mild erosive esophagitis.

The likelihood of abnormal acid exposure is the direct result of the frequency of LES failure (tLESRs). Frequency of LES failure in turn is directly related to baseline LES length. In the study by Kahrilas et al.[25] (reviewed in Chapter 9), gastric distention caused significantly higher episodes of LES failure (tLESR) and significantly higher acid exposure in the esophagus in patients with decreasing baseline LES length (i.e., increasing LES damage).

Lee et al.[24] (see Chapter 16 for review) reported that abdominal LES length in patients with GERD decreased highly significantly with increasing age of the patient. This decrease in abdominal LES length was in entire age groups of GERD patients divided into decades from <30 years to >70 years and correlated highly significantly with acid exposure in the pH test, which increased correspondingly with increasing age. This suggests a probability that increasing GERD with age is likely the result of decreasing abdominal LES length.

Cardiac epithelial length distal to the endoscopic GEJ has also been shown to increase with age when measured at autopsy in people without GERD during life.[26] Robertson et al.[5] showed that cardiac epithelial length was significantly greater in asymptomatic volunteers with central obesity compared with volunteers with normal waist circumference.

Unfortunately, there are no longitudinal studies where LES length or length of cardiac epithelium distal to the endoscopic GEJ was followed in individual GERD patients with sequential manometric and endoscopic assessment.

Obviously, the present inability to measure LES damage by any available method precludes there being any evidence of progression of LES damage in an individual person. There is no clinical, endoscopic, or histologic method that is presently used to predict any future event except the finding of Barrett esophagus that is known to be a risk indicator for future adenocarcinoma.

In Kahrilas et al.[25] (reviewed in Chapter 9), patients in three categories of GERD severity (asymptomatic or normal, GERD without hiatal hernia and GERD with hiatal hernia) had a progressive decrease in LES length in a highly significant manner. The acid exposure following gastric distension correlated with decreasing baseline LES length. The dynamic shortening of the LES with gastric distension was identical in the three groups. This suggests that a similar eating habit causes the same pressure on the LES irrespective of original LES length. This in turn suggests that LES damage is likely to be linear over the long term as long as the frequency of gastric distension during meals remains similar.

8.3 The Need for Proof of Concept

The availability of a test to accurately measure abdominal LES damage provides an instant ability to track progression of LES damage. I have described the methods of doing this with a two-measure algorithm and a one-measure algorithm. The test gives physicians an incredible opportunity to establish GERD progression by a new criterion, LES damage. This is a correlation between the cause of the disease and its severity. It is likely to produce much more precise correlations than the presently used symptoms, endoscopic changes, and acid exposure by pH or impedance testing.

I am certain that many investigators will avail themselves of the opportunity of generating data regarding progression of LES damage with time in patients with GERD when the test becomes available and recognized.

9. A PERSONAL LOWER ESOPHAGEAL SPHINCTER DAMAGE ASSESSMENT

When it was time for me to have my screening colonoscopy at age 56 years, I asked my gastroenterologist to add an upper endoscopy with four-quadrant biopsies taken at multiple levels at the endoscopic GEJ (which was the SCJ; I had no visible CLE) and at 5 mm intervals distal to that for 20 mm. I do not have typical GERD symptoms although I have experienced occasional noncardiac chest pain for many years. I had wondered if they could represent "noncardiac chest pain" that is caused by GERD. However, I have never had an empiric PPI test, largely because the episodes are intermittent.

Examination of the biopsies showed that I had a measured 4 mm of metaplastic columnar epithelium between the squamous epithelium and gastric oxyntic epithelium. This was predominantly oxyntocardiac epithelium with 0.2 mm of cardiac epithelium in one of the biopsies at the GEJ. There was no intestinal metaplasia.

All biopsies beyond 5 mm had normal gastric oxyntic epithelium. I have LES damage 4 mm, measured in one biopsy from the SCJ to the proximal limit of gastric oxyntic epithelium. This places me with a residual abdominal LES of 31 mm at age 56 years. I am in the phase of compensated LES damage where my LES damage is well within its reserve capacity and my LES remains highly competent. My symptoms are not caused by GERD.

With this measurement and using the simple algorithm where the date of birth, date of endoscopy, and measured LES damage are entered into a program as three data points, the exact amount of LES damage into the future becomes known. My LES damage at age 56 years was 4 mm. The computer program assumes that my LES damage began at age 15 years. The linear slope of my LES damage will show that my LES damage at age 97 years will be 8 mm. Having LES damage <15 mm at age 97 years is a guarantee that I will not develop GERD, no visible CLE, no Barrett esophagus, and no risk of esophageal adenocarcinoma in my lifetime. My LES will reach the critical damage level of 25 mm only if I live to be 270 years old!

Can anyone reading this who does not have GERD, as presently diagnosed, guarantee that he/she will not develop Barrett esophagus or adenocarcinoma in the future?

10. RESEARCH AND DEVELOPMENT OF TECHNOLOGY IN SUPPORT OF NEW METHOD

10.1 Research and Development Focus up to the Present Time

Research and development is always designed to satisfy a clinical need. When combined with commercial incentive, the medical device and pharmaceutical industry can produce miracles. But they have to be directed by physicians who define the need.

In GERD, the clinical need that has been presented to companies is a demand for better methods of acid suppression to control symptoms, better technology to diagnose GERD and its complications, and better technology to treat complications of GERD when acid suppression fails.

Research from the medical device and pharmaceutical industry has provided miraculous advancements over the past half century, to satisfy these clinical needs. When summarized, they are an impressive list that has produced the ability of physicians to:

1. improve the ability to neutralize gastric acid (acid neutralizers) and suppress gastric acid secretion (H_2 receptor antagonists; PPIs of ever-increasing effectiveness);
2. improve visualization of endoscopic changes by endoscopes that have ever-increasing resolution (narrow band imaging) and record changes with video-imaging and high-resolution photography;
3. improve the ability to identify intestinal metaplasia and dysplasia at endoscopy (confocal endomicroscopy and chromoendoscopy);
4. improve biopsy detection of intestinal metaplasia and dysplasia;
5. improve staging of tumors [endoscopic ultrasound (EUS), optical coherence tomography] and provide tissue samples from areas other than the mucosa such as lymph nodes in the mediastinum (EUS-directed needle biopsy);
6. improve treatment of Barrett esophagus (radio-frequency ablation, cryotherapy);
7. treat early cancer by minimally invasive endoscopic procedures (endoscopic mucosal resection and submucosal dissection);
8. treat late cancer with better surgical techniques, improved chemotherapy, and radiation;
9. develop methods of augmenting and repairing the damage LES, with endoscopic methods (Stretta and TIF), and surgical methods (LINX, EndoStim, and fundoplication);
10. improve methods of assessing reflux (pH and impedance testing) and pressures and motility in the esophageal muscle (standard and high-resolution manometry).

All this wonderful technology has improved the ability to diagnose and treat GERD in a manner that has *not been effective* in the big scheme of things. GERD as a disease has arguably become worse, not better. While more patients achieve symptom control today, mortality of GERD from esophageal adenocarcinoma rates are still increasing after a sevenfold increase from 1975 to 2006.[27] The number of patients whose quality of life is disrupted by symptomatic GERD is still significant.[28] The prevalence and incidence of GERD is increasing all over the world.[29] The progression from GERD to Barrett esophagus and adenocarcinoma is uncontrolled.[7] The cost associated with this failed treatment is gigantic, estimated in the billions of dollars.

10.2 Research in the Pipeline

In the pipeline, it seems like we can anticipate even more effective drugs to suppress acid secretion; newer imaging and endoscopy at ever-increasing resolution and computer analysis to allow us to diagnose cellular changes by endoscopy and imaging with more accuracy and definition; and newer technology to augment and repair the badly damaged LES.

LES augmentation and repair is presently a last resort for refractory GERD. It is not used frequently in relation to the overall number of GERD patients and has significant problems with effectiveness and complications.

10.2.1 What Are We Really Doing?

Better acid suppression will decrease the number of people who fail treatment. It will not stop progression of GERD to visible CLE, Barrett esophagus, and adenocarcinoma. It will likely make it worse, as improving acid suppression has done in the past 30 years.

Better technology is directed only at detecting cellular changes associated with failure of treatment such as Barrett esophagus, dysplasia, and adenocarcinoma. GERD is beyond control at that point; the progression to adenocarcinoma is largely inevitable when the cellular change in the thoracic esophagus progresses to visible CLE. Endoscopic surveillance, radiofrequency ablation and endoscopic mucosal resection are unnecessary if Barrett esophagus is eradicated.

Visible CLE→intestinal metaplasia→increasing dysplasia and adenocarcinoma are not prevented by medical therapy. Cancer rates are increasing even as acid suppression becomes more powerful.[29] Nissen fundoplication, which may lower the rate of progression, is not done as a cancer-preventive measure.[30] Surgery is also performed too infrequently to impact cancer rates even if it is cancer preventive.

We are really doing nothing except watching the incidence of GERD, Barrett esophagus, and adenocarcinoma rise. When these conditions occur, we just react, often with poor results. There is never a thought about prevention.

10.3 New Direction

The proposed new test of assessing LES damage by histology opens a door to a radically new method of investigating and treating GERD with the razor-sharp objective of preventing visible CLE. This is equivalent to preventing the progression of LES damage to the point at which it becomes incompetent and abnormal reflux is established. If this can be achieved and visible CLE is effectively prevented, Barrett esophagus and esophageal adenocarcinoma will be prevented.

This is the primary objective. When the new method is evaluated objectively, it has the potential to develop a much more ambitious end point: the complete prevention of GERD as a disease.

The difference between the proposed new method and the present method is profound in its simplicity. It is like a management technique for solving problems that is known as "root cause analysis." GERD is presently managed without paying attention to its root cause. This is doomed to failure. The new method attacks the root cause, LES damage. Understanding, measuring, and controlling the root cause will have a much higher probability of success in managing GERD and preventing its complications.

Present treatment of GERD is purely reactive, permitting the disease to progress before it is diagnosed and then using acid suppression as a one-size-fits-all treatment. The management algorithm continues to be reactive as the disease progresses, changing only when dysplastic Barrett esophagus and adenocarcinoma arise.

The new proposed management is proactive with the aim of early diagnosis by the new pathologic test for LES damage, using a new algorithm to predict progression of LES damage in the future, recognizing the point in the patient's life LES damage will be severe enough to cause GERD and produce severe reflux sufficient to cause treatment failure and visible CLE. Being able to identify the minority of people who will progress to severe GERD permits concerted preventive measures that will prevent progression to the point of severe LES damage.

I will show that the new test and algorithm will permit stratification of the entire population, whether they have or do not have symptoms of GERD at the time of the test, into the following categories (Fig. 20.1):

1. People whose abdominal LES damage will never progress to >15 mm in their natural lifetime. These patients will not develop GERD. They are 70% of the population.
2. People whose abdominal LES damage will progress to 15–25 mm at some future point in their life. These people are on the path to GERD with the severity of reflux never progressing beyond mild. They will never have sufficient reflux to cause visible CLE. These are the 70% of the 30% of GERD patients whose symptoms are easily controlled with acid-reducing agents. They can be shown their future and advised to change their diet to a GERD-friendly diet. If this does not slow progression, as evidenced by a follow-up test, they can be given the following choice:
   a. Develop GERD and take PPIs for life as needed to control symptoms with an assurance that they will not develop visible CLE and cancer or

b. Have a procedure to stop progression of LES damage and prevent the onset of symptomatic GERD. This removes the risk of complications with long-term PPI therapy.

3. People who are predicted to progress to abdominal LES damage >25 mm at some point in their lives. This is the 30% of GERD patients who are at high risk to progress to treatment failure, develop visible CLE, and adenocarcinoma. They will need to have a procedure to stop progression of LES damage. This is better done sooner than later although a trial of dietary modification and a second assessment of LES damage to test impact is possible.

10.4 Needed Research and Development

The new clinical need resulting from the ability to measure LES damage directs research and development in a completely different direction. The new needs are simple. The new test to assess LES damage can become a stand-alone test that will allow understanding of GERD from the normal state to advanced disease associated with a completely destroyed LES.

10.4.1 A New Biopsy Device to Measure the Dilated Distal Esophagus

There is no biopsy device that can procure a sample of mucosa that can measure the dilated distal esophagus accurately. The needed sample must procure a mucosal sample that is 20–25 mm long, 2 mm wide, and 1 mm deep. It must be positioned vertically along the lesser curvature with its proximal end just proximal to the SCJ and its distal end 25 mm distal. The device will be designed to take the specimen with an ease that is similar to a standard biopsy.

Initially, this will be designed for delivery through a standard endoscope. Once it is established, there will be a huge incentive to develop technology that allows the biopsy device to procure a sample without the need for endoscopy. The ultimate goal is to allow the test to be done rapidly and with low cost in a doctor's office without sedation.

10.4.2 Education of Physicians Treating Gastroesophageal Reflux Disease

This is a revolutionary new concept, but one that is very easy to understand. LES damage is the known cause of GERD. Having a new test designed to measure the fundamental cause of a disease has value that is likely to appeal to a sufficient number of physicians. Directed education of those interested should allow the concept to be quickly understood.

The only requirement to understanding this new concept is an open mind. Unfortunately, this is a rare commodity in the medical world. My experience with physicians at conferences where I explain the new concept has been disbelief except for a small number of people who understand the concept immediately and completely. It is impossible for many physicians with years of medical education to believe that some of the most concrete and hallowed facts they have learned can be wrong. Then they face the problem of a difficulty in understanding histology. Then they look at me, a not very well-known or influential person in the field. And, they conclude that this cannot be true. This is a closed mind.

The contrast was dramatic when I explained the concept to my patent attorney who had no prior knowledge of GERD. In one telephone call lasting 1 h he had perfectly understood everything I said. The next day, he sent me a summary of our conversation that I could have placed in this book. It was complete understanding. This is an open mind. Medical students and pathology residents that I teach have no problem in understanding.

A little knowledge is a dangerous thing! I am hopeful that there are a few readers who have reached this point in the book. You are my hope for a changing future in GERD. To get to this point, you have to understand and become sufficiently interested in the concept. You are the people who will push this new concept forward. I hope that by the time this happens, a new biopsy device has been developed by a medical device company and is available for you to use and confirm this new test of LES damage.

10.4.3 Testing to Validate the New Test of Lower Esophageal Sphincter Damage

Once implemented, the test will produce a large volume of data rapidly. GERD is such a common disease and upper gastrointestinal endoscopy such a common procedure that it will not be difficult to rapidly develop a database. This database will aim to correlate the measured abdominal LES damage with symptoms, duration of the disease, response to treatment, pH test, impedance, and manometry. These data will provide a completely new method of understanding GERD from the new vantage point of quantitative LES damage measured to an accuracy of micrometers.

From these data, correlations between measured LES length and severity of GERD utilizing a variety of different clinical physiological and endoscopic parameters will emerge. The tables that have been presented in this book from theoretical considerations will be replaced by actual data.

10.4.4 Pathologic Measurement

Pathologic measurement requires only a standard section stained with the routine hematoxylin and eosin stain. This is as cheap as it gets. In cases where the pathologist has difficulty in differentiating oxyntocardiac from gastric oxyntic epithelium, digested periodic acid Schiff stain can be used. However, there are many examples in the literature where pathologists have accurately differentiated between these various epithelial types.[23,31,32]

Pathologists usually respond very rapidly to a clinical demand. The training required for this is miniscule and it is likely that the measurement will have a very high precision and low interobserver variation.

10.4.5 Management of the Early Phase of Clinical Testing

There must be a careful initial assessment of the method under highly controlled conditions. This will be the duty of the medical device company that develops the biopsy device. There must be limitation of the use of the device to endoscopists who commit to taking the time and trouble to obtain a perfect specimen. There must be good ancillary support to document clinical, endoscopic, and physiologic testing data. Specimen collection, fixation, and transportation must be controlled. Ideally, a single pathologist with expertise must read the slides. These are necessary to ensure uniformity of application of the test.

10.4.6 Long-Term Follow-up

The validity of the algorithm that predicts the progression of LES damage into the future requires follow-up studies. If the test is adapted to be simple without requiring endoscopy, the test can be repeated, making accumulation of data regarding progression of LES damage relatively easy and not requiring long-term follow-up.

11. METHODS TO CONTROL PROGRESSION OF LOWER ESOPHAGEAL SPHINCTER DAMAGE

Ultimately, the only reason for performing a test to measure LES damage is to identify a subpopulation of people at high risk of progressing to GERD and, within that population, identifying a person at risk of progressing to severe GERD manifested by failure of PPI therapy to control symptoms, visible CLE, Barrett esophagus, and adenocarcinoma. There is a probability that this is possible with the new test at any stage of LES damage in persons with or without symptoms of GERD.

The ability to predict the future state of the LES opens a window of opportunity for intervention at an early stage to prevent severe LES damage in the future. This opportunity has value only if there is a method to prevent or slow progression of LES damage. There is no proven method at present to do this. This is not surprising. A method cannot exist if there has never been a demand.

The only demand that has existed for intervention in correcting LES damage has been the repair of a severely damaged LES that has caused severe symptoms that have not been controlled with maximum PPI therapy. At this point of disease, the residual abdominal LES is commonly <10 mm. All methods at present have the objective of repairing a severely damaged LES.

The new objective is to intervene at a much earlier stage of LES damage in patients predicted to progress to severe damage many years and often decades in the future. At the time the intervention is undertaken, the LES is relatively intact. The objective of the intervention is to slow down progression sufficiently to turn the line of future progression downward such that the LES never reaches a predetermined severity of damage.

For example, let us say that the test for LES damage is performed in an asymptomatic male with central obesity at age 42 years. He has abdominal LES damage measured at 15 mm (Fig. 20.11). He is at the end of the phase of compensated LES damage with a residual abdominal LES of 20 mm. Assuming onset of LES damage at age 15 years, this is a rate of progression of LES damage of 0.55 mm/year.

The future projection of LES damage for him (assuming linear progression of LES damage) shows that he will reach the critical point of 25 mm of LES damage at age 62 years, 20 years into the future. At that age, he will have severe reflux with a high and imminent risk of treatment failure and visible CLE with the onset of cancer risk in the near future if these do not already exist.

Present management will simply wait until the person develops symptoms and then treat with PPIs to control symptoms. In the early stage of the disease, his symptoms will likely be well controlled by PPI therapy. However, PPIs do nothing to stop the progression of LES damage or reflux. With this person's trajectory of LES damage progression, it is likely that he will have trouble controlling his symptoms with PPIs in the next 10 years. When treatment failure occurs at the expected 62 years, the LES damage is 25 mm (residual manometric abdominal LES length will be 10 mm). Endoscopy done at this time will likely show severe erosive esophagitis. This will heal with continued PPI treatment but, as was seen in the Pro-GERD study,[7] he will have a 19.7% risk of developing visible CLE in the next 5 years when LES damage progresses by another 2.75 mm.

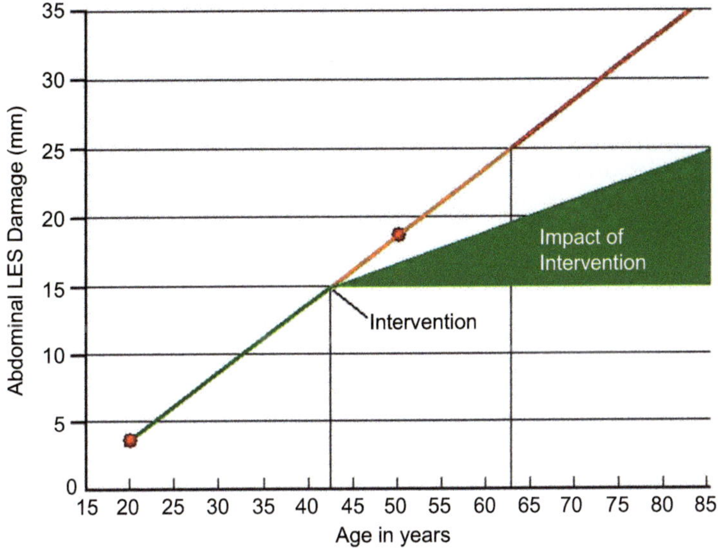

FIGURE 20.11 Progression of abdominal lower esophageal sphincter (LES) damage in a 42-year-old person whose LES damage measures 15 mm by the new histologic test. The algorithm predicts that the LES damage will progress to reach 25 mm when the person reaches age 62 years. Immediate intervention to stop or slow down progression sufficiently will prevent the person from reaching 25 mm of damage in his expected lifetime (*green zone*). Severe GERD, treatment failure, visible columnar-lined esophagus, and adenocarcinoma will be effectively prevented.

TABLE 20.2 Changes With Age in the Functional Residual Length of the Abdominal Lower Esophageal Sphincter (LES) in a Person Who Has LES Damage of 15 mm (i.e., Residual Abdominal LES Length 20 mm) by the Test at Age 42 Years, Assuming Linear Progression of LES Damage With and Without Intervention to Slow Down the Rate of Progression

Rate of LES Damage (mm/yr)	At 42 Years (mm)	At 50 Years (mm)	At 56 Years (mm)	At 62 Years (mm)	At 75 Years (mm)	At 85 Years (mm)
0.55[a]	20	15.75	12.45	9.15	2.00	0
0.23	20	18.16	16.78	15.40	12.41	10.11
0	20	20	20	20	20	20

[a]Calculated by linear LES damage of 15 mm at age 42 years from onset at 15 years.
The abdominal LES lengths in green represent lengths at which the LES is likely to be competent. The lengths in orange represent an LES that is susceptible to failure with gastric distension (i.e., at risk of postprandial reflux). The lengths in red represent an LES that is below the length at which LES failure occurs at rest.

With the new management method, the test of LES damage at age 42 years will predict progression to abdominal LES damage of 25 mm at age 62 years. There is a window of opportunity of 20 years to intervene in the attempt to prevent LES progression.

All that needs to be done is to slow the rate of progression. There is no need for repair or improvement of LES function. Obviously, if progression of LES damage can be stopped completely (i.e., to a rate of zero), the residual abdominal LES length will remain at the same 20 mm all his life and the person will never develop any significant reflux. He will not develop GERD.

However, a significant slowing of the rate of progression of LES damage can ensure that he will not reach LES damage of 25 mm until age 85 years (Fig. 20.11). A simple calculation indicates that if the progression of LES damage can be decreased from the existing 0.55–0.23 mm/year, he will not reach the critical LES damage level of 25 years until age 85 years. Between 0 and 0.23 mm/year, he will achieve the desired objective (green area in Figs. 20.2 and 20.9; Table 20.2).

If successful in slowing the progression of LES damage to a level between 0 and 0.23 mm/year, the person will never enter the red zone. He may develop mild GERD but is highly unlikely to have treatment failure and visible CLE. Without visible CLE, he never reaches a point where he is at risk for adenocarcinoma.

Given the lag time needed to convert visible CLE→intestinal metaplasia→increasing dysplasia→adenocarcinoma, it can be reasonably stated that adenocarcinoma has been effectively prevented in this person by slowing the rate of progression of LES damage to prevent visible CLE until the age of 85 years.

It is also important to understand that there is a long window of opportunity where the least invasive methods of slowing the rate of progression of LES damage can be tried and tested for efficacy. For example, a test of dietary adjustment with a GERD-friendly diet or a newly developed prokinetic drug can be tested for a 5-year period followed by a repeat measurement of LES damage to see if the progression curve has improved sufficiently.

At present, the only available methods to modify the LES are invasive methods requiring endoscopy (TIF, Stretta) or surgery (LINX, EndoStim, and surgical fundoplication). These are used to repair an LES that has become defective and produced such severe reflux that medical therapy has failed to control symptoms. The success of these procedures is judged by improvement of LES function and decrease in the amount of reflux by a pH test. With fundoplication, there is evidence of increased mean LES pressure and increase in abdominal length after successful surgery.

With the new method, the intervention is much less demanding. In the patient mentioned above (Fig. 20.11), the abdominal LES has a residual length of 20 mm (LES damage of 15 mm by the new test) when the patient is 42 years old. He has little or no symptoms of GERD. The intervention only requires slowing down the rate of progression of LES damage in the future to <40% of its present rate (from 0.55 to 0.23 mm/year). There is no requirement for improvement of LES function or complete stoppage of progression of LES damage to zero.

This should open the door to much less invasive methods of intervention to slow down the progression of LES damage. The long window of opportunity to achieve this can allow testing of noninvasive methods, assessing their impact by a repeat test of LES damage after a trial of 5–10 years depending on the patient.

11.1 Modification of Diet

LES damage is primarily the result of repetitive gastric overdistension caused by frequent meals of high volume, particularly if the content causes delayed gastric emptying.[33] Eating frequent high-volume meals with high fat content is likely to be associated with GERD than more moderate-volume meals.

The relationship between dietary indiscretion and LES damage is likely to be inconsistent. It would depend on the innate resistance of the distal esophageal squamous epithelium to be damaged by exposure to gastric contents as well as the composition of gastric contents (bile, pepsin, acid, *Helicobacter pylori*, etc.).

If a patient with early GERD on acid-reducing drug therapy is shown a computer algorithm that projects a high risk of failure to control the disease in the future necessitating surgery, he/she may be persuaded to adjust the diet.

If the patient's predicted age of severe LES damage is >10 years in the future, a trial of dietary modification with repeat assessment of LES damage in 5 years is reasonable. If the slope of future LES damage has been turned downward with dietary modification to a sufficient extent, more invasive procedures may be avoided.

11.2 Drugs That Stimulate the Neuromuscular Complex of the Lower Esophageal Sphincter

One line of research that has been of great interest but with little in the way of success in the past has been the attempt to develop drugs that can have a positive impact on the LES. These are usually aimed at modulating the local neuromuscular mechanism that maintains LES tone. The problem is that all the drugs that have been produced have not produced an improvement in LES function in the patient with GERD.

The prokinetic drugs that possibly improve LES function are baclofen, bethanechol, cisapride (a serotonin receptor agonist), metoclopramide, and domperidone. These all have had limited benefit in improving symptom control in GERD patients when used clinically. Cossentino et al.[34] in a randomized clinical trial, showed that baclofen, a GABA agonist, reduced the number of tLESRs in patients with GERD.

However, these drugs are presently considered as the last resort effort in the treatment of GERD only in patients who have failed PPI therapy. At this advanced point in the course of the disease, the LES is largely destroyed. It is unreasonable to expect prokinetic agents to stimulate a sphincter that no longer exists.

It is possible that these prokinetic drugs may produce a positive effect if used early in the disease with the objective of decreasing the rate of progression of LES damage. In such people, the drugs will have an intact sphincter to act upon. The objective is not to stimulate a dead sphincter, it is to prevent or slow down the rate of progression of damage of an existing sphincter.

There is a potential new treasure trove of drugs that may be developed in the future to satisfy this entirely new objective of GERD management. Like dietary modification, these drugs can be given a trial of a few years if the patients have a long window between the test and the predicted time of severe LES damage.

11.3 Lower Esophageal Sphincter Augmentation/Repair Procedures

Numerous LES augmentation procedures are available. These include endoscopic procedures such as Stretta[35] and TIF[36] as well as laparoscopic procedures such as LINX,[37] EndoStim,[38] and fundoplication. These are only used sparingly at the present time and largely in those patients who have failed acid-suppressive therapy for GERD.

Selection criteria for these procedures usually require an abnormal acid exposure in the pH test (i.e., pH<4 for >4.5% of the test period). The LES is severely damaged at this point of the disease and these procedures are really attempts at repairing or augmenting a severely damaged LES. They are generally effective in reducing symptoms and decreasing acid exposure, but they have significant failure rates with significant complications in this role.

All these procedures have been shown to decrease reflux by objective testing, usually proved by a significant decrease in acid exposure in a pH test done before and after the procedure. This would suggest that they augment LES function even in the setting of severe LES damage.

Nissen fundoplication is the most used and best studied of the sphincter repair/augmentation procedures. In expert hands, this procedure has a nearly 90% success rate wherein it controls symptoms and normalizes acid exposure in the esophagus. It has significant complications, including a 15% incidence of dysphagia and a high incidence of gas bloating and inability to belch and vomit. In the hands of general surgeons whose expertise is less, poor selection of patients for surgery, lack of technical experience in terms of number of procedures performed per year, and failure of good follow-up all lead to outcomes that are less than satisfactory.

Mason et al.[39] from our unit, showed that Nissen fundoplication has a direct impact on changing the dynamics associated with gastric distention and can therefore impact the progression of LES damage very effectively.[39]

They introduce the aim of their study: "It is accepted based on clinical experience that Nissen fundoplication can prevent gastroesophageal reflux in patients with defective mechanical properties of their sphincters. The effectiveness of the procedure under these conditions lies in its ability to restore the mechanical properties of the sphincter (LES pressure, overall sphincter length, and length of sphincter exposed to abdominal pressure) to normal. The advent of laparoscopic Nissen fundoplication has made surgery more acceptable to the patients in whom the disease is in its early stages. These patients can have mechanically normal sphincters. The cause of reflux in these patients has been ascribed to neurologically mediated transient LES relaxation (tLESRs) that commonly occur when the stomach is distended after a meal. The Nissen fundoplication has also been shown to be effective in preventing reflux in these patients.... Our hypothesis was that reconstruction of the sphincter with a Nissen fundoplication completely eliminates tLESRs by some effect on the dynamics of sphincter function in response to gastric distension."

This study is an experiment on 10 adult chacma baboons chosen because of their anatomic and physiologic similarity to humans. LES length, common cavity episodes, and resistance of the sphincter in anesthetized animals were determined during a pressure–volume test before and after Nissen fundoplication. The Nissen fundoplication significantly increased the resting LES pressure, overall sphincter length, and abdominal sphincter length (Table 20.3).

Infusion of water into the stomach caused a progressive increase in intragastric pressure. This followed a phase of initial active gastric relaxation that allowed for a large volume to be accommodated with relatively small increases in pressure (i.e., the reservoir function of the stomach). After this phase, LES length decreased by 0.88 ± 0.14 mm for every 50 mL increment in gastric volume. There was a significant correlation between LES length and common cavity episodes.

After fundoplication, there was a significant reduction of the effect of gastric distension on LES length, which decreased by only 0.16 ± 0.08 mm for every 50 mL volume increment. This allowed the infusion of a large gastric volume with minimal effect on LES length. Because of this, there was a concomitant decrease in the frequency of common cavity episodes.

TABLE 20.3 Characteristics of Resting Lower Esophageal Sphincter (LES) Before and After Fundoplication. Data Are Given as the Median (Interquartile Range)

Characteristic	Before Fundoplication	After Fundoplication	*P* Value
Total LES length (mm)	23.6 (21.0, 26.5)	39.0 (34.5, 42.2)	.006
Abdominal LES length (mm)	16.1 (15.2, 19.7)	30.8 (23.1, 38.1)	.006
LES pressure (mmHg)	9.2 (5.9, 12.2)	14.1 (9.6, 18.4)	.03

The authors conclude, in their discussion: "This study has provided a rational basis for the use of a Nissen fundoplication in patients in whom the disease is in its early stages or in patients with reflux disease who have mechanically normal sphincter properties. In these patients, the Nissen fundoplication prevents shortening of the sphincter during gastric distension and interrupts the underlying pathophysiological basis for reflux in the early stage of the disease."

At present, Nissen fundoplication is performed in patients with a defective LES or a person with proven reflux by a pH test that shows abnormal acid exposure (i.e., pH<4 in the esophagus for >4.5% of the time). This level of acid exposure correlates with a significantly damaged LES.

When LES damage is measured by the new histologic test and the algorithm for future LES damage predicts the person will progress to 25 mm of LES damage in the future, the LES length at the time of measurement will be higher (i.e., LES damage is at an earlier stage). This study shows that a Nissen fundoplication, if done at this time, will greatly reduce the basic cause of progression of LES damage, which is dynamic shortening of the LES with gastric overdistension. The procedure will therefore be highly effective in preventing or at least slowing down the rate of progression of LES damage.

It is likely that Nissen fundoplication, which has the requirement of laparoscopic surgery, is a procedure that is unnecessarily invasive to produce the desired effect of preventing progression of LES damage. Even simpler procedures such as LINX, EndoStim, TIF, and Stretta, which have been shown to be effective in reducing reflux, are likely to be too complicated.

There is no method at present to identify early in the course of disease the patients who will progress to treatment failure and complications in the future. The new method of assessment of LES damage and identification of the minority of patients who will develop severe LES damage in the future will be intervention aimed at preventing progression of LES damage rather than repairing a severely damaged LES. The intervention can be undertaken while the LES is still relatively intact.

The ability to select patients by an accurate prediction of future severe LES damage will provide an impetus to develop new and more effective techniques of intervention aimed at preventing progression of LES damage rather than LES repair and augmentation. Certainly, the sheer number of people who suffer from GERD is large enough for this to have significant commercial appeal. The ideal solution will be either dietary modification or the development of an effective prokinetic drug that will be capable of slowing down the rate of progression of LES damage when used early in the course of the disease when the LES is largely intact.

This ability to predict the occurrence of severe GERD/LES damage will open a window where this minority of high-risk patients can be treated at an early stage to prevent severe LES damage in the future. If this is successful, there is the probability of a decline in the incidence of complications such as Barrett esophagus and adenocarcinoma that correlate with severe LES damage and severe reflux.

REFERENCES

1. Chandrasoma P. Controversies of the cardiac mucosa and Barrett's esophagus. *Histopathology* 2005;**46**:361–73.
2. Chandrasoma PT. Histologic definition of gastro-esophageal reflux disease. *Curr Opin Gastroenterol* 2013;**29**:460–7.
3. Chandrasoma P, Makarewicz K, Wickramasinghe K, Ma YL, DeMeester TR. A proposal for a new validated histologic definition of the gastroesophageal junction. *Hum Pathol* 2006;**37**:40–7.
4. Ayazi S, Tamhankar A, DeMeester SR, et al. The impact of gastric distension on the lower esophageal sphincter and its exposure to acid gastric juice. *Ann Surg* 2010;**252**:57–62.
5. Robertson EV, Derakhshan MH, Wirz AA, Lee YY, Seenan JP, Ballantyne SA, Hanvey SL, Kelman AW, Going JJ, McColl KE. Central obesity in asymptomatic volunteers is associated with increased intrasphincteric acid reflux and lengthening of the cardiac mucosa. *Gastroenterology* 2013;**145**:730–9.
6. Chandrasoma P, Wijetunge S, Ma Y, DeMeester S, Hagen J, DeMeester T. The dilated distal esophagus: a new entity that is the pathologic basis of early gastroesophageal reflux disease. *Am J Surg Pathol* 2011;**35**:1873–81.
7. Malfertheiner P, Nocon M, Vieth M, Stolte M, Jasperson D, Keolz HR, Labenz J, Leodolter A, Lind T, Richter K, Willich SN. Evolution of gastro-oesophageal reflux disease over 5 years under routine medical care – the ProGERD study. *Aliment Pharmacol Ther* 2012;**35**:154–64.
8. Clark GWB, Ireland AP, Chandrasoma P, DeMeester TR, Peters JH, Bremner CG. Inflammation and metaplasia in the transitional mucosa of the epithelium of the gastroesophageal junction: a new marker for gastroesophageal reflux disease. *Gastroenterology* 1994;**106**:A63.
9. Chandrasoma P. Pathophysiology of Barrett's esophagus. *Semin Thorac Cardiovasc Surg* 1997;**9**:270–8.
10. Chandrasoma PT, DeMeester TR. *GERD: from reflux to esophageal adenocarcinoma.* San Diego: Academic Press; 2006.
11. Hayward J. The lower end of the oesophagus. *Thorax* 1961;**16**:36–41.
12. Odze RD. Unraveling the mystery of the gastroesophageal junction: a pathologist's perspective. *Am J Gastroenterol* 2005;**100**:1853–67.
13. Association of Directors of Anatomic and Surgical Pathology. Recommendations for reporting of resected esophageal adenocarcinomas. *Am J Surg Pathol* 2000;**31**:1188–90.

14. McClave SA, Boyce Jr HW, Gottfried MR. Early diagnosis of columnar lined esophagus: a new endoscopic diagnostic criterion. *Gastrointest Endosc* 1987;**33**:413–6.
15. Sharma P, McQuaid K, Dent J, Fennerty B, Sampliner R, Spechler S, Cameron A, Corley D, Falk G, Goldblum J, Hunter J, Jankowski J, Lundell L, Reid B, Shaheen N, Sonnenberg A, Wang K, Weinstein W. A critical review of the diagnosis and management of Barrett's esophagus: the AGA Chicago workshop. *Gastroenterology* 2004;**127**:310–30.
16. Allison PR, Johnstone AS. The oesophagus lined with gastric mucous membrane. *Thorax* 1953;**8**:87–101.
17. Barrett NR. Chronic peptic ulcer of the oesophagus and 'oesophagitis'. *Br J Surg* 1950;**38**:175–82.
18. Barrett NR. The lower esophagus lined by columnar epithelium. *Surgery* 1957;**41**:881–94.
19. Spechler SJ, Sharma P, Souza RF, Inadomi JM, Shaheen NJ. American Gastroenterological Association medical position statement on the management of Barrett's esophagus. *Gastroenterology* 2011;**140**:1084–91.
20. Korn O, Csendes A, Burdiles P, et al. Anatomic dilatation of the cardia and competence of the lower esophageal sphincter: a clinical and experimental study. *J Gastrointest Surg* 2000;**4**:398–406.
21. Leodolter A, Nocon M, Vieth M, Lind T, Jasperson D, Richter K, Willich S, Stolte M, Malfertheiner P, Labenz J. Progression of specialized intestinal metaplasia at the cardia to macroscopically evident Barrett's esophagus: an entity of concern in the Pro-GERD study. *Scand J Gastroenterol* 2012;**47**:1429–35.
22. Jain R, Aquino D, Harford WV, Lee E, Spechler SJ. Cardiac epithelium is found infrequently in the gastric cardia. *Gastroenterology* 1998;**114**:A160. [Abstract].
23. Ringhofer C, Lenglinger J, Izay B, Kolarik K, Zacherl J, Fisler M, Wrba F, Chandrasoma PT, Cosentini EP, Prager G, Riegler M. Histopathology of the endoscopic esophagogastric junction in patients with gastroesophageal reflux disease. *Wien Klin Wochenschr* 2008;**120**:350–9.
24. Lee J, Anggiansah A, Anggiansah R, et al. Effects of age on the gastroesophageal junction, esophageal motility and reflux disease. *Clin Gastroenterol Hepatol* 2007;**5**:1392–8.
25. Kahrilas PJ, Shi G, Manka M, Joehl RJ. Increased frequency of transient lower esophageal sphincter relaxation induced by gastric distension in reflux patients with hiatal hernia. *Gastroenterology* 2000;**118**:688–95.
26. Chandrasoma PT, Der R, Ma Y, et al. Histology of the gastroesophageal junction: an autopsy study. *Am J Surg Pathol* 2000;**24**:402–9.
27. Pohl H, Sirovich B, Welch HG. Esophageal adenocarcinoma incidence: are we reaching the peak? *Cancer Epidemiol Biomarkers Prev* 2010;**19**:1468–70.
28. Toghanian S, Wahlqvist P, Johnson DA, Bolge SC, Liljas B. The burden of disrupting gastro-esophageal disease; a database study in US and European cohorts. *Clin Drug Investig* 2010;**30**:167–78.
29. El-Serag HB, Sweet S, Winchester CC, et al. Update on the epidemiology of gastro-oesophageal reflux disease: a systematic review. *Gut* 2014;**63**:871–80.
30. Chang EY, Morris CD, Seltman AK, et al. The effect of antireflux surgery on esophageal carcinogenesis in patients with Barrett's esophagus. a systematic review. *Ann Surg* 2007;**246**:11–21.
31. Marsman WA, van Sandyck JW, Tytgat GNJ, ten Kate FJW, van Lanschot JJB. The presence and mucin histochemistry of cardiac type mucosa at the esophagogastric junction. *Am J Gastroenterol* 2004;**99**:212–7.
32. Glickman JN, Fox V, Antonioli DA, Wang HH, Odze RD. Morphology of the cardia and significance of carditis in pediatric patients. *Am J Surg Pathol* 2002;**26**:1032–9.
33. Tamhankar AP, DeMeester TR, Peters JH, et al. The effect of meal content, gastric emptying and gastric pH on the postprandial acid exposure of the lower esophageal sphincter (LES). *Gastroenterology* 2004;**126**(Suppl. 2):A-495.
34. Cossentino MJ, Mann K, Ambruster SP, et al. Randomised clinical trial: the effect of baclofen in patients with gastroesophageal reflux. *Aliment Pharmacol Ther* 2012;**35**:1036–44.
35. Noar M, Squires P, Noar E, et al. Long-term maintenance effect of radiofrequency energy delivery for refractory GERD: a decade later. *Surg Endosc* 2014;**28**:2323–33.
36. Trad KS, Turgeon DG, Deljkich E. Long-term outcomes after transoral incisionless fundoplication in patients with GERD and LPR symptoms. *Surg Endosc* 2012;**26**:650–60.
37. Ganz RA, Peters JH, Morgan S, et al. Esophageal sphincter device for gastroesophageal reflux disease. *N Engl J Med* 2013;**368**:719–27.
38. Rodriguez L, Rodriguez P, Gomez B, et al. Long-term results of electrical stimulation of the lower esophageal sphincter for the treatment of gastroesophageal reflux disease. *Endoscopy* 2013;**45**:595–604.
39. Mason RJ, DeMeester TR, Lund RJ, et al. Nissen fundoplication prevents shortening of the sphincter during gastric distension. *Arch Surg* 1997;**132**:719–24.

Chapter 21

Application of the New Method to Present and Future Management of GERD

What I have presented in this book is a revolutionary new method of understanding the pathophysiology of gastroesophageal reflux disease (GERD) viewed from the sole perspective of damage to the abdominal segment of the lower esophageal sphincter (LES). This is based on a new histologic test that can accurately measure abdominal LES damage by a mucosal biopsy of the dilated distal esophagus.

1. THE NEW HISTOLOGIC TEST OF ABDOMINAL LOWER ESOPHAGEAL SPHINCTER DAMAGE

Application of the new histologic test depends on the new understanding of the pathogenesis of LES damage that I have described in detail in the earlier chapters of the book. It is based on strong evidence that gastric overdistension during heavy meals causes dynamic shortening of the LES at its distal end (Chapters 8 and 9).[1,2] This exposes the distal esophageal squamous epithelium to gastric contents. In time, with repetitive exposure, the squamous epithelium undergoes cardiac metaplasia.

Cardiac epithelium between the endoscopic gastroesophageal junction (GEJ) [= squamocolumnar junction (SCJ) in the person without visible columnar-lined esophagus (CLE)] and the true GEJ, which is the proximal limit of gastric oxyntic epithelium, defines the dilated distal esophagus. The length of the dilated distal esophagus (= length of cardiac epithelium distal to the SCJ in the person without visible CLE) is a measure of abdominal LES damage (Fig. 21.1).

The accurate measurement of the length of the dilated distal esophagus is not feasible with present endoscopic biopsy tools. A new biopsy device that can obtain a 20- to 25-mm-long sample of mucosa must be developed for the test to be put into practice. The length of cardiac epithelium can be measured to an accuracy of micrometers if an appropriate mucosal biopsy sample can be obtained. This is the amount of LES damage.

To put this revolutionary concept into perspective, this area that is the basis of the new test of LES damage is presently considered normal stomach by gastroenterologists and pathologists.[3] The guidelines for management of GERD recommend that no biopsies be taken from this area.[4] It is difficult to imagine a greater mistake in the history of modern medicine.

The new test can measure LES damage at any point in the adult life of a person, whether or not the patient has any clinical evidence of GERD. Although the test can be done at any time after age 20 years, it will have the highest accuracy if done over the age of 30 years.

The measured length of LES damage can be fed into a new simple algorithm with the patient's date of birth and date of the test to produce a linear slope of the person's future abdominal LES status for his/her entire lifetime (Fig. 21.2).

The test is therefore a revolutionary new concept for the diagnosis of GERD. Whether it will work as well as I have suggested is to be established by appropriate clinical testing after the new medical device becomes available. How it will impact the diagnosis and management of GERD is up to the medical community at large.

This is the first glimmer of hope for a method that has the potential to prevent esophageal adenocarcinoma and even eradicate GERD as a disease. The medical establishment, faced with a disease out of control with significant failure of treatment to control symptoms and an increasing incidence of Barrett esophagus and GERD-induced adenocarcinoma, should embrace this new opportunity aggressively. Time will tell if this proves to be true.

2. DRAWING A LINE IN THE SAND: VISIBLE COLUMNAR-LINED ESOPHAGUS

Visible CLE is the line drawn in the sand in the new management of GERD that is based on the availability of the new test of LES damage (Fig. 21.3). This is based on the certainty that if visible CLE can be prevented, a person has no risk of developing adenocarcinoma of the thoracic esophagus. It is also based on the reality that once visible

GERD. http://dx.doi.org/10.1016/B978-0-12-809855-4.00021-X

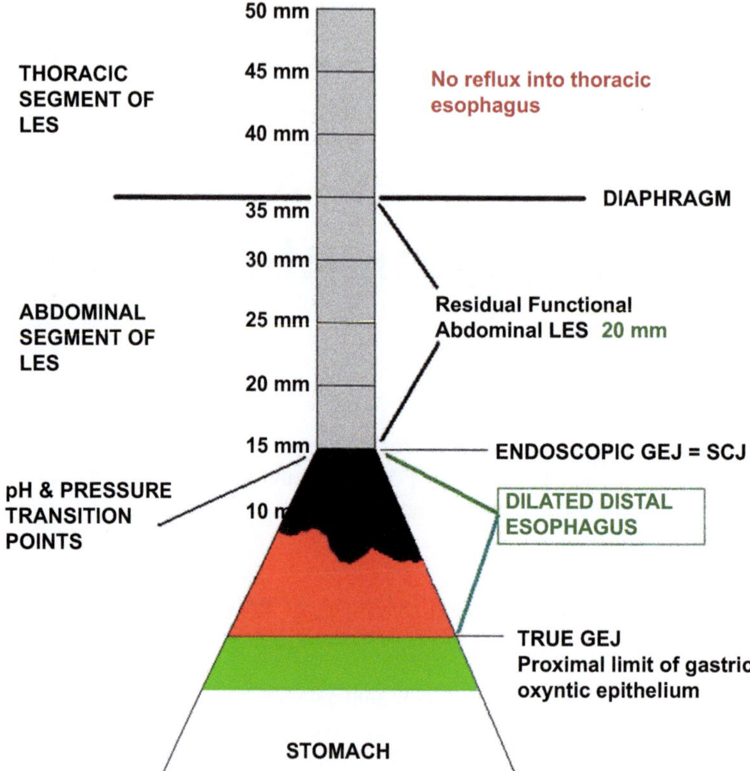

FIGURE 21.1 The dilated distal esophagus is the area between the endoscopic gastroesophageal junction (GEJ), the distal limit of squamous epithelium (gray) in the person with no visible columnar-lined esophagus, and the true GEJ, which is the proximal limit of gastric oxyntic epithelium (green). It is composed of cardiac (black) and oxyntocardiac epithelium in this person who has lower esophageal sphincter (LES) damage of 15 mm (residual abdominal LES of 20 mm).

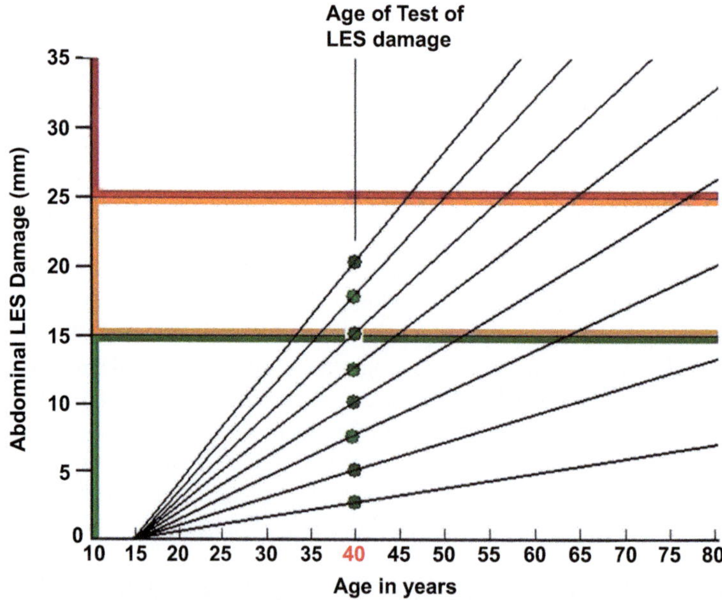

FIGURE 21.2 New test for lower esophageal sphincter (LES) damage done at age 40 years shows LES damage varying from 2.5 to 20 mm at increments of 2.5 mm. The projected LES damage into the future is shown by the multiple lines based on the measured LES damage. These show that people with 10 mm or more LES damage at this age will progress to LES damage of 25 mm by ages ranging from 47 to 75 years; those with less LES damage will remain at <25 mm LES damage throughout their life.

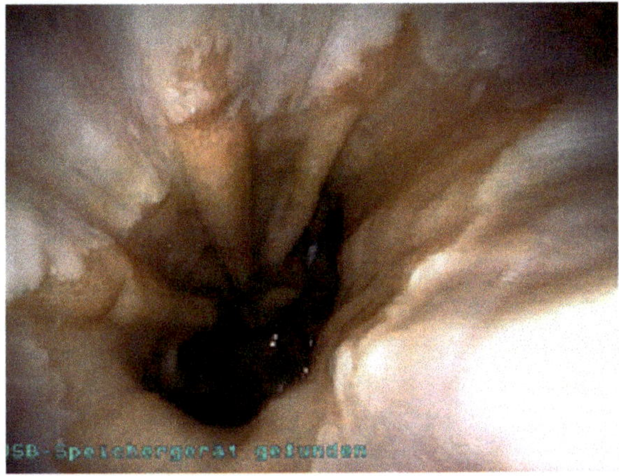

FIGURE 21.3 Antegrade endoscopic view of the squamocolumnar junction (SCJ), showing a 1 cm segment of flat, pink epithelium between the proximal limit of rugal folds and the serrated SCJ, representing short segment visible columnar-lined esophagus (CLE). Irrespective of its length, visible CLE is the line in the sand for the new method. It is the cellular change that must be prevented to successfully eradicate esophageal adenocarcinoma. *Photograph courtesy of Dr. Martin Riegler, Reflux Medical, Vienna, Austria.*

CLE occurs, medical treatment cannot stop progression in the sequence visible CLE→intestinal metaplasia→increasing dysplasia→adenocarcinoma.

Visible CLE results from reflux of sufficient severity to produce columnar metaplasia in the thoracic esophagus. It has the highest association of any cellular change in the esophagus with a defective LES and abnormal acid exposure.[5] In the Pro-GERD study,[6] the sequence of cellular abnormality in patients who were prospectively evaluated by endoscopy while under routine acid-suppressive therapy, was erosive esophagitis of increasing severity→visible CLE.

There is evidence that visible CLE and Barrett esophagus can be prevented at any stage before it arises if reflux into the thoracic esophagus is controlled. This can be achieved by antireflux surgery, but likely by any intervention that can decrease the severity of exposure to cumulative reflux over a long period of time.

In their review of the literature on the impact of antireflux surgery in GERD, Gutschow et al. addressed the question[7]: *Does antireflux surgery prevent the development of Barrett esophagus?* They defined Barrett esophagus as any visible CLE in the lower segment of the esophageal tube at endoscopy with intestinal metaplasia on biopsy.

The study used a computerized literature search using Medline and PubMed databases, review of published work, and the references cited therein. The literature cited was graded as to whether the study was prospective and randomized, prospective and controlled, prospective or retrospective series, or case reports.

For the question of prophylaxis of Barrett esophagus by antireflux surgery, five references focusing on endoscopic follow-up after surgery for uncomplicated GERD were found. Two studies were prospective and randomized, two were prospective and controlled, and one was prospective. Four studies with a total of 228 patients undergoing antireflux surgery showed no new cases of Barrett esophagus after 3.5–5 years of follow-up. The single study reporting Barrett esophagus after antireflux surgery had 15 patients who underwent surgery with two developing new Barrett esophagus after a 20-year median follow-up. This study, unlike the others, had not had preoperative biopsies and could therefore be discounted. The authors conclude: "In summary, there is no report in the present literature that confirms the development of Barrett esophagus after successful anti-reflux surgery."

The study suggests that prevention of visible CLE and Barrett esophagus is feasible in a person whose endoscopy is negative for visible CLE and who is treated with the objective of stopping reflux. This can be done with a procedure that corrects a severely damaged LES. The relationship of visible CLE and reflux into the esophagus is absolute. This contrasts with the data in the Pro-GERD study where PPI therapy for GERD resulted in 9.7% of patients progressing from no visible CLE to visible CLE within 5 years.[6]

Visible CLE has a relatively low but significant prevalence in the population. Ronkainen et al.[8] in a population study that performed screening endoscopy in a randomly selected sample, reported a prevalence of 10.3% of visible CLE. Rex et al.[9] reported a prevalence of 18.3% with visible CLE in patients over 50 years of age who presented for screening colonoscopy. Both these studies included people with a history of heartburn; the prevalence of visible CLE was greater in those with heartburn but not significantly. Kulig et al.[10] in the initial report of the Pro-GERD study of 6215 patients with significant GERD reported a visible CLE prevalence of 11.3% at the index endoscopy.

With present management guidelines for endoscopy, therefore, it is likely that approximately 10%–20% of patients will already have crossed the line in the sand that I have drawn. These will represent unacceptable failures in the suggested new management algorithm. It therefore demands that endoscopy be pushed to an earlier stage in the disease than in the above studies[9,10] to ensure the lowest prevalence of visible CLE at the index endoscopy.

I will show how the availability of the new test of LES damage can drive the point of endoscopy to an earlier stage of the disease.

Ultimately, though, if visible CLE is to be prevented in its entirety, it must be recognized that completely asymptomatic persons have a significant prevalence of visible CLE. Screening of asymptomatic persons represents the only method of preventing visible CLE in these people. This will become feasible only if the new test is expanded to become a practical screening test (see below).

3. CORRELATION BETWEEN ABDOMINAL LOWER ESOPHAGEAL SPHINCTER DAMAGE AND VISIBLE COLUMNAR-LINED ESOPHAGUS

There is a good correlation between the amount of LES damage and the competence of the LES. In turn, this determines frequency of LES failure and the occurrence of reflux episodes into the thoracic esophagus. Abnormal reflux is defined as acid exposure (pH<4) in the thoracic esophagus for >4.5% of a 24-h period of continuous monitoring. The presence of an abnormal pH test correlates with a defective LES at manometry, defined by mean LES pressure<6mmHg, total LES length<20mm, and abdominal length<10mm.[11]

Present criteria only establish correlations between a defective LES as defined above with cellular changes in the thoracic esophagus in patients with symptomatic GERD. Stein et al.[5] reported that 93% of patients with visible CLE have a defective CLE as defined above.

The question of what happens between a normal initial abdominal LES length of 35mm and the <10mm that defines a defective LES has never been addressed. The reason for this is that a test to measure LES damage has never existed.

The new test of LES damage provides the ability to establish more exact correlations between the amount of LES damage and the point at which the patient is at risk for developing visible CLE. This will, for the first time, provide a correlation between a diagnostic test (the new test for LES damage) and the point at which the patient develops a risk for developing visible CLE. This permits converting the line in the sand from a cellular criterion to a measurable objective amount of LES damage as determined by the new test.

In this book, I have, with the scarce available evidence, suggested that this amount of LES damage at which the risk of visible CLE begins is 25mm. It is very likely that this number will change downward, hopefully by a small amount, when actual data emerge in clinical validation of the test, if and when this happens.

4. TEMPORAL PROGRESSION OF ABDOMINAL LOWER ESOPHAGEAL SPHINCTER DAMAGE

LES damage is an inexorably progressive change. Being a battle between the LES and an eating habit that puts pressure on the LES from below, it is probable that the progression of damage is linear as long as the eating habit does not change. In Chapter 19, I presented evidence supporting a linear progression of LES damage with increasing age. Lee et al. have shown a progressive increase in acid exposure and decrease in length of the abdominal LES with increasing age in patients with GERD that supports linear progression.[12]

The entire temporal progression of damage of the normal abdominal LES (35mm in length) to its complete destruction (0mm in length) can be followed by this histologic test with an accuracy of micrometers. This can be done by the new test to measure LES damage at two points in the person's life or by one test during adult life (see Chapter 19). Both of these will permit calculation of the rate of progression of LES damage in any individual to two decimal points. Table 21.1 shows the temporal progression for 1mm/year increments of the rate of progression of LES damage.

Table 21.1 shows three separate zones, depicted in green, orange, and red. These zones indicate the probability and severity of GERD at different point in the person's life based on different rates of progression to age 75 years. The calculation can be extended into older ages if necessary.

1. **The green zone** indicates LES damage up to 15mm (a residual abdominal LES length of 20mm or greater) to 75 years of age. This person will remain in the phase of compensated LES damage till age 75 years and will not develop severe GERD. These people will have a rate of progression of LES damage of 2mm/decade or less.
2. **The orange zone** is the stage of mild GERD with LES damage of >15–25mm (a residual abdominal LES length of <20–10mm) at or below age 75 years. Persons in this group will have a competent LES at rest but becomes increasingly susceptible to LES failure with stresses placed on the LES with increasing age. This may impair normal eating. The person is likely to remain well controlled with proton pump inhibitors (PPIs) throughout life and will have no risk of

TABLE 21.1 Changes With Age of the Functional Residual Length of the Abdominal Lower Esophageal Sphincter (LES) With Different Rates of Progression of LES Damage. This Assumes That the Original Length of the Abdominal LES at Maturity Is 35 mm, That LES Damage Begins at Age 15 Years, and That LES Damage Has a Linear Progression Over the Long Term

Rate of LES Damage (mm/decade)	At 25 Years (mm)	At 35 Years (mm)	At 45 Years (mm)	At 55 Years (mm)	At 65 Years (mm)	At 75 Years (mm)
zero	35	35	35	35	35	35
1	34	33	32	31	30	29
2	33	31	29	27	25	23
3	32	29	26	23	20	17
4	31	27	23	19	15	11
5	30	25	20	15	10	5
6	29	23	17	11	5	0
7	28	21	14	7	0	0
8	27	19	11	3	0	0
9	26	17	8	0	0	0
10	25	15	5	0	0	0

The abdominal LES lengths in green represent lengths (20–35 mm) at which the LES is likely to be competent. The lengths in orange represent an LES that is susceptible to failure with gastric distension (i.e., at risk of postprandial reflux). The lengths in red represent an LES that is below the length (10 mm) at which LES failure occurs at rest.

developing visible CLE at an age that, given the lag phase for carcinogenesis in visible CLE, will result in adenocarcinoma of the thoracic esophagus.

3. **The red zone** represents LES damage > 25 mm (a residual abdominal LES length of < 10 mm) at some point in life before age 75 years. Persons in this group are predicted to fail PPI therapy and be at risk for developing visible CLE. They are the group at risk for adenocarcinoma. Persons with a rate of progression of LES damage > 5 mm/decade will be in this group. Persons with 10 mm/decade of LES damage may develop visible CLE around age 40 years; such patients are not rare at the present time.

These numbers are for increments of 1 mm/decade for rate of progression of LES damage. For the individual person, the test will provide an individualized slope of future LES damage that is much more accurate based on the exact age and exactly measured LES damage to two decimal points (Fig. 21.2).

A sobering thought is that present management defines GERD only when a person progresses to the red zone. The new method being proposed here has the goal of preventing progression of the disease as defined by LES damage in the entire population to the point at which they are diagnosed as having the disease at the present time.

5. POTENTIAL IMPACT OF THE NEW DIAGNOSTIC METHOD

At present, the new diagnostic method that I have proposed is one that is supported by all available evidence. It has a logic to it that is alluring because it concentrates on the cause of GERD. Also, it is based on histologic interpretation of biopsies, a time-honored and proven method of diagnosis in most diseases of the gastrointestinal tract. It simply makes sense to everyone who has heard it.

However, it will go nowhere unless the theory of the new method is believed, new devices necessary for accurate deployment of the test are developed, and the necessary clinical testing is done to validate the method.

I have provided theoretical measurements of LES damage that correlate with reflux and cellular changes in this book based on data available from measurements made in autopsy and resected specimens. The data in existence is scanty, based on a relatively small number of studies that followed my suggestion that cardiac epithelium was a metaplastic esophageal epithelium rather than the normal proximal gastric lining.[13] These will be replaced by more accurate measurements of LES damage made by the new test in live patients with intact anatomy.

There is much to be done.

5.1 Acceptance of the New Method

It is impossible to predict how this new diagnostic method will be accepted. If its acceptance was dependent on the medical establishment, I would suggest that the time frame for acceptance would likely be measured in decades.

This new method will get no traction with people who fail to accept that cardiac epithelium is esophageal and not gastric, that the true GEJ is the proximal limit of gastric oxyntic epithelium and not the end of the tube or the proximal limit of rugal folds, and that the gap between the present endoscopic GEJ and true GEJ is the dilated distal esophagus and not the proximal stomach.

The power of the various vested interests in the medical establishment that will resist the acceptance of this new method is enormous. PPIs are the largest source of revenue from GERD to a variety of interest groups. These groups will find ways to suppress studies that support the method and produce contrary evidence by flawed studies. The studies necessary to prove the new method, if they depended on the establishment, will likely not be performed. Without proof of concept, the method will likely disappear.

However, this new diagnostic test has the lure of commercial incentive. There is the need to develop a new biopsy device and build a simple algorithm to predict future LES damage based on the measured LES damage. These medical devices and algorithm, given the size of the market, have potential financial value to the developers. If acceptance of this method depends on good American capitalism, it has hope. The new biopsy device and algorithm have a fair chance of being developed.

If that happens, even the remote possibility of controlling GERD in this new way will likely stimulate a small but sufficient number of people to do the needed clinical testing to validate the method. This will happen because there will invariably be idealistic physicians who recognize that the present management of GERD is problematic and are looking for ways to improve the outcomes for their patients. In particular, the very idea that the new method has a realistic goal of preventing esophageal adenocarcinoma will be enticing. I know many such people who will participate in clinical trials of the new method with enthusiasm and vigor.

The one thing that is clear to me is that a scientist capable of garnering the support of a commercial entity with resources is far more likely to take his/her ideas forward quickly than a scientist trying to promote his/her ideas by the traditional methods of writing papers and giving lectures. It is so easy for those papers and lectures to be suppressed, ignored and rejected.

The pen may be mightier than the sword, but it cannot hold a candle to money.

5.2 Assume That the New Diagnostic Method Has Been Validated

For purposes of discussion as to how the new diagnostic method can impact the present and future management of GERD, let us assume that we are 5 years in the future. A biopsy device that can measure LES damage has been developed and validated as being able to secure a 20 × 2 × 1 mm sample of mucosa from the SCJ vertically down for 20 mm. Clinical trials have been done and show an excellent correlation between the measured LES damage and the GERD status at a cellular level.

Accurate data have been produced by these clinical trials that have divided all people into normal (zero LES damage); the phase of compensated LES damage (>0–15 mm/or other; no significant LES failure and reflux); mild GERD (>15–25 mm/or other; increasing LES failure and reflux but no visible CLE); and severe GERD (>25 mm/or other; severe LES failure and reflux with risk of visible CLE).

Furthermore, adequate albeit short-term, follow-up data are available that have validated the algorithm that predicts future LES damage in a patient based on the measurement of LES damage by the new test.

At this point in the discussion, let us also assume that the biopsy device requires endoscopy for placement at the appropriate location to secure the sample. This requires a gastroenterologist, an endoscopy suite, preparation, heavy sedation, small but significant risk of complications, postprocedure recovery, and high cost.

In the next section, we will look to 10 years in the future.

6. USE OF THE NEW DIAGNOSTIC METHOD IN THE PRESENT MANAGEMENT ALGORITHM OF GASTROESOPHAGEAL REFLUX DISEASE

The present management algorithm for GERD has the following problems:

1. There is no reliable and sensitive way to diagnose early GERD.[4] The diagnostic criteria as defined at endoscopy, pH testing, and manometric assessment of the LES are designed for specificity.
2. Patients are placed on long-term unnecessary PPI therapy with a small but significant false-positive rate of diagnosis of GERD by a flawed empiric PPI test.[14]
3. There is no present method to adequately assess severity of cellular changes of GERD. There is a significant discordance between symptoms and symptom control with treatment on the one hand, and acid exposure and endoscopic and cellular changes on the other. Asymptomatic people and patients whose symptoms are well controlled with acid-suppressive drugs can progress all the way to adenocarcinoma.[15]

4. There is no ability to recognize cellular changes before the point at which troublesome symptoms develop. This is when GERD is presently defined. There is no GERD before this point in the disease per the Montreal definition. As a result, there is a failure to detect serious cellular changes that are discordant with symptom severity or control of symptoms with PPI therapy.
5. There is no method to accurately predict future complications. As a result, the management of GERD is entirely reactive without any serious attempt at preventing disease progression. Visible CLE, which is the line in the sand in the new method, is simply allowed to happen with no real concern or recognition that it is an irreversible cellular change that is never reversed with medical therapy. In the US, visible CLE is ignored when biopsies are negative for intestinal metaplasia. As a result, adenocarcinoma in the thoracic esophagus is also simply allowed to happen.

The new diagnostic test for LES damage has the potential to correct all of these problems.

6.1 Exclusion of Gastroesophageal Reflux Disease as a Cause of Symptoms

There are data that as many as 30% of patients under long-term PPI therapy under the present management algorithm may not have objective evidence of GERD.[14] Use of PPIs in these people has the combined impact of needless inconvenience, unnecessary cost estimated at billions of dollars, and exposure to the increasingly recognized risks of long-term PPI therapy.

The new diagnostic test for LES damage provides a simple and highly effective method of solving this problem. If the person has abdominal LES damage of < 15 mm by the new test, the symptoms are not the result of GERD. The exact amount of LES damage that will become the criterion for reliable definition of the absence of GERD will be determined by the validation trials; the present < 15 mm number is based on best available evidence.

The test for LES damage can be used before PPI treatment is given, essentially replacing the empiric PPI test with an objective measure of LES damage. The problem of people being treated needlessly with PPIs long term with a false-positive empiric PPI test will disappear. This is important. Unnecessary PPI therapy that exposes a person to the risk of adverse reactions to the drugs violates the fundamental "do-no-harm" principle of physicians. If the new test of LES damage is proven to be effective in preventing unnecessary PPI therapy, endoscopy will move to a point close to the onset of symptoms suspected of being caused by GERD.

In Chapter 18, I suggested a comprehensive assessment of LES damage that includes histologic measurement of LES damage, manometric measurement of the residual LES, and a pH study with a heavy meal as a stress test during the period of pH monitoring. Whether these expensive additions are necessary is unknown at this time. Ideally, the simple biopsy measurement of LES damage will be adequate as a stand-alone test except in a few patients with a marginal result.

6.2 Earlier Endoscopy in Patients With Gastroesophageal Reflux Disease

The reason why endoscopy is not recommended at present in patients with symptoms of GERD is that endoscopy or ancillary testing with pH monitoring and manometry are not sufficiently sensitive for a diagnosis of GERD.[4] Normal endoscopy, normal manometry (mean pressure > 6 mmHg, abdominal LES length > 10 mm), or a normal pH test (pH < 4 for < 4.5% of the period or a DeMeester score < 14) does not mean that the person does not have GERD. The pH test and manometry are designed to have high specificity, not high sensitivity. The rationale for empiric PPI therapy is that there is no diagnostic test of adequate sensitivity to exclude GERD.

The effect of having a new diagnostic test to accurately define the absence of GERD (LES damage < 15 mm by histology) is that the point at which endoscopy is indicated will likely move to an early stage of the disease. No longer will treatment failure or dysphagia be the indication for endoscopy in the GERD patient. It will be the need to confirm the diagnosis before long-term PPI therapy is undertaken.

If the new test provides a reliable answer to the question: "Are the symptoms caused by GERD or not?" it will be impossible to justify beginning long-term PPI treatment without the test. It is highly likely that the new test will provide the answer to that question with great reliability. It is certain that a person with LES damage of < 10 mm will have a competent LES. These are the lengths of cardiac epithelium seen at autopsy in persons without GERD.[16,17]

The length I have suggested (15 mm of LES damage, which leaves a residual abdominal LES length of 20 mm) is likely to be accurate and an excellent predictor of lack of sufficient reflux to cause GERD. In Zaninotto et al.[11] the median abdominal LES length of their asymptomatic volunteers was ~22 mm. The median 24-h pH exposure in this group was pH < 4 for 1.1%. It seems reasonable to accept that a 1.1% exposure of the thoracic esophagus to a pH < 4 is not likely to be the reason for symptomatic GERD.

Endoscopy, which is presently done many years and even decades after starting PPI therapy in patients with GERD, now moves to the point where a decision needs to be made about the need for PPI therapy. The specter of lawsuits for adverse

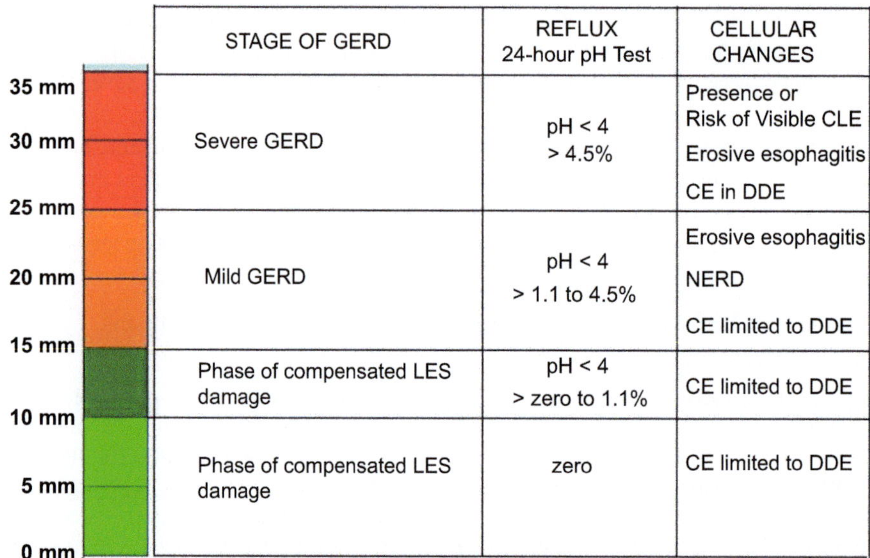

STAGE OF GERD	REFLUX 24-hour pH Test	CELLULAR CHANGES
Severe GERD	pH < 4 > 4.5%	Presence or Risk of Visible CLE Erosive esophagitis CE in DDE
Mild GERD	pH < 4 > 1.1 to 4.5%	Erosive esophagitis NERD CE limited to DDE
Phase of compensated LES damage	pH < 4 > zero to 1.1%	CE limited to DDE
Phase of compensated LES damage	zero	CE limited to DDE

FIGURE 21.4 Correlation between abdominal lower esophageal sphincter (LES) damage (shortening) and reflux (as indicated by pH monitoring) and likely cellular changes in the esophagus. *CE*, cardiac epithelium; *CLE*, columnar-lined esophagus; *DDE*, dilated distal esophagus; *GERD*, gastroesophageal reflux disease; *NERD*, nonerosive reflux disease.

side effects resulting from long-term PPI use when a simple test was available to avoid that problem will drive the utilization of the test at this earlier point in the management algorithm.

With a disease that progresses by a slow and inexorable increase in LES damage, endoscopy performed many years and even decades earlier than at present is likely to shift the prevalence of cellular change observed at the index endoscopy to a lesser stage in all objective parameters of severity of GERD:

1. The measured length of LES damage by the new test will be at an earlier stage, i.e., less LES damage, than if the endoscopy were delayed until the point of treatment failure.
2. Fewer patients will have severe (Los Angeles grade C/D) erosive esophagitis, visible CLE, Barrett esophagus, and neoplasia.
3. If biopsies are taken from the SCJ in endoscopically normal persons, as they are required in the new test, fewer patients will have intestinal metaplasia.

The migration of endoscopy to an earlier stage of GERD will therefore provide new opportunities to prevent progression to visible CLE, which is the line in the sand in the new management method. It increases dramatically the probability of successfully preventing esophageal adenocarcinoma. The reason for this is that the endoscopy is not merely looking for Barrett esophagus, which is the present objective; it is also looking for a measure of LES damage that will predict the occurrence of visible CLE decades into the future.

6.3 Accurate Assessment of Severity of Gastroesophageal Reflux Disease

At present, the complete dependence on symptoms for diagnosis and the treatment objective of symptom control completely mask the underlying progression of the esophageal epithelium at a cellular and molecular level.

The fundamental cellular progression is primarily dependent on the amount of reflux, which in turn depends on the amount of LES damage. The pH of the refluxate is of secondary importance.

The new diagnostic test for GERD is likely to provide a much more precise and objective assessment of severity of GERD in terms of the structure and function of the LES (Fig. 21.4). When compared with symptoms, which have a poor correlation with reflux severity and cellular changes, measured LES damage is almost certain to be a much superior assessment of severity of GERD in the individual patient. The exact correlations between LES damage and cellular changes will be determined by future clinical testing that will supplant the theoretical numbers I have proposed.

6.4 Detection of Early Cellular Changes of Gastroesophageal Reflux Disease

The greatest advantage of using pathologic measurement of LES damage in assessing severity of GERD is that the changes in the LES can be quantitated before any cellular changes are seen in the thoracic esophagus. This is because the normal state, where there is no LES damage as evidenced by the complete lack of cardiac epithelium (with and without parietal and/or goblet cells), can be accurately defined.

The cellular changes associated with early progression of LES damage in the phase of compensated LES damage and progression to clinical GERD with and without troublesome symptoms that presently define GERD can be recognized with the new test of LES damage. The progressive elongation of the dilated distal esophagus that is presently ignored and misinterpreted as normal proximal stomach becomes the center of attention in the new method.[18]

6.5 Accurate Prediction of Progression of Gastroesophageal Reflux Disease

The most important management tool provided by the new test for LES damage is the ability to predict future LES damage by the newly developed algorithm (see Section 4). Future LES damage levels are predicted through the entire life of the individual (Table 21.1 and Fig. 21.2).

LES damage can be tracked into the future from the time the test of LES damage is done. This allows identification of people who will never get GERD (the green zone), those who are predicted to get mild GERD that has no risk of visible CLE or adenocarcinoma that will likely remain under good control with PPI therapy all their lives (the orange zone), and those who are predicted to get into trouble at some point in the future (the red zone). Trouble is failure to control symptoms with maximum PPI therapy, the need for surgical intervention, and the occurrence of visible CLE, which is the line in the sand for preventing adenocarcinoma.

These projected predictions will show that there is a large window of opportunity to attack the disease by methods to prevent progression of LES damage. Preventing progression of LES damage is not a concept that exists at the present time. It will only emerge when the new diagnostic test becomes available. I trust the ingenuity of people given the task of protecting the LES by new simple technologies to come up with methodology to achieve this. To surgeons and medical device makers that have produced transoral incisionless fundoplication (TIF),[19] Stretta,[20] LINX,[21] EndoStim,[22] and various types of fundoplication that have significant success in controlling a severely damaged LES, this new task should be relatively simple.

7. NEW MANAGEMENT OF DIFFERENT PATIENT GROUPS DEFINED BY THE NEW TEST

The new test of LES damage has divided the population tested into a green, an orange, and a red zone (see Table 21.1). This gives us information regarding the status of the person from the time of the test to the end of his/her natural life span.

In Table 21.1, I have set the end point of predicted LES damage at an age of 75 years. This is not to suggest that the natural life span is 75 years. It is based on the fact that if visible CLE is prevented until age 75 years, the lag phase for the remaining steps needed for adenocarcinoma (i.e., visible CLE→intestinal metaplasia→increasing dysplasia→adenocarcinoma) will probably take another 25 years. As such, if visible CLE is prevented to age 75 years, the likelihood is that adenocarcinoma is effectively prevented. Visible CLE and Barrett esophagus are irrelevant conditions if they cannot progress to adenocarcinoma within the life span of the person.

Lee et al.[12] reported that elderly patients with GERD tend not to have escalation of symptoms even as their esophageal acid exposure continues to increase and their abdominal LES length continues to decrease. This suggests that the probability of failure of PPIs to control symptoms likely decreases after age 75 years.

The dual objectives of preventing adenocarcinoma and preventing failure of PPIs to control symptoms are likely to be met by the objective of preventing the patient from reaching the red zone of severe GERD by age 75 years. I believe this is a realistic target to aim for. If desired, this age can be 85 years for added safety.

I will consider the impact of the new method on the management of patients projected to be in the three zones during their lifetime. The reader should step back at this point and recognize that we are now contemplating addressing future LES damage and future GERD in our patients. At the point the test is done, they may or may not have symptoms of GERD. Their LES will be relatively intact.

Present management never contemplates this. It simply treats people with PPIs and reacts to endpoints such as treatment failure, visible CLE, Barrett esophagus, and adenocarcinoma when they arise. The new method is proactive in that it provides a pathway to preventing those endpoints.

7.1 The Green Zone

These are persons who will remain in the green zone for their natural life span. These people will never develop GERD of any significant consequence. This is 70% of the population whose battle between their LES and their eating habits is won for the entirety of their life by their LES. Their LES damage is not sufficient to progress beyond the reserve capacity of the LES. They never move out of the phase of compensated LES damage.

I am one of this group. At 4 mm of LES damage at age 58 years, I am predicted to have LES damage of 8 mm at age 97 years. I will never develop any manifestation of GERD in my lifetime. My LES will enter the red zone only if I live to age 270 years. The normal LES is an amazingly powerful structure in resisting damage.

The green zone is not equivalent to the asymptomatic GERD patient. The absence of symptoms of GERD is no guarantee of absence of cellular changes of GERD. Most importantly, Lagergren et al.[15] reported that 40% of patients with esophageal adenocarcinoma and 71% of patients with adenocarcinoma of the "gastric cardia" had no symptoms of GERD by their definition. A person in the green zone by the new test has a guarantee he/she will not develop clinical GERD let alone esophageal adenocarcinoma in his/her lifetime. If they develop symptoms of any kind, they are almost certainly caused by something other than GERD. If an empiric PPI test controls the symptoms, it is a false-positive empiric PPI test.

7.2 The Orange Zone

These are persons who will be predicted to enter the orange zone at some point in their life, as defined by moderate abdominal LES damage (<25 mm) by age 75 years. Even if they reach the critical <25 mm point after age 75 years, they are highly unlikely to progress to visible CLE→intestinal metaplasia→dysplasia→adenocarcinoma within their remaining life span. They are therefore not at risk for adenocarcinoma in their lifetime.

These people will have a choice that is made with a guarantee that they will not progress to Barrett esophagus and adenocarcinoma. They are likely to maintain control of their symptoms reasonably well with PPIs for their lifetime. Some amount of failure of symptom control and disruption of quality of life is possible at the upper region of this zone.

This is the population that will benefit most from intensive education regarding the mechanism and predicted course of their LES damage. The ability to have them see the display of their individualized progression of LES damage from the green to the orange zone 15–25 years in the future will be dramatic.

This is a powerful teaching moment. It has the highest probability that these patients will follow advice regarding modification of their diet to become more GERD-friendly. With the detection of LES damage at an early stage with projection of future trouble, dietary modification can have a significant impact in reducing the rate of progression of LES damage.

The alternative is that some pharmaceutical company will develop a new prokinetic drug that increases the resistance of the LES to damage. This can also have the effect of slowing down the progression of LES damage.

If simple dietary modification or a prokinetic drug proves even slightly effective in reducing the rate of progression of LES damage, it may reduce the need for PPI therapy in the future and remove any possibility that the patient's symptoms will not be controlled with PPIs. Ideally, a repeat test of LES damage 5–10 years later will show that these noninvasive treatment modalities have shifted the patient to the green zone for their lifetime.

7.3 The Red Zone

This final group of people will be those whose LES damage algorithm predicts they will enter the red zone at some future point within their expected life span. This is presently defined by LES damage that reaches 25 mm during their lifetime.

25 mm is an amazing number. It shows the power of the LES in resisting damage. If one assumes that the onset of LES damage is at 15 years of age, the rate of LES damage required to produce 25 mm of shortening by age 75 years is 0.42 mm/year. A squamous epithelial cell measures ~40 μm. If less than 10 squamous cells transform to cardiac epithelium per year, the person will not reach the red zone by age 75 years. The rate of LES damage required to reach 25 mm of damage by age 50 years is 0.71 mm/year. This means that if 18 squamous cells undergo cardiac metaplasia per year, the patient is in the red zone at age 50 years.

This is almost the entire spectrum of GERD. People whose eating habit causes cardiac metaplasia in squamous cells at a rate of < 10 cells/year will not enter the red zone until age 75 and will likely never develop treatment failure or adenocarcinoma. In contrast, the person who transforms > 18 squamous cells per year to cardiac epithelium is at risk for visible CLE and cancer risk at age 50 years.

The choices available to people projected to reach the red zone during their lifetime are fewer. They must at all costs avoid entering the red zone by age 75 years if they are to avoid entering the point at which visible CLE occurs and lack of risk of adenocarcinoma cannot be guaranteed (Fig. 21.5).

The exact decisions that will be made by a person with the future prospect of entering the red zone will have the following logic:

1. If the projected date of entry into the red zone is within 10 years, he/she should immediately consider having a procedure that prevents LES damage from progressing. This can be a presently available endoscopic [TIF, Stretta] or surgical

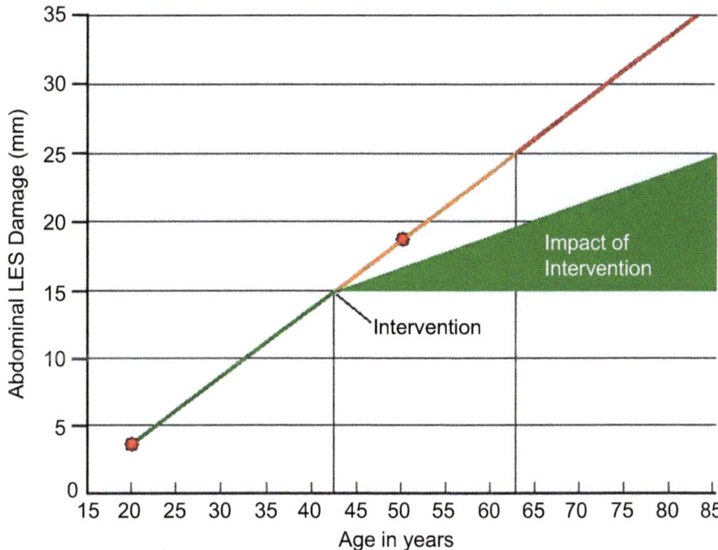

FIGURE 21.5 Progression of lower esophageal sphincter (LES) damage in a 42-year-old man with 15 mm of abdominal LES damage by the new histologic measurement. This person is projected to reach 25 mm of LES damage at age 62 years. Intervention performed at age 42 to stop or slow down the progression of LES damage can prevent him from reaching 25 mm of abdominal LES damage until age 85 years. Visible columnar-lined esophagus and risk of adenocarcinoma are prevented with successful intervention.

(LINX, Endo-Stim, Nissen fundoplication) LES augmentation procedure. These are all likely to be highly effective (Fig. 21.5). This decision becomes easier if there is a prokinetic drug that is effective or a new and simpler endoscopic interventional procedure that does not require surgery has been developed.

2. If the date of entry into the red zone is over 10 years in the future, a trial of dietary modification and/or a prokinetic drug can be tried. A repeat test in 5 years to see whether the slope of progression of LES damage has improved would create a new scenario of choice. The decision at that time will be based on the test result.

Assuming that some innovative surgeon in conjunction with an enterprising medical device company has developed an effective method of preventing progression of LES damage, the path is clear for prevention of esophageal adenocarcinoma.

The new method permits identification of the small minority of people (the red zone) who are at risk for visible CLE, which is the necessary precursor of adenocarcinoma. This determination is made at a time that depends only on the age of the patient at endoscopy. The earlier the age, the lesser the LES damage at the time the test, and the greater the likelihood that LES progression can be slowed down effectively, either by dietary modification, a new prokinetic drug that has been developed, or by a procedure.

8. DIFFERENCES IN MANAGEMENT BETWEEN THE PRESENT AND NEW METHOD

The main advantage of the newly proposed management is that it is proactive compared to the reactive nature of the present management.

The new management separates GERD by the new test of LES damage into three groups whose risk of adenocarcinoma can be reliably assessed. This allows identification of the relatively small percentage of the population who are at risk of progressing to severe GERD defined by the onset of risk for development of visible CLE. This allows concentration of all treatment effort at this red zone group. The persons in the green zone can be ignored with the knowledge that they will never develop GERD or adenocarcinoma. The persons in the orange group can be managed with the knowledge that they will likely be well controlled by PPI therapy and never develop adenocarcinoma.

In contrast, the present management of GERD treats everyone with troublesome symptoms of GERD with PPI therapy after the diagnosis is made by a flawed empiric PPI test.[14] The sole objective of treatment is symptom control. Because PPI therapy does nothing to reduce reflux[23] or the rate of progression of LES damage, the slope of progression of LES damage remains unaltered (Figs. 21.5 and 21.6).

There is no thought paid to the prevention of complications that are known to occur with this management method: failure of symptom control with PPI therapy, the occurrence of extraesophageal complications of GERD including pulmonary fibrosis, increasing incidence of Barrett esophagus, and increasing incidence of esophageal adenocarcinoma.

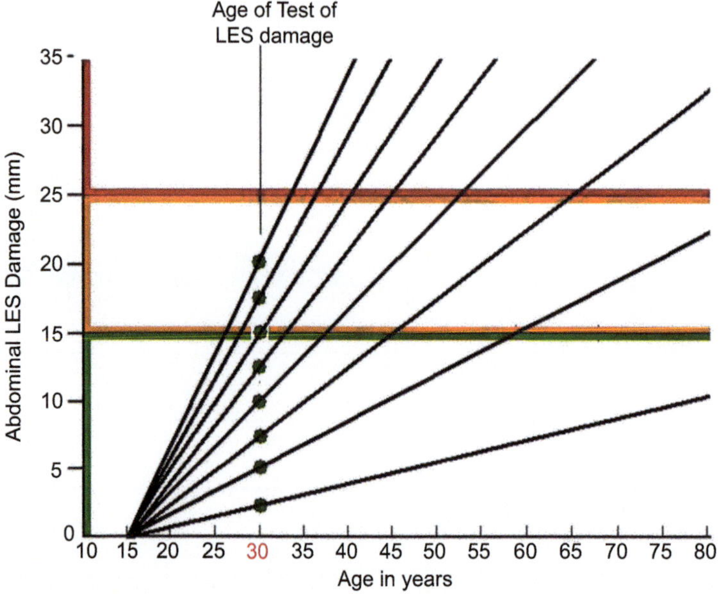

FIGURE 21.6 New test for lower esophageal sphincter (LES) damage done at age 30 years shows LES damage varying from 2.5 to 20 mm at increments of 2.5 mm. The projected LES damage into the future is shown by the multiple lines. These show that people with 7.5 mm of abdominal LES damage at this age will progress to LES damage of 25 mm by age 62 years; those with <5 mm LES damage will remain at <25 mm LES damage throughout their life.

The management is purely reactive with hope and prayer only that complicated GERD does not occur in the individual being treated. This is a futile hope because all the evidence points to a certainty that 30% of people so treated will progress to a state where their lives are disrupted by GERD to a varying extent. And, a certainty that ~20,000 people per year will go on to develop adenocarcinoma.

The lack of logic in the present management protocol is shown by the fact that less than 15% of peoples who develop adenocarcinoma have never had an endoscopy. Many of these people have never had significant symptoms and another large number have had their symptoms well controlled with PPIs.

9. FUTURE DIRECTIONS IN MANAGEMENT OF GASTROESOPHAGEAL REFLUX DISEASE

The limitation of the new test in the initial phase will be the need to perform endoscopy to position the new biopsy device in the correct place and with the correct orientation to take the biopsy. The main reason for endoscopy in the initial phase will be to correlate the LES damage, as measured by the new device, with endoscopic changes, pH testing (a Bravo capsule can be inserted for this at the same time as the endoscopy and coupled with a GERD stress test as described in Chapter 17), and manometry.

9.1 A New Assumption of Future Technologic Development

Let us assume that we are 10 years into the future. The new test has been validated and shown to be a stand-alone test that can assess both the severity of GERD by LES damage and the severity of reflux and cellular changes in the esophagus and predict the occurrence of these factors at specific times in the future of the patient's life.

This has made obsolete the need for endoscopy, pH testing, and manometry in the assessment of patients with GERD. The new measurement of LES damage coupled with the algorithm that predicts future damage will be the only test that is needed.

Endoscopy is only needed as a method of placing the biopsy device at the right place and orientation to take the biopsy sample for the test. For this simple task, endoscopy is far too complicated and expensive. Surely, innovation will produce a new method of making this measurement without the need for endoscopy.

9.2 Innovation to Remove the Need of Endoscopy for the New Diagnostic Test

The stage is now set for the final innovation that is needed to move the diagnostic test from the expensive arena that is occupied by the gastroenterologist. This arena is a high-cost environment that requires an endoscopy suite with postprocedure recovery, special expertise, anesthesiology, personnel, and equipment. It is also associated with a small but defined risk of perforation. As long as the performance of the new diagnostic test requires this arena, the ability to perform the test will be limited by the availability of resources and cost.

The impact of this limitation will inevitably be to delay the performance of the test to a later point where the person has become older, the LES damage has progressed to a further point, and the highest probability of a positive outcome has been compromised.

What will be needed is the ability to perform the biopsy without the need for endoscopy. This can be achieved by a new method of introducing the new biopsy device that does not require formal endoscopy.

Lesser endoscopy techniques such as transnasal endoscopy that can be performed rapidly, safely, and cheaply in a doctor's office will suffice if it can be made to accommodate the biopsy device. A simple guide wire with a miniature camera to provide visualization of anatomy purely for recognizing the SCJ and positioning the device vertically as needed is all that is needed to procure the mucosal biopsy.

Nonhistologic methods may also be developed in the future. These would permit the measurement of the mucosa lined with cardiac epithelium (with and without parietal and goblet cells) without the need for biopsy. The probability that such nonhistologic methods will be successful in differentiating oxyntocardiac from gastric oxyntic epithelium accurately is low at this point in time, but may change in the future.

High-resolution photography with the use of dyes and sophisticated image analysis is a possible solution. Optical coherence tomography, which can identify differences in the location, thickness, and regularity of the muscularis mucosae in squamous, cardiac, and gastric oxyntic epithelia may provide a possible solution. The optical device can theoretically be inserted into the stomach with a wire and then provide a longitudinal image as it is withdrawn across the LES zone into the esophagus. New technologies not yet known may provide a solution. Again, the commercial incentive is immense based on the market and obvious demand for such technology.

9.3 Screening for Gastroesophageal Reflux Disease

If new technology permits measurement of the dilated distal esophagus (cardiac epithelium with and without parietal and/or goblet cells distal to the normal SCJ) without endoscopy by a cheap, safe, and easily applied device, there develops a sudden ability to screen asymptomatic persons for LES damage.

The test can potentially be easier than a Pap smear or mammogram and certainly a lesser procedure than colonoscopy screening. Unlike those screening tests, the test needs to be done only once except in marginal cases where repeat testing may be needed after 5 and 10 years to provide a second assessment.

If this is done in everyone at age 30 years, the test will provide an individualized projection of LES damage into the future up to the end of their lives (Fig. 21.6). Persons can be told, with the results of the test and prediction algorithm, the following interpretation:

1. They will never progress out of the green zone and they need not worry about getting GERD or esophageal adenocarcinoma (70% of the population).
2. They will enter the orange zone at some point in their lives. They will develop GERD, but will never be at risk for esophageal adenocarcinoma (20%–25% of the population).
3. They will enter the red zone and be at risk of developing visible CLE (5%–10% of the population). They will, if untreated, develop GERD, be at risk for disruption of their lives by the failure of drugs to control symptoms, and be at risk for adenocarcinoma in the distant future. At present, 20,000 people per year in the United States will develop adenocarcinoma annually.

These are predictions that are made while the person is asymptomatic. (GERD is rare before 30 years of age.) Examination of Fig. 21.6 shows that persons with abdominal LES damage that is >5 mm at age 30 years are likely to progress into the red zone. At LES damage of 7.5 mm, a 30-year-old person is predicted to reach 25 mm of LES damage at age 65 years and 10 mm at age 50 years. This suggests that the vast majority of people will have a length of cardiac epithelium < 7.5 mm at age 30 years. This is concordant with the autopsy population in this age group.[16,17]

All decisions on management are made proactively based on future predictions. They are made while the LES is largely intact. The ability to prevent progression of LES damage is likely to be much easier than the present proposition of waiting for LES destruction and trying to repair it.

Screening persons who are asymptomatic provides the ultimate solution to GERD. By identifying the 30% of people who will develop GERD in the future by a screening test done at age 30 years, the person can be given the following choice:

1. Preventing the occurrence of GERD by opting to undergo the procedure to prevent progression of LES damage into the orange zone or red zone. The likelihood of a patient opting for the procedure will increase greatly if the procedure is easy, either noninvasive with dietary modification or prokinetic drugs, or performed by endoscopy without the need for

surgery, is safe without complications, and is relatively inexpensive. Even if the person does not buy into the procedure then, the probability of acceptance at the onset of the symptoms that have been predicted by the test is likely to result in acceptance of the procedure.

2. In the person predicted to enter the red zone in the future, a method or procedure to control the rate of progression of LES damage is essential. It is better to have it done sooner than later, possibly after a short trial of diet and prokinetic drugs. This will prevent the onset of GERD in this person.

The possibility of eradicating GERD as a human disease now becomes feasible.

10. COST COMPARISON BETWEEN NEW AND PRESENT METHODS OF MANAGEMENT OF GASTROESOPHAGEAL REFLUX DISEASE

The unbelievable reality about the new method is that it can result in a massive cost savings to health care as it pertains to GERD.

The ultimate total expenditure of applying the new method can be summarized as follows:

1. Cost of procuring the biopsy sample: If the biopsy sample can be obtained in a doctor's office by a nurse or physician's assistant quickly, safely, and without sedation, its cost will be minimal. If endoscopy is necessary, the cost will be higher, the application of the test will be more limited, and GERD will not be completely eradicated. The cost of treating residual GERD patients with PPIs remains.
2. Cost of the biopsy device, which is likely to be low, at least after a patent period has expired.
3. Cost of pathologic interpretation of the biopsy to provide an accurate measurement of LES damage requires basic processing, a routine microscopic slide, a pathologist, and a standard microscope fitted with an ocular micrometer for measurement. This will have the cost of one biopsy sample (CPT code 88305). There is no capital expenditure.
4. Cost of the algorithm used to calculate future LES damage. This is simple and, possibly after a patent period, essentially without cost.
5. Cost of the procedure to limit progression of LES damage in the 30% of the population destined to develop GERD by the projection made by the test. If this can be achieved by dietary modification, the cost is essentially zero. If a prokinetic drug can do the trick, the cost will be minimal. If a procedure is necessary, the cost will be higher and will depend on whether it is performed by endoscopy or surgery.

The potential cost savings if the new method achieves its ultimate potential can be summarized as follows:

1. The cost of human suffering resulting from preventing esophageal adenocarcinoma. This is inestimable in monetary terms.
2. The cost of human suffering from "disruptive GERD," i.e., GERD that disrupts life because PPI therapy fails to control symptoms adequately, which has been estimated to cost $20 billion annually to businesses in the United States from lost productivity and increase in health-care cost.[24]
3. The cost of PPIs and other acid-reducing drugs used in the treatment of GERD, again billions of dollars per year.[25]
4. The cost of treating patients with symptoms believed to be GERD based on a false-positive empiric PPI test.[14] The monetary cost is estimated to be billions of dollars per year. The tragedy and possibility of litigation when these patients prescribed PPIs unnecessarily develop complications of PPI therapy is inestimable in both human and monetary terms.
5. The cost of endoscopy, pH and impedance testing, and manometry in the diagnosis of GERD.
6. The cost of regular endoscopic surveillance of patients found to have Barrett esophagus at endoscopy.
7. The cost of treating dysplasia in Barrett esophagus with endoscopic mucosal resection and radio-frequency ablation.
8. The cost of treating patients with esophageal adenocarcinoma, including staging procedures, esophagectomy and its complications, chemotherapy, molecular testing for personalized oncologic treatment, radiation therapy, follow-up, and hospice care.

The replacement of the old method by the new method will achieve the impossible: eradicating (or at least decreasing dramatically) a common human disease while decreasing the cost associated with treating that disease. The only comparable situation in the history of medicine regarding cancer has been the control of uterine cervical squamous carcinoma with screening Pap smears.

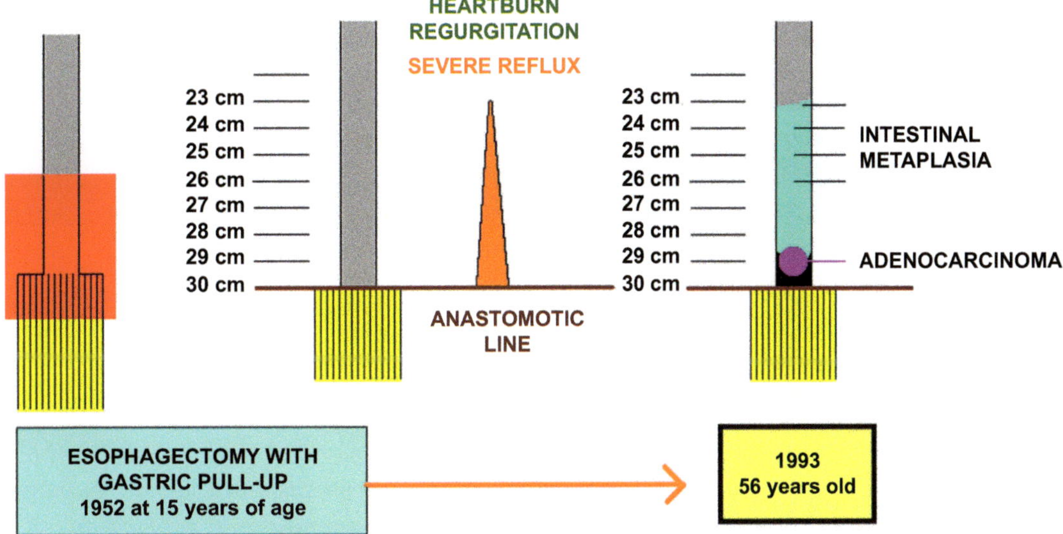

FIGURE 21.7 Summary of changes in case study. The patient had the distal esophagus and proximal stomach (*red square*) removed at age 15 years for a stricture. His reflux (*orange cone*) required proton pump inhibitors (PPIs) for control soon after surgery. Symptom control with PPIs failed 41 years after his surgery when endoscopy was performed. This showed a long segment of columnar-lined esophagus from 23 to 29 cm with an invasive adenocarcinoma at 29 cm.

11. THE LARGE WINDOW OF OPPORTUNITY FOR PREVENTING ESOPHAGEAL ADENOCARCINOMA

11.1 A Case Study

The patient was a 56-year-old man who presented with a long and eventful history. As a 9-month-old baby, his mother saw him swallow a penny. After some acute trouble with gagging and coughing, the baby appeared to be fine. The mother assumed that the baby had successfully swallowed the penny and the problem was over. She saw no reason to take him to a physician.

When he was 7 years old, he developed dysphagia. A chest X-ray showed the penny stuck in his thoracic esophagus. The penny was removed endoscopically. An area of stricture at the site was dilated. He continued to experience dysphagia for solids, resulting in multiple and increasingly frequent dilatations of the stricture. Finally, at age 15, the stricture became resistant to dilatation and it was decided that surgery was the only viable option.

He underwent a resection of the stricture by an esophagogastrectomy that removed the distal esophagus and proximal stomach, including the stricture, the entire LES, and a part of the proximal stomach. An anastomosis was created between the esophagus and pulled-up stomach at 30 cm from the incisor teeth (Fig. 21.7).

After recovery from the surgery, his dysphagia resolved. However, very soon after, he began experiencing heartburn and regurgitation. He was treated with acid-suppressive drugs (initially H_2 receptor antagonists and then PPIs), which controlled his heartburn completely with partial control of regurgitation.

With the passage of the next few decades, there was a progression of disease clinically. Control of symptoms required escalation to PPIs in increasing dosage and frequency. He developed aspiration events that were producing a chronic cough associated with slowly worsening pulmonary function.

At age 56 years (41 years after the esophagectomy), his symptoms were finally deemed to have failed control with medical therapy. He had an endoscopy performed by his treating physician. Based on the findings, he was referred to the USC Foregut Surgery Unit. Repeat endoscopy showed that the surgically created esophagogastric anastomosis was intrathoracic and at 30 cm from the incisors. There was a long segment of CLE extending from the anastomosis to 23 cm. A small nodule was present at 29 cm.

Biopsies reported by me showed the following:

1. 29 cm, bx: Invasive moderately differentiated adenocarcinoma.

2. 29 cm nodule, bx: Invasive moderately differentiated adenocarcinoma.
3. 26 cm, bx: Reflux carditis with intestinal metaplasia (Barrett esophagus).
4. 25 cm, bx: Intestinal (Barrett) metaplasia with low-grade dysplasia.
5. 24 cm, bx: Intestinal (Barrett) metaplasia with low-grade dysplasia.
6. 23 cm, bx: Reflux esophagitis with intestinal metaplasia (Barrett esophagus).
7. Brushings at 29 cm: Malignant cells present consistent with adenocarcinoma.

Back then, endoscopic mucosal resection and radio-frequency ablation was not available. The patient underwent esophagectomy. He had a T2N0 adenocarcinoma of the esophagus with no evidence of distant metastases.

Several things about the pathology are notable:

1. The long segment of Barrett esophagus consisted of epithelium with almost complete intestinal metaplasia in cardiac epithelium in the biopsies from 23 to 29 cm. Although not reported, the density of goblet cells progressively decreased from 23 to 29 cm. At 29 cm, between the anastomosis and the distal edge of the carcinoma, there was cardiac epithelium without intestinal metaplasia.
   This is the usual pattern of intestinal metaplasia in visible CLE. In Chapter 14, I have suggested that this pattern of intestinal metaplasia in CLE strongly suggests that alkalinization of the esophagus with acid-reducing drugs is the cause of intestinal metaplasia in visible CLE.
2. The adenocarcinoma was immediately above the anastomotic line, i.e., the new GEJ. The preferred location of neoplasia in a patient with Barrett esophagus is the lowest point at which intestinal metaplasia is present, i.e., at the junction of intestinal metaplasia and nonintestinalized cardiac epithelium. This is to be expected because the concentration of carcinogen is highest at the distal end of the column of gastric contents that reflux into the esophagus.

This patient had his entire LES removed surgically at age 15 years. He is therefore a human model that begins the disease at the point of complete destruction of the LES at age 15 years. He had, as expected, an acute onset of severe reflux that is associated with a zero LES length. The fact that it took 41 years for this person to progress from that point to adenocarcinoma shows that there was an incredibly long window of opportunity for preventing esophageal adenocarcinoma even with a residual LES of zero length.

This patient's esophagus above the anastomosis must have progressed through the following sequential cellular changes while he was being treated with PPIs: squamous epithelium→erosive esophagitis (controlled with PPIs)→visible CLE from 23 to 30 cm→intestinal metaplasia in visible CLE (Barrett esophagus) likely beginning at the SCJ at 23 cm and progressively extending toward the anastomic line, reaching 29 cm at the time of endoscopy→increasing dysplasia→adenocarcinoma at 29 cm, 1 cm above the anastomosis.

From the perspective of diagnosis and management, the patient developed GERD symptoms soon after the surgery. These were well controlled symptomatically for four decades while this entire sequence of cellular change was occurring. When PPIs failed to control symptoms sufficiently, he had an endoscopy, which showed a relatively early adenocarcinoma.

It is important to define the elements of this sequence that are predictable. It is predictable that he will develop severe reflux immediately after complete removal of the LES at the esophagectomy. It is predictable that he will, after a period of time that depends on a variety of factors (amount and frequency of reflux episodes, resistance of the squamous epithelium, etc.), very likely develop visible CLE with or without PPI therapy. Without endoscopy, the time at which he developed visible CLE and intestinal metaplasia is unknown.

Based on the data in the study of Dresner et al.[26] who followed the mucosal changes above the anastomotic line in patients undergoing esophagogastrectomy, it can be postulated that this patient likely developed cardiac epithelium within the first 5 years, i.e., at age 20 years and intestinal metaplasia within 15 years, i.e., at age 30 years. The process of carcinogenesis, which began when intestinal metaplasia occurred, therefore took 25 years assuming that his early, small cancer arose 1 year before presentation at age 56 years.

The least predictable part of his course is the time from the occurrence of intestinal metaplasia to the onset of adenocarcinoma. This is primarily dependent on the carcinogen level in his gastric contents and secondarily with the severity of reflux, the latter being the method of delivery of the carcinogen into the thoracic esophagus that contains the target cell (in the epithelium marked by the presence of goblet cells).

11.2 Time Frame for Progression to Cancer in the Typical Gastroesophageal Reflux Disease Patient

The age that I have assumed in this book when LES damage begins in a normal person is 15 years. Starting with an abdominal LES of 35 mm, all people progress through a phase of compensated LES damage, where the LES degrades by 15 mm. This ends, in those people who reach the end of this phase at some point in time, with the onset of symptoms of GERD.

Let us assume that the person under consideration has the onset of GERD symptoms around age 35 years. This is likely the end of the phase of compensated LES damage in the index person under consideration. This indicates a rate of progression of LES damage of 7.5 mm/decade, assuming onset of LES damage at age 15 years and that the end of the compensated phase is 15 mm of abdominal LES damage.

With this rate of progression of LES damage, the patient will reach the point of severe GERD where LES damage is 25 mm at age 48 years. This is the common median age for established troublesome GERD in many studies. At this age, the person has enough reflux to be at significant risk for developing visible CLE. Let us assume that this person develops visible CLE and Barrett esophagus in the next 5 years, i.e., around 53 years. This was shown to be the fate of 10% of GERD patients under treatment with PPIs in the Pro-GERD study.[6]

All changes in the GERD sequence until the occurrence of Barrett esophagus correlate with the severity of reflux and will therefore correlate with LES damage. It is critically important in management to understand that the window of opportunity closes as soon as visible CLE develops.

The reason for this is that the transformation of visible CLE→Barrett esophagus→adenocarcinoma is very unpredictable. This progression becomes primarily related to the level of carcinogen in the gastric contents, not necessarily LES damage and severity of reflux although these have secondary importance in that it is reflux that delivers the carcinogen to the target cell in the esophagus.

Without knowledge of the exact nature of the carcinogen or the ability to measure carcinogenicity, the risk of cancer in this person is unpredictable. People with zero carcinogen levels will never progress to neoplasia even if they have Barrett esophagus. People with low carcinogen levels will probably die of some other cause before they develop esophageal adenocarcinoma. At the other extreme, patients with Barrett esophagus with high carcinogen levels will progress rapidly.

Let us assume that this index person under consideration has a high carcinogen level in his gastric contents. The patient can rapidly progress from Barrett esophagus to adenocarcinoma by age 65–70 years. This is the common age range for adenocarcinoma in the typical GERD patient.

Because of the unpredictability of the Barrett esophagus→adenocarcinoma progression, the only way that one can be certain that adenocarcinoma is prevented is to prevent visible CLE. The window of opportunity to do this is measurable in decades if the histologic test of LES damage is performed at age 30–35 years. Once visible CLE occurs, the ability to control progression disappears and is in the hands of destiny, measured largely by the unknown level of carcinogen in gastric contents.

The new method described in this book is the first that has provided a pathway to the goal of preventing GERD, visible CLE, and adenocarcinoma. It marks a potential revolution in the diagnosis and management of GERD.

REFERENCES

1. DeMeester TR, Peters JH, Bremner CG, Chandrasoma P. Biology of gastroesophageal reflux disease: pathophysiology relating to medical and surgical treatment. *Annu Rev Med* 1999;**50**:469–506.
2. Robertson EV, Derakhshan MH, Wirz AA, Lee YY, Seenan JP, Ballantyne SA, Hanvey SL, Kelman AW, Going JJ, McColl KE. Central obesity in asymptomatic volunteers is associated with increased intrasphincteric acid reflux and lengthening of the cardiac mucosa. *Gastroenterology* 2013;**145**:730–9.
3. Odze RD. Unraveling the mystery of the gastroesophageal junction: a pathologist's perspective. *Am J Gastroenterol* 2005;**100**:1853–67.
4. Kahrilas PJ, Shaheen NJ, Vaezi MF. American Gastroenterological association medical position statement on the management of gastroesophageal reflux disease. *Gastroenterology* 2008;**135**:1380–2.
5. Stein HJ, Barlow AP, DeMeester TR, Hinder RA. Complications of gastroesophageal reflux disease. Role of the lower esophageal sphincter, esophageal acid and acid/alkaline exposure, and duodenogastric reflux. *Ann Surg* 1992;**216**:35–43.
6. Malfertheiner P, Nocon M, Vieth M, Stolte M, Jasperson D, Keolz HR, Labenz J, Leodolter A, Lind T, Richter K, Willich SN. Evolution of gastro-oesophageal reflux disease over 5 years under routine medical care – the ProGERD study. *Aliment Pharmacol Ther* 2012;**35**:154–64.
7. Gutschow CA, Schroder W, Prenzel K, et al. Impact of antireflux surgery on Barrett's esophagus. *Langenbeck's Arch Surg* 2002;**387**:138–45.
8. Ronkainen J, Aro P, Storskrubb T, Johansson S-E, Lind T, Bolling-Sternevald E, Vieth M, Stolte M, Talley NJ, Agreus L. Prevalence of Barrett's esophagus in the general population: an endoscopic study. *Gastroenterology* 2005;**129**:1825–31.
9. Rex DK, Cummings OW, Shaw M, Cumings MD, Wong RKH, Vasudeva RS, Dunne D, Rahmani EY, Helper DJ. Screening for Barrett's esophagus in colonoscopy patients with and without heartburn. *Gastroenterology* 2003;**125**:1670–7.
10. Kulig M, Nocon M, Vieth M, Leodolter A, Jaspersen D, Labenz J, Meyer-Sabellek W, Stolte M, Lind T, Malfertheimer P, Willich SN. Risk factors of gastroesophageal reflux disease: methodology and first epidemiological results of the ProGERD study. *J Clin Invest* 2004;**57**:580–9.
11. Zaninotto G, DeMeester TR, Schwizer W, Johansson KE, Cheng SC. The lower esophageal sphincter in health and disease. *Am J Surg* 1988;**155**:104–11.
12. Lee J, Anggiansah A, Anggiansah R, et al. Effects of age on the gastroesophageal junction, esophageal motility and reflux disease. *Clin Gastroenterol Hepatol* 2007;**5**:1392–8.

13. Chandrasoma P. Pathophysiology of Barrett's esophagus. *Semin Thorac Cardiovasc Surg* 1997;**9**:270–8.
14. Bytzer P, Jones R, Vakil N, Junghard O, Ling T, Wernersson B, Dent J. Limited ability of the proton-pump-inhibitor test to identify patients with gastroesophageal reflux disease. *Clin Gastroenterol Hepatol* 2012;**10**:1360–6.
15. Lagergren J, Bergstrom R, Lindgren A, Nyren O. Symptomatic gastroesophageal reflux as a risk factor for esophageal adenocarcinoma. *N Engl J Med* 1999;**340**:825–31.
16. Chandrasoma PT, Der R, Ma Y, et al. Histology of the gastroesophageal junction: an autopsy study. *Am J Surg Pathol* 2000;**24**:402–9.
17. Kilgore SP, Ormsby AH, Gramlich TL, et al. The gastric cardia: fact or fiction? *Am J Gastroenterol* 2000;**95**:921–4.
18. Chandrasoma P, Wijetunge S, Ma Y, DeMeester S, Hagen J, DeMeester T. The dilated distal esophagus: a new entity that is the pathologic basis of early gastroesophageal reflux disease. *Am J Surg Pathol* 2011;**35**:1873–81.
19. Trad KS, Turgeon DG, Deljkich E. Long-term outcomes after transoral incisionless fundoplication in patients with GERD and LPR symptoms. *Surg Endosc* 2012;**26**:650–60.
20. Noar M, Squires P, Noar E, et al. Long-term maintenance effect of radiofrequency energy delivery for refractory GERD: a decade later. *Surg Endosc* 2014;**28**:2323–33.
21. Ganz RA, Peters JH, Morgan S, et al. Esophageal sphincter device for gastroesophageal reflux disease. *N Engl J Med* 2013;**368**:719–27.
22. Rodriguez L, Rodriguez P, Gomez B, et al. Long-term results of electrical stimulation of the lower esophageal sphincter for the treatment of gastroesophageal reflux disease. *Endoscopy* 2013;**45**:595–604.
23. Blonski W, Vela MF, Castell DO. Comparison of reflux frequency during prolonged multichannel intraluminal impedance and pH monitoring on and off acid suppression therapy. *J Clin Gastroenterol* 2009;**43**:816–20.
24. Toghanian S, Wahlqvist P, Johnson DA, Bolge SC, Liljas B. The burden of disrupting gastro-esophageal disease; a database study in US and European cohorts. *Clin Drug Investig* 2010;**30**:167–78.
25. Greenberger NJ. Update in gastroenterology. *Ann Intern Med* 1998;**129**:309–16.
26. Dresner SM, Griffin SM, Wayman J, Bennett MK, Hayes N, Raimes SA. Human model of duodenogastro-oesophageal reflux in the development of Barrett's metaplasia. *Br J Surg* 2003;**90**:1120–8.

Author Index

Note: Page numbers followed by "f" indicate figures, "t" indicate tables.

D

E

F

G

T

U

V

W

X

Y

Z

Subject Index

Note: Page numbers followed by "f" indicate figures, "t" indicate tables.

D

F

G

Q

R

S

T

CPI Antony Rowe
Chippenham, UK
2018-01-05 17:40